Herpetology

Herpetology

3rd Edition

Laurie J. Vitt and Janalee P. Caldwell
Sam Noble Oklahoma Museum of Natural History
Department of Zoology
University of Oklahoma
Norman, Oklahoma

AMSTERDAM • BOSTON • HEIDELBERG • LONDON • NEW YORK • OXFORD • PARIS
SAN DIEGO • SAN FRANCISCO • SINGAPORE • SYDNEY • TOKYO

Academic Press is an Imprint of Elsevier

Academic Press is an imprint of Elsevier
30 Corporate Drive, Suite 400, Burlington, MA 01803, USA
525 B Street, Suite 1900, San Diego, California 92101-4495, USA
84 Theobald's Road, London WC1X 8RR, UK

This book is printed on acid-free paper.

Library of Congress Cataloging-in-Publication Data
Vitt, Laurie J.
Herpetology/Laurie J. Vitt, Janalee P. Caldwell. -- 3rd ed.
p. cm.
Includes bibliographical references and index.
ISBN 978-0-12-374346-6 (hard cover : alk. paper) 1. Herpetology. I. Caldwell, Janalee P. II. Title.
QL641.Z84 2009
597.9--dc22

2008018399

British Library Cataloguing-in-Publication Data
A catalogue record for this book is available from the British Library.

ISBN: 978-0-12-374346-6

For information on all Academic Press publications
visit our Web site at www.books.elsevier.com

Printed in China
09 10 11 12 9 8 7 6 5 4 3 2

Dedication

First and foremost, we dedicate this book to George R. Zug, our former coauthor (see *Herpetology: An Introductory Biology of Amphibians and Reptiles,* Second Edition), who, like us, had a vision of a herpetology textbook that would capture the interest of students who want to learn about what we consider to be the most fascinating organisms on Earth. We also dedicate this book to Coleman and Olive Goin, pioneers in North American herpetology whose original herpetology textbook led to this entire series. Finally, we dedicate this book to all researchers, teachers, students, and amateurs who share a common interest in amphibians and reptiles (even though these two groups are not closely related phylogenetically), and who share the common goal of conserving these remarkable organisms and their natural habitats.

L. J. V. and J. P. C.

Contents

Foreword

The diversity of living creatures on our planet is extraordinary—and thus, trying to understand how those organisms function, and how and why they do the things they do, is an awesome challenge. To make the challenge a bit more manageable, we traditionally divide the study of biology into many categories, some based on methodology (e.g., "microscopy" or "molecular biology"), some on function (e.g., "ecology" or "physiology"), and some on relatedness among the things that are to be studied (e.g., "ornithology" or "herpetology"). At first sight, this last way of slicing the cake seems a bit old-fashioned—surely we can simply ask the same questions and use the same methods, regardless of what kind of organism we might be studying? If so, are traditional taxonomy-based divisions just historical relics of the early naturalists, doomed to eventual extinction by the rise of powerful conceptual and methodological advances?

Nothing could be further from the truth. Entrancing as the new approaches and conceptual divisions are, the reality of life on Earth is that organisms do fall into instantly recognizable types. Few people would mistake a tree for a lizard, or a whale for an insect. The reason is simple: Evolution is an historical process that creates biodiversity by the accumulation of small changes along genealogies, with the vast majority of species becoming extinct during that process. So the end result at any time in Earth's history is a series of terminal branches from the great tree of life—terminal branches that form larger branches, that in turn coalesce to form even larger branches, and so forth. All the species within each of those larger branches share common ancestors not shared by any species on the other branches, and as a result, the species within each branch resemble each other in many ways. For example, no amphibian embryo grows up with an amniotic membrane around it in the egg, whereas every reptile embryo has one.

The evolutionary conservatism of major characteristics such as metabolic rates, reproductive modes, feeding structures, and the like in turn have imposed evolutionary pressures on myriad other features—and the end result is that the diversity of life is packaged into a meaningful set of categories. That is the reason why most of us can easily distinguish a frog from any other kind of animal and can even tell the difference between a crocodile and a lizard. And it is a major reason why there is immense value in defining a scientific field based on evolutionary relatedness of the creatures being studied, not just on methods or concepts. So "herpetology" *is* a useful category: If we really want to understand what animals do, we can't ignore the history behind each type of organism. Many of its features will be determined by that history, not by current forces. Because of that historical underpinning, the most effective way to answer general questions in biology may be to work within one or more of those major branches in the tree of life. Starting from common ancestors, we can see with much greater clarity how evolutionary forces have created rapid change in some cases (why are chameleons so incredibly weird compared with other lizards?), have produced remarkably little change over vast timescales in others (can it really be true that crocodiles are more closely related to birds than to lizards?), and have even generated convergent solutions in distantly related species exposed to similar adaptive challenges (like horned lizards in the deserts of North America compared with thorny devils in the deserts of Australia).

Allied to the greater clarity that comes from comparing like to like, and including genealogy in our thinking, are other great advantages to taxon-based categories like "herpetology." Organisms are composites of many traits, and these need to work together for the creature to function effectively. So we can't really look at metabolic rate separately from foraging behavior, or social systems separately from rates of water loss. Biology forges functional links between systems that our conceptual and methodological classification systems would treat in isolation from each other, ignoring their need for integration within a functioning individual. And there are many other advantages also. In a purely pragmatic sense, the methods that we use to study animals—such as the ways we observe them, catch them, handle them, mark them, and follow them around—depend enormously on many of the traits that differ so conspicuously between major vertebrate lineages. A textbook of herpetology can thus teach us more about *how* to study these animals than can a textbook focused on any single functional topic. And lastly, the conservation challenges facing reptiles and amphibians also are massively affected by their small body sizes, low rates of energy use, primarily tropical distributions, and the like—so that if we are to preserve these magnificent animals for future generations, we need a new generation of biologists who can comprehend the sophisticated functioning of these threatened creatures. This marvelous book captures the excitement of herpetology and will do much to instill that appreciation.

Rick Shine
School of Biological Sciences
University of Sydney
Sydney, Australia

Acknowledgments

We first acknowledge all herpetologists who have published results of their research, thus providing the basis for our textbook. Several students and colleagues in our laboratory have provided continual help and insight, often serving as crucial critics. In particular, Gabriel C. Costa, Donald B. Shepard, Adrian A. Garda, and Christina A. Wolfe provided continual input.

The following friends and colleagues provided photographs, graphics, or information or read portions of the text: Ronald Altig, J. Pedro do Amaral, Chris Austin, Teresa Cristina S. Avila-Pires, R. W. Barbour, Richard D. Bartlett, Dirk Bauwens, Daniel Blackburn, James Bogart, Franky Bossuyt, William R. Branch, A. Britton, Edmond D. Brodie III, Edmond D. Brodie Jr., Rafe M. Brown, Samuel (Buddy) Brown, Andrew Campbell, Jonathan A. Campbell, David Cannatella, Karen Carr, R. S. Clarke, Guarino R. Colli, Suzanne L. Collins, Tim Colston, Justin D. Congdon, William E. Cooper Jr., Gabriel C. Costa, E. G. Crespo, Orlando Cuellar, I. Das, C. Ken Dodd Jr., William E. Duellman, Carl H. Ernst, Danté B. Fenolio, April Fink, Darrel Frost, Chris Funk, Adrian A. Garda, Luis Gasparini, D. J. Gower, Lee Grismer, W. Grossman, Blair Hedges, W. Ron Heyer, David Hillis, Walter Hödl, Jeffrey M. Howland, Raymond B. Huey, Victor H. Hutchison, Karl-Heinz Jungfer, Ken Kardong, D. R. Karns, Michael Kearney, Jeffrey W. Lang, William Leonard, Randy Lewis, Albertina Lima, Jonathan B. Losos, Michael A. Mares, Otavio A. V. Marques, Brad Maryan, Chris Mattison, Roy W. McDiarmid, James McGuire, Phil A. Medica, Peter Meylan, Ken Miyata, Edward O. Moll, Donald Moll, Robert W. Murphy, D. Nelson, K. Nemuras, Brice Noonan, Ronald A. Nussbaum, Mark T. O'Shea, David Pearson, David Pfennig, Eric R. Pianka, Michael Polcyn, Louis W. Porras, Jennifer Pramuk, F. Rauschenbach, Chris Raxworthy, Steve M. Reilly, R. P. Reynolds, Stephen J. Richards, Gordon H. Rodda, Santiago Ron, James Rorabaugh, Herbert I. Rosenberg, C. A. Ross, Rodolfo Ruibal, Anthony. P. Russell, Paddy Ryan, Ivan Sazima, Wade Sherbrooke, D. Schmidt, Cecil Schwalbe, Terry Schwaner, Antonio Sebben, Stephen C. Secor, Bradley Shaffer, Donald B. Shepard, Rick Shine, Cameron Siler, Koen Stein, James R. Stewart, Stanley E. Trauth, Linda Trueb, R. G. Tuck Jr., H. I. Uible, Wayne Van Devender, Miguel Vences, Vidal, J. Visser, Richard Wassersug, Graham Webb, R. Whitaker, John J. Wiens, Steve Wilson, Christina A. Wolfe, Z. Zheng, and George R. Zug.

Organizations permitting us to use their illustrative materials include Academic Press, American Association for the Advancement of Science, American Museum of Natural History, American Society of Ichthyologists and Herpetologists, American Society of Integrative Biology, Blackwell Science Inc., Cambridge University Press, Charles University Press, Chelonian Research Foundation, Cornell University Press, CRC Press Inc., Ecological Society of America, Elsevier Science Ltd. (TREE), Ethology Ecology & Evolution, Herpetological Natural History, Harvard University Press, The Herpetologist's League, Kluwer Academic Publisher, The McGraw-Hill Companies, Muséum National d'Historie Naturelle, Paris, Museum of Natural History, University of Kansas, Division of Amphibians and Reptiles, National Museum of Natural History, Princeton University Press, Smithsonian Institution, National Research Council of Canada, Savannah River Ecology Laboratory, University of Georgia, Museum of Comparative Zoology (Harvard University), Selva, Smithsonian Institution Press, Society for the Study of Amphibians and Reptiles, Society for the Study of Evolution, Society of Systematic Biologists, Springer Verlag, University of Chicago Press, John Wiley & Sons Inc., Cambridge University Press, National Academy of Sciences (USA), and others.

Introduction

It is an admirable feature of herpetologists that they are able to cross the boundaries between different aspects of their subject, which remains, perhaps more than other branches of zoology, a single coherent discipline.

A. d'A. Bellairs and C. B. Cox, 1976.

In 2001, we joined George R. Zug to publish the second edition of his highly successful textbook, *Herpetology: An Introductory Biology of Amphibians and Reptiles*. For us, and for George, it was an exciting collaboration, and we fueled each other's interest and enthusiasm as we worked together on the book. George made it clear at the time that the second edition would be his last involvement in the textbook, not because he had lost interest (herpetologists never lose interest!), but rather because he was at a point in his career where he had other goals to accomplish.

Herpetology is a rapidly evolving field, and although it is a taxonomically delimited field, research on amphibians and reptiles has set new directions, defined new fields, and led to major discoveries in all conceptual areas of biology—discoveries that have changed the way we think about life on Earth. We know more now than we ever did, and we will continue to know and understand more as innovative technologies allow us to explore new ideas in ways never before thought possible. At the same time, we are losing species and habitats at a rate unparalleled in the history of life, and much of it can be tied directly to human activity and indirectly to human population growth. When Coleman and Olive Goin published *Introduction to Herpetology* in 1962, population of the Earth was nearly 3 billion; when George Zug published the first edition of *Herpetology: An Introductory Biology of Amphibians and Reptiles* in 1993, the population was 5.4 billion; today, the world population approaches 7 billion! Not only has the world population reached an unprecedented high, but the exponential rate of population increase is reflected in the exponential increase in environmental effects. We consider it imperative that students understand the basis for life around them and the connections between our survival and the survival of other species. The biology of amphibians and reptiles provides a unique opportunity to achieve that goal, for several rather obvious reasons. Amphibians (frogs in particular) have gained enormous popularity in the arts and crafts trade, partly because they are colorful and diverse, and partly because they are nonthreatening. The pet trade has brought amphibians and reptiles into the homes of millions of people and sparked their interest in these remarkable animals. It is our hope that we can use that interest to draw students into understanding general biological concepts, all of which apply to the biodiversity surrounding us that helps sustain life on Earth.

Our primary goals in revising *Herpetology: An Introductory Biology of Amphibians and Reptiles* are to (1) update the text to reflect some of the truly exciting discoveries that have been made since about 2000 when we completed the second edition, published in 2001; (2) update the taxonomy, which in some cases has changed radically as the result of much more sophisticated evolutionary analyses (e.g., anurans); and (3) introduce the reader to some of the leading herpetological researchers by featuring their work throughout the book. In doing the latter, we emphasize that many truly phenomenal researchers make major discoveries every day—we have selected a few from the many, and with future editions, our selections will vary. Our intent is not to slight any researcher by noninclusion but rather to highlight a few of the many in an attempt to make research discovery a little more personal. After all, successful herpetologists are really just normal people driven by their interest in herpetology, just as rock stars are normal people driven by their interest in music and the performing arts.

We have explicitly tried to keep the text at a level that will be of use to undergraduates with only a basic background in biology, as well as those with a much broader background. We have added color throughout the text, which we believe aids significantly in showcasing how special these animals are. Color is also very useful in chapters in which we discuss crypsis, aposomatic coloration, and social behaviors mediated by visual displays. We remind the reader that not only are amphibians and reptiles part of our own evolutionary history, but they also are an integral part of our natural heritage. They, along with all other animal and plant species, compose life on Earth.

Classification and nomenclature continue to change, and if anything, the rate of change is greater than it ever has been. New fossils, new techniques for obtaining and interpreting phylogenetic data, and the beginnings of a truly phylogenetic taxonomy and its associated nomenclature are changing amphibian and reptilian classification monthly. The ability to recover relationships among taxa at all levels based on combinations of morphological, gene sequence, behavioral, physiological, and ecological data (total evidence) demonstrates the complexity of the evolutionary history of amphibians and reptiles. At the same time, it brings us much closer to constructing phylogenetic hypotheses that accurately reflect relationships. Most striking is the observation that classical Linnean taxonomy presents a false impression about relationships of taxa. For example, Linnean taxonomy implies that all families are equal age, that all orders are equal age, and so on. Although some elements of Linnean taxonomy are useful in allowing us to talk about amphibians and reptiles, the basic notion that organisms can be placed in arbitrary groups and given names is highly misleading. Our classification contains a mix of lower-taxonomic-level Linnean taxonomy (to facilitate discussion) and phylogenetic taxonomy (to reflect relationships).We use species, genus, subfamily, and family as labels, emphasizing that each does not correspond to a given phylogenetic distance or evolutionary time period (e.g., not only are different "families" different ages, they also are nested within each other). We have attempted to be as current as possible, and our classification sections reflect published interpretations through December 2007. Numerous phylogenetic hypotheses exist for most groups of amphibians and reptiles, resulting in different classifications, sometimes strikingly different. We have selected a single cladistic interpretation for each group or combined the results of two interpretations when a single cladistic analysis for all members of the group (clade) was not available. We discuss other interpretations and analyses, but not necessarily all available studies, to ensure that readers are aware that other interpretations exist. We use Latinized familial and subfamilial group names for monophyletic groups and Anglicized or Latinized names in quotes for groups that are of uncertain monophyly. Some authors have not assigned family names to some species and groups of species that represent a sister taxon to another family; where Latinized familial names are available, we have used the available name or elevated a subfamilial name if that latter taxon includes the same set of species. Distributions are an important component of an organism's biology; our maps show the natural (nonhuman dispersed) distribution as best we were able to determine it.

Part I

Evolutionary History

Chapter 1

Tetrapod Relationships and Evolutionary Systematics

Herpetology is the study of amphibians and reptiles. Living amphibians and reptiles are representatives of a small number of the many historical tetrapod radiations (Table 1.1; Fig. 1.1). Amphibians were the first truly terrestrial vertebrates. Their ancestors were lobe-finned fishes (Sarcopterygii), a group of bony fishes (Osteichtyes). These fishes appeared in the Devonian Period (more than 380 million years before present [mybp]) and radiated in fresh and salt water. The earliest fossils assigned to Tetrapoda (from Greek, tetra = four, poda = foot) included *Acanthostega* and *Ichthyostega*, both of which were completely aquatic but had four distinct limbs. They appeared in the late Devonian (about 360 mybp) and are in a group of tetrapods referred to as ichthyostegalians (Fig. 1.1). Amphibians have successfully exploited most terrestrial environments while remaining closely tied to water or moist microhabitats for reproduction. Most amphibians experience rapid desiccation in dry environments, but some species have evolved spectacular adaptations that permit existence in extreme habitats.

During the Carboniferous, about 320 mybp, the anthracosaurs appeared (Fig. 1.1). They not only were able to reproduce on land in the absence of water but also had an effective skin barrier that presumably reduced rapid and excessive water loss. Extant reptiles (including birds) and mammals descended from anthracosaurs. The study of birds and mammals, formally called Ornithology and Mammalogy, respectively, are beyond the scope of this book.

Amphibians and reptiles (collectively, herps) are not each other's closest relatives evolutionarily, yet they have traditionally been treated as though they are related (e.g., "herpetology" does not include birds and mammals). Nevertheless, many aspects of the lives and biology of amphibians and reptiles are complementary and allow zoologists to study them together using the same or similar techniques. Biological similarities between amphibians and reptiles and the ease of field and laboratory manipulation of many species have made them model animals for the study of ecology. Amphibians and reptiles have played prominent roles in research on ecology (e.g., tadpoles, salamander larvae, lizards, the turtle *Trachemys scripta*), behavior (e.g., the frogs *Engystomops* [*Physalaemus*] and *Lithobates* [*Rana*] *catesbeianus*), phylogeography (e.g., the lizard genus *Crotaphytus*, plethodontid salamanders), genetics (*Xenopus*), developmental biology (e.g., *Xenopus*, plethodontid salamanders, reptiles), viviparity (squamates), and evolutionary biology (e.g., *Anolis*, *Lepidodactylus*).

AMPHIBIANS AND REPTILES—EVOLUTIONARY HISTORY

Living amphibians are represented by three clades: Gymnophiona (caecilians), Caudata (salamanders), and Anura (frogs) (Table 1.2). Detailed characterizations and taxonomy of living amphibians and reptiles are given in

TABLE 1.1 A Hierarchial Classification of Vertebrates Showing the Position of the Tetrapoda and Its Subgroups as Members of the Bony Fish Clade.

Vertebrata
- Gnathostomata—jawed vertebrates
  - Osteichthyes (Teleostomi)—bony fishes
    - Actinopterygii—rayfinned fishes
      - Teleostei
    - Sarcopterygii—fleshy-finned fishes
      - Coelacanthiformes (Actinistia)—coelacanths
      - Dipnoi—lungfishes
      - Ostelepiformes—*Eusthenopteron* and relatives
      - Porolepiformes
      - Tetrapoda—tetrapods
        - Ichthyostegalia—*Ichthyostega* and relatives
        - Amphibia—amphibians
          - Colosteidae—*Greererpeton* and relatives
          - Temnospondyli—temnospondyls
            - Lissamphibia—extant amphibians
        - Anthracosauria—anthracosaurs
          - Amniota—amniotes
            - Reptilia (Sauropsida)—reptiles
            - Synapsida—synapsids

Note: The origin of the tetrapods among the sarcopterygians is presented as unresolved. Category titles are not assigned to the hierarchical ranks, and some ranks or nodes are absent. Alternate group names are in parentheses; these alternates are nearly equivalent but not identical in taxa content. Differences between this classification and that derived from Fig. 1.1 result from a combination of different sets of taxa, characters, and analyses.

Part VII. **Caecilians** superficially resemble earthworms (Fig. 1.2). All extant caecilians lack limbs, are strongly annulated, and have wedge-shaped, heavily ossified heads and blunt tails. This morphology reflects the burrowing lifestyle of these tropical amphibians. **Salamanders** have cylindrical bodies, long tails, distinct heads and necks, and well-developed limbs, although a few salamanders have greatly reduced limbs or even have lost the hindlimbs (Fig. 1.2). Salamanders are ecologically diverse: Some are totally aquatic, some burrow, many are terrestrial, and many others are arboreal, living in epiphytes in forest canopy. **Frogs** are unlike other vertebrates in having robust, tailless bodies with a continuous head and body and well-developed limbs (Fig. 1.2). The hindlimbs typically are nearly twice the length of the body, and their morphology reflects their bipedal saltatory locomotion. Not all frogs jump or even hop; some are totally aquatic and use a synchronous hindlimb kick for propulsion, whereas others simply walk in their terrestrial and arboreal habitats. Among amphibians, frogs are the most speciose and widely distributed group; in addition, they are morphologically, physiologically, and ecologically diverse.

Living reptiles are represented by the clades Archosauria (turtles, crocodilians, and birds), and Lepidosauria (tuataras and squamates) (Table 1.2). Until recently, turtles were considered as the outgroup to all other reptiles because their skulls have no fenestre (openings), which placed them with the anapsids, an extinct and very old group of reptiles. Recent nuclear DNA data indicate that their "anapsid" skull condition may be derived from a diapsid skull and thus they belong in with crocodilians and birds. **Turtles,** like frogs, cannot be mistaken for any other animal (Fig. 1.3). The body is encased within upper and lower bony shells (carapace and plastron, respectively). In some species, the upper and lower shells fit tightly together, completely protecting the limbs and head. Although turtles are only moderately speciose, they are ecologically diverse, with some fully aquatic (except for egg deposition) and others fully terrestrial. Some are tiny in size whereas others are gigantic, and some are herbivores and others are carnivores. Other living archosaurs include the closely related crocodilians and birds. Birds are reptiles because they originated within Archosauria, but they have traditionally been treated as a separate group of vertebrates. **Crocodilians** are predaceous, semiaquatic reptiles that swim with strong undulatory strokes of a powerful tail and are armored by thick epidermal plates underlain dorsally by bone. The head, body, and tail are elongate, and the limbs are short and strong. The limbs allow mobility on land, although terrestrial activities are usually limited to basking and nesting.

Tuataras and the squamates comprise the Lepidosauria. Represented by only two species on islands off the coast of New Zealand, the lizard-like **tuataras** diverged early within the lepidosaurian clade (Fig. 1.3). Lizards, snakes, and amphisbaenians comprise the Squamata. These three groups are easily recognized and, as a result, are often treated in popular literature and field guides as though they are sister taxa or at least equal-rank clades. They are not. Snakes and amphisbaenians are nested within lizards (see Chapters 20 and 21). Squamates are the most diverse and speciose of living reptiles, occupying habitats ranging from tropical oceans to temperate mountaintops. Body forms and sizes vary considerably. Most are terrestrial or arboreal, though many snakes are semiaquatic, spending much of their lives in or immediately adjacent to freshwater, or, less commonly, in estuaries and seawater. The term "lizard" is usually used to refer to all squamates that are not snakes or amphisbaenians. Thus "lizards" are highly variable morphologically and ecologically, but most have four well-developed limbs and an elongate tail. Amphisbaenians are elongate with short, stubby tails, scales arranged in rings around the body, and mostly limbless (the exception is *Bipes*, which has two mole-like front limbs). They are subterranean. They are a monophyletic group of lizards. Snakes are the most speciose of several groups of limbless or reduced-limbed lizards. A few snakes are totally aquatic and some are even totally subterranean. Like amphisbaenians, snakes are a monophyletic group of lizards.

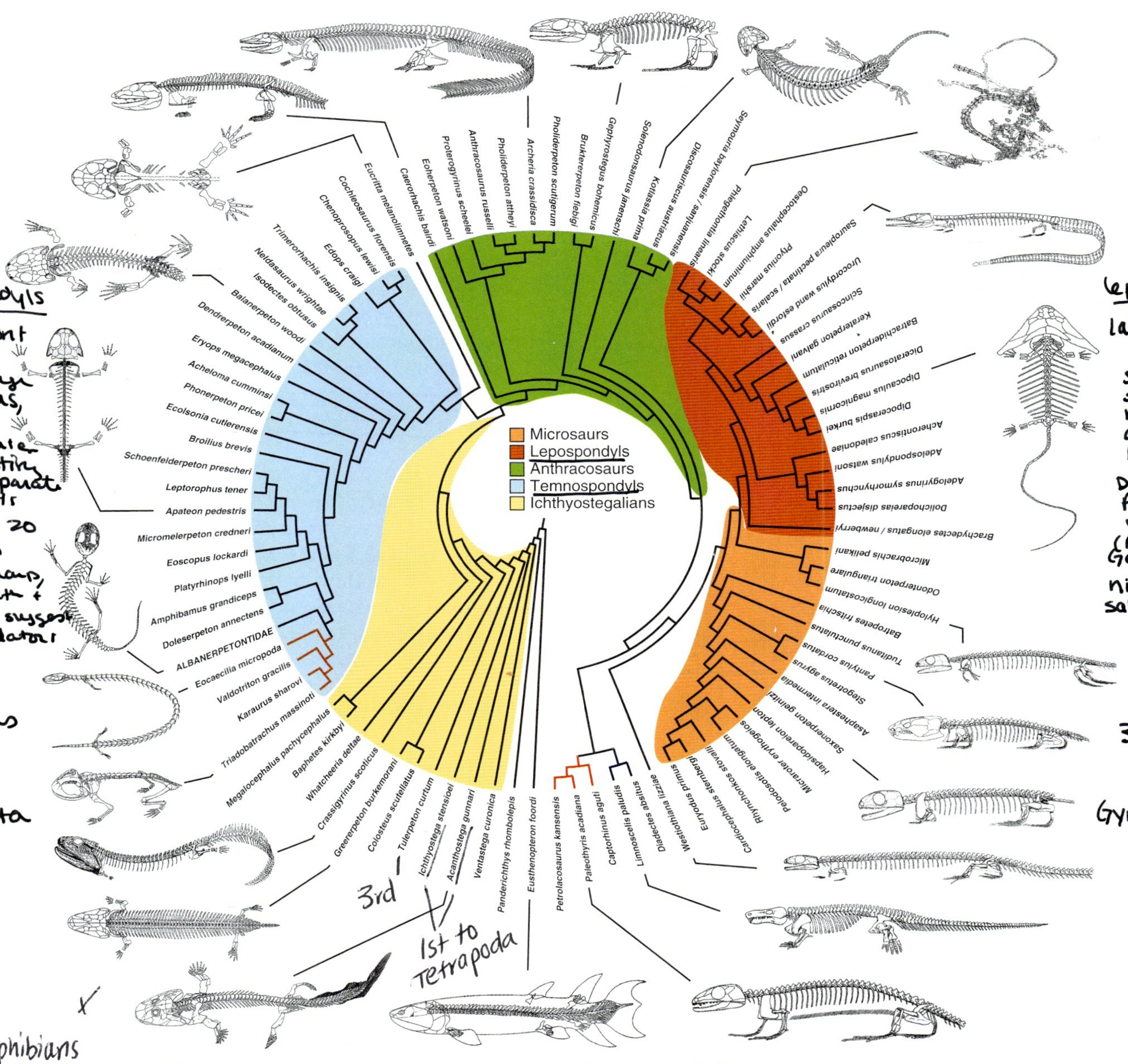

FIGURE 1.1 A super-tree of relationships among early (fossil) tetrapods. To aid in interpreting the structure of the tree, we have color-coded major groups that are discussed in the text. In addition, color-coded lines indicate clades from which extant ampibians (orange) and extant reptiles (red and blue) arose. Orange lines indicate the Lissamphibia, the group from which all extant amphibians originated. Blue lines indicate the Parareptilia, the group from which turtles were once believed to have originated. Although modern turtles have historically been placed in the Parareptilia based on their anapsid skull, recent molecular data indicate that they are nested within the Eureptilia. Red lines indicate the Eureptilia, the group from which all modern reptiles originated. It is useful to refer back to this graphic as you read through the history of tetrapod evolution in order to tie group or fossil names with appropriate evolutionary groups. Adapted from Ruta and Coates, 2003, and Ruta et al., 2003.

RELATIONSHIPS AMONG VERTEBRATES

Origin of Tetrapods

The transition from fish to tetrapod set the stage for one of the most spectacular radiations in the evolutionary history of life, ultimately allowing vertebrates to invade nearly all of Earth's terrestrial environments. Understanding the complexity of the early evolutionary history of tetrapods has been a challenge for paleontologists because

TABLE 1.2 A Hierarchical Classification for Living Amphibians and Reptiles.

Tetrapoda
- Amphibia
  - Microsauria
  - Temnospondylia
    - Lissamphibia
      - Gymnophiona—caecilians
      - Caudata—salamanders
      - Anura—frogs
- Anthracosauria
  - Amniota
    - Synapsida
    - Reptilia
      - Parareptilia
      - Eureptilia
        - Diapsida
          - Sauria
            - Archosauria
              - Testudines—turtles
              - Crocodylia—crocodylians
              - Aves—birds
            - Lepidosauria
              - Sphenodontia—tuataras
              - Squamata—lizards (including amphisbaenians and snakes

Sources: Carroll, 2007; Gauthier et al., 1988a, 1989.

many fossil taxa are represented only by fragments of jaws or limbs, making it difficult to determine phylogenetic relationships. To help orient readers, we recommend that you repeatedly examine Figure 1.1 while reading the text. The first (but not the most primitive) tetrapod found was *Ichthyostega* (Ichthyo = fish; stega = roof). For many years, this abundant fossil and another fossil, *Acanthostega*, represented by a few skull fragments, were the only known tetrapods. In 1985, *Tulerpeton* was discovered in Russia. The next discoveries of tetrapods were made because of a fortuitous event. In 1971, a graduate student conducting a sedimentology project in Greenland collected tetrapods that were placed in a museum but never studied. When these specimens were examined more closely, they were recognized as *Acanthostega*. This discovery led to a resurgence of interest in early tetrapods, and many other fossils present in museums from previous work were reexamined and studied in detail. Additional material of various species made it easier to identify fragments that had not previously been recognized as tetrapods. In addition, new techniques such as CT (computed tomography) scanning allowed reinterpretations of previously collected material. The result of the study of this material led to discarding the original idea that tetrapods evolved from lobe-finned fishes (sarcopterygians) that were forced onto land because of major droughts during the Devonian. The idea was that only those fish that could evolve limbs for terrestrial movement on land survived. Although various scientists challenged this idea, it was not until the discovery of well-preserved material of *Acanthostega* in the late 1980s that a new paradigm of tetrapod evolution became widely accepted. *Acanthostega* was clearly a tetrapod but was not a land animal. It had four limbs with digits, but no wrists and could not have supported itself on land. This realization and a reinterpretation of *Ichthyostega* as a fish with limbs led to the currently accepted idea that tetrapod limbs functioned for locomotion in shallow, vegetated Devonian swamps. Only later did their descendants emerge onto land.

An increase in exploration of Devonian sites has provided new material in recent years, and a much clearer picture of the evolution of this group is emerging (Fig. 1.1). To date, 17 distinct Devonian tetrapods from nine localities worldwide have been discovered, and 12 genera have been described. Other significant discoveries include several new proto-tetrapods and other tetrapods from the Early Carboniferous. The localities and named tetrapod genera include Pennsylvania (*Hynerpeton, Densignathus*); Scotland (*Elginerpeton*); Greenland (*Ichythostega, Acanthostega*); Latvia (*Obruchevichthys, Ventastega*); Tula, Russia (*Tulerpeton*); Livny, Russia (*Jakubsonia*); New South Wales (*Metaxygnathus*); China (*Sinostega*); and Canada (*Tiktaalik*). Most early tetrapods are known from Euramerica, where, in Late Devonian, this land mass was separate from Gondwana. Two species, *Metaxygnathus* from Australia and *Sinostega* from China, are known from Gondwana. It is probable that additional discoveries in northern Gondwana and China will support a global distribution of early tetrapods.

About 30–40 million years (a short time, geologically speaking) after the first tetrapods appeared, two lineages, amphibians and anthracosaurs, gave rise to all extant tetrapods. Reptiles evolved from one descendent lineage of the early anthracosaurs. These evolutionary events occurred in landscapes that appeared alien compared to the familiar landscapes of today. Plants, like animals, were only beginning to radiate into terrestrial environments from a completely aquatic existence. Upland deserts consisted of bare rock and soil. Plants grew only in valleys and along the coasts where water was abundant. Early diversification of terrestrial arthropods was underway, which clearly affected amphibian and reptile diversification by providing a rich and abundant food supply.

We first examine what some of the key fossils tell us. We then summarize some of the morphological, and sensory, respiratory, and feeding system changes that were associated with the invasion of land.

FIGURE 1.2 A sampling of adult body forms in living amphibians.

Although many details are uncertain, five to seven well-known key fossils illustrate the transition from fish to tetrapod (Fig. 1.4). Tetrapods arose from osteolepiform lobe-finned fishes represented in this figure by *Eusthenopteron*. *Panderichthys* and *Tiktaalik* were large, flat predatory fish considered transitional forms between osteolepiform fishes and tetrapods. They had strong limb-like pectoral fins that enabled them to support their bodies and possibly move out of water. *Acanthostega* and *Ichthyostega* were primitive tetrapods. All of these species ranged in size from 0.75 to 1.5 m in length. Many other important fossils from this period exist (e.g., Fig. 1.1), each with its own place in the story of tetrapod evolution, and we refer the interested reader to the paleontological literature for more details on these.

Key Fossils

Because of their importance in reconstructing the evolutionary history of tetrapods, we comment briefly on seven of the key fossil genera, *Eusthenopteron, Panderichthys, Elpistostege, Tiktaalik, Acanthostega, Ichthyostega*, and *Tulerpeton*.

Eusthenopteron—A tristichopterid fish, more or less contemporary with *Acanthostega, Eusthenopteron* is a member of the tetrapod stem group. It is convergent with tetrapods in many respects, including having enlarged pectoral fins, and a flat, elongate snout (Fig. 1.4). As a whole, fishes in this group (also including rhizodontids and osteolepidids) were ambush predators that lived in shallow waters.

Panderichthys—This large Middle Devonian elpistostegalian sarcopterygian fish from Latvia that lived 385 million years ago is the best-known transitional proto-tetrapod. Complete specimens are available from the Middle to Late Devonian. It had a flat head, long snout, and dorsally situated eyes (Fig. 1.4). The tetrapod-like humerus was dorsoventrally flattened, presumably lending strength for support of the body, although the fins have fin rays, not digits. A midline

FIGURE 1.3 A sampling of adult body forms in living reptiles.

fin is present only on the tail. *Panderichthys* was a predatory fish that may have used its fins to "walk" in shallow freshwater swamps.

Elpistostege—This elpistostegalian sarcopterygian fish from the early Late Devonian of Canada is most closely related to *Tiktaalik* (Fig. 1.4). It is known only from skull and backbone fragments, but has long been recognized as an intermediate form. *Elpistostege*, unlike *Tiktaalik*, appears to have occurred in an estuarine habitat, possibly indicating that these fishes as a group were exploiting a variety of habitats.

Tiktaalik—The recent discovery of many specimens of this elpistostegalian sarcopterygian from a single Late Devonian locality in Arctic Canada greatly improved our understanding of the transition to tetrapods within fishes. This species may prove as significant as the well-known *Archaeopteryx*, a fossil that represents the divergence of birds within reptiles. Phylogenetically, *Tiktaalik*, with *Elpistostege*, is apparently sister to *Acanthostega* + *Ichthyostega*. In many ways, *Tiktaalik* was like *Panderichthys*—both had small pelvic fins with fin rays and well-developed gill arches, evidence that both were aquatic (Fig. 1.4). *Tiktaalik* had a combination of primitive and derived features. Primitive features included rhombic, overlapping scales like *Panderichthys*, lack of a dorsal fin, paired pectoral and pelvic fins with lepidotrichia (fin rays), and a generalized lower jaw. Derived features in *Tiktaalik* included a flat body with raised, dorsal eyes, a wide skull, and a mobile neck. The robust forefin and pectoral girdle indicated that it was capable of supporting itself on the substrate. These features represent a radical departure from previously known, more primitive sarcopterygian fishes. Discovery of an intermediate fossil such as *Tiktaalik* helps to visualize the mosaic pattern of morphological changes that occurred during the transition from sarcopterygian fishes to the earliest tetrapods. In fish, breathing and feeding are coupled because taking water in over the gills in a sucking motion also pulls in food. These features became separated in *Tiktaalik*. The longer skull and mobile neck allowed a quick snap of the head to capture prey.

Acanthostega—This primitive transitional Late Devonian tetrapod from Greenland lived 365 million years ago. Study of this best-known tetrapod changed our understanding of early tetrapod evolution. The forelimb clearly had eight digits, but the limb had no wrist

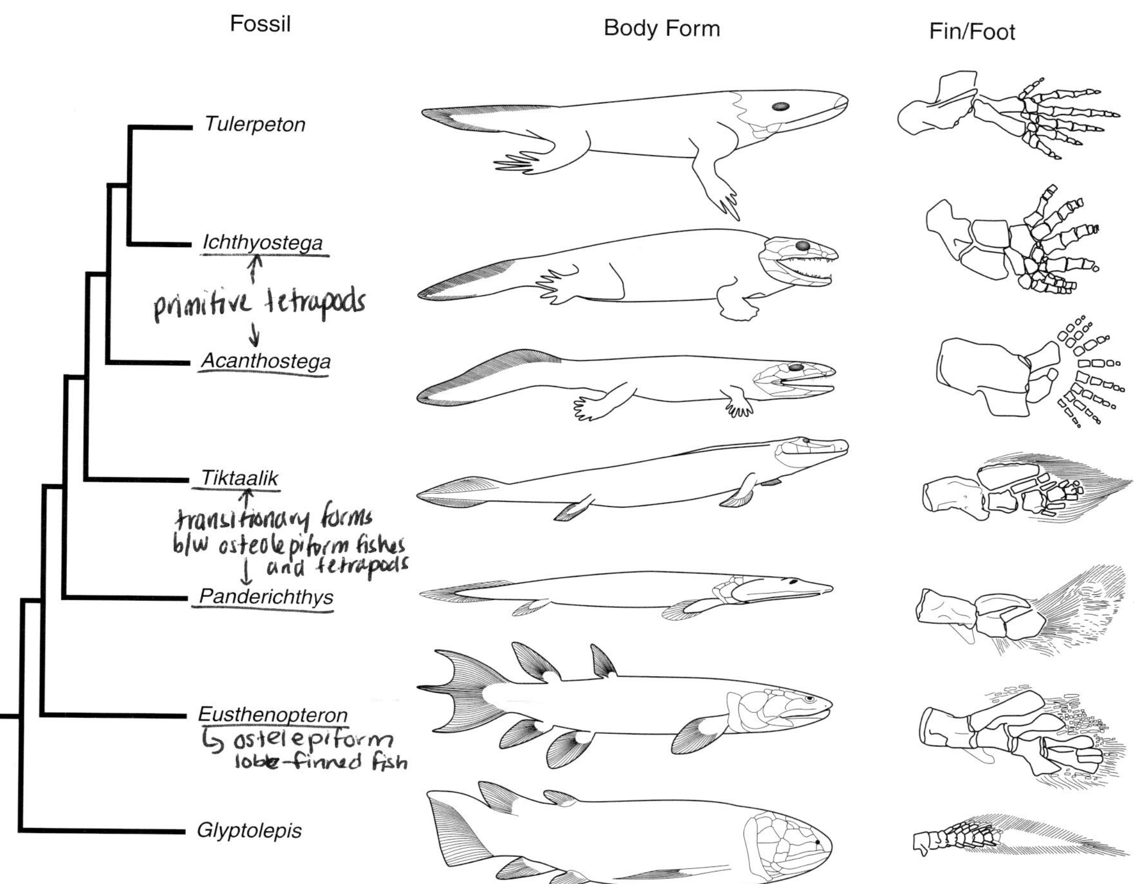

FIGURE 1.4 Relationships, body forms, and limb structure of the seven key fossil vertebrates used to recover the evolution of supportive limbs in tetrapods. *Glyptolepis* is the outgroup. Adapted from Ahlberg and Clack, 2006; Clack 2006; Daeschler et al., 2006; and Schubin et al., 2006.

bones or weight-bearing joints, thus showing that limbs with digits evolved while these animals lived in water and that they most likely did not have the ability to walk (Fig. 1.4). Because the limb is similar to the fish *Eusthenopteron*, it is considered to be primitive. *Acanthostega* had 30 presacral ribs; the fish-like ribs were short and straight and did not enclose the body. It had a true fish tail with fin rays; the tail was long and deep, an indication that it was a powerful swimmer, and it had fish-like gills. Of 41 features unique to tetrapods, *Acanthostega* had two-thirds of them. It had a large stapes that remains as part of the auditory system of more recent tetrapods. The lower jaw of *Acanthostega* bore the inner tooth row on the coronoid bone, a feature indicative of a tetrapod and not a fish. This finding led to a close study of other jaw fragments already present in museums; these jaw fragments could now be distinguished as either fish or tetrapod. *Acanthostega* most likely lived in freshwater rivers.

Ichthyostega—A primitive Late Devonian tetrapod from Greenland, *Ichthyostega* lived 365 million years ago. It had a forelimb with seven digits in a unique pattern. Four main digits formed a paddle bound together by stiff webbing, and three smaller digits formed a leading edge (Fig. 1.4). Twenty-six presacral imbricate ribs were present. It had a true fish tail with fin rays but may have had some ability to move about on land. Based on overall skeletal morphology, *Ichthyostega* likely had some ability for dorsoventral flexion of the spine, and the limbs may have moved together rather than alternately. Preparation of recently collected material revealed that the auditory apparatus is adapted for underwater hearing. *Ichthyostega* may have lived in freshwater streams and may have been able to move about on land to some extent.

Tulerpeton—This primitive Devonian tetrapod from Russia was described in 1984. Both the forelimb and hindlimb had six digits (Fig. 1.4). The robust shoulder joint and slender digits indicate that *Tulerpeton* was less aquatic than either *Acanthostega* or *Ichthyostega*.

Major Features of Early Tetrapod Evolution

The radiation of elpistostegalian fish (*Panderichthys, Elpistostege*, and *Tiktaalik*) indicates that the tetrapod origin was within the Euramerican landmass. *Panderichthys* and *Elpistostege* are found in deltaic and estuarine settings, but *Tiktaalik* is found in nonmarine sediment, indicating that the elpistostegalian fishes were exploiting new habitats. Although changes occurred in nearly all systems during the transition from water to land, it remains difficult to determine which changes preadapted tetrapod ancestors to move to land (exaptation) and which represent true responses (adaptations) to the transition.

Respiration

Lungs appeared early in the evolution of bony fishes, long before any group of fishes had other terrestrial adaptations. Indeed, lungs are the structural predecessors of swim bladders in the advanced fishes. Lungs may have developed as accessory respiratory structures for gaseous exchange in anoxic or low-oxygen waters. The lung structure of the fish–tetrapod ancestor and the earliest tetrapods is unknown because soft tissue does not readily fossilize. Presumably lungs formed as ventral outpocketings of the pharynx, probably with a short trachea leading to either an elongated or a bilobed sac. The internal surface may have been only lightly vascularized because some cutaneous respiration was also possible. Respiration (i.e., ventilation) depended upon water pressure. A fish generally rose to the surface, gulped air, and dived (Fig. 1.5). With the head lower than the body, water pressure compressed the buccal cavity and forced the air rearward into the lungs, since water pressure was lower on the part of the body higher in the water column. Reverse airflow occurred as the fish surfaced headfirst. This mechanism is still used by most air-breathing fish for exhalation. The fossil record provides additional insight on this. Gills were present in the fish–tetrapod ancestor but presumably absent in adult ichthyostegalians. The tetrapodomorph fish *Tiktaalik* occurred in shallow, wandering streams in tropical or subtropical climates. This type of habitat selected for respiratory advances such as the buccal and costal pumping mechanisms employed by tetrapods. The broad skull allows space for the buccal pumping. An enlarged spiracular tract led to respiratory modifications that allowed breathing in aquatic–terrestrial habitats. The buccal-force pump replaced a passive pump mechanism. Air entered through the mouth with the floor depressed, the mouth closed, the floor contracted (elevated) and drove air into the lungs, and the glottis closed, holding the pulmonary air at supra-atmospheric pressure. Exhalation resulted from the elastic recoil of the body wall, driving air outward. Thus respiratory precursors for invasion of land were present in aquatic tetrapod ancestors.

Movement

The transformation of fins to limbs was well underway before early tetrapods moved to land. The cause remains debatable, but fleshy fins seem a prerequisite. The fleshy fins of sarcopterygian fish project outward from the body wall and contain internal skeletal and muscular elements that permit each to serve as a strut or prop. Because limbs evolved for locomotion in water, presumably initially for slow progression along the bottom, they did not support heavy loads because buoyancy reduced body weight. The fin-limbs probably acted like oars, rowing the body forward with the fin tips pushing against the bottom. Shifting from a rowing function to a bottom-walking function required bending of the fin-limb to allow the tip to make broader contact with the substrate (Fig. 1.6). The underlying skeletal structure for this is evident in *Tiktaalik* (Fig. 1.4). Bends or joints would be the sites of the future elbow–knee and wrist–ankle joints. As flexibility of the joints increased, limb segments developed increased mobility and their skeletal and muscular components lost the simple architecture of the fin elements. Perhaps at this stage, fin rays were lost and replaced by short, robust digits, and the pectoral girdle lost its connection with the skull and allowed the head to be lifted while retaining a forward orientation as the limbs extended and retracted. The ichthyostegalians represent this stage. Their limb movements, although in water, must have matched the basic terrestrial walking pattern of extant salamanders, i.e., extension–retraction and rotation of the proximal segment, rotation of the middle segment (forearm and crus), and flexure of the distal segment (feet). As tetrapods became increasingly terrestrial, the vertebral column became a sturdier arch with stronger intervertebral links, muscular as well as skeletal. The limb girdles also became supportive—the pelvic girdle by a direct connection to the vertebral column and the pectoral girdle through a strong muscular sling connected to the skin and vertebral column. Although stem tetrapods (e.g., *Acanthostega*) had more than five digits, pentadactyly (five digits) predominates in descendent clades. The evolution of pentadactyly and terrestriality appear closely linked. The recently discovered *Pederpes finneyae*, a terrestrial tetrapod from the end of the Early Carboniferous, probably has the earliest hindlimb capable of walking.

Feeding

The presence of a functional neck in *Tiktaalik* provides some insight into the early evolution of inertial feeding, in which the mouth–head of the tetrapod must move forward over the food. While in the water, the fluidity and resistance of water assisted in grasping and swallowing food. In shallow water or out of water, the ability to move the head would provide a substantial advantage

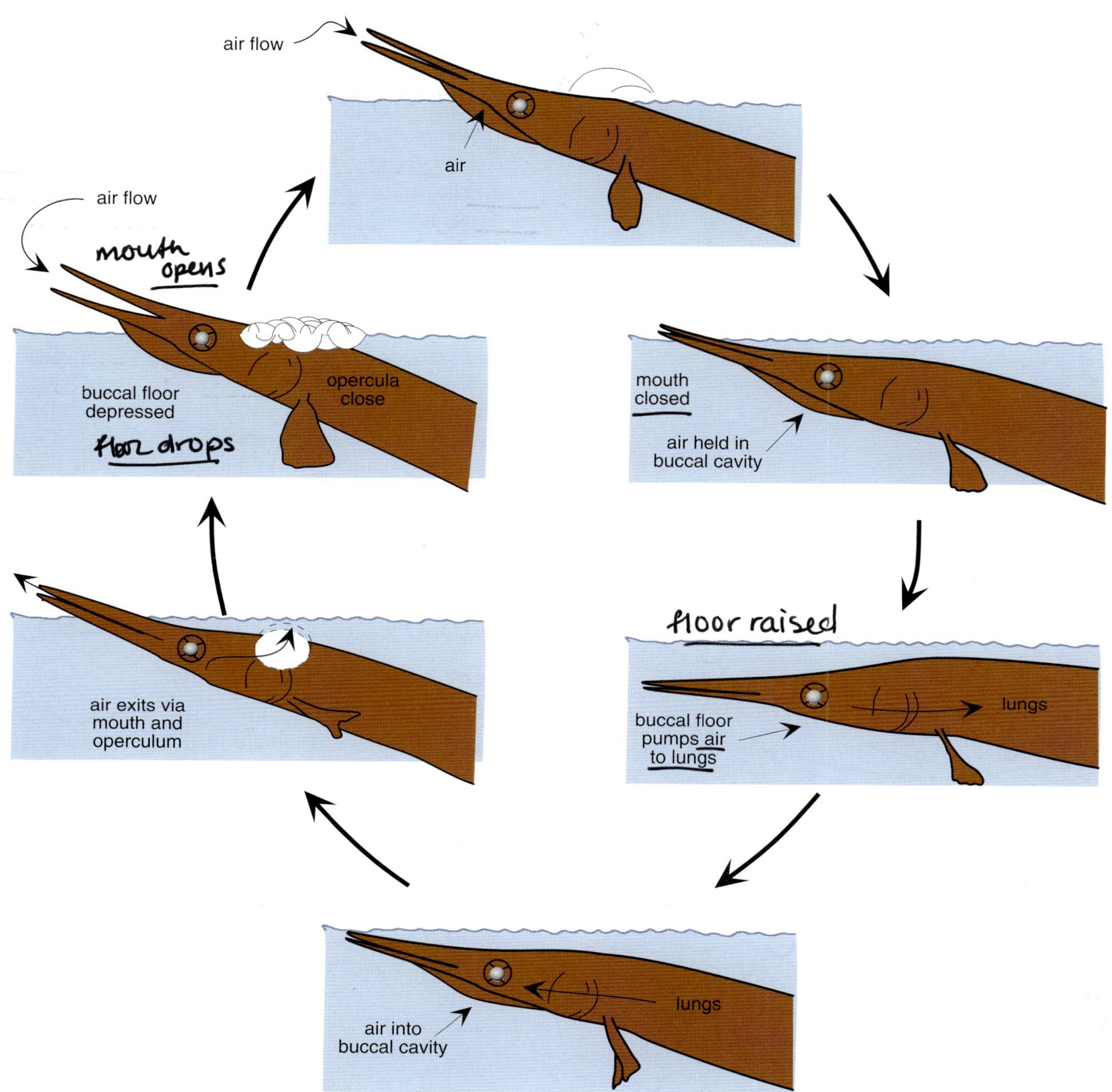

FIGURE 1.5 Air-breathing cycle of the longnosed gar (*Lepisosteus osseus*). As the gar approaches the surface at an angle, it drops its buccal floor and opens its glottis so air can escape from the lungs (bottom center, clockwise). By depressing the buccal floor, the gar flushes additional air from the opercular chamber. Once flushed, the gar extends its snout further out of the water, opens its mouth, depresses the buccal floor drawing air into the buccal cavity, and shuts the opercula. The mouth remains open and the floor is depressed further; then closing its mouth, the gar sinks below the surface. Air is pumped into the lungs by elevating the buccal floor. Adapted from Smatresk, 1994.

in capturing prey. Several modifications of the skull may have been associated with this feeding behavior. With the independence of the pectoral girdle and skull, the skull could move left and right, and up and down on the occipital condyles–atlas articulation. The snout and jaws elongated. The intracranial joint locked and the primary palate became a broader and solid bony plate.

Skin

The skin of larval amphibians and fish is similar. The epidermis is two to three layers thick and protected by a mucous coat secreted by numerous unicellular mucous cells (Chapter 2). The skin of adult amphibians differs from that of fish ancestors. The epidermis increased in thickness to five to seven layers; the basal two layers are composed of

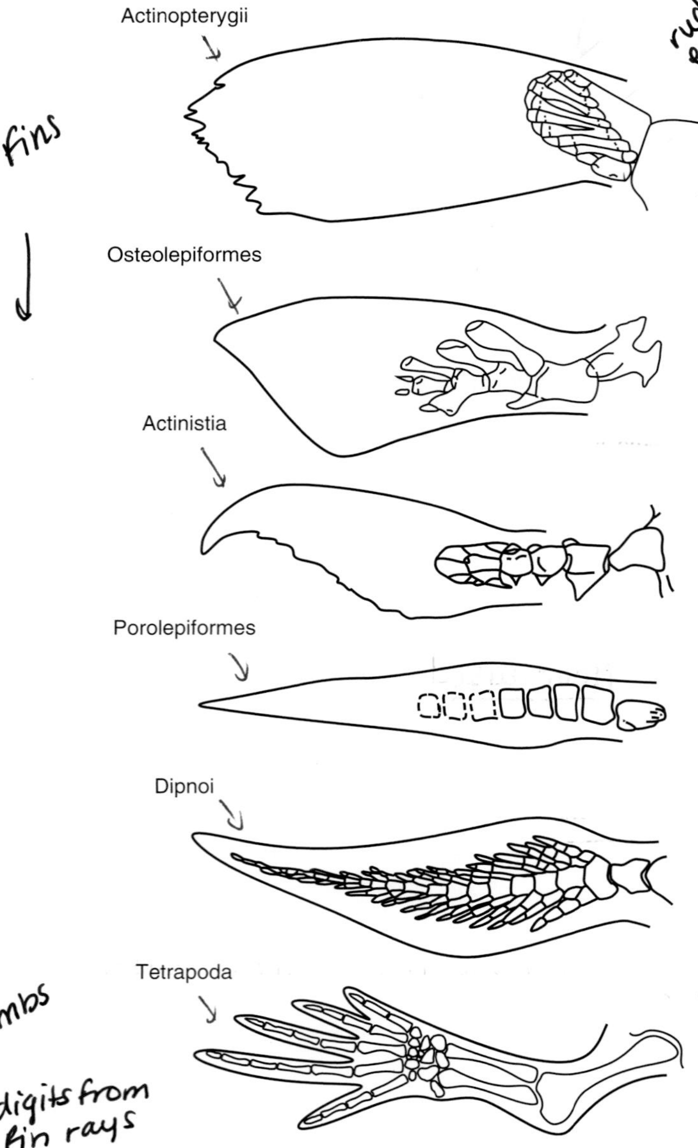

FIGURE 1.6 Fin and limb skeletons of some representative fishes and tetrapods. Top to bottom, ray-finned or actinopterygian fin, osteolepiform lobed fin, actinistian lobed fin, porolepiform lobed fin, lungfish or dipnoan lobed fin, and a tetrapod limb. Adapted from Schultze, 1991.

living cells and are equivalent to fish or larval epidermis. The external layers undergo keratinization and the mucoid cuticle persists between the basal and keratinized layers. Increased keratinization may have appeared as a protection against abrasion, because terrestrial habitats and the low body posture of the early tetrapods exposed the body to constant contact with the substrate and the probability of greater and frequent surface damage.

Sense Organs

As tetrapods became more terrestrial, sense organs shifted from aquatic to aerial perception. Lateral line and electric organs function only in water and occur only in the aquatic phase of the life cycle or in aquatic species. Hearing and middle ear structures appeared. The early tetrapod *Ichthyostega* has a unique specialized ear, and the middle ear was modified in early tetrapods. Changes in eye structure evolved in early tetrapods sharpening their focus for aerial vision. The nasal passages became a dual channel, with air passages for respiration and areas on the surfaces modified for olfaction.

The preceding summarizes the major anatomical alterations that occurred in the transition to tetrapods within fishes. Many physiological modifications also occurred; some of these are described in Chapter 6. Some aspects, like reproduction, remained fishlike: external fertilization, eggs encased in gelatinous capsules, and larvae with gills. Metamorphosis from the aquatic larval to a semiaquatic adult stage was a new developmental feature. The unique morphological innovations in the stem tetrapods illustrate the divergent morphology and presumably diverse ecology of these species. This diversification was a major feature of the transition from water to land.

EVOLUTION OF EARLY ANAMNIOTES

Ancient Amphibians

Given the existing fossil record, clearly defining Amphibia is a challenge. The monophyly of living amphibians, the Lissamphibia (caecilians, frogs, and salamanders) seems highly probable, and they are members of the temnospondyl clade. The assignment of extinct taxa to the Temnospondyli is more controversial. *Edops* (Fig. 1.7) and relatives, *Eryops* and relatives, trimerorhachoids, and a diverse assortment of taxa labeled dissorophoids make up the major groups of extinct temnospondyls. Aistopods, baphetids (= Loxommatidae), microsaurs, and nectrides

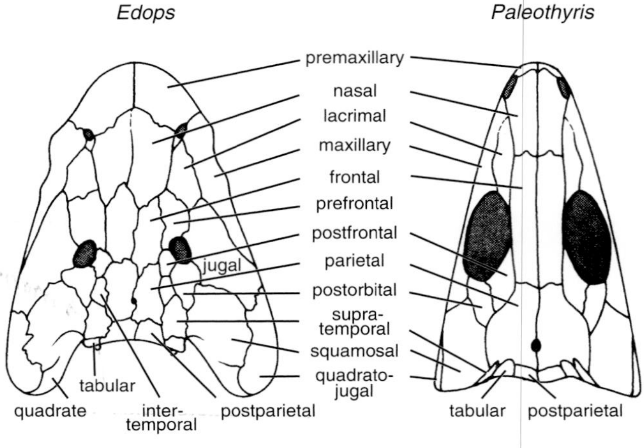

FIGURE 1.7 Comparison of the skulls of an early amphibian *Edops* and an early reptile *Paleothyris*. Scale: bar = 1 cm. Reproduced, with permission, from Museum of Comparative Zoology, Harvard University.

have been identified as amphibians, although their relationships remain controversial. The baphetids are not amphibians; presumably they are an early offshoot of the early protoamniotes and possibly the sister group of the anthracosaurs. Details on the appearance and presumed lifestyles of these extinct groups are provided in Chapter 3. All these groups may have had their origins in the Early Carboniferous, and only a few lineages survived and prospered into the Permian. As an aside, the lepospondyls and labyrinthodonts were once widely recognized groups of extinct amphibians. The members of the lepospondyls (= Aistopoda + Microsauria + Nectridea) shared features associated with small body size and aquatic behavior, but not features of phylogenetic relatedness that would support the monophyly of lepospondyls (Fig. 1.1). The labyrinthodonts encompassed phylogenetically unrelated taxa united by shared primitive (ancestral) characters. Thus, the groups were polyphyletic and their use has been largely discontinued.

By defining Amphibia by its members, it is possible to identify unique characters shared by this group. These characters are surprisingly few: (1) the articular surface of the atlas (cervical vertebra) is convex; (2) the exoccipital bones have a suture articulation to the dermal roofing bones; and (3) the hand (manus) has four digits and the foot (pes) five digits. Other features commonly used to characterize amphibians apply specifically to the lissamphibians, although some of them may apply to all Amphibia but are untestable because they are soft anatomical structures that have left no fossil record.

Modern Amphibians—The Lissamphibia

The living amphibians are generally thought to share a common ancestor and, hence, to represent a monophyletic group. Numerous patterns of relationship have been proposed, but only three patterns continue to have time-tested evidence and advocates. The proposed patterns are (1) frogs arose from a different ancestor than salamanders and caecilians; (2) frogs and salamanders are a sister group, and caecilians are a sister group to their clade; (3) caecilians and salamanders are a sister group, and frogs are a sister group to their clade. The preponderance of evidence argues for monophyly and probably a dissorophid ancestry. Remaining uncertainty derives from the long time gap between the potential temnospondyl ancestors in the Upper Carboniferous and the occurrence of all three groups in the Early to Middle Jurassic. These Jurassic forms show the major characteristics of their extant descendants but few traits of ancient amphibians. Only the Lower Triassic frog *Triadobatrachus massinoti*, from Madagascar, shows a possible link to the dissorophid temnospondyls. *T. massinoti* shares with them a large lacuna in the squamosal bone that may have housed a tympanum. Neither salamanders nor caecilians have tympana, although they have greatly reduced middle ears, suggesting the loss of the outer ear structures.

A number of other unique traits argue strongly for the monophyly of the Lissamphibia. All share a reliance on cutaneous respiration, a pair of sensory papillae in the inner ear, two sound transmission channels in the inner ear, specialized visual cells in the retina, pedicellate teeth, the presence of two types of skin glands, and several other unique traits.

Three structures, gills, lungs, and skin, serve as respiratory surfaces in lissamphibians; two of them frequently function simultaneously. Aquatic amphibians, particularly larvae, use gills; terrestrial forms use lungs. In both air and water, the skin plays a major role in transfer of oxygen and carbon dioxide. One group of terrestrial amphibians, the plethodontid salamanders, has lost lungs, and some aquatic taxa also have lost lungs or have greatly reduced ones; these amphibians rely entirely on cutaneous respiration. All lunged species use a force-pump mechanism for moving air in and out of the lungs. Two types of skin glands are present in all living amphibians: mucous and granular (poison) glands. The mucous glands continuously secrete mucopolysaccharides, which keep the skin surface moist for cutaneous respiration. Although the structure of the poison glands is identical in all amphibians, the toxicity of the diverse secretions produced is highly variable, ranging from barely irritating to lethal to predators.

The auditory system of amphibians has one channel that is common to all tetrapods, the stapes–basilar papilla channel. The other channel, the opercular–amphibian papilla, allows the reception of low-frequency sounds (< 1000 Hz). The possession of two types of receptors may not seem peculiar for frogs because they are vocal animals. For the largely mute salamanders, a dual hearing system seems peculiar and redundant. Salamanders and frogs have green rods in the retina; these structures are presumably absent in the degenerate-eyed caecilians. Green rods are found only in amphibians, and their particular function remains unknown.

The teeth of modern amphibians are two-part structures: an elongate base (pedicel) is anchored in the jawbone and a crown protrudes above the gum. Each tooth is usually constricted where the crown attaches to the pedicel. As the crowns wear down, they break free at the constriction and are replaced by a new crown emerging from within the pedicel. Only a few living amphibians lack pedicellate teeth. Among extinct amphibians, pedicellate teeth occur in only a few dissorophids.

Living amphibians possess other unique traits. All have fat bodies that develop from the germinal ridge of the embryo and retain an association with the gonads in adults. Frogs and salamanders are the only vertebrates able to raise and lower their eyes. The bony orbit of all amphibians opens into the roof of the mouth. A special muscle stretched across this opening elevates the eye. The ribs of amphibians do not encircle the body.

This large number of unique similarities argues strongly for the shared ancestry of the living amphibian groups. Whether salamanders and frogs or salamanders and caecilians are sister groups remains unresolved; different data sets and analyses support one or the other of these pairs, but not a frog–caecilian sister relationship.

EVOLUTION OF EARLY AMNIOTES

Early Tetrapods and Terrestriality

Fully terrestrial tetrapods presumably arose in the Early to Middle Mississippian period (360–340 mybp; Lower Carboniferous). Uncertainty arises because few tetrapod fossils are known from this period. Tetrapod fossils appear with high diversity in the Late Mississippian and Early Pennsylvanian (340–320 mybp). The diversity includes the first radiation of the amphibians and the appearance of the anthracosaurs and the earliest amniotes. This interval saw the emergence of waterside from shallow-water forms and to increasingly abundant and diverse terrestrial forms. Unlike the largely barren landscape of the Late Devonian during the transition from fish to tetrapod, Carboniferous forests were widespread, composed of trees 10 m, and taller, probably with dense understories. Plant communities were beginning to move into upland areas. While plants diversified on land, invertebrates and vertebrates were also evolving terrestrial residents.

The evolution of terrestrial forms required modifications in anatomy, physiology, behavior, and a host of other characteristics. True terrestriality required major reorganizations of lifestyle and life processes. The shifts from eggs that required water or moisture for deposition to those that could withstand dry conditions and from free-living embryos to direct development was critical in the move to land, but other adaptations were also required. Movement and support without the support of water required adjustments in the musculoskeletal system. Feeding in air required behavioral and morphological shifts, as did the use of different prey and plant materials for food. Gravity, friction, abrasion, and evaporation obligated modification of the integument for protection and support and internal mechanisms to regulate water gain and loss. Modification was not confined to the preceding anatomical and physiological systems. These changes did not occur synchronously; some were linked, others were not; some required little modification because of exaptation ("preadaptation"), and others required major reorganization. The diversity of changes is reflected in the diversity of Lower Carboniferous amphibians and the polyphyletic anthracosaurs.

Amphibians remained associated with aquatic habitats, and several independent clades moved at least partially to land. Many of these were successful in terms of high abundance or diversity and geologic longevity. Nevertheless, amphibians remained tied to moisture. As amphibians diversified in association with aquatic habitats, an increasing number of anthracosaurs and their descendants shifted to terrestriality in all phases of their life (Fig. 1.8; Table 1.3). These are represented today by the amniotes (Amniota).

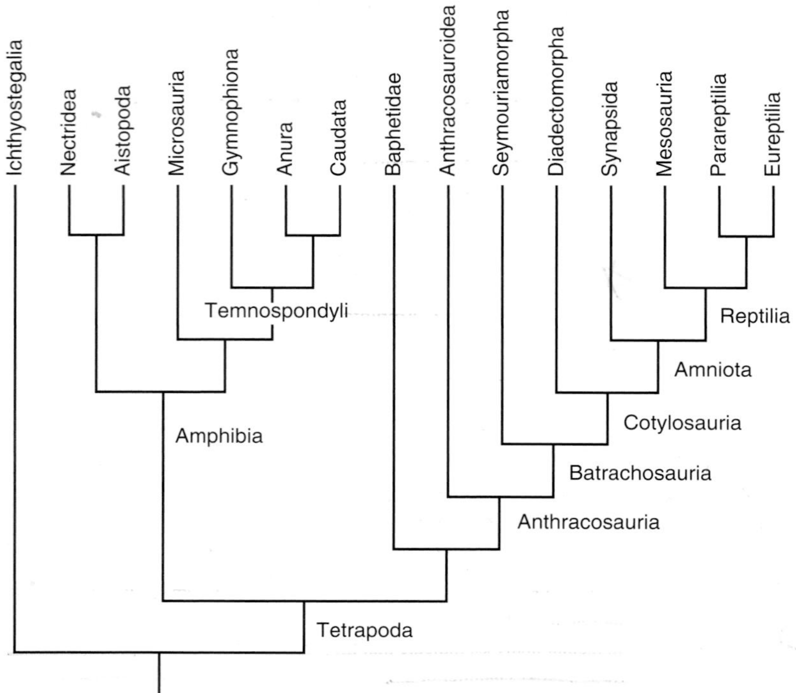

FIGURE 1.8 A branching diagram of the evolution within the Tetrapoda, based on sister group relationships. The diagram has no time axis, and each name represents a formal clade-group name. After Clack, 1998; Gauthier et al., 1988a,b, 1989; and Lombard and Sumida, 1992; a strikingly different pattern is suggested by Laurin and Reisz, 1997.

TABLE 1.3 A Hierarchical Classification of Anthracosaur Descendents.

Tetrapoda
- Amphibia
- Anthracosauria
  - Anthracosauroidea
  - Batrachosauria
    - Seymouriamorpha
    - Cotylosauria
      - Diadectomorpha
      - Amniota
        - Synapsida
        - Reptilia

Note: This classification derives from the sister-group relationships displayed in Figure 1.7. Because of the hierarchial arrangement, a reptile or mammal is an anthracosaur, although paleontologists commonly use anthracosaur to refer to the extinct tetrapod groups that are not Amphibia but also likely not Amniota.

Full terrestriality required that organisms have the ability to reproduce and develop without freestanding water. The evolution of the amniotic egg, which could be deposited on land and could resist dehydration, solved this problem (see Chapter 2 for anatomical details; note that many reptilian eggs still must absorb moisture to complete development). Internal fertilization set the stage for production of closed (shelled) eggs. By enclosing an embryo in a sealed chamber (shelled egg), the evolution of extraembryonic membranes not only provided embryos with protection from the physical environment, but also provided a reservoir for metabolic waste products.

Internal fertilization is not a prerequisite for direct development, nor does direct development free the parents from seeking an aquatic or permanently moist site for egg deposition. Among extant amphibians, internal fertilization predominates in caecilians and salamanders, but only a few anurans with direct development have internal fertilization. When an egg is encased in a protective envelope, the encasing process must be done inside the female's reproductive tract, and if sperm is to reach the egg–ovum surface, the sperm must be placed within the female's reproductive tract as well. Sperm delivery and fertilization must precede egg encasement.

Because internal fertilization has arisen independently in the three extant amphibian clades, it is reasonable to assume that internal fertilization could easily arise in protoamniote anthracosaurs. One problem with the fossil record for early tetrapods is that anamniotic eggs do not readily fossilize (there are no hard parts), and as a consequence it is difficult to reconstruct events leading to the evolution of internal fertilization and the shift to shelled eggs. The common scenario suggests that naked amniotic eggs with direct development were laid first in moist areas. Selection to reduce predation by microorganisms drove the replacement of gelatinous capsules by the deposition of an increasingly thicker calcareous shell and the shift of egg deposition to drier sites. Recent modification of this hypothesis has placed more emphasis on the development of the fibrous envelope precursor to the shell and the supportive role of such an envelope for a large-yolked egg. Other scenarios, such as the "private pool" theory, have directed attention to the development of the extraembryonic membranes and their encapsulation of the egg or embryo. Each hypothesis has a facet that reflects an aspect of the actual evolutionary history, but none provides a full explanation. Lacking historical data (fossils), we cannot determine whether the amniotic membranes evolved in embryos held within the female's oviduct or whether they evolved in externally shed eggs. Either explanation is equally parsimonious from available information on extant vertebrates (Fig. 1.9). Similarly, we cannot determine when and how a fibrous envelope replaced the sarcopterygian's gelatinous envelope, although a fibrous "shell" likely preceded a calcareous one because calcium crystals are deposited in a fibrous matrix in all living reptiles.

Juveniles and adults also required a protective envelope because of the desiccative effect of terrestrial life.

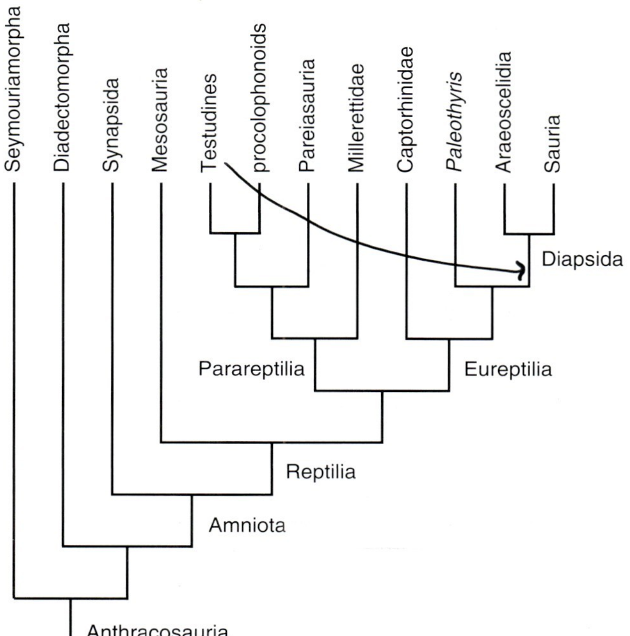

FIGURE 1.9 A branching diagram of the evolution of basal Amniota and early reptiles, based on sister group relationships. The diagram has no time axis, and each capitalized name represents a formal clade-group name. Opinion varies on whether the mesosaurs are members of the Reptilia clade or the sister group of Reptilia. If the latter hypothesis is accepted, the Mesosauria and Reptilia comprise the Sauropsida. Turtles (Testudines) are shown here as nested within the Parareptilia based on morphology. More recent molecular analyses indicate that they are nested in the Eureptilia (see Chapter 18). After Gauthier et al., 1989; Laurin and Reisz, 1995; and Lee, 1997; a strikingly different pattern is suggested by deBraga and Rieppel, 1997.

Changes in skin structure are invisible in the fossil record, but the skin of present-day amphibians suggests that the initial evolutionary steps were an increase in skin thickness by adding more cell layers and keratinization of the external-most layer(s). Keratinization of skin effectively reduces frictional damage and the penetration of foreign objects but appears to be ineffectual in reducing water loss. Early modifications of the integument were also driven by its increased role in the support of internal organs to compensate for the loss of buoyancy and compression of water. These changes occurred in deep dermal layers and involved altering fiber direction and layering.

Associated with increasingly impermeable skin (effectively reducing cutaneous respiration) was the shift to more effective pulmonary respiration. The first modifications of lungs were probably an increase in size and internal partitioning. The latter is commonly associated with increased vascularization. Once again, these modifications apparently occurred in the protoamniotes. When and where they occurred can be partially identified by examining rib structure and the appearance of a complete rib cage. A rib cage (thoracic basket) signals the use of a thoracic respiratory pump for ventilation of the lungs. The rib cage appears incomplete in most anthracosaurs and seymouriamorphs, so those groups probably were still largely dependent on the buccal force pump. The rib cage of diadectomorphs extends further ventrally; although it still appears incomplete, this condition may mark the transition from buccal to thoracic ventilation.

Anthracosaurs and early amniotes lacked otic notches, denoting the absence of eardrums. Although not deaf, they were certainly insensitive to high-frequency sounds. It is doubtful that their olfactory sense was as limited. Well-developed nasal passages in fossils and the presence of highly developed olfactory organs in living reptiles indicate that this sense was well developed in the earliest amniotes. Nasal passages contained conchae, which may have aided in the reduction of water loss. Eyes were also likely well developed at this stage, for vision is extremely important in foraging and avoiding predators in an aerial environment.

Locomotory and postural changes for a terrestrial life are reflected in numerous changes in the postcranial skeleton. Vertebral structure changed to produce a more robust supporting arch. The pleurocentrum became the main component of the vertebral body, displacing the intercentrum forward and upward. Neural arches became broader, zygapophyses tilted, and regionalization of neural spine height occurred, yielding differential regional flexibility with an overall strengthening of the vertebral column. Modification of the two anterior-most cervical vertebrae (atlas–axis complex) stabilized lateral head movement during walking and running. Modifications in the limb and girdle skeletons are not as evident in the early anthracosaurs as those appearing in later amniotes. The humerus remained a robust polyhedral element that had a screwlike articulation with the glenoid fossa. The shoulder or pectoral girdle lost dermal bone elements but remained large. The iliosacral articulation was variable and depended upon the size and robustness of the species, although two sacral ribs usually attached to each ilium. Hindlimbs commonly were larger and sturdier, demonstrating their increasing role in propulsion.

The skull became more compact and tightly linked, although it was still massive in many anthracosaurs and early amniotes (Fig. 1.7). A major trend was the reduction of the otic capsule in early tetrapods, without the concurrent development of structural struts; thus, the skull roof and braincase became weakly linked. Different strengthening mechanisms appeared in different lineages. The diadectomorphs and reptiles shared the unique development of a large supraoccipital bone to link the braincase and skull roof. The cheek to braincase solidification occurred in three general patterns within the amniotes. The anapsids developed a strong attachment of the parietal (skull roof) to the squamosal (cheek) along with a broad and rigid supraoccipital attachment. In the diapsids, the opisthotic extended laterally to link the braincase to the cheek. A lateral expansion of the opisthotic also occurred in the synapsids but in a different manner.

The robust stapes with its broad foot plate was a critical strut in the strengthening of the skull. This role as a supportive strut precluded its function as an impedance matching system (see the discussion of ears in Chapter 2). Later, the opisthotic became the supportive unit, and the stapes (columella) became smaller and took on its auditory role. This change occurred independently in several reptilian lineages; although the results are the same, the evolutionary route to the middle ear of turtles differed from that of the archosaurs and lepidosaurs. The synapsids followed an entirely different route and evolved the unique three-element middle ear seen today in mammals.

Early Amniotes

The Amniota derives its name from the amniotic egg, a synapomorphy shared by all members (Fig. 1.9 and Fig. 1.10). Other anthracosaurs may have had amniotic eggs, although they are not classified as amniotes. A fossil taxon cannot be identified as an amniote or anamniote by structure of its egg, because few fossil eggs of anthracosaurs have been found. Further, no eggs have been found in association with an adult's skeleton or with a fossil embryo showing extra-embryonic membranes. Bony traits must be used to determine which taxa are amniotes and which ones are not, and there is no unanimity in which bony traits define an amniote. Indeed, amniotes are commonly defined by content; for example, Amniota comprise the most recent common ancestor of mammals and reptiles and all of its descendants.

Unquestionably, anthracosaurs are the ancestral stock that gave rise to the amniotes (Fig. 1.1 and Fig. 1.9). They

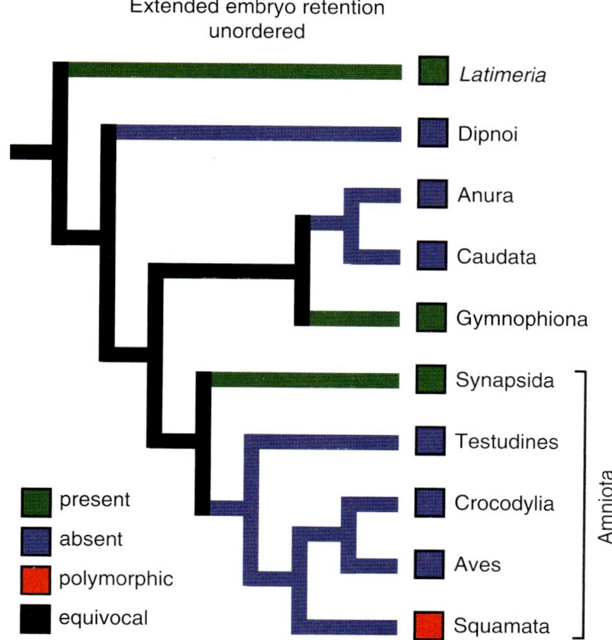

FIGURE 1.10 Presence of the amnion defines the Amniota. Viviparity is not necessarily associated with presence of an amnion. This distribution of egg-retention based on extant species does not permit the identification of the condition in basal amniotes. The origin of terrestrial amniotic eggs as an intermediate stage is equally parsimonious with the evolution of amniotic eggs within the oviduct to facilitate extended egg retention. After Laurin and Reisz, 1997.

possess features present in amniotes but not in Paleozoic or later amphibians. Anthracosaurs and amniotes share such features as a multipartite atlas–axis complex in which the pleurocentral element provides the major support. Both possess five-toed forefeet with a phalangeal formula of 2,3,4,5,3 and a single, large pleurocentrum for each vertebra. These traits are also present in the seymouriamorphs and diadectomorphs.

The seymouriamorphs are an early divergent group of anthracosaurs, although their fossil history does not begin until the Late Pennsylvanian. These small tetrapods, sometimes incorrectly called amphibians, probably had external development and required water for reproduction. Significantly, neither the seymouriamorphs nor the diadectomorphs are amniotes (Fig. 1.9).

The diadectomorphs shared a number of specialized (derived) features with early amniotes—traits that are not present in their predecessors. For example, both groups lost temporal notches from their skulls, have a fully differentiated atlas–axis complex with fusion of the two centra in adults, and possess a pair of sacral vertebrae. They share a large, platelike supraoccipital bone and a number of small cranial bones (supratemporal, tabulars, and postparietals) that are lost in advanced reptiles. The stapes of both were stout bones with large foot plates, and apparently eardrums (tympana) were absent. These latter features do not suggest that they were deaf, but that their hearing was restricted to low frequencies, probably less than 1000 Hz, much like modern-day snakes and other reptiles without eardrums. Possibly their development included preamniotic changes, such as partitioning of the fertilized egg into embryonic and extraembryonic regions, or even a full amniotic state.

The first amniote fossils are from the Middle Pennsylvanian, but they are not primitive amniotes in the sense of displaying numerous transitional traits. These first amniotes are *Archaeothyris* (a synapsid), *Hylonomus* (a reptile), and *Paleothyris* (a reptile; Fig. 1.9); already the divergence of the synapsids and reptilian stocks was evident. The Synapsida is the clade represented today by mammals; they are commonly called the mammal-like reptiles, an inappropriate and misleading name. The pelycosaurs were the first major radiation of synapsids and perhaps gave rise to the ancestor of the Therapsida, the lineage leading to modern mammals.

Divergence among the basal reptiles apparently occurred soon after the origin of the synapsids, and again because of the absence of early forms and the later appearance of highly derived reptilian clades, there is uncertainty and controversy about the early evolutionary history of the reptiles. The Mesosauria of the Lower Permian are considered a sister group to all other reptiles or a sister group to all other parareptiles (Fig. 1.9). Mesosaurs were specialized marine predators, and their specializations have provided few clues to their relationships to other early reptiles.

Controversy surrounds the origin of turtles and whether the Parareptilia is paraphyletic or monophyletic. Recent discoveries and better preparation of old and new fossils have led to a redefinition of the Parareptilia and to its recognition as a clade including the millerettids, pareiasaurs, procolophonoids, and turtles. The latter two taxa are considered to be sister groups. However, another interpretation recognizes pareiasaurs and turtles as sister groups. A strikingly different interpretation considers the turtles as diapsids and further suggests a moderately close relationship to lepidosaurs. Molecular data support the diapsid relationship by yielding a turtle–archosaur (crocodylian + bird) sister-group relationship or a turtle–crocodylian one. These data support the idea that turtles are more closely related to other living reptiles than to living mammals, but they do not provide information on the early history of reptile evolution. As noted earlier in the discussion of fish–tetrapod relationships, molecular data yield a simple phylogeny of living taxa only. Relationships of extinct taxa and their sequence of divergence based strictly on morphology add complexity to phylogenies and often reveal relationships different from molecular-based phylogenies. One difficulty with molecular studies is that, for early divergences, few taxa are used. As new taxa are added to the analyses, proposed relationships can change greatly. Nevertheless, it appears that the best current data suggest that turtles are nested within diapsids, which we adopt here.

Prior to the preceding studies, turtles were considered a sister group to the captorhinids, and these two taxa were the main members of the Anapsida, the presumed sister group of the Diapsida. The parareptiles were considered to be paraphyletic. In spite of the different placement of turtles, the preceding studies agree on the monophyly of the parareptiles and the sister-group relationship of captorhinids to all other eureptiles (Fig. 1.9). *Paleothyris* (Fig. 1.9) is one of the oldest eureptiles, although already structurally derived from, and the potential sister group to, all diapsid reptiles.

RADIATION OF DIAPSIDS

Diapsida is a diverse clade of reptiles. It has a long taxonomic history and its member content is generally accepted with only minor controversy, excluding the current disagreement about inclusion of turtles. Modern diapsids include lizards, snakes, turtles, birds, and crocodylians; extinct diapsids include dinosaurs, pterosaurs, ichthyosaurs, and many other familiar extinct taxa. The stem-based name Diapsida is derived from the presence of a pair of fenestrae in the temporal region of the skull; diapsids are also diagnosed by a suborbital fenestra, an occipital condyle lacking an exoccipital component, and a ridged-grooved tibioastragalar joint.

The earliest known divergence yielded the araeoscelidians, a short-lived group, and the saurians (Fig. 1.11, Table 1.4). The araeoscelidians were small (about 40 cm total length) diapsids of the Late Carboniferous and were an evolutionary dead end. In contrast, the saurian lineage gave rise to all subsequent diapsid reptiles. Members of the Sauria share over a dozen unique osteological features, including a reduced lacrimal with nasal–maxillary contact, no caniniform maxillary teeth, an interclavicle with distinct lateral processes, and a short, stout fifth metatarsal.

TABLE 1.4 A Hierarchical Classification of the Early Reptilia.

Amniota
- Synapsida
- Reptilia
  - Parareptilia
    - Millerettidae
    - unnamed
      - Pareiasauria
      - unnamed
        - procolophonoids
  - Eureptilia
    - Captorhinidae
    - unnamed
      - *Paleothyris*
      - Diapsida
        - Araeoscelidia
        - Sauria
          - Archosauromorpha
            - Archosauria
              - Pseudosuchia
                - Testudines
                  - Crocodylia
                  - Ornithodira
                    - Aves
          - Lepidosauromorpha
            - Lepidosauria
              - Sphendontida
              - Squamata

Note: This classification derives from the sister-group relationships in Figures 1.8 and 1.10.

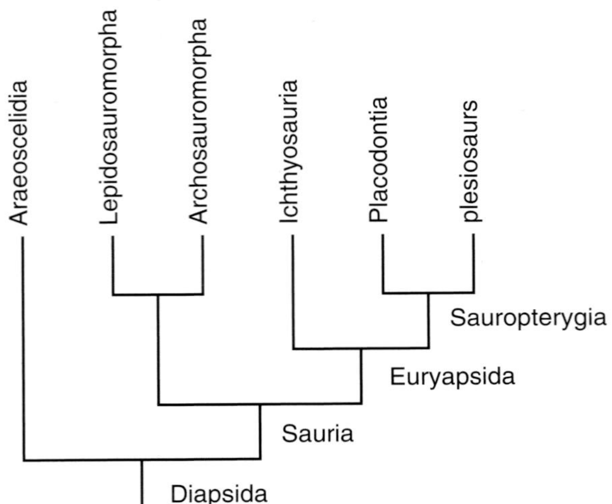

FIGURE 1.11 A branching diagram of the evolution of basal reptile clades, based on sister-group relationships. The diagram has no time axis, and each capitalized name represents a formal clade-group name. Plesiosaurs is used as a vernacular name and is equivalent to Storrs's (1993) Nothosauriformes. After Caldwell, M., 1996; Gauthier et al., 1989.

The Euryapsida apparently arose from an early split in the Sauria clade (Fig. 1.11). They comprise a diverse group of mainly aquatic (marine) reptiles, ranging from the fishlike ichthyosaurs to the walruslike placodonts and the "sea-serpent" plesiosaurs. Individually these taxa and collectively the Euryapsida have had a long history of uncertainty in their position within the phylogeny of reptiles. Only since the late 1980s has their diapsid affinity gained a consensus among zoologists, although different interpretations about basal relationships remain. For example, are they a sister group of the lepidosauromorphs or a sister group of the lepidosauromorph–archosauromorph clade? Is the Ichthyosauria a basal divergence of the euryapsids or perhaps not an euryapsid? The monophyletic clade interpretation rests on sharing six or more derived characters, such as a lacrimal bone entering the external nares, an anterior shift of the pineal foramen, and clavicles lying anteroventral to the interclavicle.

The Archosauromorpha and the Lepidosauromorpha are the other two clades of the Sauria (Fig. 1.11) with living representatives, including turtles, crocodylians and birds in the former, and tuataras and squamates (lizards, including amphisbaenians and snakes) in the latter. Both clades have had high diversity in the deep past, although the dinosaurs focus attention on the diversity within archosauromorphs, specifically on the archosaurs. However, the Archosauria had earlier relatives (e.g., rhynchosaurs, protorosaurs, and proterosuchids; Fig. 1.12), and furthermore, the archosaurs are much more than just dinosaurs. The archosaurs encompass two main lineages, the Crocodylotarsi (or Crurotarsia) and the Ornithodira; they share a rotary cruruotarsal ankle, an antorbital fenestra, no ectepicondylar groove or foramen on the humerus, a fourth trochanter on the femur, and other traits. Aside from the two main groups, archosaurs include some early divergent taxa, for example, Erythrosuchidae, *Doswellia*, and *Euparkeria*. These taxa appear to have been carnivores and ranged in size from the 0.5-m *Euparkeria* to the 5-m erythrosuchid *Vjushkovia*. These basal lineages were relatively short lived. The Ornithodira and Crocodylotarsi radiated broadly and have modern-day representatives.

The Ornithodira includes the Pterosauria and Dinosauria (Fig. 1.12). The pterosaurs were an early and successful divergence from the lineage leading to the dinosaurs. The leathery-winged pterosaurs seemingly never attained the diversity of modern birds or bats but were a constant aerial presence over tropical seashores from the Late Triassic to the end of the Cretaceous. The dinosaurs attained a diversity that was unequaled by any other Mesozoic group of tetrapods. Their size and diversity fan our imaginations; nonetheless, numerous other reptile groups (e.g., phytosaurs, prestosuchians) were highly diverse, and some of these were just as remarkable as the ornithischian and saurischian dinosaurs.

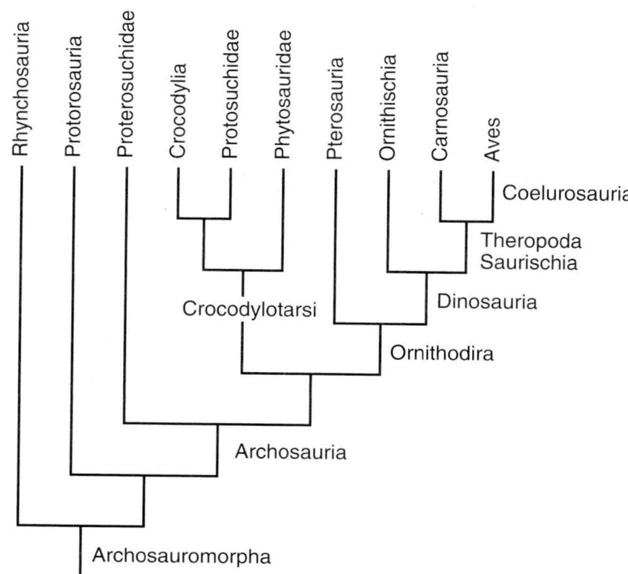

FIGURE 1.12 A branching diagram of the evolution within the Archosauromorpha, based on sister-group relationships. The diagram has no time axis; numerous clades and branching events are excluded; and each capitalized name represents a formal clade-group name. After Benton and Clark, 1988; Gauthier et al., 1989; Gower and Wilkinson, 1996.

Dinosaur evolution is well studied and outside the province of herpetology but relevant to the evolution of the living reptiles. Birds (Aves) are feathered reptiles, and *Archaeopteryx* is a well-known "missing link" that has a mixture of reptilian and avian characteristics. Although no one would argue that *Archaeopteryx* is not a bird, a controversy exists over the origin of birds. The current consensus places the origin of birds among the theropod dinosaurs (Fig. 1.12); however, three other hypotheses have current advocates, although all hypotheses place the origin of birds within the Archosauria. The theropod dinosaur hypothesis has the weight of cladistic evidence in its support. The other proposed bird ancestors are an early crocodyliform, among the basal ornithodiran archosaurs, and *Megalanocosaurus*, another basal archosaur taxon. Although these latter interpretations represent minority positions, the cladistic near relatives (birdlike theropods) of birds occur much later (>25 mybp) in the geological record than *Archaeopteryx*.

Crocodylotarsi, the other major clade of archosaurs, has an abundance of taxa and a broad radiation in the Mesozoic and Early Tertiary. The Crocodylia, a crown group including the most recent common ancestor of the extant Alligatoridae and Crocodylidae and its descendants, remains a successful group but shows only one aspect of crocodylotarsian radiation. The earliest radiations in the middle and Late Triassic included phytosaurs, aetosaurs, and rauisuchids. The phytosaurs were long-snouted crocodylian-like reptiles, and the position of their nostrils on a hump in front of the eyes suggests a similar aquatic ambush behavior on terrestrial prey. The aetosaurs were armored terrestrial herbivores, and the rauisuchids were terrestrial predators that developed an erect, vertical limb posture and reduced dermal armor. Another lineage, the Crocodyliformes, which includes the later-appearing Crocodylia, also appeared in the Middle Triassic and yielded the diversity of Jurassic and Cretaceous taxa. The crocodyliforms had members that were small wolflike, large bipedal tyrannosaurus-like, giant marine crocodylian-like, and a variety of other body forms.

The Lepidosauromorpha, the archosauromorph's sister group, consists of several basal groups and the lepidosaurs (Fig. 1.13). All share derived traits such as a lateral ridge of the quadrate supporting a large typanum, no cleithrum in the pectoral girdle, an ectepicondylar foramen rather than a groove in the humerus, and a large medial centrale

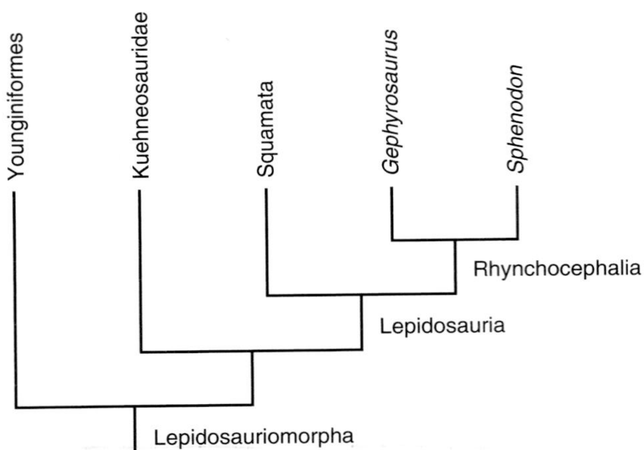

FIGURE 1.13 A branching diagram of the evolution within the Lepidosauromorpha, based on sister-group relationships. The diagram has no time axis; numerous clades and branching events are excluded; and each capitalized name represents a formal clade-group name. After Gauthier et al., 1989; Rieppel, 1994; Caldwell (1996) and deBraga and Rieppel (1997) provide different interpretations of lepidosauromorph relationships.

in the forefoot. The earliest known and basal group is the Younginiformes from the Upper Permian and Lower Triassic. They were aquatic, and adaptation to an aquatic life is a recurrent theme in the evolution and radiation of lepidosauromorphs. Another basal group with a highly specialized lifestyle was the Kuehneosauridae. They had elongate thoracic ribs that probably supported an aerofoil membrane and permitted them to glide from tree to tree or to the ground, as in the extant gliding lizard *Draco*. The kuehneosaurids are the sister group to the Lepidosauria. The Lepidosauria is a strongly supported clade with a wealth of derived features that are shared. Some of these features are teeth attached loosely to the tooth-bearing bones, fusion of the pelvic bones late in development, hooked fifth metatarsals, and paired copulatory organs (hemipenes; rudimentary in *Sphenodon*). Of the two sister groups within the Lepidosauria, only two species of tuataras (sphenodontidans) survive. The Sphenodontida has acrodont dentition and a premaxillary enameled beak. Sphenodontidans were moderately diverse and abundant in Late Triassic and Jurassic, and largely disappeared from the fossil record thereafter. From the beginning, the terrestrial taxa had the body form still seen in the tuataras. *Glevosaurus* is their sister taxon and shared a similar habitus; however, it had triangular teeth with a shearing bite. The squamates are the sister group of the sphenodontidans (Fig. 1.13) and are more abundant and speciose than the latter group from their first appearance in the Late Jurassic to today. In an all-inclusive sense, the squamates (lizards and snakes) were and are predominantly small-bodied (<0.5 m) carnivores. Based on fossil, morphological, ecological, and behavioral data, they apparently split early into the two major lineages, Iguania and Scleroglossa, that dominate the world's herpetofaunas. Recent nuclear DNA studies suggest that the Iguania are nested within Autarchoglossa (a subclade of the former Scleroglossa), which would eliminate Scleroglossa as a squamate clade. This result is intriguing and, if supported by a preponderance of data, will result in reconsideration of many interpretations of the evolution of ecology, morphology, behavior, and physiology that assume an Iguania–Scleroglossa sister relationship. Until these issues are resolved, we retain the two-clade arrangement (Iguania–Scleroglossa) of the Squamata. The fossil history of the Squamata and the other extant reptilian and amphibian groups is detailed in Chapter 3; similarly, the phylogenetic relationships of the major groups are examined in the Overview sections of each chapter of Part VI.

LINNEAN VERSUS EVOLUTIONARY TAXONOMY

Taxonomy is the naming of organisms and groups of similar organisms. Classifying objects is part of human nature and has its origins deep in prehistory. The earliest human societies began to name and recognize plants and animals for practical reasons, such as what is good or bad to eat, or what will or will not eat humans. This partitioning of objects places them into conceptual groups and is practiced daily by all of us. This may seem straightforward on the surface, but the degree to which we now understand the evolution of life on Earth has shaken the very foundations of our thinking on naming organisms and groups of organisms. In most introductory biology courses, we learn Linnean taxonomy, a formal system of classification that dates from Linneaus's tenth edition of *Systema Naturae* in 1758. This catalogue gave a concise diagnosis of all known species of plants and animals and arranged them in a hierarchical classification of genus, order, and class. Categories (taxa) were based on overall similarity. Linneaus's catalogue was the first publication to use consistently a two-part name (a binomial of genus and species). Scientific names of plants and animals remain binomials and are given in Latin (the language of scholars in the eighteenth century). The botanical and zoological communities separately developed codes for the practice of nomenclature. The most recent code for zoologists is the *International Code of Zoological Nomenclature, Fourth Edition* (the Code), effective January 2000. The *International Code of Zoological Nomenclature* can now be found online at http://www.iczn.org/iczn/index.jsp. The Linnean taxonomy system implies that taxonomic categories (genera, orders, classes, etc.) provide information about similarity (e.g., all species in a genus share something) and that this similarity reflects evolutionary history. Evolutionary taxonomy rests on the assumption that similarity reflects homology (e.g., the characteristics that species in a genus share have a common origin) and results in evolutionary "trees" that reflect both degrees of relatedness and time (Fig. 1.14). The resulting problem is that what we traditionally think of as taxonomic categories (e.g., the

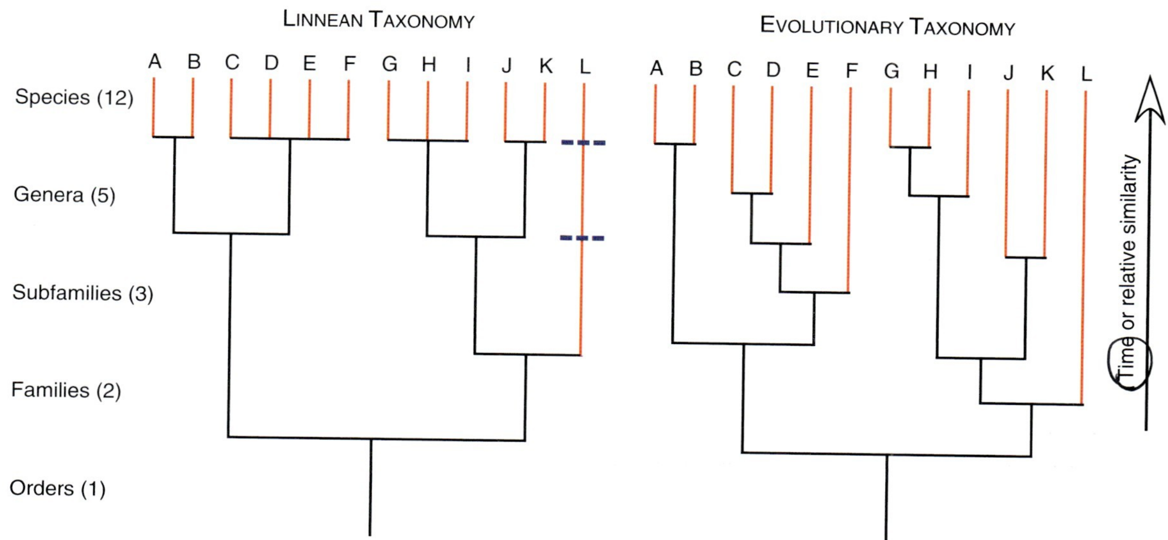

FIGURE 1.14 Linnean taxonomy places organisms in categories based on overall similarity. Evolutionary taxonomy places organisms in clades based on relatedness (homologies), which has a clear time component. A dendogram based on Linnean taxonomy (a) contains many polytomies because categories are discreet, (b) can contain some "species" (A–F and G–K) that are "equal" in rank with similar hierarchical organization to the subfamily level and others (L in particular) that contain this structure only in name, and (c) has no time component. Thus species L is in L subfamily. Dashed lines indicate where the taxonomic categories would occur for species L. A dendogram of evolutionary relationships has no clear genus, subfamily, or family structure but presents a relatively accurate hypothesis of known relationships and relative divergence times. Species are endpoints of divergences. Because of the implicit lack of a time element, individual taxonomic groups in the Linnean system often do not have comparable evolutionary histories across taxa. For example, a family of scorpions might have a much deeper (older) evolutionary history than a family of snakes.

families Colubridae [snake] and Ranidae [frog]) are not the same age in terms of their evolutionary histories. Each genus within each family has a different evolutionary history and thus the "Linnean" categories fall apart. The *International Code of Zoological Nomenclature* fails to break from the Linnean typological paradigm and consequently does not reflect evolutionary history. Changes have been proposed, and some heated discussion has followed. Throughout this book, we continue to use several lower categories of Linnean taxonomy (genera, subfamilies, and families) simply to make it possible to talk about groups of amphibians and reptiles. We do not assign taxonomic categories to higher-level clades.

Rules and Practice

The *International Code of Zoological Nomenclature* is a legal document for the practice of classification, specifically for the selection and assignment of names to animals from species through family groups. Unlike our civil law, there are no enforcement officers. Enforcement occurs through the biological community's acceptance of a scholar's nomenclatural decisions. If the rules and recommendations are followed, the scholar's decisions are accepted; if the rules are not followed, the decisions are invalid and not accepted by the community. Where an interpretation of the Code is unclear or a scholar's decision uncertain relative to the Code, the matter is presented to the International Commission for Zoological Nomenclature (a panel of systematic zoologists), which, like the U.S. Supreme Court, provides an interpretation of the Code and selects or rejects the decision, thereby establishing a precedent for similar cases in the future.

The Code has six major tenets:

1. All animals extant or extinct are classified identically, using the same rules, classificatory hierarchies, and names where applicable. This practice avoids dual and conflicting terminology for living species that may have a fossil record. Further, extant and fossil taxa share evolutionary histories and are properly classified together.
2. Although the Code applies only to the naming of taxa at the family-group rank and below, all classificatory ranks have Latinized formal names. All except the specific and subspecific epithets are capitalized when used formally; these latter two are never capitalized. For example, the major rank or category names (phylum, class, order, family, genus, species) for the green iguana of Central America are Chordata, Vertebrata, Tetrapoda, Iguanidae,

Iguana iguana. The names may derive from any language, although the word must be transliterated into the Roman alphabet and converted to a Latin form.

3. To ensure that a name will be associated correctly with a taxon, a type is designated—type genus for a family, type species for a genus, and a type specimen for a species. Such a designation permits other systematists to confirm that what they are calling taxon X matches what the original author recognized as taxon X. Comparison of specimens to the type is critical in determining the specific identity of a population. Although the designation of a single specimen to represent a species is typological, a single specimen as the name-bearer unequivocally links a particular name to a single population of animals.

Of these three levels of types, only the type of the species is an actual specimen; nonetheless, this specimen serves conceptually and physically to delimit the genus and family. A family is linked to a single genus by the designation of a type genus, which in turn is linked to a single species by a type species, and hence to the type specimen of a particular species. The characterization at each level thus includes traits possessed or potentially possessed by the type specimen. An example of such a nomenclatural chain follows: *Xantusia* Baird, 1859 is the type genus of the family Xantusiidae Baird, 1859; *Xantusia vigilis* Baird, 1859 is the type species of *Xantusia*; and three specimens, USNM 3063 (in the United States National Museum of Natural History) are syntypes of *Xantusia vigilis*.

Several kinds of types are recognized by the Code. The holotype is the single specimen designated as the name-bearer in the original description of the new species or subspecies, or the single specimen on which a taxon was based when no type was designated. In many nineteenth-century descriptions, several specimens were designated as a type series; these specimens were syntypes. Often syntypic series contain individuals of more than one species, and sometimes to avoid confusion, a single specimen, a lectotype, is selected from the syntypic series. Partially because of this kind of problem, more recent Codes do not approve the designation of syntypes. If the holotype or syntypes are lost or destroyed, a new specimen, a neotype, can be designated as the name-bearer for the species. Other types (paratypes, topotypes, etc.) are used in taxonomic publications; however, they have no official status under the Code.

4. Only one name may be used for each species. Yet commonly, a species has been recognized and described independently by different authors at different times. These multiple names for the same animal are known as *synonyms* and arise because different life history stages, geographically distant populations, or males and females were described separately, or because an author is unaware of another author's publication. Whatever the reason, the use of multiple names for the same animal would cause confusion, hence only one name is correct.

Systematists have selected the simplest way to determine which of many names is correct, namely by using the oldest name that was published in concordance with rules of the Code. The concept of the first published name being the correct name is known as the Principle of Priority. The oldest name is the primary (senior) synonym, and all names published subsequently are secondary (junior) synonyms (Table 1.5). Although simple in concept, the implementation of the Principle may not promote stability, especially so when the oldest name of a common species has been unknown for many decades and then is rediscovered. Should *viridisquamosa* Lacépéde, 1788 replace the widely used *kempii* Garman, 1880 for the widely known Kemp's ridley seaturtle *Lepidochelys kempii*? No. The goal of the Code is to promote stability of taxonomic names, so the Code has a 50-year rule that allows commonly used and widely known secondary synonyms to be conserved and the primary synonym suppressed. The difficulty with deviating from priority is deciding when a name is commonly used and widely known—the extremes are easy to recognize, but the middle ground is broad. In these circumstances, the case must be decided by the international commission.

In deciding whether one name should replace another name, a researcher determines whether a name is "available" prior to deciding which of the names is "valid." The concept of availability depends upon a taxonomic description of a new name obeying all the tenets of the Code in force at the time of the description. Some basic tenets are as follows: published subsequent to 1758 (tenth edition of *Systema Naturae*), a binomial name for a species-group taxon, name in Roman alphabet, appearing in a permissible publication, and description differentiates the new taxon from existing ones. If the presentation of a new name meets these criteria and others, the name is available. Failure to meet even one of the criteria, such as publication in a mimeographed (not printed) newsletter, prevents the name from becoming available. Even if available, a name may not be valid. Only a single name is valid, no matter how many other names are available. Usually, the valid name is the primary synonym. The valid name is the only one that should be used in scientific publications.

5. Just as for a species, only one name is valid for each genus or family. Further, a taxonomic name may be used only once for an animal taxon. A homonym (the same name for different animals) creates confusion and is also eliminated by the Principle of Priority. The oldest name is the senior homonym and the valid one. The same names (identical spelling) published subsequently are junior homonyms and invalid names. Two types of homonyms are possible. Primary homonyms are the same names published for the same taxon, for example, *Natrix viperina bilineata* Bonaparte, 1840 and *Tropidonotus viperina bilineata* Jan, 1863. Secondary homonyms are the same

TABLE 1.5 Abbreviated Synonymies of the European Viperine Snake and the Cosmopolitian Green Seaturtle.

***Natrix maura* (Linnaeus)**
1758 *Coluber maurus* Linnaeus, Syst. Nat., ed. 10, 1:219. Type locality, Algeria. [original description; primary synonym]
1802 *Coluber viperinus* Sonnini and Latreille, Hist. nat. Rept. 4:47, fig. 4. Type locality, France. [description of French population, considered to be distinct from Algerian population]
1824 *Natrix cherseoides* Wagler in Spix, Serp. brasil. Spec. nov. :29, fig. 1. Type locality, Brazil. [geographically mislabeled specimen mistaken as a new species]
1840 *Coluber terstriatus* Duméril in Bonaparte, Mem. Accad. Sci. Torino, Sci. fis. mat. (2) 1:437. Type locality, Yugoslavia. Nomen nudum. [= naked name; name proposed without a description so *terstriatus* is not available]
1840 *Natrix viperina* var. *bilineata* Bonaparte, Op. cit. (2) 1:437. Type locality, Yugoslavia. Non *Coluber bilineata* Bibron and Bory 1833, non-*Tropidonotus viperinus* var. *bilineata* Jan 1863, non-*Tropidonus natrix* var. *bilineata* Jan 1864. [recognition of a distinct population of *viperina*; potential homonyms listed to avoid confusion of Bonaparte's description with other description using *bilineata* as a species epithet]
1929 *Natix maura*, Lindholm, Zool. Anz. 81:81. [first appearance of current usage]
***Chelonia mydas* (Linneaus)**
1758 *Testudo mydas* Linnaeus, Syst. Nat., ed. 10, 1:197. Type locality, Ascension Island. [original description; primary synonym]
1782 *Testudo macropus* Wallbaum, Chelonogr. :112. Type locality, not stated. Nomen nudum.
1788 *Testudo marina vulgaris* Lacédè, Hist. nat. Quadrup. ovip. 1: Synops. method., 54. Substitute name for *Testudo mydas* Linnaeus.
1798 *T. mydas minor* Suckow, Anfangsg. theor. Naturg. Thiere. 3, Amphibien :30. Type locality, not stated. Nomen oblitum, nomen dubium. [forgotten name, not used for many years then rediscovered; name of uncertain attribution, tentatively assign to *mydas*]
1812 *Chelonia mydas*, Schweigger, Königsber. Arch. Naturgesch. Math. 1:291. [present usage but many variants appeared after this]
1868 *Chelonia agassizii* Bocourt, Ann. Sci. nat., Paris 10:122. Type locality, Guatemala. [description of Pacific Guatemalan population as distinct species]
1962 *Chelonia mydas carrinegra* Caldwell, Los Angeles Co. Mus. Contrib. Sci. (61): 4. Type locality, Baja California. [description of Baja population as a subspecies]

[Modified from Mertens and Wermuth 1960, and Catalogue of American Amphibians and Reptiles, respectively.]
Note: The general format of each synonym is as follows: original date of publication; name as originally proposed; author; abbreviation of publication; volume number and first page of description; and type locality. Explanations of the synonyms are presented in brackets.

names for different taxa, e.g., the insect family Caeciliidae Kolbe, 1880 and the amphibian family Caeciliidae Gray, 1825.

6. When a revised Code is approved and published, its rules immediately replace those of the previous edition. This action could be disruptive if the new Code differed greatly from the preceding one, but most rules remain largely unchanged. Such stasis is not surprising, for the major goal of the code is to establish and maintain a stable nomenclature. Rules tested by long use and found functional are not discarded. Those with ambiguities are modified to clarify the meaning. When a rule requires major alteration and the replacement rule results in an entirely different action, a qualifying statement is added so actions correctly executed under previous rules remain valid. For example, the first edition of the Code required that a family-group name be replaced if the generic name on which it was based was a secondary synonym; the second and third editions do not require such a replacement; thus, the latter two editions permit the retention of the replacement name proposed prior to 1960 if the replacement has won general acceptance by the systematic community. Such exceptions promote nomenclature stability.

Evolution-Based Taxonomy

The preceding rules illustrate the typological approach of Linnean taxonomy, especially the emphasis on named categories and fixed levels within the hierarchy. The adoption of cladistics as the major practice and conceptual base of current systematics has increased the advocacy for a taxonomy and nomenclature that are based on the principle of descent (homology). Hierarchies can represent the basic evolutionary concept that organisms are related through common descent, but the rigid structure of the Linnean hierarchy system fails to accomplish that (Fig. 1.14). Advocates for an evolution-based taxonomy argue that the taxonomic system should directly reflect phylogeny and retain only those elements that do not interfere with the accurate and efficient depiction of this phylogeny. A consequence of this demand is a change in how a taxon is named. In the Linnean system, a taxon is defined in terms of its assumed category or hierarchical position; in contrast, the evolution-based system defines a taxon in terms of its content, i.e., the clade containing the most recent ancestor of X and all of its descendants. A result of the latter practice is a classification in which a species can have a hierarchical position equivalent to a clade with dozens of species in several lower "level" clades (Fig. 1.14). Another consequence is the abandonment of category labels, such as family, order, or class, resulting in the development of the *PhyloCode*, which, like the *International Code of Zoological Nomenclature*, is a set of rules for nomenclature, in this case, entirely based on the hierarchical reality of evolutionary trees. If all scientists were to switch to a *PhyloCode* taxonomy and totally abandon Linnean taxonomy, not only would most scientists be confused for a long period of time (only a relatively small number of scientists working with organisms are systematists), but also the most basic understanding of "groups" of organisms would be lost to the public. Homology-based phylogenetic relationships are real, classifications are not. What this means is that, as Charles Darwin pointed out in 1859, a single evolutionary tree links all organisms that have ever lived. Phylogenies are our best approximation of what happened historically, and they improve as techniques and sampling improve. Although we construct classifications, no true "classification" exists in nature; rather, classifications are hierarchically ordered lists of organisms that allow us to talk about them in a reasonable fashion. When we say "the family Viperidae," most of us form a mental image of the vipers and pitvipers. To say "the clade comprised of the first snake (ancestor) to have only a left carotid artery, edentulous premaxillaries, blocklike, rotating maxillaries with hollow teeth ... and all of its descendents" is a bit abstract for most of us. Even within evolutionary systematics, nomenclature is confusing because clades can be node-, stem-, or apomorphy-based (Fig. 1.15). We adhere to a combination of a Linnean classification system to the family level for ease of discussion, but a phylogenetic system at higher levels and the recognition that a phylogenetic system underlies our use of Linnean taxonomy.

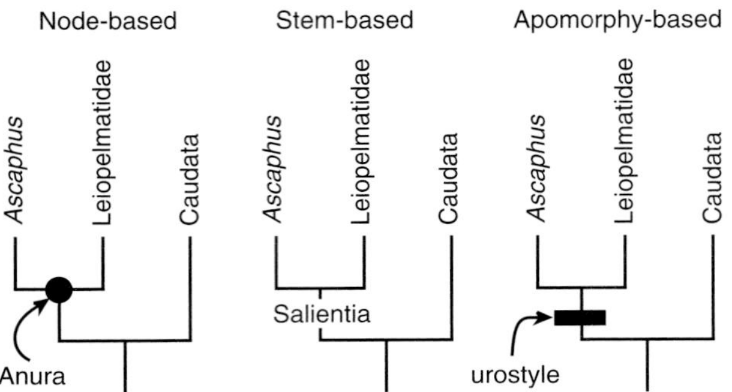

FIGURE 1.15 In evolutionary taxonomy, names of evolutionary groups of organisms (clades) can be confusing. Node-based clades are defined as the most recent common ancestor (the black circle) and all descendents. For example, Anura is the most recent common ancestor of *Ascaphus* and Leiopelmatidae. Stem-based clades are defined as those species sharing a more recent common ancestor with a particular organism (the stem) than with another. Thus Salientia is all taxa (in this case *Ascaphus* and Leiopelmatidae) more closely related to Anura than to Caudata. Apomorphy-based clades share a particular unique character (the bar in the graphic on the right). Thus Anura would be the clade stemming from the first amphibian to have a urostyle (a skeletal feature unique to frogs).

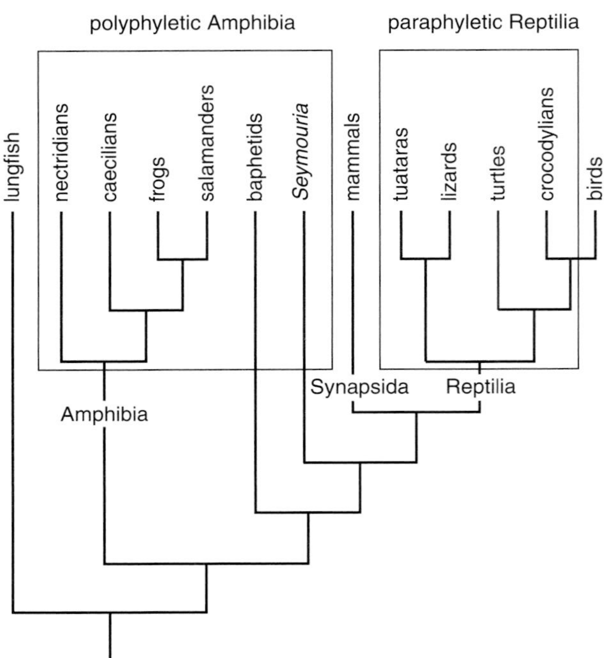

FIGURE 1.16 An abbreviated cladogram of tetrapods illustrating monophyly, paraphyly, and polyphyly. The heavier lines and capitalized group names depict the monophyletic groups of Amphibia and Reptilia recognized in the text. The boxes define earlier concepts of Amphibia (polyphyletic) and Reptilia (paraphyletic).

An example of problems that can arise from classification systems that are not based on relationships appears in Figure 1.16. The branching diagram shows evolutionary relationships as we currently understand them for extant tetrapods and three extinct groups. The group that we typically have called "amphibians" contains three groups (clades) with independent origins (polyphyletic), and the group that we typically call "reptiles" does not contain one of the members of the Reptilia clade, birds (paraphyletic). Homology-based classification systems that get away from Linnean systems present a much more realistic representation of the evolution of life.

Species are the basic units of our classifications and the only real units, existing not as artificial categories but as real entities. Typically, a species is defined as a set of unique, genetically cohesive populations of organisms, reproductively linked to past, present, and future populations as a single evolutionary lineage. Our hierarchical classification places closely related species together in the same genus and combines related genera into the same subfamily, and related subfamilies into the same family. At each level, we proceed backward in time to points of evolutionary divergence—specifically to a speciation event that gave rise to new lineages (Fig. 1.14). As we learn more about the genetics of populations, our definitions become a bit less clear; nevertheless, species are usually the end points of our phylogenies (some interesting exceptions exist—for example, "species" produced by hybridization are "end points" originating from other extant "end points"—see Chapter 4).

SYSTEMATICS—THEORY AND PRACTICE

Systematics is the practice and theory of biological classification. Thus modern systematics centers on discovering and describing the full diversity of life, understanding the processes resulting in this diversity, and classifying the diversity in a manner consistent with phylogenetic relationships (i.e., evolutionary history). Systematics has never been as relevant as it is today. Whether unraveling the interworkings of a cell, tracing the transmission route of a disease, or conserving a fragment of natural habitat, we must know the organisms with which we are working. Correct identification provides immediate access to previously published information on a particular species. Just knowing what they are is only a first step. Knowing where they came from (evolutionary history) and the underlying mechanisms allowing them to adapt (change) has taken center stage in the fight to combat infectious disease (e.g., AIDS, bird flu, ebola virus) and our attempts to maintain biodiversity (conservation strategies). Knowledge of a species' evolutionary relationships opens a wider store of information because related species likely function similarly.

Most importantly, our ability to recover the evolutionary history of extant species by using the tools of modern evolutionary systematics has changed the way we approach all areas of organismal biology. "Comparative" historically meant comparing two or more species, often species living in the same kind of habitat. Today, "comparative" means restricting species comparisons to variance in biological traits not explained by common ancestry. For example, two desert lizards might be similar ecologically because (1) they independently evolved sets of traits allowing existence in xeric environments or (2) they share a common ancestor that was adapted to xeric environments. These competing hypotheses can be tested only by knowing the structure of evolutionary relationships among the species, an approach that is becoming known as "tree thinking." Throughout this text you will encounter phylogenetic analyses applied to ecology, behavior, physiology, biogeography, and morphology, and it should become clear that this powerful conceptual approach is leading to a much better understanding of the natural world than we have ever experienced.

Systematic Analysis

Systematic research is a search for evolutionary patterns. Investigations span the spectrum from analyses of intraspecific variation to the deepest phylogenetic levels. At one end, the researcher examines species through the analysis and definition of variation within and among populations and/or closely related species. At the opposite end, research is directed at the resolution of genealogical relationships among species, genera, and higher taxonomic groups.

Species and their relationships are discerned by examining individuals. An individual's attributes provide a means to infer its affinities to another individual (or larger group). Such inferences of relationships provide a framework to examine evolutionary processes and the origin of diversity. Diversity occurs at many levels, from the variety of genotypes (individuals) within a deme (local interbreeding population) to the number of species within a genus (or higher group) or within a habitat or geographical area. Only through the recognition of which group of individuals is a particular species and which ones are something else can we address other biological questions.

Types of Characters

Any inheritable attribute of an organism can serve as a character. A character can be anatomical (e.g., a process or foramen on a bone, number of scales around midbody, snout-vent length), physiological (resting metabolic rate, thyroxine-sensitive metamorphism), biochemical–molecular (composition of venom, DNA sequence), behavioral (courtship head-bobbing sequence), or ecological (aquatic versus terrestrial). Types of characters used depend upon the nature of the questions being asked. Only a sampling of characters can be presented here.

Systematic study involves the comparison of two or more samples of organisms through their characters. This comparison involves two procedural concepts: The OTU, operational taxonomic unit; and character states. OTUs are the units being compared and can be an individual, population, species, or higher taxonomic group. The actual conditions of a character are its states, for example, an eye iris being blue or green, or a body length of 25 or 50 mm. The assumption of homology is implicit in comparison of character states; that is, all states of a character derive from the same ancestral state. Characters can be either qualitative (descriptive) or quantitative (numeric). Qualitative characters have discrete states, that is, "either/or" states: vomerine teeth present or absent, and the number of upper lip scales. Quantitative characters have continuous states: Head length of an individual can be recorded as 2, 2.3, 2.34, or 2.339 cm.

To be useful for systematics, a character's states generally have lower variation within samples than among samples. A character with a single state (invariant condition) in all OTUs lacks discriminatory power among the samples. A highly variable character with numerous states in one or more samples adds confusion to an analysis and should be examined more closely to identify the cause of the high variability (e.g., lack of homology) and be excluded if necessary.

Knowledge of the sex and state of maturity of each specimen is critical for recognition of variation between females and males, and among ontogenetic stages. Both must be considered whether the characters are anatomical, behavioral, or molecular in order to avoid confounding intraspecific variation with variation at the interspecific or higher level.

Morphology

Three discrete classes of anatomical characters are recognized: (1) mensural or morphometric characters are measurements or numeric derivatives (e.g., ratios, regression residuals) that convey information on size and shape of a structure or anatomical complex; (2) meristic characters are those anatomical features that can be counted, such as number of dorsal scale rows or toes on the forefoot; and (3) qualitative characters describe appearance; for example, a structure's presence or absence, color, location, or shape.

1. The most common morphometric character in herpetology is snout-vent length (SVL). This measurement gives the overall body size of all amphibians, squamates, and crocodylians, and how it is measured differs only slightly from group to group depending on the orientation of the vent, transverse or longitudinal. Because of their shells, carapace length and plastron length are the standard body size measurements in turtles. Numerous other measurements are possible and have been employed to characterize differences in size and shape. Mensural characters are not confined to aspects of external morphology but are equally useful in quantifying features of internal anatomy, for example, skeletal, visceral, or muscular characters.

As in all characters, the utility of measurements depends on the care and accuracy with which they are taken. Consistency is of utmost importance, so each measurement must be defined precisely, and each act of measuring performed identically from specimen to specimen. The quality of the specimen and nature of the measurement also affect the accuracy of the measurement. Length (SVL) of the same specimen differs whether it is alive (struggling or relaxed) or preserved (shrunk by preservative; positioned properly or not); thus, a researcher may wish to avoid mixing data from such specimens. Similarly, a skeletal measurement usually will be more accurate than a visceral one because soft tissue compresses when measured or the end points often are not as sharply defined. Differences can also occur when different researchers measure the same characters on the same set of animals. Thus within a sample, variation of each character includes "natural" differences between individuals and the researcher's measurement "error." Measurement error is usually not serious and is encompassed within the natural variation if the researcher practiced a modicum of care while taking data. The use of adequate samples (usually >20 individuals) and central tendency statistics subsumes this "error" into the character's variation and further offers the opportunity to assess the differences among samples and to test the significance of the differences, as well as providing single, summary values for each character.

2. Meristic characters are discontinuous (= discrete). Each character has two or more states, and the states do not grade into one another. The premaxillary bone can have 2, 3, or 4 teeth, not 2.5 or 3.75 teeth. Meristic characters encompass any anatomical feature (external or internal) that can be counted. Researcher measurement error is possible with meristic characters. These characters are examined and summarized by basic statistical analyses.

3. Qualitative characters encompass a broad range of external and internal features, but unlike mensural or meristic characters, they are categorized in descriptive classes. Often a single word or phrase is adequate to distinguish among various discontinuous states, for example, pupil vertical or horizontal, coronoid process present or absent, carotid foramen in occipital or in quadrate, or bicolor or tricolor bands at midbody. Qualitative characters can have multiple states (>2), not just binary states. Even though these characters are not mensural or meristic, they can be made numeric, simply by the arbitrary assignment of numbers to the different states or by size comparison (e.g., 1x width versus 3x height).

The preceding characters emphasize aspects of gross anatomy, but microscopic characters may also be obtained. One of the more notable and widely used microscopic (cytological) characters is karyotype or chromosome structure. The most basic level is the description of chromosome number and size: diploid (2n) or haploid (n) number of chromosomes, and number of macro- and microchromosomes. A slightly more detailed level identifies the location of the centromere (metacentric, the centromere is in the center of the chromosome; acrocentric, the centromere is near the end; and telocentric, the centromere is at the end) and the number of chromosomes of each type or the total number (NF, nombre fundamental) of chromosome arms (segments on each side of the centromere). Special staining techniques allow the researcher to recognize specific regions (bands) on chromosomes and to more accurately match homologous pairs of chromosomes within an individual and between individuals.

Molecular Structure

The preceding characters are largely visible to the unaided eye or with the assistance of a microscope. Chemical and molecular structures also offer suites of characters for systematic analysis. The nature of these characters can involve the actual structure of the compounds (e.g., chemical composition of the toxic skin secretions in the poison frogs or nucleotide sequences of DNA fragments) or comparative estimates of relative similarity of compounds (e.g., immunological assays).

Many systematists have widely and enthusiastically adopted techniques from molecular biology. Their use in systematics rests on the premise that a researcher can assess and compare the structure of genes among individuals to assess relationships among species and higher taxa through examination of molecular structure of proteins and other compounds that are only a few steps removed from the gene. Molecular data offer a different perspective, sometimes yield new insights, and in many instances permit us to answer questions that cannot be addressed with other kinds of characters. Importantly, whatever the nature of a character, the fundamental assumption is that the character being compared between two or more OTUs is homologous, and this requirement applies to molecular characters as well as gross anatomical ones.

The techniques of molecular systematics are varied and complex. Different techniques are selected to investigate different levels of relationships. Electrophoresis is especially good (and relatively inexpensive) for examining genetic relationships of individuals within and among populations and of populations within and among species. Other techniques (e.g., DNA sequencing) may be used at this level of comparison, but often they are used more effectively in the examination of higher-level relationships, which represent older divergences and speciation. The next three sections offer a glimpse of molecular techniques in systematics.

Electrophoresis is a useful molecular technique for determining patterns of variation within populations. Proteins are the major structural components and chemical regulators of cells. Their structure is a direct reflection of the DNA sequence of genes that code their formation—a gene's DNA sequence is transcribed to make messenger RNA (mRNA), and mRNA is translated into a chain of amino acids, the protein. Enzymes, one group of proteins, catalyze the cell's chemical reactions. This high specificity makes them critical and key components in cellular metabolism and potentially useful systematic characters. This potential is further enhanced by the occurrence of structurally different forms (allozymes) of the same enzyme. The different allozymes arise from gene mutations that have altered the structure of the DNA, and this alteration is translated directly into an altered structure of the enzyme.

Electrophoresis can identify different allozymes (also other chemical compounds) via their different mobility in an electrical field. Each enzyme is a chain of amino acids and has a specific size, shape, and net charge (positive, negative, or neutral). Mobility is a function of the allozyme's charge, shape, and size; minor mutations may alter one or more of these facets of allozymic structure and affect the allozyme's speed of migration.

In biological studies, the electrophoretic apparatus consists of an electric-power pack, positive and negative electrodes to a buffer tray, and a sheet of gel matrix stretching between the positive- and negative-charged buffer trays. Tissue samples (fluid homogenates, often of muscle or liver tissue) are placed in a row at one end of the gel, and when the power is turned on, the allozymes begin to migrate toward the positive or negative electrodes. After several hours, the gel is removed and stained (using specific biological dyes for each type of enzyme) to reveal the position of the allozymes for each sample (Fig. 1.17). Most current studies assay for 15–30 enzymes.

Each enzyme is a character, and its allozymes are its states. The stained gel (zymogram) comprises the raw data and shows whether each sample shares the same allozymes. Electrophoretic data are variously coded and can be simple counts of allozyme matches and mismatches, frequencies, or the presence or absence of heterozygotes. The manner of coding and analysis depends on the problem addressed, whether examining intrapopulational or intergeneric relationships.

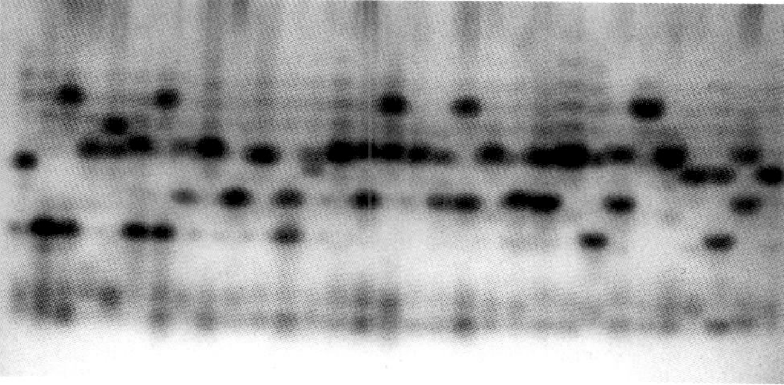

FIGURE 1.17 An electrophoretic gel (zygogram) of an esterase stain for numerous individuals of the salamanders *Plethodon cinereus* and *P. shenandoah*. Each vertical bar is an individual salamander, whereas each dark crossbar is a stained enzyme. Courtesy of A.Wynn.

Immunology (antigen–antibody or immunological reaction) provides a mechanism for estimating the genetic affinities of species. The concept is simple. Homologous proteins of closely related species are structurally similar, and as relationships become distant (increasing divergence from a common ancestor), protein structure becomes increasingly different. When a foreign protein (the antigen) is introduced into a host animal, the host's normal reaction is to produce antibodies specifically constructed to intercept and deactivate the antigen. By using these antibodies to test the level of the immunological reaction with the antigens of many different species, the researcher obtains an estimate of the similarity of each test antigen to the antigen of the donor OTU. A strong reaction indicates a high similarity in protein structure, and a weak reaction a low similarity. The antibody "recognizes" (attaches to) specific amino acid sequences of the donor's antigen, and fewer and fewer sequences are recognized as the structural differences of the test antigens increase. The basic protocol requires the introduction of an antigen into a host animal (rabbit, goat, etc.), time for the host's immunological system to produce antibodies, removal of blood serum from host and purification of antiserum, and performance of *in vitro* comparisons of the antisera reactions of the antigens from a series of test OTUs.

Of the several immunological tests, immunodiffusion, immunoelectrophoresis, and micro-complement fixation (MCF) are used in systematic studies, and blood albumins are the usual proteins compared. All three tests translate the level of antigen–antibody reaction into a numerical estimate of protein similarities and hence the relative similarity of the OTUs.

Examination of nucleic acids (DNA and RNA structure) may be the most powerful technique for recovering evolutionary relationships among organisms. The attractiveness of nucleic acids for inferring phylogenetic relationships is that their nucleotide sequences are the basic informational units encoding and regulating all of life's processes, and a huge number of nucleic acids (characters in this case) can be examined. Examination of nucleic acid sequences began in the 1980s as advances in methodology and equipment made the techniques more accessible and affordable to systematists. It has now become an indispensable part of systematics and is applied in most major fields of biology. A major feature of nucleic acid analyses is their broad comparative power and spectrum, ranging from the ability to examine and identify individual and familial affinities (e.g., DNA fingerprinting) to tracing matriarchal lineages (mitochondrial DNA, or mtDNA) and estimating phylogenetic relationships across diverse taxonomic groups (nuclear DNA). While extremely valuable for systematic studies, nucleic acid characters are not a panacea and have their own set of difficulties in analysis and interpretation.

Several techniques are available for comparing nucleotide sequences among different taxa. DNA–DNA hybridization was the first technique used on a large scale in vertebrate systematics although not in herpetology. It takes advantage of the disassociation of the two strands of DNA at high temperatures and the reassociation of complementary strands as the temperature drops. DNA strands from two OTUs are combined, disassociated, and then allowed to reassociate. Complementary strands from the same as well as different OTUs will reassociate. The number of mismatched base pairs (nucleotides) in the hybrid DNA molecules increases with evolutionary time of divergence, which is reflected by a depression of the melting or disassociation temperature. The difference in melting temperature between pure and hybrid DNA provides a measure for assessing the level of relationship.

More recently, the technology for determining the sequence of nucleotides (base pairs; see Table 1.6) has become increasingly accessible and is generally preferred, because sequence data provide discrete character information rather than estimates of relative similarity between nucleic acids (e.g., as in DNA hybridization) or their products (immunological tests). Several sequencing protocols are available, and it is necessary to select or target a specific segment of a particular nucleic acid owing to the enormous number of available sequences within the cell and its organelles. First, the nucleic acid to be examined is selected (e.g., mitochondrial or nuclear DNA, ribosomal RNA) and then specific sequences within this molecule are targeted. The target sequence is then amplified using a polymerase chain reaction (PCR) to produce multiple copies of the sequence for each OTU being compared. The sequence copies are isolated and purified for sequencing. Sequence determination relies on site-specific cleavage

TABLE 1.6 Sample of mtDNA Sequence Data for Select Iguania.

Anolis	CAATT TCTCC CAATT ACTTT AGCTT TATGC CTATG ACACA CAACA
Basiliscus	CAATT TTTAC CAATC ACCCT AGCCC TCTGC CTATG ACACG TAGCC
Oplurus	CAATT TCTTC CAATC ACATT AGCCC TATGC CTATG GTATA CCTCA
Sauromalus	CAATT TCTCG CCCTC ACACT AGCCC TATGC CTATG TCTCA CTTTC
Chamaeleo	CAATT TCTAC CCCAT ACCCT AGCCA TATGC CTACT CTACA CTGCC
Uromastyx	CAATT CCTAC CCCTG ACCTT AGCCA TATGC CTATT ATACA CAAAC

Note: The sequences represent the 401st to 445th positions on the ND2 gene. They are presented here in sets of five to permit ease of comparison. Abbreviations are as follows: A, adenine; C, cytosine; G, guanine; T, thymine.
Source: Macey et al., 1997: Fig. 1.

of the target sequence into fragments of known nucleotide sequences and the separation and identification of these fragments by electrophoresis. The homologous sequences are then aligned and provide the data for analyzing the phylogenetic relationships among the OTUs. The process is summarized in Figure 1.18.

METHODS OF ANALYSIS

The opportunities for analysis are as varied as the characters, and this field is rapidly evolving. Choice of analytical methods depends on the nature of the question(s) asked and should be made at the beginning of a systematic study, not after the data are collected. With the breadth of systematic studies ranging from investigations of intrapopulational variation to the relationships of higher taxonomic groups, the need for a carefully designed research plan seems obvious.

Systematic research often begins when a biologist discovers a potentially new species, notes an anomalous distribution pattern of a species or a character complex, or wishes to examine the evolution of a structure, behavior, or other biological aspect, and thus, requires a phylogenetic framework. With a research objective formulated, a preliminary study will explore the adequacy of the characters and data collection and analysis protocols for solving the research question.

A small set of available analytical techniques follows. These techniques segregate into numeric and phylogenetic ones. Numeric analyses offer a wide choice of methods to describe and compare the variation of OTUs and/or their similarity to one another. Phylogenetic analyses address common ancestry relationships of OTUs, specifically attempting to uncover the evolutionary divergence of taxa.

Numeric Analyses

Any study of variation requires the examination of multiple characters scored over numerous individuals. The resulting data cannot be presented en masse but must be summarized and condensed. Numeric analyses provide this service. The initial analysis examines the variation of single characters within each sample using univariate statistics. The next phase compares individual characters within subsamples (e.g., females to males), the relationship of characters to one another within samples, and character states of one sample to those of another sample using bivariate statistics. The final phase usually is the comparison of multiple characters within and among samples using multivariate analysis. Each phase yields a different level of data reduction and asks different questions of the data, for example: (1) What is the variability of each character? (2) What is the difference in the means and variance between sexes or among samples? (3) What is the

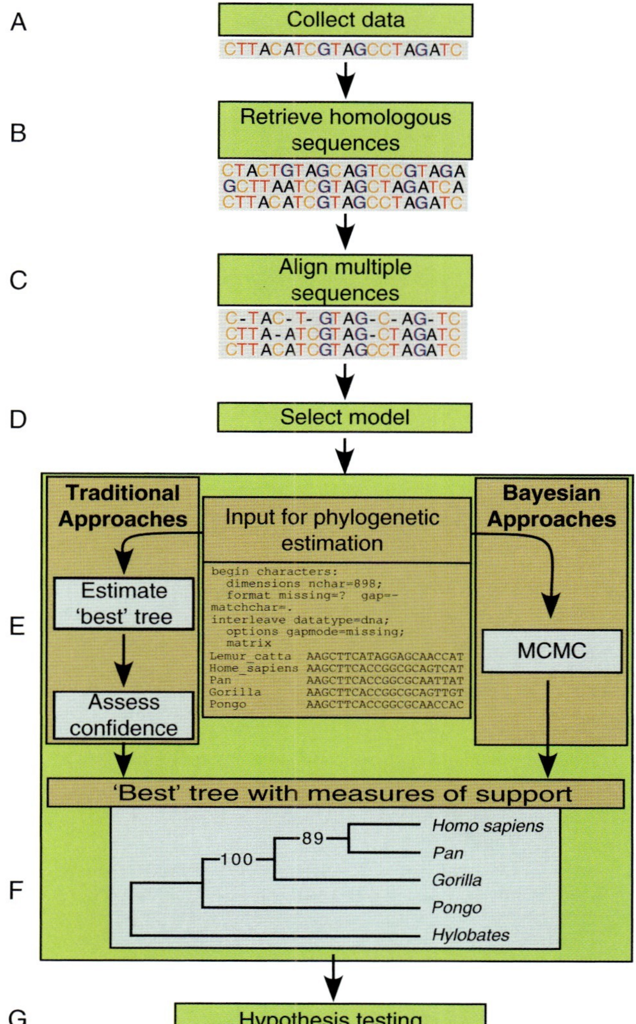

FIGURE 1.18 The production of phylogenetic trees from gene sequence data is a relatively easy process, at least conceptually. Gene sequences are assembled from the organisms of interest (A). These can be obtained from animals collected, tissues borrowed, or sequences already available (see GenBank). Typically, at least one outgroup (distantly related taxon) is included to root the tree (determine oldest nodes within the group of interest). Homologous sequences are then assembled from the various samples (B). All sample sequences are then aligned (homologous nucleotides in columns) to identify insertions and deletions (different nucleotides than expected based on homology) (C). These indicate evolutionary change for a particular sample sequence. Models of sequence evolution for analyses are then chosen (D) based on data available and model complexity. Traditional analyses and/or Bayesian analyses are then applied to data to reconstruct evolutionary trees from the data (E). A number of traditional approaches exist (Table 1.8) that are based on analyses of bootstrapped data (a subsample of data used to define models to test with remaining data) (E). The relatively newly applied Bayesian approaches use a Markov chain Monte Carlo (MCMC) analysis, a randomization procedure that has much stricter rules (E, and see Holder and Lewis, 2003). Both of these produce numerous trees that differ slightly in structure. A "best" tree is selected based on a set of criteria, or in some cases, several "best" trees are reported if the analyses provide support for more than one (F). Because all phylogenetic trees are hypotheses, they can then be tested with additional data (G).

TABLE 1.7 Examples and Definitions of Numeric Analytical Tools.

Univariate
Frequency distributions. Presentation techniques to show frequency of occurrence of different data classes or character states. Frequency tables, histograms, pie charts, and other techniques permit easy visual inspection of the data to determine normality of distribution, range of variation, single or multiple composition, etc.
Central tendency statistics. Data reduction to reveal midpoint of sample for each character and variation around the midpoint. Mean (average value), mode (most frequent value), and median (value in middle of ranked values); variance, standard deviation, standard errors (numeric estimates of sample's relative deviation from mean); kurtosis and skewness (numeric estimates of the shape of a sample's distribution).
Bivariate
Ratios and proportions. Simple comparisons (A:B, % = B/A × 100) of the state of one character to that of another character in the same specimen.
Regression and correlation. Numeric descriptions (equation and value, respectively) of the linear relationship and association of one character set to another.
Tests of similarities between samples. A variety of statistical models ($\chi 2$, Students' *t*, ANOVA/analysis of variance) test the similarity of the data between samples.
Nonparametric statistics. Statistical models containing no implicit assumption of particular form of data distribution. All other statistics in this table are parametric, and most assume a normal distribution.
Multivariate
Principal components analysis/PCA. Manipulation of original characters to produce new uncorrelated composite variables/characters ordered by decreasing variance.
Canonical correlation. Comparison of the correlation between the linear functions of two exclusive sets of characters from the same sample.
Discriminant function analysis/DFA. Data manipulation to identify a set of characters and assign weights (functions) to each character within the set in order to separate previously established groups within the sample.
Cluster analysis. A variety of algorithms for the groupings of OTUs on the basis of pairwise measures of distance or similarity.

(*In part, modified from* James *and* McCullough, 1985, 1990.)

covariance of characters within and among samples (Table 1.7)?

Even the briefest species description requires univariate statistics. A new species is seldom described from a single specimen, so univariate analysis shows the variation of each character within the sample and provides an estimate of the actual variation within the species. Means, minima, and maxima, and standard deviations are the usual statistics presented. An in-depth study of a group of species typically uses univariate and bivariate statistics to examine the variation within each species and one or more multivariate techniques to examine the variation of characters among the species and the similarities of species to one another.

Multivariate analysis has become increasingly important in the analysis of systematic data, particularly mensural and meristic data sets (Table 1.7). Multivariate analysis allows the researcher to examine all characters and all OTUs simultaneously and to identify patterns of variation and association within the characters, and/or similarities of OTUs within and among samples. For example, principal component analysis is often used in an exploratory manner to recognize sets of characters with maximum discriminatory potential or to identify preliminary OTU groups. These observations can then be used in a discriminant function analysis to test the reliability of the OTU groupings. Because these techniques are included in most statistical software, use without an awareness of their limitations and mathematical assumptions may occur. Users should be aware that combining meristic and mensural characters, using differently scaled mensural characters, or comparing data sets of unequal variance can yield meaningless results.

Cluster analysis is another multivariate technique, although it is not strictly statistical in the sense of being inferential or predictive. The numerous clustering algorithms use distance or similarity matrices and create a branching diagram or dendrogram. These matrices derive from a pairwise comparison of each OTU for every character to every other OTU in the sample (Fig. 1.19). The raw data in an OTU x Character matrix are converted to an OTU x OTU matrix in which each matrix cell contains a distance or

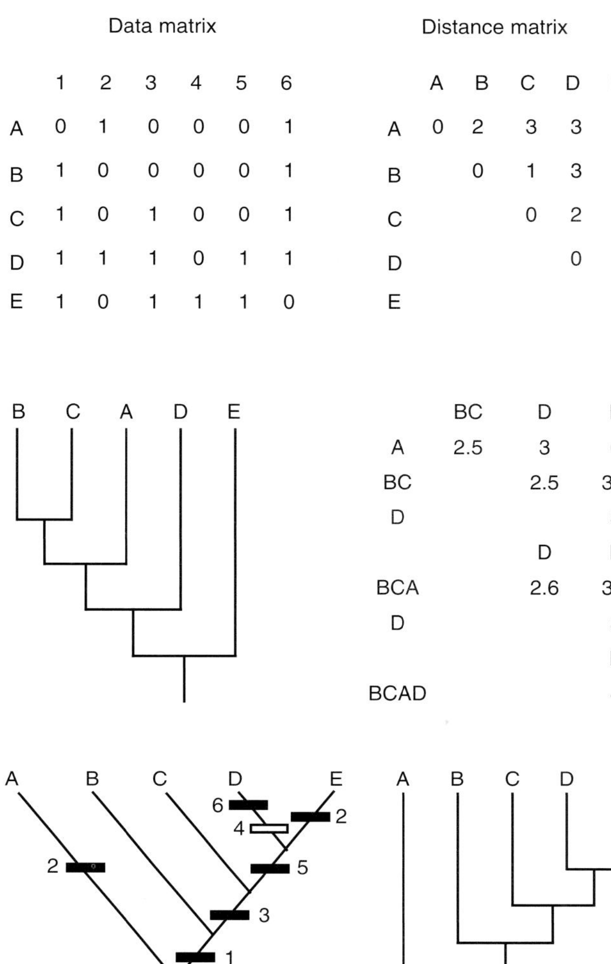

FIGURE 1.19 Construction of branching diagrams by two methods: phenetics and cladistics. The OTU x Character matrix (upper left) contains five OTUs (A–E) and six characters (1–6). Each character has two states, 0 or 1 (e.g., absent or present, small or large, etc.). Pairwise comparison of OTUs creates an OTU x OTU matrix. The distance values are the sums of the absolute difference between states for all six characters. Zeros fill the diagonal because each OTU is compared to itself; only half of the matrix is filled with the results of a single analysis because the two halves are mirror images of one another. An unweighted pair-group method (UPGM) clustering protocol produces a phenetic dendrogram (phenogram, middle left); in UPGM, the most similar OTUs are linked sequentially with a recalculation (middle right) of the OTU x OTU matrix after each linkage. The cladogram (lower left) derives directly from the OTU x Character matrix. The solid bars denote a shared-derived (synapomorphic) character state, the open bars an evolutionarily reversed state, and the character numbers. For comparison with the UPGM phenogram, the cladogram is present in a different style without the depiction of character state information.

similarity value. The clustering algorithm uses these values to link similar OTUs and OTU groups to one another, proceeding from the most similar to the least similar.

The preceding numeric techniques do not provide estimates of phylogenetic relationships; rather, they summarize the level of similarity. Overall similarity has been argued as an estimate of phylogenetic relationship. This concept is the basic tenet of the phenetic school of systematics, which came into prominence in the late 1950s and then rapidly was replaced by phylogenetic systematics. Phenetics as a classification method has largely disappeared (although many of its analytical algorithms remain) because its basic premise of "similarity equals genealogical relationship" is demonstrably false in many instances, and the resulting classifications do not reflect accurately the evolutionary history of the organisms being studied. Another basic premise of the phenetic school was that large character sets produce more robust and stable classifications; unfortunately, the addition of more characters usually changes the position of OTUs on the dendrogram and yields a dissimilar classification. This instability of OTU clustering arises from the use of unweighted characters and the swamping of useful characters by ancestral (= primitive) and nonhomologous (= homoplasic) ones.

Phylogenetic Analyses

Phylogenetic analysis has been variously practiced since the publication of Darwin's *Origin of Species*. However in the mid-1960s, with the publication of the English language edition of Willi Hennig's *Phylogenetic Systematics*, systematists began more rigorous and explicit character analyses and the reconstruction of phylogenies (taxa genealogies). This approach gives repeatability to systematic practices and is broadly known as *cladistics*. The basic tenets of phylogenetic systematics are as follows: (1) only shared similarities that are derived are useful in deducing phylogenetic relationships; (2) speciation produces two sister species; (3) speciation is recognizable only if the divergence of two populations is accompanied by the origin of a derived character state.

Character analysis plays a major role in phylogenetic reconstruction, because it is necessary to determine the ancestral or derived status for each character state. A special terminology is associated with the determination of character state polarity: plesiomorphic, the same state as in the ancestral species; apomorphic, a derived or modified state relative to the ancestral condition; autapomorphic, a derived state occurring in a single descendant or lineage; synapomorphic, a shared-derived state in two or more species. Sister groups are taxa uniquely sharing the same ancestor; synapomorphic characters identify sister groups. We reiterate that characters can be anything from gene sequences to morphology to ecology.

Determination of character state polarity can use one or more protocols. Outgroup comparison is generally considered the most reliable method. Operationally, the researcher identifies a candidate sister group(s) (outgroup) of the group being studied (ingroup) and then examines

TABLE 1.8 Comparison of Methods for Analyzing Phylogenetic Data. (From Holder and Lewis, 2003).

Method	Advantages	Disadvantages	Software
Neighbor joining	Fast	Information is lost in compressing sequences into distances; reliable estimates of pairwise distances can be hard to obtain for divergent sequences.	PAUP MEGA PHYLIP
Parsimony	Fast enough for the analysis of hundreds of sequences; robust if branches are short (closely related sequences or dense sampling).	Can perform poorly if substantial variation in branch lengths exists.	PAUP NONA MEGA PHYLIP
Minimum evolution	Uses models to correct for unseen changes.	Distance corrections can break down when distances are large.	PAUP MEGA PHYLIP
Maximum likelihood	The likelihood fully captures what the data tell us about the phylogeny under a given model.	Can be prohibitively slow (depending on the thoroughness of the search and access to computational resources).	PAUP PAML PHYLIP
Bayesian	Has a strong connection to the maximum likelihood method; might be a faster way to assess support for trees than maximum likelihood bootstrapping.	The prior distributions for parameters must be specified; it can be difficult to determine whether the Markov chain Monte Carlo (MCMC) approximation has run for long enough.	MrBayes BAMBE

the distribution of character states for each character in these two groups. If a state occurs only in the ingroup (but not necessarily in all members of the group), it is hypothesized to be apomorphic, and if present in both in- and outgroups, it is considered plesiomorphic. Ontogenetic analysis, commonality, and geological precedence are supplementary methodologies and are rarely used now owing to their low reliability.

Once the characters have been polarized, the researcher can construct a cladogram by examining the distribution of apomorphic states. Numerous computer algorithms are available for the evaluation of character state distributions and cladogram construction. For complex or large data sets, computer analysis is required. However, the following protocol demonstrates some fundamentals of cladogram construction. Figure 1.19 uses the OTU x Character matrix for the sequential linkage of sister groups, and all "1" states are considered apomorphic. Linkage proceeds as follows: D and E are sister taxa, synapomorphic for character 5; C and D–E are sister groups, synapomorphic for character 3; B and C–D–E are sister groups, synapomorphic for character 1; A and B–C–D are sister groups, synapomorphic for character 6. Taxon E shows the plesiomorphic state for character 6, which might suggest that E is not a member of the ABCD clade; however, it does share three other apomorphic characters, and the most parsimonious assumption is that character 6 underwent an evolutionary reversal in E. Similarly, the most parsimonious assumption for the synapomorphy of character 2 in taxa A and D is convergent evolution. These shared character states of independent origin are nonhomologous or homoplasic. Phylogenetic inference experienced major advances in theory and application during the 1990s. Inferring phylogeny from large data sets and particularly molecular ones is complex, often requiring days or weeks of analyses using the best computers available. The most frequently used analyses are summarized in Table 1.8.

QUESTIONS

1. Define the following terms in a phrase or a sentence.
   OTU—
   Clade—
   Sister taxa—
   Synapomorphy—
   Type specimen (holotype)—
   Paraphyly—
   Polyphyly—
2. Why was the recently discovered fossil of the tetrapodomorph fish *Tiktaalik* such an important find?
3. Why was the amniotic egg such an important innovation in the evolution of tetrapods?
4. Explain the difference between evolutionary taxonomy and Linnean taxonomy.
5. Are amphibians more closely related to fishes or mammals?
6. Describe in detail how the transition from water to land occurred and what the major morphological preadaptations were that facilitated this transition.

ADDITIONAL READING

Benton, M. J. (1997). *Vertebrate Paleontology* (2nd ed.). Chapman & Hall, London.

Carroll, R. L. (1988). *Vertebrate Paleontology and Evolution.* W. H. Freeman & Co., New York.

de Queiroz, K. (1997). The Linnean hierarchy and the evolutionization of taxonomy, with emphasis on the problem of nomenclature. *Aliso* **15:** 125–144.

Hillis, D. M., Moritz, C., and Mable, B. K. (Eds.). (1996). *Molecular Systematics.* Sinauer Associates, Inc., Sunderland, MA.

Kardong, K. V. (2006). *Vertebrates: Comparative Anatomy, Function, Evolution* (4th ed.). McGraw Hill, Boston, MA.

Schultze, H. P. and Trueb, L. (Eds.). (1991). *Origins of the Higher Groups of Tetrapods. Controversy and Consensus.* Comstock Publishing Associates, Ithaca, NY.

Sumida S. S., and Martin, K. L. M. (2000). *Amniote Origins: Completing the Transition to Land.* Academic Press, San Diego, CA.

REFERENCES

General

Blackburn, 2006a,b; Flintoft and Skipper, 2006; Gibbons, 1990; Hertz et al., 1993; Janzen and Phillips, 2006; Jenssen and Nunez, 1998; McGuire et al., 2007; Radtkey et al., 1995; Reilley et al., 2007; Roughgarden, 1995; Ryan and Rand, 1995; Tinsley and Kobel, 1996; Vitt and Pianka, 1994; Wake and Hanken, 1966; Wilbur, 1997; Wright and Vitt, 1993.

Relationships Among Vertebrates: Origin of Tetrapods

Coates and Clack, 1995; Coates et al., 2000; Coates et al., 2002; Cote et al., 2002; Daeschler et al., 2006; Edwards, 1989; Fox, 1985; Fritzsch, 1990; Gensel and Andrews, 1987; Graham, 1997; Hughes, 1976; Lauder and Gillis, 1997; Little, 1990; Romer, 1960; Ruta et al., 2003a; Ruta et al., 2006; Schmalhausen, 1968; Schultze, 1991; Scott, 1980; Thomson, 1980, 1993; Whitear, 1977; Zhu et al., 2002.

Key Fossils

Ahlberg et al., 2000; Alhberg et al., 2002; Ahlberg and Clack, 2006; Ahlberg et al, 2005; Blom, 2005; Carroll, 2007; Clack, 1998, 2006; Daeschler et al., 2006; Eernisse and Kluge, 1993; Gauthier et al., 1988c, 1989; Hanken, 1986; Hedges et al., 1993; Hedges and Maxson, 1996; Panchen, 1991; Panchen and Smithson, 1988; Schmalhausen, 1968; Schubin et al., 2006; Schultze, 1991, 1994; Schwenk, 2003; Vorobyeva and Schultze, 1991.

Major Features of Early Tetrapod Evolution

Ahlberg, et al., 2005; Brainerd and Owerkowicz, 2006; Lebedev, 1997; Ruta et al., 2003b; Shubin et al., 2004.

Evolution of Early Anamniotes: Ancient Amphibians

Ahlberg, 1995; Ahlberg and Milner, 1994; Benton, 1997a,b,c; Carroll, 1988, 1995, 2007; Garland et al., 1997; Panchen and Smithson, 1988; Ruta and Coates, 2007; Ruta et al., 2001; Ruta et al., 2003; Trueb and Clouthier, 1991a,b; Wake, 1989.

Modern Amphibians

Bolt, 1991; Carroll, 1988, 1997, 2007; Laurin, 2002; Trueb and Clouthier, 1991b; Zardoya and Meyer, 2001.

Evolution of Early Amniotes: Early Tetrapods and Terrestriality

Benton, 1990, 1997a; Carroll, 1988; Clack, 2006; Frolich, 1997; Goin, 1960; Hotton et al., 1997; Lauder and Gillis, 1997; Laurin and Reisz, 1997; Martin and Nagy, 1997; Packard and Kilgore, 1968; Packard and Packard, 1980; Packard and Seymour, 1997; Sumida, 1997.

Early Amniotes

Carroll, 1988; deBraga and Rieppel, 1997; Gaffney, 1980; Gauthier et al., 1988b, 1989; Hedges and Poling, 1999; Laurin and Reisz, 1995; Lee, 1993, 1995, 1997; Reisz and Laurin, 1991; Rieppel and deBraga, 1996; Zardoya and Meyer, 1998.

Radiation of Diapsids

Benton, 1997; Benton and Clark, 1988; M. Caldwell, 1996; Callaway, 1997; Clark, 1994; Evans, 1988; Farlow and Brett-Surman, 1997; Feduccia, 1999; Feduccia and Wild, 1993; Gauthier, 1986, 1988c; Martin, 1991; Ostrom, 1991; Rieppel, 1993, 1994; Storrs, 1993; Tarsitano and Hecht, 1980; Witmer, 1991.

Linnean Versus Evolutionary Taxonomy

Avise, 2006; de Queiroz, 1997; de Queiroz and Gauthier, 1992; Hillis, 2007; Mayr and Ashlock, 1991.

Rules and Practice

Mayr and Ashlock, 1991; Salemi and Vandamme, 2003.

Evolution-Based Taxonomy

Benton, 2000; de Queiroz 1997; de Queiroz and Gauthier, 1992, 1994; Schuh, 1999.

Systematics—Theory and Practice

de Queiroz and Gauthier, 1994; Hall, 2007; Nielsen, 2005; Oosterbrock, 1987; Simpson, 1961.

Systematic Analysis

Avise, 2004; Felsenstein, 2004; Goodman et al., 1987; Swofford et al., 1996.

Types of Characters

Buth, 1984; Green and Sessions, 1991; Guttman, 1985; Hillis and Moritz, 1990; Hillis et al, 1996; Holder and Lewis, 2003; Maxson and Maxson, 1986, 1990; Moritz and Hillis, 1996; Murphy et al., 1996; Palumbi, 1996; Richardson et al., 1986; Sessions, 1996.

Methods of Analysis

Gascuel, 2007; Hennig, 1966; James and McCullough, 1985, 1990; Kachigan, 1982; Kluge and Farris, 1969; Nei and Kumar, 2000; Schmitt, 1989; Semple and Steel, 2003; Sokal, 1986; Swofford et al., 1996; Thorpe, 1976, 1987; Watrous and Wheeler, 1981; Wiley, 1981; Wiley et al., 1991.

Chapter 2

Anatomy of Amphibians and Reptiles

DEVELOPMENT AND GROWTH

Ova, Sperm, and Fertilization

All vertebrate life begins with a single cell, the zygote. For most amphibians and reptiles, this single cell results from the fusion of an ovum and a spermatozoan, the female and the male sex cells, a process called *fertilization*. Fertilization occurs predominantly outside the female's body and reproductive system (external fertilization) in frogs and inside the female's reproductive system (internal fertilization) in all caecilians, most salamanders, and all reptiles. Sex cells or gametes are unlike any other cells in the body because they have one-half the number of chromosomes (a haploid condition, 1N) of the typical body cell. Their sole role is fusion and creation of a new individual. They, of course, differ in structure as well; these details and those of the subsequent aspects of gametogenesis and fertilization are presented in Chapter 4.

Cells that will produce gametes differentiate early in development and migrate from their origin along the neural tube to the gonadal area of the embryo. The surrounding cell mass differentiates into the gonadal tissues and structures that support and nourish these precursors of the sex cells. The precursor cells can produce additional cells by the usual mode of cell division (mitosis); however, gamete production requires a special mechanism (meiosis) that halves the number of chromosomes. Consequently, each spermatozoan and ovum has the haploid number of chromosomes, and upon fertilization, the chromosome number is restored to diploid, or 2N. The series of steps in this meiotic or reductive cell division is known generally as *gametogenesis*. Oogenesis is the production of ova, and spermatogenesis is the production of spermatozoa. Gametogenesis defines an individual's sexual maturity, and as you will see later in this chapter, sexual maturity and what we think of as morphological maturity can often be offset (i.e., they may not occur together).

EARLY DEVELOPMENT

Embryogenesis

Development consists of control of cell growth and differentiation (embryogenesis) and morphogenesis (see following section). The zygote is the single cell resulting from the fusion of the nuclei of the spermatozoan and ovum. The zygote undergoes successive divisions (cleavage) that result in formation of a blastula, a ball of cells. Cleavage is a progressive division of the larger zygote cell into smaller and smaller cells. Cleavage continues until the cells of the blastula reach the size of normal tissue cells. No overall change in size or mass of the original zygote occurs; however, the amount of yolk in the zygote greatly affects the manner of cleavage, the resulting blastula, and the blastula's subsequent development. Because of their differing yolk content, the transformation of amphibian and reptilian zygotes into embryos is not identical and, therefore, the term *development* has two different but overlapping meanings. Development usually refers to all embryological processes and the growth (enlargement) of the embryo. Development can also refer to just the embryological processes, including embryogenesis (the formation of the embryo and its embryology through metamorphosis, hatching, or birth), organogenesis (the formation of organs), and histogenesis (the formation of tissues).

Ova are categorized by their yolk content. Isolecithal ova have a small amount of yolk evenly distributed throughout the cell. Mammals have isolecithal ova, but amphibians and reptiles do not. These two clades have mesolecithal (moderately yolked) and macrolecithal (heavily yolked) ova, respectively (Table 2.1); the ova of most direct-developing amphibians tend toward macrolecithal. The latter situation highlights the developmental modes of the two yolk classes. Moderate amounts of yolk permit only partial development of an embryo within the egg and its protective capsules before it must hatch and become free-living, at which time it is called a larva. Large amounts of yolk permit complete development of an embryo within an egg or within or on one of its parents; when a "macrolecithal" embryo hatches, its development is largely complete and it is a miniature replicate of its parents.

Hatching occurs long after the zygote is formed, and the developmental routes are varied. Cleavage of the mesolecithal ovum is complete or holoblastic, that is, the first cleavage furrow divides the zygote into two equal halves, the second furrow into four equal-sized cells, and so on. Yolk concentration is greater in the bottom half of the zygote, and cell division is slower there; nonetheless, the result is a blastula—a ball of cells with a small cavity in the upper half. In contrast, cleavage of a macrolecithal ovum is incomplete or meroblastic because the mass of yolk allows only a superficial penetration of the cleavage furrow (Table 2.1). These furrows are confined to a small area on the top of the zygote, and the resulting blastula is a flat disc of cells covering about one-third of the surface of the original ovum. The entire mesolecithal blastula becomes the embryo, whereas only the disc-blastula of a macrolecithal ovum becomes the embryo and associated extra-embryonic membranes.

The next phase, gastrulation, includes cell movement and cell division, and results in the formation of the three embryonic tissue layers. These layers (ectoderm, mesoderm, and endoderm) are the precursor tissues to all subsequent tissues. They still consist of undifferentiated cells at the conclusion of gastrulation, but by the cells' segregation into layers, their fate is determined. Ectoderm

TABLE 2.1 Summary of Development in Extant Amphibians and Reptiles.

	Amphibia	Reptilia
Ovum size (diameter)	1-10 mm	6-100+ mm
Yolk content	Moderate to great	Great
Fertilization	External or internal	Internal
Cleavage	Holoblastic[1]	Meroblastic
Embryo	Ovum-zygote elongating to pharyngula	Cleavage-cell disk folding to pharyngula
Fate of ovum-zygote	Zygote becomes entire embryo	Cell disk forms embryo and extra-embryonic structures
Mode of development	Indirect or direct	Direct

[1] *In amphibians with large, yolked eggs and direct development, meroblastic cleavage has been reported only for the salamanders of the genus* Ensatina (Hanken and Wake, 1996).
Source: In part, after Ellison, 1987.

becomes epidermal and neural tissues; mesoderm forms skeletal, muscular, circulatory, and associated tissues, and endoderm forms the digestive system tissues. In amphibian gastrulation, an indentation appears on the upper surface of the blastula. The indentation marks the major area of cell movement as the cells migrate inwardly to form the embryonic gut tube with the mesoderm lying between this tube and the external (ectoderm) layer. At the completion of gastrulation, the embryo is still largely a sphere. In reptilian gastrulation, cell movement creates an elongate, but unopened, indentation (the primitive streak) along the future anteroposterior axis of the embryonic disc. A cavity does not form and the endoderm appears by a delamination of the underside of the embryonic disc. This delamination typically precedes the formation of the primitive streak.

Before gastrulation concludes, a new set of cell movements and proliferation begins. This embryonic process is neurulation and, as the name suggests, establishes the neural tube, the precursor of the brain and spinal cord. Neurulation is accompanied by an elongation of the embryo. The embryo begins to take on form. Simultaneously, endodermal and mesodermal layers are proliferating, moving, and continuing their differentiation. The fate of these cells is determined at this point; they are committed to specific cell and tissue types. These processes in amphibians and reptiles result in a "pharyngula" stage in which the basic organ systems are established. However, amphibian and reptilian pharyngulae have strikingly different appearances and futures. The amphibian pharyngula contains all the yolk within its body as part of the digestive system. It will soon hatch from its gelatinous egg capsule and become a free-living larva; of course, direct-developing amphibian embryos follow a different development pathway, although their anatomy is largely the same as that of typical amphibians with larvae.

The reptilian pharyngula lies on top of a huge yolk mass, and this yolk mass is extra-embryonic; it is not part of the pharyngula. It becomes part of the embryo only through conversion of the yolk for nutrition. The endodermal tissue continues to grow outward and will eventually encompass the yolk mass, thereby forming the yolk sac (Fig. 2.1 and Fig. 4.3). While the reptilian pharyngula develops, the cells of superficial layers (ectoderm and mesoderm) of the extra-embryonic disc also proliferate and move. They grow upward and over the pharyngula and enclose it in an amniotic sheath (Fig. 2.1). The overgrowth begins at the anterior end of the embryo and proceeds in a wavelike manner to enclose the embryo. Because this up-and-over growth is a fold of tissue, the resulting sheath consists of four layers around a cavity: ectoderm, mesoderm, cavity mesoderm, and ectoderm. The outer two layers form the chorion, the cavity is the amniotic cavity, and the inner two layers become the amnion. Eventually, the chorion grows to encase the entire zygotic mass including the yolk sac (Fig. 4.4), whereas the amnion encloses only the embryo (Fig. 2.1). The allantois is the third "extra-embryonic" membrane, but unlike the amnion and chorion, it is an outpocketing of the hindgut. The allantois consists of endoderm and mesoderm and grows outward into the amniotic cavity, in many instances filling the entire cavity, with its outward wall merging with the amnion. This amniotic complex forms a soft "shell" within the leathery or hard shell of the typical reptilian egg.

MORPHOGENESIS

Developing Form and Function

Morphogenesis is the unfolding of form and structure. *Unfolding* refers to the differentiation of undifferentiated (unprogrammed) cells and the organization of these differentiated cells into tissues (histogenesis), organ systems (organogenesis), and a functional organism (embryogenesis). *Growth* simply refers to the enlargement of an organism and/or its component parts. While cells differentiate and take on specific functions, they also multiply. This multiplication can yield an increase in size (growth) of an organ or organism; however, cell multiplication can also produce migratory cells, such as neural crest cells, which migrate elsewhere in the embryo before forming a specialized tissue or organ, or cells with special functions such as blood cells, some of which transport oxygen and others that fight infections. These two phenomena and related ones are not considered as growth.

Morphogenesis has its beginning in the pharyngula, and subsequent development focuses on organogenesis and histogenesis. Within many amphibians, these two processes proceed rapidly to produce structures that enable the embryo to live outside the egg. Most larvae have full sensory capabilities for finding food and escaping predators as well as other necessary structures to perform the full range of life processes for survival. Hatching in a typical amphibian embryo occurs when specialized epidermal cells secrete a gelatinous substance to dissolve the egg capsule. Direct-developing amphibians and reptiles remain within the egg capsules or shells until embryogenesis is complete, hatching as miniature replicates of adults. The details of organogenesis are available in embryological textbooks, but one aspect, the timing of ontogenetic events, is an essential element of amphibian development and evolution, and indeed is critical to the evolution of new lifestyles and body forms in all organisms. Changes in developmental timing at any stage of an organism's ontogeny have the potential to create a structurally and physiologically different organism. In additon, structures and functions within an individual can vary in developmental timing independently.

Heterochrony

Shape arises from differential growth within a structure. If a ball of cells multiples uniformly throughout, the result is an ever-enlarging sphere; however, if the cells in one

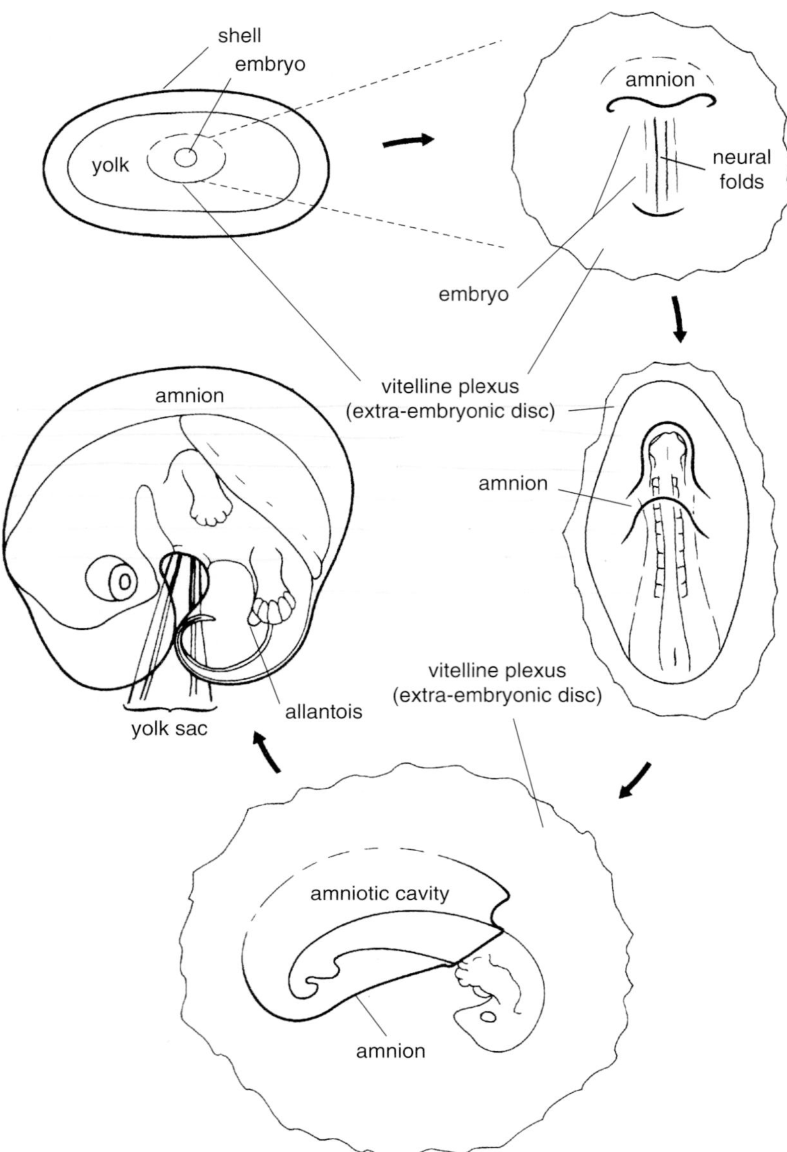

FIGURE 2.1 Selected developmental states of a turtle embryo showing the formation of the extra-embryonic membranes. Clockwise from upper left: shelled egg showing early embryogenesis; embryonic disc during neural tube formation and initiation of amniotic folds; embryo during early morphogenesis as somites form showing rearward growth of the amniotic fold as it envelopes the embryo; embryo in early organogenesis with initial outgrowth of the allantois; near-term embryo encased in amnion showing the yolk-sac attachment protruding ventrally. Adapted from Agassiz, 1857.

area grow more slowly than surrounding cells, the sphere will form a dimple of slow-growing cells. Such differential growth is a regular process of development, and each pattern of differential growth is usually genetically programmed so that every individual of a species has the same, or at least similar, body form, although environmental factors can alter the pattern. Timing and rate of growth are the essential ingredients for the production of specific shapes and structures, and shape and structure affect the function of tissues, organs, or organisms.

Changes in timing and/or rate of growth (i.e., heterochrony) have been a common feature in the evolution of amphibians and reptiles, and especially in salamanders. The recognition of heterochrony as a concept arose from the observation that differences in the morphology of some species could be explained by changes in their ontogeny. Ontogenetic processes can begin earlier (pre-displacement) or later (post-displacement) or can end earlier (hypomorphosis) or later (hypermorphosis) than in an ancestor (Table 2.2). These alterations are measured relative to the normal onset (beginning) or offset (termination) times; they refer specifically to the development of a trait or feature of an organism, such as foot structure or head shape. Alterations of the ontogeny also occur when the speed of the developmental rate is shifted either faster (acceleration) or slower (deceleration); either of these shifts can result in a different morphology. The final condition of the trait relative to its condition in the ancestor determines the pattern of heterochrony. A trait might not develop fully (truncation), it might develop beyond the ancestral condition (extension), or it might remain the same as the ancestral trait even though the developmental path differs. A single or related

TABLE 2.2 Patterns and Processes of Heterochrony.

Pattern	Simple pertubations (process)	Pattern: Interspecific (process)	Pattern: Intraspecific (process)
Truncation of trait offset shape	Decelerated (deceleration) Hypomorphic (hypomorphosis) Post-dispaced (post-displacement)	Paedomorphic (paedomorphosis)	Paedotypic (paedogenesis)
Extension of trait offset shape	Accelerated (acceleration) Hypermorphic (hypermorphosis) Pre-displaced (pre-displacement)	Peramorphic (peramorphosis)	Peratypic (peragenesis)
No change in trait offset shape	Must involve more than one pure perturbation	Isomorphic (isomorphosis)	Isotypic (isogenesis)

Source: Reilly *et al.*, 1997.

set of traits can change in descendants without affecting the developmental timing and rates of other traits; paedogenesis (Table 2.2) is a common heterochronic event in amphibians. These processes and the resulting patterns occur at two different scales, intraspecific and interspecific. Changes in a trait within populations (intraspecific) or a species result in different morphs within the same population, such as carnivorous morphs of spadefoot tadpoles. Differences in a trait's development among species (interspecific) reflect phylogenesis. These two levels of heterochrony and the complex interplay of heterochronic processes have led to confusion and an inconsistent use of terms. Dr. Steve Reilly and his colleagues constructed a model that demonstrates some of this complexity and applies a set of terms making the process of heterochrony relatively easy to understand (Table 2.2). By understanding this simple model, much developmental variation within and among species can be attributed to heterochrony.

This simple model centers on developmental patterns in an ambystomatid salamander in which individuals with larval morphology as well as individuals with adult morphology can reproduce. *Paedomorphosis* and *paedogenesis* refer to a developmental process in which a trait fails to develop to the point observed in the ancestral species or individuals, respectively. The axolotl (*Ambystoma mexicanum*) is a paedomorphic species. Morphological development of certain traits in the axolotl is truncated relative to that in its ancestral species *Ambystoma tigrinum*. Intraspecifically, morphs of *Ambystoma talpoideum* with larval traits can reproduce, hence their morphological development is truncated relative to their reproductive development and thus they exhibit paedogenesis (Fig. 2.2). Many other examples exist. For example, the tiny head relative to body size in New World microhylids frogs likely represents truncatation of head development (Fig. 2.3).

Peramorphosis and *peragenesis* refer to a developmental process in which a trait develops beyond the state or condition of that trait in the ancestral species or individuals, respectively. The male *Plestiodon* [*Eumeces*] *laticeps* develops a very large head relative to head size in its sister species *P. fasciatus*, which presumably represents the ancestral condition. The larger head is an example of peramorphosis; however, individuals within populations of *P. laticeps* have variable head size. This intraspecific variation likely arises from sexual selection and represents peragenesis, assuming that a smaller head size is the population's ancestral condition, a reasonable assumption considering that females and juveniles have relatively small heads.

Isomorphosis and *isogenesis* refer to a developmental process in which a trait is identical to the trait in the ancestral species or individuals, respectively, but the developmental pathway is different. For isomorphy or isogenesis to occur, development must undergo two or more heterochronic processes in order to "counteract" differences in developmental timing and speed. The various species of the salamander *Desmognathus* display direct and indirect development with variable durations of embryogenesis, yet adult morphology (head shape, skull, and hypobranchial architecture) is nearly identical, exemplifying isomorphosis. Isogenesis occurs in *Ambystoma talpoideum* where adult terrestrial morphology is identical in those individuals that underwent a typical developmental pattern and in those individuals that were paedogenic (Fig. 2.2).

EMBRYONIC LIFESTYLES

Protective Barriers

Tetrapod zygotes have barriers to protect them from predation by micro- and macroorganisms, from physiological challenges, and from abiotic physical threats. For those amphibians and reptiles with internal development, whether intra- or extrauterine, the parent's body provides the shield;

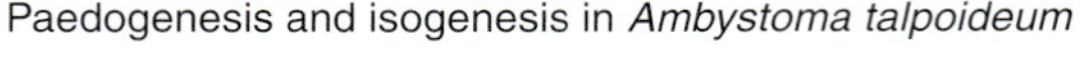

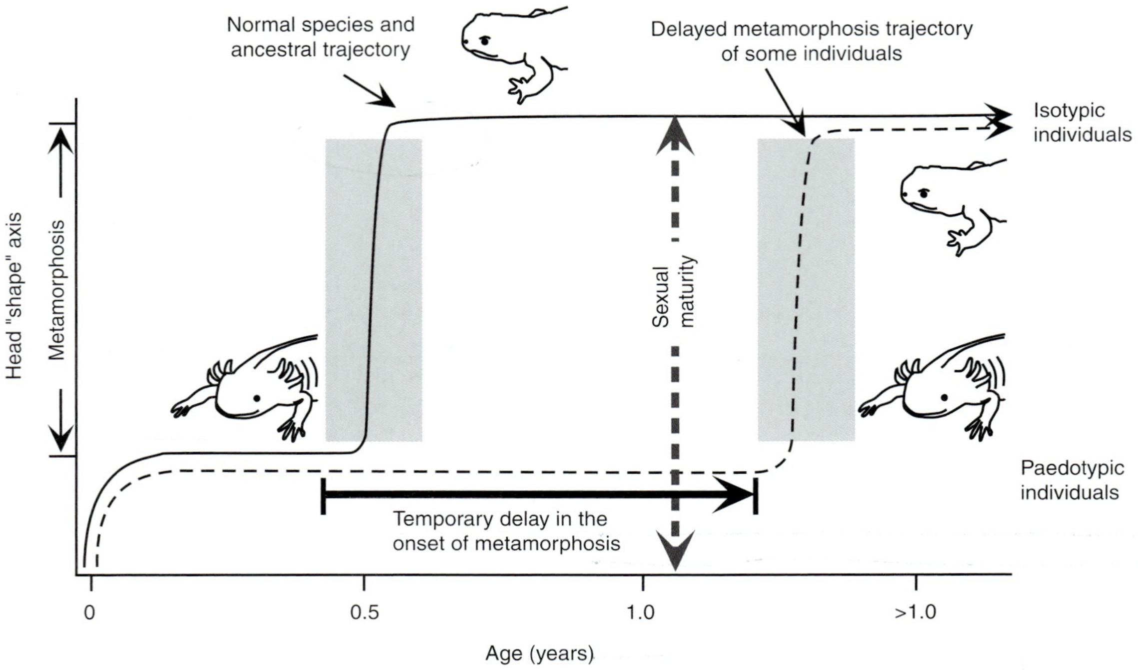

FIGURE 2.2 Paedogenesis and isogenesis in *Ambystoma talpoideum*. The life history of *A. talpoideum* demonstrates the complexities of trait development patterns. The ancestral condition for this species is metamorphosis into a terrestrial salamander in less than one year. Under certain environmental conditions, paedogenesis occurs when metamorphosis is delayed and results in sexual maturation of the individual with retention of larval traits (i.e., the larval morphology) producing paedotypic individuals. Isogenesis occurs when similar early larvae follow different developmental trajectories but ultimately produce similar adults. The adults are termed isotypic individuals. Figure courtesy of S. M. Reilly.

however, for externally deposited zygotes (eggs), a protective barrier must be deposited around the ova before they are released to the outside. Amphibians encase their ova in several mucoprotein and mucopolysaccharide layers that can be penetrated by a sperm in the cloaca or immediately upon release of eggs into the external environment. These layers form the gelatinous capsules and egg masses of amphibians (Fig. 4.4). Reptiles, which have internal fertilization, can encase their zygotes in a fibrous capsule that is made even more durable by the addition of calcium salts, thereby producing calcareous shells. Additional details of protective barriers are in Chapter 4.

FIGURE 2.3 The concept of heterochrony can be applied to a wide variety of traits. The New World microhylid, *Dermatonotus muelleri*, has a tiny head relative to its body and, because other New World microhylids are similar, truncation of head development likely occurred in an ancestor to the clade of New World microhylids. (Luis Gasparini)

Larvae—Free-Living Embryos

The diversity of amphibian larval morphology equals the diversity of adult stages. Most larvae feed during their free-living developmental period; however, some do not eat and depend upon the yolk stores of the original ovum. Caecilian and salamander larvae resemble adults in general appearance and anatomical organization (Fig. 2.4). The transition (metamorphosis) from embryonic larva to nonembryonic juvenile is gradual with only minor reorganization. In contrast, the anuran larva (tadpole) undergoes a major reorganization during its metamorphosis from embryo to juvenile because the tadpole is anatomically different from the juvenile and adult.

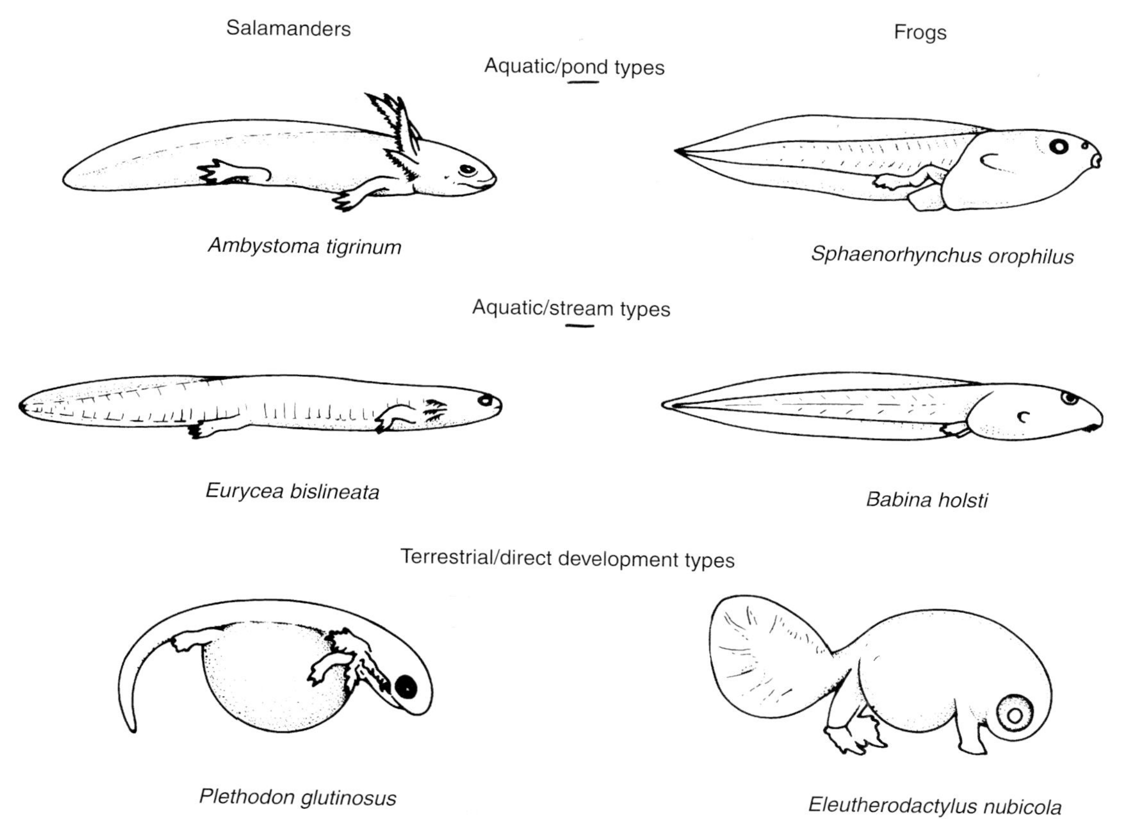

FIGURE 2.4 Body forms of some amphibian larvae arranged by habitat type.

With few exceptions, larvae of the three amphibian groups are aquatic. All share anatomical characteristics associated with an aquatic existence. They have thin, fragile skin consisting of two or three epidermal layers. The skin is heavily vascularized owing to its role as a major respiratory surface, a role shared with the gills. All amphibian larvae develop pharyngeal slits and external gills—usually three pairs that project from the outside of the pharyngeal arches. The external gills persist and function throughout the larval period in salamanders, basal anurans, and caecilians. In tadpoles of neobatrachian frogs, external gills are resorbed and replaced by internal gills, which are lamellar structures on the walls of the pharyngeal slits. All larvae have lidless eyes and large, nonvalvular nares. They have muscular trunks and tails for undulatory swimming, and the tails have dorsal and ventral fins. The skeleton is entirely or mainly cartilaginous. All have well-developed lateral line systems.

Caecilian and salamander larvae are miniature adult replicates, differing mainly by their smaller size, pharyngeal slits and gills, tail fins, a rudimentary tongue, and specialized larval dentition. In contrast the body plan of the anuran tadpole bears little similarity to the adult's. In general, tadpoles are well designed for consuming food and growing; the most salient feature of the body is a large coiled intestine. Mouth and eyes are situated anteriorly, the centrally located body is spherical, and a muscular tail provides the thrust that results in tadpole movement. Functional limbs do not appear until late in larval life and then only the hindlimbs are visible externally. The front limbs develop at the same time as the hindlimbs, but they are enclosed within the operculum and emerge only at metamorphosis.

The general tadpole body form has been modified into hundreds of different shapes and sizes, each adapted to a specific aquatic or semiterrestrial habitat and feeding behavior. This diversity has been variously partitioned. In the 1950s, Dr. Grace Orton recognized four basic body plans; her morphotypes defined the evolutionary grade of tadpoles and to some extent their phylogenetic relationships. Another approach is to examine the relationship between tadpole morphology and ecological niches. One such an analysis defined 18 guilds based on ecomorphology, which, with their subcategories, included 33 body types. Although morphotypes can define adaptive zones of tadpoles, they do not necessarily reflect phylogenetic relationships because considerable convergence has occurred. Both classifications emphasize external, oral, and pharyngeal morphology.

Most tadpoles have a large, fleshy disc encircling their mouth. Depending on the manner of feeding and the type of food, the oral disc ranges in position from ventral (suctorial, to anchor in swift water and scrape food off rocks) to dorsal (grazing on surface film in calm water) and in shape from round to dumbbell. The margin of the disc is variously covered with papillae, and these have a variety of shapes. Their actual function remains uncertain, although chemosensory, tactile, and current detection are some possibilities. Tadpoles lack teeth on their jaws; instead many tadpoles have keratinous jaw sheaths and parallel rows of keratinous labial teeth on the oral disc above and below the mouth. The labial teeth are not homologous with teeth of other tetrapods. The jaw sheaths cut large food items into smaller pieces; the rows of labial teeth act as scrapers or raspers to remove food from rocks and plant surfaces. The oral–pharyngeal cavity is large. Its structures trap and guide food into the esophagus, as well as pump water through the cavity and across the gills. The gills are initially visible externally, but at hatching or shortly thereafter, an operculum grows posteriorly from the back of the head to fuse to the trunk, enclosing the gills and the developing forelimbs. To permit water flow, a single spiracle or pair of spiracles remains open on the posterior margin of the operculum. Because the operculum covers the gill region, the head and body form a single globular mass. Adhesive glands are transient structures present near the mouth in early embryonic stages at the time of hatching. The glands secrete a sticky substance that tadpoles use to adhere to their disintegrating egg mass or to some structure in the environment. Because of the fragility of the newly hatched larvae, adherence provides stability for the larva until the oral disc and tail musculature develop fully and locomotion becomes possible.

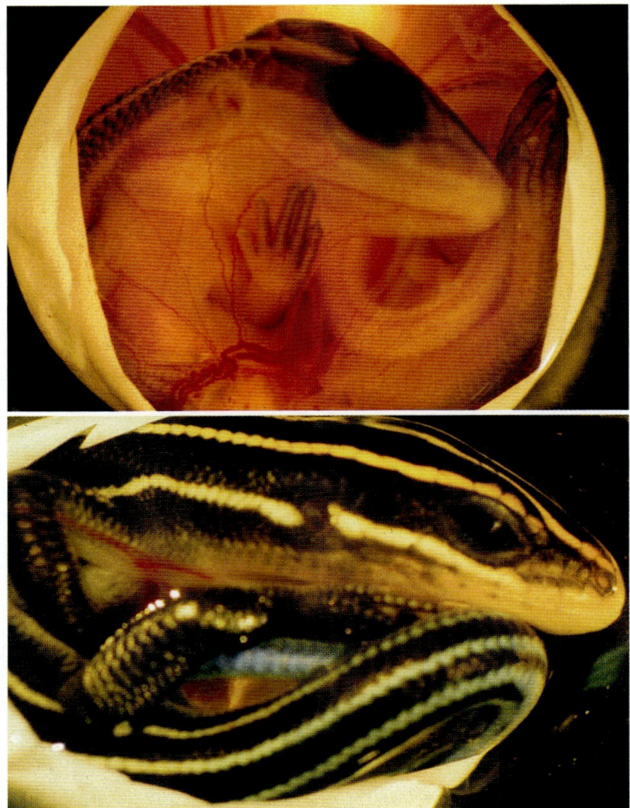

FIGURE 2.5 Reptiles are tightly coiled inside of eggs prior to hatching. Embryos of *Plestiodon fasciatus* inside of eggs. Developmental stage 39 (upper); stage 40 (lower). (James R. Stewart)

Life in an Eggshell

The eggshell protects the reptile embryo, but in so doing, it imposes special costs on embryo growth and physiology. An amphibian larva can grow to near adult size before metamorphosing, although most do not. A reptile in an eggshell cannot grow in size within the shell but must undergo complete development prior to hatching. By folding and curling, a reptile embryo can attain surprising lengths, but it is still smaller than would be possible outside of a shell (Fig. 2.5). Determinants of offspring size are complex and discussed elsewhere (see "Life Histories" in Chapter 5). Most reptile hatchlings are, however, heavier than the mass of the original ovum. Metabolism of the yolk uses water absorbed through the shell, and the embryo grows beyond the original ovum.

Just as temperature, water availability, and gas exchange affect the physiological processes of juveniles and adults, they also have the greatest impact on developing eggs. Eggs are not laid randomly in the environment; females select sites that offer the greatest potential for egg and hatchling survival. Oviposition site selection has been honed by natural selection over generations of females. Nevertheless, the abiotic and biotic environments are extremely variable, and eggs and their enclosed embryos must tolerate and respond to these varying conditions. A few examples illustrate the breadth of nesting environments and egg–embryo physiological responses.

Temperature tolerances of embryos lie typically within the tolerance range of the juveniles and adults of their species, but because the rate of development is temperature dependent and eggs lack the mobility to avoid extremes, exposure to extremes is likely to be fatal. At low temperatures, development slows down and hatching is delayed, resulting in emergence at suboptimal times or embryos that never complete development. At high temperatures, the embryo's metabolism increases exponentially so that yolk stores are depleted before development is completed, and of course, either extreme can be directly lethal by damaging cells and/or disrupting biochemical activity. The selection of protected oviposition sites potentially avoids the extremes of temperature and provides a stable temperature environment. But temperatures do fluctuate within and among nests, and in some reptiles with temperature-dependent sex determination, skewed sex ratios among hatchlings can result (see Chapter 5).

Moisture is no less critical for the proper development and survival of reptile embryos than for amphibians; however, amphibians typically require immersion in water, whereas immersion of most reptile eggs results in suffocation of embryos. Embryos do not drown; rather, the surrounding water creates a gaseous-exchange barrier at the shell–water interface, and the small amounts of gases that cross are inadequate to support cellular metabolism. The Australian sideneck turtle *Chelodina rugosa* avoids this dilemma, even though females lay their eggs in submerged nests. Once the eggs are laid, development stops. Developmental arrest typically occurs in the gastrulation phase, and embryogenesis begins only when the water disappears and the soil dries, permitting the eggs and/or the embryos to respire. The relative availability of water affects the rate of development and absolute size of the hatchlings, at least in turtles. *Chrysemys picta* eggs in high-moisture nests hatch sooner and produce larger hatchlings than those from nests with lower moisture. Developmental abnormalities can also result if hatchlings experience dehydration as embryos.

Adequate gas exchange is an unlikely problem for species that lay or attach their eggs openly in cavities or crevices (e.g., many geckos), but for the majority of reptiles that bury their eggs, adequate gas exchange can be critical. Changes in soil permeability affect the diffusion of air, drier soils having the highest diffusion rates and wet soils the lowest. Similarly, the friability of soils and associated aspects of particle size and adhesiveness influence the movement of gas through the soil. In selecting a nesting site, a female must balance her ability to dig an egg chamber with the presence of adequate moisture to prevent desiccation and yet not retard gaseous diffusion, as well as a multitude of other factors.

CHANGING WORLDS—METAMORPHOSIS, HATCHING, AND BIRTH

Metamorphosis

Metamorphosis in amphibians is the transformation of the larva to a miniature adult replicate, and usually from an aquatic to a terrestrial or semiterrestrial lifestyle. It signals the completion of embryogenesis. Some developmental processes, such as maturation of gonads, continue through the juvenile stage, but the major structural and physiological features are in place at the conclusion of metamorphosis. Metamorphosis is nearly imperceptible in caecilians and salamanders but dramatic in frogs (Table 2.3). Anuran larvae require major structural and physiological reorganization because of the striking differences between the larval and the juvenile/adult stages. Change does not occur all at once but gradually, each step leading to next level of transformation. Unlike insect pupae, metamorphosing tadpoles remain active, capable of avoiding predators and environmental stresses.

During much of larval life, growth is emphasized over morphogenesis. Morphogenesis is greatest in the early stages and then slows for caecilian and salamander larvae; frog larvae similarly undergo major development changes in their early stages, but they also display distinct structural changes throughout larval life (Fig. 2.6). Larval life span is variable—from less than 20 days in some spadefoot (frogs in the family Scaphiopodidae) populations to several years in other frogs and salamanders. The duration is species specific and genetically fixed, but not rigidly so. torrent
Metamorphosis marks the beginning of the end of larval life; once begun, metamorphosis usually proceeds rapidly. Rapidity is necessary to reduce the transforming amphibian's exposure to predation or other potential stresses when it is neither fully aquatic nor fully terrestrial.

Metamorphosis is initiated internally by the hormone thryroxine, but environmental factors can initiate early thyroxine release if a larva has completed certain morphogenic events. For example, crowding, reduced food or oxygen, drying of water bodies, or increased predation can result in thyroxine release. Although thyroxine and its derivatives promote metamorphosis, they do not operate alone. The thyroid is present early in larval life, but its secretory activity is apparently inhibited by corticoid hormones, such as corticosterone. Furthermore, prolactin is abundant in early larval stages and makes the body tissues insensitive to thryroxin. When these inhibitions are removed, the thyroid secretes thyroxin, effecting transformation.

Hatching and Birth

In amphibians, the timing of hatching depends upon the life history. For those species with larvae, hatching occurs early in embryogenesis typically at Gosner stage 17, and for those species with direct-developing embryos, hatching occurs at the completion of development. Direct-developing embryos do not pass through a major metamorphic event. Exit from the egg in either situation requires penetration of the gelatinous egg capsules. The actual hatching mechanism is known only for a few species, but because these all share "hatching" glands on the snout and head of the larvae, the mechanism is probably common to most other amphibians. These glands secrete proteolytic enzymes that weaken and dissolve the capsules, allowing the larva or juvenile to escape. Froglets in the genus *Eleutherodactylus* are assisted by an egg tooth, a bicuspid structure located on the upper lip. Stage 15 embryos use the structure to slice through the tough outer egg capsules. The structure sloughs off within two days after hatching. Birth, whether from an intra-uterine or extra-uterine situation and whether as a larval or juvenile neonate, appears to be triggered by a combination of maternal hormonal activity and embryonic/fetal secretions.

TABLE 2.3 Anatomical Changes in Frogs and Salamanders Accompanying Metamorphosis.

FROGS	SALAMANDERS
Buccal region	
Major remodeling	Slight remodeling
Oral disc with papillae and keratinous denticles and beak disappears	
Jaws elongate, enlarging mouth, and teeth develop	Teeth change from bicuspid to monocuspid
Buccal musculature reorganized	
Tongue muscles develop	Tongue muscles develop
Pharyngeal region	
Remodeling with shortening of the pharynx	
Gills and pharyngeal slits disappear	Gills and pharyngeal slits disappear
Rearrangement of aortic arches	Rearrangement of aortic arches
Modification of hyoid and segments of the branchial skeleton for tongue support	Modification of hyoid and segments of the branchial skeleton for tongue support
Viscera	
Lung development completed	Lung development completed
Stomach develops	Digestive tube modified slightly
Reduction of intestine and change of digestive epithelium	
Reduction of pancreas	
Pronephros kidney disappears	Pronephros kidney disappears
Skin	
Number of epidermal cell layers increases	Number of epidermal cell layers increases
Pigmentation and pattern change	Pigmentation and pattern change
Skeleton	
Ossification moderate to strong	Ossification slight to moderate
Major remodeling of cranial skeleton	Little change in cranial skeleton
Loss of tail; development of urostyle	
Sense organs	
Protrusion of eyes with development of eyelids	Protrusion of eyes with development of eyelids
Remodeling of eye and growth of eye muscles	
Development of stapes in middle ear	

Note: These structural changes represent only a portion of anatomical changes occurring during metamorphsis.
Source: Hourdry and Beaumont, 1985.

Birth in reptiles appears to be triggered largely by maternal hormonal activity, although a maternal–fetal feedback mechanism plays an essential role in the female's hormonal cycles. Hatching in reptiles requires the penetration of the amnionic membranes and the eggshell. Reptiles use a projection on the tip of the snout to break through these two enclosures. In turtles, crocodylians, and *Sphenodon*, the projection is a keratinous protuberance, the egg caruncle, which slices through the encasing layers (Fig. 2.7). Crocodylian and turtle embryos extract calcium from the eggshell during their embryogenesis, and this weakening of the eggshell makes it easier to rupture. Squamates presumably lost the caruncle and replaced it with an egg tooth that projects outward from the premaxillary bone. Hatching can be extended, requiring

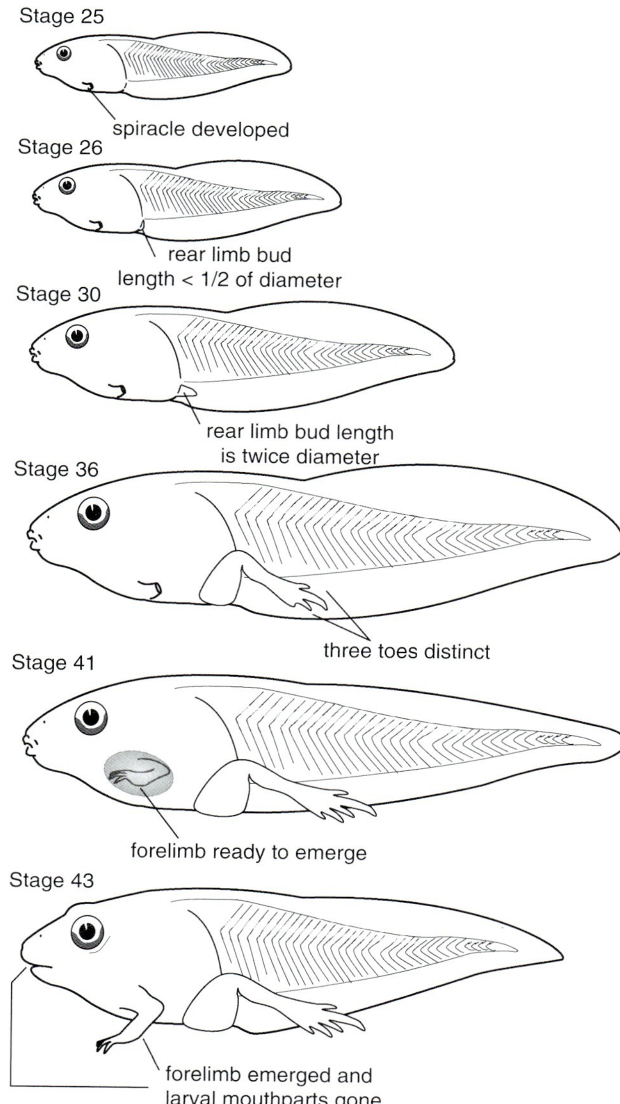

FIGURE 2.6 Selected larval stages of a typical anuran. Stage terminology from Gosner (1960).

several hours to a day for complete emergence, but can also be rapid, with near synchrony of hatching among eggs in the same nest. A few turtles have delayed emergence, hatching in autumn but not emerging from the nest until spring. This situation alerts us to the possibility that hatching and nest emergence are potentially separate events in other reptile species as well. Generally, parents are not involved in the hatching and emergence process. Nevertheless, parental crocodylians aid their young during hatching and emergence and some skinks (New World *Mabuya*) remove embryonic membranes from neonates when they are born. The possibility exists that many more species aid in the hatching and emergence process but are simply difficult to observe and thus are unreported.

FIGURE 2.7 Photograph of the egg of a *Geochelone sulcata* just beginning to hatch. The arrow points to the emerging egg tooth as it begins to slice through the leathery shell. (Tim Colston)

GROWTH

Growth is the addition of new tissue in excess of that required for the replacement of worn-out or damaged tissue. As a cellular process, growth rate in ectotherms depends on temperature, slowing and ceasing as temperature declines; excessive temperature also slows or halts growth because maintenance and metabolic costs exceed energy procurement. Growth is influenced by the availability and quality of food. In this respect, ectotherms have an advantage over endotherms by ceasing to grow during food shortages and renewing growth when food becomes available. This is one reason why reptiles and amphibians often persist in large numbers in extreme environments such as deserts when resources become low, either seasonally or as the result of extended drought. Metabolic demands of endothermy in mammals and birds render them vulnerable to starvation when resources are low.

Growth occurs primarily in the embryonic and juvenile stages of amphibians and reptiles. Embryonic growth usually is proportionately greater than juvenile growth, because embryos possess abundant, high-quality energy resources in the form of the yolk that require little energy expenditure for acquiring and processing. Juveniles and free-living amphibian larvae face variable food supplies, often with low energy content, and must expend energy to obtain and process food, while simultaneously avoiding predation and environmental hazards. From hatching or birth, most reptiles and amphibians will increase 3- to 20-fold in length, but some species may increase over 100-fold in mass. Growth may or may not continue indefinitely throughout life; unlike mammals and birds, in which growth slows dramatically or stops at sexual maturity (determinate growth), sexually mature amphibians and reptiles generally continue growing, giving the impression that growth is indeterminate.

Mechanics of Growth

All tissues grow during juvenile life, although the rate varies among tissues. Growth can be measured by changes in overall size, most often in length. Mass is more variable owing to numerous factors, such as hydration, gut contents, and reproductive state, each of which can change an animal's weight without changing its overall length. Skeletal growth is the ultimate determinant of size because the skeleton is the animal's supportive framework. Skeletal elements of amphibians and reptiles usually lack epiphyses and grow by apposition, a process in which one layer forms on top of another. Because of these attributes, extended growth is possible and leads to the assumption of indeterminate growth in these animals. Other reasons for assuming that indeterminate growth occurs are the large sizes of individuals in some species and the continuation of growth long after sexual maturity.

Both indeterminate (attenuated) and relatively determinate (asymptotic) growth exist in amphibians and reptiles, but the evidence for one or the other is lacking for most species (Fig. 2.8). Indeterminate growth may be less frequent than commonly assumed. Adult size for most species lies within a narrow range, suggesting that growth ceases. Older adults of some reptilian species have fused epiphyses. The two growth patterns in natural populations are difficult to distinguish because a narrow adult size range may indicate only that high mortality truncates the growth or size potential of the species or population.

Whatever the end point, juvenile growth is rapid and slows as sexual maturity is approached. Most juvenile growth fits one of two curvilinear patterns: parabolic growth, which may begin rapidly and remain rapid for most of juvenile life, or sigmoid growth, which is initially slow, becomes rapid, and then slows again. Both patterns show a plateau associated with maturity, a result of the reallocation of energy resources to reproduction. Individual growth curves are not smooth (Fig. 2.8), particularly in ectotherms. Growth proceeds fast or slow depending on the abundance of food, and it may halt for months at a time in species in seasonal environments, including tropical wet and dry seasons. Growth may proceed ratchetlike for the first few years of sexual maturity because energy is alternately allocated to reproduction and then to growth. Ratchetlike growth curves also result from seasonal variation in resource availability.

The ultimate size of an individual depends on its genetic potential, size at hatching, abundance and quality of food during juvenile growth, and its sex. Heredity determines the potential range of growth rate and size or age at sexual maturity. Beginning with hatchlings of the same size, species, and sex, faster growth or longer juvenile life yields larger adults. These factors and others yield the variations in adult size within and between species.

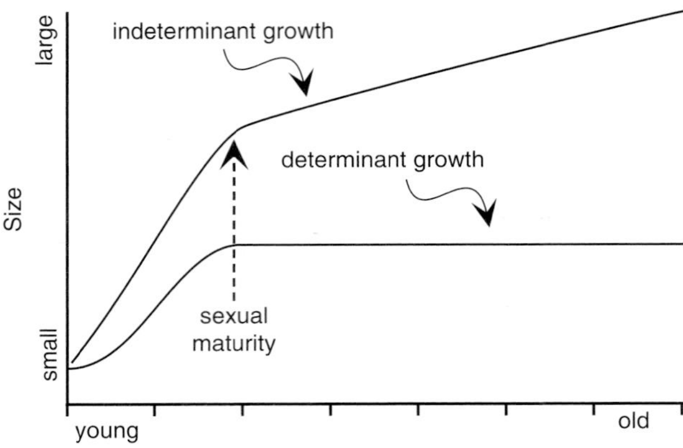

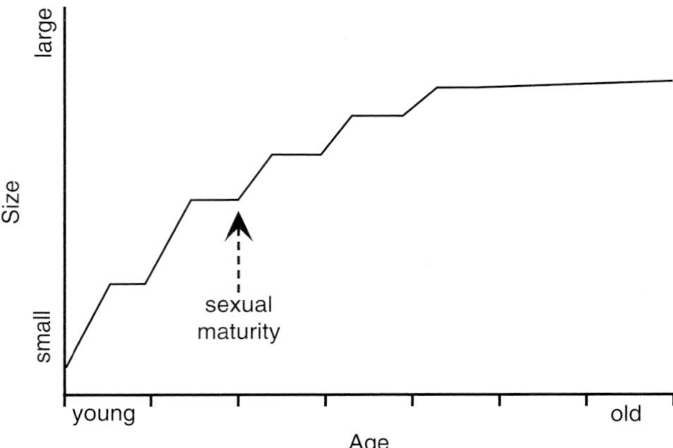

FIGURE 2.8 General growth pattern trends for amphibians and reptiles. Top: comparison of indeterminant and determinant growth. When growth is relatively indeterminant, constant growth rate as a juvenile is followed by slower, but continuous growth once sexual maturity is reached. When growth is determinant or asymptotic, a sigmoid pattern during juvenile stages is followed by slower growth after sexual maturity and finally curtailment of growth. Bottom graph: hypothetical growth for an ectotherm in a seasonal environment follows a pattern of rapid growth during equable seasons and greatly reduced or no growth during adverse seasons.

Age

The length of time an individual lives is not as critical as the time required to reach the major life-history events of hatching or birth, sexual maturity, and reproductive senility. Reproductive periodicity, the time interval between episodes of the production of offspring, is another critical age-related aspect of an individual's life history. In amphibians with a larval stage, two intervals are critical: embryogenesis within the egg and larval period to metamorphosis. All of these events are regularly subjected to selection within a population, and the modal condition within a population can shift.

Age at sexual maturity ranges from 4 to 6 months (*Arthroleptis poecilonotus*, an artholeptid frog) to 7 years (*Cryptobranchus*, hellbender salamander) for amphibians and from 2 to 4 months (*Anolis poecilopus*, a polychrotine lizard) to 40+ years (*Chelonia mydas*, green seaturtle) for reptiles. These marked extremes reflect differences in adult size only in part, because not all small species mature so quickly or large ones so slowly (Table 2.4). Age of maturity is a compromise among many variables on which selection may operate to maximize an individual's contribution to the next generation. Maturing and reproducing quickly is one strategy, but small body size reduces the number and/or size of offspring and smaller adults tend to experience higher predation. Maturing later at a larger body size permits the production of more and/or larger offspring but increases the probability of death prior to reproducing, and may yield a smaller total lifetime output of offspring. The resulting diversity in size and age at sexual maturity, number and size of offspring, and the frequency of reproduction illustrate the numerous options molded by natural selection for attaining reproductive success.

Longevity often indicates a long reproductive life span of an individual or species. The reproductive life span of some species (e.g., *Uta stansburiana*) is a single reproductive season, and most individuals disappear from the population within a year of hatching. Longevity in a few surviving individuals of *Uta stansburiana* can exceed 3 years in natural populations. For other species the reproductive life span can be a decade or longer, and individuals may live more than half a century (e.g., *Geochelone gigantea*). Annual or biennial species have little time for growth, so these species typically are small; the opposite is not true for the long-lived species. Although many long-lived species are large, some, such as the yucca ghost-lizard, *Xantusia vigilis*, are tiny yet long-lived. Often small-bodied long-lived reptiles or amphibians have secretive lifestyles.

INTEGUMENT—THE EXTERNAL ENVELOPE

The skin is the cellular envelope that forms the boundary between the animal and its external environment, and as such, serves multiple roles. Foremost are its roles in support and protection. The skin holds the other tissues and organs in place, and yet it is sufficiently elastic and flexible to permit expansion, movement, and growth. As a protective barrier, it prevents the invasion of microbes and inhibits access by potential parasites, resists mechanical invasion and abrasion, and buffers the internal environment from the extremes of the external environment. The skin also serves in physiological regulation (e.g., heat and osmotic regulation), sensory detection (chemo- and mechanoreception), respiration, and coloration.

TABLE 2.4 Natural Longevity of Select Amphibians and Reptiles.

Taxon	Adult size (mm)	Age at maturity (mo)	Maximum age (mo)
Cryptobranchus alleganiensis	330	84	300
Desmognathus quadramaculatus	73	84	124
Eurycea wilderae	34	48	96
Anaxyrus americanus	72	36	60
Lithobates catesbeianus	116	36	96
Chrysemys picta	119	72	360
Geochelone gigantea	400	132	840±
Trachemys scripta	195	50	288
Sphenodon punctatus	180	132	420+
Aspiciscelis tigris	80	21	94
Gallotia stehlini	120	48	132+
Uta stansburiana	42	9	58
Diadophis punctatus	235	32	180+
Pituophis melanoleucus	790	34	180+

Note: Body size is for females at sexual maturity (mm, snout-vent length except carapace length for turtles); age of maturity for females (months); maximum age of either sex (mo).
Sources: Salamanders – Ca, Peterson *et al.*, 1983; Dq, Bruce, 1988*b*; Organ, 1961; Ew, Bruce, 1988*a*. Frogs – Aa, Kalb and Zug, 1990; Lc, Howard, 1978. Turtles – Cp, Wilbur, 1975; Gg, Bourne and Coe, 1978; Grubb, 1971; Ts, Frazer *et al.*, 1990. Tuataras – Sp, Castanet *et al.*, 1988. Lizard – At, Turner *et al.*, 1969; Gs, Castanet and Baez, 1991; Us, Tinkle, 1967; Medica and Turner, 1984. Snakes – Dp, Fitch, 1975; Pmd, Parker and Brown, 1980.

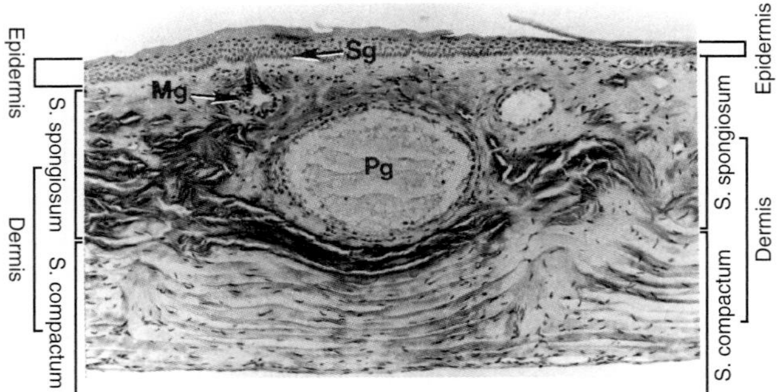

FIGURE 2.9 Amphibian skin. Cross section through the ventral skin of a marine toad *Rhinella* [*Bufo*] *marina*. Abbreviations: Mg, mucus gland; Pg, poison or granular gland; Sc, stratum compactum; Sg, stratum germinativum; Ss, stratum spongiosum.

Amphibian skin consists of an external layer, the epidermis, which is separated from the internal layer, the dermis, by a thin basement membrane (Fig. 2.9). The epidermis is typically two to three cell layers thick in larvae and five to seven layers thick in juvenile and adult amphibians. The innermost layer of cells (stratum germinativum) divides continuously to replace the worn outer layer of epidermal cells. The outer cell layer is alive in larvae, but in most juveniles and adults, cells slowly flatten, keratinize, and die as they are pushed outward. This layer of dead, keratinous cells (stratum corneum) shields the inner layers of living cells from injury. The dermis is a thicker layer, containing many cell types and structures, including pigment cells, mucous and granular glands, blood vessels, and nerves, embedded in a connective tissue matrix (Fig. 2.9). The innermost layer of dermis is a densely knit connective tissue (stratum compactum), and the outer layer (stratum spongiosum) is a looser matrix of connective tissue, blood vessels, nerve endings, glands, and other cellular structures. In caecilians and salamanders, the stratum compactum is tightly linked with the connective tissue sheaths of the muscles and bones. In contrast, much of the body skin is loosely attached in frogs.

The skin of reptiles has the same cellular organization as in amphibians. Notably, the epidermis is thicker with numerous differentiated layers above the stratum germinativum. Differentiation produces an increasingly thick, keratinous cell membrane and the eventual death of each cell. This basic pattern is variously modified among reptilian clades and occasionally among different parts of the body of the same individual. Reptiles uniquely produce ß-keratin as well as α-keratin, which they share with other vertebrates. ß-keratin is a hard and brittle compound, whereas α-keratin is elastic pliable.

On all or most of the body, skin is modified into scales. The scales are termed *plates, scutes, shields, laminae, lamellae, scansors,* or *tubercles,* depending upon the taxonomic group, the size and shape of the scales, and the location of the scales on the body. Some names are interchangeable, whereas others refer to specific structures. For example, scutes are the same as shields, but scansors are scales or lamellae beneath the digits that allow geckos to cling to nonhorizontal surfaces. All reptilian scales are keratinized epidermal structures, but those of the lepidosaurs are not homologues of crocodylian and turtle scales. Scales commonly overlap in squamates but seldom do in crocodylians and turtles.

Two patterns of epidermal growth occur. In crocodylians and turtles, the cells of the stratum germinativum divide continuously throughout an individual's life, stopping only during hibernation or torpor. This pattern is shared with most other vertebrates, from fishes to mammals. A second pattern, in which growth is discontinuous but cyclic, occurs in lepidosaurs (see the later section "Ecdysis"). Upon shedding of the outer epidermal sheath (Oberhautchen), the germinative cells enter a resting phase with no mitotic division. The renewal phase begins with the synchronous division of the germinative cells and the differentiation of the upward-moving epithelium into two distinct layers separated by a narrow layer of cell secretions.

The surface of each reptilian scale is composed entirely of ß-keratin, and the interscalar space or suture is composed of α-keratin. This distribution of keratin produces a durable and protective scale surface with junctures between the scales that allow flexibility and expansion of the skin. Although the preceding pattern is typical, the scales on the limbs of some turtles have surfaces composed of α-keratin, and in softshell and leatherback turtles, the surface of the shell is composed of α-keratin. In most of the hard-shelled turtles, the scutes and sutures contain only ß-keratin. The two-layered epidermis of the lepidosaurs has an α-keratin inner layer and a ß-keratin Oberhautchen.

An anomaly of special interest is the occurrence of individual snakes that are nearly scaleless in several species of colubrids and viperids. Only the labial and ventral scales are usually present. The remainder of the skin is a smooth sheet of soft, keratinous epidermis. Genetically, scalelessness appears to be a simple Mendelian homozygous recessive trait.

INTEGUMENTARY STRUCTURES

Amphibian Glands and Skin Structures

Amphibians have several types of epidermal glands. Mucous and granular (poison) glands occur in all postmetamorphic (i.e., juvenile and adult) amphibians and are numerous and widespread on the head, body, and limbs (Fig. 2.9). Both types are multicellular, flask-shaped glands with the bulbous, secretory portion lying within the stratum spongiosum of the dermis; their narrow necks extend through the epidermis and open on its surface. Although occurring over the entire body, the glands are not evenly distributed; their role determines their density and location. Mucous glands are the most abundant; about 10 of them are present for every granular gland. The mucous glands are especially dense dorsally, and they continuously secrete clear, slimy mucus that maintains a thin, moist film over the skin. The granular glands tend to be concentrated on the head and shoulders. Presumably, predators that attack these vulnerable parts of the body would be deterred when encountering poisonous or noxious secretions produced by the glands. The granular glands are often aggregated into macroglands, such as the parotoid glands of some frogs and salamanders (Fig. 2.10). Usually these macroglands contain more complex individual glands.

Larvae have a greater variety of epidermal glands. Most are single-celled (unicellular), although many can be concentrated in a single region. For example, the hatching glands are clustered on the dorsal forepart of the head. Unicellular mucous glands are widespread and secrete a protective mucous coat over the surface of the living epidermis. This mucous coat also serves as a lubricant to enhance the flow of water over the larva when swimming. Merkel and flask cells are scattered throughout the larval epidermis, but they are not abundant in any region. Their functions are uncertain. Merkel cells might be mechanoreceptors, and flask cells may be involved in salt and water balance.

The skin of amphibians ranges from smooth to rough. Some of the integumentary projections are epidermal, but most involve both the epidermis and the dermis. Integumentary annuli of caecilians and costal grooves of salamanders match the segmentation of the axial musculature and vertebral column. Each primary annulus and each costal groove lies directly over the myosepta (connective tissue sheet) between the muscle masses; thus, the number of annuli equals the number of trunk vertebrae. In caecilians,

FIGURE 2.10 The tropical toad *Rhaebo guttatus* has enlarged paratoid glands behind the head as well as many other glands over the body surface. Secretions from the paratoid glands are toxic. (Janalee P. Caldwell)

this annular pattern can be complicated by the development of secondary and tertiary grooves; the secondary ones appear directly above the myosepta. The warts, papillae, flaps, tubercles, and ridges in frogs and salamanders can be aggregations of glands or simply thickenings in the underlying dermis and epidermis.

Although amphibians lack epidermal scales, they do have keratinous structures. Clawlike toe tips of pipid frogs, spades of scaphiopodid frogs, and rough, spiny skin of some frogs and salamanders are keratinous. These structures persist year-round. Other keratinous structures are seasonal and usually associated with reproduction. Many male salamanders and frogs have keratinous nuptial pads on their thumbs at the beginning of the mating season; some even develop keratinous spines or tubercles on their arms or chests. At the end of the mating season, these specialized mating structures are typically shed, and they redevelop in subsequent breeding seasons.

Dermal scales exist only in caecilians, although not in all species. These scales are flat, bony plates that are buried deeply in pockets within the annular grooves. Whether these scales are homologues of fish scales remains uncertain. Some frogs, such as *Ceratophrys* and *Megophrys*, have osteoderms (bony plates) embedded in or immediately adjacent to the dermis. In some other species of frogs, the dorsal skin of the head is compacted and the connective tissue of the dermis is co-ossified with the skull bones, a condition known as exostosis.

REPTILIAN SCALES, GLANDS, AND SKIN STRUCTURES

Scales of crocodylians, turtles, and some lizards (e.g., anguids, cordylids, scincids) are underlain by bony plates, called *osteoderms* or *osteoscutes*, in the dermis. Organization of osteoderms aligns with organization of the dermis. The outer layer of osteoderms is spongy, porous bone; the inner layer is compact, dense bone. Usually osteoderms are confined to the back and sides of the animal and attach loosely to one another in symmetrical rows and columns to permit flexibility while maintaining a protective bony armor. In crocodylians and a few lizards (*Heloderma*), the osteoderms fuse with the dorsal skull elements, forming a rigid skull cap. In turtles, the carapace (upper shell) arose from the fusion of osteoderms with vertebrae and ribs dorsally, whereas the plastron (lower shell) arose from the fusion of osteoderms and the sternum ventrally.

Reptiles have a variety of skin glands. Although common over the body, the multicellular glands are typically small and inconspicuous. Their secretions are mainly lipid- and wax-based compounds that serve as waterproofing, surfactant, and pheromonal agents.

Aggregations of glandular tissues occur in many reptiles. Musk or Rathke's glands are present in all turtles except tortoises (Testudinidae) and in some emydid turtles. These glands are usually bilaterally paired and lie within the bridge between the top and bottom shells, opening to the outside through individual ducts in the axilla and inguinal region or on the bridge. Male tortoises have a mental gland just behind the tip of the lower jaw. Both male and female crocodylians have paired mandibular and cloacal glands. The occurrence of large glands is more erratic in lepidosaurs. Some geckos and iguanians have a series of secretory pores on the underside of the thighs and pubis (Fig. 2.11). Each pore arises from the center of an enlarged scale and produces a waxy compound containing cell fragments. These femoral and precloacal (pubic) pores do not open until the lizards attain sexual maturity and often occur only in males. They may function as sexual scent glands. Snakes and some autarchoglossan lizards have paired scent glands at the base of the tail; each gland opens at the outer edge of the cloacal opening. These saclike glands release copious amounts of semisolid, bad-smelling fluids. For some species, the fluid may serve in defense, whereas in other situations, they may function for sexual recognition. Other glandular aggregations occur but are limited to a few reptiles. For example, a few Australian geckos have specialized squirting glands in their tails, and some marine and desert species of turtles, crocodylians, and lepidosaurs have salt glands.

Specialized keratinous structures are common in reptiles. All limbed species with functional digits have claws, which are keratinous sheaths that encase the tips of the terminal phalanges. The sheaths have three layers. The outermost layer is formed of hard ß-keratin. The claws form either as full keratinous cones, as in crocodylians and turtles, or as partial cones, as in lepidosaurs. The upper and lower jaw sheaths of turtles are also keratinous structures and replace the teeth as the cutting and crushing surfaces. Hatchling turtles, crocodylians, and *Sphenodon* have an egg caruncle on the snout to assist in hatching.

FIGURE 2.11 Femoral pores of the male of the lizard *Sceloporus undulatus* are located along the posterior edge of the underside of the thighs. They appear as lines of black spots. (Laurie Vitt)

A dozen or more types of small, epidermal sense organs occur in reptiles, particularly in lepidosaurs. Most are barely visible, appearing as tiny pits or projections. These epidermal structures are not shed during the sloughing cycle. Presumably, most of these structures respond to tactile stimuli; however, the presence of a light-sensitive region on the tail of a seasnake suggests a broader range of receptors and sensitivities. These organs are often concentrated on the head but are also widespread on the body, limbs, and tail.

ECDYSIS

Adult amphibians shed their skin in a cyclic pattern of several days to a few weeks. This shedding, called *ecdysis*, *sloughing*, or *molting*, involves only the stratum corneum and is commonly divided into several phases. At its simplest, the shedding cycle consists of epidermal germination and maturation phases, pre-ecdysis, and actual ecdysis. These phases are controlled hormonally, although timing and mechanisms differ between species and amphibian groups. The stratum germinativum produces new cells that move outward and upward in a conveyer-belt-like fashion as new cells are produced beneath them. Once these new cells lose contact with the basement membrane, they cease dividing and begin to mature, losing their subcellular organelles. Pre-ecdysis is signaled by appearance of mucous lakes between the maturing cells and the stratum corneum. The lakes expand and coalesce, and the cellular connections between the dead cells of the stratum corneum and the underlying, maturing cells break. Externally, the skin commonly splits middorsally first over the head and then continues down the back. Using its limbs, the frog or salamander emerges from the old skin, which is often consumed. During the pre-ecdytic and/or the ecdytic phase, the epidermal cells beneath the mucous lakes complete their keratinization and die.

The shedding process of larval amphibians is not well known. In the mudpuppy *Necturus maculosus* and probably in most other larvae, the skin is shed as single cells or in small pieces. The shed skin is not keratinized and may be alive when shed. The epidermal cells mature as they are pushed to the surface, but keratinization is not part of maturation.

In reptiles, different epidermal organizations and growth patterns produce different shedding or sloughing patterns. In the epidermis of crocodylians and the nonshell epidermis of turtles, cell growth is continuous and the portions of the outer surface of the skin are shed continuously in flakes and small sheets. Depending on species, the scutes of hard-shelled turtles are either retained or shed seasonally. When retained, successive scutes form a flattened pyramid stack, because an entire new scute develops beneath the older scute at the beginning of each growing season. Scute growth is not confined to the margins, although each new scute is thickest there and much compressed beneath the older scutes.

The shedding pattern in lepidosaurs is more complex and intimately tied to the unique epidermal growth pattern. In the tuataras and most lizards, the skin is shed in large patches, whereas in snakes the skin is usually shed as a single piece. But in all lepidosaurs, the sequence of epidermal growth and shedding is identical (Fig. 2.12). During the resting stage, the epidermis has a basal germinative layer of cells, a narrow band of α precursor cells, a thin meso-layer of mucus and other cell secretions, and externally the beginnings of an outer-generation layer capped by the Oberhautchen. The resting stage ends as cell proliferation and differentiation begins in the outer-generation layer. Then the germinative cells begin to divide. As each newly formed layer of cells is pushed upward and outward by cell division below them, the cells differentiate and produce the inner-generation layer. This inner-generation layer forms the precursor of the scales (outer-generation layer) for the next epidermal cycle. As the Oberhautchen nears completion, the outer-generation layer separates from the inner layer and is shed, completing the shedding or sloughing cycle (Fig. 2.13). This cycle is repeated at regular intervals when food is abundant. This growth–shedding (renewal) phase requires about 14 days. The resting phase may last from a few days to many months.

FIGURE 2.13 *Anolis punctatus* shedding its skin. Note that the old skin separates in several places from the new skin (Laurie Vitt).

COLORATION

The color of amphibians is affected by the presence of pigment cells (chromatophores) in the dermal layer of the skin. Three classes of chromatophores are melanophores, iridophores, and xanthophores. The primary pigment in melanophores is eumelanin, which imparts black, brown, or red coloration. Pigments in iridophores are purines such as guanine; these cells reflect light because of pigment-containing organelles arranged in stacks. Xanthopores impart yellow, orange, or red coloration because they contain pteridine pigments. In addition to containing different pigments, each of the three cell types is structurally different. The three classes of chromatophores are arranged as a unit and produce an animal's external coloration (Fig. 2.14). For example, the blue color of iridophores combined with the yellow color of xanthophores produces a green-colored skin.

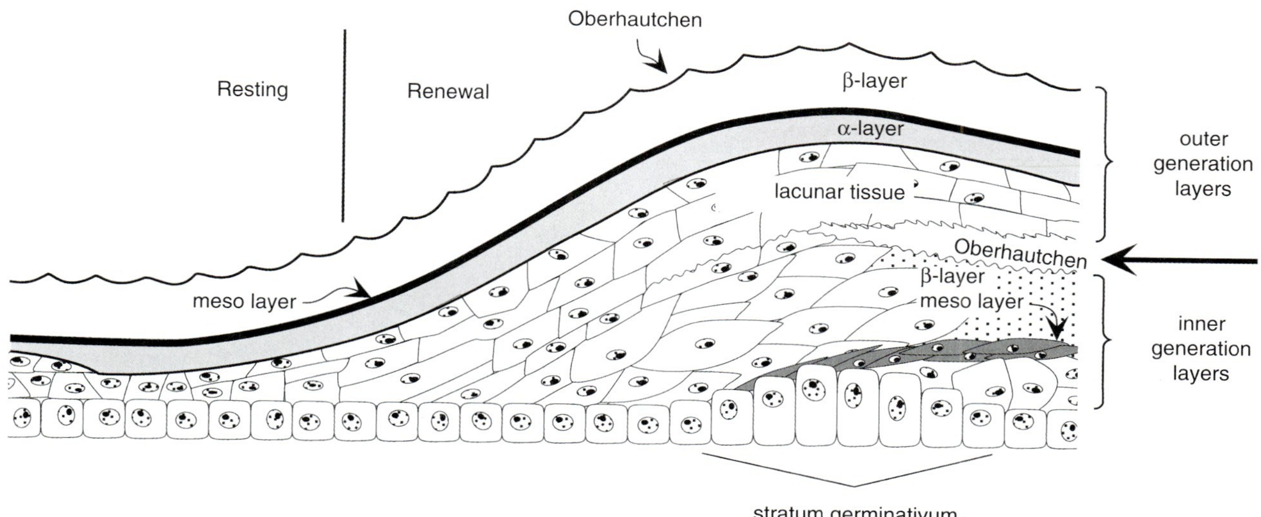

FIGURE 2.12 Diagram of the sequential cellular changes during a single shedding cycle in squamate epidermis. Adapted from Landmann (1986).

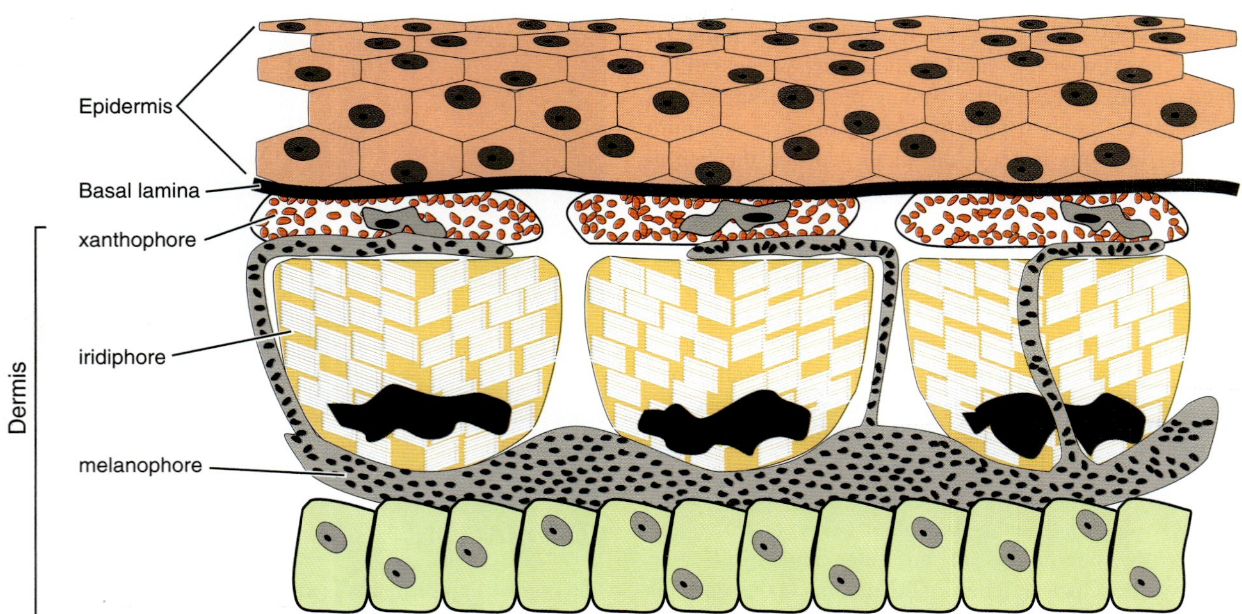

FIGURE 2.14 The arrangement of chromatophores in amphibian skin, called the *dermal chromatophore unit*. The unit consists of xanthophores, which give yellow, orange, or red coloration; the iridophores, which reflect light and cause bright colors; and the basal melanophores, which have dendritic processes that extend between the xanthophores and the iridophores.

Melanophores have a central cell body with long, attenuated processes radiating outward. Melanophores occur individually in the epidermis or as part of the dermal chromatophore unit. Epidermal melanophores are common in larvae and are often lost or reduced in their number at metamorphosis. The dermal chromatophore unit contains a basal melanophore, an iridophore, and a terminal xanthophore. Dendritic processes of the melanophore extend upward and over the iridophore, which is then overlain by a xanthophore (Fig. 2.14). The color produced by the unit depends largely upon the color of pigment in the xanthophore and the reflectivity of the iridophore. Melanophores are largely responsible for lightening or darkening of the color produced in the other two chromatophores.

Color changes can occur quickly, in less than a minute, by dispersal or reduction of the eumelanin within the melanophores' processes. Increased eumelanin darkens the observable color of the skin, while reduced eumelanin allows colors produced by the iridophores and xanthophores to predominate. Slow color changes may take weeks to months and occur when pigment concentration increases or decreases within the chromatophores or when pigment is in adjacent cells. Short-term color changes are controlled by hormonal or nervous stimulation. Some species have spectacular coloration and patterns that aid in crypsis (Fig. 2.15).

Reptiles generally have two types of color-producing cells. Melanophores are scattered throughout the basal layers of the epidermis. During the renewal phase of epidermal growth, the melanophores send out pseudopodia that transfer melanin into the differentiating keratocytes. The melanin-bearing keratocytes occur in the ß-layer of crocodylians, iguanian lizards, and snakes, and in the α- and ß-layers in many other lizards.

The second type of cell that produces color is the chromatophore, which is structurally similar to that in amphibians. Different types of chromatophores are stacked in the outer portion of the dermis. A single layer of xanthophores (= lipophores and erythrophores) lies beneath the

FIGURE 2.15 Frog skin contains a variety of pigments that often result in bizarre intricate patterns, as in this Amazonian *Ceratophrys cornuta* (Janalee P. Caldwell).

basal membrane of the epidermis. Beneath the xanthrophores are two to four layers of iridophores (= guanophores and leukophores), and at the bottom are large melanophores. This organization may represent the general pattern for all reptiles that change color, because stacked chromatophores are absent in some species that do not change color. The presence, density, and distribution of chromatophores within each layer vary within an individual and among species to produce the different colors and color patterns.

SKELETON AND MUSCLES—SUPPORT, MOVEMENT, AND FORM

The evolutionary transition from a fishlike ancestor to amphibians was accompanied by major reorganizations within the musculoskeletal system. As ancestral tetrapods shifted their activities from an aquatic to a terrestrial environment, the buoyant support of water disappeared, and the pull of gravity required a strengthening of the vertebral column to support the viscera. Simultaneously, these ancient tetrapods were shifting from undulatory locomotion to limbed locomotion. The new functions and demands on the musculoskeletal system required a more tightly linked vertebral column, elaboration of the limbs and girdles, and modification of the cranium for capture and ingestion of terrestrial food. As in amphibians, the reptilian musculoskeletal system is adapted primarily for terrestrial limbed locomotion, and some species are secondarily modified for aquatic or terrestrial limbless locomotion. With the exception of turtles, reptiles retain considerable lateral flexure of the body, and only in archosaurs does dorsoventral flexure become an important component of locomotion.

Each extant amphibian group has had a long and independent evolutionary history. Many structural differences appeared during this long divergence, and these differences are nowhere more apparent than in the composition and organization of the musculoskeletal system. Similarly, the long independent evolution of each reptilian group is strongly evident in all aspects of their musculoskeletal system. This great diversity permits us to present only a general survey of he musculoskeletal systems of amphibians and reptiles.

HEAD AND HYOID

The cranial skeleton of vertebrates contains elements from three units: the chondrocranium, the splanchnocranium, and the dermocranium. The chondrocranium (neurocranium) comprises the skeleton surrounding the brain and the sense organs, that is, the olfactory, optic, and otic capsules. The splanchnocranium is the branchial or visceral arch skeleton and includes the upper and lower jaws, the hyobranchium, and the gill arches and their derivatives. Most elements from these two cranial skeletons appear first as cartilage. Cartilaginous precursors define the position of the later developing bony element. Bone formed by replacement of cartilage is called *replacement* or *endochondral bone*. The dermocranium contains the roofing elements that lie external to the chondro- and splanchnocranial elements. These roofing elements have no cartilaginous precursors; instead, ossification centers develop in the dermis and form dermal or membrane bones.

All three crania are represented by numerous skeletal elements in fish and in the fish ancestors of amphibians. The earliest amphibians showed a loss of elements from each of the crania and a firmer articulation of the remaining elements. The reduction has continued in modern tetrapod clades, which have lost additional, but often different, elements in each group. Fewer elements have been lost in the caecilians, in which the skull is a major digging tool and must remain sturdy and firmly knit, often by the fusion of adjacent elements (see Fig. 15.1).

In extant amphibians, much of the chondrocranium remains cartilaginous throughout life (Fig. 2.16). Only the sphenoethmoid (orbitosphenoid in salamanders), which forms the inner wall of the orbit, and the fused prootic and exoccipital, which form the rear of the skull, ossify. Within the skull proper, the bony elements of the splanchnocranium are the stapes (ear) and the quadrate (upper jaw). Meckel's cartilage forms the core of the mandible (lower jaw), and ossification in its anterior and posterior ends form the mentomeckelian bone and articular, respectively. The dermal bones form the major portion of the adult skull, linking the various cranial elements and forming a protective sheath over the cartilaginous elements, the brain, and the sense organs. The skull is roofed from anterior to posterior by the premaxillae, nasals, frontals, and parietals; each side of the skull contains the maxilla, septomaxilla, prefrontal, and squamosal. Dermal bones also sheath the skull ventrally, creating the primary palate (roof of mouth). The palate consists of vomers, palatines, pterygoids, quadratojugals, and a parasphenoid, which is the only unpaired dermal bone in the amphibian skull. The dermal bones of the mandible are the dentary, angular, and prearticular, which encases Meckel's cartilage. Teeth occur commonly on the premaxillae, maxillae, vomers, palatines, and dentaries.

The jaws of vertebrates arose evolutionarily from the first visceral or branchial arch. The second visceral, hyomandibular, arch supported the jaws and bore gills, and the third and subsequent visceral arches comprised the major gill arches. Remnants of these arches remain in modern amphibians. The jaws consist mostly of dermal bones; only the mentomeckelian, articular, and quadrate are bony remnants of the first arch. The quadrate becomes part of the skull proper, and the dorsalmost element of the hyomandibular arch becomes the stapes for transmission of sound waves from the external eardrum, the tympanum,

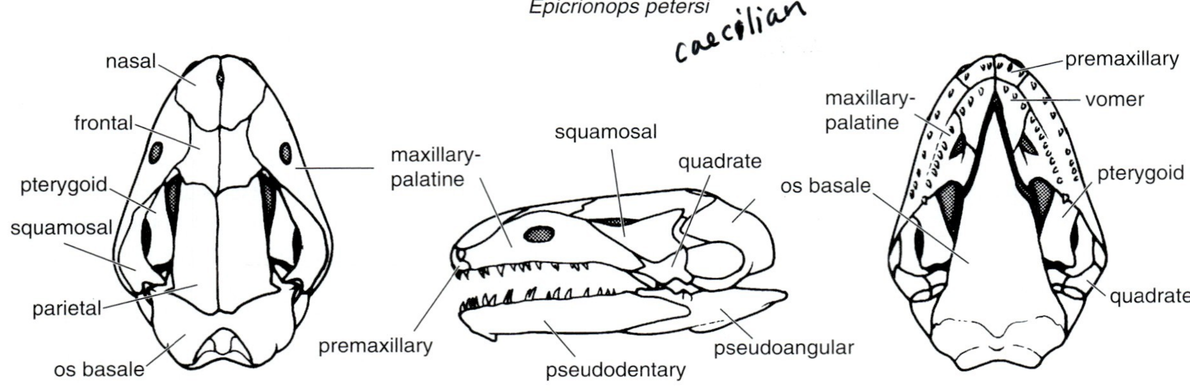

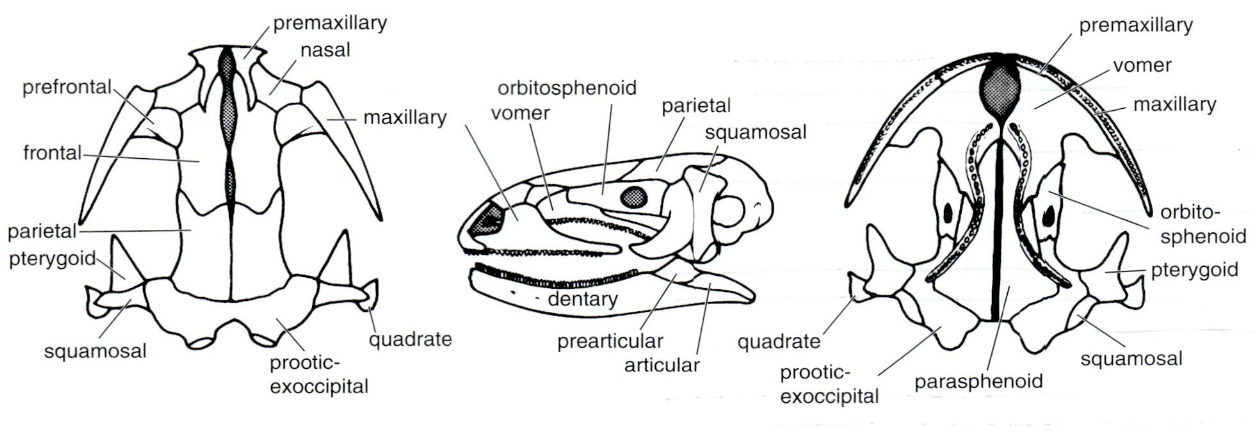

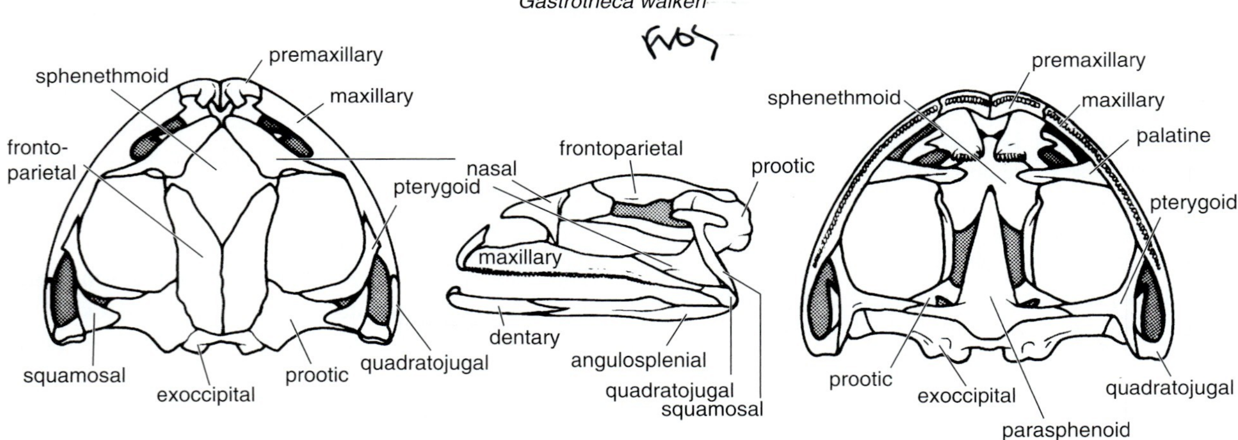

FIGURE 2.16 Cranial skeletons of representatives of the three clades of extant amphibians. Dorsal, lateral, and ventral views (left to right) of the caecilian *Epicrionops petersi*, the salamander *Salamandra salamandra*, and the frog *Gastrotheca walkeri*. Reproduced, with permission, from Duellman and Trueb (1986).

to the inner ear. The ventral portion of the second arch persists as part of the hyoid apparatus. The subsequent two to four visceral arches may persist, at least in part, as gill arches in larvae and in some gilled adults (e.g., Proteidae), and also as elements of the hyoid in juveniles and adults. Some elements from the more posterior visceral arch become structural supports in the glottis, larynx, and trachea.

The composition and architecture of the hyoid is highly variable within and between each group of living amphibians. In all, the hyoid lies in the floor of the mouth and forms the structural support for the tongue. In some species, the components of the hyoid can be traced accurately to their visceral arch origin; in others species, their origin from a specific arch element is uncertain. The hyoid elements in primitive salamanders retain an architecture similar to that of the visceral arches of fishes, but with the loss of arch elements (Fig. 2.17). In more advanced salamanders, the number of hyoid elements is further reduced. The hyoid remains cartilaginous in caecilians without segmentation of hyoid arms into individual elements. The anuran hyoid is a single cartilaginous plate with two to four processes and has little resemblance to its visceral arch precursor.

The cranial musculature contains one functional group for jaw movement and another for respiring and swallowing. The jaw muscles fill the temporal area of the skull, extending from the area of the parietal, prootic, and squamosal to the mandible. The muscles that attach to the dorsal surface of the mandible close the mouth, and those that attach to the lateral and ventral surface of the mandible open the mouth. The muscles that function in respiration and swallowing form the floor of the mouth, throat, and neck. These muscles move and support the gills and/or the hyoid and the tongue.

In reptiles, the anterior portion of the chondrocranium remains cartilaginous, even in adults, and consists mainly of continuous internasal and interorbital septa and a pair of nasal conchae that support olfactory tissue. Between the eyes and ears, the chondrocranium ossifies as the basisphenoid, and further posteriorly, the basioccipital, a pair of exoccipitals, and the supraoccipital bones develop below and behind the brain (Fig. 2.18). The occipital elements encircle the foramen magnum, the site at which the spinal cord exits the skull. Below the foramen magnum, the exoccipitals and the basioccipital join to form a single occipital condyle, which bears the articular surface between the first cervical vertebra, the atlas, and the skull.

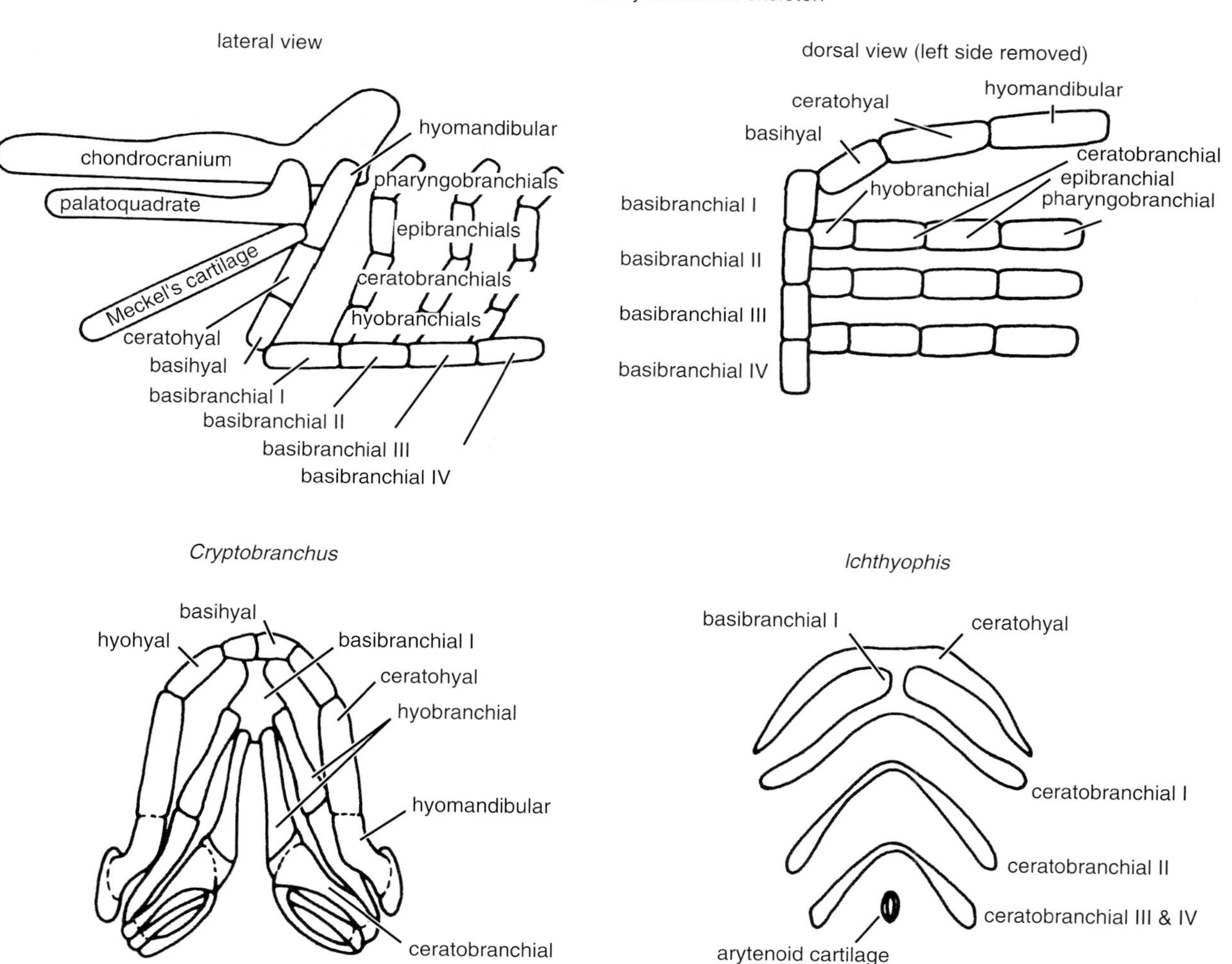

FIGURE 2.17 The hyobranchial skeleton of a typical vertebrate, the salamander *Cryptobranchus* (dorsal view), and the caecilian *Ichythyophis* (ventral view). Reproduced, with permission, from Duellman and Trueb, 1986.

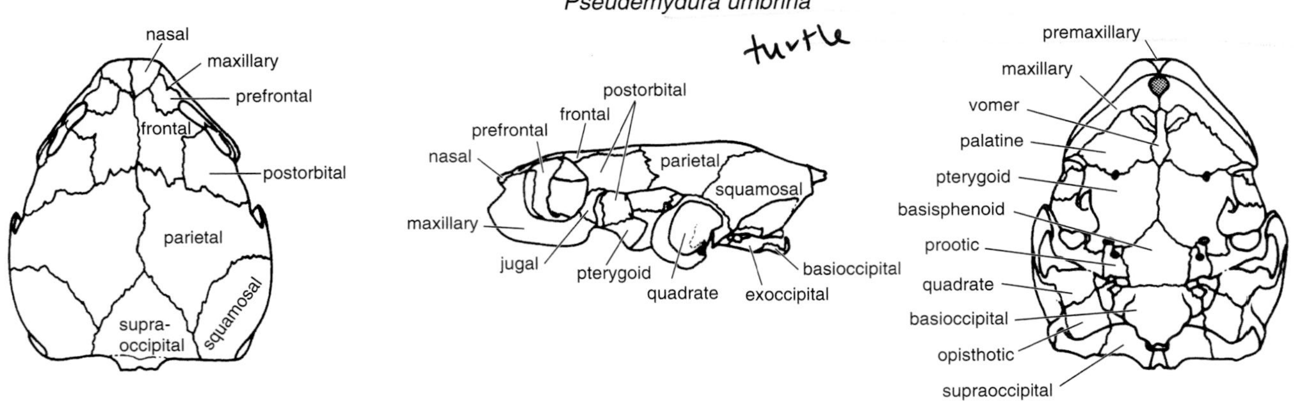

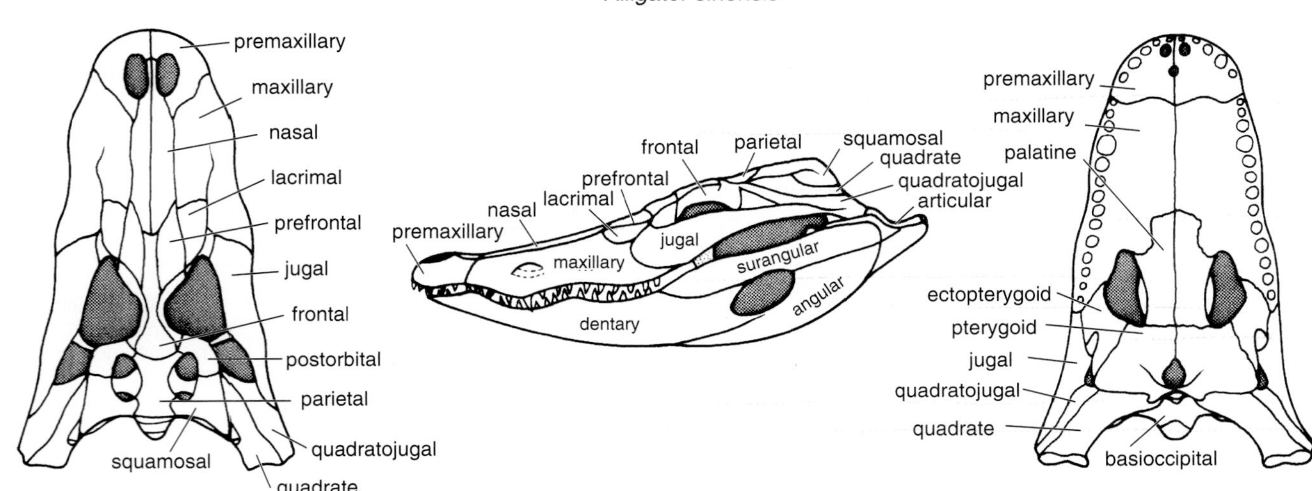

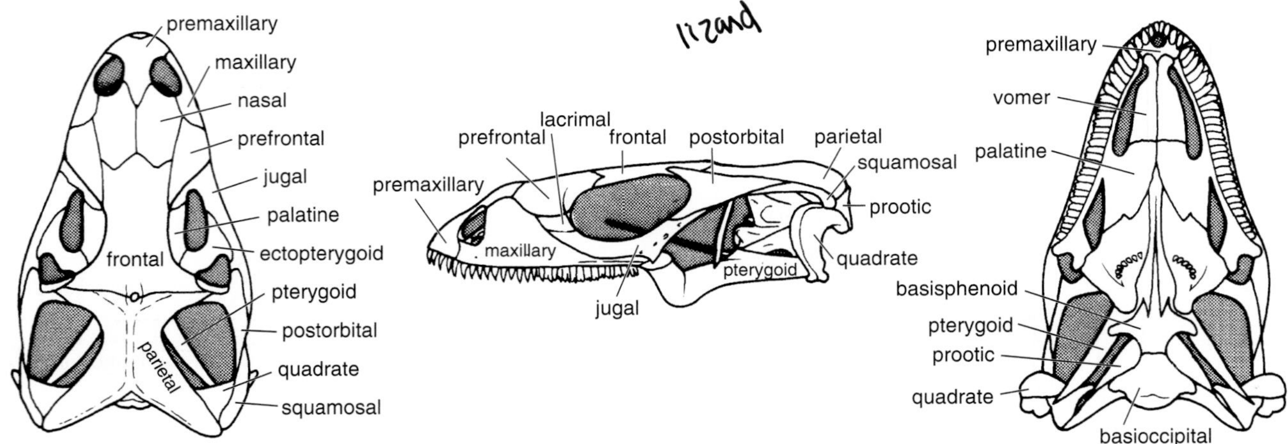

FIGURE 2.18 Cranial skeletons of representatives of the three clades of living reptiles. Dorsal, lateral, and ventral views (left to right) of the turtle *Pseudemydura umbrina*, the crocodylian *Alligator sinensis*, and the lizard *Ctenosaura pectinata*. Adapted from Gaffney (1979), Iordansky (1973), and Oelrich (1956), respectively.

Regions of each otic capsule remain cartilaginous, although much of the capsule becomes the epiotic, prootic, and opisthotic bones.

The stapes of the middle ear is a splanchnocranial element, as are the quadrate and the epipterygoid; the latter is small in lizards and turtles and is lost in snakes and archosaurs. The quadrate is a large bone on the posterolateral margin of each side of the skull. It bears the articular surface for the lower jaw. On the mandible, the articular bone provides the opposing articular surface and is the only splanchnocranial element of the lower jaw. The reptilian hyoid arch is reduced and consists of a large midventral plate, usually with three processes that extend upward and posteriorly.

Dermal bones compose the major portion of the reptilian skull and mandible, forming over and around the endochondral bones. From anterior to posterior, the roof of the dermocranium contains the nasals, prefrontals, frontals, and parietals, all of which are paired. The upper jaws, the premaxillae, and the maxillae join the roofing bones directly. The cheek and temporal areas contain a postorbital, postfrontal, jugal, quadratojugal, and squamosal bone on each side. The primary palate or roof of the mouth consists of premaxillae and maxillae anteriorly, and a median vomer that is bordered laterally by the palatines and posteriorly by the pterygoids and occasionally a parasphenoid. When a secondary palate forms as in crocodylians, it derives largely from the premaxillae and maxillae. A few other dermal bones, for example, the septomaxilla and the lacrimal, are present in some extant reptiles. The jugals, the quadratojugals, prefrontals, postfrontals, and squamosals are absent individually or in various combinations in some taxa.

The mandible or lower jaw contains numerous paired dermal bones including dentaries, splenials, angulars, surangulars, coronoids, and prearticulars (Fig. 2.18). Only the dentary bears teeth, and in the upper jaw, only the maxilla, premaxilla, palatine, and pterygoid bear teeth. Teeth can be absent on one or more of these teeth-bearing bones. In turtles, teeth are entirely absent; their cutting and crushing functions are performed by the keratinous jaw sheaths.

Typical reptilian teeth are cone-shaped and arranged in a single, longitudinal row. This basic shape has been variously modified. For example, the teeth are laterally compressed and have serrated edges in some herbivorous lizards and are elongated and posteriorly curved in snakes. When the teeth attach to the bone by sitting in sockets as in crocodylians, they are referred to as thecodont (Fig. 2.19). Pleurodont teeth found in most lepidosaurs arise from a one-sided groove in the jaw. Acrodont teeth, which attach directly to the bone surface, occur in two lizard clades. Tooth replacement is continuous throughout life, except in most acrodont forms, in which teeth are replaced in juveniles.

The skulls of turtles and all other extant reptiles are distinct (Fig. 2.18). In the turtle skull, the bony temporal

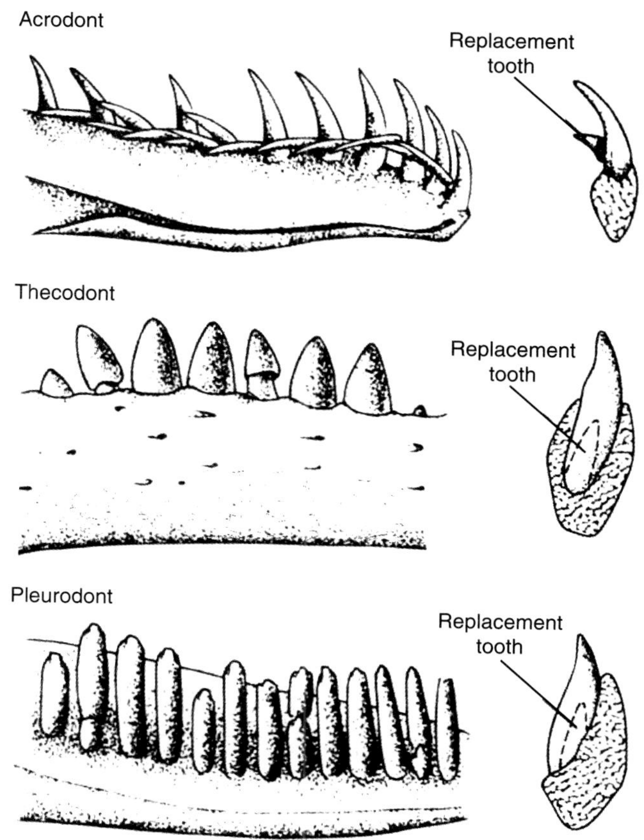

FIGURE 2.19 Reptile teeth can sit on top of the jaw (acrodont), embedded in the jaw (thecadont), or on the side of the jaw (pleurodont). Tooth location is one of the many important taxonomic characters used to separate major taxa. Adapted from Kardong, 2006.

arcade composed of parietals, squamosals, postorbitals, jugals, and quadratojugal lacks openings (anapsid). Although the lack of openings in the quadratojugal has historically placed turtles with anapsids, this condition appears to be secondarily derived from a diapsid ancestor in turtles (Fig. 2.20). In the typical diapsid skull, the temporal area has two openings called *fenestrae*, an upper one between the parietal and the postorbital–squamosal, and a lower one between the squamosal and jugal–quadratojugal. Most living turtles have emarginated temporal arcades, leaving a small arch of bone behind each eye. Only a few turtles, such as the seaturtles, retain a nearly complete arcade. Crocodylians retain the basic diapsid architecture, although the upper or superior temporal fenestra is small (Fig. 2.18). In lepidosaurs, only *Sphenodon* retains the two fenestrae. Squamates have only one upper fenestra or none at all. In squamates with only one upper fenestra, the lower temporal arch (composed of the squamosal, quadratojugal, and jugal) has been lost. In squamates with no fenestrae, the upper arch (composed of the squamosal and parietal) *or* the upper and middle arches (composed of the squamosal and postorbital) have been lost.

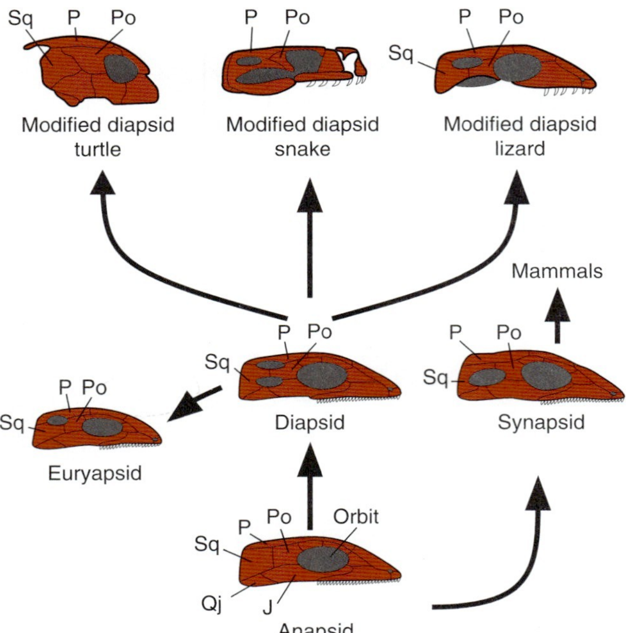

FIGURE 2.20 Evolution of skull openings (fenestre) in modern reptiles. Variation exists in the openings (fenestre) behind the orbit and the position of the postorbital (Po) and squamosal (Sq) bones that form the arch from the orbit to the back of the skull. The anapsid (closed) condition is thought to be ancestral. Lizards (including snakes) clearly have modified diapsid (two fenestre) skulls. Turtles, which have been placed historically in the Parareptilia based on the absence of a second fenestra, more likely have a highly modified diapsid skull in which both fenestre have closed. Other bones shown include the quadratojugal and the jugal. Adapted from Kardong, 2006.

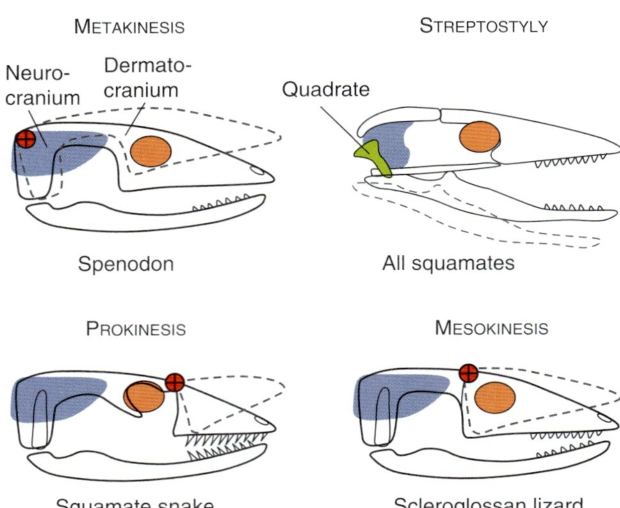

FIGURE 2.21 Evolution of jaw structure and function in squamates. Clockwise from upper left, ancestors of squamates had rigid jaws and skulls such that the skull lifted as a unit when opening the mouth (metakinesis). The "hanging jaw" of squamates (streptostyly) allowed rotation of the lower jaws on the quadrate bone. Scleroglossans have kinetic joints in the skull located behind the eyes (mesokinesis), and snakes have an extra joint located anterior to the eyes (prokinesis). Increased flexibility of the skull allows greater prey-handling ability. The red circle with a cross indicates focal point of rotation.

The loss of arches and fenestrae in the diapsid skull is associated with increased flexibility of the skull. Hinges between various sections of the skull allow the skull to flex, a process known as kinesis (Fig. 2.21). A hinge can occur in the back of the skull (a metakinetic joint) between the dermal skull and the braincase at the parietal–supraoccipital junction; this hinge is the oldest kinetic joint and occurred early in reptilian evolution and today occurs in *Sphenodon*. Two other joints developed in the dermal roofing bones. A dorsal mesokinetic joint lies between the frontals and parietals in many lizards, and in many snakes, a prokinetic joint occurs at the contact between the nasals and the prefrontals or frontals. The most striking kinesis of the lepidosaurs, particularly in snakes, is streptostyly or quadrate rotation; each quadrate is loosely attached to the dermocranium and has a free ventral end. This loose ligamentous attachment allows the quadrates to rotate and to swing forward and backward, and inward and outward. Streptostyly enhances the jaw's grasping ability and increases the gape.

The complexity in the arrangement and subdivision of muscles mirrors the diversity of the bony architecture of the head. Reptiles lack facial muscles, but the diversity of jaw and tongue muscles permits a wide range of feeding and defense behaviors. The jaw's depressor and adductor muscles arise from within the temporal arcade and attach to the inside and outside of the mandible. In highly kinetic skulls, muscles are more finely subdivided and permit a wider range of movements of the individual bones, including those of the upper jaw. Throat muscles are typically flat sheets of muscles that extend onto the neck. Beneath these muscles, the hyoid muscles are thicker sheets and longer bundles that attach the hyoid plate and processes to the mandible and to the rear of the skull and the cervical vertebrae.

VERTEBRAL COLUMN

The amphibian vertebral column combines rigidity and strength to support the head, limb girdles, and viscera, and yet it allows enough flexibility to permit lateral and dorsoventral flexure of the column. These seemingly conflicting roles are facilitated by the presence of sliding and rotating articular facets on the ends of each vertebra and by overlapping sets of muscular slips linking adjacent vertebrae.

Each vertebra consists of a ventral cylinder, the centrum, and a dorsal neural arch that may have a dorsal projection, the neural spine (Fig. 2.22). The anterior end of the centrum articulates with the posterior end of the preceding centrum. These central articular surfaces are variously shaped. In opisthocoelous vertebrae, the anterior surface is convex and the posterior surface is concave.

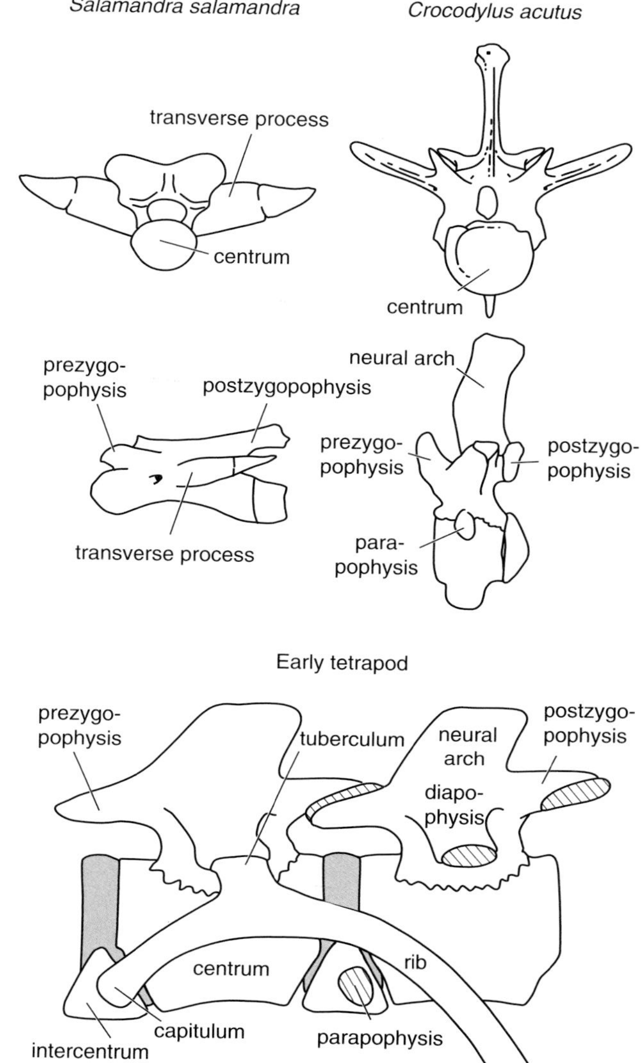

FIGURE 2.22 Anterior and lateral views of vertebral morphology of the tetrapods *Salamandra salamandra* and *Crocodylus acutus* and a schematic lateral view of an early tetrapod. Adapted in part from Francis (1934), Mook (1921), and Goodrich (1930).

In procoelous vertebrae, the anterior surface is concave and the posterior surface is convex; in amphicoelous vertebrae, both surfaces are concave. Intervertebral discs, usually of fibrocartilage, lie between central surfaces of adjacent vertebrae. A pair of flat processes extends from the prezygapophyses and postzygapophyses that form the anterior and posterior edges of the neural arch, respectively (Fig. 2.22). These processes form another set of articulations between adjacent vertebrae. Articular surfaces for the ribs lie on the sides of each vertebra; a diapophysis lies dorsal to the base of the neural arch and a parapophysis lies on the side of the centrum. Ribs are much shorter in amphibians than in the other tetrapods, such as reptiles and mammals, and do not extend more than halfway down the sides.

The first postcranial vertebra, the atlas, is modified to create a mobile attachment between the skull and the vertebral column. The atlantal condyles on the anterior surface articulate with the paired occipital condyles of the skull. The succeeding vertebrae of the trunk match the general pattern previously described. The number and shape of the vertebrae differ in the three amphibian groups. Salamanders have 10 to 60 presacral vertebrae, including a single atlas or cervical vertebra and a variable number of trunk vertebrae. The trunk vertebrae are all similar and have well-developed zygapophyses, neural spines, and usually bicapitate, or two-headed ribs. Rather than exiting intervertebrally between neural arches of adjacent vertebrae as in other vertebrates, the spinal nerves of salamanders often exit through foramina in the neural arches. Postsacral vertebrae are always present in variable numbers and are differentiated into two to four precaudal (cloacal) and numerous caudal vertebrae. Caecilians have 60 to 285 vertebrae, including a single atlas, numerous trunk vertebrae, no sacral vertebrae, and a few irregular bony nodules representing precaudal vertebrae. The trunk vertebrae are robust with large centra and neural spines; most bear bicapitate ribs. Frogs have 5 to 8 presacral vertebrae. The atlas (presacral I) lacks transverse processes, which are usually present on all other presacral vertebrae. Ribs are absent in most frogs but are present on presacrals II through IV only in *Ascaphus*, *Leiopelma*, discoglossids, bombinatorids, and pipids. Each sacral vertebra has large transverse processes called *sacral diapophyses,* although whether they are true diapophyses is uncertain. The sacral vertebra articulates posteriorly with an elongate urostyle, which represents a rod of fused postsacral vertebrae (Fig. 2.23).

The musculature of the vertebral column consists of epaxial (dorsal trunk) muscles and hypaxial (flank or ventral trunk) muscles. The epaxial muscles consist largely of longitudinal slips that link various combinations of adjacent vertebrae. These muscles lie principally above rib attachments (apophyses) and attach to the neural arches and spines. These muscular components provide rigidity and strength to the vertebral column. The hypaxial muscles support the viscera and contain the oblique muscle series that occurs on the flanks and the rectus muscle series that occurs midventrally along the abdomen.

The trend for increased rigidity of the vertebral column that began in early tetrapods is further elaborated in reptiles. The vertebrae form a firmly linked series and additionally elaborated intervertebral articular surfaces interwoven with a complex fragmentation of the intervertebral muscles. In reptiles, vertebral rigidity is augmented by regional differentiation of the vertebrae. This regionalization permits different segments of the column to have different directions and degrees of movement and is reflected in the architecture of both bones and muscles.

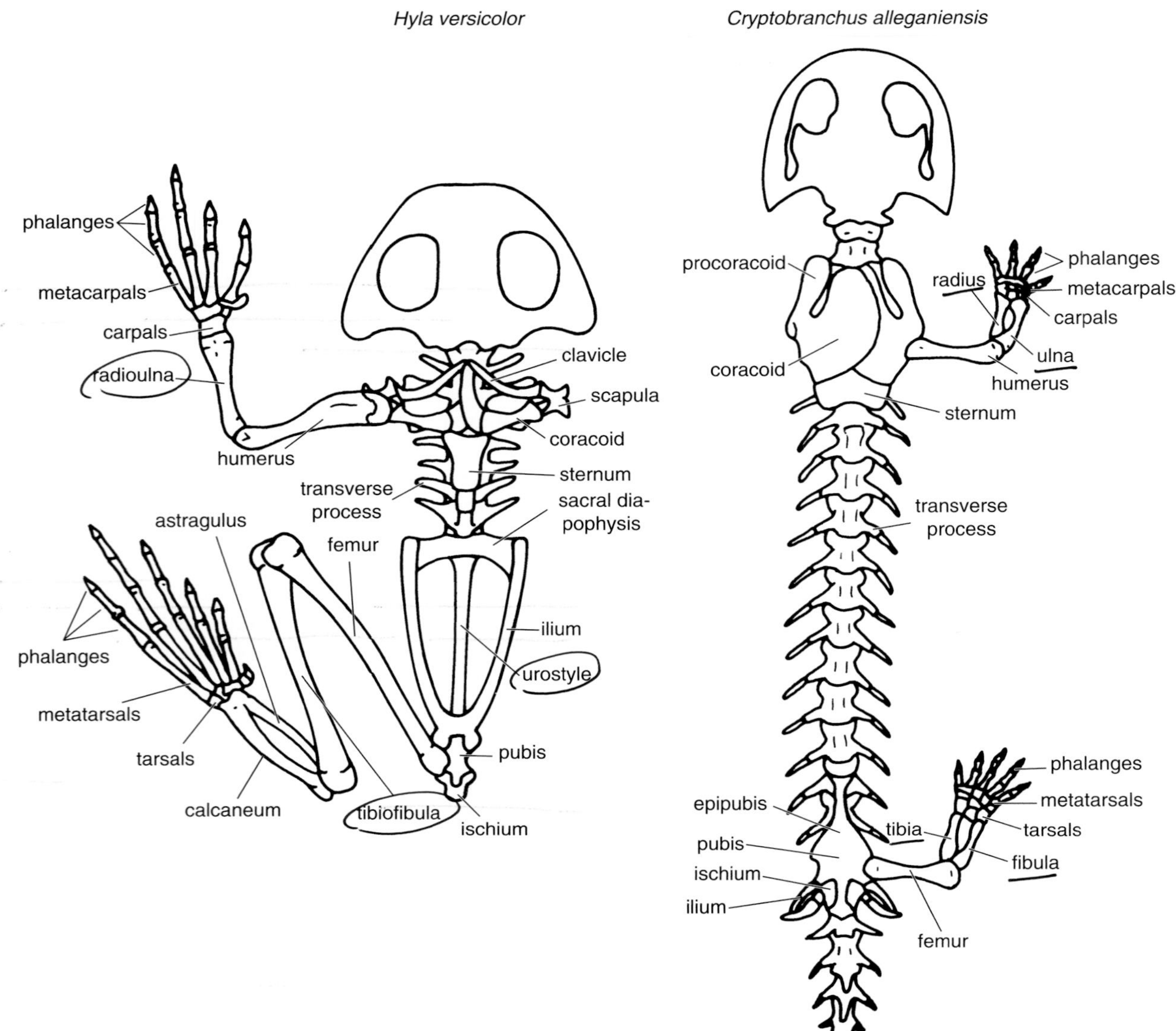

FIGURE 2.23 Postcranial skeletons (ventral view) of a gray treefrog (*Hyla versicolor*) and a hellbender (*Cryptobranchus alleganiensis*). Adapted from Cope (1898).

Reptilian vertebrae and vertebral columns are variable across taxa, but some features are shared by most reptiles (Fig. 2.22). The centra are the weight-bearing units of the vertebral column. Each centrum is typically a solid spool-shaped bone, but in *Sphenodon* and some geckos, the notochord persists and perforates each centrum. A neural arch sits astride the spinal cord on each centrum. The legs or pedicels of each arch fuse to the centrum or insert into notches on the centrum. Neural spines vary from short to long, and wide to narrow, depending upon the position within the column and the type of reptile. The intervertebral articular surfaces, or zygapophyses, consist of an anterior and a posterior pair on each vertebra and arise from the top of pedicels. The articular surfaces of the anterior zygapophyses flare outward and upward, and the posterior surfaces are inward and downward. The angle of these articular surfaces determines the amount of lateral flexibility. When the articular surfaces are angled toward the horizontal plane, flexibility between adjacent vertebrae increases, but if the surfaces are angled toward the vertical plane, rigidity increases. The pedicels also bear the articular surfaces for the ribs. For two-headed ribs, the upper surface is the transverse process or diapophysis, and the lower surface is the parapophysis. The ribs of extant reptiles are single-headed and articulate with the transverse process in all lineages except crocodylians. In many lepidosaurs, accessory articular surfaces occur at the base of the neural spine; a zygosphene projects from the front of the arch into a pocket, the zygantrum, on the rear of the preceding vertebra. The articular surfaces between the centra are variable, but the procoelous ball-and-socket condition is widespread, occurring in all extant crocodylians and most lepidosaurs.

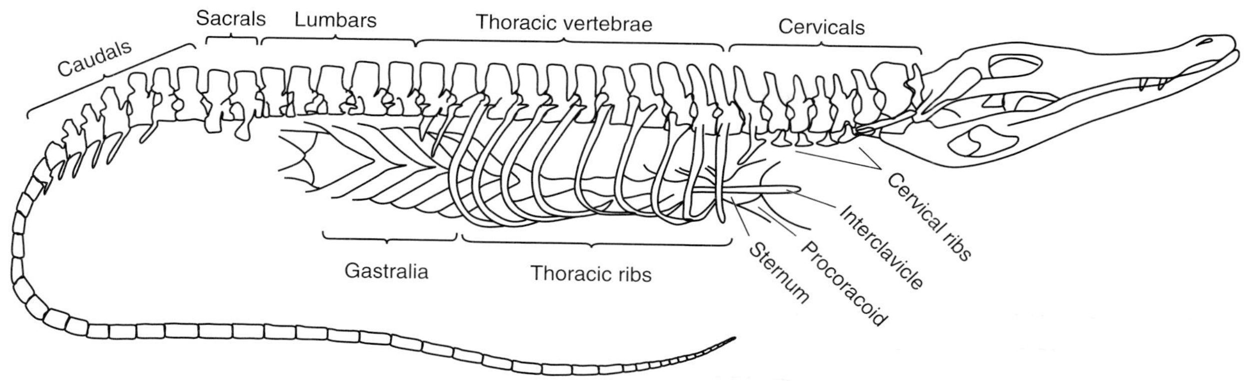

FIGURE 2.24 Partial skeleton of a crocodylian showing the variation in structure of vertebrae. The vertebral column is divided into five regions. Note the location of the gastralia (floating "ribs"). Redrawn from Kardong, 2006.

The most variable central articular patterns occur in the cervical vertebrae, where, for example, procoelous, opisthocoelous, and biconvex centra exist in the neck of an individual turtle.

Regional differentiation of the vertebrae is characteristic of crocodylians (Fig. 2.24). They have 9 cervical, 15 trunk, 2 sacral, and numerous caudal vertebrae. The first 2 cervical vertebrae, the atlas and axis, are constructed of several unfused components. The atlas bears a single anterior surface for articulation with the occipital condyle of the skull. The axis and subsequent cervical vertebrae bear two-headed ribs that become progressively longer toward the trunk. The first 8 or 9 trunk vertebrae have ribs that extend ventrally to join the sternum and form the thoracic basket. The remaining thoracic vertebrae have progressively shorter ribs. The ribs of the sacral vertebrae anchor the vertebral column to the ilia of the pelvic girdle. The caudal or postsacral vertebrae become sequentially smaller and laterally compressed, and progressively lose their processes posteriorly.

The limbed lepidosaurs have the same regional differentiation pattern as crocodylians. Vertebral number is much more variable, although all have a pair of sacral vertebrae. Generally, 8 cervical vertebrae and ribs exist only on the posterior 4 or 5 vertebrae; however, *Varanus* has 9 and chameleons have 3 to 5 cervical vertebrae. Trunk vertebrae are even more variable in number; 16 to 18 vertebrae appear to be the primitive condition, but the vertebral number can be fewer than 11 in chameleons and considerably more in elongated lizards, particularly in limbless and reduced-limbed anguids and skinks. Caudal vertebrae are similarly variable in number. In limbless squamates, differentiation is limited; the atlas and axis are present, followed by 100 to 300 trunk or precloacal vertebrae, several cloacal vertebrae, and 10 to 120 caudal vertebrae.

In contrast, vertebral number is nearly invariable in turtles (Fig. 2.25). All living turtles have 8 cervical vertebrae; when present, cervical ribs are rudimentary and confined to the posteriormost vertebrae. The variable neck lengths of different turtle species arise from the elongation or shortening of vertebrae. Of 10 trunk or dorsal vertebrae, the first and last are attached but not fused to the carapace. The middle 8 are firmly fused or co-ossified with the neural bones of the carapace. The trunk ribs extend outward and fuse with the costal bones of the shell. The 2 sacral vertebrae link the pelvic girdle to the vertebral column by short, stout ribs. Caudal number is variable but less than 24 in most species.

The division of the vertebral column muscles into epaxial and hypaxial bundles persists in reptiles, although the distinctiveness of the two types is not obvious. Similarly, the segmental division largely disappears in reptiles. Most axial muscles span two or more vertebral segments and often have attachments to several vertebrae. The

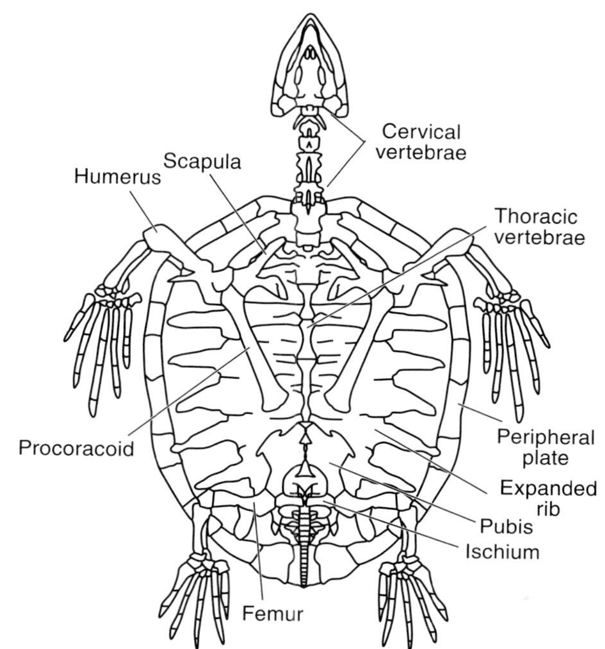

FIGURE 2.25 Skeleton of a modern turtle showing fusion of vertebrae to the shell. Adapted from Bellairs, 1969.

complexity of the intervertebral muscles is greatest in the limbless taxa. Unlike fish, their undulatory locomotion is not a uniform wave of contraction but requires individualized contraction patterns, depending upon which part of the body is pushing against the substrate. Turtles lack trunk musculature. Epaxial and hypaxial muscles, however, do extend inward from the neck and tail to attach to the carapace and dorsal vertebrae.

GIRDLES AND LIMBS

The limbs of amphibians and other tetrapods have evolved for terrestrial locomotion from the fins of fish ancestors. The girdle and limb components, the appendicular muscles and skeleton, of tetrapod vertebrates derive from the girdle and fin components of their fish ancestors. Several opposite trends are evident in the evolution of limbs from fins. The anterior (pectoral) girdle loses its articulation with the skull and has a reduced number of elements. In contrast, the posterior (pelvic) girdle becomes elaborated and enlarged; it articulates with the vertebral column. Within the limbs, the number of skeletal elements is reduced, and a series of highly flexible joints appears along the limb. On forelimbs, the first joint is formed where the propodial segment of the humerus meets the epipodial segment of the radius and ulna. On the hindlimbs, the first joint is formed where the femur meets the fibula and tibia. Additional joints are formed where the mesopodial segments of the carpal (front limbs) or tarsal (hindlimbs) elements meet the metapodial segment of the metacarpals (front limbs) or metatarsals (hindlimbs), and the phalanges (front and hindlimbs) (Fig. 2.23). These morphological specializations largely reflect the change in function of the appendages from steering and stability in fish locomotion to support and propulsion in tetrapod locomotion (Fig. 1.4).

The girdles provide internal support for the limbs and translate limb movement into locomotion. Primitively, the amphibian pectoral girdle contained dermal and endochondral elements. Dermal elements originate in the dermis. However, endochondral elements originate when hyaline cartilage is replaced by bone. The endochondral coracoid and scapula form the two arms of a V-shaped strut that has a concave facet, the glenoid fossa, at their juncture; the glenoid fossa is the articular surface for the head of the humerus. The dermal elements, including the cleithral elements and a clavicle, strengthen the endochondral girdle. A dermal interclavicle—the only unpaired pectoral element—provides midventral strengthening to the articulation of the left and right clavicles and coracoids. This midventral articulation includes the sternum posteriorly. The pelvic girdle, forelimbs, and hindlimbs contain only endochondral elements. Three paired elements form the pelvic girdle. A ventral plate contains the pubes anteriorly and the ishia posteriorly; an ilium projects upward on each side from the edge of the puboishial plate and articulates with the diapophyses of the sacral vertebra. A concave facet, the acetabulum, lies at the juncture of the three pelvic elements and is the articular surface for the head of the femur.

The girdles are anchored to the trunk by axial muscles. Because the pectoral girdle lacks an attachment to the axial skeleton, a series of muscles forms a sling that extends from the back of the skull across the anterior trunk vertebrae to insert on the scapula and humerus. The pelvic girdle has a bony attachment to the vertebral column, and its muscular sling is less extensive. The muscles of the limbs divide into a dorsal extensor and a ventral flexor unit. Within each unit, most of the muscles cross only a single joint, such as from the girdle to the humerus or from the humerus to the ulna.

Caecilians have lost all components of the appendicular skeleton and musculature. Limbs and girdles are present in most salamanders, although they may be reduced in size and have lost distal elements, as in the dwarf siren. All frogs possess well-developed limbs and girdles. Salamanders and frogs have only four, or sometimes fewer, digits on the forefeet. The missing digit in frogs and salamanders is the fifth or postaxial (outer) digit. The hindfeet of anurans and salamanders usually retain all digits, but if one is lost, it is also the fifth digit.

Reduction and loss are common features of the salamander skeleton. The pectoral girdle is largely cartilaginous and contains only the scapula, procoracoid, and coracoid. These three elements are regularly indistinguishably fused and ossified only in the area of the glenoid fossa. The left and right halves of the girdle overlap but do not articulate with one another. A small, diamond-shaped, cartilaginous sternum lies on the ventral midline posterior to the girdle halves and is grooved anteriorly for a sliding articulation with the edges of the coracoids. The humerus, the radius, and the ulna have ossified shafts, but their ends remain cartilaginous. The carpals are often entirely cartilaginous or have a small ossification node in the center of larger cartilaginous elements. Reduction by loss and fusion of adjacent carpals is common in salamanders. The phalanges ossify, but their number in each digit is reduced. The common phalangeal formula for most modern amphibians is 1–2–3–2 or 2–2–3–3, compared to the 2–3–4–5–4 formula of ancestral amphibians.

The salamander pelvic girdle has a more robust appearance than the pectoral girdle. The ilia and ischia are ossified, although the pubes remain largely cartilaginous. The two halves of the girdle are firmly articulated, and a Y-shaped cartilaginous rod, the ypsiloid cartilage, extends forward and likely supports the viscera. The hindlimb elements show the same pattern of ossification as those of the forelimbs; the hindfoot is typically 1–2–3–3–2 and the loss of the fifth toe is common, for example, in *Hemidactylium*.

The appendicular skeleton of frogs is robust and well ossified. The saltatory locomotion of anurans, both in jumping and landing, requires a strong skeleton. The pectoral girdle contains a scapula capped by a bony cleithrum and a cartilaginous suprascapula and, ventrally, a clavicle and a coracoid; an omosternum (or episternum) and a sternum extend anteriorly and posteriorly, respectively, from the midline of the girdle. Two types of girdles, arciferal and firmisternal, occur in anurans. In both types, the clavicles articulate firmly on the midline. In the firmisternal girdle, the coracoids are joined firmly through the fusion of their epicoracoid caps. In contrast, the epicoracoid caps overlap in arciferal girdles and can slide past one another. The two girdle types are quite distinct in many species, although in others, the girdle structure is intermediate. The humerus is entirely ossified and has an elevated, spherical head. The epipodial elements fuse into a single bony element, the radioulna. The carpal elements are bony and reduced in number by fusion. The phalangeal formula is rarely reduced from 2–2–3–3.

The anuran pelvic girdle is unlike that of any other tetrapod (Fig. 2.23). A plate, formed by the pubis and ischium, is compressed into a bony, vertical semicircular block on the midline; the ischia lie posterodorsally and the pubes form the ventral edge. The ilia complete the anterior portion of the pelvic block, and each ilium also projects forward as an elongate blade that attaches to the sacral diapophysis. The hindlimb elements are elongate and proportionately much longer than the forelimb. The epipodial elements are also fused into a single bone, the tibiofibula, which is typically as long or longer than the femur. Two mesopodial elements, the fibulare (astragalus) and the tibiale (calcaneum), are greatly elongate, giving frogs a long ankle. Most of the other mesopodial elements are lost or greatly reduced in size. With the exception of a few species, frogs have five toes and seldom deviate from a 2–2–3–4–3 phalangeal formula.

The limb and girdle skeletons of extant reptiles share many components with that of extant amphibians; nonetheless, the morphology and function of the muscular and skeletal components are different. Little of the reptilian endochondral skeleton remains unossified. The reptilian rib or thoracic cage is linked to the pectoral girdle through the sternum. A shift in limb posture occurred with the development of a less sprawled locomotion. Salamanders and lizards have similar gait patterns and considerable lateral body undulation when walking or running (Fig. 2.26). Lizards differ from salamanders in that they have more elevated postures and a greater range of limb movement. No reptile has a musculoskeletal system so tightly linked to saltatory locomotion as that of frogs, although some lizards can catapult themselves using thrust from the tail.

FIGURE 2.26 Graphic showing primitive lateral-sequence gait of a salamander. The center of mass (red circle) remains within the triangle of support (dashed line), and three of the four limbs meet the ground at the same time. During a trot gait (not shown), diagonal limbs meet the ground at the same time and the center of gravity falls on a line connecting those limbs. Often, the tail is used to stabilize the trot gait, which forms a triangle of support. Redrawn from Kardong, 2006.

Early reptiles had a pectoral girdle composed of five dermal components—including paired clavicles and cleithra, and an episternum (interclavicle)—and the paired, endochondral scapulocoracoids, each with two or three ossification centers, the scapula, the coracoid or the anterior and posterior coracoids. A cleithrum lies on the anterolateral edge of each scapula. Cleithra disappeared early in reptilian evolution and do not exist in extant reptiles. The episternum is a new girdle element, lying ventromedial and superficial to the sternum (Fig. 2.27). The clavicles extend medially along the base of the scapulae to articulate with the anterior ends of the episternum. The endochondral components lie deep to the dermal ones. The scapula is vertical element, and the coracoid is horizontal; at their junction, they support the glenoid fossa for the articulation of the humerus. The coracoids of the left and right sides meet medially and are usually narrowly separated by a cartilaginous band, which is continuous posteriorly with the broader, cartilaginous sternum. The sternum bears the attachments for the anterior sternal (thoracic) ribs and often a pair of posterior processes that receive the attachments for additional ribs. Posterior to the thoracic ribs, a series of dermal ribs, the gastralia, may support the ventral abdominal wall (see Fig. 2.24). These abdominal ribs are superficial to, and are not joined to, the thoracic ribs or any sternal processes, although the connective tissue sheath of the gastralia may attach to the epipubis of the pelvic girdle.

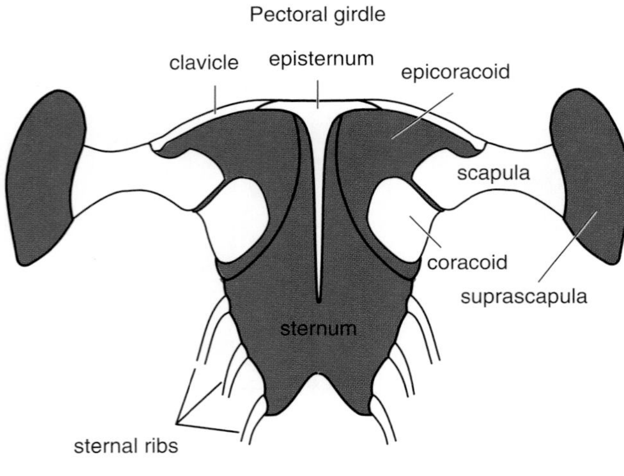

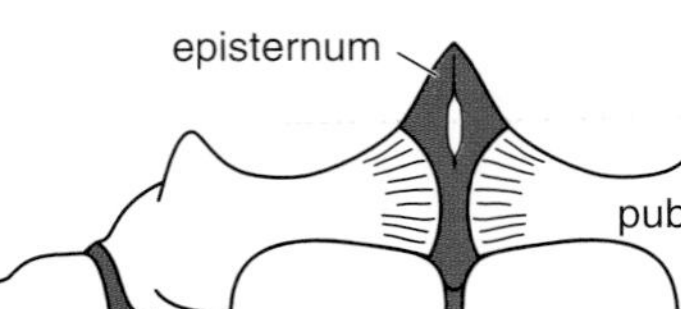

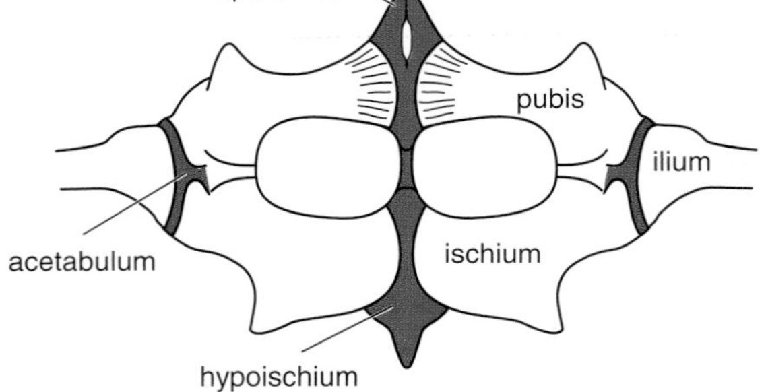

FIGURE 2.27 Ventral views of the pectoral (upper) and pelvic (lower) girdles of a juvenile tuatara (*Sphenodon punctatus*). Adapted from Schauinsland, 1903.

Crocodylians, *Sphenodon*, and some lizards have gastralia, although the gastralia and sternum are absent snakes and turtles. The ventral shell of turtles, the plastron, is largely a bony neomorph, defined as a novel and unique structure; only the clavicles and the episternum appear to have become part of the plastron. Snakes have lost all pectoral girdle elements, and many limbless lizards have greatly reduced endochondral elements; occasionally, the dermal elements are lost. Even limbed lizards show a reduction of dermal elements; the episternum is reduced to a thin cruciform rod of bone in most. Chameleons lack the clavicles and the episternum. Clavicles are absent in crocodylians, but the episternum remains as a median rod.

The reptilian pelvic girdle contains three pairs of endochondral elements: the vertical ilia that attach to the sacral vertebrae dorsally, and the horizontal pubes (anterior) and ischia (posterior). The elements form a ventral plate that joins the left and right sides of the girdle (Fig. 2.27). An acetabulum occurs on each side at the juncture of the three bones. These elements persist in all living reptiles, with the exception of most snakes. In all, the puboischiac plate develops a pair of fenestrae that often fuse into a single large opening encircled by the pubes and ischia. The plate becomes V-shaped as the girdle deepens and narrows. In most reptiles, the ilia are rodlike. In a few primitive snake families, a rod-shaped pelvic bone remains on each side. Its precise homologues are unknown, but it does bear an acetabulum and usually processes that are labeled as ilial, ischial, and pubic processes. The femur is vestigial and externally covered by a keratinous spur.

Early reptiles had short, robust limb bones with numerous processes. In modern species, the propodial elements, the humerus and femur, are generally smooth, long, and columnar with a slight curve; their heads are little more than rounded ends of the bony element. Only in turtles are the heads elevated and tilted from the shaft as distinct articular surfaces. The epipodial pairs are of unequal size, with the ulna or tibia the longer, more robust weight-supporting element of the pair. With the rotation of the epipodium, the ulna developed a proximal olecranon process and a sigmoid notch for articulation with the humerus. The tibia lacks an elevated process but has a broad proximal surface for femoral articulation. The mesopodial elements consist of numerous small block-like bones. The arrangement, fusion, and loss of these elements are highly variable, and the wrist or ankle flexure usually lies within the mesopodium. The metapodial elements are elongate and form the base of the digits. The basic phalangeal formula for the reptilian forefoot (manus) is 2–3–4–5–3 and that for the hindfoot (pes) 2–3–4–5–4. Most extant reptiles have lost phalanges within digits or occasionally entire digits.

The pectoral girdle and forelimbs attach to the axial skeleton by muscles that extend from the vertebrae to the interior of the girdle or to the humerus. A similar pattern of muscular attachment exists for the pelvic girdle and hindlimbs, although this girdle attaches firmly and directly to the vertebral column through the sacral ribs–ilia buttress. Within the limbs, the single-jointed muscles serve mainly as rotators, and the multiple-jointed muscles serve as extensors and flexors, many of which extend from the distal end of the propodium to the manus or pes.

NERVES AND SENSE ORGANS—COORDINATION AND PRECEPTION

The nervous system of vertebrates has four morphologically distinct, but integrated, units: the central nervous system, the peripheral nervous system, the autonomic nervous system, and various sense organs. The first three of these units are composed principally of neurons or nerve cells, each of which consists of a cell body and one or more axons and dendrites of varying lengths. The appearance of nervous system structures depends upon the organization of various parts of the neurons within the structure. For example, nerves are bundles of axons, and the gray matter of the brain results from concentrations of cell bodies. The sense organs show a greater diversity of structure and organization, ranging from single-cell units for mechanoreception to multicellular eyes and ears. Neurons or parts of neurons are important components of sense organs, but most sense organs require and contain a variety of other cell and tissue types to become functional organs.

NERVOUS SYSTEMS

The central nervous system includes the brain and the spinal cord. Both derive embryologically and evolutionarily from a middorsal neural tube. The anterior end of this tube enlarges to form the brain, which serves as the major center for the coordination of neuromuscular activity and for the integration of, and response to, all sensory input. The brain is divided during development by a flexure into the forebrain and hindbrain. The forebrain and hindbrain are each further partitioned, structurally and functionally, into distinct units (Fig. 2.28). From anterior to posterior, the forebrain consists of the telencephalon and the diencephalon, the midbrain consists of the mesencephalon, and the hindbrain consists of the cerebellum and medulla oblongata. Twelve pairs (10 in extant amphibians) of cranial nerves arise from the brain as follows: olfactory (I) from the telencephalon; optic (II) from the diencephalon; oculomotor (III), trochlear (IV), and abducens (VI) from the mesencephalon; and trigeminal (V), facial (VII), auditory (VIII), glossopharyngeal (IX), and vagus (X) from the medulla. The accessory (XI) and hypoglossal (XII) cranial nerves also originate from the medulla in other vertebrates, but apparently a shortening of the cranium places them outside the skull in amphibians; hence, they become spinal nerves.

The embryonic flexure disappears in amphibians as subsequent embryonic growth straightens the brain. The morphology of the brain is similar in the three living groups, although the brain is shortened in frogs and more elongate in salamanders and caecilians. The telencephalon contains elongate and swollen cerebral hemispheres dorsally encompassing the ventral olfactory lobes. The cerebral hemispheres compose half of the total amphibian brain (Fig. 2.28). The small, unpaired diencephalon lies behind the hemispheres and merges smoothly into the mesencephalon's bulbous optic lobes. Internally, the diencephalon is divided into the epithalamus, thalamus, and hypothalamus. A small pineal organ, the epiphysis, projects dorsally from the epithalamus; a parietal process, lying anterior to the epiphysis, is absent in extant amphibians. The anterior part of the ventral hypothalamus holds the optic chiasma where the optic nerves cross as they enter the brain, and the posterior part holds the infundibular area, from which the hypophysis or pituitary gland projects. Behind the optic lobes, the hindbrain is a flattened triangular area tapering gradually into the spinal cord. Neither the cerebellum, the base of the triangle abutting the optic lobes, or the medulla is enlarged.

Brain size and morphology vary considerably among reptile clades. In all reptiles, the basic vertebrate plan of two regions, the forebrain and the hindbrain, is maintained, and flexure of the brain stem is limited. The brain case is commonly larger than the brain, so that its size and shape do not accurately reflect dimensions and morphology of the brain. The forebrain of adult reptiles contains the cerebral hemispheres, the thalamic segment, and the optic tectum, and the hindbrain contains the cerebellum and medulla oblongata. The cerebral hemispheres are pear-shaped with olfactory lobes that project anteriorly and end in olfactory bulbs. These lobes range from long, narrow stalks with tiny bulbs in many iguanian lizards to short, stout stalks and bulbs in tortoises. Their sizes reflect the reliance on olfaction for many functions in amphibians and reptiles. The thalamic area is a thick-walled tube compressed and hidden by the cerebral lobes and the optic tectum. The dorsal, epithalamic portion has two dorsal projections. The anteriormost projection is the parietal (parapineal) body; in many lizards and *Sphenodon*, it penetrates the skull and forms a parietal eye. The posterior projection, the epiphysis, is the pineal organ and is typically glandular in turtles, snakes, and most lizards, although in some lizards and *Sphenodon*, it is a composite with a

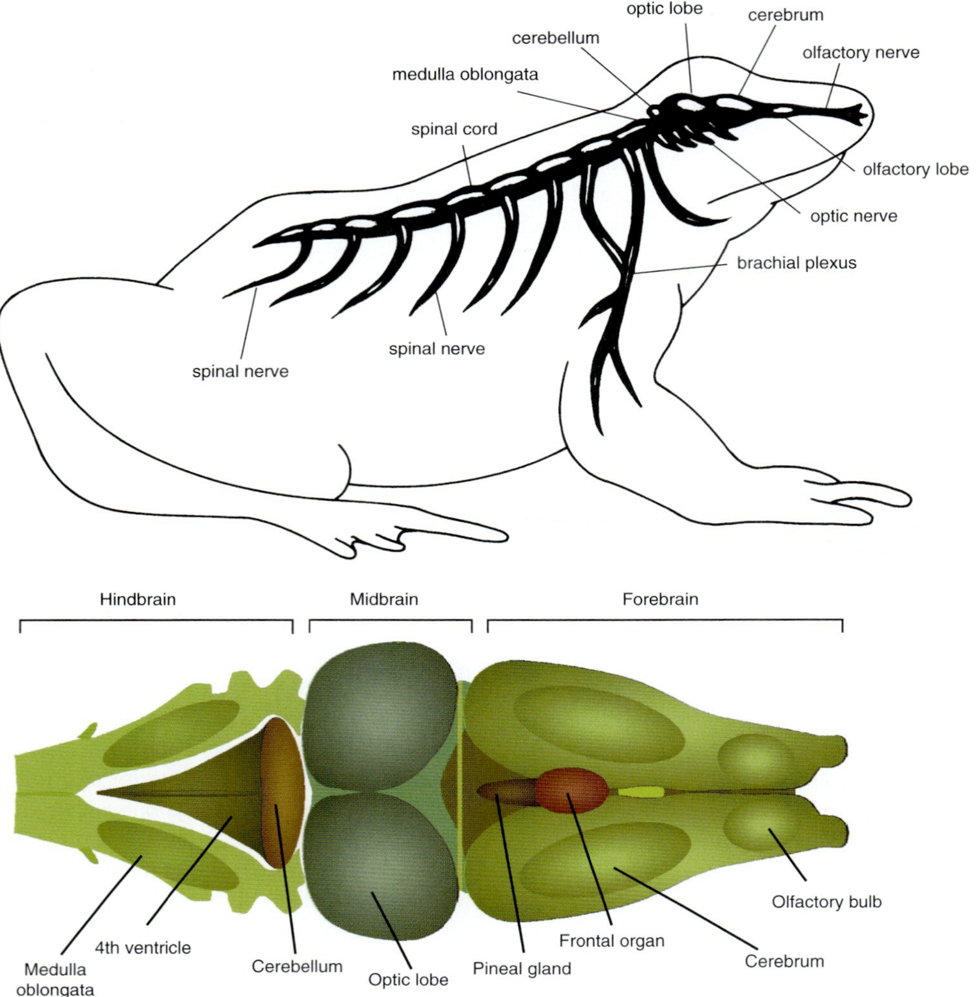

FIGURE 2.28 Above: A diagrammatic lateral view of the brain and spinal cord of a frog. Below: Structure of the frog brain.

rudimentary retinal structure like the parietal body and glandular tissue. Crocodylians lack a parietal–pineal complex. The ventral portion of thalamic area is the hypothalamus. In addition to its nervous function, the thalamus, the hypothalamus, and the adjacent pituitary gland function together as a major endocrine organ. The dorsal part of the posterior portion of the forebrain is the optical tectum and the ventral portion is the optic chiasma. The cerebellum and medulla are small in extant reptiles.

The spinal cord is a flattened cylinder of nerve cells that extends caudad through the vertebrae. A bilateral pair of spinal nerves arises segmentally in association with each vertebra for the entire length of the cord. Each spinal nerve has a dorsal sensory and a ventral motor root that fuse near their origins and soon divide into dorsal, ventral, and communicating nerve branches. The neurons of the first two branches innervate the body wall, as well as the skin, muscle, and skeleton; the neurons of the communicating branches join the central nervous system and the autonomic system to innervate the viscera, including the digestive, urogenital, circulatory, endocrine, and respiratory organs.

The spinal cord extends to the end of the vertebral column in salamanders and caecilians, but in anurans, the cord ends at the level of the sixth or seventh vertebrae, and a bundle of spinal nerves, the cauda equina, continues caudad through the neural canal. In all reptiles, the spinal cord extends from the medulla posteriorly to the end of the vertebral column. The diameter of the cord is nearly uniform from brain to base of tail, except for a slight expansion in the region of the limbs. The organization of the spinal nerves is similar in all living amphibians and reptiles. The dorsal root contains somatic and visceral sensory neurons and some visceral motor neurons; the somatic motor and some visceral motor neurons compose the ventral root.

The nerves and their ganglia aggregations of neuron cell bodies, exclusive of the skull and vertebral column, compose the peripheral and autonomic nervous systems. The peripheral system contains the somatic sensory neurons and axons of motor neurons; the autonomic system contains the visceral sensory and some motor neurons. The latter are generally associated with the involuntary activity of the smooth muscles and glands of the viscera. Both the peripheral and autonomic systems are similar in the three amphibian groups, but neither system has been studied extensively, especially the autonomic system. The peripheral nerves transmit the animal's perception of the outside world to the central nervous system and then transmit messages to the appropriate organs for the animal's response.

SENSE ORGANS

Sense organs provide the animal with information about itself and its surroundings. The sense organs that monitor the internal environment and those that monitor the external environment are integrated either directly with the central nervous system or indirectly with it through the autonomic and peripheral networks. The eyes, ears, and nose are obvious external receptors. Heat and pressure receptors of the skin are less obvious, as are internal receptors, such as the proprioceptors of joints and muscles.

Cutaneous Sense Organs

The skin contains a variety of receptors that register the environment's impingement on the animal's exterior. Pain and temperature receptors consist of free and encapsulated nerve endings, most lying in the dermis but a few extending into the epidermis. Mechanoreceptors, sensitive to pressure and touch, are similarly positioned in the skin. The pressure receptors may also sense temperature.

The lateral line system of larval and a few adult amphibians is the most evident of the cutaneous sense organs. Superficially it appears as a series of pores on the head and body of aquatic larvae and some aquatic adults, such as cryptobranchid, amphiumid, proteid, and sirenid salamanders; typhlonectid caecilians; and pipid frogs. The mechanoreceptor organs or neuromasts are arranged singly or in compact linear arrays called *stitches* to form the various lines or canals that traverse the head and trunk. Each neuromast contains a small set of cilia projecting from its outer surface. The cilia bend in only one axis, thereby sensing water pressure or current changes along only that axis. They are sensitive to light currents and used to locate food. Neuromasts are reduced only in species living in rapidly flowing water.

Recently, ampullary organs were discovered on the heads of some larval salamanders and caecilians. These electroreceptors are less numerous, lying in rows parallel to the neuromasts. Like neuromasts, ampullary organs provide the larva with a sense of its surroundings, identifying both stationary and moving objects lying within the electrical field surrounding the larva.

Cutaneous sense organs are especially common in reptiles and occur in a variety of forms. In addition to pain and temperature receptors, several types of intraepithelial mechanoreceptors register pressure, tension, or stretching within the skin. Mechanoreceptors with discoid endings or terminals occur over most of the body, and mechanoreceptors with branching terminals lie within the hinges between scales of lepidosaurs. Mechanoreceptors with coiled, lanceolate, or free terminals are confined to the dermis. On the surface of the skin, tactile sense organs are abundant; they range in shapes from buttonlike and smooth to those with barbed bristles.

The pit organs of some boids, pythonids, and viperids are specialized structures in the dermis and epidermis that house infrared heat receptors. In *Boa*, these receptors, both intraepidermal and intradermal types, are scattered on unmodified supra- and infralabial scales. In *Python*, a series of pits occurs in the labial scales, and the heat receptors are concentrated on the floor of the pit. In pitvipers (crotaline snakes), a pit organ occurs on each side of the head between the naris and the eye (Fig. 2.29). The openings face forward and their receptor fields overlap, giving them stereoscopic infrared vision. The heat receptors lie within a membrane stretched across the pit.

Ears

The ears of tetrapods, including frogs, lizards, and mammals, are structurally similar and serve two functions: hearing, the reception of sound waves; and balance, the detection of the position and movement of the animal's

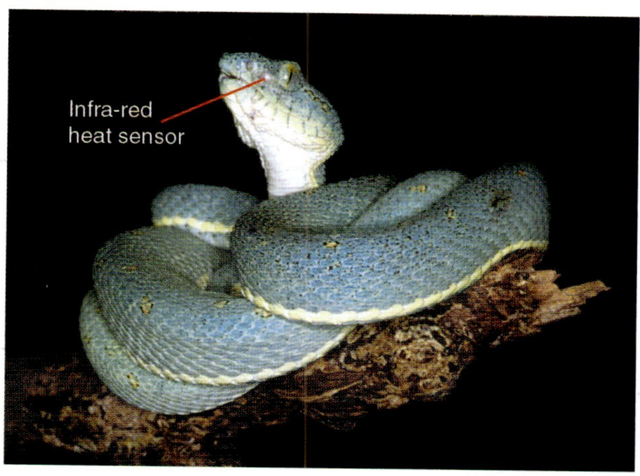

FIGURE 2.29 Infra-red heat-sensing pits are located below and posterior to the nares in pit-vipers. These sense organs detect movement across a thermal landscape based on relative temperature. The snake in the photograph is *Bothriopsis bilineata* from the Amazon rain forest. (Laurie Vitt)

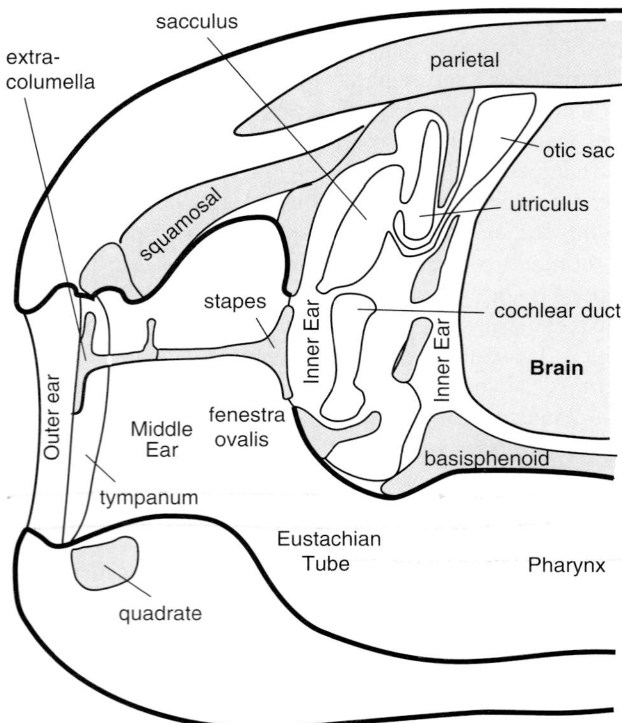

FIGURE 2.30 Lateral view of the anatomy of a lizard's ear. The otic capsule consists mainly of the opisthotic and prootic. Adapted from Baird, 1970.

head. The receptors for both functions are neuromasts located in the inner ear. These neuromasts differ somewhat from those of the lateral line system, but they similarly record fluid movements along a single axis by the deflection of terminal cilia.

Ears are paired structures, one on each side of the head just above and behind the articulation of the lower jaw. Each ear consists of an inner, middle, and outer unit (Fig. 2.30). The inner ear is a fluid-filled membranous sac, containing the sensory receptors and suspended in a fluid-filled cavity of the bony or cartilaginous otic capsule. The middle ear contains the bone and muscular links that transfer vibrations from the eardrum, the tympanum, to the inner ear. An outer ear is usually no more than a slight depression of the tympanum or may be absent. Salamanders, caecilians, and some frogs lack tympana. In these amphibians, low-frequency sounds may be transmitted via the appendicular and cranial skeleton to the inner ear. For reptiles, an outer ear occurs only in crocodylians and some lizards; tympana are flush with the surface of the head in turtles and some lizards. A special muscle allows crocodylians and most geckos to close the ear cavity.

The middle ear of reptiles contains a tympanum and two ear ossicles, the stapes and the extracolumella, within an air cavity. The tympanum receives sounds and transmits the vibrations along the extracolumella–stapes chain to the oval window of the inner ear. The middle ear cavities are large in turtles, large with left and right cavities connected in crocodylians, small and nearly continuous with the pharynx in most lizards, narrow canals in snakes, and usually absent in amphisbaenians. The stapes is typically a slender columnar bone, and its cartilaginous tip, the extracolumella, has three or four processes that reach the tympanum. In snakes, the stapes abuts against the quadrate bone for transmission of vibrations.

Unlike reptiles, the amphibian middle ear has two auditory pathways: the tympanum–stapes path for airborne sounds and the forelimb–opercular path for seismic sounds. Both pathways reach the inner ear through the fenestra ovalis of the otic capsule. The tympanum–stapes path is shared with other tetrapods. In amphibians the stapes is a single bony rod that extends between the external eardrum and the fenestra ovalis of the inner ear. In most frogs, the stapes lies within an air-filled cavity, and in salamanders and caecilians, the stapes is embedded in muscles. The limb–opercular path is unique to frogs and salamanders. Sound waves are transmitted from the ground through the forelimb skeleton onto the tensed opercular muscle that joins the shoulder girdle to the operculum lying in the fenestra ovalis.

In amphibians and reptiles, the membranous inner ear basically consists of two sacs joined by a broad passage. The dorsal sac or utriculus has three semicircular canals that project outward from it. One of these canals lies horizontally, the other two are vertical, and all three are perpendicular to one another. This orientation allows movement to be recorded in three different planes and provides information for the sense of balance. The neuromasts are clustered in patches, one patch in each semicircular canal and one or more patches in the utriculus and the ventral sac, the sacculus. In amphibians, the sacculus also contains several outpocketings, including the amphibian papilla, basilar papilla, lagena, and endolymphatic duct. The two papillae contain patches of neuromasts specialized for acoustic reception. Reptiles lack the amphibian papilla but have a cochlear duct from which the auditory sensory area projects ventrally from the sacculus and adjacent to the oval window.

Eyes

Eyes vary from large and prominent to small and inconspicuous in extant amphibians. All have a pair of eyes located laterally or dorsolaterally on the head. Most terrestrial and arboreal salamanders and frogs have moderate to large eyes, whereas fossorial and aquatic species usually have small eyes. Eyes are degenerate and lie beneath the skin in caecilians and cave-dwelling salamanders; in a few caecilians eyes lie beneath bone. The eyes of most reptiles are large and well developed. The eyes are degenerate only in a few fossorial species and groups. They

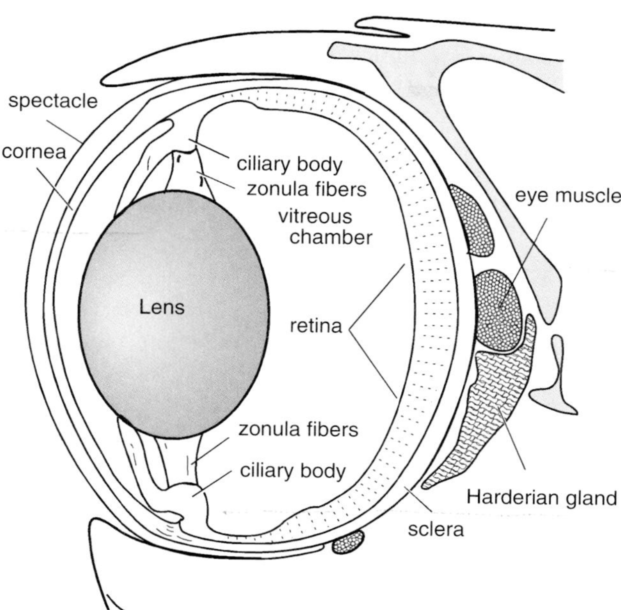

FIGURE 2.31 Cross section of the anatomy of a snake's eye. Adapted from Underwood, 1970.

have disappeared completely, leaving no pigment spot visible externally, in only a few species of scoleophidian snakes.

The structure of the eye is similar in all vertebrates (Fig. 2.31). It is a hollow sphere lined internally with a heavily pigmented sensory layer, the retina. The retina is supported by the sclera, a dense connective tissue sheath forming the outside wall of the eyeball. The cornea is the transparent part of the outer sheath lying over a gap in the retina that allows light to enter the eye. In postmetamorphic amphibians, eyelids and a nictitating membrane slide across the exposed cornea to protect and moisten it. A spherical lens lies behind the cornea and is anchored by a corona of fibers that extend peripherally to the cornea–scleral juncture. The amount of light passing through the lens and onto the retina is regulated by a delicate, pigmented iris lying behind the cornea. Its central opening, the pupil, is opened (dilated) or closed (contracted) by peripherally placed muscles. The eye retains its spherical shape by the presence of fluid, the vitreous humor in the cavity behind the lens and the aqueous humor in front of the lens.

Light enters the eye through the iris and is focused on the retina by the lens. The organization of the retina's several layers differs from what might be expected. The sensory or light-registering surfaces are not the innermost surface of the eye. Instead, the innermost layer consists of transmission axons that carry impulses to the optic nerve, and the next layer contains connector neurons that transfer impulses from the adjacent receptor cell layer. The deepest layer contains pigment cells adjacent to the sclera. The receptor surfaces of the sensory cells face inward, not outward toward the incoming light, and against and in the pigment layer. Amphibians have four kinds of light receptors: red and green rods, and single and double cones. The cones are the color receptors that possess specialized pigments sensitive to a narrow range of wavelengths. When light strikes these pigments, their chemical state is changed. Amphibians are the only vertebrates with two types of rods, and the green rods are unique to amphibians. These rods are absent in taxa with degenerate eyes. The visual pigment of the rods is sensitive to all wavelengths of light; hence, rods register only the presence or absence of light.

The eyes of reptiles, except snakes, have a ring of bony plates (scleral ossicles) embedded in the sclera and surrounding the cornea. Pupils range from round to elliptical and are usually oriented vertically, although occasionally they are horizontal is some species. The reptilian eyeball and lens are usually spherical (Fig. 2.31). Rather than moving the lens for accommodation, lens shape is changed by the contraction of radial muscles in the ciliary body encircling the lens. Crocodylians and turtles share a duplex retina (rods and cones) with other vertebrates and possess single and double cones and one type of rod. In squamates, the retina has been modified. Primitive snakes have a simplex retina consisting only of rods; advanced snakes have a duplex retina of cones and rods, although the cones are probably transformed rods. In lizards, the simplex retina contains two or three different types of cones.

Nasal Organs

Olfaction or smelling is performed by bilaterally paired nasal organs and the vomeronasal (Jacobson's) organ. Each nasal organ opens to the exterior through the external naris and internally into the buccal cavity via the choana (internal naris). Between these openings in amphibians is a large olfactory (principal) cavity and several accessory chambers that extend laterally and ventrally; the vomeronasal organ is in one of the accessory chambers. A nasolacrimal duct extends from the anterior corner of each eye to the principal cavity. The surface of the chambers contains support and mucous cells and is lined with ciliated epithelium. The ciliated neuroepithelium occurs in three patches. The largest patch occupies the roof, medial wall, and the anterior end of the principal cavity. A small, protruding patch occurs on the middle of the floor, and another small patch is present in the vomeronasal organ chamber. The neuroepithelium of the principal cavity is innervated by neurons from the olfactory bulb of the brain, and the vomeronasal organ is innervated by a separate olfactory branch. Olfaction is a chemosensory process. The actual receptor site on the cell is unknown but may be either at the base of each cilium or near the cilium's junction with the cell body.

The nasal organs of salamanders are composed of a large main cavity partially divided by a ventrolateral fold. Aquatic salamanders have the simplest and smallest nasal cavities, but they possess large vomeronasal organs. Frogs, in general, have a complex nasal cavity consisting of three chambers and a large vomeronasal organ. Caecilians have simple nasal cavities similar to salamanders but with a major modification, the sensory tentacle. The size, position, and structure of the tentacle vary among different species; however, in all, the tentacle arises from a combination of nasal and orbital tissues as a tubular evagination from the corner of the eye. The tentacle's exterior sheath is flexible but nonretractable. The tentacle proper can be extruded and retracted into its sheath. Odor particles are transported via the nasolacrimal duct to the vomeronasal organ.

In reptiles, each nasal organ consists of an external naris, a vestibule, a nasal cavity proper, a nasopharyngeal duct, and an internal naris. These structures serve as air passages and are lined with nonsensory epithelium. The sensory or olfactory epithelium lies principally on the roof and anterodorsal walls of the nasal cavity. These passages and cavities are variously modified in the different reptilian groups. The vestibule is a short tube in turtles and snakes, and is much longer and often curved in lizards. A concha covered with sensory epithelium projects into the nasal cavity from the lateral wall. *Sphenodon* has a pair of conchae, squamates and crocodylians have one, and turtles have none. The vomeronasal organ is an olfactory structure, used primarily to detect nonaerial, particulate odors (Fig. 2.32). It arises embryologically from the nasal cavity but remains connected to this cavity as well as to the oral cavity only in *Sphenodon*. In squamates, it communicates with the oral cavity by a narrow duct. Odor particles are carried to the vicinity of the duct by the tongue. Well developed in squamates, this organ is absent in crocodylians; in turtles, it lies in the main nasal chamber rather than in a separate chamber.

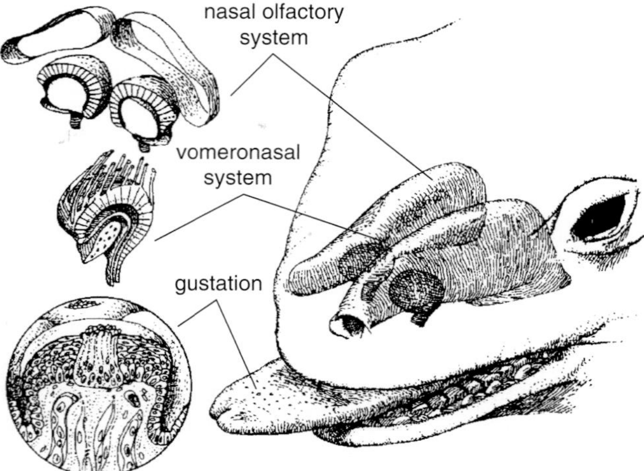

FIGURE 2.32 Lepidosaurians can have gustatory organs (taste buds), nasal olfactory systems (sense of smell), and/or vomeronasal systems (chemosensory using the tongue to transport chemicals). Adapted from Schwenk, 1995.

Internal Sense Organs

The major internal sense organs are the proprioceptor organs embedded in the muscles, tendons, ligaments, and joints. These organs record the tension and stress on the musculoskeletal system and allow the brain to coordinate the movement of limbs and body during locomotor and stationary behaviors. The proprioceptors show a structural diversity from simple nerve endings and netlike endings to specialized corpuscles. Structurally, the proprioceptors of reptiles are similar to those of amphibians.

Taste buds or gustatory organs are present in all amphibians, although they have been little studied and nearly exclusively in frogs. There are two types: papillary organs, located on fingiform papillae on the outer surface of the tongue, and nonpapillary organs, located throughout the buccal cavity, except on the tongue. Each type of taste bud is a composite of receptor and support cells. The buds are highly sensitive to salts, acids, quinine (bitter), and pure water. In many reptiles, taste buds occur on the tongue and are scattered in the oral epithelium (Fig. 2.32). Structurally, they appear similar to those of amphibians and share the same sensory responses. In squamates, taste buds are abundant in fleshy-tongued taxa and are greatly reduced or absent in taxa (e.g., most snakes) with heavily keratinized tongue surfaces.

HEART AND VASCULAR NETWORK—INTERNAL TRANSPORT

The circulatory system is a transport system that carries nutrients and oxygen to all body tissues and removes waste products and carbon dioxide from them. This system contains four components: blood, the transport medium; vascular and lymphatic vessels, the distribution networks; and the heart, the pump or propulsive mechanism.

Blood

Amphibian blood plasma is a colorless fluid, and it contains three major types of blood cells, erythrocytes, leucocytes, and thrombocytes. The blood cells are typically nucleated, although in salamanders a small number of each of the three types lacks nuclei. Erythrocytes carry oxygen to and carbon dioxide from the tissues; both gases attach to the respiratory pigment hemoglobin. Erythrocytes vary in size among amphibian species, but, in general, amphibians have the largest erythrocytes known among vertebrates. Leucocytes consist of a variety of cell types, most of which are involved in maintenance duties such as removing cell debris and bacteria or producing antibodies. The thrombocytes serve as clotting agents. Only the erythrocytes are confined to vascular vessels; the other blood cells and the plasma leak through the walls of the vascular vessels and bathe the cells of all tissues. The plasma and cells reenter the vascular vessels directly or collect in the lymphatic vessels that empty into the vascular system.

Blood plasma is colorless or nearly so in most reptiles. A few skinks and crotaline snakes have green or greenish-yellow blood. In addition to dissolved salts, proteins, and other physiological compounds, the plasma transports three types of cells: erythrocytes, leucocytes, and thrombocytes, all of which have nuclei in reptiles.

Arterial and Venous Circulation

The vascular vessels form a closed network of ducts that transports the blood. Blood leaves the heart through the arteries that divide into smaller and smaller vessels, the arterioles. The smallest vessels, the capillaries, are only slightly larger than the blood cells flowing through them. Within the capillary beds, the plasma and some leucocytes and thrombocytes leak through to the lymphatic system. Beyond the capillaries, the vessels become progressively larger. Venules, comparable to arterioles in size, lead to the larger veins, which return blood to the heart.

In amphibians, blood leaves the heart through the conus arteriosus, which soon divides into three aortic arches, the pulmocutaneous arch, the systemic arch, and the carotid arch (Fig. 2.33). The position and number of aortic arches are highly variable in amphibians. The pulmocutaneous arch divides into cutaneous arteries that serve the skin and into pulmonary arteries that lead to the respiratory surfaces where gaseous exchange occurs. The systemic arch curves dorsally and fuses on the midline with its bilateral counterpart to form the dorsal aorta. Vessels that branch from the dorsal aorta as it extends posteriorly provide blood to all viscera and limbs. The branches of the carotid arch carry blood to the tissues and organs of the head and neck. The venous system has a comparable distributional pattern of vessels but in reverse. A pair of common jugular veins drains the numerous veins of the head and neck; the subclavian veins gather blood from the smaller veins of the forelimbs and skin; and the pulmonary veins drain the lungs. A single postcaval vein is the major efferent vessel for the viscera and hindlimbs. All these veins, except the pulmonary vein, empty into the sinus venosus, which opens directly into the heart (Fig. 2.33). The sizes, shapes, and branching patterns within the vascular network are nearly as variable within a taxon as they are between unrelated taxa. The visceral arches of amphibian larvae give rise to the aortic arches of adults, although adults lose the first two arches. Of the remaining arches, some salamanders retain all, whereas anurans retain three, and caecilians retain two.

The arterial and venous networks of reptiles are similar to those of adult amphibians, but, like amphibians, the reptilian groups differ from each other. For example, the pattern of vessels to and from the trunk of snakes and turtles is not the same. The major trunk vessels leading from the heart to the viscera, head, and limbs and those vessels returning the blood to the heart are more similar among species and groups than they are different.

In reptiles, the pulmonary artery typically arises as a single trunk from the cavum pulmonale of the right ventricle and bifurcates into the right and left branches above and in front of the heart (Fig. 2.34). The systemic arteries

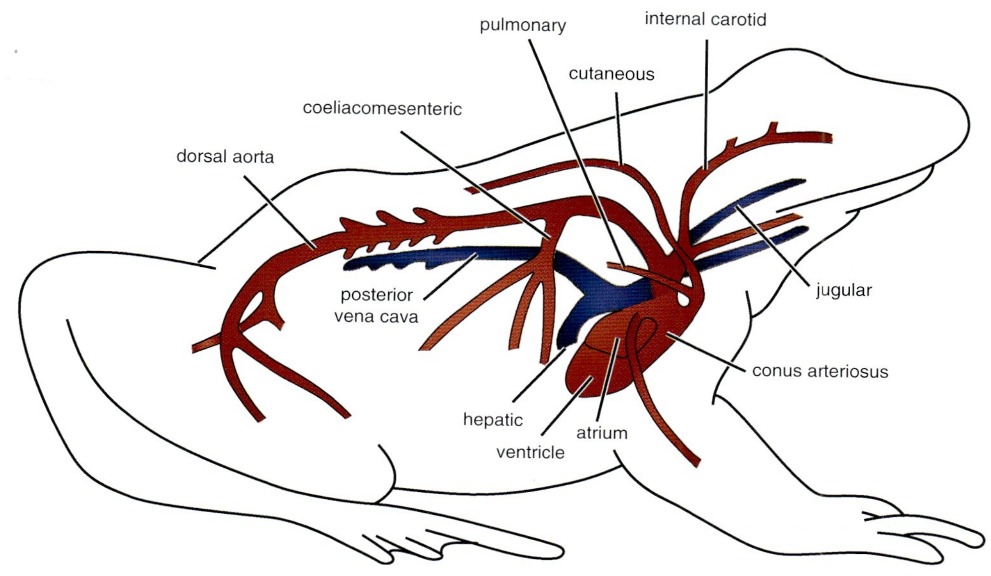

FIGURE 2.33 Lateral view of the circulatory system of a frog.

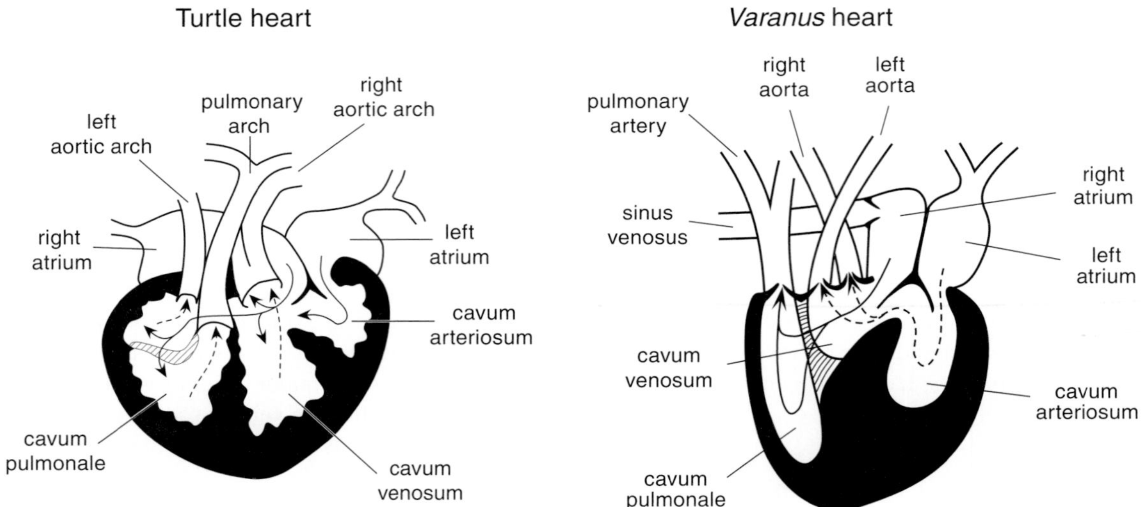

FIGURE 2.34 Heart anatomy of a turtle and a varanid lizard; diagrammatic ventral views of frontal sections. The arrows indicate only the general pathway of blood flow through the ventricle into the aortic arches. Adapted from Burggren, 1987.

(aortas) arise separately but side by side from the cavum venosum of the left ventricle. The left systemic artery curves dorsally and bifurcates into a small ductus caroticus and the larger systemic branch. The right systemic artery bifurcates in front of the heart; the cranial branch forms the major carotid network, and the systemic branch curves dorsally to join the left systemic branch. This combined aorta (dorsal aorta) extends posteriorly and its branches serve the limbs and the viscera. The major venous vessels are the jugular veins that drain the head and the postcaval vein that receives vessels from the limbs and viscera. The jugular and postcaval trunks join into a common sinus venosus; in turn, it empties into the right atrium.

Lymphatic Network

The lymphatic network is an open system, containing both vessels and open cavities or sinuses within the muscles, in the visceral mesenteries, and beneath the skin. It is a one-way network, collecting the plasma and other blood cells that have leaked out of the capillaries and returning them to the vascular system. The lymph sinuses are the major collection sites, and the subcutaneous sinuses are especially large in frogs. The sinuses are drained by lymphatic vessels that empty into veins. In amphibians and fishes, lymph hearts lie at venous junctions and are contractile structures with valves that prevent backflow and thereby speed the flow of lymph into the veins. Frogs and salamanders have 10 to 20 lymph hearts; the elongate caecilians have more than a hundred.

The lymphatic system of reptiles is an elaborate drainage network with vessels throughout the body. This network of microvessels gathers plasma (lymph) from throughout the body, and the smaller vessels merge into increasingly larger ones that in turn empty into the main lymphatic trunk vessels and their associated sinuses. The trunk, vessels, and sinuses empty into veins. Major trunks collect plasma from the limbs, head, and viscera, forming a network of vessels that outlines the shape of the reptile's body. The occurrence of valves is irregular, and plasma flow can be bidirectional; however, the major flow in all trunks is toward the pericardial sinus and into the venous system. A single pair of lymphatic hearts but no lymph nodes occur in the pelvic area.

Heart

Heart structure is highly variable in amphibians. All have a three-chambered heart composed of two atria and one ventricle, but the morphology of the chambers and the pattern of blood flow through the chambers vary (Fig. 2.33). The differences are associated with the relative importance of cutaneous and pulmonary respiration. Even differences in an amphibian's physiological state modify the flow pattern—a hibernating frog might have a flow pattern that mixes pulmonary and systemic blood in the ventricle, whereas an active frog does not. The atria are thin-walled sacs separated by an interatrial septum. The sinus venosus empties into the right atrium, and the pulmonary veins empty into the left atrium. Both atria empty into the thick, muscular-walled ventricle, which pumps the blood into the conus arteriosus. Although the ventricle is not divided by a septum, oxygenated and unoxygenated blood can be directed into different arterial pathways. Such segregation is possible owing to the volume and position of the blood in the ventricle, the nature of the ventricular contractions, the spiral fold of the conus arteriosus, the branching pattern of the arteries from the conus, and the relative resistance of the pulmonary and systemic pathways.

No single model represents a generalized reptilian heart. Heart size, shape, structure, and position are linked to other aspects of each species' anatomy and physiology. The animal's physiology is a major determinant of heart structure and function, but phylogeny and behavior also play determining roles. In snakes, heart position is correlated with arboreal, terrestrial, and aquatic habits. Among these variables, three general morphological patterns are recognized.

The typical reptilian heart of turtles and squamates (Fig. 2.34) is three-chambered, with two atria and a ventricle with three chambers or cava. From left to right, the cava are called the cava arteriosum, the cava venosum, and the cava pulmonale. The right atrium receives unoxygenated venous blood from the sinus venosus and empties into the cavum venosum of the ventricle. The left atrium receives oxygenated blood from the lungs via the pulmonary veins and empties into the cavum arteriosum. Because the three ventricular cava communicate and the muscular contraction of the ventricle is single-phased, oxygenated and unoxygenated blood mix, and blood exits simultaneously through all arterial trunks. Blood in the cavum pulmonale flows into the pulmonary trunk, and blood in the cavum venosum into the aortas.

Monitor lizards (varanids) possess a higher metabolic rate than other lizards and also have differences in the architecture of the ventricular cava, which communicate with one another (Fig. 2.34). The cavum venosum is small—little more than a narrow channel linking the cavum pulmonale with a greatly enlarged cavum arteriosum. Ventricular contraction is two-phased so that the pumping cycle creates a functionally four-chambered heart. Although mixing of unoxygenated and oxygenated blood can occur and probably does in some circumstances, the cavum pulmonale is isolated during systole (contraction), and unoxygenated blood is pumped from the right atrium to the lungs. Within crocodylians, the ventricle is divided into separate right and left muscle components. Uniquely, the two aortas in crocodylians arise from different ventricular chambers, the left aorta from the right chamber and right aorta from the left chamber. This arrangement provides an opportunity for unoxygenated blood to bypass the lungs in special physiological circumstances, such as during diving, by altering the pattern of ventricular contraction.

DIGESTIVE AND RESPIRATORY ORGANS—ENERGY ACQUISITION AND PROCESSING

The digestive and pulmonary systems are linked by a common embryological origin, similar functions, and shared passageways. The lungs and respiratory tubes form as an outpocketing of the principal regions. Both systems are intake ports and processors for the fuels needed to sustain life: oxygen for use in respiration, and water and food for use in digestion (see Chapter 6, "Water Balance and Gas Exchange," and Chapter 10, "Foraging Ecology and Diets").

Digestive Structures

The digestive system of amphibians has two major components, a digestive tube that has specialized regions and various digestive glands. The digestive tube or tract extends from the mouth to the anus, which empties into the cloaca. From beginning to end, the regions are the buccal (oral) cavity, the pharynx, esophagus, stomach, and small and large intestines. The general morphology of these regions is similar within amphibians, although the digestive tract is short in anurans and long in caecilians.

The mouth opens directly into the buccal cavity and is bordered by flexible, immobile lips. The buccal cavity is continuous posteriorly at the angle of the jaw with the pharynx. The primary palate forms the roof of the buccal cavity, and the tongue lies on its floor. The tongue is variously developed in amphibians. In its least-developed form, the tongue is a small muscular pad lying on a simple hyoid skeleton, as seen in pipid frogs. Some salamanders and many advanced frogs have tongues that can be projected very rapidly for long distances in order to capture prey. These projectile tongues have a more elaborate hyoid skeleton and associated musculature with a glandular pad attached to the muscular base.

Amphibian teeth are typically simple structures; each tooth has an exposed bicuspid crown anchored to a base, or pedicel in the jaw. Caecilians and a few frogs have unicuspid curved teeth. Salamanders and caecilians have teeth on all the jawbones; most frogs lack teeth on the lower jaw and a few lack teeth on the upper jaw.

The pharynx is the antechamber for directing food into the esophagus and air into the lungs. A muscular sphincter controls the movement of food in the thin-walled esophagus, and peristaltic movement propels food downward into the stomach. The stomach is an enlarged and expandable region of the digestive tube. Its thick muscular walls and secretory lining initiate the first major digestive breakdown of food. The food bolus passes from the stomach through the pyloric valve into the narrower and thin-walled small intestine. The forepart of the small intestine is the duodenum, which receives the digestive juices from the liver and pancreas. The small intestine of amphibians has only a small amount of internal folding and has villi to increase surface area for nutrient absorption. It is continuous with a slightly broader large intestine in caecilians, salamanders, and some frogs. In advanced frogs,

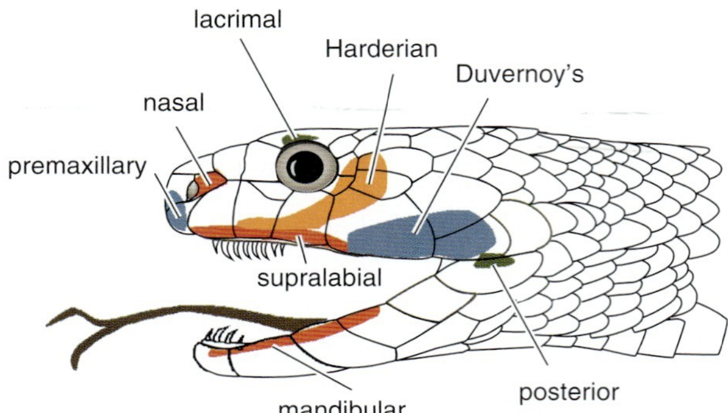

FIGURE 2.35 A variety of glands occur in the oral region of the head of reptiles, although not all reptiles have all glands shown. Premaxillary, nasal, and palatine glands secrete mucous to lubricate the mouth. Lacrimal and Harderian glands secrete fluids that wet the vomeronasal region and the eyes. The Duvernoy's gland occurs in venomous snakes and produces venom.

a valve separates the large and small intestines. The large intestine empties into the cloaca, which is a sac-like cavity that receives the products and by-products of the digestive, urinary, and reproductive systems. The cloaca exits to the outside through the vent.

The mouth of reptiles opens directly into the buccal cavity, and a variety of glands are situated in the head region (Fig. 2.35). The lips bordering the mouth are flexible skin folds, but they are not movable in lepidosaurs. Lips are absent in crocodylians and turtles. Tooth rows on the upper and lower jaws of most reptiles form a continuous border along the internal edge of the mouth. Turtles lack teeth and have keratinous jaw sheaths. In reptiles, teeth typically serve for grasping, piercing, and fragmenting food items. In many squamate reptiles (e.g., snakes), teeth aid in prey manipulation during swallowing. Only in a few species do the teeth cut and slice (e.g., *Varanus*) or crush (*Dracaena*). A well-developed tongue usually occupies the floor of the mouth. Tongue morphology varies in association with a variety of feeding behaviors; chameleons have projectile tongues, and varanoid lizards and snakes have telescoping tongues. The roof of the buccal cavity is formed by the primary palate. Two pairs of structures open anteriorly in the roof of the buccal cavity; the small Jacobson's organ opens just inside the mouth and is immediately followed by the larger internal nares. Crocodylians have a secondary palate that creates a separate respiratory passage from the internal nares on the primary palate to the beginning of the pharynx. This passage allows air to enter and exit the respiratory system while food is held in the mouth. A few turtles and snakes (aniliids) have developed partial secondary palates.

The pharynx is a small antechamber behind the buccal cavity. A valvular glottis on its floor is the entrance to the trachea. On the rear wall of the pharynx above the glottis, a muscular sphincter controls the opening into the esophagus. The eustachian tubes, one on each side, open onto the roof of the pharynx. Each tube is continuous with the middle-ear chamber to permit the adjustment of air pressure on the tympanum. Middle ears and Eustachian tubes are absent in snakes.

The esophagus is a distensible, muscular walled tube of variable length between the buccal cavity and the stomach. In snakes and turtles, the esophagus may be one-quarter to one-half of the body length (Fig. 2.36). It is proportionately shorter in reptiles with shorter necks. The stomach is a heavy muscular and distensible tube, usually J-shaped and largest in the curved area. The stomach narrows to a thick muscular sphincter, the pylorus or pyloric valve. This valve controls the movement of the food bolus from the stomach into the small intestine. The small intestine is a long narrow tube with little regional differentiation externally or internally; the pancreatic and hepatic ducts empty into its forepart. The transition between the small and large intestine is abrupt. The diameter of the latter is several times larger than the former, and often a small outpocketing, the caecum, lies adjacent to the juncture of the two intestines. The large intestine or colon is a straight or C-shaped tube that empties into the cloaca. The large intestine is the least muscular and most thin-walled structure in the digestive tract.

The cloaca is part of the digestive tract and is derived from the embryonic hindgut. A muscular sphincter, the anus, lies between the large intestine and the cloaca. The dorsal portion of the cloaca is the coprodaeum and is the route for the exit of feces. The urodaeum or urogenital sinus is a ventral outpocket of the cloaca and extends a short distant anterior to and beneath the large intestine. Digestive, urinary, and genital products exit via the vent, a transverse slit in turtles and lepidosaurs and a longitudinal slit in crocodylians.

Digestive Glands

A variety of glands occurs within the digestive tract. The lining of the buccal cavity contains unicellular and multicellular glands. Multicellular glands secrete mucus that lubricates the surface, and although numerous and widespread in terrestrial amphibians, they are less abundant in aquatic taxa such as pipid frogs and aquatic salamanders. The intermaxillary gland opens in the middle of the palate and secretes a sticky compound that helps prey adhere to the tip of the tongue. Numerous unicellular and multicellular glands are present in the lining of the remainder of the digestive tract; most secrete mucus and a few secrete digestive enzymes and acid into the stomach.

The liver and pancreas are major secretory structures that lie astride the stomach and duodendum and are derived from the embryonic gut. The liver is the largest of the digestive glands, serving as a nutrient storage organ and producer of bile. The bile drains from the liver into the gallbladder and then moves via the bile duct into the

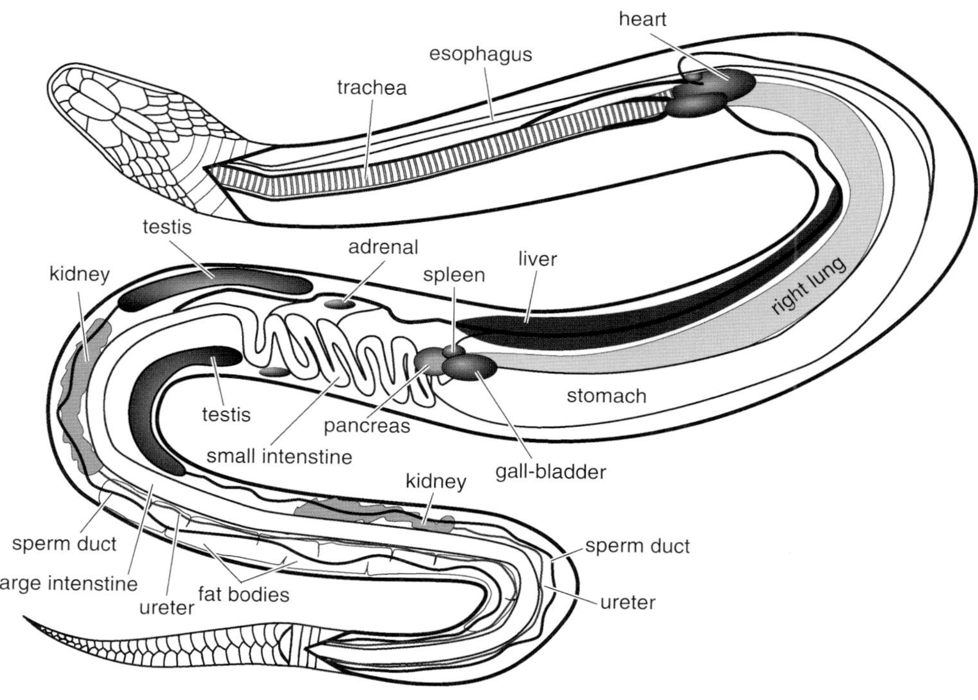

FIGURE 2.36 Visceral anatomy of a generalized male snake; a ventral view.

duodenum, where it assists in the breakdown of food. The pancreas is a smaller, diffuse gland. It secretes digestive fluids into the duodenum and also produces the hormone insulin.

Similarly, the oral cavity of reptiles contains numerous glands. The small, multicellular mucous glands are a common component of the epithelial lining and compose much of the tissue on the surface of the tongue. Larger aggregations of glandular tissue, both mucous and serous, form five kinds of salivary glands: labial, lingual, sublingual, palatine, and dental. In venomous snakes, the venom glands are modified salivary glands. Mucous glands occur throughout the digestive tract. The stomach lining is largely glandular and has several types of gastric glands. The small intestine has many small glands within its epithelial lining. The liver, usually the largest single organ in the visceral cavity, and pancreas produce secretions that assist in digestion. The pancreas is a smaller, more diffuse structure that lies within the visceral peritoneum.

Respiratory Structures

Lungs

The respiratory passage includes the external nares, olfactory chambers, internal nares, buccopharyngeal cavity, glottis, larynx, trachea, bronchial tubes, and lungs. The glottis, a slit-like opening on the floor of the pharynx, is a valve that controls airflow in and out of the respiratory passages. The glottis opens directly into a boxlike larynx. This voice box occurs in all amphibians but is anatomically most complex in frogs. The larynx exits into the trachea; the latter bifurcates into the bronchi and then into the lungs. Bronchi are absent in all frogs except the pipids. Amphibian lungs are highly vascularized, thin-walled sacs. Internally, they are weakly partitioned by thin septa composed of connective tissue. This weak partitioning and the small size, or even absence, of the lungs emphasizes the use of multiple respiratory surfaces in amphibians. Lung ventilation is triphasic by means of a buccopharyngeal force pump mechanism. Inhalation begins with nares open, glottis closed, and depression of the buccopharyngeal floor, which draws air into this cavity. The glottis then opens, and elastic recoil of the lungs forces the pulmonary air out and over the new air in the buccopharyngeal pocket. The nares close, and the buccopharyngeal floor contracts and pumps air into the lungs as the glottis closes to keep air in the lungs under supra-atmospheric pressure. Similar, but faster and shallower throat movements occur regularly in frogs and salamanders, rapidly flushing air in and out of the olfactory chambers.

Reptiles have an identical respiratory pathway. Air exits and enters the trachea through the glottis at the rear of the pharynx. The glottis and two or three other cartilages form the larynx, a simple tubular structure in most reptiles. The larynx is the beginning of the trachea, a rigid tube of closely spaced cartilaginous rings within its walls (the rings are incomplete dorsally in squamates). The trachea extends down the neck beneath the esophagus and forks into a pair of bronchi, each of which enters a lung.

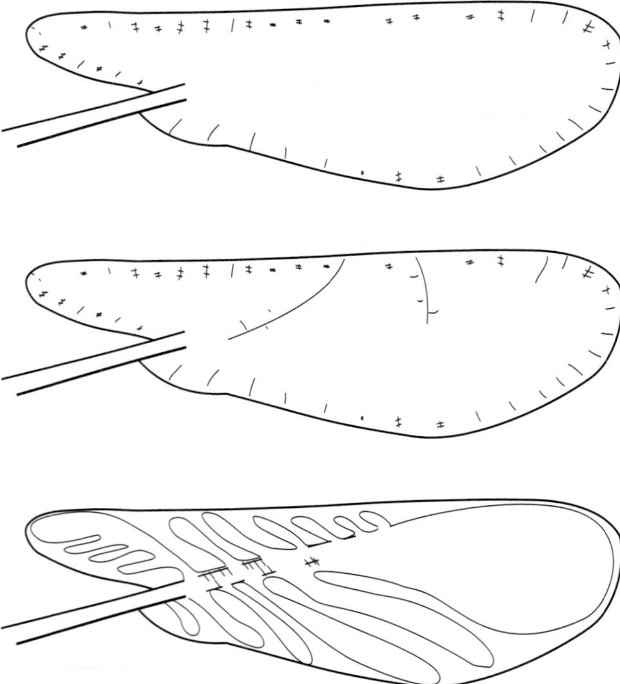

FIGURE 2.37 Internal morphology of generalized reptilian lungs; schematic cross sections of a single-chambered lung (top), a transitional lung (middle), and a multichambered lung (lower). The central chamber of a single-chambered lung is not divided by a major septum, although small niches are commonly present along the wall. The transitional lung has a central lumen partially divided by large septum. The multichambered lung is partitioned into numerous chambers of various sizes; all chambers communicate with the intrapulmonary bronchus via an airway. Adapted from Perry, 1983.

Lung structure is variable among reptiles (Fig. 2.37). Most lepidosaurs have simple saclike lungs. Each bronchus empties into a large central chamber of the lung. Numerous faveoli (small sacs) radiate outward in all directions, forming a porous wall around the central chamber. The walls of the faveoli are richly supplied with blood and provide the major surface for gaseous exchange. Iguanians have the central chamber of each lung divided by a few large septae. These septae partition the lung into a series of smaller chambers, each of which possesses porous faveolar walls. Varanids, crocodylians, and turtles also have multichambered lungs; a bronchus extends into each lung and subdivides into many bronchioles, each ending in a faveolus. In some lizards, smooth-walled tubes project from the chamber beyond the surface of the lung. No gas exchange occurs in these air sacs; rather, the sacs may permit the lizard to hold a larger volume of air. The sacs are used by some species to inflate their bodies to intimidate predators.

Development of air sacs is even more extensive in snakes because of their highly modified lungs. A single functional right lung and a small, nonfunctional left lung are the common condition (Fig. 2.36). A functional left lung occurs only in a few snakes (e.g., *Loxocemus*), and in these snakes, it is distinctly smaller than the right lung. The trachea and right bronchus extend into the lung and empty into a chamber with a faveoli-filled wall as in most lizards. Snake lungs are typically long, one-half or more of the snake's body length. Usually the posterior one-third or more is an air sac.

Many snakes also possess a tracheal lung. This lung is a vascular, faveoli-dense sac that extends outward from where the tracheal rings are incomplete dorsally; posteriorly, it abuts the right lung. Breathing occurs by the expansion and contraction of the body cavity. Among squamates, the thoracic cavity is enlarged during inhalation by the contraction of the intercostal muscles drawing the ribs forward and upward. Compression of the cavity during exhalation occurs when the muscles relax and the weight of the body wall and adjacent organs squeeze the lungs. In crocodylians, the diaphram contracts and enlarges the thoracic cavity for inhalation; abdominal muscles contract and drive the liver forward for exhalation. In turtles with rigid shells, the posterior abdominal muscles and several pectoral girdle muscles expand and compress the body cavity for breathing.

Other Respiratory Surfaces

Lungs are only one of several respiratory structures in amphibians. A few caecilians have a small third lung budding off the trachea. The buccopharyngeal cavity is heavily vascularized in many amphibians and is a minor gas exchange surface.

Gills are the major respiratory structures in larvae and a few adult salamanders. Three pairs of external gills, which develop and project from the outside of the pharyngeal arches, occur in salamanders and caecilians. External and internal gills occur sequentially in anuran larvae; the former arise early, remain largely rudimentary, and are replaced quickly by the latter.

In most adults and larvae, the skin is the major respiratory surface and is highly vascularized. Gas exchange in all vertebrates requires a moist surface; drying alters the cell surfaces and prevents diffusion across cell membranes.

Reptiles are dependent upon their lungs for aerial respiration. None of the aquatic species has developed a successful substitute for surfacing and breathing air. Long-term submergence in reptiles is possible owing to a high tolerance to anoxia, a greatly suppressed metabolism, and varying degrees of cutaneous respiration. Softshell turtles are purported to obtain more than 50% of their respiratory needs by cutaneous and buccopharyngeal respiration when submerged, but experimental results of different investigators are conflicting. The accessory cloacal bladders of turtles have also been proposed as auxiliary respiratory structures; however, their walls are smooth and lightly vascularized, unlike most respiratory surfaces.

URINARY AND REPRODUCTIVE ORGANS—WASTE REMOVAL AND PROPOGATION

The urinary and reproductive systems are intimately related in their location along the midline of the dorsal body wall and by a shared evolutionary history. Through generations of vertebrates, the male gonads have usurped the urinary ducts of primitive kidneys for transportation of sperm. Most adult amphibians have opisthonephric kidneys, whereas amniotes have metanephric kidneys. The development of these two kidney types is different, but both pass through a transient embryonic stage, the mesonephros. In amniotes, ducts from the ancestral opisthonephric kidney system have been usurped by the reproductive system, and the opisthonephric kidney system, including the ducts, has been replaced by the metanephric kidney system and ducts. The structures of each system are paired.

Kidneys and Urinary Ducts

Kidneys remove nitrogenous waste from the bloodstream and maintain water balance by regulating the removal or retention of water and salts. The functional unit of the kidney is the nephron or kidney tubule. Each nephron consists of a renal corpuscle and a convoluted tubule of three segments, each of variable length in different species. The corpuscle encloses a ball of capillaries, and most filtration occurs here. Filtration (selective secretion) may also occur in the tubule, but resorption of salts and water to the blood is the major activity as the filtrate passes through the tubule. The tubules of adjacent nephrons empty into collecting ducts, which in turn empty into larger ducts and eventually into the urinary duct that drains each kidney.

Primitively and embryologically, the kidney developed from a ridge of mesomeric tissue along the entire length of the body cavity. In modern amphibians, a holonephric kidney exists embryologically but never becomes functional. Instead the functional kidney (pronephros) of embryos and larvae arises from the anterior part of the "holonephric" ridge. The pronephros begins to degenerate as the larva approaches metamorphosis, and a new kidney, the opisthonephros, develops from the posterior part of the ridge. The tubules of the anterior end of the male's opisthonephric kidney take on the additional role of sperm transport. In primitive salamanders, this new role causes the anterior end of the kidney to narrow and the tubules to lose their filtration role. In caecilians, the kidney remains unchanged, and in anurans and advanced salamanders, the kidney shortens into a compact, ellipsoidal organ as a result of the loss of the anterior end. A single urinary duct, the archinephric duct, receives urine from the collecting ducts of each kidney and empties into the cloaca (Fig. 2.38). Two principal patterns characterize urinary drainage in amphibians. Only the archinephric duct drains the kidney in caecilians and primitive salamanders, whereas in frogs and advanced salamanders, the archinephric duct drains the anterior portion of the kidney, and an accessory duct drains the posterior one-half. The bladder has a single, separate duct, the urethra, that empties into the cloaca; fluids enter and exit the bladder through this duct.

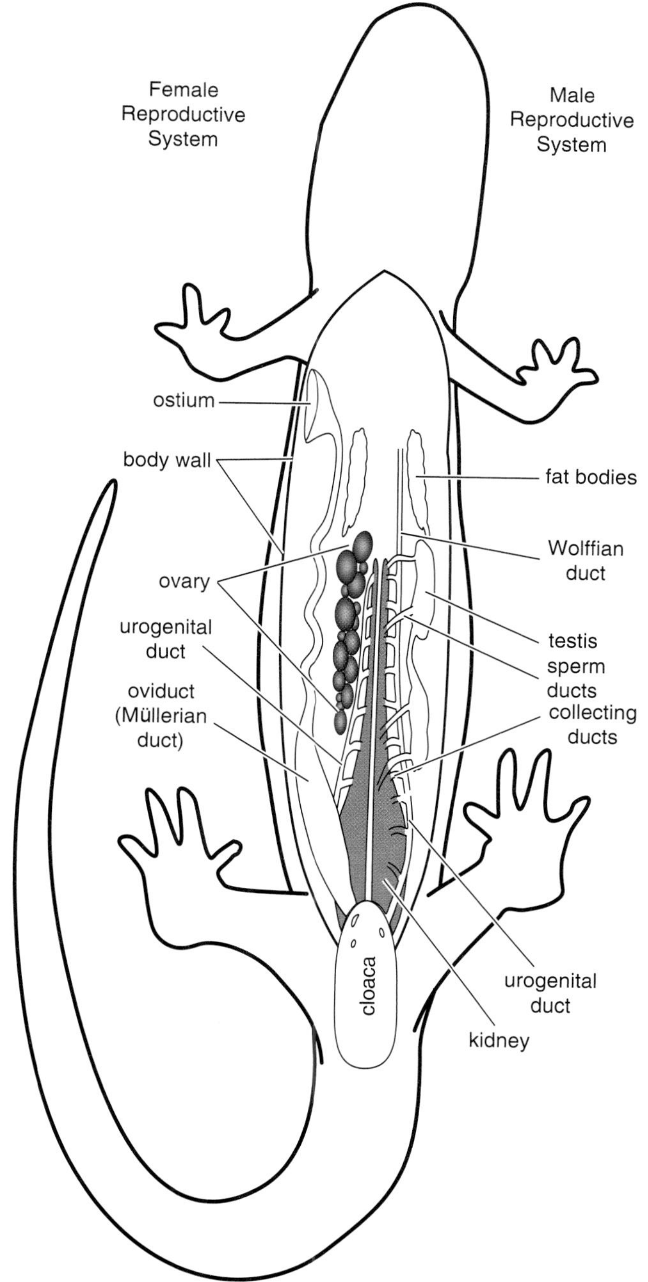

FIGURE 2.38 Ventral view of the reproductive tracts of a female (left side) and male (right side) salamander.

The metanephric kidneys of reptiles vary in size and shape. They are smooth, equal-sized, and nearly spherical in some lizards (Fig. 2.39), and smooth or rugose,

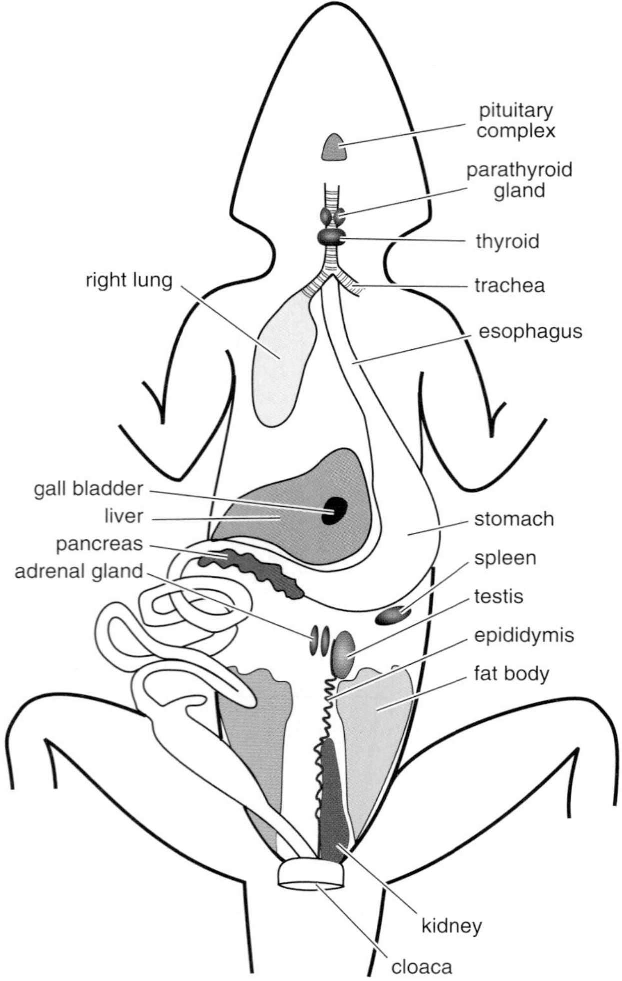

FIGURE 2.39 Schematic lizard showing the location of some digestive and endocrine glands.

elongated cylinders in snakes (Fig. 2.36). The kidneys are lobate spheroids in crocodylians and turtles. In all forms, the kidneys lie side by side on the dorsal body wall in front of the cloaca, and in all, a ureter drains each kidney and empties independently into the cloaca (Fig. 2.38). An elastic-walled urinary bladder is present in turtles and most lizards but absent in snakes and crocodylians. The bladder joins the cloaca through a single median duct, the urethra, through which urine enters and exits.

Gonads and Genital Ducts

In both amphibians and reptiles, the female and male gonads (ovaries and testes, respectively) develop from the same embryological organs. The undifferentiated organs arise on the body wall between the middle of the kidneys. Germ cells or gametes migrate into each organ and initiate the reorganization and consolidation of the pregonadal tissue into an external cortex and internal medulla. Later, when sexual differentiation occurs, the cortex is elaborated into an ovary in females, and the medulla into a testis in males.

Structurally, the male and female gonads are quite different. The ovary is a thin-walled sac with the germ cells sandwiched between the inner and outer ovarian walls. The germ cells divide, duplicate themselves, and produce ova. A single layer of follicle cells in the epithelium of the ovarian wall encases each ovum, providing support and nourishment. This unit, the follicle, which consists of the ovum and follicle cells, grows into the ovarian lumen. Numerous developing follicles form the visible portion of the ovaries in gravid females. The testis is a mass of convoluted seminiferous tubules encased in a thin-walled sac. Small amounts of interstitial tissue fill the spaces between the tubules. The developmental cycle (gametogenesis) of the ova and spermatozoa is presented in Chapter 4.

In amphibians, spermatozoa collect in the lumen of the seminiferous tubules and then move sequentially through progressively larger collecting ducts into the kidney collecting ducts before emptying into the archinephric duct. Because of its dual role in urine and sperm transport, the archinephric duct is called the urogenital or Wolffian duct. The oviducts (Müllerian ducts) are paired tubes, one on each side of the dorsal body wall, lateral to each ovary. Each arises *de novo* as a fold of the peritoneum or, in salamanders, by a splitting of the archinephric duct. The anterior end of the oviduct remains open as an ostium; ova are shed into the body cavity and move to and through the ostium into the oviduct. The posterior part of the oviduct is expanded into an ovisac, which empties into the cloaca; after ovulation, eggs remain briefly in the ovisac prior to amplexus and egg laying. Oviducts form in both males and females, degenerating although not disappearing in many male amphibians, where this nonfunctional duct is called Bidder's duct. Similarly, some males retain a part of the gonadal cortex attached to the anterior end of the testis. This structure, common in bufonids, is Bidder's organ.

In reptiles, a pair of ovaries occupies the same location as the testes of the males, and the right ovary precedes the left in squamates. Each ovary is an aggregation of epithelial cells, connective tissue, nerves, blood vessels, and one or more germinal cell beds encased in an elastic tunic. Depending upon the stage of oogenesis, each ovary can be a small, granular-appearing structure or a large lobular sac filled with spherical or ellipsoidal follicles. An oviduct is adjacent to but not continuous with each ovary. The ostium (mouth) of the oviduct lies beside the anterior part of the ovary; it enlarges during ovulation to entrap the ova. The body of the oviduct has an albumin-secreting portion followed by a thicker shell-secreting portion. The oviducts open independently into the urogenital sinus of the cloaca.

The testis is a mass of seminiferous tubules, interstitial cells, and blood vessels encased in a connective tissue sheath. The walls of seminiferous tubules are lined with germinal tissue. The sperm produced by these tubules empties through the efferent duct into the epididymis on the medial face of the testis. The ductuli coalesce into the ductus epididymis that runs to the cloaca as the vas (ductus) deferens. In shape, testes vary from ovoid to spindle-shaped. The testes are usually adjacent to each other, although the right testis lies anteriorly, especially in snakes and most lizards.

All living reptiles have copulatory organs, which are rudimentary in *Sphenodon*. Crocodylians and turtles have a single median penis that originates in the floor of the cloaca. Squamates have a pair of hemipenes, each of which originates at the junction of the cloacal vent and base of the tail.

ENDOCRINE GLANDS—CHEMICAL REGULATORS AND INITIATORS

The endocrine system is comprised of numerous glands scattered throughout the body. The glands are an integrative system, initiating and coordinating the body's reactions to internal and external stimuli. Unlike the nervous system, the glands do not communicate directly with one another and their target organs. Instead, they rely on vascular and neural pathways to transmit their chemical messengers. Unlike other organ systems, the endocrine system is a composite of unrelated anatomical structures from other systems; for example, the pituitary of the nervous and digestive systems, the gonads of the reproductive system, or the pancreas of the digestive system. Only a few of the many glands and their functions are mentioned here, and these are described only superficially. The commonality of all endocrine organs is their secretion of one or more chemical messengers, hormones, that stimulate or arrest the action of one or more target organs, including other endocrine glands or tissues. Hormones work in both short-term cycles and continuously to maintain a stable internal environment and in the long term and cyclically to control periodic behaviors, such as reproduction.

Pituitary Gland

The pituitary gland or hypophysis is the master gland of the body. Structurally, it consists of two parts: the neuropophysis, which arises from the ventral portion of the diencephalons, and the adenopophysis, which is derived from the roof of the buccal cavity. The neuropophysis and adenopophysis interdigitate and are joined by neural and vascular connections. The brain receives stimuli that trigger the release of neurohormones by the brain cells. These hormones reach the neuropophysis through blood vessels or secretory axons of neurons ending in the neuropophysis. In turn, the neuropophysis produces hormones that stimulate the adenopophysis (e.g., GnRH, gonadotropin-releasing hormone) or act directly on the target organs (ADH, antidiuretic hormone; MSH, melanophore-stimulating hormone). The adenopophysis secretes six major hormones: adrenocorticotropin, two gonadotropins (FSH, LH), prolatin, somatotropin, and thyrotropin. These hormones control growth, metamorphosis, reproduction, water balance, and a variety of other life processes.

Pineal Complex

The pineal complex consists of a pineal (epiphysis) and a frontal (parapineal) organ, each arising embryologically from the roof of the diencephalon (Fig. 2.39). These two organs are light receptors as well as endocrine glands. As light receptors, they record the presence or absence of light, and, as glands, they produce and release melatonin. These two functions are associated with cyclic activities, including both daily cycles or circadian rhythms and seasonal cycles. Frogs possess both a pineal organ lying inside the skull and a frontal organ piercing the skull and lying beneath the skin on top of the head. Caecilians and salamanders have only the pineal organ, which may extend upward to, but does not pierce, the skull roof. All reptiles except crocodylians have pineal organs that lie on the brain but do not exit the skull. Some lizards (e.g., iguanians) have pineal organs that pass through the skull and form a parietal eye.

Thyroid and Parathyroid Glands

These two glands are linked because of their shared location in the throat adjacent to the larynx and trachea (Fig. 2.39). Although both arise embryologically as outpocketings of pharyngeal pouches, they have quite dissimilar functions. The parathyroid hormones regulate calcium levels in the blood, and hence control bone growth and remodeling. The thyroid is well known for its accumulation of iodine and the importance of its hormones in controlling development, metamorphosis, and growth. Amphibians typically have a bilobular thyroid and a pair of parathyroids. In reptiles, the thyroid assumes a variety of forms. It is a single, nearly spherical organ in turtles and snakes. In crocodylians, it is an H-shaped, bilobular organ, which has a lobe on each side of the trachea connected by a narrow isthmus. Some lizards share this bilobular condition, others have a lobe on each side but no isthmus, and still others have a single median gland. In *Sphenodon*, the gland is transversely elongated. The reptilian parathyroid appears as one or two pairs of granular glands, usually at the base of the throat adjacent to the carotid arteries.

Pancreas

The pancreas is composed of both exocrine and endocrine tissues. The exocrine portion secretes digestive enzymes; clusters of cells, the Islets of Langerhans, secrete the hormone insulin. Insulin is critical for regulating carbohydrate metabolism; it stimulates the liver and adipose tissue to remove glucose from the bloodstream through glycogen production and fat synthesis, respectively. Insulin facilitates striated muscle activity by increasing the movement of glycogen into the muscle cells. In amphibians, the pancreas is a diffuse gland that lies within the mesentery between the stomach and duodenum. The reptilian pancreas is a compact organ that lies in the mesentery adjacent to the duodenum (Fig. 2.39).

Gonads

Aside from producing gametes, the gonads also produce sex hormones. The maturation and production of gametes are closely regulated by the brain, through the production of hypothalamohypophyseal hormones, and the pituitary by the production of gonadotropins. In turn, the hormonal response of the gonads influences the secretory cycles of these two organs. In addition to initiating gametogenesis, the gonadotropins stimulate the production of estrogens and androgens, the female and male sex hormones, by gonadal tissues. Estrogens and androgens are steroids, and several closely related estrogens or androgens are produced in each sex. Stimulation and inhibition of the reproductive structures are obvious actions of the sex hormones, but they interact also with a variety of other tissues. They induce the skin to produce secondary sexual characteristics, and they provide a feedback mechanism to the hypothalamic–pituitary complex. Estrogens are produced largely by the follicle cells in the ovarian follicles and the corpus lutea. Androgens are derived principally from the cells of Leydig that lie in the interstitial tissue between the seminiferous tubules. The Sertoli cells also produce minor amounts of androgens.

Adrenals

The adrenals are bilaterally paired glands that lie anterior to the kidneys in reptiles (Fig. 2.39) and elongate glands that lie on the ventral surface of the kidneys in amphibians. Each adrenal is an admixture of two tissues: The interrenal (cortical) cells form the main matrix of the gland, and adrenal (medullary) cells form strands and islets within the interrenal matrix. These two tissues have different embryological origins and distinctly different functions. The chromaffin cells produce adrenaline and noradrenaline, both of which affect blood flow to the brain, kidney, liver, and striated muscles, mainly during stress reactions. The interrenal tissue produces a variety of steroid hormones. One group of interrenal hormones affects sodium and potassium metabolism, another group affects carbohydrate metabolism, and a third group (androgens) affects reproductive processes.

QUESTIONS

1. With what you know about determinant and indeterminant growth, describe growth in a frog from the time that the animal hatches from an egg until it dies of old age. Indicate how food supply and temperature might affect growth.
2. Describe the differences between morphological and physiological color change in amphibians.
3. Describe and compare the morphology of frog larvae (tadpoles) and salamander larvae. How do these differences relate to their general ecology?
4. Describe the key differences in skeletal structure between adult frogs and adult salamanders. How might these translate into ecological differences?
5. How do amphibians and reptiles differ in terms of their early development (eggs)? How do reptiles dispose of metabolic waste products while inside of a shelled egg?
6. What is heterochrony and how does it work both within species and among species?
7. What is the difference between paedogenesis and isogenesis in a salamander?

ADDITIONAL READING

Benton, M. J. (1997). *Vertebrate Paleontology* (2nd ed.). Chapman & Hall, London.

Carroll, R. L. (1988). *Vertebrate Paleontology and Evolution.* W. H. Freeman & Co., New York.

de Queiroz, K. (1997). The Linnean hierarchy and the evolutionization of taxonomy, with emphasis on the problem of nomenclature. *Aliso* **15:** 125-144.

Hillis, D. M., Moritz, C., and Mable, B. K. (Eds.). (1996). *Molecular Systematics.* (2nd ed.). Sinauer Associates, Inc., Sunderland, MA.

Kardong, K. V. (2006). *Vertebrates: Comparative Anatomy, Function, Evolution* (4th ed.). McGraw Hill, Boston, MA.

Schultze, H.-P. and Trueb, L. (Eds.). (1991). *Origins of the Higher Groups of Tetrapods. Controversy and Consensus.* Comstock Publishers, Inc., Ithaca, NY.

Schwenk, K. (2000). *Feeding, Form, Function, and Evolution in Tetrapod Vertebrates* Academic Press, San Diego, CA.

REFERENCES

General

Bellairs, 1969; Clarke, 1997; Duellman and Trueb, 1986; Gans and coeditors, 1969 et seq.; Hanken and Hall, 1993: Vol. 1, Chapt. 8, Vol. 2, Chapt. 6-7; Hayes, 1997a.

Development and Growth: Ova, Sperm, and Fertilization
Wassarman, 1987.

Early Development
Elinson, 1987.

Embryogenesis
Buchholz et al., 2007; Nelsen, 1953; Rugh, 1951.

Morphogenesis: Developing Form and Function
Fox, 1983; Hanken, 1989.

Heterochrony
Reilly et al., 1997.

Embryonic Lifestyles: Protective Barriers
M. Packard and DeMarco, 1991; Seymour and Bradford, 1995.

Larvae—Free-Living Embryos
Altig and Johnston, 1986, 1989; Exbrayat, 2006a, 2006b; Orton, 1953; Townsend and Stewart, 1985.

Life in an Eggshell
Ewert, 1985; Kennett et al., 1993a, 1993b; G. Packard et al., 1977, 1991; G. Packard and M. Packard, 1988; M. Packard et al., 1982.

Changing Worlds: Metamorphosis
Denver, 1997; Fox, 1986; Hayes, 1997b; Hourdry and Beaumont, 1985; Wakahara and Yamaguchi, 2001; Werner, 1986.

Hatching and Birth
Ewert, 1985; M. Packard, 1994; Touchon et al., 2006; Townsend and Stewart, 1985.

Growth: Mechanics of Growth
Andrews, 1982a; Seben, 1987.

Age
Andrews, 1976; Barbault and Rodrigues, 1979; Grubb, 1971; Peterson et al., 1983.

Integument
Alibardi, 2003; Bechtel and Bechtel, 1991; Elias and Shapiro, 1957; Fox, 1986; Heatwole and Barthalmus, 1994.

Integumentary Structures
Komnick, 1986; Landmann, 1975, 1986; Lillywhite and Maderson, 1982; Nokhbatolfoghahai and Downie, 2007; Quay, 1986; Ruibal and Shoemaker, 1984; Zimmerman and Heatwole, 1990.

Ecdysis
Irish et al., 1988; Maderson, 1965; Maderson et al., 1998.

Coloration
Bagnara, 1986.

Skeleton and Muscles: Head and Hyoid
Bellairs and Kamal, 1981; Edmund, 1969; Frazzetta, 1986; Gaffney, 1979; Haas, 1973; Iordansky, 1973; Mueller, 2006; Oelrich, 1956; Romer, 1956; Trueb, 1973.

Vertebral Column
Carroll et al., 1999; Gasc, 1981; Hoffstetter and Gasc, 1969; Romer, 1956.

Girdles and Limbs
Burke, 1989; Romer, 1956.

Nerves and Sense Organs: Nervous System
Dicke and Roth, 2007; Llinás and Precht, 1976.

Sense Organs: Cutaneous Sense Organs
Catton, 1976; During and Miller, 1979; Landmann, 1975; Lannoo, 1987a,b; Spray, 1976; Whimster, 1980.

Ears
Baird, 1970; Capranica, 1986; Fritzsch and M. Wake, 1988; Hetherington, 1985; Hetherington et al., 1986; Manley, 1990; Rose and Gooler, 2007; Wever, 1978, 1985.

Eyes
Schaeffel and de Queiroz, 1990; Underwood, 1970.

Nasal Organs
Halpern, 1992; Parsons, 1970; Scalia, 1976.

Internal Sense Organs
During and Miller, 1979; Schwenk, 1985.

Heart and Vascular Network: Blood
Saint-Girons, 1970.

Arterial and Venous Circulation
Burggren, 1987; Saint-Aubain, 1985; Young, 1988.

Lymphatic Network
Ottaviana and Tazzi, 1977.

Heart
Farrell et al., 1998; Jones, 1996.

Digestive and Respiratory Organs: Digestive Structures
Guard, 1979; Junqueira et al., 1999; Luppa, 1977; Schwenk, 1986; Young, 1997.

Digestive Glands
Jaeger and Hillman, 1976; Kochva, 1978; Komnick, 1986; Miller and Lagios, 1970; Saint-Girons, 1988; Schaffner, 1998.

Respiratory Structures: Lungs
Kinkead, 1997; Perry, 1983, 1998; Ruben et al., 1996; Wallach, 1998.

Other Respiratory Surfaces
Warburg et al., 1994.

Urinary and Reproductive Organs: Kidneys and Urinary Ducts
Fox, 1977.

Gonads and Gonadal Ducts
Fox, 1977.

Endocrine Glands
Pang and Schreibman, 1986.

Pituitary Gland
Saint Girons, 1970; Schreibman, 1986.

Pineal Complex
Engbretson, 1992; Quay, 1979; Ralph, 1983.

Thyroid and Parathyroid Glands
Lynn, 1970.

Pancreas
Epple and Brinn, 1986; Miller and Lagios, 1970.

Gonads
Fox, 1970.

Adrenals
Gabe, 1970.

Chapter 3

Evolution of Ancient and Modern Amphibians and Reptiles

Tetrapods adapted first to a shallow-water existence and then to a totally terrestrial one. Some taxa remained associated with water whereas some terrestrial groups later returned to the water (e.g., many turtles). The origin of terrestriality was followed quickly by an eruption of new species with new lifestyles and body forms. As portrayed in Chapter 1, this adaptive radiation was not confined to amphibians and reptiles but occurred in other ancient tetrapods that left no living descendants. Although amphibian (anamniote) diversification began earlier, amniotes were the dominant group by the mid-Permian in terms of number of species and individuals, based on the fossil record. The history of these adaptive radiations is complex and extensive. We introduce some extinct amphibian and reptilian taxa and discuss the history of the clades that compose the modern herpetofauna.

HISTORY OF AMPHIBIANS

Radiation Among Early Anamniotes

Tetrapods in the Late Devonian were aquatic, but adaptations had appeared that would permit them to become terrestrial. Vegetation completely covered lowland coastal areas and floodplains, and plants were no longer confined to water or the margins of streams, lakes, and seas. Herbs and shrubs were the dominant plants, but trees had appeared and may have formed forests in some places. Plants were even beginning to invade the upland areas. Some arthropods likely were fully terrestrial, but the few vertebrates were, at best, semi-aquatic. These early tetrapods (e.g., *Acanthostega*, *Ichthyostega*) lived in the heavily vegetated, shallow water. Their large size (0.5 to 1.2 m total length [TL]), large heads, and tooth-filled jaws suggest that they were formidable predators and fed on large prey. They had fusiform bodies and strong tails (Fig. 1.4), suggesting that they were capable of fast burst swimming. They also had short and stout fore- and hindlimbs, perhaps permitting them to "walk" slowly and stalk prey in dense aquatic vegetation (Fig. 1.4). Unlike subsequent tetrapods, all known early tetrapods had more than five digits; *Acanthostega* had eight digits on its forefeet.

Amphibians of the Late Paleozoic

Tetrapods largely disappeared from the fossil record at the end of the Devonian. They next appeared en masse in the Upper Mississippian and Lower Pennsylvanian when fossils representing lowland lake and swamp assemblages reappeared. More than a dozen clades are recognized and include several groups of anthracosaurs, at least three amphibian groups, and the enigmatic *Crassigyrinus* (Fig. 3.1). Most of these tetrapods were aquatic, although the lowlands were inhabited by an even greater diversity of plants and plant communities. Amphibians certainly lived there as well. The anthracosaurs *Proterogyrinus* and *Eoherpeton* were large (1 m TL) aquatic predators. The baphetids (= loxommatids) were moderate-sized, reduced-limbed animals, which are known principally from skulls with numerous pointed teeth, some enlarged and fanglike. Baphetids are recognized by an anterior elongation of each orbit that probably housed a large gland. Two other anthracosaur groups, the embolomeres and eoherpetontids, were dominant members of the early tetrapod community. Both groups were mainly aquatic.

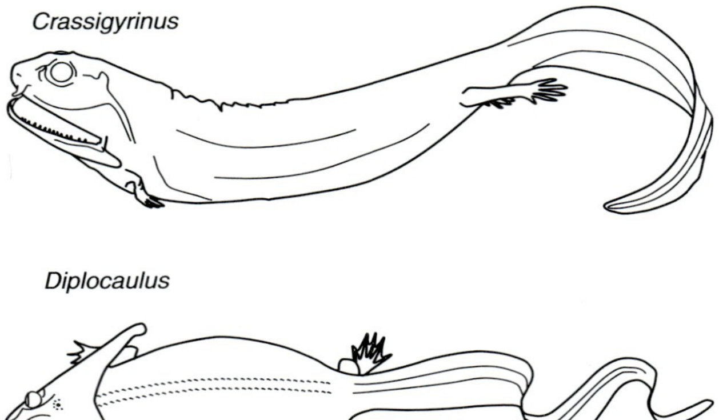

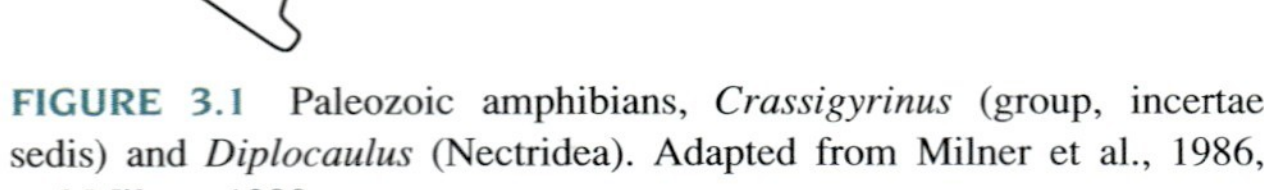

FIGURE 3.1 Paleozoic amphibians, *Crassigyrinus* (group, incertae sedis) and *Diplocaulus* (Nectridea). Adapted from Milner et al., 1986, and Milner, 1980.

The embolomeres were the largest tetrapods, ranging from 1 to 4 m (TL). All had heavy crocodile-like skulls and short, robust limbs. The limbless aistopodans were delicate eel-like amphibians; none exceeded 70 cm (TL). Presumably they were aquatic and semiaquatic, because they had fragile skulls unlike those of burrowing animals. The colosteids were small, aquatic amphibians that had elongated bodies and small, well-developed limbs. The temnospondyl trimerorhachoids included limbed and reduced-limbed forms and likely contained both semiaquatic and aquatic members. They were present in the fossil record from the mid-Mississippian to the end of the Permian (Fig. 3.2). Edopoids, another temnospondyl group, also persisted through the Carboniferous into the Early Permian. They were larger, more robust amphibians. The earlier edopoids were mainly aquatic, but the later ones became increasingly terrestrial.

In the Pennsylvanian, many new tetrapod groups appeared: geophyrostegid and limnoscelid anthracosaurs, eryopoid amphibians, nectrideans (Fig. 3.1), and three groups of microsaurs. This fauna lived predominantly in the lowlands. The climate was generally hot and wet and supported diverse and dense plant communities. By this time the uplands also bore a thick plant cover. These tetrapods were primarily aquatic forms, although a few had become terrestrial. For example, the terrestrial limnoscelids were moderate-sized (1–2 m TL), with robust bodies and limbs and long tails. Structurally, they had some features that would appear in reptiles. The eryopoids included aquatic to terrestrial, small to large amphibians. The heavy

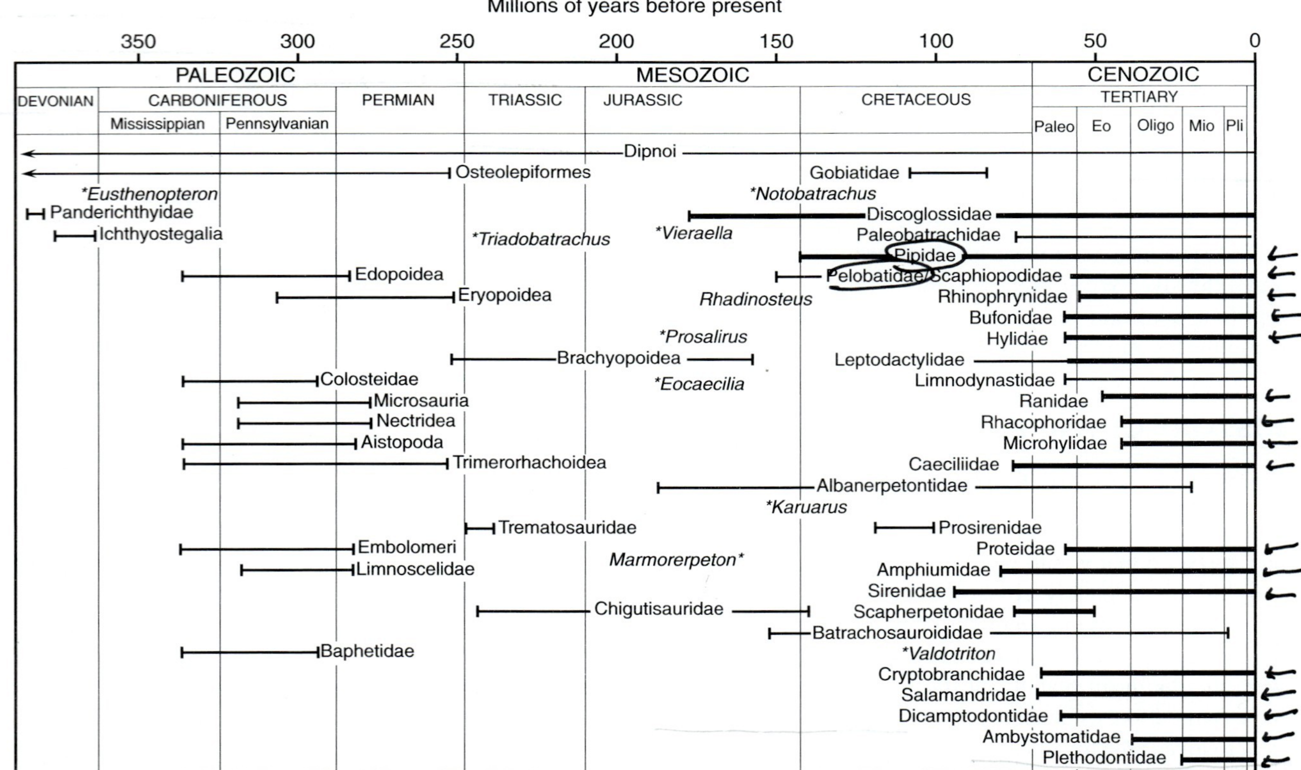

FIGURE 3.2 Geological occurrence of some early tetrapods, and extinct and living amphibians. Abbreviations for Cenozoic epochs: Paleo, Paleocene; Eo, Eocene; Oligo, Oligocene; Mio, Miocene; Pli, Pliocene; Pleistocene is the narrow, unlabeled epoch on the far right side of the chart. The Dicamptodontidae is now included in Ambystomatidae.

bodied *Eryops* is characteristic of this group, although it was larger (nearly 2 m TL) than most eryopoids. The nectrideans were moderate-sized, newtlike amphibians, all less than 0.5 m TL. The heads of some were arrow-shaped with large, laterally projecting horns. This head shape appears to facilitate rapid opening of the mouth for suction–gape feeding. Other nectrideans had more typically shaped heads with strong dentition for snap-and-grasp feeding. The microsaurs were small (most <50 cm TL), salamander-like tetrapods. They commonly had long bodies and tails, with short limbs. Presumably they were predominantly aquatic and semiaquatic.

Amniote fossils appeared in the Late Carboniferous (Fig. 3.2). They had already diverged into reptiles and synapsids. Many ancient anamniote groups persisted into the Early Permian, but not beyond. Only the trimerorhachoids and eryopoids survived into the Late Permian. Seymouriamorphs (Fig. 3.3) appeared in the early Permian and persisted through nearly the entire Permian. A few small anthracosaurs appeared briefly. The loss of amphibian and anthracosaurian diversity occurred concurrently with diversification of amniotes and a shift to an arid climate. Aquatic habitats shrunk and disappeared. Plant cover was reduced, and the drier upland vegetation spread into the lowlands. This changing climate and landscape favored terrestrial adaptations of amniotes.

The trimerorhachoids had greatly reduced limbs and were probably highly aquatic. The eryopoids as a group remained diverse in size, body form, and habits (aquatic to terrestrial). The surviving embolomeres had become shallow-water inhabitants in contrast to their deep-water ancestors of the Carboniferous. They did not survive beyond the Early Permian. The seymouriamorphs were a much more successful group of semiterrestrial and terrestrial anthracosaurs. Generally, they had large heads with well-developed jaws, robust bodies, and strong limbs.

FIGURE 3.3 *Seymouria*, an Early Permian anthracosaur from Texas. Scale: bar = 5 cm. (R. S. Clarke)

Amphibians of the Early Mesozoic

In the Triassic, reptiles and synapsids had become the dominant terrestrial vertebrates. A few anthracosaur groups survived into the earliest Triassic but soon disappeared. In contrast, amphibians experienced a minor diversity explosion with the appearance of at least seven different groups of presumed temnospondyls, including the first lissamphibian. The radiation included small to large temnospondyls with several groups having species in the 1.5-3 m range (e.g., capitosauroids, chigutisauroids, and metoposaurids) and some mastodonsaurids to 6 m TL. All large species appear to have been highly aquatic, and most had crocodile-like body forms (Fig. 3.4). The mastodonsaurids were a short-lived group found only in Lower Triassic sediments of northern Eurasia. The 2-m (TL) trematosaurs were another Lower Triassic taxon with triangular to gharial-like heads; some were marine, an anomaly for amphibians. Three temnospondyl groups (brachyopoids, capitosauroids, and plagiosaurids) occurred throughout the Triassic (Fig. 3.2). Although never common in the fossil record, they persisted throughout this period. The plagiosaurus were peculiar amphibians with broad flattened bodies and heads, and a back armored with numerous small, bony pustules. The brachyopoids were the most diverse group and appeared in the Late Permian and survived into the Lower Jurassic. The Chigutisauridae were the longest lasting of the extinct temnospondyls, surviving into the Early Cretaceous. One group of temnospondyls, the Lissamphibia, still survives. The first lissamphibian to appear in the fossil record is the Lower Triassic frog, *Triadobatrachus massinoti*. Its occurrence attests to at least an Early Mesozoic divergence among lissamphibians. However, frogs vanished from the fossil record for another 50 million years, and salamanders appeared before the frogs' reappearance. *Triadobatrachus massinoti* is unquestionably a frog (Fig. 3.5), although it had 14 body vertebrae and a short tail of 6 vertebrae. Its pelvic girdle and skull are similar to those of modern frogs.

History of the Lissamphibia

The origin of the lissamphibians likely occurred in the mid-Permian, possibly even earlier. The fossil evidence now suggests that the ancestor was a temnospondyl dissorophid. Presumably, this ancestral stock was composed of small, semiaquatic salamander-like amphibians with external fertilization, a larval developmental stage, and many other physiological and anatomical features shared by today's lissamphibians. No fossils are available to show the timing and manner of divergence of the three modern groups. Caecilians are often depicted as diverging first because of their extreme structural divergence from frogs and salamanders. But if occurrence of fossils is used as an indicator, frogs diverged first. This issue remains

FIGURE 3.4 Triassic landscape showing early reptiles including the dycinodont *Placerias* (left), a group of theropods in the genus *Coelophysis* (right), several phytosaurs (crocodile-like), and a group of metaposaurs (labyrinthodont amphibians). By Karen Carr, with permission of the Sam Noble Oklahoma Museum of Natural History.

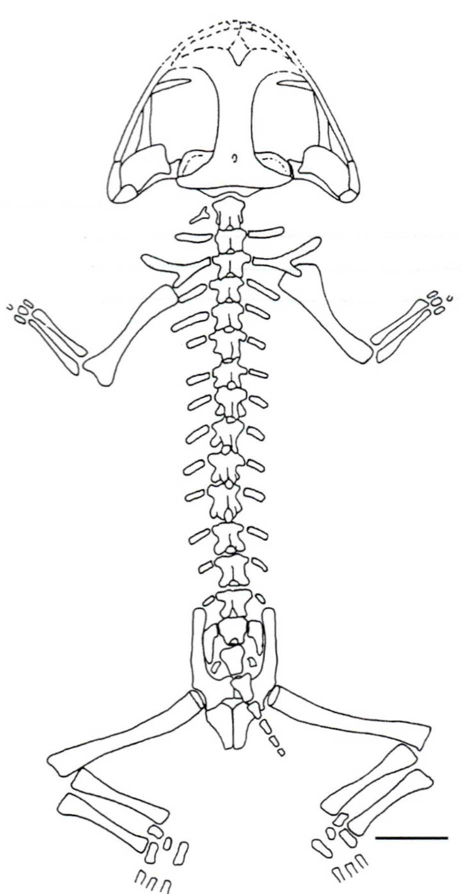

FIGURE 3.5 *Triadobatrachus massinoti*, the earliest known frog, from the Triassic of Madagascar. Adapted as a partial reconstruction from Estes and Reig, 1973. Scale bar = 1 cm.

unresolved; however, it is certain that by the mid-Jurassic, only lissamphibians and chigutisaurids remained of the previously numerous amphibian clades (Fig. 3.2).

Caecilians

Caecilians are poorly represented by fossils, and this thin evidence has kept their origin and evolution controversial. Until recently, they were known by a single Paleocene fossil vertebra from Brazil and a Late Cretaceous vertebra from Bolivia. The discovery of an Early Jurassic caecilian in the southwestern United States is significant because it extends the history of the group deep into the Mesozoic and closer to its potential ancestors of the Upper Permian or Lower Triassic. This caecilian, *Eocaecilia micropodia* (Eocaeciliaidae), is represented by most of the skeleton, including limb and girdle elements and the skull. The former elements alone demonstrate that it is not an aistopodan, although they do not resolve the question of lissamphibian monophyly. *Eocaecilia micropodia,* however, does answer questions of skull and limb evolution in the Apoda. The Apoda is the clade (stem-based) encompassing the fossil taxa and the ancestor and all descendants of the extant gymnophionans (Table 3.1).

A single vertebra from each of two South American caecilians and the recent find of four vertebrae in the Upper Cretaceous of the Sudan help define the geological and geographic occurrence of caecilians but assist little in understanding their evolutionary history. The Brazilian fossil is most similar to the vertebrae of the African *Geotrypetes* (Caeciliidae) and has been named *Apodops*. If this similarity denotes actual relationship, it provides another

TABLE 3.1 A Hierarchical Classification of the Extant Caecilians (Gymnophiona).

Gymnophiona
- Rhinatrematidae
- Neocaecilia
  - Diatrata
    - Ichthyophiidae
    - Uraeotyphlidae
  - Teresomata
    - Typhlonectidae
    - "Caeciliidae"
    - Scolecomorphidae

Note: This classification is based on the phylogenetic relationships depicted in Fig. 15.1. Category titles are not assigned to the hierarchical ranks. A name in quotation marks indicates that the group is not monophyletic.

example of Gondwanan affinities among African and South American amphibians.

Albanerpetontids

Albanerpetontids are a group of salamander-like lissamphibians that were linked to prosirenid salamanders until recently. They are moderately abundant as microfossils from Middle Jurassic to Early Miocene deposits of North America, Europe, and Central Asia. Although abundant, they are represented largely by disassociated skeletal elements, but even these fragments show albanerpetontids to be very different "salamanders." They had a unique peg-and-socket symphyseal joint in the mandible, a two-part craniovertebral joint, and sculptured osteoderms dorsally from snout to tail. They were small lissamphibians, <15 cm TL.

The discovery of a complete and fully articulated specimen (*Celtedens*; Fig. 3.6) permitted the recognition of the albanerpetontids as a separate clade of lissamphibians, likely the sister group of the salamander–frog clade. Albanerpetontids must also have an equally ancient origin in the earliest Triassic or before and are absent or unrecognized in the fossil record until the Jurassic.

Salamanders

Extant salamanders comprise two clades: Cryptobranchoidei (including Sirenidae) and Diadectosalamandroidei (Table 3.2). Both clades occur as fossils, and several other clades (e.g., karaurids and prosirenids) are known only from fossils. The extinct and extant salamanders form the Urodela (stem-based clade) with a history extending from the Middle Jurassic, about 165 mybp (million years before present) (Fig. 3.2). Urodelan history is linked mainly to the northern hemisphere (Holarctic) and to

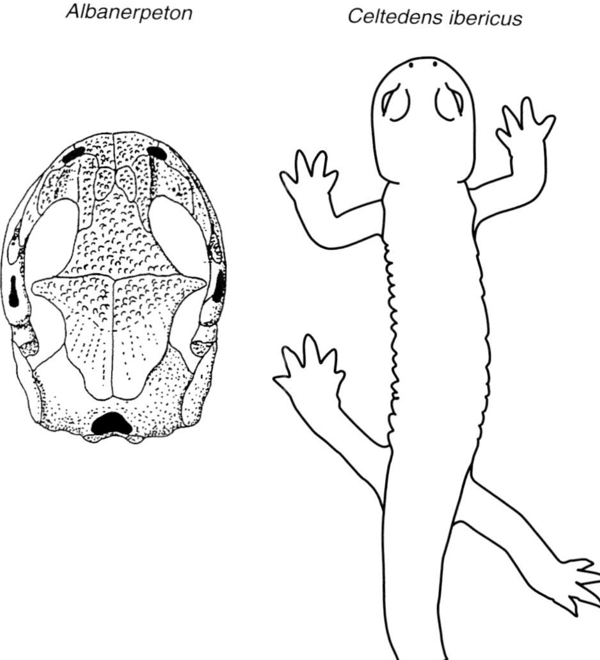

FIGURE 3.6 Albanerpetontidae, salamander-like lissamphibians from the Cretaceous and Tertiary. Skull of *Albanerpeton* and morphology of *Celtedens ibericus*. After Estes and Hofstetter, 1976, and as suggested by skeleton in McGowan and Evans, 1995, respectively.

TABLE 3.2 A Hierarchical Classification of the Extant Salamanders (Caudata).

Caudata
- Cryptobranchoidei
  - Cryptobranchidae
  - Hynobiidae
- Unnamed clade
  - Sirenidae
  - Diadectosalamandroidei
    - Hydatinosalamandroidei
      - Proteidae
      - Treptobranchia
        - Ambystomatidae
        - Salamandridae
    - Plethosalamandroidei
      - Rhyacotritonidae
      - Xenosalamandroidei
        - Amphiumidae
        - Plethodontidae

Note: This classification is based on the phylogenetic relationships displayed in Fig. 16.1, based on Frost et al. (2006), Roelants et al. (2007), and Wiens et al. (2005).

the ancient continent of Laurasia; nonetheless, recent fossil salamander discoveries in Africa and South America show that the relatively recent dispersal of plethodontids southward is not the first occurrence of salamanders on Gondwanan-derived continents.

The earliest salamanders are two species of *Marmorerpeton* from a Middle Jurassic deposit in central England. They were moderate-sized (<30 cm TL), presumably totally aquatic salamanders. Their relationships are uncertain, in part because they are represented by only a few vertebrae, a humerus, and miscellaneous skull elements. They appear to be related to the extinct scapherpetotids, but they also have some primitive features suggesting a possible sister-group relationship to all other Urodela. The earliest crown-group salamander was the cryptobranchid *Chunerpeton tianyiensis*, which was discovered in the Middle Jurassic of Inner Mongolia, dated at 161 mybp.

The karaurids are another ancient group of salamanders (see Fig. 1.1). They are known presently from a few fossils from the Upper Jurassic of Kazakhstan. The fossil of *Karaurus sharovi* is fortunately nearly complete (Fig. 3.7). Its primitive morphology indicates that the karaurids are a sister group of the Caudata. *Karaurus* was small (about 120 mm SVL) and terrestrial, judging from its body form and the dermal sculpturing (skin fused to bone) on the skull bones.

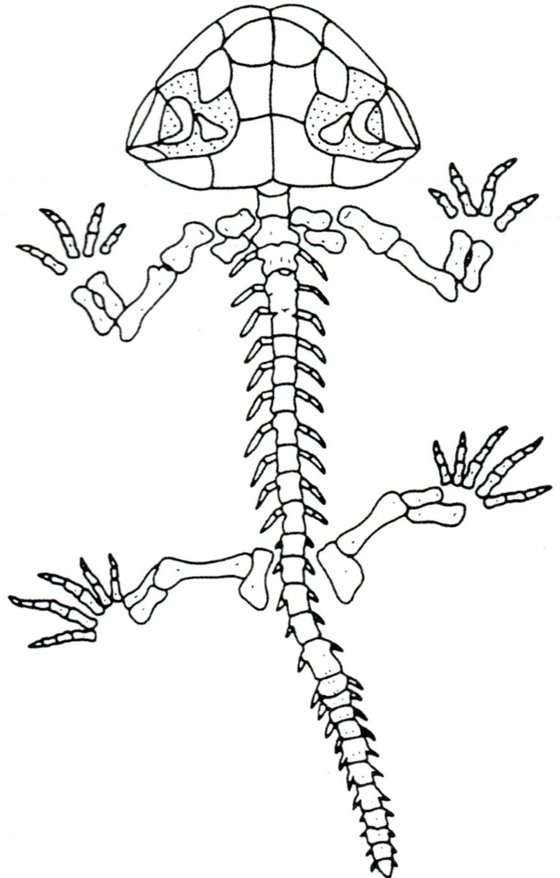

FIGURE 3.7 *Karuarus sharovi* (about 15 cm TL), the earliest known salamander, from the Late Jurassic of Russia. Adapted as a partial reconstruction from Carroll, 1988.

The first batrachosauroidids appeared soon after the karaurids in the Late Jurassic (Fig. 3.2), but unlike the latter, they persisted as an occasional member of freshwater assemblages until the Early Pliocene and are found only in North American deposits. They are similar to proteids; however, it is uncertain whether this similarity is related to the retention of a larval morphology as adults (heterochrony) or an indication of phylogenetic relationship. An assortment of other salamander fossils has been found from the Late Jurassic. Most are too fragmentary or incomplete, such as the Wyoming *Comonecturiodes marshi*, to indicate their affinities.

Salamanders are largely absent from Cretaceous deposits until the Late Cretaceous. The exceptions are the batrachosauroidids, prosirenids, and a diadectosalamandroid. The prosirenids consists of two species, *Prosiren elinorae* of Texas and *Ramonellus longispinus* of Israel. Both are assumed to share the sirenid morphology, with elongate bodies and presence of only forelimbs. Other characteristics suggest that they are not sirenids, and that they might not even be closely related. The late Lower Cretaceous *Valdotriton* is anatomically a modern salamander and a diadectosalamandroid. Because it is represented by six complete skeletons, its proposed inclusion in the diadectosalamandroid clade is robust, but it is not a member of any currently named family. It appears "intermediate" between the proteids and all other diadectosalamandroids.

Two extant families, Amphiumidae and Sirenidae, and the extinct scapherpetonids make their first appearance in the Upper Cretaceous. *Proamphiuma* from a Montana Cretaceous deposit is the first fossil amphiumid. Like many fossils with "pro" in their names, *Proamphiuma* is a structural precursor to *Amphiuma* (Paleocene to Recent), and the relationship actually may be ancestor to descendant. The amphiumids have remained a strictly North American group throughout their 60+ million-year history.

Sirenids first appeared in the North American Cretaceous as the giant *Habrosaurus*, which survived into the Early Paleocene. This siren looked much like its living relatives, except for specialized shovel-shaped teeth. Other Cretaceous sirenids are *Kababisha humarensi* and *K. sudanensis* from Africa and *Notoerpeton bolivianum* from South America. Another somewhat younger sirenid also occurred in Africa. Sirenids are unknown then until the Middle Eocene when *Siren* appears in North America, where the remainder of the sirenid fossil history is found. *Pseudobranchus* occurred first in Pliocene deposits of Florida.

The extinct scapherpetontids were a group of moderate-sized salamanders living from the Late Cretaceous to the Early Eocene in North America. These salamanders are related to the present-day dicamptodontines (currently placed within Ambystomatidae), and *Scapherpeton* and *Piceoerpeton* share the *Dicamptodon* body form. *Lisserpeton* appears to have had an elongate body and reduced limbs. Interestingly, one species of *Piceoerpeton* occurred on Ellesmere Island within the present Arctic Circle. Fossil dicamptodontines made their first appearance in North America during the Eocene but somewhat later than the last scapherpetontid. However, fossil dicamptodontines appeared first in the Upper Paleocene of Europe and again in the Middle Miocene. Upper Paleocene trackways in western North America are attributed to a dicamptodontid because of the unique bilobate palm impressions. Furthermore, the trackways are associated with a redwood flora, an association occurring today in *Dicamptodon*. Subsequent North American fossil occurrence is in the Middle Miocene.

Other modern salamanders (Cryptobranchidae, Proteidae, and Salamandridae) appeared in the Paleocene (Fig. 3.2). The cryptobranchoid *Cryptobranchus* occurred first in the Paleocene of Saskatchewan and again in the Appalachian and Ozark Pleistocene assemblages. *Andrias* has a much more extensive history. The oldest *Andrias* fossils are from the European Upper Oligocene, and *Andrias* persisted there at least through the Pliocene and in the North American Miocene. Within its present range, *Andrias* has been found only in Japanese Pleistocene deposits. The fossil forms were also giant salamanders, one with a TL of more than 2 m. The only other salamanders that might have attained such lengths were some fossil sirenids, but it is difficult to confirm because all fossil sirenids are known only from a single or short series of vertebrae. Hynobiidae, the other cryptobranchoid lineage, has no fossil record.

Proteids occurred first in the Late Paleocene of North America and the Middle Miocene of Europe. These fossils represent the extant *Necturus* and *Proteus*, as well as two extinct genera from the Miocene of Europe. All were small, perennibranchiate salamanders (gill-bearing as larvae and adults). *Ambystoma* appeared in the Eocene of North America and is moderately common in Pleistocene deposits.

Of living salamanders, salamandrids have the most speciose fossil record, with representatives of 18 genera and more than 50 species. Living genera, such as *Notophthalmus,* extend as far back as the Miocene, *Taricha* and *Triturus* to the Oligocene, and *Salamandra* and *Tylotrito*n to the Eocene. The extinct genera derive principally from the Paleocene to Oligocene. The fossil species of the extinct and extant genera match the extant species in size and body form and probably shared the diversity of behaviors and ecology seen in modern species.

Today, the plethodontids are the most speciose of the salamanders, and yet they have a meager fossil record. Half a dozen genera are represented, and four of these occur no earlier than the Pleistocene. A few vertebrae attributable to *Aneides* have been found in an Early Miocene deposit in Montana, and a fossil trackway from the Early Pliocene of California has been referred to *Batrachoseps*.

Frogs

The Salientia encompasses all taxa of extinct and living frogs, and the Anura, a crown-group clade, contains the ancestor of all living taxa and its descendant taxa. The "proanurans" is an informal name for the earliest and structurally most primitive frogs. Proanuran taxa include *Triadobatrachus* and other extinct frogs that have sister-group relationships to one another or to the Anura clade; in most instances the relationships are uncertain. Anurans previously have been divided into three subgroups: a grade of early frogs (extinct), and the extant Mesobatrachia and Neobatrachia. These subgroups appear more or less sequentially and chronologically in the fossil record relative to their branching or cladistic pattern (Fig. 3.2). However, categorizing extant frogs into Mesobatrachia and Neobatrachia is inconsistent with recent phylogenetic analyses based on sister-group relationships. Modern frog taxonomy is much more complex and hierarchical in structure (Table 3.3; Fig. 17.1). The first frog fossil is from Madagascar, suggesting a Gondwanan origin for frogs. However, the next frog fossil was found in North America. These two occurrences and the ancientness of the lissamphibians suggest that the groups giving rise to modern lissamphibians were widespread on the megacontinent of Pangaea. Subsequent fragmentation of this megacontinent could have yielded modern families of both Gondwanan and Laurasian origins.

The fossil record reflects a higher diversity of frogs than of salamanders and caecilians, similar to that observed among the modern lissamphibians. Only frogs are known from the Triassic. At least six frog taxa have been found in Jurassic deposits compared to three salamanders and one caecilian. In the Cretaceous (Fig. 3.2), salamanders and frogs are equally represented, and in the Tertiary, the extant families for both salamanders and frogs appear, establishing the diversity seen today.

The first frog is *Triadobatrachus massinoti* from the Early Triassic of Madagascar (Fig. 3.5). Although it had more vertebrae (± 14 presacral vertebrae) than later frogs and a short stumpy tail, it was clearly a frog. *T. massinoti* is unlikely to be the ancestor of later frogs; nonetheless, it provides a glimpse of the divergence in anatomy of frogs away from early temnospondyls. Its body size of about 10 cm SVL and the lack of any large frog fossils suggest that frogs remained relatively small throughout their evolutionary history, unlike many other earlier amphibian groups. Only a single fossil exists for *Triadobatrachus*, and it may represent a juvenile of an aquatic form or a metamorphosing individual of a semiterrestrial one.

TABLE 3.3 A Hierarchical Classification of the Extant Frogs (Anura).

- Salientia
  - *Triadobatrachus*
  - Anura
    - Leiopelmatidae
    - Lalagobatrachia
      - Xenoanura
        - Pipidae
        - Rhinophrynidae
      - Sokolanura
        - Costata
          - Alytidae
          - Bombinatoridae
        - Acosmanura
          - Anomocoela
            - Pelobatoidea
              - Pelobatidae
              - Megophryidae
            - Pelodytoidea
              - Pelodytidae
              - Scaphiopodidae
          - Neobatrachia
            - Heleophrynidae
            - Phthanobatrachia
              - Hyloides
                - Sooglossidae
                - Notogaeanura
                  - Australobatrachia
                    - Calyptocephalellidae
                    - Myobatrachoidea
                      - Limnodynastidae
                      - Myobatrachidae
                  - Nobleobatrachia
                    - Hemiphractidae
                    - Meridianura
                      - Brachycephalidae
                      - Cladophrynia
                        - Cryptobatrachidae
                        - Tinctanura
                          - Amphignathodontidae
                          - Athesphatanura
                            - Hylidae
                            - Leptodactyliformes
                              - Centrolenidae
                              - Cruciabatrachia
                                - Leptodactylidae
                                - Chthonobatrachia
                                  - Ceratophryidae
                                  - Hesticobatrachia
                                    - Cycloramphidae
                                    - Calamitophrynia
                                      - Leiuperidae
                                      - Agastorophrynia
                                        - Bufonidae
                                        - Nobleobatia
                                          - Hylodidae
                                          - Dendrobatoidea
                                            - Aromobatidae
                                            - Dendrobatidae

TABLE 3.3 *(continued)*

- Ranoides
  - Allodapanura
    - Microhylidae
    - Afrobatrachia
      - Xenosyneunitanura
        - Brevicipitidae
        - Hemisotidae
      - Laurentobatrachia
        - Arthroleptidae
        - Hyperoliidae
  - Natatanura
    - Ptychadenidae
    - Victorana
      - Ceratobatrachidae
      - Telmatobatrachia
        - Micrixalidae
        - Ranixalidae
          - Ametrobatrachia
            - Africanura
              - Phrynobatrachidae
              - Pyxicephaloidea
                - Petropedetidae
                - Pyxicephalidae
            - Saukrobatrachia
              - Dicroglossidae
              - Aglaioanura
                - Rhacophoroidea
                  - Mantellidae
                  - Rhacophoridae
                - Ranoidea
                  - Nyctibatrachidae
                  - Ranidae

Note: This classification is based on the phylogenetic relationships shown in Fig. 18.1, taken from Frost et al., 2006, and Grant et al., 2006.

The next proanuran, *Prosalirus bitis*, derives from the mid-Lower Jurassic (151–154 mybp) of Arizona and from the same deposits as *Eocaecilia*. Its limb and girdle morphology is essentially modern and indicates that *P. bitis* was a jumping frog. The body was truncated, although the actual number of presacral vertebrae is unknown. Similarly its affinities to other Jurassic frogs and extant families are not clear. The Patagonian *Vieraella herbstii* is potentially a contemporary of *P. bitis*. However, its fossil origin has not been precisely dated. Structurally, *V. herbstii* and the later Patagonian *Notobatrachus degustori* are even more modern in occurrence (158–172 mybp; Fig. 3.8). They have a suite of primitive characteristics, such as nine presacral vertebrae, free ribs, and a partially fused astragalus–calcaneum, all traits shared with *Ascaphus* and *Leiopelma*. As a result, these ancient Patagonian frogs have been considered representatives of the extant Leiopelmatidae. Their similarity is a reflection of primitiveness, not phylogenetic relatedness. They are best considered the sister group to modern anurans. *Vieraella herbstii* was a small frog (about 28 mm SVL). *Notobatrachus degustori* was much larger (120–150 mm SVL), roughly three times the size of modern leiopelmatids.

The next group of frogs to appear was the Alytidae [Discoglossidae]. This extant group appeared regularly in fossil assemblages during the last 170 my. *Eodiscoglossus* appeared in the Late Jurassic of Spain and persisted into the Early Cretaceous. At least skeletally, it seems nearly identical with today's *Discoglossus*. Two other genera appeared in the Late Cretaceous of western North America, and one of them (*Scotiophryne*) survived into the Paleocene. Alytids are absent throughout the Eocene. One alytid, *Latonia*, reappeared in the Oligocene of Europe. Modern *Discoglossus* and *Alytes* are found in the European Miocene and Pleistocene, respectively.

The assignment of some taxa to the Alytidae or Bombinatoridae clades is uncertain because of the recency of the recognition of these clades and the continued use of the older alytid concept by anuran paleontologists. *Bombina* is known since the Early Miocene in Europe. If either

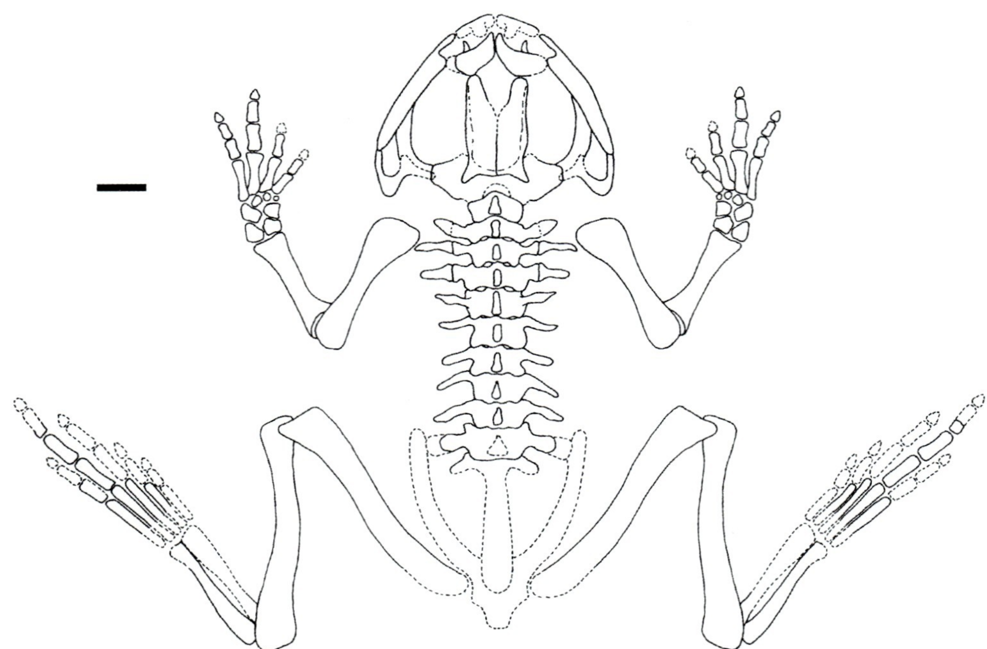

FIGURE 3.8 *Vieraella herbstii*, an ancient frog from the Jurassic of Patagonia. Scale bar = 2 mm. Adapted from Estes and Reig, 1973.

Enneabatrachus or *Scotiphyrne* are bombinatorids, then this group has a history extending from the late Upper Jurassic or earliest Lower Cretaceous.

Pelobatidae and/or Scaphiopodidae, two modern clades of anatomically conservative and similar frogs, appeared in the Late Jurassic of Asia and North America. Fossils from the western North American Morrison Formation cannot be assigned to a particular genus but are unquestionably either pelobatids or scaphiopodids. The next appearance was in the Cretaceous; *Eopelobates* and *Kizylkuma* differ sufficiently from their later-appearing relatives to be placed in a separate clade (Eopelobatinae). *Eopelobates* had a long existence from the Late Cretaceous to the Middle Miocene and an equally broad geographic occurrence from western North America through temperate Asia to Europe. The eopelobatine species were generally moderate-sized (50–60 mm SVL), terrestrial frogs. They lacked spades on the heels, a prominent characteristic of modern pelobatids/scaphiopodids but presumably shared many features of their natural history. The pelobatids appear in the European basal Miocene (*Pelobates*), and the scaphiopodids appear in the Early Oligocene of North America (*Scaphiopus*). A closely related group, the Asian Megophryidae, is unknown as fossils. The related pelodytid frogs had a brief appearance in the Eocene of central Europe and the Miocene of western North America.

Gobiates, a Cretaceous frog from Central Asia, was initially considered a near relative to *Eopelobates*, but it is morphologically quite distinct. It is another basal or proanuran group, even though it is presently known only from the mid-Cretaceous and is now recognized as a distinct lineage (Fig. 3.2). *Gobiates* was moderately speciose with about a dozen species.

The recently extinct paleobatrachid frogs were a long-lived clade. They appeared first in the Upper Cretaceous and went extinct in the early Pleistocene. Throughout their entire history, they were confined to Europe, with one questionable Cretaceous occurrence in North America. Although apparently abundant, they were only moderately speciose, with less than two dozen species recognized throughout their 120-million-year history. All paleobatrachids were moderate to small frogs, generally less than 50 mm SVL, and strictly aquatic. They had long, robust hindlimbs and long digits on both the fore- and hindfeet. *Neusibatrachus*, the oldest paleobatrachid, occurred first in the Late Jurassic but then is unknown in the fossil record until the Miocene. *Paleobatrachus* (Fig. 3.9), with 12 species, spanned the Eocene to Pliocene period. Fossils of this taxon are abundant in a series of freshwater deposits in eastern Czech Republic. In this area, volcanic gases apparently poisoned the waters of streams and ponds, periodically causing massive die-offs of all aquatic animals. These gases also stimulated diatom blooms, and the diatom skeletons buried frogs and even tadpoles. Burial was rapid, and imprints of soft parts remain to help paleontologists reconstruct the anatomy and life histories of the paleobatrachid frogs.

The paleobatrachids and pipids are sister groups, and all paleobatrachids resembled the modern clawed frogs (*Xenopus*). Pipids did not appear in the fossil record until the

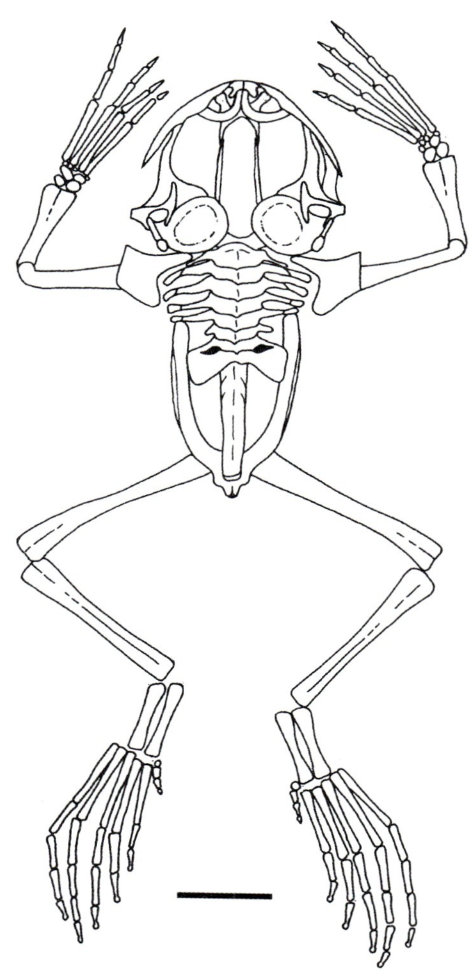

FIGURE 3.9 *Paleobatrachus grandiceps*, a representative of the extinct Paleobatrachidae, from the Oligocene of eastern Europe. Scale bar = 10 mm. Adapted from Estes and Reig, 1973.

Early Cretaceous, but they are more likely ancestors rather than descendants of paleobatrachids. The paleobatrachid's restricted distribution in Europe throughout their history contrasts sharply to the presence of pipids in South America and Africa since the Cretaceous. The Upper Jurassic pipoid *Rhadinosteus* may resolve this dilemma. Three definite pipids occurred in the Early Cretaceous of the eastern Mediterranean, suggesting an early radiation of the African pipids. *Xenopus* occurred early in Africa, from the Late Cretaceous of Nigeria and the Oligocene of Libya. It is a remarkably adaptable frog genus, and even today it is the most speciose of the pipid clade. The ancient pipids (*Saltenia* and *Shelania*) of the South American Paleocene derive from the southern portion of that continent. *Shelania* fossils are frequently found as complete or nearly complete skeletons in Patagonian sediments. These fossils provide valuable insights into the evolution of pipid frogs.

Although fossorial rather than aquatic, the Rhinophrynidae is the sister group of the paleobatrachid–pipid clade. The first rhinophrynids occurred in the Lower Eocene of western North America. Others occurred in the Oligocene but thereafter disappeared from the fossil record. The Jurassic *Rhadinosteus* represents an early pipoid and structurally is most similar to the rhinophrynids, likely indicating the divergence of the extant pipoid families.

More advanced frogs also began to appear in the Early Tertiary, even somewhat earlier than rhinophrynids. Surprisingly, considering their present diversity, neobatrachians are neither abundant nor diverse throughout much of the Tertiary. Only in the Pliocene and Pleistocene do they become more common in fossil beds. Excluding fossil records from the Pliocene, only the bufonids, hylids, leptodactylids, limnodynastids, microhylids, ranids, and rhacophorids have Tertiary representatives. Leptodactylids are definitely known from the Upper Cretaceous of South America, and if an Indian fossil's identity is confirmed, hylids will likewise have a Late Cretaceous occurrence. A nearly continuous record exists for budonids in South America from their first occurrence in the Late Paleocene. They also were present in North America and Europe from the mid-Tertiary onward. Although all fossil bufonids have been assigned to the genus *Bufo*, recent reorganization of the former *Bufo* into numerous genera will require a reexamination of fossil material in order to place fossil taxa in the appropriate new genera. Hylids (described as *Hyla*) appeared in the Oligocene in North America and in the Miocene in Europe. The only other fossil hyline hylid is *Proacris* from the Miocene of Florida. The Miocene *Australobatrachus* is the first fossil representing pelodryadine hylids and was contemporaneous in the Late Miocene with the still extant *Litoria*. Leptodactylidae has a broader and more diverse fossil history, and, although most fossils have been found in the New World, some have been found in the European Eocene. The ceratophryd *Wawelia* occurred in the Miocene of Argentina, and a Cretaceous fossil is potentially a ceratophryd. Two genera of telmatobine ceratohpryds are represented in the Oligocene and Miocene. A few *Eleutherodactylus* (Brachycephalidae) and *Leptodactylus* (Leptodactylidae) species occured in the Pleistocene. An *Eleutherodactylus* in amber from the Hispaniolan Eocene and its amber-associated biota provide important insights into the early distribution of the Mesoamerica biota and landmass movements. The widespread and diverse ranids are represented in the fossil record only by *Ptychadena* in the Moroccan Miocene and an assortment of nearly 50 species of *Rana* from the Oligocene onward of Europe and the Miocene through Pleistocene of North and Central America.

HISTORY OF REPTILES

The first tetrapod is known from the Late Devonian, the first amphibian from the Middle Mississippian, and the first amniotes from the Middle Pennsylvanian (Fig. 3.10).

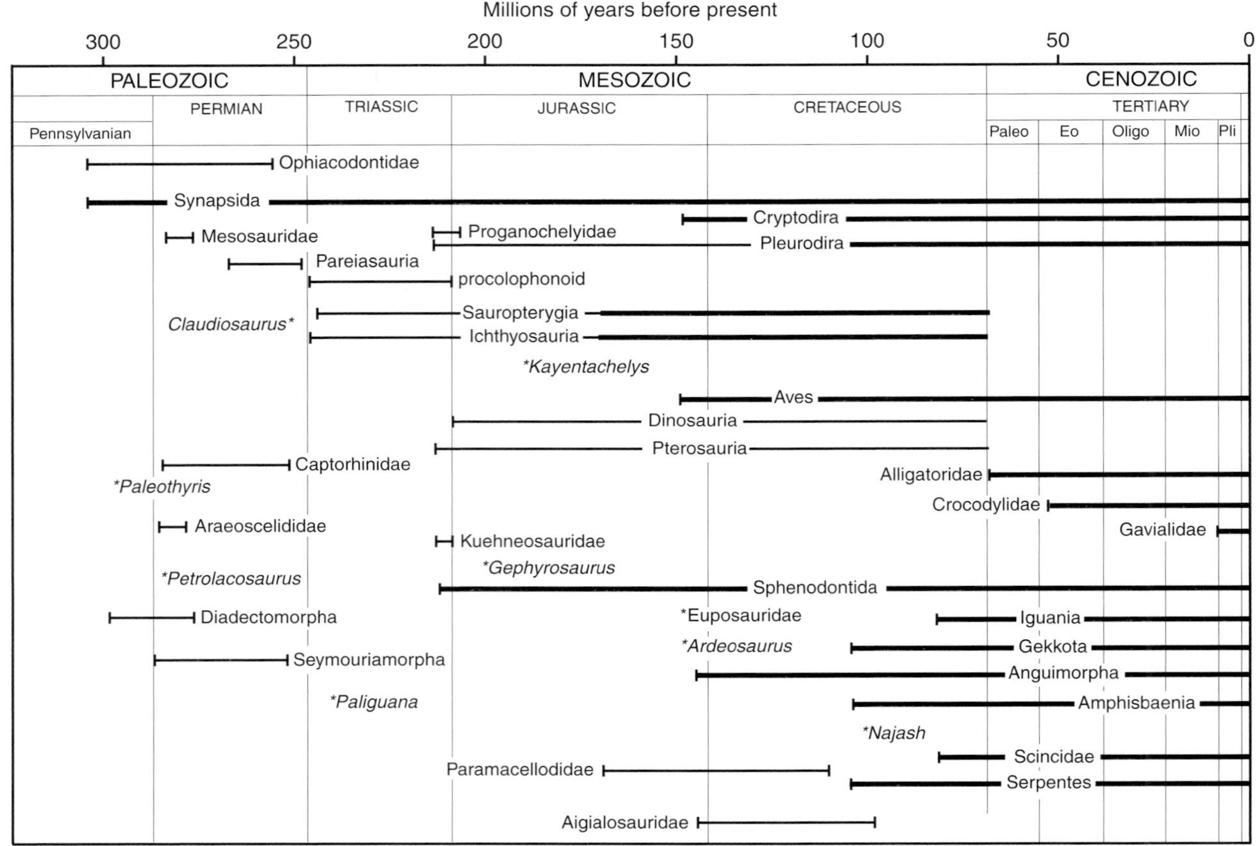

FIGURE 3.10 Geological occurrence of some early anthracosaurs and amniotes, and extinct and living reptiles. Abbreviations for Cenozoic epochs: Paleo, Paleocene; Eo, Eocene; Oligo, Oligocene; Mio, Miocene; Pli, Pliocene; Pleistocene is the narrow, unlabeled epoch at the top of the chart. Asterisk indicates insuffficient fossil material to depict how long the taxon persisted.

These first amniotes were *Archaeothyris* (synapsid), *Hylonomus* and *Paleothyris* (reptiles), showing that clades that ultimately would produce mammals and modern reptiles were already established in the Late Carboniferous. These three amniotes were small and lizardlike, but structurally quite distinct from modern lizards.

Many of the anthracosaurs were aquatic tetrapods (see Chapter 1), although some, such as *Proterogyrinus*, were definitely terrestrial. Anthracosaurs lived from the Late Devonian into the Late Permian. Thus, amniote ancestors diverged early in the history of anthracosaurs. The seymouriamorph anthracosaurs (Fig. 3.3; see also Fig. 1.1) diverged later. Their fossil history begins in the Early Permian at a time when amniotes were beginning to establish their dominance on land. Diversity of these moderate-sized (25–100 cm TL) tetrapods was low, although their fossil remains are moderately abundant in the Early Permian. They too were terrestrial. Terrestrial seymouriamorphs disappeared from the fossil record in the mid-Permian, but aquatic seymouriamorph fossils appeared in the late-Permian. Another group of anthracosaurs, the diadectomorphs, is structurally more primitive than early amniotes, although they appeared in the Late Pennsylvanian subsequent to the origin of amniotes (Fig. 3.10). Although primitive because of their early occurrence, they were specialized tetrapods. *Diadectes* was large (3 m TL) and had a partial secondary palate and molariform cheek teeth suggesting an herbivorous diet. Like many early reptiles, this group was short lived evolutionarily.

Radiation Among Early Amniotes

Several contemporaneous taxa of reptiles and synapsids from a buried forest of the Middle Pennsylvanian in Nova Scotia, Canada, are the earliest known amniotes. They apparently lived in hollow, upright trunks of buried trees and were entombed when the forest was periodically flooded. These (*Archaeothyris*, *Hylonomus*, and *Paleothyris*) were small, approximately 15 cm long (SVL). Many of the later Paleozoic amniotes were quite large, particularly in comparison to most living reptiles. The explosive radiation of reptiles was still millions of years away in the future Mesozoic. Nonetheless, amniotes, particularly pelycosaurs (synapsids), began to assume a dominant role in terrestrial vertebrate communities of the Permian.

Protomammals: The Synapsids

Archaeothyris is an ophiacodontid. Ophiacodontids had only a modest history with low diversity, perhaps surviving into the Late Permian. They are the basal members and potential ancestors of the pelycosaurs. Pelycosaurs diversified into two dozen genera and numerous species in six or more clades. They became the major tetrapods of the Early Permian in both abundance and number of species. The earliest pelycosaurs were small (ca. 30 cm SVL) and lizardlike. They had large heads with big, widely spaced teeth, suggesting that they were effective carnivores of large prey. This basal stock radiated into several groups of medium to large carnivores and at least two groups of herbivores. Two clades, *Edaphosaurus* (herbivorous edaphosaurids) and *Dimetrodon* (carnivorous sphenacotontids), had members with a dorsal "sail" of elongated neural spines on the trunk vertebrae. Both pelycosaurs were large (*Dimetrodon* to >3 m TL). The sail was likely a thermoregulation mechanism. In *Dimetrodon*, for example, surface area of the sail scales with body mass in a typical volume-to-area relationship that is associated with thermoregulation in extant reptiles (see Chapter 7). Some other pelycosaurs were varanid-like and probably were agile and carnivorous, similar to present-day varanids. Pelycosaurs began to disappear in the middle of the Late Permian. Their decline might have been brought about by the success of another early synapsid lineage that gave rise to the therapsid radiation of the Upper Permian. Later in the Triassic, mammals arose within the therapsids.

Paleozoic Reptiles

Many early reptiles have skulls with a solid bony temporal area (i.e., no temporal fenestrae; see Fig. 2.20). Taxa without temporal fenestrae were, at one time, considered to be closely related and called the Anapsida. While this relationship is no longer accepted, "anapsids" remains a vernacular name for early reptiles sharing the anapsid skull. Other clades defined originally on the nature of temporal fenestration persist, for example, Diapsida and Synapsida. Captorhinids define the Eureptilia, and fossil reptiles lacking temporal fenestrae define the Parareptilia (Fig. 1.9). Modern and fossil turtles lack temporal fenestrae ("anapsid"), but appear to have secondarily lost temporal fenestrae and thus are now considered nested within Eureptilia. Only the eureptiles have a fossil presence in the Late Pennsylvanian. *Hylonomus* (Fig. 3.11) and *Paleothyris* (Fig. 1.9) are two of these eureptiles, and a third is *Petrolacosaurus*. *Petrolacosaurus* was a moderate-sized (ca. 40 cm TL) terrestrial reptile, iguanalike with enlarged upper canines. It is typically linked to the short-lived *Araeoscelis* clade (Araeoscelidia) of the Lower Permian. Araeoscelidans are basal diapsids and the sister group to the Sauria. All were lizardlike in head and body proportions, but their limbs were gracile and elongate with fore- and hindlimbs of nearly equal length. Their dentition was simple and indicates a general carnivorous diet.

Thereafter, no other diapsids or saurians are found until *Claudiosaurus* and *Paliguana* of the Upper Permian. These two diapsids were not contemporaries (Fig. 3.10). The former is a long-necked, marine reptile that has been considered a plesiosaur or, at least, a basal sauropterygian. Evidence now suggests that the body form of *Claudiosaurus* is independently evolved and that *Claudiosaurus* arose prior to the archosauromorph–lepidosauromorph divergence. *Paliguana* has similarly been linked to a later appearing group, the Squamata. This relationship is uncertain, although *Paliguana* certainly is a diapsid and may be a squamate.

The Captorhinidae represents a primitive group of eureptiles, and some features suggest an origin prior to that of *Paleothyris*. The captorhinids were medium-sized, lizardlike reptiles, although the broad-jowled head was proportionately larger than that of most lizards. The teeth showed regional differentiation with large, pointed incisors in front and double to triple rows of short, cone-shaped teeth in the rear. Their bodies were slender and limbs moderately long, suggesting that they were agile carnivores.

Several groups of Permian reptiles, the Mesosauridae, Millerettidae, procolophonoids, and Pareiasauria (Fig. 3.12), have proven exceedingly difficult to classify, and for lack of a better name, were called the parareptiles and presumed to be unrelated. Additional fossils, improved preparation, and new analytical techniques now indicate that the Parareptilia, excluding Mesosauridae, is a monophyletic clade (Fig. 1.9).

The mesosaurs (Early Permian) were miniature (ca. 1 m TL), marine, gharial-like reptiles. They had long,

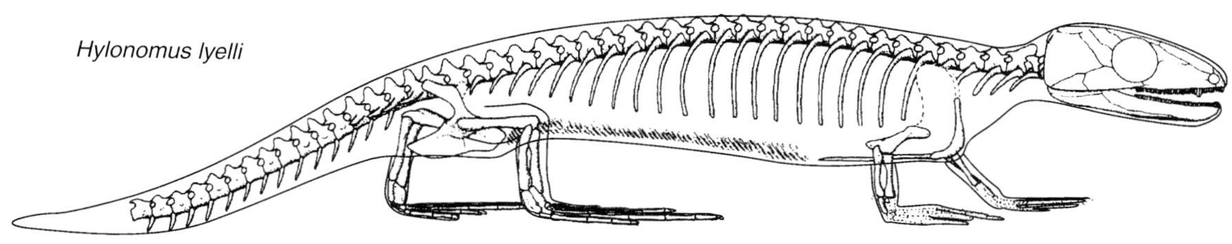

FIGURE 3.11 *Hylonomus lyelli*, the earliest known reptile, from the Early Permian of Nova Scotia. Size, about 42 cm SVL. Adapted from Carroll and Baird, 1972.

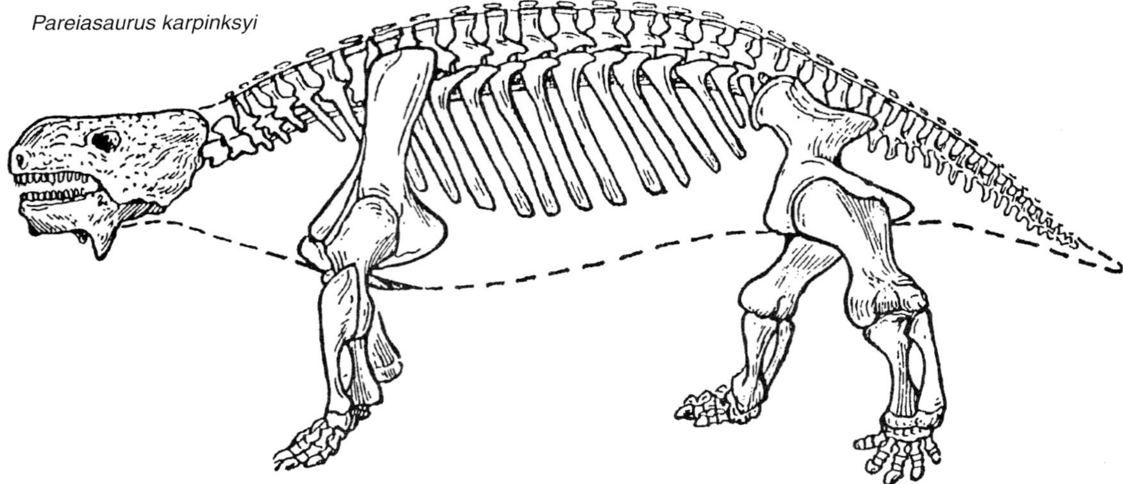

FIGURE 3.12 *Pareiasaurus karpinksyi*, a pareiasaur from the Late Permian of Russia (about 3 m TL). Adapted from Gregory, 1951.

narrow-snouted skulls, and the long, thin teeth of the upper and lower jaw curved outward and interdigitated when the jaws were closed. Such jaws are effective for catching fish with a sideward sweep of the head. The body and tail were similarly elongated and the tail laterally compressed for undulatory swimming. Nonetheless, the limbs were well developed, and the hindlimbs and feet were large, perhaps used as rudders.

The parareptiles are a diverse group of small to large reptiles, mainly of Middle Permian to Lower Triassic age. The procolophonoids are strictly Mesozoic, occurring throughout the Triassic. The millerettids were small, lizard-like reptiles. Their small heads and simple conical teeth match the appearance of many iguanians living today, and they probably shared a diet of insects. The pareiasaurs and the procolophonoids were more diverse. The pareiasaurs were the giants of the parareptiles with some taxa to 3 m (TL) (Fig. 3.12). They had large barrel-shaped bodies, elephantine limbs, and proportionately small, broad-jawed heads capped with thick bone and numerous projections. The teeth were closely spaced with laterally compressed leaf-shaped crowns. By all indications, the pareiasaurs were slow, lumbering herbivores. The procolophonoids were small to medium-sized lizardlike reptiles. Their stocky bodies, short limbs, and broad-jowled heads gave them the appearance of modern *Uromastyx* or *Sauromalus*, and they may have shared the herbivorous habits of these extant lizards. Unlike the pareiasaurs, their widely spaced, thick, bulbous-crowned teeth were probably used for crushing rather than mincing. Numerous complete skeletons of *Owenetta* (a procolophonoid) show that this small reptile of the Late Permian shares many features with the oldest known turtle, *Proganochelys*; however, it is unlikely that procolophonoids contained the ancestors of turtles based on recent nuclear gene analyses that place turtles in with diapsids.

Like *Claudiosaurus*, *Eunotosaurus* is another enigmatic Permian reptile. This small (20 cm SVL) lizardlike creature from the Middle Permian was once considered the link between the basal reptiles and turtles because it had eight pairs of broadly expanded ribs on the trunk. However, the pectoral girdle lies external to the ribs, and the skull is strongly divergent from the cranial morphology of any early turtles.

Age of Reptiles—Radiation in the Mesozoic

Reptiles dominate the fossil beds of the Mesozoic. They are both the most numerous and the largest fossils. This high species richness reflects the diversity of Mesozoic reptiles and their amazingly broad radiation. They were the dominant terrestrial and aerial animals, and although not the dominant marine ones, many were major predators in marine environments. The following summaries touch only briefly on this diversity.

Marine Reptiles

The sauropterygians and ichthyosaurs are a presumed clade (Euryapsida) of marine reptiles, whose origin has been uncertain until recently. Evidence now suggests that these exclusively Mesozoic reptiles diverged early from the saurian line prior to origin of the lepidosauromorphs and archosauromorphs (Fig. 1.11). The sauropterygians were immensely successful aquatic reptiles that appeared early in the Triassic and remained abundant until the end of Cretaceous. The sauropterygians consist of two distinct but related groups, the placodonts (Middle and Upper Triassic) and the "plesiosaurs" in the broadest sense.

The placodonts, although presumably aquatic, did not have a strongly aquatic-designed morphology. They had short, broad heads, stout bodies, and long, laterally compressed tails. Their limbs were short and well developed with a terrestrial front and hind foot anatomy. Most were 1–2 m TL, and some had dermal carapaces resembling turtle shells. The broad heads and tooth morphology suggest herbivory or a diet of shelled invertebrates, gathered in coastal and shallow-water environments. The "plesiosaurs" had a body form unlike that of any other aquatic tetrapods. Although streamlined, the body was large and stocky with a long, flexible neck and large flipperlike limbs. The Triassic nothosaurs were small to moderate-sized (20 cm to 4 m TL) reptiles with the tail extending one-third to nearly one-half of the total length. This morphology suggests that they swam by undulatory movements of the tail and posterior half of the body, using the limbs as rudders. The subsequent plesiosaurs appeared in the mid-Triassic and were abundant in the Jurassic through the Middle Cretaceous. They were generally large creatures from 10 to 13 m in total length. The body was barrel-shaped with a short tail, less than body length, and very large flipperlike limbs. In one group, the neck was very long ending in a tiny head, and in another group the neck was shorter with a large, elongated head. How they swam is uncertain. The two most likely possibilities are aquatic flight like penguins and seaturtles whose limbs move in a figure-8 stroke as in flying birds, or alternatively, with the more paddlelike stroke of seals. No matter how they swam, they were probably excellent and fast swimmers.

The ichthyosaurs were also a successful group of marine reptiles, although they declined greatly in abundance in the Early Cretaceous and disappeared by the mid-Cretaceous. As their name implies, the ichthyosaurs were fishlike reptiles (Fig. 3.13), with morphology similar to that of mackerel and tunas. They ranged in size from about 1.5 to 15 m. Their fishlike form and the presence of fetuses within the body cavity of some individuals indicate that they were viviparous (live-bearing). Most other Mesozoic marine reptiles probably were oviparous and had to return to land like modern seaturtles to deposit their eggs.

FIGURE 3.13 The Ichthyosaur *Ichthyosaurus intermedius* was one of the large marine reptiles present during the Jurassic. Photograph by Sarah Riebolt, courtesy of the Museum of Paleontology, University of California, Berkeley.

Among early crocodyliforms, several groups became highly aquatic and perhaps totally so. The most specialized group was the metriorhynchids (Middle Jurassic to Lower Cretaceous). At least 15 species have been recognized. All were about 3 m long with heavy, streamlined heads, bodies, and tails. The tail had a sharklike downward bend at its tip (heterocercal), and the limbs were flippers. The head was long-snouted and strongly toothed. By all appearances, they were excellent swimmers and successful fish predators. The marine metriorhynchid crocodiliform *Geosaurus* from Patagonia had a pair of lobulated protuberances (nasals) in the skull suggesting that they already had salt glands. Thus, as early as 140 million years ago, an extra-renal osmoregulatory system existed, which may partially explain the success of this group in marine environments. Other marine crocodyliformes included the telosaurids (late Lower Jurassic to early Lower Cretaceous), dyrosaurids (Middle Cretaceous to Eocene), and a few more Mesozoic families of brief geologic occurrence. These taxa were more typically crocodilian in appearance, although with a tendency toward streamlining and reduction of dorsal armoring.

In the Middle Cretaceous, the first marine turtles appeared. They already had streamlined shells and flipper-forelimbs, indicating a much earlier origin. Three clades are evident in these seaturtles, the Cheloniidae, Protostegidae, and Dermochelyidae. Cheloniids and protostegids were moderately abundant and widespread throughout the Upper Cretaceous and had a modest radiation. The protostegid *Archelon ischyros* was the largest of the seaturtles and had a carapace length of nearly 3 m. Today's giants, the dermochelyids, did not appear until late in the Cretaceous.

A clade of aquatic lizards split early from the evolutionary line leading to the extant varanoid groups. The dolichosaurs (Middle to Late Cretaceous) were long-necked plesiosaur-like lizards with low diversity. Their relationship to the mosasaurs is unclear. The earliest mosasaurs were the small (1–2.5 m TL) aigialosaurs, monitor-like in general appearance, although they had shorter necks, reduced but not structurally reorganized limbs, and a laterally compressed, heterocercal tail. They lived in the Late Jurassic to Middle Cretaceous seas. The Late Cretaceous mosasaurs (Fig. 3.14) had a moderate adaptive radiation that produced a variety of different sizes and feeding morphologies (e.g., at least 16 different body forms are recognized). These body forms remained somewhat lizardlike, even though the mosasaurs were highly aquatic animals. The head was elongate and

FIGURE 3.14 Cretaceous sea showing several typical reptiles, including the turtle *Protostega* (left), the mosasaur *Platecarpus* (largest reptile), and a plesiosaur (top). The extinct bony fish *Xiphactinus* (bottom right) and the aquatic bird *Hesperornis* (center right) are also shown. By Karen Carr, with permission of the Sam Noble Oklahoma Museum of Natural History.

narrow, joined by a short neck to an elongate trunk and tail. Their limbs were modified into flippers by a shortening of the pro- and epipodial elements and an elongation (i.e., hyperphalangy) of the meso- and metapodial elements and phalanges. The sinuous body and tail were both used in undulatory swimming, with flippers serving as rudders. Terrestrial locomotion would have been most difficult. Nonetheless, they were likely oviparous, and females would have had to come ashore to lay eggs. This difficult task was compounded further by their size, as the smallest genus was 2.5 m (TL), and the largest reached nearly 12.5 m. Some mosasaurs were surface creatures; others probably dove regularly to depths of several hundred meters for food. All were carnivorous predators.

Gliders and Fliers

Most airborne animals develop flight surfaces by modifying anterior appendages or by stretching membranes between anterior and posterior appendages. Several groups of diapsid reptiles independently had modified ribs and associated muscles that formed an airfoil. This ribcage adaptation is unique to diapsids and exists today in *Draco*, a group of Indomalaysian agamid lizards (Fig. 3.15). The thoracic ribs are greatly elongated and for more than one-half of their length are free of the body cavity and attached to each other by a thin web of skin. Limbs are well developed, and *Draco* can run nimbly up and down tree trunks, with the elongated ribs folded tightly against the body. When pursued, they jump into the air. The elongated ribs unfold like a fan and create an airfoil that allows them to glide long distances at a gentle angle of descent.

The first flying reptile appeared in the Late Permian. *Coelurosaurus* was a moderately large diapsid (ca. 18 cm SVL) with membranes arising from each side of the trunk and creating an airfoil of nearly 30 cm width. The original description suggested that this airfoil was supported by the ribcage as in *Draco*. Subsequent examinations show the airfoil to be supported by dermal rods that would have appeared *Draco*-like in gliding flight. Although highly specialized as a glider, *Coelurosaurus* had many primitive diapsid features and is a basal member of the neodiapsid clade.

The Late Triassic kuehneosaurids were also gliders. They had ribcage airfoils like that of *Draco* (Fig. 3.15). They are an early divergent lineage and the sister group of the lepidosaurs. Another Late Triassic glider, *Sharovipteryx*, known from a single fossil, had large membranes extending from each hindlimb to the base of the tail and perhaps small ones from the forelimbs to the trunk, creating a stealth-bomber profile with a long, thin tail projecting posteriorly. *Sharovipteryx* is a small (<10 cm SVL) diapsid of uncertain affinities.

The typical vertebrate airfoil of modified forelimb wings was used for flight by two groups of ornithodiran archosaurs—pterosaurs and birds. Both of these aerial reptiles were capable of self-propulsive, "flapping" flight. Some proponents, however, still argue for only gliding flight in pterosaurs. The pterosaurs developed a membranous wing that stretched from the posterior edge of the forelimb to the body. The proximal skeletal elements were shortened and robust for the attachment of flight muscles. Most of the wing's span attached to a greatly elongated fourth digit, that is, elongation of metacarpal IV and especially the phalanges, each of which was longer than the

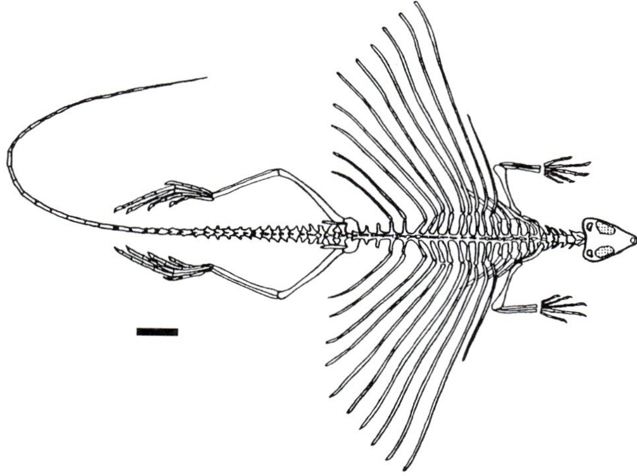

FIGURE 3.15 Top: The sauropsid reptile *Mecistotrachelos apeoros* was one of several gliding reptiles in the Triassic. Its large limbs suggest that it may have been arboreal. It had a much longer neck than that of other gliding reptiles such as *Kuehneosaurus* and *Icarosaurus* (by Karen Carr, with permission of the Virginia Museum of Natural History). Middle: Skeleton of *Kuehneosaurus*, a diapsid glider, from the late Upper Triassic showing ribs modified to support the airfoil. Scale: bar = 4 cm (adapted from Robinson, in Romer, 1966). Bottom: *Draco jareckii*, an Agamid lizard that glides using a rib-supported airfoil (R. M. Brown).

humerus. The birds modified their specialized scales (feathers) to produce an airfoil surface. The forelimb provided the support for the feathers and the anterior edge of the airfoil. In birds, the humerus is short, and the radius and ulna elongate along with elongate metacarpals and phalanges of the first three digits.

The pterosaurs appeared in the Late Triassic as full-winged fliers and persisted as a group throughout the remainder of the Mesozoic (Fig. 3.16). Nearly a hundred species of pterosaurs are recognized—from small species (15 cm wingspan) to the aerial giants, *Pteranodon* (7 m wingspan) and *Quetzalcoatlus* (11–12 m wingspan). Some were scavengers, insectivores, piscivores, carnivores, and even filter-feeders. Their distant relatives, the birds, did not appear until the Late Jurassic (*Archaeopteryx*), and bird diversity either remained low throughout the remainder of the Mesozoic or, alternatively, only a few kinds were fossilized.

The present controversy concerning the origin of birds from within dinosaurs or from other and earlier archosauromorphs is based on how flight evolved. The nondinosaur proponents suggest flight arose from gliding down; in contrast, the dinosaur proponents advocate that flight arose from running and jumping up. The gliding-down advocates point to the small forelimbs of the proposed dinosaur–bird ancestors and the low probability of such limbs becoming wings. The running–jumping advocates note that limb evolution can proceed in either direction and feathers were present to provide lift.

Archosauromorphs

The archosaurs, the so-called "Ruling Reptiles" of the Mesozoic, are a monophyletic group represented today only by turtles, crocodilians, and birds. Three groups likely represent the two major clades of archosaurs, Testudines and Crocodylotarsi + Ornithodira. Prior to recent molecular data placing turtles in the Eureptilia, turtles were not included in the Archosauromorpha. The Crocodylotarsi includes a diverse group of crocodylians and relatives. The Ornithodira contains the dinosaurs, pterosaurs, and their relatives. The divergence of these two groups is evident by the Middle Triassic.

The rhynchosaurs, proterosuchids, erythrosuchids, and *Euparkeria* were early offshoots of the diapsid lineage that led to the archosaurs. They show a sequential alteration of the skeleton toward the archosaurian mode and a trend toward increasing size. Proterosuchids (Late Permian to Early Triassic) were moderate-sized, varanid-like reptiles with a sprawling gait. The erythrosuchids, present from Early to Middle Triassic, were large (ca. 5 m), heavy-bodied reptiles with the beginnings of a more erect limb posture and the archosaurian triradiate pelvic girdle. *Euparkeria*, however, was less than 1 m TL, and it likely was quadrapedal, walking on all four limbs.

Euparkeria, from the Early Triassic, is variously considered the most primitive or the sister group of archosaurs. It appeared much like a short-necked monitor lizard and is

FIGURE 3.16 Cretaceous coastal scene showing several reptiles characteristic of the period, including the carnivorous *Dienonychus* (left; some restorations show *Dienonychus* with feathers), the coelurosaurian *Ornithodesmus* (in the air), and a group of the ornithopod dinosaurs *Tenontosaurus*. By Karen Carr, with permission of the Sam Noble Oklahoma Museum of Natural History.

the first of this clade with dermal bony armor, a trait that occurs in numerous subsequent archosaurs. Of the archosaurian lineages, the crocodylotarsians radiated broadly beginning in the Middle Triassic. The ornithodirans (pterosaurs and dinosaurs), did not appear until later, with the first definite dinosaur fossils from the Triassic–Jurassic boundary. These first fossils contain representatives of three taxa, and all three were lightweight, bipedal saurischian dinosaurs, demonstrating that the saurischian–ornithischian divergence had occurred. The diversity of dinosaurs was great (Fig. 3.17). They ranged in all sizes from 1 to 25 m (TL) and had an enormous variety of shapes. They had equally varied diets and occupied a wide range of habitats. Because this diversity and their evolution are so broadly covered elsewhere, that literature is recommended to the reader.

The Crocodylotarsi includes a large number of families, most of which had a general crocodylian body form that was variously modified. The diversity of this group does not match that of the ornithodiran archosaurs. Nonetheless, nearly two dozen families and numerous species are known from the Mesozoic. Until recently, the classification emphasized levels (grades) of specialization or divergence from the basic pseudosuchian stock. These grades, such as the protosuchian (Fig. 3.18) or mesosuchian, contained multiple groups. That classification is now being replaced by monophyletic groupings. However, the new classification is not yet firmly established, in part because the fragmentary nature of some of the extinct species and genera does not permit reliable determination of relationships.

FIGURE 3.17 Jurassic scene showing typical reptiles including a *Stegosaurus* (lower left), an *Apatosaurus* (largest), the carnivorous *Saurophaganax* (bipedal), a group of *Camptosaurus* (right), and two *Archeopteryx* (flying). By Karen Carr, with permission of the Sam Noble Oklahoma Museum of Natural History.

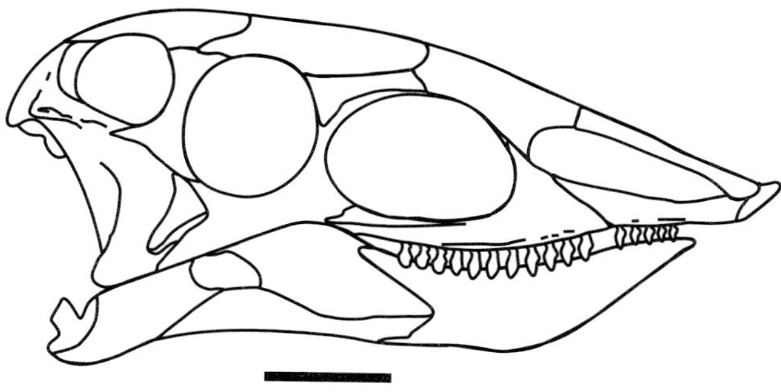

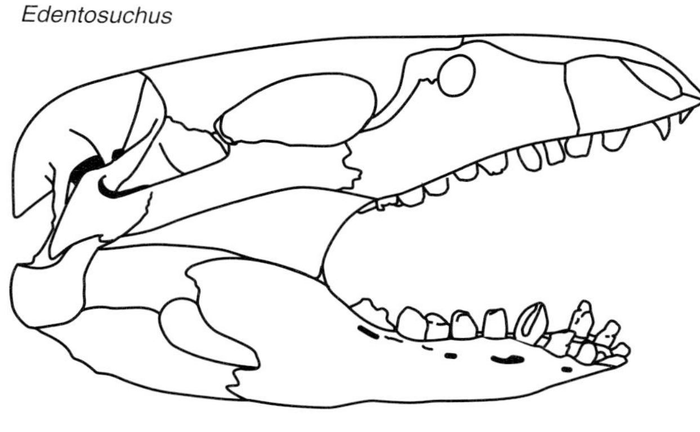

FIGURE 3.18 Cranial structure of ancient crocodilians: the aetosaur *Stegonolepis* (above) of the Upper Triassic and an unnamed *Edentosuchus*-like protosuchid of the Early Jurassic. Scale bar = 1 cm. Adapted from Walker, 1961 and Seus et al., 1994, respectively.

The phytosaurs from the Late Triassic are the most primitive crocodylotarsians and an early offshoot of the main crocodylian lineage. They were 2–4 m (TL) gharial-like animals. However, their teeth were small and remained inside the mouth when closed, and their nostrils were on a raised bony mound at the base of the long, narrow snout. The aetosaurs of the Late Triassic are another early evolutionary side branch. They had a small, piglike head (Fig. 3.18) on a heavily armor-plated crocodylian body and tail. Their small, leaf-shaped teeth suggest an herbivorous diet, which would make them the earliest herbivorous archosaurs.

Several other divergent groups appeared and disappeared in the Triassic. The main crocodylian clade, Crocodyliformes, was represented by a few subclades (e.g., teleosaurids) in the Early Jurassic, but the diversity of this group did not arise until the Late Jurassic and Early Cretaceous. The low Jurassic diversity results from the presence of only a few terrestrial and freshwater fossil deposits, the habitats in which crocodyliforms were radiating. A marine radiation of crocodyliforms is evident from the late Lower Jurassic through the Middle Cretaceous, and one group, the dyrosaurs, persisted into the mid-Tertiary. All were highly aquatic. The teleosaurids from the Early Jurassic through the Early Cretaceous were gharial-like crocodyliforms (1–9.5 m TL) of estuarine and near-shore habitats. The forelimbs of the teleosaurids were greatly reduced, and swimming probably was accomplished through the undulatory movement of the body and tail. The hindlimbs remained large and likely served as rudders. Another clade included the monstrous (>11 m TL), semiaquatic *Sarcosuchus*, an Early Cretaceous pholidosaurid. Other members of this marine radiation were metriorhynchids and *Pelagosaurus*.

The neosuchians, the lineage leading to the modern crocodylians, consist of much more than the sole surviving Crocodylia and include several Cretaceous groups, such as *Bernissartia*, a small alligator-like, molluscivorous form. The Crocodylia, the modern crocodylian clade, presumably arose in the Early Cretaceous. Members of the extant families did not appear until the Late Cretaceous, and they have been the prominent semiaquatic crocodylians since then (see the later section "History of Extant Reptiles"). A few species became terrestrial, and the pristichampsines had hooflike feet.

Extinct Lepidosauromorphs

The lepidosauromorphs are the second major diapsid lineage. The first appearance of this group occurred in the Late Permian. The Younginiformes, including *Youngina*, *Acerosodontosaurus*, and Tangsauridae, are basal members of this early radiation that survived into the Early Triassic. *Youngina* was a slender diapsid that would have been easily mistaken for many modern lizards and was likely an agile, terrestrial insectivore. The tangsaurids were similar but had laterally compressed tails and probably an aquatic lifestyle. Another group of Upper Permian–Lower Triassic lepidosauromorphs includes *Paliguana*, *Saurosternon*, and *Palaeagama*; the relationships of these eolacertilians are uncertain and debated. They were medium-sized (<20 cm TL) lizardlike diapsids.

Thereafter, the lepidosauromorphs are largely absent from the fossil record until the Late Triassic when the Sphenodontida and the kuehneosaurids appeared (Fig. 3.10). Kuehneosauridae is the sister group to the Lepidosauria (Fig. 1.13). Kuehneosaurids (Fig. 3.15) and the eolacertilians are similar in size.

The first sphenodontidan was *Brachyrhinodon taylori* from the Upper Triassic of Virginia and a likely contemporary of the first kuehneosaurid. Sphenodontidans were never an exceptionally diverse group, and most appeared much like the living tuataras, *Sphenodon*. The exception is a small group of aquatic genera, the pleurosaurines, which have elongated bodies and tails, and usually a barracuda-like head (Fig. 3.19). A sphenodontidan mini-radiation occurred from the Late Triassic to the Late Jurassic, during which this group was moderately abundant. Thereafter, the fossil presence of sphenodontidans declined through the Cretaceous, and no Tertiary forms have been found.

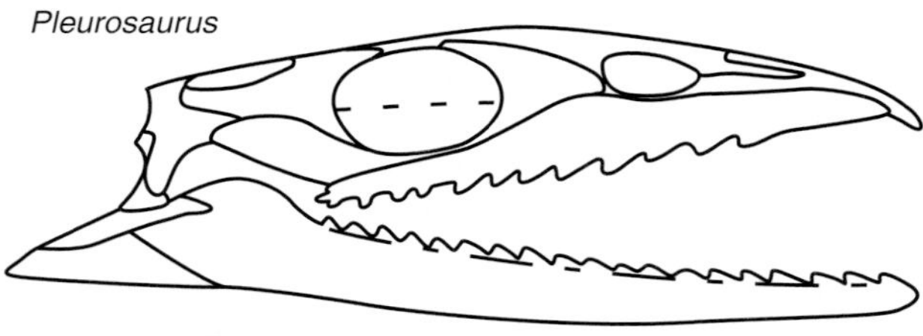

FIGURE 3.19 Cranial structure of the marine sphenodontidan *Pleurosaurus* from the Late Jurassic. Scale bar = 1 cm. Adapted from Carroll and Wild, 1994.

Lepidosaurs that are unquestionably squamates do not appear until the Middle Jurassic (Fig. 3.10). The Paramacellodidae, which are often considered scincomorphs, had a broad history from the Middle Jurassic into the Middle Cretaceous. Four other presumably more basal squamate clades, Ardeosauridae, Bavarisauridae, Dorsetisauridae, and Euposauridae, appeared in the Late Jurassic and apparently all became extinct in the Early Cretaceous. The ardeosaurids contain three genera, *Ardeosaurus*, *Eichstaettisaurus,* and *Yabeinosaurus*, which appear gecko-like in some features and have been considered gekkotans. This gekkotan relationship is now questioned. The bavarisaurids contain two genera, *Bavarisaurus* and *Palaeolacerta*, and similarly share some features with gekkotans. The other two families have been linked with extant lizard families, but these relationships also are uncertain. The euposaurids resemble agamids, but other evidence suggests that they are sphenodontidans. The dorsetisaurids resemble anguimorphs although not convincingly so. The Early Cretaceous *Scandnesia* is another basal squamate, whose affinities lie basal to the Iguania and possibly with *Eichstaettisaurus.*

Not all Upper Jurassic squamates are of uncertain affinities. *Parviraptor estesi* is a medium-sized anguimorph (ca. 15 cm SVL) and appears to be the sister group of the varanoids. The Cretaceous marine lizards (aigialosaurids, mosasaurs, and others) are strikingly similar to the varanoids, and this similarity includes a number of derived traits that are shared, suggesting a close relationship. The Necrosauridae, occurring from the Early Cretaceous to the Oligocene, also have some uniquely varanoid traits and have been proposed as a sister group of the helodermatids.

Aside from the preceding fossil representatives, the extant squamate families lack a fossil presence until the Middle Cretaceous or later. These taxa are discussed in the following section.

History of Extant Reptiles

Crocodylians

The Crocodylia, as now defined, is a clade consisting of the ancestor of extant crocodylians and all its descendants. Members of this clade, vernacularly the crocodylians, appeared first in the Late Cretaceous, although no members of the extant families occur in the fossil record until the Tertiary (Fig. 3.10). The older and broader definition of Crocodylia included protosuchians, eusuchians, and other groups and extends the history into the Lower Jurassic. A few members of these older clades survived into the mid-Tertiary; however, the Tertiary belongs to the crocodylians. The higher clades (gavialoids, alligatoroids, and crocodyloids) include many fossil taxa, and these reveal a Cretaceous divergence of gaviaoloids from the other crocodylians (Table 3.4).

TABLE 3.4 A Hierarchical Classification of the Extant Crocodilians (Crocodylia).

Reptilia
- Diapsida
  - Archosauria
    - Crocodylotarsi
      - Crocodyliformes
        - Crocodylia
          - Gavialoidea
            - Gavialidae
          - Brevirostres
            - Alligatoroidea
              - Alligatoridae
            - Crocodyloidea
              - Crocodylidae

Note: This classification derives from the phylogenetic relationships proposed in Brochu (1997*a,b*, 2001, 2004).

Gavialis has only a Pliocene and Recent occurrence. Extinct gharial or gavialoid fossils occur in the Late Cretaceous and were geographically widespread. Taxa occurred in North America (Cretaceous to Pliocene), South America (Oligocene to Pliocene), Europe (Oligocene to Miocene), Australia (Pliocene), and southern Asia (Eocene to Recent). All had the long, narrow snout associated with a specialized diet of fish. Most extinct gharial species equaled the size of the living species, but a Pliocene *Gavialis* from India apparently reached total lengths of 15–18 m.

The clade containing *Borealosuchus* and the pristichampsines are sister groups to the alligatoroid–crocodyloid clade, and both likely arose in Late Cretaceous. *Borealosuchus* was broad-snouted and alligator-like. It appeared at the end of the Cretaceous and survived into the Paleocene of North America and Europe. The pristichampsines must also have arisen in the Cretaceous; however, they appeared only briefly in the Middle Eocene of Europe. They were peculiar crocodylians with heavy dorsal and lateral armor and hooflike terminal phalanges.

The earliest alligatoroid and crocodyloid fossils are also Late Cretaceous. The Cretaceous alligatoroids include *Brachychampsa* and *Stangerochampsa*. Several other lineages arose and disappeared in the Early Tertiary. The alligatorines appeared first in the Early Oligocene, although the group certainly arose much earlier because the caimans were present in the Early Tertiary, represented by *Eocaiman* from the Middle Paleocene to Middle Miocene and the nettosuchids from the mid-Eocene to the Pliocene of South America. The nettosuchids had a unique jaw articulation and typically a broad, elongate snout. Their ducklike snout suggests a mud-noodling behavior for buried prey. *Melanosuchus* and *Caiman* appear only in the Neotropic Late Miocene and Pleistocene, respectively. In contrast, *Alligator* ranges from the Early Oligocene to the present in North America and Asia.

Crocodyloids similarly had a moderate diversity in the Late Cretaceous and Early Tertiary. The crocodylids first appeared in the lowest Eocene. The tomostomines occurred in the Middle Eocene of Egypt and China, then intermittently in northern Africa and Europe from the Oligocene to the Middle Miocene and then not again until the Late Pliocene in Asia. All shared the narrow, elongate skull. The crocodylines include a variety of lineages of which the "true" *Crocodylus* is of only recent origin from the Pliocene to the present. The Australian–New Caledonian Tertiary crocodylids appear to represent a separate evolutionary stock, the mekosuchines, that likely were displaced in the Pleistocene by the arrival of *Crocodylus* from Asia. The mekosuchines had a variety of body and head forms, ranging from narrow elongate skulls like gharials to short, broad-headed species. *Quinkana* was pristichampsine-like in having hooflike terminal phalanges. *Mekosuchus* survived into the Recent era in New Caledonia and apparently was hunted to extinction by the first humans to arrive there.

Turtles

Turtles have a good fossil record. Their bony shells are durable structures—in life and in death. The history of turtles extends back nearly 220–210 mybp to the Late Triassic where the most primitive turtle, *Proganochelys,* occurred. *Proganochelys quenstedti* was unquestionably a turtle (Fig. 3.20). Osteoderms were present and the axial skeleton was modified into a true shell. The ribs and vertebrae were fused to dermal bones to form a carapace, and some pectoral girdle elements and dermal bones fused to form a plastron. *P. quenstedti* also had a number of early amniote characteristics that were lost in later turtles. Teeth were present on the palatines but absent from the upper and lower jaws. It had a large carapace with a length of 90 cm (CL), and it was a semiaquatic turtle, well protected by its bony shell and bony neck spines (Fig. 3.20). *P. quenstedti* is not a "transitional" turtle. Rather, it is a member of the Casichelydia, which is the sister-group to the pleurodire–cryptodire clade within Testudines.

A pleurodire, *Proterochersis*, was contemporaneous and sympatric in Europe with *Proganochelys*. It was somewhat smaller (ca. 50 cm CL) and likely terrestrial.

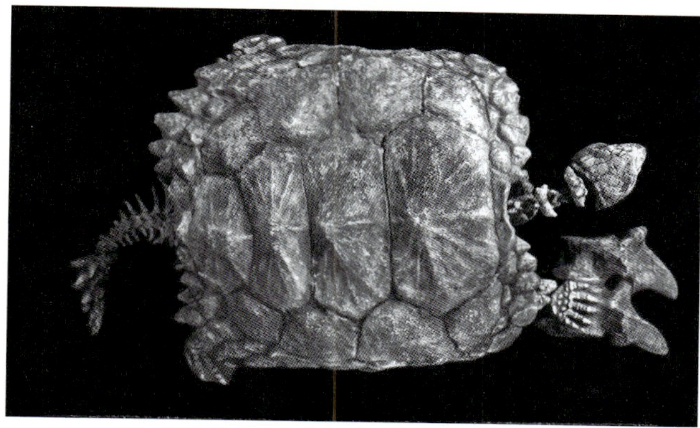

FIGURE 3.20 *Proganochelys quenstedti*, the most ancient turtle, from the Lower Triassic of Germany; approximately 15 cm CL. From Gaffney, 1990; courtesy of the American Museum of Natural History.

The pelvic girdle was fused to the plastron, indicating that it was the earliest pleurodire and confirming that the divergence of cryptodires and pleurodires had occurred. Two other contemporaries are *Australochelys* from the Late Triassic–Early Jurassic of Africa and South America and *Paleochersis* from the Late Triassic of Africa and South America. All subsequent fossil turtles are either cryptodires or pleurodires.

After *Proterochersis*, pleurodires are absent until the brief appearance of *Platychelys* in the Late Jurassic. Pleurodires do not occur again until the Early Cretaceous, but from then to the present, they are represented in many fossil faunas, particularly those of the Southern Hemisphere. Although now confined to the southern continents, a few pleurodires occurred in the Northern Hemisphere at least through the Miocene. Some Tertiary pleurodires were marine or estuarine and reached the size of modern seaturtles, although they did not develop the morphology and locomotor mode of the cryptodiran seaturtles. Chelids do not appear until the Oligocene or Miocene and only in South America and Australia. In contrast, the fossil history of the extant pelomedusoids begins in the Early Cretaceous. Pelomedusid sidenecks occur first in the Late Cretaceous with all subsequent fossils confined to Africa. Podocnemids had a much broader distribution in Africa (Late Cretaceous to Eocene), southern Asia (Late Cretaceous to Pliocene–Pleistocene), Europe (Eocene), and South America (Late Cretaceous onward) (Table 3.5).

The oldest turtle in North America and the first cryptodire is *Kayentachelys aprix,* from the late Early Jurassic (185 mybp) of western North America. It was a moderate-sized (30 cm CL), semiterrestrial turtle. Structurally, *K. aprix* was a cryptodire, although it had a number of features not seen in modern turtles, such as small teeth on the roof of the mouth. Thereafter, fossil cryptodires are absent until the appearance of the Pleisochelyidae and Pleurosternidae in the lower Late Jurassic; subsequently, cryptodires

TABLE 3.5 A Hierarchical Classification of the Extant Turtles (Testudines).

- Testudines
  - Pleurodira
    - Chelidae
    - Pelomedusoides
      - Pelomedusidae
      - Podocnemididae
  - Cryptodira
    - Trionychoidea
      - Carettochelyidae
      - Trionychidae
    - Unnamed clade
      - Kinosternoidae
        - Dermatemydidae
        - Kinosternidae
      - Chelydridae
      - Chelonioidea
        - Cheloniidae
        - Dermochelyidae
      - Testudinoidea
        - Emydidae
        - Platysternidae
        - Testudinoidae
          - Bataguridae
          - Testudinidae

Note: This classification derives from the phylogenetic relationships shown in Fig. 18.1.

remained part of the reptilian fauna. Both fossil families contained moderate-sized, aquatic turtles, and neither is related to any of the later-appearing turtle groups. Pleurosternids are the sister group to all subsequent cryptodires. The pleisochelyids are structurally more advanced turtles and the sister group to the meiolaniids and all extant groups of cryptodires. In origin, the baenoids likely arose between the pleurosternids and pleisochelyids; however, the first fossil baenids did not appear until the Middle Cretaceous and persisted into the mid-Tertiary. These heavy-shelled, moderate-sized turtles were strictly North American and probably aquatic to semiaquatic.

Extant or recently extinct clades of cryptodires began to appear in the Cretaceous. The meiolaniids arose prior to the origin of the chelydrids, yet neither has the temporal depth of the chelonioids, which appeared early in the Lower Cretaceous. The meiolaniids, or horned tortoises, do not occur in the fossil record (Australia and South America) until the Eocene and probably survived into prehistoric times. Most were large (1 m CL), high dome-shelled species. They had large heads with a bizarre arrangement of horns or spines projecting from the posterior margin of the skull. The first fossil of chelydrids (*Chelydropsis*) occurred in the Oligocene and the first snapping turtles, *Chelydra* and *Macrochelys*, in the Miocene.

The oldest known seaturtle, *Santanachelys gaffneyi*, occurred in the Middle Cretaceous (ca. 112 mybp). *S. gaffneyi* was a large (1.5 m CL), protostegid seaturtle. It and other protostegids had all of the typical features that are seen in extant seaturtles, such as streamlined shells and forelimb flippers. They are the sister group to the extant leatherback seaturtles but probably did have keratinous scutes on their shells. The leatherbacks (Dermochelyidae) did not appear until the Eocene and thereafter experienced a modest radiation of several genera and a dozen species. The other group of chelonioids includes the typical hard-shelled seaturtles, which, depending on whose opinion is followed, include the toxochelyids, osteopygids, and cheloniids or just the cheloniids including all the preceding as subfamilies. The toxochelyids and osteopygids appeared near the end of the Cretaceous. Toxochelyids did not survive into the Tertiary, and osteopygids persisted into the Oligocene. The extant cheloniid genera likely arose in the Late Miocene, although fossils identified as *Chelonia* and *Caretta* have been reported from Eocene and Oligocene sediments.

The trionychoids and testudinoids also occur in the Cretaceous, and both are represented by extant genera (Table 3.5). Fossil geomydid–testudinoids might be incorrectly identified, thereby shifting the first appearance of the testudinoids to the Eocene. The modern genera of these turtles began to appear in the Miocene, concurrently with the disappearance of the Early Tertiary genera, although a few of the latter remained into the Pliocene.

Lepidosaurs

Sphenodon guentheri and *S. punctatus* are the only surviving members of an old (220+ million years), conservative lineage (Fig. 3.10). Although this clade extends deep in time, *Sphenodon* has no fossil presence beyond subrecent records, and with few fragmentary exceptions, the sphenodontidans disappear from the fossil record after the Late Cretaceous.

In contrast, the geological history of the extant squamate families and near relatives begins in the Late Jurassic (ca. 150 mybp), and squamate diversity is evident in the Late Cretaceous (ca. 70–65 mybp; Fig. 3.10; Tables 3.6 and 3.7). Even though the assignment of Middle Jurassic squamates to modern taxa is debated, the numerous groups of Cretaceous squamates and the structural similarity of Jurassic squamates to them argue for a mid-Mesozic or earlier radiation. The chronology of first geological occurrence appears in Table 3.8.

The broader hierarchical groupings, such as Iguania, Scleroglossa, and Anguimorpha, have earlier occurrences than the families previously listed because these clades include species of uncertain familial assignment and those of extinct families. Few Cretaceous squamates represent

TABLE 3.6 A Hierarchical Classification of the Extant Lepidosauria, Exclusive of Snakes.

- Lepidosauria
  - Sphenodontida
    - Sphenodontidae
  - Squamata
    - Iguania
      - Iguanidae
      - Acrodonta
        - Agamidae
        - Chamaeleonidae
    - Scleroglossa
      - Nyctisaura
        - Xantusiidae
        - Unnamed clade
          - Gekkota
            - Gekkonidae
          - Annulata
            - Dibamidae
            - Amphisbaenia
              - Amphisbaenidae
              - Bipedidae
              - Rhineuridae
              - Trogonophidae
      - Autarchoglossa
        - Lacertiformes
          - Lacertidae
          - Teioidea
            - Teiidae
            - Gymnophthalmidae
        - Diploglossa
          - Cordylidae
          - Scincidae
          - Anguimorpha
            - Anguidae
            - Unnamed clade
              - Xenosauridae
              - Varanoidea
                - Helodermatidae
              - Thecoglossa
                - Varanidae
                - Pythonomorpha
                  - Mosasaurs
                  - Serpentes

Note: The squamate classification derives from the phylogenetic relationships depicted in Fig. 20.2. Squamate phylogeny remains a challenge, and other interpretations abound (see Chapter 20).

TABLE 3.7 A Hierarchical Classification of the Extant Snakes.

- Serpentes
  - Scolecophidia
    - Leptotyphlopidae
    - Unnamed clade
      - Anomalepididae
      - Typhlopidae
  - Alethinophidia
    - Unnamed clade
      - Tropidophiidae
      - Aniliidae
    - Unnamed clade
      - Unnamed clade
        - Cylindrophiidae
        - Uropeltidae
      - Macrostomata
        - Henophidia
          - Bolyeriidae
          - Unnamed clade
            - Unnamed clade
              - Pythonidae
              - Unnamed clade
                - Loxocemidae
                - Xenopeltidae
            - Unnamed clade
              - Calabariidae
              - Unnamed clade
                - "Boiidae"
                - Erycidae
        - Caenophidia
          - Acrochordidae
          - Colubroidea
            - Viperidae
            - Unnamed clade
              - Colubridae
              - Unnamed clade
                - Atractaspididae
                - Elapidae

Note: This classification derives from the phylogenetic relationships shown in the right-hand graphic in Fig. 21.1. This interpretation is different from that in the Second Edition and is depicted only to show how phylogenetic rankings are constructed. Squamate phylogeny remains a challenge (see Chapter 21). Category titles are not assigned to the hierarchical ranks. A name in quotation marks indicates that the group may not be monophyletic.

recent genera and species, or even subfamilies, and as noted in the section on "Extinct Lepidosauromorphs," most Jurassic and Lower Cretaceous squamates have debatable relationships to modern families and higher clades. This situation is highlighted by the presumed Cretaceous iguanid *Pristiguana* and the agamid *Priscagama*, both of which are likely correctly placed, but their primitive or generalized nature has made even an Iguania assignment suspect. Furthermore, based on fossil and morphological data, Iguania is cladistically basal to Gekkota and Anguimorpha. Nevertheless, gekkotans and anguimorphs occur earlier in the fossil record, which is consistent with the hypothesis that the Iguania are derived within squamates (Fig. 20.2; see Chapter 20).

The transition from the Cretaceous squamate fauna to a modern one begins in early Tertiary with a mix of extant and extinct genera and a few extinct subfamilies or families. Extant genera become prominent in the Miocene,

TABLE 3.8 The Chronology of First Geological Occurrence for Squamates.

Family	Period/Epoch	Time of Occurrence
Gekkonidae	Middle Cretaceous	112–100 mybp
Iguanidae	Middle Cretaceous	98–94 mybp
Agamidae	Late Cretaceous	98–94 mybp
Anguidae	Late Cretaceous	98–94 mybp
Xenosauridae	Middle Cretaceous	98–94 mybp
Helodermatidae	Late Cretaceous	98–94 mybp
Varanidae	Late Cretaceous	98–94 mybp
Aniliidae	Cretaceous	98–94 mybp
Scincidae	Late Cretaceous	88–84 mybp
Teiidae	Late Cretaceous	82–72 mybp
Xantusiidae	Middle Paleocene	62–60 mybp
Amphisbaenidae	Late Paleocene	56–54 mybp
Rhineuridae	Early Eocene	52–50 mybp
Boidae	Early Eocene	52–48 mybp
Trophidophiidae	Eocene	52–40 mybp
Typhlopidae	Eocene	50–45 mybp
Lacertidae	Eocene	45–40 mybp
Cordylidae	Oligocene	36–34 mybp
Colubridae	Oligocene	35–30 mybp
Elapidae	Early Miocene	24–20 mybp
Chamaeleonidae	Middle Miocene	20–15 mybp
Viperidae	Middle Miocene	20–15 mybp
Acrochordidae	Miocene	20–10 mybp

although extinct ones were still numerous. By the Pliocene, modern squamate genera and even a few extant species compose more than 90% of the fauna. Nonetheless, a few ancient taxa lingered into the latest Tertiary or Quaternary. A spectacular example is the Australian varanid *Megalania*, a huge goanna. Its average size was about 1.5–1.6 m (SVL), but some individuals reached total lengths of nearly 7 m (4–4.5 m SVL). These giants, probably weighing more than 600 kg, must have been formidable predators, equivalent to lions or tigers.

The earliest presumed iguanian is represented by a dorsal skull fragment from the Middle Cretaceous of Central Asia. Even though it appears unquestionably iguanian, the fossil lacks characteristics for familial assignment. Fossils from the Late Cretaceous sites in the Gobi Desert and western North America represent four or more genera of Iguanidae and the same for the Agamidae. These iguanids appear most similar to modern crotaphytines. Iguanids occur subsequently in most Tertiary periods, with the first definite iguanine, *Armandisaurus*, from the Lower Miocene of New Mexico, although the *Aciprion* fragment from the Late Eocene may also be an iguanine. While the precise status of *Pristiguana* from the Brazilian Cretaceous remains unclear, *Priscagama* and others, such as *Mimeosaurus* and *Flaviagama*, are certainly agamids. Agamids also appear regularly, if not abundantly, in most Tertiary periods. Leiolepidines appear in Early Eocene deposits in Central Asia, and Australian Miocene deposits contain both extant and extinct agamid genera. Chamaeleonids are known from the European and African Miocene and questionably from the Chinese Paleocene.

Hoburogecko from the Middle Cretaceous of Mongolia is the first gekkotan. Gekkotans are not abundant or frequent as fossils. Furthermore the assignment of pre-Pliocene fossil gekkotans to the currently recognized subfamilies is difficult. Their presumed sister group, the Annulata, has a much older and more extensive record.

Xantusiidae are the proposed sister group to the Gekkota–Annulata clade. Xantusiids appeared in the Middle Paleocene as the primitive *Palaeoxantusia,* which persisted into the Eocene. Modern *Xantusia* appeared first in the Late Miocene.

The first amphisbaenian is the Middle Cretaceous *Hodzhakulia* from Central Asia. Although represented only by maxillary and dentary fragments, these bones exhibit features that confirm their amphisbaenian identity. The more complete *Sineoamphisbaena* was found recently in a Late Cretaceous deposit of Mongolia. Its skull and the presence of forelimbs indicate that it was a primitive amphisbaenian and suggest that this taxon is the sister group to all other amphisbaenians. The next amphisbaenian was a shovel-headed form, *Oligodontosaurus*, from the Late Paleocene of western North America. Although similar to rhineurids, which appeared first in the Early Eocene of the American West, *Oligodontosaurus* had a distinct jaw structure and is placed in its own lineage. The rhineurids are abundant in the Oligocene of the American West and are remarkably similar to the single species surviving today in Florida. *Hyporhina*, another Oligocene shovel-nosed amphisbaenian from the West, represents another lineage. It is probable that these shovel-headed lineages comprise a single monophyletic group. The amphisbaenids have a fossil history beginning in the Late Eocene.

The Lacertiformes contains two families with a fossil history, although that of the Lacertidae is poor with a spotty history from the Eocene onward. In contrast, the Teiidae have a longer history. Fossil teiines and tupinambines occurred first in the Late Cretaceous, and concurrently with the polyglyphanodontines. The latter are structurally similar to extant *Dicrodon* and *Teius*, although more primitive in some features. The polyglyphanodontines were moderately diverse and abundant and occurred in western North America and Central Asia. In spite of their abundance, they disappeared after the Cretaceous.

The "scincomorphs," the cordylids and scincids, have a modest fossil history. Cordylids have an uncertain occurrence in the Late Cretaceous of western North America. These Cretaceous fossils are inadequate for taxonomic designation, although they have enough traits to indicate that they most likely are cordylids. The next cordylid occurrences were in the Oligocene of France and the Miocene of Kenya. A fossil jaw from the Lower Cretaceous of Spain has been identified as a scincid, although definite scincid fossils are confirmed only from Late Cretaceous assemblages of western North America. Scincids did not appear again until the Oligocene in North America and the Miocene in Asia and Australia.

The anguimorphs, represented by *Parviraptor estesi*, occurred in the Upper Jurassic, but the first anguid was the Late Cretaceous glyptosaurine *Odaxosaurus* from the American West. Glyptosaurines were heavy-bodied, broad-headed lizards with an armor of tubercular sculptured osteoderms covering the head and body (Fig. 3.21). This group, common through the early Tertiary of Eurasia and North America, disappeared in the Middle Miocene. The anguines appeared first in the Middle Eocene of the Northern Hemisphere.

The Xenosauridae, another basal group of anguimorphs, is the sister group to the varanoids. Xenosaurids occurred in the Middle Cretaceous of Central Asia and in the Late Cretaceous and the Upper Paleocene to Lower Eocene of western North America. Thereafter, they disappeared from the fossil record and today occur as one species in China and several species in Mexico.

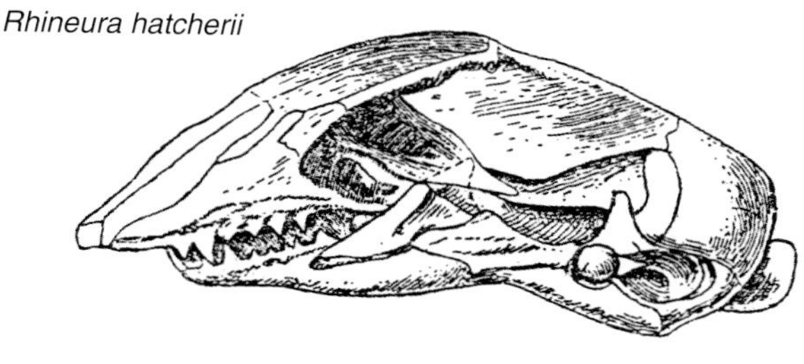

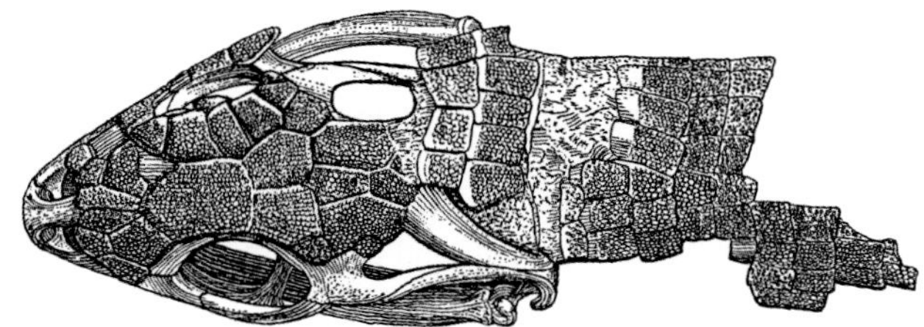

FIGURE 3.21 Skulls of two extinct taxa of North American lizards, the Middle Oligocene wormlizard *Rhineura hatcheri* (top; lateral view) and the Middle Oligocene glyptosaurine *Peltosaurus granulosus* (bottom; dorsal view). Adapted from Gilmore, 1928.

Varanoids are broadly and abundantly present in the fossil record owing to the great diversity that encompasses snakes, mosasaurs, aigailosaurs, helodermatids, necrosaurids, and varanids. The mosasaurs and aigailosaurs were briefly reviewed in the earlier section "Marine Reptiles." The Necrosauridae includes an assortment of primitive terrestrial varanoids whose history extended from the Late Cretaceous to the Eocene of North America and to the Oligocene of Eurasia. Helodermatids have a much more extensive history than their modern distribution and diversity indicate. *Paraderma bogerti* was one of two or three Upper Cretaceous beaded lizards in North America. These early helodermatids and the Mongolian *Estesia* had grooved teeth, suggesting that use of venom has a long history in these lizards. Later records of helodermatids from the Eocene and Oligocene of Europe, the Oligocene and Miocene of south-central North America, and the Pleistocene of the American southwest desert indicate that these lizards were much more widespread in the past. The earliest varanid is *Palaeosaniwa canadensis* from the Late Cretaceous of Alberta. The Mongolian Late Cretaceous also had several lizards that may be varanids. Subsequently, the varanid *Saniwa* occurred in the Late Paleocene to the Oligocene of North America and Europe, and *Iberovaranus* occurred in the Spanish Miocene. The first known *Varanus* is from the Lower Miocene of Kenya, and subsequent *Varanus* fossils occur within the distribution of the extant varanids.

Whether the snakes arose from a varanoid ancestor, deeper within anguimorphs, or even within iguanians, their fossil record dates to as early as the iguanians. The oldest known snake is represented by two vertebrae from the Early Cretaceous (127–121 mybp). Although two vertebrae might seem an inadequate base on which to recognize a snake, snake vertebrae have several unique features that easily separate them from other squamates, and yet they retain features that are typical lepidosaurian. The vertebrae alone are, however, inadequate to determine the relationship of this fossil to other snakes. *Lapparentophis defrennei* from the Middle Cretaceous (100–96 mybp) is known only from three trunk vertebrae. *L. defrennei* is an alethinophidian and presumably was a terrestrial snake. Two other snakes of equal antiquity, *Simoliophis* and *Pouitella*, are apparently not closely related to one another or to *Lapparentophis*, other than being primitive snakes. These three snake genera also do not seem to be related to any of the living families of snakes, and they or their descendants do not occur later in time.

The remarkable discovery of *Najash*, a "limbed" snake that appears to be the earliest limbed snake from a terrestrial deposit (others like *Pachyrhachis* were marine), significantly changes how we view limb loss in snake evolution. The fossils, from the Cenomanian–Turonian (Upper Cretaceous), about 90–95 mybp, have a sacrum that supports a pelvic girdle and functional limbs situated outside of the rib cage. Thus it differs from other "limbed" snake fossils in having the sacral elements. Reconstruction of the skull using computed tomography shows that *Najash* is clearly a snake (Fig. 3.22). Phylogenetic analysis place *Najash* as sister to all known snakes (Fig. 3.23), and based on skull and other characteristics, *Najash* was likely a terrestrial/subterranean species. Consequently, snakes likely arose from a terrestrial/subterranean ancestor rather than a marine ancestor. The marine hypothesis for the origin of snakes has been based largely on *Pachyrhachis*. *Pachyrhachis* was recognized as a peculiar long-bodied varanoid of the Middle Cretaceous. It had small limbs and was apparently marine, but aspects of its skull and vertebrae were snakelike. The initial discoverer proposed that it was a mosasaur or relative of a mosasaur. However, when reexamined, it was declared to be a limbed snake and the sister group to all subsequent snakes. Although this proposition remains controversial, recent analysis of skull morphology

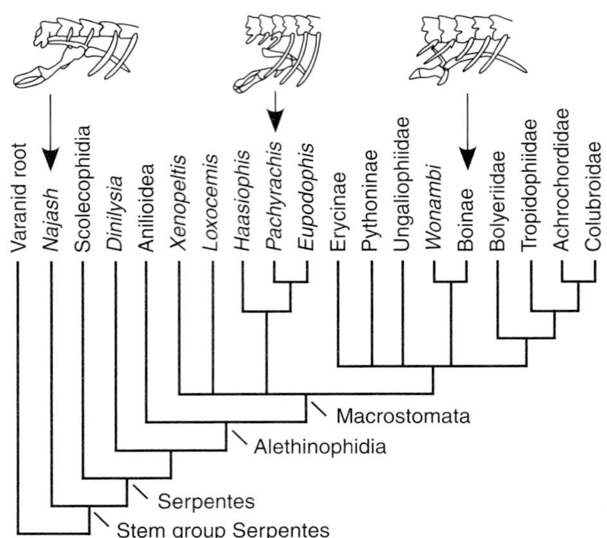

FIGURE 3.23 The newly described fossil snake *Najash* not only has bony elements of the sacrum and hindlimbs but was also terrestrial/subterranean. Combined with other skeletal features, *Najash* appears to be sister to all known snakes, suggesting that snakes had a terrestrial origin rather than a marine one. Elements of the pelvis and hindlimbs are shown for *Najash, Pachyrhachis,* and the Boinae for comparison. Adapted from Apesteguía and Zaher, 2006.

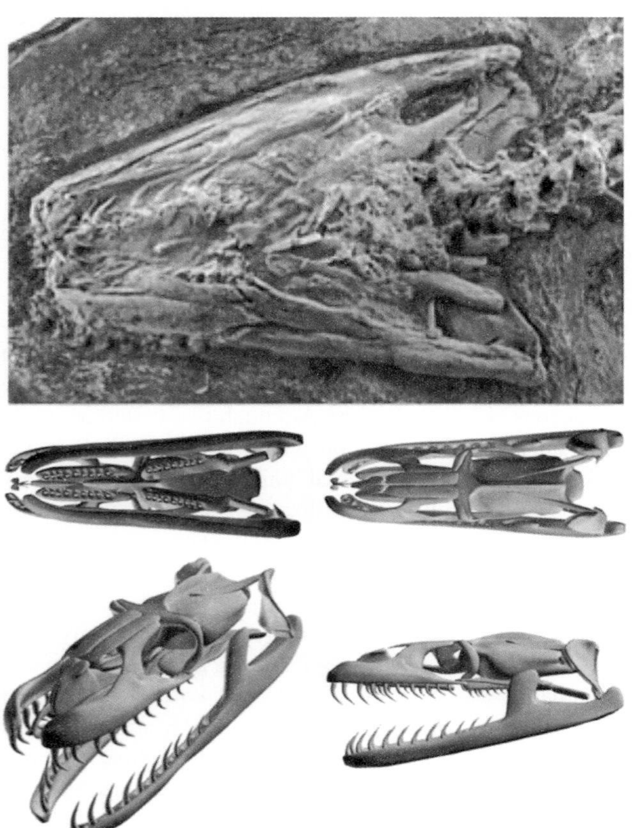

FIGURE 3.22 The structure of the head of the fossil snake *Pachyrhachis problematicus* (upper) was reconstructed using X-ray computed tomography (lower image), showing that the skull is indeed that of a basal macrostomatan snake, which means that limb loss occurred independently in different snake clades. Adapted from Polcyn et al., 2005.

place it within macrostomatan snakes, which indicates first that it is not sister to all other snakes, and second, when combined with other data, suggests that limb loss must have evolved several times within snakes.

Other snakes appeared in the Late Cretaceous. One of these, *Coniophis,* was initially considered an aniliid; however, it might be a boid. *Gigantophis* and *Madtsoia* were large snakes equal in size to the largest extant boids and initially considered a lineage within boids. As a group, madtsoiids are Gondwanan and occur in fossil assemblages from Australia (Early Eocene to Pleistocene), Madagascar (Cretaceous), Africa (Cretaceous to Late Eocene), and South America (Cretaceous to Early Eocene); recently, one was discovered in a Spanish Cretaceous deposit. In Australia, the madtsoiids (*Wonambi*, *Yurlunggur*, and several undescribed taxa) were a major group of snakes throughout the Tertiary. Was their disappearance linked to an increasing diversity of pythons in the Late Tertiary and Quaternary?

The unique *Dinilysia* (Dinilysiidae; Fig. 3.24) is known only from the Late Cretaceous of Patagonia. It was also a large snake, roughly equal in size and appearance to *Boa constrictor*. It is one of the rare fossil snake finds, consisting of a nearly complete skull and part of the vertebral column. In spite of the completeness of its skeleton, the relationships of *Dinilysia* remain uncertain, although it appears to be an alethinophidian.

Additional booids (a vernacular label for alethinophidian snakes that are not caenophidians) appeared in the

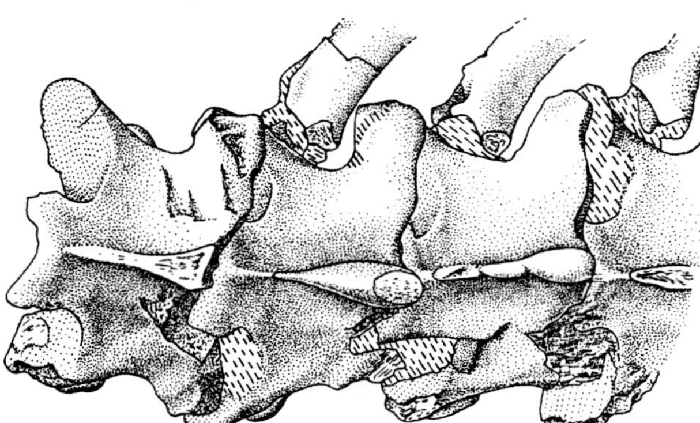

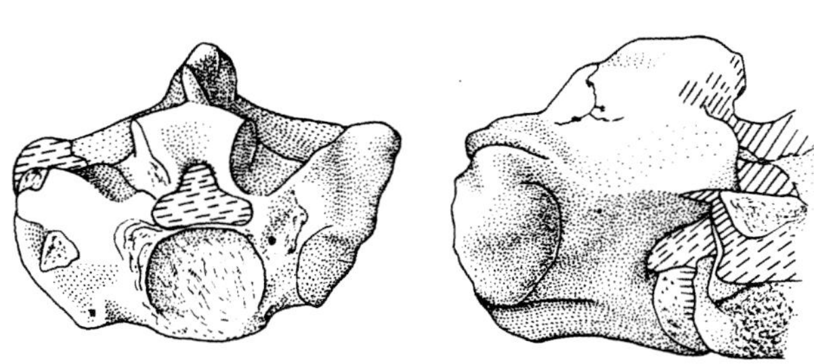

FIGURE 3.24 Trunk vertebrae from the Upper Cretaceous snake, *Dinilysia patagonica*; dorsal view of a series of four vertebrae (top), anterior view (bottom left) and lateral view (bottom right) of individual vertebrae. Adapted from Rage and Albino, 1989.

Early Tertiary and seemed to be the dominant snakes through the Eocene. Apparently climatic events caused major snake extinctions at the Eocene–Oligocene boundary. Snake diversity remained low through the Oligocene, and dominance in the snake faunas shifted to the caenophidians. Some of the booids were related to modern species. *Charina brevispondylus* from the Middle Eocene of Wyoming, for example, is the sister species of *Charina trivirgata*. A variety of boines and erycines were present in the Eocene. *Coniophis* also occurred in the Eocene of North America and Europe and was accompanied by other aniliids. Scolecophidians have an extremely poor fossil history. Only a few fossils have been found, and the earliest are from the Eocene. These fossils have been assigned tentatively to the typhlopids.

The first caenophidians appear in the Eocene and include acrochordoid and colubroid relatives. These caenophidians include extinct families and genera, none with clear affinities to modern taxa. The Oligocene presented the first colubrids, for example "*Coluber*" and *Texasophis*. Thereafter, colubrids occur with increasing frequency. Acrochordids appeared first in the Middle Miocene, but two earlier Paleocene and Eocene fossils are of a related but extinct group. The first elapid was the European *Palaeonaja* from the Early Miocene; subsequently in the Miocene, elapids occurred in Eurasia and North America. Viperids also appeared first in the Miocene. As with lizards, fossil snake faunas become increasingly modern in appearance through the Pliocene, and by the Middle Pleistocene, most snake faunas are composed solely of modern taxa.

QUESTIONS

1. Describe the early evolution of caecilians, salamanders, and frogs. What are the key fossils that tie each modern group to extinct groups? In addition, provide evidence that all three modern groups most likely are lissamphibians.
2. Describe in detail the reptile fauna of the Mesozoic. Can you speculate why the apparently diverse marine reptile fauna of the Mesozoic disappeared?
3. Gliding reptiles have evolved several times during the evolutionary history of reptiles. Describe at least three different gliding reptiles (extinct or extant) and provide evidence that each was an independent origin of gliding.
4. What is the oldest turtle fossil and why has it been so difficult to trace the origin of turtles in the fossil record?
5. What are some of the reasons for the discontinuities in the fossil records of amphibians and reptiles and how do these discontinuities affect reconstruction of the evolutionary histories of these tetrapods?

ADDITIONAL READING

Ahlberg P. E., and Milner, A. R. (1994). The origin and the early diversification of tetrapods. *Nature* **368:** 507-514.

Benton, M. J. (1997). *Vertebrate Paleontology* (2nd ed.). Chapman & Hall, London.

Carroll, R. L. (1988). *Vertebrate Paleontology and Evolution* W. H. Freeman & Co, New York.

Carroll, R. L. (2007). The Paleozoic ancestry of salamanders, frogs and caecilians. *Zoological Journal of the Linnean Society* **150** (suppl. 1), 1–140.

Farlow J. O., and Brett-Surman, M. K. (Eds.) (1997). *The Complete Dinosaur* Indiana University Press, Bloomington, IN.

REFERENCES

General

Benton, 1997a; Carroll, 1988.

Amphibians of the Late Paleozoic

Ahlberg and Milner, 1994; Milner, 1980, 1988, 1993.

Amphibians of the Early Mesozoic

Milner, 1994; Milner et al., 1986.

History of Lissamphibia: Caecilians

Estes, 1981; Evans et al., 1996b; Gao and Shubin, 2003; Jenkins and Walsh, 1993; Rage, 1986; Wilkinson and Nussbaum, 2006.

Albanerpetontids

McGowan and Evans, 1995; Milner, 1994.

Salamanders

Adler, 2003; Estes, 1981; Evans and Milner, 1994, 1996; Evans et al., 1988, 1996; Holman, 1995; Hutchinson, 1992; Rage et al., 1993; Rocek, 1994a.

Frogs

Báez, 1996; Báez and Basso, 1996; Báez and Pugener, 1998; Báez and Trueb, 1997; Duellman, 2003; Estes and Reig, 1973; Evans and Milner, 1993; Evans et al., 1990; Henrici, 1994, 1998; Holman, 1995; Hutchinson, 1992; Milner, 1994; Pledge, 1992; Poinar and Cannatella, 1987; Prasad and Rage, 1991, 1995; Rage and Rocek, 1989; Rocek, 1994b; Rocek and Lamaud, 1995; Rocek and Nessov, 1993; Sánchiz and Rocek, 1996; Shubin and Jenkins, 1995; Spinar, 1972; Tyler, 1991a,b.

Radiation Among Early Amniotes: Protomammals: The Synapsids

Hotton et al., 1986.

Paleozoic Reptiles

Benton, 1991; Carroll, 1991; Laurin and Reisz, 1995.

Age of Reptiles: Marine Reptiles

Fernández and Gasparini, 2000; Hirayama, 1997; Hua and Buffetaut, 1997; Martin and Rothschild, 1989; McGowan, 1991; Rieppel, 1993; Tarsitano and Riess, 1982.

Gliders and Fliers

Colbert, 1967, 1970; Dyke et al., 2006a,b; Frey et al., 1997; Gans et al., 1987; Stein et al., in press; Robinson, 1962.

Archosauromorphs

Benton, 1997b; Buffetaut, 1989; Parrish, 1997; Russell and Wu, 1997; Seus et al., 1994; Walker, 1961.

Extinct Lepidosauromorphs

Benton, 1994, 1997a; Durand, 2004; Estes, 1983a,b; Evans, 1993, 1994; Evans and Barbadillo, 1998; Evans and Milner, 1994; Kardong, 1997; Rieppel, 1994.

History of Extant Reptiles: Crocodylians

Brochu, 1997a, 1999, 2001, 2003, 2004; Buffetaut, 1989; Fernández and Gasparini 2000; Gasparini, 1996; Harshman et al., 2003; Markwick, 1995; Norell, 1989; Norell et al., 1994; Russell and Wu, 1997; Steel, 1973; Tchernov, 1986; Willis, 1997; Wu et al., 1996a,b.

Turtles

Fujita et al., 2004; Gaffney, 1986, 1990, 1991; Gaffney et al., 1987; Gaffney and Kitching, 1994; Gaffney and Schleich, 1994; Gardner et al., 1995; Hirayama, 1997, 1998; Joyce et al., 2004; Meylan, 1996; Meylan and Gaffney, 1989; Mlynarski, 1976; Parham and Fastovsky, 1998; Parham et al., 2006; Shaffer et al., 1997; Tong et al., 1998.

Lepidosaurs

Albino, 1996; Alifanov, 1996; Averianov and Danilov, 1996; Beck, 2005; Caldwell and Lee, 1997; Cifelli and Nydam, 1995; Clos, 1995; Denton and O'Neill, 1995; Estes, 1983a,b; Evans, 1993, 1994; Evans and Chure, 1998; Fry et al., 2005; Gao, 1997; Gao and Hou, 1995; Gao and Nessov, 1998; Greene, 1997; Holman, 1995; Kardong, 1997; Kearney and Stuart, 2004; Lee, 2005; Lee et al., 2006; Norell and de Queiroz, 1991; Parmley and Holman, 1995; Polcyn et al., 2005, 2006; Pianka and King, 2004; Pianka and Vitt, 2003; Rage, 1984, 1987, 1996; Rage and Richter, 1994; Scanlon, 1992; Sullivan and Estes, 1997; Sullivan and Holman, 1996; Wu et al., 1997a,b; Zaher, 1998.

Part II

Reproduction and Reproductive Modes

The ability of organisms to reproduce and send their genes into future generations separates living from nonliving things. Although unisexual reproduction maximizes reproductive rates (no investment in males), sexual reproduction provides the raw material on which natural selection operates: variation among individuals. Organisms cannot predict the environment of their offspring; consequently, production of numerous, slightly different offspring that results from reshuffling of genes during reproduction provides the opportunity for adaptation to changing environments. Individuals best able to survive and reproduce given the abiotic and biotic environments at the time will send the most descendents in the next generation.

Amphibians and reptiles enhance reproductive output and offspring survival in many ways. Fertilization can occur inside or outside the body of the female, and development can be direct or indirect. These and other characteristics define the *modes of reproduction*. Amphibians exhibit a spectacular diversity of reproductive modes. Their complex life history, which often includes a larval stage and radical metamorphosis, no doubt set the stage for the evolution of the great diversity of reproductive modes observed today. In reptiles, two major modes are generally recognized: oviparity, the deposition of eggs, and viviparity, the birth of fully formed individuals. However, much variation occurs within oviparous species in terms of egg retention and development prior to egg deposition. Viviparity is complex, with some species having no placentae, others having simple placentae, and yet others having placentae that rival those of eutherian mammals. Although most amphibians and reptiles reproduce sexually, some species consist entirely of females that reproduce unisexually. In some cases, females "steal" the genomes of sexual species with which they live but do not pass them on; in others, females produce identical daughters clonally, eliminating involvement of males entirely. Parental care is widespread in amphibians and reptiles, varying from attendance of eggs to protection and/or feeding of offspring.

Chapter 4

Reproduction and Life Histories

The transition from a totally aquatic life to living at least part of the time on land presented a major challenge in vertebrate evolution and led to an explosion of reproductive adaptations. Because external fertilization was the ancestral condition of the first amphibians, standing water was required for reproduction. The evolution of internal fertilization allowed some amphibians independence from standing water for breeding. Direct development (no free-living larval stage) or attendance of eggs in moist microhabitats permitted development away from water. The evolution of the amniotic egg characterized one clade of tetrapod vertebrates, the Amniota (reptiles [including birds] and mammals). Amniotic structures allow respiration and storage of nitrogenous waste within the egg, making it possible for development to occur on land in "dry," although not desiccating, egg deposition sites. These factors, among others, ultimately led to the successful and broad diversification of tetrapod vertebrates.

GAMETOGENESIS AND FERTILIZATION

In most amphibians and reptiles, a female and a male are necessary for reproduction, although some remarkable exceptions exist (see the later section "Sexual Versus Unisexual Reproduction"). Within species, reproductive activity is usually synchronous, although some interesting exceptions are known. Internal (hormonal) controls mediate reproductive timing, but ultimately reproduction is triggered directly or indirectly by environmental cues, such as temperature, rainfall, or photoperiod (Fig. 4.1). Hormonal changes cause gametogenesis, the production of the sex cells or gametes (ova in females, sperm in males), a process that is similar in all vertebrates. In addition to gamete production, the gonads produce hormones that feed back on the brain, the pituitary, and other organs, and ultimately influence the physiology and behavior of reproduction.

Gamete Structure and Production

Male gametes (spermatozoa) are produced by cells (spermatogonia) in the seminiferous tubules of the testes during spermatogenesis (Fig. 4.2). Spermatogonia undergo mitotic divisions to produce additional spermatogonia, which differentiate into primary spermatocytes. In turn, these cells undergo two meiotic divisions to produce spermatids. Differentiation of the spermatids produces spermatozoa that receive nutrition from Sertoli cells. Each spermatozoon is a highly modified cell with three sections: a head, a midpiece packed with mitochondria for the cell's energy needs, and a filamentous tail for locomotion (Fig. 4.3). The head contains the cell nucleus capped by an acrosome. The acrosome produces proteolytic enzymes that digest the egg capsule and allow the spermatozoon to penetrate into an egg. Among amphibians and reptiles, spermatozoan morphology is highly variable. Whether sperm morphology will prove to be a useful character in phylogenetic analyses remains to be seen, and attempts to correlate sperm morphology with breeding habits have even proven equivocal.

Herpetology, 3rd Ed.

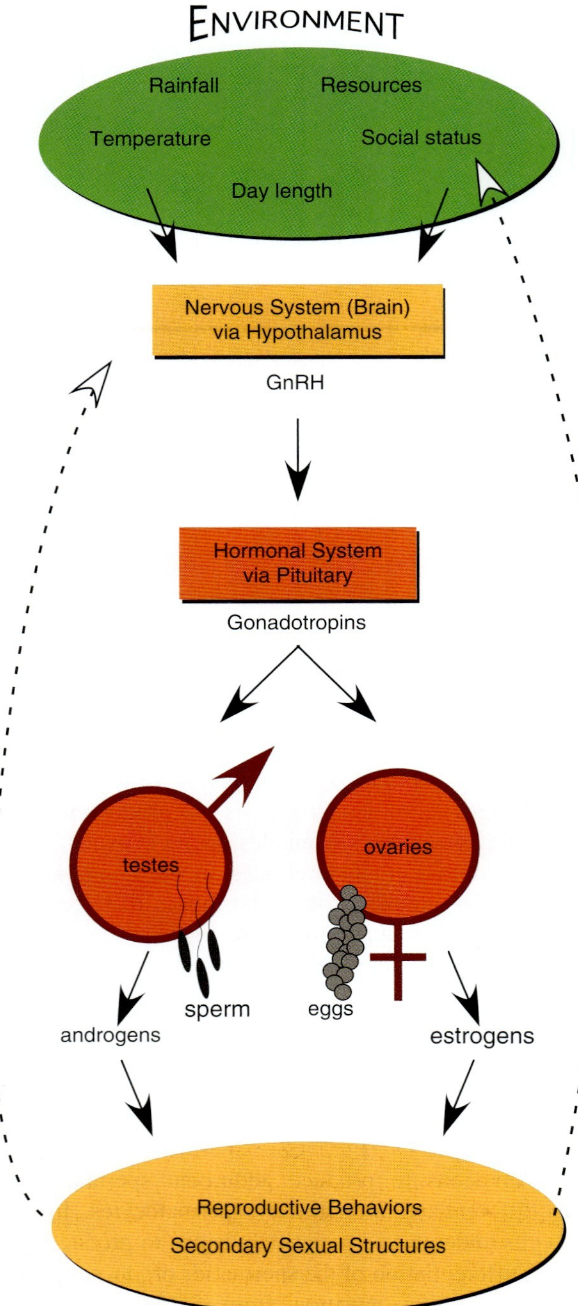

FIGURE 4.1 Sexual behavior and ultimately reproduction are mediated by interactions between environmental factors, the nervous system (brain), and the hormonal system. Gonadotropin-releasing hormone (GnRH) stimulates the pituitary to produce gonadotropins (lutenizing hormone and follicle-stimulating hormone), which, in turn, stimulate testes or ovaries to produce mature gametes and androgens. Androgens not only effect development of secondary sexual structures but also feed back on sexual behavior and the brain.

In females, the gametes or ova are produced in the ovary (Fig. 4.4). Primordial gonocytes occur in capsules of nonsex cells known as follicles, which are located in the wall of the ovary. Primordial gonocytes divide by mitosis to produce

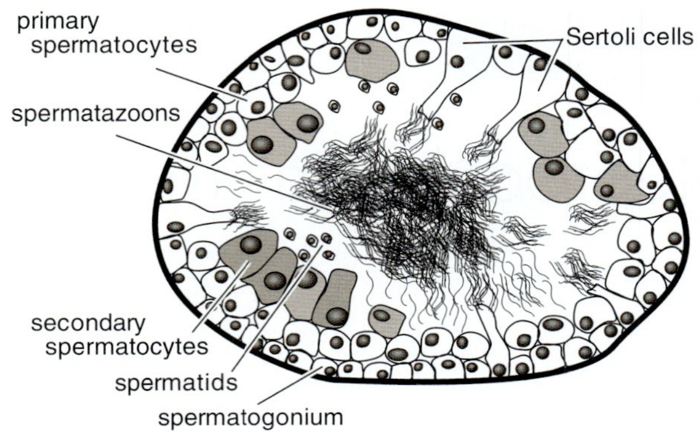

FIGURE 4.2 Spermatogenesis. Diagrammatic representation of a cross section through a seminiferous tubule in a reptile testis.

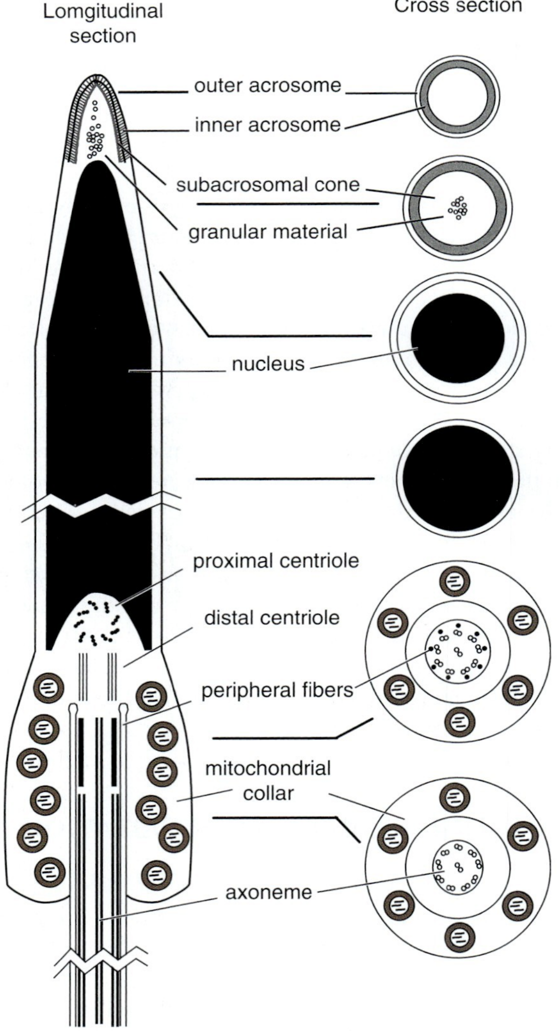

FIGURE 4.3 Structure of spermatozoan of a hylid frog. Only the base of the tail is shown, and the head of the sperm has been shortened. Redrawn from Costa et al., 2004.

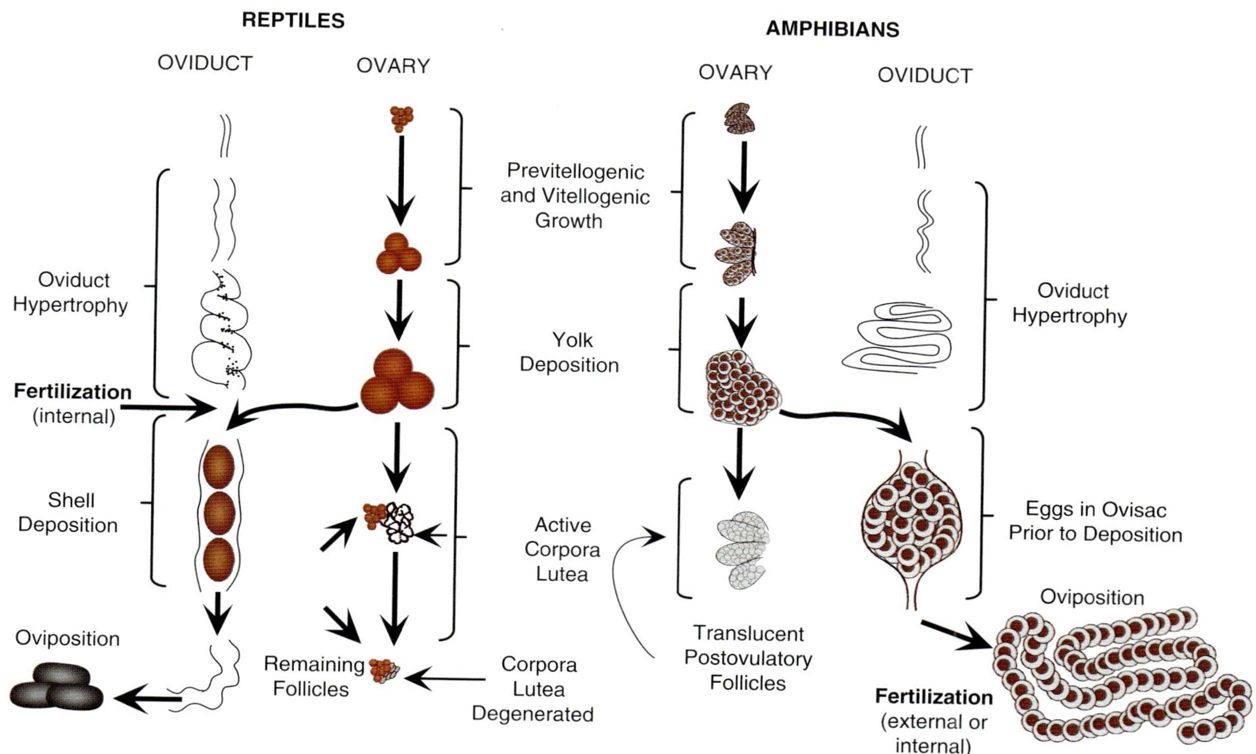

FIGURE 4.4 Development of eggs in amphibians and reptiles. Fertilization occurs internally in all reptiles after eggs are ovulated into the oviducts. Fertilization occurs externally in most amphibians. Corpora lutea are often prominent in reptiles but rare in amphibians. Following production of the clutch, the process is repeated as unused ovarian follicles mobilize lipids for production of the subsequent clutch. Subsequent clutches may be produced within the same season or in the following season, depending upon species and the environment.

oogonia (cells that will produce eggs). Oogonia undergo mitotic divisions and enlarge to produce primary oocytes, which then undergo two meiotic divisions. The first meiotic division produces a secondary oocyte and the first polar body, a nonfunctional cell; the second meiotic division produces the ovum and a secondary polar body. Each oogonium thus yields only one ovum.

Nutrients accumulate in the cytoplasm of the ovum by a process known as vitellogenesis. Vitellogenin is a precursor of yolk proteins and is synthesized in the liver in amphibians. It is transported in the bloodstream to the ovary, where it is sequestered by growing oocytes. In the oocytes, it is cleaved into the yolk proteins lipovitellin and phosvitin. These compounds are stored as yolk in the ovum until needed during embryogenesis.

At metamorphosis, the number of nonvitellogenic oocytes in the ovary of a female amphibian increases rapidly. Evidence from studies on the toad *Bufo bufo* indicates that the total number of oocytes to be used during the lifetime of the female is reached early in the juvenile stage. *Bufo bufo* can produce from 30,000 to 40,000 oocytes during this time. Species producing smaller clutches of eggs have fewer nonvitellogenic oocytes.

Vitellogenic growth and maintenance of small oocytes are initiated by the hormone gonadotropin and signal the beginning of an ovarian cycle. In mature amphibians, the ovaries contain a set of small, nonvitellogenic oocytes that are not responsive to gonadotropin, and a set of larger oocytes that are responsive to gonadotropin. Apparently, once vitellogenesis begins for one set of oocytes, intraovarian regulatory mechanisms prevent additional small oocytes from responding to gonadotropin.

In all amphibians, fat bodies are discrete structures located adjacent to the gonads. The complex relationship between the gonads and fat bodies has been debated for many years. Experimental evidence regarding the role of fat bodies is contradictory. In many species, fat bodies are large in juvenile females and in those females with ovaries undergoing vitellogenesis. Other species, however, show no correlation between fat body size and the ovarian cycle. Lipids are stored in other organs, including the liver and gonads, as well as fat bodies. For example, in newly metamorphosed *Ambystoma opacum*, 36% of lipids were stored in fat bodies, compared to 17% in *Ambystoma talpoideum*. Increased lipid levels may increase survivorship of these salamanders when they enter the terrestrial environment after metamorphosis.

In reptiles, vitellogenin is selectively absorbed during a process called *pinocytosis* by oocytes and enzymatically converted to the yolk platelet proteins lipovitellin and phosvitin. The first phase of vitellogenesis is usually slow, with little observable growth in the ova. During the last phase of vitellogenesis, ovum growth is rapid. Prior to ovulation (release of ova from ovaries), a mature ovum is 10–100 times its original size.

In both amphibians and reptiles, ovulation occurs when the follicular and ovarian walls rupture, releasing ova into the body cavity where they migrate into the infundibulum of each oviduct. In most amphibians, the postovulatory follicles exist only for a short time and do not secrete hormones. In viviparous amphibians and reptiles, walls of the follicle transform into corpora lutea (Fig. 4.5). The corpora lutea produce progesterone, and this hormone prevents the expulsion of developing embryos.

As ova pass through the oviduct, protective membranes are deposited around them. In amphibians, the ovum is already enclosed in a vitelline membrane that was produced by the ovary. Each ovum is coated with layers of mucoproteins and mucopolysaccharides as it moves through the oviduct. The number of layers or capsules around the ovum is species specific. In some salamanders, as many as eight capsules surround the ovum. Anurans typically have fewer capsules than salamanders. Amphibian eggs are anamniotic because they lack the extraembryonic membranes characteristic of reptiles and mammals. In both viviparous and oviparous reptiles, three extraembryonic membranes, the allantois, amnion, and chorion, develop during embryogenesis (Fig. 4.6; also see Chapter 2). The allantois serves as a respiratory surface for the developing embryo and storage sac for nitrogenous wastes.

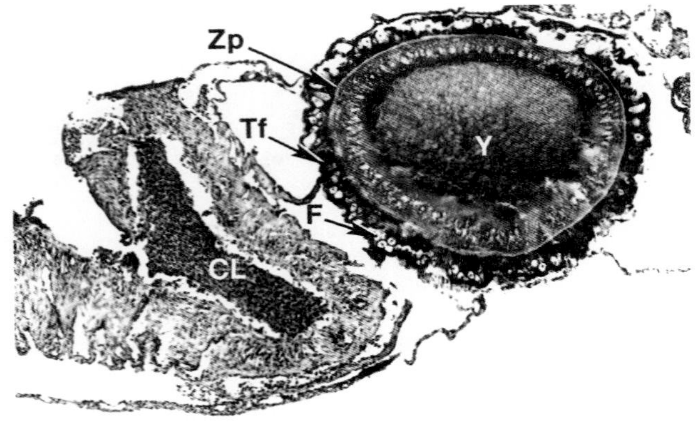

FIGURE 4.5 Oogenesis. Cross section through the ovary of the skink *Carlia bicarinata*, showing a corpus luteum (left) and a maturing follicle (right) with its ovum. Abbreviations: CL, corpus luteum; F, follicular cells; Tf, theca folliculi; Y, yolk; Zp, zona pellucida. (D. Schmidt)

In egg-laying reptiles, the ovum is ultimately encased in a durable and resistant shell. While in the upper portion of the oviduct, the ovum is sequentially coated with albumin and several thin layers of protein fibers. The fiber layer is impregnated with calcite crystals in crocodilians and squamates, and argonite crystals in turtles. Shortly after ovulation and fertilization (12 hours or less in *Sceloporus woodi*), endometrial glands in the oviduct produce the proteinaceous fibers that constitute the support structure of the eggshell (Fig. 4.7). Shell structure varies considerably among species of oviparous reptiles, but all shells provide some protection from desiccation and entry of small organisms.

Fertilization—Transfer and Fusion of Gametes

Fertilization occurs when a spermatozoon and an ovum unite to form a diploid zygote. External fertilization occurs when this union occurs outside the bodies of the

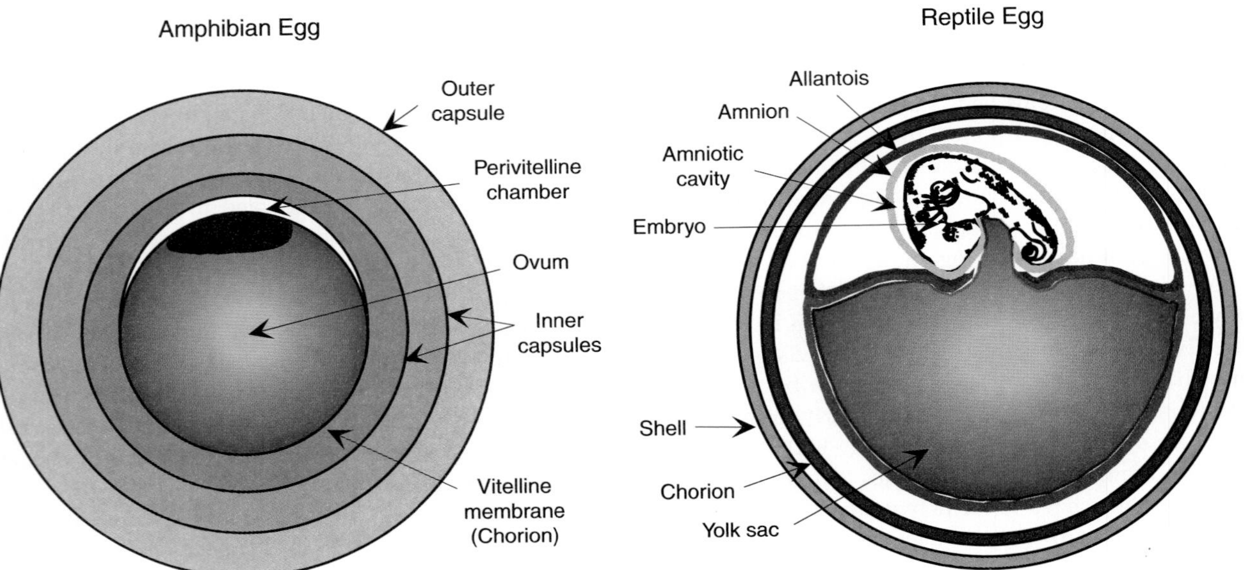

FIGURE 4.6 Comparison of anatomy of the anamniotic amphibian egg and the amniotic reptile egg.

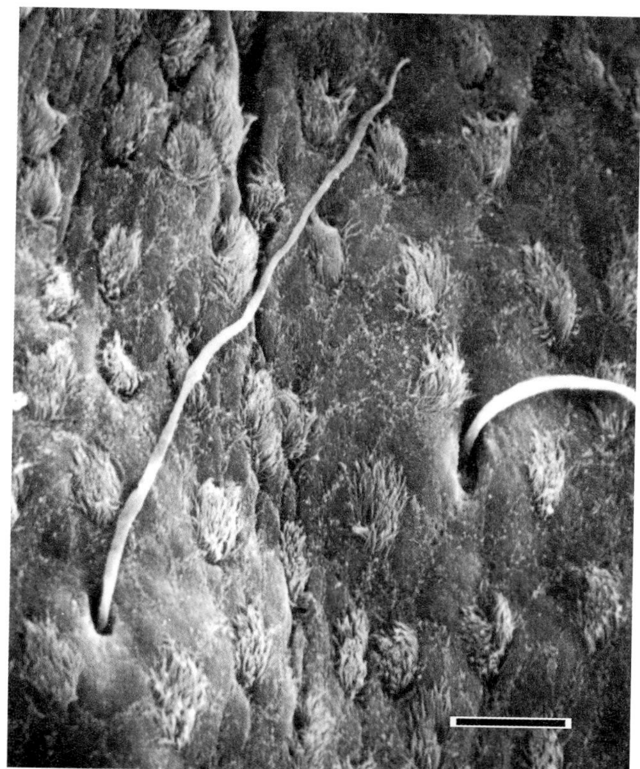

FIGURE 4.7 Wall of the oviduct of the lizard *Sceloporus woodi* during shell production. Two proteinaceous fibers are emerging from the endometrial glands of the oviduct. Scale bar = 5 μm. Adapted from Palmer et al., 1993.

male and female, and internal fertilization occurs when the union occurs within the female's body, almost always in the oviducts.

Males produce millions of tiny spermatozoa, whereas females produce relatively few eggs. Even though the eggs of some amphibians are small, they are orders of magnitude larger than spermatozoa. During mating, many sperm reach the surface of an egg but only one penetrates the cell membrane of the ovum to fertilize it. When sperm first arrive at the egg, a few adhere to the surface. Enzymes produced by the acrosome digest a tiny hole in the egg capsule, bringing the sperm head into contact with the plasma membrane. The enzymes break down receptors binding the sperm pronucleus to the surface of the egg, and the sperm pronucleus moves into the cytoplasm of the ovum. In response to the entry of the sperm pronucleus, the vitelline membrane separates and elevates, lifting all other sperm from the ovum's surface. As the successful sperm pronucleus moves to the ovum pronucleus, the ovum pronucleus completes its final meiotic division. The fusion of the two pronuclei is the final stage of fertilization and restores the diploid (2N) condition to the fertilized ovum, which is thereafter called the zygote. The zygote soon begins development via typical cell division—mitosis. Embryonic development continues in externally fertilized eggs (amphibians), but developmental arrest occurs in internally fertilized eggs (reptiles) after development to a gastrula stage. Salamanders are unusual because they have polyspermic fertilization, in which more than one sperm pronucleus enters the ovum's cytoplasm, but only one sperm pronucleus fuses with the egg pronucleus.

Reproductive Behaviors Associated with Mating

Courtship and mating behaviors vary greatly among species of amphibians and reptiles. Vocal (auditory), visual, tactile, or chemical signals used during courtship not only bring individuals together for reproductive purposes but also provide opportunities for mate choice. Reproductive behaviors are influenced by hormones (Fig. 4.1). Males, but not always females, have mature gametes when mating occurs. In females of some species, sperm can be stored and used to fertilize eggs long after mating.

Sperm are transferred to females in a variety of ways. In most frogs and cryptobranchoid salamanders, external fertilization occurs; the male releases sperm on the eggs as they exit from the female's cloaca. In most frogs, the male grasps the female so that his cloaca is positioned just above the female's cloaca. This behavior is called amplexus, and the exact positioning of the male with respect to the female varies among species (Fig. 4.8). In the two frogs that have internal fertilization (see following text), the mating behavior is termed coplexus. In salamanders with external fertilization, amplexus can occur, or the male can follow the female and deposit his sperm directly on the egg mass during or after deposition.

Relatively few frogs, including the two species of *Ascaphus*, possibly some of the 14 species of the bufonid genus *Mertensophryne*, presumably all of the 13 species of the bufonid genus *Nectophrynoides*, 1 species in the bufonid genus *Altiphrynoides*, 2 species of the bufonid genus *Nimbaphrynoides*, and 2 species of *Eleutherodactylus*, all diadectosalamandroid salamanders, and all reptiles have internal fertilization. Internal fertilization usually requires morphological structures to deliver sperm, and complex mating rituals often are found in these species. All frogs with internal fertilization except *Ascaphus* and *Mertensophryne* use cloacal apposition to transfer sperm. Although the tuatara *Sphenodon* has rudimentary hemipenes, cloacal apposition is used to transfer sperm. Males of other reptiles, the frogs *Ascaphus*, and caecilians have intromittent organs that deposit sperm into the cloaca adjacent to the oviductal openings. The intromittent organ in *Ascaphus* is modified from the cloaca; vascularization of the tissue permits engorgement of the organ with blood, facilitating deposition of sperm into the female's cloaca.

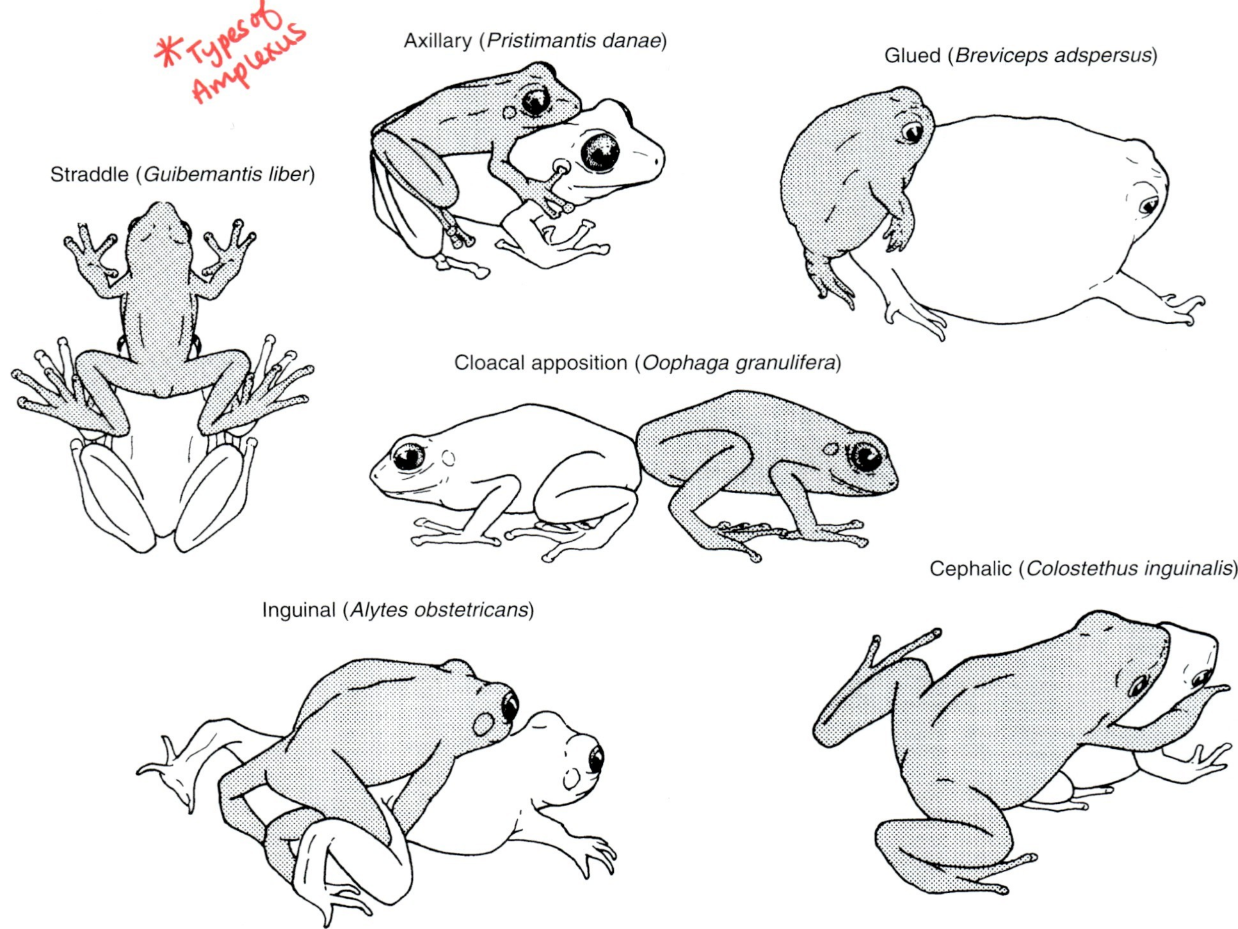

FIGURE 4.8 Positions used by frogs during amplexus. Adapted from Duellman and Trueb, 1986.

Mertensophryne micranotis has a protruding spiny vent, which may be used to transfer sperm to the female's cloaca. The male reproductive structure of caecilians, the phallodeum, is a pouch in the cloacal wall that is everted into the female's cloaca through a combination of muscular contractions and vascular hydraulic pressure and is withdrawn by a retractor muscle.

Males of salamanders with internal fertilization produce spermatophores that are deposited externally. The spermatophore consists of a proteinaceous pedicel capped by a sperm packet; the structure is produced from secretions of various glands in the male's cloaca. Male salamanders have elaborate courtships that rely on secretions from various types of glands to stimulate females to move over the spermatophores and pick up the sperm packets with the lips of the cloaca (Fig. 4.9). In turtles and crocodilians, a penis of spongy connective tissue becomes erect and retracts depending on vascular pressure; it is structurally similar to and probably homologous with the mammalian penis. A hemipenis is used for intromission in male squamates. Hemipenes are paired structures located in the base of the tail that are everted from openings in the posterior part of the cloaca by vascular pressure. Hemipenes of squamates are not homologous with intromittent organs of turtles and crocodylians. Usually only one hemipenis is everted and used during copulation. A retractor muscle withdraws the hemipenis following copulation.

In reptiles, fertilization occurs in the upper region of the oviducts prior to eggshell deposition (Fig. 4.4). Fertilization also occurs in the upper region of the oviducts in caecilians. In contrast, fertilization occurs in the cloaca in salamanders. The exact timing of fertilization varies among species. It can occur immediately after copulation (most lizards) or be delayed (salamanders, turtles, and snakes) for a few hours to years after copulation. Sperm storage structures, which occur in salamanders, turtles, and squamates, facilitate retention of sperm for long periods of time. Delayed fertilization permits females to mate with more than one male and can result in multiple paternity among the resulting offspring (see Chapter 9). For example, female spotted salamanders (*Ambystoma maculatum*) can and often do pick up sperm packets from more than one male, store them, and fertilize their eggs with sperm from multiple males.

The sperm storage structure in salamanders, the spermatheca, is located in the roof of the cloaca. The spermatheca is

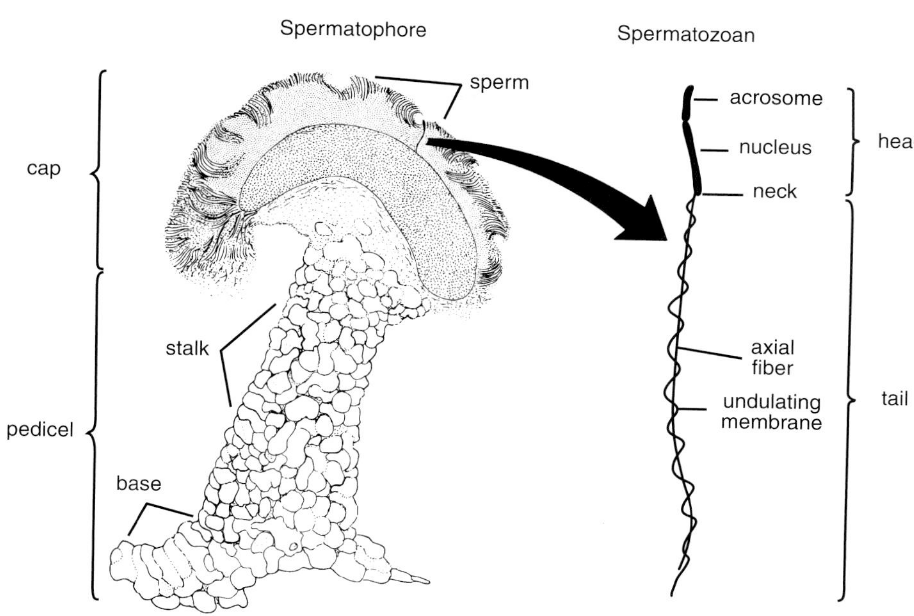

FIGURE 4.9 Diagrammatic representations of a spermatophore and a single spermatozoan of the salamander *Ambystoma texanum*. Sperm are located on the periphery of the cap of the spermatophore; the sperm heads point outward and tails are directed inward. Adapted from Kardong, 1992.

composed of either simple tubes, each of which opens independently into the cloaca, or a cluster of tubules that opens by a common duct into the main cloacal chamber. Stored sperm are expelled by muscular contraction as the eggs enter from the oviducts. Sperm storage tubules typically do not unite to form a common duct in reptiles. They are confined to the upper-middle section of the oviducts between the infundibulum and the shell-secreting area in turtles, and to the base of infundibulum and lower end of the shell-secreting area in squamates. Because of their location in squamates, their function for long-term storage of sperm has been questioned. The mechanism for expelling sperm from the tubules is unknown.

REPRODUCTIVE ECOLOGY

Ecology of Nesting

Amphibians

A nest is a discrete structure constructed by a reproductive adult for egg deposition. Many amphibians deposit eggs in water, and consequently, a nest is not commonly built. Similarly, most frogs and salamanders laying eggs on land do not construct nests but rely on preexisting sites under leaf litter (e.g., *Eleutherodactylus*), on top of leaves (e.g., *Phyllomedusa*), or on top of soil under surface objects (e.g., plethodontid salamanders). Amphibians with terrestrial nests are limited to humid environments. Frogs in several families (e.g., Leptodactylidae, Myobatrachidae, Rhacophoridae) construct foam nests in which the eggs reside (see Chapter 5). Foam nests are constructed on the surface of water (e.g., *Leptodactylus ocellatus, Physalaemus*) or in shallow depressions on land (e.g., *Leptodactylus mystaceus*). The foam ultimately dissolves, and tadpoles drop into the water below and continue development. Larvae from terrestrial foam nests are washed into small, nearby streams or ponds during rainstorms or can develop entirely in the nest and emerge as froglets. The tadpoles of some frogs with terrestrial foam nests (e.g., *Leptodactylus mystaceus*) can generate their own foam should the foam generated by the female begin to dissolve. One leptodactylid, *Lithodytes lineatus*, calls from the entrances of leaf-cutter ants (*Atta*) and constructs foam nests in underground ant chambers. Because some of these contain water, the tadpoles can develop there. Gladiator frogs (*Hypsiboas rosenbergi* and *H. boans*) construct water-filled basins that isolate the eggs from streams; the eggs are deposited as a surface film on water in the basins (Fig. 4.10). A few African frogs deposit eggs underground near water (e.g., *Leptopelis*). Subsequently the tadpoles emerge and enter the water. Other frogs construct underground nests, attend the eggs, and tunnel from the nest to the water (e.g., *Hemisus*). The nest of salamanders with parental care is simply a cavity in the ground or beneath vegetation (see "Parental Care"). Typically, the female salamander coils around its egg clutches (e.g., *Hemidactylium scutatum*).

In amphibians with aquatic larvae, females must select an egg deposition site either in water or in a place from which the larvae can get to water. In amphibians with terrestrial or arboreal clutches, high humidity is necessary to prevent desiccation. Potentially high predation risks of different kinds are associated with different egg deposition sites. Temporary ponds typically harbor predaceous insect larvae of dragonflies, damselflies, caddisflies, diving beetles, and crustaceans that can feed on amphibian eggs and larvae. Tadpoles are sensitive to chemical cues emitted by some insect larvae and respond to these larvae by

FIGURE 4.10 Nest of the Gladiator frog, *Hypsiboas boans*, from western Brazil. (J. P. Caldwell)

FIGURE 4.12 Indian python (*Python molurus*) brooding clutch of eggs. (M. T. O'Shea)

decreasing activity or by remaining in hiding places for long periods of time. Clutches of *Centrolenella* and *Agalychnis* deposited in arboreal microhabitats are subject to predation by grapsid crabs, cat-eyed snakes, and various insects. Eggs in streams and permanent ponds or lakes are subject to additional predation by fish.

Reptiles

Most oviparous reptiles construct nests for egg deposition. Because a majority of reptile eggs require at least some water for development, nest sites usually occur in moist soil; inside of rotting logs or piles of humic material; inside rotted areas of standing trees; under logs, rocks, or other surface items; or on the surface in relatively closed spaces, such as crevices, where humidity is high. Among crocodylians, most species construct aboveground nests that isolate the eggs from water (Fig. 4.11; e.g., *Crocodylus porosus* and *Alligator mississippiensis*). *Crocodylus johnsoni*, however, places its eggs in burrows in sand. Most species of turtles dig nests in the ground (e.g., *Gopherus berlandieri*, *Malaclemys terrapin*, *Emydoidea blandingi*, *Chelydra serpentina*, *Kinosternon flavescens*, *Apalone mutica*). At least one species, *Chelodina rugosa*, deposits its eggs in sand underwater during the wet season. In this case, development is arrested and begins when the sand dries during the dry season. Pythons deposit eggs inside holes within vegetation and coil around the eggs. This behavior reduces water loss and provides heat by shivering thermogenesis (Fig. 4.12). Most lizards and snakes deposit eggs in damp soil or rotting logs and humus (e.g., various species of *Plestiodon* [*Eumeces*], *Crotaphytus collaris*, *Ameiva ameiva*, *Farancia abacura*, *Pituophis melanoleucus*, *Plica plica*, and *Sceloporus aeneus*). Many snakes and lizards and some turtles deposit eggs in ant or termite nests, and still others deposit eggs in crevices in rocks (e.g., *Tropidurus*, *Platysaurus intermedius*, *Phyllopezus pollicaris*) or under the bark of trees (e.g., *Gonatodes humeralis*).

FIGURE 4.11 Nest of the saltwater crocodile (*Crocodylus porosus*). (R. Whitaker)

Egg placement greatly influences the survival and growth rates of the embryos. For most reptiles, mortality is greatest in the egg stage. Amphibians also suffer high egg mortality, but proportionally, mortality is greatest in the larval stage. In both amphibians and reptiles, the female's selection of a site for her clutch will influence the survivorship of her offspring. Good site selection yields high survivorship; poor site selection results in low survivorship or even a total loss of the clutch. The site selected must have the appropriate biophysical environment for the proper development of the embryos and must provide some protection from predation and the vagaries of environmental fluctuations, such as avoiding pond drying or excessive temperatures.

The biophysical environment of the nest site influences the duration of incubation, developmental rate, hatching success, and even the size of offspring (Fig. 4.13). Short incubation time should be advantageous because it reduces

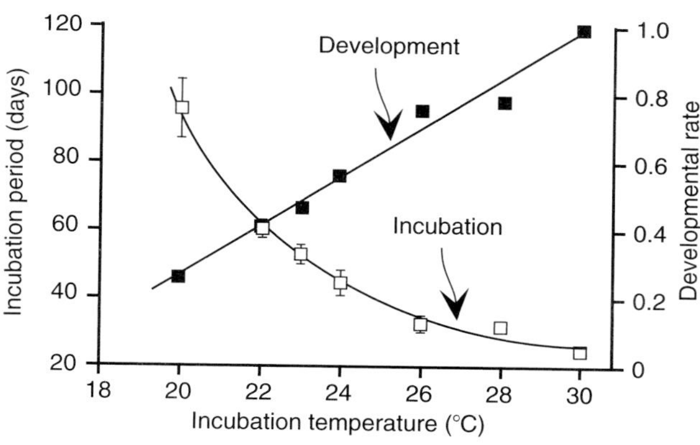

FIGURE 4.13 Effects of temperature on incubation period and developmental rate in eggs of the Australian skink *Bassiana duperreryi*. Developmental rate is the inverse of the observed incubation period divided by the shortest incubation period in the laboratory. Adapted from Shine and Harlow, 1996.

the time that eggs are exposed to mortality factors. However, incubation times are often quite long. Apparently reducing developmental time can have high costs in terms of hatching success and offspring quality. For example, hatching success is high at temperatures varying from 24°C to 28°C and much lower at temperatures exceeding 32°C in the European lizard, *Podarcis muralis*. Moreover, the hatchlings from the eggs incubated at lower temperatures are larger in body size (length and mass), grow faster, and perform better in sprint speed trials than hatchlings incubated at higher temperatures, even though incubation time is shorter (i.e., growth lower but development faster) at higher temperatures. The best balance between incubation time and offspring quality in *P. muralis* occurs at temperatures around 28°C, even though this temperature is lower than optimal temperatures for adult performance. These results support the hypothesis that some, perhaps many, species have multiple optima. In this case, one optimum temperature exists for embryonic development and another for adult performance.

Size of turtles determines to some extent where eggs are deposited, because larger turtles have longer hindlimbs for digging nests. Striped mudturtles (*Kinosternon baurii*) select nest sites close to vegetation (grass tussocks and other herbaceous plants) with little open ground. Temperatures in the nests are lower than the soil in more exposed sites. Hatching success in eggs experimentally placed in nests close to vegetation is substantially greater than those in more exposed areas. Mudturtles dig shallow nests, and as a result, nests near vegetation avoid detrimentally high incubation temperatures. Larger turtle species deposit their eggs deep enough in exposed areas to avoid extreme temperatures.

Most oviparous reptiles in temperate-zone environments deposit eggs in spring or early summer, and the eggs hatch in late summer or fall. These hatchlings (neonates) must immediately begin to feed in order to grow and store energy for overwintering. In some species, however, eggs hatch in the nest in fall, but the neonates remain in the nest through the winter (Fig. 4.14) and emerge in the spring. This phenomenon is much more widespread than commonly recognized. Among turtles worldwide, delayed emergence occurs in at least 12 genera. In the painted turtle (*Chrysemys picta*), neonates emerge in fall or spring depending on locality, and in some areas, either may occur. Presumably, overwintering neonates emerge at a time (spring) when resources are most abundant and potential predation is reduced. Warming temperatures of spring might be the cue predicting the arrival of good conditions. Spending the winter in the nest has associated costs. In winters with little or no snow cover, nests freeze, killing the neonates, but in winters with snow cover, neonates do not freeze because snow insulates the nest. A 5-year study on the nesting ecology of *C. picta* showed that winter mortality due to freezing was significant, varying up to as much as 80% in a given year.

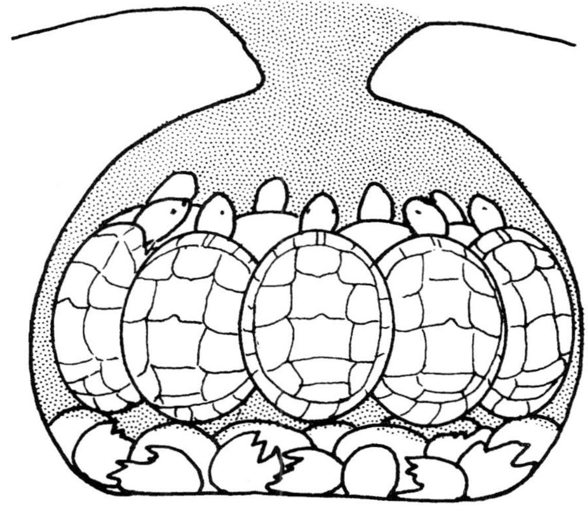

FIGURE 4.14 Spatial arrangement of hatchlings of *Chrysemys picta* in the nest during winter. From Breitenbach et al., 1984.

Temperature-Dependent Sex Determination

An amazing discovery in 1971 revealed that incubation temperatures influenced the sex of hatchlings in two turtle species, *Testudo graeca* and *Emys orbicularis*. This discovery was surprising because the assumption was that sex was genetically controlled in all vertebrates. Sex reversal occurs in larval amphibians and fishes, but this is not the same. Subsequent studies have shown that temperature-dependent sex determination (TSD; more generally referred to as *environmental sex determination,* or ESD) is widespread in reptiles, e.g., tuataras—both species of *Sphenodon*; crocodylians—confirmed for 12 species in 3 families, likely occurs in all species; turtles—confirmed for 64 of 79 species assayed, with representatives of 12 families; squamates—confirmed for 5 species in one genus of Diplodactylidae, 5 species in three genera of Eublepharidae, 15 species in eight genera of Agamidae, and 3 species in three genera of Scincidae. Because sex chromosomes are not involved and TSD occurs early during development, it is clearly sex determination and not sex reversal. TSD is usually associated with a lack of heteromorphic chromosomes, but this does not necessarily cause TSD. For example, the squamate clades Teiidae and Scincidae lack heteromorphic chromosomes but there is no evidence that TSD occurs in Teiidae, and a vast majority of skinks appear to have GSD. The most common types of genetic sex determination (GSD) in reptiles and amphibians are male/female heterogamety as XY/XX (male heteromorphic), ZZ/ZW (female heteromorphic), or homomorphic sex chromosomes (sex chromosomes undifferentiated, but sex determination as in heterogametic forms). A recent phylogenetic analysis of origins of sex determination in vertebrates revealed that GSD was the ancestral condition in sauropods, and that not only did TSD arise independently several times, it also was lost several times! Among species with TSD, the temperature range over which sex is determined is relatively small and varies somewhat among species (Fig. 4.15; Table 4.1).

In studied species with TSD, sex determination occurs in the second trimester of development, and the "average" temperature during that period regulates the direction of gonad differentiation. At the threshold temperature range, the gonads can become either ovaries or testes. In most crocodylians and lizards, males result from high temperatures, whereas females result from low temperatures. In turtles, females develop at high temperatures and males at low ones; in a few crocodylians, turtles, and lizards,

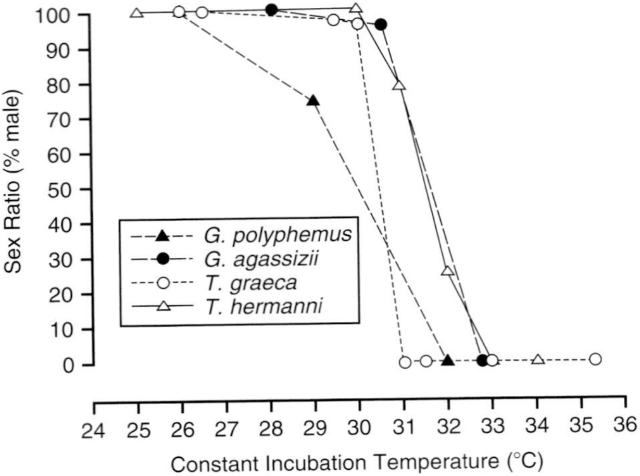

FIGURE 4.15 Sex ratios for four tortoise species (*Gopherus polyphemus, G. agassizii, Testudo graeca, T. hermanni*) raised at different incubation temperatures showing that males are produced at low developmental temperatures and females are produced at high developmental temperatures. Adapted from Burke et al., 1996.

females develop at high and low temperatures, males at intermediate ones. The physiological mechanism of TSD is just beginning to be understood. At temperatures appropriate for the production of one sex over the other, the enzyme aromatase is produced in individuals that will become females and 5-reductase is produced in those that will become males. These enzymes induce the conversion of testosterone to estradiol to initiate ovary differentiation or dihydrotestosterone to initiate testes differentiation, respectively. Genes that code for the production of aromatase or 5-reductase are turned on or off depending on temperature. Yolk steroid hormones, which can affect embryo development and growth, may also influence sex determination.

The ecological implications and consequences of TSD are fascinating and complex, and hypotheses range from TSD having no adaptive value at all (neutral) to TSD affecting maternal behavior, survival, fecundity, and sex ratios. Genetic, maternal, and environmental factors contribute to determination of sex phenotypes (Fig. 4.16).

The many hypotheses on the evolution of TSD are summarized in Figure 4.17. What emerges is the realization that endogenous and environmental factors that favor shifts in sex phenotype are complex, and no single hypothesis is likely to explain all cases of TSD. The ability to manipulate offspring sex should result when the fitness return for the female parent varies depending on sex of the offspring. The most robust model for the evolution of environmental sex determination (TSD in particular) was proposed by Drs. Eric Charnov and Mike Bull in 1977 and is now called the Charnov–Bull model. Their model specifies that TSD will be favored if three conditions are met: (1) The environment is patchy (spatially or temporally) such that one sex produced in a particular patch has higher fitness than it would have if produced in another patch, (2) patches cannot be chosen either by offspring or parents, and (3) mating is random with respect to patch. Until recently none of the many studies on TSD have unequivocally supported this model.

In the Australian Jacky Dragon lizard (*Amphibolurus muricatus*), nutrient-deprived females produce eggs double the size of eggs produced by females with high-quality diets, and the sex ratio of these offspring is highly male-biased even though yolk steroid levels are similar for male and female eggs. If large body size of male offspring translates into higher reproductive success (likely in a polygynous mating system), then females may be able to enhance their fitness by producing not only larger and presumably more competitive offspring in response to low resource levels, but also sexes (males) likely to contribute more to future generations. Additional experiments address the issue of whether producing sons or daughters would pay off in the expected manner based on operational sex ratios (OSR; see Chapter 9). Theory predicts that when a shortage of breeding males exists, producing male offspring has a

TABLE 4.1 Mechanisms of Sex Determination in Amphibians and Reptiles.

	Genetic sex determination			
	Heterogamety in males	Heterogamety in females	Homogamety	Temperature-dependent sex determination
Amphibians				
Salamanders	Plethodontidae, Proteidae, Salamandridae	Plethodontidae, Ambystomatidae, Sirenidae	None	None
Frogs	Bombinatoridae, Hylidae, Leptodactylidae, Pelodytidae, Ranidae	Bufonidae, Discoglossidae, Leiopelmatidae, Pipidae, Ranidae	None	None
Reptiles				
Turtles	Chelidae, Bataguridae, Staurotypidae	Bataguridae	Chelidae	Pelomedusidae, Podocnemidae, Bataguridae, Carettochelyidae, Cheloniidae, Chelydridae, Dermatemydidae, Dermochelyidae, Emydidae, Kinosternonidae, Testudinidae, Trionychidae
Crocodilians	None	None	None	Alligatoridae, Crocodylidae, Gavialidae
Tuataras	None	None	None	Sphenodontidae
Squamates	Iguania, Gekkonoidea, Teiidae, Scincidae	Gekkonoidea, Lacertidae, Amphisbaenia, Varanidae, Boidae, Colubridae, Elapidae, Viperidae	Iguania, Gekkonoidea, Lacertidae, Teiidae, Scincidae, Colubridae, Elapidae	Agamidae, Diplodactylidae, Eublepharidae, Gekkonidae, Scincidae

Note: Taxa for which the mechanism remains unknown are not included. Taxa may appear more than once if different sex determining mechanisms occur in different species.
From Cree et al., 1995; Deeming, 2004; Ewert et al., 2004; Harlow, 2004; Hillis and Green, 1990; Janzen and Paukstis, 1991; Lang and Andrews, 1994; Nelson et al., 2004; Viets et al., 1994.

potentially higher payoff than producing female offspring, and vice-versa. However, females in the male-biased experimental enclosures produced more male than female offspring in their first clutch of the season, exactly the opposite of what theory predicted (Fig. 4.18). Why might this be the case? Jacky Dragon habitat varies spatially and temporally, and as a result, the likelihood that OSR is predictable at a given place and time is low. Rather, females may be adjusting their offspring sex ratio to match offspring sex to the sex that has been most successful in that particular habitat patch for juveniles. To complicate matters, this occurs only in offspring produced in the first clutch of the season. First-clutch offspring have higher survival rates and are more likely to reach sexual maturity earlier than hatchlings produced later in the season.

Viviparous skinks with TSD provide a particularly ideal experimental system for examining the relationship between OSR and sex allocation because females can regulate developmental temperatures of their offspring by behavioral thermoregulation of their own bodies during pregnancy. The Southern Water Skink (*Eulamprus tympanum*) of Australia has a gestation period of 3–4 months, during which the potential exists for females to manipulate the sex ratio of their offspring. Field and laboratory experiments show that females do behaviorally thermoregulate differently depending on the OSR. When the sex ratio is female-biased, pregnant females maintain higher temperatures than they do when the sex ratio is male-biased. Even though body temperatures differ, the difference in temperature is not enough to result in differences in the sex ratio of neonatal lizards. Thus, even though the potential exists for maternal control of sex allocation to offspring in response to OSR, the lizards do not appear to do it. These examples bring us back to an earlier point—determinants of sex allocation are highly complex and may not be the same in different species. Carefully designed experiments placed in the context of what occurs in natural populations are necessary to determine which abiotic and biotic factors determine sex allocation in reptiles with TSD.

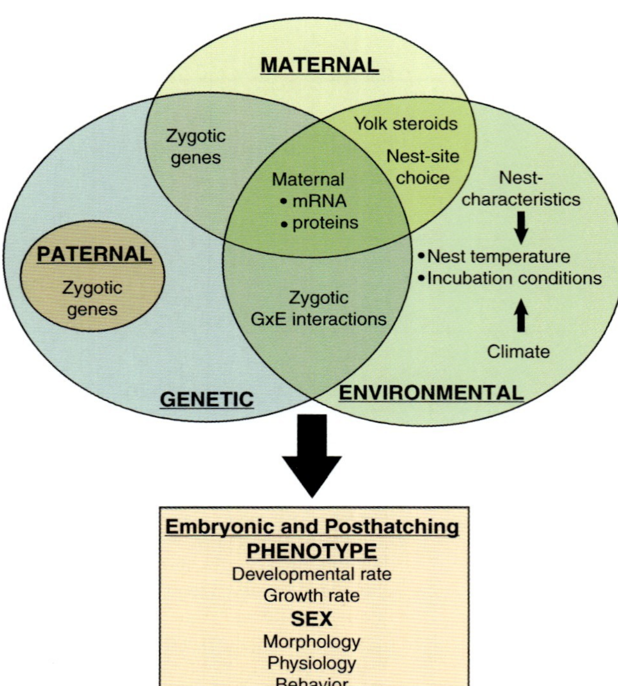

FIGURE 4.16 Genetic and environmental factors affect embryo and hatchling phenotypes and can affect the sex of offspring in species that have temperature-dependent sex determination (TSD). Maternal effects cut across genetic and environmental effects, whereas paternal effects are only genetic. Adapted from Valenzuela, 2004.

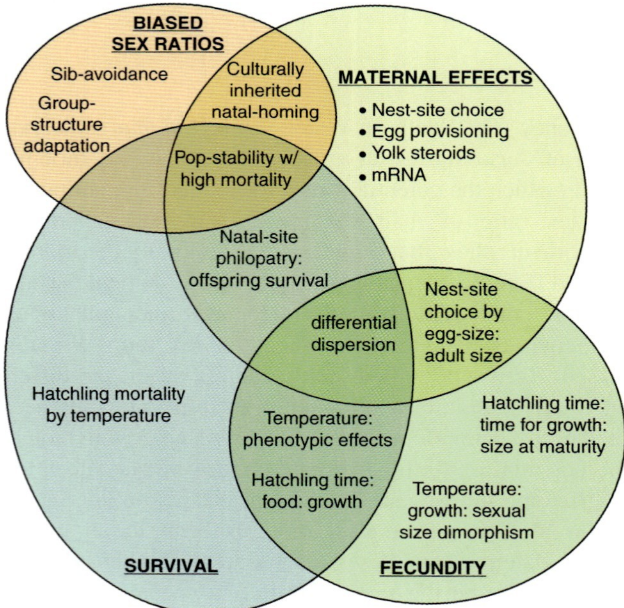

FIGURE 4.17 Evolutionary hypotheses to explain TSD center on sex ratios, maternal effects, fecundity, and survival, none of which is mutually exclusive. Most hypotheses can be categorized by the fitness component that they address. Adapted from Valenzuela, 2004.

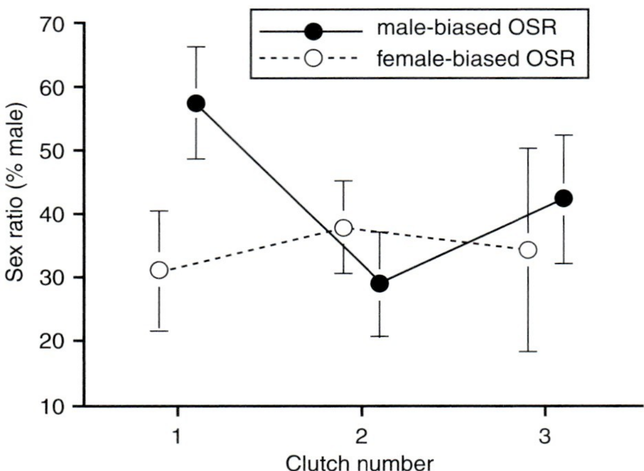

FIGURE 4.18 Offspring sex ratios differ in offspring produced in the first clutch of the season for Jacky Dragons in Australia in response to differing operational sex ratios (OSR) in experimental arenas of the mother. The response is exactly the opposite from what theory predicts. In successive clutches (2–3) sex ratios did not differ as a result of varying OSR, but the sex ratio of hatchlings was biased toward females. Adapted from Warner and Shine, 2007.

As climates change, either due to natural cyclical events or human-induced global warming, changes in temperatures in nesting habitats alter population sex ratios and, ultimately, the survival of species. Major sex ratio biases have already been observed in populations of *Alligator mississippiensis* and *Caretta caretta* based on nest location. In both cases, the sex ratios were highly biased toward females. Finally, any efforts to manage populations of sensitive species in which TSD occurs must consider the potential long-term effects of variation in nest temperatures, either under natural conditions or when eggs are reared in the laboratory for release into the wild.

Brood Size and Size of Young

Assuming that energy is limited, a given reproductive effort (clutch mass or energy) can be expended either by the production of a few large offspring or by many small ones. The identification and measurement of the trade-offs between size and number of offspring are difficult and generally relate to natural selection operating on eggs, larvae (amphibians), or juveniles. In most instances, offspring size within a population is relatively constant. Natural selection should favor the offspring size yielding the highest probability of survival and future reproduction success. As a result, energy expenditure theoretically favors either a clutch of numerous small offspring or one of fewer larger offspring (often referred to as *r*- versus *K*-selection). The number and relative size of eggs varies greatly in amphibians and reptiles. Many bufonids, for example, produce thousands of tiny eggs, whereas *Eleutherodactylus* and dendrobatid frogs produce a few large eggs or offspring. The maximum number of eggs produced by any reptile is much smaller than the maximum numbers produced by some frogs and salamanders; nevertheless, great variation exists among reptile species as well. Historical constraints render most ecologically based generalizations about size and numbers of young subject to further examination.

Field data and laboratory studies show that a trade-off exists between the number and size of offspring produced in sand lizards (*Lacerta agilis*). Further, the total reproductive investment is determined by resource availability. Resource levels also influence the allocation of energy to individual offspring, which are larger when resources are most abundant. Independent of this source of variation in hatchling size, hatchling mass is greatest in small clutches and lowest in large clutches, demonstrating the trade-off between offspring numbers and size (Fig. 4.19). In the Australian water python (*Liasis fuscus*), clutch size increases with maternal body size and is associated with the physical condition of the females. Healthier females produce larger clutches, independent of the effect of body size, and they also produce larger eggs, confirming the trade-off between offspring size and number within a population of snakes.

Females of the salamander *Ambystoma talpoideum* have increased clutch size and egg size as they grow larger. The increase in body size of females and potentially the greater energy available to adults as a function of their body size account for the increase in both clutch size and egg size. The number of offspring is maximized, and relatively large offspring presumably hatch earlier and metamorphose at a larger size.

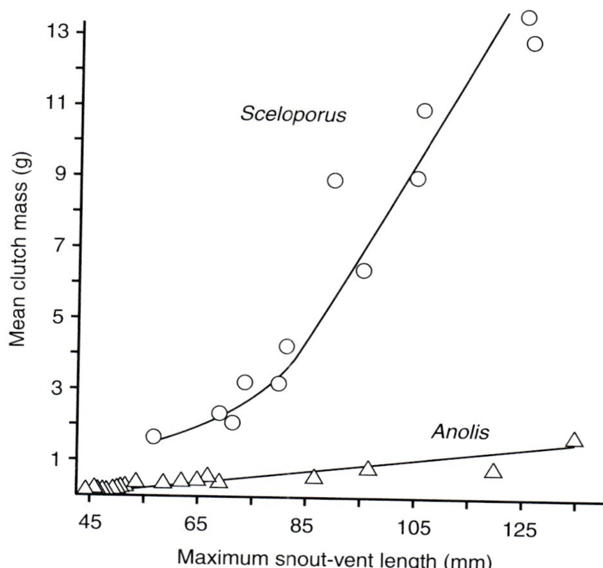

FIGURE 4.20 Species and populations of *Sceloporus* lizards with variable clutch size have relatively massive clutches of eggs at any given body size when compared with *Anolis* lizards that have fixed clutch sizes of a single egg. In addition, clutch mass increases linearly with body size in *Anolis* but exponentially in *Sceloporus*. Adapted from Andrews and Rand, 1974. Refer to the original paper for species identifications.

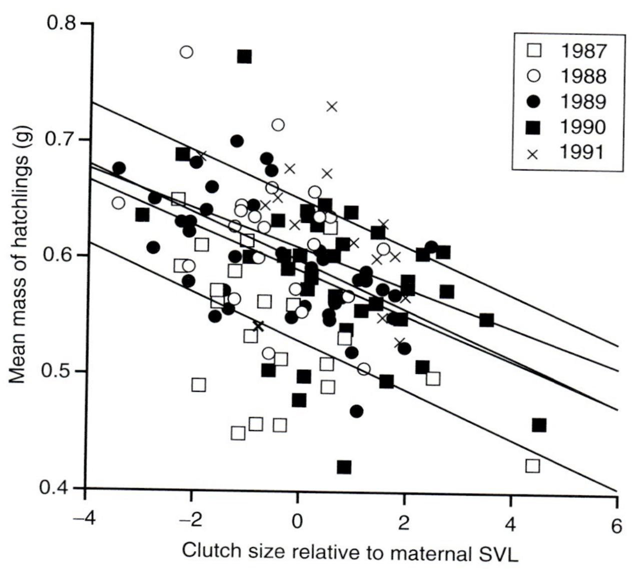

FIGURE 4.19 Annual variation in the trade-off between number of eggs and size of eggs in *Lacerta agilis*. The influence of body size on clutch size has been removed by expressing clutch size as residuals from the common regression. Adapted from Olsson and Shine, 1997.

Some lizards always produce clutches with the same number of eggs, one or two, depending on species, and no variation exists among females in the number of offspring produced in any single reproductive episode. Selection cannot operate on clutch size in these species, and as a result, there can be no trade-off between offspring size and number as in species with variable clutch size. This issue was considered indirectly in the comparison of clutch mass of *Sceloporus* (variable clutch sizes) with *Anolis* (no variation in clutch size) (Fig. 4.20). Gekkonid lizards have a clutch size of one or two eggs, depending upon species. They produce relatively larger offspring than lizards with variable clutch size, and egg size is relatively constant within a population even though egg quality may vary. When clutch size is invariant at a low value, some potential energy savings of producing few offspring can be transferred to the production of larger offspring. Presumably, females produce the largest offspring possible because dividing the clutch into numerous packets is not an option. A tight linear relationship between female and offspring size across species with invariant low clutch size adds support to this hypothesis (Fig. 4.20). Moreover, this observation suggests the possibility that selection on optimal offspring size influences the evolution of female body size in species with low, invariant clutch size.

Several nonexclusive hypotheses can explain low clutch size (and clutch mass) in anoles, geckos, gymnophthalmids, and some other lizards. One hypothesis is that low clutch size is favored because anoles have adhesive toe pads for locomotion on smooth surfaces in arboreal environments and are unable to bear the extra load associated with carrying eggs. If the ancestor of all *Anolis* lived on smooth surfaces where toe pad lamellae determined load-bearing capacity and there was a functional limit on how large adhesive toe pads could be in *Anolis*, then a single evolutionary event could explain low clutch size in all anoles. This hypothesis might also be applicable to gekkonine geckos in which at least two evolutionary events are necessary. Most gekkonines have a clutch size of two eggs, but some (e.g., *Thecadactylus rapicauda*) have a clutch size of one egg. Many gekkonines have evolved very large adhesive toe pads, and it seems unlikely that the load bearing associated with their small clutches is limited by the size of their toe pads. Both anoles and gekkonines have claws as well, so the load-bearing hypothesis is tenable only if the ancestor in each instance lived on smooth surfaces where adhesive toe pads were necessary. A large number of present-day arboreal lizards have large clutches (e.g., *Polychrus marmoratus*); they do not have adhesive toe pads and do not live on smooth surfaces. Few present-day anoles live on smooth surfaces. The load-bearing hypothesis does not apply to sphaerodactyline geckos, gymnophthalmids, or other lizards (some scincids, tropidurines, and others) that also lack adhesive toe pads.

Another hypothesis is that reduced clutch size allows more frequent clutch production, thereby providing the opportunity to distribute offspring in time and space and reduce the probability that all eggs will be lost. Similar to the load-bearing hypothesis, this hypothesis is difficult to test, especially in lineages with invariant clutch size, because a single evolutionary event in the ancestor of each lineage can account for the low and invariant clutch size in the entire lineage. Nevertheless, this might be the reason low and invariant clutch size evolved in the ancestor.

Yet another hypothesis is that low and invariant clutch size allows the production of relatively larger and presumably more competitive offspring, and females are simply producing the largest offspring possible given energetic and morphological constraints. The overall negative relationship between clutch size and offspring size in squamates lends some support to this hypothesis.

A final hypothesis is that low clutch size results from morphological constraints on females for the use of specific microhabitats. Species conforming to the load-bearing hypothesis might be included here as a special case. Low fixed clutch size in dorsoventrally flattened lizards that use narrow crevices for escape might be another example (e.g., *Platysaurus*).

Even though egg size tends to be constant (i.e., optimal) in many species, this constancy is not universal. Data on three turtle species raise questions concerning the generality of the optimal offspring size theory. Turtles are potentially constrained in egg size because their pelvic girdle is less flexible than that in many other reptiles. Eggs cannot be larger in width than the diameter of the pelvic aperture. In the chicken turtle, *Deirochelys reticularia*, small females (135 mm plastron length) have narrow pelvic apertures and produce small eggs. As they grow larger, they produce bigger eggs (Fig. 4.21). In the painted turtle

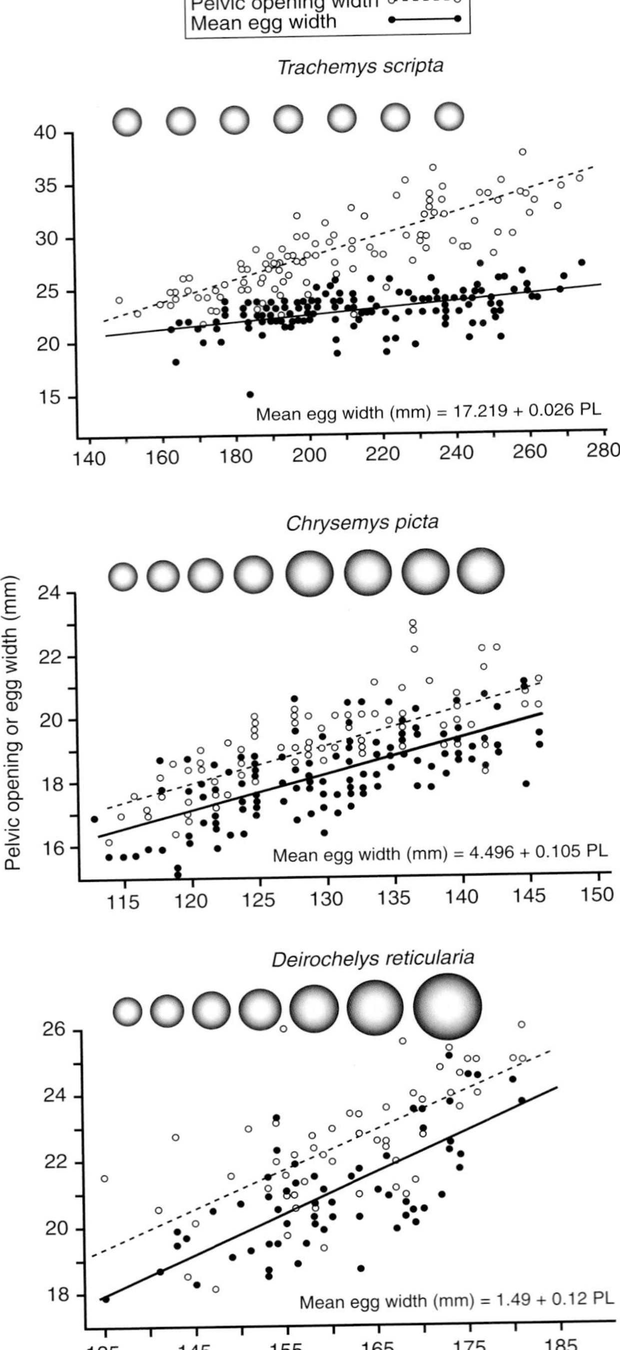

FIGURE 4.21 Variation in the size of the pelvic opening of turtles and width of eggs associated with increasing body size in three species of emydid turtles. Adapted from Congdon and Gibbons, 1987.

(*Chrysemys picta*), small females (115 mm plastron length) have narrow pelvic apertures and produce small eggs. At about 125–130 mm plastron length, even though the size of the pelvic aperture continues to enlarge with increasing female size, egg diameter levels off. In the slider (*Trachemys scripta*), all females produce eggs of about the same diameter regardless of their body size or the size of their pelvic apertures. Lack of variation in egg size in sliders indicates that selection has resulted in an optimal offspring size that is smaller than could be produced relative to the pelvic aperture diameter. In painted turtles, the optimal egg size (i.e., egg size at which there is no further increase) is constrained in small females owing to their narrow pelvic apertures; however, egg production by small females is advantageous even if the eggs and resulting hatchlings are below optimal size. Apparently, there is no optimal egg size in chicken turtles because egg size is directly associated with female body size across the entire body-size range of adults.

Seasonality in Reproduction

Reproduction among amphibians and reptiles varies from highly seasonal to aseasonal, and no single generalization explains the observed variation. A majority of temperate-zone species are seasonal in reproduction, but among tropical species, species reproduce in the wet season, dry season, over extended periods, or even nearly continuously. Reproduction in all species is hormonally mediated, and androgen production reaches a peak just prior to mating.

Amphibians

Temperature and rainfall no doubt are the major determinants of timing of reproduction, but the asynchrony of reproduction among species of amphibians occurring at single localities confirms that temperature and rainfall alone are not the sole determinants of reproductive timing. In temperate zones, some amphibians breed in late winter (e.g., *Pseudacris ornata* and *P. nigrita*—but not synchronously; *Plethodon websteri*), in spring (e.g., *Siren intermedia* and *Hydromantes ambrosii*), early to mid-summer (e.g., *Hyla arborea*), fall (e.g., *Ambystoma opacum*), and both spring and fall (e.g., *Lithobates sphenocephala*). The salamander *Rhyacotriton olympicus* has an extended breeding season in western Oregon, and females contain sperm in October through July. The Carolina gopher frog (*Lithobates capito capito*) breeds for only a few days sometime between January and April. Long-term studies on the salamander *Ambystoma talpoideum* show that breeding migrations of adults occur from September through January, always during the coldest month; however, the number of breeding adults is correlated with the cumulative rainfall.

Although most temperate-zone amphibians appear to breed annually, some reproduce biennially (every other year). In Louisiana, approximately 35% of female *Amphiuma tridactylum* reproduce each year, suggesting that most individuals reproduce every other year or even less often. The proximate explanation is that vitellogenesis requires nearly a full year, eggs are deposited in midsummer (July), and females attend the eggs until November. Thus, the complete cycle requires more than a year.

Males of many salamanders have testicular cycles that coincide with ovarian cycles of sexually mature females in the population. Male *Ambystoma talpoideum* in the southeastern United States have enlarged testes from September through January, coincident with the presence of enlarged ova in females. However, in the salamander *Plethodon kentucki*, males breed annually but females breed biennually or even less frequently. Presumably the inability of individual females to reproduce each year results from energy-accumulation limitation associated with season length. Females of species of *Plethodon* in environments with extended seasons for foraging, as in the southern United States (e.g., *Plethodon websteri*), breed annually, whereas species like *Plethodon kentucki* in environments with short activity seasons reproduce biennially.

In seasonal tropical environments, most amphibians breed during the wet season, although exceptions are known. During the dry season in northeastern Costa Rica, none of eight species of hylid frogs reproduce even though some males of several species vocalize year-round. Hylid species with explosive breeding patterns (*Smilisca baudinii* and *Scinax elaeochrous*) reproduce early in the wet season, whereas other hylids (*Dendrosophus ebraccatus* and *Agalychnis callidryas*) reproduce throughout the wet season. In Rondônia, Brazil, even though most frog species breed during the wet season, the gladiator frogs (*Hypsiboas boans*) breed during the dry season, constructing nests in sand at stream edges.

In relatively aseasonal tropical environments, many amphibians breed year-round or at least appear to have extended breeding seasons. Six species of frogs in an aseasonal rain forest in Borneo bred throughout the year. Among frog species at Santa Cecilia, Ecuador, a relatively aseasonal tropical environment, many frogs (e.g., *Scinax ruber*, *Lithobates palmipes*, *Dendrosophus sarayacuensis*, *Ameerega parvula*) reproduce throughout most of the year, whereas others reproduce during periods varying from 3 to 5 months (e.g., *Phyllomedusa vaillantii*, *Leptodactylus wagneri*). The timing and intensity of rainfall appear to determine exactly when breeding occurs. It remains unknown whether individuals breed throughout the year or whether breeding at different times of the year involves different individuals.

Reptiles

A vast majority of reptiles are seasonal in reproduction, but when continuous reproduction occurs, it is in tropical species. Nearly all temperate-zone reptiles worldwide reproduce seasonally. For most, ovulation of eggs occurs in spring, egg deposition occurs in early to midsummer, and hatching occurs in late summer. In most temperate-zone viviparous species, ovulation occurs in spring with parturition in late summer. Additional studies corroborate the preponderance of this pattern for temperate-zone reptiles in general (Table 4.2). Cold winter temperatures are a major constraint on reproductive seasonality of temperate reptiles. Soil temperatures and insolation are high enough to allow rapid embryonic development only in summer. Nevertheless, some exceptions exist. In the high elevation, viviparous species of *Sceloporus*, ovulation and fertilization occur in late fall or early winter, gestation occurs during winter and spring, and offspring are born in early or midsummer (Table 4.2).

The length of the cold season is a major constraint on the duration of the reproductive season, independent of latitude. This constraint has been neatly demonstrated for the seasonally breeding, tropical montane lizard *Sceloporus variabilis*. At high elevations, gravid females (i.e., containing oviductal eggs) occur from December to July,

TABLE 4.2 Selected Examples of Temperate-Zone Reptiles with Seasonal Breeding Patterns.

Species	Family	Country	Source
Spring Breeding			
Alligator mississippiensis	Alligatoridae	United States	Joanen, 1969
Malaclemys terrapin	Emydidae	United States	Reid, 1955
Kinosternon flavescens	Kinosternidae	United States	Christiansen and Dunham, 1972
Chelydra serpentina	Chelydridae	United States	Congdon et al., 1987
Sternotherus odoratus	Kinosternidae	United States	McPherson and Marion, 1981
Apalone muticus	Trionychidae	United States	Plummer, 1977a
Chelodina longicollis	Chelidae	Australia	Parmenter, 1985
Cophosaurus texanus	Iguanidae	United States	Howland, 1992
Sceloporus undulatus	Iguanidae	United States	Gillis and Ballinger, 1992
Japalura brevipes	Agamidae	Taiwan	Huang, 1997b
Cordylus polyzonus	Cordylidae	South Africa	Flemming and van Wyk, 1992
Sphenomorphus taiwanensis	Scincidae	Taiwan	Huang, 1997a
Ctenotus (7 species)	Scincidae	Australia	James, 1991
Tupinambis rufescens	Teiidae	Argentina	Fitzgerald et al., 1993
Takydromus hsuehshanensis	Lacertidae	Taiwan	Huang, 1998
Mabuya capensis	Scincidae	South Africa	Flemming, 1994
Mehelya capensis	Colubridae	South Africa	Shine et al., 1996a
Mehelya nyassae	Colubridae	South Africa	Shine et al., 1996a
Coronella austriaca	Colubridae	Italy	Luiselli et al., 1996
Seminatrix pygaea	Colubridae	United States	Seigel et al., 1995
Thelotornis capensis	Colubridae	South Africa	Shine et al., 1996c
Natrix natrix	Colubridae	Italy	Luiselli et al., 1997
Aspidelaps scutatus	Elapidae	South Africa	Shine et al., 1996b
Sistrurus miliarius	Viperidae	United States	Farrell et al., 1995
Fall Breeding			
Sceloporus jarrovi	Iguanidae	United States	Goldberg, 1971
Sceloporus grammicus	Iguanidae	Mexico	Guillette et al., 1980
Sceloporus torquatus	Iguanidae	Mexico	Guillette and Cruz, 1993

whereas at low elevations gravid females are found from January to September. The elevation between the two sites differs by 955 m, enough to shorten the high-elevation reproductive season by at least a month. The few high-elevation species that produce offspring in spring or early summer are pregnant during the winter; they bask regularly thereby elevating their body temperature and speeding embryonic development. This temperature constraint is relaxed in lowland tropical environments.

Spermatogenetic cycles usually coincide with ovarian cycles in temperate reptiles. The male cycle in these species is considered prenuptial because mating takes place prior to the production of eggs. The terminology has been changed recently to reflect hormonal and gonadal events in the reproductive cycle. When gonadal and hormonal events in males and females coincide, the cycle is called associated. Associated reproduction (Fig. 4.22) does not always occur, particularly in snakes. In some species (e.g., *Tropidoclonion lineatum*), sperm production and mating occur in fall and the fall-mated females store sperm. Fertilization occurs the following spring, and offspring are produced in late summer. Sperm storage appears obligatory in some species. In these cases, the male's sperm production is out of phase with the female's ovulation and fertilization, hence a postnuptial or dissociated cycle. In some tropical species, reproduction is nearly continuous in the population, but in most instances, whether individual males or females are continually sexually receptive is not well known. As a result, seasonal patterns are much more obscure among tropical reptiles. At one time, it was believed that tropical squamates had continuous reproduction in aseasonal tropical environments or reproduced during the wet season in wet–dry seasonal tropical environments. The currently known diversity of seasonal patterns of tropical squamate reproduction suggests that no single explanation is sufficient.

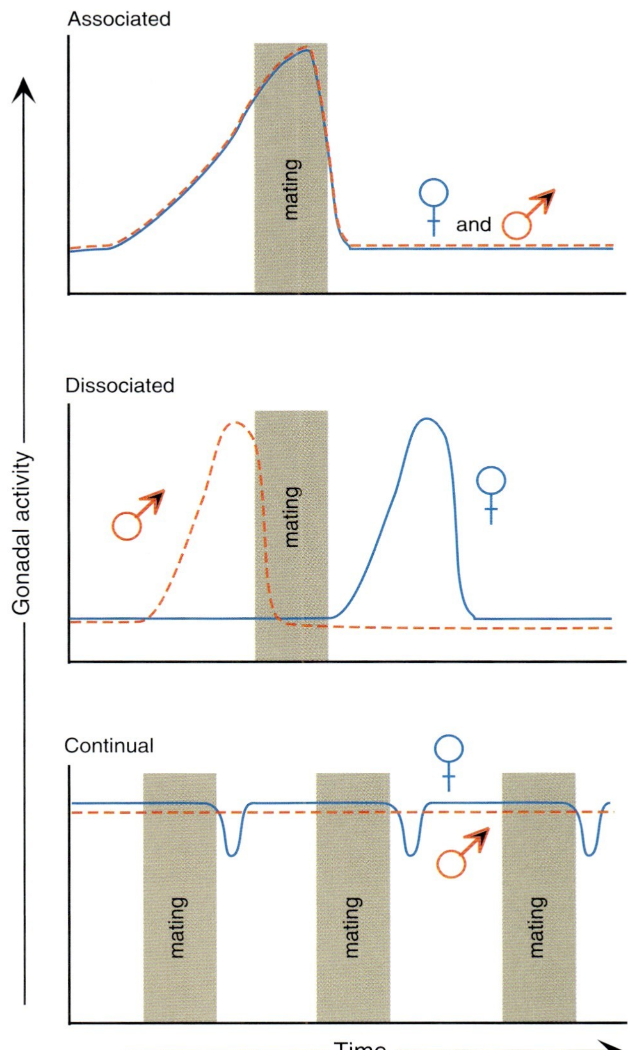

FIGURE 4.22 Schematic diagrams of sex steroid production in relation to gametogenic cycle of a spring-breeding temperate-zone reptile. Steroid levels match the peaks of gametogenesis; androgen production begins simultaneously with spermiogenesis and continues until the testes regress; estrogen production occurs during final maturation of ovarian follicles, stopping at their maturation and ovulation. Corpora lutea produce progesterone, which continues while ova remain in the oviducts; production declines and corpora lutea degenerate with egg-laying, but in viviparous taxa, progesterone is produced throughout pregnancy. Adapted from Whittier and Crews, 1987.

Among tropical Australian crocodylians, *Crocodylus johnsoni* produces eggs during the dry season, whereas *C. porosus* produces eggs at the beginning of the wet season. Among tropical snakes, some species reproduce nearly year-round (e.g., *Styporhynchus mairii*, *Liophis poecilogyrus*, *L. miliarius*, *L. viridis*), others reproduce mostly in the dry season (e.g., *Liophis dilepis*, *Philodryas nattereri*, *Waglerophis merremii*), and still others reproduce in the wet season (e.g., *Oxybelis aeneus*, *Oxyrhopus trigeminus*).

Among lizards, two studies from different continents demonstrate the diversity of reproductive patterns in seasonal tropical environments. At one highly seasonal site in Caatinga of northeast Brazil where the entire lizard fauna was studied, four gekkonids, one gymnophthalmid, and two teiids reproduced nearly continuously; two tropidurines, one scincid, one teiid, and one anguid reproduced primarily during the dry season; and one polychrotine reproduced during the wet season. At a tropical site in the Alligator Rivers Region in the Northern Territory of Australia, lizard species also varied with respect to reproductive seasonality. Among the skinks, one species (*Cryptoblepharus plagiocephalus*) reproduced year-round, five species of *Carlia* and three species of *Sphenomorphus* reproduced during the wet season, and one species of *Lerista*, two species of *Morethia*, and most *Ctenotus* reproduced during the dry season. Among the agamid lizards, *Diporiphera* and *Gemmatophora* reproduced during the wet season, and *Chelosania* reproduced during the dry season. Such high diversity in the reproductive timing at a single site with similar environmental variables demonstrates that seasonality in rainfall is only one of several determinants of reproductive seasonality in lizards.

As in temperate reptiles, the male spermatogenetic cycle may or may not coincide with the female reproductive cycle in tropical species, and presumably, sperm storage occurs in species with dissociated cycles. In some instances, spermiogenesis may occur year-round regardless of whether females are seasonal or aseasonal in reproduction. In species with continual reproduction, individual males presumably produce sperm throughout the year and females produce successive clutches; however, individual females might be cyclic but the female population is continuous, because there are always some females in the population that are preovulatory.

We now return to the question, "Why do most reptiles, regardless of where they live, reproduce seasonally?" We know from temperate-zone studies that abiotic factors (season length and temperature) restrict reproduction in most species to spring, summer, or fall. We also know that most tropical reptiles reproduce seasonally (*Anolis* lizards and geckos are the most striking exceptions—but even these have peak periods). Studies on the Australian keelback snake (*Tropidonophis mairii*), a seasonally reproducing tropical species, reveal that nesting occurs after monsoon rains stop, when soils are best for embryogenesis (damp but not waterlogged). Waterlogged soil is lethal to developing embryos, but damp soil provides hydric conditions that result in offspring survival and large offspring size. Biotic factors, such as predation on eggs or hatchlings, or resource availability for hatchlings, may be less important because they do not vary in a corresponding way with reproduction. Storage of fat typically cycles with reproduction; fat stores are mobilized to produce eggs in females and, because fat stores become depleted during the mating season in males, they are apparently used to supply at least part of the energy necessary for reproductive-related behaviors. In seasonally reproducing reptiles (temperate or tropical), fat stores are at their lowest in males just prior to mating and in females just as eggs are being produced.

The search for a general explanation of seasonality in reptile reproduction must center on tropical species for two reasons: (1) many reptile species are tropical, and (2) the extended period of cold temperatures associated with winter in temperate environments is not a constraint in tropical environments. Dr. Rick Shine has proposed possible phylogenetic conservatism for tropical Australian lizards and snakes in stating, "The observed seasonal timing of reproduction in squamates may reflect the ancestry of the lineage: for example, many of the dry season breeders belong to genera that are characteristic of the arid zone (e.g., *Ctenotus, Lerista*), whereas the wet-season breeders tend to be species characteristic of more mesic habitats (e.g. *Carlia*)." Consequently, the evolutionary histories of species may partially determine seasonality of reproduction as well.

Sexual Versus Unisexual Reproduction

A majority of amphibians and reptiles reproduce sexually, with males and females contributing genetic material to offspring. In a few taxa, reproduction occurs without the male's genetic contribution (Table 4.3), and in fewer yet, populations reproduce clonally. Three general types of unisexual reproduction have been classically recognized in reptiles and amphibians: hybridogenesis, gynogenesis, and parthenogenesis. What was recognized previously as gynogenesis in certain unisexual *Ambystoma* is probably better described as kleptogenesis. Hybridogenesis is the production of all-hybrid populations from two parental species. Kleptogenesis is unisexual reproduction in which females have a common cytoplasm but "steal" genomes from males of sexual species, which are not passed on to the next generation. Parthenogenesis is cloning, in which each female produces identical daughters with no interaction with males of other species.

Hybridogenesis

Hybridogenesis occurs when half of the initial genome is passed on but the other half is not. Hybridogenetic females originate as crosses from two different sexual species and cannot reproduce without mating with the male of a sexual species. These females produce only female offspring, all containing only the genome of the mother. During gametogenesis, the male genome is not included and the female genome is duplicated, reconstituting a diploid zygote that develops into a hybridogenetic female. Because only the genome of the female is passed on, hybridogenesis is hemiclonal.

In Europe, two closely related frogs, *Pelophylax* [*Rana*] *lessonae* (L genome) and *R. ridibunda* (R genome), hybridize over a wide geographic area, resulting in the formation of a complex of hybrids referred to collectively as *Pelophylax esculenta*. *Pelophylax esculenta* is widespread in Europe (from France to central Russia) and is usually sympatric with *P. lessonae*. Hybridogenesis is more complex in this system because both diploid (LR) and triploid (LLR) *P. esculenta* exist. Although hybridization between *P. lessonae* and *P. ridibunda* continues producing some *P. esculenta*, most are produced when *P. esculenta* hybridize with either *P. lessonae* or *P. ridibunda*, and some are produced by hybridization between two *P. esculenta* (Fig. 4.23). Where *P. esculenta* occurs with *P. lessonae*, the L (*P. lessonae*) genome is lost in the germ line (but not in the soma) of *P. esculenta* during meiosis, and only

TABLE 4.3 Genera of Unisexual Amphibians and Reptiles.

Genus	Number of species	Mode of reproduction	Representative species
Ambystomatidae			
Ambystoma	3±	K&H	*platineum*
Ranidae			
Pelophylax [*Rana*]	5	H&P	*esculenta*
Agamidae			
Leiolepis	1	P	*triploidea*
Chamaeleonidae			
Brookesia	1	P	*affinis*
Gekkonidae			
Hemidactylus	3+	P	*garnotii*
Hemiphyllodactylus	1	P	*typus*
Heteronotia	4+	P	*binoei*
Lepidodactylus	1+	P	*lugubris*
Nactus	1	P	*pelagicus*
Gymnophthalmidae			
Gymnophthalmus	2	P	*underwoodi*
Leposoma	2	P	*percarinatum*
Teiidae			
Aspidoscelis	13+	P	*uniparens*
Cnemidophorus	2+	P	*cryptus*
Kentropyx	1	P	*borckianus*
Lacertidae			
Lacerta	5+	P	*unisexualis*
Xantusiidae			
Scincidae			
Menetia	3+	P	*greyii*
Lepidophyma	2	P	*reticulatum*
Typhlopidae			
Ramphotyphlops	1	P	*braminus*

Note: K = *kelptogenesis*, H = *hybridogenesis*, P = *parthenogenesis*.
Sources: In part Adams et al., 2003; Bogart et al. 2007; Darevsky, 1992; Menezes et al., 2004; Reeder et al., 2002; and Vrijenhock et al., 1989.

gametes with the R genome are produced. These gametes arise by a premeiotic shedding of the *P. lessonae* genome and then a duplication of the remaining *P. ridibunda* genome followed by normal meiotic division. These must combine with an L genome from *P. lessonae* to produce new hybrids, a process often referred to as *sexual parasitism* (similar to kleptogenesis; see following text). When two *P. esculenta* mate, the RR offspring usually die during development because of a high load of paired deleterious genes. In regions where *P. esculenta* occurs with *P. ridibunda*, the R (*P. ridibunda*) genome is lost in the germ line (but not in the soma) of *P. esculenta* during meiosis, and only gametes with the L genome are produced. These gametes must combine with an R genome from *P. ridibunda* to produce new hybrids. Male hybrids are produced because of XX–XY sex determination and the presence of X and Y gametes from *P. lessonae* males, but *P. esculenta* usually do not mate with hybrid males because they prefer the smaller males of the sexual species. In a few areas (Denmark, southern Sweden, and northern Germany)

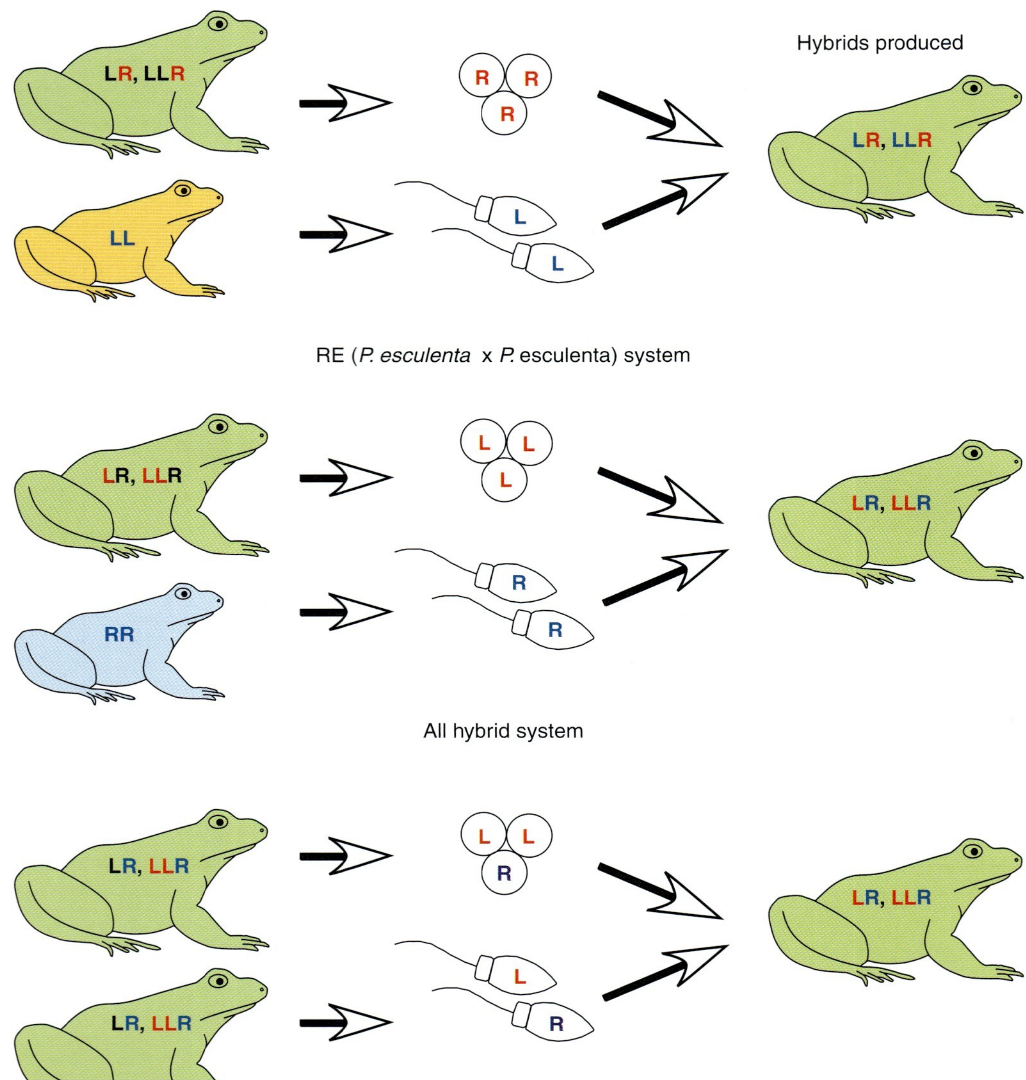

FIGURE 4.23 Hybridogenesis in the frog *Pelophylax* [*Rana*] *esculenta*. Two general breeding systems (LE and RE) exist involving sexual and unisexual species, with considerable variation within each. At three localities in Denmark, southern Sweden, and northern Germany, all-hybrid populations of *P. esculenta* occur in the absence of sexual species. Because the male-determining "y" factor is on the L genome, hybridization can and does produce male hybrids. In the RE system, male hybrids (LyRx) are more successful than female hybrids (LxRx) in reproducing with *P. ridibunda,* resulting in female hybrids being less common. Hybrid triploids are produced in some populations when a *P. lessonae* male (LL) fertilizes a *P. esculenta* (LR) egg.

in which neither sexual species exists, *P. esculenta* hybridize with each other. These populations have a high frequency of triploids (LLR) that supply the L gametes.

The all-hybrid populations of *P. esculenta* are of particular interest because the pattern of gametogenesis should lead to generation of parental genotypes (e.g., LL and RR) as well as various hybrids. However, adults containing parental genotypes do not appear in most all-hybrid populations, leading to the conclusion that they must die early in development. Parental genotypes (LL or RR) appear in some all-hybrid populations as juveniles. Consequently, it appears clear that parental genoptypes are produced, but disappear before or shortly after metamorphosis. Embryos, larvae, feeding tadpoles, and metamorphosing individuals with parental genotypes die, indicating a large hybrid load (cost to hybridogenesis) (Fig. 4.24).

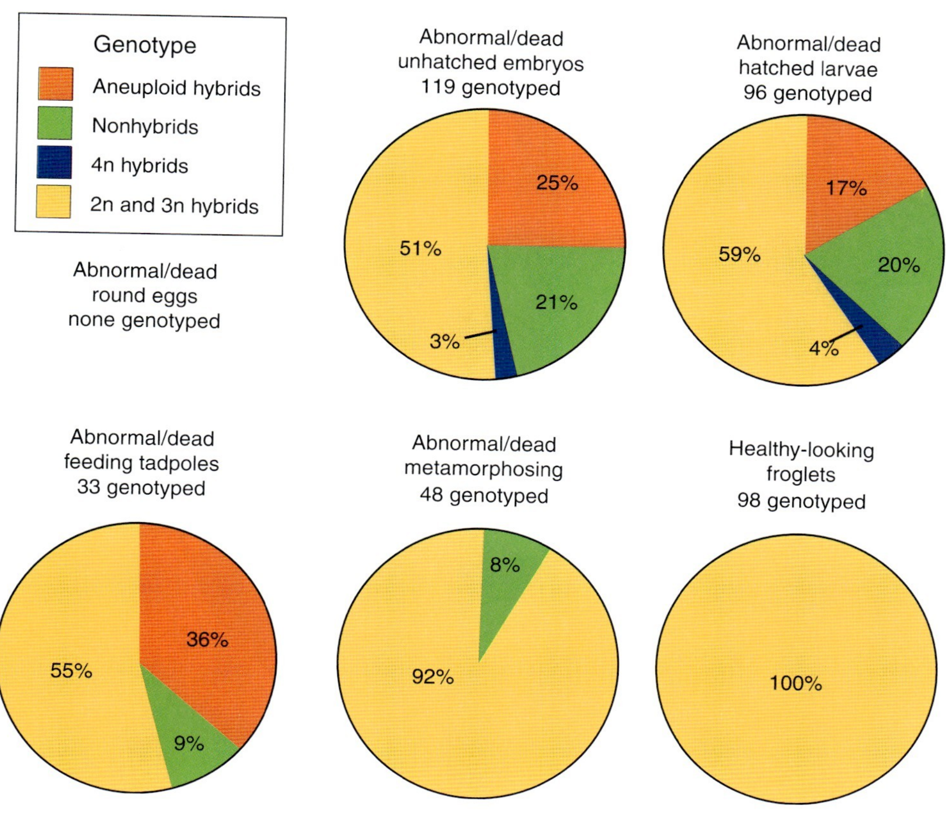

FIGURE 4.24 The cost (hybrid load) to hybridogenesis in *Pelophylax [Rana] esculenta* is high, with about 63% of offspring produced in all-hybrid populations dying before or during metamorphosis (Christiansen et al., 2005). Aneuploidy occurs when the ploidy level is not a multiple of the haploid number of chromosomes for the species.

Kleptogenesis

In parts of northeastern North America, many breeding aggregations of mole salamanders in the *Ambystoma laterale–jeffersonianum* complex consist of diploid males and females and polyploid individuals, usually females. When the composite nature of these breeding populations was first recognized, it was assumed that both the diploid individuals (*A. laterale* [genome LL] and *A. jeffersonianum* [JJ]) and the polyploid females (*A. tremblayi* [LLJ], *A. platineum* [LJJ]) were genetically distinct and reproductively isolated species, and that the unisexual polyploids were maintained by hybridogenesis or gynogenesis (a process in which sperm from a sexually reproducing male of another species is necessary to initiate egg development but the sperm genome is not incorporated into the egg). Currently, 22 distinct unisexual *Ambystoma* with chromosome numbers varying from diploid to pentaploid are recognized, and these are usually associated with one or more of the four following sexual species: *A. laterale, A. tigrinum, A. jeffersonianum*, and *A. texanum* (Fig. 4.25). Recent genetic data illustrate the complexity of the *Ambystoma laterale–jeffersonianum* complex and show that the unisexuals diverged from a sexual species, *A. barbouri*, 2.4–2.9 million years ago, a species that today does not overlap geographically with the unisexuals. Thus all unisexual populations share the same mtDNA (derived from *A. barbouri*), and depending upon where each population occurs, the local unisexuals "steal" nuclear genomes from sexual males in the breeding ponds (Fig. 4.26). This process results in individuals that derive the adaptive benefits associated with genomes of the local sexual species while retaining the ability to eliminate deleterious genes. As local conditions change through time, the gene frequencies in the sexual species change as they adapt to changing environments, and the kleptogens (unisexuals) gain the benefits by continuing to incorporate genomes of the sexual species in their soma through kleptogenesis.

FIGURE 4.25 Four sexual species of *Ambystoma* from which unisexual *Ambystoma* "steal" genomes. From left to right, *A. jeffersonianum, A. tigrinum, A. texanum*, and *A. laterale*. (J. P. Bogart).

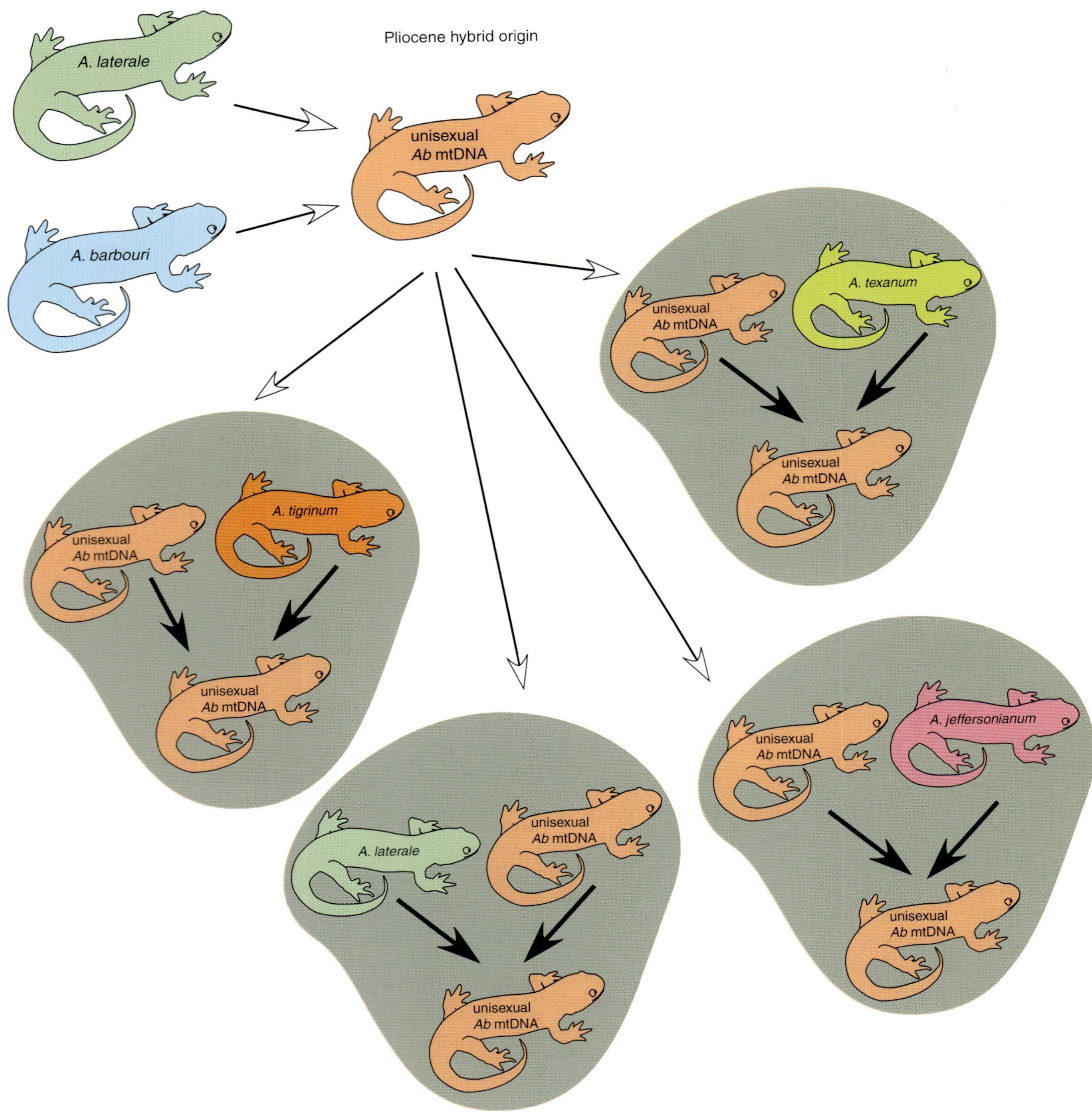

FIGURE 4.26 Kleptogenesis occurs in salamanders of the *Ambystoma laterale–jeffersonianum* complex. mtDNA has persisted unchanged since the hybrid origin of unisexual populations in the Pliocene, but unisexuals pick up and use genomes of sexual species each time they breed yet do not pass those genomes on from generation to generation. In effect, they are "stealing" genes adapted to local conditions from sexual males.

Parthenogenesis

Parthenogenesis occurs when females reproduce without the involvement of males or sperm. Inheritance is clonal, and female offspring are genetically identical to their mothers. Parthenogenesis was first discovered in the Armenian lizard *Lacerta saxicola*, and obligate parthenogenesis is now known to occur in eight lizard families and one snake family. Of the approximately 40 species of parthenogenetic squamates currently recognized (Table 4.3), all of the strictly clonal species appear to have originated as the result of hybridization of two sexual species or by backcrossing with a sexual. Confirmation that parthenogenesis was occurring in these lizards resulted from studies in which laboratory-born individuals were raised to maturity in isolation and began producing offspring. Because these parthenoforms were produced by hybridization, heterozygosity is high. Genetic variation within an individual is high, but genetic variation among individuals is nearly nonexistent. Low genomic variation within clones of parthenogenetic lizards has been demonstrated with studies on histocompatibility of skin transplants. Nearly 100% of skin grafts transplanted between individuals (two populations) of the parthenogenetic species *Aspidoscelis* [*Cnemidophorus*] *uniparens* were permanently accepted, whereas no skin grafts transplanted between individuals within a population of the sexual species *A. tigris* were accepted, suggesting that all *A. uniparens* can be traced back to a single individual (Fig. 4.27). Even though some parthenogenetic whiptail lizards appear to be uniform across their range (e. g., the widespread *A. uniparens*), others do not. Three distinct color pattern classes exist in the unisexual whiptail *A. dixoni*, whose origin can be traced to a single hybridization event between the sexual species *A. tigris marmorata* and *A. gularis septemvittata*. Histocompatibility studies based on reciprocal skin transplants confirm the single origin and suggest that pattern classes resulted from mutation or recombination occurring within the single historical unisexual.

The cytogenetic events that result in production of eggs with the same ploidy as the mother have been detailed only in *A. uniparens*. Premeiotic doubling of chromosomes yields a tetraploid oogonium, which is followed by normal meiosis that produces eggs with the same chromosome number as the female parent. Most parthenogenetic squamates are diploid, but some are triploid. The triploid condition results from backcrossing between a female of hybrid origin and a normal male of one of the original parental species (Fig. 4.28).

A rather interesting twist on parthenogenesis has recently been described for Komodo Dragons and Indian Pythons. In both of these species, captive females produced eggs with no involvement of sperm, and the offspring were all female. In the pythons, all parthenogenetically produced offspring were genetically identical to the mother, whereas in the Komodo Dragons, parthenogenetically produced females were homozygous for all loci but variation existed among offspring. In both cases, parental females also reproduced sexually. Consequently, these systems represent nonhybrid induced instances of facultative parthenogenesis and present a reproductive system that might be of particular interest for studies on the evolution of sex.

Although the cytogenetic mechanism initiating development is unknown in parthenogenetic squamates, a rather strange behavior, pseudocopulation, in which one female behaves as a male and attempts to mate with another female, occurs commonly under laboratory conditions in some parthenogens. This behavior has been observed in the field but appears to be uncommon. Comparisons of hormone levels in the courted females and the courting females of three parthenogens (*Aspidoscelis tesselatus*, *A. uniparens*, and *A. velox*) show that the courted female is preovulatory and the courting female is postovulatory or oogenetically inactive (Fig. 4.29). Courtship and pseudocopulation stimulate ovulation, indicating that the courted female is responding as though mating has occurred. Females that experience pseudocopulation

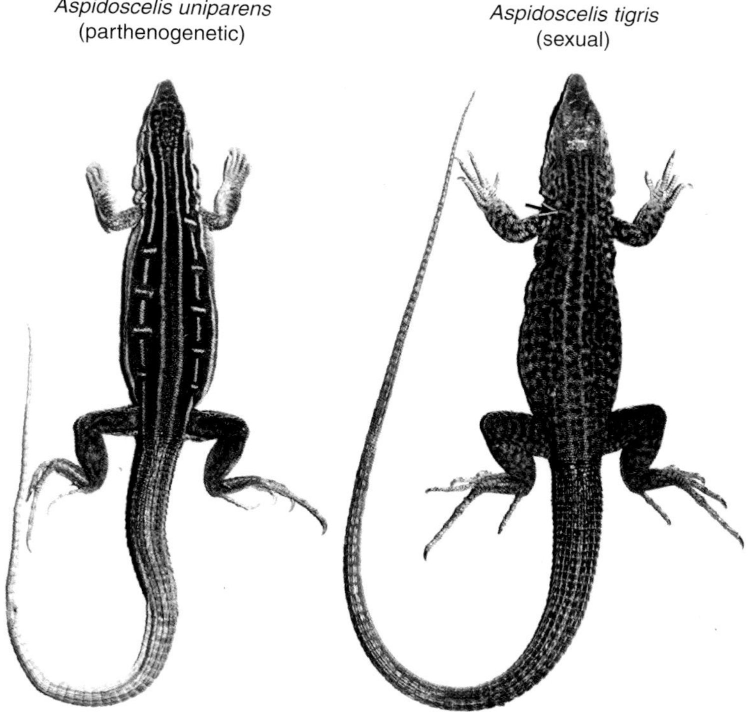

FIGURE 4.27 Skin-graft test for genetic similarity in the unisexual *Aspidoscelis uniparens* (left) and the bisexual *A. tigris* (right). Because of the clonal nature of *A. uniparens*, all 9 grafts were accepted; in contrast, all 10 grafts were rejected in *A. tigris*. Adapted from Cuellar, 1976.

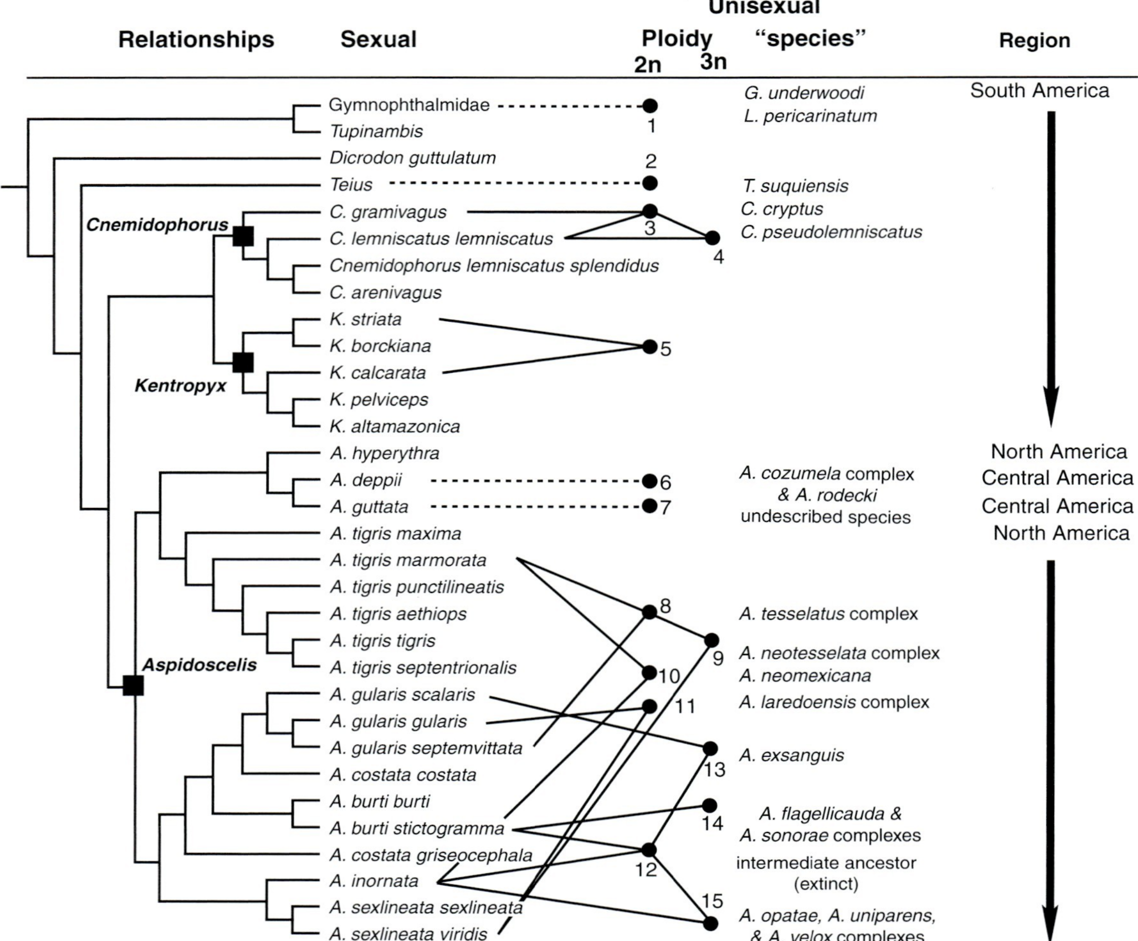

FIGURE 4.28 Genealogy of the parthenogenetic teiid and gymnophthalmid lizards. The lines originating on species names denote the parents that hybridized to create the parthenoforms/parthenogens (black circles). In many cases, a single hybridization event produced diploid parthenoforms, in others, a single hybridization produced triploid parthenoforms, and in yet others, backcrosses between a parthenoforms and a sexual species produced triploid parthenoforms. Parthenogenesis has arisen independently in the Teioidea multiple times. Adapted from Reeder et al., 2002.

appear to produce eggs at a faster rate than those that do not engage in the behavior. The evolutionary significance of pseudocopulation remains unclear because no genomic differences are known between females in the clones that participate in pseudocopulation and those that do not. Nevertheless, this system provides unique opportunities to study the role of specific behaviors (courting and copulatory behavior) on female reproduction without the added variables associated with males.

Because each female of parthenogenetic squamates produces only females, the reproductive rate in terms of potential population growth is enormous compared to that of sexually reproducing squamates (Fig. 4.30). Given this apparent advantage to unisexual reproduction, why is unisexual reproduction so rare in vertebrates and indeed in most animals? This question is revisited in the discussion on sex ratios.

LIFE HISTORIES

An organism's life history is a set of coevolved traits that affect an individual's survival and reproductive potential. Key life-history traits include age-specific survivorship (or the converse, age-specific mortality), brood size, size of young at birth or hatching, distribution of reproductive effort, interaction of reproductive effort with adult mortality, and the variation in these traits among an individual's progeny. Approaching the study of life histories from the perspective of easily measurable traits, whether they are quantitative (e.g., age-specific survivorship, offspring size) or qualitative (e.g., oviparous versus viviparous, montane versus desert), allows identification of natural patterns and provides insight into factors that can influence the evolution of life histories. Such an approach is particularly useful for generating testable hypotheses. From one

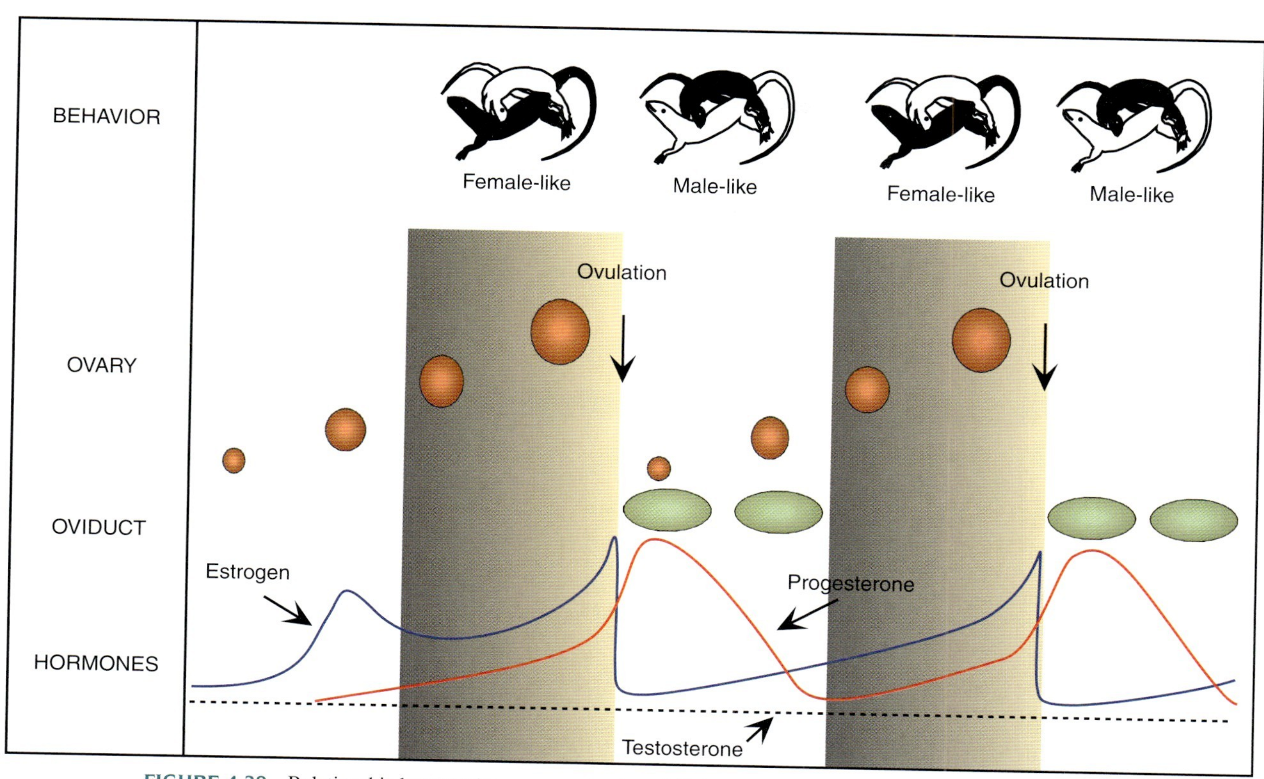

FIGURE 4.29 Relationship between hormone production, follicle development, and behavior in parthenogenetic whiptail lizards (*Aspidoscelis*) during pseudocopulation. Adapted from Crews and Moore, 1993.

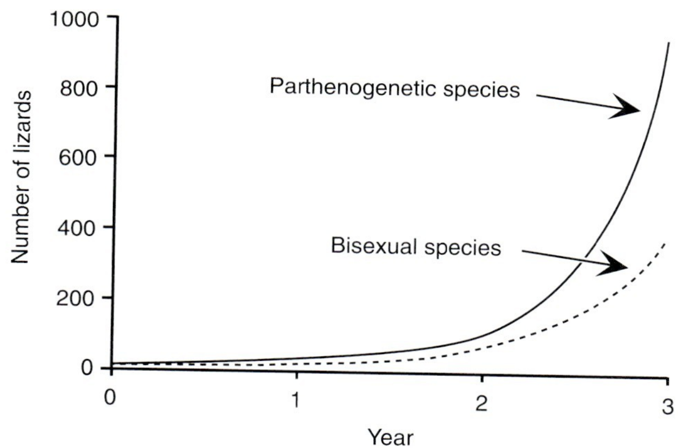

FIGURE 4.30 Hypothetical growth rates for populations of parthenogenetic and sexually reproducing *Aspidoscelis* based on laboratory data on *A. exsanguis* and assuming no mortality. The starting point on the graph represents hatching of one egg. Because 50% (males) of the sexually reproducing species do not produce eggs, population size of the parthenogenetic population is more than double that of the sexual species after only 3 years. Adapted from Cole, 1984.

perspective, a life history represents a set of rules that determine energy allocation decisions based on variation in operative environments. Operative environments include the ranges of temperatures, humidities, resources, and other variable conditions experienced by individuals throughout their lifetimes. This allocation perspective provides the potential to identify underlying evolutionary causes of observed patterns by an examination of trade-offs in energy use. Heuristically attractive, the allocation approach brings all aspects of ecology, physiology, and behavior into life histories.

Life history studies, by definition, are studies of populations. A snapshot view can be obtained by examining the age structure of a population at one or several different times. Age distribution analysis examines the size of each age class within a population at a single moment in time. Size of age classes can be the actual number of individuals in the age class or the proportion of the total population. The age distribution pattern for a population will be stable through time if its survivorship (l_x) and fecundity (m_x) schedules remain constant. In a stable age distribution, the proportion of individuals in each class remains constant. Resulting data can be presented as age (size) distributions (Fig. 4.31). Some salamander populations, such as plethodontids in climax Appalachian forest, appear to have nearly stable age distributions. Annual population loss through mortality, emigration, and aging in each age class is matched by recruitment through aging and immigration. Equilibrium population size in these salamanders derives from a longevity greater than 10 years, a stable environment, low predation, and occupation of all suitable habitat by adults. Time series of age distributions can provide estimates of cyclical patterns in populations

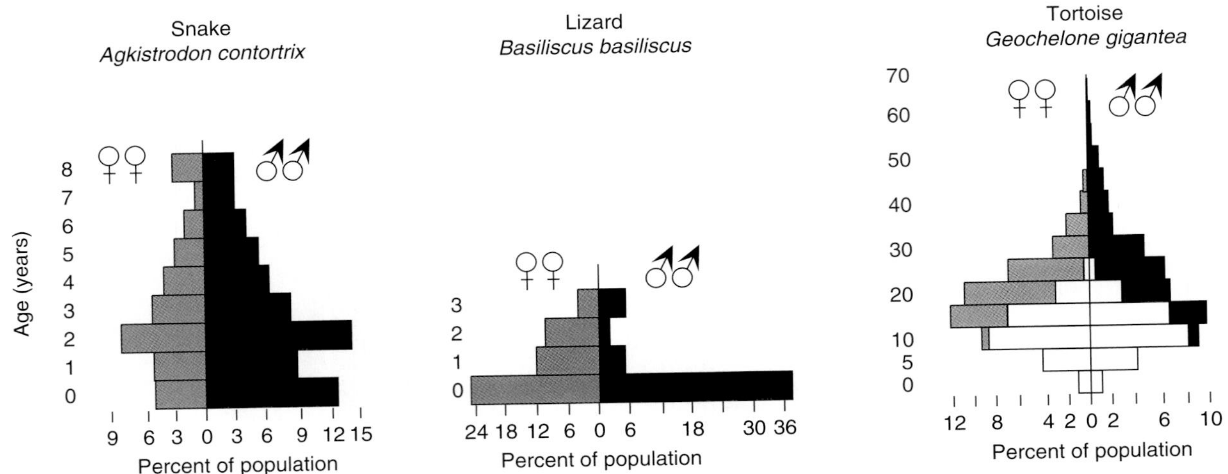

FIGURE 4.31 Age distribution patterns of a snake, lizard, and tortoise population. Point-in-time patterns differ between a moderate-lived snake, *Agkistrodon contortrix*; a short-lived lizard, *Basiliscus basiliscus*; and a long-lived tortoise, *Geochelone gigantea*. The bars denote the percent (of total population) of males or females present in each age class; open bars, unsexed individuals; shaded bars, females; solid bars, males. Adapted from Vial et al., 1977; Van Devender, 1982; and Bourn and Coe, 1978, respectively.

(e.g., timing of reproduction, relative densities of different age/size classes) and are very useful in designing long-term studies. Age distributions usually include all individuals of both sexes. What age distributions do not yield are accurate measures of age-specific survivorship or mortality, even though relative numbers of different categories can be tracked through time.

Capture–recapture studies that follow cohorts (groups of animals of the same age) provide much more detailed information and can be used to determine whether populations are remaining stable, decreasing, or increasing through time. In addition, they can be used to construct survivorship curves, which differ considerably depending upon where in the life history most mortality occurs. To estimate survivorship curves, an individual cohort (e.g., all eggs laid in one season) is followed through time until all individuals disappear from the population. The first age group might be egg to hatchling, and if 90% of eggs are destroyed by predators, then survivorship (l_x) for that cohort is 0.1 (10% survived). At each subsequent age group, the proportion surviving can be calculated as the number surviving divided by the original number. These data can be used to construct a life table (which is usually based just on females, because they are the reproducing individuals). Life tables provide a summary of a population's current state and can suggest whether the population is likely to persist. Life tables also permit intra- and interspecific comparisons of populations. Primary components of life tables are the average age of sexual maturity (i.e., age at which individuals first begin to reproduce) and age-specific mortality (l_x) and fecundity (m_x). By multiplying l_x by m_x, the age-group fecundity can be calculated ($l_x m_x$). A number of other measures of a population's state can be derived from these data, including mean generation time (T), net reproductive rate (R_o, also called the replacement rate), reproductive value (v_x), intrinsic rate of natural increase (r), and others. R_o is especially informative; it ranges between 0 and $\approx$ 10 for vertebrates. A value of 1.0 indicates that the population is stable (births = deaths). Declining populations have $R_o < 1.0$ and increasing populations have $R_o > 1.0$. An example of a typical life table appears in Table 4.4.

Survivorship (l_x) and mortality (d_x and q_x) are different aspects of the same population phenomenon, the rate of mortality of a cohort. Survivorship (l_x) maps the cohort's decline from its first appearance to the death of its last member. Age-specific mortality (d_x or q_x) records the probability of death for the surviving cohort members during each time interval. The pattern of a cohort decline is often shown by plotting survivorship against time. Four hypothetical survivorship curves represent the extremes and medians of possible survivorship patterns (Fig. 4.32). In Type I (rectangular convex curve), survivorship is high (i.e., early mortality low, $q_x < 0.01$) through juvenile and adult life, and then all cohort members die nearly simultaneously ($q_x = 1.0$). Type III is the opposite pattern (rectangular concave curve), where mortality is extremely high ($q_x > 0.9$) in the early life stages and then abruptly reverses to almost no mortality ($q_x < 0.01$) for the remainder of the cohort existence. The Type II patterns occupy the middle ground, either with a constant number of deaths (d_x) or a constant death rate (q_x). In Type II patterns with a constant death rate (q_x), survivorship declines more rapidly because the actual number of deaths at any time is based on a

TABLE 4.4A Two Examples of Life Tables.
A. Life table for a French population of Wall Lizards, *Podarcis muralis*.

Age	Survivors		Mortality		Life Expectancy
x	n_x	l_x	d_x	q_x	e_x
0-1	570	1.000	376	0.66	1.01
1-2	194	0.340	146	0.75	0.99
2-3	48	0.084	23	0.48	1.48
3-4	25	0.044	13	0.52	1.36
4-5	12	0.021	6	0.50	1.31
5-6	6	0.011	3	0.50	1.05
6-7	3	0.005	2	0.50	0.70
7-8	1	0.002	1	1.00	

Note: Abbreviations and explanations: x, *age interval (1 yr);* n_x, *actual number of members alive at beginning of age interval;* l_x, *proportion of cohort alive at beginning of interval;* d_x, *number of cohort members dying during age interval;* q_x, *age-specific death rate (proportion of individuals dying during interval that were alive at beginning of interval);* e_x, *average life expectancy (yr) for members alive at beginning of interval;* m_x, *age-specific fecundity rate (average number of offspring produced by surviving cohort during each interval;* $l_x m_x$, *total fecundity of surviving cohort members in each interval;* R_o, *replacement rate or net reproductive rate (average lifetime fecundity for each cohort member).*
Data from Barbault & Mou, 1988; *italized values are hypothetical, assuming constant* q_x *for adults.*

TABLE 4.4B Survivorship and fecundity schedule for a North Carolina population of Applachian dusky salamanders, *Desmognathus ocoee* (formerly *D. ochrophaeus*).

x	l_x	m_x	$l_x m_x$	
0	1.000	0.0	0.000	
.				
4	0.087	4.5	0.392	
5	0.055	4.5	0.248	
6	0.034	4.5	0.153	
.				
8	0.013	4.5	0.058	
.				
10	0.005	4.5	0.022	
.				
12	0.002	4.5	0.009	$R_o = 1.026$

Data from Tilley, 1980.

percentage of the remaining population. Although these idealized patterns are never matched precisely by natural populations, the patterns offer a convenient descriptive shorthand for comparing population data.

Most amphibians with indirect development, crocodylians, and turtles have Type III survivorship. Amphibian eggs and larvae commonly experience high predation. Increased size resulting from growth may temporarily render older larvae less subject to predation by aquatic predators, but predation is again high during metamorphosis and early terrestrial life. For those species that breed in temporary ponds, death of entire cohorts is a regular threat because ponds may dry prior to metamorphosis. Many turtle populations suffer high nest predation; freshwater species and seaturtles often have 80–90% of their nests destroyed within a day or two of egg deposition. The majority of the remaining amphibians and reptiles have Type II-like patterns with moderate and fluctuating mortality during early life and then a moderate to low and constant death rate during late juvenile and adult life stages. Weather (e.g., too wet or too dry) appears to be the major cause of juvenile mortality in many Type II species. No amphibian or reptile attains a close match to a Type I survivorship. Among reptiles, *Xantusia vigilis* approaches this pattern. Species with parental care may have an initial low mortality, but even crocodylians cease parental care well before their offspring are fully predator- and weatherproof. Populations do not have fixed survivorship patterns. Annual patterns are most similar in populations with a nearly constant age structure, but even these populations can shift from one pattern to another due to a catastrophic event or an exceptional year of light or heavy mortality. Males and females in the same cohort may have different survivorship curves, and if the difference is great, the resulting population will have an unequal sex ratio.

Approaches to the study of life histories of amphibians and reptiles have been quite different from the outset because most amphibians have complex life cycles involving a larval stage, and reptiles do not. Many amphibians can be observed only during their breeding season, which places further limitations on the study of their life histories. Consequently, life history studies in amphibians and reptiles have historically emphasized different variables and/or focused on different stages of the life history. Recent appreciation of the role of history in determining life history traits has helped to identify evolutionary origins of traits that are carried through to all members of various clades (e.g., matrotrophic viviparity in squamates).

Reproductive Effort and Costs of Reproduction

Reproductive effort was originally defined in terms of energy allocation, emphasizing conditions that might cause organisms to divert more or less energy to reproduction. Reproductive effort is usually viewed in terms of the total amount of energy spent in reproduction during a defined time period, such as one reproductive episode or season. This approach is particularly useful because it provides the opportunity to examine the effects of the timing

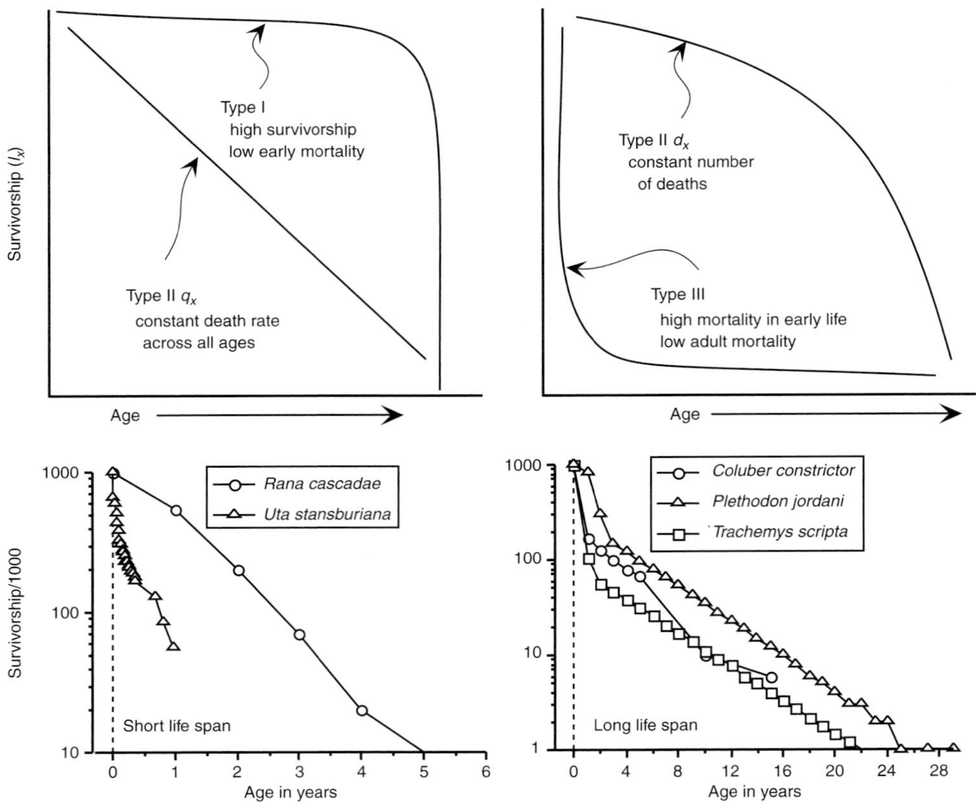

FIGURE 4.32 Top: Hypothetical survivorship curves for animal populations (see text). Bottom: Representative survivorship curves for amphibians and reptiles with short life spans (left) and long life spans (right). Although the lower graphs are superficially similar, note the great difference in age scale. Data from the following: Amphibians—*Pj*, Hairston, 1983; *Rc*, Briggs and Storm, 1970; Reptiles—*Cc*, Brown and Parker, 1984; *Ts*, Frazer et al., 1990; *Us*, Tinkle, 1967.

and intensity of reproductive investment on other life history traits during the animal's lifetime. Reproductive effort has two components: the energy invested by the female, and the way that energy is proportioned into individual offspring. Investing heavily in reproduction at one time has numerous costs. If the total energy investment (potential expenditure) in an individual's lifetime reproductive effort is the individual's reproductive value at the beginning of adulthood, an individual's reproductive value will decline with each successive reproductive event. More explicitly, any current reproductive investment decreases an individual's reproductive value, and what remains (of the total potential investment) is called residual reproductive value. An individual of a species that reproduces only once in its lifetime expends its entire reproductive value in that single event, in contrast to declining reproductive value for species that reproduce repeatedly. In the latter case, natural selection should favor age-specific reproductive efforts that maximize reproductive value at each age. Reproductive effort, thus, is defined in terms of costs and reflects a trade-off between energy allocated to reproduction and its effect on future fecundity and survival. Theoretically, a species investing heavily in reproduction early in life (high reproductive effort during each episode) should have relatively short life expectancy, and a species that invests little in reproduction (low reproductive effort in each episode) should have high life expectancy.

In general, amphibians and reptiles meet these predictions. In relative terms, amphibians typically have high reproductive efforts and relatively short life expectancies compared to reptiles. Most frogs and salamanders lay clutches of eggs comprising a large portion of their body mass and thus constitute a large portion of their overall energy budget. Some reptiles, such as crocodylians and turtles, deposit relatively small clutches (in terms of total energy) during each reproductive episode but do so year after year.

A tortoise that spreads its reproductive effort over a long life and short-lived *Anolis* lizards that episodically invest in reproduction during a short life, represent two extremes in the patterns of energy allocation to reproduction. Alhambra tortoises (*Geochelone gigantea*) reach sexual maturity in 13–17 years, live to 65–90 years, reach body sizes varying from, 19–120 kg, and reproduce repeatedly. Among three isolated island populations, reproductive effort varies depending on density, which reflects per capita resource availability. Reproductive effort is greatest in the population with the lowest density and reaches its minimum in the population with the highest density. When resources increase from one year to the next, reproductive effort increases more in the high-density population than in the low-density population, suggesting that low per capita resource availability constrains reproductive effort in the high-density population.

Anoles (*Anolis*) are early maturing, live relatively short lives, are relatively small in body size, and reproduce repeatedly. Each reproductive episode results in the production of a single egg. Because all anoles produce a single egg clutch, this reduced, fixed clutch size likely evolved in an ancestor to the anole clade. Compared with similar-sized lizards with variable clutch size, anoles have low reproductive efforts per episode (Fig. 4.20). Although the evolutionary cause of the low reproductive effort in anoles (and other lizards with low, invariant clutch size) remains unknown, it does allow frequent egg deposition and deposition of eggs at different sites. These lizards are likely hedging their bets in an environment in which egg mortality is high and unpredictable. Effectively, they place their eggs in many baskets, each of which might experience a different set of conditions. When examined on an annual basis, which for many anoles *is* the total natural life span, reproductive effort is actually high, constituting about 25% of the total annual lifetime energy budget. These short-lived lizards divide their lifetime reproductive efforts into numerous episodes, each of which represents a relatively low investment in reproduction. The primary cost of repeated reproduction over short time intervals is a short life span for these lizards.

Costs of reproduction can be divided into two major categories: potential fecundity costs and survival costs. Fecundity costs represent the energetic expenditure of reproduction. The tortoise and *Anolis* patterns previously described and many others center on these costs. Energy invested in reproduction is energy that is not available for growth or maintenance. Survival costs are more complex but center around the increased vulnerability of females that are carrying eggs or embryos, either directly from the effect of clutch mass on mobility or indirectly due to reduced physical condition following parturition and its effect on escape behavior or overwintering. Survival costs appear to be much more important as determinants of reproductive effort in relatively short-lived organisms.

Survival costs can be indirectly estimated by comparing performance or behaviors of animals with and without eggs. The ratio of clutch weight to body weight (relative clutch mass) provides an operational estimate of the burden of a clutch on a female. Gravid Australian skinks exhibit reduced performance as measured by running speed, and females of some skinks bask more when gravid than when not gravid. As relative clutch mass increases, females become progressively slower in running trials, suggesting that survival costs increase proportionately with increased reproductive effort.

The implication from these kinds of studies is that reduced performance could increase risk of predation, but actually measuring increased predation risk under natural conditions is difficult. Some studies have shown that running speed (a common measure of performance) may not affect vulnerability to some predators, and other studies have shown that females can offset their risk by reducing activity or remaining close to retreat sites. A particularly enlightening study on Australian Garden Skinks (*Lampropholis guichenoti*) compared the effects of decreased body temperature, eating a large meal, and losing the tail, all of which can occur repeatedly during the life of the lizard, on performance of pregnant and nonpregnant females. All of these decreased performance, but decreased body temperature, eating a large meal, and losing the tail had nearly double the effect that pregnancy had. When taken in the context of the total behavioral repertoire of an individual over its lifetime, performance effects of pregnancy are relatively minor. More importantly, these results suggest that performance reduction due to pregnancy may not have been a strong selective force in the evolution of reproductive investment.

Costs can also be incurred by ecological constraints on reproductive investment. Dorsoventrally flattened lizards in South America (*Tropidurus semitaeniatus*) and Africa (*Platysaurus* species) have reduced clutch size and low relative clutch mass, presumably as part of a coevolved set of morphological and reproductive traits designed to enhance use of narrow crevices for escape. In the small-bodied Australian skink, *Lampropholis delicata*, body shape varies geographically with relative clutch volume, suggesting either that morphology constrains reproductive investment per episode below optimal levels or that life history trade-offs have resulted in the coevolution of morphology and reproductive investment. Comparison between oviparous and viviparous populations of the Australian skink *Lerista bougainvillii* reveals that body volume increases as the result of a combination of increased female size and increased relative clutch mass associated with viviparity, even though the number and size of offspring remains relatively constant. In this example, the added clutch mass represents a survival cost of viviparity.

Although most attention has been given to females in assessing costs of reproduction, some evidence suggests that males incur high costs as well. In male European adders (*Vipera berus*), sperm production is a major reproductive expenditure. Total body mass decreases during spermatogenesis and prior to the initiation of reproductive behaviors associated with finding and courting females.

Life-History Variation

Amphibians

Life histories of most amphibians consist of egg, larval, juvenile, and adult stages. Because of the distinct morphological, physiological, and behavioral changes that occur during metamorphosis and the change in habitat between larval and juvenile stages, amphibian life cycles are complex. In species with direct development, the larval stage is absent. In some species, individuals with larval morphology become sexually mature and reproduce, and the "typical" adult morphology is never achieved. Other interesting variations in life histories of amphibians occur as well. Life history studies of amphibians have concentrated either on the dynamics of the larvae, which are relatively sedentary and constitute a primary growth stage, or on adults, which are relatively mobile and are the dispersal and reproductive stage. Additionally, numerous experimental studies have focused on larvae because the larval period likely regulates amphibian population size.

No long-term life history studies exist on caecilians. Compared with other amphibians, they produce relatively small clutches of large eggs or small broods of large offspring. It would not be surprising if most species are late-maturing and long-lived, but long-term studies are necessary to determine this, and the secretive habits of caecilians have prevented such studies. However, a recent study on *Ichthyophis kohtaoensis* in the Mekong Valley of northeastern Thailand demonstrates that even the most cyptozoic species can be studied if proper techniques are developed. These caecilians mate and deposit eggs at the beginning of the monsoon season. Nests are deposited terrestrially and females remain with the eggs until they hatch during the peak or near the end of the rainy season. The larvae are aquatic and become terrestrial when they metamorphose into juveniles at the end of the dry season. The terrestrial juveniles and adults live in a variety of habitats where they spend most of the time in the soil (dry season) or under leaf litter and decaying vegetation (wet season). Based on size distributions, they appear to reach sexual maturity in 3 years. Their densities are low, with only about 0.08 individuals per square meter. Caecilians in general have potentially diverse and interesting life histories, and because they live either in the soil or in water, increased understanding of their life histories is necessary to determine impacts of human activity on these elusive amphibians.

Life history characteristics vary greatly among salamander species. Among the species with aquatic larvae, salamanders differ from frogs in that the larval morphology is similar to that of the adult except gills are present and limbs may be less well developed than in the adults. Frogs exhibit the greatest diversity in life histories among tetrapod vertebrates. Among species with aquatic larvae, the larval morphology is entirely different from that of the adult. Larval morphology changes to adult morphology as a consequence of a major metamorphosis during which the tail is resorbed, larval mouthparts are replaced by adult mouthparts, fore- and hindlimbs emerge from the body, and major changes occur in the physiology and morphology of the digestive system. In species with direct development, hatchlings are nearly identical morphologically to adults but much smaller in body size.

The complexity of amphibian life histories is evident through the factors influencing survival at each stage. Amphibian eggs experience mortality from desiccation due to drying of egg deposition sites and predation by insects, fish, reptiles, birds, and even other amphibians. Terrestrial-breeding amphibians and those that place their eggs on vegetation above water have eliminated sources of egg mortality associated with the aquatic habitat. Survival of bullfrog (*Lithobates catesbeianus*) eggs, for example, varies from 10–100%; predation by leeches and developmental abnormalities are major sources of mortality. The quality of the male territory appears to be the primary determinant of egg survival in these frogs. In woodfrogs (*Lithobates sylvaticus*), survival of eggs is extremely high (96.6%). Amphibian larvae experience some of the same sources of mortality, but because of their mobility, rapid growth rates, and in some instances, production of noxious chemicals for defense, they are able to offset some mortality. Amphibian larvae of many species are capable of rapid growth as a result of their ability to respond to rapid increases in food availability typically occurring in breeding sites. For larvae, the environment rapidly changes from one in which resources are abundant and predators are scarce just after ponds fill, to environments rich in predators (mostly aquatic insects) and relatively low in resources as larval density increases. Larger larvae are less susceptible to predation and metamorphose at a larger body size. Survival rates of larvae vary considerably. Bullfrog (*Lithobates catesbeianus*) tadpoles in Kentucky have a survival rate varying from 11.8–17.6% among ponds. In the salamander *Ambystoma talpoideum*, survival to metamorphosis varies among ponds and among years within particular ponds. In one pond in South Carolina, no larvae metamorphosed over a 4-year period. In another pond, survival varied from 0.01% to 4.09% over a 6-year period. The length of time that ponds held water (hydroperiod) accounted for much of the variation in larval survival.

The juvenile stage is also a rapid growth stage, and because recently metamorphosed amphibians are inexperienced in their new environment, mortality due to predation is likely high. Experienced adults likely face their greatest threat of mortality during breeding events. High and localized densities of amphibians during breeding provide opportunities that do not exist during much of the year for predators. In some frog species, male vocalizations actually attract predators such as the frog-eating bat, *Trachops,* which orients on the call and captures calling males.

Reptiles

Life histories of all reptiles include egg (or embryo, in viviparous species), juvenile, and adult stage. Reptile life histories are much simpler than those of amphibians because there is no larval stage.

Crocodylians and Turtles

All crocodylians and turtles, when compared with squamates, are late maturing, reproduce over extended time periods (many years), and are long-lived. Most mortality occurs in early life history stages, the eggs and juveniles. Clutch size varies from 6–60 eggs among species of crocodylians (19 species) and is largest in *Crocodylus niloticus* (60) and *Crocodylus porosus* (59). The largest clutch size for an individual was a *C. porosus* with 150 eggs. Larger species and individuals tend to produce larger clutches. The Philippine crocodile, *Crocodylus mindorensis,* exemplifies a typical crocodylian life history. Females produce multiple clutches of 7–25 eggs that hatch after 77–85 days. Females guard the nest, and vocalizations of pipped young cue the females to open the nest and transport juveniles to water.

Squamates

Considerable variation exists in life history traits of squamates. Many small lizards, such as *Uta stansburiana,* are early maturing (9 months), reproduce repeatedly, and have short life spans. Others, such as *Cyclura carinata,* are late maturing (78 months), produce a single brood per year, and are relatively long-lived. Among snakes, similar life-history variation exists. *Sibon sanniola* reaches maturity in 8 months and produces a single clutch per year, whereas *Crotalus horridus* reaches sexual maturity in 72 months and produces a brood every other year. Early attempts to determine relationships among the life history characteristics of squamates were based on a limited set of data. Nevertheless, it was clear that lizard life histories could be grouped into species that mature at large size, produce larger broods, and reach sexual maturity at a relatively late age and those that mature at small size, produce smaller broods, and reach sexual maturity early in life. More sophisticated analyses based on more extensive data sets and inclusion of additional variables confirm some of these generalizations and refute others. Lizard life histories can be categorized primarily on the basis of brood frequency (Fig. 4.33). Single-brood species are subdivided into three categories: (1) oviparous species with delayed maturity and large brood size; (2) oviparous species with small broods; and (3) viviparous species. Multiple-brooded species include the following: (1) small-bodied,

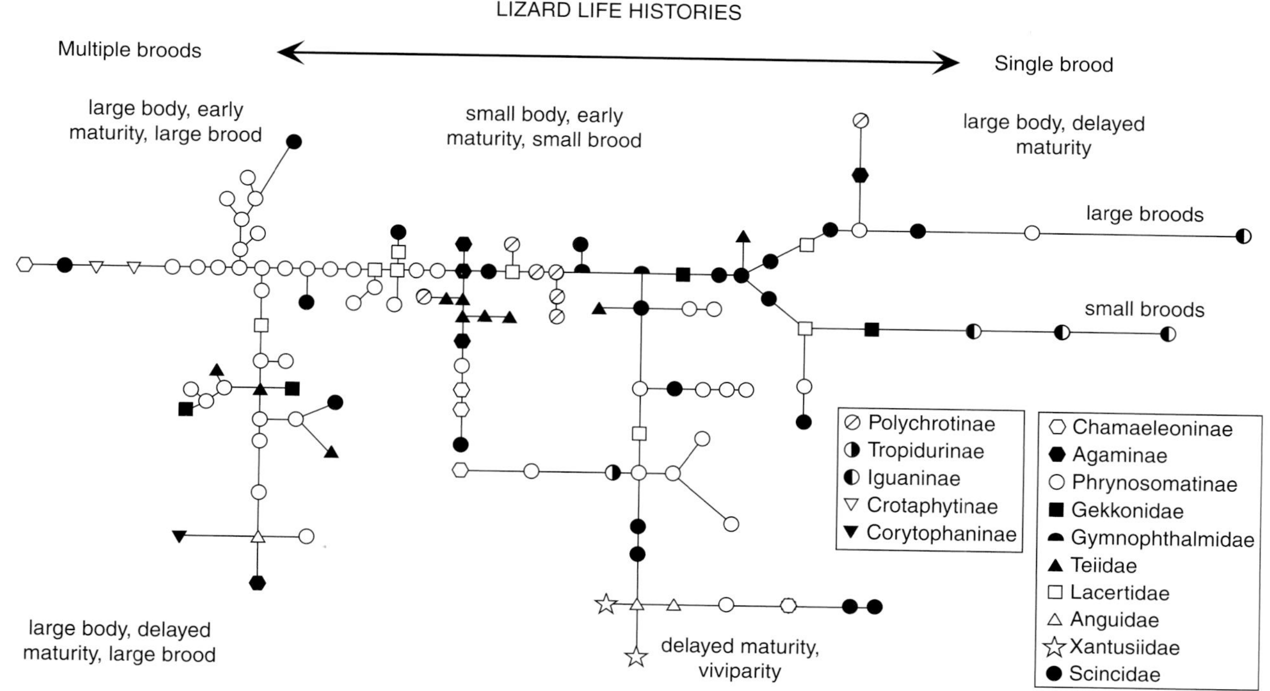

FIGURE 4.33 Prim diagram showing axes of variation in life history traits of lizards. Adapted from Dunham et al., 1988, with taxonomy for the Iguania updated.

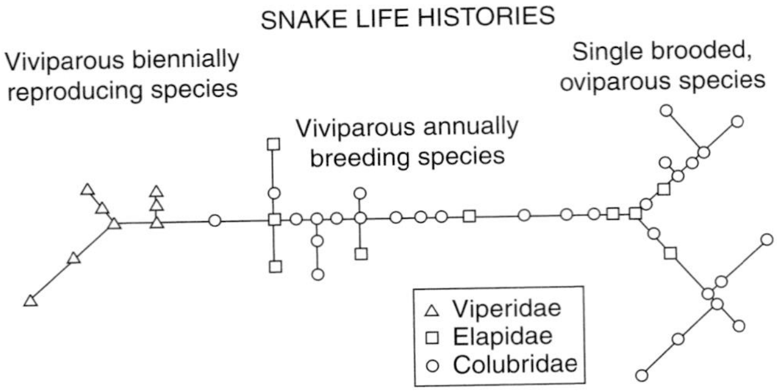

FIGURE 4.34 Prim diagram showing axes of variation in life history traits of snakes. Only three snake families are included, so the analysis must be considered preliminary. Nevertheless, coadapted sets of life history traits appear evident. Adapted from Dunham et al., 1988, with errors corrected.

early-maturing species with small broods; and (2) larger-bodied species, with early maturity and large broods. Snake life histories fall into three categories (Fig. 4.34). The first includes oviparous and single-brooded species (mostly colubrids) that have increased body size, clutch size, and delayed maturity. The second category is comprised of viviparous species that breed annually (some elapids and colubrids). The third group consists of viviparous species that reproduce biennially (all of the viperids and the garter snake, *Thamnophis sirtalis*, from the northern part of its range).

Seasonal Versus Aseasonal Environments

Early theory suggested that life histories of organisms living in seasonal environments, particularly temperate zones, should be different from those living in aseasonal environments, particularly the wet tropics. Seasonal environments were considered to be less resource limited than aseasonal environments (*r*- versus *K*-selection). Species in aseasonal environments should spread out their reproductive investment temporally and thus produce more clutches with fewer and larger eggs in each clutch. Larger offspring would presumably have a competitive advantage in such resource-limited environments. In seasonal environments, reproductive investment is constrained to fewer, larger clutches because the season is short and competition among offspring would be relaxed due to high resource availability. Some support for these ideas is evident in the first insightful analysis of life history data for lizards, albeit the data used for analysis were limited.

Testing these ideas with squamates has been difficult partly because different evolutionary lineages are involved in most comparisons, and partly because reproductive variables used are not always comparable. For example, some lineages (anoles, geckos, gymnophthalmids, and others) have clutch sizes that do not vary, whereas other lineages (most iguanians and autarchoglossans) have clutch sizes that vary with body size. Hypotheses should be tested within taxonomic groups that have species in both seasonal and aseasonal environments. A comparison of life history characteristics among Australian lizards revealed that congeneric tropical and temperate-zone species do not differ in clutch size when the effects of body size are eliminated. Body size of egg-laying skinks does not differ between tropical and temperate environments. Tropical skinks, however, have lower clutch sizes and lower relative clutch masses than temperate-zone species. Also, greater numbers of species with low, invariant clutch size occur in the tropical environment. Consistent variation in life history traits exists that is not attributable to seasonal versus aseasonal environments. Egg volume, for example, increases with female body size in *Cryptoblepharus* but not in species of *Carlia*.

Phenotypic Plasticity in Life History Traits

Variation in life-history traits includes timing of reproduction, size and number of offspring, number of clutches, and individual growth rates. This variation is well known within amphibian and reptile populations and most often appears associated with variable energy resource availability. This within-population variation is termed phenotypic plasticity and has been examined in both field and laboratory studies.

Both the number of eggs produced and the frequency of clutch production decline in New Mexico populations of the lizard *Urosaurus ornatus* as the result of reduced resources. Lower rainfall during 1 year reduces prey populations, and that in turn limits the lizard's ability to obtain adequate energy for reproduction. Variation in prey availability associated with rainfall also accounts for variation in growth rates in the lizard *Sceloporus merriami*.

In laboratory studies where energy intake has been precisely controlled, both garter snakes (*Thamnophis marcianus*) and ratsnakes (*Pantherophis guttata*) respond to increased resource availability by increasing clutch mass and the number of offspring produced. *Pantherophis guttata* also increases relative clutch mass in response to increased resource availability. In both of these snakes, individual offspring size does not respond to resource levels, indicating that offspring size is optimized within narrow limits in these snakes.

Variation of biophysical regimes within reptile nests can influence the phenotypic variability of offspring. Female skinks (*Bassiana duperreyi*) living in a mountainous region of southeastern Australia deposit clutches under logs or rocks at different depths, resulting in nearly identical patterns of temperature fluctuation. Experiments reveal that incubation periods decrease and developmental rates increase with increasing temperature. Hatchlings from clutches incubated at 22°C are larger in snout-vent length, have lower running speeds, and spend less time basking 1 month after birth than hatchlings from clutches incubated at 30°C. High variance associated with fluctuating nest temperatures also influences hatchling phenotypes (morphology, running speed, activity levels, and basking behavior), as does identity of the mother (maternal effects). Similar phenotypic responses to nest conditions occur in pythons (*Liasis fuscus*) that facultatively brood their eggs in tropical Australia. Python nest temperatures are influenced by both nest site selection and whether the female broods the eggs.

Until recently, most experimental studies on the influence of temperature or moisture on reptile development and ultimately phenotypes of offspring produced have been performed under constant temperature conditions. Few, if any, amphibian or reptile eggs experience constant temperatures throughout development. Carefully designed field and laboratory experiments that attempt to mimic natural conditions offer the opportunity to determine the phenotypic consequences of variation in the developmental environment of amphibians and reptiles and will contribute considerably to our understanding of the evolution of life histories.

QUESTIONS

1. What are advantages and disadvantages of external and internal fertilization in amphibians and reptiles?
2. Why are sea turtles poor examples of ***r***- versus ***K***-selected species?
3. What is the key difference in female reproductive systems between amphibians and reptiles?
4. Explain how the transfer of genomes occurs in unisexual ambystomatid salamanders and why unisexual populations are able to persist.
5. What is the significance of a close relationship between egg size and size of the pelvic opening in the Chicken turtle (*Deirochelys reticularia*)?
6. Describe the differences between the following unisexual methods of reproduction and give an amphibian or reptile example (species) of each.
   Kleptogenesis—
   Hybridogenesis—
   Parthenogenesis—

ADDITIONAL READING

Bull, J. J. (1980). Sex determination in reptiles. *Q. Rev. Biol.* **55:** 3–21.

Duellman, W. E., and Trueb, L. (1994). *Biology of Amphibians.* McGraw-Hill Book Co., New York.

Dunham, A. E., Miles, D. B., and Reznick, D. N. (1988). Life history patterns in squamate reptiles. In *Biology of the Reptilia. Volume 16. Ecology B. Defense and Life History.* C. Gans and R. B. Huey (Eds.). pp. 441–552. A. R. Liss, New York.

Exbrayat, J.-M. (2006a). *Reproductive Biology and Phylogeny of Gymnophiona (Caecilians).* Science Publishers, Enfield, NH.

Jamieson, B. G. M. (Ed.). (2003). *Reproductive Biology and Phylogeny of Urodela.* Volume 1 of the series *Reproductive Biology and Phylogeny,* series Editor, B. G. M. Jemieson. Science Publishers, Inc., Enfield, NH, USA.

Pianka, E. R., and Vitt, L. J. (2003). *Lizards: Windows to the Evolution of Diversity.* University of California Press, Berkeley, CA.

Sever, D. M. (Ed.). (2003). *Reproductive Biology and Phylogeny of Anura.* Vol. 2 of the series *Reproductive Biology and Phylogeny.* B. G. M. Jemieson (series Ed.). Science Publishers, Inc., Enfield, NH.

Stearns, S. C. (1976). Life-history tactics: a review of the ideas. *Quarterly Review of Biology* **51:** 3–47.

Stearns, S. C. (1977). The evolution of life history traits: a critique of the theory and a review of the data. *Annual Review of Ecology and Systematics* **8:** 145–171.

Stearns, S. C. (1992). *The Evolution of Life Histories.* Oxford University Press, Oxford.

Tinkle, D. W. (1967). The life and demography of the side-blotched lizard, *Uta stansburiana. Miscellaneous Publications of the Museum of Zoology, University of Michigan* **132:** 1–182.

Valenzuela, N., and Lance, V. (2004). *Temperature-Dependent Sex Determination in Vertebates.* Smithsonian Books, Washington, D. C.

Wells, K. D. (2007). *The Ecology and Behavior of Amphibians.* University of Chicago Press, Chicago.

Wilson E. O., and Bossert, W. H. (1971). *A Primer of Population Biology.* Sinauer Assoc., Inc., Stamford, CT.

REFERENCES

Gametogenesis and Fertilization: Gamete Structure and Production

Aranzábal, 2003a, 2003b; Costa et al., 2004; Duellman and Trueb, 1986; Elinson and Fang, 1998; Exbrayat, 2006; Follett and Redshaw, 1974; Fouquette and Delahousaye, 1977; Garda et al., 2004; Houck and Arnold, 2003; Jorgensen, 1992; Kardong, 1992; Kitimasak et al., 2003; Newton and Trauth, 1992; Osborne and Thompson, 2005; Oterino et al., 2006; Palmer et al., 1993; Scheltinga and Jamieson, 2003a,b, 2006; Scott et al., 2007; Sever, 2003; Toranzo et al., 2007; Villecco et al., 1999; Wake and Dickie, 1998.

Reproductive Behaviors Associated with Mating

Duellman and Trueb, 1986; Myers and Zamudio, 2004.

Reproductive Ecology: Ecology of Nesting

Auffenberg and Weaver, 1969; Brandão and Vanzolini, 1985; Breitenbach, 1982; Breitenbach et al., 1984; Broadley, 1974; Burger and Zappalorti, 1986; Burke et al., 1993; Caldwell, 1992; Caldwell and Lopez, 1989; Caldwell et al., 1980; Congdon et al., 1983 1987; Downie, 1984; Duellman and Trueb, 1986; Ernst, 1971; Fitch, 1954, 1955; Gibbons and Nelson, 1978; Goodwin and Marion, 1978; Guillette and Gongora, 1986; Hayes, 1983; Huey, 1982; Hunt and Ogden, 1991; Hutchison et al., 1966; Iverson, 1990; Joanen, 1969; Kennett et al., 1993; Kluge, 1981; Lawler, 1989; Legler and Fitch, 1957; Magnusson, 1980; Overall, 1994; Packard and Packard, 1985; Packard et al., 1977, 1981; Petranka et al., 1987; Plummer, 1976; Reid, 1955; Riemer, 1957; Riley et al., 1985; Rowe and Dunson, 1995; Shine, 1985a; Shine and Fitzgerald, 1996; Shine and Harlow, 1996; Snell and Tracy, 1985; Van Damme et al., 1992; Vaz-Ferreira et al., 1970; Vitt and Colli, 1994; Vitt et al., 1997; Vitt, 1986, 1991, 1993; Webb et al., 1977; Wilbur, 1975; Wilson, 1998.

Temperature-Dependent Sex Determination

Allsop et al., 2006; Bull, 1980; Burke et al., 1996; Cree et al., 1995; Crews et al., 1994; Deeming, 2004; Ewert et al., 2004; Ferguson and Joanen, 1982; Harlow, 2004; Hillis and Green, 1990; Janzen and Krenz, 2004; Janzen and Paukstis, 1991; Lang and Andrews, 1994; Lovern and Passek, 2002; Lovern and Wade, 2003; Morreale et al., 1982; Mrosovsky, 1994; Nelson et al., 2004; Pieau, 1971, 1975; Pough et al., 1998; Rhen and Lang, 2004; Valenzuela, 2004; Vogt and Flores-Villela, 1992; Viets et al., 1994; Warner et al., 2007; Warner and Shine, 2005, 2007, 2008; Wibbels et al., 1994.

Brood Size and Size of Young

Andrews and Rand, 1974; Broadley, 1974; Christian and Bedford, 1993; Congdon and Gibbons, 1987; Fitch, 1970; Madsen and Shine, 1996; Olsson and Shine, 1997; Pianka, 1986; Rand, 1982; Selcer, 1990; Semlitsch, 1985b; Smith and Fretwell, 1974; Vitt, 1981, 1986.

Seasonality in Reproduction

Asana, 1931; Barbault, 1975, 1976; Barbault and Rodrigues, 1978; Benabib, 1994; Brown and Shine, 2006; Caldwell, 1987, 1992; Christiansen and Dunham, 1972; Congdon et al., 1987; Derickson, 1976; Donnelly and Guyer, 1994; Doody and Young, 1995; Duellman, 1978; Duellman and Trueb, 1986; Farrell et al., 1995; Fitch, 1970, 1982; Fitzgerald et al., 1993; Flemming, 1994; Flemming and van Wyk, 1992; Fontenot, 1999; Friedl and Klump, 1997; Gillis and Ballinger, 1992; Goldberg, 1971; Guillette and Cruz, 1993; Guillette et al., 1980; Hahn and Tinkle, 1965; Hahn, 1967; Hartmann et al., 2005; Howland, 1992; Huang, 1997a, 1997b, 1998; Inger and Bacon, 1968; Inger and Greenberg, 1966; James, 1991; James and Shine, 1985; Joanen, 1969; Kupfer et al., 2005; Luiselli et al., 1996, 1997; Marvin, 1996; McPherson and Marion, 1981; Noble and Bradley, 1933; Nussbaum and Tait, 1977; Parmenter, 1985; Plummer, 1977; Reid, 1955; Salvidio, 1993; Seigel and Ford, 1987; Seigel et al., 1995; Semlitsch, 1985a,b, 1987; Semlitsch and West, 1983; Semlitsch et al., 1995; Sever et al., 1996; Sexton et al., 1971; Shine, 1985a; Shine et al., 1996a, 1996b, 1996c.; Vitt, 1983, 1992; Volsøe, 1944; Webb, 1977; Webb et al., 1977; Whittier and Crews, 1987; Wilhoft, 1963.

Sexual Versus Unisexual Reproduction

Adams et al., 2003; Avise et al. 1992; Berger, 1977; Bi and Bogart, 2006; Bi et al., 2007; Bogart, 2003; Bogart et al., 1989, 2007; Christiansen et al., 2005; Cole, 1984; Cole and Townsend, 1983, 1990; Cordes and Walker, 2006; Crews and Moore, 1993; Cuellar, 1976; Darevsky, 1958, 1992; Ezaz et al., 2006; Groot et al., 2003; Kearney and Shine, 2004; Kearney et al., 2006; Lenk et al., 2005; Menezes et al., 2004; Milius, 2006; Normark et al., 2003; Parker and Selander, 1976; Ragghianti et al., 2007; Robertson et al., 2006; Reeder et al., 2002; Som and Reyer, 2007; Vorburger, 2006; Vrijenhoek et al., 1989; Watts et al., 2006.

Life Histories: General

Barbault and Mou, 1988; Bruce, 2005; Dunham, 1993; Dunham and Miles, 1985; Dunham et al., 1988; Shine, 2003a; Stearns, 1976; Tilley, 1980; Tinkle et al., 1970.

Reproductive Effort and Costs of Reproduction

Andrews, 1979; Andrews and Rand, 1974; Bell, 1980; Cooper et al., 1990; Downes and Shine, 2001; Fisher, 1930; Hirschfield and Tinkle, 1975; Nagy, 1983a,b; Olsson and Madsen, 1996; Olsson et al., 1996, 1997; Qualls and Shine, 1995; Rocha, 1990; Seigel et al., 1987; Shine, 1980, 2003b; Shine and Schwarzkopf, 1992; Shine et al., 1996d; Swingland, 1977; Tinkle and Hadley, 1973, 1975; Vitt, 1981; Vitt and Congdon, 1978; Warner and Shine, 2006; Weiss, 2001; Williams, 1966.

Life-History Variation

Alcala and Brown, 1987; Andrews, 1982; Bourn and Coe, 1978; Briggs and Storm, 1983; Brown and Parker, 1984; Caldwell, 1993; Cecil and Just, 1979; Congdon et al., 1983, 1987; Duellman and Trueb, 1986; Frazer et al., 1990; Gibbons, 1972; Greer, 1975; Hairston, 1983; Howard, 1978a,b; Iverson, 1979; Kofron, 1983; Reagan, 1991; Pechmann et al., 1989; Seigel, 1983; Semlitsch, 1987; Semlitsch and Gibbons, 1988; Shine, 2003a,b, 2005; Shine and Olsson, 2003; Slade and Wassersug, 1975; Tinkle, 1967; Tinkle et al., 1970; Tinkle et al., 1981; Tuttle and Ryan, 1981; Van Devender, 1982; Vial et al., 1977; Wassersug, 1974; Webb et al., 1983; Wilbur, 1980; Wilbur and Collins, 1973; Woolbright, 1991.

Seasonal Versus Aseasonal Environments

Dobzhansky, 1950; Dunham et al., 1988; Fitch, 1970, 1982, 1985; James and Shine, 1988; Pianka, 1970; Tinkle et al., 1970.

Phenotypic Plasticity

Ballinger, 1977; Bull, 1987; Dunham, 1978; Ford and Seigel, 1989; Seigel and Ford, 1991; Shine and Harlow, 1996; Shine et al., 1997a,b; Stearns, 1976.

Chapter 5

Reproductive Modes

The most fundamental way to describe reproductive modes in extant tetrapods, if not most animals, is by identifying the reproductive product (egg versus live young) and sources of nutrients for development (yolk versus mother). Even though this bipartite classification of reproductive modes can be easily applied across broad taxonomic groups, the term reproductive mode is used differently in describing how amphibians and reptiles reproduce. With few exceptions, embryos of oviparous amphibians and reptiles receive all their fetal nutrition from yolk within the egg, a process known as lecithotrophy (lecitho = yolk; trophy = food). The exception is one (and possibly more) caecilian species in which the embryos peel and eat the lipid-rich skin of the mother, a process termed dermatophagy. Embryos of viviparous species can receive nutrition entirely from yolk, by oviductal secretions, by feeding on eggs or siblings embryos in the uterus, or by a simple or complex placenta. When the mother provides at least some nutrients, via either secretions or a placenta, the process is called matrotrophy (matro = mother), and many different types of matrotrophy have been identified (Table 5.1). Patrotrophy (patro = father) is provision of some nutrients by the father. These terms are used in this and related chapters.

DEFINING REPRODUCTIVE MODES

Amphibians

Amphibian reproductive modes are defined by a combination of characteristics, including breeding site, clutch structure, location of egg deposition (terrestrial or aquatic), larval development site, and parental care, if present. This complex suite of characters is needed because of the rich diversity of reproductive behaviors and life histories among anurans. In contrast, caecilian and salamander reproductive modes are less diverse, although no less complex and interesting.

The ancestral reproductive mode in amphibians is assumed to include external fertilization, oviparity, and no parental care. Within salamanders and frogs, some species have external fertilization and some internal fertilization, although the latter is rare in frogs. Caecilians are the exception among living amphibians because all known species have internal fertilization. Oviparity occurs in nearly all salamanders and frogs, and about one-half of caecilians. Parental care in caecilians and salamanders includes egg attendance. Parental care is more diverse in frogs, occurring in about 6% of known species. The three groups of amphibians will be discussed individually because of major differences in their reproductive modes (Table 5.2).

Caecilians

Because caecilians are fossorial and secretive, less is known about their reproduction and life history than about those of either frogs or salamanders. All male caecilians have a copulatory organ, the phallodeum, and presumably all have internal fertilization. More than one-half of caecilian species are viviparous. In these species, development occurs in the oviduct, and some form of maternal nutrition is provided. Fully metamorphosed young caecilians are eventually born, although the duration of pregnancy is known for only a few species. The remaining species are oviparous, depositing eggs on land. The oviparous species have either direct development or eggs that hatch as free--living aquatic larvae. In at least one species, *Boulengerula taitanus*, and perhaps other direct-developing oviparous species, the young remain with the mother and feed on

Herpetology, 3rd Ed.

TABLE 5.1 Fetal Nutritional Adaptations in Amphibians and Reptiles.

Nutritional pattern	Definition	Occurrence
Lecithotrophy	All nutrients for development to hatching or birth contained in egg as yolk when it is ovulated	All amphibians and reptiles that deposit eggs and viviparous species in which there is no matrotrophy
Matrotrophy	Some or all nutrients for developing fetuses provided by female during gestation	
Oophagy	Developing fetuses feed on sibling ova	Known only in *Salamandra atra*
Adelphophagy	Developing fetuses feed on developing siblings (also called *uterine cannibalism*)	May occur in *Salamandra atra*
Histophagy	Developing embryos feed on maternal secretions	Some viviparous caecilians, frogs, and salamanders
Histotrophy	Developing fetuses absorb maternal secretions	May occur through large saclike gills in typhlonectid caecilians and through fine papillae around the mouths of *Nectophrynoides occidentalis*
Placentotrophy	Developing embryos receive nutrients from the mother by placental transfer	Squamate reptiles with a placenta
Patrotrophy	Male provides some nutrients for developing tadpoles	Tadpoles carried in vocal sacs of *Rhinoderma darwinii* may absorb nutrients from male

Note: Individual species can utilize more than one nutritional adaptation. Three other types of matrotrophy occur in fishes but are not shown.
Adapted in part from Blackburn *et al.*, 1985.

TABLE 5.2 Reproductive Modes in Amphibians and Reptiles.

Amphibians

Caecilians
- I. Fertilization internal
  - A. Oviparity
    - 1. Eggs terrestrial; development direct
    - 2. Eggs terrestrial; development indirect (larvae)
  - B. Viviparity
    - 1. Birth and neonates terrestrial
    - 2. Birth and neonates aquatic

Salamanders
- I. Fertilization external
  - A. Oviparity
    - 1. Eggs aquatic; development indirect, larvae aquatic
- II. Fertilization internal
  - A. Oviparity
    - 1. Eggs aquatic; development indirect, larvae aquatic
    - 2. Eggs terrestrial; development indirect, larvae aquatic
    - 3. Eggs terrestrial; development indirect, larvae terrestrial and nonfeeding
    - 4. Eggs terrestrial; development direct
  - B. Viviparity
    - 1. Birth and neonates terrestrial
      - a. lecithotrophy
      - b. matrotrophy
    - 2. Oviductal histophagy
    - 3. Oophagy or adelphophagy

Frogs
- I. Eggs aquatic
  - A. Eggs deposited in water
    - 1. Eggs and feeding tadpoles in lentic water
    - 2. Eggs and feeding tadpoles in lotic waters
    - 3. Eggs and early larval stages in constructed subaquatic chambers; feeding tadpoles in streams
    - 4. Eggs and early larval stages in natural or constructed basins; subsequent to flooding, feeding tadpoles in natural ponds or streams
    - 5. Eggs and early larval stages in subterranean constructed nests; subsequent to flooding, feeding tadpoles in ponds or streams
    - 6. Eggs and feeding tadpoles in water in tree holes or aerial plants
    - 7. Eggs and nonfeeding tadpoles in water-filled depressions
    - 8. Eggs and nonfeeding tadpoles in water in tree holes or aerial plants
    - 9. Eggs deposited in stream and swallowed by female; eggs and tadpoles complete development in stomach
  - B. Eggs in bubble nest
    - 1. Bubble nest floating on pond; feeding tadpoles in ponds
  - C. Eggs in foam nest (aquatic)

1. Foam nest floating on pond; feeding tadpoles in ponds
2. Foam nest floating on pond; feeding tadpoles in streams
3. Foam nest floating on water accumulated in constructed basins; feeding tadpoles in ponds
4. Foam nest floating on water accumulated in axils of terrestrial bromeliads; feeding tadpoles in ponds

D. Eggs embedded in dorsum of aquatic female
1. Eggs hatch into feeding tadpoles
2. Eggs hatch into froglets

II. Eggs terrestrial or arboreal (not in water)
A. Eggs on ground, on rocks, or in burrows
1. Eggs and early tadpoles in excavated nests; subsequent to flooding, feeding tadpoles in streams or ponds
2. Eggs on ground or rock above water; upon hatching, feeding tadpoles move to water
3. Eggs on humid rocks, in rock crevices, or on tree roots above water; upon hatching, feeding semiterrestrial tadpoles live on rocks or in rock crevices in a water film
4. Eggs hatch into feeding tadpoles carried to water by adult
5. Eggs hatch into nonfeeding tadpoles that complete their development in nest
6. Eggs hatch into nonfeeding tadpoles that complete their development on dorsum or in pouches of adult
7. Terrestrial eggs hatch into froglets by direct development

B. Eggs arboreal
1. Eggs hatch into feeding tadpoles that drop into lentic water
2. Eggs hatch into feeding tadpoles that drop into lotic water
3. Eggs hatch into feeding tadpoles that develop in water-filled cavities in trees
4. Eggs hatch into froglets by direct development

C. Eggs in foam nest (terrestrial or arboreal)
1. Form nest on humid forest floor; subsequent to flooding, feeding tadpoles in ponds
2. Foam nest with eggs and early larval stages in basins; subsequent to flooding, feeding tadpoles in ponds or streams
3. Foam nest with eggs and early larval stages in subterranean constructed nests; subsequent to flooding, feeding tadpoles in ponds
4. Foam nest with eggs and early larval stages in subterranean constructed nests; subsequent to flooding, feeding tadpoles in streams
5. Foam nest in subterranean constructed chambers; nonfeeding tadpoles complete development in nest
6. Foam nest arboreal; tadpoles drop into ponds or streams

D. Eggs carried by adult
1. Eggs carried on legs of male; feeding tadpoles in ponds
2. Eggs carried in dorsal pouch of female; feeding tadpoles in ponds
3. Eggs carried on dorsum or in dorsal pouch of female; nonfeeding tadpoles in bromeliads or bamboo internodes
4. Eggs carried on dorsum or in dorsal pouch of female; direct development into froglets

III. Eggs retained in oviducts

IV. Fetal nutrition either lecithotrophic or matrotrophic via histophagy.
A. Ovoviviparity; fetal nutrition provided by yolk.
B. Viviparity; fetal nutrition provided by oviductal secretions.

Reptiles

Crocodylians
I. Fertilization internal
A. Oviparity
1. Parental care, nest attendance
2. Parental care, nest and hatchling attendance

Turtles
I. Fertilization internal
A. Oviparity
1. Parental care, none
2. Parental care, possibly digging nests when eggs hatch

Tuataras
I. Fertilization internal
A. Oviparity
1. Parental care, none

Lizards & Snakes
I. Fertilization internal
A. Oviparity
1. Parental care, none
2. Parental care, nest attendance

B. Viviparity
1. Lecithotrophy
2. Matrotrophy
a. Minimal nutrient transfer to primarily lecithotrophic embros
b. Placentotrophy; nutrient transfer variable, but in some cases, nearly 100%

Note: Terminology for fetal nutrition follows Blackburn et al., 1985. *All oviparity involves lecithotrophic nutrition except in some species of caecilians. Amphibian modes are modified from* Duellman and Trueb (1994) *and* Haddad and Prado (2005).

her modified, lipid-rich skin for at least a week. During one week of feeding, the young of *B. taitanus* increased about 11% in total length, while the mother lost 14% of her weight.

Viviparity in caecilians has evolved several times, based on distribution of this character on a recent caecilian phylogeny. As in other viviparous amphibians, the number and size of ova of these species are smaller than in oviparous species; egg size is from 1 to 2 mm in diameter, and egg number is from 10 to 50. Initially, the yolk provides nutrients for development, but the yolk is soon exhausted, and the fetus scrapes nutrient-rich secretions from the walls of the oviduct. This type of matrotrophy is known as histophagy and is accomplished using specialized fetal dentition (Table 4.1). The fetal dentition is lost at birth and replaced by the typical caecilian dentition of juveniles and adults.

Nutritional investment in young of primitive oviparous caecilians by yolk only was previously thought to be the ancestral condition. Recent studies have shown that fetal or fetal-like dentition is present in some oviparous species in addition to viviparous species, including species of *Boulengerula*, *Siphonops*, and *Caecilia*. The young of some other African species of caecilians have fetal dentition, but observations of the use of this dentition in feeding has yet to be observed in any of these species. The distribution of this type of dentition in both oviparous and viviparous species indicates that this character is homologous and may be ancestral. Independently derived viviparous lineages of caecilians may have evolved from ancestors that already had specialized fetal dentition (similar to the dematotrophic species *B. taitanus*) and thus were preadapted for evolving young that used fetal dentition to feed on oviductal secretions.

The ovarian cycle and oviductal morphology are known in only a few species of caecilians. The gestation period is approximately 11 months in one species, *Dermophis mexicanus*. Corpora lutea are large in pregnant females of the few species that have been studied. Corresponding high levels of progesterone are found in the blood, and as in other vertebrates, the production of progesterone by the corpora lutea apparently functions to prevent expulsion of the fetuses prior to birth. Proliferation of the epithelial layer of the oviduct begins about the second or third month of pregnancy. The content of the secretion changes throughout the gestation period; initially, the contents are mainly free amino acids and carbohydrates that gradually become rich in lipids near the end of gestation.

Oviparous caecilians have free-living larvae or direct development. Ovum size of species with free-living larvae ranges from 8 to 10 mm, the largest among all reproductive modes. Oviparous species with direct development have eggs ranging in size from 3 to 6 mm. Size of these eggs contrasts with oviparous salamanders and frogs, in which direct-developing species have the largest eggs with greater amounts of yolk than species with free-living larvae. Specialized feeding on the mother's skin after birth in at least one species may be advantageous in that a few large offspring may be produced, but investment in offspring can be diverted if conditions are unfavorable. Clutch size in oviparous species ranges from 6 to 50 eggs. Length of the larval stage is unknown for most caecilians, but in some Old World taxa, the larval period is about one year, and in *Ichthyophis kohtaoensis*, it is about 6 months.

Gill structure of the viviparous typhlonectids differs from the free-living larval caecilians that have the typical triramous gills of other larval lissamphibians (Fig. 5.1). Typhlonectid gills are large saclike structures. They appear to function as pseudoplacentae, allowing gas and nutrient exchange between the parent and fetus. These gills are lost soon after birth.

Salamanders

Salamanders in the families Hynobiidae and Cryptobranchidae, and presumably Sirenidae, have external fertilization. All other salamanders have internal fertilization. Hynobiid salamanders deposit paired egg sacs, which are then fertilized by the male. Clutch size in one species varies from 24 to 109. Cryptobranchids deposit paired strings of eggs. Reproduction has not been observed in the four sirenid species, but two nests of *Siren intermedia* had 206 and 362 eggs, each attended by a female. Studies of oviductal anatomy of the two species of *Siren* revealed no sperm in the oviducts. The absence of a sperm storage organ and of spermatozoa in the oviducts at the time of oviposition provides strong evidence that external fertilization occurs in sirenids. In all other salamanders, eggs are fertilized by sperm held in sperm storage structures as they pass through the oviduct, and in all species studied, sperm occur in the spermathecae prior to and after oviposition.

All other lineages of salamanders have internal fertilization by means of the male spermatophore (see Chapter 4 and Fig. 4.9). After the female picks up spermatophores, sperm are then stored in the female's spermatheca, a storage organ in the roof of the cloaca, and eggs are fertilized inside the female's cloaca as they pass by the spermatheca.

Several modes of reproduction are found among the families of salamanders with internal fertilization (Table 5.2). Eggs and larvae may be aquatic, or eggs may be terrestrial and larvae may be either aquatic or terrestrial. Terrestrial eggs with direct development are common in one lineage of salamanders, the Plethodontidae. Eleven species of salamanders in two genera, *Salamandra* and *Lyciasalamandra*, are viviparous; all are in the family Salamandridae.

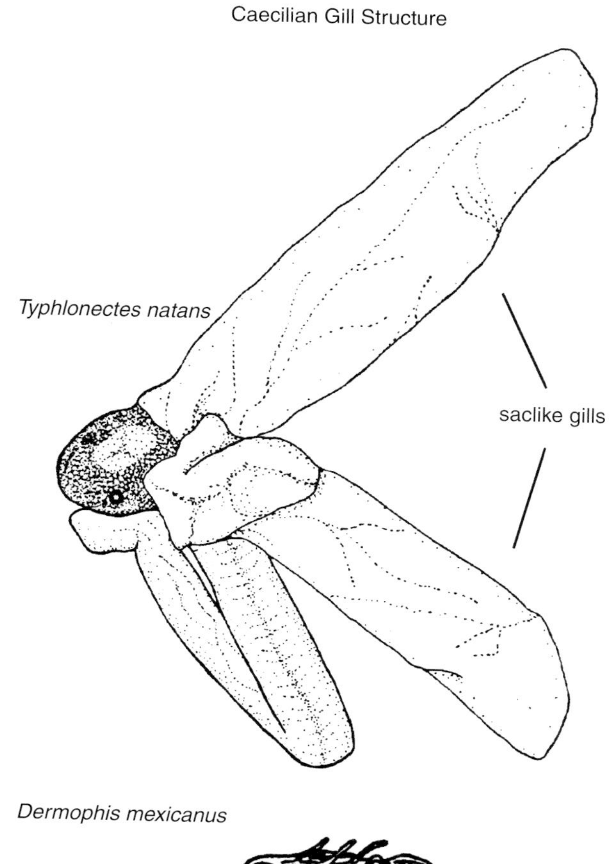

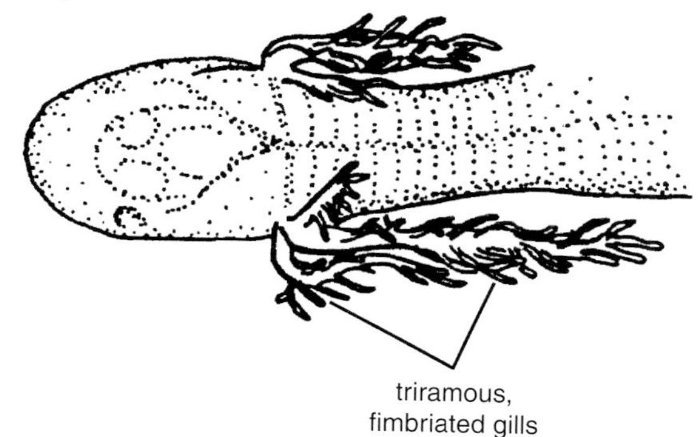

FIGURE 5.1 Gill structure in the larvae of viviparous caecilians. Bottom, *Dermophis mexicanus* (Caeciliidae) with triramous, fimbriated gills; top, *Typhlonectes natans* (Typhlonectidae) with enlarged, saclike gills; these highly vascularized gills may absorb nutrients from the parent. Adapted from M. Wake, 1993.

Anurans

Frogs have the greatest diversity of reproductive modes among vertebrates (with the possible exception of teleost fishes), but most are oviparous and lecithotrophic. The ancestral reproductive mode in amphibians includes deposition of eggs in water, but many extant species have partial or fully terrestrial modes of reproduction. Amphibian eggs are permeable and require water to prevent desiccation. Many terrestrial reproductive modes occur in tropical regions where humidity and temperature are high.

Reproductive modes in amphibians are categorized primarily by the three major situations in which eggs are placed for development (Table 5.2). These are (1) eggs deposited in aquatic habitats, such as ponds, streams, water-holding plants or tree holes, or small basins of water constructed by individuals of certain species of frogs; (2) eggs deposited in arboreal or terrestrial habitats, such as leaves above pools or streams, burrows on land, or on the body of the male or female; and (3) eggs retained in the frog's body. Within each of these three major categories, further subtypes are found; in all, 39 modes of reproduction have been described and new ones continue to appear as more detailed observations are made. Some examples of each of these three major categories serve to illustrate the complexity and, in some cases, the bizarre aspects of frog reproduction.

The ancestral mode of reproduction in which deposition of aquatic eggs that hatch into free-living larvae that complete development in standing or flowing water is common (Modes 1 and 2 under "Frogs" in Table 5.2). The gladiator frogs, large hylids that occur in parts of Central and South America, are examples of frogs that construct basins in which eggs are deposited (Mode 4; see Fig. 4.10). Basins are built commonly at the edges of streams with sand or mud substrates. The male frog constructs the basin by pivoting on his body and pushing the substrate out with his limbs. In *Hypsiboas boans*, the males call from sites above the basin or nest, whereas in *Hypsiboas rosenbergi*, males call from small platforms at the edge of the nest. Upon arrival of the female, eggs are deposited as a surface film in the nest; subsequent rains break down the edges (ramparts) of the nest, releasing the tadpoles into the main body of the stream. In Sichuan Province, China, females of the megophryid frog *Leptobatrachium* (*Vibrissaphora*) *boringiae* deposit their eggs in a doughnut-shaped mass on the underside of submerged rocks in fast-moving streams (Fig. 5.2). Using an asymmetric inguinal amplexus, the male pushes eggs up to the bottom surface of the rock with its right hindlimb as the amplexing pair rotates horizontally and counterclockwise. Three limbs of the female provide support and power for rotation as she lifts off the substrate. The result is a circular egg mass with a doughnut hole in the middle.

Several species of hylid frogs (e.g., *Anotheca spinosa*, *Osteocephalus oophagus*, *Trachycephalus resinifictrix*) deposit eggs in water in arboreal microhabitats such as bromeliads or tree holes; their larvae are either omnivorous or are fed unfertilized eggs by the female parent who periodically returns to the deposition site (Mode 6). The Mesoamerican *A. spinosa* deposits eggs in bromeliads, bamboo internodes, or tree holes. After amplexus and egg deposition, the male disappears, but the female continues to visit the developing tadpoles about every 4.85 days and deposits nonfertile (nutritive) eggs for the tadpoles to eat. Metamorphosis requires 60–136 days,

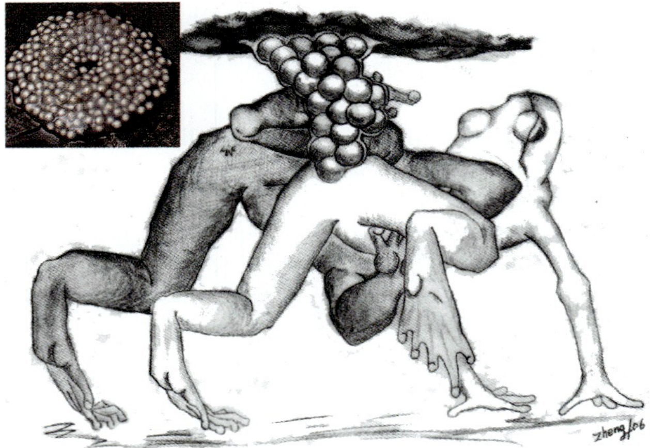

FIGURE 5.2 Females of the frog *Leptobatrachium boringiae* position themselves on the substrate under submerged rocks during asymmetrical inguinal aplexus while the male (top) pushes the eggs to the undersurface of the rock with his right hindleg. The end result is a mass of eggs that looks like a doughnut (insert). Adapted from Cheng and Fu, 2007, drawing by Z. Zheng.

FIGURE 5.3 Production of a foam nest by a paired male and female *Leptodactylus knudseni*. These large leptodactylids may deposit eggs in the same nest more than once. Tadpoles develop in the foam and are washed into a nearby pond if heavy rains occur. (W. Hödl, 1990)

after which the female approaches a calling male, and a new fertile clutch is deposited. Similar behavior has been observed in the South America species, *O. oophagus*, but periodically the female and male return together in amplexus and deposit a new clutch of fertilized eggs. If tadpoles are present, they consume the newly deposited fertile eggs; otherwise, the new eggs hatch and begin developing. In contrast, tadpoles of *P. resinifictrix* are omnivorous, feeding on both detritus and on conspecific eggs.

One of the most unusual reproductive modes occurs in the Australian gastric brooding frog, *Rheobatrachus silus*. The female deposits aquatic eggs and then swallows them. The eggs develop in her stomach (Mode 9). It is thought that prostaglandin E_2 produced by the developing young inhibits the production of gastric secretions during the gestation period. In several months, fully formed froglets emerge from the mother's mouth. This frog is thought to be extinct.

Aquatic eggs can also be placed in a foam nest that floats on the surface of small ponds or other aquatic habitats (Modes 11–13; Fig. 5.3). Many leptodactylid frogs construct foam nests, including *Leptodactylus* and *Physalaemus*. In *P. ephippifer*, the foam is produced from cloacal secretions of the amplexing male and female. The male rotates his legs in a circular motion, whipping the cloacal secretions into a froth. Egg expulsion and fertilization begin once a substantial foam mass has been produced (Fig. 5.4). Each pair of frogs produces a floating nest of 300 to 400 eggs. In *Leptodactylus labyrinthicus*, only about 10% of the eggs in the foam nest constructed by a pair are fertilized. Nests are frequently constructed prior to the rainy season, and tadpoles remain in the nest for prolonged periods, feeding on the unfertilized eggs (Fig. 5.5). When rains begin, the tadpoles are flooded into the pond, where they feed on newly deposited eggs of small hylids.

An unusual reproductive mode is found in some species of the aquatic *Pipa* (Modes 15–16). Eggs are embedded into dorsum of the female during a complicated mating ritual in which the male and female undergo turnovers under water (Fig. 5.6). While upside down, eggs are extruded from the female's cloaca and are pressed against her dorsum by the male. They become embedded in the female's skin, where they develop into tadpoles (e.g., *P. carvalhoi*) or froglets (e.g., *P. pipa*) in about 2 months.

The second major category of reproductive modes includes frogs that deposit their eggs in arboreal or terrestrial sites. Throughout tropical regions of the world many species deposit eggs on land. In the Amazonian region, for example, more than one-half of all species have terrestrial eggs.

Males and females of many species of frogs in the aromobatid genus *Allobates* court on land, after which relatively small clutches of eggs (up to 30; Mode 20) are deposited in leaf litter on the forest floor. During the initial period of development, the male (in most species, the female in some) attends the eggs. When the eggs hatch after about a week, the parent frog wriggles down among the tadpoles and they move up onto its back. The parent frog then transports the tadpoles to water, often a small stream or pool in the forest, where the tadpoles swim free and complete their development without further parental care. Terrestrial nests occur in nearly all aromobatids and dendrobatids.

Four species of aromobatids, *Anomaloglossus stepheni*, *Anomaloglossus degranvillei*, *Allobates chalcopis*, and *Allobates nidicola*, have nonfeeding tadpoles. Of these, three complete their development in terrestrial nests

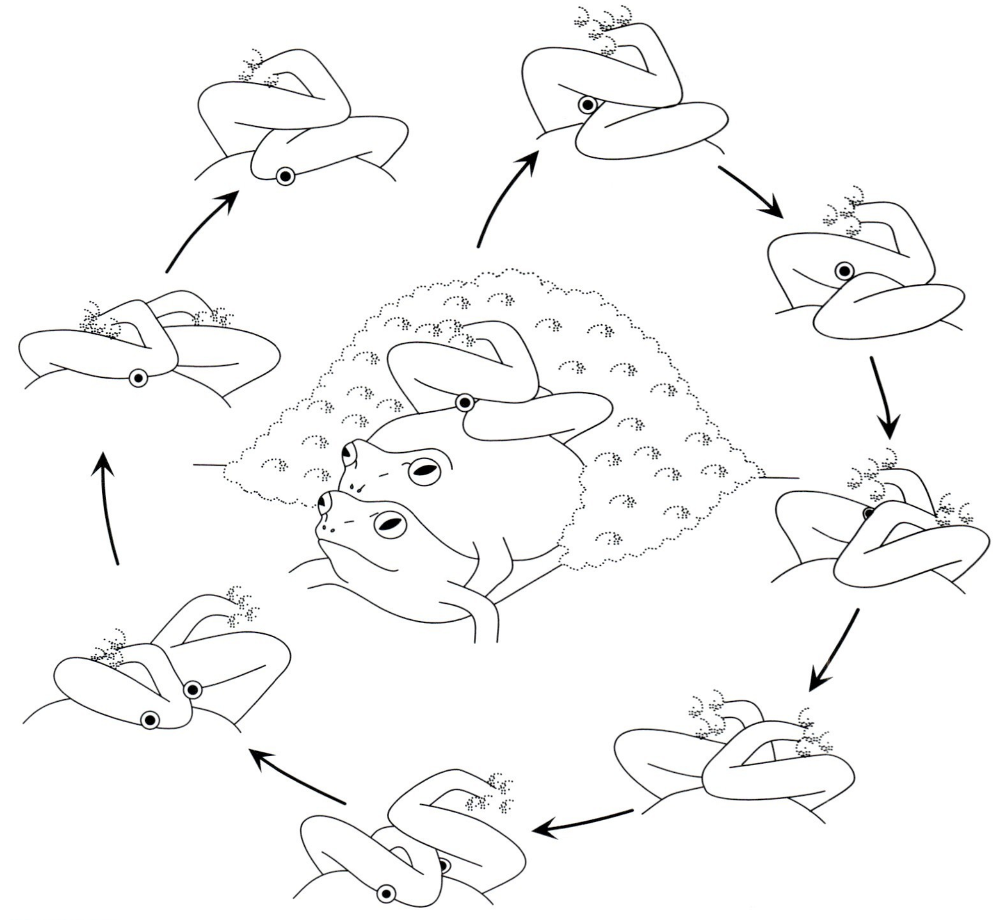

FIGURE 5.4 Secretions from a male and female are whipped by rapid leg movements into a foam nest by the Brazilian leiuperid, *Physalaemus ephippifer*. At the same time, eggs are deposited and fertilized. The black circles represent the path of an egg as it is extruded from the female and pushed into the growing mound of foam; several hundred eggs will be deposited in a single nest. Adapted from Hödl, 1990.

FIGURE. 5.5 Larva of *Leptodactylus labyrinthicus* after eating eggs of other frog species inhabiting ponds in Goiás, Brazil. (A. Sebben)

(Mode 21). These nidicolous tadpoles remain in the nest in the forest leaf litter about 30 days prior to metamorphosis. Tadpoles of *Anomaloglossus degranvillei* complete their development while carried on the parent's back (Mode 22). Mouthparts of three of these tadpoles are reduced to varying degrees, although mouthparts of *A. chalcopis* are fairly well developed.

Males of the Australian frog *Assa darlingtoni* (Myobatrachidae) have inguinal pouches for tadpole transport (Mode 22). After an extended amplexus lasting up to 9 hours, a terrestrial clutch of eggs is produced. The clutch is guarded by the female, and after about 11 days, the egg mass begins to liquefy and the tadpoles hatch. The male returns and performs a complex series of movements to guide the larvae to his inguinal pouches; the movements can include using his feet to scoop and tuck the tadpoles under him. The larvae use their tails to move onto the male and into the pouches. The male continues to feed and call while carrying the developing tadpoles. After 59–80 days in the pouch, the froglets emerge fully formed, having more than doubled their weight.

Other unrelated species of frogs carry tadpoles in pouches on their backs or, in one genus (*Rhinoderma*), in the vocal sacs of the male. An experiment with *R. darwinii* from Argentina suggested that the male provides

FIGURE 5.6 From top to bottom: Mating ritual of *Pipa parva*. The pair somersaults in the water as eggs are released and fertilized; the male presses the eggs into the female's dorsum, where they embed in her skin. A female *Pipa parva* with freshly deposited eggs on her dorsum. Tadpoles emerging from pockets on the back of a female *Pipa carvalhoi*. (K.-H. Jungfer, 1996)

nutrients for the larvae. Radioactive material injected into the lymphatic sacs of males carrying an average of 11 larvae in their vocal sacs appeared in the tissues of the larvae, suggesting patrotrophy. If nutrients were indeed transferred from the male to the larvae, this would

FIGURE 5.7 Direct-developing eggs of *Pristimantis* sp. In this species, eggs are deposited in leaf litter in a tropical forest. Note the well-developed back legs of the embryos. (J. P. Caldwell)

represent the first case of patrotrophy in amphibians. It may also occur in *Assa* and other species in which males carry larvae for extended time periods.

Direct development has evolved repeatedly in anurans. Eggs are deposited in terrestrial nests and embryos develop entirely within the eggs, emerging as froglets (Mode 23). Just over one-quarter of all New World tropical frogs compose an assemblage referred to as *eleutherodactyline frogs*; all have direct development (Fig. 5.7). Male *Eleutherodactylus cooki* in Puerto Rican caves guard clutches of about 16 eggs. Occasionally, a male guards a nest with double and triple clutches. Froglets emerge from the nest within 22–29 days after egg deposition.

Many frogs have arboreal clutches of eggs that are attached to leaves or tree branches above water. When the tadpoles hatch, they drop into ponds (Mode 24), streams (Mode 25), or water-filled cavities or tree holes (Mode 26). These modes occur in many species of hylid frogs in the genera *Dendropsophus*, *Phyllomedusa*, and *Agalychnis* and in most Centrolenidae (Fig. 5.8).

Two closely related species of *Agalychnis*, *A. craspedopus* and *A. calcarifer*, deposit eggs above pools formed in buttresses of large fallen trees. Their courtship may last up to 12 hours. About mid-morning, the amplexing pair deposits a small clutch of eggs on the side of the tree or on hanging vegetation above the water. *Agalychnis calcarifer* lays 20 to 28 eggs in a clutch, and *A. craspedopus* 14 to 21 eggs per clutch. Their eggs are relatively large (9–12 mm diameter) and heavily laden with yolk; ovum diameter is 4 mm. In 7–15 days, the eggs hatch and the larvae drop into the water, where they complete their development in several months.

FIGURE 5.8 Frogs in the family Centrolenidae deposit their eggs on the undersides of leaves over moving water, where they develop into tiny tadpoles that drop into the stream. (J. P. Caldwell)

Many New Guinea microhylids have large eggs that undergo direct development (Mode 27). The eggs are deposited in arboreal sites such as leaf axils or hollow stems, and a parent remains with the eggs. One species of *Oreophryne* deposits about 10 eggs on the upper surfaces of leaves. The egg mass is enclosed in a membrane that is distinct from the egg capsules. Presumably, this membrane adds an extra degree of protection. A male attends the eggs during part or all of their development.

Many species of tropical *Leptodactylus* produce foam nests either in burrows or in small depressions. Developing tadpoles are subsequently washed into nearby pools or streams that form with the onset of heavy rains, or remain in the nest and are fed unfertilized eggs by the female (Modes 29–31). *Leptodactylus mystaceus* males call from small depressions to attract a female. During amplexus, the pair constructs a foam nest in the depression (Fig. 5.9), and upon completion of the nest and egg deposition, the pair separates and both parents depart. No further parental care occurs. If rains are delayed, the original foam produced by the parents begins to dissipate. However, the tadpoles generate new foam by vigorously wriggling their bodies together. During this time, tadpole development is arrested until rains begin. The leptodactylid frog, *Leptodactylus fallax*, is highly unusual in that males construct and fight over nesting burrows to which they attract females by calling, and after a foam nest is constructed and eggs deposited, the female not only remains with the nest and defends it but also feeds the larvae unfertilized eggs. Males also remain with the nest to defend it.

Reptiles

Reptile reproductive modes are defined on the basis of whether they lay eggs (oviparity) or produce live young (viviparity) and whether nutrition is provided exclusively by the yolk (lecithotrophy) or at least partially by the mother (matrotrophy) or father (patrotrophy) (Table 5.1). All crocodylians, turtles, the tuatara, and a majority of snakes and lizards lay eggs. In most of these, hatching of eggs appears to be synchronous (Fig. 5.10). About 20% of squamates are viviparous. In oviparous reptiles, embryo nourishment comes from the yolk (lecithotrophy). Females of some oviparous species, such as the snake *Opheodrys vernalis* and the lizard *Lacerta agilis,* retain eggs until the embryos are within only a few days of hatching. Among those species that bear live young, maternal contribution of nutrients (matrotrophy) to development varies considerably. In some viviparous species, development of embryos is supported entirely by yolk in the egg (lecithotrophy), just as in oviparous species. Examples include the live-bearing horned lizard *Phrynosoma douglassi* and all

FIGURE 5.9 Left: Nest construction by a male *Leptodactylus mystaceus*; male calls from the depression to attract a female. Right: The male and a female produce a foam nest in which they deposit eggs. The nest is abandoned, and tadpoles are flooded from the nest when heavy rains occur. (J. P. Caldwell)

FIGURE 5.10 Synchronous hatching occurs when eggs of the Amazonian lizard, *Plica plica*, are disturbed. (L. J. Vitt)

snakes in the Boinae. In others, such as *Mabuya heathi*, developmental nutrition derives entirely from the mother via a placenta.

VIVIPARITY

Viviparity has evolved independently at least 113 times in amphibians and reptiles, with most origins (at least 108) occurring in squamates. Indeed, the evolution of viviparity from oviparity has occurred more frequently in squamate reptiles than in all other vertebrate groups combined (a mere 37 nonsquamate origins). Viviparity provides parents more control over development of offspring than does oviparity because the female carries the offspring inside her body. Consequently, predation on eggs in the nest is not a threat, although costs of carrying offspring may be considerable (see Chapter 4). Female performance (i.e., escaping from predators) can be reduced during pregnancy due to the large size of developing young, and females carry their offspring for a longer time period than do oviparous species.

The geographical distribution of viviparous species raises additional questions. Viviparous species might be expected to occur in temperate zones or at high elevations where temperatures are low enough or seasons short enough that eggs in nests would never hatch. In these areas, the female could regulate her body temperature behaviorally and thus regulate the temperature of developing embryos (the Cold-Climate Hypothesis). Although some species of squamate reptiles occur in cold climates, this explanation or hypothesis does not apply to all amphibians and reptiles. For example, all viviparous caecilians and most viviparous frogs are tropical, negating temperature as a likely explanation for these taxa. The same is true for viviparous squamates; many have temperate zone distributions (e.g., *Elgaria coerulea*) or live at high elevations (e.g., *Sceloporus jarrovi, S. aeneus*), but some are tropical (e.g., *Mabuya nigropunctata, M. heathi*) or live in deserts (e.g., *Xantusia vigilis*) and did not have an ancestor living in cold environments. Moreover, the evolutionary events leading to viviparity in amphibians differ from those in reptiles (Fig. 5.11).

Viviparity in amphibians, which occurs in about 3% of known taxa, has arisen relatively infrequently compared with reptiles. Apparently all caecilians have internal fertilization, and many species are viviparous. All but a few frogs have external fertilization, thereby predisposing them for oviparity. Seemingly this predisposition also exists for salamanders, because fertilization occurs in the cloaca as the eggs are released and not in the oviducts.

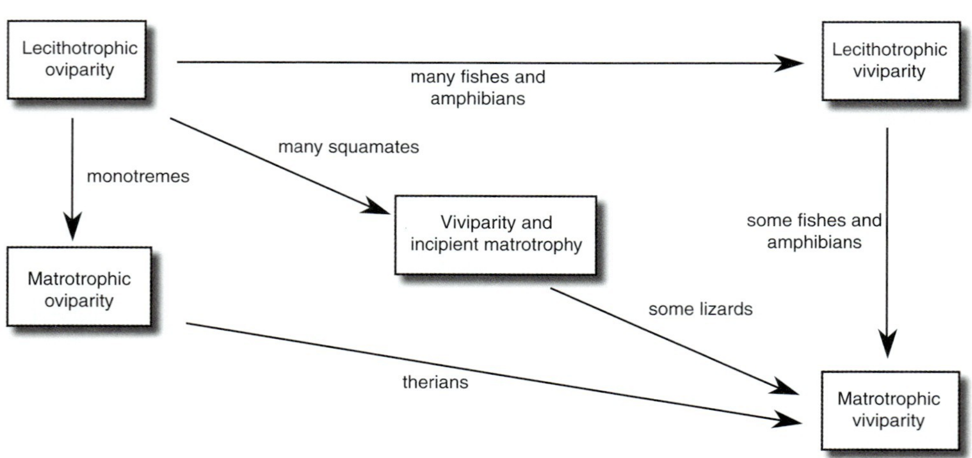

FIGURE 5.11 Evolutionary events leading to vivparity and matrotrophy in vertebrates. Adapted from Blackburn, 2006.

Nevertheless, viviparous species have evolved within a few clades of frogs and salamanders. Among viviparous amphibians, nutrition for developing embryos is either lecithotrophic or matrotrophic but without a placenta. Fetuses most commonly ingest or absorb nutrient-rich secretions from the female's oviducts.

Among salamanders, viviparity has evolved independently four times in 11 viviparous species in two salamandrid sister genera, *Salamandra* and *Lyciasalamandra*. Within the genus *Salamandra*, presumably 4 species are viviparous, 2 of which have been studied in detail. *Salamandra salamandra* is unique in that some populations have lecithotrophic viviparity, whereas others have matrotrophic viviparity. Comparison of the developmental sequence in the two types of viviparity in *S. salamandra* revealed major differences. In the matrotrophic populations, the female ovulates about 20–60 eggs, but about 50% of the eggs undergo developmental arrest almost immediately. The developing embryos quickly use their yolk and undergo rapid development of the gills, limbs, and digestive system. In addition, rate of development among the remaining embryos is asynchronous. In the lecithotrophic viviparous populations, the female produces about the same number of eggs, but the yolk is consumed slowly, and the embryos do not have accelerated development of the digestive system. In the matrotropic populations, the mouth opens early in development and the embryos begin to consume the arrested eggs (oophagy) and smaller embryos (adelphophagy) in utero. From 1 to 15 embryos continue to develop and grow and are born as fully terrestrial, metamorphosed juveniles in about 3 months. In the lecithotrophic viviparous populations, embryos also remain in the uterus about 3 months, but they remain in the egg in the uterus, hatching only right before they are released. They are released into water as gilled larvae and require another 1 to 3 months in water to metamorphose into terrestrial juveniles. *Salamandra atra* differs from *Salamandra salamandra* in that only two young are produced, one in each oviduct, and gestation can last for 3–4 years, depending on climatic conditions. Nutrition is provided by secretions from giant epithelial cells that develop in the uterus during the second year of gestation (histophagy) and adelphophagy. Although *Salamandra atra* ovulates from 28–104 eggs, all but 2 have incomplete gelatinous coats and disintegrate into a yolky mass that is consumed by the developing embryos.

All seven species in the genus *Lyciasalamandra* are viviparous. *L. luschani*, like *Salamandra atra*, produces only two offspring, one from each oviduct. Gestation extends from 5 to 8 months, and nutrition is also provided by oophagy in the uterus.

Only five frog species, one in the Caribbean eleutherodactyline assemblage, *Eleutherodactylus jasperi*, and two each in the closely related African bufonid genera *Nectophrynoides* and *Nimbaphrynoides*, are viviparous. *Eleutherodactylus jasperi* of Puerto Rico has lecithotrophic viviparity, in which eggs develop inside the fused lower portions of the oviducts. The female retains the eggs for about 33 days from the time of amplexus to birth of three to five froglets. Small-bodied females give birth to relatively large-bodied young. This production of relatively large offspring is typical of other viviparous amphibians (e.g., caecilians). No morphological evidence exists for transfer of nutrients to the embryos; in addition, some yolk remains in the intestines of the froglets when they are born.

Nectophrynoides tornieri and *Ne. viviparous* have lecithotrophic viviparity. They produce large, yolk-filled eggs ranging from 3 to 4 mm in diameter. *Nimbaphrynoides occidentalis* and *Ni. liberiensis* have matrotrophic viviparity and produce small eggs from 0.5 to 0.6 mm in diameter. Gestation lasts about 9 months for *Ni. occidentalis*, and during the last 2 months of gestation, the oviducts produce a concentrated mucopolysaccharide secretion to nourish the embryos. The embryos or larvae have a ring of large papillae around the mouth that may aid in absorption of nutrients. Birth of 4–35 froglets occurs in the early rainy season. Formerly, the two species of *Nimbaphrynoides* were considered to be in the genus *Nectophrynoides*, but new information has led to their separation as different genera. In addition, recent work has led to the discovery of previously unknown species of *Nectophrynoides*, which now contains 13 species. Life history and reproductive details are unknown for many of the newly described species.

Respiratory adaptations of viviparous amphibians are not the same (or homologous) as those in viviparous reptiles. Preexisting respiratory structures of embryonic amphibians (gills, skin) increase their vascularization and, when juxtaposed against the oviduct lining, enhance gas exchange during development (Fig. 5.1).

The possibility exists that viviparity evolved in some amphibians with internal fertilization because of the potential competitive advantage of large offspring. Viviparity might also have arisen during a transition to terrestrial breeding when coupled with internal fertilization. By retaining fertilized eggs within the oviduct until hatching occurs, females prevent their eggs from desiccation by moving to a more humid microenvironment. Nevertheless, these scenarios do not answer why viviparity is so rare in salamanders and frogs. If a primary advantage of viviparity among amphibians is the production of large offspring, then this advantage can be achieved in salamanders by other means. Dividing the clutch into fewer larger eggs and attending the eggs in terrestrial nests can offset high mortality associated with the production of many small eggs deposited in aquatic environments. Nest attendance combined with direct development, as occurs in most plethodontid salamanders, eliminates the costs associated with placing eggs in high mortality environments, relaxes selection on the numbers of offspring, and increases the selective advantages of producing fewer, larger, and presumably more competitive offspring. Because amphibians do not require the higher developmental temperatures of reptiles, little thermal advantage to carrying offspring in the body of the female exists. Egg attendance and direct development obtain the same result in amphibians as viviparity does in squamates—elimination of a high-mortality stage of the life history.

Among squamate reptiles, temperature and specifically cold climates, appears to be the primary factor promoting the evolution of viviparity. The basic arguments are (1) females carrying offspring can behaviorally obtain and maintain body temperatures above substrate temperatures, whereas eggs experience the vagaries of environmental temperatures; (2) development is more rapid in embryos at higher temperatures; and (3) neonate survival is higher because accelerated development allows them to enter the environment earlier and become established prior to cold weather. Evidence supporting the Cold-Climate Hypothesis comes from a variety of sources. Squamate reptiles occurring at the highest latitudes and elevations are all viviparous, and recently evolved viviparous squamates tend to inhabit cold environments.

A species need not currently live in a cold environment for cold climate to have been the selective factor leading to viviparity. Once viviparity has arisen within a clade, the viviparous species could have dispersed into warmer areas, so their current distributions might be quite different from those in the past. For example, the Australian snake genus *Pseudechis* (Elapidae), contains five oviparous and one viviparous species. Only the viviparous species, *P. porphyriacus,* inhabits a cold habitat. Alternative ecologically based hypotheses for the evolution of viviparity are rejected on the basis of comparative data, suggesting that the Cold-Climate Hypothesis is the only viable one explaining viviparity in *P. porphyriacus.*

The Cold-Climate Hypothesis addresses the conditions under which viviparity might evolve but does not directly address the adaptive significance of viviparity. Viviparity allows females to manipulate thermal conditions that embryos experience during development because they carry the offspring with them (Maternal-Manipulation Hypothesis). Experiments on Death Adders (*Acanthophis praelongus*) from tropical Australia reveal that pregnant females maintain relatively more constant body temperatures than nonpregnant females. Offspring of females allowed to select their body temperatures within a 25°–31°C diel cycle produced larger offspring that survived better compared with females that produced their offspring under normal conditions similar to those experienced by nonpregnant snakes (23°–33°C diel cycle). Thus females can manipulate development in a way that enhances not only their offspring fitness, but theirs as well. The Maternal-Manipulation Hypothesis may provide a much more general explanation for the evolution and maintenance of viviparity under a wide variety of conditions compared with the Cold-Climate Hypothesis and should also be applicable to amphibians.

Our understanding of morphological aspects of squamate viviparity, in terms of placental development, dates back to a review of placentation in reptiles in 1935 by H. C. Weekes. Functionally, the transition from oviparity to viviparity involved simultaneous egg retention and nutrient transfer (Fig. 5.12). Studies on a facultative placentotrophic snake reveal some clues on the functional transition from lecithotrophy to matrotrophy via placentotrophy. Embryos of the colubrid snake *Virginia striatula* can develop exclusively on yolk reserves as in typical lecithotrophic reptiles, or they can receive some nutrients (particularly calcium) from the female's oviducts. Calcium passes across the oviductal lining to the embryo's yolk sac, which is pressed against the oviduct, thus establishing a functional relationship between the maternal and fetal tissues. In a sense, the mechanism is not very different from the production of eggshells in oviparous species, except that female tissue transfers calcium to fetal tissue rather than to a fibrous matrix that becomes the shell. Once a transfer mechanism arises, the transfer of other nutrients can follow and a reduction in yolk can occur. Females no longer need to invest all their energy in offspring at one time. Rather, they can spread their nutritional commitment to offspring over a more extended time period. These results suggest that viviparity and matrotrophy evolved simultaneously in squamates. This kind of change can ultimately lead to obligate placentotrophy.

Among the relatively few squamate species that have functional placentae, the transition from producing eggs with shells to producing live offspring that receive

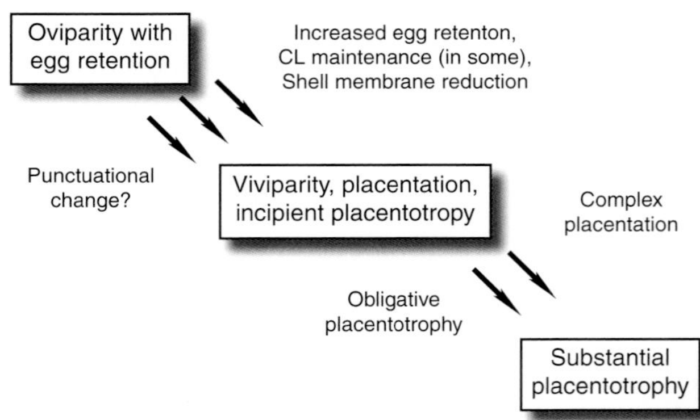

FIGURE 5.12 Evolutionary events leading to viviparity and matrotrophy in squamates. CL refers to corpora lutea. Adapted from Blackburn, 2006.

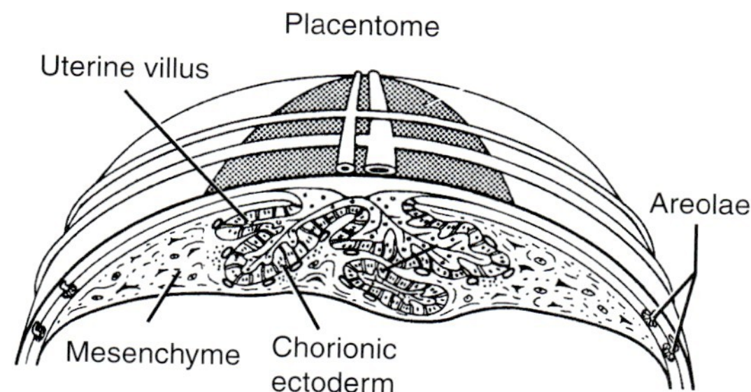

FIGURE 5.13 Diagrammatic representation of the chorioallantoic placenta in *Mabuya heathi*. The placenta lies above the embryo and consists of hypertrophied uterine (maternal) and chorionic (fetal) tissue forming the placentome, the joint structure for nutrient transfer to the embryo, waste transfer to the female, and gaseous exchange. The interdigitating structures are the chorionic areolae, the site of transfer and exchange. Adapted from Blackburn and Vitt, 1992.

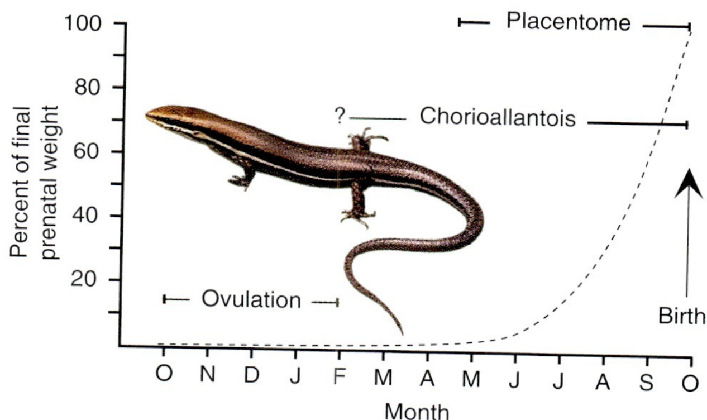

FIGURE 5.14 Generalized pattern of growth in embryos of the viviparous New World skink, *Mabuya heathi*. The embryo increases more than 74,000% of its freshly ovulated mass as the result of nutrient uptake from the female. Adapted from Blackburn and Vitt, 1992.

nutrients from the mother while in the oviduct was much more complex, requiring respiratory, hormonal, and nutritive specializations. Embryonic membranes, including the yolk sac or the chorioallantois in viviparous placentotrophic squamates, became highly vascularized and interdigitated with the wall of the oviduct to accommodate gas and nutrient exchange (Fig. 5.13).

The timing of hormone production and release were required adjustments to avoid the expulsion of embryos prior to their complete development. In oviparous vertebrates, the corpora lutea, which typically degenerate rapidly following ovulation, produce progesterone. Because progesterone inhibits oviduct contraction, a decreasing level of progesterone results in oviductal contraction and expulsion of eggs or embryos. In viviparous species and those with egg retention, the corpora lutea persist following ovulation and continue to produce progesterone so that embryos are not expelled.

Most New World skinks in the genus *Mabuya* receive all nutrients for development through a highly specialized and unique placenta that is functionally similar to the placenta of eutherian mammals (Fig. 5.13). The complex type of matrotrophy that occurs in all New World *Mabuya* certainly did not arise independently in each species. Rather, an ancestor that colonized the New World was already viviparous. Surprisingly, juvenile-sized females 3 months or less in age ovulate tiny ova similar to those ovulated by adult-sized females. Because little growth of the embryos occurs during the first 4–7 months of gestation, the body size of these "juvenile" females increases sufficiently to accommodate the developing embryos by the time rapid embryonic growth begins (Fig. 5.14).

PARENTAL CARE

Parental care is defined as any form of post-ovipositional parental behavior that increases the survival of the offspring at some expense to the parent. We do not include matrotrophic provisioning of young that occurs in some viviparous species because this parental contribution to offspring survival occurs prior to birth. Parental care occurs in a diversity of taxa (Table 5.3), indicating that it has arisen independently many times within amphibians and reptiles. The number of evolutionary origins of parental care in amphibians is much lower than the number of species with parental care. Most amphibians and reptiles show no parental care other than nest construction for egg deposition.

General Categories of Parental Care

Parental care is represented in amphibians and reptiles by a variety of behaviors, and not all apply to both amphibians and reptiles. They can be summarized in general as the following:

1. Nest or egg attendance—A parent remains with the nest or eggs but without detectable nest defense.
2. Nest or egg guarding—A parent remains with the nest or eggs and actively defends against conspecifics or predators.
3. Egg brooding—Defined slightly differently for amphibians and reptiles. In amphibians, brooding is used for species that retain the embryos somewhere on or in the body but not in the oviducts. In reptiles, it refers only to a parent facilitating incubation by raising the temperature of the eggs.
4. Egg, larval, or hatchling transport—A parent carries offspring from one place to another.
5. Feeding of young—A parent brings food to offspring, e.g., tadpole feeding.
6. Guarding or attending young—A parent stays with young after the eggs hatch.

Nest or Egg Attendance

Egg attendance occurs in caecilians (females), salamanders (either or both sexes), frogs (either or both sexes), crocodylians (either or both sexes), a few turtles (females),

TABLE 5.3 Known Taxonomic Distribution of Parental Care in Amphibians and Reptiles.

Group	Care provider	Families	Species	Percent
Caecilians*	Female	2/6 families	8/162 species	5
Salamanders*	M or F	8/9 families	72/354 species	20
Frogs*	M or F	20/45 families	±325/5450	6
Turtles†	F	2/14 families	3/260	1
Crocodylians‡	M or F	All familes	All species	100
Amphisbaenians		Unknown		
Lizards§	Female	6/15 families	41/3000	1.3
Snakes§†	Female	6/11 families	47/1700	2.8

Note: Viviparous species are not included. The numbers of evolutionary origins for each taxonomic group are lower than the number of species exhibiting parental care.
Sources: * Crump (1995); † Iverson (1990); ‡ Greer (1970); § Shine (1988); †§ Somma (2003).

and many squamates (females). Functions of egg attendance vary. In amphibians they include aeration of aquatic eggs, hydration of terrestrial eggs, protection from pathogens or predators, or manipulation to prevent development adhesions. Attending females of the salamander *Necturus maculosus* aerate their aquatic eggs by rapid gill movements. Glass frogs, Centrolenidae, deposit eggs on leaves above streams and small rivers in Neotropical rain forests. In species in which males are territorial, such as *Hyalinobatrachium fleishmanni*, males attend the nests (Fig. 5.15). However, in at least one species with nonterritorial males, *H. prosoblepon*, females attend the nest. Other frogs carry their eggs with them, either on or in their back (Figs. 5.6 and 5.15). Males of the tropical leptodactylid frog *Eleutherodactylus coqui* provide water to eggs by direct transfer across their skin. Fungus attacks eggs of the New Guinea frog *Cophixalus parkeri*, and developmental abnormalities occur when attending females are removed. In reptiles, nest attendance may aid in hydration of eggs. The attending female of the skink, *Plestiodon septentrionalis*, regulates egg water exchange by moving the eggs, coiling around the eggs, or expanding the nest cavity thus exposing different proportions of the egg surface to substrate and air. Similar functions have been suggested for crocodiles (e.g., *Crocodylus porosus*). Nest attendance in reptiles may prevent drowning of eggs (e.g., *Opisthotropis latouchi*), deter fungal infection (e.g., *Plestiodon fasciatus*, *Gerrhonotus liocephalus*), or aid in keeping eggs hidden (e.g., *Iguana iguana*).

Nest or Egg Guarding

Nest or egg guarding occurs in salamanders (either sex), frogs (either sex), crocodylians (either sex), and squamates (females). Attending females of the salamander *Plethodon cinereus* and males of the tropical hylid frog, *Hypsiboas rosenbergi*, aggressively attack conspecifics that approach the nest. Arthropod predators attack nests of the frog *Cophixalus parkeri* after removal of the parent. Following oviposition, female *Iguana iguana* interact aggressively with other females that attempt to use the same nest sites. Female *Plestiodon* and a number of snake species including *Naja naja* aggressively attack when disturbed while guarding eggs (Fig. 5.16). Removal of female *Mabuya longicauda* from their nests results in a 70% reduction in hatching success; nearly all predation was by the oophagous snake *Oligodon formosanus*. Female *Mabuya* aggressively attacked the oophagous snakes, although they abandoned their nests if confronted with a predatory snake. Females of the Nile crocodile (*Crocodylus niloticus*) aggressively defend their nests against monitor lizards (*Varanus*) that attempt to prey on the eggs.

Egg, Larval, or Hatchling–Froglet Transport

Transport of early life-history stages is widespread in frogs (either sex) and crocodylians (either sex). In many frog species, eggs are carried, usually by the female, while they develop (see the section "Defining Reproductive Modes," as well as Figs. 5.6 and 5.15). In some instances, transport includes brooding (see following text). Transport of tadpoles is common, occurring in seven frog families. Most frequently, tadpoles are carried on the back of one parent as in most species of aromobatids and dendrobatids (Fig. 5.17). Tadpoles are carried from a terrestrial nest site to water in the vocal sacs of male *Rhinoderma rufum*. Males of New Guinea microhylids guard eggs of their direct-developing offspring; in addition, males of at least two species, *Liophryne schlaginhaufeni* and *Sphenophryne cornuta*, transport their froglets on their back after they have hatched. Froglets jump off the parent's back at regular intervals and thus are dispersed from the nest site, possibly reducing competition among them or lowering predation risks. Females of most crocodylians (e.g., *Crocodylus mindorensis*, *Crocodylus niloticus*) carry the hatchlings in their mouth to water (Fig. 5.18).

FIGURE 5.15 Clockwise from top left: Male *Hyalinobatrachium valerioi* attending three clutches of eggs of different ages; female of *Stefania evansi* brooding exposed eggs on its back; female *Flectonotus fitzgeraldi* brooding five eggs in dorsal pouches; an amplexing pair of *Gastrotheca walkeri*. Large, pale yellow eggs are expelled singly from the female's cloaca, fertilized by the male, and manipulated into the brooding pouch on the female's back. Photographs: *H. valerioi*, W. Hödl; all others, K.-H. Jungfer.

FIGURE 5.16 Female of the skink *Plestiodon fasciatus* attending her clutch of eggs. (L. J. Vitt)

Egg Brooding

Brooding in anurans involves retaining the eggs and/or larvae on the body of the parent for a longer period of time than that required to simply transport the larvae from a nest site to an aquatic site. A variety of behaviors can be observed among the many species that exhibit brooding. Eggs may be carried only until they develop into larvae, or they may be carried until they metamorphose into froglets. In aromobatids and dendrobatids, the eggs are not carried but hatch in a terrestrial nest. In most species, tadpoles are quickly transported to an aquatic site, but in a few species, the tadpoles may be retained on the dorsum of the parent for a few days to a week or more, or they may be carried on the dorsum until metamorphosis. Although it is difficult to categorize all species, brooding includes sequestering the offspring on or in the body for

FIGURE 5.17 Events leading to deposition of tadpoles of the dendrobatid frog *Epipedobates tricolor*. From top to bottom, amplexus, tadpole attendance, tadpole transport, and release of tadpoles in water. (K.-H. Jungfer)

FIGURE 5.18 *Crocodylus palustris* carrying newly hatched offspring to water. (J. W. Lang)

some period of time, whereas transport involves moving the eggs or larvae from one site to another.

The male parent in *Assa darlingtoni* picks up its tadpoles from a terrestrial nest and carries them in inguinal pouches for the remainder of their development until metamorphosis. Eggs are placed in a dorsal pouch in the hylid lineage *Gastrotheca*; in some species, they are carried until they develop into tadpoles, whereas in others, they are carried until metamorphosis (Fig. 5.15). A few large eggs are carried in an exposed position on the back of the hylid *Stefania*, where they remain until they develop into froglets (Figs. 5.15 and 5.19). In contrast, *Flectonotus*, another hylid, broods a few large eggs in a dorsal pouch that opens by splitting down the midline (Fig. 5.15). In the gastric brooding frogs, *Rheobatrachus silus* and *R. vitellinus* (Rheobatrachidae), brooding of eggs and/or larvae occurs in the stomach of the female; in one species, froglets emerge after metamorphosis, whereas in the other species, the female releases tadpoles. Development in these frogs is supported entirely by yolk contained in the eggs. In contrast to *Rhinoderma rufus*, male *Rhinoderma darwinii* brood their tadpoles in their vocal sacs until metamorphosis occurs.

Brooding in reptiles is known only in oviparous boids, and it may be ubiquitous in pythons. The primary advantage of brooding is faster development of embryos by maintaining a higher temperature. Shivering thermogenesis provides the heat. Pythons generate their own heat while brooding the eggs; this behavior raises the temperature of the clutch and increases developmental rates of the embryos (see Fig. 4.12). In at least one python species, *Liasis fuscus*, brooding of eggs is facultative and initiated by low nest temperatures. In another species, *Python molurus*, brooding appears obligatory because nonbrooded eggs have a high incidence of abnormal embryos.

FIGURE 5.19 Froglets that have nearly completed their development on the back of a brooding female *Stefania evansi*. (K.-H. Jungfer)

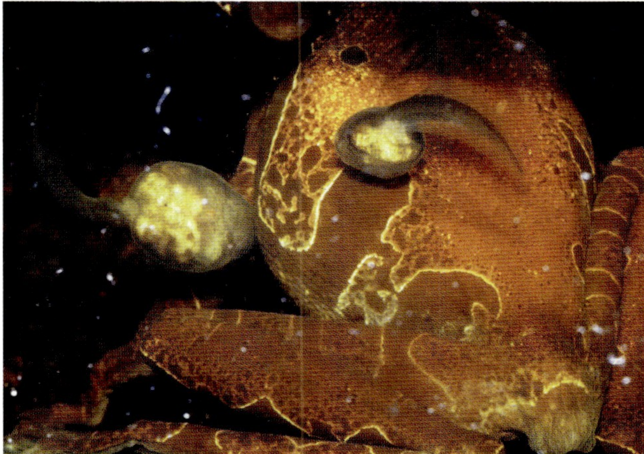

FIGURE 5.20 Top: A female *Anotheca spinosa* feeding trophic eggs to her tadpoles. Begging behavior of the tadpoles may stimulate egg laying. Bottom: Trophic eggs consumed by a tadpole of *Anotheca spinosa* are visible through the transparent skin. (K.-H. Jungfer)

Feeding of Young

Some frogs have evolved the ability to feed trophic eggs to their developing tadpoles. This behavior has evolved in several clades of hylid, aromobatid, dendrobatid, and rhacophorid frogs. In all of these species the tadpoles develop in restricted microhabitats that have little or no food available. Typically, these developmental sites include tree holes, bamboo segments, bromeliad axils, or other types of water-holding plants. Females of several species of dendrobatid frogs (e.g., *Oophaga pumilio*, *Ranitomeya ventrimaculata*, *R. vanzolinii*) deposit trophic eggs in the tadpole's aquatic microhabitat; in *R. vanzolinii*, both the male and female play a role in feeding the tadpoles. Trophic eggs may be fertilized or unfertilized, depending on the behavior of the species and whether courtship with the male is necessary to induce egg deposition in the female. After initially mating and depositing fertilized eggs, a female *Anotheca spinosa* (Hylidae) returns periodically to deposit unfertilized eggs for the tadpoles (Fig. 5.20). Physical contact by the tadpoles with the female's cloaca appears to stimulate release of the eggs. In contrast, pairs of the Amazonian hylid frog, *Osteocephalus oophagus*, return repeatedly to the same microhabitat to mate and deposit eggs. The first clutch deposited in an unused site develops into tadpoles, and later clutches serve as food for the tadpoles. After metamorphosis of the tadpoles, the original pair continues to deposit eggs, and more tadpoles develop. Tadpoles not provided with eggs die.

Guarding or Attending Young

Attending or guarding young (including tadpoles) occurs in frogs, viviparous lizards, and crocodylians. Although widespread taxonomically (i.e., Leiopelmatidae, Leptodactylidae, Hemisotidae, Microhylidae, Pyxicephalidae, and Ranidae), attendance and guarding of tadpoles has been verified in relatively few species. In some instances, the parent (usually the female) remains with the tadpoles and aggressively attacks animals that disturb the tadpole aggregation (e.g., *Leptodactylus ocellatus;* Fig. 5.21). Parental frogs have been observed to accompany the tadpole schools as they move around in ponds, and some terrestrially breeding frogs remain with the foam nest or tadpoles. In ranids, parental attendance includes species that dig channels that allow tadpoles to move from one body of water to another

FIGURE 5.21 Top: Adult female of *Leptodactylus ocellatus* situated at the edge of her tadpole school. For perspective, the tadpoles just below the frog in the top panel are about 50 mm in total length. The female remains with the tadpole school and aggressively attacks intruders. Bottom: Tadpole school of *Leptodactylus ocellatus* from central Brazil. (J. P. Caldwell)

(e.g., *Pyxicephalus adspersus*) or dig tunnels from terrestrial nest sites to water (e.g., *Hemisus*). Juveniles of the scincid lizard *Egernia saxatilis* remain in the territory of the family group to which they are related. Juveniles are indirectly protected from unrelated adults of the same species, which attack, kill, and eat juveniles, because family groups defend their territories against intrusion by other groups. Adults tolerate the presence of juveniles to which they are related. In a few viviparous squamates, females aid offspring emerging from placental membranes following birth (e.g., South American *Mabuya*, North American *Xantusia*, Neotropical *Epicrates*). Among crocodylians (observed in four genera: *Crocodylus*, *Alligator*, *Caiman*, and *Paleosuchus*), adults approach eggs in which juveniles have begun vocalizing prior to hatching and crack open the eggs with their mouths. The parents help free the hatchlings and often pick them up in their mouths and carry them to water. Juveniles of all studied species emit distress calls that elicit approach of adults, suggesting a protective function.

Evolution of Parental Care

Several behaviors associated with parental care appear obligatory, and their evolution is readily understood. For example, if live-bearing skinks tear open the placental membranes, more neonates survive. Similarly, more neonates survive when the female eats the membranes, reducing the likelihood that chemical cues from the membranes attract chemosensory-oriented predators. If frogs with terrestrial eggs that hatch into tadpoles did not transport their tadpoles to water, no descendents would pass on that particular behavior. A selective advantage accrues to frogs that move their nests farther and farther from water if intensity of predation decreases with distance from water. Simultaneous selection favors the evolution of obligatory larval transport. Clearly, the primary benefit of all forms of parental care is the increased probability of offspring survival. The diversity of parental care behaviors in amphibians and reptiles suggests that a variety of evolutionary trajectories achieve that end, and a single explanation is inadequate to explain the origin of the numerous and different types of parental care.

The majority of parental care in amphibians occurs in nonaquatic species with terrestrial modes of reproduction. In aquatic amphibians with parental care, the driving force behind the evolution of parental care appears to be physiological. For example, in *Cryptobranchus alleganiensis*, females increase oxygen availability to the developing offspring by moving the eggs around; a similar behavior occurs in *Necturus maculosus*. Development and survival depend on oxygen, and mechanisms that favor increased oxygen availability, especially in low-oxygen situations, should be favored. Similar physiological arguments could be made for at least some terrestrial amphibians that exhibit parental care.

In amphibians, parental care is associated with increased terrestriality (Fig. 5.22). Removing the egg and larval stages from water presumably confers a selective advantage because the life-history stages with the highest mortality are either eliminated (as in direct-developing species) or shifted to sites with lower mortality. In many amphibians with parental care, offspring size increases with increasing terrestriality—increased size apparently increases larval survivorship. The evolutionary cost of increasing offspring size is a reduction in offspring number, but if parental care effectively reduces mortality, this cost should be offset by the increased survival of protected offspring. Among reptiles, parental care appears associated primarily with the protection of eggs from predators or fungus, but in some instances, such as in brooding pythons, parental care provides a physiological function. As in amphibians, different evolutionary scenarios explain parental care in different species.

Parental care, by definition, has costs to the parent(s). Although parental care is relatively easy to observe and document, measuring the costs of parental care is much more difficult. For many examples provided in preceding text,

FIGURE 5.22 Adult female of the salamander *Plethodon albagula* attending her egg clutch. (S. E. Trauth)

costs have not been measured. Costs to amphibians and reptiles may include a decrease in future survival, possibly because of increased predation or a reduction of time available for food gathering. Even if survival is not affected, a decrease in investment of future offspring may occur because of the time invested in the current offspring. The benefit to the parents in terms of increased survivorship of offspring must outweigh the costs, or parental care would not evolve. As an extension of this concept, biparental care, which is rare in amphibians, would not evolve unless offspring survival were higher when both parents are involved in care than if only one of them provided care.

Synthesis

The diversity of reproductive patterns, life histories, and reproductive modes of amphibians and reptiles offers nearly unlimited opportunities for testing ecological and evolutionary theory. The decisions females make when selecting nest sites and constructing nests can have profound effects on survival and development of eggs as well as on the morphology, performance, and in some instances, even the sex of offspring. Amphibians and reptiles can reproduce within very short time periods, over extended time periods, or may even skip years between reproductive episodes. Investment in reproduction is costly in terms of both energy and survival. The interaction between age-specific reproductive effort and its survival costs has produced an impressive diversity of life-history patterns in amphibians and reptiles. Life histories vary from species with high reproductive efforts, early attainment of sexual maturity, and short life spans to low reproductive efforts, late maturity, and long life spans. Life history patterns are constrained by phylogeny, with some lineages comprised of species having life histories quite different from species in other lineages. Similar constraints due to morphology and foraging behavior exist. Among species with variable clutch size, a trade-off exists between the size of offspring and the number of offspring produced. Species producing many offspring typically produce small offspring, whereas species producing few offspring usually produce relatively large offspring. The large number of species in which offspring size does not appear to vary supports the idea that offspring size is optimized. However, offspring size variation in some species appears related to resource availability, morphological constraints, or even the possibility that more than a single optimum exists.

Viviparity has released many squamates and some amphibians from mortality associated with clutch deposition and prolonged, unprotected incubation periods, but not without associated costs. Performance of females carrying offspring can be reduced, and behavioral modifications are associated with carrying offspring over extended time periods. Viviparity in squamates has evolved independently more times than in all other vertebrates combined, rendering squamates an ideal system for examining the evolution of viviparity.

QUESTIONS

1. Why are reproductive modes defined so differently for amphibians and reptiles?
2. What are the advantages of viviparity in reptiles and to what does the Maternal-Manipulation Hypothesis refer?
3. Describe examples of each of the following using an amphibian or reptile species:
   egg attendance
   egg brooding
   egg guarding
   larval transport
4. What is the difference between lecithotrophy and matrotrophy and what are some clear examples of each?

ADDITIONAL READING

Blackburn, D. G. (1993). Standardized criteria for the recognition of reproductive modes in squamate reptiles. *Herpetologica* **49:** 118–132.

Blackburn, D. G. (2000a). Classification of the reproductive patterns of amniotes. *Herpetological Monographs* **14:** 371–377.

Buckley, D., Alcobendas, M., Garcia-Paris, M., and Wake, M. H. (2007). Heterochrony, cannibalism, and the evolution of viviparity in *Salamandra salamandra*. *Evolution and Development* **9:** 105–115.

Crump, M. L. (1995). Parental care. In *Amphibian Biology. Volume 2, Social Behaviour*. H. Heatwole and B. K. Sullivan (Eds.), pp. 518–567. Surrey Beatty & Sons Pty Ltd., Chipping Norton Pty Ltd, N.S.W.

Duellman, W. E., and Trueb, L. (1986). *Biology of Amphibians*. McGraw-Hill Book Co., New York.

Greene, H. W., May, P. G. D. L., Hardy, S., Sciturro, J. M., and Farrell, T. M. (2002). Parental behavior by vipers. In *Biology of the Vipers*. G. W. Schuett, M. Hoggren, M. E. Douglas, and H. W. Greene (Eds.), pp. 179–225. Eagle Mountain Publications, Eagle Mountain, Utah.

Guillette, L. J., Jr. (1987). The evolution of viviparity in fishes, amphibians and reptiles: An endocrine approach. In *Hormones and Reproduction in Fishes, Amphibians, and Reptiles*. D. O. Norris and R. E. Jones (Eds.), pp. 523–569. Plenum, New York.

Shine, R. (1985b). The evolution of viviparity in reptiles: an ecological analysis. In *Biology of the Reptilia. Volume 15, Developmental Biology B*. C. Gans and F. Billett (Eds.), pp. 677–680. John Wiley & Sons, Inc., New York.

Shine, R. (1988). Parental care in reptiles. In *Biology of the Reptilia, Vol. 16. Ecology B. Defense and Life History*. C. Gans and R. D. Huey (Eds.), pp. 275–329. Alan R. Liss, Inc., New York.

Somma, L. A. (2003). *Parental Behavior in Lepidosaurian and Testudinian Reptiles*. Krieger Publishing Company, Malabar, FL.

Stearns, S. C. (1976). Life-history tactics: a review of the ideas. *Quarterly Review of Biology* **51:** 3–47.

Wake, M. H. (1992a). Reproduction in caecilians. In *Reproductive Biology of South American Vertebrates*. W. C. Hamlett (Ed.), pp. 112–120. Springer-Verlag, New York.

Wells, K. D. (2007). *The Ecology and Behavior of Amphibians*. University of Chicago Press, Chicago, IL.

REFERENCES

Defining Reproductive Modes

Amphibians

Blackburn, 1999a, 2000a,b; Caldwell, 1992, 1994; Caldwell and Lima, 2003; Crump, 1995; Delsol et al., 1981, 1983, 1986; Duellman and Trueb, 1986; Exbrayat, 2006a, 2006b; Gibson and Buley, 2004; Godley, 1983; Goicoechea et al., 1986; Haddad and Prado, 2005; Hödl, 1990; Hoogmoed and Cadle, 1990; Joglar et al., 1996; Johnston and Richards, 1993; Juncá, 1998; Juncá et al., 1994; Jungfer, 1996; Jungfer and Schiesari, 1995; Kluge, 1981; Kokubum and Giaretta, 2005; Kupfer et al., 2006; Kusano, 1980; Nickerson and Mays, 1973; Prado et al, 2004, 2005; Salthe, 1969; Schiesari et al., 1996; Sever et al., 1996a; Shepard and Caldwell, 2005; Wake, 1980, 1992a, 1993a, 1993b.

Reptiles

Blackburn, 1982, 1985; Blackburn et al., 1984; Salthe and Duellman, 1973; Shine, 1985b; Thompson and Speake, 2006; Wake, 1992a.

Viviparity

Blackburn, 1982, 1985, 1991, 1993a,b, 1998a,b, 1999a,b,c,d, 2000, 2002, 2005, 2006; Blackburn and Fleming, 2008; Blackburn and Vitt, 1992, 2002; Blackburn et al., 1984, 1985; Exbrayat, 2006; Exbrayat and Hraoui-Bloquet, 2006; Guillette, 1987; Heulin, 1990; Miaud et al, 2001; Qualls, 1996; Shine, 1983, 1985a,b, 1987, 2002; Shine and Bull, 1979; Stewart and Blackburn, 1988; Stewart and Thompson, 2002, 2003; Stewart, 1989; Thompson and Speake, 2006; Tinkle and Gibbons, 1977; Wake, 1978, 1980, 1982, 1992; Webb et al., 2006; Weekes 1935.

PARENTAL CARE

General

Alcala et al., 1987, 1988; Bachmann, 1984; Bell, 1985; Bickford, 2002; Brust, 1993; Burrowes, 2000; Caldwell and Oliveira, 1999; Campbell and Quinn, 1975; Corben et al., 1974; Crump, 1995; Duellman and Trueb, 1986; Ehmann and Swan, 1985; Garrick and Lang, 1977; Garrick et al., 1978; Gross and Shine, 1981; Groves, 1981; Huang, 2006; Jacobson, 1985; Jungfer, 1996; Jungfer and Schiesari, 1995; Junger and Weygolt, 1999; Kluge, 1981; Kok et al., 1989; Lescure, 1979; McDonald and Tyler, 1984; Miller, 1954; Modha, 1967; Noble and Mason, 1933; O'Connor and Shine, 2004; Pooley, 1974, 1977; Rand and Rand, 1976; Rebouças-Spieker and Vanzolini, 1978; Salthe and Mecham, 1974; Shine, 1988; Shine et al., 1997; Simon, 1983; Somma, 2003; Somma and Fawcett, 1989; Summers and Amos, 1997; Taigen et al., 1984; Vaz-Ferreira and Gehrau, 1975; Vinegar et al., 1970; Webb et al., 1977; Wells and Bard, 1988.

Evolution of Parental Care

Bishop, 1941; Crump, 1995; Shine, 1988; Wells, 1981.

Part III

Physiological Ecology

Life ultimately depends on chemical reactions that occur within cells of individual organisms. Physiological processes operate within a narrow environmental range and function best in an even narrower range. The environment must be neither too hot nor too cold, neither too wet nor too dry, and it must have the proper proportions of gases, especially oxygen. Cellular chemistry and function are closely integrated with osmotic balance, the maintenance of specific ionic concentrations within cells and tissues. Chemical reactions in turn require energy that is produced by oxidation of fuel to power life processes, and the efficiency of these reactions depends upon temperature. Osmoregulation, respiration, thermoregulation, and energetics make up the most important physiological processes.

Amphibians and reptiles live in diverse environments that vary greatly in solute concentrations, temperature, oxygen availability, and fuel and nutrient resources. An individual's behavior and physiological homeostatic mechanisms interact to maintain its internal environment within tolerance limits, thereby ensuring the animal's survival and ultimately its ability to reproduce. For ectotherms, temperature may be the single most important physiological variable because all cellular processes are temperature dependent. Nonetheless, all environmental variables affect life and how it is lived. Physiological ecology examines the complex interplay between physiological processes and the organism's physical and chemical environments. It integrates behavioral and ecological phenomena in seeking explanations for the evolution of physiological traits.

Chapter 6

Water Balance and Gas Exchange

In an active amphibian or reptile, thousands of cellular reactions occur every second. These reactions require an aqueous medium, and further, water and oxygen are required to convert fuel to usable energy. These metabolic reactions power the chemistry of digestion, absorption, waste removal, cell repair and division, reproduction, and a multitude of other functions. To survive, amphibians and reptiles must maintain internal body fluids that provide a stable environment for the cells. The concentration of body fluids typically differs greatly from concentrations of solutes in the external environment and continually challenges their internal balance. Water loss and ion gain are the primary osmoregulatory challenges to amphibians and reptiles in saltwater; water loss and the resultant increased concentrations of ions are the major challenges for terrestrial species, and water gain and ion loss or decreasing concentrations of ions are the primary challenges faced by freshwater species.

WATER AND SALT BALANCE

Osmoregulation—Maintaining Homeostasis

Osmoregulation, the control of water and salt balance, presents different challenges to organisms living in freshwater, saltwater, and aerial or terrestrial environments (Fig. 6.1). Many structures and organs are involved in osmoregulation, including the skin, gills, digestive tract, kidneys, and bladder. In freshwater, an amphibian or reptile is hyperosmotic. The ionic concentration of the body is greater than that of the environment, and if not regulated, water moves in, cells swell and possibly burst, and ions become too dilute. Excessive hydration can be avoided in several ways. Permeability of the skin can be decreased or urinary output can be increased, although salts must be conserved. Marine or brackish species face the opposite challenge. They are hyposmotic in relation to their environment. The ionic concentration of the body is less than the environment; thus, water moves out if unregulated, causing dehydration and a concentration of salts in the body fluid. Dehydration can be circumvented by decreasing permeability of skin and reducing the amount of water in urine, although nitrogenous waste must still be removed before reaching toxic levels. Terrestrial species are also at risk of dehydration, but from evaporation rather than osmotic loss of water. They counteract this problem physiologically in a manner similar to marine species. The basic physics of water loss and gain is rather simple, but the mechanisms by which amphibians and reptiles accomplish osmoregulation are varied and often complex (Fig. 6.2).

Kidney Function

Kidneys play a major role in osmoregulation in both amphibians and reptiles. In the two groups, kidneys are morphologically and functionally similar (Fig. 6.3).

Herpetology, 3rd Ed.

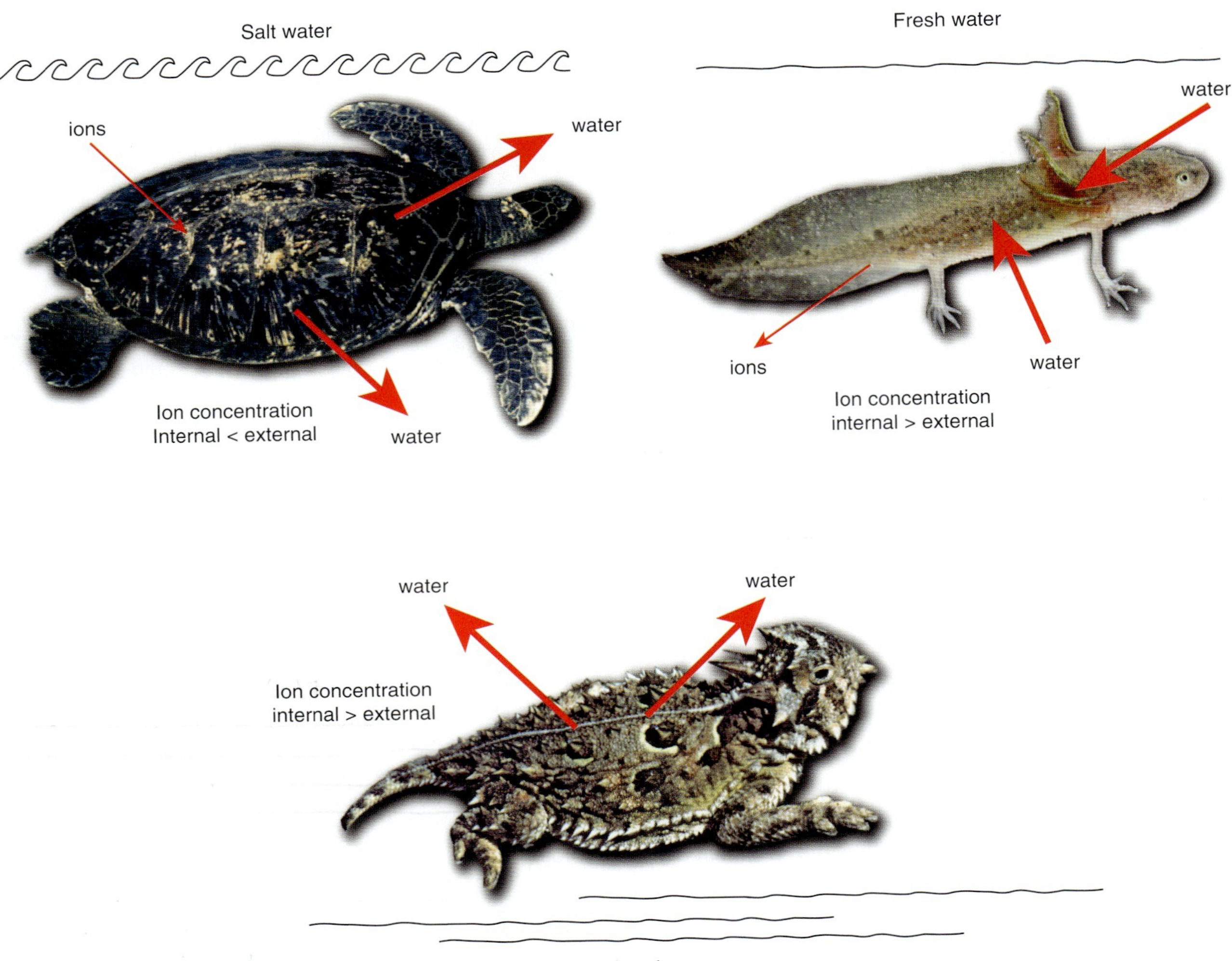

FIGURE 6.1 Osmotic challenges of amphibians and reptiles in saltwater, freshwater, and on land. In saltwater, the animal is hyposmotic compared to its environment, and because its internal ion concentration is less than that of the surrounding environment (internal < external), water moves outward. In freshwater, the animal is hyperosmotic to its environment, and the greater internal ion concentration (internal > external) causes water to move inward. On land, the animal is a container of water and ions, but because the animal is not in an aqueous environment, internal fluctuations in ionic balance result from water loss to the relatively drier environment. The animal actually has much higher ion concentrations (internal > external) than surrounding air, and if ionic concentrations reach high levels, as they do in some desert reptiles, ion transfer can occur via salt glands, usually in the nasal or lacrimal region.

Metabolic by-products and water diffuse into the kidney tubules from the circulatory system via the glomeruli, where capillaries interdigitate with the kidney tubules. In the proximal tubules, glucose, amino acids, Na^+, Cl^-, and water are resorbed. Nitrogenous waste products and other ions are retained in the urine, and additional water and Na^+ are removed in the distal tubules. In amphibians, due to a high filtration rate, about one-half of the primary filtrate enters the bladder even though more than 99% of filtered ions have been resorbed. As a consequence, urine produced by most amphibians is dilute. Some striking exceptions include African reedfrogs (*Hyperolius*), which exhibit increased levels of urea in plasma during dry periods, and the frogs *Phyllomedusa* and *Chiromantis*, which are uricotelic.

In reptiles, the filtration rate is lower than that of amphibians, and resorption of solutes and water is greater. Between 30 and 50% of water that enters the glomeruli of reptiles is resorbed in the proximal tubules alone. Urine generally empties into the large intestine in reptiles, but some have urinary bladders. In all cases, whether amphibian or reptile, urine flows from the urinary ducts into the cloaca and then into the bladder or the large intestine. Additional absorption of Na+ by active transport can occur in some freshwater reptiles from water in the bladder. Most reptiles produce

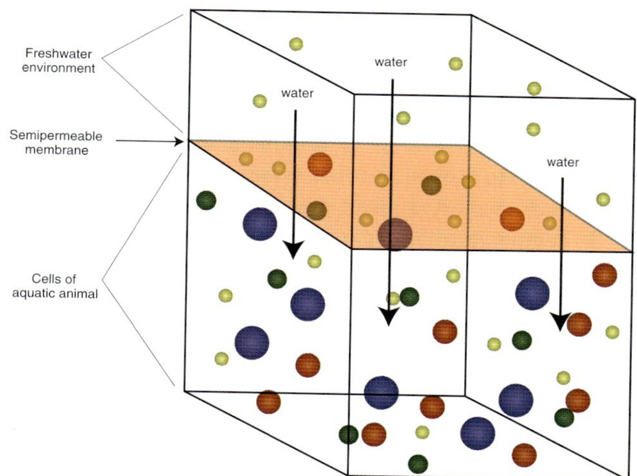

FIGURE 6.2 Model depicting how transfer of water occurs in cells based on a freshwater system. Water moves by the process of osmosis across the semipermeable membrane of the cell. The direction of water movement depends upon ionic gradients. If ion concentrations are higher inside the cell than outside (as in this example), then water moves in to balance concentration of ions. Semipermeable membranes do not allow all molecules to pass through. Rather, some do and some do not. In addition, cells have the capability to actively transport molecules across membranes. Amphibians and reptiles use a variety of behavioral and physiological mechanisms to maintain water and ionic balance because few natural environments are isotonic with their body fluids.

relatively concentrated urine, which minimizes water loss. In some species that live in deserts or marine environments, salt glands and other structures are involved in the control of Na^+ excretion.

Kidney structure differs somewhat between amphibians and reptiles, partially as a result of different embryonic origins. The opisthonephros of amphibians develops from posterior extensions of the pronephric kidney, whereas the metanephros of reptiles develops from the posterior lumbar mass of nephrogenic tissue. The opisthonephric kidneys of adult amphibians have two types of nephrons. In addition to fluids that are filtered from plasma in the glomeruli in the ventral nephrons, dorsally located nephrons collect fluid directly from the coelomic cavity. All filtration in reptiles occurs through glomeruli in the metanephric kidneys (Fig. 6.3).

Gaining and Losing Water

The body of an amphibian or reptile is composed of about 70–80% water, in which various ions necessary for proper physiological function are dissolved. Sodium, magnesium, calcium, potassium, and chloride are critical ions for normal physiological functions. Amphibians and reptiles live in environments varying from xeric deserts to

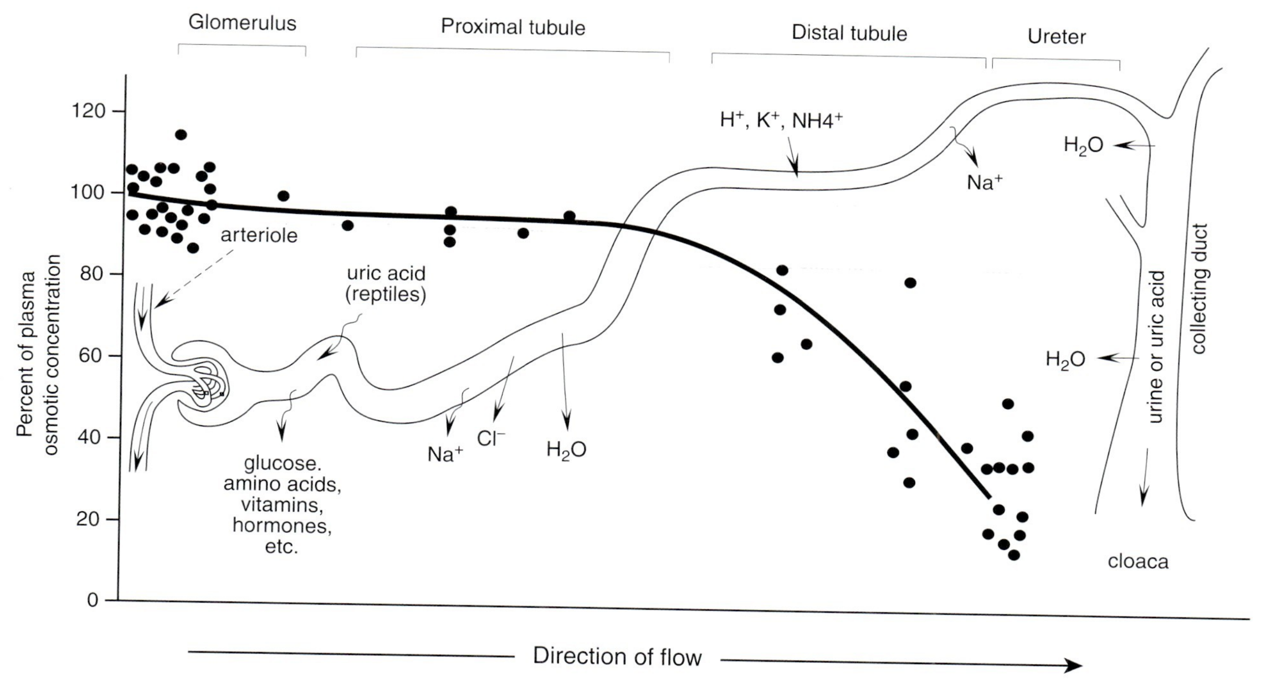

FIGURE 6.3 Diagrammatic representation of the functional kidney in amphibians and reptiles (see text for differences). Solid circles and the heavy line represent the reduction in osmotic concentration of urine for an amphibian. The line would be lower for a reptile. Adapted in part from Withers, 1992.

montane cloud forests and from fresh to saltwater, each of which presents special challenges for the maintenance of osmotic balance. For an organism to function normally, the ionic concentration of intra- and extracellular fluids must remain within certain specific limits, and the nitrogenous by-products of metabolism must be excreted from the body to avoid poisoning the organism. Most amphibians and reptiles maintain homeostasis, but a few species can tolerate high plasma solute concentrations for extended time periods (anhomeostasis). For those living in water, the external environment contains a complement of dissolved ions similar to their bodies, but in different proportions.

Water enters and exits the body in a variety of ways (Table 6.1). Many aspects of water gain and loss differ in amphibians and reptiles, primarily because of the structure and permeability of the skin. Amphibian skin is unique among vertebrates because it is highly permeable and lacks any kind of structures—scales, feathers, or fur—to make it less permeable. Consequently, water balance is the major physiological issue for amphibians, and evaporative water loss is one of the most important mechanisms for thermoregulation. In contrast, the epidermis of the skin of most reptiles is covered with scales, which dramatically reduces water gain and loss. Thus, water loss is less of an issue for reptiles, allowing them to maintain activity at higher body temperatures, often during daytime and in dry environments (Chapter 7). Because maintenance of osmotic balance is different in amphibians compared to other vertebrates, including reptiles, we discuss the two groups separately.

Amphibians

The skin of amphibians is highly permeable to water. It also functions as a major respiratory organ, through which they obtain oxygen and expel carbon dioxide, and the skin must be kept moist for exchange of gases to occur. Aquatic species take in water easily and must deal with an overabundance of water, whereas terrestrial or arboreal species often face the opposite problem of losing water rapidly and risking dehydration. This rapid evaporative water loss limits the time of activity for terrestrial species. Amphibians have evolved numerous physiological mechanisms and behavioral responses to deal with water loss or gain and thus to maintain osmotic balance. Temperature affects many functions of amphibians, including metabolic rate, locomotion, digestion, developmental rate, and calling rate, and is intertwined with water balance (see Chapter 7).

TABLE 6.1 Routes of Water Gain and Loss in Amphibians and Reptiles.

Gain	Loss
Food (preformed water)	Excretion
Drinking	Feces
Integument	Urine
Metabolism	Salt Glands Respiration Integument

Note: Some routes are specific to only amphibians or reptiles; see text.
Source: Adapted from Minnich, 1982.

Amphibians acquire water primarily through the skin, a process sometimes referred to as *cutaneous drinking*. They also acquire some water from food (called *preformed water*), and they gain a limited amount of water through metabolic processes when food is digested. Unlike reptiles and most other vertebrates, amphibians do not drink water orally.

Evaporation can be a significant source of water loss in terrestrial and even in semiaquatic species of amphibians. The skin of amphibians does not deter evaporative water loss as it typically does in reptiles. Experiments performed in the early 20th century revealed that skinned and normal frogs lose water at the same rate, and both lose water at the same rate as freely evaporating models of the same size and shape. Under arid conditions with no ability to regulate their water loss, most amphibians would not survive longer than 1 day. Water is lost not only through evaporation but also during respiration and excretion.

In aquatic amphibians, in which water is continually taken in through the skin, excretion of dilute urine aids in maintaining osmotic balance. Aquatic and semiaquatic amphibians are capable of producing urine at high rates to offset the high water influx through their permeable skin. Very dilute urine is produced to conserve salts. In contrast, when terrestrial amphibians begin to dehydrate, urine production declines rapidly in order to conserve water. Glomerular filtration rate decreases within 30 minutes to 1 hour after a frog or toad begins to dehydrate. Toads may have cutaneous osmotic sensors that detect changes in extracellular fluid volume.

In addition to its role in osmoregulation, the highly distensible amphibian bladder functions as a water-storage organ. Terrestrial species of frogs and salamanders can hold as much as 20–50% of their body mass as bladder water, whereas aquatic species such as *Xenopus* have small bladders capable of holding only 1–5% of their mass. An Australian desert frog, *Litoria* (= *Cyclorana*) *platycephala*, can hold as much as 130% of its normal body mass in bladder water, and not surprisingly, this species is called the "water-holding" frog. Many species of frogs are capable of reabsorbing their bladder water to maintain suitable levels of plasma solutes. Bladder water extends the survival time of amphibians in environments in which they are losing water.

The antidiuretic hormone, arginine vasotocin, acts on skin, kidneys, and bladder and enables terrestrial species, particularly those that live in arid areas, to preserve water. When dehydration begins to occur, the rate of urine formation is decreased, and water begins to be reabsorbed from the dilute urine in the bladder. The hormone increases the permeability of the skin and the bladder, so that rehydration can occur more rapidly when the animal encounters water. Hormonal control of cutaneous water uptake appears to be similar to that which regulates the drinking response in other tetrapods.

Morphological modifications of the skin in amphibians aid in water uptake. Different regions of the body have different degrees of permeability. Skin varies from smooth to granular. In general, aquatic amphibians, semi-terrestrial species that live near water, such as *Lithobates* and *Rana*, or rain forest species such as *Leptodactylus* and dendrobatids, have smooth ventral skin, whereas terrestrial species have granular ventral skin. Granular skin is more highly vascularized and enhances water absorption; thus, toads and tree frogs typically have granular venters. The granular skin surface of frogs, especially toads, also creates narrow grooves that serve as water channels to keep the skin on the dorsal surface of the body moist. Evaporation from the back pulls water onto the back via molecular adhesion, and capillary action pulls water from the venter, which is in contact with the soil. Salamanders have numerous vertical body grooves, the largest of which are the costal grooves that channel water from the salamander's underside to its back.

A specialized area of the skin, the "pelvic patch," is present on the posterior region of the venter and on the ventral surfaces of the thighs in many species of anurans (Fig. 6.4). This region is more highly vascularized than any other skin surfaces. By adpressing these surfaces to moist soil or to water on leaves, for example, a frog can quickly absorb water. This behavior is called the water absorption response. This mechanism provides as much as 70–80% of total water uptake in some species of toads. In the red-spotted toad, *Anaxyrus punctatus*, the hormone arginine vasotocin greatly increases the ability of the pelvic patch to absorb water. Toads vary in how dehydrated they become before beginning to show this response. The red-spotted toad begins to show the water absorption response after losing only 1–3.6% of its body weight, although other species of toads can lose much more body weight before showing the response. Body size and whether the habitat is arid or mesic have been proposed as determinants of when the water absorption response is initiated.

Behavioral adjustments are the overriding mechanisms for water retention in terrestrial amphibians. Most species adjust daily and seasonal activity to minimize water loss, and they seek humid or enclosed retreats such as crevices or burrows when inactive. Dehydrated or resting amphibians typically adopt water-conserving postures. These postures include folding the arms and legs tightly beneath the body and flattening the ventral surface close to the substrate (Fig. 6.5). When emerging (at night, for example), many species seek damp substrates from which they absorb water. In the water-absorbing posture, frogs and toads hold the hind limbs away from the body and press the ventral surfaces onto the substrate (Fig. 6.6). If the substrate contains renewable water (a pond edge, for example), the frog remains in the same position; however, if the substrate is nonporous, the frog continually readjusts its position to take up additional water.

The tropical rain frog *Eleutherodactylus coqui* provides an example of how frogs use adjustments in posture and activity to regulate water flux. By resting the chin on the substrate and drawing the limbs up underneath the

FIGURE 6.4 Photo of *Anaxyrus punctatus* (Bufonidae) on glass showing the ventral water absorption patch. (L. J. Vitt and J. P. Caldwell)

FIGURE 6.5 Water-conserving posture in the hylid frog *Hyla chrysoscelis*. The posture minimizes surface area exposed, thus reducing evaporative water loss through the skin. (J. P. Caldwell)

FIGURE 6.6 Typical posture of frog absorbing water from surface. The frog (*Chiasmocleis albopunctata*; Microhylidae) maximizes contact of the ventral body surface with the damp substrate. (J. P. Caldwell)

body during periods of inactivity, a minimum amount of surface is exposed, and cutaneous water loss is reduced (Fig. 6.7). While calling, males expose a maximum amount of surface area, which results in increased water loss, and in addition, expansion and contraction of the body and vocal sac during calling causes the boundary air surrounding the frog to mix with environmental air, increasing the rate of water loss even more. A threefold difference can occur between frogs in water-conserving postures versus nonwater-conserving postures. Because of water loss during calling, males experience increased solute concentrations that negatively impact metabolism and potentially result in reduced calling performance, at least as measured by jumping experiments. The ability of these frogs to absorb water from damp surfaces partially offsets the osmoregulatory costs of calling activity, and the payoff for calling is an increase in reproductive success.

Crowding, piling together, or remaining in tight retreats can minimize exposed surface and reduce water loss as well. The physical process underlying these behaviors is quite simple. For example, a salamander (*Ensatina*) in a small rock crevice would have its dorsal and ventral surfaces pressed against the rocks, thus greatly reducing water loss from those regions of the body.

In some frogs, modifications of skin or use of glands or other structures in skin reduce water loss. Several groups of arboreal frogs, collectively called waterproof frogs, have independently evolved specialized mechanisms for withstanding arid conditions by decreasing permeability of the skin. The mechanism for water loss reduction in certain species of *Phyllomedusa*, a genus of South American hylids, involves secretion of lipids from specialized skin glands. In *P. sauvagii*, the glands secrete a variety of lipids, with wax esters most abundant. The frogs have an associated stereotypic behavior, in which they systematically use their arms and legs to wipe the lipids evenly over all surfaces of their bodies (Fig. 6.8). The skin becomes shiny and impermeable to water, and the frogs reduce evaporative water loss while in their arboreal

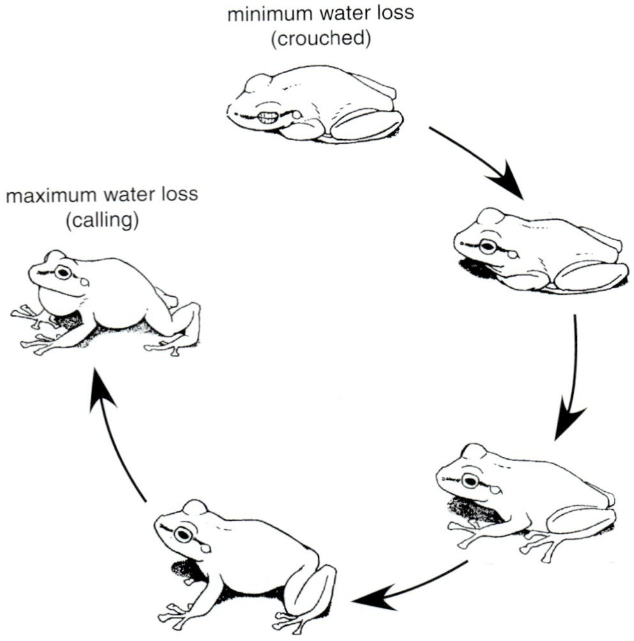

minimum water loss (crouched)

maximum water loss (calling)

FIGURE 6.7 *Eleutherodactylus coqui* uses different postures to regulate water loss. The chin-down posture with legs underneath the body minimizes water loss. Water loss is greatest during bouts of calling by males when the greatest amount of skin surface area is exposed. Adapted from Pough et al., 1983.

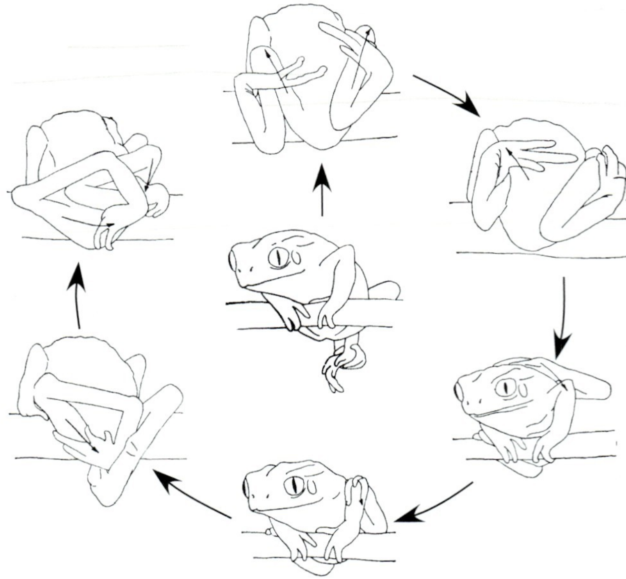

FIGURE 6.8 *Phyllomedusa sauvagii* (Hylidae) spreads lipids from lipid glands in the skin by a series of stereotyped movements using the feet. Arrows indicate direction of foot movement. Adapted from Blaylock et al., 1976.

perches. In waxed frogs, the rate of water loss is low at low temperatures, but above 35°C, the wax melts and no longer forms an evaporation barrier. *Phyllomedusa sauvagii* lives in arid areas of Bolivia and adjacent countries, where the highest environmental temperatures occur during the rainy season when the frogs reproduce; thus, dehydration is not a problem at that time. A smaller species, *Phyllomedusa hypochondrialis*, is similar in lipid production and wiping behavior although it is from two to four times more effective at reducing evaporative water loss than that of other species of *Phyllomedusa*, and it is the only species that has a specific arrangement of dorsal lipid-secreting glands. *Phyllomedusa hypochondrialis* lives in open habitats in South America that experience high daytime temperatures and have an extended dry season (Fig. 6.9). Ecological studies on this and other species of *Phyllomedusa* are needed to interpret physiological and morphological differences among these species.

FIGURE 6.9 *Phyllomedusa hypochondrialis* (Hylidae), a common frog in semiarid and savanna areas of Brazil, has the ability to reduce water loss considerably more than other *Phyllomedusa* species by waxing its skin when active and exposing the maximum amount of its small body (upper). During dry season, these frogs minimize water loss by minimizing surface area exposed. The frog in the lower panel is just beginning to emerge from a nearly balled-up state. It was found under a dry pile of vegetation at the peak of the dry season with several others. The skin surface was dry, and it took the frog nearly 10 minutes to come out of an apparent state of torpor. (J. P. Caldwell and L. J. Vitt)

Other species in unrelated clades have evolved similar behavior, indicating that frogs living in arid conditions are under strong selective pressure to conserve water. *Litoria caerulea*, an Australian tree frog, secretes lipids from its skin and uses its arms to wipe the secretion over its dorsal surfaces to provide a barrier to water loss. An Indian rhacophorid, *Polypedates maculatus*, lives in semiarid habitat and wipes skin sections over its body. Unlike *Phyllomedusa sauvagii*, the skin secretions do not reduce water loss completely, and *Polypedates* seeks moist habitats after wiping behavior. Several species of *Hyla* from Florida (USA) also secrete lipids and perform simple wiping behaviors; however, these species generally live in mesic environments that are only periodically dry. Overall, the Florida species exhibit a higher evaporative water loss than true waterproof frogs.

South African waterproof frogs in the genera *Chiromantis* (Rhacophoridae) and *Hyperolius* (Hyperoliidae) lose water at the same rate as expected for reptiles when exposed to arid conditions. Their mechanism for prevention of cutaneous water loss differs from *Phyllomedusa*, and their skin does not contain wax glands. These frogs live in semiarid areas in Africa where temperature can exceed 40°C; thus, heat gain is more of a challenge for these frogs than for *P. sauvagii*. The waterproofing mechanism lies in the structure of the dermal layer of the skin. In all frogs, the dermis contains various types of chromatophores arranged in layers (Fig. 2.14; see Chapter 2). In *Chiromantis petersii* and *C. xerampelina*, the iridophores are several layers thick. In the dry season, the iridophores increase in number, filling the stratum spongiosum. The iridophores function in part to lower internal temperature by lowering radiation absorption, thereby reducing the rate of water loss. In *Hyperolius viridiflavus*, an African species that estivates in exposed areas with high temperatures and low relative humidity, the number of iridophores present exceeds that necessary for radiation reflectance. Instead, accumulation of additional iridophores aids in elimination of nitrogen. Iridophores contain mainly the purines guanine and hypoxanthine, which contain nitrogen. In addition to the skin, the liver epithelium and other internal connective tissues fill with iridophores, supporting the interpretation that the iridophores function in a capacity other than radiation reflectance.

Another mechanism used by some amphibians to survive extended dry seasons or drought includes the formation of an impermeable encasement called a cocoon. The ability to from a cocoon has evolved independently in numerous taxa of frogs, including *Limnodynastes* and *Neobatrachus* (Limnodynastidae), *Litoria* (Fig. 6.10) and *Smilisca* (Hylidae), *Ceratophrys* and *Lepidobatrachus* (Ceratophryidae), *Leptopelis* (Arthroleptidae), and *Pyxicephalus* (Pyxicephalidae), and in a few salamanders, including *Siren*. Cocoons form in individuals that burrow in soil during dry periods. Even though evolved in unrelated species, cocoon formation appears to be similar among these species. The cocoon forms from the

FIGURE 6.10 The hylid frog *Litoria novaehollandiae* (Hylidae), encased in its cocoon, emerges after a rainstorm and begins to eat the cocoon. The cocoon consists of retained layers of shed skin. (S. J. Richards)

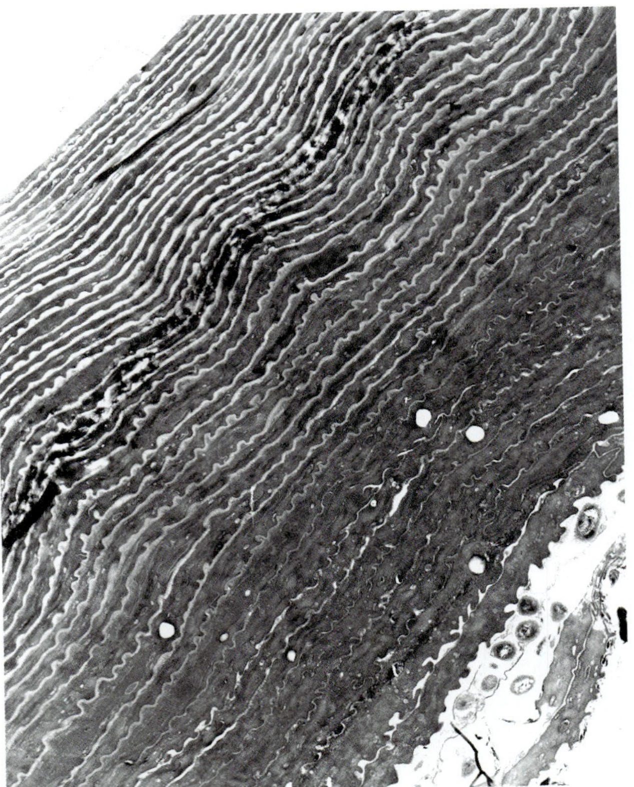

FIGURE 6.11 Photomicrograph showing 39 layers of stratum corneum forming the cocoon of the South American frog, *Lepidobatrachus llanensis* (Ceratophryidae). (R. Ruibal)

accumulation of multiple layers of shed epidermal skin. The layers of skin are not truly shed but remain attached to the frog. With each ecdysis event, another epidermal layer lifts off the new skin and fuses to the previous layer (Fig. 6.11). In the Argentinian *Lepidobatrachus llanensis*, the cocoon accumulates at the rate of 1 layer a day, and in the Australian hylid *Litoria alboguttata*, a 24-layer cocoon formed in 21 days after water was withheld in a laboratory experiment. In a study of field-excavated *Neobatrachus aquilonius* (Limnodynastidae), frogs were encased in cocoons ranging from 81–106 layers thick; one individual had a cocoon composed of 229 layers. The multilayered cocoon creates an impermeable sac around the frog that opens only at the nares, allowing the frog to breathe. Apparently individuals enter the soil while it is still damp and begin to form the cocoon only when they begin to dehydrate. After the cocoon is formed, the frogs cease voiding urine and remain hydrated from water they have stored. The cocoon prevents respiration through the skin, but the need for ventilation is reduced because of metabolic depression.

Similarly, the salamanders *Siren intermedia* and *S. lacertina* form cocoons when ponds in which they live dry. As the ponds dry, the sirens begin to aestivate in the bottom mud to avoid dehydration, and cocoon formation begins as the habitat continues to dry. Initially, the cocoons were reported to form from dried mucous gland secretions, but subsequent studies revealed that salamander cocoons consist of epidermal layers, just as in frogs.

Reptiles

Unlike amphibians, most reptiles gain or lose almost no water through their impermeable skin, which is largely resistant to movement of water or ions. Water loss and gain must remain in balance, and reptiles lose and gain water in several ways. Drinking freshwater is an important source of water gain, and reptiles accomplish this behavior in a variety of ways. Some desert lizards (e.g., *Coleonyx variegatus* and *Xantusia vigilis*) drink water that condenses on their skin when they enter cool burrows. Some South African tortoises collect water in their shells during rainfall. By elevating the posterior carapace higher than the anterior region, an individual can cause the water to run along the edges of the ridged carapace toward its head. This behavior may be more common than currently known. For example, the two tropical turtles, *Kinosternon scorpiones* and *Platemys platycephala*, spend much of their time on land and have deeply grooved carapaces. Both experience extended dry seasons. The possibility exists that they use their shells to capture water, but this behavior has not yet been reported.

Some desert lizards living in xeric environments are capable of acquiring water from their skin by assuming stereotyped behaviors that result in capillary transport of water toward the mouth through channels between scales. This behavior has been observed in the laboratory for *Moloch horridus* and *Phrynocephalus helioscopus* and in the field for *Phrynosoma cornutum*. Typically, the body is arched during rainstorms in *P. cornutum*, and water moves from the back to

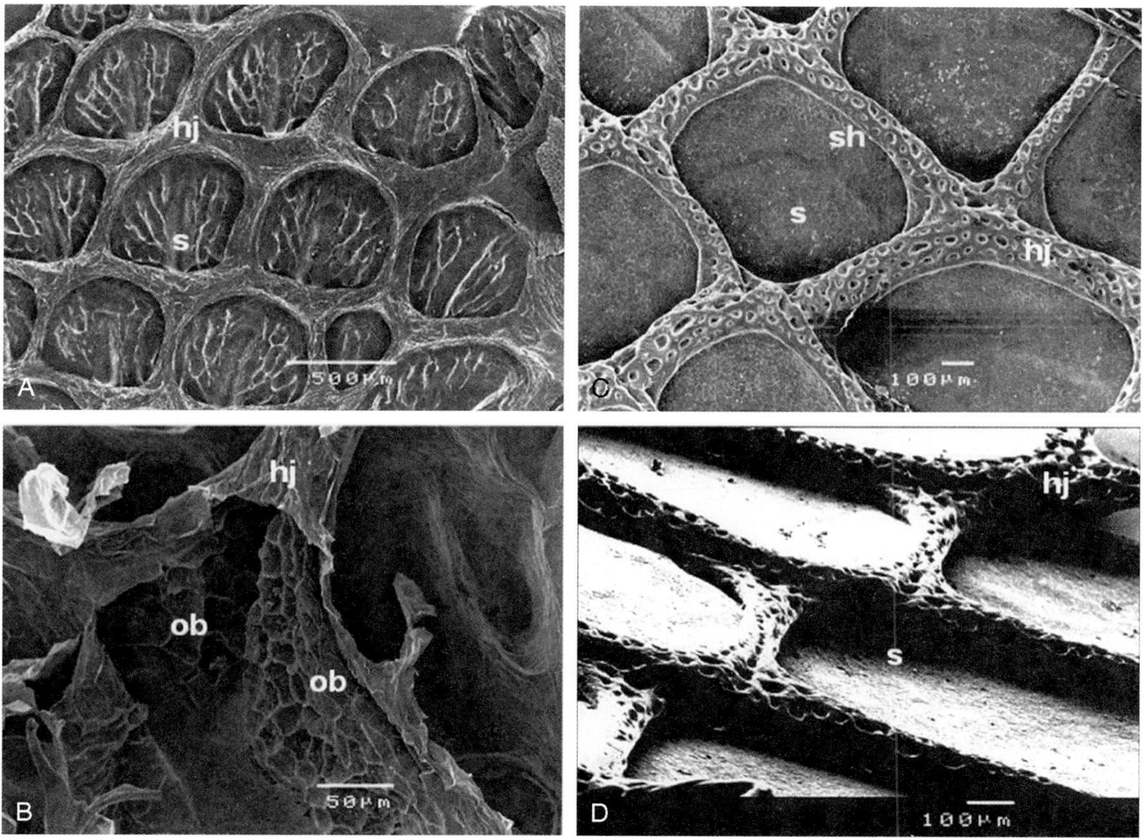

FIGURE 6.12 Scale microstructure in the two desert lizards *Moloch horridus* (Agamidae: A and B) and *Phrynosoma cornutum* (Phrynosomatinae: C and D). The medial surface of ventral scale epidermis (β-layer) is shown at two magnifications for each species, showing the interior bracing support of each scale and the scale-defining interconnections of scale hinges. Hinge joints (hj), deep portions of scale hinges that spread laterally, form a continuously connected surface. The epidermis is ruptured in B (medial side of β-layer) showing Oberhäutchen (ob) cover on walls of the scale hinge. The β-level epidermis in *P. cornutum* (D) shows surface pitting on the different levels of the hinge joint.

the mouth. The mechanism for moving water across the skin in both *Moloch* and *Phrynosoma* has only recently been investigated in detail by Dr. Wade Sherbrooke and his collaborators. Although it has been known for some time that these lizards could move water from the skin to the mouth, it was thought that water simply moved along spaces between scales, ultimately reaching the mouth. However, the interscalar spaces are along scale hinges, and each of the scale hinges has an expanded base and a channel that is nearly closed, sort of like a tiny straw (Fig. 6.12). Scale hinges on the body of the lizard are interconnected to form a complex network of tiny channels through which water flows by capillary action. The β-level keratin of the skin is very thin along the scale hinges, and the walls of the hinge joints have a complex topography that effectively increases surface area of the channels to facilitate water transfer. Scale surfaces at the rear of the jaw in both lizard species are also modified to allow the jaw to function as a buccal pumping mechanism. By creating a water pressure gradient at the edge of the mouth, water moves through the interconnected capillary system of water channels in the skin, providing the ability to drink water that has been captured on the skin surface and transported through the capillary system (Fig. 6.13). Because *Moloch* and *Phrynosoma* are in different lizard clades (Agamidae and Phrynosomatinae, respectively) and live on different continents (Australia and North America, respectively) but in similar habitats (deserts), similarity in structure and function of the water-capturing system is an example of convergent evolution.

Variable amounts of water are obtained from food, but the impact of this water depends on the electrolyte concentration of the food. Diet choice or feeding rates can be influenced to some degree by the concentration of electrolytes in a particular food item. Free or preformed water in the insect prey of many desert lizard species exposed to extreme heat and prolonged periods of low humidity may be the sole source of water over extended dry periods. *Urosaurus graciosus*, a small lizard in the Sonoran Desert, forages in the canopy of relatively small trees and shrubs during morning and late afternoon and remains inactive

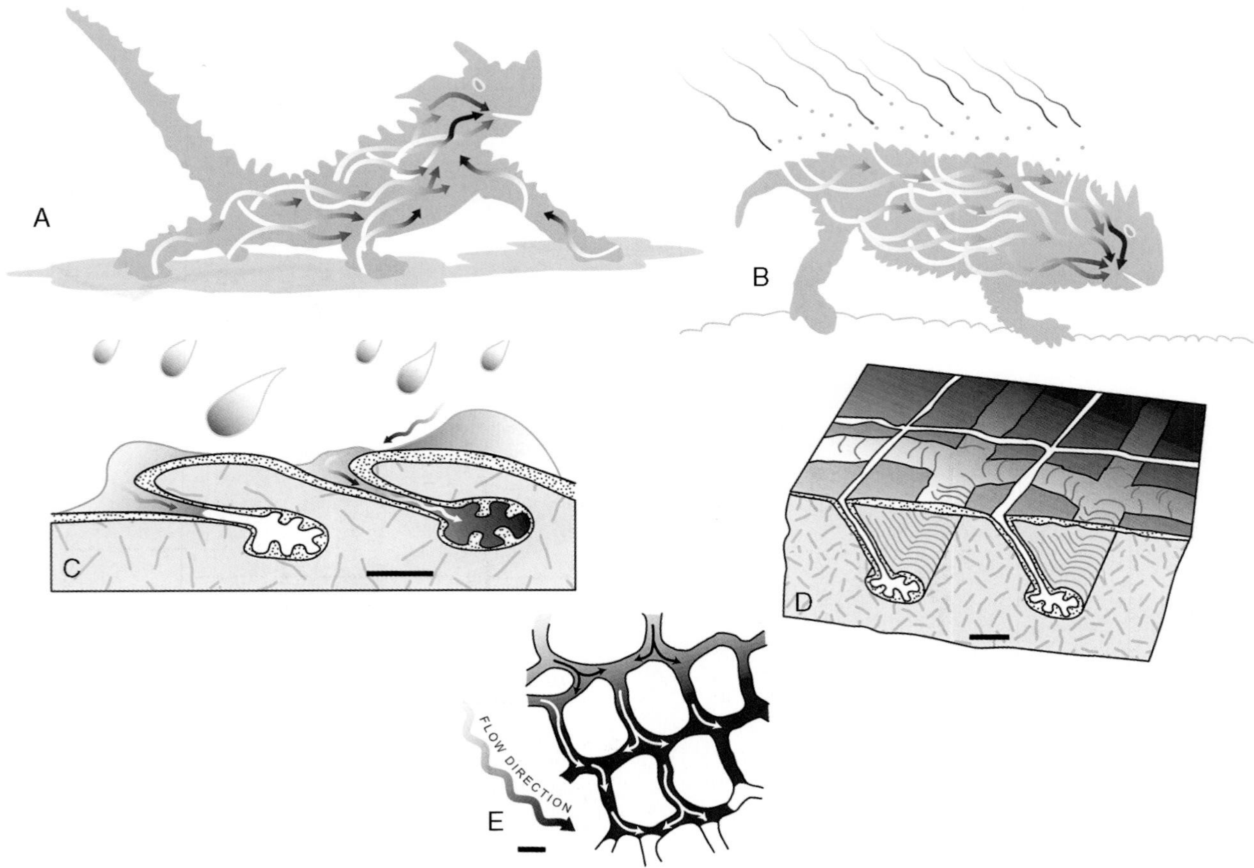

FIGURE 6.13 Schematic summary of the mechanism for cutaneous water collecting, transport, and drinking in *Moloch horridus* (A) and *Phrynosoma cornutum* (B). Arrows indicate directional movement of water. C through E are generalized models of the morphology of the water transport system in the two lizards. In C and D, narrow passageways below each scale expand into scale hinge joints. Water moves through the channels and collects in the scale hinges, which are interconnected (D). The scale hinge-joint channel system consists of a continuous floor of channels, all directed so that water ultimately flows to the corner of the lizard's mouth. Adapted from Sherbrooke et al., 2007.

in shady sites on tree trunks during the hottest part of the day. Due to a lack of cool retreats, body temperature increases from 35°C while foraging to more than 38°C while inactive in the afternoon. Water loss is high, 38.5 ml/kg/day. In the same habitat, a closely related and similar-sized species, *Urosaurus ornatus*, lives in larger trees adjacent to rivers where afternoon temperatures are lower as a result of the shading effect of the canopy, and it maintains body temperatures at 36°C or lower throughout the day. Water loss in *U. ornatus* is lower, 27.7 ml/kg/day. Differences in thermal ecology account for differences in water loss between the two species. Both species gain water primarily from the insects they eat; infrequent rainfall and dew are the only other water sources. *Urosaurus graciosus* eats an average of 11.5 prey items per day and has an average stomach volume of 0.129 cm^2, whereas *U. ornatus* eats an average of 7.7 prey per day and has an average stomach volume of only 0.066 cm^2. *Urosaurus graciosus* offsets its high rate of water loss by ingesting substantially more insect prey.

Water loss occurs in reptiles through a combination of metabolic and evaporative processes. Water is lost in feces and relatively concentrated urine. Water lost in feces ranges from 8 to 70% of that taken in by food. Terrestrial reptiles lose some water during respiration and, in many cases, use respiratory water loss to aid in thermoregulation (Chapter 7). Production of metabolic water contributes to osmoregulation in some species of reptiles. For example, metabolic water contributes 12% of total water gain in *Dipsosaurus dorsalis*. However, reptiles cannot produce metabolic water at a rate that exceeds their evaporative water loss. The temperature at which a reptile digests its food affects how much metabolic water is produced.

Reptiles have a variety of water-storage sites that help offset water loss. The bladder is a common site of water storage. For example, the bladders of desert tortoises can occupy more than one-half of the peritoneal cavity. Other sites of water storage include the stomach in the lizard *Meroles* (formerly *Aporosaura*) *anchietae*. The accessory lymph sac in the lateral abdominal folds of chuckwallas and the baggy folds of skin around the legs in diamondback terrapins also hold water. Nevertheless, behavior plays an important role in water retention in reptiles, and it is tightly interwoven with thermoregulation. Many reptiles use daily and seasonal activity patterns to regulate body temperatures, and they seek humid or enclosed crevices or burrows when inactive to reduce water loss.

Freshwater crocodiles often experience prolonged periods of drought in which no surface water is available. Aestivating crocodiles (*Crocodylus johnsoni*) in Australia spend 3 to 4 months inactive underground with no access to water. Body temperatures increase with time as a result of increasing environmental temperatures, but water loss rate is only about 23% of the rate prior to aestivation. The crocodiles do not dehydrate and appear to have no physiological mechanisms specifically associated with aestivation. Refuges used for aestivation appear to adequately accommodate homeostasis.

Chuckwallas (*Sauromalus obesus*) in the Mojave Desert maintain relatively constant solute concentrations even though they live in environments in which water is highly seasonal and unpredictable. When the vegetation that comprises their diet is abundant, they obtain sufficient water from their plant food, and the excess water is excreted. When vegetation is dry, the lizards do not eat and remain inactive inside crevices where temperatures are relatively low. Their water loss rates are low in this situation. Although the plants that they eat are always hyperosmotic, primarily because of high K^+ concentrations, excretion of potassium urate by nasal salt glands removes electrolytes with little associated water loss. Effectively, these lizards separate electrolyte excretion from water excretion, thereby maintaining homeostasis.

Some terrestrial species are able to withstand fairly large fluctuations in body water. For example, the tropical lizard *Sceloporus variabilis* has higher levels of water and metabolic flux than most similar-sized temperate sceloporine lizards. Physical, biotic, and behavioral differences between *S. variabilis* and its temperate-zone relatives account for increased rates of water and energy exchange. These lizards move more and are active longer, and both water and food are more readily available to them than to most temperate-zone sceloporines.

Just as some amphibians aggregate to reduce evaporative water loss of individuals, evidence exists suggesting that lizards do the same. Banded geckos, *Coleonyx variegatus*, provide an example. Many herpetologists believed that these geckos aggregated for social purposes, and although early experiments demonstrated that they aggregate, reasons for aggregations were at best speculative. These nocturnal desert geckos have evaporative water loss rates two to three times higher than those of diurnal desert lizards living in the same habitats. Recent studies by Jennifer Lancaster and colleagues reveal that evaporative water loss rates are nearly double for individual geckos when placed in containers alone than they

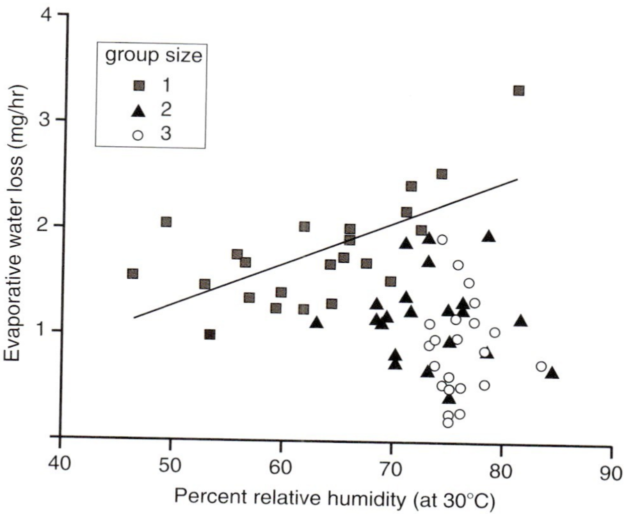

FIGURE 6.14 Rates of evaporative water loss (EWL) are lower for individual geckos (*Coleonyx variegatus*) when they are aggregated in groups of two or three in retreats than they are when geckos are alone in retreats. Adapted from Lancaster et al., 2006.

are when three are in a container (Fig. 6.14). Similarly, water loss rates are much lower when geckos are aggregated in retreats than when they are alone. Thus, aggregation in banded geckos may reduce evaporative water loss by increasing relative humidity within retreats. Interestingly, and different from what some amphibians do, water loss rates were not reduced concordant with an increase in the number of geckos in a retreat. That is, three geckos in a retreat did not gain substantially over two geckos in a retreat in terms of individual reductions in water loss. This suggests that reducing exposed surface area of their bodies is not the primary mechanism providing the physiological payoff. Rather, some of the water lost through respiration and evaporation is regained. This study exemplifies the sometimes complex interactions between physiology, social behavior, and ecology. Skinks in the genus *Plestiodon* that brood eggs likely regain some of their water lost through respiration, but this has not been investigated. Their eggs are known to gain water lost by brooding females during respiration (Chapter 5).

Nitrogen Excretion

Digestion of food and catabolism of protein result in the production of wastes, including various nitrogen-containing products, particularly ammonia, urea, and uric acid. Prolonged dehydration leads to accumulation of nitrogen waste, which causes death if not removed or diluted. Organisms that primarily excrete ammonia are called ammonotelic; those that excrete urea, ureotelic; and those that excrete uric acid, uricotelic. Among amphibians, reptiles, and other vertebrates, patterns of nitrogen excretion appear generally related more to habitat than to phylogeny, but this idea has not been explored in the context of modern phylogenetics.

Aquatic animals in general excrete ammonia. Ammonia is a small molecule that readily diffuses across skin and gills if sufficient water is available, but the kidneys inefficiently excrete ammonia. Ammonia is highly toxic, and animals cannot survive even moderate ammonia concentrations in their body fluids. For this reason during the evolution of terrestriality, selection favored the excretion of a less toxic form of nitrogen, such as urea or uric acid. Uric acid is the least toxic nitrogenous by-product. For example, three species formerly considered to be in the genus *Rana* (now *Limnonectes kuhlii*, family Dicroglossidae, and *Hylarana signata* and *H. chalconota*, family Ranidae) are obligatory ammonotelics, and individuals of these species die when deprived of water. Most ranids are ureotelic and can tolerate moderate dehydration without dying. Totally aquatic amphibians, such as *Xenopus* and nearly all larvae, excrete ammonia. At metamorphosis, the larvae of most species switch from excreting ammonia to excreting urea. Under normal conditions, *Xenopus* continues to excrete ammonia throughout its life, but it is physiologically adaptable. When its aquatic habitats dry, *Xenopus* aestivates in the mud and physiologically shifts to urea excretion, thereby avoiding the toxic effects of ammonia accumulation. Urea is soluble in water and has relatively low toxicity compared to ammonia. In many amphibians, urea is the primary excretory product. All terrestrial species produce urea. Certain liver enzymes that function in urea production are widespread in aquatic and terrestrial amphibians, suggesting that this method of excretion appeared early in the evolutionary history of tetrapods.

Uric acid has a low solubility and requires very little water for excretion. Most snakes and lizards excrete uric acid, which serves to conserve water in species living in arid areas. Uricotelism appears to have evolved independently in a few lineages of waterproof frogs. *Phyllomedusa sauvagii* and some species of *Chiromantis* produce urates, salts of uric acid, even when ample water is available. Ninety percent of the water filtered by the kidney is reabsorbed in *Phyllomedusa sauvagii*.

The saltwater crocodile, *Crocodylus porosus*, takes in substantial amounts of saltwater during feeding and has no freshwater available. As a result, while they are in saltwater, a net loss in body water occurs. Most sodium (55%) is excreted through lingual salt glands, but a considerable amount (42%) is excreted across the cephalic epithelium. Loss of water occurs primarily across the skin (55%) and epithelia of the head (36%).

Reptiles have little difficulty with osmoregulation in freshwater. Because of their relatively impermeable skin, water influx and solute efflux across the skin are relatively low. Aquatic species that take in significant amounts of water produce dilute urine and reabsorb solutes in the kidney, urinary bladder, and colon.

The Terrestrial Transition

Most amphibian larvae are aquatic and must undergo a transition to terrestrial life. Because larvae are hyperosmotic in relation to their aquatic environment and adults are hyposmotic in relation to their terrestrial environment, the osmoregulatory challenges are reversed and require different behavioral, morphological, and physiological solutions (see Fig. 6.1). This change in lifestyle sets amphibians apart from all other vertebrates and reflects part of the transition from water to land that led to the diversification of terrestrial tetrapods.

With few exceptions, anuran larvae live in aquatic habitats. Thus, behavioral adjustments to water gain or loss are not possible. Because excess water influx is a problem, amphibian larvae would be predicted not to take in water through the mouth. However, studies have shown that larvae ingest large quantities of water when feeding. Water turnover decreases during metamorphosis, but whether this change results from a decrease in ingestion of water during feeding is unknown.

At metamorphosis, the organs responsible for osmoregulation undergo extreme morphological and physiological changes. Larval skin has a simpler structure than adult skin, for example, and gills are replaced by lungs in many species. Whereas adults can regulate ion exchange across the skin by active transport of solutes, anuran larvae are incapable of this type of regulation, apparently because they lack the proper enzymes to carry out the reactions. Instead, active transport of solutes occurs in the gills of anuran larvae. In contrast to tadpoles, active transport of solutes occurs across the skin in salamander larvae.

Marine and Xeric Environments

Marine Environments

No amphibian is truly marine. Nevertheless, 61 species of frogs and 13 species of salamanders are tolerant of hypersaline environments to some degree (Table 6.2). Three species of frogs (*Fejervarya* [= *Rana*] *cancrivora*, *Pseudepidalea* [= *Bufo*] *viridis*, and *Xenopus laevis*) live in habitats with unusually high salinity, and a few species of the salamander *Batrachoceps* live near saltwater in tidal areas. *Fejervarya cancrivora* inhabits estuaries in Southeast Asia, where it feeds predominately on marine crabs and crustaceans. This frog, in addition to other brackish species, remains in osmotic balance with saltwater by maintaining a high level of urea in the blood. To create these high levels, urea is retained, and, in addition, urea synthesis is increased. The enzymes responsible for these reactions are found at higher levels in frogs that inhabit the most saline environments.

Sea turtles, sea snakes, diamondback terrapins (*Malaclemys*), and some species of *Crocodylus* are found

TABLE 6.2 Amphibians Known to Inhabit or Tolerate Brackish Water

Ambystomatidae	Brachycephalidae
Ambystoma subsalsum	*Eleutherodactylus martinicensis*
Dicamptodon ensatus	
	Leiuperidae
Plethodontidae	*Pleurodema tucumanum*
Batrochoseps major	
Plethodon dunni	Microhylidae
	Gastrophyrne carolinensis
Salamandridae	
Taricha granulosa	Pelodytidae
Lissotriton (= *Triturus*) *vulgaris*	*Pelodytes punctatus*
Sirenidae	Pelobatidae
Siren lacertina	*Pelobates cultripes*
Alytidae	Scaphiopodidae
Discoglossus sardus	*Spea hammondii*
Bombinatoridae	Pipidae
Bombina variegata	*Xenopus laevis*
Bufonidae	Dicroglossidae
Anaxyrus (= *Bufo*) *boreas*	*Fejervarya* (= *Rana*) *cancrivora*
Pseudepidalea (= *Bufo*) *viridis*	*Euphlyctis* (= *Rana*) *cyanophlyctis*
Hylidae	Ranidae
Acris gryllus	*Lithobates* (= *Rana*) *clamitans*
Pseudacris regilla	

Note: *List includes only selected species.*
Source: *Adapted from Balinsky, 1981. Scientific names and families updated.*

TABLE 6.3 Occurrence of Salt Glands in Reptiles

Lineage	Salt-secreting gland	Homologies
Turtles		
Chelonids, dermochelids, and *Malaclemys terrapin*	Lacrymal gland	Lacrymal salt gland of birds
Lizards		
Agamids, iguanids, lacertids, scincids, teiids, varanids, xantusiids	Nasal gland	None
Snakes		
Hydrophines, *Acrochordus granulatus*,	Posterior sublingual gland	None
Cerberus rhynchops	Premaxillary gland	None
Crocodylians		
Crocodylus porosus	Lingual glands	None

in water of varying degrees of salinity. The ionic concentration of body fluids in these species is maintained at higher levels than in freshwater species. Much of the increase in solutes is due to higher levels of sodium, chloride, and urea. This response also typically occurs when freshwater species are experimentally placed in saltwater.

Reptiles in saline habitats tend to accumulate solutes as the salinity level increases. Numerous species have independently evolved salt glands that aid in the removal of salt (Table 6.3). Other species survive in saltwater because of behavioral adjustments. The mud turtle, *Kinosternon baurii*, inhabits freshwater sites that are often flooded by saltwater, but when salinities reach 50% of saltwater, the turtle leaves water and remains on land. One important key to the survival of reptiles in marine environments is that they do not drink saltwater. Experiments with freshwater and estuarine species of *Nerodia* revealed that drinking is triggered in freshwater species experimentally placed in saltwater, presumably because of dehydration and sodium influx. These snakes continue to drink saltwater, which leads to their eventual death. In contrast, estuarine species in the same genus are not triggered to drink saltwater, presumably because their skin is not permeable to saltwater and they do not become dehydrated.

Xeric Environments

Some reptiles living in extreme environments can withstand extreme fluctuations in body water and solute concentrations. Desert tortoises (*Gopherus agassizii*) inhabit a range of environments in deserts of southwestern North America. By storing wastes in their large urinary bladder and reabsorbing water, they minimize water loss during droughts. Nevertheless, during extended droughts, they can lose as much as 40% of their initial body mass, and the mean volume of total body water can decrease to less than 60% of body mass. Rather than maintaining homeostasis in the normal sense, concentrations of solutes in the body increase with increasing dehydration (anhomeostasis), often to the highest levels known in vertebrates, but the most dramatic increase occurs in plasma urea concentrations. When rainfall occurs, increases in solute concentrations are reversed when tortoises drink water from depressions that serve as water basins (Fig. 6.15). Following the ingestion of water, they void the bladder contents, and plasma levels of solutes and urea return to levels normally seen in reptiles in general. They then store large amounts of water in the bladder, and as conditions dry out, the dilute urine remains hyposmotic to plasma for long periods, during which homeostasis is maintained. When the

FIGURE 6.15 The desert tortoise, *Gopherus agassizii* (Testudinidae), either drinks from natural depressions or constructs shallow water-catchment basins in the desert floor following periodic rainstorms (Medica, et al., 1980). (P. A. Medica; photo not included in published paper)

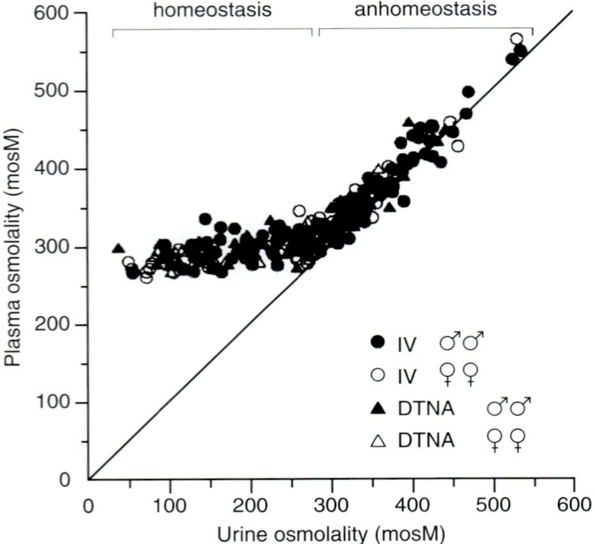

FIGURE 6.16 Concentration of plasma solutes remains stable (homeostasis) in desert tortoises as long as urine stored in the bladder is hyposmotic to plasma. When solute concentration of the plasma reaches that of the urine (isosmotic), solute concentrations increase in both (anhomeostasis). Data from two populations are included (IV = Ivanpah Valley; DTNA = Desert Tortoise Natural Area, both in the Mojave Desert). Adapted from Peterson, 1996.

urine reaches an isosmotic state, solute concentrations in both plasma and urine increase (Fig. 6.16).

RESPIRATORY GAS EXCHANGE

Respiration is the process by which animals acquire oxygen. Oxygen is essential for cellular metabolism, during which food is converted to energy by oxidation. By-products of this process are carbon dioxide and water, which must be eliminated. External respiration refers to the transfer of oxygen from the environment across the surface of the respiratory organ to the blood and to the reverse flow of carbon dioxide from the blood to the environment. Internal respiration refers to gas exchange between the blood and the cells of the body tissues. At the cellular level, this transfer of oxygen and carbon dioxide occurs by passive diffusion, and like water, gases flow from areas of high concentration to areas of low concentration.

Differences in the physical properties of water and air determine the available oxygen supply for all animals, including amphibians and reptiles. Water is denser than air and holds much less oxygen, and the solubility of both oxygen and carbon dioxide decreases as temperature increases. Both water and air contain very little carbon dioxide; thus, the diffusion gradient for carbon dioxide out of an animal is high. The high viscosity of water relative to air encourages concentration-gradient stagnation at the boundary layer of the respiratory surfaces. This problem is overcome by mixing the boundary layer through increased ventilation, stirring and moving the boundary layer by ciliary action, or similar mechanisms that prevent the stagnation effect. Ventilation that involves moving water is energetically expensive because the density of water provides resistance to movement and water generally has a low oxygen concentration. In air, the flow of oxygen into and out of the lungs is less energetically expensive because air has a high concentration (21%) of oxygen, and the low density of air offers little resistance to ventilation movements. The major disadvantage of air breathing is the loss of water from the respiratory surfaces, which must be kept moist to function properly. Gas exchange other than that which occurs from the skin of amphibians takes place across surfaces that are not exposed directly to air. These surfaces are found in protected cavities inside the body (e.g., lungs, the cloaca) where they can be kept moist and water loss can be minimized.

Respiratory structures in amphibians include the skin, gills, lungs, and the buccopharyngeal cavity. No reptiles have gills, and cutaneous respiration is rare because of their impermeable skin. A few species of reptiles (e.g., *Apalone*) respire with the cloaca in addition to using the lungs. Gills are used only for aqueous respiration, lungs are used primarily for aerial respiration with some exceptions, and the skin and buccopharynx are used for aquatic and aerial respiration in different species. Most amphibians and many reptiles rely on more than one respiratory surface, using them simultaneously in some situations and alternately in others. Although the respiratory surfaces are derived from different anatomical systems, they share several traits because efficient gas exchange requires a steep concentration gradient and thin membranes between the two exchange media. Thus, respiratory surfaces are heavily vascularized and have one or only a few cell layers between the capillaries and the exchange medium.

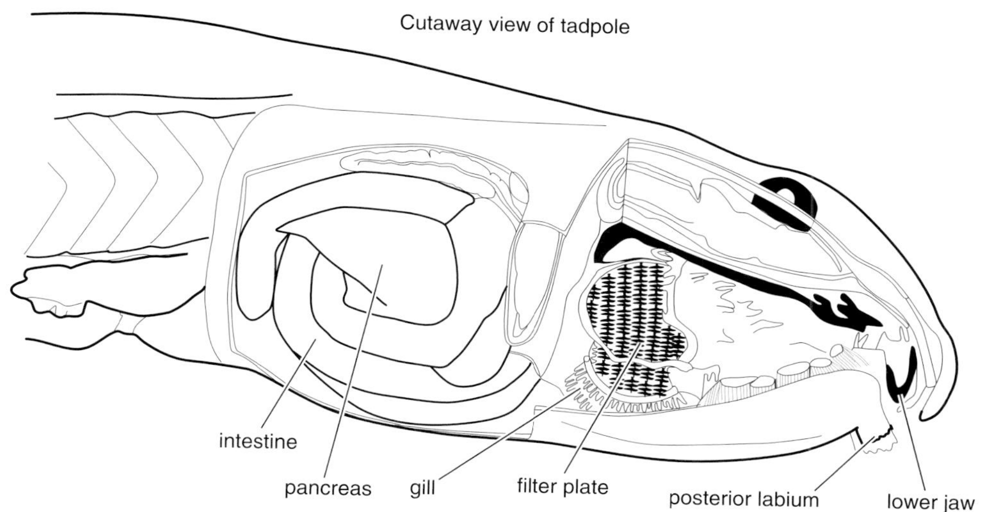

FIGURE 6.17 Longitudinal section through a tadpole, showing the placement of the internal gills beneath the operculum. Adapted from Viertel and Richter, 1999.

A variety of mechanisms increase movement of water or air across the exchange surfaces to prevent gradient stagnation at the interface.

Respiratory Surfaces

Gills

Gills are evaginated respiratory surfaces used for breathing in water. Gills are present in all amphibian larvae and in some aquatic salamanders. They are typically highly branched structures. The numerous branches increase the available surface area for gas exchange, but owing to this branchiate structure and the absence of skeletal support, gills are strictly aquatic respiratory organs. Water is necessary to support the gills and to spread open all surfaces for gas exchange. During the early developmental stages of anuran larvae, transient, external gills develop but soon atrophy. Internal gills remain and are enclosed by a fold of skin called the operculum (Fig. 6.17). In egg-brooding hylids that retain embryos in cavities or pouches on their backs (i.e., *Gastrotheca* and *Flectonotus*), large, thin bell-shaped gills encase all or part of the embryo, providing a surface for gas exchange (Fig. 6.18).

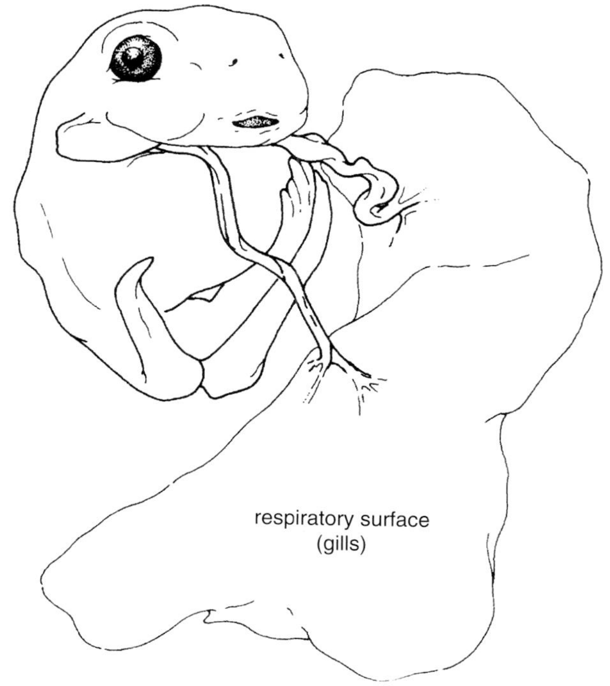

FIGURE 6.18 Direct-developing young of the hylid frog *Gastrotheca cornuta* (Amphignathodontidae). Offspring develop in the dorsal pouch of the female, and oxygen diffuses from the female across the thin, bell-shaped gills of the froglet. Adapted from Duellman and Trueb, 1986.

Larval salamanders have gills that vary in size and structure depending on the nature of the aquatic environment (Fig. 6.19). Salamanders (larvae or adults) that live in ponds have large, feathery gills, whereas those that live in streams or other habitats with moving water have smaller, less filamentous gills. Nonmoving water has a lower amount of dissolved oxygen, and larger gills with an increased surface area permit salamanders to survive in these habitats. Salamanders that retain gills as adults include proteids, such as *Necturus*, cryptobranchids, and paedomorphic plethodontids and ambystomatids. Gills are extensively vascularized and account for up to 60% of the oxygen intake in *Necturus*.

In still water, a boundary layer forms around the gills and must be disrupted so that oxygenated water will be available to the animal. Some salamanders gently move the gills back and forth to raise the diffusive conductance

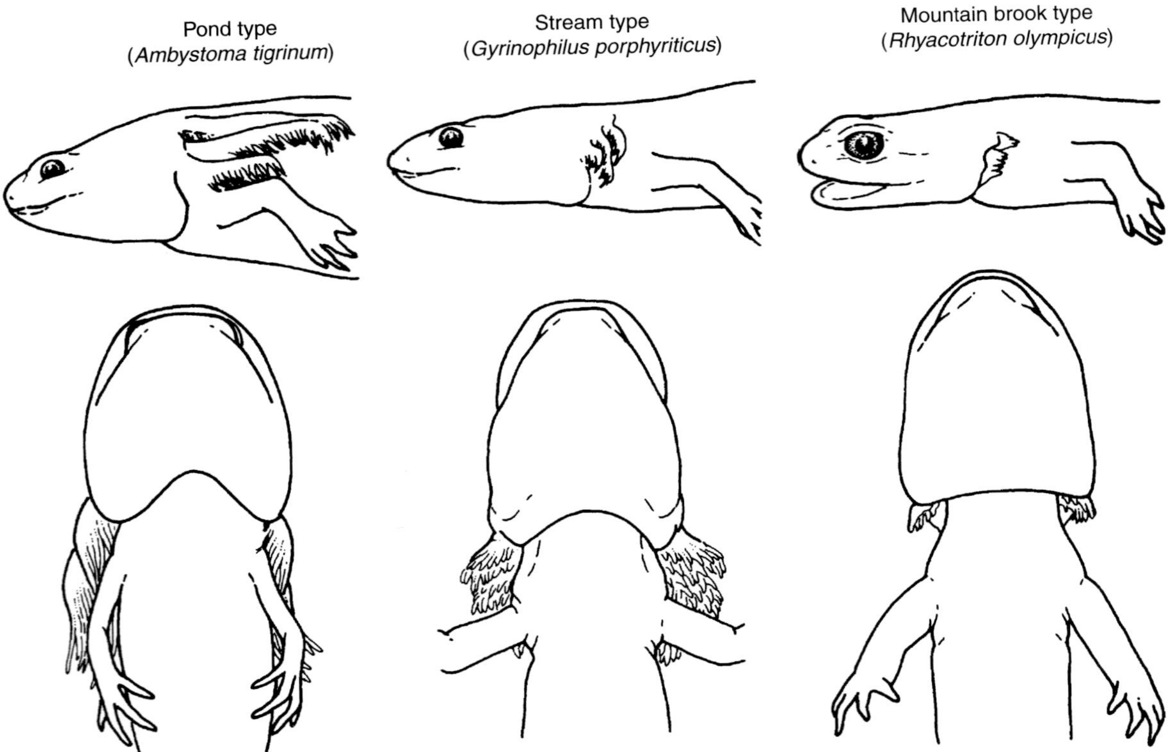

FIGURE 6.19 Adaptive types of salamander larvae, or in some cases, paedotypic adults. Adapted from Duellman and Trueb, 1986.

for oxygen. The internal gills in anuran larvae are perfused by a buccal-pump mechanism, during which water enters the mouth, passes over the gills, and exits through a single spiracle or a pair of spiracles. The relative size of gills and other respiratory surfaces varies in response to the availability of oxygen in aquatic environments.

Buccal Cavity and Pharynx

The buccopharyngeal membranes serve as a respiratory surface in a wide variety of amphibians and reptiles. In this type of respiration, the membranes in the mouth and throat are permeable to oxygen and carbon dioxide. In some species that remain submerged in water for long periods, gas exchange by this route can be significant. Respiration across the buccopharyngeal cavity provides a small percentage of gas exchange in lungless plethodontid salamanders. Some turtles (*Apalone*, *Sternotherus*) can extract sufficient oxygen by buccopharyngeal and cutaneous exchange for survival during long-term submergence, such as during hibernation. Because of low temperatures during hibernation, oxygen requirements for metabolism are reduced.

Skin

The highly permeable skin of amphibians is a major site of gas exchange in terrestrial, semiaquatic, and aquatic species. Cutaneous respiration accounts for some gas exchange in certain species of reptiles (Fig. 6.20). Exchange of respiratory gases occurs by diffusion and is facilitated by a relatively thin layer of keratin and a rich supply of capillaries in the skin. Exchange of gases across the skin in water is limited by the same physical factors as exchange across other respiratory surfaces.

Ventilation of skin, as with gills and other respiratory surfaces, is required to disrupt the boundary layer that can develop. *Xenopus* has been observed to remain submerged longer and to move less frequently in moving compared to still water. Most plethodontid salamanders have neither lungs nor gills and are largely terrestrial (Fig. 6.21). The majority of their gas exchange occurs through the skin. In these salamanders, in contrast to others, there is no partial separation of the oxygenated and venous blood in the heart. Many species of this diverse group, because of their mode of respiration, are limited to cool, oxygenated habitats and to nonvigorous activity. Their oxygen uptake is only one-third that of frogs under similar conditions. Plethodontids that inhabit tropical habitats where temperatures can be high, such as *Bolitoglossus* in tropical rain forests, are active primarily on rainy nights. Waterproof frogs sacrifice their ability to undergo cutaneous respiration in exchange for the skin resistance to water loss.

Some amphibians increase their capacity for cutaneous respiration by having capillaries that penetrate into the

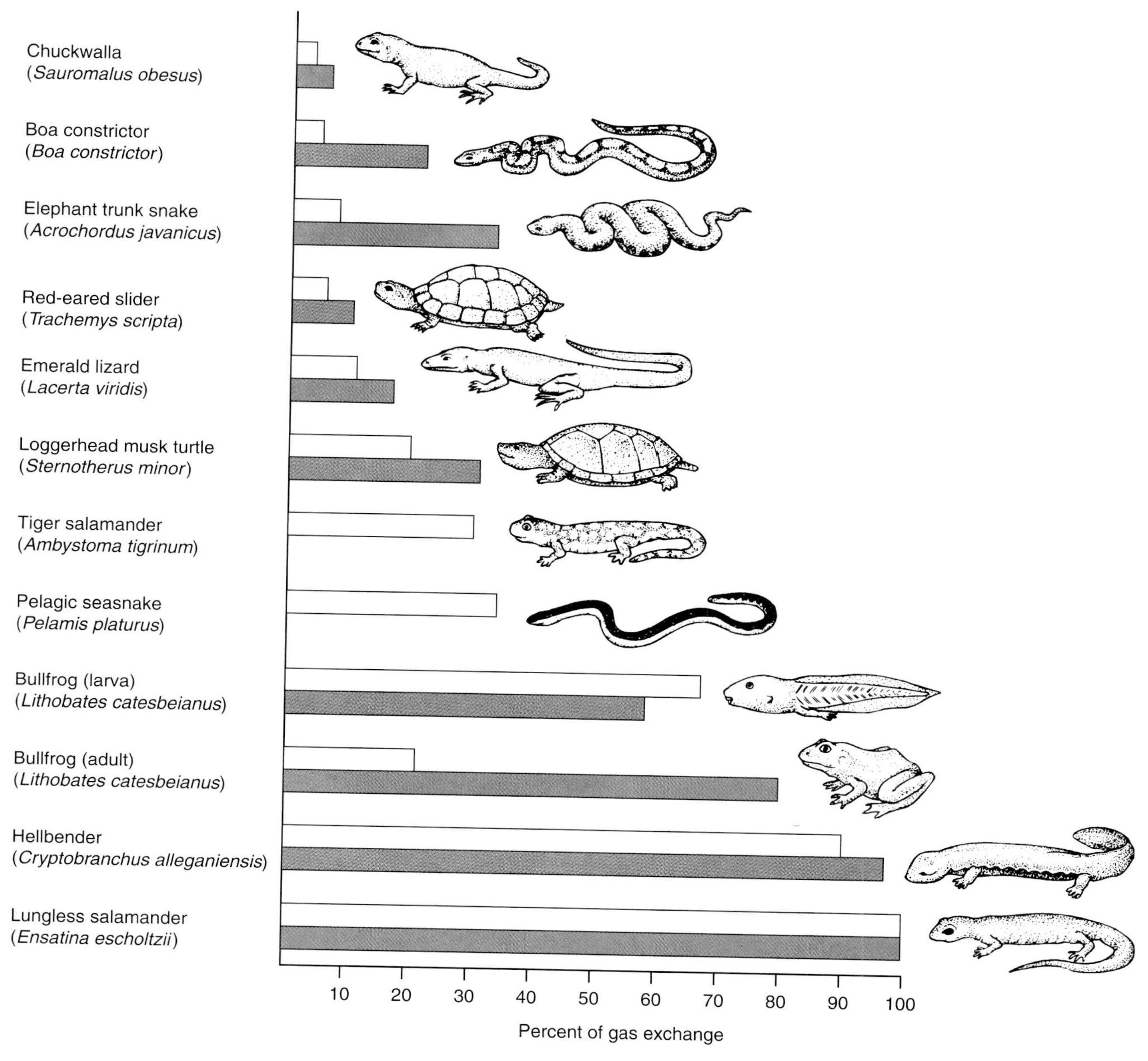

FIGURE 6.20 Cutaneous exchange of gases in amphibians and reptiles. Open bars indicate uptake of oxygen; shaded bars indicate excretion of carbon dioxide. Values represent the percent of total gas exchange occurring through the skin. Adapted from Kardong, 1995.

epidermal layer of skin. This modification is carried to an extreme in *Trichobatrachus robustus*, the "hairy frog," which has dense epidermal projections on its thighs and flanks. These projections increase the surface area for gaseous exchange. Hellbenders, *Cryptobranchus alleganiensis*, live in mountain streams in the eastern United States. These large salamanders have extensive highly vascularized folds of skin on the sides of the body, through which 90% of oxygen uptake and 97% of carbon dioxide release occurs. Lungs are used for buoyancy rather than gas exchange. The Titicaca frog, *Telmatobius culeus*, which inhabits deep waters in the high-elevation Lake Titicaca in the southern Andes, has reduced lungs and does not surface from the depths of the lake to breathe. The highly vascularized skin hangs in great folds from its body and legs (Fig. 6.22). If the oxygen content is very low, the frog ventilates its skin by bobbing. Other genera of frogs, salamanders, and caecilians (typhlonectines) have epidermal capillaries that facilitate gas exchange.

Gas exchange in tadpoles occurs across the skin to some degree in all species. Tadpole skin is highly permeable, similar to that of adults. Gas exchange across the

FIGURE 6.21 Plethodontid salamanders, like this *Plethodon angusticlavius*, have no lungs. All respiration occurs across other skin surfaces. Consequently, all live in wet or moist habitats, most are secretive and/or nocturnal, and most are small in body size. (L.J. Vitt)

FIGURE 6.22 The Titicaca frog, *Telmatobius coleus* (Ceratophryidae), lives at great depths in Lake Titicaca and does not surface to breathe. The large folds of skin greatly increase the surface area of the skin, facilitating cutaneous respiration. (V. H. Hutchison)

skin is prevalent in bufonids and some torrent-dwelling species that do not develop lungs until metamorphosis. Microhylids, some leptodactylids, and some pipids have reduced gills, thus increasing their reliance on cutaneous respiration.

Recent studies show that some reptiles, once thought not to exchange gases through the skin, actually may use cutaneous respiration for as much as 20–30% of total gas exchange. In some aquatic species, such as *Acrochordus* and *Sternotherus*, gas exchange across the skin is especially significant for carbon dioxide (Fig. 6.20). Even in terrestrial taxa such as *Lacerta* and *Boa*, measurable amounts of gas exchange occur cutaneously. A sea snake, *Pelamis platurus*, frequently dives and remains submerged. During these dives, oxygen uptake equals 33% of the total, and 94% of the carbon dioxide loss is through the skin. Exchange does not occur through scales but rather through the skin at the interscalar spaces.

Lungs

Lungs are the principal respiratory surface in many terrestrial amphibians and all reptiles. All extant amphibians with lungs utilize a positive-pressure buccal pump mechanism (Fig. 6.23); in contrast, reptiles (and mammals) use thoracic aspiration. In amphibians, the floor of the mouth is alternately raised and depressed. When depressed, the nostrils are open and air is taken into the buccal cavity, where it is temporarily stored. When the floor of the mouth is elevated, the nostrils close. Buccal pumping is a continual process and is a separate function from lung ventilation. At periodic intervals, the glottis is opened and deoxygenated air in the lungs is quickly expelled. The airstream passes rapidly over the oxygenated air in the buccal cavity, and the two air masses mix very little if any. The oxygenated air is then forced into the lungs.

Thoracic aspiration is used to ventilate the lungs in reptiles. The walls of the lungs can change shape, forcing air in or out of them. In lizards, intercostal muscles between the ribs contract and force the ribs forward and outward. In turn, this movement enlarges the pleural cavity around the lungs, causing them to enlarge and fill with air. Other intercostal muscles then contract, bringing the ribs backward and inward, decreasing the size of the pleural cavity and forcing air out of the lungs.

The left lung of advanced snakes is greatly reduced. The faveoli, compartments that open into the central portion of the lung and contain the actual respiratory surfaces, are abundant in the anterior portion of the lung but gradually decrease and are absent in the posterior portion. Respiration, therefore, occurs only in the anterior part of the lung. Ribs and their associated intercostal muscles extend the entire length of the snake's body and control inflation and deflation of the lungs as in lizards; however, different regions of the body can move independently. The posterior part of the lung serves in a special capacity when the anterior part of the body cannot be used for ventilation. Because of the long, narrow body form of snakes and because they engulf prey much larger than their body, the ribs in the forward part of the body cannot move as prey is being swallowed. Instead, the posterior ribs move in and out, causing the sac-like posterior part of the lung to inflate and deflate and function as a bellows. Cartilaginous rings hold the trachea open, and air is thus forced in and out of the respiratory part of the lung by the action of the posterior lung.

Crocodylians use the liver to press against the lungs and force air in and out. Certain muscles cause the liver, which is located posterior to the lungs, to move. Turtles and tortoises have a special problem in that their lungs are contained inside immobile shells. The lungs and other

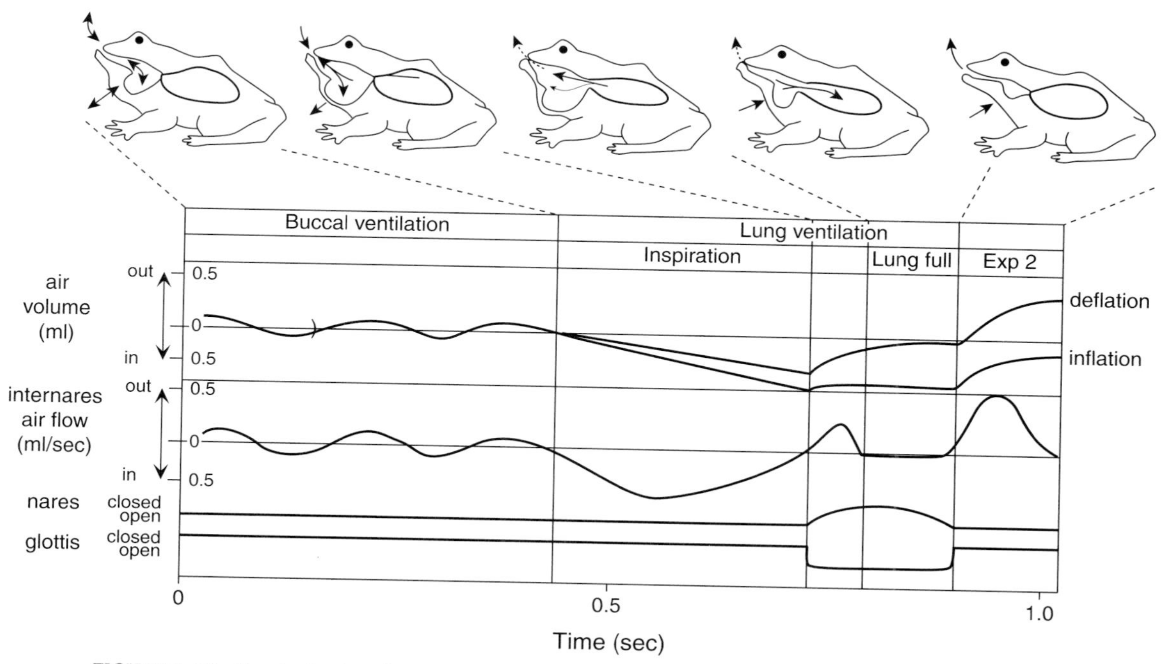

FIGURE 6.23 Respiration in a frog. Oxygenated air is taken into the buccal cavity through the nares. Deoxygenated air in the lungs is rapidly expelled and does not mix with the air in the buccal cavity. Elevation of the buccal cavity (the buccal pump) forces the new air into the lungs. The glottis is then closed to hold the oxygenated air in the lungs, and the remaining air in the buccal cavity is expired by further elevation of the buccal cavity. Adapted from Withers, 1992.

viscera are located in a single cavity, so pressure on any part of the cavity will affect the lungs. In many species, breathing is facilitated by moving the legs in and out of the shell, which decreases or increases the body cavity, causing the lungs to fill and empty.

Some anurans have aquatic larvae that develop lungs and breathe air as tadpoles. This mechanism may account for a significant amount of oxygen uptake, but it is not the only source in any species. As much as 30% of oxygen uptake may be via the lungs in some species. Tadpoles do not appear to be dependent on air breathing. Development and survivorship is not affected in bullfrog tadpoles if they are forcibly submerged.

Lungs also play a role in buoyancy regulation in tadpoles and adults of some aquatic species. Tadpoles occupy different positions in the water column; some are benthic, spending most of their time grazing on bottom substrate, whereas others float and feed in midwater or hang at the water surface. Specific gravity is controlled by the amount of air in the lungs; tadpoles that are prevented from gulping air sink to the bottom of experimental chambers and are unable to maintain their position in the water column.

RESPIRATION AND METABOLISM

Gas exchange is a direct function of metabolism. Metabolic activities, whether anabolic or catabolic, require the energy derived from oxidation, so oxygen is required even in a resting or hibernating state. Metabolism can occur in the temporary absence of oxygen, but an oxygen debt develops that must be repaid. Metabolic rate is measured by oxygen consumption or carbon dioxide production; metabolism and gas exchange are inseparable.

Body size and temperature influence gas exchange. As mass increases, oxygen consumption and carbon dioxide production increases, although the consumption rate declines with increasing mass. This mass-specific relationship reflects the general physical principal that mass increases as a cube of length whereas surface area increases as a square of length. The respiratory surface area may be unable to meet metabolic needs without modifications. Modifications include increasing surface by additional folds (skin) or partitions (lungs), increasing vascularization and/or placing blood vessels closer to surface, and increasing gas transport capacity of blood and increasing flow rate. Such changes can occur ontogenetically, but they also occur across taxa of different sizes.

Aerobic metabolism is strongly temperature dependent, and oxygen consumption increases two to three times for every 10°C increase in body temperature. Metabolic activity is similar in amphibians and reptiles, but different groups have different basal metabolic rates. For example, anurans typically have higher and more temperature-sensitive rates than salamanders. The temperature–metabolism relationship is linear in the majority of ectotherms (see Chapter 7), but a few snakes and lizards have decoupled metabolism from temperature over narrow temperature ranges, usually within their preferred activity temperatures, and metabolism remains constant for 3–5°C range. Gas exchange and metabolism are influenced in varying amounts by a host of other factors. Some species show daily and/or seasonal fluctuations of the basal rate, indicating an endogenous rhythm. Metabolism in temperate species of amphibians can be acclimated and adjusted to seasonal temperature changes. Health and physiological state can modify basal metabolic rates. In alligators, for example, metabolic rate is two times higher in an animal that has fasted 1 day compared to one that has fasted 3 to 4 days.

The wide altitudinal distribution of amphibians (sea level to more than 4500 m) suggests that some rather obvious respiratory adaptations to high altitude should exist to maintain aerobic metabolism. Because the partial pressure of oxygen is lower at high altitudes than at low altitudes, one expectation is that high-altitude species might have low individual erythrocyte volume and high erythrocyte counts as well as reduced hemoglobin content and hemoglobin oxygen affinity (P_{50}—the partial pressure of oxygen at which 50% of the hemoglobin is saturated). Although a considerable amount of variation exists among species in these physiological variables, it does appear to be the case that the size of erythrocytes is reduced with increasing elevation (Fig. 6.24). Because data exist for only a few species, the amount of variation in physiological traits associated with respiration attributable to elevation (i.e., adaptation), history (phylogeny), or other factors remains poorly known. Nevertheless, because frogs in several different clades show decreased erythrocyte size with increased elevation, some physiological responses related to respiration are apparent.

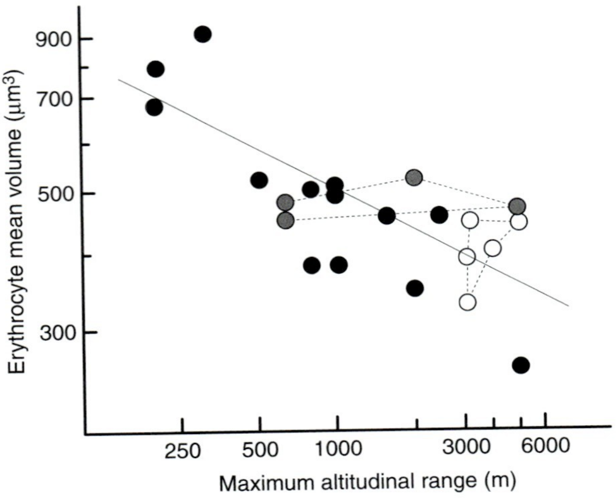

FIGURE 6.24 Mean volume of erythrocytes decreases with increasing elevation in several frog species. Shaded circles are bufonids (originally reported as in the genus *Bufo,* now in the genera *Duttaphrynus* and *Rhinella*), open circles are *Telmatobius* (Ceratophryidae), and closed circles represent much earlier data from 22 anuran species in eight genera from Chile. Dashed polygons enclose data for bufonids and species of *Telmatobius*. Adapted from Navas and Chauí-Berlinck, 2007.

Aerobic and Anaerobic Metabolism

An animal's normal activities are fueled by energy from aerobic metabolism, a process also called cellular respiration and that requires oxygen. Cellular respiration, the chemical transformation of food substrate in biochemical pathways, should not be confused with respiration in the sense of exchange of gases across membranes. Although aerobic metabolism generates most of the energy used by an organism, energy can also be obtained by anaerobic metabolism when oxygen is not available. Anaerobiosis is a vital process for animals because it allows rapid conversion of muscle glycogen to glucose, thus releasing energy quickly for a rapid burst of activity, such as escaping from a predator, or for surviving an anoxic event, such as prolonged submergence under water by an animal that normally breathes air. Although vital for survival, anaerobiosis is energetically costly, and prolonged use of anaerobiosis is debilitating. However, some activities, such as movement of lizard tails after autotomy, are sustained anaerobically and the oxygen debt is not repaid (see Chapter 11). During burst activity, anaerobiosis provides energy at 5 to 10 times the aerobic level, but the process rapidly depletes energy stores. Also, lactic acid accumulates in the muscle within a few minutes, causing the animal to become visibly fatigued. Recovery may require hours or even days, although the oxygen debt and lactic acid removal can proceed rapidly if anaerobiosis is not excessive. Anaerobic metabolism is highly inefficient, requiring as much as 10 times the food input for an equivalent amount of aerobic work, and total nutrient and energy replacement requires much longer. Anaerobiosis is temperature independent within much of a species' temperature activity range, thus permitting an escape response equally as rapid at a low temperature as at a high one. We examine whole-organism energetics in the next chapter.

QUESTIONS

1. How does the desert tortoise survive extended periods of drought and how does the mechanism it uses differ from that of other desert species?
2. Describe the morphology and function of the pelvic patch in frogs.
3. Why would a calling frog lose more water than a crouched frog?
4. What is a "cocoon" in frogs, what is its function, and in what kinds of environments would you expect to find cocooned frogs?

5. Because of the permeability of their skin, amphibians in general experience high rates of water loss but skin permeability also functions in amphibian respiration. Discuss the key elements of this apparent trade-off.
6. Based on what you know about cutaneous respiration in amphibians, what differences in skin architecture would you expect between frogs and salamanders living in deep water compared to those living on land?
7. Explain how the Thorny devil and the Texas horned lizard acquire water in their extreme desert environments.

ADDITIONAL READING

Duellman, W. E., and Trueb, L. (1986). *Biology of Amphibia*. McGraw-Hill, New York.

Dunson, W. A. (1979). Control mechanisms in reptiles. In R. Gilles (Ed.), *Mechanisms of Osmoregulation in Animals: Regulation of Cell Volume* (pp. 273–322). Wiley Interscience, New York.

Hillyard, S. D. (1999). Behavioral, molecular, and integrative mechanisms of amphibian osmoregulation. *J. Exp. Zool.* **283:** 662–674.

Jorgensen, C. B. (1997). 200 years of amphibian water economy: From Robert Townson to the present. *Biol. Rev.* **72:** 153–237.

Minnich, J. E. (1982). The use of water. In C. Gans and F. H. Pough (Eds.), *Biology of the Reptilia: Physiology* (pp. 325–395). Academic Press, London.

Shoemaker, V. H. (1992). Exchange of water, ions, and respiratory gases in terrestrial amphibians. In M. E. Feder and W. W. Burggren (Eds.), *Environmental Physiology of the Amphibians* (pp. 125–150). University of Chicago Press, Chicago.

Shoemaker, V. H., and Nagy, K. A. (1977). Osmoregulation in amphibians and reptiles. *Ann. Rev. Physiol* **39:** 449–471.

Spotila, J. R., O'Connor, M. P., and Bakken, G. S. (1992). Biophysics of heat and mass transfer. In M. E. Feder and W. W. Burggren (Eds.), *Environmental Physiology of the Amphibians* (pp. 59–80). University of Chicago Press, Chicago.

Ultsch, G. R., Bradford, D. F., and Freda, J. (1999). Physiology: Coping with the environment. In R. W. McDiarmid and R. Altig (Eds.), *Tadpoles: The Biology of Anuran Larvae* (pp. 189–214). University of Chicago Press, Chicago.

Wells, K. D. (2007). *The Ecology and Behavior of Amphibians*. University of Chicago Press, Chicago.

REFERENCES

Water and Salt Balance

Osmoregulation—Maintaining Homeostasis and Kidney Function

Bentley, 1966; Garland et al., 1997; Kardong, 1995; Nagy, 1972; Petriella et al., 1989; Shoemaker and Bickler, 1979; Shoemaker and McClanahan, 1980; Shoemaker and Nagy, 1977; van Beurden, 1984; Withers, 1992.

Gaining and Losing Water

Adolph, 1932; Barbeau and Lillywhite, 2005; Benabib and Congdon, 1992; Bennett and Licht, 1975; Bentley and Blumer, 1962; Cartledge et al., 2006; Christian and Parry, 1997; Christian et al., 1996; Congdon et al., 1982; Cree, 1988; Czopek, 1965; Dunson, 1970; Etheridge, 1990; Geise and Linsenmair, 1986; Gomez et al., 2006; Hillyard et al., 1998; Hoff and Hillyard, 1991; Lasiewski and Bartholomew, 1969; Lopez and Brodie, 1977; Louw and Holm, 1972; Mautz, 1982; McClanahan and Baldwin, 1969; McClanahan et al., 1976, 1978, 1983; Minnich, 1979; Nagy, 1982; Norris and Dawson, 1964; Pough et al., 1983; Reno et al., 1972; Schmuck and Linsemair, 1988; Schwenk and Greene, 1987; Sherbrooke, 1990; Shoemaker, 1992; Shoemaker and Nagy, 1977; Tracy and Christian, 2005; Withers and Richards, 1995; Wilson et al., 2001; Withers et al., 1982; Young et al., 2005, 2006.

Nitrogen Excretion

Mazzoti and Dunson, 1989; Shoemaker and McClanahan, 1980; Taplin et al., 1982; Withers, 1998.

The Terrestrial Transition

Martin and Nagy, 1997; Martin and Sumida, 1997.

Marine and Xeric Environments

Dunson, 1978, 1979, 1980; Katz, 1989; Nagy and Medica, 1986; Peterson, 1996.

Respiratory Gas Exchange

Feder and Burggren, 1985; Mautz, 1982.

Respiratory Surfaces

Crowder et al., 1998; Czopek, 1961; Foxon, 1964; Full, 1986; Guimond and Hutchison, 1972, 1976; Hutchison et al., 1976; Kardong, 1995; Spotila et al., 1992; Tu et al., 1999; Ultsch, Bradford and Freda, 1999; West and VanVliet, 1983.

Respiration and Metabolism

Bartholomew, 1982; Bennett, 1982, 1991; Jorgensen, 1997; Navas and Chauí-Berlinck, 2007; Seymour, 1982.

Chapter 7

Thermoregulation, Performance, and Energetics

Globally, temperature is the master limiting factor in the distributional and diversity patterns of amphibians and reptiles. No amphibian or reptile can survive in the frigid environment of Antarctica, and only a few occur marginally within the limits of the Arctic. Their greatest diversity lies within the tropics and warm temperate areas. Even at a regional scale or, smaller yet, in a single habitat, the spatial occurrence and temporal activity pattern of each amphibian or reptilian species is related one way or another to temperature. Because amphibians and reptiles are ectothermic and rely on environmental sources for heat gain, their options for activity are more limited than those for endothermic tetrapods, which maintain elevated body temperature from metabolic heat. Metabolic heat arises from cellular or mitochondrial metabolism. All amphibians and reptiles produce metabolic heat but at a level far below that of mammals and birds, and few have the necessary insulation to prevent its rapid loss. Nonetheless, many reptiles and some amphibians regulate their body temperatures within relatively narrow ranges by taking advantage of the sun and warm surfaces in the environment for heat gain and shade, retreats, water, and cool surfaces for heat loss.

The sun is the ultimate heat source for all amphibians and reptiles, but they also gain heat indirectly by conduction and convection. Amphibians in general operate at lower body temperatures than most reptiles, are more often nocturnal, and may limit activity to periods when humidity is high or rainfall occurs. Differences between amphibians and reptiles in water and ion exchange resulting from differences in skin permeability account for much of the difference in activity periods. Amphibians have permeable skin and experience rapid water loss at high temperatures or low humidities (Chapter 6). The highly impermeable integument of reptiles permits direct exposure to sunlight without excessive water loss. Basking is the most observable heat-gain behavior in reptiles (Fig. 7.1), even though most amphibians and many reptiles gain heat indirectly from surfaces they come in contact with. For amphibian and reptile species living in arid habitats or open tropical habitats, environmental temperatures may be too high during much of the day for sustained activity. As a result, activity is shifted to cooler microhabitats or cooler times of day. Patches in the environment or physical structures serve as heat sinks for such species. Exactly how an individual species responds to the thermal complexity of its environment is influenced by a diversity of abiotic and biotic factors, some of which are extremely difficult to measure directly. A phylogenetic component to thermoregulation also exists, which is just beginning to be explored in detail.

All physiological processes in ectotherms are temperature dependent. The most obvious among these for anyone who has observed or maintained amphibians is water balance; as temperature increases, rates of water loss increase. For reptiles, the most apparent process affected by temperature is behavior; a cold reptile is not as active as a warm one. Of course, the ability to perform reflects the effect of temperature on a multitude of physiological processes. Behavior of amphibians is also strongly influenced by temperature. The differences in response to temperature and moisture between amphibians and reptiles are nicely illustrated by comparing behavioral and thermal responses by an amphibian and a reptile to the daily progression of

FIGURE 7.1 Like many lizards, *Sceloporus poinsetti* basks on boulders in direct sunlight to gain heat. (L. J. Vitt)

environmental temperatures at a high elevation site in Peru (Fig. 7.2). Both the lizard *Liolaemus multiformis* and the toad *Rhinella spinulosa* emerge in morning and bask in sun to gain heat. The lizard basks and feeds at a body temperature of about 30°C until rain occurs. The lizard then retreats for the remainder of the day. The toad ceases activity and enters a retreat at mid-morning when its body temperature exceeds 20°C. When its body temperature falls to about 12°C, it emerges again to bask and gain heat. Once warm, it enters the retreat again. When rain occurs, the toad emerges and remains active the remainder of the day at a relatively low body temperature (± 12°C.). At some localities in Peru, these toads reach body temperatures as high as 32°C on sunny days, but high body temperatures are experienced only during a short time period in morning.

Rates of oxygen consumption and consequently metabolic processes are temperature dependent; hence, all life's processes including development, growth, and reproduction are temperature dependent. Most aspects of behavior and an individual's resistance and reaction to disease vary with temperature as well. The challenge for an individual amphibian or reptile is to center its activity within a range of temperatures that optimizes behavioral and physiological function while concurrently minimizing the risk of mortality. In general, all these processes are components of performance, and how an individual performs on an instantaneous, daily, and seasonal basis determines its survival and, consequently, its fitness, that is, the number of offspring contributed to the next generation. Behavior and physiology related to thermoregulation in ectotherms, particularly reptiles, may be mediated by regulation of brain temperature rather than body temperature directly. This is suggested by the observation that brain temperature is the most sensitive regulator for panting (respiratory water loss resulting in cooling) rather than sensors in other regions of the body.

Because the study of thermoregulation in ectothermic vertebrates has received so much attention during the past 50 years, a complex and often controversial terminology has developed. We restrict our discussion to terms that are currently widely accepted (Table 7.1). Under conditions of

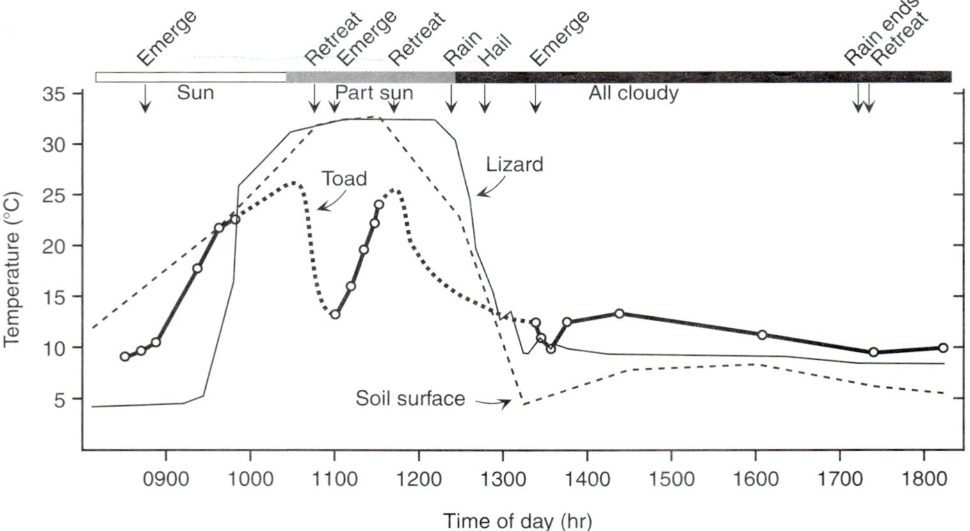

FIGURE 7.2 Lizards and frogs occurring at high elevations in Peru maintain different body temperatures and behave differently in response to temperature at the same locality. The lizard *Liolaemus multiformis* maintains a relatively high and constant body temperature throughout the day by basking. The toad *Rhinella spinulosa* reaches about 23°C in the morning by basking in sun and then retreats under a boulder at about 1050 hr. By repeated emergences and retreats, the toad maintains body temperatures below surface soil temperatures. Adapted from Pearson and Bradford, 1976.

TABLE 7.1 Terminology in Studies of Amphibian and Reptile Thermoregulation.

Term	Definition
Activity temperature range	Normal range of temperatures in which activity occurs
Mean activity temperature (T_b)	The mean of all temperatures of active animals (T_b = body temperature)
Preferred temperature	The temperature selected by individuals in a thermal gradient when all external influences have been removed
Set point	The range of temperatures or temperature at which animals attempt to regulate T_b
Operative temperatures (T_e)	Equilibrium temperature for an animal in a particular environment
Voluntary minimum	The lowest temperature tolerated voluntarily in the lab
Voluntary maximum	The highest temperature tolerated voluntarily in the lab
Critical thermal minimum	The low temperature that produces cold narcosis thus preventing locomotion and escape
Critical thermal maximum	The high temperature that at which locomotion becomes uncoordinated and the animal loses its ability to escape conditions that will lead to its death
Poikilothermy	Wide variation in T_b in response to environmental temperature
Homeothermy	Constant T_b (within ± 2°C) even with greater environmental temperature fluctuations
Ectothermy	Condition in which the external environment is the source of heat
Endothermy	Condition in which heat is produced metabolically (internal)
Heliothermy	Gaining heat by basking in sun
Thigmothermy	Gaining heat by conduction (e.g., lying on a warm rock not exposed to sun)
Acclimation	Functional compensation (relatively short time periods) to experimentally induced environmental change
*Thermoregulation	Maintenance of a relatively constant T_b even though environmental temperatures vary
*Thermal conformity	T_b varies directly with environmental temperature; there is no attempt to thermoregulate.

Note: Effective use of these terms requires context. For example, to simply say that a lizard is a thermoregulator is meaningless without a time component—it may thermoregulate behaviorally while active in the daytime but actually be a thermal conformer at night while in a refuge.
Sources: Cowles and Bogert, 1944; Hutchison and Dupré, 1992; and Pough and Gans, 1982.

normal activity, amphibians and reptiles usually cease activity when they cannot maintain body temperatures within a specific range. The activity range is bounded by the voluntary minimum and voluntary maximum temperatures (Fig. 7.3). If body temperature is allowed to rise or fall outside these bounds, the possibility exists that critical thermal maximum or critical thermal minimum will be reached. At these temperatures, the individual cannot escape conditions if they worsen, and they will ultimately reach the lethal maximum or minimum and die. It is unlikely that individuals would ever reach critical temperatures under normal circumstances because of their ability to behaviorally thermoregulate. Many trade-offs exist between maintaining body temperature near the preferred temperature and other behaviors. Individuals often engage in behaviors at temperatures that could place them at risk. For example, Gila monsters (*Heloderma suspectum*) have been observed

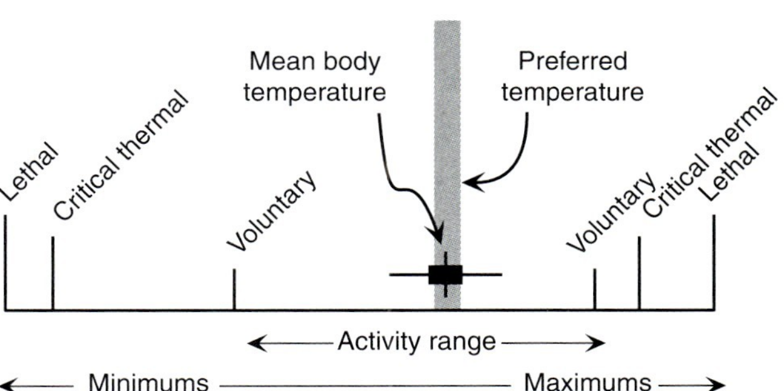

FIGURE 7.3 Profile of body temperature characteristics of an ectotherm. Mean body temperature is based on field data taken on active animals. Set temperature is based on temperatures selected by individuals with external influences eliminated. See Table 7.1 for definitions.

engaged in male–male combat so intensively that they allowed their body temperatures to fall to 17°C, which is 13°C below their mean body temperature in the field.

All animals have a set-point temperature or a set-point temperature range regulated by the hypothalmus, a region of the brain that controls temperature. The set-point temperature is essentially the thermostat setting that signals when an animal should initiate body temperature regulation. For mammals and birds (endotherms), the response is primarily physiological and involves the initiation or curtailment of metabolic heat production. In ectotherms, the response is usually behavioral and, to a lesser degree, physiological. As an ectotherm's body temperature shifts away from the set-point temperature, the animal moves, changes orientation, or changes posture to effect heat gain or loss. However, some evaporative cooling occurs in reptiles, usually via panting and respiratory water loss. In amphibians, water loss reduces heat gain via evaporative cooling. Evaporative cooling is an effective temperature control mechanism only if the amphibian has ready access to water in order to avoid dehydration. When ectotherms are brought into the laboratory and placed in thermal gradients, they tend to select a rather narrow range of temperatures as long as all external cues that might influence thermoregulatory behavior have been eliminated. The mean of these selected temperatures is the preferred temperature. Assuming that the animals are under no physiological stress, the preferred body temperature approximates the set-point temperature.

THERMOREGULATION

Heat Exchange with the Environment and Performance

Heat exchange with the environment occurs via radiation, convection, and conduction (Fig. 7.4). A terrestrial or arboreal ectotherm receives radiant energy from the sun directly or indirectly from reflected solar radiation and heat transfer from substrate and air. Sunlight striking a surface is variously absorbed and reflected; the absorbed solar radiation converts to heat and raises the temperature of the object. No natural object totally absorbs or reflects solar radiation, and most organisms have a mixture of absorptive and reflective surfaces. Many can change the absorptive–reflective nature of their surfaces by color change. Dark surfaces are strongly absorptive, light ones reflective; an animal's colors and pattern, and the ability to change color, reflect a balance between thermal requirements, social advertisement, and crypsis. Subterranean ectotherms

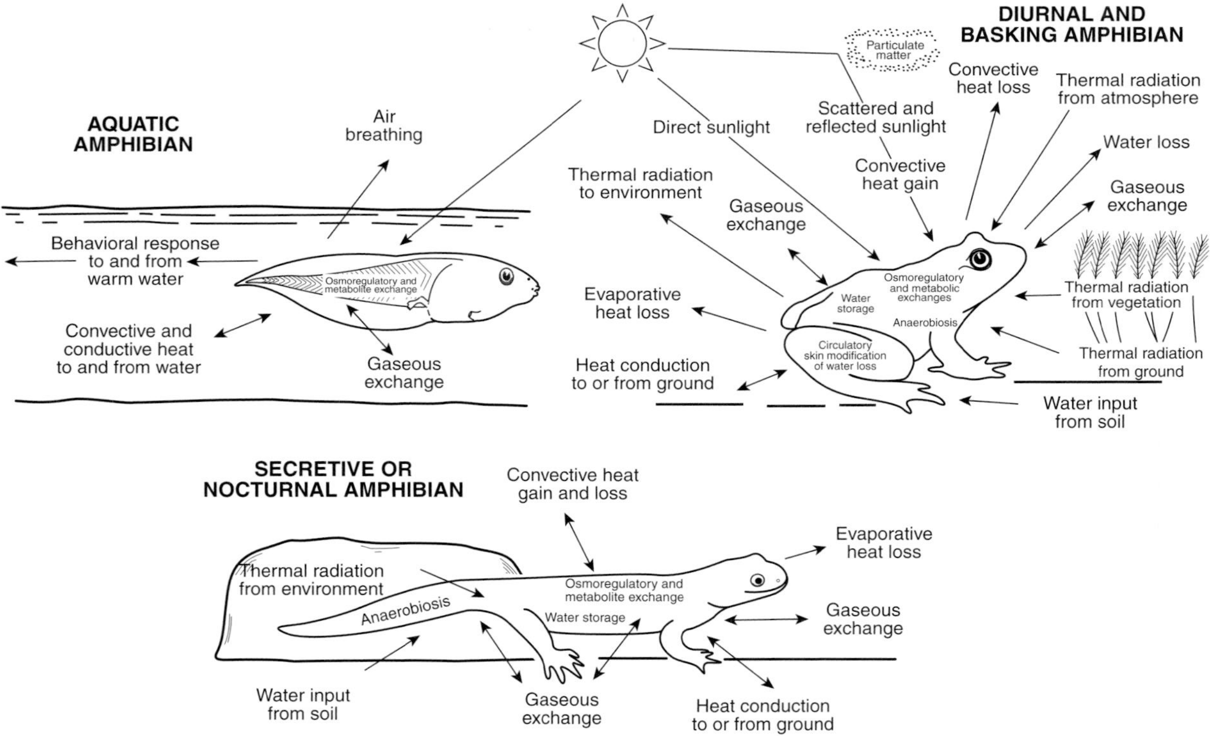

FIGURE 7.4 The environments in which individual amphibians and reptiles live provide different opportunities for heat exchange based on the medium and the physical structure of the habitat. A reptile differs from the amphibians shown because water loss is much lower and influences body temperature much less. The reptile also is limited in its ability to absorb water directly from the environment. Adapted from Brattstrom, 1979.

gain heat by conduction from their microhabitat or by coming into contact with the undersides of warm surfaces (e.g., rocks) that are exposed to direct sunlight. For amphibians, smaller body size not only translates into potentially higher rates of heat exchange but, for the same reason (high surface to volume ratio), translates into high rates of water loss. For both small amphibians and reptiles, physiological control of heat exchange (other than by evaporative cooling) is mediated for the most part by its effect on behavior. Interestingly, body size of anurans increases in cold climates as the result of their thermoregulatory abilities, whereas body size of salamanders decreases, indicating that their thermoregulatory abilities are minimal at best. For anurans, optimizing the trade-off between heating and cooling rates allows them to reach larger size in cold climates.

Most amphibians and reptiles control body temperatures when possible because most life processes vary with temperature. These processes have been fine-tuned by natural selection to be optimal within the activity range of individual species (Fig. 7.5). The activity range itself is influenced by a myriad of physical and biological factors and differs among species as well as within species (Fig. 7.6).

In frogs, the ability to jump is critical for escape from terrestrial predators. Effective escape involves both a trajectory and an escape distance. Escape distance is a function of the distance moved with each jump and the number of consecutive jumps. Green frogs, *Lithobates clamitans*, can move more than 100 cm in a single jump. The distance moved, however, is temperature dependent. At body temperatures below 10°C and above 25°C, jumps are shorter than those between 10 and 25°C (Fig. 7.7). Presumably, cold frogs are less able to escape than frogs within their activity temperature range.

Because of the thermal sensitivity of active escape behaviors, the lizard *Agama savignyi* alters its escape behavior to offset the effects of temperature on specific escape behaviors. At higher temperatures, lizards are more likely to run than at lower temperatures. In addition, a shift from flight behavior to threat behavior not involving flight occurs with decreasing temperatures (Fig. 7.8). The levels of threat also vary from lunging to bite at relatively high temperatures to attacking, lashing with the tail, and leaping off the substrate to bite at lower temperatures.

Field studies confirm that temperature influences performance in natural situations. The South American lizard *Tropidurus oreadicus* flees greater distances when body temperatures are low than when body temperatures are high. This result suggests that the lizards behaviorally respond to reductions in physiological performance associated with low body temperatures by running farther and presumably minimizing risk.

Nocturnal ectotherms are active at lower temperatures than most diurnal ectotherms, and their body temperatures tend to be more variable. Based on the hypothesis that physiological performance should be optimized at the normal activity temperatures of ectotherms, performance in nocturnal ectotherms should be greatest at the low temperatures at which activity occurs. However, nocturnal geckos perform better at temperatures above their normal activity temperatures. Their best performance temperatures are often similar to temperatures associated with maximal performance in diurnal lizards. Low body temperatures of nocturnal geckos are a consequence of nocturnal activity. Nocturnal activity results in suboptimal performance, at least as measured in the laboratory, and potentially affects escape from predators, limits feeding success, and has other consequences. Thermal physiology may reflect evolutionary conservatism, and consequently, no physiological adjustments have been made to enhance

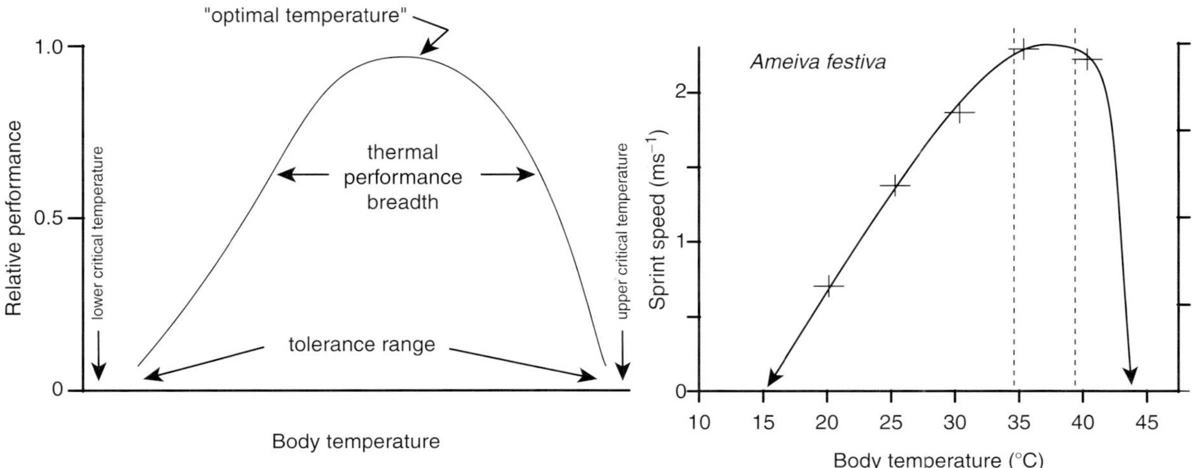

FIGURE 7.5 Theoretically, physiological and behavioral performances are maximized across a relatively narrow range of body temperatures in ectothermic vertebrates (left). Empirical data on *Ameiva festiva* demonstrate that performance is constrained by temperature. Adapted from Huey and Stevenson, 1979, and Van Berkum et al., 1986.

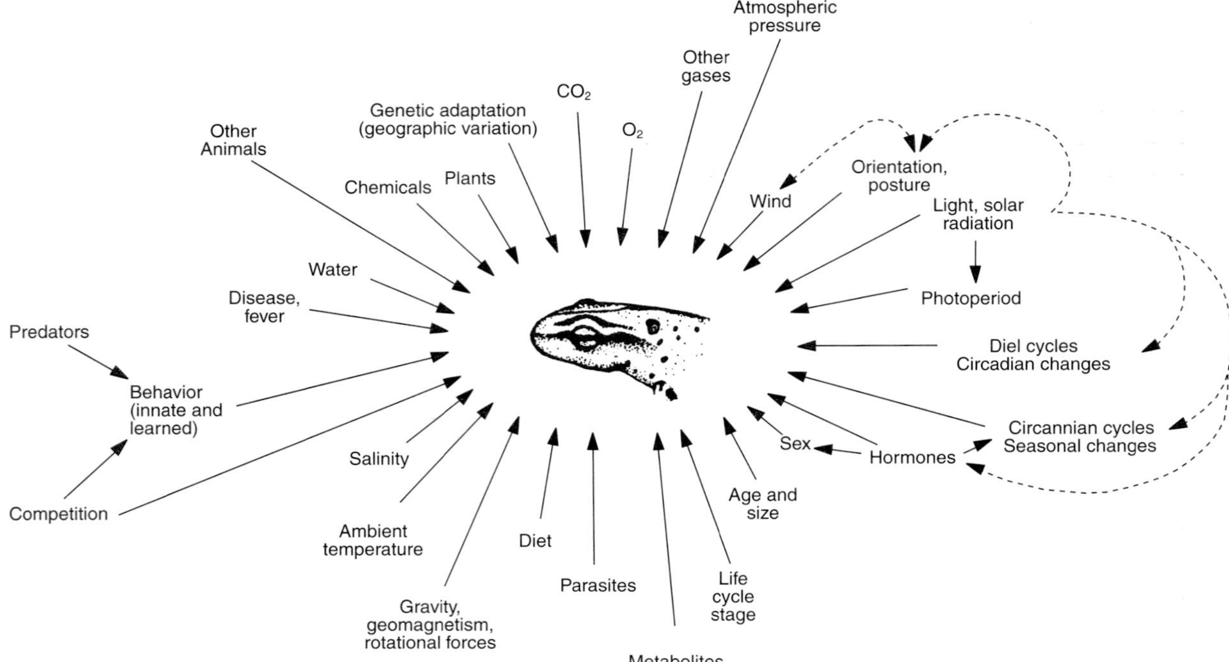

FIGURE 7.6 A multitude of factors influence heat exchange and thus body temperatures and thermal ecology of amphibians and reptiles. The effects of some variables are direct, such as orientation and exposure during basking. Others are indirect. Predators, for example, can interfere with an amphibian or reptile's ability to use basking sites, thereby forcing it to maintain activity at suboptimal temperatures. Adapted from Hutchison and Dupré, 1992.

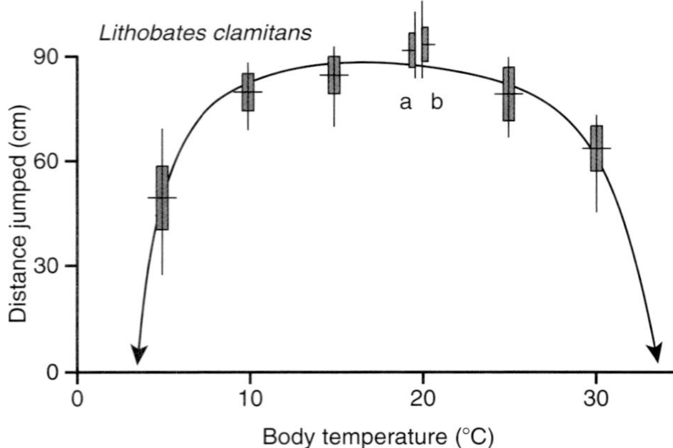

FIGURE 7.7 Green frogs, *Lithobates clamitans*, jump varying distances depending upon their body temperature but do most poorly at low and high temperatures. "a" and "b" refer to different sets of samples at 20°C. Adapted from Huey and Stevenson, 1979.

nocturnal activity. Nevertheless, a trade-off exists between nocturnal and diurnal activities. The cost to survival and fitness of nocturnality is sufficiently low that shifting diurnal activities, such as digestion, that operate most efficiently at high temperatures, to lower temperatures does not incur a fitness cost. Moreover, many geckos spend the day in retreats (e.g., crevices) that are warmer than the temperatures that they experience while active at night. Even though some aspects of performance appear reduced due to nighttime activity (lower body temperatures), an evolutionary shift has occurred resulting in a lower minimum cost of locomotion (C_{min}) as compared with diurnal lizards. C_{min} is simply the amount of energy required to move 1 g of body mass a distance of 1 km. Consequently, locomotion in these lizards is more energy efficient.

Evolutionary studies suggest that thermal physiology and performance have evolved together in 11 species of lacertid lizards. Species that have narrow distributions of preferred body temperatures and can achieve near-maximum sprint speeds across a wide range of body temperatures have the highest levels of performance (Fig. 7.9). As relative hindlimb length has increased evolutionarily, so has maximum sprint speed. The optimal temperature for sprinting has also evolved with maximum sprint speed. As morphology has changed evolutionarily, so has physiology and behavior. In some instances, thermal physiology does not appear to have kept pace evolutionarily with performance. Two populations of the lacertid *Podarcis tiliguerta* are separated by a 1450 m elevational gradient. Set temperatures based on laboratory studies are identical, suggesting that their thermal physiology is similar. At high elevations, lizard body temperatures average 25.4°C, whereas at low elevations they average 30.2°C. High-elevation lizards have lower sprint velocities than low-elevation lizards and do not achieve the absolute speeds reached by low-elevation lizards. The possibility exists that lizards at higher elevations do not face the same risks as lizards at lower elevations, and thus selection on

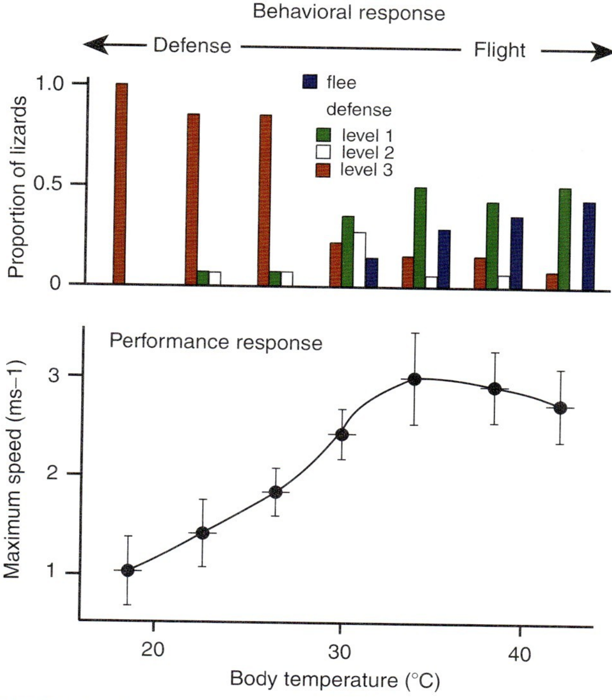

FIGURE 7.8 Body temperatures determine escape behaviors of active *Agama savignyi*. At high body temperatures, lizards rely on running (flight) for escape, reflecting the optimization of running speeds at high body temperatures. At lower body temperatures, alternative escape behaviors become more frequent. Defense occurs at three levels: level 1, gape and lunge; level 2, upright stance, body inflated, and tongue protruding; and level 3, all of the preceding plus attack, lashing with tail, and leaping off substrate to bite. Adapted from Hertz et al., 1982.

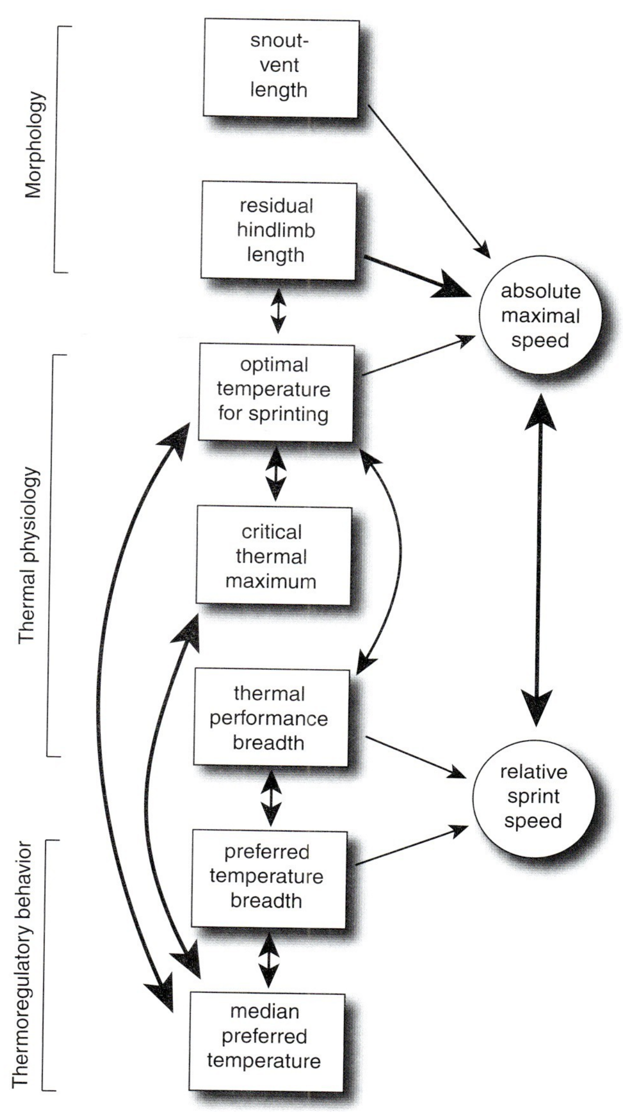

FIGURE 7.9 Morphology, physiology, and behavior have evolved together among 13 species of lacertid lizards. Relative hindlimb length (residual hindlimb length) is correlated with absolute maximal running speed, which correlates with relative sprint speed. Attributes of thermal physiology are also correlated. Heavy lines indicate significantly correlated traits and, because the analysis is based on phylogenetically independent contrasts, the effects of phylogenetic relatedness have been accounted for. Adapted from Bauwens et al., 1995.

temperature-related performance characteristics has been relaxed at high elevations.

The temperatures that ectotherm eggs experience can have cascading effects on performance of individuals. In addition to increasing developmental rates, higher temperatures during development can affect relative size and performance of hatchlings. In some species, the effects appear to be carried through at least part of the juvenile life history stage and possibly through life. Studies conducted by Dr. Rick Shine have addressed these issues with a combination of field studies and experiments. Field-collected and laboratory-hatched eggs of the Australian skink *Bassiana duperreyi* produce hatchlings that differ in size and performance depending upon incubation regime. Eggs incubated on a cycle around 20°C ("cold incubated") hatch later and produce smaller hatchlings relative to original egg size than those incubated on a cycle around 27°C ("hot incubated"). The hot-incubated lizards are relatively heavier and have longer tails than those from the cold incubator. Hot-incubated lizards perform better in sprint speed trials and maintain their superior performance for at least 20 weeks (the entire study). Whether these incubation-mediated phenotypic differences among offspring translate into differences in individual fitness remains to be determined. However, some details of the study suggest that they do. Survival among all study animals was low, but hot-incubated lizards survived better than cold-incubated lizards in the laboratory trials. Relative size and performance in natural populations likely influence individual fitness through effects on time to sexual maturity, body size at sexual maturity, and ability to compete for high-quality mates. However, natural nests of these lizards do

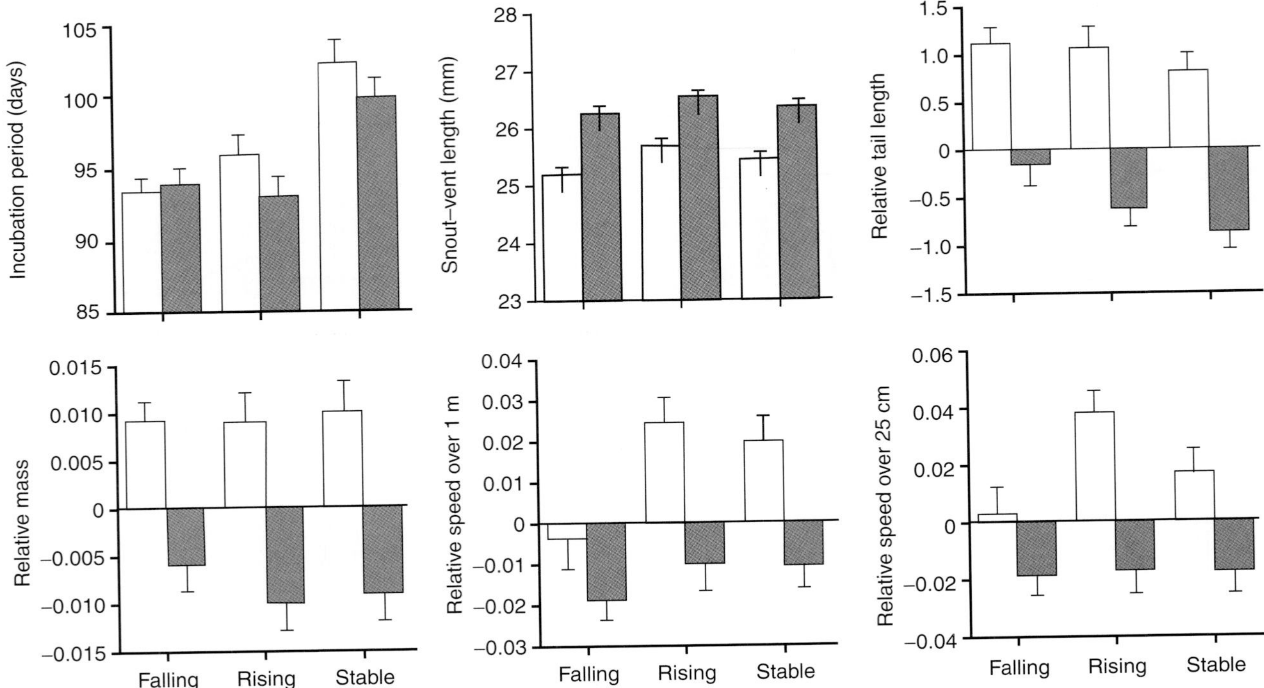

FIGURE 7.10 Hatchling sex and incubation treatment (falling, rising, or stable temperatures during development) influence incubation periods and phenotypic traits of hatchling *Bassiana duperreyi*. Some traits (tail length, speed, and mass) were correlated with size (snout–vent length), so values shown are adjusted for the effect of size. Open bars, males; Shaded bars, females. Adapted from Shine, 2004.

not experience cold-versus-hot incubation. Rather, a single clutch deposited by a female experiences an increasing set of developmental temperatures early in the season at lower elevations or a decreasing set of temperatures at higher elevations. Some nests experience relatively stable (but fluctuating) temperatures through development because of the timing of egg deposition. Laboratory experiments designed to mimic natural conditions produced interesting results. In addition to affecting development time (hatching delayed under stable temperatures), these differences in temperature regimes modified hatchling phenotypes independent of overall mean incubation temperature (Fig. 7.10). Deformities occurred more frequently in hatchlings from eggs experiencing falling temperatures, and deformed hatchlings performed worse and died earlier than their nondeforming siblings. Differences in body size and performance existed between sexes (males weighed more and ran faster), but the most interesting differences were associated with temperature regimes during development. Hatchlings from eggs experiencing falling temperatures were smaller and slower than those from eggs experiencing rising temperatures. These studies not only point to the importance of mimicking natural conditions in designing experiments, but also show that effects of incubation on development are extremely complex, affecting not only development time and probability of producing deformities but also hatchling phenotypes.

Control of Body Temperature

Most amphibians and reptiles control their body temperatures within relatively narrow ranges while active (Table 7.2). Much of this control is behavioral, either directly as the result of short-term movements or posturing to maximize heat gain or loss. Among amphibians, behavioral thermoregulation is difficult to separate from behavioral mechanisms for water conservation because they lose water readily through the skin (Chapter 6). In many salamanders, for example, behavioral control of temperature does not occur. In amphibians without cutaneous control of evaporative water loss, body temperatures are only slightly above environmental temperatures because of evaporative cooling. Most reptiles and many frogs rely on behavioral mechanisms to thermoregulate. Because sun exposure and temperatures of natural environments vary spatially and temporally, behaviors resulting in thermoregulation vary accordingly (e.g., Fig. 7.2). Microhabitat selection and adjustments in time of activity account for much of the control of body temperature (Table 7.2).

Many salamanders, particularly plethodontids, do not thermoregulate behaviorally, at least in the same ways that many frogs and most reptiles do. Body temperatures of tropical plethodontids along an elevational gradient parallel environmental temperatures and change seasonally with environmental temperatures. Sympatric tropical species of plethodontids do not differ in body temperatures, indicating that niche segregation by temperature does not exist.

TABLE 7.2 Examples of Body Temperatures of Amphibians and Reptiles in °C.

Species	Minimum voluntary	Maximum voluntary	Mean	No. species
Salamanders				
Cryptobranchidae	9.8	28.0	—	1
Sirenidae	8.0	26.0	24.0	3
Amphiumidae	—	—	24.0	1
Salamandridae	4.5	28.4	16.0	4
Temperate ambystomatids	1.0	26.7	14.5	9
Tropical ambystomatids	10.5	30.0	19.0	12
Temperate aquatic plethodontids	2.0	22.0	11.3	9
Temperate terrestrial plethodontids	−2.0	26.3	13.5	28
Tropical plethodontids	1.8	30.0	14.2	43
Frogs				
Ascaphus	4.4	14.0	10.0	1
Pelobatidae	12.2	25.0	21.4	2
Leptodactylidae	22.0	28.0	24.7	5
Rhinella	3.0	33.7	24.0	17
Hylidae	3.8	33.7	23.7	14
Gastrophryne	15.5	35.7	26.5	2
Lithobates	4.0	34.7	21.3	12
Lizards				
Anguidae	11.0	34.7	23.0	3
Anniellidae	13.8	28.3	21.0	1
Chamaeleonidae	21.0	36.5	21.0	2
Gekkonidae	15.0	34.0	24.9	3
Gerrhosaurinae	19.0	41.0	33.3	1
Helodermatidae	24.2	33.7	27.2	1
Iguanidae	18.0	46.4	36.7	50
Lacertidae	35.0	41.5	38.4	3
Scincidae	13.2	39.5	30.4	16
Teiidae	27.0	45.0	40.5	9
Xantusiidae	11.5	32.2	23.1	4
Snakes				
Boidae	12.2	34.0	25.1	3
Colubridae	9.0	38.0	26.8	41
Viperidae	17.5	34.5	27.0	12
Pelamus platurus	—	—	24.9	1
Turtles				
Chelydridae	5.0	24.5	—	1
Emydidae	8.0	35.2	26.7	6
Kinosternidae	16.2	28.8	*23.0	2
Testudinidae	15.0	37.8	30.6	3

Note: *Estimated.
Sources: Brattstrom, 1963, 1965; Duellman and Trueb, 1986; and Feder et al., 1982.

Maximum body temperatures of tropical and temperate-zone plethodontids are similar, but temperate-zone species have lower minimum body temperatures because temperate-zone microhabitats experience cool periods that tropical ones do not experience. One explanation for the apparent lack of behavioral thermoregulation in these salamanders is that they cannot exploit warm terrestrial microhabitats because they are too dry, except possibly in the moist tropical lowlands. Likewise, they would have difficulty exploiting warm aquatic microenvironments such as those in the moist tropical lowlands, because of the oxygen deficiency of warm water. Because they lack lungs, gulping air to gain oxygen while in low oxygen water is not an alternative. They are thus restricted to relatively cool terrestrial microhabitats and maintain temperatures similar to those microhabitats. Most other salamanders have body temperatures similar to their microhabitats most of the time. Ambystomatid salamanders, for example, spend most of their lives underground. When they migrate to ponds to breed, migrations take place during rainy nights that offer no opportunities for behavioral thermoregulation.

Some frogs regulate body temperatures by basking in sun (e.g., Fig. 7.2). Bullfrogs (*Lithobates catesbeianus*) vary in body temperatures from 26–33°C while active, even though environmental temperatures vary more widely. During the day, they gain heat by basking in sun and lose heat by a combination of postural adjustments and use of the cold pond water as a heat sink. At night when water temperatures are low, bullfrogs move from shallow areas to the center of the pond where water is relatively warmer. In the morning, they return to the pond edge to bask and gain heat. Although bullfrogs clearly cannot maintain high body temperatures at night, they behaviorally select the warmest patches in a relatively cool mosaic of the nighttime thermal landscape, thereby exercising some control over their body temperatures. Similar observations have been made on other frog species.

An alternative to moving between microsites to gain and lose heat is to use water absorption and evaporative water loss to moderate body temperatures. By having part of the body against moist substrate, a frog can absorb water to replace water lost by evaporative cooling, thereby maintaining thermal stability even though environmental temperatures may be relatively high (see Fig. 6.6). Likewise, by regulating evaporative water loss, some frogs are capable of maintaining body temperatures in cooling environments by reducing evaporative water loss.

Control of evaporative cooling to stabilize body temperatures during periods of high ambient temperatures occurs in other ways as well. The best-known examples are the waterproof frogs *Phyllomedusa* and *Chiromantis*, which allow body temperatures to track environmental temperatures until body temperatures reach 38–40°C. Skin glands then begin secretion and evaporative cooling allows the frog to maintain a stable body temperature even if environmental temperature reaches 44–45°C. Some Australian hylid frogs in the genus *Litoria* are able to abruptly decrease water loss across the skin in response to high body temperatures and thus avoid desiccation. Other species of *Litoria* with low skin resistance to water loss are able to reduce their body temperature by evaporative water loss and thus avoid reaching potentially lethal body temperatures. Several *Litoria* species have independently evolved high resistance to water loss, and it is usually associated with an increase in their critical thermal maximum temperatures.

Control of body temperatures in diurnal lizards and snakes often involves behavioral shifts throughout the day. Some Amazon rain forest populations of the South American tropidurid lizard *Tropidurus hispidus* live on isolated granitic rock outcrops that receive direct sunlight. The rain forest acts as a distribution barrier; the lizards do not enter the shaded forest. During the day direct sunlight causes the rock surfaces to heat up to nearly 50°C, which is above the critical thermal maximum for most animals. The lizards forage and interact socially on the rock surfaces, maintaining relatively constant body temperatures throughout the day by moving between rock patches exposed to sun and shady areas (Fig. 7.11). During morning, lizards bask on relatively cool rocks to gain heat. During afternoon, lizards use relatively cool rocks in shade as heat sinks to maintain activity temperatures.

Even though many laboratory and field studies of reptilian thermal ecology and physiology maintain that the species under study thermoregulate with some degree of accuracy, comparisons of environmental temperatures and set temperatures are necessary to reach that conclusion. The high-elevation Puerto Rican lizard *Anolis cristatellus* lives in open habitats and basks in sun to gain heat. During summer and winter, body temperatures of the lizards are higher than environmental temperatures, which indicates that they thermoregulate (Fig. 7.12). However, in summer, environmental temperatures are higher than in winter, and as a result, the lizards are able to achieve higher body

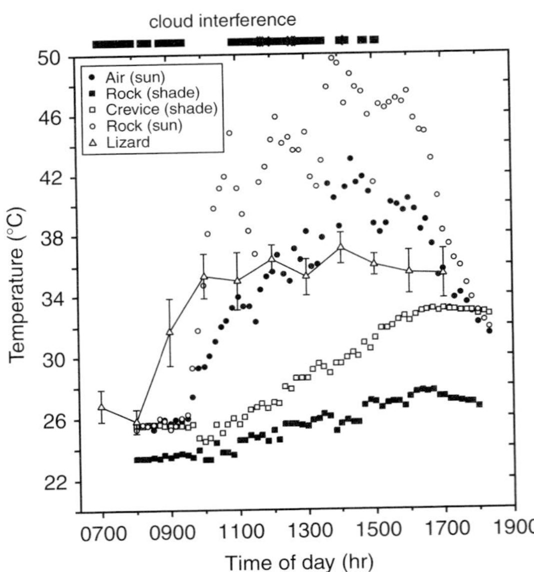

FIGURE 7.11 Many lizards regulate their body temperatures within a relatively narrow range by behavioral adjustments. During morning, when rock surfaces are relatively cool, *Tropidurus hispidus* basks in sun to gain heat. As rock temperatures increase during the day, the lizards spend more time on shaded rock surfaces or in crevices using cool portions of their habitat as heat sinks. Late in the day, when exposed rock surfaces cool, lizards shift most activity to open rock surfaces that remain warm and allow the lizards to maintain high body temperatures longer. Adapted from Vitt et al., 1996.

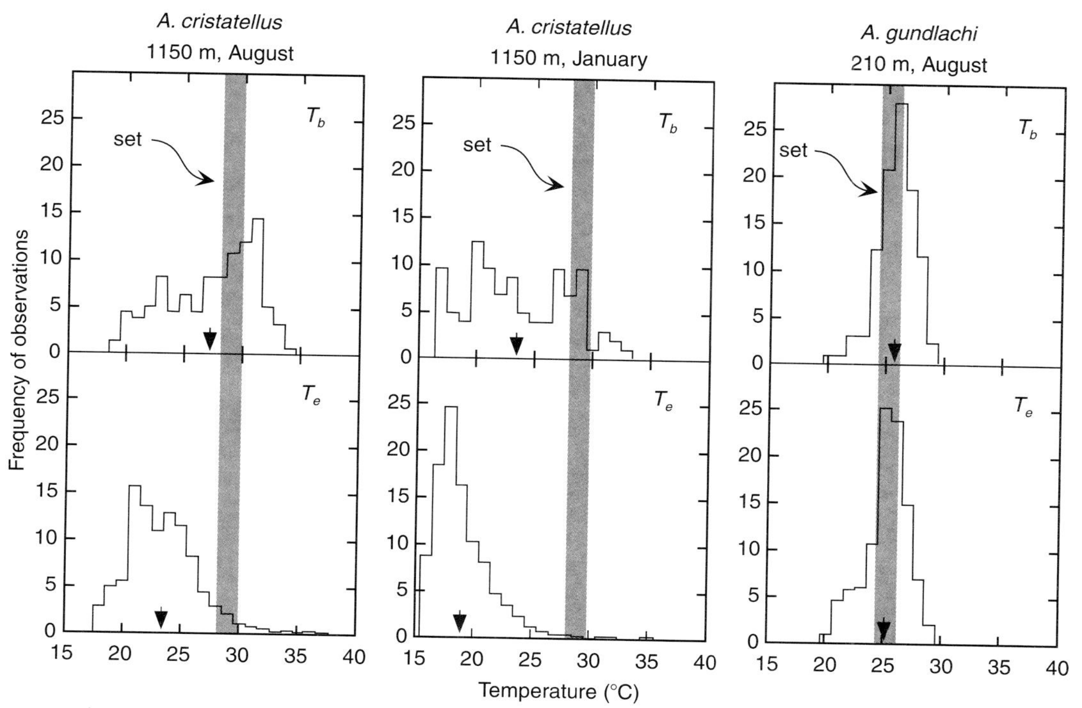

FIGURE 7.12 Precision of thermoregulation in *Anolis* lizards. The upper panels show body temperatures of lizards; the lower panel shows operative environmental temperatures; and the shaded bars show set or preferred temperatures of the lizards. Arrows indicate means. Similarity of body temperatures to set temperatures indicates accuracy of thermoregulation. The accuracy of thermoregulation varies with season in *Anolis cristatellus* as the result of shifts in environmental temperatures. The accuracy of thermoregulation is high in *A. gundlachi,* even though it does not thermoregulate. In this case, the environmental temperatures are nearly identical to set temperatures of the lizards. Adapted from Hertz et al., 1993.

temperatures much more easily than they do in the same environment during winter. Even though their body temperatures are variable in both summer and winter, in summer their body temperatures more closely approximate set temperatures. *Anolis gundlachi* lives in a forest environment at low elevations where opportunities to bask and gain heat are limited. Its body temperature is much lower than that of *A. cristatellus*. However, because its set temperature is nearly identical to environmental temperatures, no thermoregulation is necessary for lizards to maintain body temperatures near their set temperature, and for the same reason, body temperatures vary little from set temperatures; they are maintained with a high degree of accuracy.

Teiid lizards offer interesting comparisons because body size varies considerably among species; most are active at high body temperatures (37–39°C), most have short daily activity periods, and most are carnivorous, but a few are herbivorous and have extended activity periods. The relatively large-bodied tropical whiptail, *Cnemidophorus murinus*, averages 37.2°C when active in an environment in which hourly temperatures in most microhabitats exceed 40°C during midday (1100–1600 hr). Because these lizards are herbivorous, long activity periods appear necessary to digest plants that they eat. They remain active during the entire day, avoiding extreme temperatures by shifting from open microhabitats in morning to shaded microhabitats from midday on. Nevertheless, they are unable to maintain body temperatures within their preferred (T_{sel}) range because the habitat offers limited refuges from extreme temperatures (Fig. 7.13). The scrubby plants under which they seek shade allow enough sunlight so that about 63% of the lizards' time is spent in filtered sun. The lizards experience body temperatures very near temperatures known to be lethal for many lizard species. In this species, benefits of being able to digest low-energy plant material outweigh risks associated with operating at near-lethal body temperatures.

Snakes differ from most lizards and all other reptiles in that their surface-to-volume ratios are high as the result of their elongate and relatively slender morphology. Morphology, and more specifically, how morphology is used, can have large effects on heat exchange. Postural adjustments change exposure of body surfaces; a stretched-out snake has much more surface area exposed than a coiled snake. Behaviorally, snakes seek out microhabitats with appropriate temperatures. For example, file snakes (*Acrochordus arafurae*) are able to maintain body temperatures within a range of 24–35°C by selecting microhabitats within that range and avoiding other microhabitats.

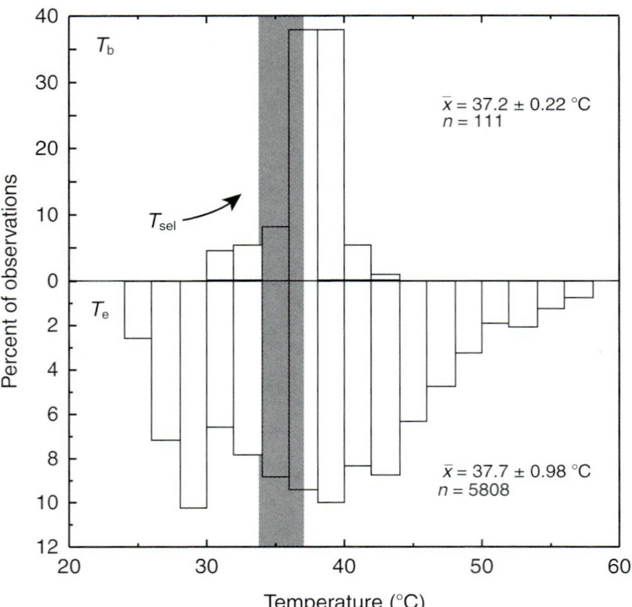

FIGURE 7.13 Thermal ecology of the herbivorous teiid lizard *Cnemidophorus murinus* on Bonaire. The lizards maintain body temperatures (T_b) during most of the day that exceed their preferred body temperatures (T_{sel}). Although environmental temperatures (T_e) vary considerably, they are highest from 1100–1600 hr (not shown), providing limited microhabitats with low temperatures. Adapted from Vitt et al., 2005.

Although the preceding kinds of studies provide accurate snapshots of thermoregulation in reptiles against a background of available microhabitat temperatures (null distributions), they do not include effects of body size on heating and cooling rates (thermal inertia), nor do they explicitly include the set of potential movements that an individual might make during its activity, and the effects of those movements on temperature as part of the null models. A recent study by Dr. Keith Christian and collaborators proposes a null model incorporating thermal inertia (by including body mass and heating and cooling rates) and behavior (via generating a random set of movements within the habitat). Such an approach takes into account the effects of temperature at both past and present locations in the environment. The two relatively large varanid lizards, *V. gouldii* and *V. panoptes,* occupy different habitats (sand and floodplains, respectively) in northern Australia. Their ecology is fairly well known, and much data exist on their thermal ecology. Null models of microhabitat temperatures indicate that both species should be able to easily maintain body temperatures within their set-point ranges. *Varanus gouldi* does maintain its body temperature within its set-point range during its entire daily activity period, but *V. panoptes* does so for only one time period during the day (Fig. 7.14). These results demonstrate that *V. gouldii* and *V. panoptes* interact differently with their thermal environments even though they are relatively similar in size and morphology, as well as being closely related.

Because many reptiles gape and pant when overheated, they cool as a result of evaporative water loss. Thus respiration and thermoregulation are linked, but this is an area in which relatively little attention has been directed. For example, some turtles and snakes gape, but its role in thermoregulation has been unknown until recently. Lizards and crocodylians gape, and studies have shown that gaping and panting results in some evaporative water loss and cooling. Gaping and panting thresholds change in response to dehydration, hypoxia, time of day, and season.

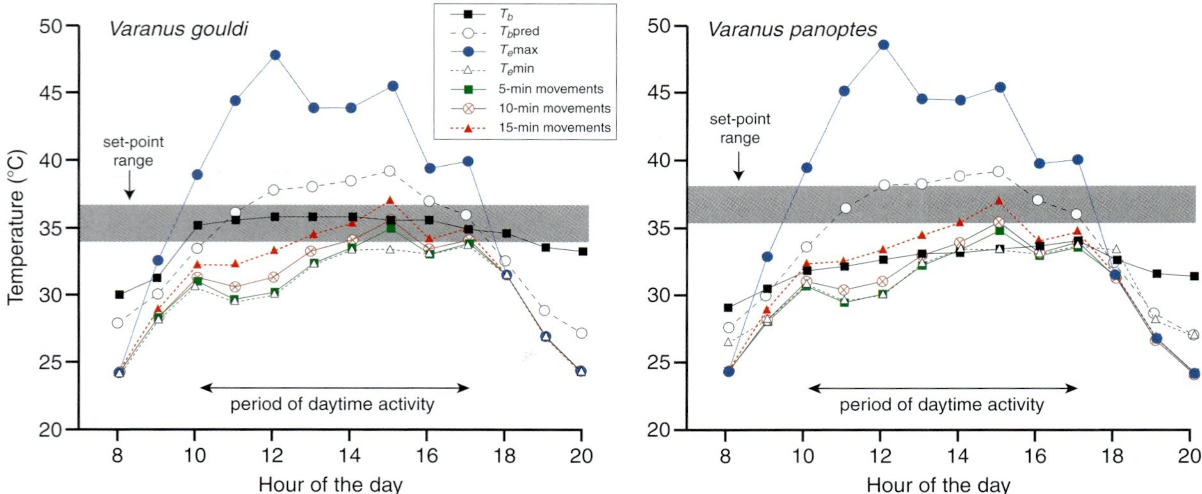

FIGURE 7.14 Temperatures during a typical wet-season day in northern Australia for the two monitor lizards, *V. gouldii* (left) and *V. panoptes* (right). Temperatures included are the following: telemetry-measured body temperatures (T_b), mean-predicted body temperatures based on models ($T_b$pred), minimal and maximal environmental temperatures ($T_e$min and $T_e$max). Three available body temperature lines are shown based on movements between thermal environments every 5, 15, and 30 minutes. Adapted from Christian et al., 2006.

In addition, sexual differences exist in gaping and panting thresholds. Recent studies on snakes show that evaporative water loss through respiration does occur in response to feeding and activity resulting in a lowering of head temperature (Fig. 7.15). Most intriguing is the suggestion that rapid temperature change in the head might indicate that the primary function of respiratory cooling is to regulate brain temperature, an effective strategy considering that the brain controls, directly or indirectly, all behaviors.

Physiological processes that facilitate heat gain and loss occur in the skin, cardiovascular system, and excretory system (Table 7.3). Limited heat production can occur by muscular activity, and the hormone thyroxin can cause heat production through its effect on metabolism. Peripheral and central temperature sensors determine physiological and whole animal responses to temperature change, and species are variously tolerant to high and low temperatures.

Temperature control by heat production is rare in amphibians and reptiles. Among the best examples of heat production are in brooding pythons and leatherback sea turtles. Female pythons increase their body temperatures during brooding eggs by contracting muscles of the body. Some of the heat is transferred to the developing clutch, which then develops at a faster rate. Body temperatures, muscle-contraction rates, and oxygen consumption rates of female *Python molurus bivittatus* increase above that of nonbrooding snakes during the time that the snakes brood their eggs (Fig. 7.16). Higher temperature-mediated development rates during incubation translate into earlier hatching. Combined laboratory and field studies on Australian pythons (*Morelia spilota*) show that brooding influences not only developmental rates but also offspring quality.

Leatherback sea turtles (*Dermochelys coriacea*) approach mammalian endothermy on a diet of jellyfish. They maintain body temperatures of 25–26°C in 8°C seawater by a combination of elevated metabolism, large body size, thick insulation, and thermally efficient regulation of blood flow to skin and appendages. Their dark skin may permit some heat gain through solar radiation, but their primary heat source is from muscular activity. Metabolic rates of leatherbacks are higher than what would be predicted on the basis of body size alone. Heat is retained by a thick, oil-filled skin (an equivalent insulator to blubbery skin of whales) and countercurrent heat exchange in the circulatory system of the limbs. As a consequence, they can enter much cooler marine environments than most reptiles. The same mechanism permits the turtles

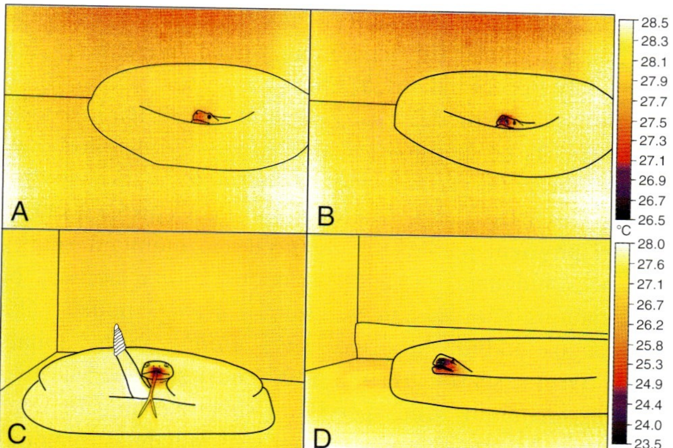

FIGURE 7.15 Thermal images of snakes showing that the head region differs dramatically from the body in temperature during cooling. (A) The head of the tropical rattlesnake *Crotalus durissus* barely cools just following apnea. (B) Four seconds later, the head of the same snake is much cooler following inspiration. (C) A different rattlesnake just after a high level of activity (tail rattling), showing significant respiratory head cooling. (D) A python showing whole-head cooling following gaping behavior and rapid respiratory rates, leading to high rates of evaporative water loss. Adapted from Tattersall et al., 2006.

TABLE 7.3 Behavioral, Morphological, and Physiological Factors Influence Heat Exchange in Amphibians and Reptiles.

1. Behavior
Microhabitat selection
Temporal adjustments of activity
Postural adjustments
Huddling or aggregation
Burrowing
2. Integument
Modification of reflectance by color change
3. Cardiovascular system
Capacity for vasomotor activity including peripheral vasomotion
Vascular shunts
Cardiac shunts
Countercurrent systems
Temperature-independent control of cardiac output
4. Evaporative cooling
Water loss from skin
Panting; water loss from oral or buccal surfaces
Respiratory water loss
Salivation; increase buccal evaporation
Urination on self to increase surface evaporation
5. Heat production
Shivering thermogenesis
Increase in cellular metabolism by hormonal stimulation
6. Temperature sensors
Peripheral
Central
7. Tolerance of hyperthermia or hypothermia
Marked capacity for hypothermia
Modest tolerance of hyperthermia

Source: Bartholomew, 1982.

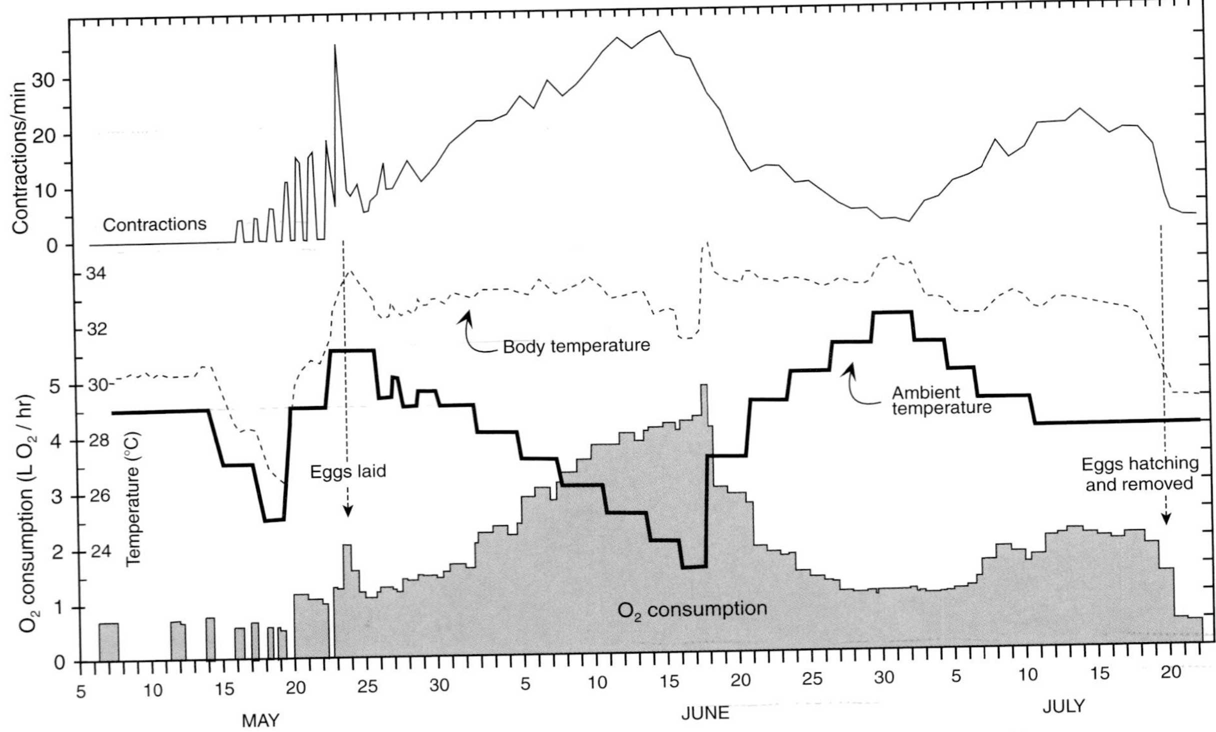

FIGURE 7.16 Brooding temperature in a python. *Python molurus bivittatus* generates heat by rapid contractions of skeletal muscle while brooding eggs. The rate of muscle contractions increases, oxygen intake increases, and CO_2 production increases during egg brooding. Adapted from Van Mierop and Barnard, 1978.

to lose heat when in warm waters. Other sea turtles are not nearly as efficient at thermoregulation. Green sea turtles and loggerheads typically maintain body temperatures only 1–2°C above temperatures of surrounding water.

Costs and Constraints of Thermoregulation

Even though physiological and behavioral processes are maximized within relatively narrow ranges of temperatures in amphibians and reptiles, individuals may not maintain activity at the optimum temperatures for performance because of the costs associated with doing so. Alternatively, activity can occur at suboptimal temperatures even when the costs are great. Theoretically, costs of activity at suboptimal temperatures must be balanced by gains of being active. Costs are varied and not well understood; they include risk of predation, reduced performance, and reduced foraging success.

The desert lizard *Sceloporus merriami* is active during the morning at relatively low body temperatures (33.3°C), inactive during midday when external temperatures are extreme, and active in the evening at body temperatures of 37.0°C. Although the lizards engage in similar behavior (e.g., in morning and afternoon, social displays, movements, and feeding), metabolic rates and water loss are greater and sprint speed is lower in the evening when body temperatures are high. Thus, the metabolic and performance costs of activity occur in the evening when lizards have high body temperatures. However, males that are active late in the day apparently have a higher mating success resulting from their prolonged social encounters. The costs of activity at temperatures beyond those optimal for performance are offset by the advantages gained by maximizing social interactions that ultimately impact individual fitness.

Biophysical models can be useful for evaluating costs and benefits of thermoregulation in ectotherms, and several of the previous examples of thermoregulation studies in reptiles used them. Microclimate data can be used to model physical parameters necessary to maintain energy balance, and data on body temperatures and activity of free-ranging animals can be used to test the models. The Galapagos land iguana *Conolophus pallidus* shifts habitats from the hot season to the garua (cool) season. The two habitats differ in wind speed and substrate and air temperatures. Temperatures in the hot season habitat (plateau) are high during the day but decline earlier in the day than those of the garua season habitat, the cliff face. The lizards select cooler microhabitats and maintain cooler body temperatures during the garua season than during the hot season, even though microhabitats with warmer temperatures are available during the garua season (Fig. 7.17). Lower cool season temperatures reflect a change in thermoregulation because warm temperatures

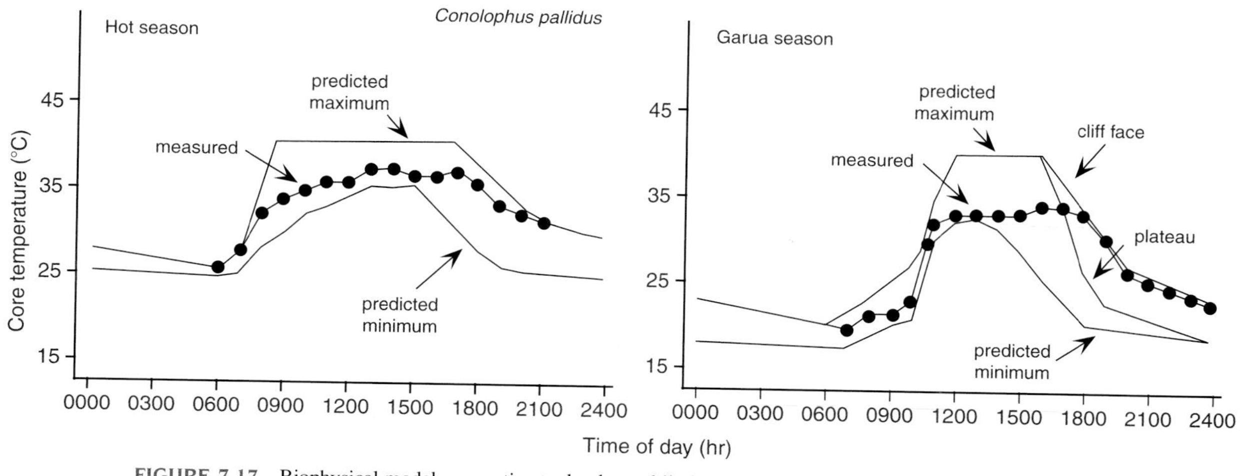

FIGURE 7.17 Biophysical models can estimate the thermal limits available to an ectotherm in a specific habitat. When coupled with temperature data from free-ranging Galapagos land iguanas (*Conolophus pallidus*), trade-offs between achieving optimal temperatures and remaining active for longer periods at suboptimal temperatures demonstrate that the physical environment can determine patterns of space use. Adapted from Christian et al., 1983.

are available year round. What the lizards gain by operating at suboptimal body temperatures during the garua season is more time at relatively high, but not the highest, temperatures. In both seasons, their body temperatures are maintained at levels that allow the longest period of constant temperature.

Costs and constraints on thermoregulation can also be examined with large data sets. Lizards that are active thermoregulators tend to have low slopes and high intercepts in the relationship of body temperatures to environmental temperatures (Fig. 7.18). These lizards attain high body temperatures early in the day and maintain high and constant temperatures throughout the day (see Fig. 7.11 for an example). Lizards that are passive thermoregulators tend to have high slopes in the relationship of body-to-environmental temperatures (their body temperatures change with that of the surrounding environment), and the intercept of the relationship is low (they operate at low body temperatures). The cost of maintaining high body temperatures throughout the day is that energy is used more rapidly and extra time and energy spent searching for prey to maintain energy use constrains other behaviors. The cost of maintaining low temperatures and fluctuating with temperatures in the environment is greater reliance on cool microhabitats or nocturnal activity and a reduced ability to transform energy into offspring.

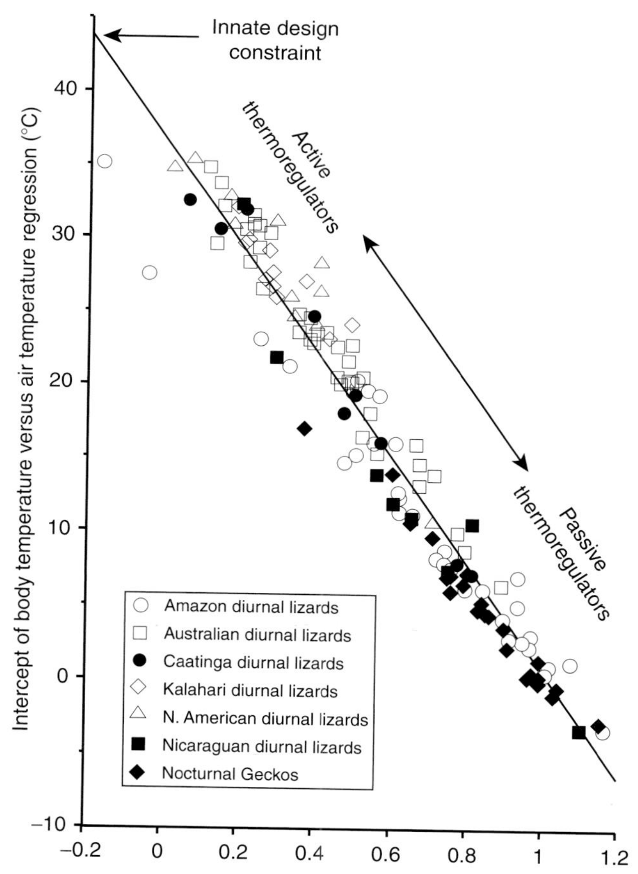

FIGURE 7.18 Thermoregulatory tactics vary among lizard species, but each habitat tends to have a mixture of active thermoregulators and passive thermoregulators. Species that are active thermoregulators tend to have high and relatively constant body temperatures throughout the day, whereas species that are passive thermoregulators tend to have lower body temperatures that fluctuate with environmental temperatures. Adapted from Pianka and Vitt, 2003.

Body Size

Rates of heat gain or loss decrease with increasing body size because, in larger animals, proportionately less surface area is available for heat exchange. In terrestrial lizards, for example, the critical mass is about 20–25 g, a size at which physiological mechanisms can have some effect on temperature control. This is one of the many reasons that behavioral control of thermal interactions assumes such importance in small ectotherms.

DORMANCY

When environmental conditions exceed an individual's capacity for homeostasis, retreat and inactivity offer an avenue for survival. Regular cycles of dormancy are major features in the lives of many amphibians and reptiles. Climatic fluctuations are the principal force for cyclic dormancy—hot and dry conditions in desert regions and near or below freezing temperatures in temperate-zone areas are examples. Seasonal fluctuations in food resources may drive dormancy in some tropical areas, although this remains unproven. Dormancy behaviors are commonly segregated into hibernation for avoidance of winter cold and aestivation for all others, including acyclic drought-caused dormancy. Depending upon the geographic range, individuals of some species may be dormant longer than they are active. For example, Arizona *Spea* appear to be active for about 1 month per year and Manitoba *Thamnophis* are active for less than 4 months each year.

Physiological studies of amphibian and reptilian dormancy indicate that many species alter cardiovascular function and suppress metabolic activities to conserve energy and ensure adequate oxygen to vital organs during extended periods of inactivity. Metabolic rates are lower than if rates were slowed just by temperature effects. The physiology of aestivation is less clear; metabolic rates generally do not drop below rates expected on the basis of temperature alone, although water loss rates are variously reduced.

Hibernation

Hibernation is a behavioral response to changing seasons. Although hibernation in mammals is often associated with changes in resource availability caused by cold temperatures, hibernation in amphibians and reptiles most likely is a direct response to cold temperatures and secondarily to changes in resource availability. Hibernation removes the animal from environments that are likely to experience temperatures low enough to kill amphibians and reptiles. Hibernacula can be underground, under water, inside of rock outcrops, or inside of hollow trees; virtually any cavity providing temperatures warmer than external temperatures can serve as hibernacula, and for many people in southern states, this means that their attics may serve as hibernacula for ratsnakes (*Pantherophis*). During hibernation, activity ceases for the most part, temperatures in the hibernation site determine body temperatures, and physiological processes are reduced to levels the same or lower than those predicted on the basis of temperature. Limited activity can occur, depending on immediate thermal conditions in the environment. In most instances, limited activity during hibernation does not involve feeding, mating, gestation, or other important life processes.

As winter approaches in temperate zones, most amphibians and reptiles seek shelter where the minimum environmental temperatures will not fall below freezing. Some amphibians and turtles avoid subfreezing temperatures by hibernating on the bottoms of lakes and streams. Because water reaches its greatest density at 4°C and sinks, animals resting on or in the bottom usually will not experience temperatures less than 4°C. Amphibians and reptiles hibernating on land are less well insulated by soil and must select sites below the frost line or be capable of moving deeper as the frost line approaches them. The few terrestrial hibernators that have been followed do move during hibernation. Box turtles (*Terrapene carolina*) begin hibernation near the soil surface, reach nearly 0.5 m deep during the coldest periods, and then inch toward the surface as environmental temperatures moderate. Hibernating snakes (*Pantherophis, Crotalus*) move along thermal gradients in their denning caves/crevices, always staying at the warmest point. The onset of hibernation in Desert tortoises (*Gopherus agassizii*) appears associated with endogenous cues rather than exogenous ones. Considerable variation exists among individuals with respect to time of entering hibernacula, but overall temperature and day length are not tightly associated with hibernation. Mid-hibernation temperatures for desert tortoise vary from about 8.9°C to 16.3°C across different years and sites.

Many aquatic hibernators rest on the bottom of ponds or streams rather than buried in the bottom. While such sites might expose them to predation, hibernation in open water permits aquatic respiration (extrapulmonary) and apparently is sufficient to meet some or all of the oxygen expenditures during dormancy in both amphibians and reptiles. In normoxic water, the oxygen demands of lunged anurans and salamanders are easily met by cutaneous respiration during hibernation. Cutaneous respiration also provides sufficient oxygen for some hibernating reptiles (*Chrysemys picta, Sternotherus odoratus, Thamnophis sirtalis*). Experiments on garter snakes hibernating submerged in a water-filled hibernaculum demonstrate that the submerged snakes use aerobic metabolism but at a more energy-conservative rate than terrestrial hibernating conspecifics. In normoxic waters, turtles also remain aerobic through cutaneous and perhaps buccopharyngeal respiration; however, if buried in anoxic or hypoxic environments such as mud, the hibernating animals switch to anaerobiosis. Survival is possible because of high tolerance for lactic acid buildup, and in some instances, submerged turtles may shuttle between normoxic and anoxic sites. When in the normoxic ones, they can shift to aerobiosis and to some extent flush the excess lactic acid. This might explain observations of turtles swimming below the ice of a frozen pond.

Freeze Tolerance

Most temperate-zone amphibians and reptiles are able to survive brief periods of supercooling (−1 to −2°C). Freezing (formation of ice crystals within the body) is lethal to all but a few species because ice crystals physically damage cells and tissue. Intracellular freezing destroys cytoplasmic structures and cell metabolism. Extracellular freezing also causes physical damage, but the critical factor is osmotic imbalance. As body fluids freeze, pure water freezes first, increases the extracellular osmotic concentration, and dehydrates the cells. Intracellular dehydration disrupts cell structure and, if extreme, causes cell death. Extracellular freezing also blocks fluid circulation and the delivery of oxygen and nutrients to the cells. The damage from freezing causes the animal's death upon thawing.

Intracellular freezing is lethal for all animals. A few species of turtles (*Terrapene carolina*, hatchling *Chrysemys picta*) and frogs (*Pseudacris crucifer*, *H. versicolor*, *Pseudacris triseriata*, *Lithobates sylvaticus*) are "freeze tolerant" and survive extracellular freezing. The frogs hibernate in shallow shelters, and although snow may insulate them, body temperatures still drop to −5 to −7°C, causing them to freeze. Ice crystals appear beneath the skin and interspersed among the skeletal muscles; a large mass of ice develops in the body cavity. As much as 35–45% of the total body water may become ice and yet the frogs survive. When frozen, a frog's life processes are suspended; breathing, blood flow, and heartbeat stop. These frogs tolerate the large volume of body ice by producing and accumulating cryoprotectants (= antifreeze) within the cells. The cryoprotectants are either glycerol (*Hyla versicolor*) or glucose (the three other species), which protect and stabilize cellular function and structure by preventing intracellular freezing and dehydration. These freeze-tolerant species also possess specialized proteins that control extracellular freezing and adjust cellular metabolism to function at low temperatures and under anaerobic conditions.

The frogs do not physiologically anticipate winter and begin to produce the cryoprotectants. Ice forms peripherally and triggers synthesis of cryoprotectants. The rate of freezing is slow, permitting the production and distribution of cryoprotectants throughout the body before any freeze damage can occur. As soon as the body begins to thaw, the cryoprotectants are removed from general circulation. Freeze tolerance extends into early spring at the time when the frogs begin reproductive activities. For the early spring breeders such as the spring peeper (*Pseudacris crucifer*) and the wood frog (*Lithobates sylvaticus*), this extended tolerance permits survival under the highly variable and occasionally subzero temperatures that occur during their late winter to early spring breeding season. Freeze tolerance appears to be lost gradually in association with the beginning of feeding.

Aestivation

Amphibians in desert and semidesert habitats face long periods of low humidity and no rain. To remain active is impossible for all but a few species; death by dehydration occurs quickly. Aridland species retreat to deep burrows with high humidity and moist soils, become inactive, and reduce their metabolism. Inactivity may dominate an anuran's life. *Spea hammondii* in the deserts of southwestern North America spend >90% of their life inactive; they appear explosively and breed with the first heavy summer rains then feed for 2–3 weeks before becoming inactive for another year. Where retreats become dehydrating, some anuran species (e.g., *Litoria* [formerly *Cyclorana*], *Neobatrachus*, *Lepidobatrachus*, *Smilisca* [formerly *Pternohyla*], *Pyxicephalus*) produce epidermal cocoons. The cocoon forms by a daily shedding of the stratum corneum; the successive layers form an increasingly impermeable cocoon, completely encasing the frog except around the nostrils (see Chapter 6). Some salamanders (e.g., *Siren*) burrow into the mud of drying ponds and produce similar epidermal cocoons.

ENERGETICS

The acquisition of energy in the natural world involves a complex interaction between the biophysical environment in which an animal lives, resources available and their distribution, the social system and how it might constrain access to resources and consequently mating success, and the risk involved in acquiring resources (Fig. 7.19). Energy available to amphibians and reptiles is limited by a combination of resource availability and the costs or risks of harvesting it. Once acquired, energy is used for three primary life processes: growth, maintenance, and reproduction. Energy can also be allocated to storage to be used at a later time. All other aspects of energy use (e.g., energetic support for performance, physiological processes) fall within these broad categories (Fig. 7.20). Compartmentalizing energy use makes it relatively easy to understand how various behaviors or processes contribute to the overall energy budget of an organism.

With the exception of brooding pythons, amphibians and reptiles generally do not use energy produced during metabolism to maintain body temperature, and their body temperatures are low during at least some periods of the day and season. Consequently, they have relatively low energetic costs of maintenance. Approximately 40–80% of energy ingested in food is invested in body tissue in ectotherms, whereas about 98% of energy ingested in food of birds and mammals (endotherms) is invested in temperature regulation and activity. The high densities and biomass that amphibians and reptiles achieve, even in low-resource environments, can be attributed to this.

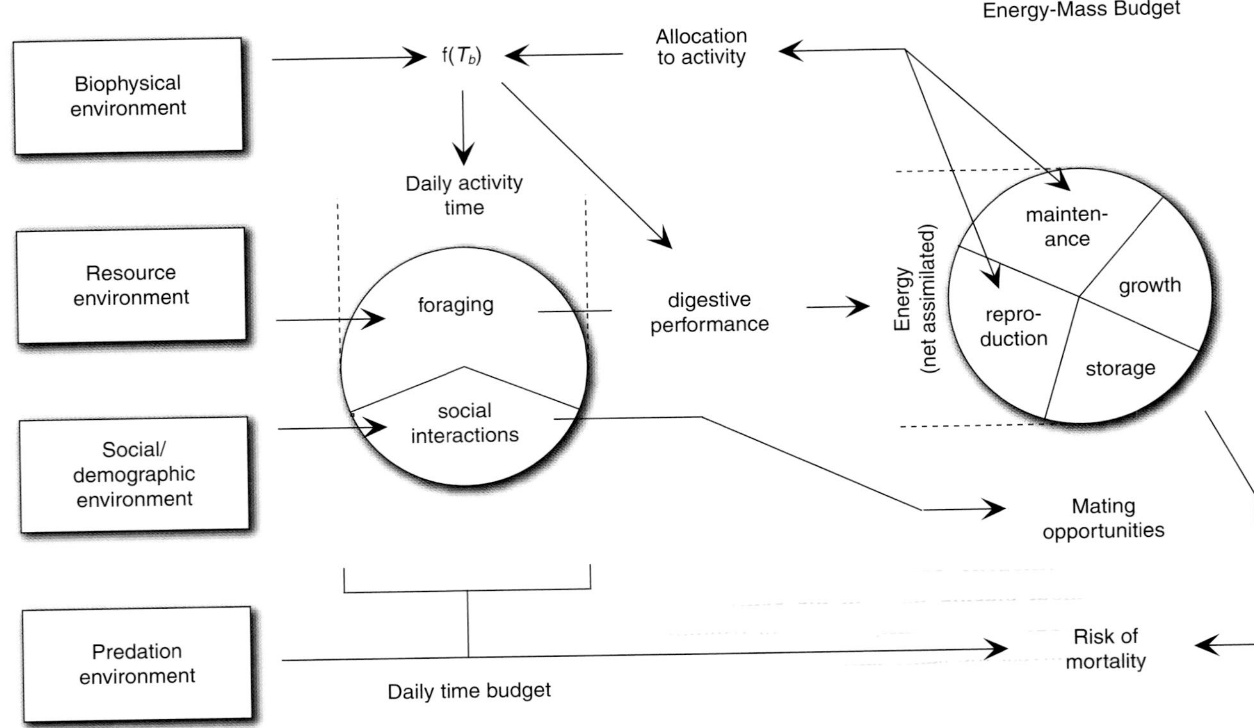

FIGURE 7.19 The ability of amphibians and reptiles to acquire energy necessary to support life is determined by a combination of abiotic and biotic factors. Adapted from Dunham et al., 1989, and Niewiarowski, 1994.

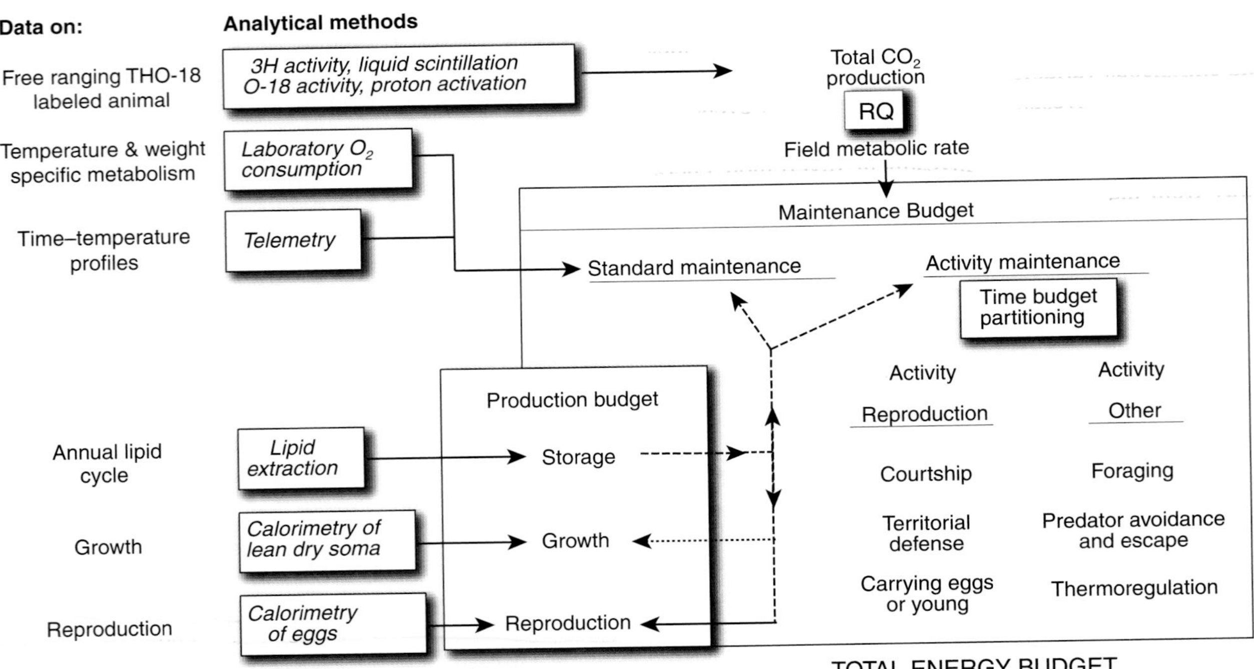

FIGURE 7.20 Schematic diagram showing the protocol for developing ecologically meaningful energy budgets for amphibians and reptiles. Analytical methods can vary depending on the species studied and the specific question asked. Complete energy budgets require partitioning of energy into growth, maintenance, storage, and reproduction. All activities belong to one of these four compartments. Adapted from Congdon et al., 1982.

Because the conversion of food (resources) into usable energy is an oxidative process, energy can be measured by measuring the rate of use of oxygen both in the laboratory and in the field. Energetic studies typically refer to oxygen consumption for a given body mass per unit time as VO_2. To standardize units for comparisons, oxygen consumption is generally presented as milliliters or liters of O_2 per gram or kilogram of body mass per hour. Standard metabolism is the minimum rate of energy consumption necessary to stay alive (usually measured when an animal is completely at rest). Resting metabolism is the rate of energy consumption of postabsorptive (not digesting food) animals when not moving but at a time of day when the animals would normally be active. Maximum metabolism is energy consumption at a high level of activity. Because rates of energy use are temperature dependent, data on metabolic rates usually contain a temperature component.

Comparing the energetic cost of specific behaviors across species without placing them in the context of a complete energy budget can be misleading. For example, if two frog species invest the same amounts of energy in reproduction as measured by the energy content or mass of their clutches, it does not follow that their reproductive investments are equal. One might be a large-bodied species that invests very little of its annual energy budget into a single reproductive event. The other might be a very small species that invests a major portion of its annual energy into a single reproductive event. Comparisons of energy use among individuals within species can be much more illuminating because trade-offs will be more evident and extraneous variables (e.g., size) can be minimized. An individual that invests more in a reproductive event than other individuals must harvest more resources to support the additional reproductive investment, divert more energy away from maintenance, or use stored energy that would otherwise be available for maintenance at a later time.

Similar to other physiological processes, the use of energy is related to temperature and body size (Fig. 7.21). Because metabolism in most animals is supported by oxygen, oxygen consumption can be used as a measure of metabolism. Not surprisingly, level of activity influences metabolic rate independent of temperature. Thus warmer ectotherms use more energy, as do more active ones. Body size also influences energy use; larger ectotherms in general use more energy than smaller ones. The energetic cost for a wide variety of behaviors has been studied in many amphibians and reptiles. These include locomotion, prey handling, foraging, and social interactions.

When resting metabolism is known, it is relatively easy to measure the energetic cost of various behaviors. The difference between the metabolic rate associated with the behavior and the rate of resting metabolism estimates the energetic cost of the activity. Because rates are temperature dependent, temperature must be controlled. Examples of activities of amphibians and reptiles that require significant amounts of energy include reproductive-related behaviors, prey acquisition, escape, foraging, and locomotion.

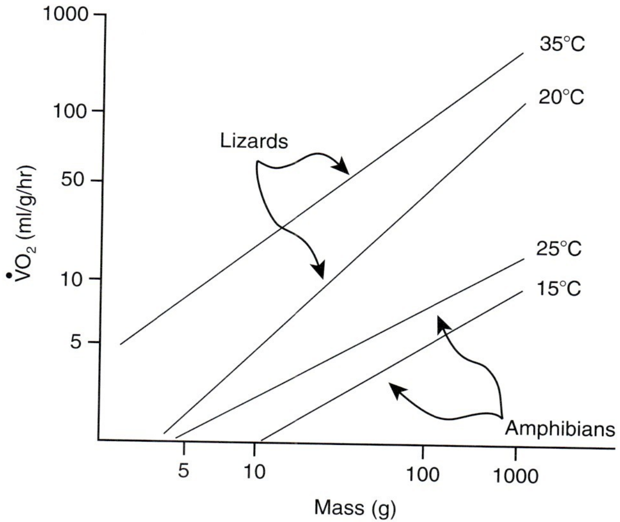

FIGURE 7.21 Effects of body mass and temperature on the rate of oxygen uptake (metabolic rate) in a reptile and an amphibian under two different thermal regimes. Amphibians have lower metabolic rates than reptiles even after the effects of size and temperature are removed. Data from Bennett, 1982, and Whitford, 1973.

Like most frogs, males of the spring peeper (*Pseudacris crucifer*) call to attract females. While at rest, males use 0.108 ml of oxygen per gram of body mass per hour, a rate similar to that of females at rest. During forced exercise, males have higher metabolic rates than females (0.110 ml/[g•hr] versus 0.91 ml/[g•hr], respectively). Males use more energy while calling (1.51 ml/[g•hr]) than they do while exercising; energy used for calling by males can be considered a cost of reproduction. The energy used for call production increases with the rate of calls produced (Fig. 7.22). The rate of call production is

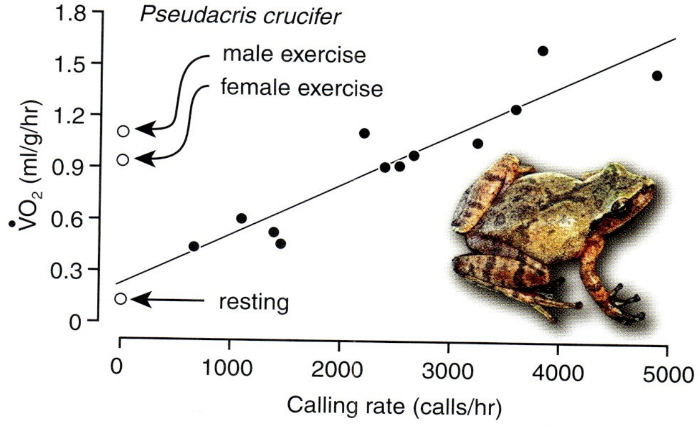

FIGURE 7.22 Male spring peepers (*Pseudacris crucifer*) expend considerable energy calling to attract females during the breeding season. The rate of calling is related to reproductive success, which explains why males expend extra energy to call at higher rates. Adapted from Taigen et al., 1985.

the primary determinant of mating success in males; females are attracted to males with the highest calling rates. Thus the high energetic cost of calling in spring peepers has a high payoff in terms of reproductive success.

Moving from place to place requires the use of energy, and different animals have different ways of moving. In general, body mass alone explains much of the variation in energetic costs of locomotion for obvious reasons. Because locomotion involves distance moved, the energetic cost of locomotion, which is called the net cost of transport, is measured as oxygen used per unit body mass per kilometer (e.g., O_2•[g•km]). In general, amphibians have lower costs of transport than reptiles but great variation exists among species, some of which is tied to the specific type of locomotion. Snakes provide a nice example of the cost of transport because morphology is relatively conservative and there are no limbs to consider. The four kinds of locomotion used by most snakes, lateral undulation, concertina, sidewinding, and rectilinear, differ considerably in terms of energy requirements. Concertina locomotion requires seven times more energy than lateral undulation. Sidewinding, which appears to have a high level of activity associated with it, requires much less energy than lateral undulation or concertina locomotion (Fig. 7.23). The snake moves by arching its body and moving its body through the arch; a relatively small part of the body touches the substrate at any one time, resulting in little resistance (Fig. 7.24).

Most energetic studies of behavior in amphibians and reptiles were conducted in the laboratory until the development of a technique using doubly labeled water. Animals are injected in the field with water that has a heavy oxygen atom (^{18}O) and a heavy hydrogen atom (^{3}H). By sampling blood periodically and examining the decay in the ^{18}O, rates of energy use can be calculated. The decay in ^{3}H provides an estimate of water flux. A number of particularly interesting estimates of energy use by free-ranging reptiles has provided new insights into trade-offs in energy use.

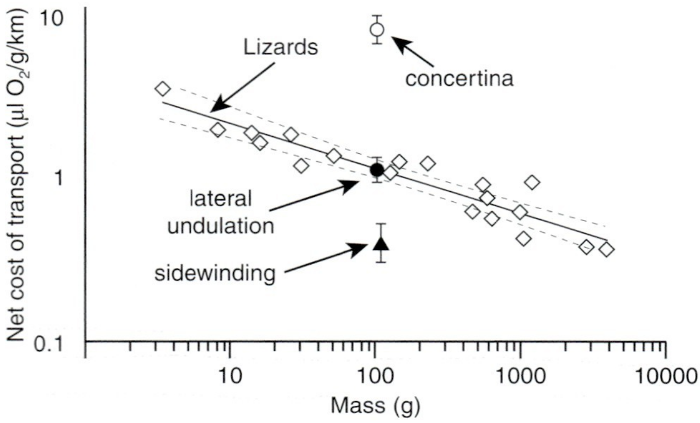

FIGURE 7.23 The net cost of transport for snakes using lateral undulation is similar to that of lizards during locomotion. Concertina locomotion is much more expensive energetically, but sidewinding has a low energetic cost. Adapted from Secor et al. 1992.

FIGURE 7.24 The sidewinder *Crotalus cerastes* during locomotion on a sand dune. (C. Mattison)

Two snakes, the sidewinder (*Crotalus cerastes*) and the coachwhip (*Masticophis flagellum*), occur together over a large part of the Mojave and Sonoran Deserts of western North America. The sidewinder is a sit-and-wait or ambush forager that remains for extended time periods in a single place waiting for potential prey to pass by. When a prey item passes, the snake strikes, envenomates, kills, tracks, and swallows the prey. The coachwhip is an active or wide forager that moves considerable distances during the day in search of prey, which it captures and swallows, usually alive (Fig. 7.25). Daily energy expenditure in both species varies with season, partially as the result of seasonal changes in temperature. Even though slight differences are apparent between the two species in standard metabolic rates (coachwhip higher at all temperatures), large differences are apparent in energy used for other activities, much of which can be attributed to foraging (Fig. 7.26). The energetically expensive foraging of coachwhips is offset by increased rates of energy acquisition. Sidewinders feed primarily on small rodents and lizards, whereas coachwhips feed on a wide variety of vertebrates, including sidewinders! Coachwhips spend more time foraging, move more frequently, and have higher prey capture rates than sidewinders, accounting for differences in energy uptake and use. Nevertheless, it is important to keep in mind that coachwhips are colubrid snakes and sidewinders are viperid snakes. Differences in energy metabolism, even though associated with foraging behavior, may reflect much more general physiological differences between major snake clades rather than direct responses to foraging mode differences between species.

SYNTHESIS

Water balance, respiration, thermoregulation, and energetics are tightly linked in ectothermic vertebrates. For amphibians, rates of water loss can be extremely high, and most species

FIGURE 7.25 The sidewinder, *Crotalus cerastes* (left), is a sit-and-wait predator investing little energy in prey search, whereas the coachwhip, *Masticophis flagellum* (right), is an active forager that spends considerable energy searching for prey. The coachwhip is eating an adult *Dipsosaurus dorsalis*. (*C. c.*, S. C. Secor; *M. f.*, J. M. Howland)

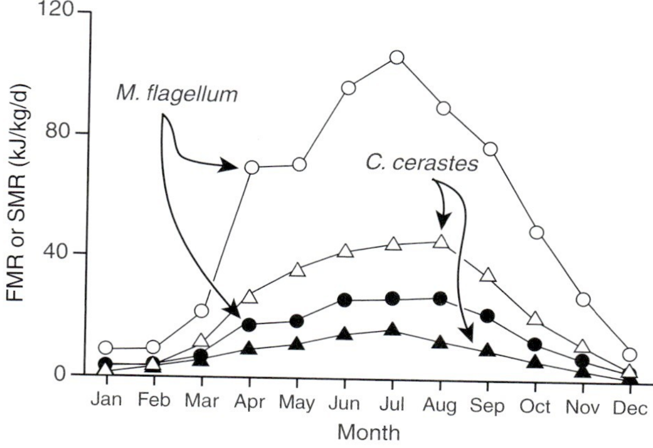

FIGURE 7.26 The sidewinder, a sit-and-wait predator, expends considerably less energy under natural field conditions than the coachwhip, an actively foraging predator. Year-long profiles of daily energy expenditures (averaged by month) are illustrated. Open symbols indicate field metabolic rates based on doubly labeled water measurements, and closed symbols indicate standard metabolic rates. Adapted from Secor and Nagy, 1994.

select microhabitats that minimize water loss. Such microhabitats are usually relatively cool or enclosed. Most amphibians take in large amounts of water and produce dilute urine, although there are some notable exceptions. One consequence of activity at low temperatures and of ectothermy in general is that metabolic rates are low (no metabolic cost of heat production). For many reptiles, activity occurs at high body temperatures, but during periods of inactivity, body temperatures are much lower. Reptiles in general take in much less water than amphibians and are capable of retaining more of what they take in. As a result, they produce relatively concentrated urine, often including uric acid as a concentrated waste product. Like amphibians, metabolic rates of reptiles are low because there is no cost of heat production (with a few exceptions); however, overall, reptilian metabolic rates are higher than those of amphibians. Because nearly all energy acquired is directed into low-cost maintenance, growth, reproduction, and storage, amphibians and reptiles can occur at high densities in environments that limit densities of homeothermic vertebrates that expend much of their ingested energy on heat production. Amphibians and reptiles can also persist through long periods of energy shortages.

Although the interplay between temperature, water economy, and energetics is well documented from a physiological perspective, the correlated evolution of these important physiological traits is only beginning to be appreciated. The evolutionary history of geckos in the genus *Coleonyx* exemplifies the possibilities an evolutionary approach to the interplay between water economy, temperature, and metabolism can have in understanding physiological processes. The ancestor of *Coleonyx* in North America appears to have had relatively low body temperature (26°C), high evaporative water-loss rate (2.5 mg/g/hr), and a low standard metabolic rate (0.07 mg/g/hr) and lived in a relatively moist, forested habitat. Two extant species, *C. mitratus* and *C. elegans*, retain these characteristics, and they are members of the earliest lineage (Fig. 7.27). During the evolutionary history of *Coleonyx*, species moved into more arid environments, ultimately into the deserts of North America. Correlated with that shift are increases in body temperatures (above 31.0°C), reductions in evaporative water loss (< 0.1 mg/g/hr), and increases in standard metabolic rate (> 0.15 mg/g/hr). In this example, the set of predictions based on a shift from mesic to xeric habitats holds true, indicating that these are indeed adaptations to life in specific environments. Finally, this example points to the importance of maintaining physiological homeostasis for amphibians and reptiles occupying diverse environments.

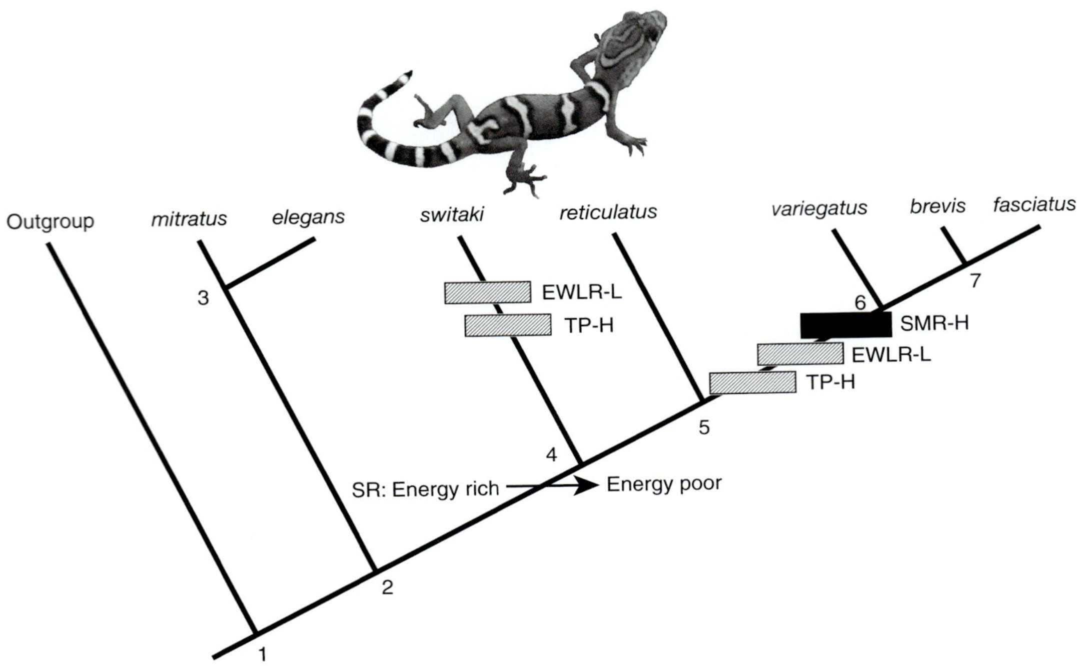

FIGURE 7.27 An hypothesis of physiological–ecological character state evolution in lizards in the genus *Coleonyx*. Four equally parsimonious hypotheses were found based on physiological data alone, but when coupled with biogeographic data, the other three were rejected. EWLR = evaporative water-loss rate, TP = temperature preference, SMR = standard metabolic rate, H = high, L = low. Solid bars indicate acquisition of a new state, and crosshatched bars indicate independent evolution of a derived state. The genera *Eublepharis, Hemitheconyx*, and *Holodactylus* make up the outgroup. Presumably, a shift occurred in the selective regime (SR) from an energy-rich to an energy-poor microhabitat during the evolutionary history of *Coleonyx*. Adapted from Dial and Grismer, 1992.

QUESTIONS

1. Explain how the use of a phylogeny unraveled the evolutionary history of temperature preferences, standard metabolic rate, and evaporative water loss in gecko species in the genus *Coleonyx*.
2. How does the energy budget of a juvenile snake differ from that of an adult female snake?
3. What does "thermal performance breadth" mean?
4. What is different in pythons in terms of how they brood their eggs compared to other reptiles?
5. Why would a sidewinder rattlesnake (*Crotalus cerastes*) expend less energy in a 24-hour period than a coachwhip (*Masticophis flagellum*)?
6. Why does *Varanus gouldii* maintain its body temperature within the set-point range whereas *Varanus panoptes* does not?

ADDITIONAL READING

Bartholomew, G. A. (1982). Physiological control of body temperature. In C. Gans and F. H. Pough (Eds.), *Biology of the Reptilia, Vol. 12. Physiology C. Physiological Ecology* (pp. 167–211). Academic Press, New York.

Bennett, A. F. (1980). The metabolic foundations of vertebrate behavior. *BioScience* **30:** 452–456.

Gregory, P. T. (1982). Reptilian hibernation. In C. Gans and F. H. Pough (Eds.), *Biology of the Reptilia, Vol. 13. Physiology D, Physiological Ecology* (pp. 53–154). Academic Press, New York.

Huey, R. B., and Stevenson, R. D. (1979). Integrating thermal physiology and ecology of ectotherms: A discussion of approaches. *Amer. Zool* **19:** 357–366.

Hutchison, V. H., and Dupré, R. K. (1992). Thermoregulation. In M. E. Feder and W. W. Burggren (Eds.), *Environmental Physiology of the Amphibia* (pp. 206–249). Univ. of Chicago Press, Chicago.

Lillywhite, H. B. (1987). Temperature, energetics, and physiological ecology. In R.A. Seigel, J.T. Collins, and S.S. Novak (Eds.), *Snakes: Ecology and Evolutionary Biology* (pp. 422–477). MacMillan, New York.

REFERENCES

General

Bartholomew, 1982; Bauwens et al., 1995; Beck, 2005; Bennett, 1980; Brattstrom, 1963, 1965; Cowles and Bogert, 1944; Crawford and Barber, 1974; Dawson, 1975; Garland, 1994; Huey and Slatkin, 1976; Hutchison and Dupré, 1992; McCue, 2004; Pearson and Bradford, 1976; Rome et al., 1992; Sinsch, 1989; Tattersall et al. 2006.

Thermoregulation

Heat Exchange with the Environment and Performance

Atsatt, 1939; Autumn, 1999; Autumn et al., 1994, 1997, 1999; Bauwens et al., 1995; Christian, 1996; Christian and Weavers, 1996; Christian et al., 1995, 2006; Elphick and Shine, 1998; Farley, 1997; Farley and Emshwiller, 1993, 1996; Hertz et al., 1982; Huey and Stevenson, 1979; Huey et al., 1989a,b; Hutchison and Larimer, 1960; Olalla-Tárraga and Rodríguez, 2007; Pearson, 1977; Rocha and Bergallo, 1990; Seebacher and Shine, 2004; Shine, 2004; Shine and Harlow, 1996; Shine et al., 1997; Van Damme et al., 1992.

Control of Body Temperature

Chong et al., 1973; Crawford and Kampe, 1971; Crawford et al., 1977; Dupré et al., 1986; Feder and Lynch, 1982; Firth and Heatwole, 1976; Frair et al., 1972; Heatwole et al., 1973, 1975; Hutchison and Dupré, 1992; Hutchison et al., 1966; Jacobson and Whitford, 1971; Lillywhite, 1970, 1987; Moll and Legler, 1971; Paladino et al., 1990; Parmenter and Heatwole, 1975; Shine and Fitzgerald, 1996; Shine and Lambreck, 1985; Shoemaker et al., 1989; Spotila et al., 1977, 1997; Tattersall and Gerlach, 2005; Tattersall et al., 2006; Tracy et al., 2006; Vitt et al., 2005; Withers et al., 1982.

Costs and Constraints of Thermoregulation

Christian et al., 1983; Grant, 1990; Huey and Slatkin, 1976; Pianka and Vitt, 2003; Porter and Gates, 1969.

Dormancy (Hibernation and Aestivation)

Blouin-Demers et al., 2000; Case, 1976b; Gregory, 1982; Litzgus et al., 1999; Nussear et al., 2007; Plummer, 2004; Wone and Beauchamp, 2003.

Energetics

Forester and Czarnowsky, 1985; Nagy, 1983a; Secor and Nagy, 1994; Secor et al., 1992; Taigen et al., 1985.

Synthesis

Dial and Grismer, 1992.

Part IV

Behavioral Ecology

Behavioral ecology is an enormous field that includes but is not limited to the ecology of movement, social interactions, foraging, and escape from predators. All these areas require behavior of one kind or another. Behavioral decisions ultimately influence individual fitness because they determine whether an individual will be able to compete within the social system of its own species, avoid predators, or successfully find food. We first consider the distribution of individuals in their environments and in relation to other individuals within their populations. The mechanisms that individuals use to navigate within and between the habitats they use are briefly summarized. We follow by examining the complexities of social behavior, centering on how individuals interact with other individuals within local populations. Individual amphibians and reptiles balance the primary benefit of social behavior, which is increased individual fitness, against the costs of acquiring the resources required to maintain activity and the potential risks of predation while carrying on these activities.

Chapter 8

Spacing, Movements, and Orientation

Movements and home ranges of reptiles and amphibians vary considerably, both among and within species. Within species, both intrinsic (age, life history stage, size, sex, reproductive status) and extrinsic (environmental quality, season, temperature, humidity) factors contribute to patterns of movement. Exactly where individual animals live is determined by complex interactions between physiological requirements of individuals and physical characteristics of their habitats. The location of other individuals can constrain spacing patterns within bounds set by the physical environment. Movements are critical for locating food and mates and avoiding environmental extremes and predators. The ability to return to high-quality microhabitats, overwintering sites, and breeding sites requires systems for orientation. Animals are not distributed randomly, because some places are better than others in terms of resource availability when balanced against the risk or costs of acquiring the resource. Animals do not remain in the same place, because they are sensitive to gradients in resources and because risk varies with location. Because of their sensitivity to resource gradients, animals orient themselves and direct their patterns of movement in organized ways, often toward resources. The payoff for making these choices is clear. Individuals that are better able to access resources while minimizing risk grow more rapidly, reproduce earlier, and if their body size is larger as a result of their resource-accruing abilities, produce more or larger offspring (see Chapters 5 and 9). If all resources were spread uniformly in the environment or even across environments, it would be difficult for individuals or species to segregate spatially, but resources are not distributed uniformly. As a result, nearly all environments are patchy in one way or another. Even if a single resource is distributed uniformly across habitat patches, other resources likely are distributed in other ways. As the number of potential resource categories increases, the likelihood that two species or two individuals would use all resources in the same way rapidly declines.

Global patterns of amphibian and reptile distribution indicate that the physical environment places limits on the spatial distribution of species. This is particularly obvious with respect to temperature (Chapter 7), because amphibians and reptiles are ectothermic. Sea snakes, for example, are largely distributed in the shallow, warm seas of southeastern Asia and northern Australia partly because they originated there and partly because the broad continental shelf areas offer a thermally appropriate habitat with a high diversity and abundance of potential prey. The absence of sea snakes in the Atlantic Ocean reflects their inability to cross cold polar currents and deep expanses of open ocean. Only a single species, *Pelamis platurus*, has traversed the Pacific and successfully colonized the coastal waters of western tropical America. This sea snake probably arrived in the eastern Pacific after the closing of the Panamanian gap (4 mybp) and hence was unable to continue its westward dispersal into the Caribbean and the Atlantic. For many crocodylians, a combination of fresh—or, in some instances, brackish—water, combined with the absence of freezing weather delimits their geographic distribution. For many amphibians, the spatial distribution of appropriate breeding sites sets limits on their distributions. Historical factors also play a role, as pointed out in Chapter 12. Past and present locations of dispersal barriers, including mountain ranges, rivers, and oceans, have excluded many species from invading areas where climatically they could survive and flourish.

Geographic distributions of species or populations can be limited by microhabitat distributions, presence or absence of competitors or predators, or even the availability of prey. Microhabitat specialists such as *Xantusia henshawi* are restricted to areas with exfoliating rock; flat lizards in the genera *Tropidurus* and *Platysaurus* are restricted to granitic outcrops in South America and South Africa, respectively. *Anolis* lizards on Caribbean islands have evolved microhabitat specialization (see Chapter 12) in response to competition with other *Anolis* and thus are limited to specific microhabitats within the same habitat. Dietary specialists occur only in microhabitats containing the prey that they eat. Horned lizards, *Phrynosoma*, which are ant specialists, do not occur in habitats lacking edible ants, and some, like *Phrynosoma cornutum*, may move very little while active because they sit along trails of harvester ants. No single factor explains the geographic distribution of any species. At a local level, a multitude of factors influences spatial distributions of individuals.

LOCAL DISTRIBUTION OF INDIVIDUALS

Distribution of individuals occurs at a number of levels. In the context of community ecology, species tend to be associated with specific microhabitat patches (the "place" resource or niche discussed in Chapter 12). A relatively easy and informative exercise is to walk through a natural habitat and list the animal species and the microhabitats where they were first observed. A tabulation of these data reveals that each species tends to be associated with different microhabitats. Because microhabitats interdigitate, any given habitat can contain a large number of species that spatially overlap, but each occurs predominately in specific microhabitats. Selection of microhabitats is enforced by competitive interactions and the risk associated with activity in unfamiliar places or patches. An individual may no longer be cryptic or may be unable to escape predators in unfamiliar patches (see Chapter 11).

Within species, individuals often move within an area that they do not defend from conspecifics, called the *home range*. Foraging and social activities occur in this area. Adjacent home ranges can overlap or they can be completely exclusive. Part or all of the home range might be defended, usually against conspecifics but occasionally against other species. This defended area is the territory, which is introduced in this chapter and discussed in the context of social behavior in Chapter 9. Spacing typically implies the spatial distribution of individuals within a species and, more specifically, within a local population. As a result, spacing usually focuses on home ranges and territories.

Home Ranges

Home ranges of amphibians and reptiles usually are associated with one or more resources. The resources include food, shelter, mates, thermoregulation sites, escape routes, and a host of other things. Home range size can vary between sexes, is often associated with body size, and is influenced by population density. For species living in two-dimensional habitats, such as fringe-toed lizards on sand dunes in southern California or plethodontid salamanders in the Great Smoky Mountains, the home range can easily be measured as the area that encompasses all the outer points within which an individual occurs. This technique is called the minimum polygon method of home range determination and does not take into consideration the amount of time or the relative frequency with which an individual might use different parts of the home range. Nevertheless, it is the most widely used method of calculating home range and has many advantages. In particular, it can be calculated easily in the field, the measurements are fairly accurate if samples are adequate, and it is based on actual observations of animal occurrences. Moreover, the amount of overlap in home ranges between individuals in the population can be easily calculated. The variation in sizes of amphibian and reptile home ranges is impressive (Table 8.1). An association between body size and home range size exists across many species, but some exceptions exist. For arboreal amphibians and reptiles, measuring home range is much more difficult, and even defining it is not easy. The Amazonian lizard *Anolis transversalis*, for example, spends much of its life in the canopy of a single or a few trees. The home range is three dimensional and thus is a volume rather than an area. Moreover, because the lizard can only move on the branches and leaves within the canopy, many gaps or unusable areas exist. Nevertheless, conceptually, a three-dimensional home range is no different from a two-dimensional one—they both represent regular use of space by individuals.

Home ranges can and often do vary through time or space; they can change radically following single events. For some species they may not even exist. Home ranges are not defended, and other individuals may use parts of them. Overlap in home ranges among individuals can be considerable. During the nonbreeding season, many terrestrial amphibians (e.g., *Ambystoma maculatum, Plethodon cinereus, Rhinella marina, Rana temporaria*) have small to moderate-sized home ranges away from water. An individual can have one or more resting and feeding sites (activity centers) within its home range, but it might use a single site for a day, a week, or longer before shifting to another site. An amphibian may not visit all sites each day or even each week, but the periodic occurrence at sites and the persistent occupancy of the total area adjacent to these sites delimit the individual's home range. For species that reproduce in ponds or streams, the home range breaks down during breeding events because adults breed in aquatic sites that are not within the home range. Terrestrial-breeding amphibians (e.g., dendrobatid, mantellid, and brachycephalid frogs, *Plethodon*) generally deposit eggs within their home ranges. Some spend their entire lives in a single home range.

TABLE 8.1 Home Range and Resource Defense in Select Amphibians and Reptiles.

Taxon	Area	Female area	Male size (mm)	Defense		Habits
				Terr	S-S	
Batrachoseps pacificus	3.6	?	42	?	?	Terrestrial
Desmognathus fuscus	1.4	?	45	?	?	Semiaquatic
Salamandra salamandra	10	>	82	?	?	Terrestrial
Atelopus varius	<20	=	25	+		Terrestrial
Lithobates clamitans	65	=	60	+		Semiaquatic
Eleutherodactylus marnockii	328	=	20	+		Terrestrial
Terrapene c. triungis	52,000	=	115	−	−	Terrestrial
Trachemys scripta	397,500	<	200[a]	−	±	Aquatic
Crocodylus niloticus	7990	<	2100[b]	+		Aquatic
Sceloporus merriami	535	<	45	+		Terrestrial
Varanus olivaceus	20,500	<	450	±	+	Arboreal
Xantusia riversiana	17	=	65	−	+	Terrestrial
Acrochordus arafurae	15,000	?	900	−	?	Aquatic
Carphophis amoenus	253	?	215	?	?	Semifossorial
Natrix natrix	99,000	>	700	?	?	Terrestrial

Note: Terr = territorial, S-S = site or resource specific.
[a]*Plastron length*
[b]*Total length*
Sources: Salamanders—Bp, Cunningham, 1960; Df, Ashton, 1975; Ss, Joly, 1968. Frogs—Av, Crump, 1986; Rc, Martof, 1953; Sm, Jameson, 1955. Turtles—Tct, Schwartz et al., 1984; Ts, Schubauer et al., 1990. Crocodilians—Cn, Hutton, 1989. Lizards—Sm, Ruby and Dunham, 1987; Vo, Auffenberg, 1988; Xr, Fellers and Drost, 1991a. Snakes—Aa, Shine and Lambreck, 1985; Ca, Barbour et al., 1969; Nn, Madsen, 1984.

Shape of home ranges varies considerably and is often related to the microhabitat specificity of a species and the physical structure of the microhabitat. Semiaquatic or aquatic species (*Desmognathus monticola, Limnonectes macrodon*, mud turtles) are linearly distributed along streams and lakeshores. As a consequence, individuals within these populations tend to have elongate, narrow home ranges along the stream edge or lakeshore, or within the watercourse. The watersnake *Nerodia taxispilota* is linearly distributed along rivers, streams, and edges of ponds and lakes in the southeastern United States. Along part of the Savannah River that divides Georgia and South Carolina, these water snakes are most common adjacent to backwater areas, such as oxbow lakes and areas along the outside banks of curves in the river where water velocity is highest. They commonly are found on steep riverbanks or perched on logs and roots out of the water. During a 2-year period, each individual *N. taxispilota* moved an average of 270 m, although most individuals moved very little. These snakes are capable of long movements but often remain in a relatively small area because of the availability of good perch sites. Additionally, steep banks and overhanging logs and roots provide some protection from aquatic and terrestrial predators.

The smooth softshell turtle, *Apalone muticus*, is linearly distributed in rivers throughout the southern and central United States. These rivers experience drastic and unpredictable fluctuations in water level, and flooding can cause major changes in the physical structure of the river channel. As a result, the home ranges of softshelled turtles are short lived. Home ranges are associated with sandbars that change periodically due to erosion, but some softshells change the location of their home ranges without any apparent change of habitat (Fig. 8.1). Some individuals often move long distances from their home ranges and then return, presumably assessing the quality of other areas in the river. Other individuals maintain approximately the same home range year after year, even when the habitat structure changes.

Home range size often varies among sexes and with reproductive state. Home ranges of male *Sceloporus jarrovi* are twice the size of female home ranges and increase in average size as the fall breeding season commences. In contrast, female home range size remains the same (Fig. 8.2). The increase in male home range size is due partially to a 50% reduction in male density from summer to fall and an increase in the proportion of the home range defended by reproductive males. By the peak

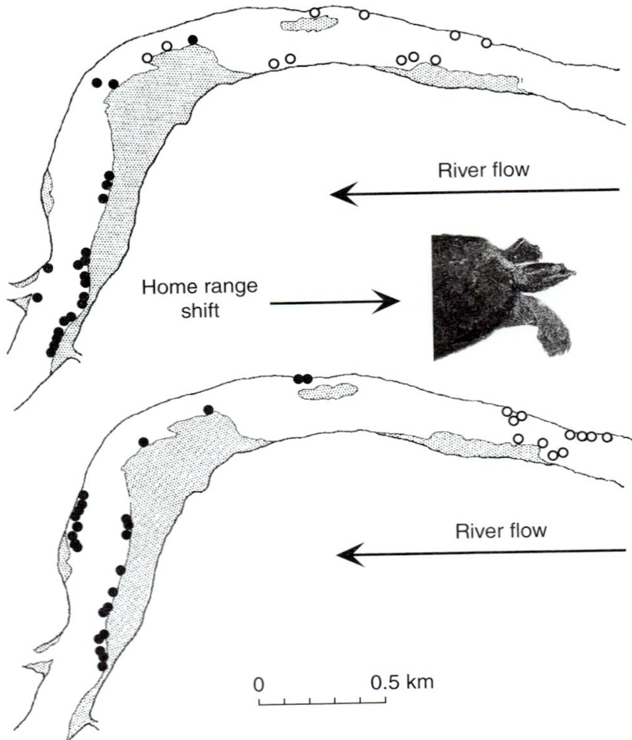

FIGURE 8.1 Shifts in the home ranges of two female *Apalone muticus* in the Kansas River. Two time periods are represented: Closed circles represent early sightings during summer, and open circles represent sighting approximately 1–2 months later (time periods are not the same for each turtle). The upper panel is a subadult that shifted its home range 1363 m upstream. The lower panel is an adult female that shifted its home range 1534 m upstream. Because the turtles are aquatic and live in rivers and streams, their distribution is linear. Adapted from Plummer and Shirer, 1975.

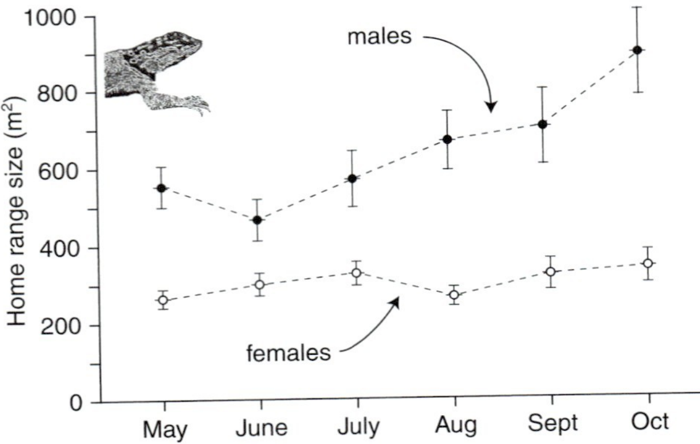

FIGURE 8.2 Seasonal variation in home range size for male and female *Sceloporus jarrovi*. Breeding occurs in fall, at which time male home ranges increase in size. Adapted from Ruby, 1978.

of the breeding season, males defend the entire home range; during this time, the home range and territory are the same.

In most species, home range size generally decreases as food availability or density increases (Fig. 8.3). In at

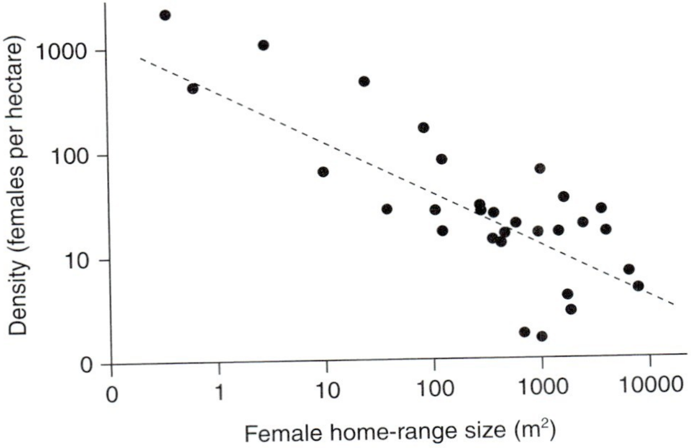

FIGURE 8.3 As female density increases, home range size decreases for most amphibians and reptiles, as shown here for territorial and nonterritorial female lizards. Adapted from Stamps, 1983.

least one instance, the local climate places constraints on lizard activity that feed back on the amount of space used by individuals. Based on long-term capture–recapture studies, Arthur Dunham and his collaborators were able to demonstrate that variation in home range size results from complex interactions between resource availability, microclimate, and physical structure of the habitat. As in *Sceloporus jarrovi*, home ranges of male *Sceloporus merriami* are larger than those of females, but geographically close populations vary greatly in home range size (Fig. 8.4). This lizard occurs across an elevational gradient in the Chisos Mountains of west Texas. Populations at higher elevations experience a much more mesic environment than those at low elevations. Males and females at the lowest elevations at Boquillas have much smaller home ranges than individuals at higher elevations, even though food availability is lowest and lizard density is highest at Boquillas. Although it appears paradoxical that lizard density could be high with low food availability, an interaction between reproductive, microhabitat, and energetic requirements accounts for the small home ranges. The environment at Boquillas is the most extreme (high temperatures, low rainfall) along the elevational gradient, and as a result, the amount of time available to each lizard for activity is reduced. Feeding rates of Boquillas lizards are low, suggesting that energy is more limited compared with higher-elevation populations. The high temperature also limits activity, and with food already in short supply, the lizards further limit their activity, which reduces home range size. The reduced activity coupled with low food availability ultimately feeds back on allocation of energy for reproduction and results in lower reproductive output. *Sceloporus merriami* is a sit-and-wait predator. In contrast, lizards that actively search for prey would be expected to have large home ranges. Actively foraging lizards, such as *Aspidoscelis* and *Cnemidophorus*, have relatively large home ranges throughout which they search for prey.

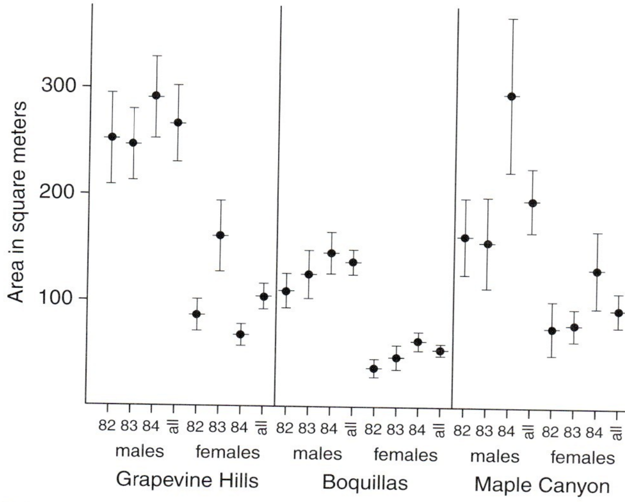

FIGURE 8.4 Home range size in *Sceloporus merriami* varies between sexes, among years, and among three different sites in the Chisos Mountains of west Texas. Boquillas, the site with the most extreme (hot and dry) environment, imposes thermal constraints on lizard activity, resulting in small home ranges. Adapted from Ruby and Dunham, 1987.

The Australian elapid, *Hoplocephalus bungaroides*, centers its home range around retreat sites in rocky outcrops and tree hollows and remains inactive most of the time. Male home ranges of *H. bungaroides* overlap very little during the breeding season, but home ranges of females are often within the home ranges of males. Females carrying eggs move less than do nonreproductive females or males and as a result have smaller home ranges. Home range size in males and females varies among years, apparently in response to the relative abundance of their mammalian prey.

Aquatic environments offer special challenges in terms of space use for amphibians and reptiles, not only because of their three-dimensional nature, but also because they fluctuate depending on rainfall or drought. Aquatic snakes and turtles often have relatively large home ranges, and their home ranges can change seasonally. During particularly dry years, their entire area of activity can shift if a pond or stream dries. Most leave their home ranges for brief periods to deposit eggs. Surprisingly, one of the larger aquatic (marine) turtles, *Chelonia mydas*, has one of the smallest home ranges once they settle in an area to feed. These turtles create a submarine pasture and focus their grazing in that small area. In contrast, another sea turtle, *Dermochelys coriaceae,* appears to move constantly, tracking the seasonal blooms of its jellyfish prey.

The Sonoran mud turtle (*Kinosternon sonoriense*) lives in rivers, streams, and man-made impoundments in the Sonoran Desert of Arizona and northern Mexico and, even though abundant in many areas, is often missed by people observing wildlife. These turtles spend much of their time under rocks or other objects under water in their habitats and tend to move at night. When streams dry during droughts or seasonally, the turtles often aestivate underground in terrestrial habitats, and they are able to withstand water deprivation for extended time periods. A long-term capture–recapture study on these turtles revealed that males move farther than females and that adults in general move farther than juveniles. Although average distance that males and juveniles moved was not associated with body size, distance moved varied with body size in females, but in a curvilinear manner. The largest (and presumably oldest) females moved less than did moderate-sized adult females but considerably more than small adult females (Fig. 8.5). Movements of these turtles occur within pools, between pools in a particular complex of pools within a stream, between complexes, and even between drainages, although the latter is a rare event. A vast majority of Sonoran mud turtle activity occurs within a single pool or its associated pool complex. Because most mud turtle species live in streams, home range length serves as a good metric for comparisons among species. Among the few mud turtles for which data exist, Sonoran mud turtles have the longest home ranges (298 m for adult males, 104 m for adult females). On average, Sonoran mud turtles also move greater distances than most other mud turtles, but not *K. flavescens*. Both *K. sonoriense* and *K. flavescens* live in habitats that experience seasonal drying, and as a result, their relatively long-distance movement patterns may be associated with finding water. Similar to many aquatic amphibians and reptiles, not only is the specific aquatic habitat of mud turtles critical to sustain natural populations, but associated terrestrial habitats and access to other aquatic habitats are as well.

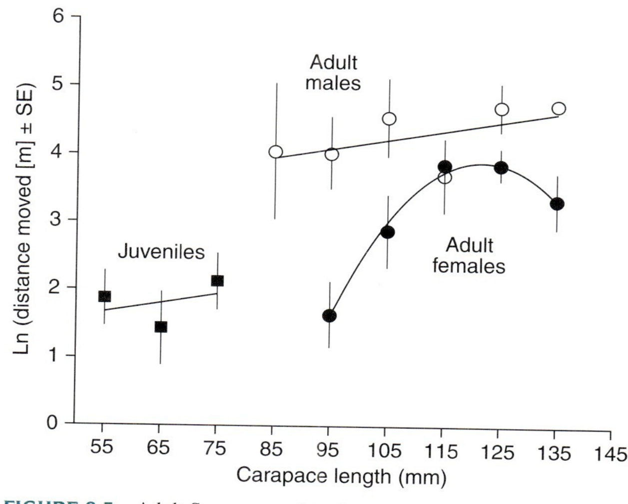

FIGURE 8.5 Adult Sonoran mud turtles move more than juveniles, and for adult males and juveniles, distance moved does not increase much with turtle size. However, in adult females, small and very large females move less than females of moderate size. Note that the *y*-axis is natural-log transformed. Adapted from Hall and Seidl, 2007.

A few other patterns of space use occur in ambush-foraging species that do not fit the typical home range model because of regular long-distance shifts in primary foraging sites. Individual prairie rattlesnakes, *Crotalus viridis*, wander until they locate an area of high prey density. They remain in that area until prey density reaches some lower threshold and prey capture becomes infrequent, after which they move to a new site. Likewise, water snakes, *Nerodia sipedon*, appear not to have traditional home ranges. Because home range size continues to increase with the number of times an individual is captured, use of space appears to consist of a series of activity centers that shift spatially. Similar use of space has been observed in other snakes.

Water pythons, *Liasis fuscus*, migrate seasonally to follow their prey, dusky rats (*Rattus colletti*), which shift their dry season distribution from soil crevices in the backswamp areas in the Northern Territory, Australia, to levee banks up to 12 km away during the wet season when the floodplain is inundated (Fig. 8.6). At the end of the wet season, the snakes return to the floodplain, even though rat density remains high on the levee. Adult male rats, which reach a larger size than females, are more abundant and reach larger size on the floodplain due to higher levels of moisture and nutrients, and these are preferred. The snakes shift their seasonal activity to coincide with

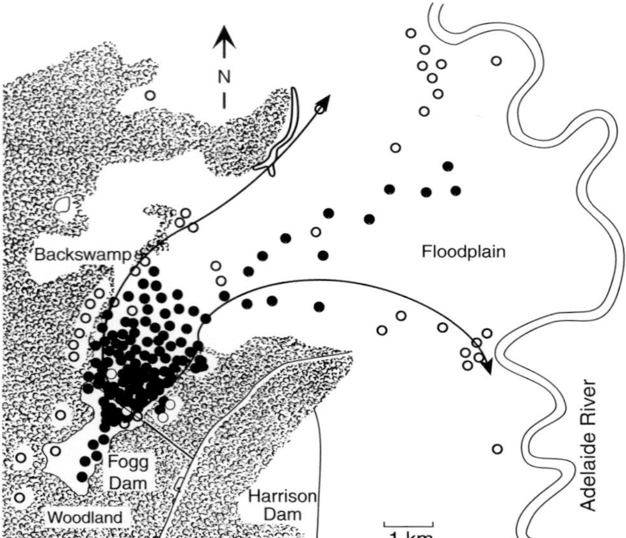

FIGURE 8.6 Locations and movements of water pythons (*Liasis fuscus*) in the Northern Territory of Australia. Solid circles indicate positions of snakes during dry season when the floodplain is dry and the backswamp contains deep crevices; open circles indicate positions of snakes during wet season when the floodplain is wet and the backswamp crevices are closed. Snakes move to high ground in wet season because rats become rare in low areas. Snakes move to the backswamp and dam during dry season because rats there are larger. Arrows show movement patterns for two radio-tracked individuals (one male and one female) showing that individuals move long distances. Adapted from Madsen and Shine, 1996.

greatest abundance of their preferred prey. The bushmaster, *Lachesis muta*, moves to microhabitats where prey capture is likely, such as along the edge of a fallen log or along trails. The snake typically remains in one spot for several weeks, rarely changing position except to raise the head at night while "searching" for passing prey. After a meal, the snake remains at the site 2–4 more weeks digesting the prey and then seeks out a new foraging site. It remains a mystery whether some sort of large, circumscribed area is involved or whether bushmasters simply move along a nonrepeating track.

The most obvious examples of age-specific differences in space use can be found in species with complex life cycles. Many larval amphibians live in aquatic environments and the adults live in terrestrial environments, so little overlap in larval and adult use of space is expected. Adults of many amphibians with complex life histories have home ranges, but whether larvae have home ranges is unclear. In arboreal lizards, juveniles use different perches than adults or disperse in response to population density. Hatchlings of the Neotropical lizard *Anolis aeneus* prefer perches averaging 1.35 cm in diameter, whereas adult females and males prefer much larger perches (8.5 and 38.6 cm diameter, respectively). Hatchling perches are closer to the ground (14.4 cm on average) than those of adult females and males (50.6 and 169.7 cm, respectively). Home range size is also much smaller for hatchlings. Because of ontogenetic differences in perch characteristics, hatchlings occupy different microhabitats than adults.

Nevertheless, some examples of age-related variation in home range size exist in reptiles, and as more studies are done, additional examples will be documented. Females of Australian sleepy lizards, *Tiliqua rugosa*, give birth within their home ranges. During spring of their first year of life, juveniles maintain home ranges that overlap much more with the home range of their mother than with home ranges of adjacent adults, even though no parental care occurs. Juvenile home ranges are about 60% of the size of home ranges of females, and juveniles move less often and for shorter distances than adult males or females. Adult males have home ranges that average about 20% larger than those of adult females.

Gopher tortoises (*Gopherus polyphemus*) were once abundant animals across the Coastal Plain of the southern United States. Harvesting of these animals for food during the last 2 centuries and, more recently, rapid loss of habitat, have resulted in drastically reduced numbers of these large tortoises. Much of the time, these conspicuous animals remain inside deep burrows in sandy soils, making relatively short forays to forage, find mates, and occasionally construct new burrows. Home ranges of adult, subadult, and juvenile gopher tortoises vary considerably, depending on sex, duration of study, and the number of movements made by individual tortoises. Home ranges vary from as small as 0.002 hectares to as large as 5.3 hectares, with most in the range of about 0.01–2.5 hectares. Using a combination of techniques (thread-spools, transmitters, and permanent marks), David Pike studied the home range and movements of hatchlings on the Atlantic Coast of central Florida. Dispersal from the nests was random with respect to direction. Hatchlings moved considerably following hatching (late summer), very little during their first winter—a time period in which yolk reserves provide most hatchling nourishment—and then resumed movements the following spring and summer. Burrow construction followed a similar pattern. Even though the number of moves was high following hatching, distance moved was low and remained low until the following spring and early summer. Home range size (minimum convex polygon) of hatchling tortoises that were radio-tracked varied from 0.014–4.81 hectares (average 1.95 ± 2.12 hectares). The considerable distances that hatchlings disperse, especially after their yolk is depleted, likely allows hatchlings to develop a spatial map of their environment and has the added benefit of moving individuals away from the nest site, spreading them out and possibly reducing predation. By the time each hatchling cohort reaches sexual maturity, dispersion of individuals would also reduce inbreeding. A key element of this study is that it brings movement patterns of hatchlings into an overall view of the ecology of gopher tortoises. As we have seen with many other animal species impacted by human activities (e.g., sea turtles), understanding ecology of hatchlings is critical to developing species-management strategies.

Territories

A territory is the area within a home range that is actively defended against intruders, usually because the area includes a defendable resource or has some other quality that is better than adjacent areas. Defense results in exclusive use of the territory by the resident. In amphibians and reptiles, when territoriality occurs, males are most often territorial and females are not. In a few species, females defend a territory as well. Most often, territories defended by males contain females whose home ranges are included within the male territory. Because territoriality allows an individual to maintain control over resources, it involves competition among individuals within species for resources that ultimately contribute to individual fitness. Natural selection favors those individuals that control and use resources in a way that positively influences their reproductive success. Discovering the connection between resource control and reproductive success is seldom easy. Every aspect of territorial behavior has costs, and obviously, the gains associated with territoriality must outweigh the costs if territorial behavior is to be maintained through time.

Imagine two individual males in a population, one that defends good places to forage from other males but allows females into those areas and breeds with them. The other male controls no resources and as a consequence does not attract females. However, he can easily find enough food to keep himself healthy by moving around. The territorial male, as the result of his territory defense behavior, might, hypothetically, be more vulnerable to predation. Nevertheless, he has many more opportunities for mating than the other male. He actually may not live very long, but long enough to reproduce, so that when predators kill him, he will have left offspring. In the meantime, the nonterritorial male remains healthy and lives a long life. Representation of his genes ends in that generation, whereas territorial genes (even with the risk attached) are passed on to the next generation. Alternatively, the long-lived, healthy male could replace a territorial male that was eliminated, shift his behavior to territorial, and achieve a high reproductive success. In this scenario, both types of male reproductive strategies are maintained in the population.

Of course, social systems and the evolution of social systems are not this simple—for example, a nonterritorial male might be able to sneak a few matings with females living within the home range of territorial males. Thus, nonterritorial genes can be passed on but at a lower frequency than territorial ones. Territoriality generally is linked with mate choice and other aspects of social systems (see Chapter 9).

Given the preceding, a territory can be defined explicitly as any defended area that meets the following three conditions: It is a fixed area, it is defended with behavioral acts that cause escape or avoidance by intruders, and these behavioral acts result in exclusive use of the area by the resident territory holder.

Territoriality is well known in some frogs and salamanders but unknown in caecilians. Because caecilians are extremely cryptozoic, territoriality may exist but be undocumented. The observations that sexual dimorphism exists in head size and is not related to sexual differences in prey, and the existence of bite marks on some individuals suggest that territoriality could occur in some caecilians. In frogs, acoustic signals serve as avoidance displays, and outright aggression can occur in threat displays. Territoriality occurs most often in frogs with extended breeding seasons and is rare or does not occur in explosive breeders or species with very restricted breeding seasons. It also occurs in frogs with extended parental care (e.g., dendrobatid frogs). In bullfrogs, males establish territories that contain good oviposition sites, which they defend with threats, displays, or wrestling matches. Large males win a majority of contests with other males, indicating that male size determines dominance. Good oviposition sites have high embryo survival. The two primary sources of mortality, developmental abnormalities and leech predation, are reduced at sites with cooler temperatures (< 32°C) and the appropriate vegetation structure to reduce leech predation on the eggs and embryos. Females are attracted to territories with a potential for low egg mortality, and because large males control these territories, they mate with more females. In this situation, the resource base for territories is high-quality egg deposition sites, and the payoff is increased reproductive success for defenders of these sites. Sneak or satellite males that are not territory holders occasionally intercept females and mate with them.

Most data on salamander territoriality is based on studies of a single clade, *Plethodon*, which is composed largely of terrestrial species. *Plethodon cinereus* marks territories with chemicals (pheromones). In the laboratory, adult male and female *P. cinereus* show "dear enemy" recognition, in which they are less aggressive toward recognized enemies than they are toward unfamiliar intruders. Evolutionarily, this reduces energy spent in continual high-level encounters with close neighbors that are unlikely to go away but will maintain distance if reminded that a territorial holder is in place. Combat, often directed at the tail, can occur, and tails can be lost as a result of encounters. Tails are important energy stores for reproduction; consequently, the loss of a tail negatively affects reproductive success. Bites during combat are also directed at the nasolabial grooves, which are important transmitters of chemical signals.

The Central American dendrobatid frog *Oophaga pumilio* lives in leaf litter on the forest floor. Males maintain territories that they aggressively defend from other males. Males call from tree bases or fallen logs, and the distribution of these structures determines inter-male distance to a large extent. Many males remain in restricted areas over long time periods and, when displaced experimentally, return to their territories. Females deposit eggs in terrestrial oviposition sites, and males use elevated perches for calling; the location of male territories must include resources.

Food available to individual animals varies both temporally and spatially and can influence space use. Males and females of the montane lizard *Sceloporus jarrovi* defend territories against conspecifics of the same size or sex. Territories that contain relatively more food tend to be smaller than territories with less food, independent of the differences in territory size associated with lizard body size. Adding food to the territory of *S. jarrovi* results in a shift in space use; the site where food is added becomes the center of the territory. In this instance, food availability appears to determine the location of the territory. In the Western fence lizard, *Sceloporus occidentalis*, home ranges of males overlap considerably, from 28–67% of space being used by at least one additional male and often several males. However, territory overlap is much smaller, ranging from 14–52%, and in most cases, overlap occurs with only a single male. Nevertheless, aggressive interactions between males are rare, even though males frequently perform push-up displays. Males remain in the same territories year after year. Most likely, males establish territories early in life, remain in those territories, and use social signals to remind neighbors that they still inhabit the territory. Because aggression is energetically expensive and potentially risky, it may not occur in *S. occidentalis*, or at least it may occur only rarely. This example shows that definitions of *home range* and *territory* are often not as clear as we might like. Not only are territories of *S. occidentalis* not defended aggressively and regularly, but males allow at least one other male to overlap in some part of their territory. Descriptors like *home range* and *territory* are conceptually useful, but it is important to keep in mind that they are just descriptors, and detailed observations frequently reveal that patterns of space use are intrinsically more complicated.

Evolution of Territoriality

Studies conducted on use of space by individuals within and between species reveal considerable variation in the proportions of home ranges that are defended. Some species defend the entire home range, others defend specific sites within the home range, and others do not appear to defend any part of the home range. Males of many species without territories aggressively attack other males that approach females either within the male's home range or while the resident is courting the female (see Chapter 9). Although adaptive scenarios can be devised to explain territorial defense in nearly every amphibian or reptile, similarities in behavior among closely related species often reflect common ancestry; individuals of many species behave the way their ancestors did.

A close examination of defense behavior in lizards suggests that evolutionary history determines a large part of behavioral patterns. Among studied lizards, defense is accomplished by direct combat, threats, or simple avoidance. Combat involves biting, wrestling, or any behaviors involving physical contact between two individuals. Threat refers to aggressive communication in which no physical contact is made. Threats most often involve push-up displays, throat expansion, or high-intensity erection and contraction of the dewlap. Avoidance defense is based on indirect displays such as chemical signals. Push-up displays are presented from a distance where the primary goal is to assert presence. Other examples of avoidance displays exist as well. The size of the area defended can range from all or part of the home range to none of it (Table 8.2). An examination of the distribution of home range defense on a lizard phylogeny shows that major shifts have occurred during the evolutionary history of lizards in the proportion of the home range defended (Fig. 8.7). This phylogenetic analysis shows that territoriality (defense of all or part of

TABLE 8.2 Ten Behavioral Categories for Lizards Based on Aggressive Defense of Resources.

	Defense area		
Defense style	All or part of home range	Specific site (Basking, Shelter)	No area (Self)
Combat	Type I	Type IV	Type VII
Threat	Type II	Type V	Type VIII
Avoidance	Type III	Type VI	Type IX
Type X - Affiliative aggregations or random distribution of animals			

Note: Each category is defined by the intersection of defense style and defense areas.
Source: Adapted from Martins, 1994.

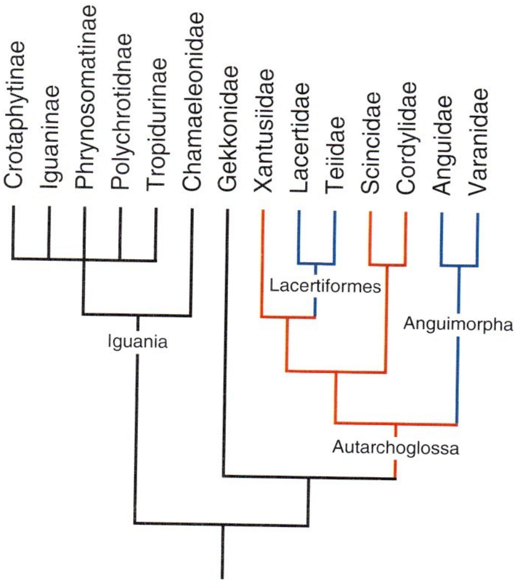

FIGURE 8.7 Phylogeny for lizards showing the evolutionary distribution of home range defense. The ancestor of all lizards presumably defended the entire home range with an overall reduction in area defended as lizards diversified, and this behavior is carried through in clades indicated by black. Site defense (clades in red) evolved in the ancestor to Autarchoglossans. A lack of home range or site defense evolved independently twice, in the ancestor to the Lacertiformes and in the ancestor to Anguimorpha. Taxonomy has been revised for consistency, but relationships to behavioral traits remain unchanged. Adapted from Martins, 1994.

the home range) is ancestral to all lizards and that adaptive scenarios are not necessary to explain territoriality in the Iguania and Gekkonidae. The loss of territoriality within the scleroglossans (particularly the Teiidae, Lacertidae, Anguidae, and Varanidae) most likely reflect the consequences of a switch from a sit-and-wait foraging mode to an active- or wide-foraging mode (see Chapter 10). We point out that these conclusions rest on the assumption that the phylogeny on which this analysis was done reflects the evolutionary history of squamate reptiles. If recent nuclear-gene-based relationships of squamate higher taxa prove to better reflect evolutionary history, then the sequence of events leading to territoriality among lizard clades will need to be reconsidered (see Chapter 20).

Other Patterns of Space Use

Many amphibians and reptiles brood or guard nests, and remain near the eggs until the eggs hatch (Fig. 8.8). The space the brooding parent uses is much smaller than the home range and is not necessarily within the home range used during the nonbrooding season. Females of the four-toed salamander, *Hemidactylium scutatum,* brood eggs in clumps of peat moss along slow-moving streams, remaining restricted to the nest for an extended time period. Lungless salamanders in the genus *Plethodon* brood egg clutches in moist areas under rocks and inside of rotting logs (Fig. 5.22). Female broad-headed skinks, *Plestiodon laticeps,* brood clutches of eggs in partially decomposed pockets within hardwood logs, rarely leaving until after the eggs hatch, and other *Plestiodon* species brood their eggs in a variety of relatively sealed chambers inside of logs, under surface objects, or in the ground (Fig. 5.16).

Aggregations occur in a wide variety of amphibians and reptiles for a number of reasons (Table 8.3). All aggregations represent nonrandom use of space, and most often are centered on scarce resources. For amphibians, the most obvious examples are aggregations of adults in ponds or other bodies of water during breeding events. Spadefoots arrive by the thousands to breed in temporary ponds, as do many other explosive-breeding frogs. Large numbers of *Physalaemus* (Leiuperidae) enter ponds that form during the early wet season in seasonally wet open areas in South America, yet locating a single individual during the dry season is difficult.

Tadpoles of a variety of anuran species form dense "schools" that move about in ponds, presumably to offset predation. In some, such as *Hypsiboas geographicus* (Hylidae), *Leptodactylus ocellatus* (Leptodactylidae), and *Lithobates heckscheri* (Ranidae) not only are the schools huge, but the tadpoles are large as well, often exceeding 60 mm in total length. Consequently, the schools appear as huge dark masses in the ponds where they occur. Schooling behavior has evolved independently many times in anurans.

A variety of species of salamanders, including *Plethodon glutinosus, Ambystoma macrodactylum,* and *Ambystoma*

FIGURE 8.8 Amphibians and reptiles usually remain in one place while brooding or attending eggs. The ceratobatrachid frog *Platymantis* (undescribed species) broods its eggs on leaves, whereas the microhylid frog *Oreophryne* (undescribed species) broods its eggs inside of hollows in branches. Photographs by Stephen J. Richards.

TABLE 8.3 Examples of Social, Nonreproductive Aggregations of Amphibians and Reptiles.

Taxon	Purpose
Salamanders, mixed, seven species	Hibernation
Plethodon glutinosus	Estivation
Salamandra salamandra	Hibernation
Rhinella tadpoles	Schooling
Hyla meridionalis, Pelodytes, Triturus, Podarcis	Hibernation
Limnodynastes juveniles	Water conservation
Xenopus laevis tadpoles	Schooling
Terrapene ornata	Hibernation
Terrapene ornata and *Kinosternon flavescens*	Hibernation
Crocodylian hatchlings	Reduce predation
Alligator mississippiensis	Feeding
Amblyrhynchus cristatus	Sleeping
Diadophis punctatus	Water conservation (?)
Pelamis platurus	Feeding
Storeria dekayi	Water conservation (?) and hibernation
Thamnophis (three species), three other snake genera, *Ambystoma*, and *Pseudacris*	Hibernation
Coleonyx variegatus	Water conservation
Typhlops richardi	Water conservation (?)

Sources: Amphibians—*S, Bell,* 1955; *Pg,* Humphries, 1956; Ss, Lescure, 1968; *R., Wassersug,* 1973; Hm, Van den Elzen, 1975; L, Johnson, 1969; Xi, Wassersug, 1973;. Reptiles—*To, Carpenter,* 1957; C and Am, *Lang,* 1989; Ac, Boersma, 1982; *Dp, Dundee and Miller,* 1968; *Pp, Kropach,* 1971; *Sd, Noble and Clausen,* 1936; T, Carpenter, 1953; Cv, Lancaster et al., 2006; Tr, Thomas, 1965.

tigrinum, aggregate in damp retreats when the terrestrial environment becomes excessively dry. In the proteiid salamander, *Proteus anguinus*, individuals aggregate in shelters in the caves in which they live, usually under stones or in crevices. Experiments show that homing in on a retreat is accomplished by use of chemical cues, which provide directional information to the salamanders as well as functioning in social behavior. Chemical cues also appear to attract other individuals, resulting in several individuals sharing shelters. Garter snakes aggregate in large numbers for both overwintering and mating, and rattlesnakes aggregate in large numbers in high latitudes and at high elevations to overwinter in dens. In fall, the lizard *Sceloporus jarrovi* aggregates along crevices

FIGURE 8.9 A basking aggregation of marine iguanas, *Amblyrhynchus cristatus*. (K. Miyata)

in mountains of southeastern Arizona to overwinter. They frequently bask in sun along the crevices to gain heat, even though they are territorial during the activity season. Snakes and lizards aggregate at talus slopes in northern Oregon because these areas are the best available nesting sites. Fifty-one lizard eggs, 294 snake eggs, and 76 snakes were found in a patch of talus within an area of 150 square feet. In tropical South America, aggregations of frogs can be found inside and under termite nests during the dry season. In addition to frogs, these aggregations often include snakes, lizards, and arthropods. The termite nests offer an environment where temperature and humidity are moderated. These are but a few examples of aggregations in amphibians and reptiles (Fig. 8.9).

MOVEMENTS, HOMING, AND MIGRATIONS

Most amphibians and reptiles move relatively little during their entire lifetime except when they are breeding. Individual box turtles, *Terrapene c. carolina*, in Maryland, for example, moved very little over 30 years or more and remained in the same home range; similar observations have been made on many other species. Individuals move to forage or change foraging positions, pursue mates, defend territories, deposit eggs, or escape predators. Most of these movements take place within the individual's home range. The benefits of moving are offset by the costs of moving (usually energy or risk of mortality). For species with cryptic morphology or coloration, moving upsets crypsis and can accrue a survival cost. Active or wide-foraging species tend to move considerably more and expend more energy, doing so within their home ranges, than do species that use the sit-and-wait foraging mode. Their alert behavior and rapid response to predators offset the cost of exposure.

Both extrinsic and intrinsic factors influence movements of amphibians and reptiles (Table 8.4). Herpetologists rapidly learn to take advantage of environmentally induced patterns of movements; amphibians, in particular, can be collected or observed in great numbers on rainy nights during spring in temperate zones and on the first rainy nights during tropical wet seasons. Rattlesnakes (particularly *Crotalus viridis* and *C. oreganus*) occur in large numbers when they aggregate for overwintering. Long-term studies on slider turtles have identified factors that cause movements in turtles (Table 8.5). These factors likely apply to most species of amphibians and reptiles. Movements outside the home range carry additional risks compared with movements within the home range, largely because traveling occurs in areas with which the individual has little or no familiarity. When these movements occur, they usually are related to breeding, finding food or water no longer available in the home range, or overwintering, or such movements are in response to catastrophes (e.g., flooding).

The most apparent dichotomy in movement patterns on a daily basis is diurnal versus nocturnal movement. Most salamanders and frogs are nocturnal, but some species such as cricket frogs (*Acris*) and striped pond frogs (*Pseudis limellum*) are both diurnal and nocturnal.

TABLE 8.4 Factors That Influence Movements of Individual Amphibians and Reptiles.

Environmental	Population	Individual
Daily temperature patterns	Density	Sex
Seasonal temperature patterns	Sex ratio	Body size
Humidity/rainfall	Age structure	Age
Habitat type or condition	Size structure	Physiological condition
Catastrophic events	Disease/ parasitism	Reproductive state
		Recent experience

Source: Adapted from Gibbons et al., 1990.

TABLE 8.5 Causes and Consequences of Movements at the Intrapopulation and Interpopulation Level for Turtles.

Category	Purpose	Primary benefits gained by moving
Intrapopulational (short-range)	Feeding	Growth; lipid storage
	Basking	Increased mobility due to body temperature increase; reduction of external parasites; enhanced digestion
	Courtship and mating (adults only)	Reproductive success
	Hiding, dormancy	Escape from predators or environmental extremes
Interpopulational (long-range)	Seasonal	
	Seeking food resources	Growth; lipid storage
	Nesting (adult females)	Direct increase in fitness
	Mate seeking (adult males)	Direct increase in fitness
	Migration (hibernation, aestivation)	Survival
	Travel from nest by juveniles	Initiation of growth
	Departure from unsuitable habitat	Survival

Adapted from Gibbons et al., 1990.

Movements of winter-breeding amphibians often occur during day and at night. The absence of daylight appears to trigger mass movements in *Pseudacris crucifer*, *P. ornata*, *P. nigrita*, and *Lithobates sphenocephala*, and both temperature and moisture determine the specific nights on which breeding will occur. On nights with low temperatures or no rainfall, breeding migrations do not occur. The risk of movement during daytime for these frogs may be tied to diurnal predators like birds. Dendrobatid frogs are diurnal and sleep at night, often perched within 0.5 m of the ground on leaves of small plants. Brightly colored species (e.g., *Dendrobates*, *Phyllobates*) offset predation by having noxious or poisonous skin secretions and advertising their toxicity with aposematic coloration, whereas other species (e.g., the closely related aromobatid frogs such as *Allobates*) offset diurnal predation by cryptic coloration and behavior (see Chapter 11, "Chemical Defense").

Depending on species, turtles can be diurnal or nocturnal. Box turtles (*Terrapene*) and tortoises (*Geochelone*) are strictly diurnal, as are many aquatic turtles (e.g., *Apalone*, *Graptemys*). Some species, like *Chelydra serpentina*, appear to be active both during the day and at night. Crocodylians are active during both day and night, but much of their diurnal activity involves basking. *Caiman crocodilus* in the Amazon of Brazil, for example, basks on sandy banks of rivers and ponds during the day and actively searches through its aquatic habitat for prey at night. When water floods the forest during the wet season, caimans enter the flooded forest in search of stranded prey. Among lizards, most are diurnal (e.g., all iguanians, teiids, gymnophthalmids), some are nocturnal (e.g., many gekkonids), and some vary their diel activity, at least on the surface, with season (e.g., helodermatids). Among snakes, nearly every possible diel pattern of activity occurs. Most desert snakes are nocturnal, but some, like *Masticophus flagellum*, are strictly diurnal. Likewise, many tropical snakes are nocturnal, but some species, including all species of whipsnakes in the genus *Chironius*, are diurnal (Fig. 8.10).

In the Mojave Desert of southern California, male sidewinders (*Crotalus cerastes*) move an average of 185 m each night while active, whereas nongravid females move only 122 m. Individuals are active on about 60% of the nights during their activity season. Greatest movements of adult males occur during spring and fall mating seasons, which suggests that they are searching for females. Activity ranges of individuals vary from 7.3–61 hectares; males, females, and juveniles have similar activity ranges. Sidewinders appear to move randomly until fall, when their movements are directed toward overwintering sites. Overwintering sites are usually located in rodent burrows at the interface between sand and alluvial habitat patches.

Freshwater turtles leave their aquatic habitats to dig nests, search for mates, overwinter, or locate new aquatic habitats when their original stream or pond dries up. Six turtle species, *Trachemys scripta*, *Kinosternon subrubrum*,

FIGURE 8.10 Some tropical colubrid snakes are diurnal, such as the tropical whipsnake, *Chironius flavolineatus* (left), but most are nocturnal, such as the burrowing snake *Apostolepis bimaculata*. (L. J. Vitt)

Pseudemys floridana, Sternotherus odoratus, Chelydra serpentina, and *Deirochelys reticularia*, are long-time residents of Ellenton Bay, a freshwater pond located approximately 2 miles from the Savannah River in South Carolina. Adults of four other species, *Pseudemys concinna*, *Clemmys gutata*, *Chrysemys picta*, and *Kinosternon bauri*, occasionally enter Ellenton Bay. Juveniles of the latter four species have never been observed at the pond, and with the exception of *P. concinna*, a majority of nonresident turtles were males (100% for *K. bauri* and *C. picta*, 80% for *C. guttata*). Only a single female *P. concinna* has entered the pond. Most of the nonresident turtles are males because long overland movements by males increases their probability of encountering females in other aquatic habitats, whereas females have less to gain by long-distance moves, particularly considering the potential costs of increased risk of predation by terrestrial predators. Overland ventures by *T. scripta* vary from 0.2 to 9 km, resulting in sightings of turtles in ponds other than their home ponds (Fig. 8.11). Many of the turtles return to their home ponds, indicating that these movements are not immigrations.

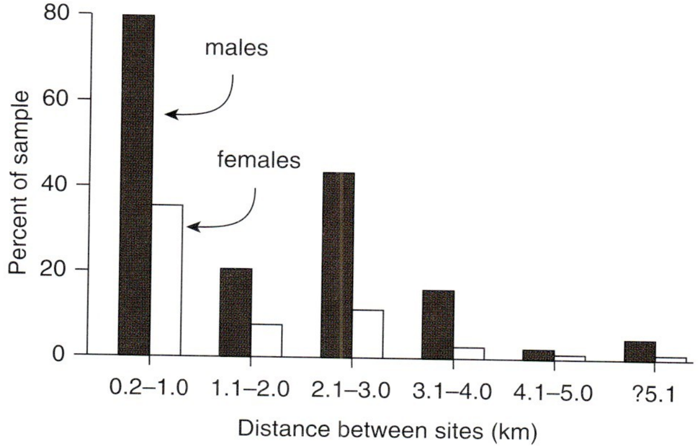

FIGURE 8.11 Long-range movements based on straight-line distances of *Trachemys scripta* between aquatic habitats in South Carolina. Travel between Ellenton Bay and Lost Lake were primarily over land. Exchanges in Par Pond could have been by a shorter overland route or a longer route through water. Adapted from Gibbons et al., 1990.

In Malaysia, the semiaquatic snake *Enhydris plumbae* occurs in water buffalo wallows, slow-moving streams, rice paddies, and a variety of other aquatic habitats. Most individuals move very little, and 44% do not move at all (Fig. 8.12). The snakes are active day and night but are observed on the surface at night. A partial explanation for the low movement in *E. plumbae* is that many occur in small, isolated bodies of water (buffalo wallows), but even those in rice paddies move very little.

Studying movement behavior of salamanders, especially terrestrial species, is logistically difficult. By inserting tiny tantalum-182 tags in the base of the tail of salamanders, individuals can be located in the habitat even though they may be buried in soil or leaf litter. A scintillation system detects radioactivity of the tags from 2 meters away. The technique appears particularly suitable for short-term studies, because the isotope has no apparent effect on salamander physical condition and the tags remain in place for about a month. This technique is not useful for longer time periods because salamanders lose body weight, suffer skin lesions, and often lose the tags after about 40 days. An early study using the technique revealed that some salamanders are capable of orientation and subsequent homing when displaced. Males of *Plethodon jordani* occupy home ranges that are about three times larger than those of females. Salamanders displaced between 22–60 meters from their home ranges return to within 7 meters of their

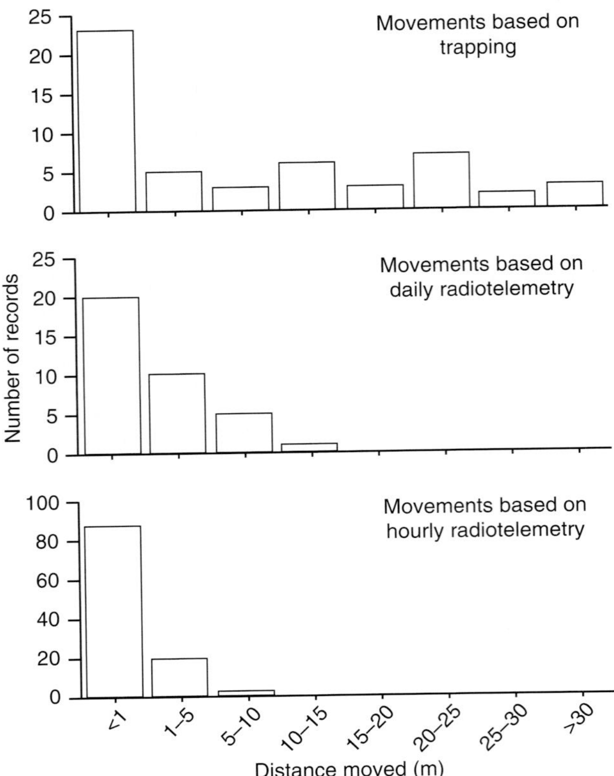

FIGURE 8.12 The snake *Enhydris plumbea* in Malaysia (Borneo) moves very little. The method of collecting movement data influences the results and might lead to misleading conclusions in species that move considerable distances. Adapted from Voris and Karns, 1996.

capture site, which indicates that they are capable of orientation. Because the displaced salamanders climbed up on vegetation, airborne chemical cues were implicated in orientation. Homing studies on *Desmognathus fuscus* in Pennsylvania add support to the hypothesis that chemical cues are involved in homing. These salamanders maintain small home ranges along a stream for extended time periods. Four groups of salamanders were displaced to discover the possible cues used in homing behavior: One group was a normal, nontreated group; the second was an anosmic (olfactory system nonfunctional) group; the third group was blind; the fourth group was a sham-treated control group. The anosmic group did not return to original home ranges, whereas varying numbers of the other treated groups did return, lending support to the hypothesis that chemical cues are involved in the orientation and homing process.

Among the most striking movements by extant amphibians and reptiles are sea turtle migrations from hatching site to feeding grounds as juveniles and, many years later, back to nesting beaches as adults. Green sea turtles, *Chelonia mydas*, emerge from eggs at Tortuguero, Costa Rica, enter the Caribbean Sea, and migrate throughout most of the Caribbean (Fig. 8.13). Their long journeys and ability to return to the beaches where they were hatched suggest a complex navigational system.

Mass Movements

Mass movements occur in some amphibians and reptiles. The use of terrestrial drift fences around amphibian breeding ponds has made it relatively easy to monitor the movements of amphibians, some of which are startling. Ambystomatid salamanders and many frogs, especially those that are explosive breeders, move en masse to and from breeding ponds. Metamorphs leaving breeding ponds often do so en masse as well. During a single year (1970), 2034 individuals of 14 species of frogs moved in or out of one permanent pond, and 3759 individuals of 13 species of frogs moved in or out of another temporary pond in South Carolina. However, the numbers of amphibians migrating into and out of ponds during breeding and metamorphosing events varies considerably among species and years. The salamander *Ambystoma opacum*, for example, did not enter or leave a small pond in South Carolina from 1970 to 1980, but in 1987, nearly 300 adult females entered and more than 800 metamorphs exited the pond. In the same pond over a 12-year period, patterns of movement among species were not concordant (Fig. 8.14). Mass movements of amphibians often result in high mortality caused by automobile traffic. Although a few parks and recreation areas now construct fencing and underground passages for migrating amphibians, the migratory biology of amphibians and other animals is usually not considered when designing roads.

Sea turtles and large freshwater turtles (*Podocnemis*) arrive at nesting beaches by the hundreds over a few nights. Garter snakes and rattlesnakes enter and leave hibernacula in large groups. Thus mass movements are common and generally appear related to breeding events or overwintering. These and the preceding examples largely represent directed and cyclic movements away from the home ranges used during the activity season.

Dispersal

Dispersal is undirected movement to locations unknown by the dispersing animal and commonly refers to juveniles leaving the home ranges of their parents to find a home of their own. Habitat instability, intraspecific competition, and inbreeding depression are considered the primary evolutionary driving forces resulting in dispersal (Fig. 8.15). Whether or not individuals should disperse is based on the relative costs and benefits of doing so. Costs to dispersal include increased predation risk associated with entering unknown and unfamiliar habitats, potential difficulties finding resources (food, shelter), and potentially

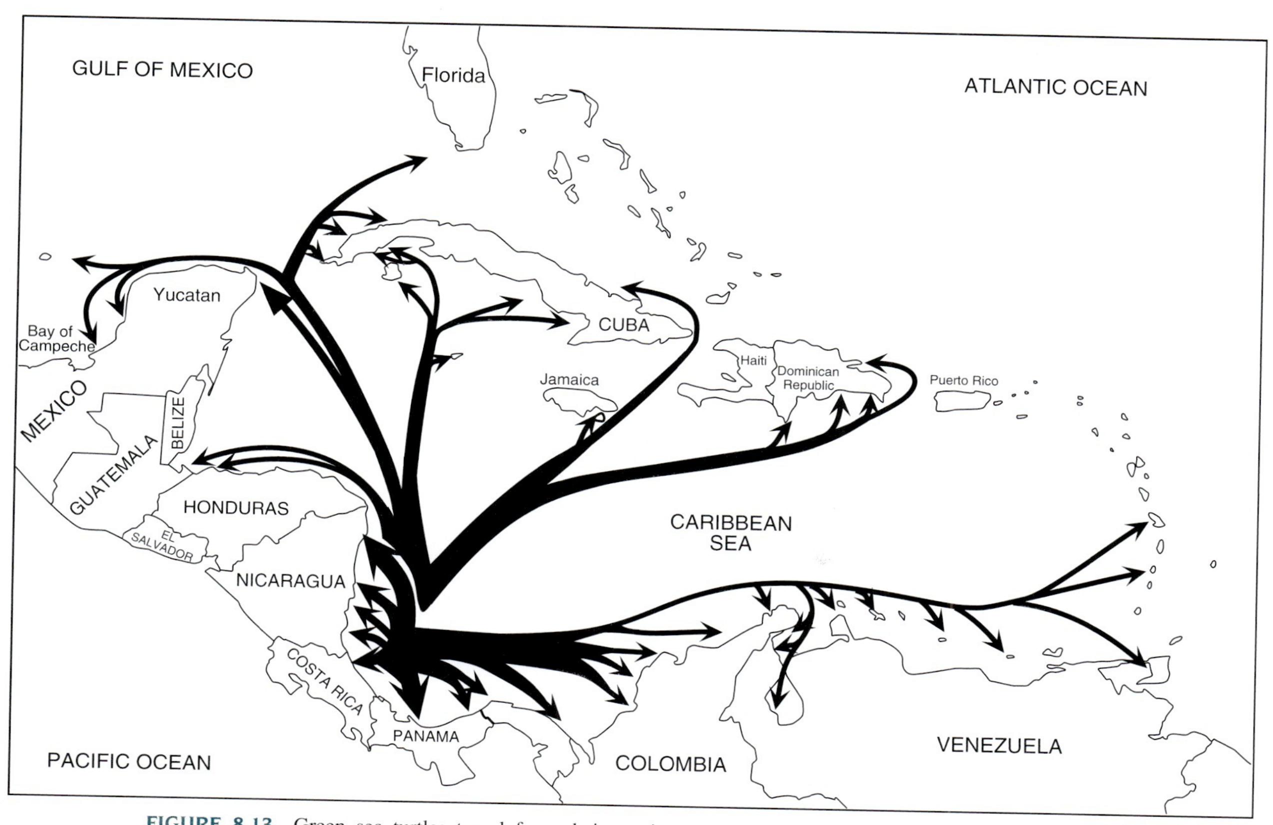

FIGURE 8.13 Green sea turtles travel from their nesting beaches throughout the Caribbean Sea to reach beaches as far north as Cuba. Adapted from Bowen and Avise, 1996.

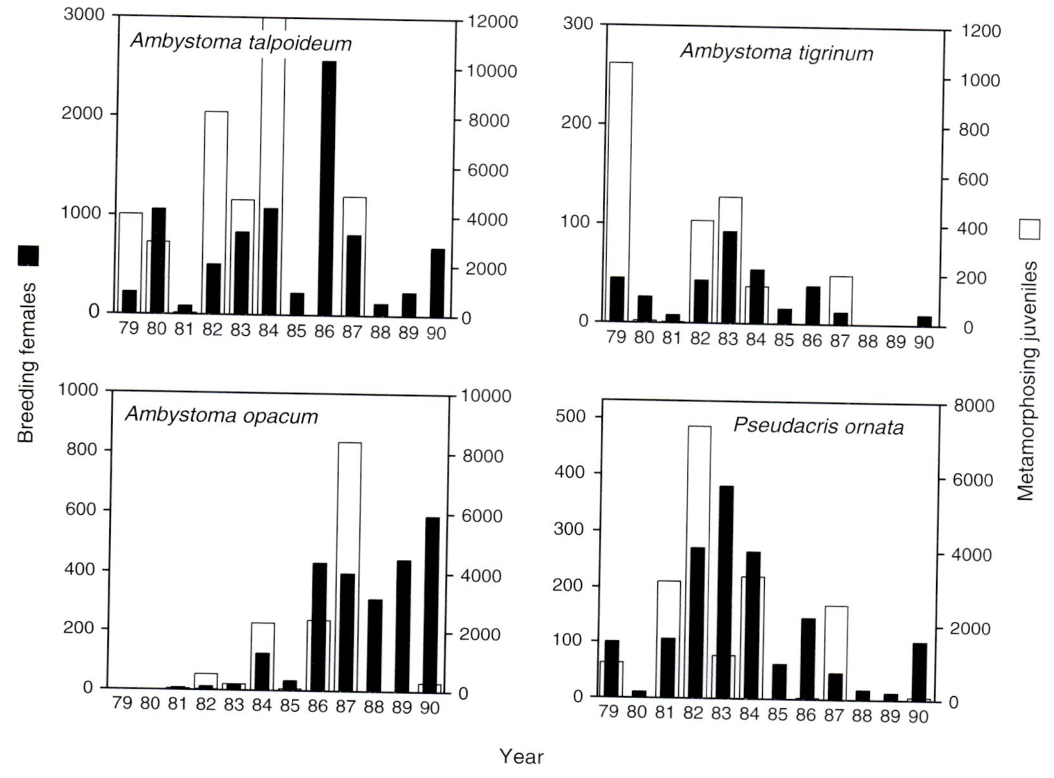

FIGURE 8.14 The number of breeding females and metamorphosing larvae of three salamander species and one frog species varies impressively from year to year in the small Carolina bay, Rainbow Bay, in South Carolina. Migration patterns of amphibians using the same breeding sites are not synchronous. Adapted from Pechmann et al., 1991.

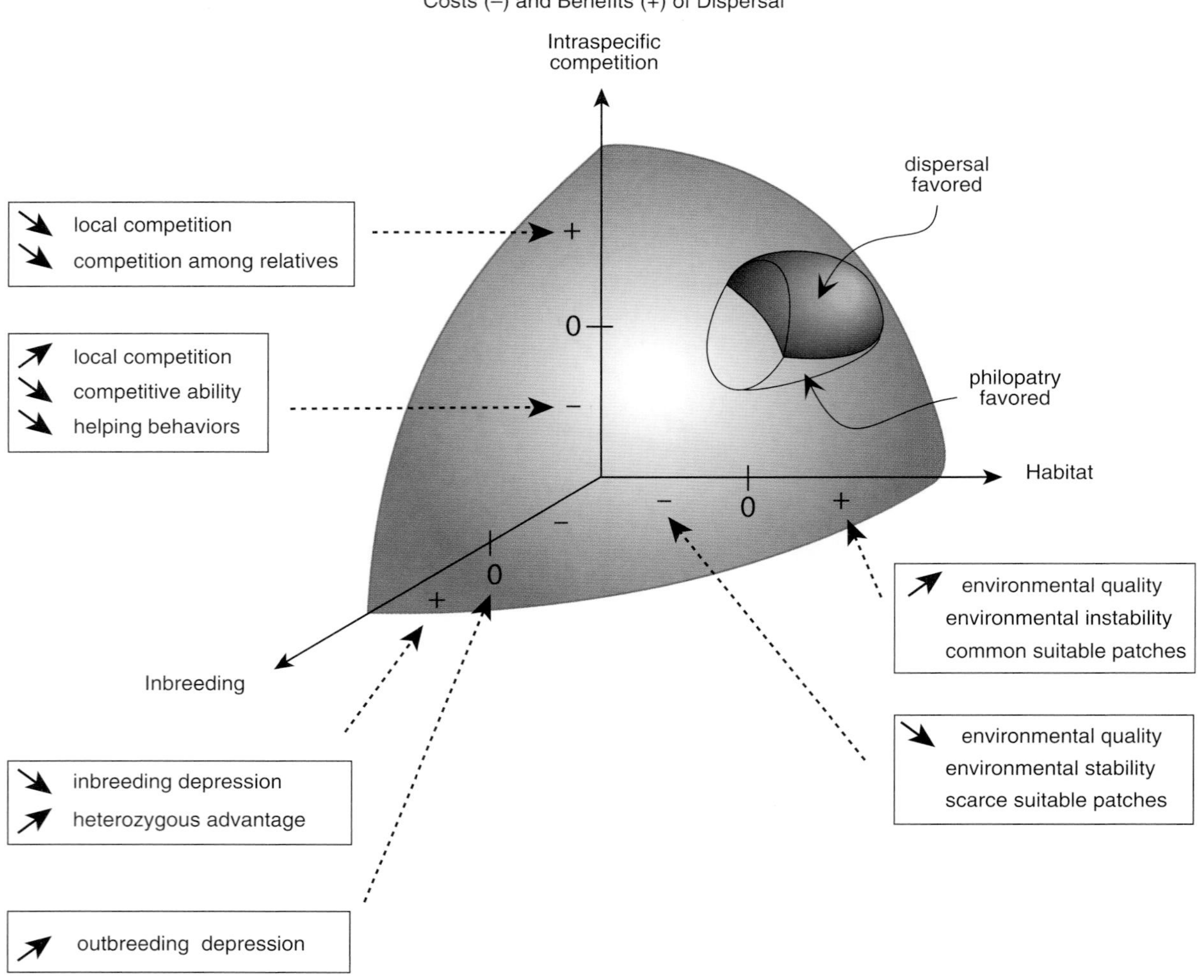

FIGURE 8.15 Model showing the relationships between costs and benefits of dispersal. The curved surface represents points where costs and benefits of dispersal are at equilibrium. Dispersal behavior will be selected above the plane, whereas philopatry will be selected below the plane. The three-dimensional volume represents a species in which some individuals (e.g., juveniles) disperse and others (e.g., adults) remain where they are. Adapted from Clobert et al., 1994.

increased aggression from unfamiliar conspecifics. Benefits include opportunities to discover better resources, increased likelihood of outbreeding, and potentially reduced local competition. In populations of the European lizard *Lacerta vivipara*, more than 50% of juveniles disperse, whereas very low numbers of yearlings or adults disperse. Dispersal of juveniles is greater when population density is high in their population of origin. High population density is an indicator of a temporally high-quality environment. High-quality environments produce offspring that are better able to compete because of relatively larger size and condition. By dispersing, these juveniles offset disadvantages associated with inbreeding. In low-quality environments, only the most competitive juveniles will survive, whether or not they disperse. These survivors will be the individuals with the best set of characteristics for the poor environment. By not dispersing and mating with other individuals that survived and thus carry traits for survival under poor conditions, individuals with traits associated with success in the poor habitat will be favored. Even though inbreeding is potentially high, the inbreeding is selective, and as a consequence, typical costs of inbreeding are relaxed compared with benefits juveniles gain by remaining in their place of origin (philopatry). In this example, a complex interaction between variation in the local environment and the costs and benefits of dispersal with respect to inbreeding determines whether juveniles should or should not disperse.

Amphibian metamorphs and hatchling sea turtles are two examples of cohorts that leave their natal sites but will return in subsequent years to breed. They do not appear to know where they are going as hatchlings, but innate navigational mechanisms will allow them to return later in life.

Metamorphosing amphibian larvae move into and through the habitat of their parents, most becoming part of the local populations. Dispersal distance usually is small, and the juveniles occupy home ranges in vacant spots among adults or in peripheral locations. Similar dispersal occurs in reptiles and direct-developing amphibians, although dispersal can occur later as large juveniles make the transition into the breeding population.

Several species of frogs in unrelated clades transport their eggs, and subsequently either their tadpoles or juveniles, on their backs. Most dendrobatids and aromobatids drop off their litter of tadpoles in one place (Fig. 5.17), but some, such as *Ranitomeya vanzolinii*, drop individual tadpoles in different places (Fig. 9.19). *Sphenophryne cornuta*, a microhylid from Australia, transports its young, dropping them off periodically (Fig. 8.16). Other frogs that carry their young on their backs, such as *Stefania evansi* (Fig. 5.19), may drop off young in different places as well. Although these behaviors are usually considered in the context of parental care and reproductive modes, they certainly play a role in dispersion as well.

Among animal species with polygynous mating systems, males generally disperse farther than females, partly because males compete for females and partly because females often disperse less as the result of their association with resources or refugia from predators. Male *Uta stansburiana* disperse during their first year of life, but in some cases, females disperse equally as far as males. Females appear to disperse until they locate good territories. Some males disperse farther than females because they have to go farther to find unoccupied territories.

FIGURE 8.16 Some amphibians carry their tadpoles or young around and aid in their dispersal. The Australian microhylid, *Sphenophryne cornuta*, drops off its young in different places. Photograph by Stephen J. Richards.

HOMING AND ORIENTATION

Homing refers to the ability of displaced individuals to return to their original location. Implicit in any discussion of homing is the idea that animals must be able to sense the direction they are moving. Amphibians and reptiles that migrate, particularly during breeding events or just before and after overwintering, generally do not move randomly. Amphibians migrating into and out of breeding ponds enter and leave by relatively predictable pathways, as do rattlesnakes moving to overwintering den sites. Orientation can involve visual, olfactory, auditory, or even magnetic cues, each of which requires a different system for reception (Fig. 8.17). Orientation requires some sort of map and a compass. If the compass is based on celestial cues such as the sun, then a clock is necessary to reset the compass as the sun's azimuth changes seasonally.

Salamanders generally cannot home for more than about 30 meters, but the newt *Taricha rivularis* in California can home for up to 2 km. Some individuals can home from about 8 km. Some turtles can home from only 0.5–1 km (*Clemmys guttata*), but others home over 500 km (sea turtles). Crocodylians can home for up to 2 or more km. In the few lizards studied, relocation to distances of about 200 m or less result in good homing ability, but at a greater distance, the lizards do not return.

Many amphibians and reptiles return to specific shelters following both short- and long-distance movements. Movements of the snake *Coluber viridiflavus* in Italy can be divided into single-day loops in which the snake leaves its shelter and returns by the end of the day, complex loops in which the snake moves greater distances over several days using temporary shelters, and large loops involving movements up to 3 kilometers and lasting up to a month. Single-day loops are primarily excursions for basking, complex loops appear to be associated with foraging, and large loops appear associated with reproductive activity (Fig. 8.18).

Landmarks

Within home ranges, most amphibians and reptiles use local landmarks. The repeated use of the same perches, foraging areas, and overnight retreats indicates that individual reptiles and amphibians recognize landmarks within their home ranges. The existence of home ranges and territories is also evidence for the ability of individuals to recognize local landmarks. On a larger spatial scale, many species appear able to recognize the kinds of habitats they live in and orient to those. Some *Anolis* lizards are known to use elevated vantage points to survey their immediate habitat. In a simple but effective experiment, three species of *Anolis* were placed on artificial

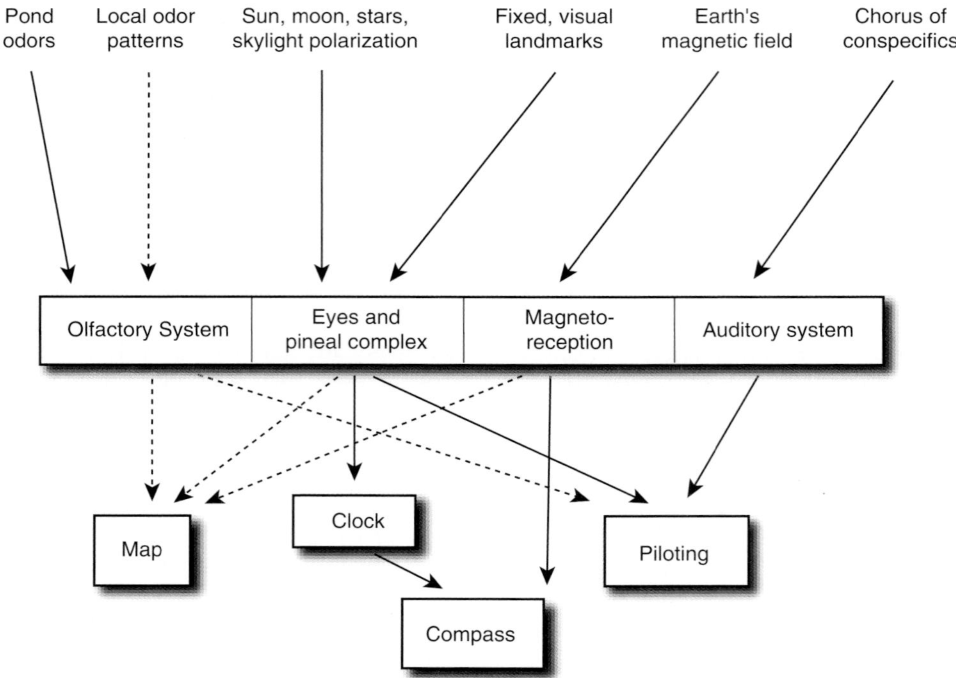

FIGURE 8.17 Relationships among cues, sensory systems, and the mechanistic basis of orientation and navigation for anurans. These relationships may be similar for most amphibians and reptiles. For terrestrial species, odors might be associated with den sites or daily retreats. Adapted from Sinch, 1990.

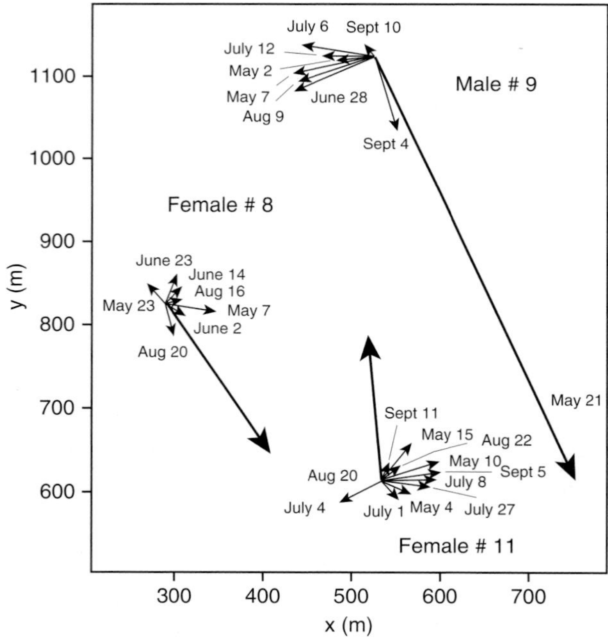

FIGURE 8.18 Movement activity of three individual *Coluber viridiflavus*. Short arrows indicate typical 1-day or complex movements, and the heavier, long arrows indicate large loops. The tip of each arrow indicates the most distant point reached by the snake during each excursion. Adapted from Ciofi and Chelazzi, 1994.

elevated posts from which they could see two vegetation types, a grass–bush habitat and a forest habitat. *Anolis auratus* and *A. pulchellus* choose the grass–shrub habitat, whereas *A. cristatellus* choose the forest habitat. Because the choices correspond with the natural habitats of the lizards, the study reveals that these species used the habitat structure as a landmark or cue to direct their movement.

Orientation and homing ability varies among lizards, even in the same habitat. In open habitats of southern Idaho, horned lizards, *Phrynosoma douglassi*, seem unable to find their original home range when displaced, yet adult sagebrush lizards, *Sceloporus graciosus*, are able to orient toward and return to their original home ranges. Horned lizards may not maintain home ranges for long because their movements follow the movements of their ant prey. Because home range and defense of all or part of the home range (territories) is ancestral in lizards, horned lizards have lost the ability to orient and return to home ranges.

When disturbed, many amphibians and most reptiles rapidly retreat along what appear to be well-known escape routes. This too demonstrates their familiarity with local landmarks. Directed long-distance movement, such as annual migrations of prairie rattlesnakes to den sites, also suggests the importance of local landmarks in orientation and navigation.

x–y Orientation

The interface between aquatic (or marine) and terrestrial environments provides a landmark for orientation by animals that use the interface. Many frogs, for example, typically jump into the water at approximately 90° to the shoreline—their jumps are nonrandom with respect to physical characteristics of the environment. The advantages to orientation toward or away from shorelines are clear. For adult amphibians that sit along the shore, escape into the water is important for avoiding terrestrial predators; for larvae facing metamorphosis, orientation toward shore is critical for emergence into the terrestrial environment; for adults during breeding migrations, orientation toward breeding sites is crucial to find aquatic environments for egg deposition. This type of orientation is termed *y*-axis orientation. Linear cliff faces, riverbanks, and a host of other physical characteristics of the environment might also serve as the basis for *x–y* orientation in terrestrial species. For aquatic amphibians, the *x* axis is the shoreline and the amphibians tend to move perpendicular (90°, the *y* axis) to it (Fig. 8.19). Of course, shorelines can face any direction. For example, a circular pond has sections that face every possible direction of the compass. Amphibians use the sun and its trajectory, which are predictable, to set their *x–y* compass based on the particular shoreline that they use. When landscape views are taken away, frogs and tadpoles retain their ability to orient perpendicular to the *x* axis as long as they can view the sky. Some evidence suggests that turtles may also use the sun to set an *x–y* compass.

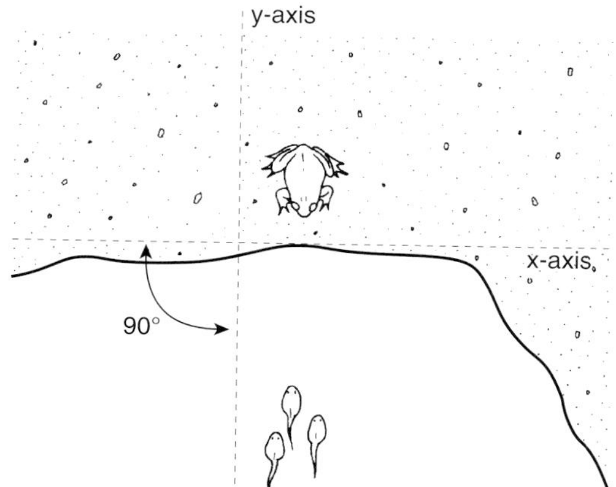

FIGURE 8.19 *Y*-axis orientation is a type of celestial orientation. The animal establishes a homing axis (*y*) perpendicular to an identifiable physical attribute of its home (e.g., shoreline, the *x* axis). Normal escape response is into the pond for the frog being approached by terrestrial predators or to shallow water for tadpoles being approached by aquatic predators; return follows the compass direction of the *y*-axis. Adapted from Adler, 1970.

Orientation by Polarized Light

Light radiates outward from the sun. As the light waves enter the earth's atmosphere, the atmosphere deflects some light waves into a plane perpendicular to the original plane of entry. This scattering or deflection is polarization, and the scattered component (i.e., polarized light) travels in a single plane along a path called the *e*-vector. Because the *e*-vector always remains perpendicular to the sunlight's entry plane rather than the earth's surface, the orientation of the *e*-vector plane relative to every spot on earth changes constantly as the earth rotates. For amphibians or reptiles that see polarized light, this changing orientation offers a directional clue. In addition, an inverse relationship exists between reflection and polarization: Over water surfaces and damp soils, reflectance is low and polarization is high; over drier soils, reflectance is high and polarization is low. Variation in polarized light over wet versus dry landscapes provides amphibians and reptiles with means to differentiate between wet and dry areas and to move to their preferred habitat.

Much indirect evidence and some clever laboratory experiments suggest that some amphibians and reptiles use polarized light in orientation and navigation. The emydid turtle *Trachemys scripta*, when displaced on sunny days to terrestrial sites 300 meters away from their home ranges in a pond, orient toward the pond even though they cannot see it. On cloudy days, turtles fail to orient, indicating that the clouds, which stop polarized light, interfere with the ability of turtles to orient. The outer segments of cones in the eyes of *T. scripta* are capable of differentially absorbing polarized light, further suggesting that the mechanism for locating ponds may be detection of polarized light reflected from aquatic habitats.

The pineal body of salamanders and possibly lizards is a polarized light receptor. Both blinded and normal-sighted *Ambystoma tigrinum* orient to a shoreline once their internal compass has been set based on a vector of polarized light. When light is blocked from the top of the head by opaque plastic, these salamanders orient incorrectly, thus implicating the pineal in orientation based on polarized light.

Although little research has been conducted on use of polarized light by lizards, some recent work on Sleepy lizards (*Tiliqua rugosa*) by Michael Freake suggests that Sleepy lizards are able to use celestial cues to orient, allowing them to determine the compass bearing of movements. Covering the parietal eye interfered with the lizard's ability to orient even though the lateral eyes were unobstructed and provided them complete access to visual cues (including celestial cues and landmarks). Consequently, it appears that Sleepy lizards use the parietal eye to detect polarized light to set a directional compass that allows them to navigate without the use of cues detected by lateral eyes. This phenomenon may be much more widespread among squamate reptiles than previously thought.

Orientation by Chemical Cues

Many habitats (e.g., ponds) and retreat sites have characteristic odors that can be used by amphibians and reptiles for orientation and navigation. In southern California, the toad *Anaxyrus boreas* breeds during spring in ponds and lakes. The toads spend the remainder of the year dispersed in the surrounding terrestrial environment. When displaced 50–200 meters from a pond on clear nights, adults orient to the pond and return; on cloudy nights they also orient to the pond but not as precisely. Blinded toads also orient to the pond, but the possibility exists that they use alternate light receptors. However, when olfactory nerves are severed and the toads rendered anosmic, the toads orient randomly on clear nights even though celestial cues are available. Thus, even in the presence of celestial cues, loss of olfactory senses removes the toads' ability to orient. Because a host of environmental factors can affect the dispersion of chemical cues in natural habitats (e.g., wind), it is likely that, once chemical cues are detected, they are used to set an internal compass. Once the compass course is set by chemical cues, frogs can use celestial cues to navigate.

Olfactory cues also appear important in orientation and navigation in some salamanders. Observations that salamanders retain the ability to home accurately without celestial cues suggest that olfactory cues are used, particularly on overcast or rainy nights. Displaced *Plethodon jordani* that are blinded return to home sites, suggesting that olfactory cues serve as orientation and navigation cues. Early studies on *Taricha rivularis,* in which salamanders were rendered anosmic by damaging the olfactory nerves, caused a reduction in the homing ability, thus demonstrating that the olfactory system is involved in orientation. The salamander *Ambystoma maculatum* migrates on cloudy and rainy nights yet locates ponds. A clever experiment, in which salamanders were placed in arenas with two paper towels, one soaked in water and mud from their home pond and the other soaked with water and mud from nonhome ponds, revealed that *A. maculatum* discriminates between the two odor sources, preferentially orienting toward the odor from their home pond. These results are consistent with field observations that when individuals of *A. maculatum* are placed in unfamiliar ponds, they often migrate back to their home pond.

Magnetic Orientation

The eastern red-spotted newt (*Notophthalmus viridescens*) is well known for its accurate homing behavior. This newt apparently detects its geographic position based on information associated with its home site (i.e., a "map") and a sense of direction ("compass"). One possible basis for such a map is the spatial variation in the magnetic field. The newts may have two different magnetoreception mechanisms that explain differences between their orientation responses to shoreline and their home pond under different conditions of light (Fig. 8.20). One mechanism involves visual centers

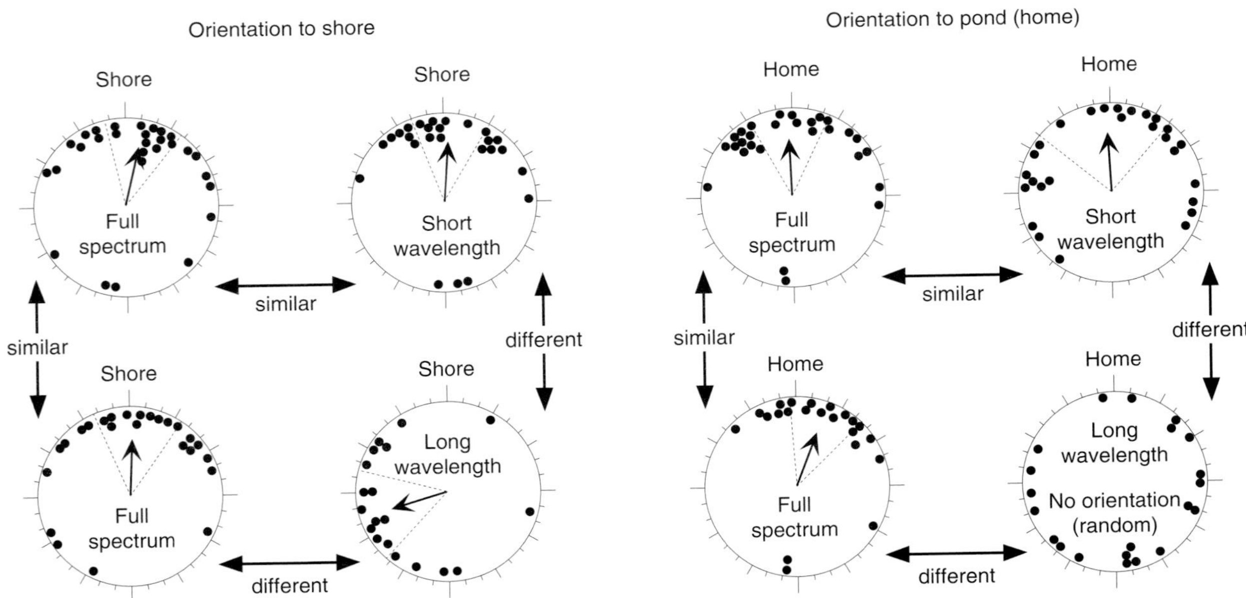

FIGURE 8.20 Diagrammatic summary of experiments on orientation toward shore and toward the home pond for eastern red-spotted newts. In both sets of experiments, controls are those with a full spectrum of light available. In the left panel, newts oriented toward shore in both of the controls and when under short wavelength light. Under long wavelength light, newts oriented approximately 90° counterclockwise from the shore, and their pattern of orientation was significantly different from both their control and the newts under short wavelength light, demonstrating the light dependency of shoreline magnetic orientation. In the right panel, newts oriented toward their home ponds in both controls and under short wavelength light but oriented randomly under long wavelength light, demonstrating the light dependency of home pond magnetic orientation. Adapted from Phillips and Borland, 1994.

in the brain that appear to respond to directional magnetic stimuli. Because visual centers are involved, this mechanism depends on light. The other mechanism involves the trigeminal nerve system, which is independent of visual input and thus does not require light. The possibility exists that a highly sensitive magnetite-based receptor responds to polarity of the magnetic field, and if present in newts, would explain their ability to home. Such receptors have been found in other vertebrates.

Alligators and sea turtles appear capable of orienting on the basis of magnetic cues as well. Sea turtles are renowned for their keen abilities to navigate, and because much of their environment is open ocean, landmarks are largely unavailable. Loggerhead sea turtles (*Caretta caretta*) that hatch in Florida, for example, appear to circle the North Atlantic Ocean and return several years later as juveniles to the American coastline. One population of green sea turtles (*Chelonia mydas*) nests on beaches of Ascension Island, more than 2200 km east of their feeding grounds off the coast of Brazil. The regular return of adults to the tiny island attests to their capability for precise orientation and navigation. Studies on mitochondrial DNA have shown that females in this population and other populations return to the beaches where they hatched. Magnetic orientation likely is involved in open ocean navigation. In laboratory experiments, hatchlings orient to magnetic fields, to wave action, and even to chemical cues. When leaving the beach following hatching, the hatchlings first orient on light from the moon and stars reflecting off the ocean, which takes them to the water. Once in the water, they orient on incoming waves and move perpendicular to them, which carry them out to sea (Fig. 8.21). When the small turtles intersect the Gulf Stream, currents carry them around the Sargasso Sea (Fig. 8.22). Magnetic cues appear to be used for

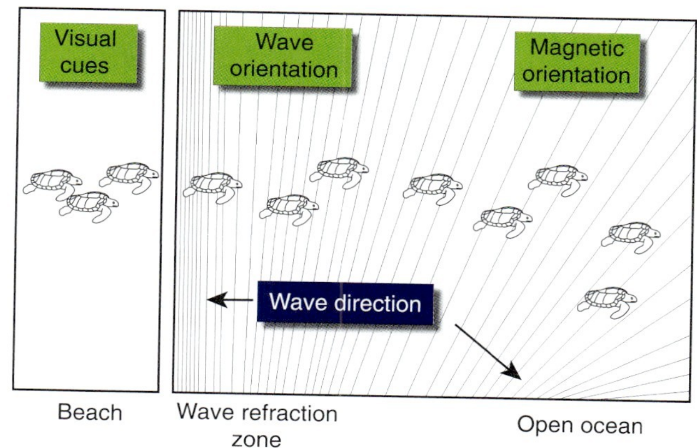

FIGURE 8.21 Different orientation cues believed to guide hatchling Loggerhead sea turtles from their nests on beaches in Florida to the open ocean. Lines indicate direction of waves. Adapted from Lohmann et al., 1997, and Russel et al., 2005.

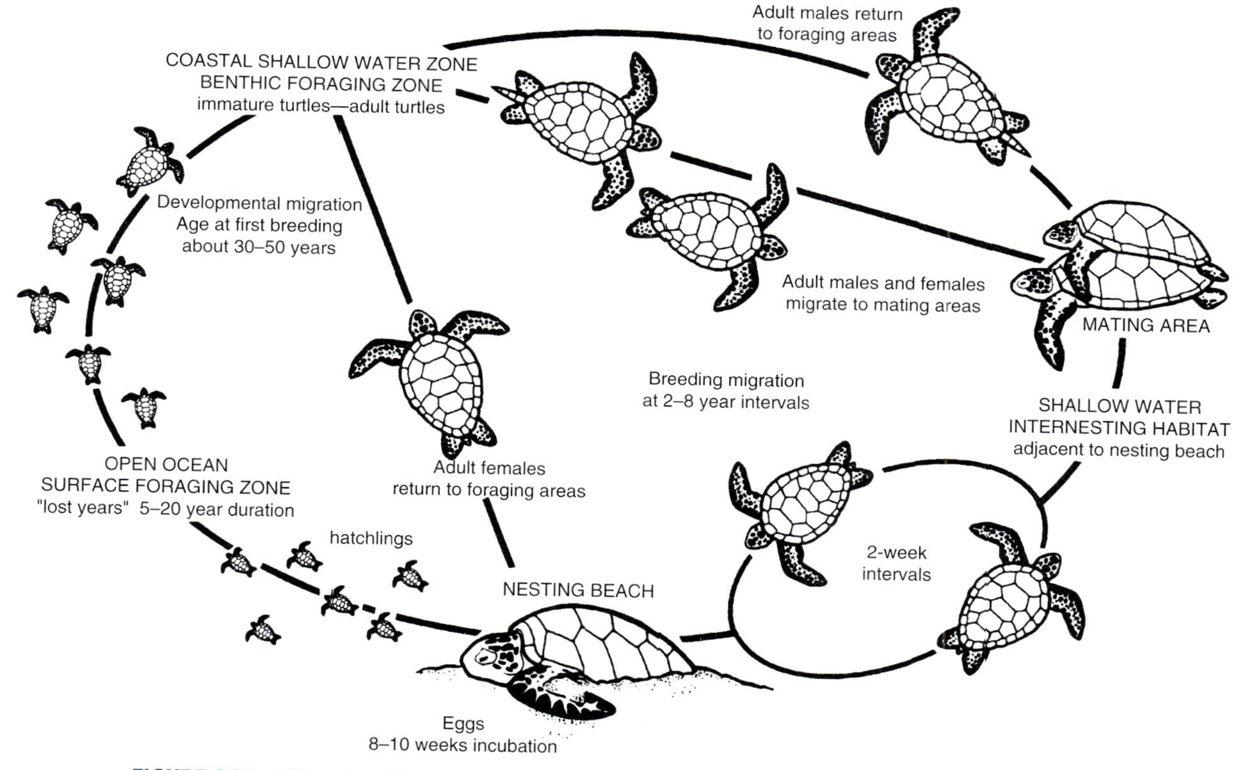

FIGURE 8.22 Life cycle of the green sea turtle showing the course of movements throughout life and possible cues used for orientation during each life history stage. Adapted from Miller, 1996.

navigation while at sea. Once they reach maturity, at an age of 30–50 years, the adult females return to beaches for nesting.

QUESTIONS

1. Using a reptile or amphibian species of your choice, discuss why you might expect the home range of a male to be larger than the home range of a female during the breeding season.
2. When different methods were used to examine movement in the Malaysian snake *Enhydris plumbae*, the results were different. What were these different methods and how do you explain the differing results?
3. Describe movements in the life cycle of the Green sea turtle and discuss orientation cues used by juveniles, immatures, and adults.
4. What is the difference between landmark orientation and *x–y* orientation and what are real examples of each?

ADDITIONAL READING

Adler, K. (1970). The role of extraoptic photoreceptors in amphibian rhythms and orientation: a review. *Journal of Herpetology* **4:** 99–112.

Bruce, R. C., Jaeger, R., and Houck, L. D. (Eds.) (2000). *The Biology of Plethodontid Salamanders.* Kluwer Academic/Plenum Publishers, New York.

Clobert, J., Massot, M., Lecomte, J., Sorci, G., Fraipont, M., and Barbault, R. (1994). Determinants of dispersal behavior: the common lizard as a case study. In *Lizard Ecology: Historical and Experimental Perspectives.* L. J. Vitt and E. R. Pianka (Eds.), pp. 183–206. Princeton University Press, Princeton, NJ.

Dingle, H. (1996). *Migration.* Oxford University Press, Oxford.

Gauthreaux, S. A. Jr. (Ed.). (1980). *Animal Migration, Orientation, and Navigation.* Academic Press, London.

Lutz, P. L., and Musick, J. A. (Eds.). (1996). *The Biology of Sea Turtles.* CRC Press, Boca Raton, FL.

Russel, A. P., Bauer, A. M. and Johnson, M. K. (2005). Migration in amphibians and reptiles: An overview of patterns and orientation mechanisms in relation to life history strategies. In *Migration of Organisms: Climate, Geography, Ecology* (A. M. T. Elewa (Ed.). Springer Publishing, New York.

Stamps, J. A. (1983). Sexual selection, sexual dimorphism, and territoriality. In *"Lizard Ecology: Studies of a Model Organism"* (R. B. Huey, E. R. Pianka, and T. W. Schoener, Eds.), pp. 169–204. Harvard University Press, Cambridge.

Wells, K. D. (2007). *The Ecology and Behavior of Amphibians.* University of Chicago Press, Chicago.

Wilson, J. (2001). A review of the modes of orientation used by amphibians during breeding migration. *J. Pennsylvania Academy of Sciences* **74:** 61–66.

REFERENCES

Introduction

Broadley, 1978; Greene, 1997; Lee, 1975; Pianka and Parker, 1975; Vanzolini et al., 1980; Whitford and Bryant, 1979; Whiting et al., 1993; Williams, 1969.

General

Galbraith et al., 1987; Gibbons, 1986; McIntyre and Wiens, 1999.

Local Distribution of Individuals

Home Ranges

Alexy et al., 2003; Anderson, 1993; Aresco, 1999; Auffenberg and Franz, 1982; Bodie and Semlitsch, 2000; Boglioli et al., 2000; Bull and Baghurst, 1998; Butler et al., 1995; Diemer, 1992; Doody et al., 2002; Duvall et al., 1985; Epperson and Heise, 2003; Eubanks et al., 2003; Hall and Steidl, 2007; Ligon and Stone, 2003; Madsen, 1984; Madsen and Shine, 1996; Mills et al., 1995; Morafka, 1994; Peterson and Stone, 2000; Pike, 2006; Plummer and Shirer, 1975; Rose, 1982; Ruby, 1978; Smith et al., 1997; Stamps, 1978, 1983; Stone, 2001; Tiebout and Gary, 1987; Van Loben Sels et al., 1997; Webb and Shine, 1997; Wilson et al., 1994.

Territories

Brown and Orians, 1970; Bunnell, 1973; Delêtre and Measey, 2004; Howard, 1978a,b; Jaeger, 1981; Jaeger et al., 1982; McVey et al., 1981; Shedahl and Martins, 2000; Simon, 1975.

Evolution of Territoriality

Martins, 1994, 2002; Stamps, 1977.

Other Patterns of Space Use

Aleksiuk and Gregory, 1974; Anderson, 1967; Brodie et al., 1969; Caldwell, 1989; Congdon et al., 1979; Duellman and Trueb, 1986; Gehlbach et al., 1969; Guillaume, 2000a,b; Marquis et al., 1986; Wells and Wells, 1976.

Movements, Homing, and Migrations

Anderson and Karasov, 1981; Barthalmus and Bellis, 1972; Bentivegna, 2002; Dodd and Byles, 2003; Godley et al., 2002; Huey and Pianka, 1981; Madison and Shoop, 1970; Pechmann and Semlitsch, 1986; Pilliod et al., 2002; Schabetsberger et al., 2004; Secor, 1994; Semlitsch, 1981; Stickel, 1950, 1989; Tuberville et al., 1996; Voris and Karns, 1996.

Mass Movements

Clevenger et al., 2002; Gibbons and Bennett, 1974; Pechmann et al., 1991.

Dispersal

Clobert et al., 1994; Doughty et al., 1994; Stamps 1983.

Homing and Orientation

Adler, 1976; Adler and Phillips, 1991; Adler and Taylor, 1973; Bowen and Karl, 1996; Ciofi and Chelazzi, 1994; Coulson, 1974; Diego-Rasilla, 2003; Dodd, 1994; Dodd and Cade, 1998; Ferguson, 1967, 1971; Freake, 1998, 1999; Gibbons and Smith, 1968; Goodyear, 1971; Goodyear and Altig, 1971; Grant et al., 1973; Grassman et al., 1984; Guyer, 1991; Irwin et al., 2004; Johnson, 2003; Kiester et al., 1975; Liebman and Granda, 1971; Lohmann, 1992; Lohmann et al., 1996, 2001, 2004; Madison, 1969; Martins, 1994; McGregor and Teska, 1989; Phillips et al., 2001; Rodda, 1984, 1985; Rodda and Phillips, 1992; Shoop, 1968; Taylor and Adler, 1973; Tracy, 1971; Tracy and Dole, 1969; Yeomans, 1995.

Chapter 9

Communication and Social Behavior

Every organism constantly interacts with other organisms. The interactions include predation, feeding, physiological responses to disease organisms, and numerous others. Social behavior is an interaction with one or more conspecifics, that is, individuals of the same species and, occasionally, between individuals of different species. Social interactions may be a regular feature of an individual's daily life, particularly for individuals living in groups or occupying adjacent territories, or they may occur once a day, once a week, and even only once a year during the reproductive season in a low-density species. Whatever their frequency, social interactions require some form of communication. Amphibians and reptiles communicate through a variety of senses: visual, chemical (nasal and vomeronasal), acoustic, and tactile. In many instances, communication involves more than one sense working together, synchronously, or sequentially.

The evolution of an organism's signal production is intimately interwoven with the evolution of its signal receptors. One system cannot change without adjustments in the other, or communication is lost and interactions fail. Frogs have an impressive array of vocalizations, most of which are used for mate attraction. Frogs have an equally impressive and sophisticated acoustic reception system that allows them to discriminate among species and among individuals. Skinks and other scleroglossan lizards can recognize conspecifics and often individuals exclusively by chemical cues. The primary benefit of high-resolution communication is the ability to identify and locate mates in a complex environment, such as a multispecies frog chorus in a densely vegetated marsh, and to discriminate critically among mates, that is, to recognize a high-quality male among the numerous calling males and select the "best" one. Signal production has an energetic cost, but further, it has a potential life-threatening cost. If a conspecific can locate another conspecific by the communication signal, so can a predator. In the Neotropics, one group of predaceous bats locates male frogs by "homing in" on the frog's advertisement call.

Social interactions are integral to an individual's survival and ultimately influence an individual's evolutionary fitness. The diversity of amphibians and reptiles has again allowed many species to serve as model organisms for the study of the evolution of communication and social behavior. The focus in this chapter is first communication and then sexual behavior because interactions and an individual's choices associated with mate choice have a more immediate and direct effect on individual fitness than a decision or interaction in the context of other types of social behavior. Other aspects of social behavior are presented in Chapters 4, 10, and 11.

COMMUNICATION

Strictly speaking, communication is defined as "the cooperative transfer of information from a signaler to a receiver." Consequently, if a male frog calls and his call is not received by another frog, or a snake produces chemical cues that are not detected by another snake, communication has not occurred. Further, most signal and reception systems of reptiles and amphibians are controlled by sex hormones and, thus, are most effective during the breeding season.

Visual communication uses either body movement or a series of movements or the flashing of a body part having a distinctive color or shape. In amphibians and reptiles, limb movements, head bobs, rapid shuttling movements, and open-mouth threats comprise the most common signals. Visual communication is best known for iguanian lizards but occurs in many other amphibians and reptiles, often in combination with other signals. Although visual displays are most often directed at specific individuals, assertion or advertisement displays can be performed by territorial males to reinforce their territory status to all males or to attract females within sight. Among reptiles, the combination of an approach with head bobs occurs in so many groups that it likely is an ancestral trait and may reflect an ancient solution to the identification of gender and conspecifics at a distance. Many reptile species are sexually dimorphic in coloration, suggesting the importance of color in species and gender recognition. Because some seasonal color changes are tied to reproductive events and under the control of androgens, color also signals an individual's reproductive condition. Most studied reptiles have color vision, which further suggests that color is used in communication.

Acoustic communication is best known in anurans, but crocodylians, some turtles, and some lizards (Gekkonidae) regularly use sound (Table 9.1). Sounds for social communication are produced by rubbing body parts together (some gekkonids, some viperids) and slapping the body against surfaces such as water (crocodylians), although vocal sounds are most prominent. These sounds (vocalizations) are produced by airflow over the vocal cords. Many frogs have vocal sacs to enhance sound transmission.

Chemical communication uses odors that are derived from glandular secretions, either volatile ones (nasal) or surface-adherent ones (vomeronasal). Chemical communication has been studied most intensely in salamanders and skinks. It is used widely by other scleroglossan lizards (includes snakes) and by some iguanians. Although few studies are available, chemical communication is probably used during reproduction and other social interactions in caecilians and in at least one clade of frogs, the Mantellidae. In amphibians and reptiles, most chemical

TABLE 9.1 Vocalizing Taxa of Amphibians and Reptiles, Exclusive of Anurans.

Taxon	Frequency	Taxon	Frequency
Ambystomatidae	+	*Sphenodontidae	+++
Ambystoma maculatum			
Amphiumidae	+	Agamidae	+
Cryptobranchidae	++	*Brachysaura minor*	
Andrias davidianus		Anguidae	+
Dicamptodontidae	++	*Ophisaurus*	
Dicamptodon ensatus		Chamaeleonidae	+
Plethodontidae	++	*Chamaeleo goetzei*	
Aneides lugubris		Cordylidae	+
Salamandridae	+	*Cordylus cordylus*	
Triturus alpestris		*Eublepharinae	+++
*Sirenidae	++	*Coleonyx variegatus*	
Siren intermedia		*Gekkoninae	+++
		Gekko gecko	
Testudines	+	*Lialis burtonis*	
*Testudinidae	++		
Geochelone gigantea		*Iguanidae	+
*Alligatoridae	+++	*Lacertidae	++
*Crocodylidae	+++	*Gallotia stehlini*	
Gavialidae	+++	Scincidae	+
		Mabuya affinis	
		Teiidae	+
		Cnemidophorus gularis	

Note: Families marked with an asterisk have one or more species presumably using vocalization for intraspecific communication. The frequency of vocalization within a family or higher group is subjectively estimated: +++, more than 50% of species; ++, moderate; +, rare, one or few species in a speciose group. Some examples of voiced species are included.
Sources: salamanders through Triturus, *Maslin, 1950;* Siren, *Gehlbach & Walker, 1970; turtles, Gans & Maderson, 1973; Geochelone, Frazier & Peters, 1981; crocodilians, Garrick et al., 1978; gharial, Whitaker & Basu, 1983; tuatara, Gans et al., 1984;* Anolis, *Milton & Jenssen, 1979; lizards, Böhme et al., 1985.*

communication relies on vomeronasal receptors, but gekkonoid lizards have well-developed nasal reception systems (olfaction) that may function in communication. Odor-bearing chemicals are picked up by the tongue or the surface of the head and transported to the nasal sac in amphibians and the roof of the mouth in reptiles and ultimately to the vomeronasal organ (Fig. 10.4). Crocodylians and turtles lack vomeronasal organs, hence this route for chemical communication is not available to them; however, both groups produce glandular secretions during the reproductive season and likely communicate chemically.

Tactile communication occurs when one individual rubs, presses, or hits a body part against another individual. Tactile communication is common in turtles and snakes (e.g., ritualized combat in viperids) but also occurs commonly in amphibians and many lizards. Often, tactile communication occurs after visual, acoustic, or chemical contact has been established. Because most species of amphibians and reptiles use a combination of signals during social communication, each group is reviewed separately.

Caecilians

Most social communication in caecilians appears to be chemically mediated. Caecilians have a specialized chemosensory organ, the tentacle (see Fig. 15.4; Chapter 2, "Sense Organs"), which evolved from elements of the orbit and nasal cavity. During metamorphosis, the eye becomes covered by skin or bone, and its nerves and muscles degenerate. Paired tentacles develop anterior to the eyes, and the lumen of each tentacle is continuous with Jacobson's organ. During burrowing, caecilians close their nostrils and use the tentacles to detect odors. Relatively little is known about caecilian reproductive behavior, but mate location may depend upon pheromones.

Salamanders

Salamander courtship relies heavily on chemical signals that are an essential component of the often elaborate and ritualized courtship behaviors of many salamanders. Visual and tactile cues are also essential in salamander courtship (Fig. 9.1). Salamanders use pheromones to distinguish between species and to locate conspecifics; additionally, odors identify the reproductive status and sex of conspecifics and stimulate sexual activity in females. The pheromones are produced by numerous types of courtship glands found only in males. Gland development is mediated by sex hormones. The courtship glands do not appear until sexual maturity, and most atrophy during the nonbreeding season.

In plethodontid salamanders, mate location is aided by "nose tapping," during which a male repeatedly touches his snout to the substrate. The snout bears a pair of nasolabial grooves; these small grooves extend from the upper lip to the nares. Odors from the substrate move along the groove by capillary action and through the nares into the vomeronasal organ. In the hemidactyliines, each groove extends to the tip of a small papilla (cirrus) that protrudes from the lip beneath each naris.

Courtship glands are most common in the Salamandridae and Plethodontidae. Males of the eastern North American newts (*Notophthalmus*) have a genial gland on each side of the head. When a male encounters a receptive female, he moves beside her and then performs a series of tail undulations that waft the pheromone toward her snout; shortly afterward, courtship continues and the female accepts the male's spermatophore. If a male finds an unreceptive female, he captures her by clasping her neck with his enlarged hindlimbs. This amplexus may last for 3 hours, and during this period, the male places his genial glands against the female's snout. The glands' secretions induce the female's sexual receptiveness and allow courtship to proceed to spermatophore transfer.

Plethodontid salamanders have two general types of courtship glands, the mental gland on the chin and the caudal glands on the back at the base of the tail. The diverse plethodontids have five types of mental glands and even more diverse secretion-delivery behaviors. In some taxa, males slap or rub the mental gland directly on the females' nares. Male *Desmognathus* have enlarged premaxillary teeth. During courtship, a male drags his enlarged teeth across the female's neck or back, lacerating her skin and simultaneously releasing secretions from his mental gland, thereby directly delivering the pheromone to her circulatory system. The secretions induce sexual activity in the female.

The caudal gland secretions maintain a female's receptivity during courtship. Caudal glands lie atop the male's tail, where their secretions are in direct contact with the female's snout during the tail-straddling walk. In this critical phase of courtship, the female straddles the male's tail as they walk in tandem. The secretions ensure that the female follows the male and is, thus, more likely to pick up his spermatophore at the end of the courtship walk. This elaborate courtship involves suites of closely integrated morphological and behavioral characters and has many variations among the more that 200 species of plethodontids, including the loss of the mental gland and associated behaviors.

Even though chemosensory cues are critical components of the elaborate courtship of salamanders, tactile signals are also essential and critical elements. Many salamanders nudge, butt, slap, or rub parts of their bodies against each other. As previously noted, these tactile

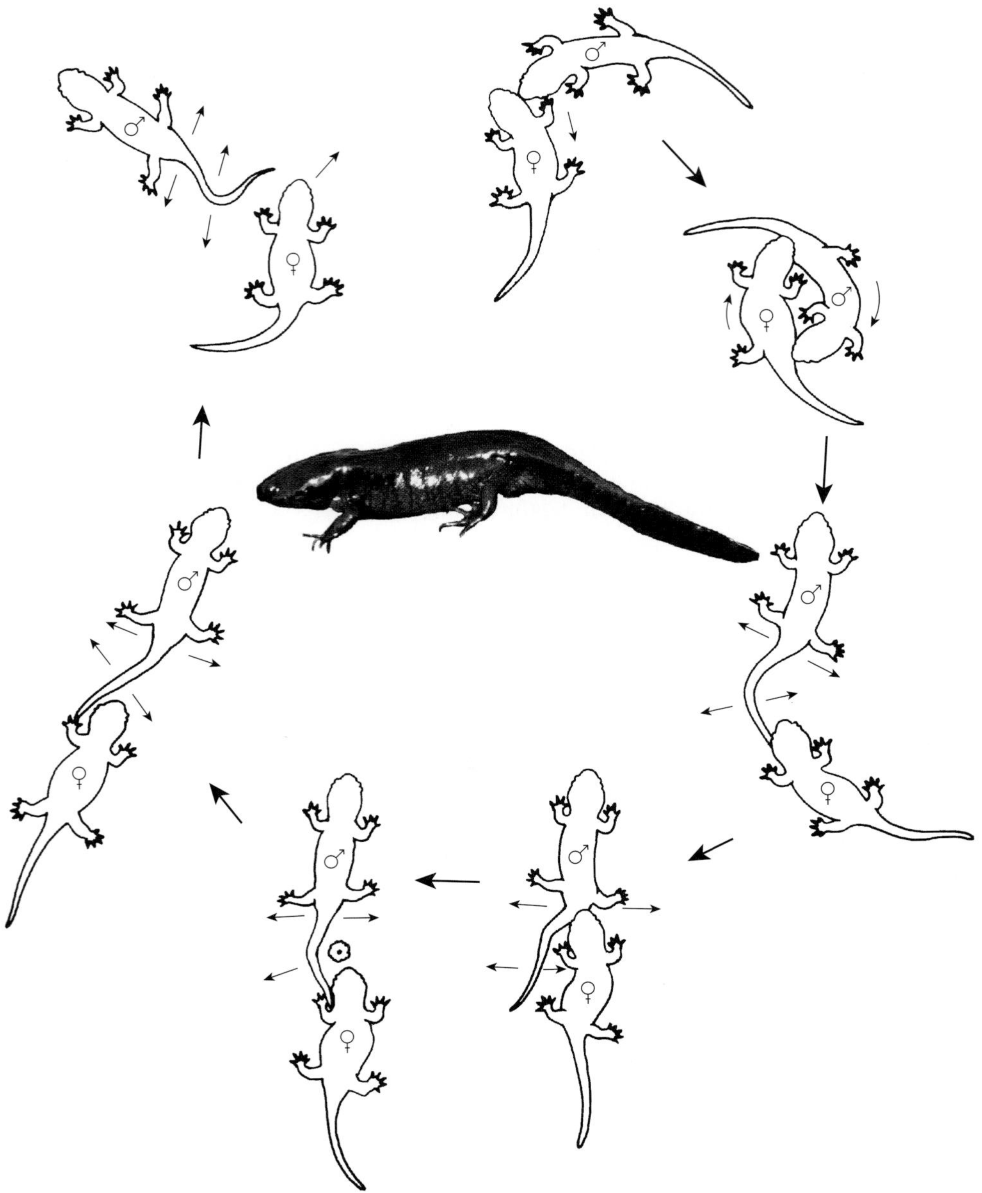

FIGURE 9.1 Courtship sequence of the mole salamander, *Ambystoma talpoideum*. The sequence begins at top center and proceeds clockwise. The male rubs the female; the female nudges the male's cloaca (bottom right), stimulating him to deposit a spermatophore (bottom left); the female briefly examines the spermatophore and then moves over it, picks up the sperm packet with her cloaca, and departs. Adapted from Shoop, 1960.

behaviors deliver pheromones to the courted female and elevate her reproductive readiness and receptiveness. Some taxa bite vigorously, and in two species, biting holds the female during courtship. In the salamandrid *Triturus*, the male whips his tail vigorously, and the force of water movement is the tactile stimulation and may even push the female away. In some cases, the tail touches the female. This tail-whipping behavior presumably increases the female's receptivity to the male.

Frogs

Acoustic signals are the primary mode of communication in frogs, many of which breed at night, although visual and chemical communication are also used by frogs. The absence of light and the anuran force-pump breathing mechanism may have been major selective factors in the evolution of vocalization. Each species of anuran has distinct vocalizations, and individual frogs produce a variety

TABLE 9.2 Broad Categories of Call Types in Frogs.

Call Type	Function
I. Advertisement call	The primary function of this type of call is the attraction of conspecific gravid females. Because the advertisement call has other functions, it is further categorized as follows:
A. Courtship call	The call a male makes to attract a conspecific female that is gravid and ready to mate.
B. Territorial call	The call produced by a male that is defending a territory when a second male vocalizes in or near or intrudes into his territory.
C. Encounter call	The call made by a male in response to the approach of another male.
II. Reciprocation call	Calls are occasionally given by a female in response to the mating call of the male; typically, female frogs of most species do not call, and these calls are rare.
III. Release call	Call given by male that is amplexed by another male; the call is usually accompanied by vibrations of the body. This kind of call is common in explosive-breeding frogs, such as *Bufo*, in situations where many males are active at one time and amplexus is nondiscriminatory.
IV. Distress call	Loud catlike scream given by females of some species of frogs when grasped by a predator. Frogs in clades not closely related, including *Hypsiboas lanciformis*, *Leptodactylus pentadactylus*, *Hemiphractus fasciatus*, *Lithobates catesbeianus*, *Lithobates sphenocephalus*, and *Hypsiboas boans* among others, produce distress calls, indicating that the ability to give these calls has evolved independently several times.

Adapted from Duellman and Trueb, 1986.

of calls, depending on the behavior in which they are engaged. Frog calls segregate into four broad categories, and of these, the advertisement call is most complex (Table 9.2). In many species, different parts of an advertisement call serve different functions. Each part of a call and the various call attributes convey specific information from the signaler. Each component can vary among individual males, and this variation forms the basis for the selection and evolution of call characteristics (Table 9.3).

Frogs produce sound by passing air over their vocal cords, as do all tetrapods. Frogs are also unusual in having vocal sacs for sound resonation. Usually only male frogs have vocal sacs, but not all species that vocalize have sacs. The shape and size of vocal sacs vary among frogs. Primitive frogs have loose folds on the sides of the mouth that are air filled during calling; these may represent primitive vocal sacs. The vocal sac is an outpocketing of the buccal cavity and communicates with it by paired vocal slits. Frogs possess three basic types: a median subgular sac, paired subgular sacs, and paired lateral sacs. The median subgular sac is the most common type and is found in many groups of frogs (Fig. 9.2).

Sound production must be coordinated with ventilation of the lungs, which is accomplished by a force-pump mechanism (see Chapter 6). The frog produces sound by passing air over the vocal cords, and the sound waves are amplified (resonated) by passage through the air in the vocal sacs (Fig. 9.3). Another function of the vocal sac is to increase the frog's calling rate. Without a vocal sac, a frog would require a few seconds to inflate the lungs using the buccal-pump mechanism. Being able to shunt air from the lungs to the vocal sac means that a shorter time

TABLE 9.3 Components of Acoustic Signals Produced by Amphibians and Reptiles.

Call Component	Description
Call or call group	A discrete acoustical signal; may be a single note in some species or a series of notes.
Call rate	The number of calls produced per minute.
Note	An individual unit of energy, such as a single pulse or a trill.
Note repetition rate	The number of notes per unit of time.
Pulses	Notes may be pulsed or unpulsed; examples of a pulse that can be heard are those forming the trill of a toad, which is made up of individual pulses.
Pulse rate	The number of pulses per second or millisecond.
Spectral frequency	The pitch of a call. In many species, a series of evenly spaced harmonics can be seen on the sound spectrogram. The harmonic with the greatest emphasis is called the dominant frequency, whereas the lowest-pitched harmonic is called the fundamental frequency.

Adapted from Duellman and Trueb, 1986.

FIGURE 9.2 A calling Graceful tree frog (*Litoria gracilenta*) from Australia. This frog has an exceptionally large median subgular vocal sac. Photograph by S. J. Richards.

interval between calls is required. This idea was tested in the frog *Engystomops* (formerly *Physalaemus) pustulosus* by Gregory Pauly and his colleagues using frame-by-frame video analysis of the first calls of males in which lungs were not inflated and comparing those calls with later calls in which air was shunted between the lungs and the vocal sac. Females in this species, and in many others, prefer to mate with males with a faster call rate, so the use of a vocal sac to produce faster calls may have a direct effect on reproductive success. However, relatively few studies have examined variation of intercall intervals between individuals within a species or between species in which males have a vocal sac and those that do not.

Sound is a type of energy that produces pressure waves, and the wave components can be depicted in a sound spectrogram (Fig. 9.3). From the spectrograms, numerous characteristics of the call can be measured and used to compare vocalizations of different species or to study variations of calls of individual males in a chorus. Call rates, note repetition rates, and the spectral frequencies are call parameters that vary among species and among individuals (Table 9.2).

Reproduction in frogs is largely dependent on male vocalizations for mate attraction, territory defense, and other male–male interactions. The importance of vocal signaling for anuran reproduction and the relatively easy access to breeding frogs have encouraged the intense and rigorous investigation of all aspects of the anuran signaling system. These studies range from the simple

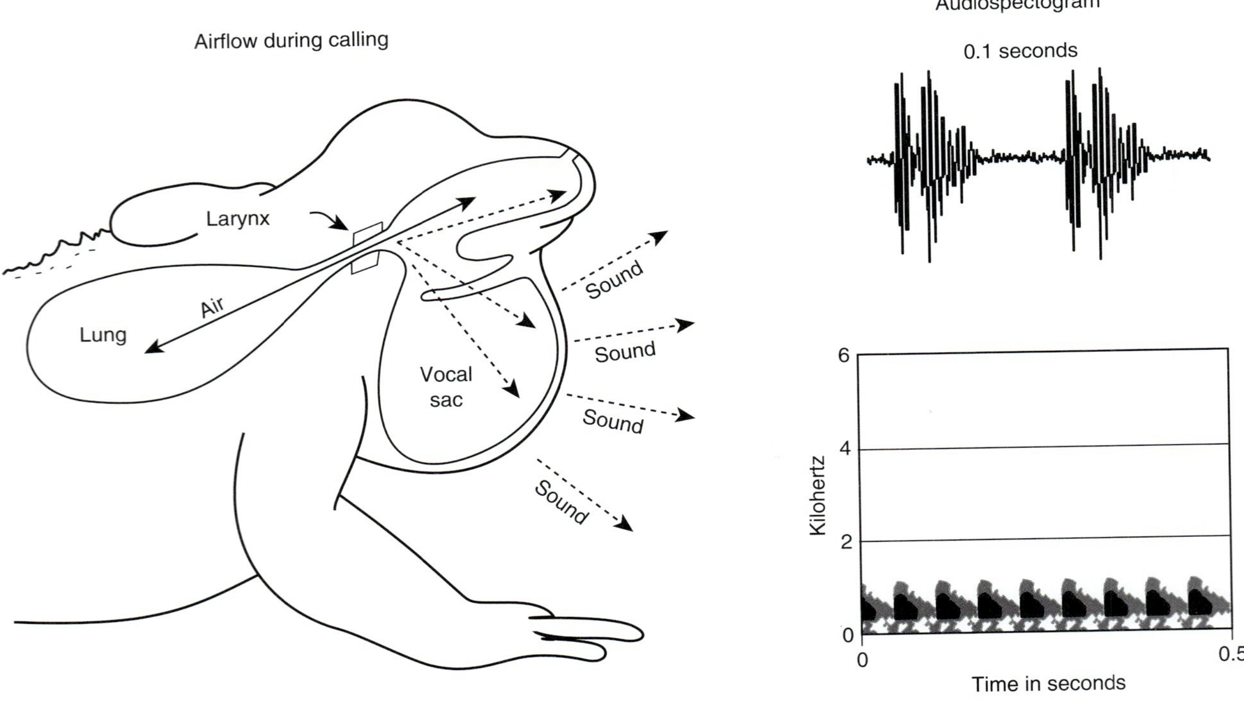

FIGURE 9.3 Sound production and call structure of the marine toad, *Rhinella marina*. Sound production (left) uses aspects of the respiratory ventilation cycle without releasing air to outside. Before calling begins, the bucco-pharyngeal force-pump inflates the lungs and vocal sacs. Then with nostrils closed, the body muscles contract, pushing a pulse of air through the larynx, vibrating the vocal cords. Sound radiates outward and is resonated by the vocal sac. The call of *R. marinus* is a deep, long trill of many continuous pulses (> 50) and lasts several seconds. The waveform (right top) and spectrogram (right bottom) show the energy envelope and pulse structure of brief segments of a call. Each pulse lasts about 0.03 second; the dominant frequency is 500–1000 kHz. Morphology adapted from Martin and Gans, 1972. Redrawn and reprinted, with permission of Wiley-Liss, a subdivision of John Wiley & Sons, Inc., © 1972. Call analysis courtesy of W. R. Heyer.

description of male calling behavior and call structure to detailed neurological investigations and behavioral experiments. Older studies emphasized the description of individual species' vocalizations and how calls serve as species-isolating mechanisms and reduce interspecific mating. The comparatively recent emphasis on mating systems and an individual's reproductive success has led to the study of those aspects of frog calls that females use to discriminate among individual males and that males use in aggressive encounters with one another.

Many studies of frog vocalizations have used playback techniques in artificial settings to learn how female frogs react to male calls. In general, females respond to conspecific calls and ignore heterospecific calls. Male vocalization, although stereotyped in some respects, is more variable than once thought. In the Neotropical *Dendropsophus ebraccatus*, females prefer males with a faster call rate and multinote calls. In male–male interactions of the same species, males produce graded aggressive calls; as males get closer to each other, the duration of the first note of the call increases. Males also show plasticity in their response to the presence of an advertisement or encounter call of conspecific males. In dense choruses, male *Pseudacris regilla* allow conspecific males to vocalize at a shorter distance before reacting with an encounter call. Females strongly prefer advertisement calls over encounter calls. Therefore in a chorus, an individual male is more likely to attract a female by producing advertisement calls and by reducing his encounter-call challenges to other males.

In some situations, males synchronize their calls, but in others, males alternate their calls. In species that migrate to ponds to breed and form dense choruses, males that call synchronously may be less obvious to predators that use acoustic cues to locate prey. For example, bats are known to locate and capture calling males of the Neotropical frog *Engystomops pustulosus*. Other species of frogs do not form choruses but are spread out in the habitat and call individually. Most brachycephalids (i.e., *Craugastor*, *Eleutherodactylus*) call from trees in rain forest habitat and usually have brief calls that make the frog difficult to locate. Most bufonids migrate to ponds to breed in large numbers, and individual calls cannot readily be distinguished. One bufonid, *Rhinella ocellata*, differs from most other bufonids in that it does not migrate to ponds to breed but calls from positions on the ground in sandy soil usually near rivers. These small toads are light colored with paired brown spots and are difficult to locate on the sandy soil. Small numbers of males call in the same general area; for example, five males called from a 150-square-mile section of a study area in central Brazil. Frequently, two males that are closest to each other precisely alternate calls (Fig. 9.4). In this situation, where density of individuals is low, alternating calls allows each male to transmit the maximum amount of information to gravid females that may be in the vicinity.

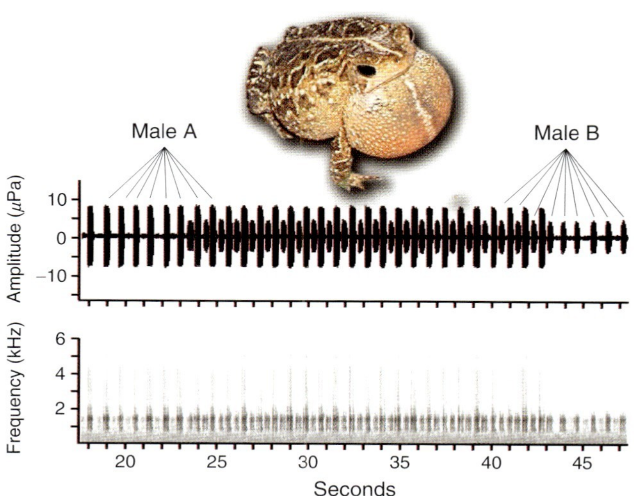

FIGURE 9.4 Waveform and spectrogram of two individuals of the toad *Rhinella ocellata*, in which calls are alternated. Calls of Male B do not overlap those of Male A.

In other studies, computer programs are used to produce synthetic calls that mimic advertisement and other vocalizations. Components of a call can be removed, changed in frequency, or otherwise modified in order to determine which components are most attractive to females. For example, male *Engystomops pustulosus* (Leiuperidae) produce calls with two parts—a whine and a chuck. Studies of marked individuals reveal that frogs can vary the complexity of their calls by producing only the whine or the whine with a variable number of chucks, up to six. Females prefer more complex calls, choosing males that give one or more chucks over those that produce only a whine. Males use complex calls only when they are in high-density choruses; when calling in isolation or in low-density choruses, males produce only the whine. The cause of this reproductive trade-off is the bat, *Trachops cirrhosus*; complex calls provide this predaceous bat with location cues. When competition among males for females is high, a male must risk predation to increase his probability of attracting a female, but when competition is low, a simple call may attract a female without increasing his risk of predation.

A number of frog species, including some exclusively nocturnal ones, use a combination of acoustic and visual communication, and some species use only visual communication. Visual signals include a variety of movements of the body and limbs. These signals include hand waving, foot raising and lowering, foot flagging, leg stretching, and toe undulations. In addition, the body can be raised and lowered, inflated, or swayed from side to side, and color changes may occur in calling males or in territorial females. Visual communication in frogs is undoubtedly more common than reported because the signals in some cases are subtle and not recognized by human observers.

To date, visual communication has been reported in 10 anuran clades. Some displays are remarkably similar among distantly related clades, suggesting independent evolution. Many of these species are diurnal, and some live in or adjacent to noisy mountain streams.

In some taxa, only males produce visual signals, but in others, both males and females use them. The hylodid *Hylodes asper* is a torrent-living frog; males vocalize and subsequently use a foot-flagging display to attract an approaching female (Fig. 9.5). Males raise a hindleg and hold it above the body; the light-colored toes are spread and the foot becomes a flag against the dark background of the frog's habitat. The female may signal a response by stretching one or both legs behind her. Males use the foot-flagging behavior and other limb movements to signal to other males that attempt to intrude into their territories. Male *Dendropsophus parviceps* use a similar foot-flagging display in response to the close approach of a conspecific male (Fig. 9.5). Male *Atelopus zeteki* use a stereotypic hand wave, presumably to signal territorial occupancy to conspecifics.

The tiny Pumpkin toadlet, *Brachycephalus ephippium* (Brachycephalidae), is a diurnal, bright orange frog that inhabits leaf litter in Brazilian coastal rain forest. Males produce an up-and-down arm display to inform other males of territorial intrusion. Visual displays are coupled with vocalizations. Most often, the intruder retreats and no physical contact ensues. Breeding occurs away from water in *B. ephippium*, and its weak advertisement call is lower than the background noise of the forest. Presumably, the low call and daytime activity patterns contributed to the evolution of its visual signals. Some dendrobatid frogs are also diurnal and brightly colored. One species, *Ameerega parvula*, uses leg-stretching displays. The cryptically colored aromobatids and dendrobatids use various types of visual displays. Both male and female *Mannophryne trinitatis* use visual displays during social interactions, and in addition, this species exhibits sex role reversals. Females of *M. trinitatis* establish territories around boulders and rocks in streambeds. Females perch on top of large boulders and challenge intruders that enter their territories by adopting an upright posture and pulsating their bright yellow throats. If the intruder ignores the visual signal, physical contact results. The frogs may stand on their hindlegs and grapple. Females attack males, males carrying tadpoles, and other females, but most aggression is directed toward other females. Males of *M. trinitatis* use rapid color change as a visual signal. Males are light brown until they begin to call; while calling, they become uniform black, losing their stripes and other markings. Two adjacent calling males, both black, may engage in grappling fights; the loser immediately becomes light brown. Females, but not males, are territorial; males use color to court females from a distance, thereby avoiding attacks from nonreceptive females. A receptive female signals her reproductive readiness by leaving her territory and approaching a calling male. Males of the cryptically colored *Allobates caeruleodactylus* have bright blue fingers (Fig. 9.6). The blue color is intense during the breeding season but fades during the nonbreeding season, indicating that the color is hormone related. Presumably the color is used as a signal indicating territorial boundaries to intruding males.

Frogs often use tactile cues to distinguish gender, particularly in explosively breeding species (Fig. 9.7). The larger body of a gravid female provides the tactile cue that identifies her gender and reproductive state to a male. In prolonged-breeding species, a female approaches a calling male, and typically, the male continues to call until touched or nudged by the female. In some poison frogs (*Ranitomeya, Dendrobates*), females follow calling males during courtship; eventually, the female strokes the male's legs, head, or chin with her forefeet, which signals her readiness to oviposit and stimulates the male to release sperm. In the hylid *Hypsiboas rosenbergi*, each male

FIGURE 9.5 Foot flagging in the Brazilian torrent frog, *Hylodes asper* (Leptodactylidae; left) and *Dendropsophus parviceps* (Hylidae; right). Photographs by W. Hödl.

FIGURE 9.6 The aromobatid frog *Allobates caeruleodactylus* has brilliant blue toes, which are probably used in visual signaling. Photograph by A. P. Lima.

constructs a basin of mud or sand at the edge of a small forest stream and calls from a platform in the basin. A female approaches and inspects the basin while the male continues to call. Only after the female touches the male does he cease calling and initiate amplexus.

Frogs in the family Mantellidae have a wide variety of femoral glands on the undersides of their thighs (Fig. 9.8). Although the structure of the glands has been studied by Miguel Vences and Frank Glaw and their colleagues, the function of the glands remains largely unknown. Supposedly, the glands function during reproduction. Three other genera of frogs in unrelated ranoid clades, *Indirana*, *Petropedetes*, and *Nyctibatrachus*, also have femoral glands, indicating that these glands may have evolved convergently. Frogs in the latter genus have reproductive behavior similar to many mantellids, which includes the lack of amplexus and positioning of the male and female on vertical leaves. In *Nyctibatrachus humayuni*, males vocalize from leaves overhanging small streams in evergreen forest in the Western Ghats of India. When approached by a female, the male moves to the side and continues vocalizing. The female deposits eggs on the exact spot from which the male had been vocalizing. If a female deposits a second group of eggs, they are placed on the spot where the male had moved to vocalize. After depositing eggs, the female departs and the male moves over the eggs and fertilizes them. Thus, in this species, the male chooses the deposition site, and fertilization is 100% successful. The male continues to call from the same location and may mate with additional females who deposit their eggs on the same leaf or a nearby leaf. Possibly the female uses the secretion from the male's femoral glands to determine where to lay eggs, although this idea has not been investigated. The glands of the mostly diurnal, terrestrial species of mantellids are composed of clusters of enlarged granular glands that have separate ducts.

FIGURE 9.7 Scramble competition in the frogs *Bufo bufo* and *Rana temporaria* during an explosive-breeding event. Photograph by W. Hödl.

FIGURE 9.8 Mantelline frogs have well-defined femoral glands on the ventral surfaces of their thighs. The left row shows species in which the glands are typically composed of single granules, each of which is a separate secretory unit. The right row shows species in which the gland is composed of granules arranged in a circle; each granule opens into a central external depression. A, *Mantella aurantiaca*; B, *Guibemantis liber*; C, *Guibemantis bicalcaratus*; D, *Gephyromantis pseudoasper*; E, *Gephyromantis cornutus*; F, *Gephyromantis luteus*; G, *Gephyromantis malagasius*; H, *Mantidactylus* cf. *ulcerosus*; I, *Mantidactylus* cf. *betsileanus*; J, *Mantidactylus albofrenatus*; K, *Mantidactylus brevipalmatus*; L, *Mantidactylus* cf. *femoralis*; M, *Mantidactylus argenteus*; N, *Mantidactylus grandidieri*. Photographs by Miguel Vences.

In contrast, glands in the most derived genus in the family, *Mantidactylus*, are arranged in a circle with the ducts leading inward to a central depression; thus, the secretory product is concentrated into one spot on the ventral thigh. Unlike the other mantellids, frogs in the genus *Mantidactylus* are semiaquatic, and concentration of the secretion from one point may allow more precise delivery in water, assuming that the glands function in reproduction. Many interesting questions remain to be investigated concerning the function of these unusual glands.

Turtles

Tortoises and turtles use combinations of visual and chemical signals during social interactions. Visual displays involve head bobs (tortoises) and displays of patterns and colors on the forelimbs, neck, and head (emydid turtles). When two tortoises interact and at least one is a male, the male first performs head bobs or sways the head back and forth. If both are male, the other one responds with a similar behavior; the interaction can escalate into butting, biting, and other aggressive acts. In desert tortoises, *Gopherus agassizii*, the interactions include all aforementioned acts, and two males, having interacted during the day, may spend the night in the same burrow only to continue the interaction the following day (Fig. 9.9). When males interact with females, the sequence begins in the same way, but when the female retreats instead of producing head bobs in response to the male's head bobs, the male continues to approach, intensifies his head bobbing, and then circles the female. After a series of behaviors including biting or ramming, the male attempts to mount the female, scratching her shell, grunting, and moving his head in and out of his shell. This behavior sequence may or may not result in copulation. Even though the initial social cues are visual, the tactile signals may ultimately initiate copulation.

In some emydid turtles, the male maneuvers around a female in the water and eventually positions himself to expose his color and striping pattern to the female. Male color patterns are species specific and presumably provide the first level of species identification. While face to face, males gently bump heads with females (a tactile cue). Following this behavior, the male attempts to position himself on the back of the female with his head above and oriented down above the head of the female. The male extends his forelimbs with the elongate claws downward and begins a rapid chewing motion with the jaws. This behavior is followed by rapid vibratory movements of the forelimbs in front of the female's head. The limbs are vibrated in a fanlike fashion but do not touch the female.

Many turtles have Rathke's glands on the bridge of the shell. These glands produce aromatic chemicals. Other turtles (e.g., testudinids, some emydids, *Platysternon*) have mental glands that are active during the breeding season. Cloacal secretions may also play a role in social communication; however, the precise function of secretions and pheromones is poorly known for turtles. Rathke's gland secretions may allow musk turtles, *Sternotherus odoratus*, to find and follow one another in the water. Anecdotal observations of turtle behavior indicate that pheromones are likely involved in many social interactions. Experimentally, reproductive behavior in *Emys orbicularis* was reduced by more than 60% by cutting the olfactory or vomeronasal nerves, indicating that the reception of chemical cues is involved in reproductive behavior.

Crocodylians

Visual signals, often in combination with auditory signals, are common in crocodylian communication. Visual signals predominate in short-distant interactions, whereas auditory cues are primarily used in long-distance communication. In alligators and some crocodiles, the behavioral sequences are similar. When an intruder enters a male's territory, the resident approaches the intruder with his head and tail partially above the surface (head emergent–tail arched posture) to signal his alertness. Chases, lunges, and real or mock fights follow. After most chases, the territorial male inflates his body (inflated posture). Depending on species, narial geysering (water forced out of the nares) occurs during male–male confrontations or, as in alligators, geysering occurs with head slaps, in which the head is raised out of the water and then slapped against the surface.

Auditory signals include bellowing (Fig. 9.10), juvenile grunts, and slapping sounds. In alligators, males and females bellow, but the duration of bellows and the time between bellows is greater in males than in females. Loud, low-frequency bellows are produced only during the breeding season and after the eggs have been deposited. Coughlike calls are used by males and females during courtship for close-range communication. Head slapping is mainly a male signal during male–male interactions. Juveniles usually grunt under conditions of distress. The grunts cause adults to orient to and move toward the young. Adult alligators can also produce grunts, and these cause juveniles to move to the adult.

Tuataras and Lizards

Tuataras rely on visual cues for male–male and male–female interactions. Males are territorial and defend their territories by first approaching the intruding male, inflating the lungs to increase the apparent size of the gular region and the trunk, elevating the dorsal crest, and darkening the skin above the shoulders and eyes. The intruder performs a similar ritual. Often, the resident performs lateral head shaking; this behavior usually causes the intruder to depart. If the intruder stays, the males approach each other. They face each other but orient their heads in opposite directions while holding their bodies parallel; then they open and rapidly snap their mouths shut. This confrontation is followed by rapid chases initiated by rapid tail whipping. Males commonly croak during the mouth-gaping phase, and during the chase, the pursuing male bites the head, body, or tail of the other tuatara. Courtship behavior is similar in the early stages (Fig. 9.11). Females perform a head nod when approached by a male; the courting male responds with what is termed the stolzer Gang, an ostentatious walk marked by frequent pauses and extremely slow

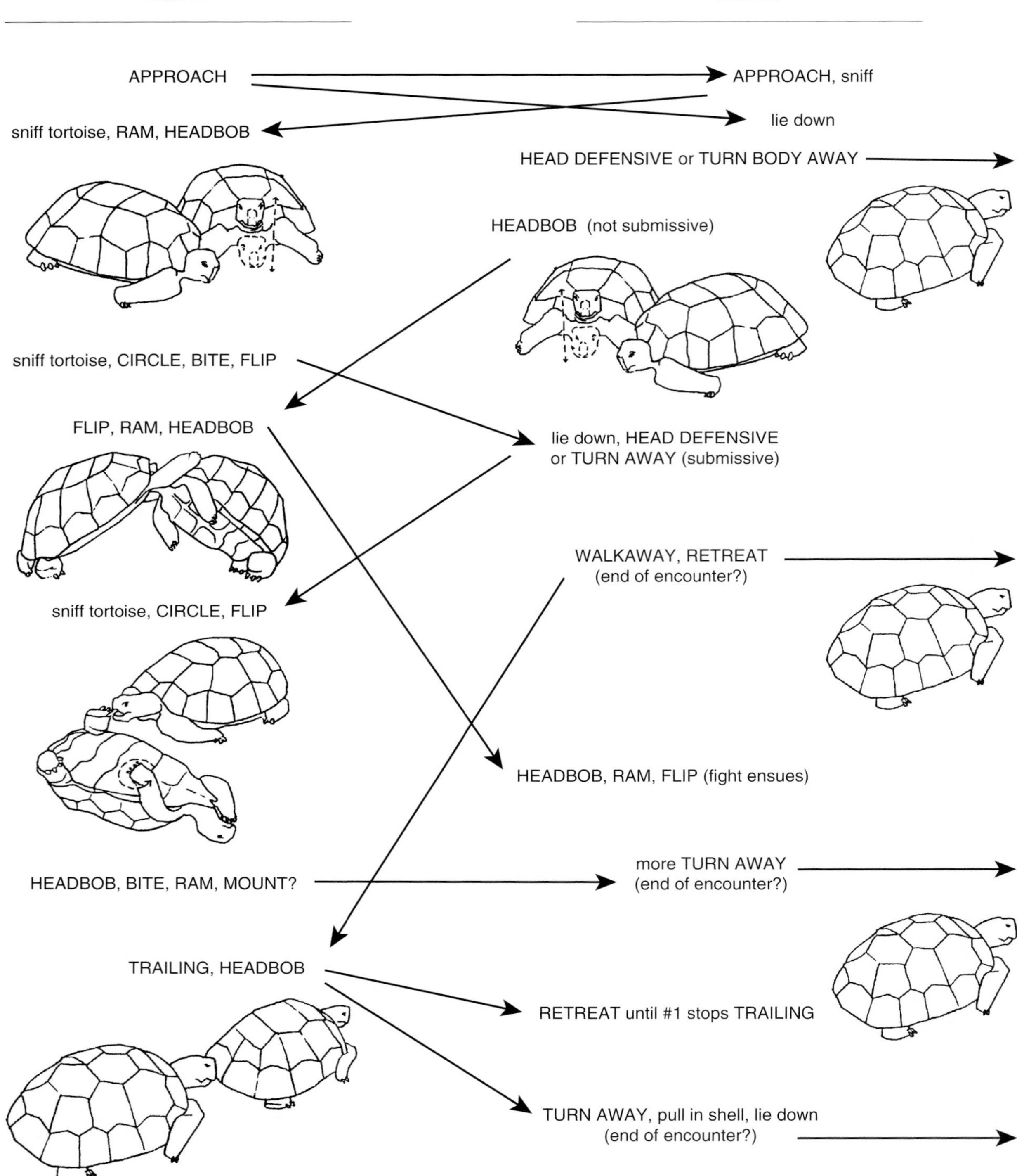

FIGURE 9.9 Sequence of behaviors that occur during an aggressive encounter between two male desert tortoises, *Gopherus agassizii*. Common alternative sequences are indicated by arrows. Terms in capital letters indicate specific behaviors that have been described. Adapted from Ruby and Niblick, 1994.

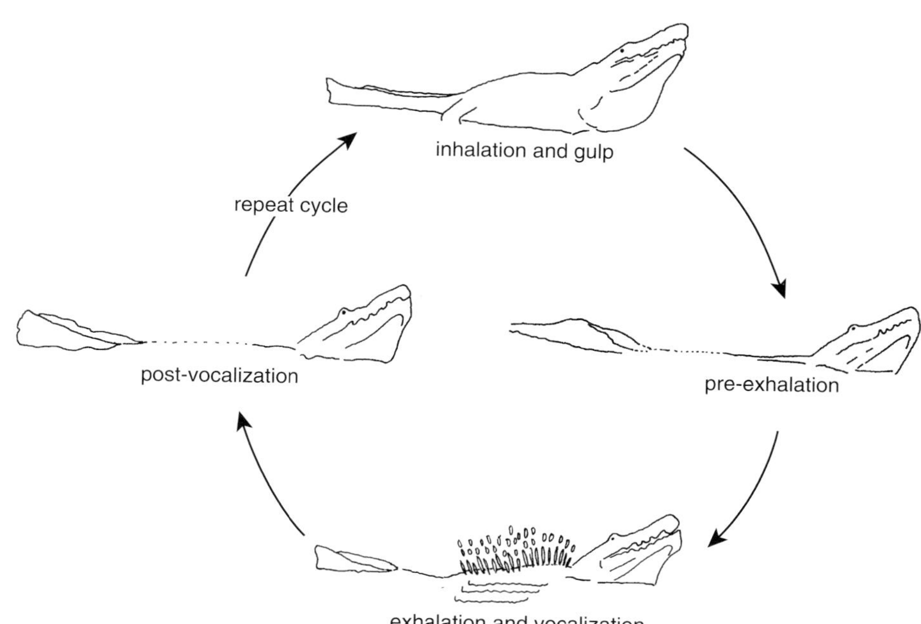

FIGURE 9.10 Sequence of events involved in the production of the bellow of an alligator, *Alligator mississippiensis*. Exhalation causes a fountain of water along the alligator's trunk and also produces a radiating series of ripples at the water surface. Adapted from Garrick and Lang, 1977.

forward progression; his limb movements are stiff legged and exaggerated.

Male lizards of many species have courting behaviors similar to tuataras. Even though stiff-legged walking can be part of the ritual, it differs by a faster forward progression and the absence of frequent pauses. This behavior is not considered to be a stolzer Gang in lizards.

Within lizards, each major clade (Iguania, Gekkota, and Autarchoglossa) emphasizes different sets of social signals. The Iguania use visual, and to a much lesser extent, chemical, and tactile signals in social communication. Nocturnal gekkotans use auditory and visual signals, whereas diurnal gekkotans use primarily visual signals. Most autarchoglossans rely primarily on chemical, signals and, to a lesser extent, on visual and tactile signals, but some interesting reversals have occurred (e.g., the Corydidae). Most iguanians and many gekkotans are territorial and sit-and-wait foragers; in contrast, most autarchoglossans are active foragers and likely not to be territorial (see Table 10.1 for a summary of traits associated with this dichotomy). The best-known examples of visual communication are in the Iguania, and the best-known examples of chemical communication are in the Autarchoglossa.

Coloration of dewlaps, heads, and patches on the lateral or ventral surfaces of the body are frequently used in visual communication. Dewlap displays of *Anolis* are combined with signature head bob displays that are species specific. These displays are categorized as simple, compound, or complex. Simple displays involve the extension of a uniformly colored dewlap and a simple head bob pattern (Fig. 9.12). Compound displays occur where the dewlap has a central color surrounded by a second color and a relatively simple head bob pattern. Complex displays result when the dewlap has an intricate pattern of two or more colors, and head bobbing and dewlap extension are relatively independent. The signature head bob display of anoles attracts females and may be the most effective cues for long-distance signaling. Considerable variation exists among individuals in signature head bob displays, dewlap color, and dewlap extension, supporting the idea that females can discriminate among individual males based on some aspect of the display. Female discrimination is confirmed by choice experiments; female anoles select males with "normal" displays over males with even slightly deviant displays. The vigor of the male's display appears to be the most important component for the attraction of females. Signature head bob displays occur in many other lizards as well (Fig. 9.13).

Even though it appears rather obvious that sexual signals (e.g., male dewlap size, color, and pattern) should communicate information allowing individuals to determine whether escalating interactions have the potential for a payoff, the relationship between signals and performance is much more complex. Simon Lailvaux and Duncan Irschick approached this problem by examining the relationship between signaling and whole-animal performance in Caribbean *Anolis* lizards that differ in degree of sexual dimorphism. Sexual dimorphism was used as a surrogate for differences in degree of territoriality (species with greater sexual dimorphism are more territorial). Not

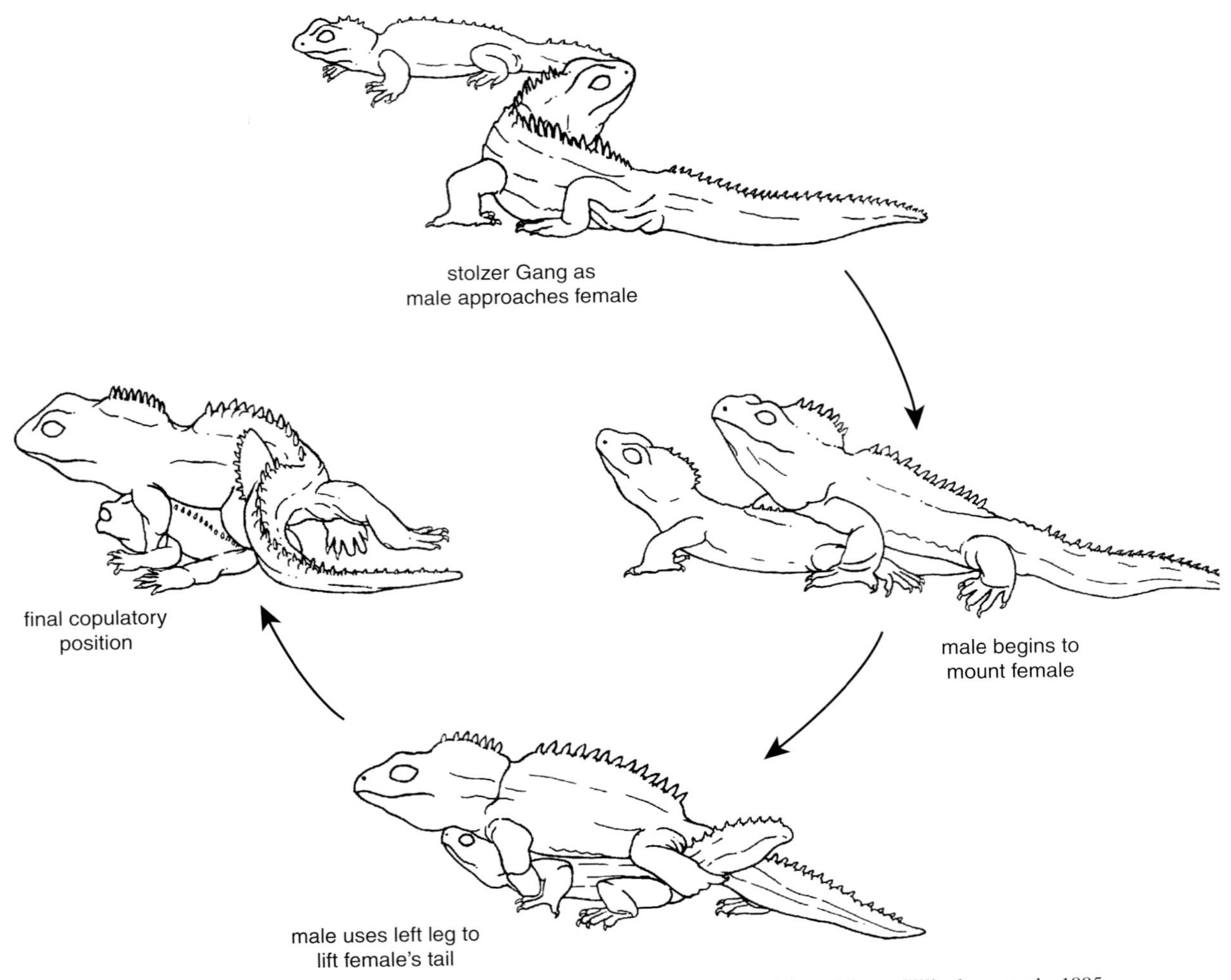

FIGURE 9.11 Mating behavior of the tuatara, *Sphenodon punctatus*. Adapted from Gillingham et al., 1995; redrawn by J. P. do Amaral.

only is a considerable amount of information available on the ecology of the different *Anolis* species, but their evolutionary relationships are known, providing the opportunity to examine performance-based fighting ability in an evolutionary and ecological context. The researchers first found that dewlap size was an honest predictor of performance (measured as bite force) in the most sexually dimorphic species, but not in the least sexually dimorphic species. Similarly, maximum bite force predicted success in male–male combat among the most sexually dimorphic species but not the least sexually dimorphic species. Surprisingly, dewlap size predicted success in male combat among the least sexually dimorphic species but not the most sexually dimorphic ones. The frequency of biting increased with increasing sexual dimorphism, but the frequency of dewlapping decreased with increasing sexual dimorphism. Thus, even though dewlap size is an honest signal, the signal is used less in species that are the most dimorphic. In these, male combat is more likely to be used to settle territorial disputes. The dewlap varies in function in relation to sexual dimorphism and is only weakly associated with other ecological, morphological, and behavioral traits. In anoles with reduced sexual dimorphism, male dewlaps play a major role in agonistic interactions with other males and are also used in social interactions with females. Female dewlaps are larger relative to those of males in these species. Predictors of fight outcomes vary depending on the degree of sexual dimorphism, and the kind of information conveyed by dewlaps varies. Finally, anoles with similar ecomorphs tend to have similar behavioral characteristics.

A number of studies have demonstrated social responses to color. Female *Anolis carolinensis* prefer males with red dewlaps over males with drab-colored dewlaps. In experiments, male *Sceloporus undulatus* attack females if they are painted with the male's ventral blue coloration; similar male aggression toward females is elicited by painting male coloration on females of *Agama agama*, *Plestiodon laticeps*, and species of *Lacerta*.

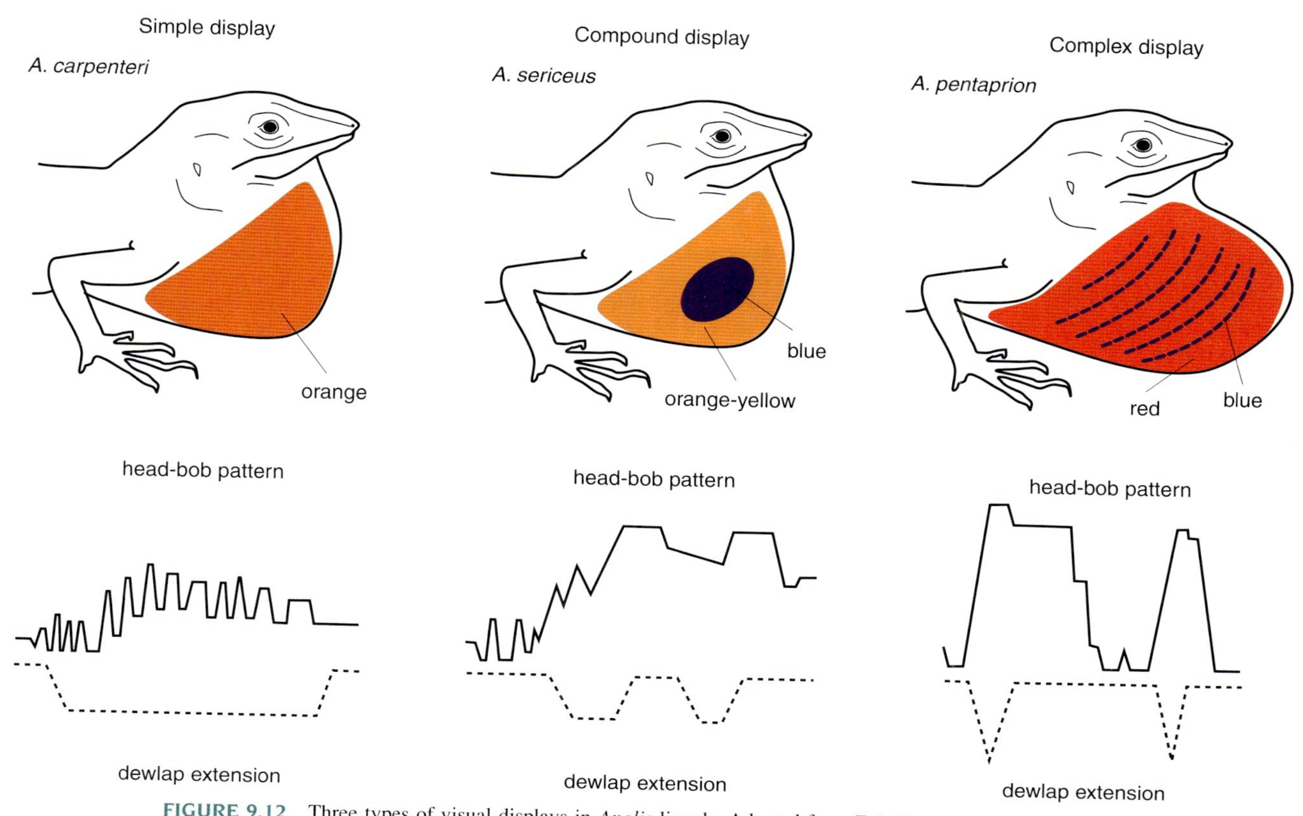

FIGURE 9.12 Three types of visual displays in *Anolis* lizards. Adapted from Echelle et al., 1971.

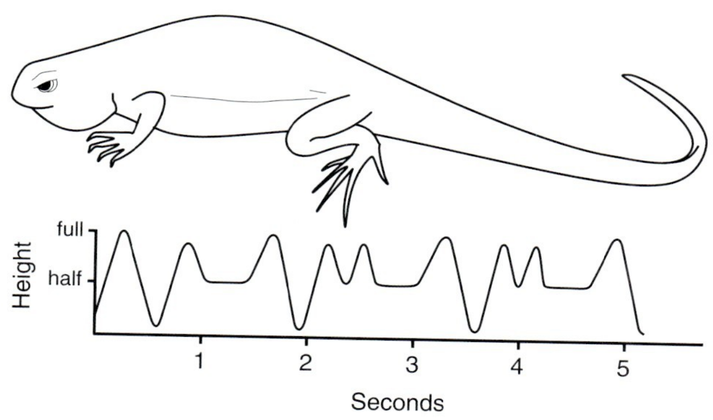

FIGURE 9.13 Display posture and movement-sequence diagram for a male desert iguana, *Dipsosaurus dorsalis*. The line in the diagram denotes the relative height of the head during a push-up defensive display sequence. Adapted from Carpenter, 1961.

In some lizards, females are brightly colored. At least seven hypothesis have been suggested to account for bright female coloration; these include sexual recognition, female signaling, aggression avoidance, sexual maturity, courtship rejection or stimulation, and conditional signaling. Conditional signaling appears best supported. Rapid color change in females signals sexual receptivity to the territorial male; additionally, the female's long-term retention of the bright colors signals her likely rejection of further courtship. Brightly colored females of keeled earless lizards, *Holbrookia propinqua*, for example, are recognized as females by males. Females in the process of undergoing rapid color change are sexually receptive, can store sperm once they mate, and are the females courted by males. Females that have completed the transition to the bright color phase aggressively reject courtship attempts by males, and bright coloration is associated with large follicles. The use of bright coloration in females as a social signal to males appears to occur only in species where males are familiar with females, suggesting that territoriality is a prerequisite for this kind of social recognition.

Although we usually associate visual signals in lizards with communication among individuals, studies on Indian rock lizards suggest that visual signals in lizards may be much more complex and serve multiple functions. These lizards live on rock outcrops and produce a complex set of signals, including push-ups, extension of legs or the gular region, tail raising, and dorsal flattening. Males dorsally flattened the body in response to birds, and because males are more exposed than females, this behavior was uncommon in females. Both males and females display to females by arching their backs and extending their gular folds, and similar behaviors are elicited in both males and females by other animals. Males extend legs in response to conspecifics, whereas females extend legs in response to other animals. Tail raising occurs in females in response to males. Taken together, these observations indicate the complexity of visual social signals in lizards.

Among lizards, the chemical communication system is best known in the clade of North American five-lined skinks (*Plestiodon laticeps*, *P. fasciatus*, and *P. inexpectatus*). Early field observations suggested that male *Plestiodon* used chemical signals to follow trails of females during the breeding season. Male broad-headed skinks, *P. laticeps*, can discriminate among species of *Plestiodon* or sex within their species, and they can determine sexual receptivity of females based on pheromonal cues alone. When an adult male first encounters another adult-sized individual, he approaches that individual. If the other skink is a female and does not respond aggressively to the approach, the male begins tongue flicking the body of the female and ultimately directs the tongue flicks toward the cloaca, where a urodeal gland produces the pheromone used for identification of species, gender, and sexual receptivity. A series of experiments in which cloacal odors were transferred to other species and sexes of skinks resulted in male *P. laticeps* attempting courtship with other species or other males containing the pheromones of sexually receptive females of *P. laticeps*. Experiments with a diversity of scleroglossan lizards have produced similar results in other chemical-signaling species.

Chemical signals seemingly permit individual recognition in some lizards. Kin recognition in lizards can be based on familiarity or kinship. Desert iguanas discriminate between their own odors and those of other desert iguanas. Similar observations have been made for the skinks *Plestiodon laticeps*, *Tiliqua rugosa*, and *Egernia stokesii*, and an amphisbaenian. Juveniles of *Egernia saxatilis* recognize kin based on chemical cues that result from familiarity. Experiments in which juveniles and mothers were separated show that ability to recognize kin is lost when they are separated. Nevertheless, chemical recognition of kin occurs in other Australian skinks. In *T. rugosa* and *E. stokesii*, mothers discriminate between their own offspring and the offspring of other females.

Monitor lizards use both chemical and tactile signals. Male Komodo dragons (*Varanus komodoensis*) tongue-flick females at various positions along the body during the initial stages of courtship. When a male nudges a female with his snout, she will either respond with an assertion display or run away. If the female runs, the male pursues her closely and attempts to court. Males always scratch females on the neck and back during courtship and may even bite the female's neck prior to copulation.

Male combat is perhaps the most spectacular example of use of tactile cues, and it occurs in a great variety of lizards and snakes. During male–male interactions, tactile cues can assume considerable importance. During the peak of breeding seasons, male *Sceloporus* engage in fights that involve bumping, biting, and even tearing of body parts, as do many skinks. It is not uncommon to observe male–male combat in which one lizard tears the tail off another. The cost of losing in male–male combat can be reduced social status or, in extreme cases, death. In some of the largest and potentially most dangerous lizards, such as varanids, male–male combat is much more ritualized and may never result in major injury to the lizard. Because the lizards are large, these wrestling matches can be spectacular events (Fig. 9.14).

FIGURE 9.14 Male–male combat in the Australian monitor lizard *Varanus panoptes*. Photograph by D. Pearson.

Auditory communication is limited in lizards. Many nocturnal geckos vocalize, and the calls undoubtedly function in communication. Nevertheless, these have not been well studied. Many geckos vocalize singularly, but the Barking gecko (*Ptenopus garrulous*) of the Kalahari calls in choruses, similar to breeding frogs. Some vocalizations are associated with aggressive interactions between males or during feeding interactions. Although geckos are best known for their vocalizations, a few other lizards vocalize. Canary Islands lacertids (*Gallotia*) may use sound in courtship, and some North American *Aspidoscelis* make sounds when picked up.

Social communication among juvenile lizards is poorly studied. Juvenile green iguanas, *Iguana iguana*, appear to recognize siblings on the basis of fecal odors. Juvenile *Anolis aeneus* defend territories and interact aggressively with other juveniles, especially when food is available. Moreover, behaviors associated with individual interactions change with age.

Snakes

Initial social communication in snakes is chemical, but tactile interactions are used as close-range signals between the sexes and, in some cases, between conspecific males. Some skin pheromones are critical for successful reproduction; they are not produced by the cloacal glands. Snakes have a diversity of glands and secretions, although the paired cloacal scent glands are best known and produce pheromones used by snakes for defense and trailing. The glands lie dorsal to hemipenes in males and in the corresponding position in females; often they are very large. Of the many explanations of cloacal gland function, defense is the most probable hypothesis because the secretions usually smell bad to humans, and some secretions repel specific snake predators. Observations of snakes returning to den sites and trailing other individuals suggest that glandular secretions are involved in these behaviors. In addition to serving as cues for locating aggregation sites, the secretions are used for discrimination during reproductive behavior.

Pheromones that attract males to females during the breeding season occur in the skin on the dorsal surface of the females. Like some scleroglossan lizards, snakes appear able to discriminate among pheromones produced by their own and other species. Garter snakes (*Thamnophis*) are best at discriminating among odors of other sympatric garter snakes, suggesting local natural selection on chemosensory abilities or the chemicals.

At middle and lower latitudes, garter snakes have an extended breeding season, and males can locate females by following pheromone trails. At northern latitudes, most garter snake breeding occurs when the snakes first emerge from the overwintering sites before they disperse. Because they overwinter in aggregations, large numbers of individuals interact. Several pheromones resembling vitellogenin are present in the skin of *Thamnophis sirtalis parietalis,* and whether on the back of a female snake or on the surface of an experimental arena, these chemicals elicit courtship behavior by males. Males generally emerge before females; they remain clustered at the den site awaiting the emergence of females. When females emerge, they are mobbed by males responding to the pheromones in their skin. Competition among males for access to the relatively few emerging females is intense, and as a result, most males do not mate. Not only can garter snakes males follow chemical trails of females, but in doing so, they obliterate the trail of the female, making it more difficult for other males to follow the female. Once close to a female, visual cues are used, but visual clues alone do not allow male snakes to discriminate between sexes, and thus they can be misled to a male based on visual cues alone.

In most snakes, tactile signals predominate in courtship once a male has determined the gender of a conspecific. Courtship and mating usually involve three discrete phases: tactile chase, tactile alignment, and intromission coitus. The tactile-chase phase includes the first contact between the snakes, including chemosensory sampling by males to determine sex. This phase usually is followed by chases or attempts to mount the female. During the tactile-chase phase, the male places his body alongside (undulation) or with a loop over the female's dorsal surface; segments of his body musculature may contract in a wavelike manner. In addition, the male often rubs his chin on the female's back or even bites her; in snakes with vestigial limbs (e.g., Boidae), the pelvic spurs scratch or titillate the female in the vicinity of her vent (Fig. 9.15). During the tactile-alignment phase, the first attempts to copulate occur. This involves rapid muscle contractions in the male's tail as it is aligned with the female's tail. These caudal vibrating movements are a tail-searching copulatory attempt. Tactile behaviors that occurred during the tactile-chase phase are often continued during the tactile-alignment phase. During the final phase, the female gapes her cloaca to allow the insertion of a single hemipenis, resulting in intromission and coitus.

Similar to lizards, male–male combat is common in snakes and has been observed in viperids, colubrids, boids, and elapids. Injury appears rare or nonexistent, likely partially a result of the fact that snakes have no weaponry (strong jaws, claws). Following gender identification by chemical cues, two males glide parallel to each other, usually with their heads raised. Although the postures vary among snake clades, male combat is generally a contest in which one male attempts to push down the head of the other male in order to establish dominance. In elapids and colubrids, the interaction is mainly horizontal, but in viperids, males lift their heads and anterior portions of their bodies off the ground, often intertwined, and push each other over, only to initiate the sequence again and again until dominance is established.

REPRODUCTIVE BEHAVIOR

Mating Systems

In general, mating systems are categorized according to the levels of polygamy within a species. Conflicting strategies between the sexes result from the differential investment of the male and female parents in the offspring. From the outset, males invest less in each individual offspring than females. Males produce millions of tiny sperm, few of which will fertilize eggs, whereas females produce relatively few eggs, each of which has a high probability of being fertilized. Each egg contains most of the energy required for development, whereas an individual sperm cell contains only genetic material and a flagellum for propelling itself to the egg. Many factors influence mating systems; these include the spatial and temporal availability of reproductively active individuals, the behavioral tactics of males and females, and numerous ecological, phylogenetic, and physiological constraints. In addition, parental care can play a significant role in mating systems. The study of mating systems of frogs and salamanders presents special challenges because many species are secretive

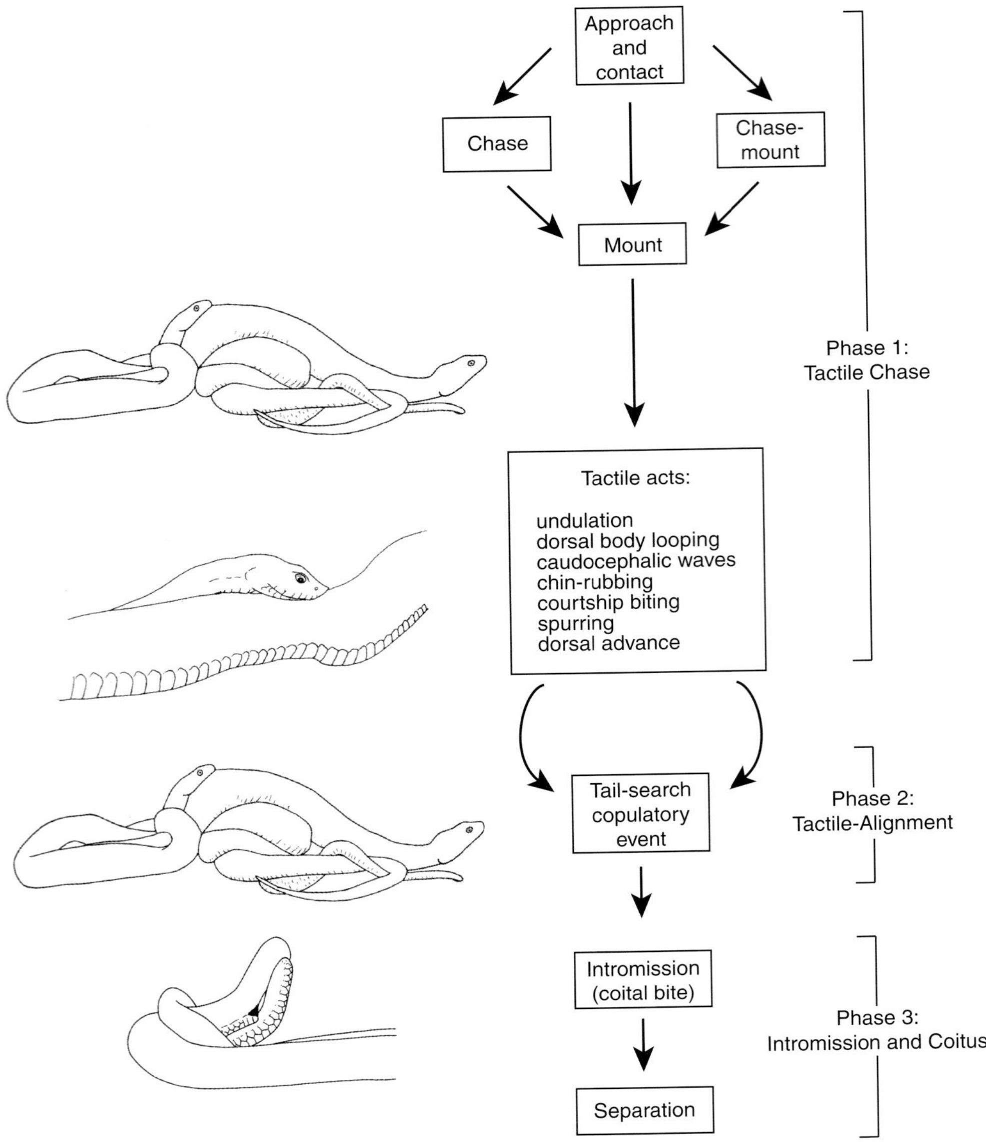

FIGURE 9.15 Three general types of tactile signals are used by snakes during courting and mating. Adapted from Gillingham, 1987, with drawings adapted from Carpenter, 1977.

or nocturnal and are thus difficult to observe. Mating systems of caecilians, for example, are largely unknown because these fossorial animals are nearly impossible to observe. The myriad behaviors in which males and females are involved and the choices each makes before and during courtship are oriented specifically toward the goal of mating and the production of offspring (Fig. 9.16).

Many aspects of amphibian and reptilian population and reproductive biology must be detailed to understand the mating system of any species. Population dynamics, mate choice and parental care behaviors, and physiological constraints on reproductive behavior are some of the factors that affect mating systems. The ratio of males and females in a breeding population is a major factor determining the structure of the mating system. If one sex is limited, the reproductive success of the other sex will be affected. Competition will occur among individuals of the abundant sex for access to individuals of the scarce (limited) sex. In most species, males compete for limited females. Determination of the limited sex cannot be made by examining the sex ratio in a large breeding aggregation, because not all individuals appear during any given breeding event. In addition, males may arrive synchronously at the breeding site, but females, even though they may be present in the environment in the same numbers as males, may arrive a few at a time over a long period. The population sex ratio may have little bearing on the sex ratio of males and females capable of

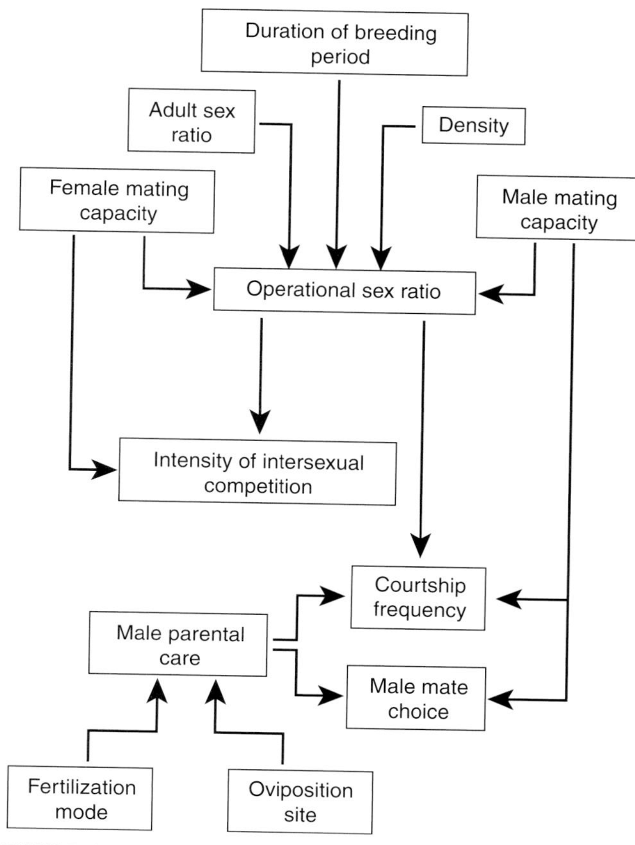

FIGURE 9.16 Determinants of the mating system in salamanders. Adapted from Verrell, 1989.

breeding at any given time. Rather, the operational sex ratio (OSR) is a critical determinant of the mating system. The OSR is the ratio of males to fertilizable females at any given time. Determining the OSR for any species presents many difficulties. In pond-breeding salamanders, for example, females may be present that are not ready to breed. In other species, females are present but breed synchronously, in effect making the OSR 1:1 for a brief period of time. Terrestrial frogs and salamanders present other problems. If males defend territories, the local OSR may be close to 1:1, depending on the amount of female movement. Among many lizards and snakes, the operational sex ratio continually changes as some females become sexually receptive and others become unavailable after fertilization occurs.

Monogamy and polygamy are the two major mating systems; polygyny and polyandry are two types of polygamy (Table 9.4). To a large extent, the number of mates acquired (mating success) by a particular sex and the number of offspring that result (fecundity) determine the kind of mating system. Relative to fecundity, if males increase their fecundity by mating with a large number of females but females have no gains by mating with more than one male, a polygynous mating system should result (Fig. 9.17). Monogamy is the likely outcome when neither males nor females gain by mating with additional individuals of the opposite sex. Monogamy also is expected in mating systems requiring both parents (biparental care) to insure the survival of offspring. Most amphibians and reptiles have polygynous mating systems, but there are many interesting exceptions. In addition, the operational sex ratio plays an important role in determining the intensity of sexual selection. Male-biased sex ratios usually result in competition among males for females, and in turn, this competition drives the evolution of sexual dimorphism.

Amphibian Mating Systems

The evolutionary framework for mating systems in amphibians considers not only OSR but also the type of breeding pattern of a species at any give time. At the extremes,

TABLE 9.4 Mating System Classification Based on Levels of Polygamy.

Mating pattern	Mating system type	Description
Polygyny	Female defense	Males defend groups of females; increased male–male competition
	Resource defense	Males defend resources required by females
	Lek	Males display at a communal site to attract females; both female choice and male–male competition intense
	Scramble competition	Males locate and mate with as many females as possible; male–male competition intense
Polyandry	Male defense	Females defend male mates in female aggregations
	Resource defense	Females defend resources required by males or by their offspring
Monogamy	Mate-guarding/ assistance	Males mate with single females and defend them against other males; OSR unity
Polygamy	Resource use	Either sex gains by multiple matings

Adapted from Sullivan et al., 1995.

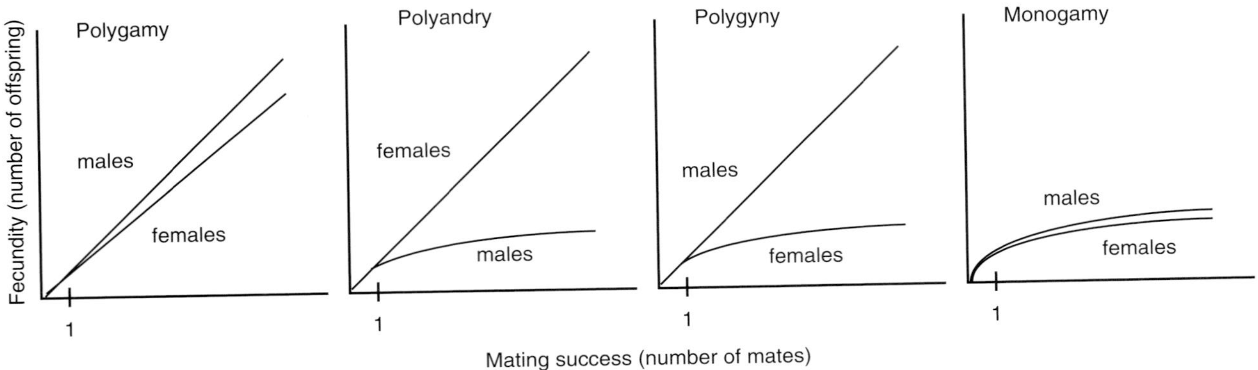

FIGURE 9.17 Mating systems as a function of the relative association between fecundity (number of offspring produced) and mating success for males and females. The line for each sex within a panel represents a sexual selection gradient. For example, in the panel in the lower left (polygyny), potential fecundity of males increases with increased number of mates, resulting in an increase in the intensity of sexual selection in males (line with a high slope), whereas females do not gain in fecundity by mating with additional males and thus there is little sexual selection operating on females (line with no slope). Adapted from Duvall et al., 1993.

some species are explosive breeders that accomplish all reproduction within 2 to 3 days when environmental conditions are suitable. Examples include the spadefoot toads, *Scaphiopus*, which breed during the first summer rains in the Southwest deserts of North America; *Lithobates sylvaticus*, which breeds during the spring thaw in ponds of northern North America; and the salamander, *Ambystoma maculatum*, which breeds in vernal ponds for about 1 month in early spring. At the opposite extreme, bullfrogs, many hylids, and many plethodontids are prolonged breeders. Males of these species establish and defend territories for several months, and females arrive gradually over a long period of time. The ratio of females to males can be very different in these contrasting circumstances.

In all species, a critical determinant of the mating system is the mode of sperm transfer. In nearly all frogs, fertilization is external and the male typically sequesters the female in amplexus during actual fertilization of the eggs. External fertilization provides opportunities for multiple males to fertilize a single clutch, but the extent to which this occurs is only beginning to be appreciated. Group spawning occurs in Spotted salamanders, *Ambystoma maculatum*, and multiple paternity can occur in as many as 70% of the clutches. Some tree frogs (*Agalychnis*) and myobatrachids (*Crinia*) have synchronous polyandry, in which more than one male amplexes with a female during spawning. The recent discovery of clutch piracy in *Rana temporaria* is yet another way in which more than one male can fertilize a clutch of eggs. Many species of salamanders have internal fertilization by means of a spermatophore that the male deposits on the substrate; the sperm packet must be picked up by the female. In most species, the male does not sequester the female in amplexus, but instead he must lead her over the spermatophore; thus, males are particularly vulnerable to interference from rival males at this critical time. Females may pick up more than one spermatophore in some species, and because most salamanders store sperm, the opportunity for sperm competition occurs, and a female's clutch may be fertilized by more than one male.

Salamanders

Mating systems in salamanders are partly established by whether the clade is aquatic or terrestrial. In general, aquatic species have shorter breeding periods than terrestrial species, partly because aquatic species depend on rainfall to establish the breeding habitat, which usually is a temporary pond. Prolonged-breeding salamanders, exemplified by the plethodontids, are typically terrestrial and usually establish and defend territories. All known species of salamanders are polygynous. Opportunities for mate choice are limited in explosive breeders but can occur in prolonged-breeding species.

Many species of mole salamanders (*Ambystoma*) are typical explosive breeders. Male spotted salamanders (*A. maculatum*) migrate to temporary ponds in early spring and often deposit spermatophores before the females arrive. Females and males engage in very little courtship; at most, some nudging occurs. Females move around the pond and pick up spermatophores from different males, resulting in multiple males fathering a single female's clutch. Competition among males is limited to the deposition of as many spermatophores as possible, and placing a spermatophore on top of another one is a common male tactic. Ultimately, a male's reproductive output is related to the number of spermatophores he can deposit.

Another pond-breeding species is the salamandrid *Notophthalmus viridescens*, the red-spotted newt of eastern

North America. Even though aquatic, these newts have an extended breeding season and a more complex reproductive behavior than ambystomatids. Females are either ready to lay eggs or become responsive to males during courtship. A male will amplex a female that is not immediately receptive, or he may attempt to induce a nonreceptive female to pick up a spermatophore without amplexus. In either case, courtship interference by other males is intense. In some populations, males mimic female behavior and nudge the cloaca of the courting male, causing him to deposit a spermatophore; later, the interfering male deposits his spermatophore to the now more receptive female.

Many studies of courtship behavior have been done under laboratory conditions. These studies have provided much information on interactions and behaviors that are difficult to obtain under natural conditions. However, the results of field studies sometimes conflict with laboratory studies. In red-spotted newts, for example, a laboratory study showed that spermatophore transfer by the courting male was successful in 60% of the amplectant pairs observed, whereas a field study of this newt revealed that spermatophore transfer was successful in only 6% of the observed pairs. Natural situations are much more complex than the laboratory, and many more factors impinge on the outcome of individual behaviors.

Terrestrial plethodontid salamanders typically establish territories that contain good food resources and reproductive sites. Their mating system is defined as resource-defense polygyny. Both male and female *Plethodon cinereus* establish territories. Male territories do not overlap each other, but female territories often overlap those of several males. Whether a female chooses among males is unknown, but females in a laboratory setting have been observed to spend more time around the territories of larger males.

Frogs

Many species of anurans exhibit explosive-breeding patterns in arid areas, as well as in forest habitats in temperate and tropical areas. The benefits of explosive breeding are obvious in deserts and semiarid areas because breeding can occur only when water is present. The advantages of explosive breeding in wetter areas seems related to the density of predators in aquatic sites, because as the length of the hydroperiod increases in ephemeral ponds, the density of aquatic predators, such as dytiscid and dragonfly larvae, also increases. Explosive breeding at the time of pond formation gives frog larvae a temporal advantage over their predators.

Explosive-breeding anurans characteristically have a high degree of male–male or scramble competition. Females usually arrive and depart the breeding site quickly, and competition for females can be intense. Males may attempt to displace amplectant males, and often, the larger male wins the contest. The OSR may not be 1:1 in all cases, and the potential for female choice may exist. Females of some species may approach specific calling males and bypass others. Explosive breeders include the spadefoot toads (*Scaphiopus*, *Spea)* of North America, *Lithobates sylvaticus* in northern North America, and *Bufo bufo* and *Rana temporaria* in Europe (Fig. 9.7).

Resource-defense polygyny occurs or has been implicated in several species of prolonged-breeding frogs. In the bullfrog *Lithobates catesbeianus*, males establish and defend territories that vary in the quality of larval habitats. Territories defended by large males have higher larval survivorship because they have lower densities of leeches that feed on the eggs and tadpoles. Whether females choose large males or some aspect of a male's territory is unknown. Relatively little is known about other species of frogs that defend territories and attend the eggs. Centrolenid frogs, for example, call from trees along streams and small rivers, and amplectant pairs deposit their eggs on leaves above the water. In *Hyalinobatrachium fleischmanni*, a species in which males attend eggs, females choose a male and initiate amplexus, but male characteristics on which choice might be based have not been determined (Fig. 9.18). In the sympatric species *Centrolene prosoblepon,* males do not attend clutches, and males initiate amplexus.

Polyandry, in which a female mates with several males, has the potential advantage of providing the female with a wider range of genetic diversity for her eggs. In the hyperoliid *Afrixalus delicatus*, the amplectant pair constructs a small nest by folding a leaf over about 35 eggs. The female does not deposit all her eggs at once, and some females break amplexus after depositing a clutch of eggs and seek another male with whom to construct another nest. Males in this system are polygynous; the behavior of females is poorly understood, and only 7% (of 100

FIGURE 9.18 A male glass frog, *Hyalinobatrachium fleischmanni*, calling from a leaf above a stream; below, a female is attracted to his call. Note that the eggs of the gravid female can be seen through the transparent venter. Photograph by W. Hödl.

observed pairs) of the pairs exhibit polyandry. True sex-role reversal in which a female mates with multiple males, each of which then cares for the resulting offspring, is unknown in frogs.

Monogamous mating systems are typically found in birds and a few mammals but are rare in other vertebrate groups. Monogamy has been widely cited to have evolved in birds because offspring survival is greater when two parents instead of only one are involved in feeding the young. Recent studies using genetic analyses have shown that monogamous relationships are more complex than was previously presumed. In birds, extra-pair fertilizations are common, even though the pair may remain socially monogamous. Males may derive increased reproductive success by fertilizing other females, even though they remain with a primary partner. In certain situations, pair-bonded females may mate with another male that may be of higher quality than the social partner but that is not available as a long-term partner. Reproductive parasitism of the other sex may occur if a male or a female can entice an unrelated individual to provide parental care for his or her offspring.

Monogamy is rare in frogs and has been implicated only in a few species of unrelated aromobatids, dendrobatids, and hylids, and like birds, the parents provide biparental care to the tadpoles. Parental care includes feeding the tadpoles trophic eggs deposited by the female parent. In all groups, the egg and larval habitat is a restricted site, such as a small tree hole or vine hole that holds water or the water-filled tanks of ground bromeliads. These sites are small, often with reduced or no light, and lack food for the tadpoles. All species of dendrobatids and aromobatids deposit terrestrial eggs and then transport their tadpoles to a small tree hole or other type of water-holding plant for development. In hylids, eggs rather than tadpoles are deposited directly into the water of these habitats. Prolonged parental care is required for tadpole survival.

In the dendrobatid *Ranitomeya vanzolinii,* males and females form pair bonds, and the pair remains together in a small territory. Clutch size is very small, about three eggs that are deposited above the waterline on the wall of a tiny tree or vine hole in the Amazonian rain forest (Fig. 9.19). Developing tadpoles are transported singly by the male to another site. Because the tadpoles are cannibalistic, they are not allowed to drop into the water in the same tree hole, where a larger tadpole may be present. No more than one tadpole occupies a small tree hole. The male and female court about every 5 days; the female then ovulates two eggs, one from each ovary. The male guides

FIGURE 9.19 A pair-bonded male and female of the spotted poison frog, *Ranitomeya vanzolinii*, emerging from a small tree hole; the male (foreground) is transporting a single tadpole on his back. On the right, an opened vine shows the cavity within used as a tadpole nursery. Pointer shows three eggs that were deposited above the waterline. When opened, a large tadpole was found in the water in the cavity. Frogs, photograph by J. P. Caldwell; vine, photograph by L. J. Vitt.

her to the tree hole containing one of their tadpoles, and the female deposits the trophic (unfertilized) eggs for the tadpole to consume. In the aromobatid *Anomaloglossus beebei*, an individual male and female form a pair bond, and both parents provide care for the tadpoles. The male parent cares for the eggs by moistening them and transports the tadpoles, whereas the female parent occasionally deposits trophic eggs for the tadpoles. The parents remain together on a small territory defended by the male.

In *Ranitomeya ventrimaculata*, which is closely related to *R. vanzolinii*, promiscuity is common. In this species, the larval habitat is a small amount of water held in the leaf axils of *Heliconia* plants. Tadpoles are not deposited singly as in *R. vanzolinii*; rather, many tadpoles from different clutches are either transported to the same axil or are allowed to slide into the pool as they develop from eggs attached just above the waterline. Cannibalism is common among the tadpoles and may provide a significant source of nutrients for the tadpoles. Indeed, the closely related *R. vanzolinii* and *R. ventrimaculata* exemplify how natural section has operated in different directions to produce two different types of mating systems. In *R. ventrimaculata*, reproductive parasitism is high, whereas it appears low in *D. vanzolinii*. The factors driving these two systems appear related to aspects of the larval habitat.

Monogamy and biparental care have also been reported in the hylid *Osteocephalus oophagus*. In this Amazonian rain forest species, an amplectant pair deposits a clutch of about 250 eggs in a tree hole. As the tadpoles develop, the same male and female return about every 5 days and deposit more fertilized eggs for the developing tadpoles to consume (Fig. 9.20). The mechanism for repeated pairing is not known. After about a month, some tadpoles metamorphose and leave the tree hole. Eggs continue to be deposited in the same tree hole, but not all of them are consumed by the older tadpoles, and these uneaten eggs hatch into more tadpoles. The result is that tadpoles of different sizes are present in a pool; generally the smaller ones are unable to obtain trophic eggs and die. Oophagy is obligatory in this species; if the parents do not regularly provide trophic eggs, the tadpoles starve.

Reptile Mating Systems

Snakes

Most snakes have polygynous mating systems, but a few are effectively monogamous. In polygynous snakes, males gain in terms of the offspring they sire by mating with more than a single female. Females maximize production of offspring by mating with a single male and investing time and energy in efficient foraging to gain the benefits associated with increased energy intake, which include a fecundity increase related to body size and condition. Prairie rattlesnakes, *Crotalus viridis*, are polygynous, and females are sexually receptive for only short periods of time, partly because they are nonreceptive during the extended gestation period. Female body size and the availability of food and heat to females influence the frequency of reproduction. Some females skip several years of breeding. Taken together, these factors result in a variable operational sex ratio; more males than females are available to breed at any one time. Thus, the OSR of these snakes depends largely on ecological factors and the peculiarities of viviparous pit viper breeding biology (Fig. 9.21).

Lizards

Most lizards have polygynous mating systems, but monogamy or at least extended pair bonding is more common than previously thought. Among iguanians, males of most

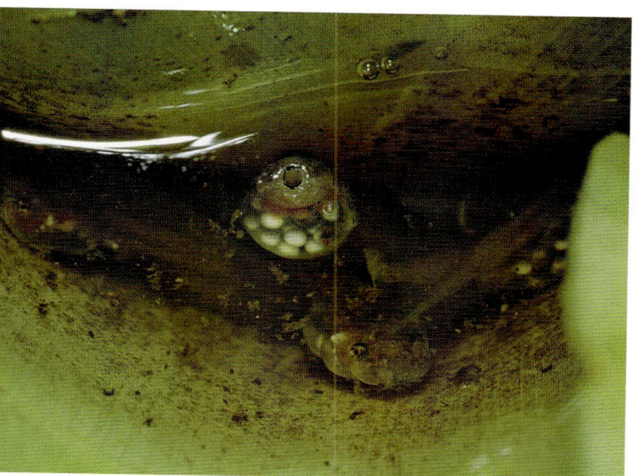

FIGURE 9.20 A marked female *Osteocephalus oophagus* (white waist band) returning to a small tree hole to deposit eggs as food for her offspring. Tadpoles nip at her cloaca to stimulate egg deposition. On the right, a tadpole that has just ingested eggs, which can be seen through the tadpole's transparent venter. Photographs by K.-H. Jungfer.

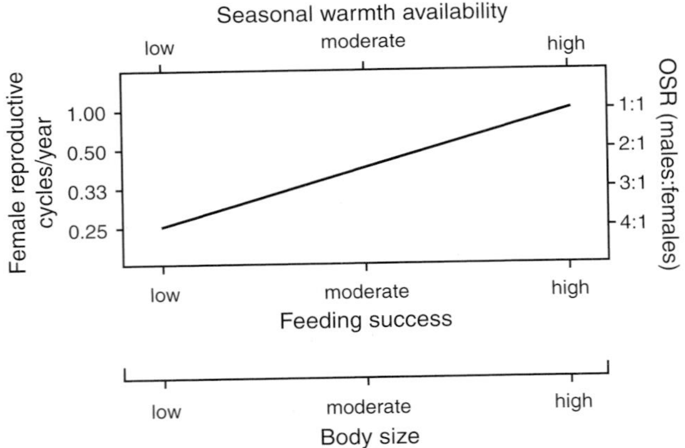

FIGURE 9.21 The operational sex ratios (OSR) of prairie rattlesnakes are intimately affected by the amount of heat available for thermoregulation and gestation, and by feeding success because it influences energy available for reproduction. Reproductive success is influenced by body size because it determines the number of offspring produced in a given season. Adapted from Duvall et al., 1992.

species defend at least part of their home range (see Chapter 8), and polygynous mating systems predominate. Male–male interactions commonly lead to sexual dimorphism; however, the degree of polygyny varies greatly. In some species, territorial males mate with only one or two females, whereas in others, individual males may mate with as many as six females. Territories or home ranges of females are often contained within the territory of the male, and male territorial boundaries are defended by males, resulting in low male home range overlap. Most communication is visual in territorial species, and a high diversity of male coloration and ornamentation occurs, perhaps as a result of intrasexual selection. Sexual dimorphism in head size is also common.

Among territorial iguanians, males appear to more vigorously pursue nonresident females that enter their territories than resident females. Males of keeled earless lizards, *Holbrookia propinqua*, and brown anoles, *Anolis sagrei*, either more intensely court nonresident females or selectively court nonresident females when offered a choice. In both these species, males appear able to recognize familiar and unfamiliar females. Vigorous courting of nonresidents might result in a nonresident female taking up residence in the male's territory or, even if the female leaves, if the male copulates with the female, the additional opportunities to sire offspring increase his individual fitness.

Even though most autarchoglossan lizards are nonterritorial, polygynous mating systems predominate. Because males usually search for females, often using a combination of visual and chemical cues, and courtship can be extended as can post-copulatory mate guarding, polygyny is often sequential. When male–male interactions occur, they are contests associated with the acquisition of a female that is being courted by one of the males. Sexual dimorphism in coloration and head size occurs in many species. In some long-lived scleroglossans, extended pair bonds and near monogamy occur.

In the sand lizard *Lacerta agilis*, females mate with many males even though they produce only one clutch of 4 to 15 eggs each season. The males with which they mate are variously related to the female. Female sand lizards appear to exert mate choice by preferentially using sperm from more distantly related males, likely as a result of intrauterine sperm competition. DNA fingerprinting studies demonstrate that males most closely related to the female sire fewer offspring that those that are more distantly related. Because several males can sire the offspring of a single female even though a single clutch is produced, the mating system is effectively polyandrous.

Most teiid lizards have sequential polygyny, in which males guard females when the females are receptive. Males often interact aggressively with other males that attempt to court the female. In *Ameiva plei*, large males win in male–male encounters and guard females during their entire sexually receptive period of 1 to 4 days. Among 21 mature males in a study site, only 6 mated with females, and the 4 largest males accounted for more than 80% of matings and sired nearly 90% (estimated) of the eggs produced. Because male body size determines success in guarding receptive females and guarding determines mating success, selection favors the evolution of large body size in males even though territoriality is not involved. In addition, females reject small males when a large courting male is removed but do not reject proxy large males, indicating that females select larger males. Presumably, females are harassed less when guarded by a large male and, as a result, can spend more time foraging.

In Australia, males and females of the long-lived, large-bodied skink *Tiliqua rugosa* form monogamous pairs that remain together up to 8 weeks prior to breeding. Males are often observed with the same female in consecutive years, suggesting long-term pair bonds as well. Unlike monogamous frogs, extended parental care does not occur, so no advantage to monogamy accrues for either sex relative to offspring survival. Moreover, the close association of males and females occurs prior to the time when the young are produced and ends after mating. Males and females are easily observed prior to and during the breeding season and, when in pairs, are often feeding. They are omnivores, with a diet dominated by plant material. A similar percentage of unpaired (69.9%) and paired (78.3%) females had food in their mouths when first observed. Thus, females feed similarly regardless of the presence of a male. The same was not true for males. Single males more frequently had food in their mouths (62.2%) than males observed with females (26.1%). A paired male follows closely behind the female while she forages, stopping when the female stops, but often not feeding. Although one reason for the male to defer feeding when the female stops is to maintain alertness for the possibility of an approaching male, another is to maintain vigilance to detect approaching predators. Both sexes gain from this behavior. The male gains by having access to a female for reproduction, and the female gains by being able to feed and gain energy for reproduction while the male watches for predators.

Although female guarding by males after copulation is generally assumed to ensure paternity, its effectiveness remains unclear. Most male *T. rugosa* are monogamous (82%), but some males sequentially pair with different females. Both females and males are occasionally observed with one or more additional partners, even though the apparent long-term bond is with only one partner. Based on microsatellite DNA analysis of females, their offspring, and their male partners, some females were found to produce offspring fathered by a second male, a finding not surprising considering the occasional extra-pair associations. Females paired with polygynous males were more likely to have extra-pair fertilizations than females paired with high-fidelity males. Females with polygynous partners have opportunities to be courted by other males.

Snow skinks (*Niveoscincus microlepidotus*) in Tasmania, are nonterritorial, but males guard females following copulations. By collecting DNA samples and using an Amplified Fragment Length Polymorphism (AFLP) procedure, Mats Olsson and collaborators were able to test the hypothesis that mate guarding ensures paternity. Nevertheless, 75% of clutches examined contained evidence that more than a single male was involved in fathering offspring within individual females' clutches.

Alternative Mating Strategies

Recently, researchers have closely studied the mating tactics of individual animals. In amphibians, external fertilization increases opportunities for alternative strategies. By marking individuals and following them for a long time, researchers have discovered that males in the same population use different strategies to obtain mates. In some cases, these alternative strategies are genetically based, but individual males of some species can switch strategies, depending on current internal or external factors.

Satellite Males

The satellite male strategy is common in frogs and occurs in some reptiles as well. In frogs, a male can adopt a calling strategy or become a satellite. Satellites do not vocalize but rather wait near a calling male to intercept females that are attracted to the calling male. Satellite male behavior occurs in numerous species of hylids, ranids, bufonids, and other clades.

Several hypotheses address the evolution of this strategy:

1. Calling sites or suitable territories are limited, and males compete for these sites and defend them by calling. Site holders are more competitive than satellites and may be larger.
2. Some males select a satellite status and become sexual parasites. This strategy includes individuals that switch back and forth between calling and satellite status, although other males are persistent satellites or persistent callers. Satellite behavior must have a payoff for it to have evolved and be maintained. In some cases, the mating success of satellites is equal to that of calling males.
3. The third and most comprehensive hypothesis predicts the adoption of satellite status because of energetic constraints mediated by hormones. Although the social and acoustic environments play important roles in determining male behavior, a male's internal physiology also dictates what strategy he adopts. Recent work has shown that vocalizing males have an increase in adrenal glucocorticoids, which in turn modifies androgen production and possibly neural mechanisms that regulate calling behavior. Chris Leary and his colleagues injected corticosteriod, a stress hormone, into two species of toads under natural conditions during breeding events and found that elevation of this hormone caused calling males to become noncalling satellites. Clearly, adoption of calling or satellite status involves interplay between the social or acoustical environment and the internal physiological status of an individual.

Experiments with satellite males of various species indicate that these hypotheses are not mutually exclusive. In *Dendropsophus minutus*, for example, some satellites begin calling when the nearby calling male is removed. This behavior supports the hypothesis of limited calling sites; however, in the same population, other satellites do not begin calling but move to another calling male and thereby support the switching hypothesis. These studies did not account for the hormonal status of individuals, which may help explain the conflicting results. In addition, Chris Leary and his colleagues discovered that satellite males in two species of toads, *Anaxyrus woodhousii* and *A. cognatus*, are smaller in size than calling males, but they are not younger. Smaller but not younger satellite males were suggested for *Pseudacris triseriata*, but other studies with bullfrogs and *Pelobates* have demonstrated that satellite males are both younger and smaller than calling males. These conflicting studies indicate that the relationship between age, body size, and adoption of calling or satellite behavior is complex. Individuals within a species may follow different growth trajectories that may influence their behavior. For example, tadpoles that metamorphose earlier in a season may have more time to feed and grow than those that result from later breeding events. It follows that energetic constraints mediated by endocrine regulation could act differently on smaller and larger individuals. More research considering both social and hormonal status of individuals is needed to understand when and why males adopt calling or satellite behavior.

For a female, interception by a satellite male may lower her fitness. Females of many species assess a potential mate's fitness based on attributes of his call. Because satellites are silent, females cannot evaluate their fitness. In at least one species, the toad *Epidalea* (formerly *Bufo*) *calamita*, females struggle to be released when amplexed by a silent male. However, a genetically superior male may behave temporarily as a satellite because of energetic or hormonal constraints. Breeding with such a male would not lower a female's fitness. Much more work is needed to understand the relationships among body size, age, and energetic and hormonal influences on satellite behavior before we can determine how fitness is affected for males or females.

A recently discovered alternative mating strategy in *Rana temporaria* has been termed *clutch piracy*. This species breeds explosively, although the operational sex ratio is strongly skewed toward males because only a small number of females arrive at the ponds each day. A female mates with one male and deposits a large spherical clutch of eggs. While an amplectant male and female are searching for a suitable egg deposition site, they can be followed by one or more males. At the moment spawning is complete, one or more of these "pirate males" will seize the clutch, clasp it, and release sperm. In one pond, 84% of clutches were clasped by one or more pirates. A male will sometimes release sperm over only the outside of the clutch, but other males actively enter the center of the clutch and release sperm into the interior. The proportion of eggs fertilized in pirated clutches fertilized only externally and nonpirated clutches was not different, but a significantly larger proportion of eggs was fertilized when pirates entered the center of the clutch. Microsatellite paternity analyses showed an average of about 26% of the embryos in the clutches was sired by pirates. In addition, one pirate that seized a clutch 1 minute after it was deposited by the parents and another that tore the clutch away from the parents just as it was deposited sired 95–100% of the eggs in the clutches. Both the pirate males and the females could have increased fitness as a result of this alternative mating strategy. Females would benefit by having their eggs sired by more than one male, which would increase genetic variation in their offspring, and pirate males would have an opportunity to sire offspring in a situation where access to females is limited. Males were observed to act as a parental male and a pirate male in the same evening, thus showing that the roles are behavioral and not genetically fixed.

Whether satellite males exist in snakes is uncertain. In the European *Vipera berus*, large males generally win male–male combat and gain access to females. Smaller males avoid interactions with large males, yet about 10% of the matings involve smaller males that "shadow" females. Even though the breeding season extends for only 3 weeks, females mate up to eight times, making it is possible for small males to mate and sire offspring even though the females have already mated. Multiple paternity of *V. berus* offspring occurs in many females, so advantages potentially exist for satellite behavior. Moreover, the operational sex ratio varies considerably through time due to seasonal weather, the availability of receptive females, and variation in survivorship. When the operational sex ratio is lowest (i.e., many receptive males compared with few receptive females), combat among males for access to females is most likely to occur, and this, in turn, intensifies sexual selection for large body size in males (Fig. 9.22).

Alternative mating strategies exist in garter snakes (*Thamnophis sirtalis parietalis*) and may ultimately be much more common than previously thought. At den sites, male–male competition is intense. Large size enhances mating success for males at den sites, and chemical cues by tongue-flicking all individual snakes are used to discriminate sex of individuals. However, not all mating occurs at den sites. Some females leave the den site before mating. Smaller males use chemical cues to follow pheromone trails of these females, and then a combination of visual and chemical cues are used when a female is found. Competition with other males is nonexistent in surrounding woodlands because most males are at the den site, and as females move away from the den site, the density of males decreases radically. Consequently, use of visual cues in woodlands does not have a high cost because misidentification is unlikely. Thus male garter snakes adjust their mating tactics to the situation at hand.

Sexual Interference

In salamanders, males do not use the satellite strategy. Because the mode of sperm transfer requires a male to entice a female to pick up his sperm packet, interference by other males is the main type of intermale competition. The major factor determining whether a male courts or interferes is whether he finds a courted or noncourted female. Males apparently adopt a courtship strategy or an interference strategy based on the circumstances.

Males have evolved at least four types of interference behaviors: interference through female mimicry, spermatophore covering, wrestling with a male already in amplexus to attempt a takeover, and overt fighting.

Some male newts (*Notophthalmus* and *Triturus*) mimic female behavior to avoid detection by a courting male. The rival male can cause the courting male to deposit his spermatophore at the wrong time by nudging his

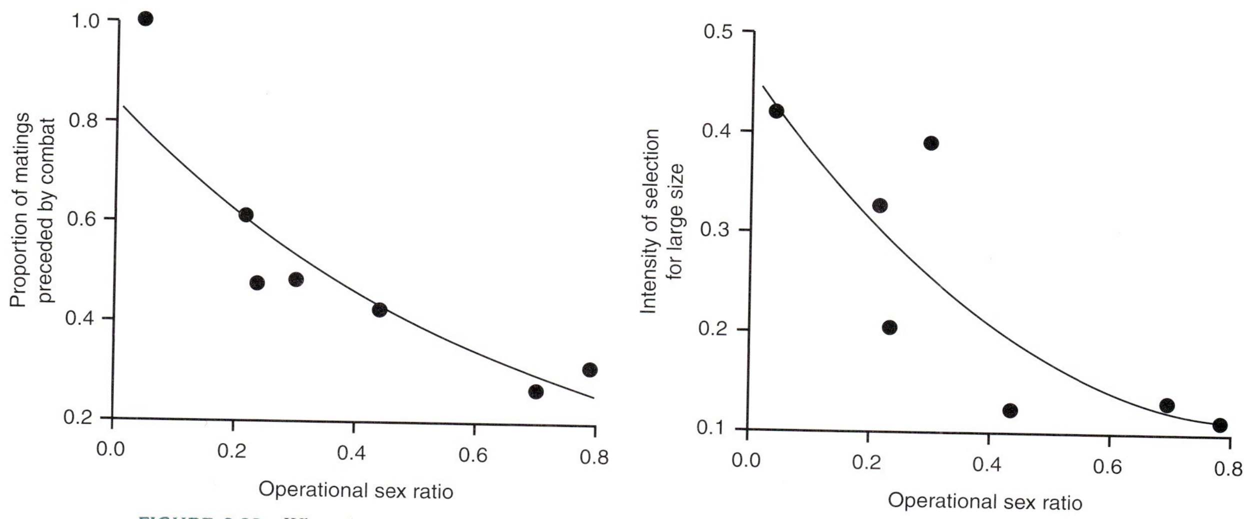

FIGURE 9.22 When the operational sex ratio in adders (*Vipera berus*) is low, male–male combat increases (left), resulting in increased sexual selection for large body size in males (right). Adapted from Madsen and Shine, 1993.

cloacal lips; subsequently, the rival deposits a spermatophore and induces the female to pick it up. In ambystomatids, an interfering male deposits his spermatophore on top of the courting male's spermatophore, thus substituting his sperm for that of the courting male. In salamandrids and ambystomatids, a courting male often amplexes a female if she is not immediately receptive. A wresting bout ensues if a rival male attempts to dislodge the amplectant male. Both size and prior ownership determine the outcome of the attempted takeover. Duration of the contest increases with increasing size of the intruder, and displacement occurs only when the amplectant male is smaller than the intruder.

Overt fighting is most common in plethodontid salamanders. Fighting includes biting, chasing, and the adoption of certain postures. One male can affect the future reproductive success of another male by inflicting physical injury. For example, damage to the tail can result in the loss of the tail and its fat reserves, whereas damage to the nasolabial grooves can interfere with a male's ability to obtain food.

Sexual interference is undoubtedly common in lizards based on numerous anecdotal observations of territory holders repeatedly chasing off smaller males and smaller males trailing courting pairs of nonterritorial lizards. In the large-bodied *Ameiva ameiva* and in broad-headed skinks *Plestiodon laticeps*, females are frequently pursued by more than one male. The courting male chases the trailing male when the latter approaches too closely. If the males are of similar size, the chase is prolonged and either delays or diverts the courting male's ability to mate.

Among the most fascinating examples of sexual interference in reptiles is the presence of "she-males" in garter snakes. Some male *Thamnophis sirtalis parietalis* elicit courtship behavior of other males in large mating aggregations. These "she-males" apparently produce an estrogen as a result of high testosterone levels; this chemical cue either causes other males to misidentify them as females or prevents their identification as males. In experiments testing the mating success of she-males and normal males, she-males were more than twice as successful as normal males, so she-males have a strong mating advantage. Behaviorally, interference works in two ways. A she-male gains because other males do not interfere during courting, or by courting a she-male, a normal male's reproductive effort is misdirected and potentially lost. While she-males may have a mating advantage, they may not have a fitness advantage because the production of estrogen potentially can result in fewer or less viable sperm.

Until recently, observations on garter snakes indicated that larger males mate more than smaller males because they have an advantage in male–male interactions at the den site. Early studies indicated that larger males are more effective at pushing the cloaca of smaller males away from the cloaca of females. Recent experiments by Rick Shine and Robert Mason reveal that larger males mate more because they are better able to coerce females, not because they have an advantage in male–male rivalry. They showed this by conducting experiments in which single males with females were compared with two males with a female. Mating occurred most often in females that were unable to resist courtship-induced hypoxia stress, and it had nothing to do with presence of other males. Thus, even though larger male size aids in male–male rivalry in garter snakes, male–female interactions are most important in providing the mating advantage to large size in males. During a breeding event, because of many males attempting to mate with the female, the female experiences hypoxia stress, which causes her cloaca to open. Females have no way to "choose" among males trying to insert a hemipene at this point. Large male body size enhances male success because they can force their hemipene into the cloaca of the female easier than smaller males during hypoxia (coercive mating).

Sexual Dimorphism and Sexual Selection

Sexual dimorphism in body size, coloration, and a variety of morphological characteristics is well known in amphibians and reptiles. Male bullfrogs, *Lithobates catesbeianus*, and green frogs, *Lithobates clamitans*, for example, have larger tympana than females because male calls are critical for territory maintenance (Fig. 9.23). In many instances, competition among individuals of the same sex (usually males) for access to individuals of the other sex is the driving force behind the evolution of sexual dimorphism. These interactions between individuals of the same sex determine reproductive success and result in intrasexual selection. Male competitive ability is at a selective advantage. Because size often dictates a male's success in contests with other males, intrasexual selection can drive the evolution of increased body size in males and result in sexual dimorphism with males larger than females. In other instances, females may choose males for mates based either on size or some other overt male trait. This female choice also can result in males being the larger sex. Sexual selection, however, is not the only factor that determines body size within each sex. For example, large female body size can be selected because size and fecundity are linked in many species. Sexual size differences can arise from differing growth trajectories, age at sexual maturity, and patterns of energy use. Differences in size between males and females more often represent a combination of the effects of sexual selection and natural selection.

In frogs, females are larger than males in about 90% of the species studied. Although sexual selection is the usual explanation for male and female size differences, other factors also are involved in anurans. For example, many species of frogs are explosive breeders, and male–male competition for mates is the rule in these species. Among prolonged breeders that maintain territories, larger males most often win in bouts with smaller intruders. In species with female choice, females choose males based on their calls, and they often prefer calls with a lower fundamental frequency, that is, those produced by larger males. All these factors typically drive selection for large size in males. Thus, sexual selection does not explain why females are larger than males.

In many species increased fecundity is correlated with large size in females. But why males do not achieve the same size as females is unknown. One explanation is that males have energetic demands associated with breeding. Males must call to attract mates, maintain territories, and compete with other males. Recent studies show that females prefer males with high calling rates or with longer or more complex calls, both of which are energetically expensive. Calling requires more energy than any other male activity. Also, males may have less time to forage, resulting in slower growth and ultimately in smaller size. In addition, sexual dimorphism in frogs is often expressed in morphological traits other than size. For example, male toads have large nuptial excrescences, and male *Leptodactylus* have huge forearms compared with females (Fig. 9.24).

In turtles, sexual dimorphism in body size and coloration is common. In some species, males are the larger sex, but in others, females are larger. Males are larger than females in most terrestrial taxa, and male combat is

FIGURE 9.23 Sexual dimorphism in the tympanum of the green frog, *Lithobates clamitans*. Female left, male right. Photographs by J. P. Caldwell.

FIGURE 9.24 Male (left) and female (right) *Leptodactylus ocellatus* showing sexual dimorphism in forelimb size. Photographs by J. P. Caldwell.

common in these species, suggesting that intrasexual selection drives the evolution of large body size in males. Males are smaller than females in most aquatic species. In these, male mobility determines reproductive success; males must locate females and court them in a high-density, three-dimensional environment, water. Increased body size in males would likely reduce their ability to gain access to females. Selection on female body size is not relaxed because the number of eggs, and in some cases, the size of individual eggs, increases with body size. Females reach large sizes even though the size of males is constrained. When life history traits are considered, the evolution of sexual size dimorphism becomes more complex. Body size and age at maturity are critical variables; they result in size differences largely because growth rates at sexual maturity decrease. If males reach sexual maturity at a younger age and growth rates are identical, males remain small relative to females even if they continue to grow. For many turtles, natural selection favors the rapid attainment of large size to deter predation, and sexual selection favors rapid maturation, particularly in males, so they can mate sooner.

In snakes, male combat is closely linked to the evolution of sexual dimorphism. In the 15% of snakes using male combat, most taxa are sexually dimorphic, and males are larger. Intrasexual selection in which relatively larger males win in male–male social interactions appears to be the ultimate cause of sexual size dimorphism in these snakes. The proximal cause appears to be the continuation of male growth after sexual maturity. In most cases where females are the larger sex, male combat does not occur.

In lizards, aggressive interactions among males appear to result in sexual dimorphism, and males are larger than females regardless of whether the lizards are territorial. In addition to males attaining larger size due to intrasexual selection, males often have larger heads or ornamentation (Fig. 9.25). In territorial species such as *Anolis* and *Sceloporus*, a male's reproductive success usually correlates with the number of females within his territory or his number of copulations. In nonterritorial species such as *Ameiva*, most *Cnemidophorus*, and *Plestiodon*, home ranges of males are large and overlap those of several females. Males not only court females for extended periods but also guard females from advances of other males, often interacting aggressively for access to females that are receptive. In both territorial and nonterritorial species, reproductive success of males is usually determined by size. Larger males are successful territory holders in territorial species or are successful at guarding females in nonterritorial species (Fig. 9.26). In both cases, larger males win aggressive encounters. In some nonterritorial species, such as *Ameiva plei*, females reject small males even in the absence of a larger male. This preference for large males allows females to continue foraging during the breeding period because the presence of a large male reduces the harassment of a female by smaller males. In *Plestiodon laticeps*, small males avoid encounters with larger males because there is a low probability of winning. Small males court females only when large males are absent, thus deferring agonistic behavior until they are larger and the probability of success is increased.

A variety of ecological factors also can influence sexual size dimorphism. In pond turtles, for example, the annual frequency of clutch production is associated with

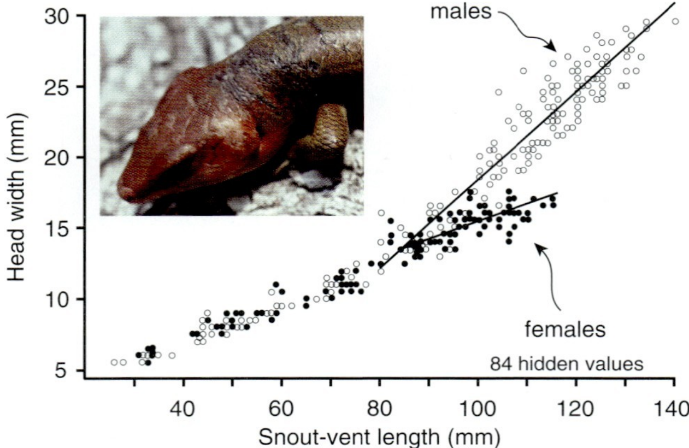

FIGURE 9.25 Sexual dimorphism in relative head size in the broad-headed skink, *Plestidon laticeps*. Note that the divergence between sexes in relative head size occurs after sexual maturity is attained (about 84 mm SVL). Scars on the head and neck of the male result from male fighting. Adapted from Vitt and Cooper, 1985.

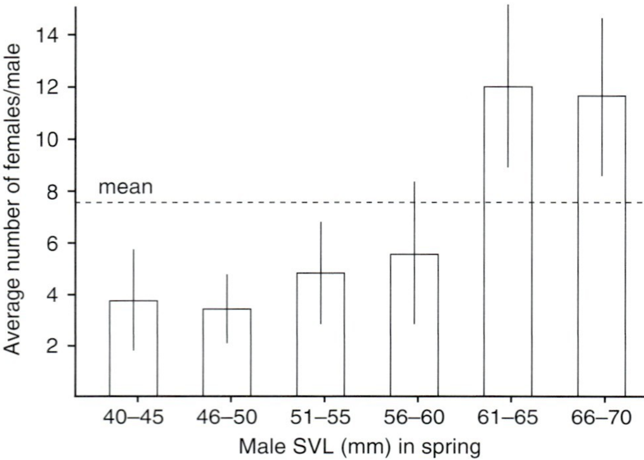

FIGURE 9.26 The number of mating opportunities (female/male) increases with body size for males of *Anolis carolinensis*. Adapted from Ruby, 1984.

sexual dimorphism. Sexual dimorphism increases with increasing number of clutches produced per season.

Snakes offer several examples of how ecology can influence body size. Because snakes swallow their prey whole, a strong association exists between the head size and maximum prey size. Furthermore, unlike lizards, snakes rarely use their heads in mating behavior, so sexual selection on relative head size does not occur. Males and females in many snake taxa have evolved differences in body size, relative size of the head, or ecology. Divergence in body size is related to reproductive differences, but the divergence in head size reflects independent adaptations of feeding behaviors in females and males. Neither sexual selection (at least directly on head size) nor resource partitioning causes sexual dimorphism in snakes. Rather, independently evolved differences in size and trophic structures account for the dietary differences between females and males.

MISCELLANEOUS SOCIAL AGGREGATIONS

Other interesting social interactions exist among amphibians and reptiles, one of which we discuss briefly. Work on salamanders (*Plethodon cinereus*) indicates that scat piles function for territorial advertisement and individual recognition. Among lizards, many species pile scats in one place, and an association exists between sociality and scat piling in many lizards. Most known lizards that pile scats are in the Australian skink genus *Egernia*. Scat piles usually consist of scats from a single individual and are placed close to basking sites. Some lizards have communal scat deposition sites (e.g., *Egernia hosmeri*). Lizards can discriminate between their own scats and those of other individuals.

The Australian gecko *Nephrurus milii* lives in social groups and also piles scats communally. However, when two or more retreats (crevices) are available, they do not pile scats in the crevice that they inhabit. Banshi Shah and collaborators conducted a clever experiment designed to test the hypothesis that these scat piles served in recognition of microhabitats for these geckos. What they found was that marking crevices with scat piles did not affect crevice use by geckos, thus falsifying the hypothesis for this species. Both aggregation and scat piling in this gecko may result from use of crevices for thermoregulation while digesting prey, rather than social communication.

Other kinds of social aggregations (e.g., overwinter denning) have been discussed elsewhere.

QUESTIONS

1. Describe parental care in the following: dendrobatid frogs (Dendrobatidae), the American alligator (*Alligator mississippiensis*), the Cane toad (*Rhinella marina*), the Five-lined skink (*Plestiodon fasciatus*), the Gastric brooding frog (*Rheobatrachus silus*), and the Western coachwhip (*Masticophis flagellum*).
2. How does mating success differ between males and females in polygynous mating systems?
3. If you found a new lizard species in which males were larger than females and had heads that were much larger than those of females, what are three possible explanations for such differences and how might those explanations cause the observed differences?
4. Prairie rattlesnakes are an example of the interaction between ecological factors and sexual selection. Define "operational sex ratio." Then, based on your definition, explain how the ecology of these snakes influences on the operational sex ratio.
5. Describe the differences between simple, compound, and complex displays in *Anolis* lizards. What kinds of information are transferred with these displays?
6. Describe the numerous ways that frogs use auditory signals in communication.

ADDITIONAL READING

Arnold, S. J., and Duvall, D. (1994). Animal mating systems: a synthesis based on selection theory. *American Naturalist* **143:** 317–348.

Birkhead, T. (2000). *Promiscuity: An Evolutionary History of Sperm Competition.* Faber and Faber Ltd., London.

Cooper, W. E., Jr., and Greenberg, N. (1992). Reptilian coloration and behavior. In *Biology of the Reptilia. Volume 18, Physiology. Hormones, Brain, and Behavior.* C. Gans and D. Crews (Eds.). pp. 298–422. University of Chicago Press, Chicago.

Duvall, D., Arnold, S. J., and Schuett, G. W. (1992). Pitviper mating systems: ecological potential, sexual selection and microevolution. In *Biology of Pitvipers.* J. A. Campbell and E. D. Brodie, Jr, (Eds.). pp 321–336. Selva, Tyler, TX.

Fox, S. F., McCoy, J. K., and Baird, T. A. (Eds.). (2003). *Lizard Social Behavior.* John Hopkins University Press, Cambridge, Mass.

Halpern, M. (1992). Nasal chemical senses in reptiles: structure and function. In *Biology of the Reptilia. Volume 18, Physiology. Hormones, Brain, and Behavior.* C. Gans and D. Crews (Eds.). pp. 422–523. University of Chicago Press, Chicago.

Heatwole, H., and Sullivan, B. K. (Eds.). (1995). *Amphibian Biology, Vol. 2, Social Behaviour.* Surrey Beaty & Sons, Chipping Norton, NSW.

Ryan, M. J. (1985). *The Túngara Frog. A study in Sexual Selection and communication.* University of Chicago Press, Chicago.

Stamps, J. A. (1983). Sexual selection, sexual dimorphism, and territoriality. *In* "Lizard Ecology. Studies of a Model Organism" (R. B. Huey, E. R. Pianka, and T. W. Schoener, Eds.), pp. 169–204. Harvard University Press, Cambridge.

Wells, K. D. (2007). *The Ecology and Behavior of Amphibians.* University of Chicago Press, Chicago.

REFERENCES

General

Halliday and Tejedo, 1995; Ryan et al., 1981.

Communication

Alcock, 1998; Ruby and Niblick, 1994.

Caecilians

Billo and Wake, 1987.

Salamanders

Halliday, 1977; Houck and Sever, 1994; Verrell, 1982, 1989.

Frogs

Blommers-Schlösser, 1975; Brenowitz and Rose, 1999; Caldwell and Shepard, 2007; Glaw et al., 2000; Haddad and Giaretta, 1999; Kluge, 1981; Krishna and Krishna, 2006; Kunte, 2004; Lima and Caldwell, 2002; Lindquist and Hetherington, 1996; Pauly et al., 2006; Pombal et al., 1994; Ryan, 1985; Ryan et al., 1982; Vences et al., 2007; Wells, 1980, 1988; Wevers, 1988.

Turtles

Auffenberg, 1965; Boiko, 1984; Halpern, 1992; Jackson and Davis, 1972; Mason, 1992.

Crocodylians

Garrick and Lang, 1977; Garrick et al., 1978.

Tuataras and Lizards

Adams and Cooper, 1988; Alberts, 1992; Auffenberg, 1981; Bull et al., 2001; Cooper, 1984, 1996; Cooper and Crews, 1988; Cooper and Greenberg, 1992; Echelle et al., 1971; Fitch, 1954; Fleishman, 1992; Gans et al., 1984; Gardner et al., 2001; Gillingham et al., 1995; Greenberg and Noble, 1944; Huyghe et al., 2005; Jenssen, 1970a,b; Jenssen and Hover, 1976; Jenssen et al., 2005; Lailvaux and Irschick, 2006, 2007; Lappin and Husak, 2005; Lena and de Fraipont, 1998; López et al., 1997; Main and Bull, 1996; Martins, 1993; Martins et al., 2004; O'Conner and Shine, 2004; Radder et al., 2006; Stamps, 1977; Tokarz, 1995; Trauth et al., 1987; Werner et al., 1987; Whiting et al., 2003; Wiens, 1999.

Snakes

Clark, 2004; Ford, 1982; Ford and Low, 1984; Ford and O'Bleness, 1986; Ford and Schofield, 1984; Garstka and Crews, 1981, 1986; Gillingham, 1987; Gregory, 1982, 1984; Halpern, 1992; LeMaster and Mason, 2002; LeMaster et al., 2001; Mason, 1992; Noble and Clausen, 1936; Quay, 1972; Shine et al., 2005a,b.

Reproductive Behavior

Mating Systems

Arnold and Duvall, 1994; Emlen and Oring, 1977; Shine et al., 2001a,b, 2004a,b, 2005a,b; Sullivan et al., 1995; Trivers, 1972; Verrell, 1989; Verrell and Halliday, 1985.

Amphibian Mating Systems

Backwell and Passmore, 1990; Birkhead and Møller, 1995; Bourne et al., 2001; Brockmann, 1993; Caldwell, 1997; Caldwell and Oliviera, 1999; Howard, 1978a,b; Jacobson, 1985; Jungfer and Schiesari, 1995; Jungfer and Weygoldt, 1999; Massey, 1988; Mathis, 1991; Mathis et al., 1995; Myers and Zamudio, 2004; Sullivan, 1989; Sullivan et al., 1995; Summers and Amos, 1997; Summers and Earn, 1999; Summers and McKeon, 2004; Verrell, 1983, 1989; Wells, 1977.

Reptile Mating Systems

Anderson and Vitt, 1990; Bull, 1988, 1994; Bull and Pamula, 1998; Bull et al., 1998; Censky, 1995, 1997; Cooper, 1985; Duvall et al., 1993; Gregory, 1974; Mason and Crews, 1985; Olsson and Madsen, 1998; Olsson et al., 1996, 2005; Sinervo and Zamudio, 2001; Stamps, 1983; Tokarz, 1992.

Alternative Mating Strategies

Byrne and Roberts, 2004; Halliday and Tejedo, 1995; Höggren and Tegelström, 1995; Jaeger, 1981; Leary et al., 2004, 2005, 2006; Madsen and Shine, 1993, 1994; Madsen et al., 1992; Mason, 1992; Mason and Crews, 1985, 1986; Shine and Mason, 2005; Shine et al., 2000a,b, 2001a,b; 2005b; Verrell, 1989.

Sexual Dimorphism and Sexual Selection

Anderson and Vitt, 1990; Berry and Shine, 1980; Blair, 1960; Censky, 1996, 1997; Congdon and Gibbons, 1987; Cooper and Vitt, 1987; Forsman and Shine, 1995; Gibbons and Lovich, 1990; Lande, 1980; Price, 1984; Ruby, 1981, 1984; Runkle et al., 1994; Savitsky, 1983; Shine 1978, 1979, 1991a, 1994; Shine et al., 2005c; Stamps, 1983; Stamps and Krishnan, 1997; Stamps et al., 1994, 1998; Trivers, 1976; Vieites et al., 2004; Vitt and Cooper, 1985; Woolbright, 1983.

Miscellaneous Social Aggregations

Bull et al., 2000; Bustard, 1970; Carpenter and Duvall, 1995; Chappel, 2003; Horne and Jaeger, 1988; Jaeger 1981, 1986; Jaeger et al., 1986; Shah et al., 2006; Stammer, 1976; Swan, 1990; White, 1976.

Chapter 10

Foraging Ecology and Diets

Amphibians and reptiles are often the most abundant terrestrial vertebrates at any locality in the warmer parts of the world, and like other animals, they must eat other organisms to survive. Given their high species diversity and abundance, their impact on other animal species—and in some instances, plants—is not trivial. Although some particularly interesting exceptions exist, caecilians generally feed on earthworms and other invertebrates, frogs and salamanders feed almost exclusively on insects (at least as adults), crocodylians feed largely on other vertebrates, turtles feed on a combination of plants and animals, and squamates feed largely on invertebrates or vertebrates, although two lizard taxa (Iguaninae and Leiolepidinae) are herbivorous. In addition, a large number of small-bodied tropidurine lizards in southern South America are also herbivorous. Many lizard squamates across taxa feed occasionally on fruits and flowers.

In nature, amphibians and reptiles have a huge diversity of food items available, yet no amphibian or reptile eats all available items. More explicitly, none samples available food randomly. Instead, an individual eats a particular subset of available food, and diets of individuals usually reflect diets of a species in a particular habitat. The preferred food can range from a variety of appropriate-sized arthropods or insects to just one prey type, such as termites. Even among species living in the same area, diets differ. Are these differences the result of competition? How much of the variation in diets that we see among species living in the same environment is historical? These issues are examined in Chapter 13. The emphasis here is how amphibians and reptiles detect, pursue, and capture their prey; the relative sizes of prey; the kinds of food they eat; and the evolution of sensory systems relative to prey choice. Diets of amphibians and reptiles are complex and influenced by many abiotic and biotic variables. As a result, methods, analyses, and interpretations of diet studies vary considerably, and no single "best" protocol exists.

FORAGING MODES

Two well-publicized foraging modes are recognized: sit-and-wait foraging (also referred to as ambush foraging) and active foraging (also referred to as wide foraging). These foraging modes were originally defined on the basis of behaviors used to locate and capture prey. Theoretically, sit-and-wait foragers invest little time and energy searching for prey. They typically remain stationary and attack mobile prey that move within their field of vision. Most foraging energy is spent in the capturing and handling of prey. Active foragers move about through the environment in search of prey, expending considerable energy in the search phase but little energy in the capture phase of foraging. Although many species of amphibians and reptiles can easily be placed into one of these two categories, some are herbivorous, and as a consequence, they do not pursue prey in the classical sense. Whether or not a "continuum" exists between sit-and-wait and wide foraging remains controversial. Early studies indicated that such a continuum should exist based on theoretical grounds. Recent studies showing that major dietary shifts (along with associated morphological and behavioral shifts) occurred deep in the evolutionary history of squamates raise questions about the reality of such a continuum. In addition, even though significant data have been collected on relevant behavioral correlates of foraging mode, we have barely begun

Herpetology, 3rd Ed.

to scratch the surface in terms of compiling data on most of the world's herpetofauna.

Our theory on foraging is based heavily on the idea that foraging behavior is evolutionarily plastic and responds to differences in prey abundance and behavior. A decade ago, discussions of foraging mode were strictly selection based. Foraging behavior was assumed to be driven in each species by a combination of competition and energetic aspects impinging upon a particular species. This interpretation was made and widely accepted in spite of the observation that specific foraging modes were shared by closely related species and groups of species. One prediction of this hypothesis is that a continuum of foraging modes should exist. The introduction of modern comparative methods that apply cladistic analyses to behavioral and ecological phenomena provides a different perspective. For example, phrynosomatine lizards are sit-and-wait foragers, whereas teiid lizards are active foragers. Mapping foraging modes on a phylogenetic dendogram reveals that sit-and-wait foraging is shared among the earliest branching squamate clades and the other lepidosaurian clade, the sphenodontans or tuataras. This observation argues for the evolution of sit-and-wait foraging in the distant past and the origin of active foraging much later or more recently in the evolution of squamates. The observation further suggests that active foraging in lizards likely arose as a single evolutionary event.

The bimodality of sit-and-wait versus active foraging appears obvious within single assemblages of species (e.g., lizards in the deserts of the southwestern United States). A synthesis of foraging data by Gad Perry shows that bimodality is evident and no continuum of foraging modes is detectable when the confounding effects of phylogeny are removed (Fig. 10.1). Phylogenetic analyses of other behaviors related to foraging also indicate that much of the variation has its origins deep within phylogeny rather than representing repeated adaptive responses to prey types, distribution, or abundance.

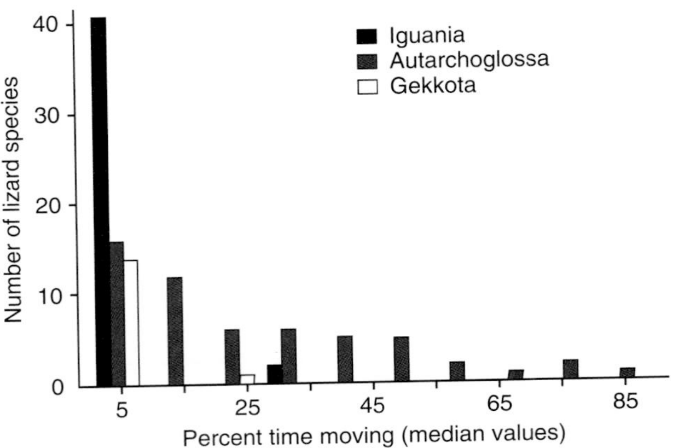

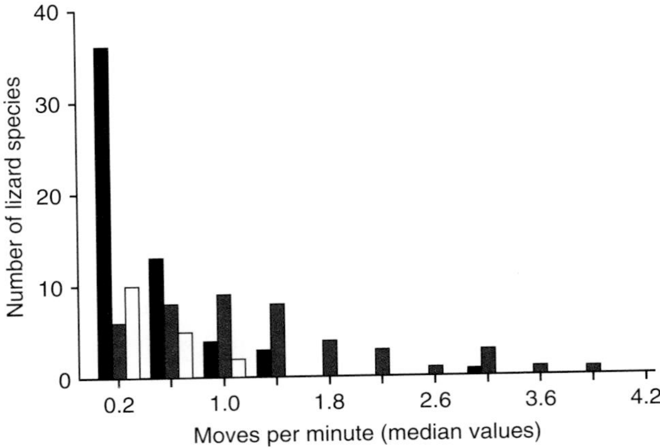

FIGURE 10.1 Two important behavioral attributes of lizard foraging, the number of moves per unit time and the percent of time spent moving, vary considerably across lizard species. Most lizard species in the Iguania, a group typically considered sit-and-wait foragers, make fewer moves and move less distance than lizards in the Autarchoglossa, a group typically considered to be active foragers. Phylogenetic analyses of percent time moving and number of moves per unit time confirm that the apparent bimodality in behavioral attributes of foraging mode have an historical basis (i.e., they reflect phylogenetic patterns rather than easily identifiable ecological patterns). Adapted from Perry, 2007.

Nevertheless, extremes in foraging behavior are apparent regardless of the number of evolutionary events causing them. Foraging behavior does not evolve in a vacuum; consequently, numerous ecological, behavioral, physiological, and life history correlates of foraging mode can be identified. Similar to time spent moving and the number of moves per unit time (behaviors associated with search behavior; Fig. 10.1), the so-called "correlates" of foraging mode likely also have a historical basis. Many correlates are intuitively obvious based on behaviors associated with prey search and capture (Table 10.1). Species that are sit-and-wait foragers typically do not move while waiting for potential prey to pass through their field of vision. They would be expected to be visually oriented or even use thermal cues (as in pit vipers), have cryptic morphology or coloration (so that neither the prey nor predators detect them), and have a physiology that results in optimal function under conditions in which little movement, other than prey attack, occurs. Actively foraging species search though a habitat for prey and are expected to use a combination of visual and chemical cues for prey detection. Because they move while foraging and have well-developed chemical senses, they can find nonmoving, clustered, or hidden prey that might not be detected by sit-and-wait foragers. Movement alone offsets crypsis to at least some degree, so active foragers would be expected to be wary because potential predators would have little problem detecting them. Rapid response would be at an advantage for these species, reducing the

TABLE 10.1 Correlates of Foraging Mode.

Character	Sit-and-Wait foraging	Active foraging
Escape behavior	Crypsis, venoms (viperids)	Flight, skin or blood toxins (*Phrynosoma* and many frogs), venoms (elapids, helodermatids)
Foraging behavior		
Movements/time	Few	Many
Movement rate	Low	High
Percent time moving	Low	High
Sensory mode	Vision	Vision and olfactory
Exploratory behavior	Low (social)	High (food)
Prey types	Mobile	Sedentary
Morphology	Associated with microhabitat	Streamlined
Head shape	Short and wide	Long and narrow
Physiological characteristics		
Endurance	Limited	High
Sprint speed	High	Intermediate to low
Aerobic metabolic capacity	Low	High
Anaerobic metabolic capacity	High	Low
Heart mass	Small	Large
Hematocrit	Low	High
Activity body temperatures	Moderate (25-37°C)	High (32-41°C)
Energetics		
Daily energy expenditure	Low	Higher
Daily energy intake	Low	Higher
Social behavior		
Home range size	Variable but smaller	Variable but larger
Territoriality	Common	Rare
Mating system	Resource-defense polygyny	Sequential-mate-defense polygyny
Social signals	Visual	Visual and chemosensory
Reproduction		
Relative clutch mass	If clutch size is variable, relatively high; if clutch size is fixed, low	Relatively low and consistent across species regardless of clutch size

Sources: Bennett and Gleeson, 1979; Brown and Nagy, 2007; Cooper, 1994a, 1995, 1999, 2000, 2002, 2007; Garland and Losos, 1994; Huey and Pianka, 1981; Huey et al., 1984; McBrayer and Corbin, 2007; Miles et al., 2007; Nagy et al., 1984; Perry, 2007; Perry et al., 1990; Perry and Pianka, 1997; Pianka, 1966; Pianka and Vitt, 2003; Pough and Taigen, 1990; Reilly and McBrayer, 2007; Schwenk, 1993, 1995; Secor and Nagy, 1994; Seigel et al., 1986; Vitt and Congdon, 1978; Vitt and Price, 1982; Werner, 1997; Werner et al., 1997; Whiting, 2007.

probability that predators could capture them. Also, because of their seemingly continual motion while foraging, their physiology should cause them to function optimally while actively searching. Support for this view of the influence of foraging ecology on other aspects of an animal's biology stems mainly from studies comparing two or a few species that differed not only in foraging behavior but also in evolutionary histories. Such analyses cannot distinguish whether the evolution of active foraging from sit-and-wait foraging caused the behavioral, physiological, and ecological differences or is just part of a complex set of coevolved traits. As compelling as foraging behavior appears to be as the driving force behind the traits listed in Table 10.1, an analysis of complete physiological, behavioral, and ecological data testing this hypothesis has not been performed. The analysis by Gad Perry is a bold step toward solving this complex puzzle and should be taken as a challenge to assemble the data set allowing such an analysis. Phylogenetic analyses by others suggest that chemosensory behavior, lingual–vomeronasal morphology, and foraging mode comprise an adaptive complex driving the evolution of lizards.

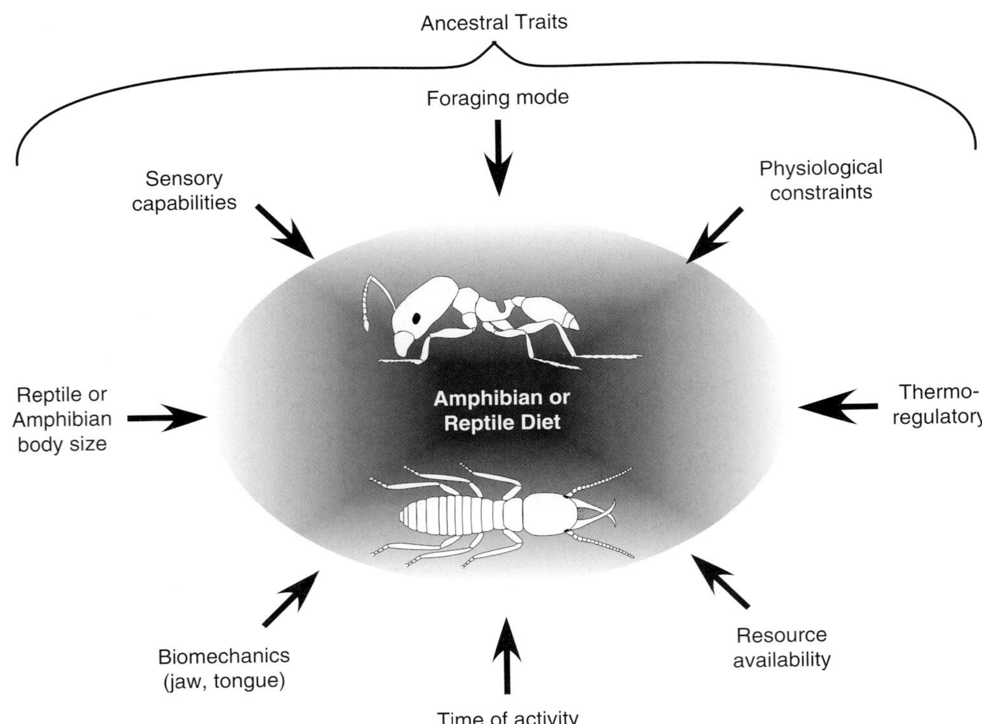

FIGURE 10.2 Diets of amphibians and reptiles are influenced by a variety of abiotic and biotic factors. In addition, the evolutionary history of each species determines a portion of prey preferences. Adapted from Vitt and Pianka, 2007.

File snakes (*Acrochordus arafurae*) offer an interesting perspective on the relationship between sensory modes and foraging behavior. Clear differences exist between males and females in foraging behavior. Male file snakes search actively for small fish in shallow water, whereas females ambush large fish in deep water. Males use chemical cues (fish scent) to detect prey, whereas females primarily use prey movement to detect prey. These differences suggest a functional relationship between foraging behavior and types of cues used for prey detection without the confounding effects of phylogeny, geography, or other variables that might account for differences.

Optimal foraging theory is a popular explanation for the evolution of foraging modes. This theory dictates that animals best able to harvest resources should be at a selective advantage when competition among individuals exists. Thus natural selection should favor the fine-tuning of resource acquisition ("optimal foraging"). Because growth, maintenance, and reproduction require energy (Chapter 7), the payoff for foraging "optimally" is presumably increased reproductive success. Although heuristically appealing, optimal foraging theory is overly simplistic, and most empirical studies fail to support most of its predictions. One prediction, however, is supported; when food is scarce, animals tend to eat a greater variety of prey types than they do when food is abundant. In natural environments, foraging is extremely complex. External, internal, and historical factors influence the ability of individual organisms to acquire food, and these factors are difficult if not impossible to model (Fig. 10.2 and Table 10.2).

Although most species of amphibians and reptiles can easily be assigned to one of the two broad foraging categories, cordylid lizards present a remarkable pattern with respect to foraging mode. Cordylines are sit-and-wait foragers; their sister taxon, the gerrhosaurines, are wide foragers; and the cordylids are nested in a clade of wide-foraging lizards (Autarchoglossa; see Fig. 20.2). Thus sit-and-wait foraging has evolved independently in the ancestor to cordylines. One cordyline, *Platysaurus broadleyi*, can vary its foraging behavior based on age, sex, and food availability. Juveniles spend nearly 10% of their time moving and thus fall on the interface between sit-and-wait and wide foraging. Adults are sit-and-wait foragers, unless figs are available. When figs are available, their foraging behavior is more like that of herbivores; the lizards move considerably, searching for figs.

TABLE 10.2 Factors Influencing Foraging Behavior.

Factors
External factors
Prey availability
Predation risk
Social interactions (e.g., competition)
Habitat structure (e.g., perch availability)
Opportunities for thermoregulation
Internal factors
Hunger
Learned experiences
Age (e.g., ontogenetic diet shifts)
Sex and reproductive state (e.g., energetic trade-offs)
Epigenetic inheritance (e.g., maternal effects)
Dietary preferences (as influenced by nutrient requirements, toxins, distasteful compounds)
Historical (phylogenetic) factors
Sensory limitations
Morphological characteristics (e.g., mouth shape, head size)
Physiological constraints (e.g., sprint speed)
Behavioral set (e.g., conservative foraging mode)

Source: Adapted from Perry and Pianka, 1997, and Vitt and Pianka, 2007.

DETECTING, CAPTURING, AND EATING PREY

Prey Detection

Prey of amphibians and reptiles can be detected by visual (usually moving prey), chemical (usually nonmoving prey), tactile (moving and nonmoving), or thermal (moving and nonmoving) cues. Many species rely on a single type of cue, but others use combinations of cues to detect prey. Caecilians appear to use their tentacles as chemosensory samplers. Salamanders and frogs primarily use visual cues to detect moving prey, and in many instances, responses to movement are so stereotypical that inanimate nonfood items can be rolled in front of some species (e.g., *Rhinella marina*) and be ingested. In other species (e.g., *Salamandra salamandra*), prey must meet a specific set of criteria to elicit attack. Certain frogs and salamanders, such as *Anaxyrus boreas* and *Plethodon cinereus*, are quite good at locating some prey items on the basis of olfactory clues alone. Prey detection in crocodylians appears to be based on a combination of tactile and visual cues, but chemical cues may also play a role. Among turtles, visual, chemical, and tactile cues can be involved in prey detection. Among squamates, the entire spectrum of cues for prey detection exists. In most iguanian and gekkotan squamates, visual cues associated with prey movement result in prey attack. In most other squamates (including snakes), chemical cues are important in prey detection and discrimination, but visual cues can also be involved, and in some (e.g., viperids and boids), thermal cues are also involved.

Visual Prey Detection

Visual prey detection is used by most amphibians and reptiles that are sit-and-wait predators and to a lesser degree by many active-foraging species. Neurophysiological studies of the anuran eye show that prey recognition derives from four aspects of a visual image: perception of sharp edges, movement of the edges, dimming of images, and curvature of the edges of dark images. Perception is greatest when the object image is smaller than the visual field. Under these conditions, anurans can determine the speed, direction of movement, and relative distance of the prey. Success in capture by visual predators depends on binocular perception in many species; most align their head or entire body axis with the prey before beginning capture behavior. Chameleons are an exception. The eyes of chameleons are independently movable, and when one eye detects a prey item, the head turns to allow both eyes to focus on the prey prior to aiming the projectile tongue. These movements give the impression that binocular vision is being used to determine the distance of the prey item (Fig. 10.3). However, accommodation (focus) is most important in coordinating prey detection and prey capture in chameleons. They can accurately

FIGURE 10.3 The eyes of chameleons, such as this *Furcifer pardalis*, move independently until a prey item is sighted. Photograph by Chris Mattison.

orient on and capture a prey item at substantial distances with only one functioning eye.

The extensive use of vision in prey capture is also apparent from the number of diurnal and nocturnal species with large, well-developed eyes. Nocturnal predators locate prey under low-light conditions and require maximum light entry into the eye for perception. Horizontal or vertical elliptical pupils allow the maximum movement of the iris and the greatest dilation of the pupil under low-light conditions (Fig. 10.4).

Nevertheless, even in species that appear to use visual cues, more than simple detection of movement is involved. For example, most phrynosomatine lizards eat a wide diversity of insects, but lizards in the genus *Phrynosoma* specialize on ants. Arguably, *Phrynosoma* do not specialize on ants because ants are usually the most abundant insects; however, by the same reasoning, other syntopic lizards are selectively not eating ants. Regardless of which species actually are the specialists, amphibians and reptiles relying on visual cues do not randomly capture all available moving prey. Prey selection demonstrates a high level of visual acuity, sufficient to discriminate based on size and shape.

Chemosensory Prey Detection

The use of chemical cues in prey detection of amphibians and reptiles is just beginning to be appreciated. Chemosensory-oriented amphibians and reptiles use one or more of three chemical senses: olfaction, vomerofaction, and taste (gustation) (see Fig. 2.32). The first two are used in prey location and identification; olfaction uses airborne odors and vomerofaction uses airborne or surface odors. The olfactory epithelium in the nasal chamber is sensitive to volatile compounds carried by the air and inspired with respiratory air or "sniffing" by rapid buccal

FIGURE 10.4 Elliptical pupils are found in some nocturnal frogs, lizards, and snakes, and in all crocodylians. The pupils, closed here, open in low light to facilitate vision at night. Clockwise from upper left: *Hemidactylus mabouia, Corallus hortulanus, Osteocephalus taurinus, and Scaphiopus hurterii*. Photographs by L. J. Vitt and J. P. Caldwell.

FIGURE 10.5 The long, flexible tongue of *Xenoxybelis argenteus* picks up particles from the air, surfaces, and potential prey. The odors are transmitted to the vomeronasal organs and allow identification and discrimination. The same sensing system is used in chemosensory-based social communication. Photograph by J. P. Caldwell.

or gular pumping. The vomeronasal (Jacobson's) organ is especially sensitive to high molecular weight compounds that are transported into the oral or nasal cavity by the snout or tongue. Olfaction acts mainly in long-distance detection, e.g., the presence of food and its general location, and triggers tongue flicking and the vomeronasal system. Vomerofaction operates as a short-range identifier and appears more important than olfaction or gustation in feeding. The vomeronasal system requires that chemicals be brought in, usually by the tongue, which can pick up volatile chemicals from the air or nonvolatile chemicals by lingual contact with surfaces (Fig. 10.5). Gustation functions during feeding as the final discriminator in those species that have taste buds.

Olfaction and vomerofaction have long been recognized as feeding senses in salamanders and scleroglossan squamates and are often used in conjunction with vision. Actively foraging predators, such as teiid lizards, use vision while moving across open-surface microhabitats but depend on vomerofaction to locate prey in dark crevices or buried in leaf litter or soil. Likewise, many salamanders probably alternate between visual and vomerofactory searching depending upon the availability of light and crypsis of the prey. Some salamanders, such as *Hydromantes italicus*, locate, orient on, and capture prey in total darkness, based on chemical cues alone. Iguanian lizards (except the Iguaninae and Leiolepidinae) and most anurans are highly visual predators, and most lack well-developed olfactory–vomerofaction systems. However, observations on *Rhinella marina* and a few other anurans that respond to chemical cues in food suggest that the role of chemoreception in prey detection by anurans may be underappreciated. Among iguanine lizards (e.g., *Dipsosaurus dorsalis*) that are herbivorous, species that have been studied are able to discriminate plants on the basis of chemicals. Historically, turtles and crocodylians were considered to be visual–tactile foragers; however, both groups produce pheromones for individual and species recognition and would seem capable of locating prey via odor or vomerodor. Experiments have shown that the American alligator can locate visually hidden food both in the water and on land, suggesting chemoreception in prey identification. Snakes are perhaps best known for their chemosensory abilities because of the often rapid sampling of the air and surfaces with their long, flexible, forked tongues (Fig. 10.5). Not only does the tongue transmit particles to the vomeronasal organs, but because it is forked and thus samples two points, directional information is also conveyed. In some garter snakes (*Thamnophis sirtalis similis*), visual cues alone do not elicit foraging even though they are important for prey capture. Foraging commences when the snakes detect chemical signals with their vomeronasal system.

Taste is a chemosensory sense but is used to discriminate rather than locate prey. When combined with the tactile sense organs of the oral epithelium, taste can serve to identify food items once in the mouth and permit rapid acceptance or rejection. Items may be rejected because of taste or because of mechanical stimulation of the tactile sense based on the presence of spines or urticating hairs.

Similar to differences among species in foraging behavior, much of the variation in use of chemical cues has a historical basis in squamates (Fig. 10.6). When scleroglossans diverged from the Iguania, chemosensory acuity for prey detection appeared and was carried into most of the taxa derived from them, including snakes. From a historical perspective, the evolution of chemosensory prey detection made a large set of prey available to scleroglossans that was not available to the visually oriented Iguania: nonmoving prey and hidden prey, many of which are larvae of insects eaten by visually oriented amphibians and reptiles. It likely also allowed scleroglossans to reject insects containing defensive chemicals (e.g., alkaloids) that might interfere with the lizard's metabolic processes. The development of the chemosensory mechanism for prey discrimination certainly played a major role in the subsequent diversification and success of scleroglossan lizards (including snakes).

Although the development of chemosensory cues for prey detection occurred in scleroglossans or their immediate ancestor, the chemosensory structures (vomeronasal organs, taste buds) were present in squamate ancestors, and many iguanians use chemical cues for social communication. Only in the iguanines, which are herbivorous, have chemical cues been demonstrated to be used for

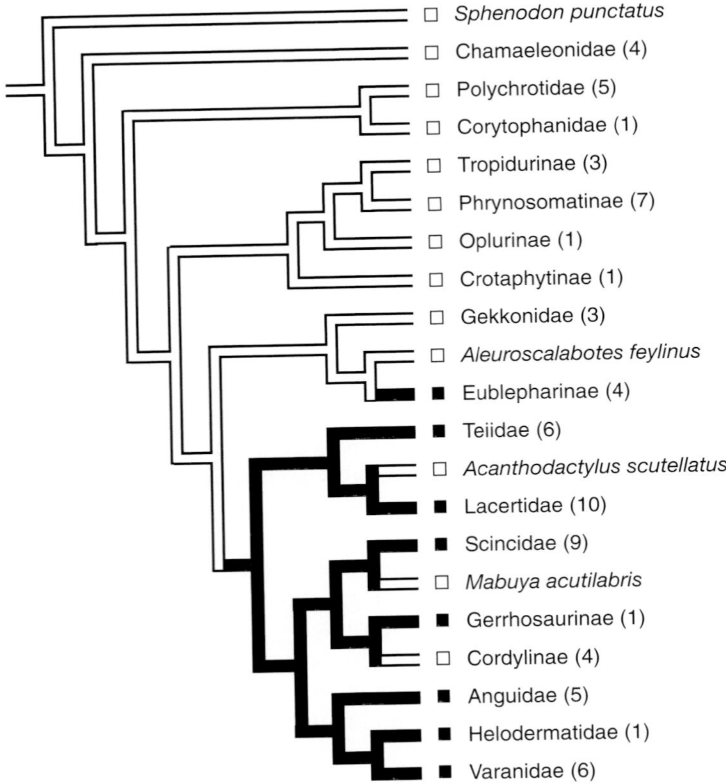

FIGURE 10.6 The evolution of prey chemical discrimination and foraging mode appears linked in squamates. Several evolutionary reversals have occurred within major clades, four of which are shown here (Eublepharinae, *Acanthodactylus scutellatus*, *Mabuya acutilabris*, and Cordylinae). In instances where reversals have occurred, chemical cues are not used for prey discrimination, even though the sensing systems are developed. Clade names have been modified to maintain consistency with those in Chapter 20. Adapted from Cooper, 2007.

discrimination among food types. However, considering the large number of tropidurine lizards now known to eat plant material, these and lizards in the Leiolepidinae need to be examined. Thus, even though the basic morphological structures for prey discrimination were in place when squamates began diversifying, most members of the arthropod-eating iguanians did not use them for prey detection and discrimination.

Auditory Prey Detection

Use of airborne sound to locate prey may occur widely in amphibians and reptiles, but it remains largely undocumented. The observations are mostly anecdotal, such as *Rhinella marina* orienting and moving toward a calling *Physalaemus pustulosus*, although a recent field experiment showed that the gecko *Hemidactylus turcicus* locates male crickets based on their calls and preys on female crickets coming to the male. The geckos *Hemidactylus frenatus* and *Cosymbotus platyurus* may also use auditory cues in combination with chemical and visual cues.

For some amphibians and reptiles, sensitivity to substrate vibrations or seismic sounds is likely a major prey-detection mechanism. Seismic sensitivity may be particularly important for fossorial (burrowing) species or those with fossorial ancestors, both for the avoidance of predators and the location of prey. Snakes, salamanders, and caecilians have no external ears, so they probably have a high sensitivity to seismic vibration, although actual tests are lacking for most species. Uniquely, both frogs and salamanders have a special pathway (opercularis system) for the transmission of vibrations from the substrate to the inner ear, and the limited data indicate that salamanders are two times more sensitive to these sounds than frogs. The opercularis system links the forelimb to the inner ear through the opercularis muscle that extends from the scapula to the opercular bone lying in the fenestra ovalis of the otic capsule. The muscle acts like a lever arm; vibrations received by the forelimb rock the tensed muscle thereby pushing or pulling the operculum and creating fluid movement in the otic capsule. These seismic vibrations are of low frequency, typically less than 200 Hz, and stimulate the neuroreceptors in the sacculus and lagena rather than those of either the basilar or amphibian papilla, although the latter may be stimulated by frequencies as low as 100 Hz. These low frequencies are made by such activities as the digging of insect prey or mammalian predators. In snakes, seismic vibrations appear to be transmitted via the lower jaw through the quadrate–columella to the inner ear. Snakes also detect seismic vibrations through mechanoreceptors in the skin, although not with the same fine-scaled resolution as with the ear. Other fossorial groups (e.g., caecilians, amphisbaenians) likely use mechanoreceptors for detection of seismic vibrations.

Thermal Prey Detection

Some snakes use thermal cues to locate and orient on prey. Infrared light (long wavelength light) is sensed by trigeminal-innervated blind nerve endings in the skin of the head. Many boas and pythons (e.g., *Corallus, Morelia, Chondropython*) and all viperid snakes in the Crotalinae (e.g., *Crotalus, Agkistrodon, Lachesis, Bothrops*) have infrared sensitive pits either along the jawline in the labial scales (boids) or in the loreal scales (crotalines) at the front of the jaw (Fig. 10.7). The pits open (face) anteriorly and provide a binocular perception field. These receptors are capable of detecting thermal objects moving within the snake's sensory thermal landscape. Temperature changes lower than 0.05°C elicit a response from some snakes. Experiments have demonstrated that snakes can accurately orient on and strike objects based on thermal cues alone. Infrared cues are putatively most effective for nocturnal snakes that feed on mammals and birds because of the large temperature differential between the background thermal landscape and the moving prey, but these cues are likely to be equally effective for a pit viper hidden in a crevice, for example, when a lizard with an elevated body temperature enters the crevice.

FIGURE 10.7 Facial heat-sensing pits between the nares and the eye on *Bothrops moojeni* and along the jaw of *Corallus hortulanus* allow these snakes to detect moving prey on the basis of their thermal image. Photographs by L. J. Vitt.

Tactile Prey Detection

Tactile prey detection is poorly understood in amphibians and reptiles, but some rather obvious examples exist. Popular films of large crocodiles appearing to come from nowhere in rivers and ponds to capture large mammals when they break the water surface likely result from tactile cues transmitted through water. The mechanism involves use of mechanoreceptors in the skin. Aquatic amphibians use the lateral line, a string of mechanoreceptors, to sense changes in water pressure reflecting from stationary or motile objects in the near vicinity to identify and locate prey. Such recognition would certainly be enhanced by a weak electric field (see lateral line in "Sense Organs," Chapter 2). Preliminary evidence from aquatic salamanders indicates that prey identification and size determination occur solely by the lateral-line system.

Alligator snapping turtles (*Macrochelys temminckii*) certainly use tactile cues when making the decision to close their mouths on an unsuspecting fish that tries to sample their wormlike tongue (Fig. 10.8). Tactile cues may also be important for turtles, such as *Chelus fimbriata*, that expand their throats rapidly to vacuum

FIGURE 10.8 The alligator snapping turtle, *Macrochelys temminckii*, lures fish into its mouth by waving its fleshy tongue. The cryptic morphology of the nonmoving turtle combined with the resemblance of the tongue to a small earthworm facilitates prey capture. Photograph by R. W. Barbour.

FIGURE 10.9 The aquatic snake *Erpeton tentaculatum* uses appendages on the front of the head to detect tactile stimuli from fish when they approach the snake. Photograph by R. D. Bartlett.

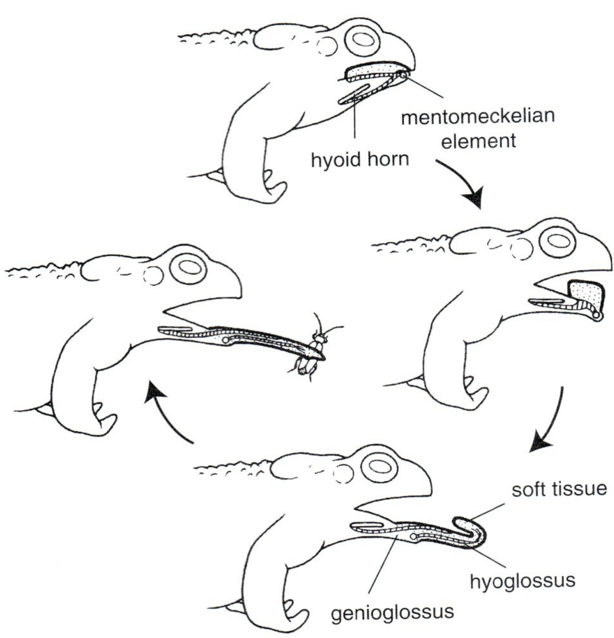

FIGURE 10.10 The anatomical mechanics of an anuran projectile tongue (*Rhinella marina*). The four schematic stages show the projection sequence from tongue at rest on the floor of the oral cavity (top) to its full extension and capture of an insect (left). Five anatomical features are highlighted: the soft tissue of the tongue (stippled); two muscles (black), the genioglossus from the hyoid to the base of the tongue and the hyoglossus from the mentomeckelian element (mm) to the base of the tongue; and two skeletal elements (white), the hyoid horn lying below the tongue and mm at the tip of the jaw. Projection begins (right) with the mouth opening; the mm snaps downward by the contraction of a transverse mandibular muscle (not shown), and the genioglossus contracts to stiffen the tongue. The tongue flips forward (bottom) from the momentum generated by the downward snap of the mentomeckelian element and the genioglossus contraction; the two tongue muscles then relax and are stretched. The tongue is fully extended and turned upside down (left), and the dorsal surface of the tongue tip encircles the prey. The genioglossus and hyoglossus muscles contract, drawing the tongue with the adhering insect back through the mouth as it closes. Adapted from Gans and Gorniak, 1982.

in fish or tadpoles moving in front of them. Flaps of skin are highly innervated and undoubtedly are involved in detection of tactile cues. Many other turtles have barbels about the jaw that are sensitive to water displacement and likely aid in feeding. The Tentacled snake, *Erpeton tentaculatum*, uses a sit-and-wait strategy to attack fish underwater (Fig. 10.9). Appendages on the head (tentacles) may provide tactile cues allowing the snake to accurately strike and capture the fish.

Prey Capture and Ingestion

Once detected, prey must be subdued and ingested in order for the amphibian or reptile to appreciate a net gain in energy from the pursuit of prey. A vast majority of amphibians and reptiles swallow their prey whole, and in most species, prey are very small relative to the size of the predator. Toads (*Rhinella* and *Anaxyrus*, for example) flick the tongue in and out at such a rapid rate that the entire event cannot be detected easily by the human eye (Fig. 10.10). At the opposite extreme are crocodylians such as *Crocodylus moreletii* in Veracruz, Mexico, that drown large prey and hold them in their mouths for as long as 3 days until they began to decompose and then dismember and eat them. Komodo Dragon lizards fatally wound moderate-sized mammals by slicing through the musculature of their body or legs with their serrated teeth. The mammals die, and monitors are attracted to the putrefying corpse, which the lizards are able to dismember, swallowing large pieces. Herbivorous lizards feed on clumped, stationary plant parts, so prey "capture" is a trivial problem. Many species of snakes kill their prey by constriction or envenomation, but some simply swallow their prey alive.

Numerous behavioral and morphological adaptations are associated with capturing and subduing prey. In catching mobile prey, motor and sensory units are finely coordinated to intercept the moving prey, and usually the strike–capture mechanism aims at the center of the mass or gravity of the prey. The center of gravity is the most stable part of the target and has the least amount of movement.

Some reptiles and amphibians use lures to attract their prey. Juvenile viperids use caudal luring enhanced by bright coloration on the tail and cryptic coloration of the body (Fig. 10.11), and lingual-appendage luring occurs in alligator snapping turtles (*Macrochelys*; Fig. 10.8). Pedal luring using the back feet occurs in some species of *Ceratophrys* frogs, and juveniles of *Ceratophrys cornuta* have white toes and webbing, possibly to enhance the outline of the foot against their leaf-litter habitat (Fig. 10.12).

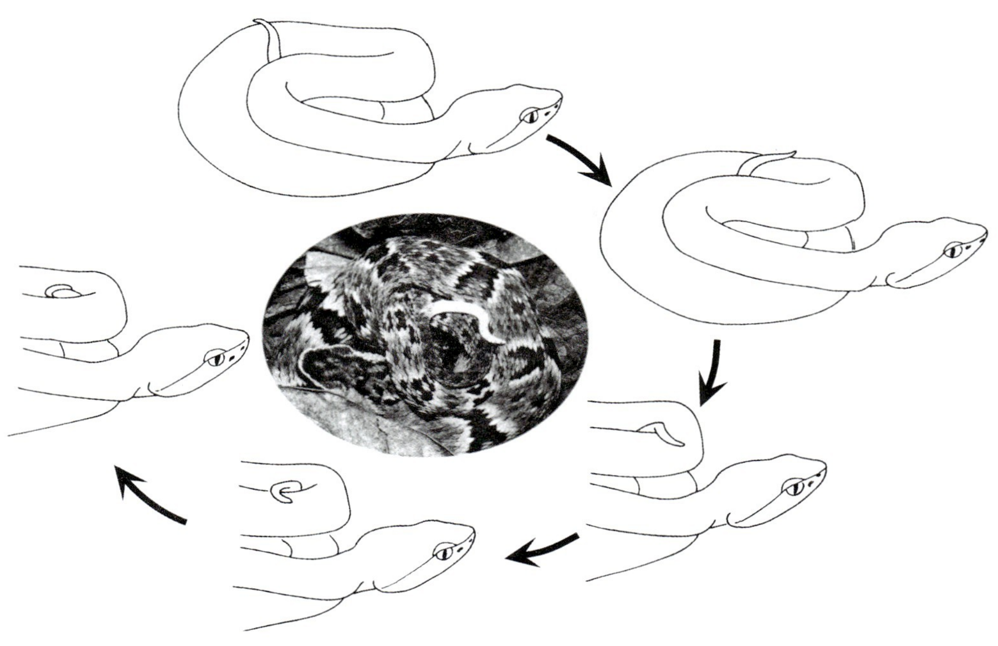

FIGURE 10.11 By waving its brightly colored tail, *Bothrops jararaca* attracts frogs and other small insectivorous animals within strike range. The insert shows the contrast between the tail color (yellow in life) and the cryptic coloration of the snake. Adapted from Sazima, 1991. Photograph by I. Sazima.

FIGURE 10.12 Some frogs, such as *Ceratophrys cornuta*, use pedal luring to attract prey. The light color of the hind toes disappears as the frogs increase in size. Photograph by J. P. Caldwell.

Biting and Grasping

Prey capture by most amphibians and reptiles involves biting and grasping. Prey are attacked, either as the result of a rapid sprint by the predator followed by biting the prey, or by a rapid movement (e.g., strike) of the head and neck from a stationary position. Reptiles or amphibians with long, flexible necks (turtles, varanid lizards) and limbless ones (amphiumas, pygopods, snakes) can and regularly do use the strike mechanism, often from ambush but also following a slow stalk of the prey. In both strikes and bites, the mouth commonly does not open until the head moves toward contact with the prey, and the bite–strike is an integrated behavior of motor and sensory units. When the open mouth contacts the prey, the tactile pressure on teeth and oral epithelium triggers rapid closure of the mouth.

Only minimal food processing occurs in the mouth of amphibians and reptiles. Teeth may crush or perforate food items, which are commonly swallowed whole. Some evidence suggests that most lizards, for example, do not swallow arthropod prey items until they have crunched the exoskeleton. If hard-bodied prey fail to crush when bitten, the broad-headed skink, *Plestiodon laticeps*, repositions the prey repeatedly and attempts to crush it. If the hard-bodied prey happens to be a female mutilid wasp (velvet "ant"), repeated biting allows the insect to use all its defense mechanisms. The powerful sting, injected deep into the tongue, causes the lizard to release the wasp. When approached by a xendontine snake, some species of *Rhinella* inflate their body by filling their lungs and tilt their back toward the snake. In response, some snakes, such as the South American snake *Waglerophis merriami*, puncture the inflated lungs of *Rhinella* with their razor-sharp and enlarged maxillary teeth. In these species, the maxillary is reduced in length and rotates forward during biting. Once deflated, the toads can be swallowed by the snake.

Fragmentation of food is limited to herbivores that bite off pieces of foliage, and large lizards, turtles, and crocodylians may use a combination of sharp jaw sheaths or

FIGURE 10.13 Juvenile Aldabran tortoises (*Geochelone gigantea*) eating a leaf from their shade tree. Photograph by G. R. Zug.

teeth and limb–body movements to break up large items. Turtles have continuously growing keratinous sheaths on upper and lower jaws; each sheath provides a uniform bladelike labial surface that is effective in cutting food (Fig. 10.13). Tooth structure in amphibians and diapsid reptiles is highly variable, ranging from simple conelike teeth to molar-like teeth or bladelike teeth with serrated edges. Specialized diets usually are associated with specialized teeth: broad and sturdy teeth for crushing mollusks are found in *Dracaena*; bladelike teeth for cutting vegetation or fragmenting large prey are found in *Iguana* and *Varanus*, respectively; long recurved teeth for feathered prey occur in *Corallus hortulanus*; and hinged teeth for capturing skinks occur in *Scaphiodontophis*.

Once captured, prey must be moved through the oral cavity into the esophagus. Three main "swallowing" mechanisms are recognized in amphibians and reptiles. Inertial feeding is mechanically the simplest and most widespread in reptiles. In its simplest form, inertial feeding involves moving the head–body over the food based on inertia alone. The food is held stationary in the mouth. Each time the mouth is slightly opened, the head is thrust forward, thereby shifting the head forward over the food (Fig. 10.14). Snakes swallow large prey in this manner by alternately advancing the left and right sides of the head over the prey using the movement of the palatoquadrate–mandibular skeletal complex. Prey are held secure by this complex on one side of the head, while the bite–grip on the opposite side of the head is relaxed with the jaws on that side of the head shifting forward and then contracting to gain a grip. The alternate forward movement of the left and right sides moves the head and body over the prey.

Manipulation of the tongue and hyoid appears to be the principal swallowing mechanism in amphibians. Some salamanders use hyoid–tongue retraction to swallow prey. After capturing a prey item and with the mouth closed, the tongue presses the prey tightly against the roof of the mouth and the vomerine and palatine teeth. The mouth opens quickly and, with the tongue still firmly holding the prey, retracts and draws the prey inward as the mouth slowly closes. This cycle is repeated until the prey move through the buccopharyngeal cavity. Swallowing in frogs also involves tongue–hyoid movement, although the actual mechanics are known in less detail. Frogs have voluminous oral cavities, and captured prey are usually completely engulfed. Apparently a similar hyoid–tongue retraction cycle without opening the mouth moves the prey inward. This movement is assisted by the compression of the palate, visible externally by the retraction of the eyes.

Constriction

Constriction is a specialized bite-and-grasp technique used by numerous snakes to hold or kill prey. A constricting

FIGURE 10.14 The mollusk-eating snake *Dipsas indica* uses inertial feeding behavior to swallow a large slug (left) and extended teeth on the lower jaw to extract a snail from its shell (right). Photographs by I. Sazima.

snake strikes its prey, and if its bite–grip is secure, a loop of the body is thrown on and around the prey. Additional loops (coils) of the body encircle the prey with continual adjustment to reduce overlapping loops. As the prey struggles and then relaxes parts of its body, the snake tightens its grip. The tightening continues, and ultimately, circulatory failure causes death. Increasing compression of the thorax stops the flow of blood to the heart. In species that have been well studied (gopher snakes and king snakes), constriction is much more controlled than generally believed. The snakes can detect muscular, ventilatory, and circulatory movements in the rodent that they are constricting and respond by tightening and loosening coils accordingly. The snakes maintain a constriction posture several minutes after the rodent stops moving, but if the snake detects circulatory, ventilatory, or muscular movement by the rodent, it reapplies pressure. When struggling ceases and the prey is dead or unconscious, the snake relaxes its coils, locates the head of the prey, and begins to swallow it. Constriction is best known in boas and pythons, but it is common in other snakes as well (Fig. 10.15).

Even some highly venomous snakes constrict their prey after biting and injecting venom. It is easy to visualize constriction in boids, where the prey typically are birds or mammals. However, snakes that constrict fish best exemplify the effectiveness of constriction as a means of subduing and killing prey. The file snake (*Acrochordus*) can attach its tail to underwater roots of mangroves as an anchor, strike a large fish, and rapidly subdue it by constriction. The rough scales on the file snake facilitate holding the fish, and the elastic body apparently serves to buffer the thrashing movements of the struggling fish. Some limbless amphibians (e.g., *Amphiuma*) may use constriction to subdue prey as well.

Injected Venoms

Venom delivery systems have evolved independently a number of times within the Squamata. Considerable variation in morphology, derivation, and effectiveness of venom-delivery systems exists. All members of the Helodermatidae, Elapidae, and Viperidae are venomous, as are several groups of colubrids. Venom subdues the prey by either anesthetizing or killing it. A nonstruggling prey is much safer and less energetically demanding to capture and swallow than a struggling one. Also, a predator can eat larger prey if they do not resist capture and consumption. Many of the viperids add a third benefit to the injection of venom by injecting proteolytic enzymes that aid in digestion.

FIGURE 10.15 Following prey detection and strike and grasp, many snakes, like this Burmese python, coil around their vertebrate prey. Not only does constriction subdue the prey, but it also causes circulatory failure, which kills the prey. Photograph by S. C. Secor.

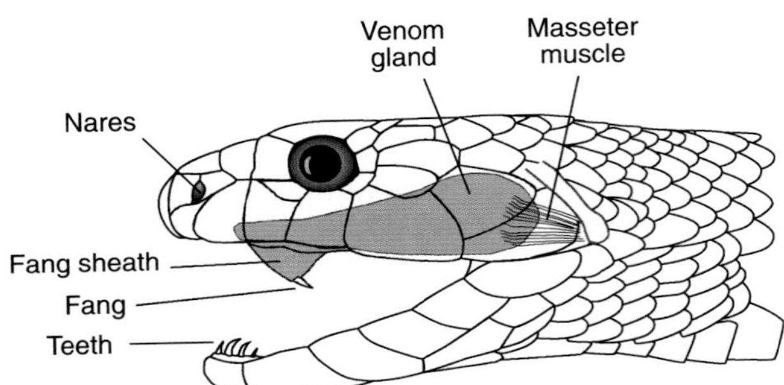

FIGURE 10.16 Venomous snakes have movable (Viperidae) or fixed (Elapidae, some Colubridae) fangs to inject venom. Venom is delivered to the fangs from the venom glands via venom ducts. Modeled after a drawing of a taipan, *Oxyuranus scutellatus*, in Shine, 1991.

A venom-delivery system contains four items: glands to produce the venom, muscles to force venom from the glands, ducts to transport venom from the gland to the injection system, and fangs (modified teeth with open or closed canals) to inject the venom into the prey (Fig. 10.16). The fangs of helodermatids and most venomous colubrids bear a single groove on one side of each enlarged tooth, whereas the fangs of elapids and viperids have closed canals. The venom is produced continuously in the venom glands and stored in venom-gland chambers. When elapids or viperids bite a prey animal, muscles over (adductor superficialis in the elapids) or around (compressor glandulas in viperids) the glands contract and squeeze a portion of the venom through the venom ducts and into the fang canals. The snake can regulate the venom dose depending on the size of the prey and possibly how much venom is available. Viperids and some elapids strike, bite, inject venom, and release the prey, whereas most elapids, colubrids, and *Heloderma* maintain their bite–grip and chew the wound to ensure deep penetration of venom. Elapids and most rear-fanged colubrids have relatively small fangs. With few exceptions, these fangs are fixed in an erect position. The greatest deviation from fixed fangs in elapids occurs in the death adder (*Acanthophis antarcticus*) of Australia, which has morphology and foraging behavior strikingly convergent with that of terrestrial viperids. The front fangs are fixed on a highly movable quadrate bone.

The venom of each species is a composite of several compounds that work synergistically to subdue the prey (Table 10.3). Typically, the venom of a species causes either tissue destruction or neurological collapse. The tissue-destruction venoms subdue the prey because the prey goes into shock, and neurological-collapse venoms prevent nerve

TABLE 10.3 Major Types of Reptilian Venoms and Some Examples of the Function of Each Type.

Enzymes	
All venoms contain several different enzymes; more than 25 enzymes occur in reptilian venoms.	
Proteolytic enzymes	digest tissue protein and peptides causing hemorrhagic necrosis and muscle lysis; also known as endopeptidases. Common in crotalines, less in viperines, absent in elapids.
Thrombin-like enzymes	interfere with normal blood clotting, either by acting as an anticoagulant or procoagulant. Common in viperids, rare in elapids.
Hyaluronidase	breaks down mucopolysaccharide links in connective tissue and enhances diffusion of venom. In all venomous snakes.
Phospholipase	modifies muscle contractibility and makes structural changes in central nervous system; also interferes with the prey's motor functions. Common in colubrids, elapids, viperids.
Acetylcholinase	interrupts ganglionic and neuromuscular transmission and eventually affects cardiac function and respiration. Common in elapids, absent in viperids.
Polypetides	
The polypetides are toxic nonenzymatic proteins of venoms. These toxins commonly act at or near the synaptic junctions and retard, modify, or stop nerve-impulse transmission.	
Crotactin	produces paralysis and respiratory distress. In rattlesnakes, crotalines.
Cobrotoxin	acts directly on heart muscle to cause paralysis. In cobras, *Naja*.
Viperatoxin	acts on medullary center in brain, resulting in vasodilation and cardiac failure. In *Vipera*.
Miscellanea	
Various ions and compounds that are found in venoms but as yet have no recognizable prey-type or taxonomic-group association.	
Inorganic ions	sodium, calcium, potassium, iron, zinc, and others; some enhance the activity of specific enzymes.
Glycoproteins	anticomplementary reactions that suppress normal immunological tissue response.
Amino acids and biogenic amines	

Note: Reptilian venoms are an admixture, consisting mainly of enzymatic and nonenzymatic proteins.

impulse transmission and interrupt all motor activity, including respiration. The immobile prey can then be eaten safely.

Projectile Tongues

Tongues are small and usually have limited or no mobility in aquatic amphibians and reptiles. Tongues became important in terrestrial animals when water was no longer present to carry food through the oral cavity into the esophagus. A protrusible tongue for sampling the environment and gathering food probably evolved early in terrestrial tetrapods, because protrusion is widespread in amphibians and reptiles. Many bite-and-grasp feeders (herbivores and carnivores) use their tongues to retrieve small items. The tongue is extended through the mouth and the item is touched by the tip or dorsal surface of the tongue. The item is held by sticky saliva and the tongue is retracted. The most dramatic tongue protrusions are the projectile tongues, which have evolved independently several times in amphibians and reptiles.

Most frogs capture prey by projecting the tongue (Fig. 10.10), but the mechanism is different from that found in salamanders (Fig. 10.17) and even differs among frogs. The frog's tongue is attached at the front of mouth and has a direct attachment to the cartilaginous symphysis joining the right and left sides of the mandible. When a prey item is identified, the frog orients its body perpendicular to the prey. The mouth opens and the lower jaw drops downward. The genioglossus muscle, which lies within the tongue, contracts, stiffening the tongue. The submentalis muscle (linking left and right mandibles beneath the middle of the tongue) contracts to form a pivot point that yanks the symphyseal cartilage downward. This movement pulls the anterior end of the tongue downward, and the momentum imparted to the tongue flicks the posterior end outward in much the same fashion as a catapult. The weight of the tongue's posterior half stretches the tongue to twice its length, and as the upper surface of the tongue hits the prey, the tip wraps over the prey. The tongue is retracted by a quick contraction of the hyoglossus muscle in the posterior region of the mouth, with the prey stuck to the tip of the tongue. For most frogs, direct orientation on the prey is necessary because the tongue flips out in direct line with the frog's head. However, the microhylid *Phrynomantis bifasciatus* is able to send its tongue out in nearly every possible direction (Fig. 10.18).

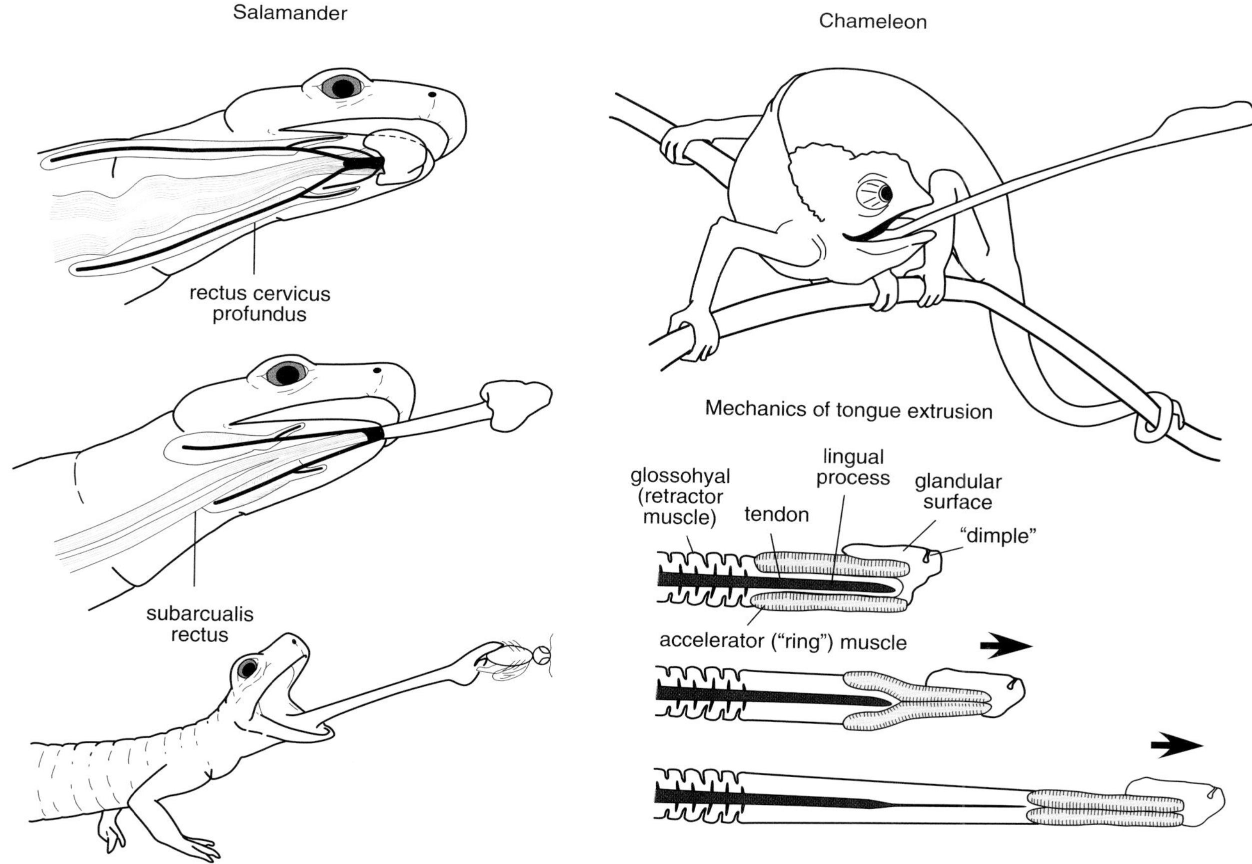

FIGURE 10.17 The anatomical mechanics of a salamander and a chameleon tongue. Salamanders redrawn from Duellman and Trueb (1986); chameleon redrawn from Kardong (1998).

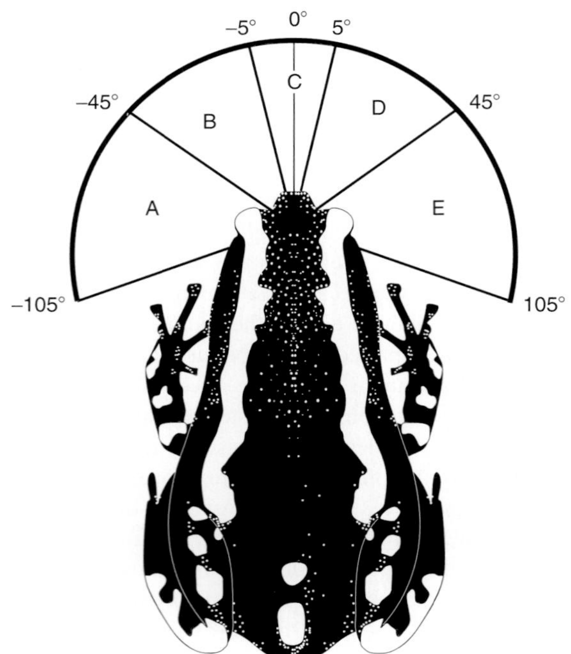

FIGURE 10.18 Unlike most frogs, the microhylid frog *Phrynomantis bifasciatus* can extend its tongue in an arc of 105° to either side of center to capture prey. It does so using hydrostatic force to push the tongue directly out of the mouth. Adapted from Meyers et al., 2004.

Rather than using muscles to pull the tongue and flip it out as in other frogs, *Phrynomantis* has a hydrostatic muscle that pushes the tongue out. The hydrostatic mechanism allows the frog to send out its tongue within a range of about 105° to either side of center. A tongue that functions in this manner should be particularly useful for frogs that feed on tiny prey, such as termites.

Terrestrial salamanders orient on prey and rapidly extrude the tongue, which, in many species, has a large pad on the tip. Mucous on the tongue tip adheres to the prey item, and longitudinal muscles retract the tongue and prey. The mechanics of tongue extrusion vary among salamander taxa. A large fleshy tongue is flopped out on a prey item in *Ambystoma*, whereas highly derived elongate tongues with fleshy tips are projected for considerably longer distance in various plethodontids. Projectile tongues appear to have evolved independently several times in salamanders, including three times within the Plethodontidae. The projectile mechanisms in salamanders derive from modifications of the hyoid apparatus, a structure that usually functions to move the floor of the buccal cavity during respiration. Salamanders with projectile tongues (plethodontids and the salamandrids *Chioglossa* and *Salamandrina*) are lungless, and as a result, the hyoid apparatus is not involved in respiration. The general mechanism of tongue extension includes the projection from the mouth of the pedestal-like tongue tip by the hyoid apparatus. The posterior, bilaterally paired hyoid arms lie in the floor of the mouth like a partially opened fan with the hinge-tip pointed anteriorly. When the hyoid muscles contract, the fan closes and drives the tip outward. The movement is rapid and the momentum, as in frogs, assists in stretching the tongue as much as 40–80% of the salamander's body length. The structure of the hyoid apparatus varies considerably among salamander species. Tongue movement in *Bolitoglossa* is so rapid that a sensory feedback system is not involved. The extensor and retractor muscles fire simultaneously, but the retractor muscle contains enough slack that it does not begin to retract the tongue until the tongue is fully extended.

Chameleons have one of the most spectacular tongue-projection systems known in vertebrates (Fig. 10.17). They can project their tongues at high speed for as much as 200% of their snout-vent length and accurately hit and capture an insect. Precise integration between the ocular system and the tongue-projection system is critical. The projectile tongue of chameleons shoots forward by a hyoid mechanism. Once a chameleon has oriented on an insect after detecting it visually, the head is extended toward the prey, the lower jaw opens, and the tongue slowly extends a short distance out of the mouth. The zygodactylus toes and prehensile tail hold the chameleon firmly to branches from which they forage. The tongue then shoots out toward the prey, the sticky tip captures the insect, and the tongue is drawn back into the mouth with the insect (Fig. 10.19). The mechanism includes a precision system of depth perception based on accommodation, a highly modified hyoid apparatus including a powerful accelerator muscle, and exceptionally contractile hyoid muscles.

Filter Feeding

No reptiles and no adult amphibians filter-feed. However, tadpoles of most frogs filter-feed. The diets of most tadpoles consist mainly of algae and protists, and hence tadpoles are microphagous ("small eating"). Comblike labial teeth that occur in rows on the oral disc scrape detritus from surfaces. Tadpoles use the movement of water in through the mouth, buccal, and pharyngeal cavity, and out through the gills (branchial arches) for both respiration and food entrapment. Microphagy requires a filter or straining mechanism to capture tiny items and direct them into the gut. A system that includes branchial food traps and gill filters in the pharynx captures smaller particles (Fig. 10.20). Buccal papillae extract large particles and funnel them directly into the esophagus.

The buccopharyngeal cavity of tadpoles is large, more than half the volume of the head of most tadpoles. The upward and downward movement of the buccal floor in association with the opening and closing of the mouth and gill filter valves (vela) moves water through this large cavity. As the mouth opens, the floor drops and draws water into the cavity. The vela prevent a major backflow

FIGURE 10.19 Ballistic tongues of some chameleons, such as this *Chamaeleo pardalis*, can extend out more than two times the length of the lizard's body. The short section of the tongue nearest the head that is directed slightly upward contains the process entoglossus, which is part of the hyglossal skeleton that is situated inside the tongue and gives it support. Photograph by M. Vences and F. Rauschenbach.

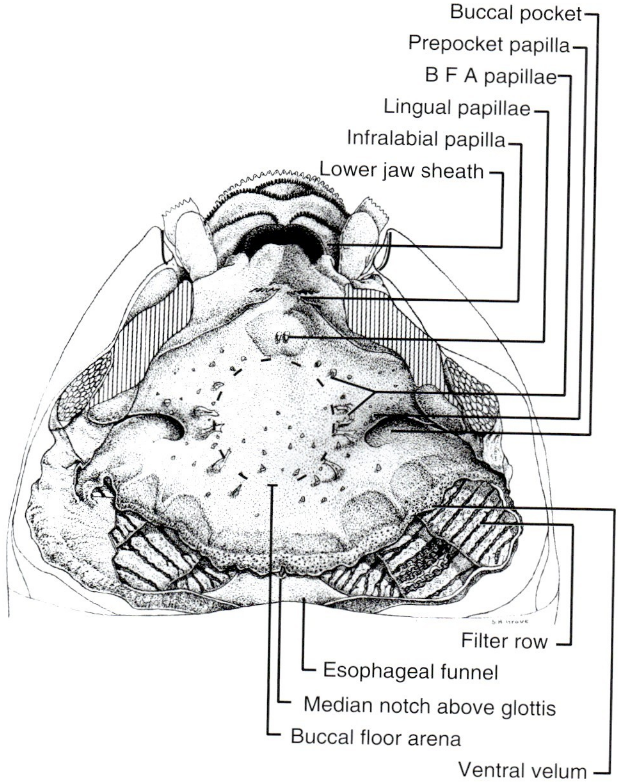

FIGURE 10.20 Floor of the mouth of the tadpole of *Pseudacris regilla*. Tadpoles have several mechanisms for filtering food particles from the water taken into their mouths. Large food particles are channeled into the esophagus by rows of papillae on the floor and roof of the mouth. Smaller particles are strained out of the water as it passes through elaborately folded filters located on the gill bars. Even smaller particles are trapped in mucous strands secreted from glands located in the mouth. Adapted from Wassersug, 1976.

through the gill openings. The mouth then closes and the floor rises, forcing the water outward through the gill slits. The flow of water brings the food particles to the rear of the cavity and in contact with the gill filter surface. Large particles cannot pass through the filter and are picked up by the papillae, which move them into the esophagus. Strings of mucus snare smaller particles touching the surface. A combination of water movement and ciliary activity moves the strings and trapped food rearward. The strings aggregate into larger clumps before passing into the esophagus with the larger food particles. The volume of food entering the mouth cavity regulates this filtering mechanism. When particle suspension density is high, the buccal pump works more slowly to prevent the gill filters and mucus traps from clogging, and conversely, if particles are sparse, the system works more rapidly.

Suction Feeding

Aquatic salamanders, pipid frogs, and some turtles capture prey by opening their mouths simultaneously with the enlargement of their buccal (mouth) cavity, creating a negative pressure gradient. Prey are literally vacuumed into the mouth by the rush of water flowing into the reduced-pressure cavity created by the enlarged buccal cavity. The Matamata turtle, *Chelus fimbriata*, offers the most vivid demonstration of suction feeding. Either from ambush or by slowly stalking or even herding prey, the Matamata moves its head so that it is aligned with the prey, usually a fish or a tadpole. The head shoots forward while the hyoid musculature simultaneously contracts, dropping the floor of the buccal cavity. With the valvular nostrils closed, a tremendous suction vacuum results. The buccal cavity may increase by three to four times its normal size. Just prior to reaching the prey, the mouth opens and prey and water surge into the buccal cavity. The mouth is shut, but not tightly. The floor of the buccal cavity rises, expelling the excess water without losing the prey. The success of this prey-capture technique depends upon accurate alignment of the head to the prey, good timing, and rapid enlargement of the buccal cavity. Matamatas respond to increased prey density by moving less in search of prey.

The hellbender, *Cryptobranchus alleganiensis*, can capture prey alongside its head in addition to prey situated in front of it. This primitive salamander is capable of asymmetrical movements of its lower jaw and hyoid apparatus, which allow it to open its mouth on only one side. The key feature is the ligamentous attachment of the left and right dentaries at the front of the mouth. The flexible attachment permits one side of the jaw to remain in place while the opposite side swings downward, accompanied by a unilateral depression of the hyoid apparatus; this series of movements results in asymmetrical suction.

PREY TYPES AND SIZES

The kinds of prey eaten by amphibians and reptiles have already been introduced in a very general way. A multitude of factors determines the kinds of prey a particular species will eat (Fig. 10.2). The spectrum of prey available in a particular habitat is certainly a major limiting factor. For example, sea turtles would not be expected to eat insects simply because there are no truly pelagic insects. Species that ingest a random sample of prey available in a particular habitat are considered generalists, whereas species that select specific portions of the prey availability spectrum are specialists. Measuring prey availability independent of predators, however, has proven difficult. Different sampling regimes produce different results, and often the sample does not contain all prey captured by the amphibians and reptiles living in the sampled habitat.

A statement by Kirk Winemiller and Eric Pianka (1990) exemplifies the problem:

> Considerable effort has been expended in grappling with the difficult problem of resource availability. Resource availabilities are not easily measured in the field. For example, when insects are sampled with sweep nets, D-vac, Tanglefoot sticky traps, and/or pitfall traps, results differ dramatically. In a study of the herpetofaunas of several sites in the high Andes, Jaime Pefaur and William Duellman fenced study plots and conducted exhaustive collections of all herps and insects encountered within the plots with the intention of using the insects as intact whole specimens for comparison standards with the stomach contents of the herps. Yet fewer than 10% of the insect species actually eaten by the herps were collected by diligent humans. . . .

Winemiller and Pianka recommended using all prey from the pooled set of consumers as a measure of resource availability. Even though the sample is not independent from the consumers, it contains only the prey eaten by the consumers and, thus, may better represent the actual prey-availability spectrum.

Most species of amphibians and reptiles eat a variety of prey types and sizes. In leaf litter habitats of the Brazilian Amazon, the frog *Leptodactylus mystaceus* relies heavily on beetles, termites, and grasshoppers. In the same microhabitat, the lizard *Anolis nitens* feeds primarily on insect larvae, roaches, and spiders (Fig. 10.21). In both species, many other prey items are eaten but to a lesser extent. Prey data based on volumetric data differ

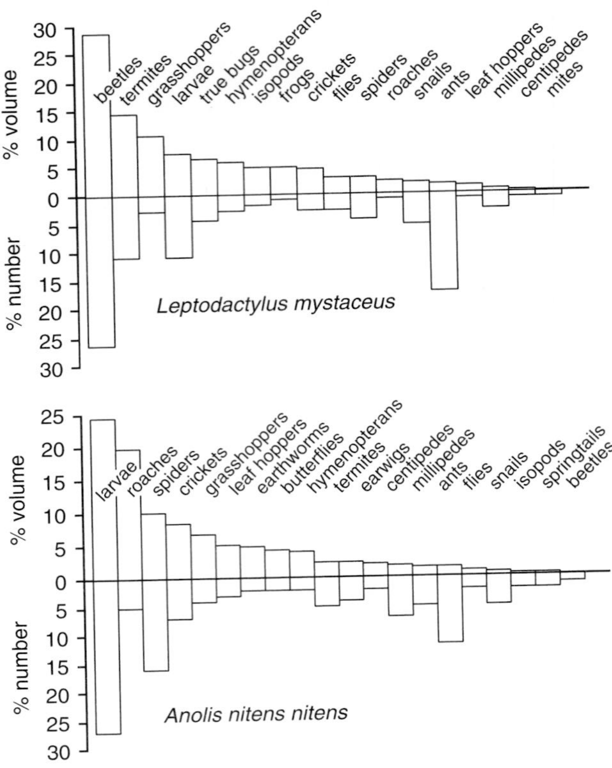

FIGURE 10.21 Representative diets of a frog, *Leptodactylus mystaceus*, and a lizard, *Anolis nitens*, that occur in the same microhabitat (leaf litter) in an Amazonian rain forest. Both species feed on a variety of arthropods and other invertebrates, but the diets are considerably different. In both species, a few prey categories dominate the diet. Volumetric data, which indicate energy gain, are not always reflected in numerical data, which indicate the cost of acquiring prey. Unpublished data from Vitt and Caldwell.

somewhat from prey data based on numeric data, largely because taxonomic groups of invertebrates vary greatly in size. Ants, for example, rank second numerically for *L. mystaceus* and third numerically for *A. nitens*, yet volumetrically, they are relatively unimportant. Because the diets of these two species are strikingly different even though they live in the same microhabitat (leaf litter), it is clear that frogs and lizards do not randomly sample available prey.

Sea turtles, sea snakes, and the marine iguana provide a different perspective on feeding in reptiles because all their foraging occurs in seawater. Green sea turtles feed on a wide variety of red, green, and brown algae, sea grasses, jellyfish, mollusk eggs, and sponges. At some localities, such as near the coast of Peru, invertebrates are much more common in green sea turtle diets, and some fish are taken. Loggerhead sea turtles feed mostly on marine invertebrates, including horseshoe crabs. Hawksbills appear to feed largely on sponges but also take other invertebrates. The diets of other species include combinations of algae and invertebrates. The leatherback sea turtle, however, feeds

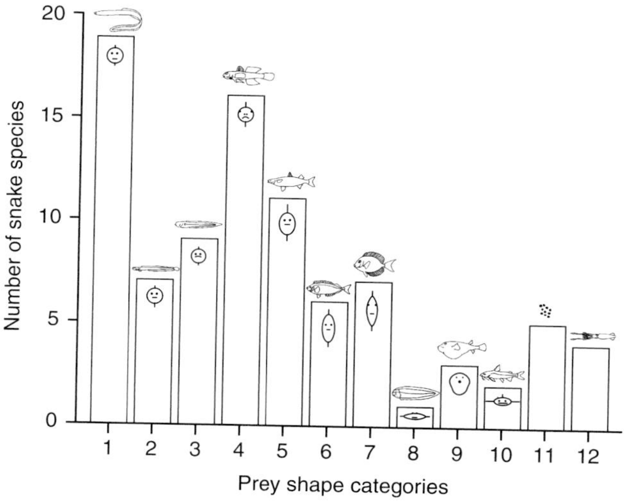

FIGURE 10.22 An examination of the shapes of prey fed on by species of sea snakes reveals that the majority of species feed primarily on fish that are elongate and nearly circular in cross section. The last two columns represent fish eggs and squids. Adapted from Voris and Voris, 1983.

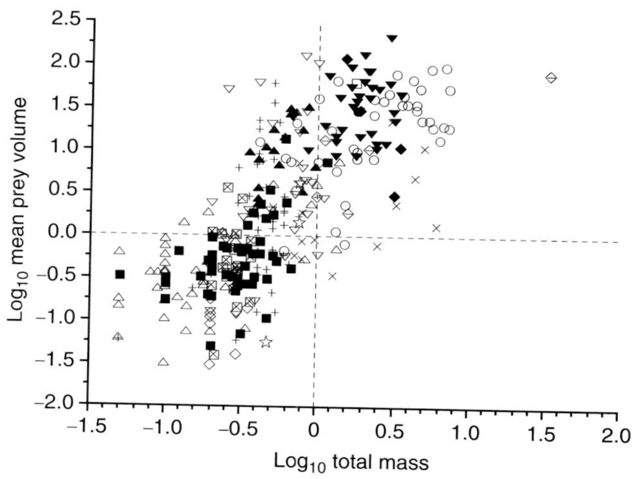

FIGURE 10.23 Both the mean size of prey eaten and the maximum prey size (not shown here) are correlated with body size of frogs and lizards. Even though a strong correlation exists with all species included, species differences in the relationship also exist. In general, species that feed on the smallest prey, mites and ants, tend to eat smaller prey and more of them than species eating other prey types. Frog species are *Elachistocleis ovalis* = x, *Leptodactylus andreae* = upright triangle, Leptodactylus bolivianus = *parallelogram with cross,* L. fuscus = *closed parallelogram,* L. mystaceus = *closed upside-down triangle, Leptodactylus* lineatus = *open star,* Physalaemus ephippifer = *closed square, and* Pseudopaludicola boliviana = *open square with cross. Lizard species are* Anolis nitens = *open circle,* Coleodactylus amazonicus = *open parallelogram,* C. septentrionalis = *cross,* Arthrosaura reticulata = *open square,* Gymnophthalmus underwoodi = *closed triangle,* Leposoma percarinatum = *upside-down open triangle, and* Tretioscincus oriximinensis = *closed circle. Adapted from Caldwell and Vitt, 1999.*

mostly on gelatinous organisms, usually scyphozoans, pelagic coelenterates, and their parasites and commensals.

Sea snakes feed on a diversity of fishes and marine invertebrates, but they mostly eat fish that are sedentary, bottom-dwelling species with fine scales or no scales at all (Fig. 10.22). Different feeding modes translate into different prey types. Marine iguanas feed exclusively on algae that they scrape off submerged rock surfaces. Marine iguanas do not forage in the terrestrial environment, but high temperatures associated with the rocks make it possible for these lizards to bask and raise their body temperatures, which aids in processing their plant diet.

Taken together, sea turtles, sea snakes, and the marine iguana sample a broad taxonomic diversity of food items available in the oceans. The overall lack of amphibians in seawater does not seem surprising because water and electrolyte balance in saltwater present major challenges to animals with permeable skin. Nevertheless, it seems surprising that such a vast and resource-rich habitat has not been exploited by more reptiles, given their ability to regulate water loss in hyperosmotic environments (Chapter 6). Of course, reptile diversity has been high in oceans in the past, and reasons for extinctions of marine clades remain unknown.

Body size of amphibians and reptiles also plays an important role in prey selection. Small species simply cannot eat prey as large as large species can. A summary of data for eight frog and seven lizard species from the northern Amazon rain forest, all living in leaf litter, shows that body size and prey size are related (Fig. 10.23). Careful examination of the data shows also that the relationship between prey size and frog or lizard body size differs among species. Frogs that are ant specialists tend to eat relatively smaller prey than species that are not ant specialists and the same is true for lizards. Not only do ant specialists eat relatively smaller prey than similar-sized non-ant specialists, but they also eat more prey items.

Small species of reptiles and amphibians often feed on some of the smallest arthropods available. Mites and tiny ants are among the smallest arthropods available in tropical rain forest leaf litter. Although many frog species eat some mites, most eat very few. However, several small species of frogs, such as the dendrobatid *Minyobates*, specialize on mites.

All blindsnakes (Leptotyphlopidae, Typhlopidae, and Anomolepididae) eat small prey, usually social insects in their nests. Even though most of these snakes are small themselves, they are large compared with their prey. Consequently skull kinesis is not necessary to successfully prey on social insect castes. A majority of snakes eat very large prey and are capable of doing so because of their feeding apparatus. The upper and lower jaws are highly kinetic, and the right and left sides of each move independently. Moreover, unlike in other reptiles and

TABLE 10.4 The Four Distinct Feeding Types of Snakes.

Type I	Extremely small prey (e.g., termites, ant larvae) that require no immobilization
Type II	Heavy, elongate prey (e.g., caecilians, other snakes) that because of their shape do not require large gapes, but because of their size require constriction or envenomation for subduction
Type III	Heavy, bulky prey (e.g., mammals, lizards) that require specializations for both subduction and swallowing
Type IV	Prey that are lightweight relative to diameter (e.g., fishes, birds) and that require gape specializations but not subduction specializations (venom or constriction)

Note: The categories are based on two measures of prey size: relative mass and relative girth.
Source: Adapted from Greene, 1997.

amphibians, in snakes the lower jaws are not fused, which allows even more freedom of movement. Taken together, these characteristics allow a large expansion of the feeding apparatus, leading to the accommodation of large prey. Based on variation in relative size and shape of prey, four distinct feeding types are recognized in snakes (Table 10.4).

Herbivory

Among amphibians, herbivory is almost totally limited to anuran tadpoles. Ingestion of plant materials has been reported in a few frogs. This limitation is due to the difficulties of digesting fiber. Tadpoles avoid the herbivory conundrum by consuming mainly the algal and bacterial scum (aufwuchs) in the water. Herbivory in tadpoles appears widespread but is poorly verified owing to few studies on tadpoles that examine which cells in the gut contents are digested and which are voided whole. Tadpoles gather their food from all levels of the water column: grazing on bottom sediments, filtering midwater phytoplankton, and skimming the surface scum. Most species specialize on a particular section of the water column and use a certain style of harvesting.

Obligate herbivory is absent in adult amphibians and uncommon in adults of reptiles even though many typically insectivorous reptiles occasionally feed on at least some plant material (Table 10.5). For example, *Tropidurus* lizards on two isolated rock outcrops in the western Amazon rain forest of Brazil eat as much as 17.6% plant materials (flowers). A population on the Rio Xingu in the eastern Amazon eats 26.5% plant materials, mostly fruits. Insects, spiders, and other invertebrates make up the remainder of the diet.

TABLE 10.5 Examples of Reptilian Herbivores, Whose Diets as Adults Are Predominantly Plant Matter.

Taxon	Food items
Turtles	
Batagur baska	Foliage, fruit, animal
Chelonia mydas	Seagrasses, algae
Melanochelys trijuga	Foliage, animal
Pseudemys nelsoni	Foliage, animal
most Testudinidae	Foliage, fruit, flowers
Geochelone carbonaria	Fruit, flowers, foliage, animal
Geochelone gigantea	Foliage
Gopherus polyphemus	Foliage, fruit
Lizards	
Gerrhosaurus skoogii	Foliage, animal
Aporosaurus anchietae	Seeds, animal
Corucia zebrata	Foliage, fruit, flowers
Dicrodon guttulatum	Fruits
Hoplodactylus pacificus	Nectar, fruit, animal
Lepidophyma smithii	Fruit, animal
all Iguaninae	Foliage, fruit, flowers
Cyclura carinata	Foliage, fruit, flowers, animal
Dipsosaurus dorsalis	Flowers, foliage, animal
Iguana iguana	Foliage, fruit, flowers
Sauromalus hispidus	Foliage, flowers, fruit
All *Phymaturus*	Foliage, flowers, fruit

Note: Some have a cellulolytic microflora in the digestive tract and/or colic modifications of the hindgut. The list does not include all well-documented cases of herbivory nor does it include the many examples of omnivory. Plant matter is arranged in order of decreasing volume in the taxon's diet.
Sources: Turtles—Bb, Moll, 1980; Cm, Bjorndal, 1980; Mt, Wirot, 1979; Pn, mT, Ernst and Barbour, 1989a; Gc, Moskovits and M. Bjorndal, 1990; Gg, Hamilton and Coe, 1982; Gp, MacDonald and Mushinsky, 1988; Lizards—As, Steyne, 1963; Aa, Robinson and Cunningham, 1978; Cz, Parker in Greer, 1976; Dg, Holmberg, 1957; Hp, Whitaker, 1968; Ls, Mautz and Lopez-Forment, 1978; al, Iverson, 1982; Cc, Auffenberg, 1982; Dd, Mautz and Nagy, 1987; Ii, Rand et al., 1990; Sh, Sylber, 1988; Phymaturus, Espinoza et al., 2004.

Herbivory poses a digestive problem for vertebrates. Vascular plants contain cellulose in the support structure of their cells. No vertebrates produce cellulase to break down cellulose. Thus, vertebrate herbivores must depend upon the presence of a gut microflora of cellulolytic bacteria to digest plant food. Without such a microflora, it is doubtful that an amphibian or reptile could eat and process enough plant matter to survive on a strictly herbivorous diet. To maintain an efficient gut microflora, a constant and elevated body temperature appears necessary. Other requirements are a constant food supply, slow passage of food items to permit adequate time for bacterial degradation, anaerobic gut environment, regulation of gut pH, and removal of fermentation waste by-products. Lowland tropical reptiles feed year-round and maintain fairly high and constant body temperatures. Once a cellulolytic microflora is obtained, it is improbable that the microflora would need to be renewed. Such microflora stability is less certain for temperate-zone reptiles because of low body-core temperatures and possible absence of a food bolus during dormancy. Low temperature and/or the purging of the digestive tract prior to hibernation or aestivation might well eliminate a specialized microflora. Only a single temperate species, the gopher tortoise (*Gopherus polyphemus*), has been closely examined, and it efficiently digests a high-fiber diet and effectively absorbs the nutrients generated by the bacterial fermentation in the hindgut. It either retains a microflora bolus or restores its microflora each spring.

The how and when of gut microflora acquisition remains unknown for many herbivorous reptiles. For *Iguana iguana*, a complex behavioral mechanism has evolved to ensure the acquisition of plant-digesting microbes. The hatchlings eat soil before emerging from the nest cavity and continue to do so after emergence as they begin to feed regularly on plants. After a few days, the young iguanas move from low shrubbery around the nesting area upward into the canopy and join older juveniles and/or adults; here they consume the feces of older individuals, and this inoculate ensures the presence of the correct microflora in their guts. Inoculation of gut microflora in hatchlings of other species from ingestion of adult feces likely occurs in other reptilian herbivores, but direct observations have not been made. *Gopherus polyphemus* defecates within its burrows and presumably eats some of its feces prior to emerging in the spring. But where do juvenile gopher tortoises and, for that matter, the young of all other reptilian herbivores obtain their fiber-digesting microfauna? In mammalian herbivores, gut microflora acquisition poses no problem, because the young and their parents are closely associated from birth through weaning. The mammalian mother regularly licks the young, and the young feed from the mother's mammary glands, so young mammals acquire the microflora early from the ingestion of the mother's saliva or fecal material. This close association of mother and offspring does not exist for any reptilian herbivore. In herbivorous Aldabra tortoises, the absence of a gut microflora leads to a low digestive efficiency (30%), in contrast to digestive efficiencies of about 65% for red-footed tortoises and 85% for green iguanas, both of which have gut microfloras.

It has long been argued that large body size is necessary for reptiles to maintain energy balance on a strictly herbivorous diet, and until recently, most known herbivorous lizards were large in body size. This idea has been challenged by an impressive data set compiled by Robert Espinoza and his collaborators. Phylogenetic analysis of diets of a monophyletic clade of tropidurine lizards in the genera *Ctenoblepharys, Phymaturus,* and *Liolaemis* revealed an estimated 18.5 independent origins of herbivory. All 10 species of *Phymaturus* are herbivorous (one origin), and the other herbivorous tropidurines are in the genus *Liolaemis*. Not only has herbivory evolved more times within these lizards than in all other lizards combined, but the rate at which the evolution of herbivory occurred is 65 times greater than that for all other lizards. Moreover, these lizards are smaller in general than all other herbivorous lizards but are well within the size range of most lizards that are not herbivorous (Fig. 10.24). After removing the effect of phylogeny, a negative correlation exists between plant consumption and environmental temperature. Thus, the evolution of herbivory is associated with low rather than high temperatures. Isolation of lizards in montane habitats that are not interconnected accounts for repeated independent origins of herbivory. Small body size appears necessary for herbivorous lizards in these habitats because they can heat rapidly, given an unpredictable thermal environment. While active, they maintain body temperatures typical of other herbivorous lizards.

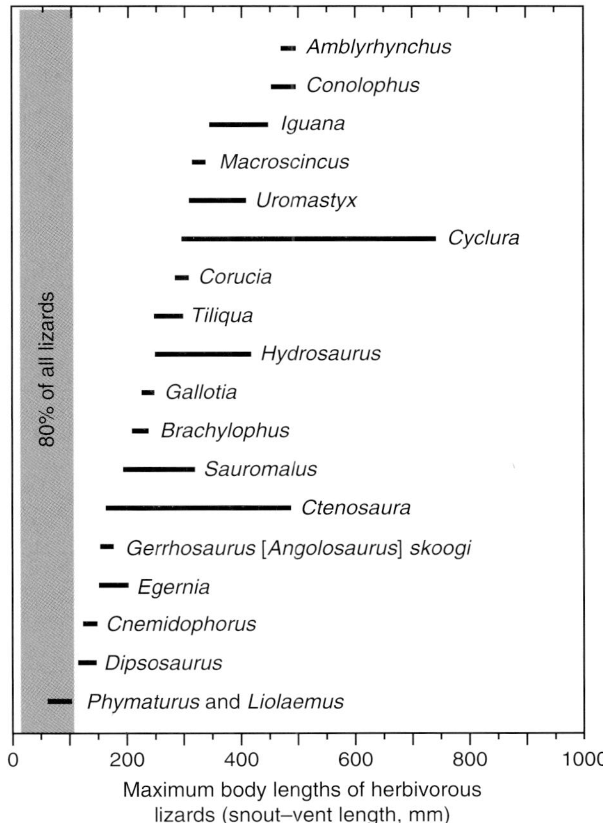

FIGURE 10.24 Body sizes of herbivorous lizards showing that herbivorous *Phymaturus* and *Liolaemus* are smaller than all other herbivorous lizards, with body sizes falling well into the size distribution for insectivorous lizards. Adapted from Espinoza et al., 2004.

In an interesting analysis of traditional large-bodied herbivorous lizards, Anthony Herrel points out that herbivorous lizards are wide foragers only because they move from plant to plant (plants are stationary). These herbivorous lizards, all in the Iguania, also discriminate chemical cues, a trait that is associated with wide-foraging lizards in the Scleroglossa. A large number of lizard species eat some plant material and can be considered omnivorous, and the most logical way to evolve herbivory is from omnivorous lizards, which already have some of the morphological adaptations necessary to feed on plants. However, most omnivorous species are in the Autarchoglossa, whereas nearly all herbivorous species are in the Iguania. Many questions remain about the evolution of herbivory and how herbivorous lizards fit in the classical foraging-mode dichotomy.

Ontogeny of Diets

Ontogenetic dietary shifts are probably common in amphibians and reptiles but are not well studied. Adult amphibians and reptiles do not necessarily eat the same prey as larvae or juveniles. The most dramatic example of a dietary shift is in amphibians with aquatic larvae and terrestrial adults. Most anurans are detritivores as larvae and insectivores as adults. Among frogs with predaceous tadpoles (e.g., *Adelphobates castaneoticus*), the dietary shift is from eating aquatic insect larvae and other tadpoles during the larval stage to eating ants during the adult stage. A dramatic example of a dietary shift occurs in semiterrestrial tadpoles of the dicroglossid frog *Nannophrys ceylonensis* of Sri Lanka. These strange tadpoles have a number of morphological adaptations that allow them to live and feed on land. They live on damp rocks where they scrape the surface film off rocks in short feeding bouts. No filter feeding occurs. Moreover, young tadpoles apparently do not feed, and the rock-scraping behavior begins at Gosner Stage 25. Tadpoles at Stage 34 coil around patches of food, apparently excluding smaller tadpoles. Although the diet consists of a variety of organisms, including protozoans, rotifers, arthropods, nematodes, and occasionally conspecific eggs and tadpoles, a dietary shift from a greater proportion of plant material to a greater proportion of animal material occurs during tadpole development, and the dietary shift is correlated with changes in the gut. Shortening of the gut in tadpoles of most frog species occurs during metamorphosis, but in *N. ceylonensis*, the timing of shortening of the gut occurs earlier and is more protracted. Consequently, a developmental shift in gut development allows these tadpoles to take advantage of animal food matter in densely shaded forests with low primary productivity. Ontogenetic dietary shifts likely occur in many tadpoles, but few species have been studied. In some frogs, such as the Chilean giant frog (*Calyptocephalella gayi* [formerly *Caudiverbera caudiverbera*], family Calyptocephalellidae), phenotypic plasticity in gut morphology and physiology suggests that adjustments necessary for dietary shifts are not uncommon in tadpoles. Interestingly, phenotypic plasticity in gut morphology and intestinal enzymes in the Chilean giant frog resulted from different temperature treatments rather than different diet treatments. The ecological significance of this result remains unstudied, but it would be interesting to examine whether changes in types of food available to these tadpoles vary with temperature, and if so, then the connection between phenotypic plasticity and diet could be made. Temperature may cause the phenotypic change, and the underlying adaptive significance may be associated with correlated changes in food supply.

Among amphibians and reptiles in which juveniles have the same morphology as adults, a large component of the dietary shift is associated with body size and thus age. Water snakes in Florida provide an example. *Nerodia erythrogaster* and *N. fasciata* feed primarily on fish as juveniles but switch to mostly frogs when they reach about 50 cm in snout-vent length (Fig. 10.25). Even though *N. rhombifera* and *N. cyclopion* feed on fish throughout their lives, the kinds of fish they eat change with snake age and size. Several factors contribute to ontogenenic diet shifts in these snakes, including the effect of snake body size on the size of prey that can be taken, differences in microhabitat use between juveniles and adults, and sexual differences (size-based) in prey types taken.

Potential ontogenetic shifts in diet can be offset by morphological variation among age groups. Juveniles and adults of the salamander *Plethodon cinereus* feed on the same prey types; small mites are among the most common prey. Prey size does not vary with head size in adults, but size of the largest prey items does vary with head size in juveniles. Consequently, size constrains the diet of juveniles in that they cannot eat the larger items that adults eat. Nevertheless, juveniles have relatively broader heads than adults, which allows them to eat all but the largest prey taken by adults (Fig. 10.26).

Evolution of Diets

Recognition that diets of amphibians and reptiles might evolve just as morphological or physiological traits is just gaining acceptance. It has long been known, for example, that within some clades, all species share a diet preference unlike that of species in closely related clades. For example, horned lizards (*Phrynosoma*) as a group eat primarily ants; all Iguaninae are herbivorous, at least as adults; dendrobatid frogs in the genera *Dendrobates, Oophaga, Ranitomeya,* and *Adelphobates* primarily eat ants; and snakes in the closely related families Typhlopidae, Leptotyphlopidae, and Anomalepidae eat eggs, larvae, and pupae of ants and termites. Indeed, insectivory in these snakes (the Scolecophidia) is one piece of evidence suggesting

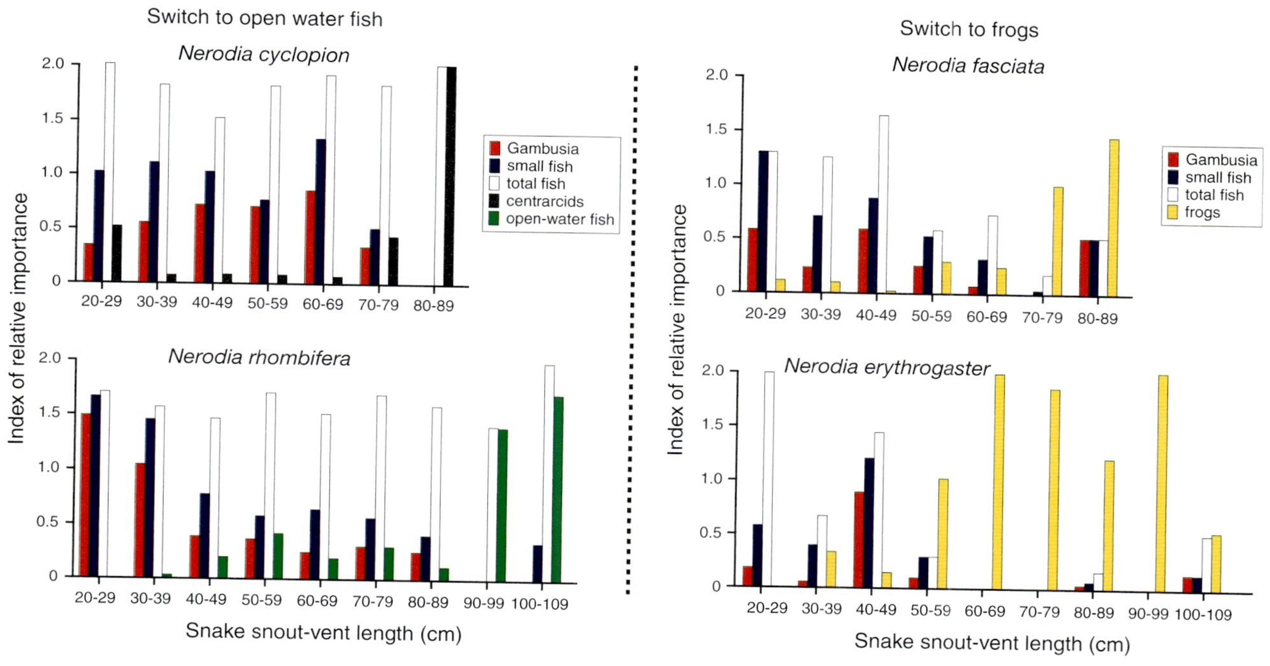

FIGURE 10.25 The diets of four species of water snakes change with age and size. Adapted from Mushinsky et al., 1982.

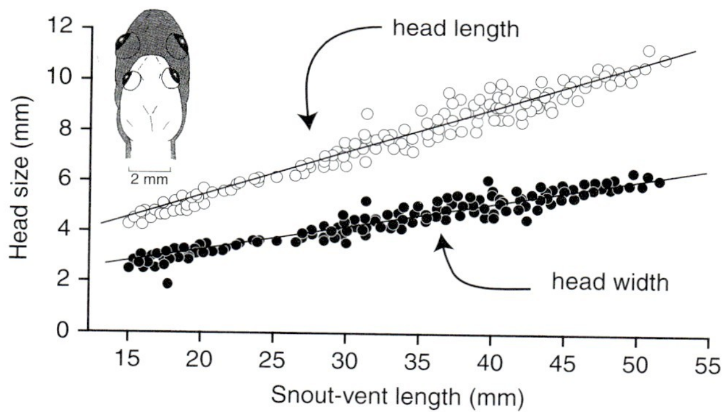

FIGURE 10.26 Although both head width and length increase with body size (snout-vent length) in *Plethodon cinereus*, head width of juveniles is proportionately greater in juveniles, which allows them to feed on relatively large prey. Adapted from Maglia, 1996.

that they are the most primitive snake clade. Species in the Sonorini clade of colubrine snakes feed on arthropods, and xenodontine colubrids feed on frogs. These and many other examples suggest that similarity in diets within particular clades reflects dietary shifts early in the evolutionary history of the clade, which, among other things, has changed the way we think about species assemblages and communities (discussed in more detail in Chapter 12).

Specialization on ants provides a particularly instructive example of the evolution of diets and exemplifies the complexity of trade-offs between foraging and predator escape strategies. Ant specialization has evolved independently in a number of families of lizards and frogs. Within the Phrynosomatinae, species in the genus *Phrynosoma* feed primarily on ants. These tanklike lizards are cryptic in morphology and coloration, move very little, and eat literally hundreds of ants each day. Most other genera of phrynosomatine lizards eat a diversity of insects, including some ants. From a strictly energetic perspective, eating ants seems to be inefficient because ants are generally small and contain a large amount of exoskeleton compared with larger insects such as caterpillars. If a lizard had to move to find each ant, the energy gain would be less than the energy required to capture the ant. Ants also often contain noxious chemicals. Consequently there are energetic costs to eating ants as well as potential metabolic-processing costs to handle ingested chemicals. Several benefits of ant eating offset the potential costs. First, ants often occur in clusters, and as a consequence, the energy involved to find a thousand ants may be the same or less than the energy to find a single large grasshopper. More importantly, the same chemicals that ants use for defense are metabolized by *Phrynosoma* and contribute to the bad taste of their blood, which appears to repel canid predators (see Chapter 11). Likewise, in dendrobatoid frogs, ants comprise most of the prey eaten by many species. Other species feed on relatively fewer ants. Most interesting is the observation that ant specialization in these small tropical frogs appears to be related to a behavioral defense complex involving toxic or bad-tasting skin secretions and aposematic coloration (Fig. 10.27). Among other things, bright coloration of numerous species warns predators that the frogs have bad-tasting or toxic skin, resulting

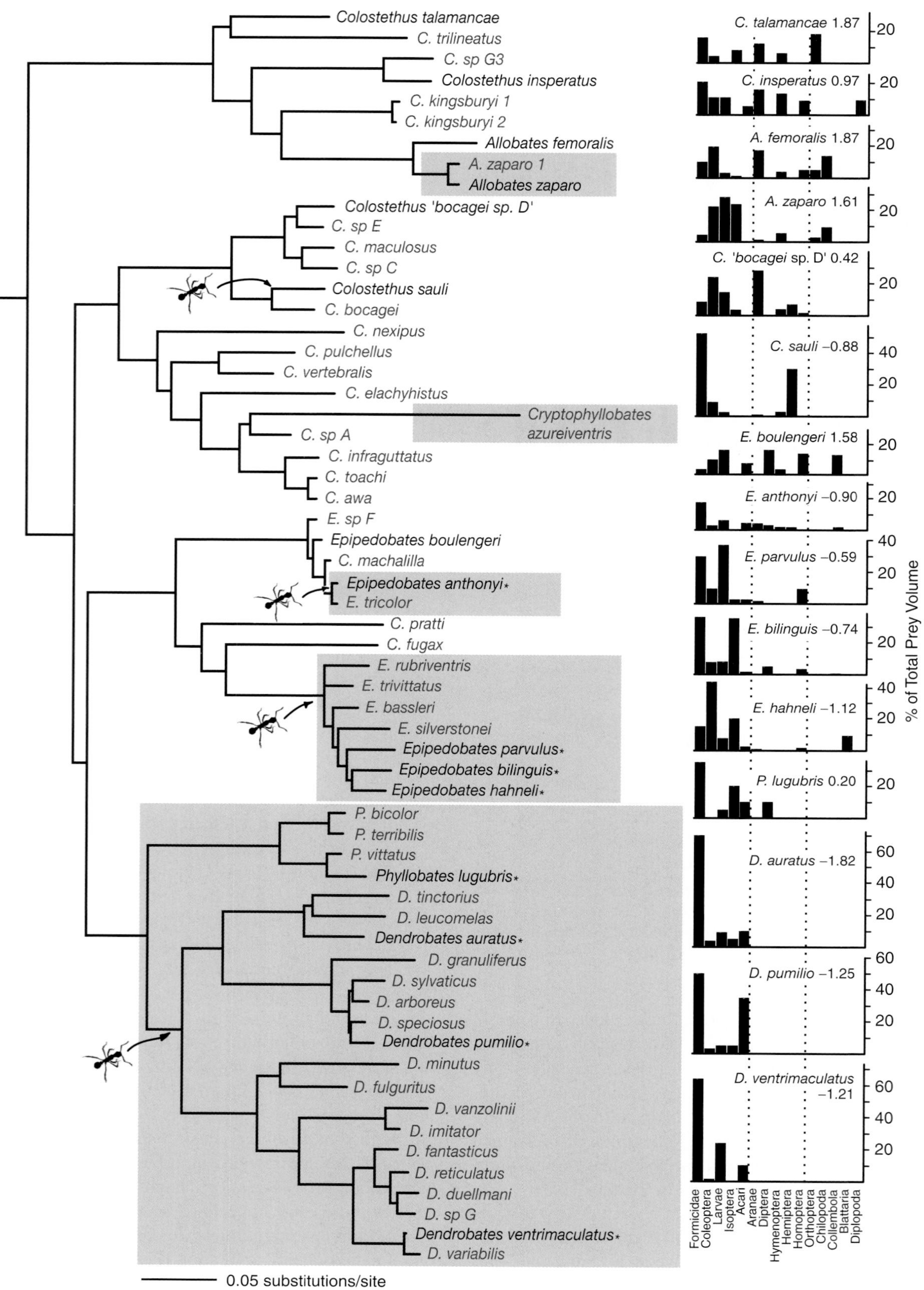

FIGURE 10.27 For legend see opposite page.

FIGURE 10.27 In dendrobatoid frogs, the evolution of specialization on ants is linked with aposematic coloration and production of skin toxins. Ants (myrmicine ants in particular) produce the alkaloids for chemical defense against predators; frogs eat the ants and are able to either move the alkaloids to the skin or combine them with other chemicals and move them to the skin and use them for predator defense. Bright coloration of these frogs usually, but not always, signals to a predator that the frog is distasteful or toxic. Ant icons indicate a dietary shift to ant specialization based on an *a priori* categorization of generalists versus specialists. Shaded boxes indicate conspicuously colored frog species, and asterisks indicate that the species are known to contain alkaloids in the skin. Frequency histograms on the right indicate relative volume contributed by the 15 most common prey types to the diet of each frog species for which dietary data were available, and these are indicated in the phylogeny by boldfaced type. Numbers to the right of frog species names in the diet panel refer to the principal components scores of dietary niche breadths, essentially ranking frogs across prey types. Note that we have retained genera and species names as in the original graphic, and thus they are inconsistent with the taxonomy that appears in Chapter 17, with the following clade names: Aromobatidae (Dendrobatidae [Hyloxalinae {Colosthethinae+Dendrobatinae}]). Nevertheless, phylogenetic relationships are the same, and as a result, interpretations regarding evolution of diets, coloration, and defensive chemicals remain unchanged. For the interested reader, we suggest tracking species names on the Web site http://research.amnh.org/herpetology/amphibia/. Adapted from Darst et al., 2005.

from the ingestion of ant chemicals as well as ingestion of chemicals from other tiny leaf litter arthropods. Brightly colored species move frequently while foraging and thus are conspicuous, whereas cryptic (non-ant specialists and nontoxic) species do not move much while foraging. Specialization on ants and the associated predator escape mechanisms have evolved repeatedly within these frogs, and in two instances (Dendrobatinae and one clade in the Colostethinae), entire clades of frogs with these coevolved traits have been generated (bottom two shaded boxes in Fig. 10.27). Additional details on predator escape in these frogs appear in Chapter 11. A similar radiation of frogs with the same set of traits (ant specialization associated with aposematic coloration and skin toxins) has evolved independently in the frog family Mantellidae in Madagascar. In addition to acquiring alkaloids from ants, some mantellids also acquire nicotine from ants that get nicotine from plants. Thus a nicotine food chain exists from plant to ant to frog! The preceding examples, from both frogs and lizards that eat ants, which are in general, small and low-energy prey, exemplify the complexity of the evolution of diets in ectothermic vertebrates. Based on energy gain alone (i.e., optimal foraging), ant specialization should be a poor strategy and selectively disadvantageous. However, because it can have added benefits in terms of sequestering chemicals for defense, energetic disadvantages are compensated for by advantages in offsetting predation.

QUESTIONS

1. How does the tongue of a toad work when it captures prey, and how does this differ from tongue use in a chameleon?
2. Some lizards and many snakes can and do eat large prey. Describe how this is possible and compare how lizards that eat large prey differ from snakes that eat large prey.
3. Reptiles and amphibians are often categorized in two broad foraging modes based on foraging behavior. What are these foraging modes and what are the behavioral and energetic bases for these different modes? List as many ecological, morphological, and physiological correlates of each foraging mode that you can think of.
4. Although the argument has been made that herbivory in lizards is associated with large body size, recent data on small-bodied South American lizards suggests that large body size is not a necessary condition of herbivory. Moreover, phylogenetic analyses show that the evolution of herbivory has occurred much more frequently in small-bodied lizards. Explain the physiological arguments for associating herbivory with body size and the phylogenetic arguments that associate herbivory with small body size.
5. Why might you expect ontogenetic dietary shifts to be more common in snakes than in lizards?

ADDITIONAL READING

Greene, H. W. (1997). *Snakes. The Evolution of Mystery in Nature.* University of California Press, Berkeley.

Huey, R. B., and Pianka, E. R. (1981a). Ecological consequences of foraging mode. *Ecology* **62:** 991–999.

MacArthur, R. H., and Pianka, E. R. (1966). On optimal use of a patchy environment. *American Naturalist* **100:** 603–609.

Perry, G., and Pianka, E. R. (1997). Animal foraging: past, present, and future. *Trends in Ecology & Evolution* **12:** 360–364.

Pianka, E. R., and Vitt, L. J. (2003). *Lizards. Windows to the Evolution of Diversity.* University of California Press, Brekeley, CA.

Reilly, S. M., McBrayer, L. B., and Miles, D. B. (Eds.) (2007a). *Lizard Ecology: The Evolutionary Consequences of Foraging Mode.* Cambridge University Press, Cambridge, Mass.

Schoener, T. W. (1971). Theory of feeding strategies. *Annual Review of Ecology and Systematics* **2:** 369–404.

Schwenk, K. (1995). Of tongues and noses: chemoreception in lizards and snakes. *Trends in Ecology and Evolution* **10:** 7–12.

Schwenk, K. (2000). Feeding: Form, Function, and Evolution in Tetrapod Vertebrates. Academic Press, San Diego, CA.

Seigel, R. A., and Collins, J. T. (Eds.) (2002). *Snakes: Ecology and Behavior*. McGraw-Hill, Inc., New York.

Wells, K. D. (2007). *The Ecology and Behavior of Amphibians*. University of Chicago Press, Chicago.

REFERENCES

Foraging Modes

Anderson, 2007; Bauer, 2007; Bonine, 2007; Cooper, 1994a,b, 1995, 1999, 2000, 2002, 2007; Cooper and Vitt, 2002b; Cooper et al., 2002; Gerritsen and Strickler, 1977; Gray, 1987; Herrel, 2007; Huey and Pianka, 1981a; MacArthur and Pianka, 1966; Perry, 1999, 2007; Perry and Pianka, 1997; Pianka, 1966; Pianka and Vitt, 2003; Reilly and McBrayer, 2007; Schoener, 1974b; Schwenk, 1993, 1995; Simon and Toft, 1991; Vanhooydonck et al., 2007; Vincent et al., 2005; Vitt and Pianka, 2004, 2005, 2007; Vitt et al., 2003; Whiting, 2007.

Prey Detection

Badenhorst, 1978; David and Jaeger, 1981; Dole et al., 1981; Luthardt and Roth, 1979.

Visual Prey Detection

Pianka and Parker, 1975; Roth, 1986; Toft, 1980.

Chemosensory Prey Detection

Cooper, 1995, 2007; Cooper and Alberts, 1990; Cooper et al., 2002; Coulson and Hernandez, 1983; Rossi, 1983; Roth, 1987; Schwenk, 1986, 1994a, 1995; Teather, 1991.

Auditory Prey Detection

Chou et al., 1988; Hetherington, 1985; Jaeger, 1976; Sakaluk and Belwood, 1984.

Thermal Prey Detection

Cock Buning, 1985; Grace et al., 2001; Greene, 1997; Shine, 1991b; Sichert et al., 2006.

Tactile Prey Detection

Hartline, 1967; Smith et al., 2002; Willi et al., 2006.

Prey Capture and Ingestion

Auffenberg, 1981; Greene, 1997; Perez-Higareda et al., 1989; Sazima, 1991.

Biting and Grasping

McBrayer and Corbin, 2007; McBrayer and Reilly, 2002; Reilly and McBrayer, 2007; Vitt and Cooper, 1988.

Constriction

Hardy, 1994; Moon, 2000; Mori, 1996; Shine, 1991b; Shine and Schwaner, 1985.

Injected Venoms

Hayes et al., 1992, 1993; Greene, 1997; Kardong and Bels, 1998; Shine, 1980a, 1991b.

Projectile Tongues

Gans and Gorniak, 1982; Larson and Gutherie, 1985; Lombard and Wake, 1976, 1977, 1986; Meyers et al., 2004; Wainwright and Bennett, 1992a,b; Wainwright et al., 1991.

Filter Feeding

Inger, 1986; Wassersug, 1972, 1980; Wells, 2007.

Suction Feeding

Cundall et al., 1987; Formanowicz et al., 1989; Holmstrom, 1978; Mahmoud and Klicka, 1979.

Prey Types and Sizes

Bjorndall, 1997; Caldwell, 1996; Caldwell and Vitt, 1999; Glodek and Voris, 1982; Pefaur and Duellman, 1980; Pianka and Vitt, 2003; Simon and Toft, 1991; Toft, 1980; Vitt and Cooper, 1986a; Vitt and Pianka, 2007; Vitt and Zani, 1996a; Voris and Voris, 1983; Wells, 2007; Winemiller and Pianka, 1990.

Herbivory

Bjorndal, 1987, 1989; Cooper and Vitt, 2002b; Das, 1996; Espinoza et al., 2004; Hamilton and Coe, 1982; Herrel, 2007; Troyer, 1984a, b; Vitt, 1993; Vitt et al., 2005.

Ontogeny of Diets

Bouchard and Bjorndal, 2006; Caldwell, 1993, 1996a; Casteñeda et al., 2006; Cooper and Lemos-Espinal, 2001; Durtsche, 1999, 2000; Hunt and Farrar, 2003; Maglia, 1996; Mushinsky et al., 1982; Wickramasinghe et al., 2007.

Evolution of Diets

Cadle and Greene, 1993; Caldwell, 1996; Clark et al., 2005; Clough and Summers, 2000; Daly et al., 2000, 2002, 2003; Darst et al., 2005; Greene, 1982, 1997; Pianka and Parker, 1975; Santos et al., 2003; Saporito et al., 2003, 2004, 2007; Savitsky, 1983; Shine, 1991b; Toft, 1980, 1995; Vences et al., 2003; Vitt and Pianka, 2007; Vitt et al., 2003.

Chapter 11

Defense and Escape

A large majority of amphibians and reptiles do not survive to reach sexual maturity, and once adults, many do not survive long enough to produce offspring. Predation is the greatest cause of mortality in natural populations and can occur in any life history stage. Eggs of amphibians are eaten by insects, leeches, fishes, other amphibians, and many reptiles. Fungus and bacteria also cause significant mortality in amphibian eggs. Eggs of reptiles are eaten by a variety of mammals, including foxes and raccoons. Many reptiles eat eggs of amphibians and reptiles, and even ants prey on reptile eggs. Small-bodied and juvenile amphibians and reptiles are prey for numerous arthropods, including insects, spiders, centipedes, and amblypygids, and nearly all vertebrate groups from fishes to mammals and birds prey on amphibians and reptiles (Fig. 11.1). Although a few invertebrates and numerous vertebrates prey on adult amphibians and reptiles, the increased body size of adults relative to juveniles and relative to the body size of predators reduces the diversity of predators that can effectively capture them. Body size of some species, such as the saltwater crocodile, Komodo monitor, anaconda, and Galapagos tortoise, renders adults virtually immune to predation by all animal species except humans.

During the evolutionary history of amphibians and reptiles, any morphological, physiological, behavioral, or ecological trait that reduced predation increased in frequency as individuals not exhibiting those traits were removed from the breeding population. The selective pressures driving the evolution of predator-escape mechanisms were and continue to be strong because as prey respond evolutionarily to predictable predation events, predators respond by evolving new or more effective ways to find and capture prey. In effect, it is an evolutionary arms race between predators and prey. The diversity of predator-escape mechanisms in amphibians and reptiles continues to surprise herpetologists; new defenses continually are being discovered. Many mechanisms are obvious, such as alert responses followed by rapid flight or the loss of tails by salamanders and lizards that allow the prey a second chance at escape. Many are much more subtle and include rapid development of amphibian eggs and tadpoles to offset predation by aquatic insect larvae or the evolution of large clutches of small eggs to offset heavy and predictable predation on early life history stages. Some involve the use of chemicals to deter or even poison predators. In several families of ant-eating frogs, chemicals obtained from the diet are mobilized and used in defense. In viperid, elapid, and some colubrid snakes, injected venoms used to acquire prey serve to fend off or even kill potentially lethal predators. Taken together, predator-escape mechanisms provide some of the most fascinating questions for biological research and lie at the center of Charles Darwin's "struggle for existence."

Escape from predation requires interference with a predator's ability to detect or identify an individual as prey or the successful escape of a potential prey once detected. In a heuristic sense, the evolution of escape mechanisms seems obvious. In nature, predator escape is much more complex because the diversity and abundance of predators is not constant in space or time, and the complement of potential predators changes depending on the life history stage of the prey and numerous other factors. Trade-offs associated with reproduction, social behavior, and activity

Herpetology, 3rd Ed.

FIGURE 11.1 Predation on amphibians and reptiles. From top: a Striped racer (*Masticophis taeniatus*) eating an adult Greater earless lizard (*Cophosaurus texanus*) (photograph by J. M. Howland); a leopard lizard (*Gambelia wislizeni*) eating a Long-tailed brush lizard (*Urosaurus graciosus*) (photograph by C. Schwalbe); a spider eating a small hylid frog (photograph by W. Hödl); a desert hairy scorpion (*Hadrurus*) eating a juvenile horned lizard (*Phrynosoma*) (photograph by J. Rorabaugh).

can influence both the evolution of escape mechanisms and the manner in which predator escape might take place. Most amphibians and reptiles employ several different predator-escape mechanisms, often using different ones during different life history stages.

ESCAPE AND EMERGENCE THEORY

As important as predator escape is in the life history of amphibians, reptiles, and animals in general, development of detailed theory on interactions between predator and prey has lagged behind theory in other areas, such as optimal foraging. Optimal escape should balance risk and cost of escape behaviors (Fig. 11.2, top). If the cost of escaping (dashed line) is high for an individual, for example, a lizard that is a long distance from refuge, then the animal should take lower risk when confronted by a predator by seeking refuge sooner or when the predator is farther away. If an individual is close to a safe refuge, then both cost to seek refuge and risk of being captured are low and the animal can allow the predator to approach more closely. The relationship of risk curves (solid lines) to cost curves should determine response of prey. In general, this relationship should hold for nearly all kinds of escape behaviors and risk factors, and theoretical predictions apply to both predator and prey behaviors. Of course, this is highly simplified compared with the natural world, in which multiple risk factors and costs are involved. Consider another example in which two cost and two risk curves are involved (Fig. 11.2, center). Intersection of these curves indicates that the relationship of risk to cost can vary considerably, and again, this is highly simplified. Finally, the first two examples show escape cost increasing linearly with distance from a predator and risk decreasing in a curvilinear fashion with increasing distance between predator and prey. However, because of the interplay between different potential risk and cost factors, risk and/or cost curves may not be monotonic, and thus several optima with respect to cost and risk may exist (Fig. 11.2, bottom). In this example, cost is shown to be nonmonotonic and three optimal solutions exist.

Although usually referred to as optimal escape theory, the same theory applies to emergence from refuges, because similar decisions must be made by animals as they begin activity and when they re-initiate activity following escape. A simple example might be a snake that by entering a crevice escaped an attempted predation event by a hawk. The cost of remaining in the crevice may be reduced foraging or access to mates, but the cost of coming out, which should decrease with time, could be sudden death by the hawk that may have remained in the area. A considerable amount of effort has been made to test optimal escape and emergence theory, mostly by William E. Cooper, Jr., and his colleagues. Because

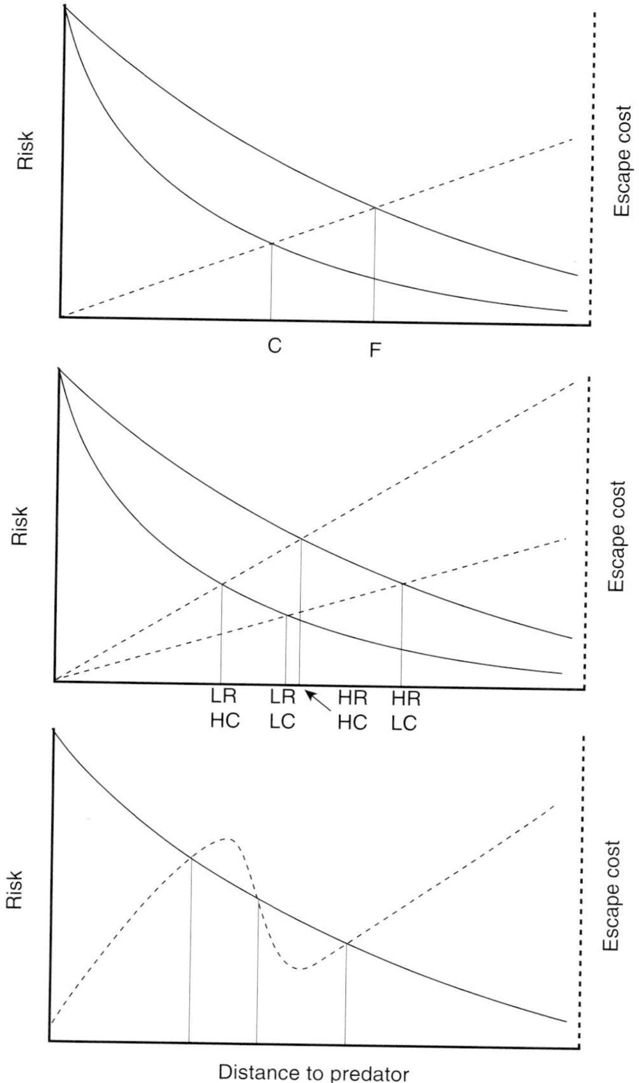

FIGURE 11.2 Theoretical models showing the relationship between risk and escape cost as a function of distance between a predator and prey. In the top panel, two prey are located at different distances from a refuge. The risk curve is higher for the one farthest from the refuge. C (close) and F (far) are optimal approach distances for each. In the center panel, multiple risk and cost curves are shown, and other options exist. Optimal approach distances are indicated by vertical lines, and each is labeled (e.g., LR = low risk, HC = high cost, and so on). In the lower panel, a monotonic risk curve intersects with a nonmonotonic escape cost curve producing several optima as indicated by vertical lines attached to curve intersections. A similar curve can be drawn for a nonmonotonic risk curve and a monotonic cost curve. Adapted from Cooper and Vitt, 2002a.

optimal escape and emergence theory centers on distance between predator and prey, the theory itself is limited, and much research points to the necessity to develop more realistic theory.

What are the costs and what are the risks? Primary costs are related to feeding opportunities and social interactions. Individuals that are hiding from predators cannot forage and cannot interact with other individuals, both of which impact individual fitness, either directly or indirectly. Risk ultimately is the probability of being captured and killed, but it has many components and they vary among species. Risk factors can include distance to refuges, structural complexity of the environment in terms of escape opportunities, light levels (affecting crypsis), perch diameters and structure of arboreal habitats, and a multitude of other variables. Speed and wariness of potential predators are risk factors that may not be easily predictable, and angle of approach can affect predation events. Another issue is body temperature and its effect on behavioral responses. Some lizards, for example, are known to flee when predators are at greater distances when their body temperatures are relatively low.

In keeled earless lizards (*Holbrookia propinqua*), lizards run faster and enter refuges more readily in response to increased speed and directness of approach, and they allow closer approach when refuges are available. Similarly, desert iguanas (*Dipsosaurus dorsalis*) flee more readily when approached rapidly and directly. Depending on time of day and temperature, they either escape into burrows or simply flee, but low temperatures cause the lizards to allow closer approach. Responses of *Anolis* lizards to simulated risk factors vary considerably and are tied to some extent to microhabitat use by the lizards. Arboreal species that escape upwards in vegetation vary approach distances inversely with perch height. Those using low perches flee when the approacher is at a greater distance than those using higher perches. Cryptic species allow closer approach than those that are not cryptic. Other studies on *Anolis* lizards have shown that anoles living on open ground or low on tree trunks venture farther from shelter and run farther when fleeing than those living in vegetation, and these behaviors appear correlated with morphology. Broad-headed skinks (*Plestiodon laticeps*) enter holes in trees to escape approaching predators and usually come back out to forage after a minimal time period in refuge. However, when a simulated risk remains near the refuge, the skinks remain in refuge longer, thus assessing risk and responding accordingly. Moreover, their latency to emergence increases with the speed at which an approach is made. These and many other experimental studies indicate that lizards respond to cost and risk of predation as expected by theory.

In the Balearic lizard (*Podarcis lilfordi*), availability of food changes the response of lizards to risk factors. Lizards with food available allow closer approach, flee less distance, and are more likely to return than those not presented with food. The amount of food available to individuals also affects escape behaviors in these lizards, with lizards willing to increase risk if the food payoff is greater.

Male broad-headed skinks take greater risks when they are guarding mates than they would otherwise. When male–female pairs are approached in the field, males are detected first but females retreat first, leaving the male

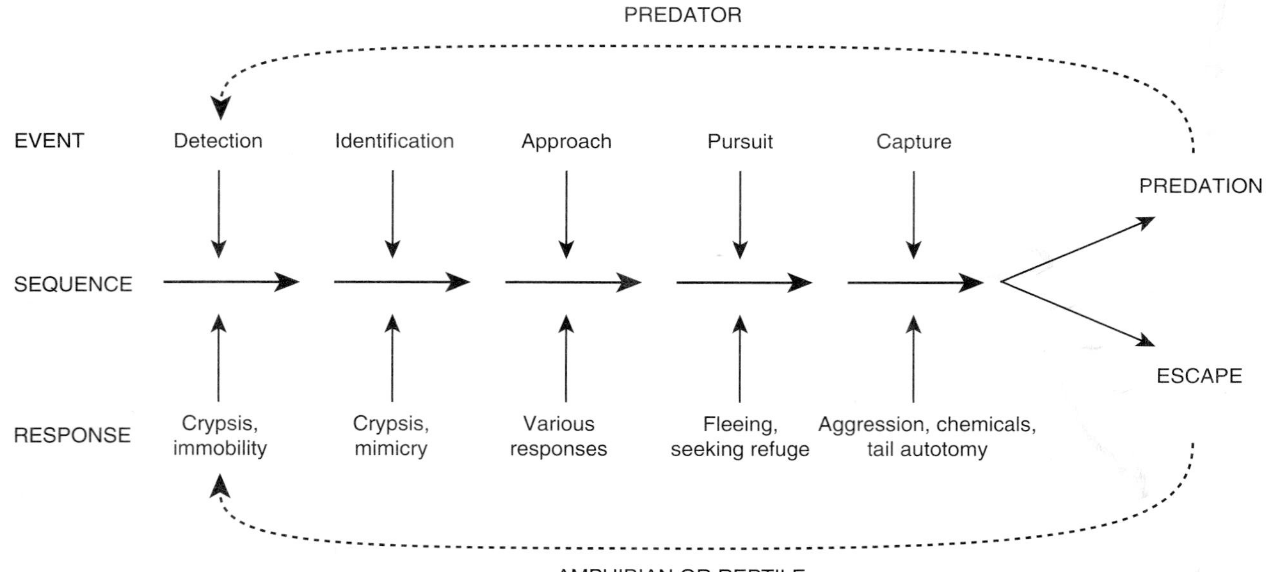

FIGURE 11.3 Sequence of events during an encounter between a predator and a prey. At any point along the sequence, the interaction can end. If the series of interactions passes through all stages, the prey either will escape or be eaten, and the process will be repeated. Adapted from Pianka and Vitt, 2003.

exposed longer. In addition, males respond to the disappearance of the female by tongue-flicking the substrate in attempts to pick up the trail of the female, whereas females never attempt to find the male. Thus males allow risk to increase because the cost of losing a mate is presumably high. When males fight during social interactions, they also expose themselves to additional risk.

We now turn our attention away from theory and testing of theory underlying escape from predators and examine some of the fascinating mechanisms used by amphibians and reptiles to escape predation. A predation or escape event can be complex and often follows a series of stages, any of which can result in continued pursuit (Fig. 11.3). Predators must detect, identify, approach, pursue, and capture a prey if they are to be successful, and a prey must avoid detection, flee if detected, or respond in a more drastic fashion if captured in order to escape. For the predator, loss of a prey means only that some energy was expended for no net gain; for the prey, failure to escape is final, resulting in death.

PREDATOR AVOIDANCE

Escaping Detection

Predators detect prey by visual, thermal, auditory, tactile, and olfactory cues. Escaping detection requires (1) interference with a predator's ability to use cues or (2) not being present when a predator might be searching for prey. Simply limiting activity to time periods when predators are unlikely to be active affords some relief from predation. The most obvious example is nocturnal activity by many amphibians and reptiles, which effectively removes them from predation by diurnal bird species. Other animals, including nocturnal snakes and bats, might be effective predators on a given species at night. Altering activity patterns involves a multitude of trade-offs. Limiting activity to night, for example, might also limit energy acquisition rates in environments where many arthropods, the primary food, are diurnal. Nighttime activity and the associated lower body temperatures of ectotherms might result in reduced performance while active, affecting both prey acquisition and escape from whatever predators might be active at night.

Crypsis and Immobility

Cryptic coloration, morphology, or both, particularly when coupled with immobility, the lack of movement, appear to be highly effective in deterring detection by visually oriented predators (Fig. 11.4). A species is cryptic if its coloration or morphology resembles a random sample of relevant aspects of the environment in which it lives. Exactly what comprises "relevant aspects" may not always be clear, but most observers have no difficulty determining that a cryptically colored species, such as a rough green snake, *Opheodrys aestivus*, matches its green leafy vegetation background. Movement offsets crypsis, and as a result, effective use of crypsis usually includes nearly total immobility. Color and pattern can vary geographically within species of amphibians and reptiles, and individuals in local populations often match the corresponding microhabitat. Rock rattlesnakes (*Crotalus lepidus*) vary dramatically in coloration across small distances in Big Bend National Park in Texas, nearly perfectly matching coloration of

FIGURE 11.4 Cryptic coloration and morphology render amphibians and reptiles nearly invisible against the appropriate background. Clockwise from the upper left: the gecko *Ptychozoon lionotum* on the trunk of a tree (photograph by L. L. Grismer); a variety of color patterns exists among individuals of the Amazonian frog *Proceratophrys*, with each cryptic against leaf litter, and the polymorphism makes it difficult for predators to form a search image (photograph by J. P. Caldwell); the frog *Theloderma corticale* assuming a balled posture on moss (photograph by D. Fenolio); and the eyelash viper *Bothriechis schlegelii* on a log (photograph by L. J. Vitt).

background soils and rocks (Fig. 11.5). Similarly, individuals of *Uta stansburiana* are various shades of gray in flatland desert habitats of Southwest deserts but nearly black on black basaltic lava flows in the eastern Mojave Desert.

Disruptive coloration can be an important component of crypsis as well. Patterns of blotches, stripes, bands, or spots break up the general outline of an individual and make it difficult to detect the whole animal, especially against a background containing a mixture of patterns and colors. In some species, such as the cycloramphid frog *Proceratophrys*, not only is coloration disruptive, but different individuals in the same place have different patterns (Fig. 11.4). Polymorphism in coloration and pattern presumably makes it difficult for predators to form a reliable search image, particularly against backgrounds that vary, such as leaf litter. It would seem particularly advantageous in species with polymorphic color patterns for individuals to select microhabitats matching their pattern. One study suggests that they do. Green and brown morphs of *Pseudacris regilla* more frequently select matching than nonmatching backgrounds. Moreover, the natural predator, *Thamnophis elegans*, has higher success at detecting the frogs when they are against nonmatching backgrounds.

Often color patterns that appear brilliant are cryptic against some backgrounds. The bright contrasting bands of the coral snake patterns (black and white or yellow; black, yellow, and red) effectively conceal snakes in forest-floor litter, particularly when patches of light filter through the canopy and reach the forest floor. The disruptive nature of banding distracts away from the overall "snake" image.

Modifications of body shape enhance the effects of color camouflage by making it difficult to find edges or by causing the animal to resemble a structural aspect of the environment. *Pipa, Phrynosoma,* trionychid turtles, viperid snakes, and many other amphibians and reptiles are dorsoventrally compressed. Flattening of the body makes it difficult to detect edges when these animals rest on a flat substrate (Fig. 11.4). Adding spines and other

FIGURE 11.5 Color-pattern diversity in rock rattlesnakes (*Crotalus lepidus*) in Big Bend National Park, Texas. Each is shown against the background where it was observed. Left, Grapevine Hills; center, Boquillas; right; Maple Canyon. Photographs by L. J. Vitt.

appendages to body edges further disrupts body shape and prevents a match with a predator's search image. Many frogs have modifications of the skin that enhance crypsis. Supraciliary processes, scalloped fringes along the outer margins of the limbs, and a variety of warts and tubercles aid in disrupting the outline of the animal. The long, thin vine snakes, *Oxybelis* and *Xenoxybelis*, are nearly impossible to detect while they are stationary because they resemble the thin branches of their habitat. Unlike most snakes that frequently extrude their tongues making them detectable by the movement, *Oxybelis* and *Xenoxybelis* even hold the tongue out for extended periods without moving it (see Fig. 10.5).

Other forms of crypsis can aid in escape from predators that use other senses to detect their prey. A small mammal might be cryptic from the perspective of infrared heat sensors of rattlesnakes that are tuned to the thermal landscape as long as it remains perfectly still. As with visually oriented predators, immobility is critical. Any movement by the small mammal would be perceived by the rattlesnake as a moving thermal signal and therefore a live animal. Relatively little is known about nonvisual cues, but it would not be surprising to discover chemicals in amphibians and reptiles that render them "cryptic" from the perspective of chemosensory-oriented predators. Some blind snakes (*Leptotyphlops*) are known to produce chemicals that protect them from attacks by the ants they prey on (see the section "Chemical Defense"), but whether these or other chemicals render them "cryptic" in the chemical landscape of a social insect nest remains poorly known. Distasteful or toxic chemicals (see following text) likely do not provide crypsis, as their actions are more direct.

Escaping or Misdirecting Identification

Once a predator has detected a potential prey item, the predator must identify it prior to attacking or attempting to eat it. The cost to a predator of misidentification potentially can be high, particularly if the prey is toxic or has other effective defenses. Prey identification can be visual, olfactory, tactile, or a combination of these cues. Prey have evolved many fascinating mechanisms to deceive predators into misidentifying them and consequently leaving them alone, the most striking of which is aposematic coloration of potentially harmful prey and mimicry by palatable prey. Some even signal the predator to let the predator know that they are ready and therefore likely to escape (pursuit-deterrence signaling).

Aposematic Coloration and Postural Warning

Many frogs and salamanders are brightly colored and produce noxious or lethal chemicals from granular glands in their skin (see "Chemical Defense"). The bright coloration, or in some instances, specific postures, warn predators of the high cost to attempted predation, causing them to discontinue approach and thus end the predation sequence. When bright coloration is associated with potentially life-threatening defense mechanisms (e. g., toxins), the coloration is considered a warning or aposematic coloration. Predators either learn or evolve recognition of these warning colors and avoid those potential prey. In this case, proper identification by the predator results in escape by the prey.

Some amphibians assume a posture known as the *unken reflex*, which warns of distasteful or toxic chemicals (Fig. 11.6). First described in the frog *Bombina*, *unken* is the German word for toad. The back is arched and the head and limbs of the body are elevated to expose bright ventral or lateral coloration while the animal remains perfectly immobile. Examples include the frog *Bombina variegata*, the eft stage of the North American salamander *Notophthalmus viridescens*, and the European salamander *Salamandrina*, all of which are red or orange on the ventral surfaces. Some frogs, such as *Pleurodema brachyops* and *Eupemphix nattereri*, assume a defensive posture with the posterior part of the body elevated to expose large

FIGURE 11.6 Some amphibian defense postures. Clockwise from top left: *Salamandrina terdigitata* assuming the unken reflex; *Rhinophrynus dorsalis* in defensive posture with head down and body inflated; *Pleurodema brachyops* presenting eyespots that are covered with high concentrations of glandular glands; *Lepidobatrachus laevis* giving an open-mouth threat display with distress call. Photographs by E. D. Brodie, Jr.

eyespots that produce noxious chemicals (Fig. 11.6). Other salamanders hide the head while lashing with their tails. Because the tail contains mucous and granular glands, this behavior presumably further deters a predator that may come in contact with the noxious secretions.

Mimicry

Although the term *mimicry* has been widely applied to nearly every situation in which one species of animal resembles another, its definition with respect to predatory escape behavior is much more explicit. Mimicry occurs when one species of animal (the mimic) resembles another species that has easily recognizable characteristics (the model) and as a result deceives a potential predator (the dupe) that might otherwise capture and eat it. The model is usually poisonous, noxious, aggressive, or otherwise protected from predation, and its colors, odors, or behaviors signal to a potential predator that it is dangerous and therefore not worth pursuing. A mimic takes advantage of an aposematically colored species that is truthfully advertising its high cost of capture. Batesian mimicry occurs when a nontoxic or otherwise nonprotected species mimics a toxic or protected species, whereas Müllerian mimicry occurs when one or more potentially dangerous species resemble each other and each is both the model and the mimic. In both types of mimicry, the assumption is that similarities in coloration, pattern, or behaviors between the mimic and the model converge. In instances where two sister taxa have the same color or pattern, mimicry probably did not evolve independently in each taxon, even though each may gain some advantage with respect to escaping predators by having similar patterns. In those species, a single evolutionary event produced the matching colors or patterns, and it occurred in their common ancestor or even further back in the evolutionary history of the group.

Many descriptive studies have identified possible mimicry systems in amphibians and reptiles, and a few experimental studies have shown that mimics of known models dupe some predators. Verifying that the models themselves are protected from predation has been much more challenging and often the evidence is indirect, partially because observation of natural predation events is uncommon. The most widely publicized and debated example of mimicry in amphibians and reptiles is coral snake mimicry, in which a number of harmless or mildly venomous snakes with various combinations of banding patterns resemble highly venomous New World coral snakes (*Micrurus* and *Micruroides*).

All species of coral snakes are highly venomous and capable of inflicting potentially lethal bites to predators. Most coral snakes have patterns of alternating, high-contrast bands, usually red, yellow, and black or at least a combination of either red and black or yellow and black. Laboratory experiments have shown that birds avoid cylindrical pieces of wood doweling with high-contrast bands, suggesting that coral snakes are models in a mimicry system. Their putative mimics (mostly colubrid snakes) have similar patterns and usually are about the same size as coral snakes. The most convincing comparative evidence that colubrid snakes mimic coral snakes is the concordant change in coloration and pattern by some colubrid snakes as coral snake patterns change geographically, described in detail by Harry Greene and Roy McDiarmid. Five species of coral snakes, *M. fulvius, M. limbatus, M. diastema, M. mipartitus*, and *M. elegans*, have distinctly different color and banding patterns through Mexico and Central America. One, *M. diastema*, has at least three distinct color patterns depending on locality. At each locality containing a specific species or color morph of coral snake, a species or color morph of the mildly venomous snake, *Pliocercus*, matches the local coral snake. Mimics not only have high-contrast banding patterns similar to coral snakes in general, but the banding patterns of the mimics vary with the banding patterns of coral snakes as they change geographically. Similar geographic matches occur between coral snakes and nonvenomous snakes throughout much of the New World, and in some instances, the "model" and "mimic," even though nearly identical to each other, do not resemble typical tribanded color patterns typical of most coral snakes (Fig. 11.7).

FIGURE 11.7 The highly venomous coral snake *Micrurus albicintus* (upper) and the nonvenomous snake *Atractus latifrons* (lower) occur together in the Brazilian Amazon. This example of a nonvenomous snake with a pattern and color that matches a highly venomous coral snake likely represents Batesian mimicry. Photographs by L. J. Vitt.

Experimental evidence also suggests that coral snake patterns provide some protection from predation in natural situations. In some clever experiments designed by Edmond D. Brodie, III, and collaborators, both plain-colored and tricolored snakelike models were placed on the forest floor in Costa Rica to determine if natural predators would disproportionately attack the plain-colored models. Because the models were made of soft plastic, predation attempts could be scored based on bite marks left by the predator. A similar experiment was conducted on a plain background to determine whether crypsis was also involved. Bird attacks (based on beak marks) were much more frequent on plain-colored models regardless of whether they were on the forest floor or on a plain background, suggesting that birds avoided the coral snake banding pattern. Further studies on models with a variety of coral snake patterns showed that attack frequency varied among models, indicating that birds can distinguish quite well among different patterns and that some patterns are more effective at deterring predation attacks. More recent experiments by David Pfennig and coworkers have shown that in areas where coral snakes do not occur, but "mimics" do occur, the response of predators to the coral snake image breaks down such that the predators do attack the banded, nonvenomous snakes. Moreover, patterns of the banded nonvenomous snakes deviate rather dramatically away from the coral snake pattern and coloration in areas where the coral snakes do not occur. These results are consistent with the original descriptions and interpretations of color and pattern matching by Greene and McDiarmid.

Numerous other examples of mimicry exist among snakes, and many have not been described in the literature. For example, the nontoxic toad-eating snake *Xenodon rhabdocephalus* varies considerably in color pattern, but in most localities, its pattern closely resembles the local pattern of either *Bothrops asper* in Central America or *B. atrox* in South America. When captured, *X. rhabdocephalus* adds to the deception by opening its mouth and erecting what appear to be large, moveable fangs, similar to species of *Bothrops*. The teeth of *Xenodon* are enlarged rear fangs mounted on a movable maxillary bone and used to puncture toads that have filled with air. The snakes do not produce venom.

Mimicry of invertebrates by amphibians and reptiles may be widespread but is only beginning to be appreciated. Juveniles of the Kalahari lizard, *Heliobolus lugubris*, are black with fine white markings and, when disturbed, arch their backs and walk stiff-legged (Fig. 11.8). Their cryptic tails are pressed against the ground to further enhance a beetle-like appearance. Overall, their coloration and behavior closely resemble that of an oogpister carabid beetle that produces noxious chemicals for defense. The adults of *H. lugubris*, which are much larger than the beetles, do not have coloration or locomotion similar to beetles. Similarly, juveniles of the Brazilian anguid lizard *Diploglossus lessonae* are similar in size, color, and pattern to an abundant rhinocricid millipede that produces a variety of noxious and toxic substances for defense (Fig. 11.9). Juveniles appear during the wet season when the millipedes are abundant, live in the same microhabitats as the millipedes, and when the dry season ends and millipedes disappear, the lizards, having reached a larger body size than the millipedes, lose the banded coloration of the millipedes. Other lizards appear to mimic scorpions, centipedes, and millipedes, but neither comparative nor experimental studies have verified that a mimicry system is involved. Considering the high density of noxious or toxic invertebrates and the fact that they were highly diversified long before the diversification of amphibians and reptiles, invertebrate mimicry by amphibians and reptiles should be common.

FIGURE 11.8 Juveniles (upper panel) of the lizard *Heliobolus lugubris* mimic the oogpister beetle (center) in the Namib Desert of Africa. The adult (lower panel), in addition to being considerably larger than the beetle, has a different color pattern and behavior. Adapted from Huey and Pianka, 1981. Photographs by R. B. Huey.

Among amphibians, mimicry is best known in salamanders. The red eft stage of *Notophthalmus viridescens* is terrestrial and unpalatable to birds because of its toxic skin. A variety of other terrestrial salamanders occurs in various parts of the range of red efts and appears to gain some benefit by resembling them. Birds avoid *Pseudotriton ruber* and red morphs of *Plethodon cinereus* based on their similarity to red efts. Likewise, *Plethodon jordani* in the southern Appalachians has brilliant red markings that warn of its distastefulness. The markings are on the cheeks or legs, depending on locality. In areas where the salamander has red cheeks, the palatable look-alike salamander *Desmognathus imitator* has red cheeks, whereas in areas where *P. jordani* has red legs, *D. imitator* also has red legs.

Many small leaf litter frogs that inhabit Amazonian rain forests have similar coloration and patterns, consisting primarily of bright white or yellow dorsolateral stripes on a dark background; some have bright yellow or orange flash colors in the groin and on the hidden surfaces of the thighs. These color patterns have evolved independently in different frog clades, and it is not uncommon to find as many as five or six species occurring in the same place. These frogs may form a geographically widespread system of models and mimics (Fig. 11.10) because some produce alkaloids in the skin that deter predators and others do not. Moreover, use of alkaloids for defense has evolved independently several times even within the group of frogs (dendrobatoids) that serve as models for the system (see Fig. 10.27).

Many species of frogs have large eyespots on the posterior surface of the body that they expose when disturbed (e.g., Fig. 11.6). Whether the eyespots represent mimicry of large, potentially dangerous animals or simply direct the attention of a predator to areas where noxious chemicals are produced is poorly studied, and both occur (see following text).

Mimicry may dupe other senses of predators, although this area remains unexplored. One possibility is auditory mimicry of the saw-toothed viper *Echis carinatus* by the gecko *Teratoscincus scincus*. Both the snake and the lizard produce a rasping sound by rubbing scales together. The

FIGURE 11.9 Juveniles of the lizard *Diploglosus lessonae* (left) mimic the toxic rhinocricid millipede (center) in northeastern Brazil. The adult lizard (right) is much larger than the millipede and has completely lost the banding pattern of the juvenile. Adapted from Vitt, 1992. Photographs by L. J. Vitt.

FIGURE 11.10 A possible mimicry complex among Amazonian leaf litter frogs. Clockwise from upper left: *Allobates femoralis* (nontoxic); *Allobates gasconi* (nontoxic); *Ameerga trivittata* (toxic); *Ameerga petersi* (toxic); *Hyloxalus chlorocraspedus* (toxic); and *Leptodactylus lineatus* (nontoxic). Photographs by J. P. Caldwell.

rasping sound would seem effective against nocturnal mammalian predators. Considering the widespread occurrence of chemical cues used in prey detection by salamanders and scleroglossan lizards (including snakes), there is every reason to suspect that mimicry systems involve chemical cues.

Mimicry of inanimate objects in the environment has also been suggested as a defense mechanism. Some horned lizards (*Phrynosoma modestum*) and the Australian agamid *Tympanocryptis cephalus* assume postures and have a morphology and coloration that mimics small rocks common in their microhabitats (Fig. 11.11). Many other amphibians and reptiles have morphologies and perform behaviors that give the impression that they are "mimicking" attributes of the physical environment. Strictly speaking, these behaviors fall into the category of crypsis in that there is no animate "model." These examples show

FIGURE 11.11 The Australian agamid lizard *Tympanocryptus cephalus* resembles a small rock in its natural habitat. Photograph by S. Wilson.

how predator-escape mechanisms sometimes cannot be easily categorized.

Escaping Approach

Species that move while foraging or have bright coloration are easily detected by predators and, as a result, rely less on crypsis and immobility. The most common response by potential prey to approaching predators, once aware that they have been detected, is locomotion away from the predator. For actively foraging lizards, this is the primary escape mechanism. The lizards often continue foraging while keeping track of approaching animals. When an animal moves within a critical distance or makes a dash at the lizard, the lizard runs to a safe distance and begins foraging again. Many aquatic snakes (e.g., *Nerodia*) and arboreal lizards (e.g., *Iguana, Uranoscodon, Crocodilurus, Physignathus*) bask on top of vegetation overhanging water and dive into the water to escape predators approaching from the land or within the vegetation. Basking crocodylians and turtles enter the water when potential predators approach. Nearly all amphibians and reptiles that use crevices or burrows rapidly enter their crevice or burrow when predators approach. Some, such as the chuckwalla (*Sauromalus obesus*), inflate their lungs with air and press their skin against the walls of the crevice to make themselves nearly impossible to extract. In the tropical arboreal lizard *Uracentron flaviceps*, all individuals in a tree often enter the same hollow when approached from within the tree. Most frog species simply jump when a predator approaches, and nearly all shoreline frogs jump into water and bury themselves in mud when approached. Some lizards (e.g., *Ptychozoon, Thecadactylus, Draco*) and frogs (e.g., *Agalychnis moreletii, Ecnomiohyla miliaria*, species of *Rhacophorus*) parachute to safety by extending their limbs and spreading their toes to stretch webbing, or by using other skin extensions as airfoils (Fig. 11.12). Some snakes (e.g., *Chrysopelia*) also parachute by using skin extensions as airfoils. Parachuting is nonrandom in that the frog, lizard, or snake can alter the trajectory while gliding. Terrestrial species, such as dendrobatids, retreat with a series of jumps that takes them well out of reach of predators. Arboreal frogs jump to other perches, which protects them from predators incapable of jumping. Some, such as *Phyllomedusa hypochondrialis*, fall to the ground when disturbed and roll into a motionless (thus cryptic) ball on the forest floor.

Numerous threat displays cause predators to discontinue approach. The rattling of rattlesnakes and hissing sounds produced by many snakes and some lizards deter approach, particularly if combined with body, neck, or head expansion. The expanded hoods of cobras, open-mouth displays of cottonmouths, and brilliant colors on the inside of lizard mouths cause many predators to keep their distance (Fig. 11.13). In many snakes, threat displays include loud hissing sounds and repeated strikes. Some frogs, including the hemiphractid *Hemiphractus* and the ceratophryid *Lepidobatrachus*, open their mouths and expose their orange or pink tongues as threat displays similar to those of lizards when disturbed. The horned frog *Ceratorphys cornuta* will quickly strike a potential predator when disturbed and hold on with its massive jaws.

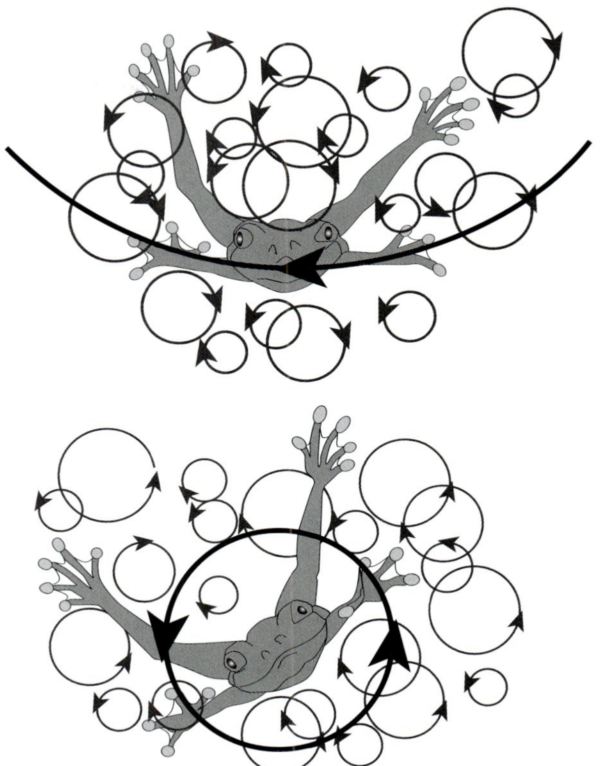

FIGURE 11.12 When jumping from trees, amphibians and reptiles experience tiny eddies of airflow that they can use in maneuvering their position, allowing them to control their gliding. Adapted from McCay, 2003.

Numerous frogs use a loud distress call to frighten predators (see Table 9.2). Usually this call is given only by females and only after the frog has been captured by the predator. The call is given with an open mouth, which is unusual in that male advertisement calls are given with closed mouths (see Fig. 11.6, *Lepidobatrachus laevis* giving distress call). The scream is frequently a loud, relatively long catlike call, and the startle effect causes certain kinds of predators to release their prey. Whether it has an effect on snakes is unknown because snakes do not pick up airborne sounds. Because of the large number of unrelated frogs that have distress calls, this trait has apparently evolved independently in many clades. This type of call is so distinctive that it is frequently used by herpetologists to track a snake–frog predatory event.

FIGURE 11.13 Threat displays by snakes and lizards. Clockwise from upper left: The Amazonian hoplocercine *Enyalioides palpebralis* faces an intruder and opens the jaws to expose the bright orange mouth and throat coloration (photograph by L. J. Vitt); the Neotropical vine snake *Oxybelis aeneus* presents an open-mouth display when disturbed (photograph by L. J. Vitt); the Chinese cobra *Naja atra* expands its hood and presents a face-on display to intruders (photograph by R. W. Murphy); the Neotropical snake *Pseustes poecolinotus* expands its throat and upper body, makes loud hissing sounds, and strikes, usually with the mouth closed (photograph by L. J. Vitt).

Some lizards use the strange behavior of waving a forelimb as they stop moving. The first impression one gets from this behavior is that it makes the lizard more rather than less conspicuous. Although several hypotheses have been proposed to explain this behavior, it appears that it signals as a pursuit deterrent. Basically, the lizard signals to potential threats that it recognizes the threat and is ready to respond by escape if necessary. From the predator's perspective, it is fruitless to waste time and energy on a prey that is likely to escape. Field experiments on the Bonaire whiptail lizard (*Cnemidophorus murinus*) falsified other hypotheses (solely an intraspecific social signal, indicates flight to follow, or is behavior used in thermoregulation) but revealed that the lizards use the behavior when the approach by the investigator is slow or at an angle, yet flee if the approach is rapid. Thus it appears that these lizards, and possibly others that perform the behavior, effectively signal the predator that it is unlikely to be successful in a predation attempt.

Escaping Subjugation and Capture

Skin, Armor, and Spines

Skin and other structures on the outside of the bodies of amphibians and reptiles can aid in resisting a predator attack. The softer, more permeable skin of amphibians has fewer structural modifications to increase its resistance to predator attacks (but see "Chemical Defenses," to follow). Aside from the assorted bony or keratinous spines that occur on the limbs and trunks of some frogs (most are associated with reproduction or digging), only the fusion of the skin to the dorsal skull roof may be defensive. This fusion provides strength to both skin and skull. For a few species, the top of the head may be used to block entry to retreats. The heavily keratinized skin of reptiles provides a durable body armor, and many modifications have evolved to give it even greater strength. The turtle shell composed of thick dermal plates is a most obvious defense structure. The ability to entirely close the shell as in *Terrapene* protects these turtles from most predators. Crocodylians, some lizards, and some amphibians have the epidermal scutes or scales underlain by bony osteoderms; this combined barrier makes penetration by a predator's teeth difficult, and both crocodylians and lizards use a spinning, thrashing movement to escape from the jaws of predators. Enlarged and spiny scales make a biting grip painful for a predator, and they render the prey difficult to hold. The horns of horned lizards (*Phrynosoma*) are longest in the areas of highest predator densities. Field observations of predators, such as the coachwhip *Masticophis flagellum*, dying from puncture wounds after swallowing horned lizards verify the effectiveness of spines. The spiny tails of *Ctenosaurus, Uromastyx*, and many other lizards strike painful blows, often cutting into flesh. Jaws and claws of large-bodied lizards and turtles can inflict painful wounds when the animals are grasped, often resulting in escape. Even hatchlings of some turtles, including *Trachemys scripta* and *Chrysemys picta*, can inflict wounds to predators substantial enough to deter predation. Hatchlings of *T. scripta* and *C. picta* have brightly colored plastrons, and bass appear to avoid them, whereas fish do not avoid dull-colored *Chelydra serpentina,* which they reject only after attempting to eat them. The bright plastral colors appear to warn fish that hatchlings are dangerous. The newt *Echinotriton*, in addition to warning predators by raising and waving its brightly colored tail, has spiny projections from the ribs that extend through the skin providing added predator deterrence (Fig. 11.14).

FIGURE 11.14 The Asian newt *Echinotriton andersoni* not only presents a display indicating that it is dangerous but also has lateral spines that can deliver poisonous secretions if a predator bites the newt. Photograph by E. D. Brodie, Jr.

Other, more subtle structural modifications protect many smaller species. The tiny chameleons *Brooksia* have the transverse processes of the vertebrae curved dorsally over the neural arches to form a shield over the spinal cord. When touched, *Brooksia* freezes, releases its grip on the branch, and falls to the ground; during the fall, it rights itself so that it always lands with the vertebral shield upright, and birds treat it as an inedible object. Many gekkonids and some scincids, in addition to autotomizing tails when grasped by predators, can lose large patches of skin when grabbed by a potential predator. As long as the body wall is not broken, the lizards heal with minimal scarring (Fig. 11.15).

FIGURE 11.15 Some lizards, such as this Amazonian gecko *Gonatodes humeralis*, can escape the grasp of predators by losing large patches of skin. The skin regenerates. Photograph by L. J. Vitt.

Chemical Defense

Amphibians and reptiles produce a wide range of antipredator chemicals. Granular skin glands of amphibians produce chemicals ranging from irritating and mildly distasteful to emetic and lethal. Granular glands can be evenly spread across the dorsal surface as in dendrobatoids or concentrated in large glandular masses as in the parotoid glands and warts of bufonids. The glandular masses are evident on many salamanders and frogs, and their locations complement their use in defense behaviors. *Rhinella, Bufo, Anaxyrus,* and *Ambystoma* have large alkaloid-producing parotoid glands on their heads that act as neurotoxins to deter predators that actually make contact. Salamanders that use tail lashing (e.g., *Bolitoglossa, Eurycea*) have heavy concentrations of glands that produce chemicals on the tail. The predator cannot approach and grab the salamander without being exposed to gland secretions. Some species, such as *Salamandra salamandra*, can spray defensive chemicals from pressurized glands up to 200 cm (Fig. 11.16).

FIGURE 11.16 Examples of defense mechanisms involving squirting or spraying of noxious or toxic substances. Left: The European fire salamander *Salamandra salamandra* squirts chemicals from skin glands when disturbed (adapted from Brodie and Smatresk, 1990; photograph by E. D. Brodie, Jr.). Upper right: The Australian gecko *Diplodactylus ciliaris* explosively sprays an unpalatable, sticky substance from glands in the tail; lower right: Secretions from the tail of *D. ciliaris* are released as droplets by cooling the lizards and prodding the skin with an electrode (adapted from Rosenberg and Russell, 1980; and Rosenberg et al., 1984; photographs by H. I. Rosenberg and A. P. Russell).

Even in species with less striking defense behaviors and glandular concentrations (e.g., *Hyla, Lithobates*), the predator receives a dose of secretions from the amphibian's granular glands as soon as it takes the prey into its mouth, and irritating secretions usually are sufficient to cause the prey to be released. Some frogs, such as *Trachycephalus venulosus*, produce a noxious skin secretion that is also gluelike, causing leaves and other debris to adhere to the jaws and mouth of the predator, facilitating escape. Many of the gluelike compounds have impressive adhesive properties. When attacked by small garter snakes (*Thamnophis couchi*), the tiny salamander *Batrachoseps attenuatus* coils around the neck of the snake, making it nearly impossible for the snake to continue swallowing it. Moreover, skin secretions from the salamander are wiped on the snake, causing nearly everything, including other parts of the snake, to adhere, and allowing the salamander to escape predation. The salamanders and frogs do not stick to their own secretions.

A remarkable number of noxious and toxic components have been identified from amphibian skin, and many of these are used for chemical defense. The known compounds fall into four groups: biogenic amines, peptides, bufodienolides, and alkaloids. The biogenic amines include serotonin, epinephrine, and dopamine; all affect the normal function of the vascular and nervous systems. The peptides comprise compounds such as bradykinin that modify cardiac function. The bufodienolides and alkaloids are similarly disruptive of normal cellular transport and metabolism and are often highly toxic.

The source of many chemicals that occur in amphibian skin appears to be the arthropods in their diets, particularly ants. Clades within the Bufonidae (e.g., *Rhinella*), Microhylidae (e.g., Microhylinae), Mantellidae (e.g., *Mantella*), and Dendrobatidae (e.g., *Dendrobates, Oophaga*, and *Phyllobates*) specialize on ants and produce some of the most toxic skin compounds. The suggestion that some frogs may optimize chemical intake for defense when selecting prey is supported by comparisons of the diets of frogs and lizards from the same microhabitats. Many leaf litter frogs of Amazonian forests feed on ants, even though more energetically profitable prey are available based on diets of lizards in the same microhabitat. The ant-eating frogs produce noxious chemicals in the skin, whereas those that eat few ants do not produce toxic skin chemicals (Fig. 11.17). The correlation between ant eating (myrmecophagy) and skin toxins is best supported for dendrobatid frogs. Ant eating, production of noxious or toxic chemicals in the skin, and aposematic coloration have evolved independently several times. Based on their presumed phylogenetic relationships, these traits have evolved together (see Fig. 10.27). A number of behavioral and life history traits have evolved concordant with myrmecophagy, including increased activity, reduced clutch size, and more extended parental care, including either prolonged feeding of tadpoles or long-term pair bonds in some lineages. The possibility exists that release from predation by visually oriented predators has relaxed some of the constraints imposed by low levels of activity in cryptic species such as *Allobates*, resulting in the evolution of complex social behaviors involving high levels of activity in other genera, such as *Dendrobates*. Species of *Allobates* eat few ants, are not aposematically colored (with one possible exception), do not produce skin toxins, and rely on crypsis for escape from detection by predators. A nearly identical set of independently derived characteristics occurs among species in the mantellid frog genus *Mantella*. Species of *Mantella* feed on very small prey, mostly ants, produce alkaloids in the skin, are diurnal, and have aposematic coloration similar to that found in dendrobatids.

Larvae of many amphibians are distasteful, which provides them some protection from predation, particularly predation by fish. Palatability varies among species within the same general habitat, as well as among closely related species. Amphibian larvae use chemical cues to detect predators and spend more time in refuges when predators are present.

With the exception of snake venoms, the chemical defenses of reptiles are more disagreeable than harmful. Turtles have musk (Rathke's) glands that open on the bridge of their shells; musk secretions have not been demonstrated as defensive, but to the human nose, the odor of kinosternid and chelid turtles is repugnant. Snakes have paired cloacal glands that are aimed at and emptied on predators. Some snakes, such as *Leptotyphlops dulcis*, produce chemicals that are effective in holding social insects at bay. Geckos also have cloacal glands that may

FIGURE 11.17 Brightly colored and toxic versus cryptically colored and nontoxic dendrobatoid frogs. Clockwise from upper left: *Oophaga pumilio*, *Adelphobates galactonotus*, *Ranitomeya ventrimaculata*, *Adelphobates quinqevittatus*, *Allobates conspicuous*, and *Allobates* sp. The last two species do not produce skin toxins for defense. Photographs by J. P. Caldwell.

or may not be used in defense; however, the squirting tail glands of the Australian gecko *Diplodactylus spinigerus* produce a sticky, odiferous compound that appears defensive against vertebrates due to its odor or taste. It may also be effective against some invertebrate predators such as spiders (Fig. 11.16), and it can be squirted up to a meter.

Most lizards do not have glands from which they can squirt chemicals for defense, but some horned lizards (*Phrynosoma*) involve their circulatory system in chemical defense. When captured by a potential predator, *P. cornutum* squirts blood from the sinuses of the eyes. At one time it was thought that blood squirted from the eyes of horned lizards gave a predator the false impression that it had been wounded by the sharp horns. However, blood of these horned lizards apparently tastes bad and causes canids to release the horned lizards. The source of bad-tasting chemicals remains unknown but may come from chemicals produced by the ants that they eat.

The glands of any amphibian or reptile can have multiple roles. Their secretions, even the most poisonous ones, probably also serve other functions, including individual and species recognition for reproductive and territorial behaviors, lubrication, waterproofing, or protection from bacteria, fungus, and parasites.

Death Feigning

Death feigning occurs in some frogs, salamanders, lizards, and snakes. In species that appear to feign death after falling from perches, the primary role of death feigning appears to be enhancing crypsis by ceasing movement. A Madagascar chameleon or an Amazonian *Phyllomedusa*, for example, that falls to the forest floor and ceases movement would seemingly disappear in the leaf litter. In North American *Heterodon* and Neotropical *Xenodon*, death feigning does not appear to enhance crypsis (Fig. 11.18). The snakes flatten their bodies, hiss, and often strike when first approached. When that threat display fails to have an effect, the snakes roll on their backs, often in a coiled or semicoiled position, open their mouths, and even drag their open mouth and tongue in the dirt. This may or may not be followed by defecation over much of the body. Exactly how this ridiculous behavior protects the snakes from predation remains unclear, but it has been suggested that the feces contain toxins from toads eaten by the snakes and thus chemical defense may be involved. Other snakes simply roll into tight balls or flatten out in a tight coil when disturbed. The tropical leaf litter snake *Xenopholis scalaris* is bright red, which suggests aposematism or possible mimicry (Fig. 11.19).

Tail Displays and Autotomy

A large number of salamanders, lizards, and snakes display their tails when first disturbed. For salamanders and a few lizards, the display is associated with the production of noxious chemicals that discourage the predator from attacking or continuing to attack (see "Chemical Defense"). In many snakes and lizards, no chemicals are produced, and it

FIGURE 11.18 Defensive display of *Heterodon platirhinos*. Photographs by L. J. Vitt.

FIGURE 11.19 When disturbed, many snakes, such as this red-bodied *Xenopholis scalaris*, coil tightly and flatten out while hiding the head and exposing bright coloration. Photograph by L. J. Vitt.

appears that the primary function of the tail displays is to distract a potential predator away from more vulnerable parts of the body (Fig. 11.20; see also Figs. 11.6 and 11.14). *Amphisbaena alba* not only raises its headlike tail off the ground when disturbed, but its head with mouth open is also raised, usually near the tail. Whether this gives a predator the impression that the animal has two aggressive heads or simply provides a 50% probability that the predator will attack the tail and allow the *Amphisbaena* to inflict a painful bite remains unknown. However, the effect is so stereotyped that nearly every rural citizen of countries where these animals live calls them "two-headed snakes." Rattlesnakes produce a loud, distinctive rattle from their specialized tail tip that not only distracts a potential predator away from the more vulnerable body as the snake crawls away to cover, but also serves to warn a potential predator that there is a high cost to any potential encounter. In this case, the cues are both visual and auditory.

Larvae of many amphibians have bright or high-contrast tail tips that redirect predator attacks from the body to the tail, thus facilitating escape. In the northern cricket frog *Acris crepitans*, larvae in temporary ponds where the primary predators are large dragonfly naiads (*Anax*) have black tail tips and suffer high rates of tail-tip damage as the result of misdirected naiad attacks, indicating the effectiveness of this defense strategy (Fig. 11.21). Larvae in lakes and streams where the primary predators are fish have translucent tails that allow the tadpoles to remain cryptic against the underwater substrate, thus reducing detection by predatory fish.

Among the most spectacular escape mechanisms in amphibians and reptiles is tail autotomy with subsequent regeneration. Tails of many salamanders, most lizards, and a few snakes can be released when grabbed by a predator, leaving the predator holding a thrashing and expendable body part while its owner flees to safety. Thus, tail loss not only allows immediate escape from a predator's grasp but also provides time for the salamander or lizard to escape while the predator is distracted by the tail. Because tails contain energy, the distracted predator does gain by continuing to devour the tail. Regenerated tails can be smaller, similar to, or larger than original tails, and in some species, tails do not regenerate at all (Fig. 11.22). When tails regenerate, vertebrae do not regenerate, and tail support is provided by a cartilaginous rod (Fig. 11.23).

In many species of lizards, coloration, size, or shape of the tail renders it conspicuous compared with the body, and although conspicuous tails can attract the attention of potential predators, the costs of attracting predators are outweighed by the benefits accrued by being able to detect the presence of a predator. For example, tails of juvenile *Plestiodon fasciatus* are brilliant blue, tails of *Vanzosaurus rubricauda* are red or orange, and regenerated tails of *Hemidactylus agrius* are bulbous. Experiments with natural predators of the banded gecko (*Coleonyx variegatus*) and the five-lined skink (*Plestiodon fasciatus*) reveal that these lizards raise the tail off of the ground, distracting the attention of snakes to the tail and away from the more vulnerable body parts. Tails are not immediately lost when grabbed by the predator. Rather, the lizards appear to allow the snake to gain a secure hold on the tail prior to releasing it. The tail is released by the lizard as the result of powerful muscle contractions in the tail. When tail autotomy occurs, segmented myomeres are

FIGURE 11.20 Reptile tail displays. Clockwise from upper left: The tiny leaf litter gecko *Coleodactylus* sp. raises its tail and waves it when disturbed; its tails are easily autotomized. The Neotropical amphisbaenid *Amphisbaena alba* waves its headlike tail above the ground while also raising its head with mouth open. The rainbow boa (*Epicrates cenchria*) coils its tail and waves it while crawling away, reflecting sunlight to produce a strobelike reflection of bluish coloration. The Amazonian coral snake *Micrurus hemprichi* hides its head in its coils and waves its short tail above the body. Photographs by L. J. Vitt.

exposed but not torn, and little bleeding or fluid loss occurs as the tail is released (Fig. 11.24). The tails immediately begin to thrash violently using anaerobic metabolism, and they continue to thrash for extended time periods. Snake predators swallow the tails, increasing the rate of ingestion as the thrashing of the tail becomes less vigorous.

Loss of tails by lizards and salamanders has potential energetic, social, and survival costs. Tails of salamanders and lizards are often used as fat-storage organs. Stored fat can be important for energetic support of reproduction and social behavior. *Coleonyx brevis* produces smaller eggs or no eggs following tail autotomy, and *Plestiodon* appears to produce smaller clutches following tail autotomy. Some lizards, such as *Uta stansburiana,* suffer reduced social status as the result of tail loss. In other species such as *Lacerta monticola*, mating success is reduced. In still others, long-term effects include reduced home range size and reduced access to females. All salamanders and lizards that lose their tails are without tail autotomy as a defense mechanism during the time period in which tail regeneration takes place.

Costs of tail loss and regeneration can also vary ontogenetically. Because juveniles do not invest directly in reproduction, tail loss in juveniles has a reproductive cost only if regeneration delays the attainment of sexual maturity or results in reduced size at sexual maturity, thus affecting clutch size. In juvenile skinks (*Plestiodon*), lizards that lose tails not only regenerate the tails, but growth rates increase enough to counter the effects of the loss of a relatively large portion of their body (Fig. 11.25). In adults, tails and their energy reserves are important for reproduction or reproductive-related activities. As a result, the cost–benefit ratio for tail autotomy changes. Tails of adult *Plestiodon* are cryptically colored similar to the body, and other predator-escape mechanisms become more important than tail autotomy.

Although an apparent cost of tail loss in salamanders and lizards might be reduced performance and hence higher risk of mortality, this is not always the case. In

FIGURE 11.21 Tadpoles of the cricket frog *Acris crepitans* have black tails that direct attacks away from the body (lower left) when they occur in pools or small ponds with predaceous dragonfly larvae. They have clear tails (lower right) when they occur in lakes or streams where maintaining crypsis is important to avoid detection by fish predators. Photographs by J. P. Caldwell.

FIGURE 11.22 Although original (left) and regenerated (right) tails of lizards are superficially similar, regenerated tails can be larger than the original, as in this tropical gecko, *Thecadactylus rapicauda*. Photographs by L. J. Vitt.

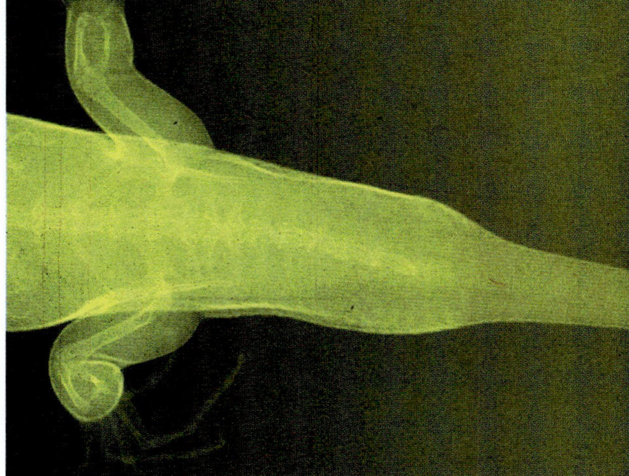

FIGURE 11.23 These X-rays of tails of *Plestiodon laticeps* show that the vertebrae of an original tail (upper panel) are replaced by a cartilaginous rod in the regenerated tail (lower panel). Photographs by L. J. Vitt.

some lizard species, individuals without tails perform better than individuals with tails intact. An experimental study in which tails were removed from hatchlings of side-blotched lizards (*Uta stansburiana*) in the field, with their subsequent growth and survival monitored, revealed reduced growth rates in lizards that lost their tails but no apparent reduction in survival when compared with hatchlings with intact tails over a 3-year period. In one year of the study, female hatchlings with broken tails survived better than those with intact tails.

The ground skink *Scincella lateralis* has taken the strategy of tail loss to the extreme. Not only do the lizards autotomize tails when attacked by a predator, but also both the skink and the autotomized tail have a high escape probability. When first autotomized, the tails jump about

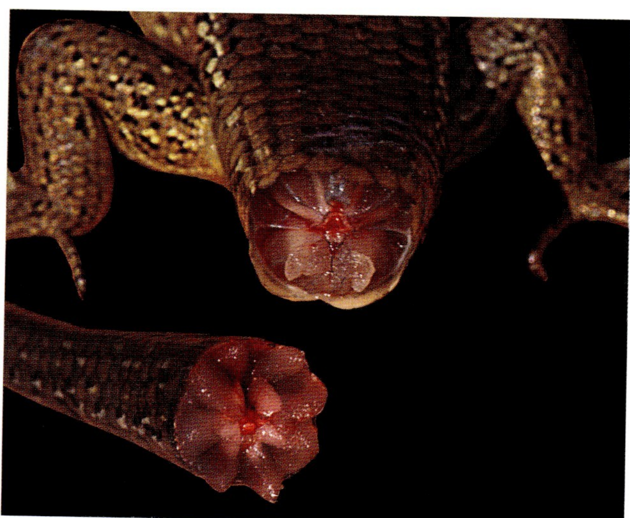

FIGURE 11.24 When tails are autotomized by lizards, myomeres separate and little bleeding occurs. Breakage occurs along cleavage planes within vertebrae, not between them. Photographs by L. J. Vitt.

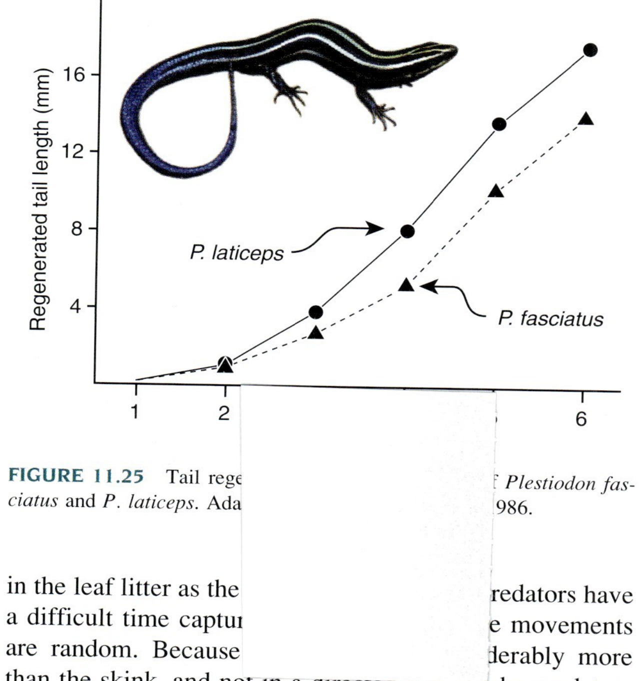

FIGURE 11.25 Tail rege f *Plestiodon fasciatus* and *P. laticeps*. Ada 986.

in the leaf litter as the redators have a difficult time captu e movements are random. Because derably more than the skink, and not in a directed way, snake predators are distracted by its movements and thus lose track of the skink. Frequently, the snake never finds the tail. Ground skinks that have lost their tails return to the site of tail loss, and if they can find their lost tails, they ingest them and regain much of the energy lost.

Although rates or frequencies of regenerated tails (indicating that the tail has been lost at least once) in natural populations have been used as relative measures of predation, tail loss rates do not necessarily estimate predation intensity because salamanders or lizards with regenerated tails are the survivors of predation attempts—there is no easy method to determine mortality among animals that did not lose their tails. Consequently, high frequency of tail loss could indicate the success of tail autotomy as a defense strategy rather than a high mortality due to predator attacks.

Similar to other mechanisms of predator escape, tail autotomy can impact other behaviors, many of which are also aimed at escaping predators. In the keeled earless lizard (*Holbrookia propinqua*), individuals remain closer to plant cover after losing their tails, and males that lose their tails flee further than they would with their tails intact. Lizards that lose their tails also feed less. Taken together, these observations indicate that keeled earless lizards adjust the levels of risk that they are willing to take based on whether they have an intact tail, which would provide them an effective escape strategy (tail autotomy) if a predator were to attack.

Offsetting Predation on Egg Clutches

Eggs of both amphibians and reptiles are susceptible to many kinds of predators and pathogens. One way in which females of many reptiles reduce predation on their eggs is by burying them in the ground or in nests of social insects. Some pond-breeding species of hylid frogs deposit their eggs singly throughout a pond or in small clumps attached to aquatic vegetation.

Developing embryos still in their eggs are not defenseless and show several kinds of responses to disturbance. In reptiles, hatching of a clutch is typically synchronous. When one embryo begins to hatch, others are apparently disturbed and also hatch. Although no studies have examined this response, hatching probably releases chemical cues that can attract predators. Thus, when one embryo hatches, selection should favor immediate hatching by the other embryos in the clutch and their quick escape from the nest site. Term embryos of the lizard *Plica plica*, when disturbed by a potential (human) predator, all hatch quickly at the same time, and the juveniles dart out of the nest in different directions, thus creating confusion (Fig. 5.10).

Recent work by Karen Warkentin has provided a detailed understanding of hatching responses to predators by developing embryos of the leaf-breeding frog *Agalychnis callidryas*. In this species, embryos hatch synchronously in response to disturbance by a predator, usually either the snake *Leptodeira annulata* or wasps, and the embryos may hatch up to 30% earlier than when they are not disturbed. When the eggs are physically disturbed, the embryos begin moving vigorously within their capsules, causing the capsules to burst, and quickly propel themselves within seconds into the water below, thus avoiding predation. Use of a miniature accelerometer enabled Dr. Warkentin to determine that vibrations made by the snake predator were longer, more widely spaced, and lower in frequency than vibrations made by rainstorms, which did not induce hatching in the embryos.

Schooling and Other Aggregations

Many amphibian larvae occur in what appear to be social groups, often referred to as schools. Tadpole schools are known from a variety of frog taxa, including the Bufonidae, Ranidae, Hylidae, and Leptodactylidae. Large numbers of larvae move around in ponds as a group, spreading out and reorganizing as the group moves about (Fig. 11.26). The possibility exists that a large school is perceived by some predators as something other than many individual tadpoles and is thus avoided. From the perspective of an individual tadpole, being a member of a large school reduces the probability that any specific individual will be the next one captured, sort of a low-probability Russian roulette. Tadpoles of the tropical frog *Leptodactylus macrosternum* form large schools that extend from the surface of the water where the tadpoles gulp air to the pond bottom where they fan out and graze for a brief period. The schools are continually reorganizing, and individuals are always moving toward the center of the school. In northern Brazil, adults of the frog *Pseudis paradoxa* repeatedly dive into the schools from the water surface and feed on these tadpoles. Larvae of dragonflies (especially *Anax*) on the pond bottom pick off tadpoles from the bottom of the school, and larvae of predaceous diving beetles (Dytiscidae) prey on tadpoles that lag behind the school. Although the schools move around throughout the pond, it is not clear whether they do so to find richer foraging sites or to avoid intense predation. In some populations of *Leptodactylus bolivianus* and *L. ocellatus*, females remain with the school of tadpoles and either direct the tadpole movements or aggressively defend the schools from potential predators.

FIGURE 11.26 Individual tadpoles are afforded some protection from predation by forming large schools. Presumably, predators do not recognize individuals as potential prey. This school of *Leptodactylus ocellatus* tadpoles was several meters in diameter. Photograph by J. P. Caldwell.

Miscellaneous Behaviors

A wide variety of other escape behaviors are used by amphibians and reptiles once predators initiate attacks. Some lizards seize parts of their own body, rendering them nearly impossible to swallow. *Cordylus cataphractus*, for example, bites and holds onto its own tail, making a loop of its body and exposing its large, armored scales to a predator. The elapid snake *Vermicella annulata* elevates loops of its body to make it difficult for predators to secure a grip on the snake. A diversity of snakes coil in a ball or hide their heads within coils.

Life History Responses to Predation

In a general way, life history responses to predation are relatively easy to visualize. For example, in species where mortality on juveniles is density dependent, production of fewer, larger, and more competitive offspring should be the evolutionary response. In species where mortality on juveniles is density independent, production of greater numbers of offspring should be the evolutionary response. Because energy for reproduction is typically limited (see Chapters 4 and 5), production of more offspring means that those offspring will be smaller. Both cases represent life history responses to predation or other mortality sources. The possible combinations of life history responses are nearly unlimited given the many variables that influence the evolution of life histories. The life histories of two species of frogs that breed in the same microhabitat exemplify the complexities of life history responses to predation. *Adelphobates castaneoticus* and *Rhinella castaneotica* breed in open, abandoned fruit capsules of the Brazil nut tree in Amazonian Brazil. The capsules fall to the forest floor, agoutis gnaw the top off the capsules and remove the Brazil nuts, and the capsules fill with water during rainstorms (Fig. 11.27). Mosquitoes, giant damselflies, and both species of frogs use the capsules for breeding. A single tadpole is transported to a capsule by *A. castaneoticus.* About 250 eggs are deposited in a capsule by *R. castaneotica.* The *Adelphobates* larva is predaceous, feeding on insect larvae and *Rhinella* tadpoles if any are in the capsule. Larvae of one mosquito species and the giant damselflies are predaceous and feed on both tadpole species if the tadpoles are small enough. The tiny *Rhinella* larvae develop rapidly in a race to metamorphose before all are eaten. The density of predators likely determines how many, if any, of the *Rhinella* tadpoles survive to metamorphosis. The relative size of mosquito, damselfly, and *Adelphobates* larvae and the order of colonization determine which of these organisms will survive to metamorphosis. For example, if a tadpole of *Adelphobates* is deposited before the insects, it feeds on all insect larvae subsequently deposited, grows, and ultimately metamorphoses. If one of the insect larvae is deposited first and grows large enough to kill a tadpole of *Adelphobates*, the insect larvae will grow and metamorphose. Experiments have shown that a

FIGURE 11.27 Life histories of two frog species using the same breeding microhabitat illustrate the evolution of complex responses to predation. After falling to the forest floor, the indehiscent fruits of the Brazil nut tree are opened by agoutis (upper left), which remove the seeds (upper right) known as Brazil nuts (center) and leave the open fruit capsule on the forest floor. After the capsule fills with water, it is colonized by two frog species and a variety of insects. The frog *Adelphobates castaneoticus* (middle right) transports a single tadpole to the capsule (bottom right), whereas the toad *Rhinella castaneotica* (lower left) deposits a small clutch of eggs (middle left). The sequence of arrival and the composition of the fauna in the capsule determine reproductive success in both frogs (see text). Adapted from Caldwell, 1993. Agouti photos, M. A. Mares; all others, J. P. Caldwell.

7-mm damselfly larva can kill a large tadpole. Thus, both relative size and sequence of deposition determine survival in this microcosm. On the one hand, *Adelphobates* has evolved a life history in which a few large and highly competitive offspring are produced to enter a competitive system. On the other hand, *Rhinella* has evolved a life history that includes a reduced clutch size compared with other species of *Rhinella*, allowing it to use the small breeding site and produce enough individual offspring to insure that at least some survive to metamorphosis.

Predators and Their Prey: The Evolutionary Arms Race

Implicit in any discussion of predators and their prey is the notion that, as prey evolve responses to predators, predators evolve responses to the changes in prey behaviors that shift predator–prey interactions. If that were not the case, then predators, prey, or both would quickly be driven to extinction. Of course it is not quite that simple because each predator has many different prey from which to select, and each prey species is influenced by more than a single predator. One of the best-documented examples of the evolutionary arms race in amphibians and reptiles involves predatory garter snakes and newts in western North America (Fig. 11.28). The newt *Taricha granulosa* has high levels of tetrodotoxin (TTX), primarily in its skin, but in some other tissues as well. TTX is highly toxic, with an individual newt containing enough of TTX to kill 10–15 humans. These salamanders are often heard about in the media when fraternity members eat one on a dare and subsequently become ill or die. TTX is a neurotoxin that blocks propagation of action potentials by

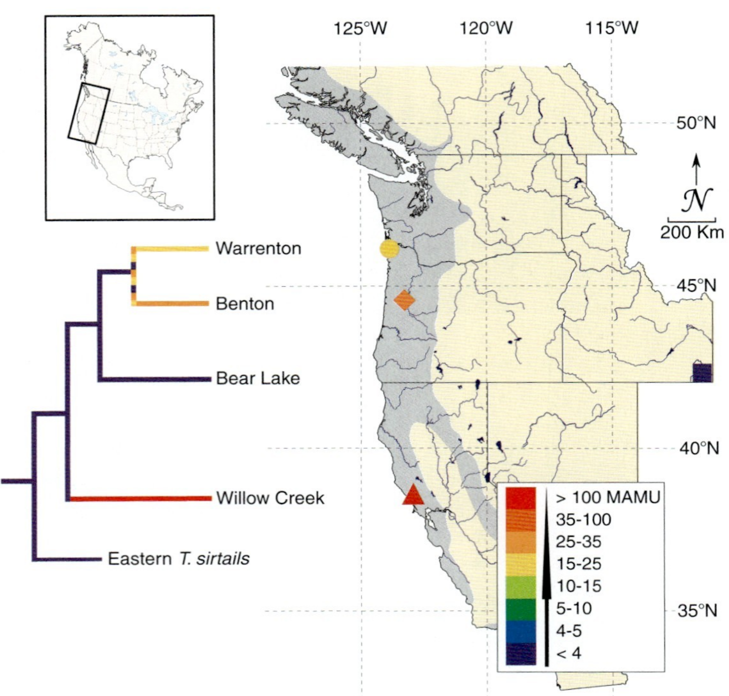

FIGURE 11.28 The geographic range of the garter snake *Thamnophis sirtalis* completely overlaps the range of newt in the genus *Taricha*. The range of *Taricha* is indicated by shading. The snakes vary in resistance to TTX, depending on locality (indicated by colors, with red indicating highest resistance). MAMU (mass adjusted mouse units) indicates the amount of TTX required to reduce a snake to 50% of its normal crawl speed. *Thamnopis sirtalis* that do not occur with the newts (eastern *T. sirtalis*) have low resistance to TTX. Not only have the garter snakes evolved high TTX resistance in response to newt toxins, but they also have evolved high resistance independently at least twice. The Bear Lake population does not occur with newts. Adapted from Geffeney et al., 2002, reprinted with permission from AAAS.

binding to sodium channels in nerves and muscles. As with skin chemicals produced by other amphibians, TTX is believed to be an effective chemical defense against most predators. The garter snake, *Thamnophis sirtalis*, occurs throughout the range of the newt and, in addition to feeding on other amphibians, feeds on *Taricha*. Snakes in the clade to which *T. sirtalis* belongs (Natricinae) have a natural (historical) resistance to TTX (many eat amphibians), but resistance in some of the populations of *T. sirtalis* in western North America (where the newts occur) is 100–1000 times that found in other natricine snakes. Among populations of *T. sirtalis*, the degree of resistance to TTX varies over three orders of magnitude, and extreme resistance to TTX has evolved independently at least twice. Among five populations in which newt toxicity and snake resistance are well documented, a nearly perfect phenotypic match exists, suggesting an ongoing evolutionary arms race. Variation exists among individual snakes in response to TTX, depending on such factors as body size (dilution effect) and perceived toxicity of individual newts (Fig. 11.29). Nevertheless, reciprocal selection seems to best explain the geographic mosaic of predator–prey interactions between these two species (Fig. 11.30). We emphasize that this interaction is only one of many for both the newts and the garter snakes, and predator–prey interactions for both are much more complex. Nevertheless, because TTX is highly toxic and garter snakes have evolved resistance to effects of TTX, it provides an ideal model for testing predictions about phenotypic and evolutionary responses in predator–prey systems.

OFFSETTING THE EFFECTS OF PARASITISM

Long-term effects of parasites on amphibians and reptiles are relatively poorly known, and new parasite species are being described at an astonishing rate. Parasites can have a nearly undetectable impact on their hosts or, in some instances, can kill their hosts. If the fitness of hosts is negatively affected by parasitism, then parasites are effectively predators because the likelihood of an infected individual's genes being represented in future generations is reduced. Potential negative effects of parasitism include anemia and reduced performance followed by reduced survival, competitiveness, social status, ability to sequester mates, and for females, reduced fecundity. Ectoparasites, such as ticks and mites, may also introduce endoparasites, such as filiarial worms and *Plasmodium* (Fig. 11.31). Parasitism is so widespread and common among amphibians and reptiles that nearly every scenario is possible. Lists of parasite species exist and new species are described continually from amphibians and reptiles, but few data are available on parasite life histories, how infestation affects an individual amphibian's or reptile's health, growth, reproductive output, or the effects on population structure and dynamics. When mass die-offs occur, such as the microsporidian epidemic of English *Bufo bufo* in the early 1960s, the decimation of *Lithobates pipiens* populations across northern North America in early 1970s, or the high incidence of viral-induced papilloma in Florida populations of *Chelonia mydas*, we are

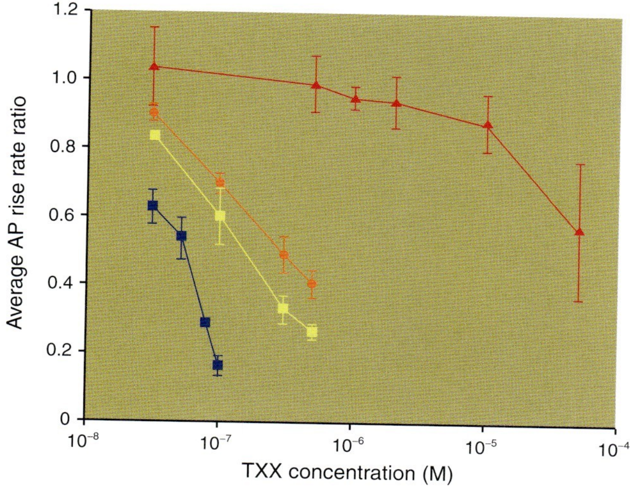

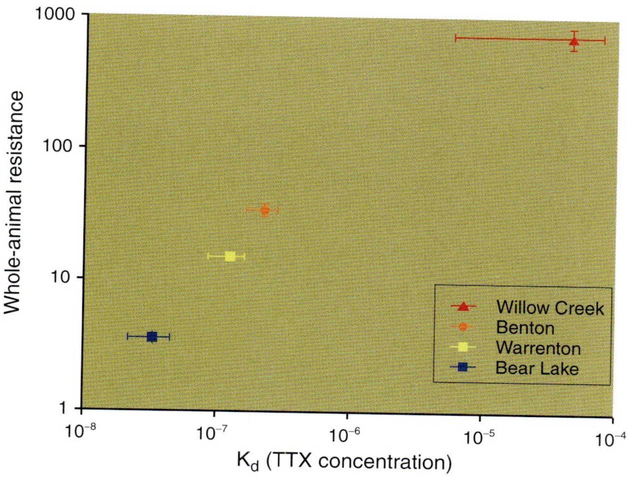

FIGURE 11.29 Variation in whole-animal and sodium-channel TTX resistance in four garter snake populations. Top: Maximum rise rate ratios as a function of TTX concentration. Bottom: Whole-animal resistance to TTX as a function of sodium channel TTX resistance. Most instructive is the Bear Lake population of garter snakes, which does not occur with newts and has low resistance to TTX. Adapted from Geffeney et al., 2002, reprinted with permission from AAAS.

reminded of the impact parasites can have on natural populations (Fig. 11.32).

Amphibians and reptiles are hosts to the usual vertebrate parasites. Internally, they include bacteria, protozoans, and various groups of parasitic "worms." External parasites include helminths and arthropods, primarily mites and ticks. All individuals likely have endoparasites of one kind or another as well as one or more ectoparasites. The level of virulence is usually unknown, but in most populations, individual amphibians and reptiles generally appear healthy, so many parasites must be benign and/or the host must be resistant to some degree, at least.

Amphibians and reptiles share many of the features of the immune system of mammals, and as a consequence, similar mechanisms modulate parasite infections. One mechanism for combating bacterial infection is elevation of body temperature, because high temperature inactivates or kills bacteria. Some lizards, snakes, and turtles behaviorally select and maintain body temperatures significantly above their normal activity temperatures. This behavioral fever mechanism appears to reduce the infection and improve the reptile's resistance. Amphibian granular and mucous glands may also function to offset parasite infection. These glands may have appeared early in amphibian evolution to protect against bacterial and fungal infections of the moist skin and still serve that function today. Magainins, isolated from the skin of *Xenopus*, have exceptional antibiotic and antifungal properties. They or related compounds likely exist in other amphibians. Other chemicals in the skin of some amphibians act as insect repellents and likely reduce exposure to insect-borne blood parasites.

Among the most common and geographically widespread parasites is malaria (*Plasmodium*), and a large number of species are known to infect amphibians and reptiles. In northern California, about 40% of the populations of *Sceloporus occidentalis* have malarial parasites. Within these populations, fewer than one-third of the individuals are infected, and males are more commonly infected than females. Performance of infected lizards is adversely affected by infection (Table 11.1), although no apparent differences in structure and dynamics between infected and noninfected populations are detectable. In Panamanian populations of *Anolis limifrons*, adult males also have the highest incidence of malarial infection during all seasons; however, no evidence of differences in general health, feeding, or reproductive behavior between noninfected and infected males exists.

Parasitism can influence the outcome of competitive interactions among species. On the island of St. Martin in the Caribbean, the lizard *Anolis gingivinus* occurs throughout the island and is a superior competitor over *A. wattsi*, but *A. wattsi* is restricted to the central hills. A malarial parasite *Plasmodium azurophilum* infects *A. gingivinus* but rarely infects *A. wattsi*. In areas where *Anolis gingivinus* is not infected, *A. wattsi* is absent, but in areas where *Anolis gingivinus* is infected, *A. wattsi* is present. The spatial distribution of the parasite in *Anolis gingivinus* is nearly identical to the spatial distribution of *A. wattsi*, which suggests that its presence renders *Anolis gingivinus* a poorer competitor when infected. In addition, *P. azurophilum* is known to reduce hemoglobin and negatively influence the immune system of *Anolis gingivinus*. Parasite-mediated competition may be common in amphibians and reptiles but is poorly documented.

Although parasitism appears to affect physiological function in some species, it does not affect others. Frillneck lizards, *Chlamydosaurus kingi*, that are infected with mosquito-transmitted filarial parasites perform equally well as uninfected lizards. Aerobic capacity, body condition,

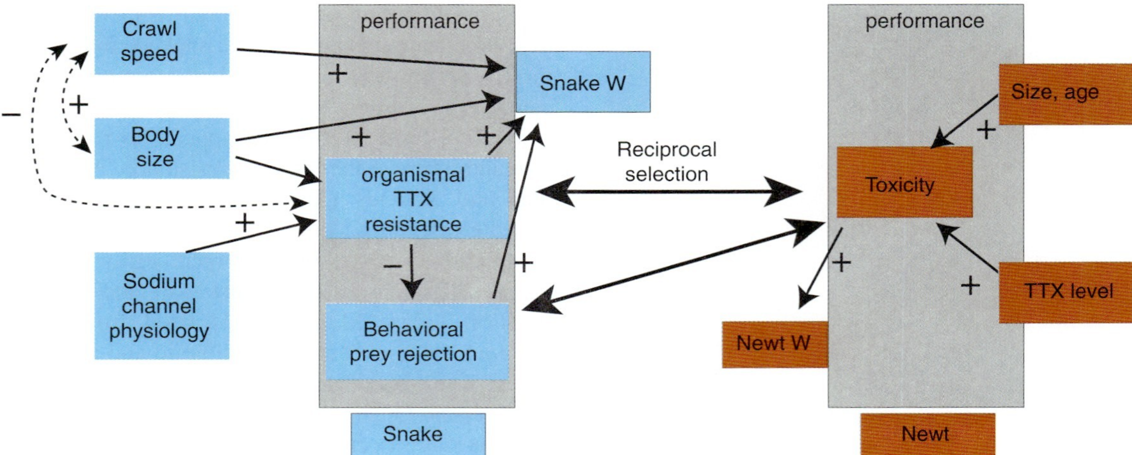

FIGURE 11.30 Summary of phenotypic interactions between garter snakes (*Thamnophis sirtalis*—blue) and newts (*Taricha granulosa*—orange). The interface of coevolution in predator and prey centers on levels of tetrodotoxins (TTX) in newts and resistance in the garter snakes. Snakes have the option of rejecting newts if their TTX levels are too high. Adapted from Brodie and Ridenhour, 2003.

FIGURE 11.31 Ticks embedded on the head of *Anolis oxylophus* (upper panel) and in the shell of *Rhinoclemmys annulata*. Ticks not only feed on the blood of their hosts but also can introduce additional parasites. Photographs by L. J. Vitt and K. Miyata, respectively.

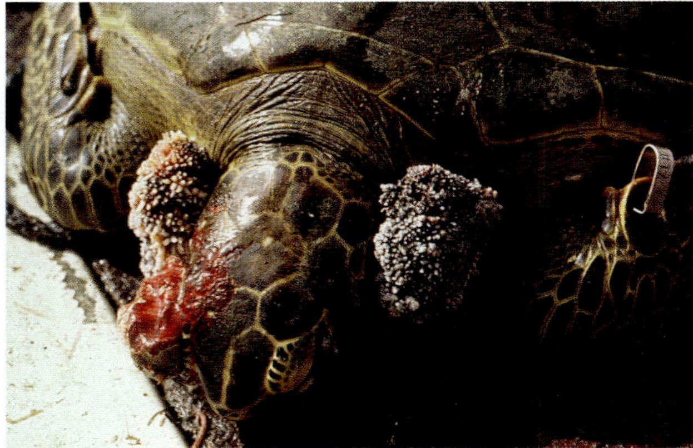

FIGURE 11.32 A subadult green sea turtle (*Chelonia mydas*) from Indian River Lagoon, Florida, with fibropapillomatosis. Photo by C. K. Dodd, Jr.

TABLE 11.1 The Effect of Malaria on the Performance of Western Fence Lizards *Sceloporus occidentalis*.

Criterion	Performance
Hemoglobin concentration	76
Metabolic rate, active	85
Burst running speed	89
Running stamina (2 min)	83
Fat stored, female	75
Clutch size	86
Growth rate	96
Mortality	114

Note: The values are the level of performance (in percent) of a sample of malaria-infected lizards compared with noninfected lizards, which are assumed to perform at 100%.
Source: Adapted from Schall, 1983.

hematocrit, and hemoglobin concentration are not related to the number of microfilariae in the blood of lizards and are not related to whether lizards are infected or not infected. Although larger lizards at the site where the parasite occurs have higher levels of infection, no effect of size (and, hence, parasite infection) is detectable on any of the performance parameters measured. Australian keelback snakes (*Tropidonophis maurii*) can be heavily infected with haemogregarine blood parasites, but similar to Frill-neck lizards, the parasites appear to have no measurable effect on various measures of performance.

A variety of mites and ticks infests reptiles and amphibians (Table 11.2). Many lizard species have mite pockets, folds of skin that often are completely packed with mites. In some lizards, folds of skin form mite pockets on the lateral surfaces of the neck anterior to the insertion of the front legs. These are often so packed with mites that large red patches are visible on the lizards from considerable distances. Exactly why lizards have mite pockets is controversial. One hypothesis is that mite pockets concentrate mites and restrict their damage to a few small areas. Another is that mite pockets reflect phylogenetic or structural constraints; the folds are present and mites use them. Implicit in the first hypothesis is the idea that overall mite loads would be reduced and thus overall damage would be less. Whether lizards actually gain anything by having mite pockets remains to be demonstrated; no apparent reason exists for mites to restrict themselves just to mite pockets. Moreover, the overall impact of tick and mite infestations remains poorly documented for amphibians and reptiles. Infestation of the mite *Hannemani dunni* is found in 100% of individuals of the salamander *Plethodon ouachitae* in some areas. In one study, each individual had an average of 20 mites on its body, and many individuals had clusters of mites on their appendages, causing deformities of the toes. The sympatric *Plethodon serratus*, which occurs in the same microhabitat as *P. ouachitae*, is not infected with mites. The reason why *P. serratus* is not attacked by mites is unknown, as is whether the mite infestation has any detrimental effects on reproduction or population structure in *P. ouachitae*. In sleepy lizards, *Tiliqua rugosa*, two species of ticks infect lizards at different localities. Long-term studies on the lizards reveal that longevity is not reduced in lizards infected with ticks. Individual lizards appear to maintain their tick loads from year to year. Lizards with the largest numbers of ticks reach the largest body size and are more likely to be in mating pairs than lizards with low tick loads. Thus, no evidence in these lizards indicates that parasite infection reduces fitness.

Elimination of a parasitic infection occurs in some species of amphibians and reptiles, although the mechanism is

TABLE 11.2 Parasites of Amphibians and Reptiles.

Parasite type	Amphibian–Reptile infected	Site of infection
Ectoparasites		
Arthropods and leeches		
Ticks	Snakes, lizards, turtles, anurans	Skin
Mites	Snakes, lizards, semiaquatic turtles	Skin, scales, cloaca
Chiggers (larva of Trombiculid mites)	Snakes, lizards, salamanders	Skin folds, joints
Myiasis (flies, mosquitoes, fleas, and gnats) (eggs are deposited)	Snakes, lizards, semiaquatic turtles, anurans	Skin, scales
Leeches (both salt- and freshwater species)	Aquatic turtles, crocodylians, aquatic amphibians	Skin, mouth
Endoparasites		
Protozoans		
Amoebiasis	Snakes, lizards	Intestinal tract
Coccidiosis	Snakes, lizards, turtles, crocodylians	Intestinal tract
Cryptosporidiosis	Snakes, lizards	Intestinal tract
Haemogregarines	Snakes, lizards, turtles, crocodylians	Bloodstream
Plasmodium and *Haemoproteus*	Snakes, lizards, turtles, crocodylians	Bloodstream
Trypanosomiasis	Snakes, lizards, turtles, crocodylians	Bloodstream
Ciliated protozoa (considered to be beneficial to some species)	Lizards, turtles	Digestive tract
Flagellated protozoa (some species not harmful to crocodylians and turtles)	Snakes, lizards, turtles, crocodylians	Digestive tract
Sarcosporidiosis	Aquatic turtles	Gallbladder

continued

TABLE 11.2 *(continued)*

Parasite type	Amphibian–Reptile infected	Site of infection
Trematodes (flukes)		
Monogenea (difficult to diagnose)	Turtles	Urinary bladder, nose, mouth, esophagus
Digenea	Crocodylians	Digestive tract
Aspidogastrea	Turtles	Alimentary tract
Spirorichidae (excluding *Digenea*)	All, especially turtles	Circulatory system
Styphlodora	Snakes	Renal tubules, cloaca, ureters
Cestodes (tapeworms)		
Pseudophyllidea	All, especially pythons	Muscles, subcutaneous tissue
Proteocephalidea	Snakes, varanid lizards	Small intestines
Mesocesttoidida	Snakes, lizards	Intestinal tract
Anoplocephalidae	Lizards, turtles, snakes	Intestinal tract
Nematotaenae	Lizards	Intestinal tract
Dilepididae	Snakes, lizards	Liver
Nematodes (roundworms)		
Ascarids	All	Gastric mucosa
Rhabditida	All, especially varanid lizards	Lungs
Strongyloids	All	Esophagus, intestine
Acanthocephalins	Lizards, turtles	Small intestine
Filarids	All	Bloodstream
Pentastomiasis	All, especially varanid lizards	Lungs, esophagus
Oxyurids	Lizards, turtles, some snakes	Lower intestine
Capillaria and *Eustrongylides*	Snakes, lizards, turtles	Liver, bile duct

unknown. For example, nearly all male spadefoots, *Scaphiopus couchii*, leave their breeding aggregations with a monogenean trematode infection, yet 50% have lost the parasites prior to hibernating.

Although much of the emphasis on amphibian and reptile parasites has historically centered on their effects or treatment, some recent studies examine origins and dispersals of parasites using modern phylogenetic methods.

QUESTIONS

1. Recall the diagram of the sequence of events that can occur during a predation attempt (detection, identification, approach, pursuit, and capture). Choose any three events and, using real examples (species), describe how the species make it through that event. Exactly how does the particular behavior result in escape?
2. Describe why tails of juvenile five-lined skinks are brilliant blue but tails of adults are cryptically colored. Frame your answer in terms of costs and benefits.
3. Why do some *Acris crepitans* tadpoles have black-tipped tails but others do not?
4. What is meant by the "evolutionary arms race" and how does this work in gartersnakes and newts?
5. Optimal escape theory makes some specific predictions with respect to risk and escape cost as a function of distance to predator. What are these, and what are some of the variables that might affect risk and cost in amphibians and reptiles?

ADDITIONAL READING

Arnold, E. N. (1988). Caudal autotomy as a defense. In *Biology of the Reptilia. Volume 16, Ecology B. Defense and Life History*. C. Gans and R. B. Huey (Eds.), pp. 235–273. Alan R. Liss, Inc., New York.

Daly, J. W. (1998). Thirty years of discovering arthropod alkaloids in amphibian skin. *Journal of Natural Products* **61:** 162–172.

Duellman, W. E., and Trueb, L. (1986). *Biology of Amphibians*. McGraw-Hill Book Co., New York.

Erspanner, V. (1994). Bioactive secretions of the amphibian integument. In *Amphibian Biology, Volume 1, The Integument*. H Heatwole and G. T. Barthalmus (Eds.). pp. 178–350. Surrvey Beaty & Sons, Chipping Norton Pty Ltd, N.S.W.

Greene, H. W. (1988). Antipredator mechanisms in reptiles. In *Biology of the Reptilia. Volume 16: Ecology B. Defense and Life History.* C. Gans and R. B. Huey (Eds.). pp. 1–152. Alan R. Liss, Inc., New York.

Pianka, E. R., and Vitt, L. J. (2003). *Lizards: Windows to the Evolution of Diversity.* University of California Press, Brekeley, CA.

Pough, F. H. (1988). Mimicry and related phenomenona. In *Biology of the Reptilia, Vol. 16, Ecology B. Defense and Life History.* C. Gans and R. B. Huey (Eds.). pp. 153–234. Alan R. Liss, Inc., New York, NY.

Wells, K. D. (2007). *The Ecology and Behavior of Amphibians.* University of Chicago Press, Chicago.

REFERENCES

General

Endler, 1986; Menin et al., 2005; Schall and Pianka, 1980; Stebbins and Cohen, 1995; Toledo, 2005; Toledo et al., 2007; Vermeij, 1982; Wells, 2007.

Escape and Emergence Theory

Brown and Shine, 2004; Cooper, 1999, 2000, 2003a, 2003b, 2005, 2006; Cooper and Peréz-Mellado, 2004; Cooper and Vitt, 2002a; Cooper et al., 2006; Martín and Lopéz, 1999; Schulte et al., 2003; Ydenberg and Dill, 1986.

Predator Avoidance

Escaping Detection

Brodie et al., 1974; Endler, 1978; Greene, 1988; Morey, 1990.

Escaping or Misdirecting Identification

Autumn and Han, 1989; Brodie, 1968, 1993; Brodie and Brodie, 1980; Brodie and Janzen, 1995; Greene, 1988; Greene and McDiarmid, 1981; Harper and Pfennig, 2007; Howard and Brodie, 1973; Huey and Pianka, 1981b; Pfennig et al., 2001, 2007; Pough, 1988; Sazima and Abe, 1991; Sherbrooke and Montanucci, 1988; Smith, 1975, 1977; Tilley et al., 1982; Warkentin, 1995, 2000, 2005; Warkentin et al., 2005; Vitt, 1992b.

Escaping Approach

Cooper, 2000; Cooper et al., 2004; Dudley et al., 2007; Duellman and Trueb, 1986; Emerson and Koehl, 1990; Greene, 1997; McCay, 2001, 2003; McGuire, 2003; McGuire and Dudley, 2005; Rossman and Williams, 1966; Shine et al., 2002; Socha, 2003; Vitt and Zani, 1996b; Vitt et al., 2002.

Escaping Subjugation and Capture

Arnold, N., 1984, 1988; Arnold, S., 1982; Bauer et al., 1989; Blouin, 1990; Branch, 1983; Britson, 1998; Britson and Gutzke, 1993; Bustard, 1967; Caldwell, 1982, 1989, 1993, 1996a, 1997; Caldwell and Vitt, 1999; Clark, 1971; Congdon et al., 1974; Cooper, 2003c; Cooper and Vitt, 1985, 1991; Daly, 1998; Daly et al., 1987, 1994, 1996, 1997, 2000, 2002, 2003; Daniels, 1983; Darst et al., 2005; Dial, 1981; Dial and Fitzpatrick, 1981, 1983; Duellman and Trueb, 1986; Erspamer, 1994; Evans and Brodie, 1994; Fox and Rostker, 1982; Fox et al., 1990; Greene, 1988; Kats et al., 1988; Maiorana, 1977; Martin and Salvador, 1993; Middendorf and Sherbrooke, 1992; Niewiarowski et al., 1997; Petranka et al., 1987; Rose, 1962; Rosenberg and Russell, 1980; Salvador et al., 1996; Schoener, 1979; Schoener and Schoener, 1982; Shine, 1980c; Summers, 2003; Summers and Clough, 2001; Toft, 1995; Vaz-Ferreira and Gehrau, 1975; Vences et al., 1997, 1998, 2003; Vitt and Cooper, 1986; Vitt et al., 1977; Wake and Dresner, 1967; Wassersug, 1971; Wassersug et al., 1981; Watkins et al., 1969; Weldon, 1990; Wells and Bard, 1988.

Life History Responses to Predation

Caldwell, 1993; Tinkle, 1967; Tinkle et al., 1970.

Predators and Their Prey: The Evolutionary Arms Race

Brodie and Brodie, 1999a,b,c; Brodie and Ridenhour, 2003; Brodie et al., 1974, 2002a,b; Geffeney et al., 2002, 2005; Hille, 1992; Janzen et al., 2002; Mosher et al., 1964; Motychak et al., 1999; Narahashi, 2001; Vermeij, 1987.

Offsetting the Effects of Parasitism

Al-Sorkhy and Amr, 2003; Arnold, N., 1986; Ayala, 1977; Bauer et al., 1990, 1993; Bentz et al., 2003; Brown et al., 2006; Bull and Burzacott, 1993; Burns et al., 1996; Christian and Bedford, 1995; Madsen et al., 2005; Monagas and Gatten, 1983; Ortega et al., 1991; Schall, 1983, 1992; Tinsley, 1990; Vaughn et al., 1974; Winter et al., 1986.

Part V

Ecology, Biogeography, and Conservation Biology

Ecology is proving to be one of the most complex fields of modern science and, from a conservation biology perspective, perhaps the most important and relevant field if we are to meet challenges resulting from habitat loss, human impacts on the environment, and global warming. Understanding how biological systems work is critical to developing management strategies that have the potential to maintain the biodiversity that supports life as we know it. Ecological studies not only require knowledge of individuals, populations, species, and communities, but also understanding of underlying physiological, behavioral, and evolutionary factors that might limit or affect organisms within these levels of organization. Ecological studies centered on species interactions (e.g., competition and predation) or systems (e.g., nutrient cycles) during much of the last century. These approaches viewed ecological characteristics of individuals, populations, and communities as relatively recent, or in the case of systems ecology, controlled by availability of critical nutrients. The rapid development of molecular systematics and analyses of phylogenetic data during the last 20 years, and its application to ecological questions, has changed the way we think about ecology, and in many respects, has brought several very broad conceptual areas together in our attempts to understand the world in which we live. Species interactions and nutrient cycles remain relevant, but in a much broader context.

The inability to predict what happens in natural ecosystems given the often clear results of controlled experimental studies attests to the complexity of ecological systems. The availability of massive data sets, relatively easy access to data as the result of the Internet, and new analytical tools (e.g., GIS [geographic information systems]) has opened new frontiers in ecology as well.

Reptiles and amphibians have played critical roles in the development and testing of ecological theory, much of which has been successfully applied to other organisms. Populations of lizards and larval frogs and salamanders, in particular, are extremely amenable to experimental manipulations. Complex, and sometimes surprising, population- and community-level responses to species interactions have been identified in studies with amphibians and reptiles. Rapid advances in phylogenetic analyses resulting in fairly robust phylogenetic hypotheses for many amphibians and reptiles have provided the basis for the rapidly developing field of historical ecology. Popular theories have been falsified using phylogenetic hypotheses in which divergence times can be estimated, and fascinating patterns have been identified, often generating new theory. Tying present-day species distributions to the history of continental movements has become much more sophisticated than it ever was, providing new and interesting insight into biogeography. Taken together, these new approaches have also changed the way we think about conservation biology. Overall, impacts of human activity have been devastating, causing the largest extinction event in the history of Earth. In some cases, species have expanded their distributions virtually overnight, particularly in species that can take advantage of man-made structures or altered ecosystems. Consequently, ecology, biogeography, and conservation biology are considered by many as among the most exciting fields in biology; all can be viewed as frontiers, and all are critical to long-term survival of amphibians, reptiles, and of course, humans.

Chapter 12

Ecology

Ecological studies seek to explain why there are so many species, why a given set of species occurs in a particular area, and how those species interact and persist. Underlying these key questions is the notion that resources are in one way or another limited (Fig. 12.1). Consequently, for a number of species to survive in a given habitat or area, those resources should be divided (partitioned), and species interactions (competition, predation, parasitism) will determine which species from the regional species pool will persist. Anyone who has done some hiking in different parts of their country or the world knows that each place seems to have its own distinct set of species. Rain forests usually have high diversity of frogs, many habitats have high snake diversity, and both deserts and rain forests have high diversity of lizards, whereas deciduous forests have low lizard diversity (Table 12.1). Reasons for these differences are complex but include abiotic factors (e.g., high latitudes contain few ectothermic terrestrial vertebrates because of temperature), biotic factors, and historical factors (e.g., no salamanders occur in Africa or Australia).

Ecological studies typically deal with assemblages of individuals of a single species (populations), the set of species in a given taxon (e.g., frogs) living in a particular area or habitat (assemblages), or all species in some defined area (communities). A population is typically defined as a group of interbreeding individuals of the same species living in the same area. Each individual is potentially able to mate with any other individual of the opposite sex. Consequently, each population represents a single gene pool, and all individuals share a recent common ancestry. Although the potential for interbreeding is seldom, if ever, realized within a single generation, complete mixing of genes (panmixis) may occur over generations in small, localized populations. Asexually reproducing organisms (e.g., parthenogenetic *Aspidoscelis*) do not fit this definition because each individual is reproductively isolated from all others. Nevertheless, they experience many of the same population phenomena as sexually reproducing species.

Populations can be variously delimited. All side-blotched lizards (*Uta stansburiana*) in western North America, western diamondback rattlesnakes (*Crotalus atrox*) in the Sonoran Desert, or cave salamanders (*Eurycea lucifuga*) in the Ozark Plateau represent populations. Although each is a biological population, the local population (= deme) is usually the unit of interest to biologists. The local population responds to local conditions:

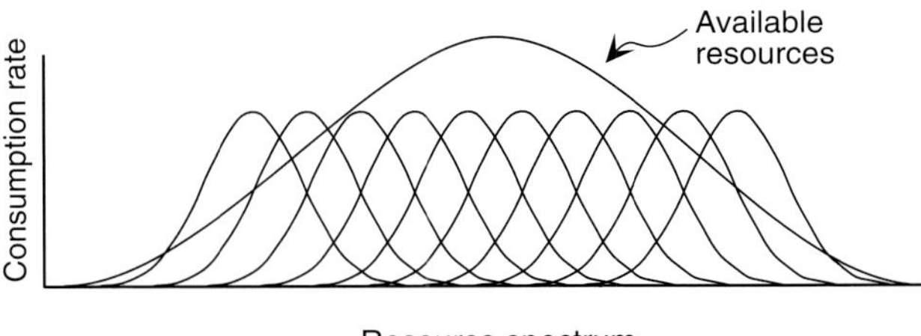

FIGURE 12.1 Graphic representation of a single resource system in which a number of consumers partition the resource. The general idea is that if resources are limited, species should divide those resources if the species are to coexist. If two species use identical resources, the one that is most effective at collecting and converting energy into offspring will outcompete the other. Adapted from Pianka, 1988.

Herpetology, 3rd Ed.

TABLE 12.1 Composition of Continental Herpetological Assemblages[1] from Different Localities, Habitats, and Climates.

Site	Caecilians	Frogs	Salamanders	Turtles	Lizards	Snakes	Crocodylians	Totals	Latitude
Andrew Exp. Forest (forest)	0	3	7	0	3	3	0	16	44°N
Barro Colorado (forest)	1	29	0	5	22	39	2	98	9°N
Big Desert (scrub)	0	4	0	0	18	2	0	24	35°S
Brazilian Pantanal (seasonally flooded savanna)	1	43	0	4	30	83	2	163	15°S
Chitwan (grassland and forest)	0	11	0	7	10	24	2	54	28°N
Jalapão National Park (cerrado)	2	32	0	2	32	46	2	116	11°S
Kivu (forest)	0	29	0	2	10	38	1	80	2°S
Lamto (savanna)	0	17	0	0	10	12	0	39	8°N
Lazo Nat. Reserve (forest)	0	6	0	0	1	6	0	13	43°N
Nanga Tekalit (forest)	1	47	0	0	40	47	0	135	3°N
Packsaddle Wildlife Management Area (short-grass prairie)	0	8	1	6	8	21	0	44	36°N
Ponmudi (forest)	2	24	0	0	16	14	0	56	9°N
Prince William (forest)	0	10	10	4	4	13	0	41	38°N
Rota (grassland and forest)	0	6	2	1	10	5	0	24	37°N
Sakaerat (fields and forests)	1	24	0	2	30	47	0	104	14°N
Santa Cecilia (forest)	3	81	2	6	28	51	2	173	0°N
Savannah R. Site (swamp and forest)	0	23	16	12	9	35	1	95	33°N

Continued

TABLE 12.1 *(continued)*

Site	Caecilians	Frogs	Salamanders	Turtles	Lizards	Snakes	Crocodylians	Totals	Latitude
Tucumán (forest)	0	16	0	1	26	24	0	67	28°S
UK Nat. Reserve (grassland and forest)	0	9	1	4	7	16	0	37	39°N
V. Crookes Reserve (grassland and forest)	0	17	0	0	8	14	0	39	30°S
Vienna (fields and forest)	0	12	5	1	5	5	0	28	48°N

[1] *Each assemblage represents the taxa likely to be present in a 25-km² area and represents multiple habitats in most cases. The data are the number of species, excluding introduced or exotic species.*
Sources: Andrew Experimental Forest, Oregon, USA, *Bury and Corn*, 1988; Barro Colorado Biological Station, Canal Zone, Panama, *Myers and Rand*, 1969; Big Desert, Victoria, Australia, *Woinarski*, 1989; Pantanal, Brazil, *Strüssmann et al.*, 2007; Royal Chitwan National Park, Nepal, *Zug and Mitchell* 1995; Jalapão National Park, Brazil, *Colli et al.*, *unpublished*; Kivu, Zaire, *Laurent*, 1954; Lamto, Ivory Coast, *Barbault*, 1972, 1975*a*, 1975*b*; Lazo State Nature Reserve, Maritime Terr., Russia, *Shaldybin*, 1981; Nanga Tekalit, Sarawak, *Lloyd et al.*, 1968 (*island; no continent at this latitude in Asia*); Packsaddle WMA, western Oklahoma, USA, *Vitt et al.*, *unpublished*; Ponmudi, India, *Inger et al.*, 1984; Prince William National Forest, Virginia, USA, *Pague and Mitchell*, *unpublished*; Rota, Spain, *Busack*, 1977; Sakaerat Experiment Station, Thailand, *Inger and Colwell*, 1977; Santa Cecilia, Ecuador, *Duellman*, 1978; Savannah River Plant, Georgia, USA, *Gibbons and Semlitsch*, 1991; Tucumán (*bosques chaqueros*), Argentina, *Laurent and Teran*, 1981; University of Kansas Natural History Reserve, Kansas, USA, *Fitch*, 1965; Vernon Crookes Nature Reserve, Natal, *Bourquin and Sowler*, 1980;Vienna, Austria, *Tiedemann*, 1990.

growing, shrinking, evolving, or even disappearing (extinction). Each local population is semi-isolated from other similar populations by minor or major habitat discontinuities, but few are totally isolated (closed), and most receive occasional immigrants from nearby or distant populations and lose members via emigration. Populations have characteristics that communities do not have, including population growth rates, survivorship schedules, birth rates, and replacement rates, to mention a few. These are discussed in more detail in Chapter 4.

Communities have identifiable characteristics that populations and species do not have. These include energy flow, nutrient cycling, and species turnover. Communities typically have structure that persists even though species composition and abundances change. Communities are composed of sets of species that are producers (plants), primary consumers (herbivores), secondary consumers (carnivores), and decomposers (bacteria, etc.). Parasites on animals are secondary consumers. At one level, the basic organization of all communities follows energy flow through the various life-forms from plants through consumers and decomposers. Life's energy derives entirely from the sun. Plants capture this radiant energy and convert it into plant tissue; herbivores eat the plants and convert the energy into animal tissue. Predators eat herbivores, and some high-order predators feed on other predators as well as on herbivores. At each step (trophic level; see inset, Fig. 12.2) in the food or energy chain, energy is lost as a by-product of metabolic activities (i.e., respiration) and because individuals are unable to assimilate all food obtained. Assimilation efficiency is typically lowest in herbivores and highest in top-order predators. Trophic pyramids reflect the sequential energy loss through trophic levels. It should be obvious that, given differences in assimilation efficiencies among trophic levels, biomass of primary producers is greatest, and biomass of each successive trophic level is lower. As adults, amphibians and reptiles are mainly primary and secondary predators, eating other consumers and in turn being eaten. Larvae of most anurans and some adult reptiles are herbivores. Consequently, herpetological assemblages occupy the middle region of the trophic chain or food web (Fig. 12.2).

No matter how broadly or narrowly defined, a community's structure is its species composition, the abundance of each species, and the interactions among species. Even though patterns of co-occurrence are evident and relatively easy to quantify, the causes for these associations are not. Abiotic, biotic, and historical factors determine the presence or absence of a species and the abundance of its members in a local community (Table 12.2). Abiotic (a = without; bio = life) factors are a function of the physical environment and each species' physiological tolerances to environmental variables (Chapters 6 and 7). Biotic factors are resource related and concern interactions with other species. These interactions may be direct (catching prey or being captured as prey) or indirect (shade from a tree or high humidity of a forest); they have positive, negative, or neutral effects on an individual's

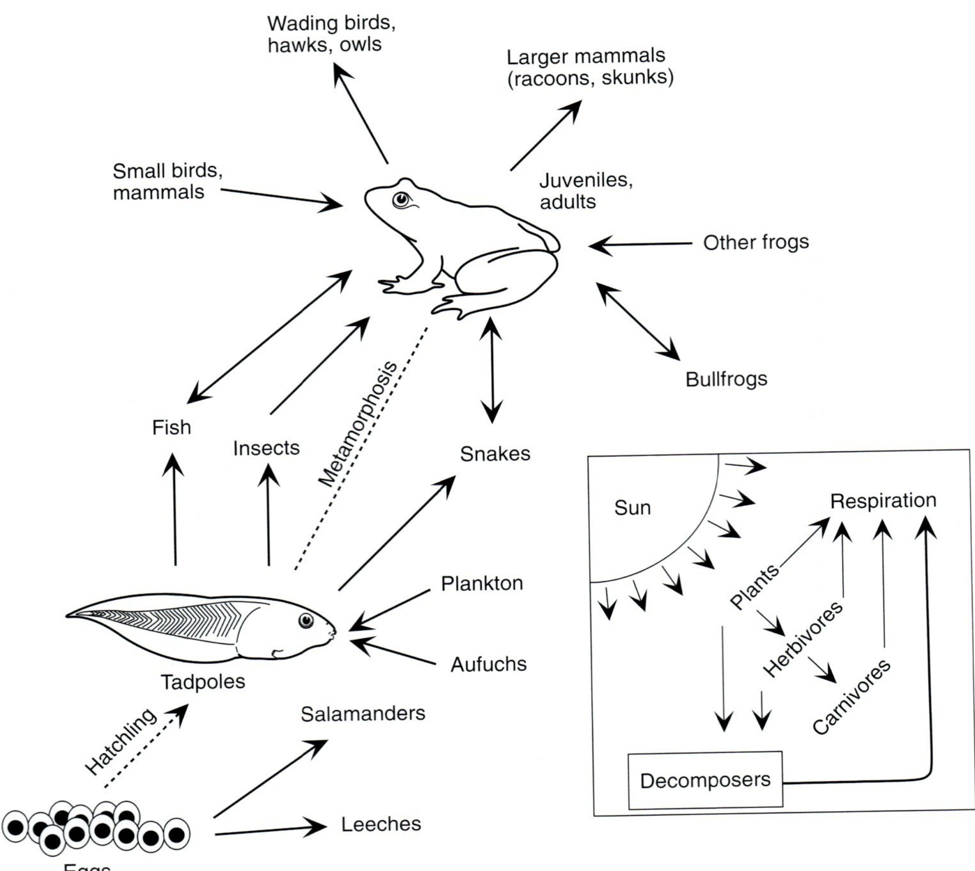

FIGURE 12.2 A generalized food web for an eastern North American pond showing the bullfrog (*Lithobates catesbeianus*) as the focal point. Arrows denote direction of energy flow, i.e., consumption. The inset (lower right) shows the general pattern of energy flow through communities. Adapted from Bury and Whelan, 1984.

TABLE 12.2 Properties Determining a Species' or Population's[1] Membership, Position, and Persistence in a Community.

Organismic	Environmental
Body size	Severity of physical environment
Diet (trophic position)	Spatial fragmentation
Mobility	Long-term climatic variation
Homeostatic ability	Resource availability
Generation time	Resource partitioning
Number of life stages	
Recruitment	

[after Schoener, 1986]
[1]*Schoener proposes these properties to examine the structure and dynamics of assemblages (e.g., intertidal algae, island anoles). These properties also highlight factors that affect an organism's survival and reproductive success, hence a population's or species' niche and community affiliations.*
Note: This summary does not include historical factors (see text).

survival and reproduction, hence influencing persistence or extinction of a population. Direct interactions include predation, mutualism, and competition, major factors that shape community structure. Historical factors include colonization and extinction events and patterns of movement. Interactions that led to present-day structure of many communities are often subtle because a long history of interactions leads to equilibrium. Species interactions that produce present-day structure may have occurred long ago. Moreover, existing differences among species at first contact might be sufficient to allow coexistence with little or no interaction. The available species pool can also have considerable impact on the structure of communities.

The terms community and assemblage are often interchanged, and most "community" studies of amphibians or reptiles are actually studies of assemblages. The term assemblage is usually applied to a taxonomic subset of species in a community. For example, one might use the term when referring to an assemblage of African savanna snakes or Madagascar rain forest chameleons. The term *ecosystem* refers to all organisms and the abiotic factors associated with those organisms in some easily definable area; for example, a Neotropical ecosystem or a freshwater stream ecosystem. Guilds are sets of species that use particular resources in the same manner; for example, there might be an ant-eating guild composed of all species of amphibians and reptiles in a given community that feed on ants. Intraguild predation and competition may occur among different stages or size classes within a guild. These terms and others are common in the ecological literature and will appear in the following sections.

We first examine some of the factors that contribute to patterns of species richness and abundance. We then provide examples of two broad types of ecological studies using amphibians and reptiles. These are experimental studies and three dramatically different types of comparative studies. Experimental studies offer the opportunity to exclude most variables and center on one or more that are of particular interest. They test specific hypotheses such that falsification eliminates explanations with no support. The disadvantage to experimental studies alone is that failure to falsify a given hypothesis does not necessarily mean that it is the correct explanation. Experiments, by their very nature, are highly simplistic compared with the natural world in which a huge number of variables interact. Nevertheless, when used in the proper perspective, experiments can be extremely powerful. Comparative studies also test hypotheses, but in a slightly different way than experimental studies. Most comparative ecological studies are based on data collected on natural populations *in situ,* that is, in the milieu of the myriad of variables that might affect individuals, populations, and communities. Resultant analyses of comparative ecological data usually produce statistical results that explain or account for a portion of the variance observed in the variables of interest. Remaining variance remains unexplained. Also, most comparative ecological studies involve very complex sets of data, and multivariate analytical methods are necessary to organize and make comparisons. Comparative studies often generate new hypotheses, some of which can be tested with experiments. The term comparative has been used differently by different investigators. Until recently, comparative simply meant comparing two or more things (a very common approach in pre-1990 physiology and ecology). Today, comparative has many meanings, but most frequently it involves use of phylogenies to sort out effects of relatedness from effects of the variables of interest. For example, two species in a community might be similar ecologically simply because they are closely related and thus have a relatively recent common ancestor, rather than because they interact. The reason that both experimental and comparative studies have persisted is partly because combinations of the two usually produce the best-supported explanations for the variation that exists in the natural world. Also, some questions of interest cannot be easily addressed with experiments, and some questions cannot be easily addressed with comparisons.

SPECIES RICHNESS AND ABUNDANCE

The striking differences in number of species at different localities and latitudes have long intrigued biologists (Table 12.1). Comparisons of species richness (diversity or density) provide geographic comparisons of community structure. Most attention has been directed at explaining the tendency for species diversity to increase from high-latitude habitats to tropical ones and, as for most other studies of communities, to examine the changing diversity within specific taxonomic groups such as frogs or lizards. The emphasis in most instances centers on just the number of species rather than a combination of the number of species and their relative densities. Accurate and comparable data on species abundance have only recently become available, and considerable variation exists in methods to determine species composition and relative abundance, making comparisons difficult (see following text).

Numerous explanations have been proposed to account for differences in species richness across landscapes and between mainland and island habitats. The primary explanations are evolutionary time, ecological time, climatic stability, climatic predictability, spatial heterogeneity, productivity, and species interactions. These explanations are not mutually exclusive, and likely multiple causes operate in different combinations at different locations and at different times.

Evolutionary and Ecological Time

Two distinct timescales are typically considered in ecological studies. Evolutionary time refers to time periods long enough for adaptation to occur. Ecological time refers to time periods short enough that adaptation has not occurred. Older communities presumably have had more time for species to adapt to the local environment. As a result, they should contain species that exhibit adaptations to various aspects of the particular environment or community, thus reflecting the influence of evolutionary time periods. Species can colonize new or modified habitats with no apparent evolutionary change (adaptation), thus these events are considered to occur over ecological time. Amphibian and reptile species of northern sites (Andrew Experimental Forest, Vienna; Table 12.1) are all wide-ranging species with distributions $> 1000\ km^2$, in contrast to tropical sites, which have many species with small distributions. Although individuals of temperate-zone amphibians and reptiles may have limited dispersal abilities, their populations are capable of expanding as their preferred habitats expand. This is evident from the reoccupation of glaciated portions of North America in the last 10-15 thousand years; the current ranges of some species (*Ambystoma laterale, Lithobates septentrionalis, Pantherophis vulpina, Emys blandingii*) are totally within formerly glaciated areas.

Climatic Stability and Predictability

Climatically stable areas have little seasonal or long-term change in temperature or rainfall. Such locations are limited to a few rain forest areas of the world, for example, Amazonian forest on the eastern slopes of the Andes. These areas generally have relatively high numbers of species of amphibians and reptiles. Santa Cecilia, Ecuador, for example, has 173 species of amphibians and reptiles, and a site near Iquitos, Peru, has about 200 species. Climatically predictable habitats with regular cycles of wet–dry or hot–cold seasons are far more numerous, but species richness in these habitats varies considerably depending upon latitude. Relative length and harshness of the cold or dry seasons are rarely considered and can be quite influential in limiting the species occurring in a particular habitat. Climatic predictability may be more imagined than real; climate records of this century emphasize the great irregularity in the beginning, end, and length of seasons. Predictability of climate may be no more regular in the tropics than in the temperate zone.

Spatial Heterogeneity

Habitats with a greater spatial or structural heterogeneity tend to have more species, within the constraints of climate. A structurally heterogeneous habitat at northern latitudes would not have a high diversity of reptiles because temperature is an overriding limiting factor for ectothermic vertebrates, but a structurally heterogeneous habitat at the same latitude as a structurally simple habitat would be expected to contain more species of ectotherms. Structurally complex forest habitats usually have more species than the structurally simpler grassland and desert habitats, but striking exceptions exist. The greatest diversity of lizards, for example, occurs in the Great Victorian Desert of Australia, not in the world's rain forests! Moreover, historical patterns of fire and its effect on habitat heterogeneity have contributed to high lizard diversity. Three habitats exist in the Sakaerat area of Thailand (Table 12.1): gardens and fields, deciduous forest, and evergreen forest; at this locality and elsewhere, herpetofaunal diversity increases proportionately with spatial heterogeneity of the habitat (54, 67, and 77 species, respectively). Productivity is often related to spatial heterogeneity. High food availability and high prey diversity allow a greater number and diversity of consumers.

A common assumption is that the abundance of each species is less in a species-rich community than in a species-poor one. This conjecture may be true for tree species in a rain forest compared with a temperate-zone deciduous forest. While probably not true for most herpetological communities, actual comparisons do not exist (see density estimates in Table 12.3). Such comparisons would examine the actual abundance (density) of each species in the area under consideration. Another abundance comparison would be to examine the relative abundance (equability or evenness) of each species within the community. However, these comparisons are confounded by differences in body size of each species, trophic position, seasonal and annual fluctuation in population densities, and widespread lack of accurate population censuses (particularly for tropical populations). Ignoring these difficulties, it is unlikely that all species are equally abundant in any community or assemblage. More likely, one or a few species have high densities, and all other species are at low densities. The Sakaerat skink assemblages show abundance patterns that are likely typical of those between common and rare species in other herpetological assemblages, whether they are from the tropics or temperate zone (Fig. 12.3).

In nearly all herpetofaunas sampled, some species are abundant and others are rare, and as pointed out earlier, sampling techniques often provide different estimates of abundance. Both species-rich and species-poor assemblages have log-normal patterns of species abundance. In the Cerrado of Brazil, a savanna-like open habitat, many species of lizards are abundant and can easily be trapped with pitfall traps. Two species, *Cnemidophorus mumbuca* and *Tropidurus oreadicus* are much more abundant than other species (Fig. 12.4), even though they are not among the smallest lizards at the site. Each species uses different microhabitats, and by quantifying characteristics of the vegetative and structural habitats at each trap site, it is possible to associate lizard species with microhabitat attributes. For example, several species are associated with presence of leaf litter, others are associated with a lack of shade (open sky), and yet others are associated with presence of fallen logs (Fig. 12.5).

Exceptions to equal abundance between species-rich and -poor assemblages occur between mainland and island anole assemblages. Island populations have densities 2 to 10 times higher than mainland populations. A few other lizards also occur at higher densities on islands, but comparisons for other amphibians and reptiles are lacking. These differential densities appear to result from differences in predation; generally, island populations experience relaxed predation rates (see Chapter 5).

Species richness also differs markedly between island and mainland assemblages. Islands have fewer species compared with comparable-sized areas on the mainland. Further, a positive relationship exists between island size and species richness, and a negative relationship exists between the distance of an island from a colonizing source (e.g., mainland) and species richness. These species-area

TABLE 12.3 Population Densities of Some Amphibians and Reptiles.

Taxon	Density	Body size	Habit–Habitat
Bolitoglossa subpalmata	4790	42	Terrestrial–trop. forest
Plethodon glutinosus	8135	63	Terrestrial–temp. forest
Arthroleptis poecilonotus	325	20	Semiaquatic–trop. savanna
Rhinella marinus	160	90	Terrestrial–trop. scrub
Eleutherodactylus coqui	100	36	Terrestrial–trop. forest
Eleutherodactylus coqui	23,000	36	Terrestrial–trop. forest
Geochelone gigantea	27	400	Terrestrial–trop. scrub
Sternotherus odoratus	194	66	Aquatic–temp. lake and river
Apalone mutica	1267	210	Aquatic–temp. lake and river
Alligator mississippiensis	0.2	1830[a]	Semiaquatic–temp. marsh
Lacerta vivipara	784	56	Terrestrial–temp. forest
Mabuya buettneri	17	78	Arboreal–trop. savannah
Uromastyx acanthinurus	0.15	110	Terrestrial–subtrop. desert
Varanus komodensis	0.09	1470	Terrestrial–trop. scrub
Xantusia riversiana	3200	70	Terrestrial–temp. scrub
Agistrodon contortrix	9	540	Terrestrial–temp. savannah
Coluber constrictor	0.3	630	Terrestrial–temp. scrub
Enhydrina schistosa	0.9	730	Aquatic–trop. tidal river
Opheodrys aestivus	429	360	Arboreal–temp. forest
Regina alleni	1289	400	Aquatic–subtrop. marsh

[a]*Total length.*

Note: Density is the mean number of individuals per hectare; body size is length (SVL; CL *for turtles; mm) of adult females.*

Sources: Salamanders—Bs, *Vial,* 1968; *Pg, Semlitsch,* 1980. *Frogs—Ap, Barbault and Rodrigues,* 1979; *Bm, Zug and Zug,* 1979; *Ec, Stewart and Pough,* 1983. *Turtles—Gg, Bourn and Coe,* 1978; *So, Mitchell,* 1988; *Tm, Plummer,* 1977*b*. *Crocodilians—Aa, Turner,* 1977. *Lizards—Lv, Pilorge,* 1987; *Vk, Auffenberg,* 1978; *Xr, Fellers and Drost,* 1991*a,b*. *Snakes—Ac, Fitch,* 1960; *Cc, Brown and Parker,* 1984; *Es, Voris,* 1985; *Oa, Plummer,* 1985; *Ra, Godley,* 1980.

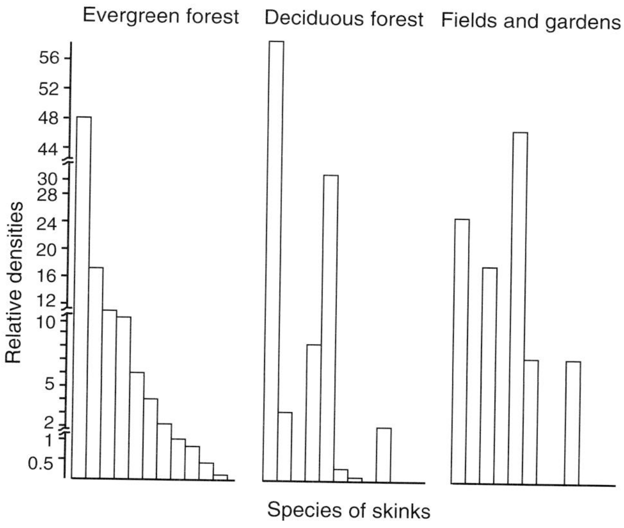

FIGURE 12.3 Relative abundance of skink assemblages in three habitats at Sakaerat, Thailand. Eleven species occur in the evergreen forest, and subsets occur in the deciduous forest and agriculture areas. The species are ranked in order of decreasing abundance for the evergreen forest assemblage, and that order is retained for the two other assemblages. Skink abundance is unequal between habitats; skinks comprise 53%, 43%, and 4% of the total lizards for the three habitats, respectively. Data from Inger and Colwell, 1977.

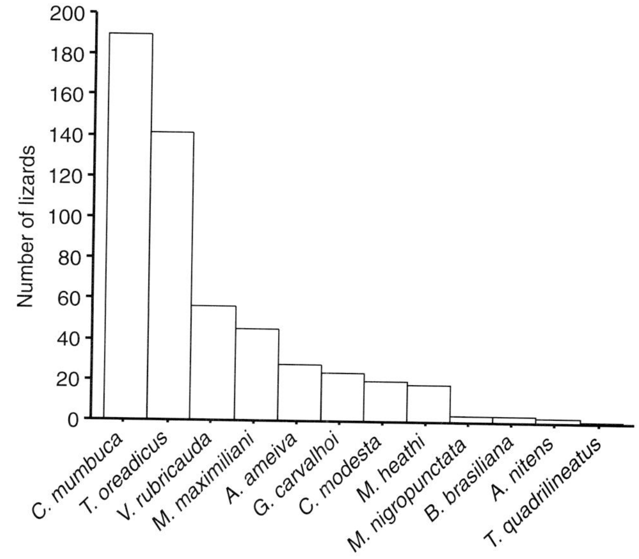

FIGURE 12.4 Abundance of lizards at Brazilian Cerrado site, based on pitfall trapping. Genera from left to right: Cnemidophorus, Tropidurus, Vanzosaura, Micrablepharus, Ameiva, Gymnodactylus, Colobosaura, Mabuya, Mabuya, Briba, Anolis, and Tupinambis. Adapted from Vitt et al., 2007b.

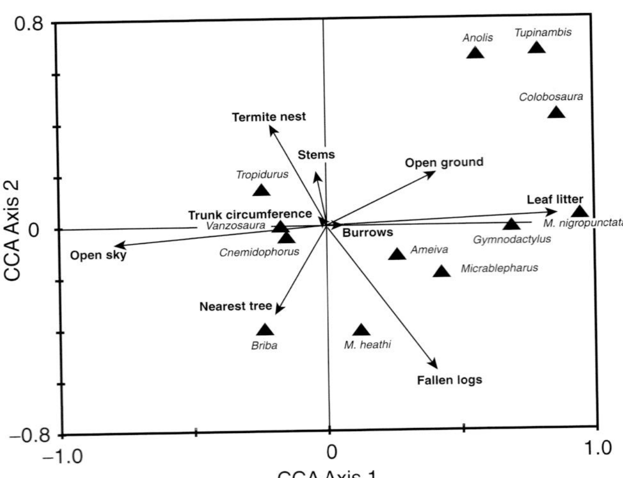

FIGURE 12.5 Relationship between structural microhabitat variables and lizard species in the Cerrado based on a Canonical Correspondence Analysis, which ordinates lizard species along two multivariate axes. Species are the same as in Figure 12.4. Adapted from Vitt et al., 2007b.

relationships led Drs. Robert MacArthur and Edward Wilson (1967) to develop an equilibrium theory of island species diversity. The equilibrium theory proposes that a balance exists between the number of species colonizing an island and the number of species going extinct. The colonization or immigration rate is a function of the island distance from a source area, and the extinction rate is a function of island size. Since immigration and extinction are assumed to be continuous processes, species number reaches an equilibrium value and remains constant even though the composition of the species assemblages changes continually.

Although a linear relationship between island size and species number exists for lizards (Fig. 12.6), few other herpetological groups have been examined. Island assemblages deviate from several predictions based on theory. Lizard assemblages commonly have higher species diversity than predicted, suggesting supersaturation. Lizard assemblages also exhibit a constant number of species over a wide range of small island sizes. These deviations

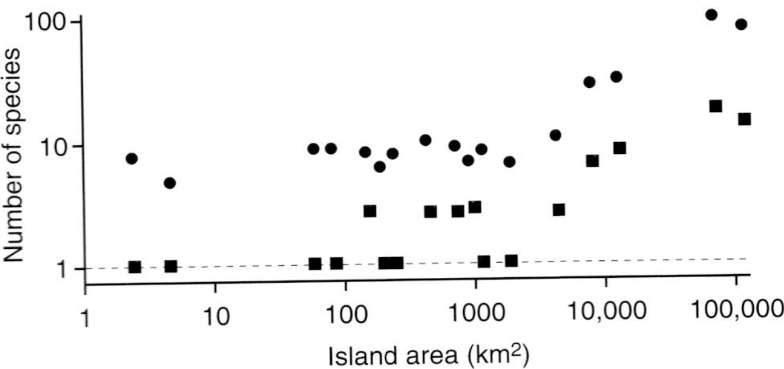

FIGURE 12.6 Relationship between island size and species richness of the *Sphaerodactylus* (squares) and lizard (circles) faunas in the West Indies. Dashed line depicts the islands with only a single species. Data from Schwartz and Henderson, 1991.

result from lower dispersal and extinction rates than the birds and insects from which the theory was developed.

A species-area effect has been proposed for peninsulas as well as islands. The peninsula effect predicts a decline in species richness from a peninsula's base to its tip. Its applicability to amphibians and reptiles remains uncertain. Species diversity does decline for some herpetofaunas (peninsular Florida) but not for others. The reptiles of Baja, California, are as diverse at the tip as at the base.

EXPERIMENTAL STUDIES

During much of its history, the study of ecology was tied closely to the study of natural history, in which detailed field observations were made, often over long time periods, to describe what was going on in nature. The key word is describe, and in a sense, natural history is like reporting; the investigator makes observations, determines which variables are worth quantifying, collects data, and puts together a story based on the information. When such studies are detailed adequately, the "story" is accurate. What is lacking in even good natural history studies are clear explanations of "how" and "why." In short, underlying mechanisms resulting in the often-neat story are speculative at best. Development of analytical technologies no doubt contributed greatly to the use of experiments in ecology, but the basic underpinnings are simple. Based on field observations, the investigator asks a question, articulates the question into a hypothesis, determines predictions based on the hypothesis, and sets up experiments to test the predictions. The experiments are designed specifically to be tested statistically, and the failure of many experiments has resulted from failure to follow the correct statistical design. Although many ecologists resisted the experimental approach in the late 1960s and early 1970s, elegant experiments designed by Dr. Henry Wilbur and others to tease apart underlying mechanisms structuring temporary pond communities using amphibian larvae and aquatic insects changed that. Since then, experimental studies have been directed at understanding nearly every aspect of aquatic and terrestrial amphibian and reptile communities. What started out as relatively simple cattle tank experiments (simulated small ponds) in which numbers of individuals of amphibian larvae and their predators–competitors could be introduced in a nested statistical design, has transformed into large-scale field experiments using enclosures in ponds or on the desert floor manipulating both biotic and abiotic factors.

In natural ponds, an unknown number of variables influence larval survival, and because natural ponds are not uniform in structure, habitat gradients exist that also introduce variation. Artificial ponds can be designed to minimize or eliminate effects of unmeasured variables, inoculated with predetermined densities of potential competitors and predators, and set up in appropriate statistical designs. They are particularly relevant to studies of amphibian larval communities, because amphibian mortality mainly occurs in the larval stage, and mortality can be density dependent, density independent, or a combination of both. Drying of ponds prior to metamorphosis, for example, is density independent, whereas the effect of competition and/or predation in ponds with long hydroperiods (the time period that the pond holds water) is usually density dependent. Moreover, size and time to metamorphosis affect fitness of amphibians because they ultimately affect adult body size.

Early experiments have examined the effects of competition and predation on survivorship, length of the larval period, and size at metamorphosis in six amphibian species. Competition among larvae of three species of *Ambystoma* was evident; each species survived better, metamorphosed more rapidly, and reached a larger size at metamorphosis in the absence of the other two salamander species. Additional experiments revealed more complex interactions when a predator (*A. tigrinum*) was added and when an alternative prey (*Lithobates sylvaticus* tadpoles) for that predator was also added. In the absence of the alternative prey, the predator was a competitor with the other *Ambystoma*. However, in the presence of *L. sylvatica* tadpoles, *A. tigrinum* fed on the frog larvae, grew rapidly, and became a predator on the other species of *Ambystoma*. In another study, increased density of the predacious salamander *Notophthalmus viridescens* in artificial ponds reduced survivorship of the gray tree frog, *Hyla versicolor,* but the surviving frog larvae were larger at metamorphosis because predation reduced larval density, and more resources were available for each individual. In the presence of a competitor, *Pseudacris crucifer*, size at metamorphosis was reduced in *H. versicolor*. These studies, and many others, demonstrate that competition and predation can have a major impact on fitness of amphibian larvae (e.g., competition can negatively influence body size at metamorphosis). Females that metamorphose at a small size will have a reduced clutch size. Competition can also increase time to metamorphosis, which increases the possibility that the pond will dry prior to metamorphosis (density-independent selection). Predation can reduce density, resulting in lower density-dependent mortality. The relatively fewer surviving metamorphs benefit because they have more food and metamorphose at a larger size.

Experimental studies that involve manipulation of natural communities are, by design, much more complex. Nevertheless, several large-scale experiments using enclosures in a natural habitat have produced results similar to artificial pond experiments. Larval *Ambystoma opacum* at high density grew more slowly, metamorphosed at smaller body size, and had lower survival than those that were enclosed at a lower density. Slower growth in the high-density enclosures also increased the probability that all larvae would die due to pond drying. Intraspecific competition in this case was dependent on hydroperiod through its effect on larval density. The intensity of competition also increased risk of predation, because larvae take greater risks to acquire resources when competition is greater. The effects of density on size at metamorphosis translated into measurable effects on adults. Females resulting from larvae that experienced low density were larger when they returned to breed than those raised at high density and, for one cohort, had larger clutch sizes.

North American spadefoot toads (*Spea*) provide a nice example of a system in which experimental manipulations can address important ecological questions. Spadefoots breed explosively when summer rains fill temporary ponds in the arid Southwest. In two species, *Spea bombifrons* and *S. multiplicata* (Fig. 12.7), tadpoles emerge from eggs as typical omnivorous tadpoles feeding primarily on detritus. However, if fairy shrimp (Anostracoda) are present in the ponds, some of the larvae change morphologically, developing a shortened gut, thickened beak, and reduced papillae and labial teeth and become carnivorous. These carnivorous phenotypes do best in ephemeral ponds with high densities of fairy shrimp, and their increased growth rates reduce time to metamorphosis, allowing them to get out of the ponds before they dry up. A series of related experiments conducted by Dr. David Pfennig and his collaborators has elucidated not only the mechanics of the transition from omnivorous to carnivorous tadpoles but also some of the underlying genetics of the system. In addition to each species having the heritable ability to produce carnivorous tadpoles, both species occur together in some ponds, which sets the stage for intense competition between the two. When the two species co-occur, a lower proportion of *S. multiplicata* exhibit the carnivorous phenotype, even though *S. multiplicata* is more abundant in the ponds than *S. bombifrons*. This suggests that *S. bombifrons* effectively reduces carnivore production by *S. multiplicata*. When tapoles of both species were reared in mixed-species experimental tanks and fed fairy shrimp, *S. multiplicata* produced fewer carnivores than expected by chance, whereas S. *bombifrons* produced more,

FIGURE 12.7 Two spadefoot toads that can produce carnivorous tadpoles in response to the presence of fairy shrimp. Left, *Spea multiplicata*; right, *S. bombifrons*.

indicating that each species responds differently not only in carnivore production from when they are reared separately, but also in opposite directions. *S. bombifrons* has a negative effect on *S. multiplicata*'s ability to produce carnivores, whereas *S. multiplicata* has a positive effect on the ability of *S. bombifrons* to produce carnivores.

Additional experiments revealed that character displacement, a shift in the proportion of carnivorous phenotypes produced, occurred in *S. multiplicata* in response to coexistence with *S. bombifrons* in the field (Fig. 12.8). In addition, the preceding feeding experiments show that phenotypic plasticity also exists, in which *S. multiplicata* enhances production of carnivore morphs in *S. bombifrons* and *S. bombifrons* suppresses carnivore production in *S. multiplicata* (Fig. 12.8). Why might both character displacement and a facultative response as the result of phenotypic plasticity occur? First, it is clear from experiments that carnivore phenotypes are produced in both species in the absence of the other only in response to fairy shrimp and that the ability to produce carnivores is heritable. Second, when the two occur together, one (*S. multiplicata*) produces mostly omnivores, even though it is capable of producing carnivores. This offsets resource use such that both species are able to reach metamorphosis; *S. bombifrons* larvae feed on fairy shrimp and *S. bombifrons* larvae feed on detritus. *S. bombifrons* larvae are better competitors for fairy shrimp. The character change (reduction of production of carnivore morphs) in *S. multiplicata* is character displacement, an evolutionary response, and the system meets all criteria of a character displacement hypothesis. Nevertheless, phenotypic plasticity results in immediate shifts in production of larval morphs in both species when they are together (a facultative response). These shifts are likely mediated by proximal cues (Fig. 12.9). The cues may be species-specific chemical cues or cues associated with rapidly changing densities of fairy shrimp. Because *S. bombifrons* is a better competitor for fairy shrimp, they would reduce the density of fairy shrimp rapidly, and *S. multiplicata* may detect this density change and respond by not producing carnivore morphs. In effect, this results in resource partitioning, with *S. multiplicata* feeding on detritus and *S. bombifrons* feeding on fairy shrimp when they occur in the same ponds.

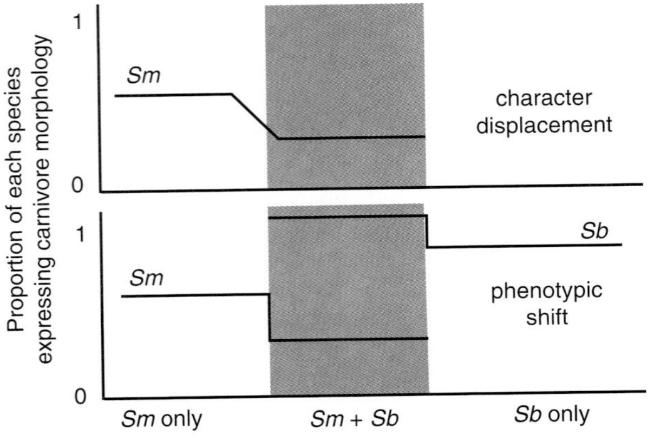

FIGURE 12.8 Character change in two species of *Spea*. In the top panel, *S. multiplicata* responds to the presence of *S. bombifrons* by producing fewer of the carnivore phenotypes, indicating that character displacement has occurred. In the lower panel, production of carnivores by *S. bombifrons* is facultatively enhanced by presence of *S. multiplicata*, and production of carnivores by *S. multiplicata* is facultatively suppressed by presence of *S. bombifrons*. Adapted from Pfennig et al., 2000.

COMPARATIVE STUDIES

Comparative ecological studies that do not involve phylogenies are highly diverse and can be broken down many ways. Often, ecological studies involve comparisons of communities in different places but in similar habitat types, an approach known as the "far-flung" approach (Fig. 12.10). Others follow a single community or assemblage through time (e.g., year after year), the "long-term" approach. Each addresses different questions. Are species assemblages in major habitat types (e.g., deserts or rain forests) structured similarly, or does community structure remain constant through time? We first describe comparative studies across continents and within a particular kind of microhabitat. Each of these illustrates different approaches to the study of communities or species assemblages, and each provides different perspectives on ecological studies in general. We then examine several comparative ecological studies in which phylogenetic hypotheses were brought into play.

Far-Flung Studies

Comparisons of contiguous communities along habitat gradients often provide insight into factors that maintain structure in undisturbed communities. For example, 105 species of reptiles and amphibians that differ in species composition, relative abundance, and microhabitat use occur across three distinct habitats in Borneo. The three environments taken together comprise a gradient from undisturbed broadleaf evergreen forest through deciduous dipterocarp forest, and into agricultural land. The evergreen forest contains the highest diversity of reptiles and amphibians (77 species), and the agricultural area contains the lowest diversity (55 species); the dipterocarp forest is intermediate (67 species). A similar trend occurs in the number of resource states (microhabitats) used by the resident species. Niche breadths are lower in evergreen forest, and the average niche overlaps, or similarities in use of resources, are higher. Niche breadths reflect the relative frequency of use of different microhabitats. Species with low niche breadths use one or a few microhabitats, whereas species with high niche

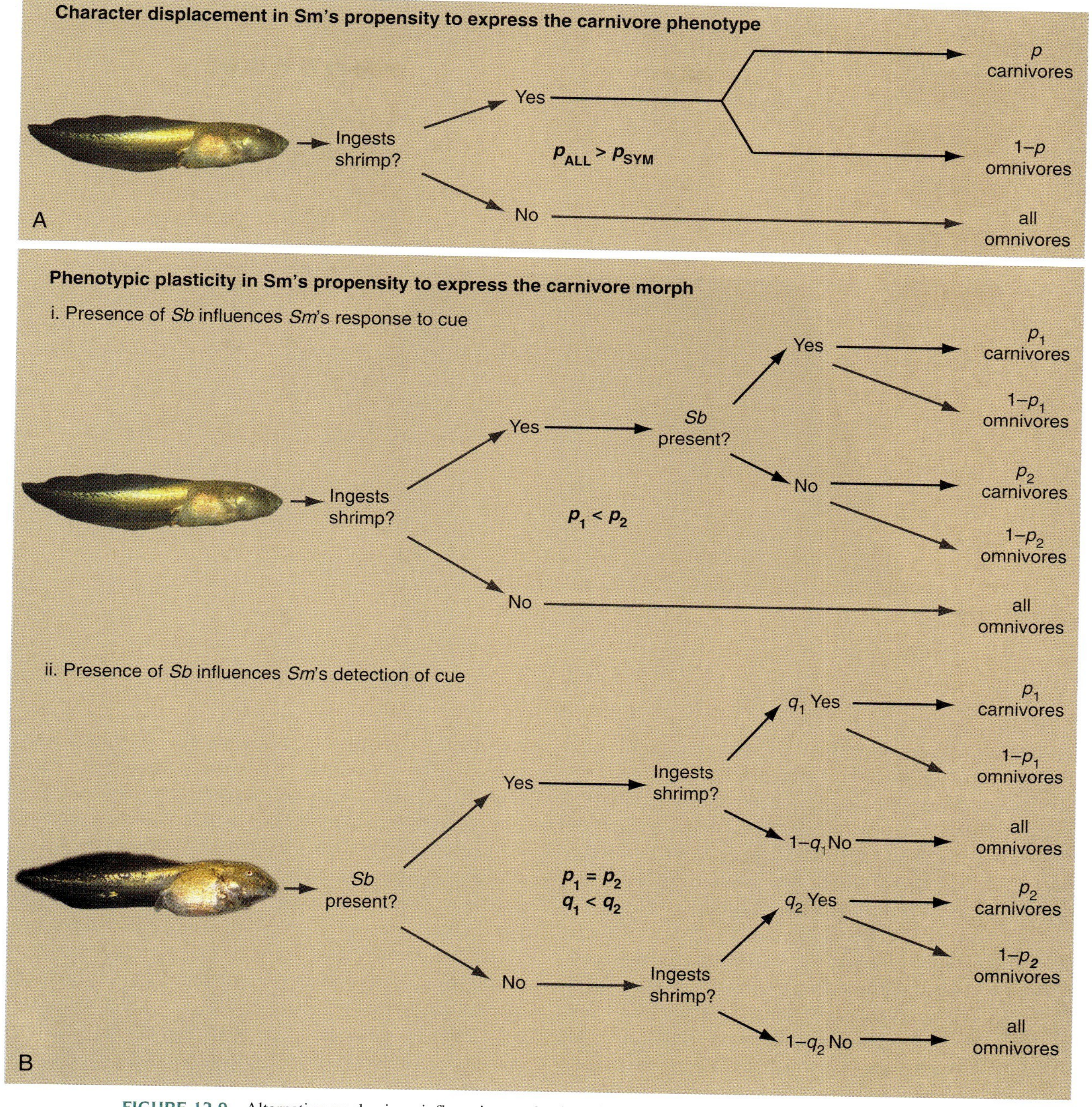

FIGURE 12.9 Alternative mechanisms influencing production of carnivore versus omnivore morphs in *Spea multiplicata* in response to presence of *S. bombifrons*. Tadpoles that ingest fairy shrimp have a probability *p* of producing carnivores and a probability 1−*p* of producing omnivores (the default morph). *S. bombifrons* modifies the morphological switch in *S. multiplicata* by character displacement (A) or phenotypic plasticity (B). In (A), *S. multiplicata* tadpoles have a fixed response to *S. bombifrons*, but the response differs in areas where they don't occur together (allopatric, *P*all) and where they do occur together (sympatry, *P*sym). Phenotypic plasticity results from the ability of *S. multiplicata* to be influenced by *S. bombifrons* in its response to shrimp chemical cues (i.e., interference competition) or ability to detect shrimp chemical cues (B. ii., exploitative competition). Adapted from Pfennig and Murphy, 2000.

breadths might use a large number of microhabitats equally. Overlaps indicate similarity between species in the use of a particular set of resources (microhabitats, in this case). The low niche breadths and higher average niche overlaps indicate that species in the evergreen forest are more similar to each other with respect to resource use than those in the other two habitats. When species are very similar in resource use, they are referred to as tightly packed. Species in the evergreen forest have high overlaps, that is, they are similar to each other in

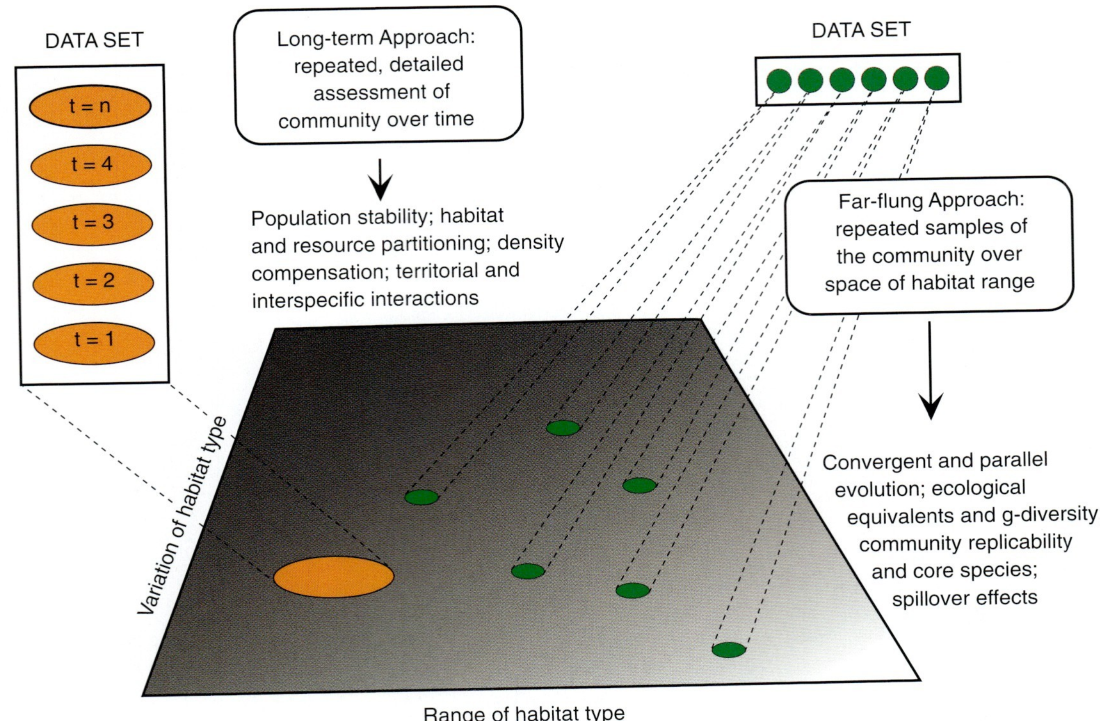

FIGURE 12.10 Graphic representation showing differences between long-term and far-flung studies in animal ecology. Abbreviation: *t* = time. Adapted from Cody, 1996.

microhabitat use, thus forming microhabitat guilds. The guilds are distinctly different from each other (low overlap). For example, one very tight terrestrial guild (low overlaps among members) contains four species of lizards, one frog, and one snake. A riparian guild contains two frogs and a turtle. The conclusion from this study, based on spatial comparisons along a habitat gradient, is that more predictable environments (evergreen forest) promote the formation of guilds that allow greater species richness. Additional information on other niche axes might make the pattern even more clear.

Deserts of the world are well known for their apparent abundance of lizard species, which makes them ideal for conducting ecological studies. Lizards are diverse ecologically, behaviorally, and taxonomically. Moreover, deserts are extreme environments because rainfall is low and highly seasonal, and thus resources are likely limited for extended time periods. Consequently one might expect species interactions to be important in structuring these lizard assemblages and convergence to occur in community structure, because deserts share many abiotic and biotic variables. Dr. Eric Pianka's studies show that climatological, historical, and resource-based differences between continents have shaped desert lizard assemblages in different ways. Nevertheless, average diet and microhabitat niche breadths of lizards are similar among the deserts, and even though the communities cannot be considered convergent, taxonomically unrelated species pairs have converged in morphology and ecology (Fig. 12.11). Differences are apparent in numbers of species, taxonomic composition of communities, and other ecological characteristics (Table 12.4). Even when species composition of identifiably similar microhabitats are compared, striking differences exist. Saltbush (Chenopodiaceae) shrub sites occur in North American, Kalahari, and Australian deserts. Six lizard species live in these sites in North American deserts, 13 in the Kalahari, and 18 in Australian deserts.

Similarly, continental comparisons of the number of species and individuals in tropical forests reveal that Costa

FIGURE 12.11 The agamid lizard *Moloch horridus* (left) of Australia is an ecological equivalent of the iguanid lizard *Phrynosoma platyrhinos* of North America. Both species are ant specialists in arid habitats. Photos by E. R. Pianka (left) and L. J. Vitt (right).

TABLE 12.4 Variation Among Continental Deserts in the Structure of Lizard Communities.[1]

Mode of life	North America			Africa			Australia		
	Mean	Range	%	Mean	Range	%	Mean	Range	%
Diurnal	6.3	4-9	86	8.2	7-10	56	18.1	9-25	60
Terrestrial	5.4	4-7	74	6.3	5.5-7.5	43	15.4	9-23.5	54
Sit-and-wait	4.4	3-6	60	2.4	1.5-2.5	16	5.3	2-7	18
Widely foraging	1.0	1	14	4.0	3-6	27	10.1	4-12	36
Arboreal	0.9	0-3	12	1.9	1.5-2.5	13	2.7	0-5.5	9
Nocturnal	1.0	0-2	14	5.1	4-6	35	10.2	8-13	36
Terrestrial	1.0	0-2	14	3.5	3-5	24	7.6	6-9	27
Arboreal	0.0	–	0	1.6	0.5-2.5	11	2.7	1-4	9
Subterranean	0.0	–	0	1.4	1-2	10	1.2	1-2	4
All terrestrial	6.4	4-8	88	9.8	9-11	67	23	15-34.5	78
All arboreal	0.9	0-3	12	3.5	2-5	24	5.4	1-9	18
Total	7.3	4-11	100	14.7	11-18	101	29.6	18-42	100

[after Pianka, 1985]

[1]*The numbers of species and their modes of life are indicated for each category and are based on multiple sites in each desert. Semi-arboreal species are assigned half to terrestrial and half to arboreal categories.*

Rican forests harbor many more individuals and species of amphibians and reptiles when compared with Bornean forests. In Costa Rica, terrestrially breeding Brachycephalid frogs form a major component of the leaf litter fauna, whereas in Borneo, viviparous skinks appear to have similar ecological roles. Subsequent work in Borneo, Thailand, and Indo-Malayan rain forests also reveals much lower amphibian and reptile densities compared with Costa Rica. Although the differences were initially attributed to differences in routes and rates of energy flow associated with differences in leaf litter, additional data suggest alternative explanations. Later studies in Borneo, Thailand, and the Indo-Malayan rain forest were conducted in areas with high leaf litter, yet amphibian and reptile density remained low. The impact of insect production related to fruiting of dipterocarp trees likely accounts for lower densities in Borneo. Climatic change that has resulted in the habitat becoming drier appears to account for reduced frog density in Thailand. The shorter hydroperiod of temporary breeding pools increases larval mortality. Data were not sufficient to suggest hypotheses to account for low densities in Indo-Malayan forests. This example shows that, on a global level, differences and similarities in community structure may have independent explanations.

Long-Term Studies

Long-term studies yield insights that are not often evident in far-flung studies. Turtles at the E. S. George Reserve in Michigan have been continually monitored since 1964—a unique investigation in which the turtle populations have experienced turnover in investigators rather than the opposite. Beginning in 1953, Dr. Owen Sexton marked and monitored turtles. As each successive researcher moved on, others took over, primarily Drs. Donald Tinkle and Justin Congdon. The turtles continued to be monitored through 2007, although data for the entire period are forthcoming. Populations of three species, *Chrysemys picta, Emys blandingii*, and *Chelydra serpentina*, comprise more than 98% of the turtle community, and these three species have remained more or less stable over 20 years of intensive monitoring (1974–1994). The size of the *Chelydra serpentina* population increased slowly during the 20 years, and *Chrysemys picta* underwent a major population decrease only to recover several years later (Fig. 12.12). A fourth species, *Sternotherus odoratus*, comprised less than 2% of the turtle community. The population of *Sternotherus odoratus*

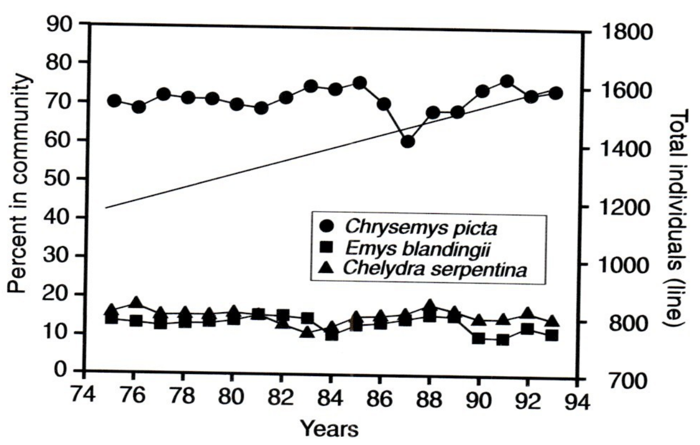

FIGURE 12.12 Annual variation in proportional representation of three turtle species in Michigan, based on capture–recapture studies. Adapted from Congdon and Gibbons, 1996.

disappeared repeatedly as a result of changes in the physical condition of marshes. No evidence suggests that species interactions are important in structuring this turtle community. Rather, environmental factors and intraspecific interactions appear to best explain patterns of population size in these and other turtles studied.

Amphibian populations were monitored continuously from 1979 through 2006 on the Savannah River Plant (SRP) in South Carolina by a large team of scientists led by Dr. J. Whitfield Gibbons. Sixteen years of data on all species that use ponds as breeding sites indicate that the length of time that ponds contain water (hydroperiod) is the primary cause of variation in population levels of the amphibian community, through either its direct effect on larval mortality or its effect on competition and predation. In the driest years, recruitment of juveniles into the population is controlled by the resulting short hydroperiod ($\leq$ 100 days). Larvae do not survive to reach metamorphosis; reproductive failure is complete or nearly complete for all species. In wetter years with longer hydroperiods ($\geq$ 200 days), both the diversity and numbers of metamorphosing juveniles increases. Not all species respond similarly to variation in the length of the hydroperiod. One frog species, *Pseudacris ornata*, actually experienced lower recruitment in years with longer hydroperiods because they were able to use temporary ponds as alternate breeding sites. Longer hydroperiods increase the number and kinds of species interactions of developing larvae. If ponds persist long enough, larval densities increase as do densities of predators, and competition and predation become major factors influencing recruitment. In this complex system, community structure appears regulated by a predictable interaction between rainfall, hydroperiod, competition, and predation.

A long-term study of the red-spotted newt, *Notophthalmus viridescens*, revealed that the population was divided into numerous subpopulations centered around a pattern of breeding ponds in Virginia. Adults are philopatric (they almost always return to their home pond) and when removed, they return to their home pond. As a result, little exchange of genes takes place between subpopulations. Moreover, because breeding success is zero in some ponds, immigrants from other ponds appear responsible for the founding of subsequent populations. Even when large numbers of newts are translocated to other ponds, a majority return to their pond of origin.

An interesting example of indirect effects of nutrient cycling on caiman populations stems from studies in tributaries of the Amazon River in Brazil. Nutrient-poor lakes contain caimans, turtles, and fish. In the forests associated with the tributaries, nutrients are held largely in vegetation and rapidly recycled into plants following decomposition; streams and lakes are often nutrient poor. Annual floods inundate low-lying forests and enlarge forest lakes. Fish that normally live in the main channel migrate into forest lakes to spawn. Unexpectedly, fish diversity and population size have declined with the increased harvest of caimans. When caimans are present, they feed on the adult fish, and their feeding and defecation nearly doubles the amounts of calcium, magnesium, phosphorus, potassium, and sodium in the water, making the system much more productive for hatchling fish and other aquatic organisms. Consequently, removing the caimans interferes with normal nutrient cycling and can negatively affect the entire system.

The existence of maintained field stations and field sites has resulted in an increasing number of studies that are based on repeated sampling, sampling across habitats, and inference from associated experimental studies. At the same site where Sexton, Tinkle, and Congdon monitored turtle populations, Dr. Earl Werner and colleagues conducted a series of studies aimed at understanding why the number of amphibian species varies across environmental gradients. Amphibians inhabiting 37 ponds were monitored over a period of 7 years. Structure (species composition) of the amphibian assemblages varied among 36 of the ponds (Fig. 12.13), and some turnover (replacement) of species occurred in ponds. Species richness was positively associated with pond area and hydroperiod and negatively associated with canopy cover. Most of the species turnover was associated with canopy versus no canopy or fish versus no fish. Additional analyses indicated that pond connectivity (a measure of relative distances to all possible sources [other ponds] of amphibians) contributed to species turnover as well. During the course of the study, an extended drought dried some of the ponds, eliminating fish and providing a natural experiment. The amphibian assemblages in those ponds responded positively when the ponds refilled, increasing in species richness (Fig. 12.14). These results demonstrate first that abiotic rather than biotic factors account for most of the variation among ponds in amphibian species richness; secondly, the natural experiment verifies an observation made in many other studies: Presence of fish negatively impacts amphibian populations.

Historical Ecology Studies

Probably the classic example of the transition of traditional comparative studies to modern comparative studies is the radiation of *Anolis* lizard ecomorphs in the Caribbean. Dr. Ernest Williams identified distinct ecomorphs of *Anolis* lizards in the Caribbean based on a long-term evaluation of patterns of morphological and ecological evolution in island lizards. He observed that morphologically similar but apparently unrelated anole species occupied similar microhabitats on different islands within the Lesser Antilles. The lizards had nearly the same body size, coloration, morphology, and behavior. The combined morphotypes and ecotypes comprised

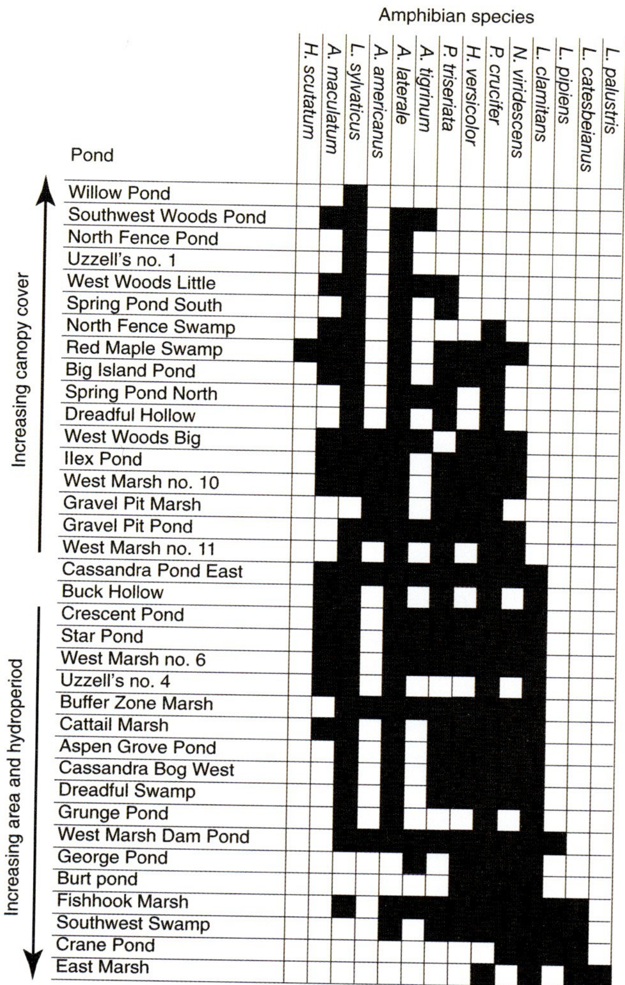

FIGURE 12.13 Incidence matrix showing occurrence of 14 amphibian species across 36 ponds on the E. S. George reserve in Michigan. Blackened squares indicate that a species was present at a particular pond at least once during the 7-year study. Ponds are rank-ordered in relation to two environmental variables: increasing canopy cover and increasing area and hydroperiod. Adapted from Werner et al., 2007.

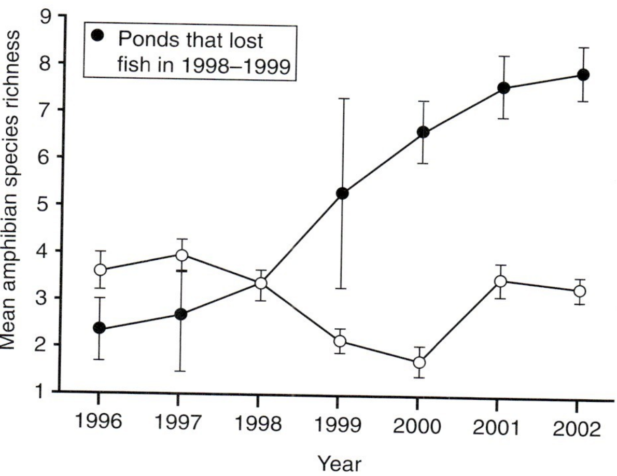

FIGURE 12.14 Annual changes in amphibian species richness for all ponds combined at the E. S. George Reserve in Michigan. Three ponds lost fish in fall of 1998 or in 1999. Species richness of these ponds is indicated by closed circles. Prior to loss of fish, these three ponds had amphibian species richness similar to that of ponds containing fish. Amphibian species richness increased dramatically following loss of fish. Adapted from Werner et al., 2007.

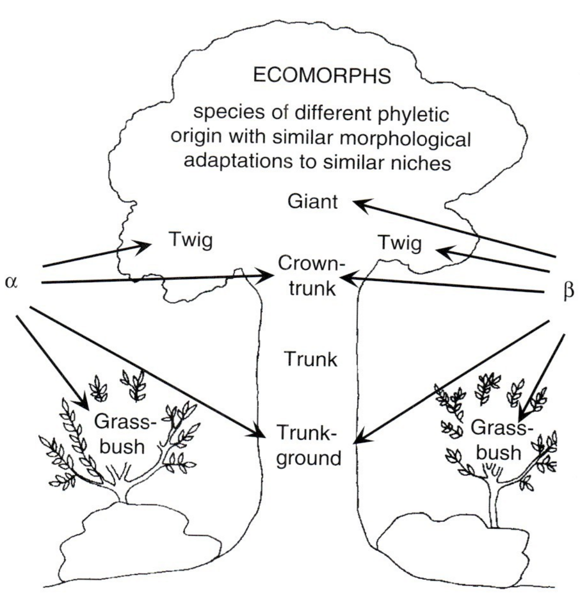

FIGURE 12.15 Ecomorphs of *Anolis* lizards in the Caribbean. α and β indicate different (independent) clades of anoles. Adapted from Williams, 1983.

what have become known as ecomorphs, which are morphologically similar animals of different species living in similar microhabitats (Fig. 12.15). For example, a species that lives in the crown of vegetation and has a specific associated morphology (large body size) is called the crown ecomorph. Most striking was the observation that similar ecomorphs on different islands were not necessarily each other's closest relative. This observation suggested that the evolution of ecomorphs on different islands was independent. Relationships of the anoles at the time were based on morphological data and not well supported. In particular, because morphology was one factor used to determine relationships, the argument that ecomorphs result from independent evolution is somewhat circular.

By examining the topology of morphological and ecological traits on independently derived cladograms based on molecular data, Dr. Jonathan Losos (1992) was able to examine patterns of community structure in both Jamaica and Puerto Rico in an evolutionary framework. As the number of anole species increased in Jamaica (Fig. 12.16), generalist species split into two specialized species, one using the trunk-ground habitat and one using

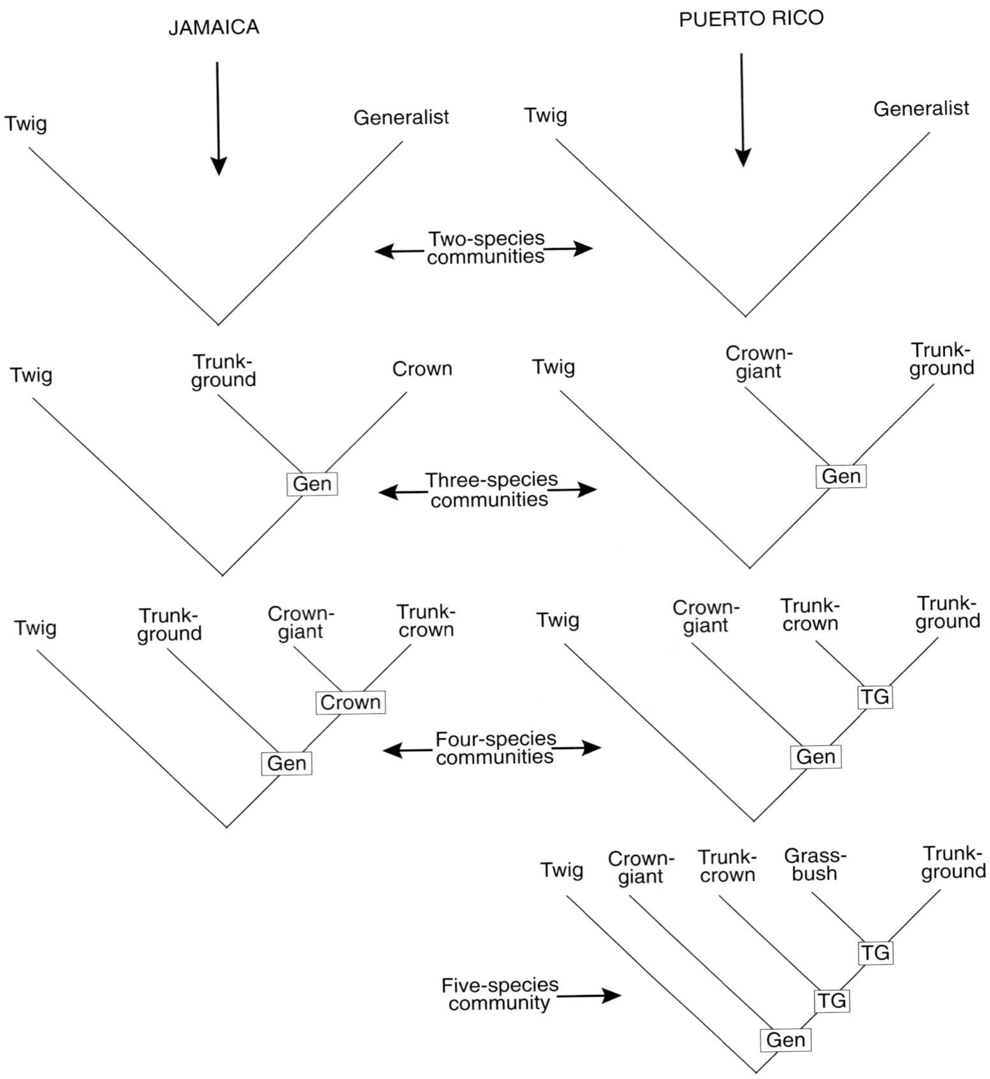

FIGURE 12.16 Patterns of community evolution in Jamaican (left) and Puerto Rican (right) *Anolis* lizards. Progression downward through communities within islands shows the evolution of anole ecomorphs. Comparison across islands (e.g., compare four-species communities) shows that evolution of ecomorphs in *Anolis* living on different islands is convergent. Adapted from Losos, 1992.

crowns of vegetation. The crown lineage then split to form one large-bodied species (crown-giant) and a smaller species that utilizes the trunk-crown interface. In Puerto Rican *Anolis*, community evolution occurred as well, but the pattern of evolution was not identical to that in Jamaica (Fig. 12.16). Similar to Jamaica, the generalist species split into two new species, but one is a crown-giant and the other is a trunk-ground species. The trunk-ground lineage then produced a trunk-crown species. Finally, the trunk-ground lineage produced yet another species, this time a grass-bush species. In both instances, morphology of the lizards is closely related to habitat use. Species that are more arboreal have longer hindlimbs and more streamlined morphology. Most striking is that similar ecomorphs were produced from two initial species (twig and generalist) that are different in Jamaica and Puerto Rico. In the four species assemblages, for example, each island has the same set of ecomorphs but no species are shared. The same ecomorphs evolved independently on each island, showing that the evolution of community structure of *Anolis* in these two islands is convergent. *Anolis* lizards on Caribbean islands are ideal for this kind of comparison because they are abundant and easy to observe and work with, and their relationships are now well known. Moreover, because as a group they share relatively similar morphology, species interactions among anoles has undoubtedly been intense. Evolutionary response should be relatively rapid because they produce individual eggs in rapid succession, most are early maturing and short lived, and generation time is low (an untested idea). Interestingly, ecomorphs are not as clear-cut in mainland Central and South American habitats as they are on islands.

Assemblages in mainland habitats that consist of species in different families and other higher order clades (e.g., those containing iguanians, gekkotans, and autarchoglossans) are much more complex. Although divergence of major anole lineages is old, it is recent compared with family-level divergences in squamates. Moreover, major squamate clades differ morphologically, physiologically, behaviorally, and ecologically from each other in dramatic ways, suggesting that at least some of the differences had origins deep in their evolutionary history. Consequently, the structure of squamate communities that we see today, particularly in mainland habitats, may largely result from events that occurred deep in their evolutionary history. Iguanians and scleroglossans, the two major clades of squamate reptiles, differ dramatically in jaw structure and function, use of the tongue, sensory systems, and foraging behaviors (Fig. 12.17). Iguanians are nearly all diurnal and most use elevated perches (rocks, tree trunks, limbs), whereas scleroglossans are both nocturnal and diurnal and some use elevated perches whereas others do not (Fig. 12.18). Within scleroglossans, gekkotans differ from autarchoglossans in several important ways as well. Both use jaw prehension for capturing prey, which frees the tongue from involvement in feeding. Gekkotans use their tongues to clean their spectacles and lips, and they discriminate prey using their olfactory chemosensory system. Autarchoglossans use their tongues and their well-developed vomeronasal system to discriminate prey, and they are in general more active lizards that gekkotans. Most gekkotans use elevated perches and are nocturnal, whereas most autarchoglossans do not use elevated perches and are diurnal (Fig. 12.18). Scleroglossans dominate lizard faunas throughout the world, suggesting the possibility that they are better competitors than iguanians, at least in terrestrial environments. The possibility exists that one of the reasons that iguanians use elevated perches and gekkotans use elevated perches and/or are nocturnal is the long history of interactions with autarchoglossans. Both may have been pushed out of terrestrial microhabitats. In addition, the combined set of traits shared by autarchoglossans clades has allowed them to diversify into subterranean microhabitats in which they have repeatedly evolved limblessness or near limblessness.

Diets of lizards differ among major clades, and at least six dietary shifts have occurred during the evolutionary history (Fig. 12.19; see also Chapter 10). The ability to discriminate prey based on chemical cues in scleroglossans may have allowed them to exclude many insects, such as ants and beetles, that contain chemical defenses that interfere with metabolism (e.g., alkaloids), or it may have allowed them to select more profitable prey with the same result. This example shows that major ecological differences among clades of squamates had their origins deep in the evolutionary history of squamates, and these rather large differences may partially account for the high species richness of many squamate assemblages throughout the world.

New World snakes reach their highest species diversity in tropical forests, and ecological diversity among tropical snakes is also high. Snake species feed on a wide variety of other organisms and live in nearly every imaginable microhabitat. Most tropical New World snakes are colubrids that are represented by three evolutionary clades, the colubrines, the South American xenodontines, and the Central American xenodontines. Central American xenodontines overall are smaller in body size than South American xenodontines, which are smaller than colubrines. The three clades differ in their relative use of microhabitats and prey types as well. Consequently, at least some of the ecological differences among species across the three clades can be attributed to historical factors. This hypothesis is strengthened by the observation that the three clades evolved in separate biogeographic centers (i.e., their evolutionary histories are different). A comparison of the structure of snake communities (colubrids only) from northern Central America to southern South America reveals that the proportion of the community represented by each of these three clades changes dramatically from north to south (Fig. 12.20). In northern localities, South American xenodontines are poorly represented, whereas Central American xenodontines are poorly represented in southern localities. Because the lineages differ considerably in ecological characteristics, Neotropical snake communities at various localities differ because of the nonequal representation of the three lineages. For example, most South American xenodontines are terrestrial, fossorial, or aquatic, whereas most Central American xenodontines are fossorial, cryptozoic, or arboreal, in that order. None of the colubrines is arboreal or aquatic and only one is fossorial. Likewise, most of the South American xenodontines eat snakes, all of the Central American xenodontines eat earthworms or mollusks, and none of the colubrines eats either of these prey categories—but all of them eat arthropods. Some similarities in diets cut across clades. For example, species from all three clades feed on lizards, amphibians, and fish. Consequently, conclusions about a specific Neotropical snake community must include a consideration of the composition of the snake community and the history of each clade. Ecological factors may play a role in determining patterns of resource use within the specific community, but only within constraints imposed by the history of the clade represented.

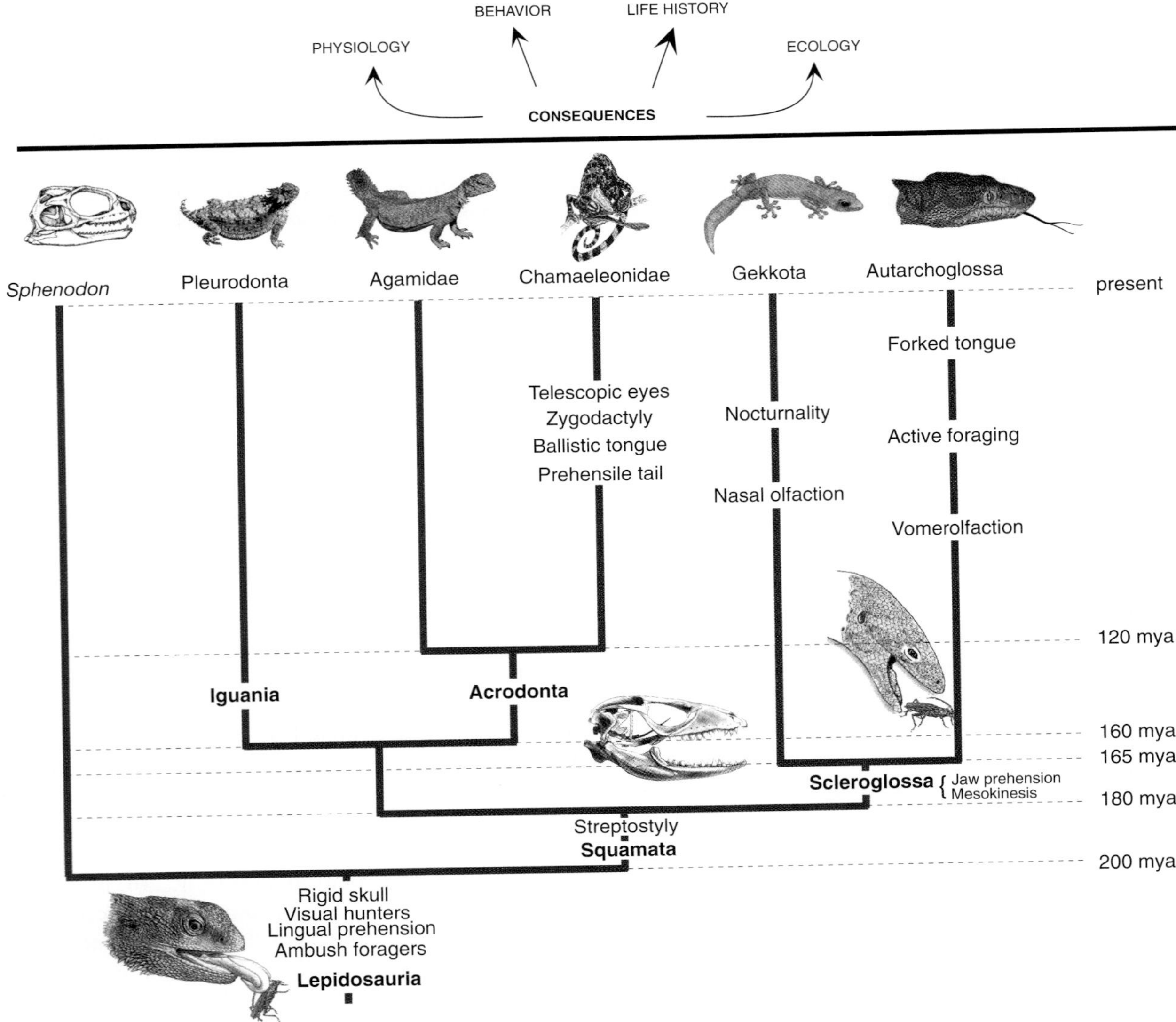

FIGURE 12.17 Diagram of the major events that occurred during the evolutionary history of squamate reptiles that affect their present-day ecology. Streptostyly, the hanging-jaw mechanisms set squamates apart from their sister taxon Rhynchocephalia. This allowed greater jaw mobility. The switch from lingual prehension (ancestral) to jaw prehension in scleroglossans freed the tongue for other uses. In gekkotans, the tongue was used as a windshield wiper for the spectacle over the eye and to clean the lips, but they also developed an effective nasal olfactory system. In autarchoglossans, the tongue and vomeronasal organ provided an effective chemosensory system, allowing prey discrimination among other things. Other correlates include differences in foraging mode, behavior, and physiology (see also Chapters 6-11). Adapted from Vitt et al., 2003.

An important point to keep in mind with respect to all ecological research involving phylogenetic hypotheses is that phylogenies are just that—hypotheses. The possibilities exist that relationships may change dramatically as new genes are used in molecular studies, new techniques appear, and better analyses are developed. For example, consider the global lizard studies previously discussed. Several recent molecular studies using nuclear genes suggest that the classical Iguania–Scleroglossa (Gekkota + Autarchoglossa) relationship of squamates may be incorrect (see Chapters 20 and 21). If these proposed relationships are established with additional data, especially independent data sets, interpretations of ecological scenarios will by necessity need to be revisited. Although major ecological differences among clades will likely hold, the pathways by which those differences evolved will change. This highlights the complexity of ecology as a science.

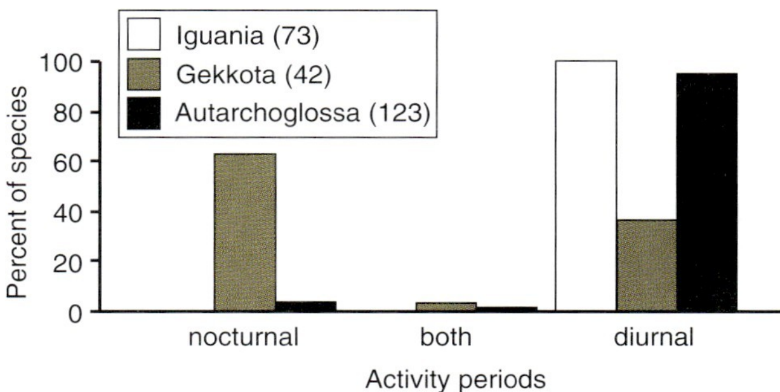

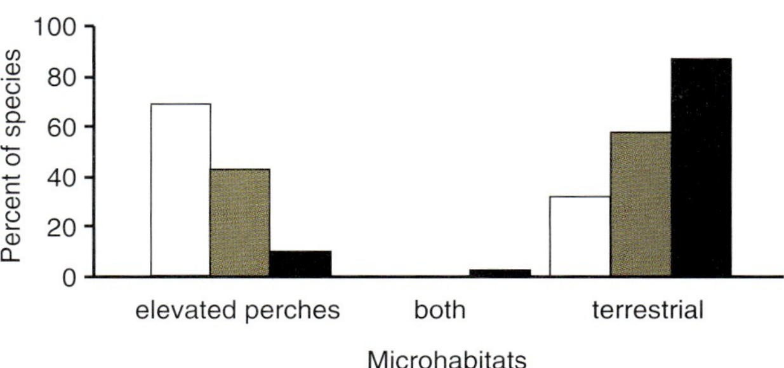

FIGURE 12.18 Differences in activity periods (top) and microhabitat use (bottom) exist between the three major squamate lizard clades. Adapted from Vitt et al., 2003.

NEW ANALYTICAL TOOLS

As the result of advances in technology and new applications of old technology, new directions have emerged recently in ecology. One of these is ecological niche modeling, which uses existing distributional data based on spot localities of species to generate climate, vegetation, or geophysical models predicting where a given species should occur. The basic idea is that the model identifies attributes of the environment that should be correlated with niche parameters of individual species. Thus the model delimits the fundamental niche of a species (i.e., the set of attributes within which the species could exist), and that is used to generate a potential distribution. In its simplest application, niche modeling generates distribution maps that fill in gaps in the known distribution. Conceptually, the process is straightforward. A data matrix is constructed consisting of GIS (Geographic Information System) layers comprised of variables of interest with values corresponding to each spot locality for the species. GIS variables can include topography, climate hydrology, land cover, or anything that might be available in digital format. Using the Genetic Algorithm for Rule-Set Prediction (GARP), which is an evolutionary computing system that learns as it computes, initial rules are set based on relationships of the variables. The analysis then proceeds through an iterative process by which rules are randomly selected, applied, perturbed, and tested with rejection and acceptance improving the genetic algorithm. Once rules converge, the evolutionary process stops and a model of the fundamental niche for the species of interest is produced. The model then locates all areas on the ecological landscape of variables that fit the model to predict the potential distribution of the species. Applications of niche modeling are just beginning to be appreciated. It can be used to predict distributions of species in fragmented landscapes, to search for biodiversity hotspots, to examine effects of habitat gradients on species distributions, and to generate hypotheses on potential species interactions. With historic climatic data, it can also be used to examine the potential effects of global change on species distributions. Amphibians and reptiles are particularly conducive to the use of ecological niche modeling because they are ectothermic and thus respond to variation in the physical environment. Moreover, use of ecological niche modeling has made museum collections much more valuable in that they contain the distributional data, and often historical distributional data, necessary to test predictions about distributions, past, present, and future. We provide several examples to follow, each of which has focused on a different application of ecological niche modeling.

Madagascar contains a unique herpetofauna with high endemicity. Deforestation has reduced natural habitats drastically, and much of the deforestation occurred before thorough surveys of the fauna and flora were conducted. Dr. Chris Raxworthy and his colleagues used distributional data on 11 species of Madagascar chameleons to generate niche models to predict their distributions across the fragmented landscape. Madagascar chameleons in the genera *Calumma, Fucifer*, and *Brookesia* are conservative in their niche requirements, and as a result, ecological niche modeling reliably predicted species distributions. More importantly, the niche models identified areas suitable for chameleons that had not been examined. When two chameleon species that do not occur together are ecologically similar, niche modeling over-predicts distributions. Comparisons of niche models for known species identified three regions in Madagascar in which over-predictions occurred. When visiting two of these, one in western and the other in northeast Madagascar, seven new locally endemic chameleons were discovered. Thus, niche modeling identified accurately regions in which new species could be found.

Steep habitat gradients offer a natural experiment for using niche modeling. Distributions of many species occur along steep habitat gradients, and in many instances, one species is replaced by another that is closely related. Along such gradients, distributions of two closely related species can be nonoverlapping or overlapping. If they are nonoverlapping, it may be because of differing niche requirements (abiotic factors), or it may be the result of interference competition (biotic factors) with one species excluding the other in the area of overlap. Use of

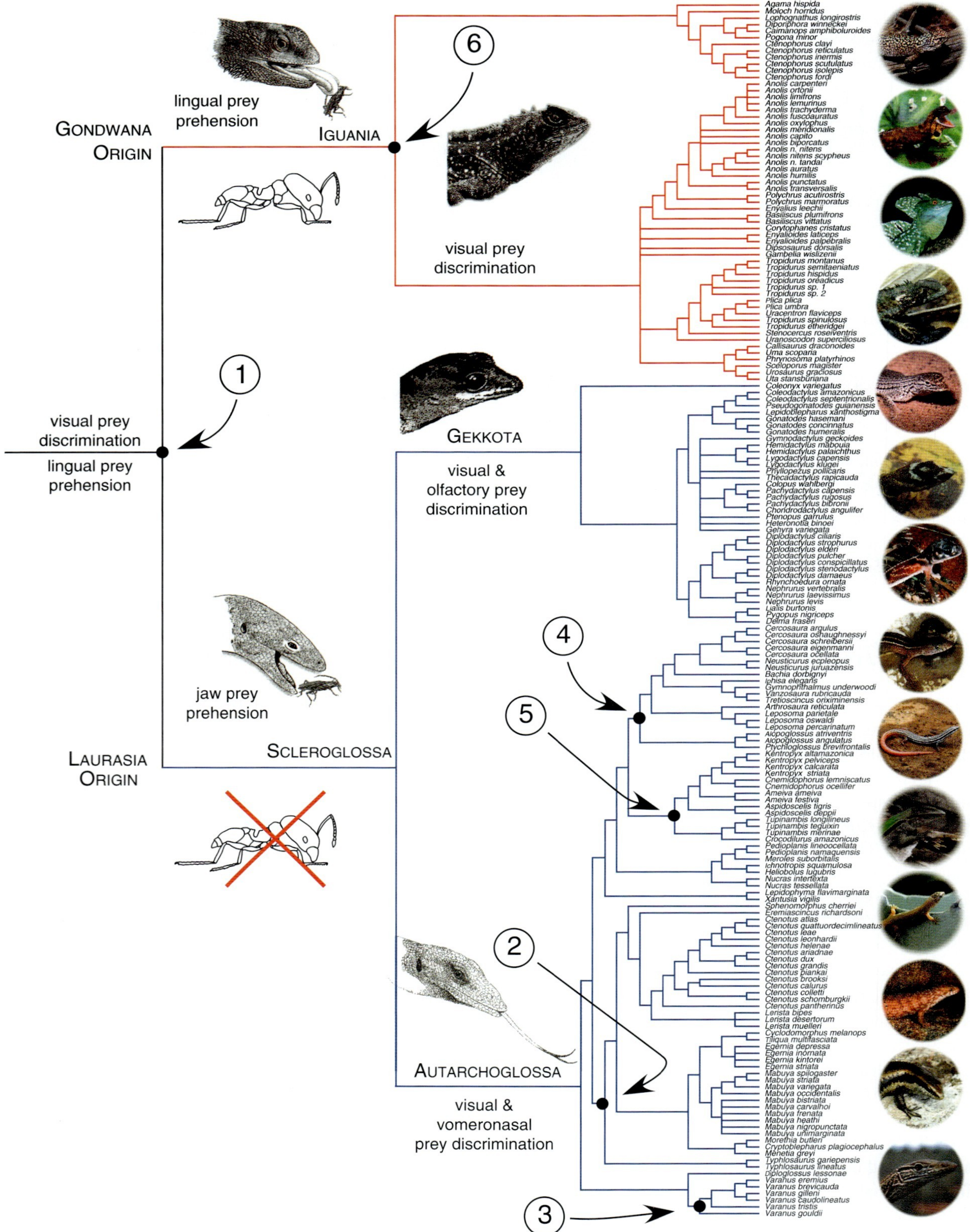

FIGURE 12.19 Major dietary shifts occurred during the evolutionary history of squamate lizards. Each numbered divergence point represents a statistically significant dietary shift, and taken together, these six divergence points account for 80% of the variation among species in diets. Adapted from Vitt and Pianka, 2005 (© 2005 National Academy of Sciences, USA), and 2007.

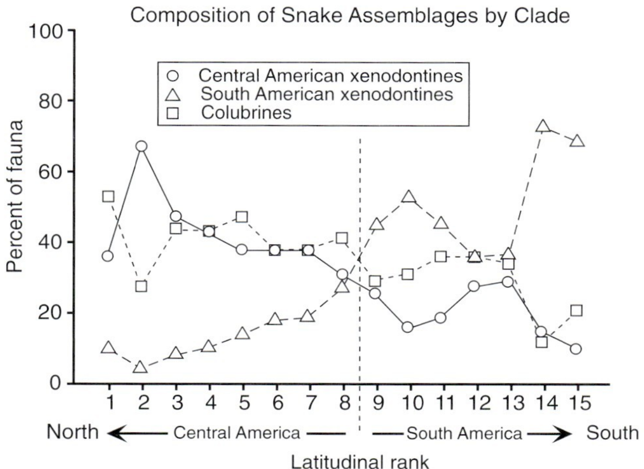

FIGURE 12.20 Changes in the proportional representation of three clades of colubrid snake species in communities across a latitudinal gradient from northern Central America to southern South America. The points at each north–south location represent specific snake communities. For example, the first set of points represents the snake fauna at Los Tuxtlas, Mexico, where 36% are Central American xenodontines (circle), 11% are South American xenodontines (triangle), and 53% are colubrines (square). Numbers along the x-axis represent rank order of localities along the north–south gradient. Values on the x-axis do not correspond to the original categories used by Cadle and Greene (1993). Ranking by latitude changed the positions of the five sets of points at the southern latitude end of the graph. The dashed line separates Central American from South American localities. Adapted from Cadle and Greene, 1993.

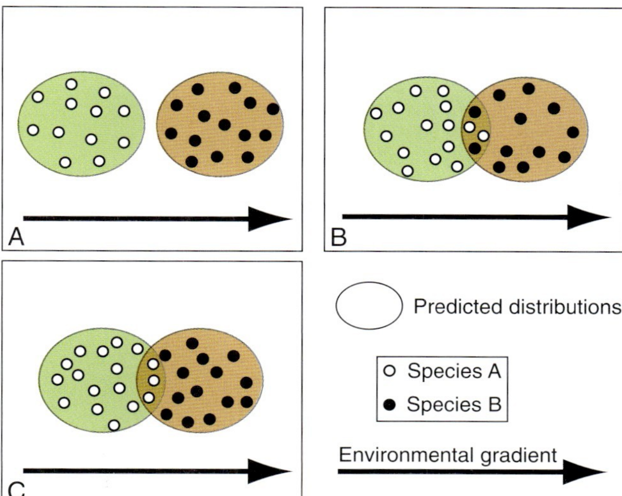

FIGURE 12.21 Three possible outcomes of using niche modeling to examine distributions of closely related species across environmental gradients. (A) No overlap in predicted distributions, (B) both species occurring in the area of predicted overlap, and (C) one of the species occurring in the area of predicted overlap. Adapted from Costa et al., 2007.

niche modeling to predict distributions provides clues about potential causes of differences in distributions and can be used to generate and ultimately test hypotheses (Fig. 12.21).

A steep environmental gradient exists in the south central United States as the eastern deciduous forest to the east gives way to the central plains. Most of the gradient occurs in Oklahoma (Fig. 12.22), and as a result, Oklahoma has high amphibian and reptile species richness with many eastern species replaced by closely related western species across the gradient. The gradient is reflected by many variables, including vegetative (plant species, richness, and structure), soil (clay in the east, sandy in the west), elevational, and climatic (rainfall, temperature, predictability of rainfall, and others). To the east and west of Oklahoma, the gradients level off. Six pairs of closely related amphibian and reptile species have distributions that end along this gradient. The west–east species pairs are *Plestiodon obsoletus* and *P. fasciatus*; *Gastrophyrne olivacea* and *G. carolinensis*; *Lithobates blairi* and *L. sphenocephala*; *Scaphiopus couchii* and *S. hurteri*; *Sistrurus catenatus* and *S. miliarius*; and *Tantilla nigriceps* and *T. gracilis*. Gabriel Costa and collaborators used GARP to generate niche models for the species of interest. To construct the models, they included distribution data from across the species ranges (including outside of Oklahoma) and started with 20 climatic variables available as GIS layers. A lower number of climatic variables contributed most to the niche models. Niche modeling accurately predicts where each species should occur and identifies zones of potential species overlaps that vary from narrow (the two species of *Sistrurus*) to very wide (the two species of *Tantilla*). Potential zones of species overlap coincided with the center of the environmental gradient, supporting the hypothesis that steep gradients influence distributions. None of the species pairs exhibited completely nonoverlapping distributions. Consequently, abiotic factors alone within the predicted zones of overlap are not likely to limit distributions (see Fig. 12.21A). Because much is known about the ecology of these species, reasons for broad overlap can be easily hypothesized. For example, the two skinks, *Plestiodon obsoletus* and *P. fasciatus*, have very different microhabitat requirements. Climatic variables may limit their distributions across Oklahoma, but within the zones of overlap, *P. obsoletus* is found in patches of prairie, usually where rocks are present, and *P. fasciatus* is found in patches of forest, with or without rocks. For *Gastrophyrne olivacea* and *G. carolinensis,* even though their predicted distributions overlap considerably, the overlap zone is dominated by *G. olivacea*, suggesting that species interactions occur. Likewise, *Scaphiopus couchii* occurs throughout the overlap zone, but *S. hurteri* has a much more restricted distribution within the zone, suggesting that species interactions may occur. Patterns of potential and actual distribution identified in this study can be used to generate and test hypotheses about mechanisms underlying overlaps

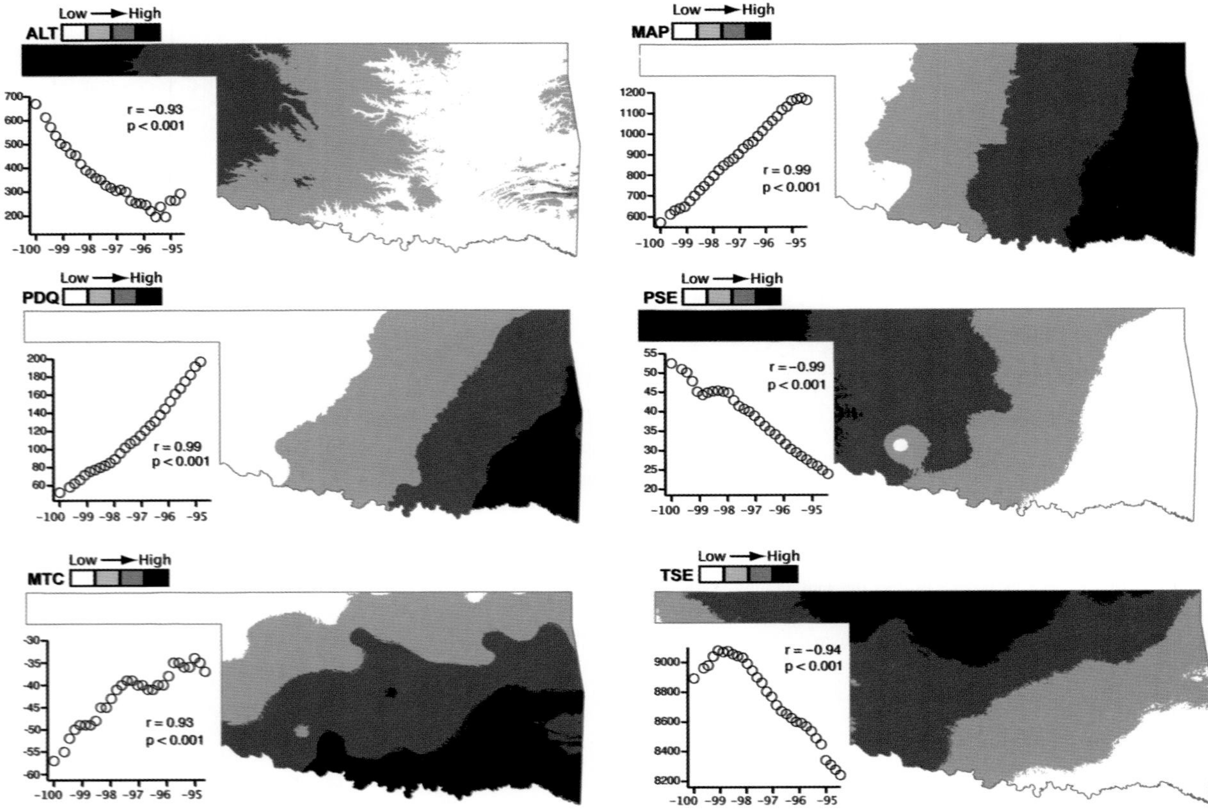

FIGURE 12.22 East–west gradient in six environmental variables used to model fundamental niches of six species pairs of amphibians and reptiles. Variables shown are (a) altitude, (b) mean annual precipitation, (c) precipitation of the driest quarter, (d) precipitation seasonality, (e) minimum temperature of the coldest month, and (f) temperature seasonality. Temperature variables are in °C, precipitation variables are in mm, and altitude is in m. Adapted from Costa et al., 2008.

in distribution. For example, does character displacement in ecological traits exist in overlap zones where each of a species pair coexist? If so, is character displacement manifested in shifts in microhabitat use, body size, prey types, or other quantifiable variables?

The kinds of data used to generate ecological niche models can influence the resultant distributions. Drs. Antoine Guisan and Ulrich Hofer used GIS data with a general linear modeling (GLM) procedure to predict distributions of 13 species of lizards and snakes in Switzerland. They compared niche models from climatic GIS data with niche models from topographic GIS data and found that climate was a much better predictor of species distributions than topography. Both kinds of data did a fairly good job of predicting distributions for 12 of the 13 species, and climatic data explained the distribution for the other one (*Coronella austriaca*). For 3 species that occur across Switzerland, *Anguis fragilis, Natrix natrix*, and *Coronella austriaca*, GLM models performed poorly, possibly indicating that biotic factors (e.g., food) might determine distributions rather than abiotic factors. The GLM procedure differs from the GARP procedure previously outlined in that it simply constructs the best-fit model from the GIS data set, whereas GARP proceeds through a learning process, adjusting rules as it eliminates relationships of variables that do not add to the model. Consequently, GARP and other methods based on iterative rule testing do a much better job of predicting distributions based on GIS data.

QUESTIONS

1. What is the significance of the occurrence of similar types of *Anolis* ecomorphs on different islands in the Caribbean?
2. Describe in some detail a reptile or amphibian example of (a) a far-flung approach to studying a community and, (b) a long-term approach to studying a community.

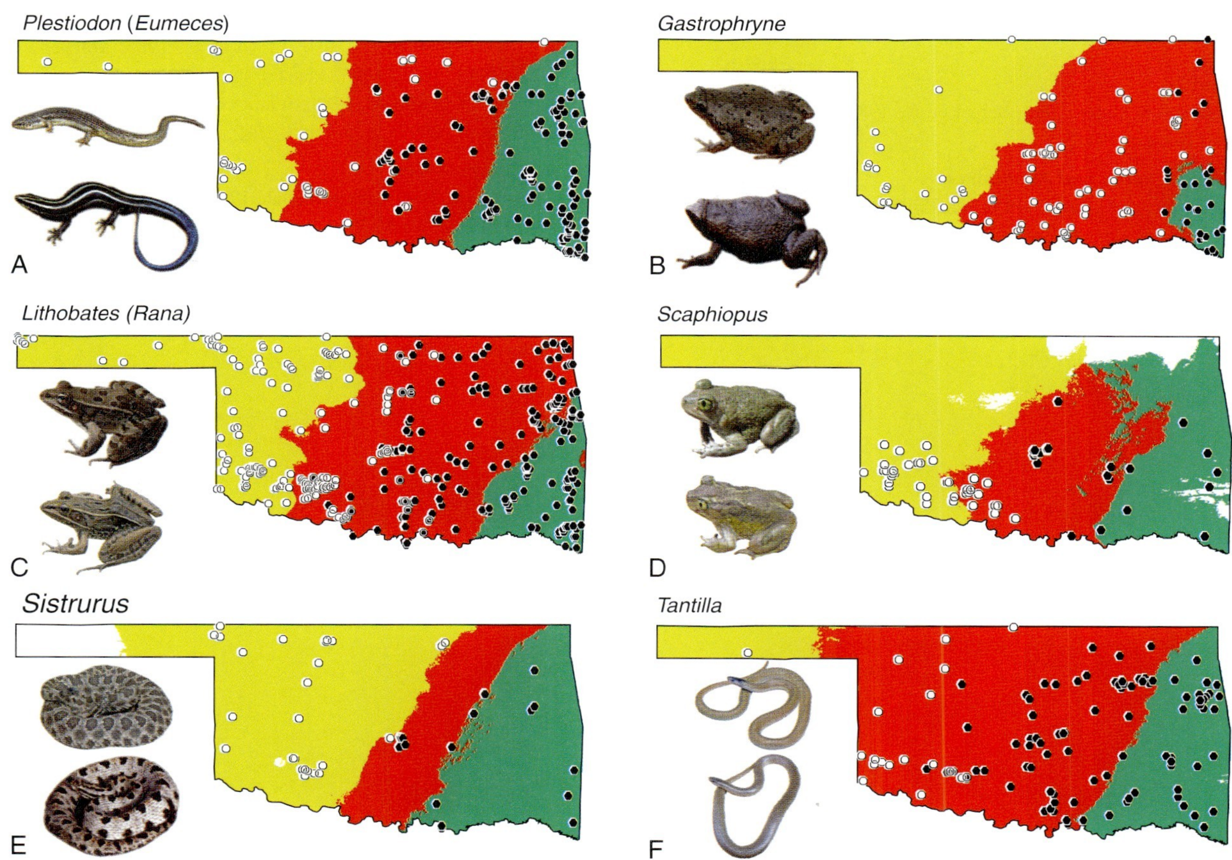

FIGURE 12.23 Distribution maps for six species pairs of amphibians and reptiles, based on niche modeling. Open circles indicate species with western distributions; closed circles represent species with eastern distributions. Yellow indicates distribution of the western species within Oklahoma, blue indicates distribution of the eastern species within Oklahoma, and red indicates an overlap zone, all based on niche modeling. Adapted from Costa et al., 2008.

3. What is it about Oklahoma that makes it an ideal place to test hypotheses about species distributions using amphibians and reptiles?
4. Explain how phylogenies can be used to understand global patterns of lizard ecology.
5. What is the role that food resources play in interactions between tadpoles of *Spea multiplicata* and *Spea bombifrons*?

ADDITIONAL READING

Cody, M. L., and Smallwood, J. A. (Eds.) (1996). *Long-term Studies of Vertebrate Communities*. Academic Press, San Diego.

Huey, R. B., Pianka, E. R., and Schoener, T. W. (Eds.) (1983). *Lizard Ecology. Studies of a Model Organism*. Harvard University Press, Cambridge, Mass.

Hutchinson, G. E. (1957). Concluding remarks. *Cold Spring Harbor Symposia on Quantitative Biology* **22:** 415–427.

MacArthur, R. H. (1984). *Geographical Ecology: Patterns in the Distribution of Species*. Princeton University Press, Princeton, NJ.

Pianka, E. R. (1986). *Ecology and Natural History of Desert Lizards. Analyses of the Ecological Niche and Community Structure*. Princeton University Press, Princeton.

Pianka, E. R., and Vitt, L. J. (2003). *Lizards. Windows to the Evolution of Diversity*. University of California Press, Brekeley, CA.

Reilly, S. M., McBrayer, L. B., and Miles, D. B. (Eds.) (2007a). *Lizard Ecology: The Evolutionary Consequences of Foraging Mode*. Cambridge University Press, Cambridge, Mass.

Seigel, R. A., and Collins, J. T. (Eds.) (1993). *Snakes: Ecology and Behavior*. McGraw-Hill, Inc., New York.

Vitt, L. J., and Pianka, E. R. (Eds.) (1994). *Lizard Ecology: Historical and Experimental Perspectives*. Princeton University Press, Princeton, NJ.

Wells, K. D. (2007). *The Ecology and Behavior of Amphibians*. University of Chicago Press, Chicago.

Williams, E. E. (1983). Ecomorphs, faunas, island size, and diverse end points in island radiations of *Anolis*. In *Lizard Ecology: Studies of a Model Organism*. R. B. Huey. E. R. Pianka, and T. W. Schoener (Eds.). pp. 326–370. Harvard University Press, Cambridge, Mass.

REFERENCES

General

Ricklefs and Schluter, 1993; Seigel et al., 1987; Wilbur, 1987; Wilson and Bossert, 1971.

Species Richness and Abundance

Andrewartha and Birch, 1954; Grismer, 2002; MacArthur, 1984; MacArthur and Wilson, 1967; Pianka, 1975, 1985, 1996; Schall and Pianka, 1978;

Experimental Studies

Morin 1987; Scott, 1990, 1994; Semlitsch et al., 1988; Wilbur, 1972; Wilbur and Fauth, 1990.

Comparative Studies

Cody, 1996; Congdon and Gibbons, 1996; Ernst, 1971; Fittkau, 1970; Gill, 1978, 1979; Inger, 1980; Lloyd et al., 1985; Mitchell, 1988; Pianka 1973, 1975, 1985; Pianka, and Parker, 1975; Scott, 1976; Semlitsch and Gibbons, 1988; Semlitsch et al., 1996; Sexton, 1959; Werner et al., 2007a, b; Zweifel, 1989.

Historical Ecology Studies

Cadle and Greene, 1993; Jackman et al., 1999, 2002; Losos, 1992; Losos et al., 1998; Schwenk, 2000; Townsend et al., 2004; Vitt and Pianka, 2005, 2007; Vitt et al., 1999, 2003; Williams 1972, 1983.

New Analytical Tools

Costa et al., 2008; Guison and Hofer, 2003; Raxworthy et al., 2003; Stockwell and Peters, 1999.

Chapter 13

Biogeography and Phylogeography

Biogeography is the study of distributions of animal and plant species across the planet and through time. Questions asked by biogeographers center on where animals and plants live, their relative abundances, and the underlying causes of their distributions and abundances. Biogeography has always been a historical science in that distributions change through time, and biogeographers have sought to explain these changes. Prior to acceptance of continental drift, most historical explanations centered on land bridges and rising and falling of oceans, at least for terrestrial organisms. The German scientist Alfred Wegener was the first scientist to use the term continental drift, but he was not the first to suggest that continents might have been connected in the distant past. Even though he understood that continents had to have drifted apart, he was unable to provide a reasonable mechanism for continental drift. In the mid-20th century, the theory of continental drift was superceded and replaced by the theory of plate tectonics, which provided a clear mechanism accounting for the drifting of continents.

Although we will not detail the theory here, the basic elements of continental drift are that continents ride on massive plates that slowly drift across the surface of the Earth. This surface, the lithosphere, consists of seven large tectonic plates as well as many smaller ones and is relatively viscous. The lithosphere rides on a denser layer, the athenosphere. Boundaries of the drifting plates converge, diverge, or transform, depending on forces underneath the plates. Convergence occurs when two plates come together, often resulting in production of mountain ranges. Divergence occurs as plates drift apart, causing formation of ridges and deep valleys on the ocean floor. Transformations occur when one plate rides up over another. Subduction is a term used to describe the process by which one of the plates moves under the other. Earthquakes and volcanic eruptions are instantaneous (geologically speaking) results of major plate movements, and major changes in the distribution of landmasses are long-term results of plate movements. During the history of the Earth, landmasses have coalesced and drifted apart many times. Because the evolution of life was occurring as continents moved, studies of biogeography changed radically when continental drift and plate tectonics were accepted by the scientific community. Land bridges were no longer the only available explanation for global patterns of distribution. Clearly some taxa had drifted with the pieces of their original turf. Land bridges of course were involved in some colonizations and recolonizations, but they were no longer the only explanations for patterns of distribution. The historical pattern of continental movements is most relevant to biogeography. Prior to the origin of ancestors of amphibians and reptiles (approximately 350 mybp), two large continents (Gondwana and Old Red Sandstone continent) existed. Old Red Sandstone continent combined with several smaller continents to form Laurussia, which drifted toward Gondwana. By approximately 250 mybp, a single supercontinent Pangaea existed, which resulted from the collision of Laurussia and Gondwana. Through time, Pangaea split into two large continental masses, Gondwana and Laurasia, each of which split further into the continents that we see today (Fig. 13.1).

Although patterns of distribution have interested scientists for at least 200 years, biogeography studies prior to about 1967 were largely descriptive. When Robert MacArthur and Edward O. Wilson published their classic book *The Theory of Island Biogeography* in 1967, the

Herpetology, 3rd Ed.

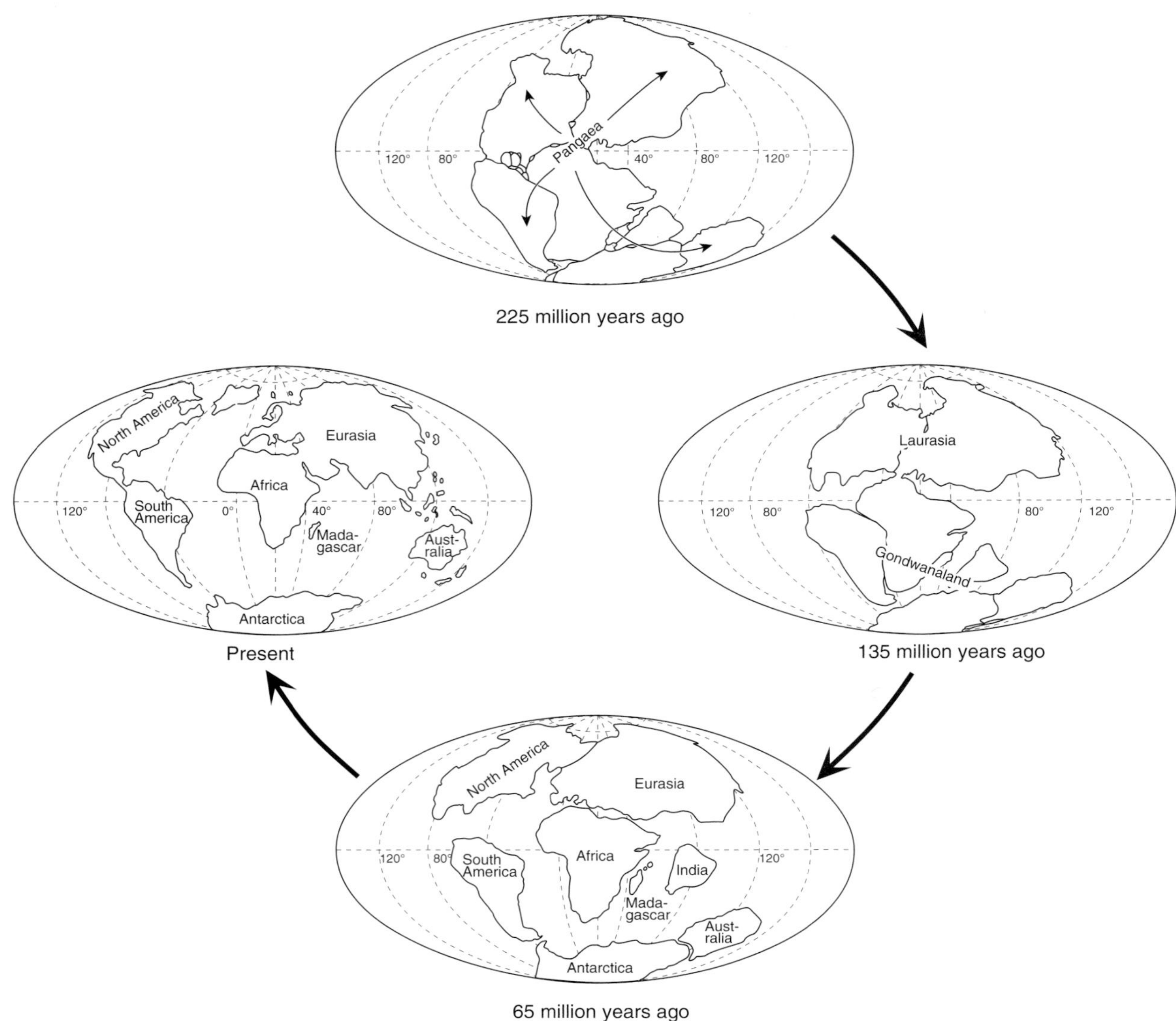

FIGURE 13.1 Diagrammatic reconstruction of the history of continental drift.

field changed dramatically, and biogeography became a predictive science rather than simply a descriptive one. We already introduced some elements of island biogeography in the last chapter, in an ecological context. To reiterate the key points, the theory of island biogeography posits that the number of species in a given area can be predicted based on a few key variables: island size, immigration rate, and extinction rate. Most studies of island biogeography reveal a close association between the number of species and island size, if all else is equal. Distance of islands from mainland habitats (sources of species) also affects number of species (richness). The farther an island is from the mainland, the fewer species are likely to be present, and this variation is largely a result of reduced immigration rates or increased extinction rates. Island biogeography has much wider application because habitats distributed across the Earth are patchy, and each patch can effectively be considered an island. Thus the underlying hypotheses associated with the theory of island biogeography can be tested in studies of aquatic, marine, and terrestrial organisms. The theory of island biogeography can be applied to ecological questions, particularly in landscape ecology and conservation biology. Unlike plate tectonics, land bridges, and changes in ocean levels, island biogeography centers on patch size and distance from sources and is most applicable to predicting numbers of species and immigration and extinction rates in a relatively short time period (ecological time). Historical biogeography (plate tectonics and its correlates) centers on long-term (geologic time) patterns of distribution and diversification and thus correlates origins and patterns of diversification of faunas with the geologic history of the Earth.

Late in the 20th century, with the rapid development of molecular systematics, historical biogeography was, for all practical purposes, reinvented. The potential existed to estimate time periods during which major evolutionary shifts occurred in a particular group of organisms based on calibrated molecular clocks and then fit patterns of divergence to independently derived estimates of time based on the fossil record and geological data. Thus emerged the field now referred to as phylogeography. Phylogeography combines gene genealogies with data on present distributions to determine what led to present-day distributions. Its impact on the fields of biogeography and ecology are just beginning to be appreciated.

In addition to the historical approaches to the study of distributions, present-day animal and plant distributions can be described on the basis of overall structure of plant communities relative to climate. Thus we see terms such as biome and biogeographic realm. Because the last chapter (12) dealt with ecology and introduced the theory of island biogeography, we start by discussing ecological determinants of present-day distributions of amphibians and reptiles. We then discuss how the history of the Earth's continents has affected amphibian and reptile distributions. Finally, we showcase several phylogeography studies to show how the combined use of well-supported phylogenetic hypotheses and geological–fossil data is rapidly improving our understanding of the history of patterns of distribution and diversification in amphibians and reptiles.

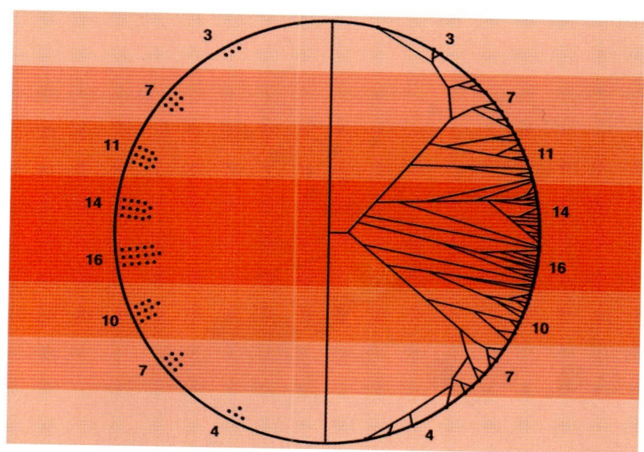

FIGURE 13.2 Graphic depicting two very different approaches to understanding global patterns of species richness. The circle represents the globe, and shades of color represent latitudinal zones with the latitude zero across the center. On the left, black points represent individual species, and clearly the number of species is correlated with latitude; tropical habitats have more species than temperate habitats. Explanations for higher diversity in tropical regions center on correlations between numbers of species and environmental variables such as temperature and moisture. On the right, evolutionary relationships of species are shown (lines) with a hypothetical monophyletic clade represented. This graphic stresses a history of diversification indicating that more clades originated in the tropics, and because of niche conservatism, few clades were able to evolve ecological traits allowing them to disperse to temperate climates. Adapted from Wiens and Donoghue, 2004.

DISTINGUISHING BETWEEN ECOLOGICAL AND HISTORICAL BIOGEOGRAPHY

Throughout the world, geographic areas contain different species of plants and animals. The flora and fauna of a given area are relatively uniform in species number and composition compared with adjacent areas. The flora and fauna can persist over large or small areas and then gradually or abruptly change to new flora and fauna. Many approaches exist to examine patterns of distribution. Geographical ecology (ecological biogeography) examines geographic patterns in the structure of different communities from a perspective of resource utilization. Island biogeography fits into this category. Ecological biogeography emphasizes overall structure of communities across space and has resulted in descriptions of biomes, biogeographic realms, and other ecologically based categories. Historical biogeography focuses on the relationships and origins of taxa, emphasizing the phylogenetic affinities of the species (their evolutionary histories) and how those tie in with the history of distributional patterns. The key difference between the ecological and historical approaches to animal distributions is that ecological approaches center on extant or relatively recent correlates of present-day distributions, whereas modern historical biogeography emphasizes the evolutionary history of the organisms of interest (Fig. 13.2). The example shown provides two alternative explanations for high diversity of organisms in the world's tropical regions. Both approaches predict similar numbers of species at various latitudes, but the biogeographical approach traces the origins of the faunas tying together phylogeny, ecology, and microevolution. The ecological approach pays little attention to the underlying evolutionary relationships of organisms at any particular place on the planet, whereas the historical biogeography approach interprets patterns of species richness in the context of evolutionary relationships. It is clear that most species and clades originated in tropical environments, with relatively few moving into colder regions. This can be nicely illustrated by examining two commonly discussed ecological parameters from a phylogenetic perspective, niche conservatism and niche evolution (Fig. 13.3). Niche conservatism refers to individual species maintaining ecological traits similar to those of their sister taxon and ancestors, whereas niche evolution refers to divergence in niche characteristics. As the graphic model demonstrates, sets of species in one region are often more similar to each other than they are to species in other regions, and when the phylogenetic relationships of the animals are examined, we find that

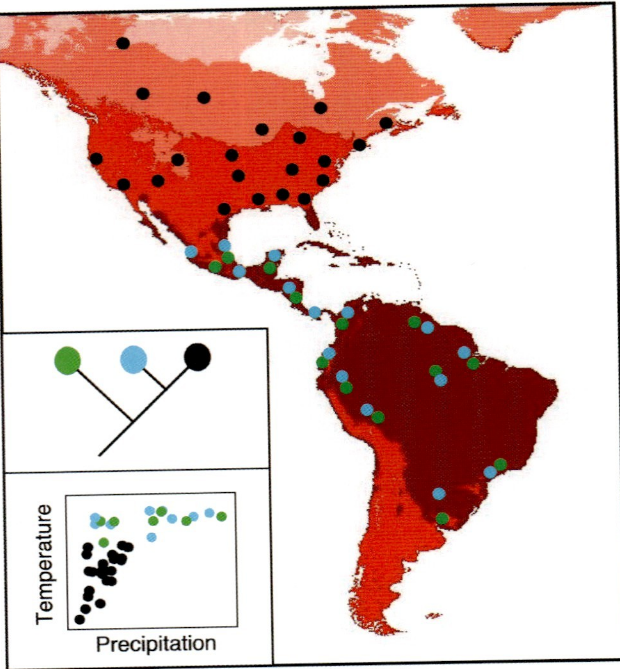

FIGURE 13.3 Graphical model showing effects of niche conservatism and niche evolution on faunas in different regions of the world. Colored circles represent different clades. Two clades (blue and green) have retained ancestral niche characteristics and have distributions restricted to tropical and subtropical environments. A third clade (black) has evolved tolerance to lower temperatures and lower precipitation and dispersed, no longer occurring in tropical regions and thus showing niche evolution. Adapted from Wiens and Donoghue, 2004.

ecologically similar species are often each others' closest relatives. Thus niches are conserved within clades, and clade niche shifts occur when divergence occurs. These sets of species (clades) often have different distributions. As discussed in the last chapter, by the time some of these sets of species (clades) come back together, they are ecologically different enough to coexist. As you will see in following text, the biogeographical approach allows us to trace historic distributions and test competing hypotheses. For example, is the presence of one clade on South America the result of tectonic events in the distant past or did their ancestors drift across the Atlantic Ocean after the breakup of Pangaea? Is high Amazonian biodiversity the result of relatively recent climatic events on speciation or are alternative hypotheses more likely? Did southern faunas that now occur in India ride continental plates to get there, and if so, when, or did they reach India and southern Asia by transpacific dispersal? How static are continental distributions, and how have relatively recent climatic or geologic events affected species distributions and the speciation process? The kinds of questions that contemporary biogeographers ask intersect ecology, biogeography, and evolution.

Ecological Determinants of Species Distributions

The ecological approach to distributions combines physiological traits of organisms with relevant environmental variables. Each organism has specific physiological tolerances and requirements. Populations represent a distribution of the requirements of their individual members. If an individual's tolerances are not exceeded and if its requirements are fulfilled, it survives. Survival of individuals does not necessarily ensure survival of the population; individuals must reproduce and so must their offspring for a population to persist. It is this latter aspect that makes age-specific mortality and age-specific fecundity the key life history traits (see Chapter 5) and, ultimately, the key determinants of a population's occurrence and persistence in any geographic area.

Because a species consists of multiple local populations, the species' distribution represents the total occurrences of its populations, and the borders of each species' distribution marks the areas where populations waver between extinction and self-perpetuation. At one season or year, conditions allow reproduction and survival of young and the population grows; in the next, reproduction could be unsuccessful and the population could drift toward extinction. Factors affecting these population cycles are climate (micro- and macro-) relative to physiological tolerances, availability and access to resources, and interspecific interactions. Many amphibians are resilient to relatively long-term (several years) environmental fluctuations, such as extended droughts. As a result, we often see drastic reductions in amphibian and reptile abundance at the local level followed by rapid increases in abundance when conditions change. The ability to withstand long time periods when resources are low is largely a consequence of physiological correlates of ectothermy. Amphibians and reptiles can persist on relatively little energy for long time periods when compared with endothermic mammals and birds. Other factors, such as historical accident and dispersal ability, determine which species are likely to occur in one area and the probability of their reaching another area.

Needless to say, amphibian and reptile species are not randomly distributed across the planet. Nevertheless, many can live in locations outside their natural distribution. Examples include increase in abundance of numerous exotic lizards in Miami, Florida; the marine toad (*Rhinella marina*) in the West Indies, Australia, and the southwest Pacific; the various species of Mediterranean geckos (*Hemidactylus*) that have colonized the New World; and the brown tree snake (*Boiga irregularis*) in Guam. Thus, physiological tolerances are not the only factors limiting and determining distributions. Normally, species that occupy an area originated from nearby areas, and a vacant habitat is "filled" by a few migrants that cross barriers (geographic or unsuitable habitats) or by the slow expansion of a population into less hospitable areas. The preceding are examples of species transported by humans across barriers (oceans) that otherwise would have restricted their abilities to reach other continents. Dispersal abilities and opportunities are variable. Small, fossorial amphibians and reptiles such as caecilians, blind snakes, and amphisbaenians should have poor dispersal abilities (but see later text), whereas large, aquatic species tend to be good dispersers. Coastal and riverside species are more likely than inland species to be transported elsewhere. Amphibians rarely cross saltwater barriers; reptiles commonly do, by drifting on floating surface objects. Many exceptions to these generalities exist. Dispersal ability and the nature of barriers are also critical in determining the level of gene flow among populations, which affects local population differentiation.

It is hardly surprising that climate affects a species' occurrence. An animal will not survive in an area where one or more of its physiological tolerances are regularly or constantly exceeded. Temperature and rainfall and their periodicity establish the climatic regimes under which individuals and populations must operate. Tolerance limits (Chapter 7) are species specific, although variation among populations exists, particularly in widespread species. Because the edges of species' ranges often closely match the isograms of rainfall and temperature, tolerance limits may define the limits of species' distributions (Fig. 13.4). Frequently, the effects of temperature and/or rainfall are greater on one life stage than on another, but the survival of each stage is critical for the survival of the population. Spring droughts may prevent temporary-pond amphibians from breeding or, if breeding does occur, larval recruitment may fail due to ponds drying early. Adults may survive (tolerate) the drought to breed again when conditions improve. Droughts extending over several years can cause local extinction, particularly in short-lived species. In turtles with temperature-dependent sex determination (Chapter 5), cooler summers may produce all-male cohorts, and if this hatchling sex ratio continues, the sex ratio may be biased toward males and result in eventual population declines.

Climate, resources, and interspecific interactions vary from area to area, and each population adjusts (adapts) to its local conditions. Because environmental and biotic conditions are nowhere the same, each population adapts differently and diverges genetically (evolving) from other populations. If this divergence continued, speciation would occur; however, speciation is a rare outcome because adjacent populations exchange individuals. Migrants bring new genes into the population from gene pools of adjacent populations. The rate of gene flow is a function of the closeness of the populations and the dispersal tendency of the species. The rate can be quite slow yet maintain the genetic continuity of distant populations. While maintaining continuity, local populations can adapt to local conditions. Often, adaptations most easily detected are traits associated with reproduction (Fig. 13.5; also see Chapters 4 and 5). Of course, environments are not static through time, and as environments change, so do distributions of species.

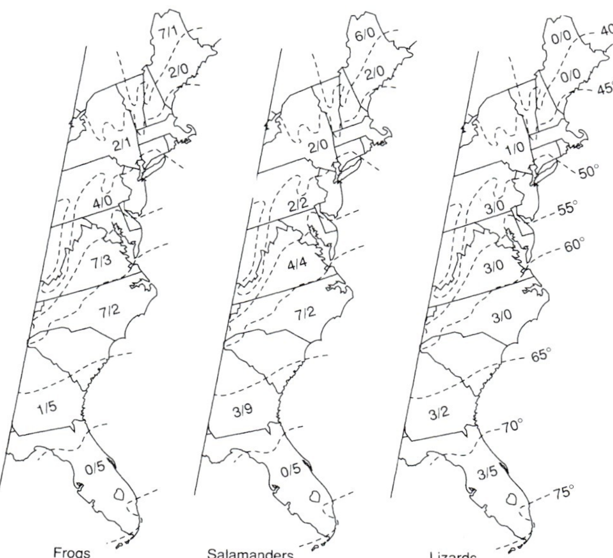

FIGURE 13.4 Temperature isotherms and the northern and southern limits of frog (30 species), salamander (26 species), and lizard (16 species) distributions in eastern North America (Piedmont and Coastal Plain). Isotherms are mean annual temperature (°F); the integers in each zone are number of species with ranges terminating in the interval (northern–southern termini). Isotherms from USDA, 1941; distributional data from Conant and Collins, 1991.

Biomes and Biogeographic Realms

Current worldwide distributions of communities form two patterns: biomes and biogeographic realms. Biomes are

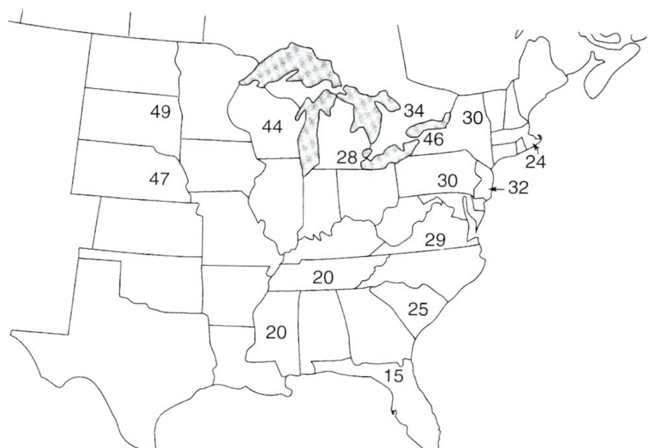

FIGURE 13.5 Geographic variation in clutch size among populations of the snapping turtle (*Chelydra serpentina*) in North America. Integers indicate mean clutch size at specific locality. Data from Fitch, 1985, and Iverson et al., 1997.

based on the similarity of the overall structure of the plant community relative to climate. Biogeographic realms (also called ecozones) are based on the evolutionary and historic distribution patterns of animals and plants. The biome concept ignores animals and recognizes communities based on plant structure (e.g., height and shape of plants, leaf structure, deciduous or evergreen vegetation) because climate is the primary determinant of vegetation. The major terrestrial biomes are tundra, boreal coniferous forest (taiga), temperate deciduous and rain forest, temperate grassland, chaparral, desert, tropical grassland and savanna, tropical scrub forest, tropical deciduous forest, and tropical rain forest. These biomes can be further subdivided in multiple ways. In all cases, biomes reflect the annual cycle of temperature and rainfall; animal distributions match the biomes in general but deviate considerably when examined in detail for amphibians and reptiles. Few amphibians or reptiles occur in the tundra, and those that do only do so marginally. Assemblages with low numbers of species are widespread in the boreal forest biome and dominated by amphibians. Northern temperate latitudes have salamanders; southern ones have none. The number of species and the diversity of the herpetofauna increase within the temperate biomes and into the tropics, but unlike plants, in which overall community structure matches climate, animal community composition is more influenced by taxonomic affinity.

Biogeographic realms (Fig. 13.6) are defined in terms of animal and plant distributions based on phylogenetic affinities. Nevertheless, they are not explicitly phylogeny-based biogeographical hypotheses. The realms derive from higher-order relationships, typically relatively large clades, and reflect past geological events (continental drift, barriers and connections for species dispersal). Indeed, the present distribution of many higher taxonomic groups of amphibians and reptiles matches the past continental connections and fragmentations proposed by the plate tectonic (drifting continents) theory (compare Fig. 13.1 with the distribution maps in Chapters 15–21). For example, salamanders occur mainly in the Holarctic (Nearctic + Palearctic) region, and frog diversity is highest in the southern hemisphere. These distributions match the Mesozoic split of the supercontinent of Pangaea into northern Laurasia and southern Gondwana. Ancient groups still retain a Laurasian or Gondwanan distribution. Pipid frogs occur in both Africa and South America. The distribution of pleurodiran turtles reflects the historical geologic links between Australia, southern Africa, and South America, a Gondwanan distribution. Cryptobranchid, plethodontid, and salamandrid salamanders occur both in North America and Eurasia, suggesting an ancient distribution throughout Laurasia. Just as these interclade relationships match ancient topographies, intergeneric and interspecific distributions reflect more recent (but still ancient) geological events and climates.

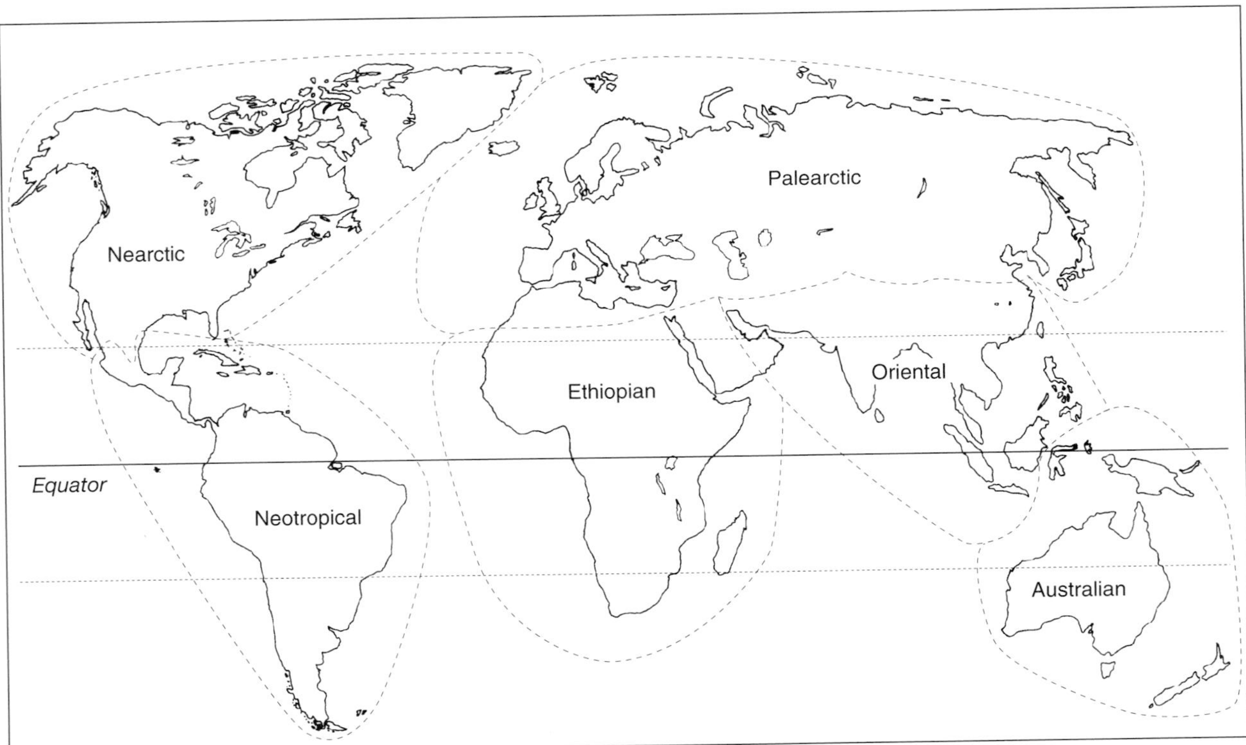

FIGURE 13.6 Biogeographic realms of the world.

Each continent has been divided into biogeographic provinces that are delimited by abrupt terminations in species distributions as one community shifts to another. These discontinuities in community structure likely reveal a prior isolation of communities and speciation that occurred within each one.

Historical Determinants of Species Distributions

In historical biogeography, the perspective shifts from the recent past of ecological biogeography to the distant past, the deep time of geological history, and from intracommunity interactions to phylogenetic relationships for reconstruction of species and higher taxa distribution patterns. Current theory and analyses are based on either a dispersal or a vicariance viewpoint, and area cladograms result (Fig. 13.7). Dispersal refers to animals moving across land or water to reach new areas, whereas vicariance refers to some kind of event (continents breaking apart, mountain ranges uplifting) physically separating populations. Many studies fall somewhere between the two extremes. Dispersal was the primary mechanism for explaining current distributional patterns and dominated biogeographic studies until the early 1970s, when the geological worldview shifted from static continents to drifting continents (plate tectonics). A vicariance-based mechanism to explain distributional patterns utilizes cladistic analyses to test distributions against phylogenetic hypotheses.

Dispersal theory rests on two basic assumptions: taxonomic groups have a center of origin, and each group disperses from its center of origin across barriers; the resulting communities or biota derive from one to several centers and dispersal events. Vicariance theory rests on the assumptions that taxonomic groups or biotas are geographically static, and that geological events produce barriers and the biota diverges subsequent to isolation. Both theoretical approaches require knowledge of phylogenetic relationships to discern the ancient dispersal routes or the areas occupied by ancient biota. Because allopatric speciation appears to be the dominant mode of speciation and the fragmentation of a biota is more likely than a biota dispersing as a single unit, vicariance interpretations are generally preferred over dispersal explanations. Vicariance explanations are also more amenable to testing. Dispersal explanations are required to account for the evolution of oceanic island biotas, such as those found in the Galápagos and Hawaiian Islands, as well as movement of taxa from patch to patch on continuous landmasses. Nevertheless, many sources of vicariance exist at the continental and regional level, including large rivers, mountain ranges, and even ecological gradients. Molecular phylogenies in which divergences can be dated provide a powerful tool for determining whether vicariance or dispersal events caused present-day distributions of

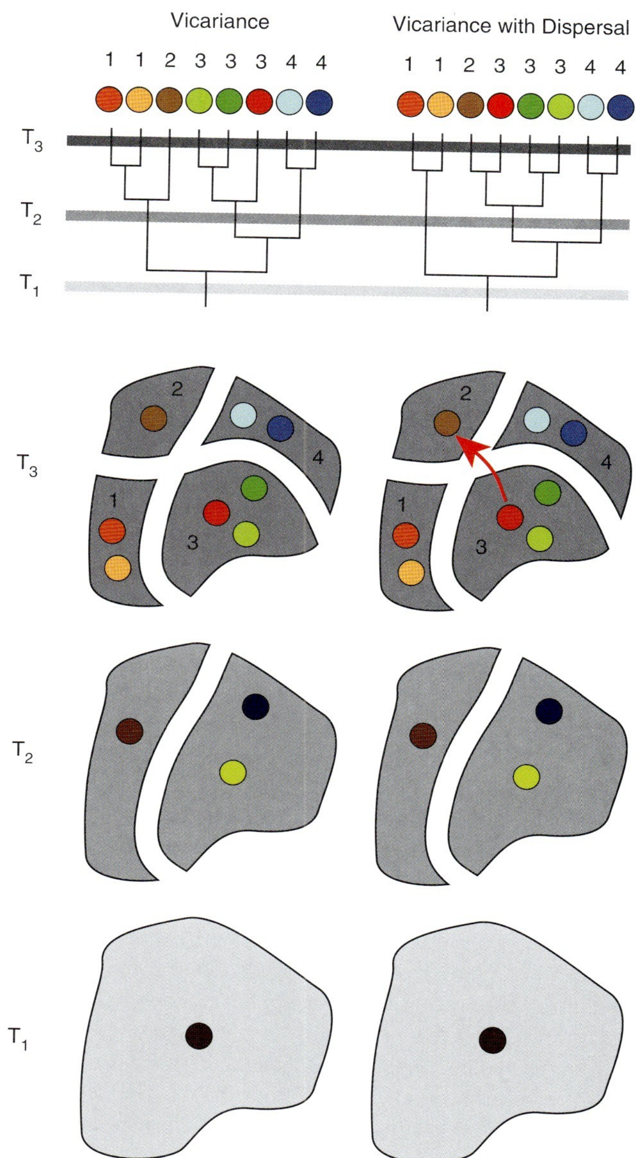

FIGURE 13.7 Graphical model showing the difference between vicariance and vicariance with a single dispersal event in the construction of area cladograms. At time 1 (T_1) the ancestor lives on a single large continent. By time 2, continents have separated, creating the first vicariant event, and additional speciation has occurred on the continent on the right. By time 3, four continents exist. On the left, no dispersal has taken place, and additional speciation events have occurred on continents 1, 3, and 4. On the right, speciation has also occurred on continents 1, 3, and 4, but one of the species on continent 3 has dispersed to continent 2 where, over time, that species has differentiated. Thus the species on continent 2 has one of the species from continent 3 as its closest relative (sister taxon). Cladograms at the top show phylogenetic relationships between the species under each scenario. Numbers across the top refer to present distribution of species on the four continents. By comparing a dated phylogeny with independently derived dates of vicariant events (in this case, continental splitting), it is possible to falsify a vicariance hypothesis. The phylogeny on the left supports a vicariance hypothesis, and the one on the right falsifies it for the species on continent 2, leaving dispersal as the only viable hypothesis for the origin of that species on continent 2. Although our example uses continents, other barriers (mountains, rivers, ecological transitions) can result in vicariance. Adapted in part from Futuyuma, 1998.

amphibians and reptiles, and some of the results that have been obtained are surprising. Combined with a knowledge of the history of the location of landmasses, dated phylogenies present the opportunity to falsify vicariance hypotheses, and when vicariant hypotheses are falsified, alternative explanations such as transoceanic dispersal must be considered. In some instances, the results of careful biogeographic hypothesis challenge what we thought we knew about distributional histories.

The geologic history of most areas and their herpetofaunas are so complex that a single theory is inadequate. The Seychelles Islands provide a case study. The herpetofauna of the Seychelles contains several levels of endemicity that strongly indicate multiple origins and suggest that components arrived at different times (Table 13.1). The oldest elements are the sooglossid frogs and caecilians. These groups have only distant (and somewhat uncertain) affinities with African taxa. Both are confined to the high granitic islands of the Seychelles that have been emergent since the Mesozoic. The granitic islands appear to be fragments of the Indian tectonic plate that broke free from the current African plate and moved northward to collide with the Asian plate. Since amphibians are noted for their inability to cross huge expanses of saltwater, these amphibians and perhaps the gecko *Ailuronyx* are derived from ancestors living on the original African–Indian plate. The rhacophorid frog *Tachycnemis* and some reptiles also appear to be derived from an early Seychellan herpetofauna, but likely from taxa that arrived via island hopping across narrow water gaps. The day-geckos (*Phelsuma*) and others are more recent arrivals that show closer affinities with Malagasian and African taxa, but presumably arrived prior to human colonization. More recent components have arrived via human transport (*Gehyra*).

A vicariance explanation has been used to explain the present-day distribution pattern of chelid turtles (Fig. 13.8). Two cladistic patterns of relationships among chelid turtle genera suggest different scenarios to account for their distributional history. One cladogram suggests that *Pseudemydura* is the sister group to all other extant chelids. The Australian *Emydura* group is the sister group of all Neotropical chelids and the Australian long-necked *Chelodina*. This pattern of cladogenic events would suggest that all modern genera arose from vicariance events in the deep past on the southern continent (Fig. 13.1) or that the ancestor of Neotropical chelids and the Australian chelids, excluding *Chelodina*, arose on the southern continent, and subsequently the ancestral *Chelodina* reached Australia and differentiated there. This latter explanation requires a dispersal event across the ocean, highly unlikely for a freshwater turtle. Another cladistic relationship offers a simpler evolutionary scenario. The ancestral chelids occurred broadly on the southern continent, and rifting of the southern continent into the South American and Australian–Antarctica continents was the vicariance event that gave rise to the ancestors for two monophyletic continental clades. While the latter offers a more parsimonious explanation, both explanations and both cladograms are hypotheses that require further testing. The strength of the vicariance model is the ability to test biogeographic hypotheses and reject those that do not match the proposed geologic or other vicariance models. A more recent phylogenetic analysis of chelid turtles fails to resolve the problem, because *Chelodina* (Australian), the South American chelids, and the remaining Australian chelids form an unresolved polytomy (Fig. 13.8). This is a nice example of how we progress in science. Two plausible hypotheses should be testable by simply resolving chelid turtle relationships. However, when the best available data are applied, we discover that the relationships between the three critical groups are not clear, and we are left with additional questions that will require more detailed data collection and interpretation.

TABLE 13.1 Relative Ages of Select Components of the Herpetofauna of the Seychelles.

Ancient	Near ancient	Near recent	Recent
(> 60 mybp)	(< 60–10 mybp)	(> 10 mybp)	(< 1000 yr)
Grandisonia			
Sooglossus	*Tachycnemis*	*Ptychadena*	
Ailuronyx	*Urocotyledon*	*Phelsuma*	*Gehyra*
	Janetscincus	*Mabuya*	
	Lycognathophis	*Boaedon*	*Ramphotyphlops*
	Pelusios seychellensis	*Pelusios subniger*	

Note: Taxa are arranged vertically: caecilians, frogs, geckos, skinks, and turtles. The age of each taxon is based on its degree of taxonomic differentiation and endemicity. The ages are arbitrary estimates beginning immediately prior to separation of the Seychelles from Gondwana (Ancient) and mark the islands' progressive isolation from faunal source areas. Data in part from Nussbaum, 1985.

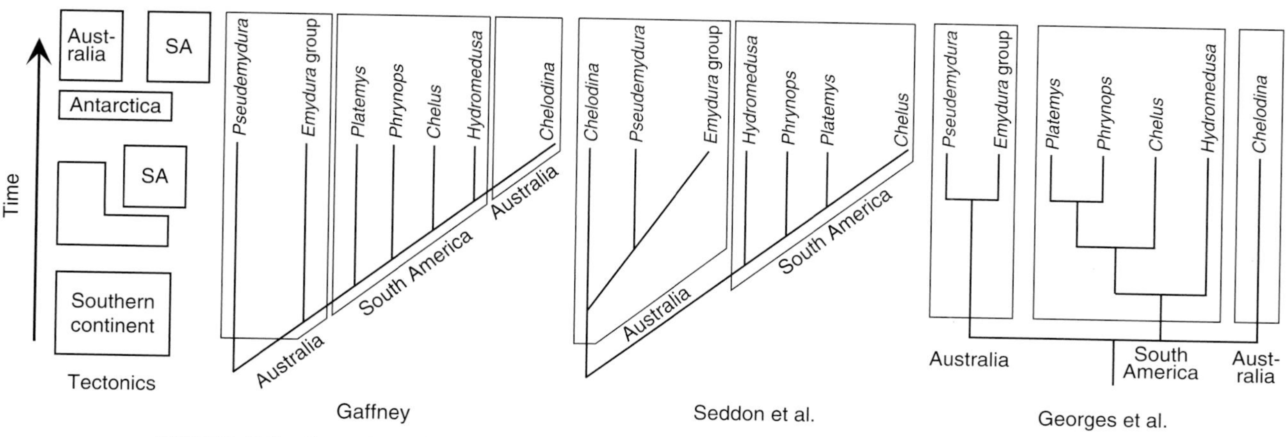

FIGURE 13.8 Comparison of phylogenetic relationships of chelid turtles and their distributions to the tectonics of southern continents. Unfortunately, the most recent cladogram (far right) fails to falsify either of the earlier hypotheses because Chelodina, South American chelids, and remaining Australian chelids form an unresolved polytomy. Cladograms adapted from Gaffney, 1977; Seddon et al., 1997; and Georges et al., 1998.

The Seychelles and chelid examples highlight the necessity of a pluralist approach to biogeographic analysis and of the need to provide explanations (hypotheses) that can be tested. Multiple levels of interpretations are likely required for the patterns of most herpetofaunas and their component species. We now examine a subset of the recent analyses that address questions of historical biogeography of amphibians and reptiles. Each of these provides new insights into old questions, and each raises additional questions.

Recovering History: Phylogenetic Approaches to Biogeography

Prior to the use of dated phylogenies, biogeographic scenarios were, for the most part, storytelling. The fossils existed, present-day distributions existed, and information on historic distribution of continents existed. What was missing was the ability to independently date divergence patterns in taxonomic groups of interest. To put it another way, distributional histories were simply fitted to the movement of continents. Dated phylogenies have changed that line of thinking dramatically, and as previously indicated, historical biogeography has transformed into phylogeography with the ability to explicitly test hypotheses. Rather than summarizing everything that is known about biogeography of amphibians and reptiles, we have selected a set of studies that make specific points about the process of distributional histories and diversification. We refer the interested reader to other sources for detailed and more complete summaries of the biogeography of amphibians and reptiles.

Amazon Biodiversity

High diversity of amphibians and reptiles in tropical rain forests is well known, and a number of hypotheses have been presented to account for this high diversity. One, the Vanishing Refuge Theory (often referred to as the Climatic Disturbance Hypothesis), which was originally applied to birds and lizards, has received considerable attention. This hypothesis basically posits that environmental fluctuations during the Pleistocene (2 million to 10 thousand years before present) resulted in repeated expansions and contractions of rain forest, resulting in repeated isolation of faunas and resultant speciation. Pollen profiles from Pleistocene deposits indicate that the rain forest was both more and less extensive in the past. Other hypotheses include (1) the Riverine Barrier Hypothesis, which suggests that the large rivers in the Amazon basin were distribution barriers for species living in terra firma forest, thus restricting gene flow resulting in divergence across rivers; (2) the Ecological Gradients Hypothesis, which suggests that habitat gradients (e.g., temperature, moisture) can serve as sufficient barriers to restrict gene flow; (3) the Historic Mountain Ridge Hypothesis, in which mountain ranges (the Andes in particular) were barriers; and (4) the Marine Incursion Hypothesis, in which influx of saltwater produced barriers. The lizard example forming the basis for the Vanishing Refuge Theory was the *Anolis nitens* (formerly *chrysolepis*) complex. At the time that this was proposed by Paulo Vanzolini and Ernest Williams, four subspecies of *A. nitens* were recognized, and for the most part, these rain forest lizards had nonoverlapping distributions. One subspecies, *A. n. brasiliensis*, was known from only a few isolated patches of dry forest south of the Amazon Basin, and these patches were believed to be remnants of a once much more widespread Amazon rain forest. Under this hypothesis, isolation had resulted in divergence of *A. n. brasiliensis*, and similar isolation events had produced other subspecies. Thus, the mechanism of speciation was expansion and contraction of forest causing isolation, genetic drift in the lizard populations that were isolated, and then when forest re-expanded, dispersion of the genetically distinct *Anolis* into surrounding rain forest. Data available at the time were convincing. The key assumption of this model was that divergence of these anoles had occurred during the Pleistocene, coincident with the period during which expansion and contraction of the rain forest had occurred. At the time that this was proposed, molecular techniques allowing reliable dating of divergences were not available. In 2001, Rich Glor and colleagues tested this hypothesis using a molecular phylogeny of the *A. nitens* complex and several outgroups, many of which are also Amazonian (Fig. 13.9). Their results unequivocally show that divergences in this and other Amazonian anoles occurred much earlier than the Pleistocene, thus falsifying the Vanishing Refuge Theory for these species.

Additional studies on another lizard complex (sphaerodactyline geckos) occurring in the Amazon also failed to support the Vanishing Refuge Theory. Although the evolutionary history of sphaerodactyline geckos is complex, revealing several very old and some recent divergences, none occurred during the Pleistocene (Fig. 13.10). Most species-level divergences in these lizards occurred during the Oligocene–Miocene, 20+ mybp, which coincides with divergence patterns observed in Amazonian anoles and a host of other vertebrates. Dramatic climate change during this time period along with orogenic events in the Andes account for some of the diversification patterns observed

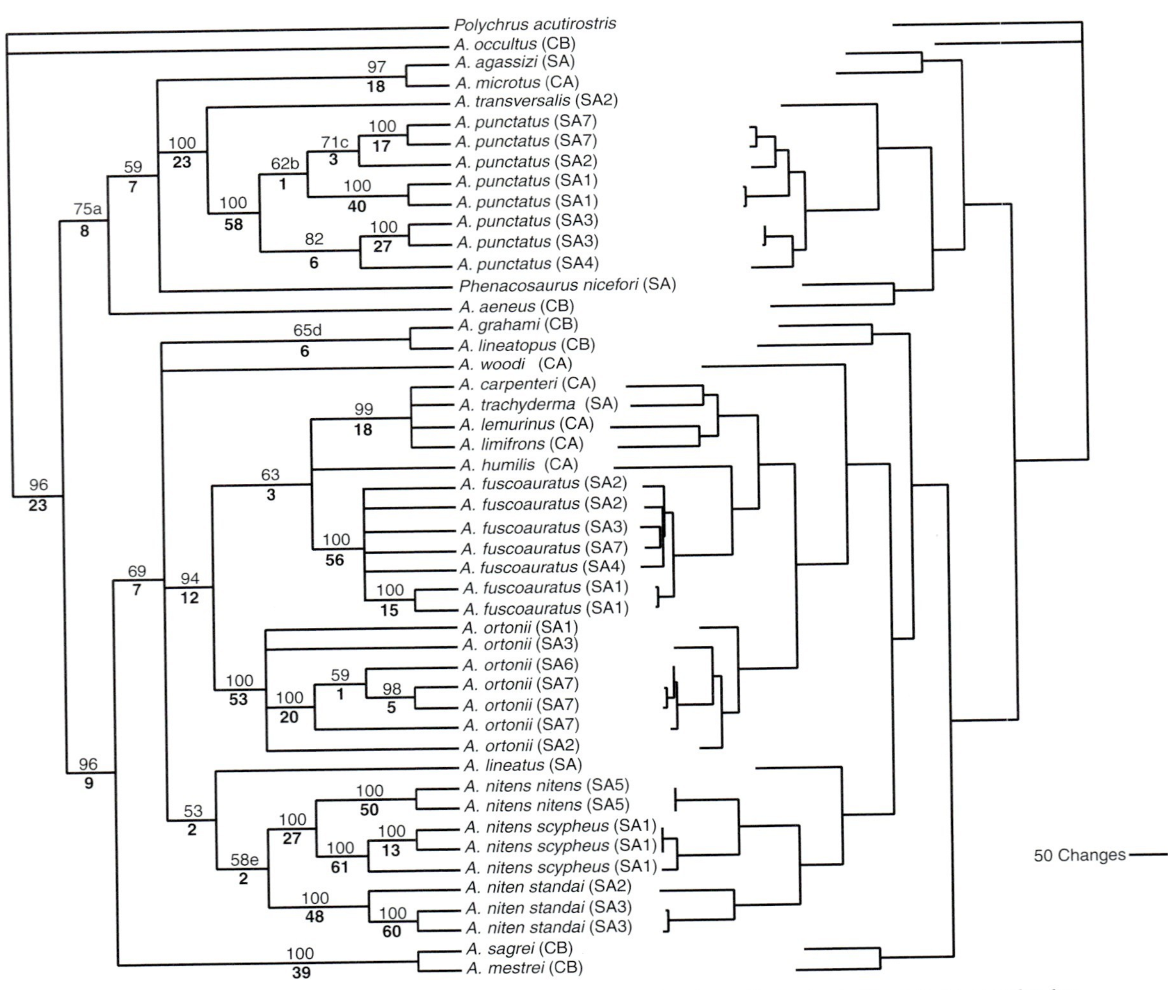

FIGURE 13.9 Although diversification of lizards in the *Anolis nitens* (formerly *A. chrysolepis*) complex has been used to support the Vanishing Refuge Theory of Amazonian diversity, molecular analysis of the group clearly shows that diversification took place much earlier. Although the described subspecies are each others' closest relatives, a deep split in haplotypes from the north (Roraima state in Brazil and Ecuador) and the south (Amazonas and Acre States in Brazil) and relatively deep splits in more recent clades placed their origins before the existence of proposed refuges. The *A. nitens* clade, at minimum, is > 15 million years old. Left, maximum parsimony bootstrap tree; right, maximum-likelihood tree. Adapted from Glor et al., 2001.

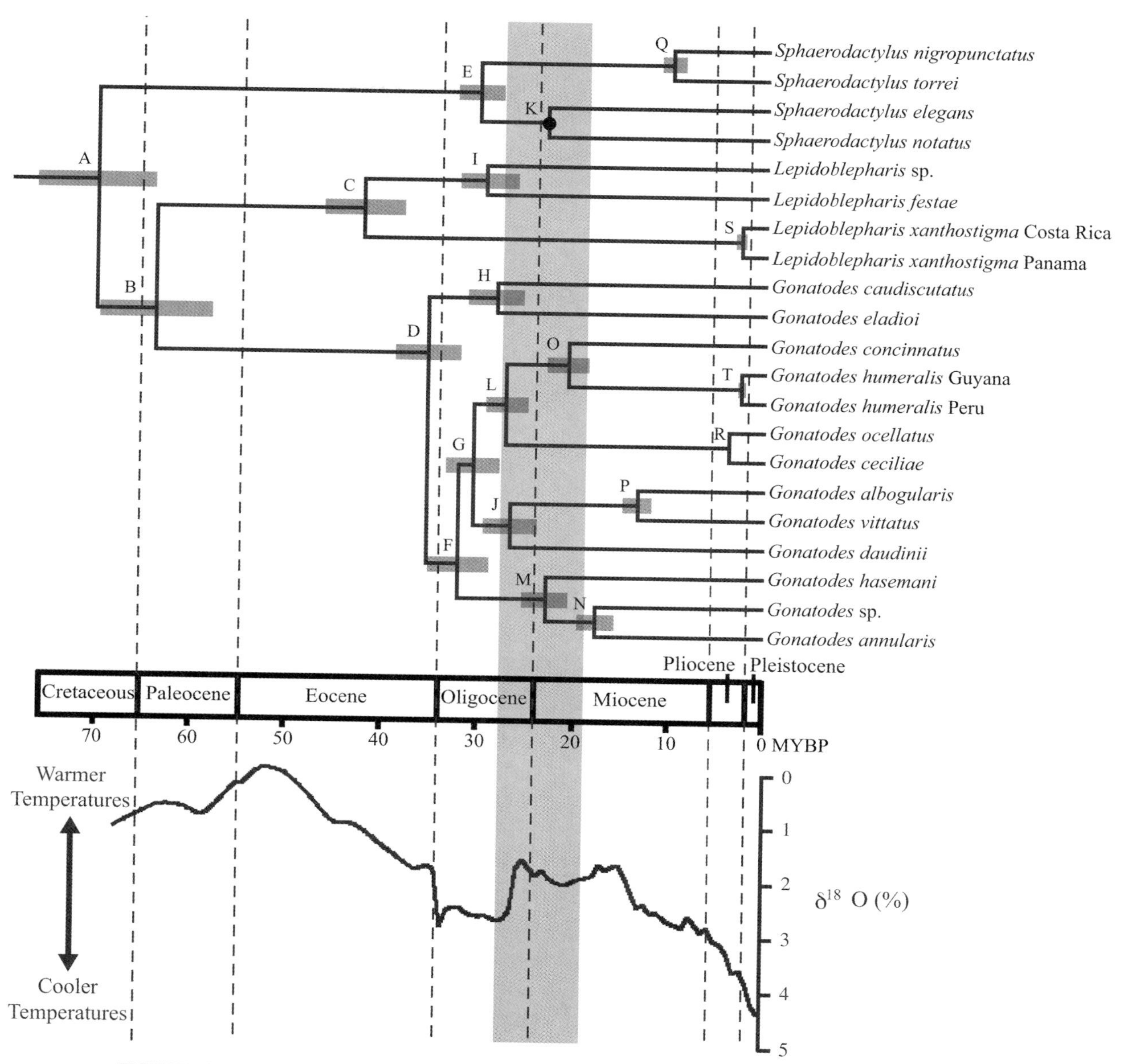

FIGURE 13.10 Phylogenetic relationships of Amazonian sphaerodactyline geckos. Based on gene sequence data, major divergences occurred during the Miocene–Pliocene, much before the Pleistocene, when Amazon refuges existed. Adapted from Gamble et al., 2008.

in both these geckos and the anoles previously discussed. Two gecko clades show an east–west distribution, *Gonatodes hasemani* versus the *G. annularis* + *G.* sp. clade, dated at about 23 mybp, and the split in the *G. humeralis* clade dated at about 1.9 mybp. Similar nonconcurrent east–west divergences occurred in Amazonian anoles (Fig. 13.9). Thus the congruent east–west biogeographic patterns within each of two very divergent lizard clades (anoles and geckos) cannot be tied to a single vicariant event because they occurred millions of years apart. Causes for some of these divergences remain unknown, but clearly diversification of lizards in the Amazon Basin is complex and cannot be explained by one event or hypothesis. Recent studies on birds and other organisms have also failed to support the Vanishing Refuge Theory.

Another recent study used a molecular phylogeny to test two competing hypotheses for the distribution of fogs in the *Physalaemus petersi* complex. This frog occurs in the western Amazon, a region that is broken up by major rivers (the Riverine Barrier Hypothesis) and also experiences an elevational gradient from west (high) to east (low) (Fig. 13.11). Because elevation gradients can influence species distribution (because of correlated ecological variables), elevation (Elevation Gradient Hypothesis)

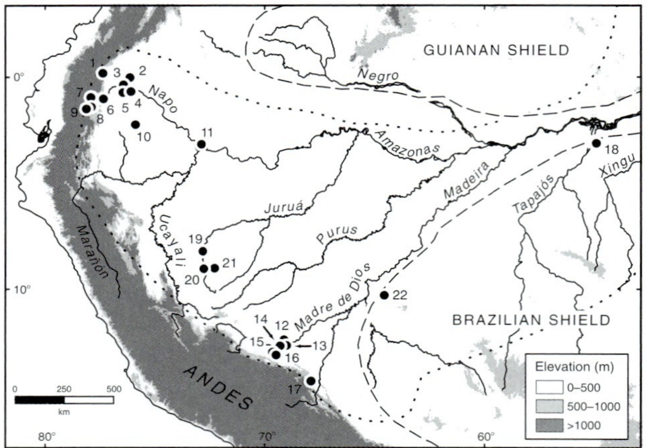

FIGURE 13.11 The frog *Physalaemus petersi* is ideal for testing biogeographic hypotheses on origin of diversity in the Amazon River basin because it occurs across an area divided by large rivers (Riverine Hypothesis) and where an elevational gradient exists associated with the Andes (Elevational Gradient Hypothesis). The dotted outline shows the approximate distribution of *P. petersi* (which extends farther to the east than shown), and the dashed lines show the locations of the Guianan (upper) and Brazilian (lower) Shields. Adapted from Funk et al., 2007.

might explain distributional patterns in these frogs. Chris Funk and colleagues used sequence data from three genes to test these hypotheses and uncovered a complex pattern of relationships. Although no evidence was found suggesting that elevation gradients played a role in diversification of these frogs, one of the rivers (the Madre de Dios) appears to have been a barrier (Fig. 13.12). Nevertheless, populations of *P. petersi* south of the Rio Madre de Dios appear to have expanded rapidly, leaving open the possibility that secondary contact of expanding lineages rather than divergence explains the pattern. Studies on dendrobatid frogs are consistent with an elevational gradient hypothesis in at least some areas, and studies on other frogs (leptodactylids and hylids) provide some evidence

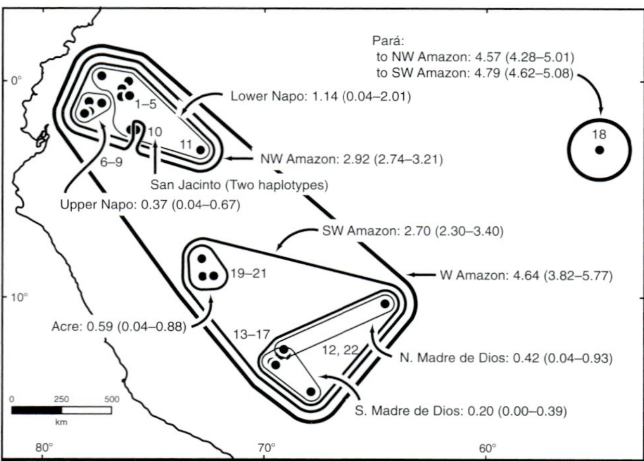

FIGURE 13.12 Geographic ranges of major clades of *Physalaemus petersi* based on haplotype divergences. Numbers indicate average percent corrected sequence divergence. Adapted from Funk et al., 2007.

for the Riverine Barrier Hypothesis. Consequently, biogeography of Amazonian frogs will not be explained by a single process. Rather, processes differ among and within groups of frogs, and although one process (e.g., riverine barriers) might explain some divergences, others (e.g., elevational gradients) may explain others.

Surprisingly, several recent studies that lack phylogenetic data and estimates of divergence times continue to invoke the Vanishing Refuge Theory to account for patterns of speciation. For example, a recent study of amphisbaenians in the *Amphisbaena fuliginosa* complex suggests that patterns of diversification in these subterranean lizards are consistent with patterns expected based on expansion and contraction of Amazon rain forest, with subsequent isolation of *A. fuliginosa* populations. A dated phylogeny for amphisbaenians will be necessary to test this hypothesis.

Historical Biogeography of Amphibians

The three major clades of extant amphibians, caecilians, frogs, and salamanders, existed by the mid-Triassic, and diversification had already been taking place in salamanders and frogs long before that. Diversification of these groups during the Mesozoic and later (more recently) has always been considered to be more or less a gradual process based on the fossil record. Although molecular phylogenies do not provide data on absolute extinction rates, dated molecular phylogenies can be used to examine net rates of diversification. Kim Roelants and colleagues assembled molecular data for 171 amphibian species of the world including all major clades. By estimating divergence times for many of the clades, they demonstrated that several episodes of amphibian diversification have occurred in the past and that the accumulation of species is not a gradual process (Fig. 13.13). Amphibian diversification either accelerated with time, or diversification of amphibians has experienced rapid extinction rates. At the end of the Cretaceous, a rapid increase in amphibian diversification took place that continued into the Eocene (Fig. 13.14). This pattern corresponded to a time period that experienced high turnover in amniote species and clades and diversification of several insect taxa (ants, beetles, hemipterans).

Until recently, most major global patterns of frog distribution have been tied to the breakup of Pangaea and subsequent drifting of continents (plate tectonics). Considering that the origin of frogs and some of the major frog clades predates the breakup of Pangaea, plate tectonics undoubtedly has played a major role in global patterns of frog diversification. Nevertheless, dated molecular data from several frog clades suggest that some of the major divergences occurred much later than the breakup of Pangaea, and that the Late Cretaceous may have experienced major diversification events inconsistent with current plate

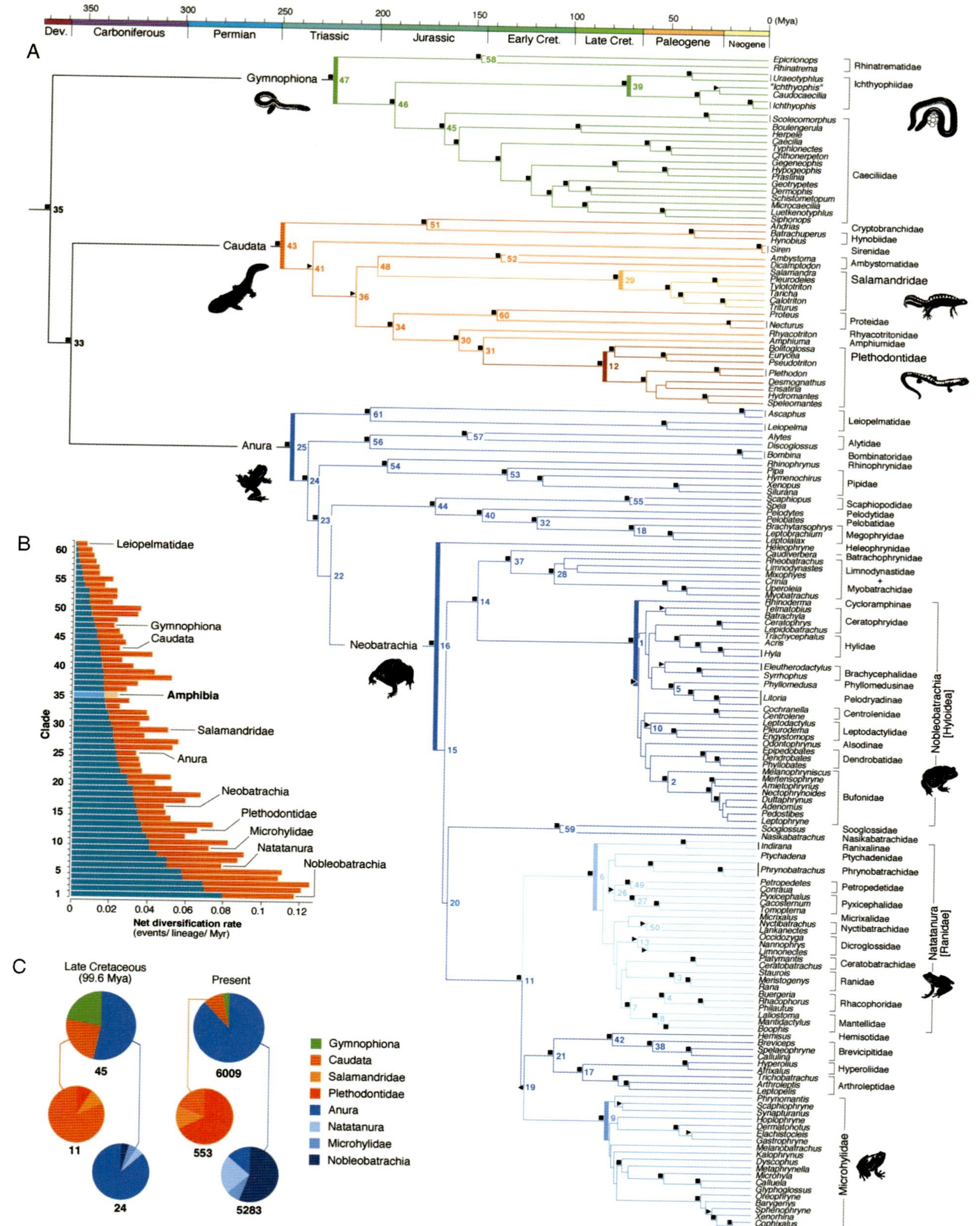

FIGURE 13.13 History of diversification of modern amphibians. (A) Phylogeny of modern amphibians with geological timescale across the top. (B) Net diversification rates for amphibian clades. Clade numbers refer to those in (A). Net diversification rates (d – b, where b = speciation rate and d is extinction rate) per clade are shown under the lowest possible relative extinction rate (red, d:b = 0) and an extremely high possible rate (blue, d:b = 0.95). (C) Comparison of proportion diversity of extant amphibian clades in the Late Cretaceous (left) and now (right). Adapted from Roelants et al., 2007 (© 2007 National Academy of Sciences, USA).

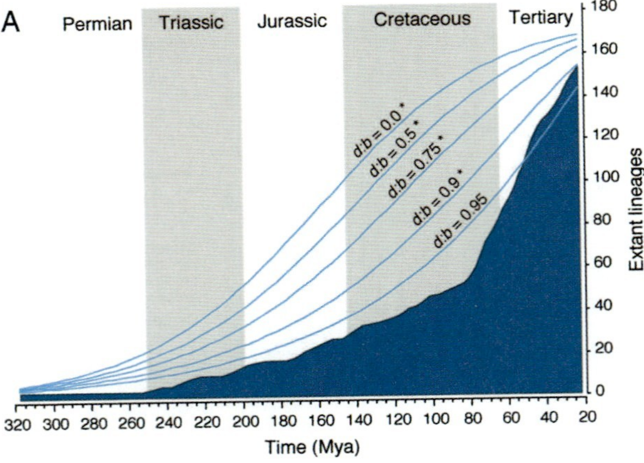

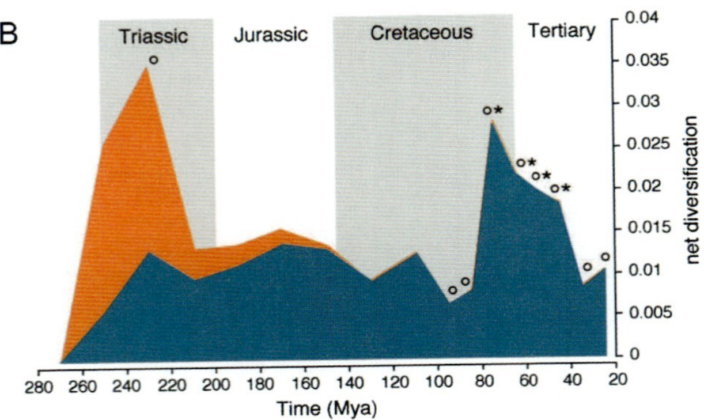

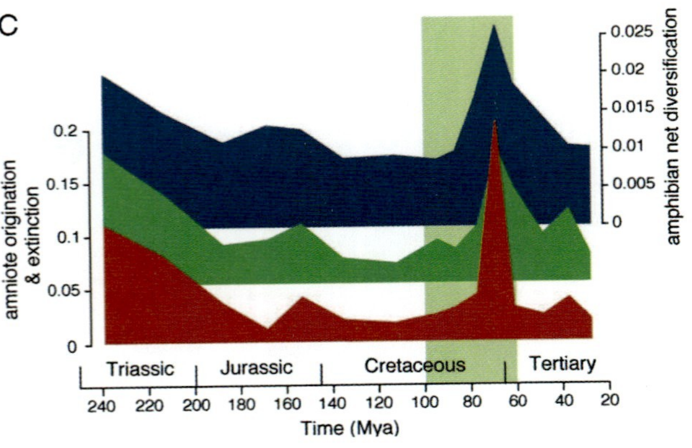

FIGURE 13.14 History of global patterns of amphibian net diversification. (A) Rate through time (RTT) plot derived from the time tree (Fig. 13.13) compared with models varying in relative extinction rates from 0 to 0.95. (B) RTT plot of net diversification rates estimated under low extinction rates (red, d:b = 0) and high extinction rates (blue, d:b = 0.95) for successive 20-million-year intervals (280–100 mybp) and 10 million year intervals (100–20 mybp). Circles and asterisks indicate estimates that differ significantly from those expected under low extinction rates (d:b = 0) and high extinction rates (d:b = 0.95), respectively. (C) Amphibian net extinction rates (blue) compared with amniote family origination (green) and extinction (red) rates based on the fossil record. Note that the amphibian data (blue) are represented on a log scale, and thus differences are even more dramatic than shown. Adapted from Roelants et al., 2007 (© 2007 National Academy of Sciences, USA).

tectonics theory, which is consistent with the preceding explanation for all amphibians. Two large frog clades, the Natanura and the Microhylidae, have always been assumed to have had a Gondwana origin. Ines Van Bocxlaer and colleagues have recently shown that divergence within microhylids and natanurans occurred during the Late Cretaceous after Gondwana had split into continents recognized today, and that because these frogs now occur on most Gondwana-derived continents, these frogs either dispersed across oceans (highly unlikely) or previously unidentified land bridges must have been present (Fig. 13.15). Because diversification events in these two major clades are congruent, a single vicariance event is postulated, providing strong evidence for land bridge connections rather than transoceanic dispersal. Several possible scenarios exist in terms of which continents retained connections to explain the observed patterns, but they all share the common element that land bridges must have existed.

Ranoid frogs currently have a nearly global distribution. Franky Bossuyt and his colleagues analyzed molecular data for all known families and subfamilies from throughout the distribution of these frogs and found that each major clade is associated with one historical Gondwanan plate (Fig. 13.16). Their phylogenetic analysis suggest that two colonization routes from Gondwana to Laurasia occurred; one group of ranoids was carried with India (Out of India Hypothesis) when it migrated north, eventually colliding with southern India. The other was along the Australia–New Guinea plate. The notion that frogs would be able to colonize and survive on the drifting Indian continent has not been without controversy. When India was isolated from other landmasses as it drifted north during the Cretaceous–Tertiary Boundary, the extensive Deccan Traps volcanism sent lava flows across the continent, rendering much of it uninhabitable. Ranoids likely survived on a part of the drifting continent that now comprises southern India and Sri Lanka. The endemic ranoid fauna that now exists in the Western Ghats of India and the central highlands of Sri Lanka are derived from the ranoids that drifted with the Indian continent as well. The genera *Micrixalis, Nyctibatrachus, Lankanectes*, and *Indirana* have no living relatives in India, live in a very restricted habitat, and have origins (Cretaceous) predating those of several much older ranoid clades (Fig. 13.17).

The recent discovery of a new frog "family," the Nasikabatrachidae, with a single burrowing species in the Western Ghats of India further attests to the importance of India in transporting very old frog clades. The strange-looking *Nasikabatrachus sahyadrensis* (Fig. 13.18) appears to be the sister taxon to the Sooglossidae, which occurs only on the Seychelles. Most likely, both the Sooglossidae and *Nasikabatrachus* diverged before the breakup of the Seychelles and India, with ancestors of *Nasikabatrachus* disappearing on the Seychelles. We include *Nasikabatrachus* in

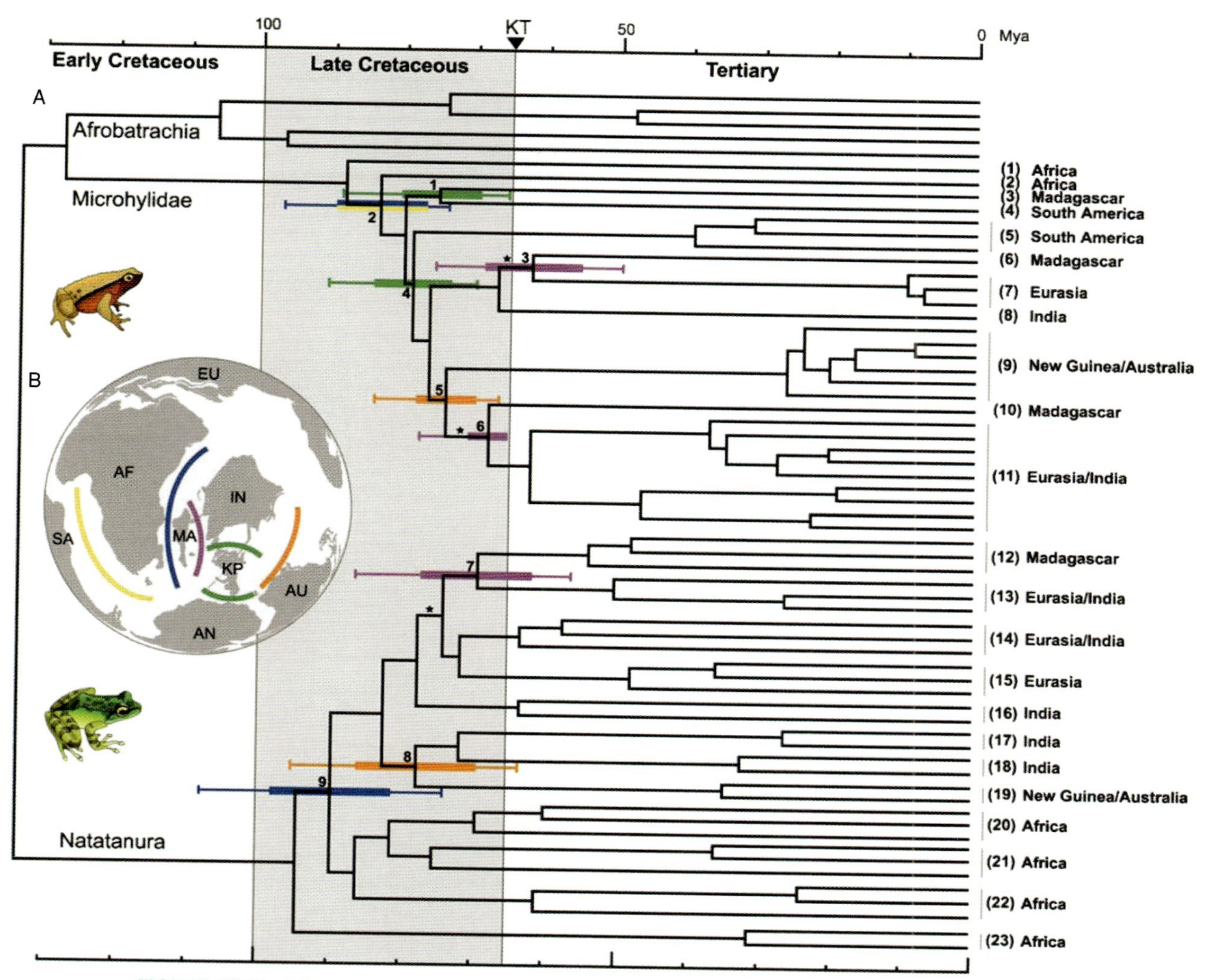

FIGURE 13.15 Divergences that most affected global distribution of Microhylidae and Natanura occurred in the Cretaceous. (A) Molecular time tree phylogeny showing divergence patterns. (B) Horizontal colored bars and lines at interval nodes (standard deviation and 95% credibility intervals) indicate vicariance events as follows: orange: Australia <–> Indo-Madagascar; yellow: Africa <–> South America; blue: Africa <–> Indo-Madagascar; purple: Madagascar <–> India (Seychelles); green: South America–Antarctica <–> Indo-Madagascar (with the Kerguelen Plateau involved). (B) Gondwana in the Late Cretaceous. Abbreviations: AF = Africa, MA = Madagascar, In = India, EU = Eurasia, SA = South America, AN = Antarctica, AU = Australia–New Guinea, KP = Kerguelen Plateau. Adapted from Van Bocxlaer et al., 2006.

the family Sooglossidae as on the *Amphibian Species of the World* Web site.

The bufonids (toads) provide another interesting example of the use of phylogenies in teasing out the distributional history of a major frog clade. With the exception of the Australia–New Guinea plate, the Antarctic, and Madagascar, bufonids have a nearly global distribution. Several alternative hypotheses have been advanced to explain their current distribution, but only recently has a dated phylogeny been used. Jennifer Pramuk and her colleagues produced a Bayesian consensus tree of relationships with time estimates based on a Bayesian algorithm calibrated with fossil data (Fig. 13.19). Bufonids originated in Upper Cretaceous, which confirms that they originated in South America after the breakup of Gondwana (Fig. 13.20). This interpretation is consistent with the lack of fossil bufonids from Madagascar, Australia, and New Guinea. Bufonids dispersed out of the New World and into Europe and Asia during the early Palaeogene. The New World clade that contains *Rhinella, Cranopsis*, and *Anaxyrus* reinvaded the New World during the Eocene on one of three possible land bridges. Divergence time estimates suggest that this most likely occurred in the early Cenozoic (65–40 mybp) on the Thulean land bridge across Iceland–Faeros just below

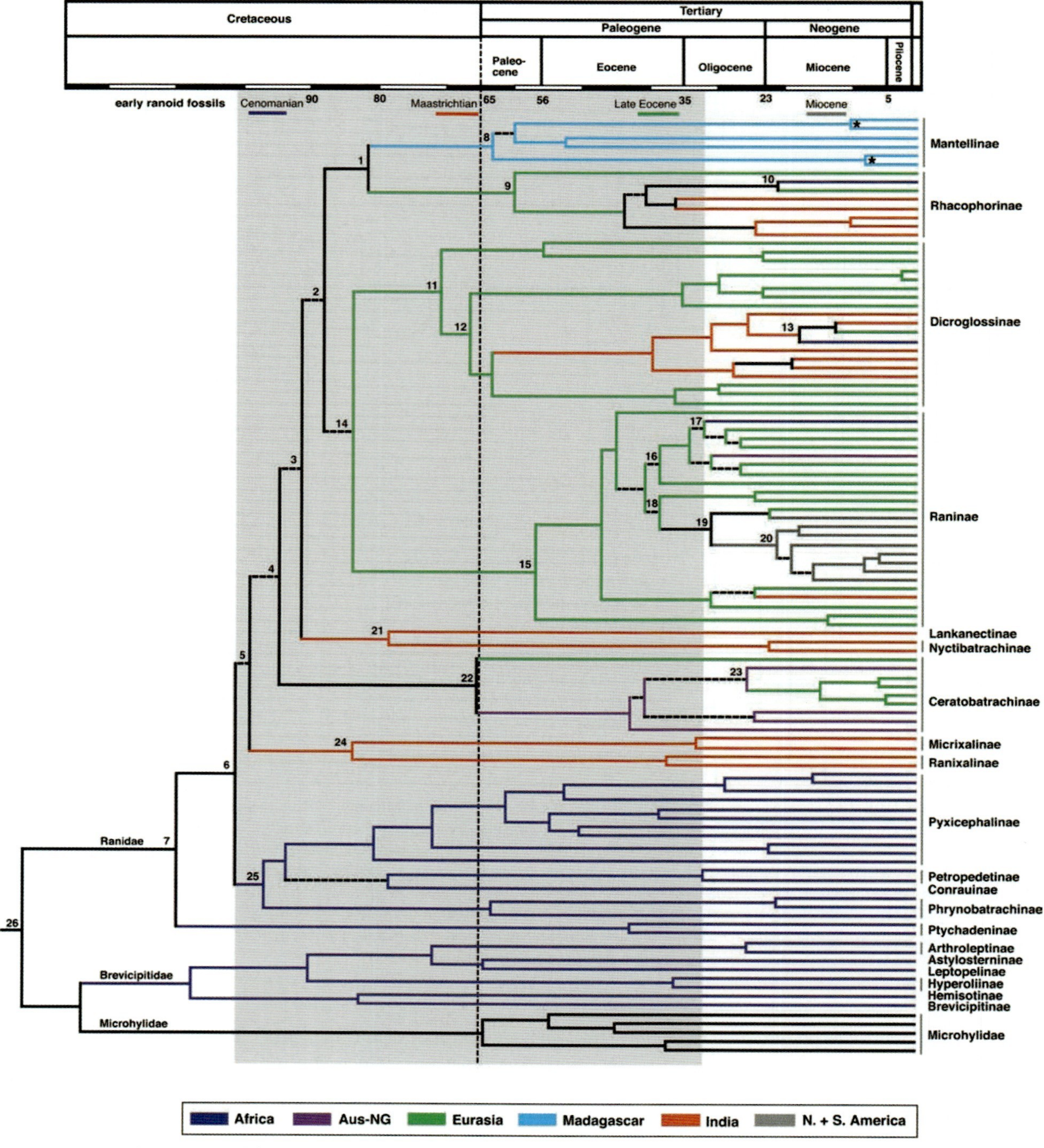

FIGURE 13.16 Biogeographic history of ranoid evolution. Dashed branches are lineages of uncertain phylogenetic position. Colored bars across the top of the phylogeny indicate age of ranoid fossils from their respective continents: (1) undetermined ranoids from the Cenomanian of Africa, (2) Ranidae from the Maastrichtian of India, (3) Raninae from the Late Eocene of Europe, and (4) Raninae from the Miocene of North America. Gray shading indicates an apparent lack of dispersal between Africa and other biogeographic units (between nodes 6 and 17) for about 70 million years. The K–T (Cretaceous–Tertiary) boundary is indicated by the vertical dashed line. Asterisks indicate calibration points. Adapted from Bossuyt et al., 2006.

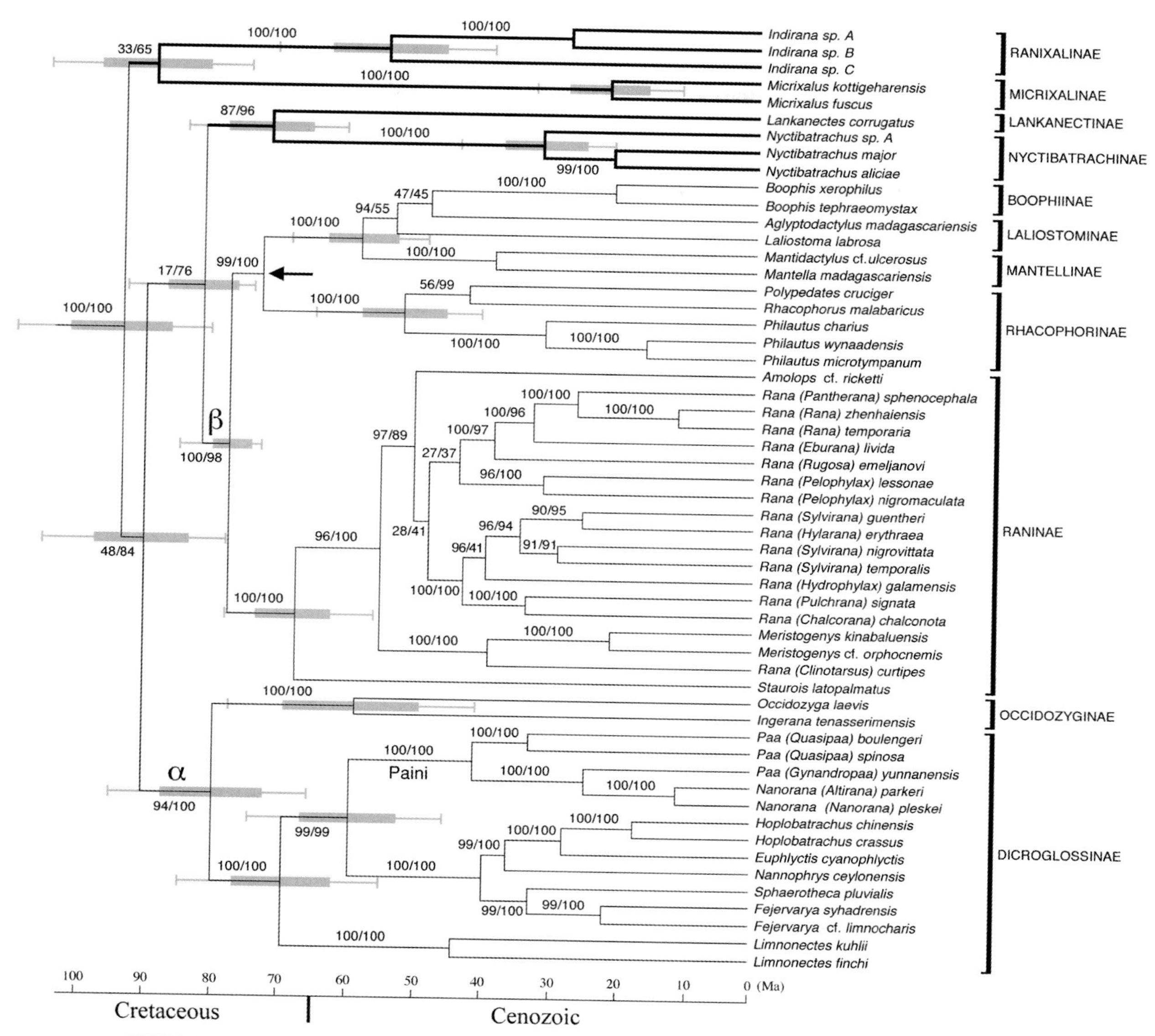

FIGURE 13.17 Dated phylogeny of ranoid frogs centering on the phylogenetic position of four families endemic to the Western Ghats of India and hills of Sri Lanka, the Ranixalinae, Micrixalinae, Lankanectinae, and Nyctibatrachinae. The phylogeny demonstrates that these clades are outside (sister to) other ranoids. Molecular dating places the origin of the clades containing these four subfamilies in the Cretaceous. Adapted from Roelants et al., 2004.

62° N latitude. This dispersal was during the latest Paleocene thermal maximum and provided a time period most suitable for long-range dispersal by ectotherms. Other dispersal patterns are evident in Figure 13.20 as well, but one of the most interesting is the origin of the Caribbean genus *Peltophryne*. The age of this clade falls out at about 51 mybp. Thus it is an old clade but not old enough to have survived the Cretaceous–Tertiary impact event (dated at 65 mybp). Although geological history of the Caribbean is complex, the age of the islands is younger than the age of the *Peltophryne* clade estimate, suggesting a dispersal event. Different from the preceding examples, much of the diversification history of bufonids seems to have occurred relatively recently, with dispersal events followed by reinvasions.

Because of its relative recency, the Caribbean presents an interesting opportunity to examine transoceanic dispersal, as seen in the preceding example with *Peltophryne*. Until recently, the frog genus *Eleutherodactylus* was believed to be the largest genus of vertebrates. Species are distributed across Middle America, South America, and Caribbean islands. A recent phylogeny of these frogs identified three

FIGURE 13.18 The recently described frog *Nasikabatrachus sahyadrensis* is among the oldest of the Neobatrachia and ties the fauna of the Seychelles to the fauna of India. Its ancestors must have been present on the Indo-Madagascan fragment of eastern Gondwana during Middle–Late Jurassic or Early Cretaceous. Photograph by S. D. Biju.

major clades tied to each of these major geographic regions. These frogs originated in South America. The Middle American clade contains about 111 species, the South American clade contains about 393 species, and the Caribbean clade contains about 171 species (Fig. 13.21). Although it has always been assumed that these frogs dispersed by land during the Cretaceous, molecular clock analysis providing approximate dates of divergence indicate that land dispersal was unlikely because the relevant landmasses were not connected. Dispersal occurred much more recently, during the early Cenozoic, with the first transoceanic dispersal event during the Middle Eocene, the second during early Oligocene, a third during early Miocene, and a fourth during the Pliocene (Fig. 13.22).

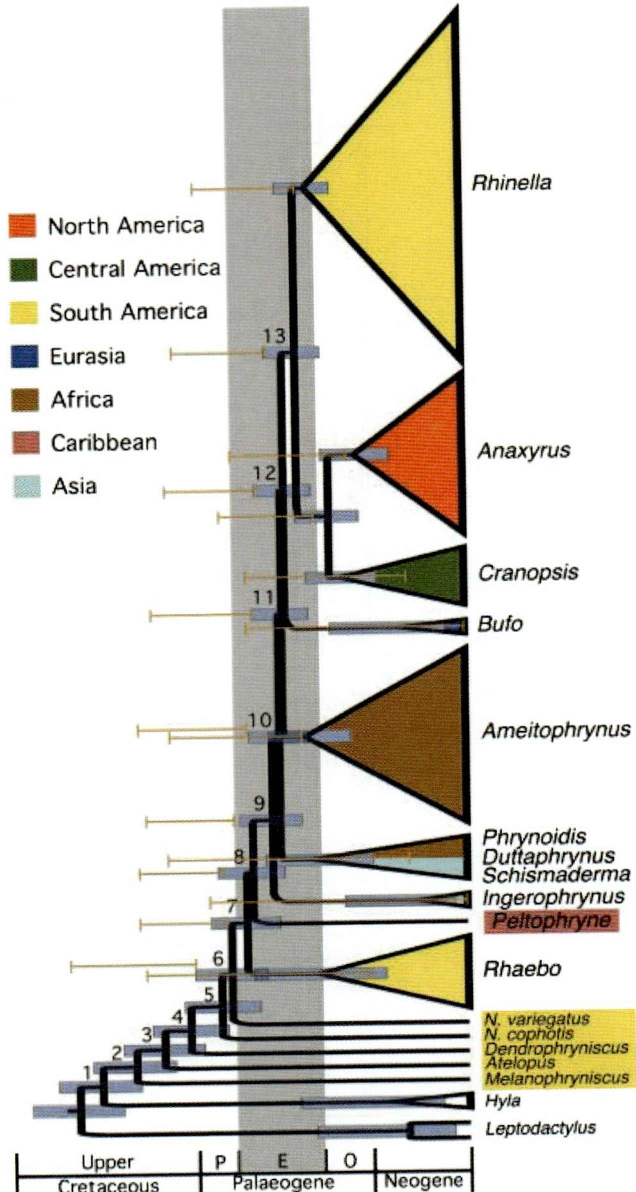

FIGURE 13.19 Early diversification of the Bufonidae occurred near the end of the Upper Cretaceous, failing to confirm a Gondwana origin of the family. Diversification into modern genera occurred later, during the mid-Paleogene. Horizontal bars and shaded rectangles indicate 95% credibility intervals of estimates of divergence times. Colors indicate geographical distributions of each lineage. Adapted from Pramuk et al., 2008.

Historical Biogeography of Caecilians

Similar to other major clades of amphibians, caecilian origins can be traced back to Pangaea, and as a result, their presence in northern and southern continents reflects a combination of very old plate tectonics (Jurassic) and relatively more recent plate tectonics (Cretaceous) (Fig. 13.23). Old World distribution of ancestors of ichthyophiids and uraeotyphlids and combined New and Old World distribution of caeciliid ancestors dates back to the breakup of Pangaea. Ancestors of ichthyophiids, uraeotyphlids, and caeciliids that occur in India today likely rode the India plate as it moved north, ultimately colliding with Asia (the "Out of India" hypothesis). Nevertheless, some interesting patterns of distribution in caecilians illustrate that nontectonic events have influenced distributions of some caecilians. In Africa two, and possibly three, sets of sister taxa have east and west representatives with disjunct distributions (Fig. 13.24). Because overall drying of Africa during the Neogene separated tropical forests in the west from those in the east, the obvious hypothesis explaining this pattern is that a

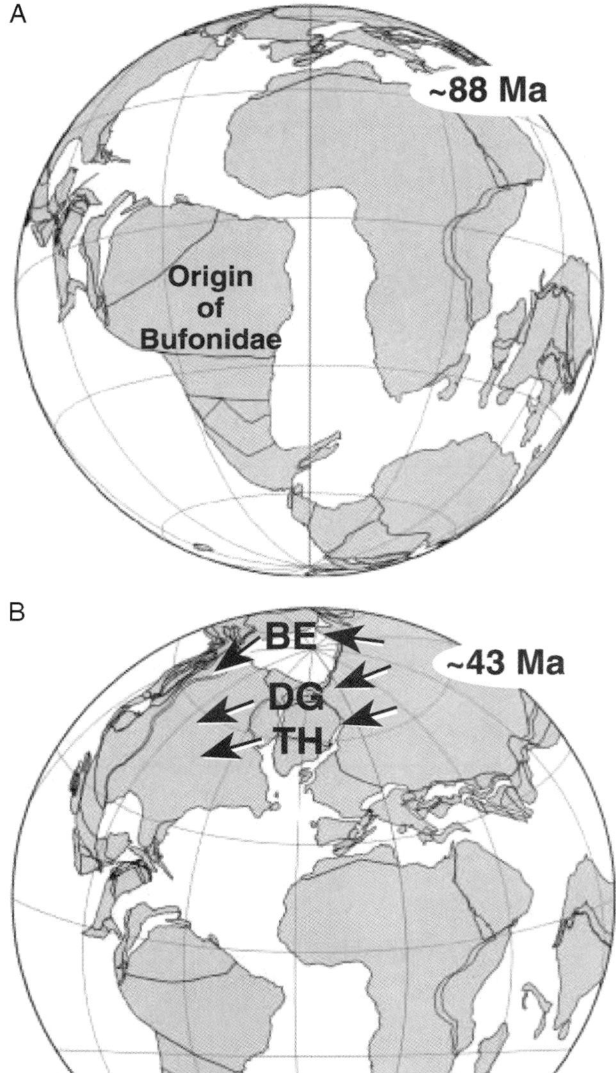

FIGURE 13.20 These maps illustrate the key geological events associated with diversification in the Bufonidae. (A) Bufonids originated in South America about 88 mybp, after the breakup of Gondwana. At some point, bufonids dispersed into the Old World and diversified into the Eurasian and African clades, likely across Beringia. (B) Approximately 43 mybp, during the Eocene, bufonids dispersed back into the New World. Although at least three possible routes existed (Berengia, DeGeer, and Thulean land bridges), the Thulean land bridge is most likely because it provided a much more mild climatic regime. Adapted from Pramuk et al., 2008.

relatively dry barrier was formed separating these species pairs spatially, or more explicitly, that a single biogeographic event accounts for the divergences in these three species pairs. However, relative dating using molecular data allows this hypothesis to be rejected because the timing of divergences is not parallel. The two species pairs *Herpele–Boulengerula* and *Scolecomorphus–Crotaphatrema* may have diverged at about the same time, but the species pair *Schistomatopem thoemse–Schistomatopem gregorii* diverged much more recently. Consequently, at least two biogeographic events must have been at play in the history of east–west divergence in African caecilians. This example shows that absolute dating is not necessary to falsify biogeographic hypotheses, but, of course, good dating would allow a test of whether drying of central Africa corresponds to specific biogeographic events in the history of African caecilians.

Historical Biogeography of Burrowing Reptiles

As we indicated earlier, oceans should be considered major barriers for burrowing species of reptiles. However, a recent study shows that some amphisbaenians crossed the Atlantic Ocean during the Eocene (40 mybp), most likely on floating rafts of land. The 165+ species of amphisbaenians presently occur in Africa, the Middle East, Europe, South America, North America, and some Caribbean islands. Because these animals live underground, the prevailing hypothesis has been that their present-day distributions date back to a Pangaea origin followed by initial separation resulting from the split of Pangaea into Gondwana (southern continent) and Laurasia (northern continent) (200 mybp), followed by the split-up of Gondwana forming Africa and South America (100 mybp). Similar to the preceding Amazonian biodiversity example, this hypothesis can be easily testable with phylogenetic data on amphisbaenians. Nicolas Vidal and colleagues used a molecular data set to demonstrate that amphisbaenian biogeography is much more complex than previously thought. The first major divergence occurred about 109 mybp and likely represents the initial split of Pangaea into Gondwana and Laurasia (Fig. 13.25). Thus Rhineuridae is now represented in North America (Laurasia origin), but all remaining amphisbaenian ancestors remained on Gondwana (southern continent). All other divergences within amphisbaenians occurred during the Cenozoic less than 65 mybp. All amphisbaenians in the New World except *Rhineura* arrived long after Pangaea had split. The divergence between Trogonophidae and Amphisbaenidae likely occurred in the Eocene (51 mybp) in Africa. South American and African Amphisbaenidae diverged in the Eocene about 40 mybp. Thus the only explanation for this divergence is transatlantic dispersal, likely on a floating island, because the distance from Africa to South America exceeded 3500 km. Ancestors of *Cadea*, which is most closely related to European *Blanus*, arrived on Cuba even later (40 mybp), either as a result of transatlantic dispersal or dispersal via Greenland. By this time, a land connection did exist, but even in this case transatlantic dispersal seems more likely than

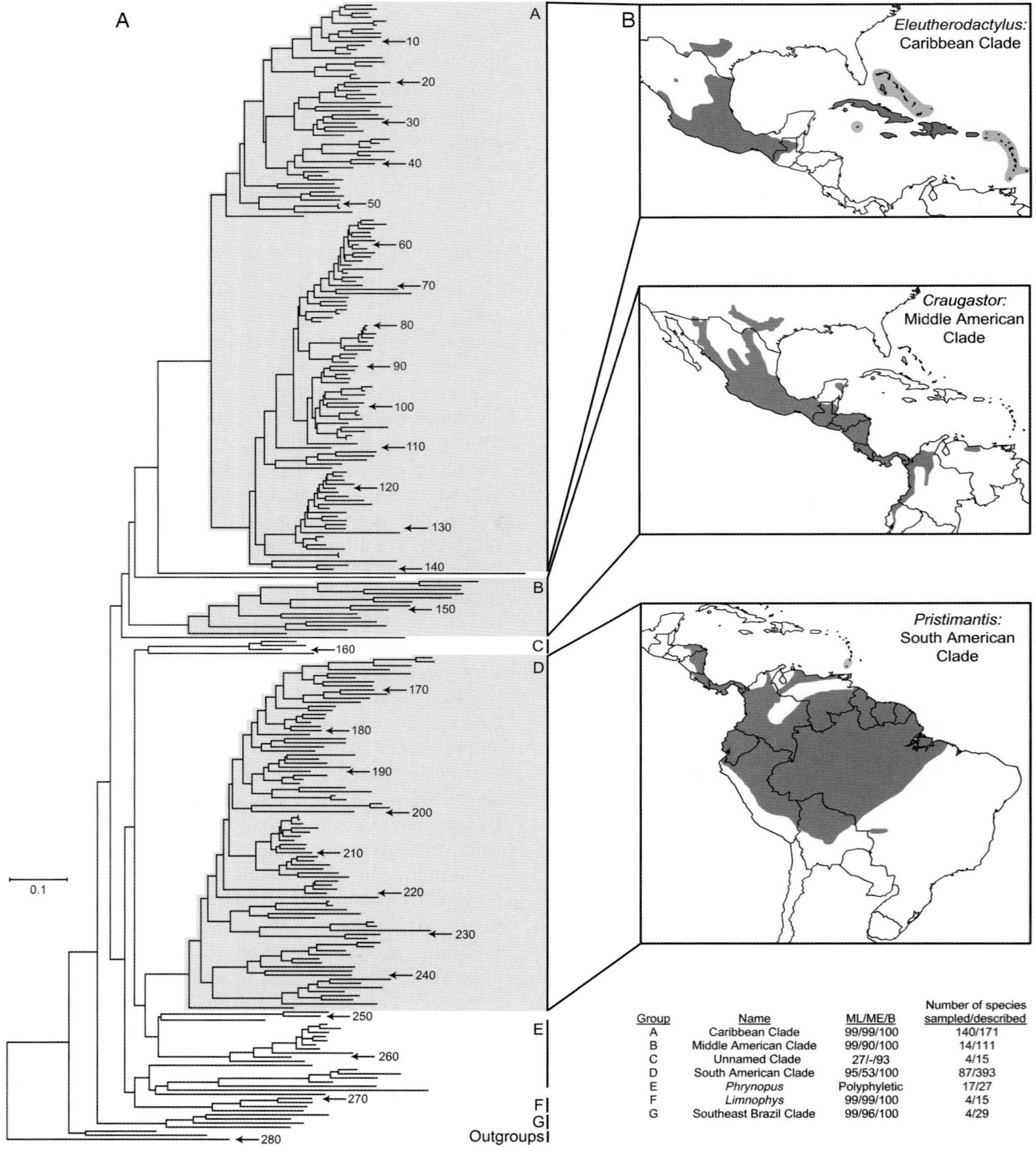

Group	Name	ML/ME/B	Number of species sampled/described
A	Caribbean Clade	99/99/100	140/171
B	Middle American Clade	99/90/100	14/111
C	Unnamed Clade	27/-/93	4/15
D	South American Clade	95/53/100	87/393
E	*Phrynopus*	Polyphyletic	17/27
F	*Limnophys*	99/99/100	4/15
G	Southeast Brazil Clade	99/96/100	4/29

FIGURE 13.21 Phylogenetic relationships of eleutherodactyline frogs showing geographical distribution for each clade. Adapted from Heinicke et al., 2007 (© 2007 National Academy of Sciences, USA).

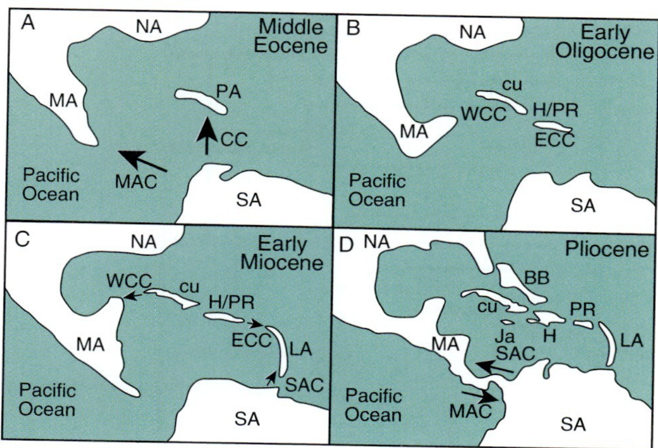

FIGURE 13.22 The origins of Middle American and Caribbean clades of eleutherodactyline frogs can be modeled based on the timing of divergences. (A) Dispersal over water from their South American origin probably occurred during the Middle Eocene (49-37 mybp), resulting in the formation of the Middle American clade (MAC) and the Caribbean clade (CC). (B) Higher sea levels led to isolation of a western Caribbean clade (WCC) on Cuba and an eastern Caribbean clade on Hispaniola and Puerto Rico during the Early Oligocene (approximately 30 mybp). (C) Dispersal from Cuba to the mainland led to the radiation of the subgenus *Syrrhopus* in southern North America during the Early Miocene (approximately 20 mybp). Concurrently, members of the ECC and South American clade (SAC) colonized the Lesser Antilles. (D) The closing of the Isthmus of Panama during the Pliocene (approximately 3 mybp) resulted in overland dispersal of MAC species to South America and SAC species to Middle America. Adapted from Heinicke et al., 2007 (© 2007 National Academy of Sciences, USA).

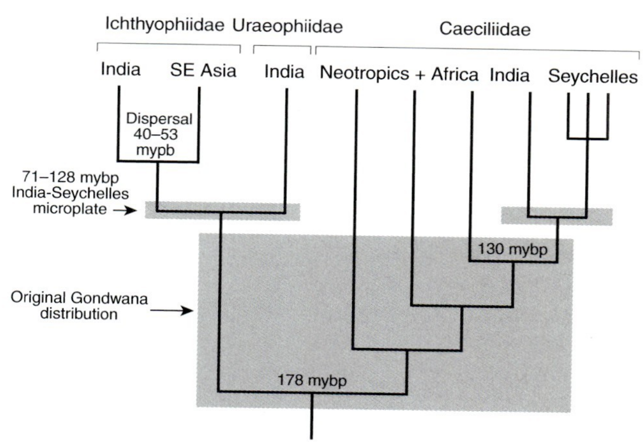

FIGURE 13.23 The first major divergence event in the history of caecilians occurred on Gondwana approximately 178 mybp with several other Gondwana divergences. This deep divergence accounts for the presence of caecilians on most southern continents today. However, an added twist is the much more recent dispersal (40-53 mybp) of members of the Ichthyophiidae into Southeast Asia. Adapted from Pough et al., 2003, and Wilkinson et al., 2002b.

dispersal across a northern land bridge. The preceding explanation is also consistent with the fossil record of amphisbaenians. Although not fossorial, the Cuban gecko *Tarentola americana* may have arrived in a similar way from the Mediterranean.

Historical Biogeography of Malagasy Reptiles

Although many recent biogeography studies reveal dispersal events that account for present-day distributions, some Malagasy reptiles appear to have biogeographic patterns consistent with Gondwana vicariance. Ninety mybp, the combined Madagascar and India plates likely had subaerial connections with Antarctica. Madagascar was connected to Antarctica via the Gunnerus Ridge, and India was connected via the Kerguelen Plateau (Fig. 13.26). And, of course, South America was connected to Antarctica to the west. Malagasy boid snakes, podocnemid turtles, and pleurodont iguanian lizards date to at least 75, 80, and 67 mybp, respectively, indicating that dispersal origins for these taxa were highly unlikely, and these taxa were likely isolated as the result of a single vicariant event during the Late Cretaceous (Fig. 13.27). Other estimates of the ages of these taxa place their divergences slightly earlier. The taxa would have to be much older (about 160 mypb) to have arisen from an African vicariant event and much younger (65 mybp) to have arisen from Laurasian vicariance. The single iguanid genus in Fiji (*Brachylophus*), which is nested within the South American iguanids, does represent a dispersal event much later from South America. If vicariance accounts for presence of boid snakes, podocnemid turtles, and pleurodont iguanian lizards on Madagascar, then they should also occur in India and Australia. Fossil iguanids and the sister group to podocnemids (the extinct Bothremydidae) did occur in India. The absence of extant podocnemids and iguanids in India has been attributed to the effect of the Deccan Traps volcanism, but keep in mind that some frog taxa were able to survive on the Indian continent during this time period.

Biogeography in the Recent Past

Although many studies of historical biogeography center on the relatively distant past, the powerful tools provided by gene sequence data allow close examination of relatively recent biogeographic events and their impact on patterns of speciation and distribution as well. Many examples exist, and similar to studies dealing with deep history, these studies are expanding rapidly because we can often tie divergence events to landscape changes. For example, mtDNA-haplotype data identify 13 independent lineages in the *Eurycea bislineata* complex (Plethodontidae) of eastern North America, indicating that species diversity is much higher than previously thought. These salamanders are tied to stream systems and can be common in many localities. Phylogeographic divergence in these salamanders is linked with historical drainage patterns (mid-Miocene and Pleistocene) rather than current ones. Shifts in the drainage patterns during glacial events split populations, resulting in the fragmentation that we see today. In this case, interruption of historic stream patterns was the vicariant event leading to separation of populations.

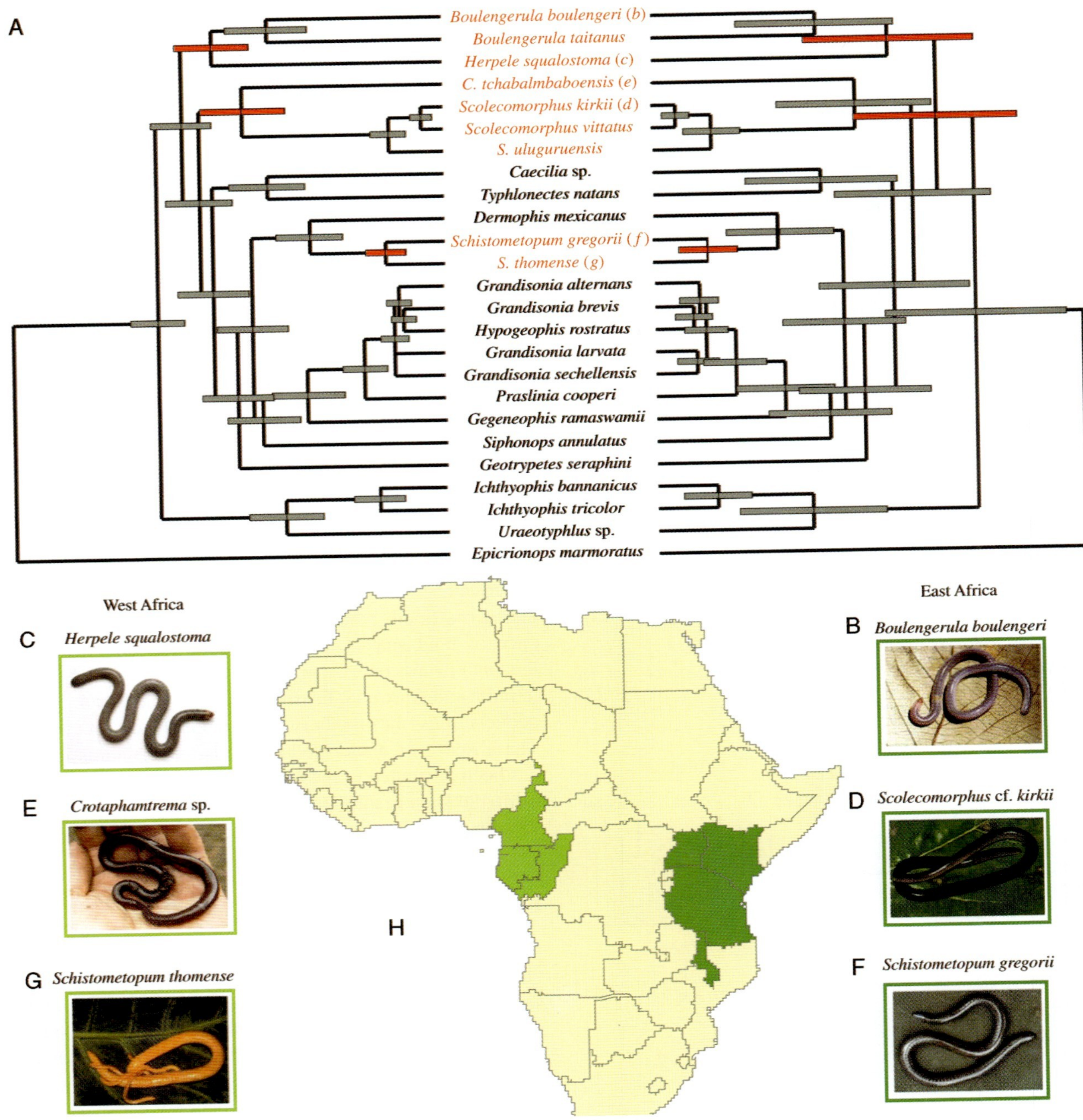

FIGURE 13.24 Divergence of African caecilians cannot be tied to a single biogeographical event. (A) Left, phylogeny based on 12S and 16S gene sequences; right, uncorrected lognormal molecular clock showing divergence times. (B through G) West (C, E, G) and East (B, D, F) African caecilians. (H) Map of Africa showing current non-overlapping distributions of West and East African caecilians. Adapted from Loader et al., 2007.

Crotaphytid lizards (Collared [*Crotaphytus*] and Leopard [*Gambelia*] lizards) provide an example of the complexities involved in relatively recent biogeographic events. Not only have populations experienced vicariance events in distant and recent past, but some populations also have come back together resulting in gene flow after considerable differentiation. These lizards occur across western and central North America extending into Baja, California, and northwestern Mexico. They are medium-sized lizards and are well known by most naturalists because of their size, conspicuousness, and ability to defend themselves when captured by inflicting painful bites that frequently break the skin. For many years, only a few species were recognized. Jimmy McGuire and his

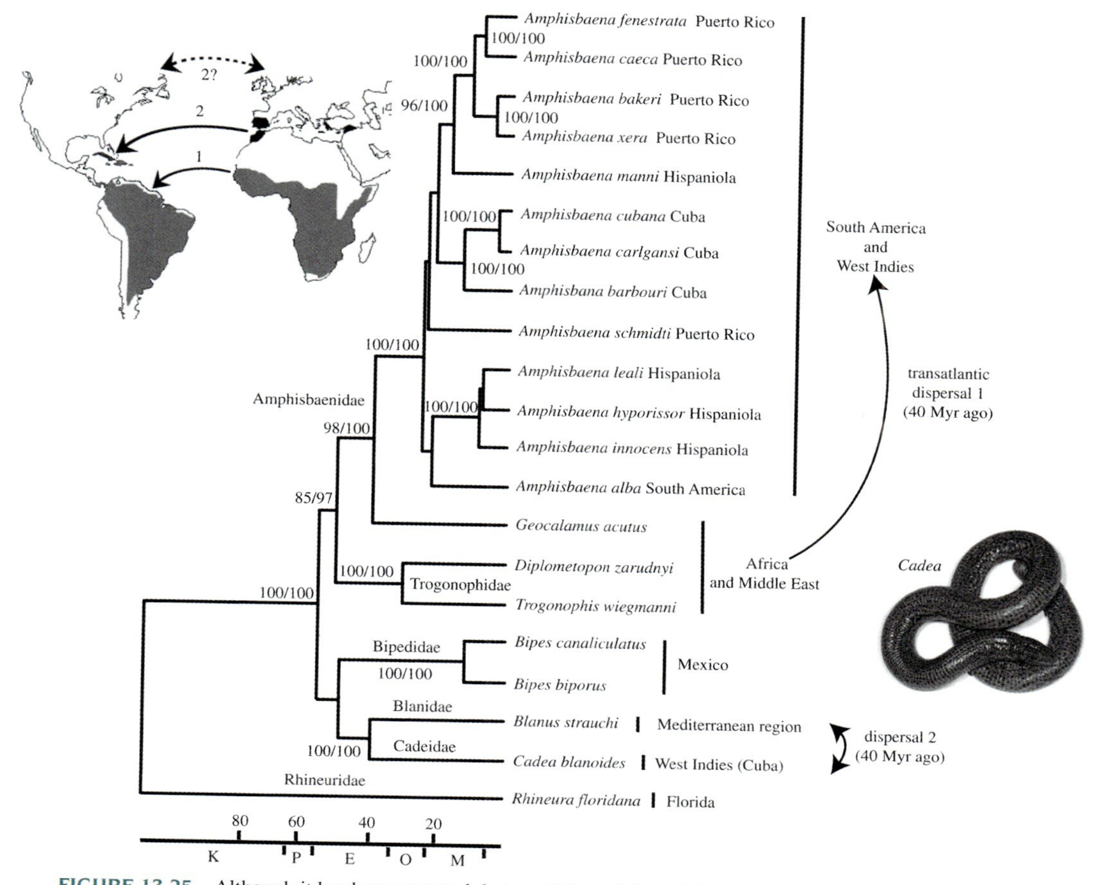

FIGURE 13.25 Although it has been assumed that small fossorial amphibians and reptiles would not be able to disperse across oceans, it appears that amphisbaenians have done just that. Based on the dated phylogeny and the position of landmasses at the time, the only supportable hypothesis for dispersal of Amphisbaenidae ancestors to the New World is transatlantic during the Eocene (arrow 1, upper left). The most likely hypothesis for dispersal of cadeids is transatlantic during the Eocene (solid arrow 2, upper left), although a complex terrestrial dispersal cannot be ruled out (dashed arrow 2, upper left). Adapted from Vidal et al., 2008.

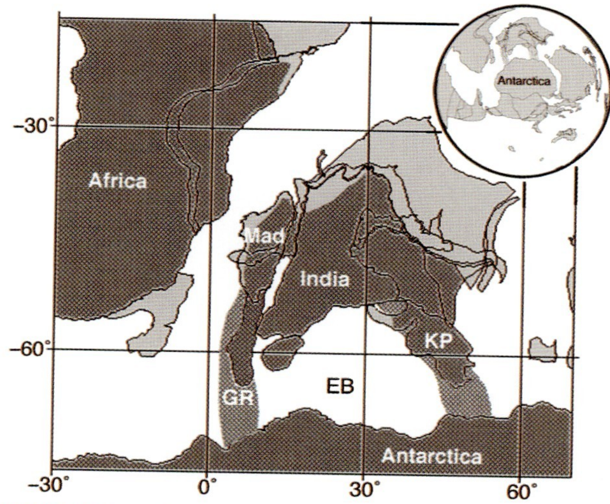

FIGURE 13.26 Subaerial (surface) connections between Madagascar and Antarctica likely existed approximately 90 mybp. Dark shading indicates submerged areas. Mad = Madagascar, KP = Kerguelen Plateau, GR = Gunnerus Ridge, and EB = Enderby Basin. Adapted from Noonen and Chippindale, 2006.

colleagues began studying collared and leopard lizards to understand their relationships by examining their morphology. This resulted in descriptions of several unrecognized species. Further work using molecular data has unraveled some of their interesting recent evolutionary history in the context of biogeography, and shows that morphological data overestimated species diversity in one clade and provided relationships based on gene sequence data inconsistent with morphological data in the other clade. The first step was to develop a phylogeny of sampled populations to determine relationships of known species. Three species of *Gambelia* had been recognized based on morphology, *G. wislizenii*, *G. sila*, and *G. copei*. The phylogenetic analysis based on mtDNA haplotypes revealed that *G. copei* is nested within *G. wislizenii* and thus does not appear to represent a distinct taxon. The situation is much more complex among the nine recognized species of *Crotaphytus* (Fig. 13.28). Based on gene sequences, northern and southern populations of both *Crotaphytus nebrius* and *C. vestigium* are not each other's closest relatives, and eastern and western populations of *C. collaris*

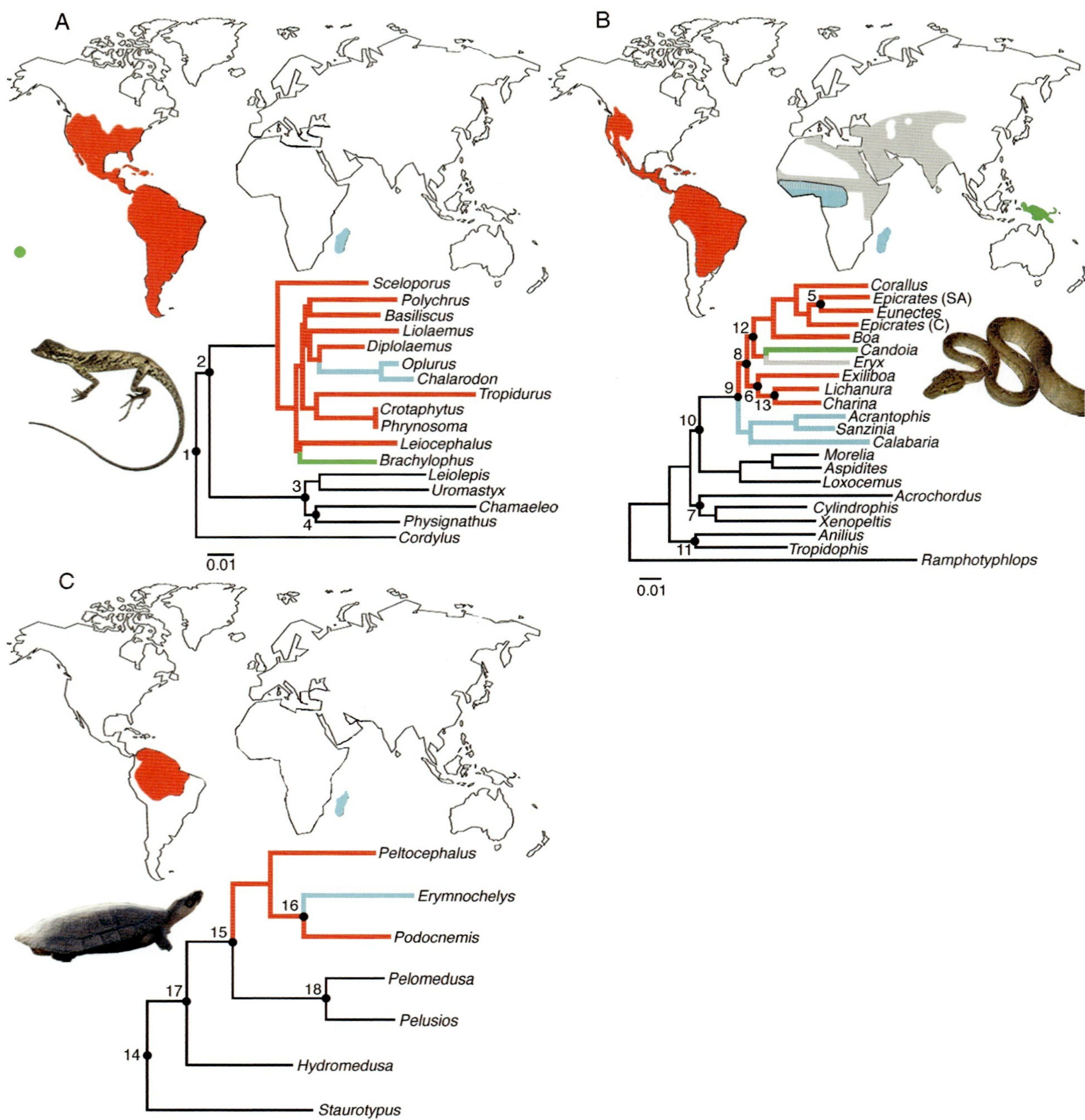

FIGURE 13.27 Phylogenetic relationships of three reptile clades: (A) pleurodont iguanid lizards, (B) boine snakes (including Ungaliophiidae and erycine genera *Eryx, Charina, Calabraria*, and *Lichanura*), and (C) Podocnemid turtles. Clades of interest are indicated by thick branches, and colors correspond to shaded geographical distributions. Adapted from Noonen and Chippindale, 2006.

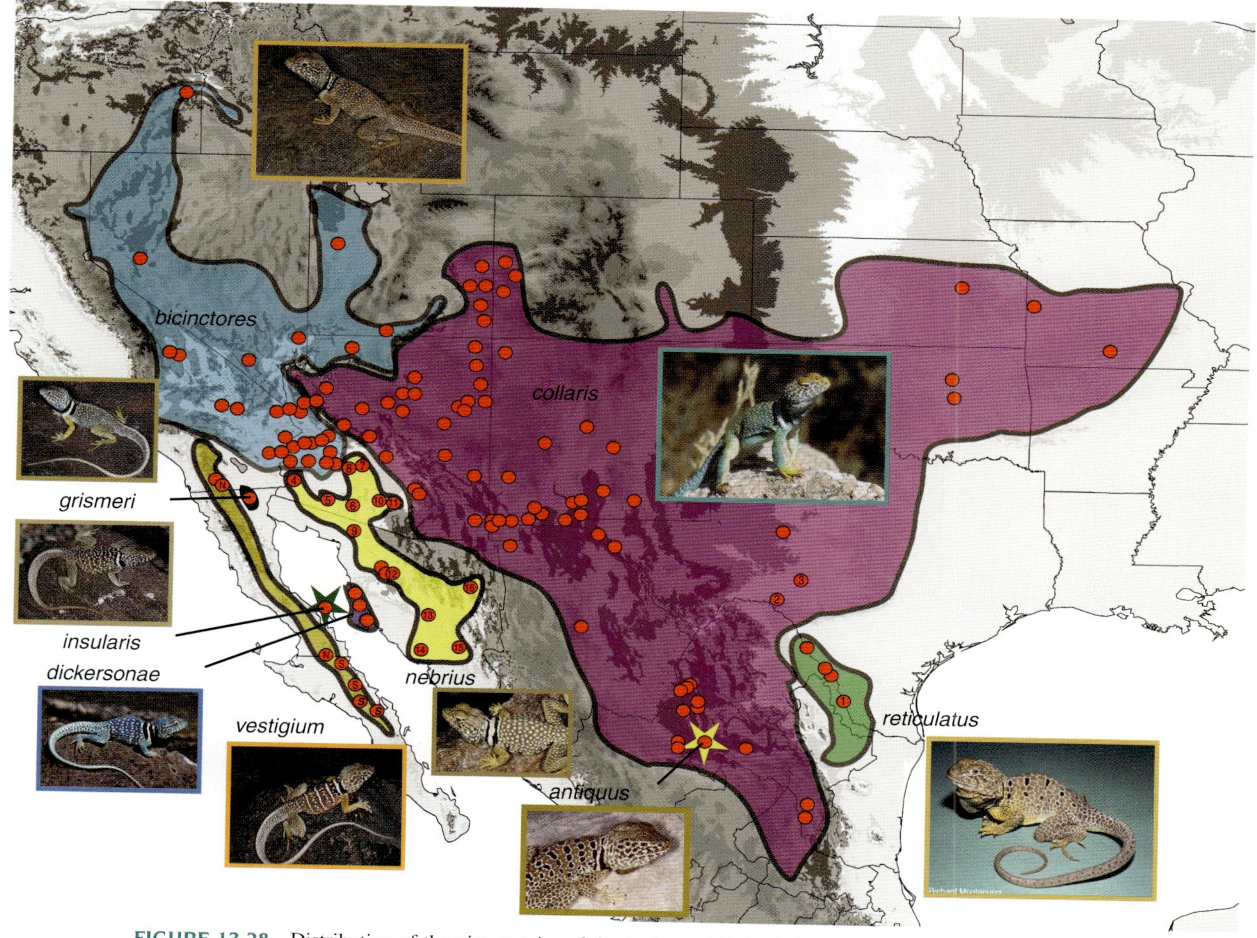

FIGURE 13.28 Distribution of the nine species of *Crotaphytus*. Circles indicate sampling localities for phylogenetic analysis. Adapted from McGuire et al., 2007.

are not each other's closest relatives (Fig. 13.29). Other apparent cases of paraphyly and polyphyly in the phylogeny exist as well. Although at first pass one might conclude that convergent morphological evolution occurred resulting in the mismatch between morphology and genes, the explanation lies in understanding the recent biogeographical history of populations of *Crotaphytus*. To examine this, Jimmy McGuire and colleagues combined recent ecological techniques with biogeographic analyses to reconstruct probable historical distributions. Using niche modeling, it is possible to determine environmental correlates of the present-day distributions of species and then use those data to model past distributions based on the history of past climates. By combining niche models with haplotype trees, the distributional history can be reconstructed. Present-day distributions of these species differ from what they were in the past, and species ranges have come in contact repeatedly, allowing introgression (movement of genes from one population into another). Consequently, *C. bicintores* appears in four different sections of the *C. collaris* topology, and other examples are apparent in the gene tree (Fig. 13.29). What makes *C. bicintores* most interesting is that introgression appears to have occurred at least three different times in about the same place during the last few million years. *Crotaphytus bicintores* has remained morphologically distinct from *C. collaris* but has picked up mitochondrial gene sequences through hybridization repeatedly with *C. collaris*. This process is described as an "introgression conveyor" (Fig. 13.30).

Although glaciation events during the Pleistocene have resulted in divergences in some lineages, they have been less important in others. Five-lined skinks occur across most of eastern North America and have always been considered a single wide-ranging and relatively uniform species. However, a recent phylogenetic analysis based on mtDNA data identifies several divergences that predate the Pleistocene and several that coincide with Pleistocene glaciation. Three main lineages are distributed from east to west (East, Central, and West clades). Three other geographically restricted lineages exist in Oklahoma, Wisconsin, and the

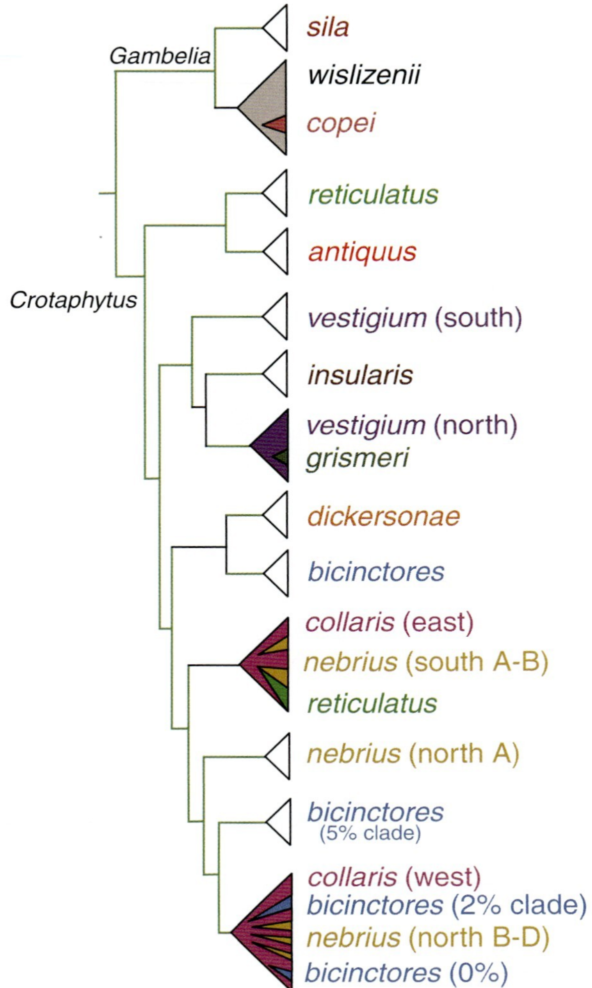

FIGURE 13.29 Phylogenetic relationships of crotaphytid lizards based on mtDNA sequence analysis. Note that species identified on the basis of morphology (names) do not sort out on the gene tree. Rather, some species (e.g., *C. bicintores*) are nested in clades with other species. For a more detailed phylogeny, see original article. Adapted from McGuire et al., 2007.

Carolinas. The Pleistocene vicariance event was caused by the Mississippi River. Glacial melt during the Pleistocene expanded the Mississippi River so that it became a barrier, splitting these skinks into east and west populations. However, populations that had been split prior to the Pleistocene formed several of the haplotype groups of these lizards. Glacial water was reduced by the end of the Pleistocene (8000 ybp), and the Mississippi was a much smaller meandering river. Skink populations that had been separated during the Pleistocene were able to disperse once again.

Summary

In this chapter, we focused on processes of biogeography, a field that is changing rapidly. Many additional examples of recent historical biogeography studies exist, and we refer the interested reader to the original literature for these. The advent of molecular-based phylogenies that can be calibrated to estimate divergence times allow testing of vicariance hypotheses, and reconstructions of historical environments based on species' niches has allowed biogeographers to pose and answer new questions about the history of amphibians and reptiles. We anticipate that biogeography, phylogenetics, and ecology will come together as one of the most powerful approaches to understanding the history of diversification and distribution of organisms. As a field that transformed from a purely descriptive science (correlating distributions with past events) to a hypothesis-testing science, biogeography is a frontier across the entire history of life.

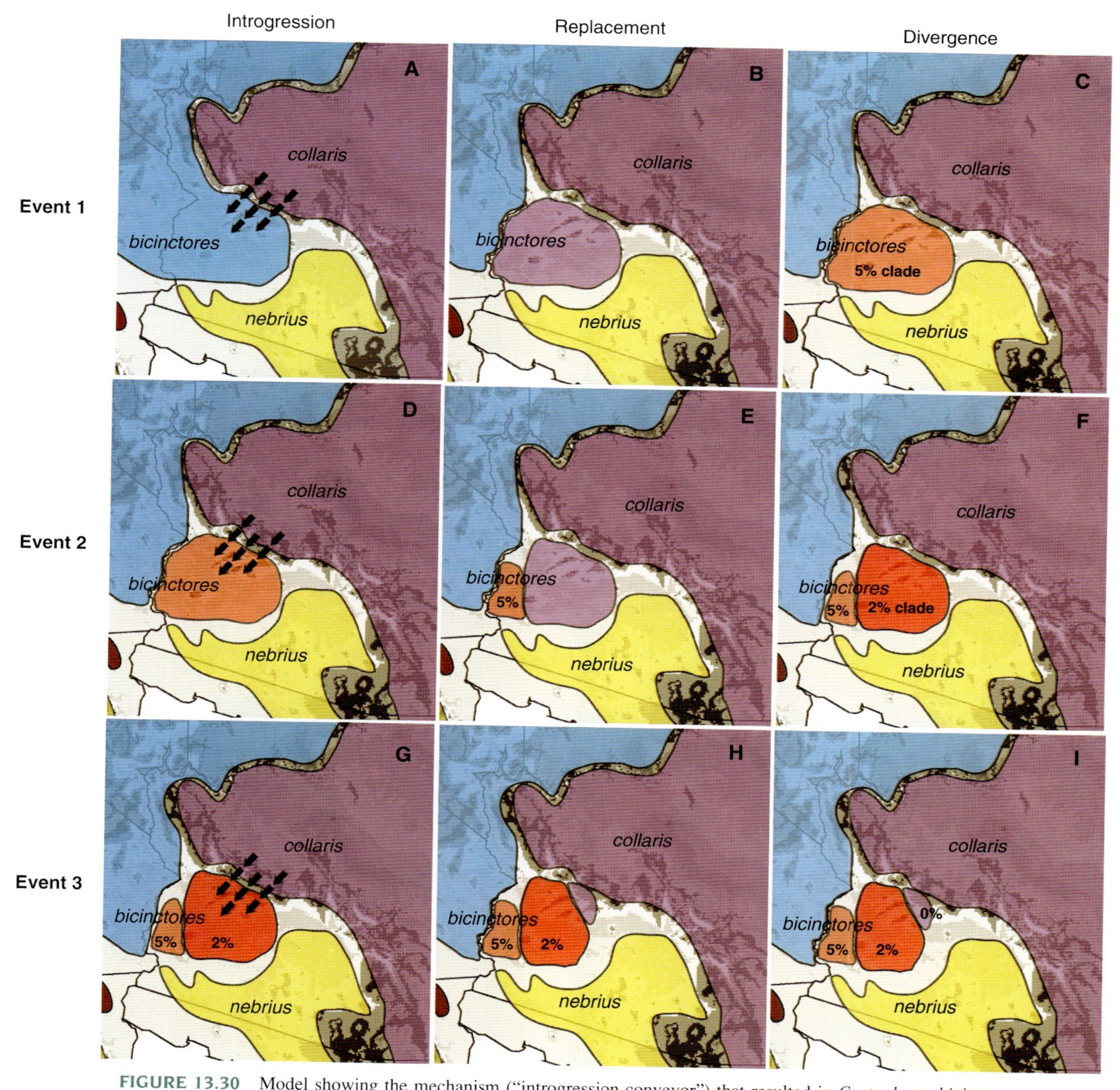

FIGURE 13.30 Model showing the mechanism ("introgression conveyor") that resulted in *Crotaphytus bicinctores* populations in southwest Arizona acquiring mitochondrial haplotypes from adjacent populations of *C. collaris*. Timing for introgression events 1, 2, and 3 are estimated at 3 mybp, 1 mybp, and recent, respectively. Adapted from McGuire et al., 2007.

QUESTIONS

1. What is phylogeography and why is it such a rapidly emerging field?
2. What is the "out of India" hypothesis and how does it relate to biogeography? Provide at least one amphibian or reptile example.
3. Describe in some detail how phylogenies have helped us understand the high diversity of frogs in tropical rain forests of South America.
4. What is an area cladogram, how does it work, and how can it be applied to testing hypotheses in biogeography?
5. What are the differences between ecological and historical biogeography?
6. How did the use of a phylogeny for lizards in the *Anolis nitens* complex resolve the issue of whether these lizards diversified in the Pleistocene?

ADDITIONAL READING

Avise, J. C. (2000). *Phylogeography: the History and Formation of Species.* Harvard University Press, Cambridge, Mass.

Brown, J. H., and Lomolino, M. V. (1998). *Biogeography* (2nd Ed.). Sinauer Associates, Inc., Sunderland MA.

Humphries, C. J., and Parenti, L. R. (1999). *Cladistic Biogeography: Interpreting Patterns of Plant and Animal Distributions* (2nd ed.). Oxford University Press, Oxford.

MacArthur, R. H., and Wilson, E. O. (1967). *The Theory of Island Biogeography.* Princeton University Press, Princeton, NJ.

Wegener, A. (1968). *The origin of continents and oceans.* (translated from the 4th German version by John Biram with an introduction by B. C. King.). Methuen, London.

Wiley, E. O. (1988). Vicariance biogeography. *Annual Review of Ecology and Systematics* **19:** 513–542.

REFERENCES

General

Brown and Lomolino, 1998; Fitch, 1985; Iverson et al., 1997; MacArthur and Wilson, 1967.

Distinguishing Between Ecological and Historical Biogeography

Brown and Lomolino, 1998; MacArthur and Wilson, 1967; Wiens and Donoghue, 2004.

Ecological Determinants of Species Distributions

Conant and Collins, 1991, 1998; Elton, 2001; Greene, 1997; Grinnell, 1917; Hutchinson, 1957; Lampo and De Leo, 1998; Phillips and Shine, 2004; Rodda et al., 1999; Urban et al., 2007.

Biomes and Biogeographical Realms

Cox and Moore, 1985; De Blij et al., 1995; Odum, 1993; Rogers and Kerstetter, 1974; Schultz, 2005; Udvardy, 1975.

Historical Determinants of Species Distributions

Brown and Lomolino, 1998; Futuyuma, 1998; Gaffney, 1977; Georges et al., 1998; Nussbaum, 1985a; Seddon et al., 1997.

Recovering History: Phylogenetic Approaches to Biogeography

Pough et al., 2003.

Amazon Biodiversity

Colinvaux, 1996; Darst and Cannatella, 2004; Endler, 1997; Funk et al., 2007; Gascon et al., 1996, 1998; Glor et al., 2001; Graham et al., 2004; Haffer, 1969, 1974, 1979; Hedges, 2001a,b; Heyer and Maxon, 1982; Kronauer et al., 2005; Moritz et al., 2000; Nores, 1999; Patten et al., 2000; Prance, 1973, 1982; Räsänan et al., 1995; Vanzolini, 1973, 2002; Vanzolini and Williams, 1970a,b; 1981; Vitt and Zani, 1996c.

Historical Biogeography of Amphibians

Bossuyt and Milinkovitch, 2000, 2001; Bossuyt et al., 2006; Van Bocxlaer et al., 2006; Darst and Cannatella, 2004; Duellman and Trueb, 1994; Inger, 1999; Roelants et al., 2004, 2007; Vences et al., 2000.

Historical Biogeography of Caecilians

Loader et al., 2007; Nussbaum, 1985a,b; Wilkinson et al., 2002, 2003.

Historical Biogeography of Malagasy Reptiles

Gaffney and Forster, 2003; Gaffney et al., 2003; Laurent, 1979; Noonan, 2000; Noonan and Chippendale, 2006; Raxworthy et al., 2002; Sites et al., 1996; Vences et al., 2001a, 2001b.

Historical Biogeography of Burrowing Reptiles

Augé, 2005; Carranza et al., 2000; Gans, 1978; Kearney, 2003; Kearney and Stuart, 2004; Raxworthy et al., 2002; Sullivan, 1985; Vidal et al., 2008.

Biogeography in the Recent Past

Howes et al., 2006; Kozak et al., 2006; McGuire, 1996; McGuire et al., 2007; Masta et al., 2002; Montanucci, 1974, 1983.

Chapter 14

Conservation Biology

I dreamed I saw the knights in armor coming, saying something about a queen
There were peasants singing and drummers drumming and the archers split the tree
There was a fanfare blowin' to the sun, there was floating on the breeze
Look at Mother Nature on the run in the 1970s
Look at Mother Nature on the run in the 1970s

Neil Young, 1970
After the Gold Rush, courtesy of Warner Brothers.

Conservation biology is no longer a fledgling subject. The primary journal in the field, *Conservation Biology*, published its 101st issue in June 2006, and the editors and former editors, Gary Meffe, David Ehrenfeld, and Reed Noss, took the opportunity to look back at their accomplishments and to the future at the urgent work that lies ahead. The accomplishments are many: The society has attracted a large membership, and the journal is recognized as one of the top scholarly journals in the field. Other journals have come into existence and are publishing excellent work in the field. Several, including *Amphibian and Reptile Conservation*, are specific to amphibians and reptiles, and some herpetological journals now include subsections under conservation. Many conservation-oriented Web sites focus on amphibians and reptiles as well. Subdisciplines within conservation biology, including conservation genetics, restoration ecology, landscape ecology, and many others, have developed in recent years. From all scholarly viewpoints, conservation biology has emerged as a true scientific discipline and has succeeded in providing an understanding of many of the underpinnings of the field, including effects of pollution on populations of plants and animals, how to approach restoration of various habitats, how to manage endangered species, and many other topics too numerous to mention.

Yet, in spite of all the successes, conservation biology has not achieved what its practitioners hold most dearly: the reversal of the tremendous loss of biodiversity, natural habitats, and even ecosystems that is occurring unabated throughout the world. Although we can find local success stories, and these should be applauded, the overall picture for most groups of plants and animals is a steady decline in number of individuals and populations and, ultimately, species. Thus, the future of conservation biology, and whether we are to succeed in reversing the depressing trends we see every day, lies in coming to terms with why the excellent scientific framework has not translated into real-world change and how new paths can be forged that will make a real difference.

The reasons for the lack of success in the real world cannot be blamed on conservation biologists. Rather, the situation in which conservation biologists find themselves is complex and multifaceted, involving governments that view growth as positive over all other considerations, the rapidly increasing human population, unsustainable harvests of many resources, and the threat of global climate change brought on by human use of products that result

Herpetology, 3rd Ed.

in greenhouse gas emissions. In recent years, as the human population expands and technology becomes more pervasive, we have seen an ever-increasing disconnect between people and nature. Children no longer play outside with abandon and make natural history discoveries on their own; instead, television, computers, cell phones, and other devices compete for leisure time. Another reason for this disconnect is that an estimated 48% of all people worldwide now live in cities, and the diversity of plants and animals with which they interact is homogenized in the sense that the same city-adapted species are seen day after day.

Solutions to these problems at a time when adults are busy with fast-paced lives and outdoor activities of children are structured at local parks and ball fields are badly needed. Children especially must be reconnected with the natural world. Children naturally love nature, and if given the chance, they will make discoveries on their own. Research has shown that children who are exposed to wild areas appreciate and value these areas as adults. But urban parks and outdoor playgrounds do not supply the appropriate habitats because of their structured settings and typically low diversity of plants and animals. One model that has been successful in helping to reverse this trend is the Chicago Wilderness area, which is a regional reserve of 225,000 acres encompassing three states and overseen by a consortium of more that 175 federal, state, and private organizations. The reserve consists of wild areas that are encompassed in federal lands, state parks, county preserves, and many other locales that include prairies, wetlands, woods, and other habitats. A great number of volunteers have made this enormously successful project work, and people are learning about their biodiversity, restoration ecology, and how best to educate children and others about conservation. Efforts such as these are necessary at a global level to save habitats and provide a connection with the natural world. With the current rapid conversion of farmland and natural habitats to residential areas in the United States and other parts of the world, the time to set aside these areas is running out.

The relationship between economics, resource use, and conservation is only beginning to be explored and debated (Fig. 14.1). Without consideration of how our consumption of material goods drives our economy and how we can make changes that affect the world, we will not be able to save our wild areas and their tremendous diversity. We must also consider the size of the human population and begin to deal openly with this subject. We explore these areas in the sections of this chapter.

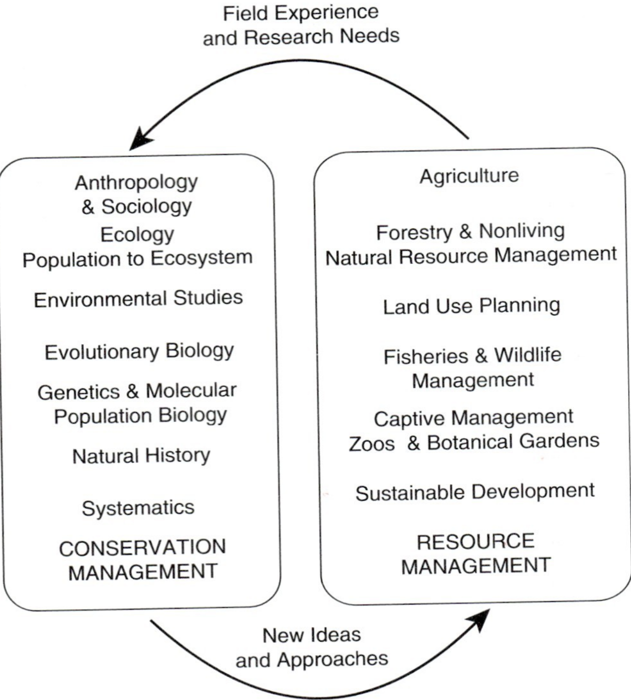

FIGURE 14.1 Conservation biology encompasses a broad range of biological and social studies to address issues and problems arising from and recognized through the use and management of natural resources. Adapted from Temple, 1991.

GENERAL PRINCIPLES

A major focus of conservation biology is the maintenance of the world's biodiversity. Biological diversity is the product of organic evolution, and biological processes from the molecular level involving DNA to the biosphere are not intelligible without reference to organic evolution. Organisms continually interact with their abiotic and biotic environments; intraspecific and interspecific interactions create the numerous local ecological theaters. Nowhere is the ecological play the same; players and conditions constantly change, bringing about a dynamic and evolving ecological world. Species and species interactions that we observe today often have a deep and complex evolutionary history, and virtually all biological communities have a history involving drifting continents, diversification events, dispersals, and climatic change. Nevertheless, the impact of a single species, *Homo sapiens*, may prove to be greater than all historic and biological interaction effects combined in terms of causing dramatic change on a global scale.

Humans have impacted natural landscapes and their living components from the time that they became organized as hunters–gatherers. More recently, human impact has increased dramatically in conjunction with exponential population growth (Fig. 14.2). In the 20th century alone, the world population size soared from less than 2 billion in the early 1900s to 6 billion in 1999. Because the growth is exponential, the time required to add another billion people to the world is now about 11 years. The Earth's human population is currently increasing by 211,090 people *per day*, and conservative estimates predict that the population will reach nearly 9 billion by 2050. Although the fertility

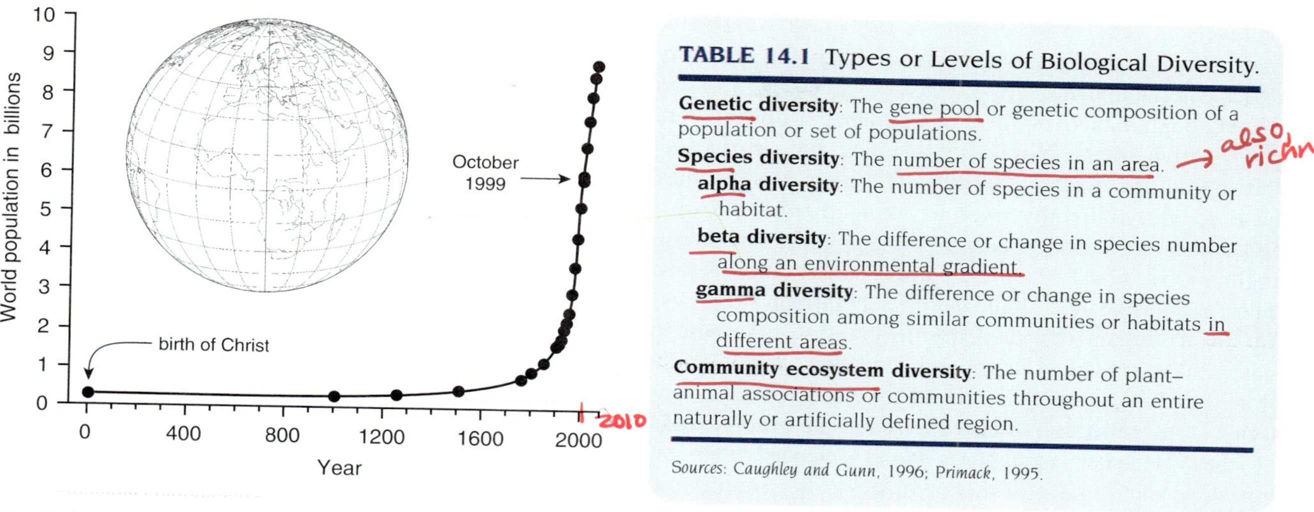

FIGURE 14.2 World population growth during the last 2000 years.

TABLE 14.1 Types or Levels of Biological Diversity.

Genetic diversity: The gene pool or genetic composition of a population or set of populations.
Species diversity: The number of species in an area.
alpha diversity: The number of species in a community or habitat.
beta diversity: The difference or change in species number along an environmental gradient.
gamma diversity: The difference or change in species composition among similar communities or habitats in different areas.
Community ecosystem diversity: The number of plant–animal associations or communities throughout an entire naturally or artificially defined region.

Sources: *Caughley and Gunn*, 1996; *Primack*, 1995.

rate is declining worldwide, the population will continue to grow until 2050, when it will reach 9.2 billion. This figure is dependent on the continued decline of the fertility rate; should this rate cease to decline by only a small amount, the world population could reach 10.8 billion by 2050.

The fertility rate is not the same throughout the world; developing countries have a higher fertility rate, and much of the growth of the population will be in these regions. However, the current fertility rate of the United States is 2.13, which is the highest since 1971. In the United States, much of the increase in population is due to immigration. Analyses by the organization Negative Population Growth, which is dedicated to achievement of a stable population that ensures a reasonable standard of living for all while protecting the environment, shows that the foreign-born population in the United States is now 31 million, a 57% increase since 1990. If the present population of the United States continues to increase on the same trajectory, the population will reach over 400 million by the year 2050. The consequences of this population size will be an increase in the loss of biodiversity, more urban sprawl, and more pressure on stressed social systems.

This tremendous increase in size of the human population has created the biodiversity crisis, as well as many other societal crises for our species. The effect of increased human populations on the biosphere is compounded by the increased rate of technological development, extraction of natural resources for energy production, conversion of natural land, and the resulting pollution from these activities.

What is biodiversity, and how does the biodiversity crisis impact the field of herpetology? At its simplest, biodiversity is the wealth of life throughout the world, including the smallest viruses and microbes to the giant whales and redwoods. Biodiversity includes the genetic diversity embodied in these organisms, and the interactions among them that form unique communities and ecosystems. The history of life on Earth is recorded in the DNA of organisms; this too is an important component of biodiversity. The study of biodiversity and its conservation require addressing diversity at several levels and in several ways (Table 14.1). Species richness, often called species diversity (see Chapter 12), is easiest to recognize conceptually simply by noting how many different species exist in a given area. Species richness is typically reported by taxonomic group. For example, 19 species of snakes occur in the metropolitan Washington area; 95 species of native frogs occur in the continental United States; and 5453 frog species occur in the world. These regions are arbitrarily defined, and the numbers offer a sense of a region's diversity but have little biological utility. A biologist or resource manager needs more precise data and asks for species-diversity metrics relative to habitats or natural plant–animal associations. The number of species in a given area is termed alpha diversity (Table 14.1). A biologist may also wish to know how species diversity changes along an environmental gradient (beta diversity) or the differences in number and kinds of species among climatically and structurally similar habitats in different geographic areas (gamma diversity). Such diversity has long fascinated herpetologists. For example, Ronald Heyer's study in 1967 of herpetofaunal change along an altitudinal gradient in Costa Rica represents a study of beta diversity, and Eric Pianka's studies in 1986 and 1994 of lizard assemblages in the deserts of Africa, Australia, and North America represent studies of gamma diversity. The fascination, however, extends beyond the number and kinds of species that occur in a region to what these occurrences reveal about the origin and interactions of an assemblage or community and the ecology of the individual species. Subsequently, these studies provide historical "snapshots" by informing us how a locality and its fauna have changed through time. It is the multitude of such biological studies and the presence of voucher specimens in the world's museums that allow scientific assessments of changing diversity and species abundance.

Although numerous, these assessments of diversity are few relative to the number of habitats and ecosystems throughout the world. Additionally, many studies may have been scientifically rigorous for their time, but they lack either the scope of data or the appropriate sampling regime to rigorously examine current conservation issues. Their proportionately small numbers highlight the necessity of continuing biodiversity inventories in order to obtain an accurate and thorough knowledge of the world's flora and fauna. Another critical aspect of these inventories is the collection of specimens and the prompt study of these voucher specimens. Good science relies on repeating and verifying observations. Verification of species occurrence relies on actual specimens because most plant and animal species cannot be reliably identified in-hand, from photographs, or from a small set of recorded measurements. Biodiversity inventories regularly identify new species, and these discoveries include amphibians and reptiles. Often these new species are common faunal members, but their uniqueness has not been recognized because they were not carefully examined. Close study of numerous groups of amphibians and reptiles is revealing that the true diversity is masked by not distinguishing among cryptic and closely related species. The biodiversity crisis has imposed urgency on the documentation of the world's biota, and the rate of discovery of new organisms seemingly has accelerated. For example, Brazil has the richest amphibian fauna in the world with 776 species, 50% of which have been described in the last 40 years; in addition, many species are undiscovered or undescribed.

Conservation biology is also concerned with other types of biotic diversities, including genetic and community–ecosystem diversity (Table 14.1). Genetic diversity is a measure of the breadth of a population's or a species' genetic constitution. Fundamentally, how variable is a population or how many genotypes does it contain? It is easy to recognize the genetic variability of human populations or those of our domestic animals because of our familiarity with them. Most populations of animals and plants are equally variable. The importance of genetic diversity is that it provides a population with broad genetic flexibility to adapt to changing conditions. Without such flexibility, a population cannot adjust to new conditions and may become extinct. Factors affecting genetic diversity include genetic bottlenecks, the founder effect, and genetic drift, which are examined in the "Extinction" section.

Community–ecosystem diversity examines the number of habitats or species assemblages throughout a broad region. Whether a community is viewed as a haphazard or an organized assemblage of organisms, the different needs and tolerances of organisms and the heterogeneity of the physical landscape can yield structurally similar assemblages of organisms. Species and genetic diversity are directly linked to ecosystem diversity, because more habitat types translate into more kinds of organisms. For each species living in multiple habitats, a broader selective landscape generates greater genetic diversity. Much of the effort of conservation biologists has focused on single-species preservation; however, the early preservation of large areas as breeding and hunt-free refuges for game species created the base for the present-day emphasis on community transition conservation. Within North America, the greatest regional herpetofaunal diversity and community segregation occurs in the southeast and southwest United States. For example, the herpetofauna of Florida's sand-ridge communities shares only a few species with the pine–flat wood or evergreen–hammock communities. In turn these terrestrial communities have few species in common with marsh to river communities. Similar sharp contrasts in species composition exist among the lizard and snake assemblages of the southwestern desert communities in the United States.

The biodiversity crisis is characterized by the loss and reduction of diversity within all three levels previously described. The most extreme loss is the extinction of a species. Extinction is a natural process and occurs continuously; however, the crisis we now face is occurring because the rate of extinction, that is the number of species lost per unit time, has greatly exceeded the normal rate. In addition, the breadth of extinctions has broadened, encompassing all sizes and types of organisms. Normally, extinction occurs at a slow pace and the number of species that disappear equals or is slightly fewer than the number of new species that appear. This gradual accumulation of species through time results in increasing diversity. The current phenomenon of rapid decline of the world's biodiversity has characteristics of a geologically ancient mass extinction event, where thousands of species are lost in a short period of time. Mass extinction is a catastrophic event; those documented from the fossil record have losses of more than 30% of the species. An estimate of 96% loss of species has been proposed for the mass extinction event at the end of the Permian. Although that estimate may be high, a 50% loss is not a high estimate for that event. If one of every two species disappears, species interactions and ecosystems change drastically. Conservationists are concerned about the loss of diversity because a high rate of extinction might lead to a cascading extinction event where the loss of one species causes the loss of multiple species. No matter how resourceful we humans are, the human species cannot be assured that it will survive a mass extinction because the complex interactions among species that support our global food supply are at risk.

In addition to assessing the diversity of amphibians and reptiles, herpetological research contributes broadly to conservation management, both in identifying causes of decline and in developing data on amphibian and reptilian biology, from genetics to natural history. Subsequent sections of this chapter examine major aspects of conservation and offer a few examples of the contribution of herpetological research.

IMPACT ON AMPHIBIAN AND REPTILE COMMUNITIES

Humans have modified the environment everywhere. Such a comment may seem to be an exaggeration, but it is not an overstatement (Fig. 14.3). Globally, our activities have resulted in a rising average annual temperature and in a rise in ultraviolet radiation at the earth's surface. These climatic effects are only one facet of our environmental alteration, which ranges from global climatic changes to the local loss of a marsh or a patch of forest. All alterations, even those occurring in polar regions, can affect amphibians and reptiles. Scientists are currently alarmed at the increasing melting rate of polar ice sheets as the result of human activity (e.g., global warming). The resultant rising sea levels will affect amphibians and reptiles in coastal and low-lying areas.

Habitat Modification, Fragmentation, and Loss

Habitat alteration and loss is the most visible human-mediated environmental change. Prehistoric human populations began the process by setting fires to catch game, thereby expanding grasslands and savannas at the expense of forest. The rise of agriculture converted grassland and forest into farms and gardens. The conversion of natural

FIGURE 14.3 Disturbed and fragmented habitats. Clockwise from upper left: Tropical Amazon rain forest during the burning season (August) of 1987 in central Rondônia, Brazil—no rain forest remains in this area today (L. J. Vitt); desertification in progress due to overgrazing by goats in northern Kenya (C. K. Dodd, Jr.); former Guatamalan cloud forest (ca. 2000 feet elevation) converted to agriculture (C. K. Dodd, Jr.); stream (Lost Creek) in Alabama degraded from coal mine runoff. The federally protected flattened musk turtle lived in the stream (C. K. Dodd, Jr.).

landscapes continues. At the end of the 20th century, the world had 24 megacities, defined as urban areas with populations that exceed 10 million. In addition, all over the world, small cities are rapidly growing and engulfing more and more natural areas. Farmland and natural areas are rapidly being converted to residential areas in many parts of the world. Although everyone is familiar with the environmentalists' plea to "save our rain forests," natural habitats of every type throughout the world are severely threatened. As one example, it is estimated that the Brazilian cerrado, a unique savanna-like biome that encompasses about 21% of Brazil's total landmass, is severely threatened (Fig. 14.4). This area is a biodiversity hot spot, with more than 420 species of amphibians and reptiles, many of them endemic, in addition to large numbers of other unique vertebrates and plant species. In the 1950s, Brazil designated a new political entity, the Distrito Federal, and built a new capital city, Brasília, in the state of Goiás, which is in the heart of the cerrado. This planned city and the accompanying highway system were built to encourage migration of people into Brazil's interior. As a result, the cerrado became a new agricultural frontier, and today, the region has been extensively converted to soybean, corn, and cattle production. Only about 20% of the original cerrado remains, and relatively little of this remnant is protected. A recent study aimed at predicting diversity of squamate reptiles demonstrates that numerous biodiversity hot spots exist or existed prior to conversion to agriculture. Gabriel Costa and his colleagues used geographical information systems (GIS) and niche modeling based on known distributions to construct a model predicting areas of high lizard and snake diversity (see discussion of niche modeling in Chapter 12). They found that many areas within the cerrado exhibit a set of environmental conditions that could maintain more than 70 squamate species (Fig. 14.5). This approach has many applications to conservation biology. For example, once potential biodiversity hot spots are identified, follow-up surveys can be concentrated in those areas. In addition, the approach allows a rapid assessment of large areas, which could identify best locations for reserves or national parks.

The direct effect of habitat loss on an amphibian or reptile species or community is obvious; they disappear from that area. The consequences, however, extend beyond the edges of the lost habitat. Clear-cutting of tropical and temperate forest affects both the abundance and presence of

FIGURE 14.4 The cerrado of central Brazil is considered a biodiversity hot spot, yet it is being converted to agriculture at an alarming rate. Clockwise from upper left: Jalapão National Park looking out from isolated sand dune area; typical undisturbed cerrado in Jalapão National Park; cattle pasture (foreground) in area that was formerly dry forest (background) in Goías State; aerial view of cerrado in area in which most of the natural vegetation has been removed for agriculture or grazing (L. J. Vitt).

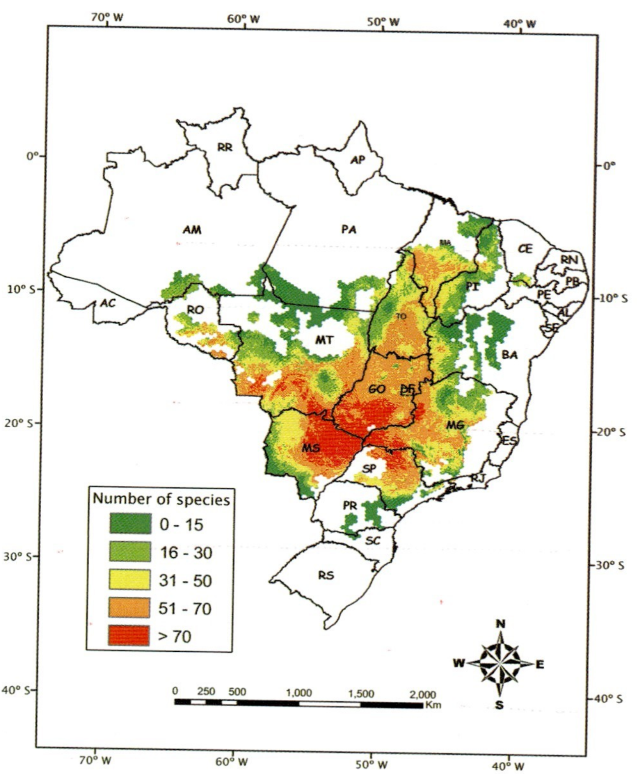

FIGURE 14.5 Niche modeling of the distributions of cerrado squamate reptiles demonstrates high potential biodiversity. Continuing field studies in patches that remain relatively undisturbed can be used to test accuracy of predicted distributions. Adapted from Costa et al., 2007.

amphibian species at the remaining forest's edge and at least to 20–30 m inside the forest. In a wet tropical forest of Amazonian Ecuador, the species richness (i.e., the number of species) of *Eleutherodactylus* frogs increased with increasing distance from the forest opening. Overall, frog diversity had only a weak linear association with distance from the opening because some species of hylid frogs may benefit from the relative openness of the forest. In a deciduous forest in Maine, the edge effect decreased the relative abundance of the native salamanders and frogs; salamanders showed the greatest sensitivity to increased light levels and reduced humidity associated with the forest edge.

Selective logging within forests has a similar effect on amphibian communities as the edge effect. Totally removing a forest by clear-cutting usually eliminates the entire amphibian community. Removal of all the trees and the associated destruction of the understory vegetation and broad disruption of the litter–ground cover expose the soil to direct sunlight. Thereafter, the soil attains significantly higher temperatures, experiences greater temperature fluctuations, and becomes drier; these microclimatic changes are lethal to amphibians. If the logged areas are left undisturbed, the forest eventually regenerates itself. The speed of the regeneration depends upon numerous factors, including, for example, size of logged area, presence of small forest stands within the logged area, species composition of the native forest, soil type and quality, and weather and climate. Temperate and tropical forests naturally develop openings because of storm damage or the death of old trees, but these gaps fill quickly with seedlings from the surrounding forest and small trees and herbaceous vegetation regenerated from rootstocks. The same process occurs in logged areas, but larger cleared areas require a longer time for the migration of seeds and seedlings throughout the area. The same principle applies to recolonization of a logged area by the amphibian community. Assuming a relatively rapid regeneration of the forest, the entire amphibian community may reassemble in 20–30 years. Again, local effects as well as logging practices are factors in reassembling the community. Data for Appalachian salamander communities suggest a range of 20 to 50 years for recolonizations; however, in managed forests, site preparation activities drastically alter the soil and other physical aspects of the site and make forest plantations uninhabitable for most species of amphibians and reptiles. Selective logging and other disturbances in tropical rain forests potentially alter community structure by changing species interactions. These activities create hotter forest openings than natural tree falls. For example, in Amazonian rain forest, canopy gaps attract large-bodied heliothermic lizards such as *Ameiva,* and these predators can reduce the population size of smaller lizard and frog species by direct predation and by interference competition for shared prey.

Natural disruptions occur regularly in all ecosystems. Floods, landslides, and fires are the usual agents. While locally devastating, the native flora and fauna have experienced such disturbances over many generations and recovery is relatively quick. Indeed, high species and community diversity of an area may be fostered by the regular occurrence of disturbances. In one sand-ridge site in the Great Victoria Desert of Western Australia, 45+ species of lizards occur; this high diversity is four times the species richness of any desert site in North America and more than double the richness in the African Kalahari. Natural wildfires are frequent but narrowly confined, thereby creating a patchwork of habitats of similar plant composition, each at a different stage of recovery from its most recent exposure to fire (see Chapter 12). Because different assemblages of lizard species are adapted to different habitats, numerous species can occur in the same area but with a reduction in competition for the same resources.

In rare cases, a natural catastrophe can decimate local populations or even eliminate entire communities. If a population within one of these communities is the single remaining population of a species, the catastrophe causes extinction. It is this latter aspect that concerns conservationists and becomes increasingly possible because of human-mediated habitat loss and alteration. Several paradigms in conservation biology arise from the problem of habitat destruction and its effects on individual species. The issues of concern are population viability and persistence, and ultimately population size, including both the absolute number of individuals and their density or number of individuals in a unit area. The metapopulation model views a population as consisting of source and sink populations. In the former, sufficient offspring are produced on average to maintain the population and produce an occasional excess of offspring that disperse because one or more critical resources are controlled by other individuals. Sink populations, on average, produce too few offspring for that population to persist and require regular migration of new individuals for their survival. Habitat destruction and alteration fragment the suitable habitat and create dispersal barriers of unsuitable habitat. The barrier may be a road, a housing development, new agriculture areas, or any of a number of other disturbances. Regardless of the barrier's size, if it significantly reduces or halts dispersal to sink populations, they soon disappear. Fragmentation can create problems for source populations and threaten their survival as well. Such factors as demographic stochasticity, inbreeding, and genetic drift can alter the genetic diversity of a population and reduce its survivability. These factors and related ones become increasingly influential in a population's survival in a habitat fragment and are of major concern in establishing reserves and refuges, which are just that, fragments of once larger natural areas.

Determining reserve size depends greatly upon the biology of the species targeted for preservation. "Bigger is better" is true, but it is an overly simplistic solution in today's world, where individuals of many species compete for space. How big does a reserve need to be to maintain genetic diversity and avoid demographic collapse? The minimum viable population (MVP) size model grew out of this debate. As initially proposed, a minimum viable population is the number of individuals necessary for a population to have a 99% chance of survival for 1000 years and to avoid extinction by natural catastrophes or the effects of demographic, genetic, and environmental stochasticity (Table 14.2). No one has attempted to derive a precise number for any amphibian or reptilian population, although modeling of turtle populations has identified the demographic features necessary for the survival of populations of these long-lived species. Some aspects of demographic, genetic, and environmental stochasticity are examined in subsequent sections.

Habitat fragmentation is such a common feature of our present landscapes that we often lose sight of its impact on natural communities and species distributions. A recent study in the Great Central Valley of California is exceptionally revealing, although not exceptional in occurrence. The total number of native amphibian species was not large, consisting of only seven species—three salamanders and four frogs—with a maximum of six species in any locality and fewer in some areas (Fig. 14.6). Breeding season surveys of over 1000 aquatic sites in the 28 counties of the Central Valley revealed that only 3 counties still retained populations of their original fauna. Species retention was greatest in hilly areas and least in flatlands, which are now largely agricultural. No county had lost its entire complement of native species, but most had lost more than one-half their species. In some areas, overall diversity has increased by the introduction of exotics; however,

TABLE 14.2 General Threats to the Persistence of Small Populations.

Demographic stochasticity: The natural fluctuation in a population's demographic characteristics over generations. It includes the following:

- Changes in population size. A population randomly increases and decreases in size through time. In a random-walk situation, a population fluctuates between highs and lows and, over many generations, the declines become more severe. If the population size fluctuates to zero, the population disappears.
- Changes in sex ratio. A random distortion of a population's sex ratio can interrupt reproductive behavior and successful juvenile recruitment.

Genetic stochasticity: The loss of genetic diversity through random events in the history of a population. The loss of diversity reduces genetic variation among individuals, hence reducing the adaptive plasticity of a population through time. Genetic stochasticity includes the following:

- Founder's effect. A population that arises from a few individuals contains only the genetic variation of the founding individuals, and this variation is likely only a small fraction of the source population.
- Genetic drift. Either through random (drift) or selective mating, allelles that occur at low frequencies in a population tend to decrease in frequency and eventually be lost.
- Inbreeding depression. Breeding with close relatives increases homozygosity.
- Bottleneck. A sudden decrease in population size with a corresponding reduction in genetic variation.

Environmental stochasticity: Unpredictable changes in the abiotic and biotic factors that affect the availability of resources and the equability of the environment. These changes include the following:

- Weather. Exceptional weather patterns may cause floods, droughts, or unseasonably hot or cold periods that disrupt feeding and reproduction or even exceed the physiological tolerance limits of a species.
- Climate. Long-term shifts in weather pattern change the seasonal rainfall, insolation, and temperature regime of a locality.
- Catastrophes. Major geological disturbances, such as landslides, volcanic eruptions, or meteor impacts, can destroy all life within an area.
- Disease and parasites. Appearance of a new disease or parasite or the change in virulence of an existing one.
- Predator. Appearance of a new predator or an improved hunting strategy by an existing one.

Sources: Caughly and Gunn, 1996; Meffe and Carroll, 1994; Primack, 1995.

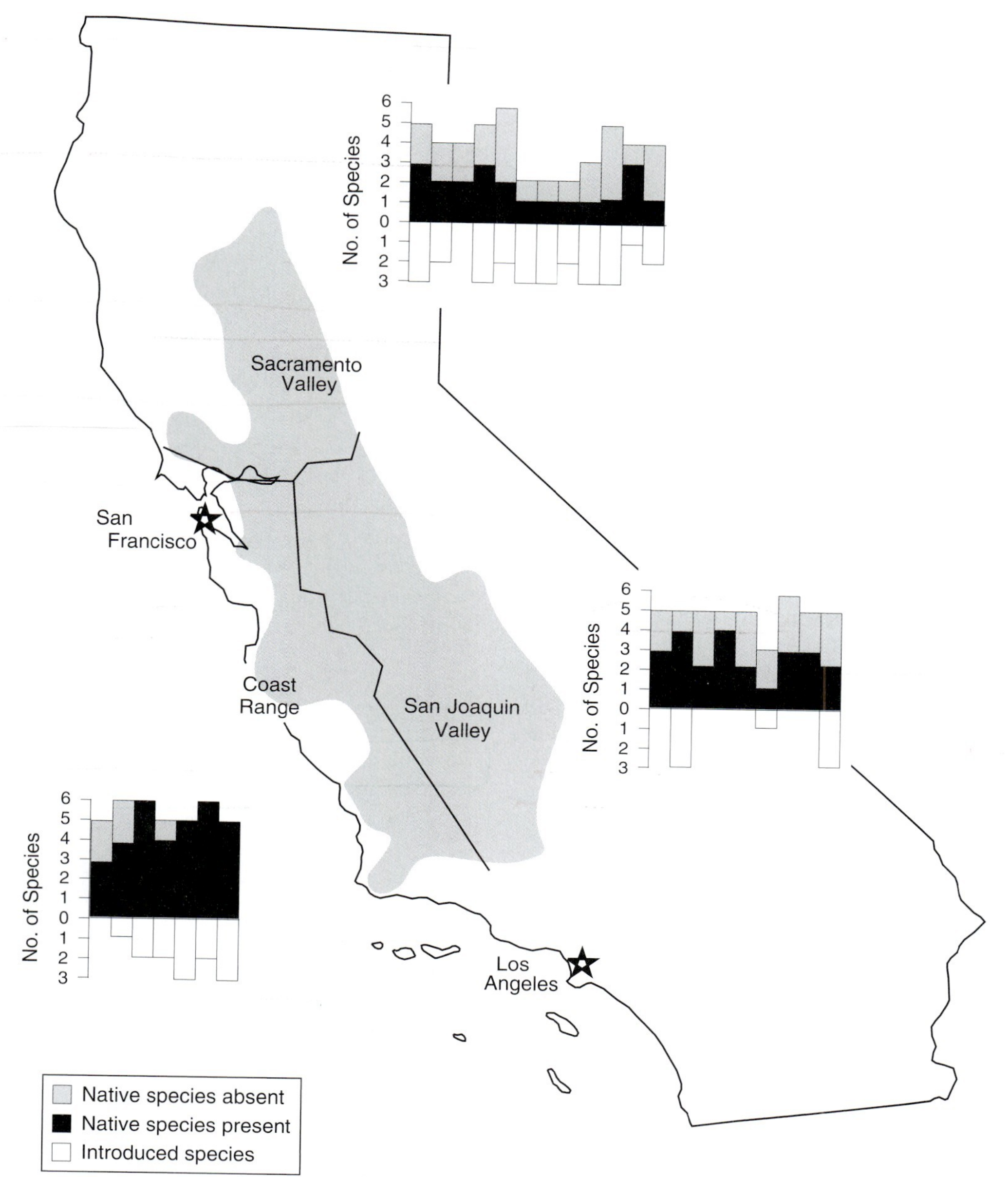

FIGURE 14.6 Amphibian faunas in the Great Central Valley of California, USA. To document the faunal changes during the 20th century, over 1000 aquatic sites were surveyed between 1990–1992. The results are summarized in three bar diagrams showing the number of exotic species present, the number of native species lost, and the number still present in each county of the three major valley provinces. Only 3 of the 21 counties surveyed still have all members of their original amphibian fauna. The lightly stippled area denotes the valley oak–grassland habitat; the dark line marks the boundary between the valley provinces. Modified from Fisher and Shaffer, 1996.

in most cases, exotics do not occur with native species, and some exotics such as the bullfrog *Lithobates catesbeianus* are partially to blame for the extirpation of native frogs.

We often fail to recognize effects of habitat modification on small amphibian and reptile species because they may be difficult to observe even when they are abundant. However, effects of human activities are usually obvious on large species. Large reptiles are particularly vulnerable to extirpation by man because they usually are good to eat, have valuable skins, are relatively easy to hunt, and have life histories that make it difficult for populations to sustain continual harvest of large (and old) animals. The giant land iguanas in the genus *Cyclura* are prime examples (Fig. 14.7). These huge lizards were once common and occurred at high density on many islands in the West Indies but have diminished to dangerously low levels recently

FIGURE 14.7 Large iguanian lizards in the genus *Cyclura*, such as this *C. nublia*, have experienced drastic population declines as a direct result of human activities. Photograph by C. Ken Dodd, Jr.

(Fig. 14.8). Indigenous peoples hunted them for food, having some impact on populations, and the influx of western Europeans and their pets and farm animals devastated local populations. Some effects were direct, such as killing them for food, skins, or removing them to send to Europe as exotic pets. Indirect effects included competition with farm animals for food (the lizards are herbivorous), destruction of nests by pigs and cattle, and predation by dogs and other human pets. As rat populations that follow colonization increased, colonists introduced mongooses and later cats to control rats. As rat populations declined, cats and mongooses ate eggs and young of the lizards. Habitat for many of the populations has been replaced by luxury hotels, shopping malls, and golf courses. Although many programs now exist in an attempt to protect these lizards, population sizes are small and the future of land iguanas appears grim. Land iguanas reach sexual maturity at an age of 6 or 7 years and can live for more than 40 years. Their mean generation time is about 20 years, and each sexually mature female produces a clutch of 2 to 10 eggs each year following attainment of sexual maturity. Removal of large individuals, especially females, has a cascading effect on future populations, especially when egg and juvenile mortality increase at the same time. Not only do large females deposit more eggs each year than smaller females, but they also are more likely to survive from natural predators simply because they are larger. Humans are not natural predators of these lizards, and their impact has been substantial.

Pollution and Disease

Everyone can recognize industrial pollution with its particle-laden smoke arising from smokestacks and its toxic

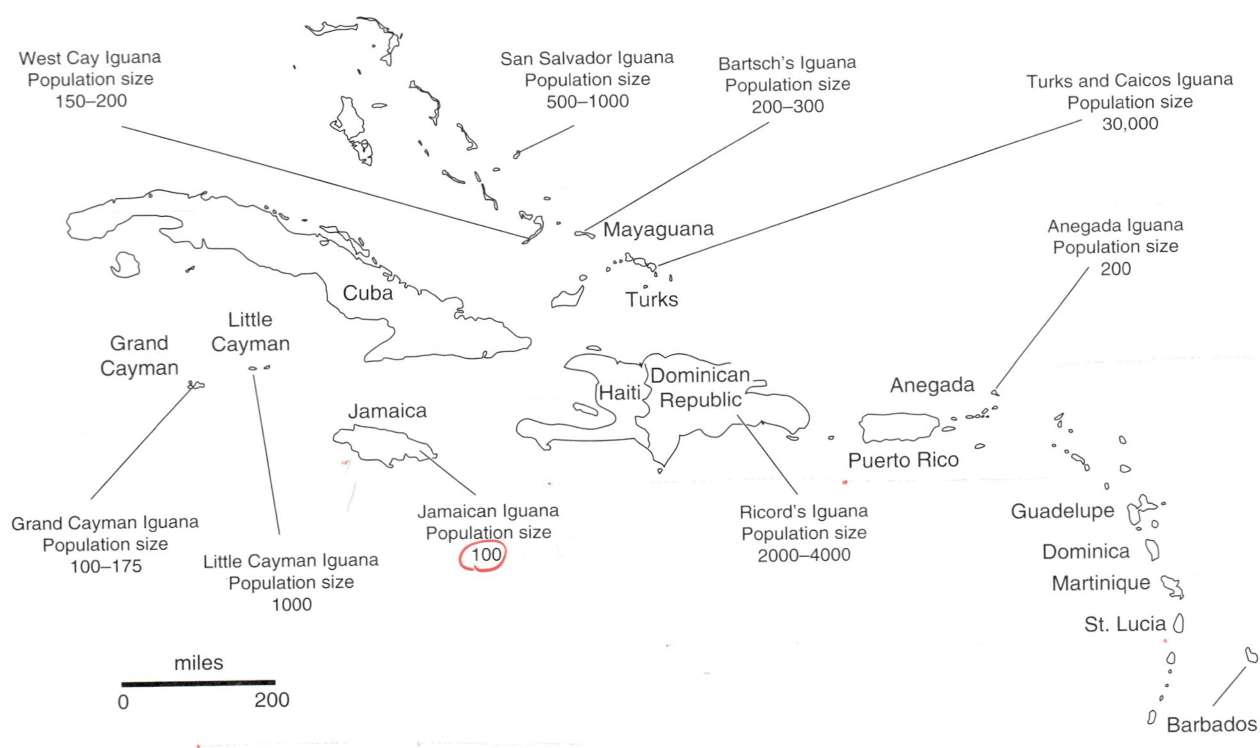

FIGURE 14.8 Map of the West Indies showing population sizes for remaining land iguanas (*Cyclura*). Adapted from Pianka and Vitt, 2003.

waste emptying into adjacent waterways; however, pollution is not always so obvious or blatantly toxic but can be lethal nonetheless. The phosphates in laundry detergents or the nitrogenous matter from a dairy farm are not toxic when diluted, but add a dozen, a hundred, or a thousand washing machines emptying their wash water into a lake, or the runoff from a dozen dairy farms into a small stream day after day, and they can have serious ecosystem effects. A single washing machine or diary farm impacts the local ecosystems by slowly altering microenvironments, making them lethal for native microfauna and microflora. As these organisms change, so does the macrofauna and macroflora. Life persists in many polluted environments, and the diversity of species and their abundance sometimes may be even greater, highlighting one of the dilemmas of conservation: When is action necessary, and what action is required?

Unfortunately, action is seldom preventative but occurs with an impending crisis or amid a full-blown one. These crises attract our attention and research efforts. Three of the most visible crises, acid rain, ecoestrogens, and sea turtle fibropapillomatosis, are briefly examined. We consider disease as a cause of worldwide amphibian declines in a separate section. These examples highlight the scope and complexity of pollution and its potential fostering or enhancement of disease in amphibians and reptiles. The pollutants, or "environmental contaminants," range from solid-waste disposal filling a breeding pool through fragmented waste (e.g., plastic bags, tar balls) to airborne or water-suspended microparticles, such as heavy metals, organic compounds from pesticides and herbicides, and PCBs. The interactions of these pollutants with life processes are understood only in a few instances and have become a major area of research. Depending upon their concentration and biochemical nature, microparticles can be lethally poisonous, carcinogenic, mutagenic, and, although less well documented, immunosuppresant.

Environmental Acidification

Acid rain has moved out of the forefront of conservation concern, in part because it has been alleviated to some extent in Europe, Canada, and the United States by the enforcement of clean-air legislation. Nevertheless, it remains a pollution problem, perhaps a low-grade one in the preceding areas but certainly a major problem in China, India, and other areas that rely mainly on coal to power their industries yet practice little pollution control. Acid rain arises from the combustion of fossil fuels and the release of sulfur and nitrogen oxides in the air. These by-products react with the moisture in the air to produce sulfuric and nitric acids and in turn are returned to the earth by snow or rainfall. While some acid rain falls locally, much of the acid pollution is carried downwind and dropped hundreds of miles from the pollution source. Normal rain is saturated with carbon dioxide and has a pH of about 5.6–5.8; acid rain commonly has a pH of 3.0 to 4.0 and is occasionally even more acidic. The first evidence of the danger of this far-removed pollution was the death of trees on mountaintops. It soon became evident that the effects were far broader, sterilizing life in seemingly unpolluted forest streams and lakes. Acid rain is most destructive when it falls in areas of hard-rock and mineral-poor soils, because the soil and water are incapable of neutralizing (buffering) the acid precipitation.

Because many amphibians are aquatic for part of their life, they are highly susceptible to the toxic effects of acid rain; however, their susceptibility is variable (Table 14.3). Some species, such as *Lithobates virgatipes* and *Hyla andersonii*, breed in acidic waters (pH < 4.0) of cedar bogs, but most amphibian species require water that is less acidic, and their eggs and larvae suffer more than 50% mortality even in water with a pH of 4.5. Acidic water affects the survivability of juveniles and adults, but its toxicity focuses on the developmental stages by disrupting the ionic balance within cells and typically killing embryos by the late gastrula stage. Less acidic water that may permit greater than 50% larval survivorship still affects development by slowing growth and morphogenesis; commonly, it produces a high

TABLE 14.3 pH Tolerance Levels of Select Species of Amphibians.

Taxon	Critical pH	Lethal pH
Salamanders		
Ambystoma jeffersonianum[2] (field)	4.0–5.0	4.2
Ambystoma jeffersonianum	4.2–4.6	4.2
Ambystoma maculatum (field)	5.5	4.2
Ambystoma maculatum	6.0–7.0	4.0–5.0
Frogs		
Acris gryllus	4.2–4.6	4.0–4.1
Hyla andersonii	3.6–3.8	3.4
Pseudacris crucifer	4.0–4.2	3.8
Hyla versicolor	3.9–4.3	3.8
Xenopus laevis	3.0	3.5
Lithobates catesbeianus	4.1–4.3	3.9
Lithobates clamitans	3.8–4.1	3.7–3.8
Lithobates pipiens	5.5–5.8	-
Lithobates sylvaticus	3.6–3.9	3.5
Lithobates virgatipes	3.6–3.8	3.4

Note: Mortality is presented as critical, denoting 50% mortality of larval sample with exposure throughout the entire development period, and lethal, denoting 100% mortality. pH exposure levels were determined in a laboratory setting unless noted otherwise.

Source: Tome and Pough, 1982.

percentage of developmental abnormalities, many of which result in death during metamorphosis. The toxic effect of acid rain is greatest on species that breed in vernal (temporary) ponds. Most of these breeding sites are dry prior to the arrival of the rains and temperatures that stimulate breeding events. The rains not only bring their acid load but also wash acid from the surrounding vegetation and land into the pond, causing pH to drop below even the tolerance levels of the most acid-tolerant species. Species living in permanent waters are buffered from these acid surges simply by the diluting effect of the large volume of water.

If acid levels are not lethal to all species, community structure shifts. For example, the glacial soils of central New York are poorly mineralized and are downwind from the heavy industry of the midwestern United States; thus, all freshwater communities are acidified. Acid precipitation differentially affects two salamanders in the amphibian communities of vernal pools. *Ambystoma jeffersonianum* is an acid-tolerant species, and its larvae can develop and metamorphose in water with a pH of <4.0. Its congener, *A. maculatum*, is less tolerant and requires water with a pH of ≥ 5.0 for successful hatching and metamorphosis. Since snowmelt and spring rainfall commonly produce breeding pools of pH 4.5, more *A. jeffersonianum* larvae metamorphose and eventually return to reproduce, slowly outnumbering the formerly dominant *A. maculatum*.

Ecoestrogens

Animals, particularly herbivores, have long experienced natural exogenous hormones. Most of these products are produced by plants and fungi as defense mechanisms to stop or reduce consumption by herbivores. Ecoestrogens (estrogen-mimicking chemicals) represent one class of these defense compounds, and interactions across generations (coevolution) result in the consumer's ability to tolerate and neutralize the ecoestrogen or to recognize and avoid its consumption. Human activity inadvertently has introduced numerous new ecoestrogens into the environment, often in excessively high levels. Some industrial pollutants, sewage effluent by-products, and pesticides and their breakdown products act as weak estrogens.

Estrogens are essential components of each animal's reproductive physiology; however, exposure to them at the inappropriate time or in excessive amounts disrupts normal reproductive behavior. Further, larvae and embryos are quite sensitive to estrogens, whose timing or concentration interrupts normal development of the reproductive system and other organ systems. The potential effects of ecoestrogens still are incompletely known, but evidence from wildlife and laboratory studies shows increasingly their effect in reducing the reproductive potential of individuals, and in causing cancer and immunosuppression.

Studies have demonstrated a striking effect of ecoestrogens on demography and reproduction in the alligator population of Lake Apopka (central Florida, USA). In 1980, this lake suffered a major pesticide spill consisting of dicofol that was contaminated with DDT and its breakdown products. The alligator population showed an immediate demographic loss of its juveniles, probably a direct result of poisoning of these age classes. Adults seemingly were unaffected; however, the population has not yet recovered. Throughout the 1980s, egg viability was 20%, compared with 80% in eggs from a Florida wildlife refuge, and it continues to remain low. The pesticides and their metabolites are persistent, requiring decades to disappear from the environment, and they continue to be present at high levels in alligator eggs. This contamination is directly toxic to many embryos, but a few survive and hatch. However, the hatchlings are not normal. Embryonic exposure to ecoestrogens has disrupted the reproductive-system development. For males, this exposure has resulted in feminization of the reproductive organs; penes are smaller and spermatogenesis is lower. In females, ovarian morphology and ovarian follicles are abnormal.

Fibropapillomatosis

Diseases are a natural phenomenon, and no plant or animal appears to be free from them. Disease becomes a concern to conservationists when it results in sudden die-offs of populations or when its frequency of occurrence increases sharply. The latter has occurred in the endangered and threatened cheloniid sea turtles, especially in the green sea turtle *Chelonia mydas*. In general, neoplasias are uncommon in wild animals and certainly so in reptile populations. However, beginning in the mid-1980s, the incidence of cutaneous papillomas, fibromas, and fibropapillomas has increased markedly in several populations of *C. mydas*. These tissue-proliferation lesions are generically labeled green turtle fibropapillomatosis (GTFP), owing to their presumed origin and highest incidence in that species. Although the lesions are not cancerous, their excessive growth internally and externally is life threatening. Externally, the growths reduce an individual's ability to escape enemies and to find and consume food (see Fig. 11.32). Internally, the papillomas enlarge and interfere with the function of the viscera, including blocking the digestive tract and disrupting kidney or lung functions.

GTFP was first reported in 1938 in a captive *C. mydas*, which had been caught 2 years earlier in the Key West area. This occurrence was to prove prophetic, because today, Florida Bay has one of the highest incidences of GTFP. Incidence levels range from 0 to 92% (Kaneohe Bay, Hawaii), and occurrence has been reported pantropically with the exception of East Atlantic coastal Africa. The highest incidence is in lagoons and bays adjacent to dense human populations. Yet, two locations only kilometers apart can have strikingly different incidences. For example, the incidence is 50% in Indian River Lagoon and 0% in the reefs off the central Florida coast. This association with human populations and waters with low circulation suggests that environmental contaminants foster GTFP; however, no matter how strong the association, no evidence presently supports a cause-and-effect association.

The etiology of GTFP remains uncertain, although viral transmission is strongly supported. The transmission of the virus from individual to individual is less certain; a marine fluke that parasitizes green sea turtles was early suggested as a vector, but evidence is inconclusive. GTFP has a distinct demographic association with juvenile turtles. The papillomas have never been reported in the youngest juveniles of the pelagic phase; however, once returning to near-shore waters, the incidence increases in some populations in the larger size classes (to about the 80–90 cm carapace length) before declining. Perhaps there is a natural remission of the disease, if infected individuals can survive the debilitating middle years. Fibropapillomas, however, are occasionally seen in nesting females, so adults are not immune to the disease.

GTFP remains a major threat to the survival of populations of *Chelonia mydas*. Throughout the 1980s and 1990s, the incidence in infected populations and the number of populations with GTFP individuals has increased. The additional threat is that GTFP now occurs in other species, particularly in those resident in habitats with a high incidence in the *C. mydas* population.

Harvesting Amphibians and Reptiles

In 1998, R. Melisch remarked, "Apart from habitat destruction and alteration, the biggest threats for wild species of plants and animals are illegal trade and unsustainable consumptive use." This remark is seemingly an overstatement of human exploitation of the world's biota, yet from Melisch's perspective as a conservationist working in southern Asia, it rings true. Further reflection supports its worldwide applicability if it encompasses all human use—legal or illegal, intentional or unintentional capture.

Many species and populations of amphibians and reptiles are negatively impacted by human commerce in the broadest sense, and this impact is as great in the developed as in developing nations (Table 14.4). For example, the European pet trade overharvested its native tortoises and those of adjacent Africa and Asia. When *Testudo* populations were decimated, these tortoises were banned from

TABLE 14.4 Wild-Caught Amphibians and Reptiles That are Most Frequently Traded on the World Market, Based on Records for 1998–2002.

Amphibians		
Category	Volume	Trade purpose
Imported amphibians		
Whole bodies (count)		
Lithobates catesbeianus NYS	3,886,546	Food
Hymenochirus curtipes	2,376,647	Pet
Cynops orientalis	1,635,362	Pet
Bombina orientalis	1,016,579	Pet
Lithobates forreri	679,937	Research
Body parts and products (count)		
Lithobates catesbeianus	293,908	Food
Limnonectes macrodon	164,591	Food
Unidentified ranoid spp.	112,289	Food
Holobatrachus tigerinus[a]	22,417	Food
Rana tigerina[a]	17,010	Food
Mass (kilograms)		
Lithobates catesbeianus	2,816,693	Food
Limnonectes macrodon	1,193,383	Food
Unidentified ranoid spp.	534,318	Food
Holobatrachus tigerinus[a]	462,763	Food
Lithobates pipiens NYS	113,050	Food, research
Exported amphibians		
Whole bodies (count)		
Hymenochirus curtipes	188,622	Pet
Cynops pyrrhogaster	112,901	Pet
Hyla cinerea	87,536	Pet
Bombina orientalis	78,606	Pet
Hymenochirus spp.	72,832	Research, pet
Body parts and products (count)		
Non-CITES entry	134	Various
Ambystoma spp. NYS	47	Pet
Limnonectes macrodon	9	Food
Ambystoma laterale NYS	9	Pet
Unidentified ranoid spp.	8	Food
Mass (kilograms)		
Rana tigerina[a]	16,330	Food
Unidentified ranoid spp.	6,000	Food
Limnonectes macrodon	1,932	Food
Lithobates catesbeianus	319	Food
Litoria spp.	50	Pet
Reptiles		
Category	Volume	Trade purpose
Imported reptiles		
Whole bodies (count)		
Hemidactylus spp.	793,591	Pet
Python regius	584,508	Pet

Continued

TABLE 14.4 *(continued)*

Reptiles		
Category	Volume	Trade purpose
Trachemys scripta[b]	305,038	Pet, food
Varanus salvator	299,447	Pet, whole skins
Iguana iguana	298,632	Pet
Body parts and products (count)		
Elaphe radiata	4,782,607	Skin products
Tupinambis tequixin	2,591,370	Skin products
Tupinambis rufescens	1,689,813	Skin products
Elaphe carinata	1,268,591	Skin products
Varanus niloticus	1,094,709	Skin products
Mass (kilograms)		
Chinemys reevesi	105,957	Traditional medicine
Elaphe radiata	8,685	Traditional medicine
Gekko gecko	8,503	Traditional medicine
Boa constrictor	8,182	Skin products
Pelodiscus (Trionyx) sinensis	5,233	Traditional medicine, food
Exported reptiles		
Whole bodies (count)		
Trachemys scripta[c]	23,655,553	Food, pet
Alligator mississipppiensis	577,440	Whole skins
Anolis carolinensis	258,284	Pet
Anolis sagrei	100,894	Pet
Pseudemys spp.	100,279	Food, pet
Body parts and products (count)		
Tupinambis rufescens	513,774	Skin products
Alligator mississipppiensis	359,734	Skin products
Python reticulatus	124,659	Skin products
Tupinambis teguixin	75,467	Skin products
Varanus salvator	54,637	Skin products
Mass (kilograms)		
Alligator mississipppiensis	101,151	Food, skin
Crotalus atrox	72,683	Food
Apalone ferox	15,007	Food
Chelydra serpentina	6,729	Food
Apalone spp.	943	Food

[a]Hoplobatrachus tigerius *and* Rana tigerina *are synonymous species names.*
[b]*Most likely contains a large number of exports accidentally labeled as imports.*
[c]*The concatenation of* Pseudemys scripta, Trachemys scripta, *and* Chrysemys scripta.
Note: Volume *refers to level of trade.*
Source: Schlaepfer et al., 2005.

commerce, and the European tortoise trade adopted the North American box turtle, *Terrapene*, as one of the replacement "tortoises," thereby setting in motion the decimation of *Terrapene* populations. The issues of harvesting plants and animals are emotionally loaded, especially concerning regulatory issues and sustainable harvest of living natural resources. Our bias is on the noncommercial, protective side.

Amphibians and reptiles are widely harvested, although the impact is focused on a relatively few species in any locality. Their harvest is largely for consumption (food and folk medicines), luxury trade (leathers, jewelry, and curios), and the pet trade. All three represent commercial exploitation, in which animals are gathered specifically for sale by collectors, and each type of harvest represents a worldwide, multimillion-dollar industry. This commercial exploitation regularly leads to overharvesting and is a principal concern of conservationists; however, local family consumption also decimates populations of the targeted species when local human populations are dependent upon wildlife as a major source of protein. The concept of sustainable harvest focuses on use developed principally as a management tool for commercially and sport-harvested species, but it is useful as well for the conservation of species overharvested for local consumption.

Sustainable use allows the limited harvest of a population, providing that the portion of the population remaining is able to reproduce and maintain itself (Fig. 14.9). Conceptually, sustainable use is easy to establish, but in practice, it is difficult to set and control harvest limits. The goal is to establish a harvest regime that garners local community support because it is commercially profitable and/or provides the local community with an adequate supply of meat. If the harvest is set too low, populations of the harvested species experience little impact and possibly grow, but the local

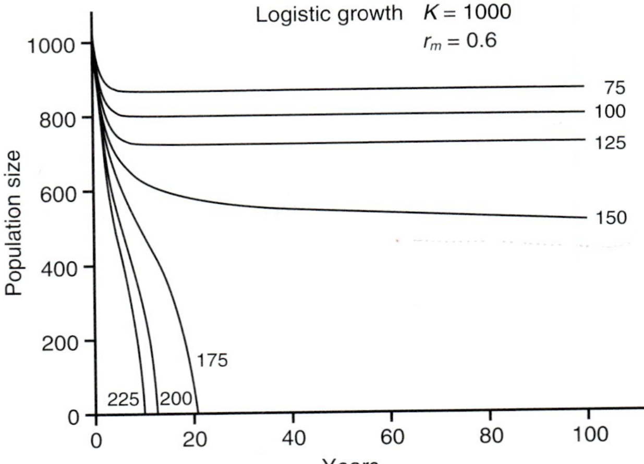

FIGURE 14.9 Hypothetical profiles of populations subjected annually to different levels of constant harvest. The values on the right indicate the number of individuals harvested annually. The profiles demonstrate the potential of sustainable-use harvesting; however, no abiotic or biotic perturbations are incorporated in the population-growth model, and natural populations would display fluctuations in population size. Modified from Caughley and Gunn, 1996.

community receives little benefit and likely will ignore the harvest limits. If the harvest is too high, the harvested population declines toward extinction. Extinction can be the actual disappearance of the population through overharvest, or it can be unsustainable-use extinction, in which population density is so low that efforts to harvest exceed the benefits to the harvester. Even in the latter situation, the population is likely to disappear because it has fallen below its minimum viable population level (for an explanation of MVP, see the section "Habitat Modification, Fragmentation, and Loss").

Turtles are presently at risk of global extinction because of the high demand for their meat for soup and their shells for traditional Chinese medicine. The situation has arisen in part because of ancient traditions combined with newfound wealth in China. Wild-caught *Cuora trifasciata* can bring US$1000/kg in China, which has instigated intense harvesting of this and many other of the 90 species of Chinese freshwater turtles and tortoises. Currently, over 1000 large turtle farms exist in China and are estimated to be worth US$1 billion, but these farms are not easing the pressure on wild turtles and tortoises. Instead, the owners of these farms are the major purchasers of wild-caught turtles. Farm-raised turtles show a marked decrease in reproductive capacity over time, so owners continually supplement their breeding stock with wild-caught animals. Turtle farming is therefore not a sustainable practice in China, and its only function is to provide short-term gain for a few people while driving many populations and species to extinction. These farms provide a method for laundering wild-caught animals as captive bred. In addition, as Chinese turtles become rarer, Chinese farmers are switching to North American turtles, such as snappers (*Chelydra, Macrochelys*) and sliders (*Trachemys*). Even if turtle farming could be made sustainable, the demand for wild-caught turtles is deeply ingrained in Chinese culture. Practitioners of traditional medicine tout the nutritional superiority of wild-caught animals over farm-raised animals, and consequently, wild-caught animals bring a much higher price. Currently, only a massive effort by the Chinese government could change turtle-farming practices, but even this could not control the black market and the desire for wild-caught animals. Chinese biologists predict that the current situation can lead to only one outcome: the extinction of China's wild turtle and tortoise populations.

Sustainable-use programs have had some successes among reptiles. Managed harvest of crocodylians began about 3 decades ago as a tool to assist the recovery of species and populations that had been devastated by unregulated hunting for their skins. The success of managed harvest and captive rearing in Papua New Guinea, Venezuela, and a few other countries stimulated other governments to begin similar programs. In most countries, managed species have shown a remarkable resilience, and populations are no longer endangered. However, with more countries producing skins, supply began to exceed demand and was then followed by a declining popularity for crocodylian leather. Ultimately, an economic depression in Asia caused the market for crocodylian skins to collapse. Market fluctuations are common for luxury items such as exotic leathers, and it now seems evident that the conservation management of a species cannot depend solely or even largely on the marketing of products from a particular species. Several other species of large reptiles (e.g., *Tupinambis, Varanus, Python, Naja*) are widely sought for the leather trade and have been examined for sustainable use; harvesting regimes and environmental education have begun in several countries. The fate of these management programs depends upon how accurately biologists have been able to assess the reproductive potential and demography of each population and upon the development of an accurate tracking of the number of individuals captured. These data are required to establish appropriate harvest quotas. Sport wildlife and fisheries depend upon quotas, which have proved to be effective management tools where special interests do not override the recommendation of the fisheries and wildlife biologists.

The commercialization of wildlife has potential negative side effects. The sustainable-use programs for crocodylians have been successful with the commercially valuable species, but the focus on these species has resulted in the neglect of truly endangered species, especially those with small distributions and less flexible demographics, such as *Alligator sinensis* and *Crocodylus mindorensis*. Further, commercialization of one group of species creates a market for all species and becomes a serious threat to endangered species. Even relatively abundant and widespread species can experience overharvesting if managers fail to distinguish between legally harvested skins and illegal ones. This situation probably occurs with the tropical American caimans, where the number of imported skins exceeds the number of skins legally exported. This concern for commercialization's fostering the uncontrolled harvest of wild populations is the reason for the sea turtle conservationists' resistance to either the farming or ranching of sea turtles. By making the marketing of all species from all areas illegal, no legal loopholes remain for the marketing of illegally harvested animals.

Human consumption of reptiles and amphibians, while relatively small compared with that of fish, birds, and mammals, is still significant. Like the skin trade, it concentrates on larger and long-lived species. Because of delayed maturity and a highly variable annual replacement rate, these species lack the demographic resilience to recover quickly from overharvesting. Among lizards, species of *Varanus*, *Tupinambis*, and certain iguanines are hunted for local consumption and in many areas have experienced sharp population declines because of overhunting. An effort to develop community-based *Iguana* farms in Belize, Honduras, and other countries for restocking wild populations and providing marketable meat was successful in identifying the proper farming protocols but has been only marginally successful in terms of enactment; widespread community support for the programs did not occur. Snakes are an important food and source of folk remedies in Asia, and their local consumption and capture for distant markets has grown greatly during the 1990s. The effect on snake populations has not been documented for most species, but it is likely to become evident by an increase in rodent populations and their devastation of grain crops. The present decline in sea turtle populations had its origins in the butchery of females and/or the harvest of eggs for human consumption (Fig. 14.10). While human consumption remains a significant threat in some regions, the incidental capture and death of sea turtles in the fisheries industry has become the major threat to sea turtle survival.

Frogs have been harvested in huge numbers from all over the world, mostly for food. Those from India and other places in Asia currently are being sent to markets in North America and Europe. The U.N. Food and Agriculture Organization estimated that, worldwide, at least 5200 tons of frogs were harvested annually from 1987 to 1998. In the United States, frogs were harvested as early as the middle 1800s. Mark Jennings and Marc Hays documented the history of *Rana aurora draytonii*, a frog that was once widespread and common in the western United States but is now almost extinct. Jennings and Hays examined numerous historical documents and anecdotal reports and determined that frog harvesting in California began about the time of the gold rush in 1849. They documented that tens of thousands of these frogs were taken annually from 1888 to 1895. Anecdotal reports indicated that large harvests were taken even before that time, indicating that the original populations must have been very large. By the early 1900s, the native frog populations were declining, most likely due to a combination of overharvesting because of demand by an increasing human population and concurrent alterations in habitat because of human activities. By the first decade of the 1900s, introductions of bullfrogs (*Lithobates catesbeianus*) were well documented, apparently as a replacement for the native *Rana aurora*. As a result of those introductions, feral populations of bullfrogs have become widespread in the western United States and, because of their large size and catholic diets, have decimated many of the native populations of frogs. Today the bullfrog has become a pest species, and numerous eradication programs have been unsuccessful. This account shows that unsustainable and unregulated harvesting can lead to decimation of native frog populations, and it also shows that introductions of exotic species with no forethought can lead to serious declines and loss of additional native species.

FIGURE 14.10 Female green sea turtle being prepared for market in Mexico. Note the large number of near-term eggs. (J. P. Caldwell)

Many amphibians and reptiles have been collected for centuries for the pet trade, for food, and for other reasons, such as use in folk medicine or for adornment (Table 14.4). This trade has had a significant impact on amphibian and reptile populations. In the United States alone, the annual trade of wild-caught amphibians and reptiles includes millions of individuals. Martin Schlaepfer and his colleagues determined that from 1998 to 2002, the United States imported 14.7 million wild-caught amphibians, in addition to 5.2 million kg of wild-caught amphibians, and 18.4 million wild-caught reptiles and reptile parts. During the same period, the United States exported 26 million wild-caught whole reptiles. Although the United States Fish and Wildlife Service has a system to keep track of this trade, it is difficult to interpret these data because most individuals are tracked only by family name and not by species (Table 14.5). These data give an idea of the huge number of amphibians and reptiles traded in the United States, and one can only imagine the numbers when the global figures are considered. No database exists to track the numbers of non-CITES species traded globally.

TABLE 14.5 Families of Amphibians and Reptiles Imported or Exported from the United States from 1998–2002 for Which More Than 100,000 Individuals Traded or More Than 50% of All Individuals Had No Species-Specific Identification.

	Individuals without species identification	
Class/Family	Number	Percentage
Amphibia		
Salamandridae	597,301	22.2
Pipidae	439,256	13.2
Ranidae	361,858	7.1
Discoglossidae	193,642	16.0
Rhacophoridae	176,949	71.5
Hylidae	171,844	35.7
Bufonidae	169,276	83.5
Hyperoliidae	12,503	67.2
Pelobatidae	7207	55.0
Plethodontidae	6513	98.7
Leptodactylidae	4321	64.1
Reptilia		
Gekkonidae	1,079,447	64.9
Lacertidae	392,743	92.5
Scincidae	206,365	61.4
Agamidae	185,168	29.1
Emydidae	166,573	27.7
Teiidae	116,922	25.7
Iguanidae	100,978	17.2
Cheloniidae	13,919	67.1
Kinosternidae	5684	87.9
Chelidae	4643	59.8

Source: Schlaepfer et al., 2005.

Exotic Species

The term exotic species refers not to species that are exciting or titillating but to introduced or nonnative species. Perhaps a chameleon in Hawaii is exciting to some people, but a seemingly innocuous introduction of an exotic species into a new habitat often has major and devastating effects on the native flora and fauna. The innocuous chameleon eats insects. If the insects are exotic also, then the chameleon causes no problem, but if it becomes established in native forests and eats native insects, many of which have small distributions, the chameleon can cause the extinction of native insects and potentially native plants that depend upon the native insects for their pollination. Generally speaking, exotic species are undesirable, and they negatively impact the native biota. Because humans have so greatly altered their immediate landscapes and bring with them (intentionally and unintentionally) many commensal species of plants and animals, they usually are unaware of the impact of an exotic species until it becomes a "plague." For example, the chestnut blight and the gypsy moth are exotics that have changed the community structure of eastern North American deciduous forests.

The impact of exotic and invasive plants on amphibians and reptiles is poorly documented. Most exotic plants accompany human alteration of native floras and habitats, making it difficult to separate the effects of habitat alteration from the dominance of exotics in human-impacted communities. Water hyacinths spread rapidly throughout southeastern United States from a small introduction in the St. Johns River, Florida, in the 1880s. The spread of this floating plant has apparently not affected any species of amphibian or reptile, and it seems to have provided a new habitat for the dwarf siren (*Pseudobranchus*), whose populations can be remarkably dense in the fibrous root masses of the plant. In turn, dwarf sirens support high densities of their predator, the black swamp snake *Seminatrix pygaea*. The preceding statement, however, is strictly anecdotal. No estimates have been made of abundance of either the salamander or the snake prior to the introduction of the water hyacinths; similarly, population data on cooter and slider turtles (*Pseudemys*) do not exist, and water hyacinths crowded out many of the native aquatic plants on which these turtles feed.

The water hyacinth story illustrates several aspects of the impact of an exotic or invading species. First, most evaluations of an exotic's impact lack preintroduction ecological data on the impacted species, and second, the impact becomes apparent only after the invading species has decimated or eliminated one or more native species. Invading species can benefit some species, and their detrimental effect on other species often is not observed because the impacted species is invisible relative to our knowledge or interest. Our evaluation of the gains or losses from an invading species depends largely on our perception of the value, either economic or aesthetic, of the impacted species. Water hyacinths were introduced for their attractiveness but rapidly became an economic burden because they clogged waterways and slowed or stopped boat traffic. At no time was there any discussion or concern for their impact on the native aquatic communities of the southeast United States. Only recently, as natural communities have declined because of human activities and the rate of invasive-species colonization has increased through the globalization of trade, have both the biological and business communities recognized the actual and potential costs of exotic species. For the former, exotics threaten the survival of native species and ecosystems; for the latter, exotics increase operation expenses, as when zebra mollusks clog water-pumping equipment or brown tree snakes cause power failures.

The most obvious impacts on native amphibians and reptiles occur on islands to which one or more vertebrate predators are introduced. Islands usually meet the criteria necessary for successful invasions by exotic species (Table 14.6). Rats have decimated bird and other reptile populations on almost every island that they have colonized—always with the transport provided by humans—and remain a major threat to many insular populations. The black rat and subsequently the Norway rat were transported worldwide by European explorers and subsequently traders and whalers, although the Polynesian rat (*Rattus exulans*) was the first to be widely introduced, probably intentionally as a food item, by people colonizing the oceanic islands of the Pacific 2000–3000 years ago. The Polynesian rat was especially destructive to bird populations but may have had a profound impact on lizard populations as well. It may have been responsible for the extinction of the tuatara from the main islands of New Zealand. The black and Norway rats remain major predators of the eggs and small juveniles of the Aldabra and Galápagos tortoises. The mongoose, introduced to control rat populations, proved unsuccessful in that role but highly successful in the elimination of ground-nesting birds and terrestrial lizards, including, the skink *Emoia nigra* from Viti Levu, Fiji. House cats, both feral and domestic, are skilled hunters and kill large numbers of amphibians and reptiles in suburban and rural areas of continents and have proved highly devastating to insular populations of lizards. Herbivores, such as goats on the Galápagos or rabbits on the California Channel Islands, change the structure and composition of vegetation, thereby affecting the availability of food for herbivorous reptiles and reducing or eliminating shelter from insolation and predators.

Mammals are not the only exotic predators that greatly affect amphibian and reptile populations. The brown tree snake *Boiga irregularis* is widely known for its consumption of the Guam avifauna, but this snake has also greatly reduced lizard density and may have caused the extinction of one or more lizard species (Fig. 14.11). In this case, it appears that the adult snakes are principally bird predators and the juvenile snakes are lizard predators.

In some cases, introduced amphibians have become major pests. Bullfrogs, *Lithobates catesbeianus*, were discussed previously. These frogs have been widely introduced as food and have become established in many areas. It is a highly carnivorous species, taking prey ranging from arthropods to mammals, including its own kind and other frogs, in its native habitat. This predaceous behavior has eliminated native frogs from many habitats. The marine or giant toad *Rhinella marina* occurs naturally in Central and South America, but it is possibly the most widely distributed exotic species. This species now occurs in the West Indies, Oceania, Philippines, Solomon Islands, New Guinea, and Australia. The Braminy blind snake *Ramphotyphlops bramineus* might have a larger exotic distribution, but it is invisible to most people because of

TABLE 14.6 General Requirements for the Successful Invasion by an Exotic Species.

I. Community–ecosystem susceptible to invasion
- A. Climatically similar to the source ecosystem of the invading species
- B. Simplified community or one stressed by human or natural disturbance
  1. Low species diversity
  2. Absence of or few predators
  3. Absence of or weak competitor species for the same resources

II. Successful invader species
- A. Broad physiological tolerances
  1. A habitat generalist
  2. Broad dietary requirements
- B. High reproductive potential
  1. Individuals mature quickly and reproduce frequently
  2. Each individual reproduces many offspring during its reproductive life
  3. Eggs and juveniles with moderate to high survivorship
- C. High genetic variability
- D. Phenotypically plastic

Note: The preceding outline includes many but not all characteristics necessary for a successful invasion. A successful invasion is the colonization and the establishment of a multigenerational and self-reproducing population of a species in an area distant from its area of natural occurrence and in a different community or ecosystem.
Source: *Modified from Meffe et al., 1994: Table 8.2.*

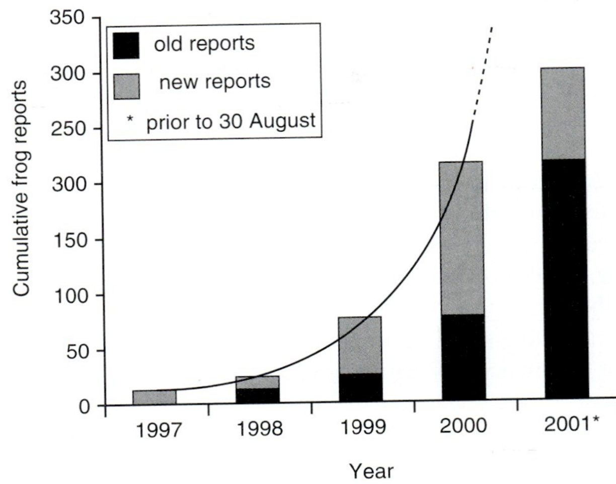

FIGURE 14.11 Increase in the abundance of the Coqui (*Eleutherodactylus coqui*) in Hawaii from 1997 to mid-2001. The rate of increase is exponential. Adapted from Kraus and Campbell, 2002.

its tiny size and secretive habits. The marine toad owes its entire exotic distribution to intentional introductions as a biological control agent, mainly for sugar cane beetle control for which it has never proved effective. Because the marine toad has been most successful in highly disturbed habitats, no evidence indicates that it has displaced any native frogs by competition or predation; however, a modicum of evidence from Australia suggests that some mammalian and avian predators have experienced population declines as the front of the toads' expanding distribution passed through new areas. Apparently, the native predators are poisoned by secretions from the toad's parotoid glands, but subsequently they either avoid the toad or learn how to kill and eat it without being poisoned. Interestingly, recent studies by Rick Shine and his colleagues indicate that some of the native Australian fauna is adapting to the presence of marine toads. Thus, even though their initial introduction had major effects on Australian reptiles and amphibians, those effects are minimized as time passes. In addition, marine toads in Australia appear to be evolving increased speed allowing them to invade new areas more rapidly!

Another frog that has become a serious pest is *Eleutherodactylus coqui*, which naturally occurs in Puerto Rico. These are relatively small frogs, 40–50 cm in SVL, and were transported to Hawaii in ornamental plants via the horticultural trade. Fred Kraus and Earl Campbell and their colleagues documented the presence of these frogs in Hawaii in 1999 and subsequently reported on their spread in 2002. It was obvious that the frogs originated from the plant trade because the initial large populations were at active nurseries. *E. coqui* cannot be mistaken or overlooked because of the male's loud, high-pitched, two-note call. Like all species in the genus, it has direct development, so reproduction proceeds year-round and frogs do not need bodies of water to reproduce. At the time that Kraus and Campbell and colleagues first reported on the presence of these frogs, they noted that populations were still small enough that they could be eradicated from Hawaii. They predicted negative ecological and societal consequences should the populations continue to expand. These frogs can reach densities of 20,000 per hectare in Puerto Rico and take 114,000 prey items per night per hectare. Not only would they have the potential to disrupt the food web in native forests and cause the demise of native birds and other insectivores, but the societal costs could be great. Guests in hotels and home owners were beginning to complain that the loud calls were disrupting their sleep, especially when a large number of frogs was present. Kraus and Campbell also noted that eradications would take cooperation of state officials, nursery owners, hotel owners, and the public. Four years later, Kraus and Campbell reported that the populations of this frog had exploded (Fig. 14.11). The earliest record of this frog on Hawaii Island was 1988. By 1998, frogs occurred at 8 sites on Hawaii Island, 12 on Maui, and one on Oahu. By 2001, frogs occurred on at least 101 sites (and possibly 124 other sites) on Hawaii Island, 36 sites on Maui, 14 sites on Oahu, and 2 sites on Kauai. This rapid increase was due to several factors, some previously mentioned, but largely because of the failure of people to act quickly. The frogs were not taken as a serious threat and the state government did not take quick action. The horticultural trade has not been regulated, and introductions continued to occur. In addition, despite educational fact sheets and public service announcements broadcasting the call, some people continued to intentionally introduce frogs to their gardens, even though transport of the frogs by humans had been illegal since 1998. Today, the species can probably never be eliminated from Hawaii. In addition to the ecological effects, infestations have led to lower property values in some areas, and some home owners have vowed to leave Hawaii rather than endure the frogs' shrill calls. The nursery industry is threatened because shipments to other places could be rejected if frogs are present in their material, and hotel owners with frogs on their property have lost business. The story of this invasion should be used as a cautionary tale and can perhaps save other islands from suffering the same fate as Hawaii. Invasive species can take a huge ecological, economic, and societal toll. Quick action is necessary when a problem is first detected.

Throughout the tropics, one or more species of geckos are human commensals, living in and on human habitation. Because of this association, they have been unintentionally transported pantropically and widely among oceanic islands. The genus *Hemidactylus* has the largest roster of exotic species, and recently the house gecko *Hemidactylus frenatus* has been spreading eastward through the Pacific Islands. Its spread has occurred at the expense of another commensal gecko, *Lepidodactylus lugubris*. *Lepidodactylus lugubris* is an all-female species that is competitively inferior to *Hemidactylus*. Both females and males of the bisexual *H. frenatus* defend their feeding sites and challenge conspecifics and any other geckos that enter their sites. This behavior drives *L. lugubris* from the prime feeding areas and through time eliminates them from most buildings, thereby providing an example of species displacement via interference competition.

Extinction

Extinction is the disappearance of a population of organisms. This natural process occurs regularly within a species, but it is much rarer for all populations of a species to become extinct. The extinction rate for sink populations within a metapopulation can be measured in a few generations or years, whereas it may take tens to

TABLE 14.7 Species Potentially at Risk for Extinction: Amphibian and Reptilian Examples.

Characteristic	Species
Only one or few populations	*Pseudemydura umbrina*, western Australian swamp turtle[9]
Small population sizes	*Alytes muletensis*, Majorcan midwife toad[3]
Small geographic ranges	*Lepidodactylus gardineri*, Rotuman forest gecko[14]
Populations in decline	*Dermochelys coriacea*, leatherback sea turtle[11]
Low population densities	*Phaeognathus hubrichti*, Red Hills salamander[4]
Low genetic variability	*Emys (Clemmys) marmorata*, western pond turtle[7]
Adults requiring large home ranges	*Lachesis muta*, bushmaster[8]
Adults of large body size	*Varanus komodoensis*, Komodo dragon[1]
Slow maturity and/or long lived	*Emys (Emydoidea) blandingii*, Blanding's turtle[2]
Low reproductive potential	*Glyptemys (Clemmys) muhlenbergi*, Muhlenberg's turtle[5]
Poor dispersal ability	*Plethodon shenandoah*, Shenandoah salamander[13]
One or more migratory life stages	Cheloniidae, hard-shelled sea turtles[10]
One or more life stages forming temporary or permanent aggregations	*Lepidochelys kempii*, Kemp's ridley sea turtle[10]
Specialized resource requirements	*Hoplocephalus bungaroides*, broad-headed snake[12]
Harvested by people	*Tupinambis nigropunctatus*, tegu[6]

Note: The taxa selected as examples are not necessarily threatened currently but represent the attribute that places them at risk of extinction.
Sources: [1]Auffenberg, 1981; [2]Congdon et al., 1994; [3]Corbett, 1989a; [4]Dodd, 1989; [5]Ernst et al., 1989a; [6]Fitzgerald, 1994b; [7]Gray, 1995; [8]Greene, 1986; [9]Kuchling, 1998; [10]Natl. Research Council, 1990; [11]Spotila et al., 1998; [12]Webb and Shine, 1998; [13]Witt, 1999; [14]Zug, 1991.

thousands of generations or years for metapopulations to become extinct, and ten thousands to millions of years for a species to become extinct. The propensity of a population or a species to become extinct usually is associated with its size or number of individuals. A list of species at risk shows the variety of vulnerabilities and their direct association with the number of individuals (Table 14.7). For example, species composed of individuals with large body size or large home ranges typically consist of populations of fewer individuals. Other factors that make species vulnerable to extinction are associated with reproduction or aggregation. Many turtles require 10 or more years to attain sexual maturity; this long interval increases the probability of mortality before an individual can reproduce for the first time. Many frogs and salamanders in seasonal environments, whether tropical wet–dry or temperate hot–cold, form breeding aggregations in temporary pools. The assemblage of most or all breeding adults in a single location at one moment in time increases the probability that a single catastrophic event can eliminate the entire population. If the breeding aggregation requires a special habitat, a string of abnormal weather patterns can temporarily eliminate the proper breeding site for the reproductive life of species with short generation times.

One or more species of amphibians and reptiles match each at-risk category (Table 14.7), and in most cases, the species have obtained their at-risk status as a result of human activities. Harvesting and natural habitat reduction and fragmentation are and have been the major factors driving amphibian and reptile species to the brink of extinction. Both factors reduce population size and genetic variation, thereby increasing the likelihood that stochastic events will cause extinction (Table 14.2). Until recently, extinction among amphibians and reptiles has been relatively low or not recognized, but the current biodiversity crisis includes numerous species and herpetofaunas (Table 14.8). Presently, amphibians (especially anurans) and turtles are at greatest risk for mass extinction. Overharvesting threatens turtles, but the cause(s) for the disappearance of frog species is less certain.

Declining Amphibians

In 1989, a small group of scientists attending the First World Congress of Herpetology in Canterbury, England, began voicing their concerns about frog disappearances in many types of habitats. These informal conversations eventually led to more formal meetings and the realization that herpetologists from all around the world were aware that frogs had disappeared from many places where they had been abundant. The Declining Amphibian Populations Task Force (DAPTF) was formed to investigate the situation and establish a worldwide communication network. The real question people wanted to know was whether frog disappearances were a global phenomenon or restricted to a smaller number of localities.

Soon, numerous scientists began pooling their data, which indicated that many populations and some species of amphibians from around the world had disappeared or were in sharp decline. The peculiarity of many of the

TABLE 14.8 Examples of Amphibian and Reptilian Extinctions During the Last 2000 Years.

Amphibians
Anura
Bufonidae
Atelopus oxyrhynchus (Venezuela)[6]
Ollotis periglenes (Costa Rica)[9]
Discoglossidae
Discoglossus nigriventer (Israel–Syria)[4]
Leiopelmatidae
Leiopelma auroraensis (New Zealand)[5]
Leiopelma markhami (New Zealand)[5]
Leiopelma waitomoensis (New Zealand)[5]
Brachycephalidae
Eleutherodactylus eneidae (Puerto Rico)[8]
Eleutherodactylus karlschmidti (Puerto Rico)[8]
Eleutherodactylus jasperi (Puerto Rico)[8]
Myobatrachidae
Rheobatrachus silus (Australia)[11]
Reptiles
Testudines
Testudinidae
Geochelone inepta (Mauritius)[1,4]
Geochelone abingtonii (Pinta, Galápagos)[7]
Crocodylia
Crocodylidae
Crocodylus raninus (Borneo)[10]
Squamata
Anguidae
Celestus occiduus (Jamaica)[2,7]
Gekkonidae
Aristelliger titan (Jamiaca, West Indies)[2]
Hoplodactylus delcourti (New Zealand)[2,5]
Phelsuma edwardnewtonii (Mascarene Islands)[2,7]
Iguanidae
Brachylopus sp. (Tonga)[2]
Cyclura collei (Jamaica)[7]
Leiocephalus eremitus (Navassa Island, West Indies)[2,7]
Leiocephalus herminieri (Martinique, West Indies)[2,7]
Lacertidae
Gallotia goliath (Canary Islands)[2]
Scincidae
Cyclodina northlandi (New Zealand)[2,4]
Leiolopisma mauritiana (Mascrene Islands)[2]
Oligosoma gracilocorpus (New Zealand)[2,4]
Colubridae
Alsophis santicrucis (St. Croix, West Indies)[7]
Dromicus cursor (Martinque, West Indies)[7]
Viperidae
Vipera bulgardaghica (Turkey)[3]

Note: Extinction is often difficult to verify for amphibians and reptiles. Many instances exist in which a species was described from one or a few voucher specimens and then is not observed for 50 or more years. The absence of observations might indicate a species of small population size, specialized habitat preferences, short or unusual seasonal activity patterns, or similar factors requiring detailed knowledge of the natural history to rediscover the species. Human expansion into and modification of natural habitats, however, increases the probability that many of these "rare" species are already extinct or soon will be.

Sources: [1]*Bour, 1984;* [2]*Case et al., 1992;* [3]*Corbett, 1989;* [4]*Day, 1981;* [5]*Gill and Whitaker, 1996;* [5]*Halliday and Heyer, 1997;* [6]*Honegger, 1981;* [7]*Joglar and Burrowes, 1996;* [8]*Pounds and Crump, 1994;* [9]*Ross, 1990;* [10]*Tyler, 1991c.*

population declines was their suddenness, their occurrence in areas presumably exposed to a minimum of human influence, and often the disappearances of some species but not others at the same locality. Amphibian populations are known to fluctuate greatly in size, so some biologists expressed concern that the declines represented natural fluctuations and that by raising a potentially false alarm, the conservation of truly threatened and endangered amphibian species would be hindered.

The initial alarm, unfortunately, has proven to be real. Since the initial concern, much research has rigorously documented the status of populations and attempted to determine the causes of declines. Attention has focused most strongly on anurans because of their greater diversity and worldwide occurrence and to a lesser extent on salamanders. When declining amphibians are discussed, however, almost nothing is mentioned about caecilians, yet these amphibians are potentially as threatened as frogs and salamanders. Caecilians are primarily tropical and frequently subterranean, which accounts for the paucity of data on their status.

David Gower and Mark Wilkinson recently summarized what is known about the conservation status of caecilians worldwide. The primary threats to these animals are habitat destruction and pollution. They noted that the habitat of two Philippine species, *Ichthyophis glandulosus* and *I. mindanaoensis*, has been cleared, and streams in which their larvae develop are polluted and nearly dry. Populations of other species in various places throughout the world could be reduced because of conversion of the land to agriculture and the concomitant use of agrochemicals, or because of urbanization. Although chytridiomycosis has caused the decline of frogs and salamanders, no studies have examined whether caecilians have this fungal infection. The fungus is found in damp soil and water, habitats in which caecilians occur, so species in high-risk areas could be infected. The fact remains, however, that almost no data exist for most species of caecilians, so their population status remains largely unknown.

Determination of the number of amphibian species that are threatened or have gone extinct is difficult because of the lack of data on population status for many species. The Global Amphibian Assessment (GAA) is a comprehensive project in which many conservation organizations and scientists have partnered to identify the scale of amphibian declines and the geographic areas affected. The GAA estimates that more than one-third of amphibians throughout the world are threatened, and possibly as many as 120 species have gone extinct since 1980.

The causes of amphibian extinctions and declines vary, and many questions remain despite the intense amount of research in this area. Habitat loss and modification are major factors in the decline of abundant and uncommon species everywhere. For example, estimates indicate that over 70% of the ponds and marshes of Great Britain have disappeared since the beginning of the 20th century and that frog and toad abundance may have been reduced by more than 90%. Habitat loss and modification is a global phenomenon. In addition, natural habitats and their herpetofaunas adjacent to and interspersed among agricultural lands experience a subtle form of poisoning from insecticides and herbicides. These chemicals and their breakdown products have a variety of effects from carcinogenic and mutagenic actions to direct poisoning and hormone mimics, thereby affecting all life stages of amphibians and reducing the survivorship of all. Environmental acidification (see the section "Pollution and Disease") is also widespread and particularly disruptive of early development. Its effects occur distant from its source. Although it does not appear to have been the direct agent for the disappearance of amphibian populations in western North America, acidification may act synergistically with other pollutants, ranging from heavy metals to ecoestrogens, to disrupt the physiology of amphibians and make them more susceptible to bacterial, fungal, and viral diseases.

Disease has become increasingly implicated in amphibian declines, particularly in those declines and disappearances occurring in presumed pristine habitats. One of the most ubiquitous pathogens is a fungus, *Batrachochytrium dendrobatidis,* which infects the skin of frogs and the mouthparts of tadpoles, causing a disease referred to as chytridiomycosis. The fungus was first identified in 1998 and then described in 1999. It has been found in museum specimens of African *Xenopus laevis* dating back as far at the 1930s. It is unclear whether the fungus has recently spread, possibly as a result of global climate change, or whether it has become more virulent. Once the fungus reaches a naïve population, however, it can cause mass mortality of an entire frog community, quickly killing nearly all frogs of most species in an area.

By 2004, the fungus had been found on every continent except Asia and Antarctica, and it had infected 14 families and 93 species of frogs and salamanders. The advancing front of the disease was particularly well documented in Costa Rica, where amphibians in the northern part of the country were affected in the middle to late 1980s, and populations in the southern part were sequentially affected in 1992, 1993, and 1996 (Fig. 14.12). Amphibians are the only host of this fungus, which attacks keratinized skin cells in adults and keratin in the mouthparts of tadpoles. The skin of adult frogs becomes rough and no longer able to function in respiration and water balance, leading to death of the infected animals. The fungus spreads easily in water by means of a flagellated zoospore, so it is readily transmitted from one individual to another. Scientists have recently documented how global warming is related to the spread of the fungus in tropical mountainous areas of Costa Rica. In this region, warm air has increased cloud cover, which in turn has provided a more moderate climate (cooler days and warmer nights), especially at midelevation localities. This change in climate has provided optimum conditions for the growth of the fungus, allowing the fungus to spread rapidly among frog populations, and appears to explain why frogs in mountainous areas from the western United States to the Andes in South America have declined precipitously in numbers.

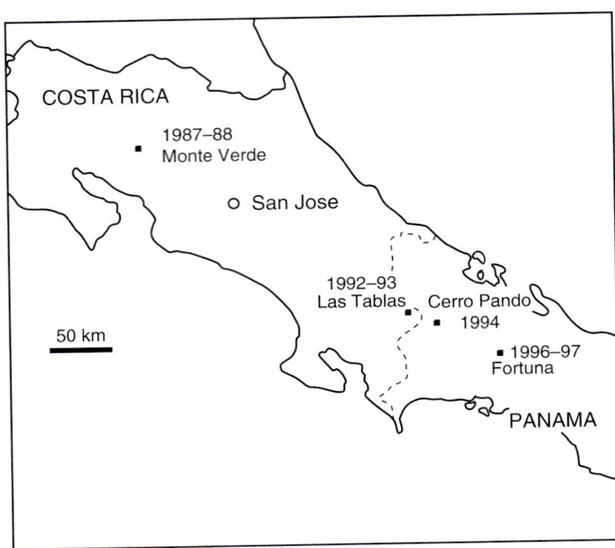

FIGURE 14.12 The timing of amphibian die-offs in Costa Rica and adjacent Panama, suggesting the spread of a virulent pathogen. The dates represent the sudden disappearance of frogs at Monte Verde and the appearance of dead and dying frogs at the other sites. Data from Lips, 1998, 1999; Pounds and Crump, 1992.

In addition to declines attributable to chytridiomycosis, deformed amphibians have begun to appear in many populations. A small number of deformed individuals in a population, usually less than 5%, is typical and can be caused by mutations, injury, or trauma. More recent reports, however, have found that 15–90% of frogs in some populations have severe deformities. Previously, deformities consisted of injured toes or feet, whereas the more recent deformities include extra legs, misshapen eyes or tails, and other whole-body deformities. In a review of these deformities, Andy Blaustein and Pieter Johnson found that at least 60 species in 46 states in the United States and in parts of Canada, Japan, and Europe have been affected.

Three possible causes of deformities, all with questions remaining, include UV-B radiation, chemical contamination, and parasitic infection. The use of chlorofluorocarbons (CFCs) and other chemicals by humans is causing a continual depletion of the protective ozone layer in the stratosphere, which causes an increase in the amount of exposure of UV-B radiation to plants and animals. Experimental studies have shown that high doses of UV-B radiation can cause deformities in amphibians; however, the types of deformities do not match those seen in wild populations. Many kinds of chemical contaminants, including herbicides, pesticides, heavy metals, and others, are prevalent in amphibian habitats, and much research has examined the effects of these chemicals on amphibians and their larvae. Many can kill larvae, and some can cause deformities. One of the problems in determining a cause-and-effect relationship is that it is difficult to isolate one particular chemical in a natural environment because so many are present. Parasite infections were first proposed as a cause of amphibian deformities as a result of observations made by Stanley Sessions and Stephen Ruth. These researchers observed that limb deformities were associated with a high incidence of cysts (metacercariae) of a trematode parasite. They conducted experiments in which resin beads the size of metacercariae were implanted into developing limb buds of *Xenopus*. These implants led to the formation of deformities similar to ones seen in the wild. Subsequent research in a number of areas has shown that infection of the trematode parasite *Ribeiroia* can cause severe deformities similar to those seen in the wild.

Questions arise with all these proposed causes of deformities, and each has multiple levels of complexity. Regarding the parasite infections, why have parasite infections become so much more prevalent in recent years? One possibility is the dramatic increase in artificial habitats such as farm ponds and catchment basins from farms where large numbers of animals are raised in enclosed buildings. These types of aquatic impoundments have high fertilizer content from cattle and other animal manure, which in turn causes a large amount of algal growth. In turn, dense algal growth causes a denser snail population, which is an intermediate host of the parasite. These artificial habitats are places where birds (a part of the parasite life cycle) and amphibians come into contact (Fig. 14.13). Other stressors, such as pesticides and other chemicals, are likely interacting to make amphibians susceptible to parasites and disease (Fig. 14.14).

Recently Joe Mendelson and colleagues called for the formation of the Amphibian Survival Alliance, which would be within the Amphibian Specialist Group of the IUCN (World Conservation Union). This organization will mount a coordinated global response to the amphibian crisis, requiring a 5-year budget of at least US$400 million. Regional centers for disease research and captive management would be established. Survey, monitoring, and habitat protection programs must be implemented, and in the most critical cases, salvage operations must be undertaken for species on the verge of extinction.

This superficial review demonstrates the numerous factors involved in amphibian declines and highlights the complexity of the problem. Considering that amphibians live at the water–land interface and thus are exposed to environmental contaminants throughout their life history, and considering that amphibians have persisted throughout the entire evolutionary history of terrestrial vertebrates, the rapid declines we see today serve as harbingers of the potential devastating effects of human activity on life on Earth.

FIGURE 14.13 The trematode *Ribeiroia odatrae* reproduces asexually inside aquatic snails (*Planorbella* sp.) and generates thousands of infectious cercaria (larvae). The cercaria burrow into developing limb buds of amphibians, forming metacercariae (cysts). These interfere with limb development, causing deformities, and it is believed that these deformed froglets may be more susceptible than normal frogs to predation by birds. When a bird eats an infected frog, the metacercariae develop into sexually reproducing parasites that release eggs back into the water, where they hatch and again infect aquatic snails, completing the life cycle. Adapted from Blaustein and Johnson, 2003.

PRESERVATION AND MANAGEMENT—IDEALS AND PROBLEMS

As natural areas shrink or are modified, species and ecosystem preservation become increasingly a management task. The ideal situation is retention of large areas of diverse habitats without management, except for their protection from the illegal extraction of natural resources, such as wildlife poaching and logging. Such areas still remain, but for many species and ecosystems, active management is required. Three major management tools are establishment of refuges, the management of animals in captivity, and the reestablishment of populations using reintroductions. None of the three is a panacea.

Reserves and Corridors—Saving Habitats

As noted in the discussion on habitat fragmentation, a key issue in establishing a reserve or refuge is how much area to preserve. Reserve size is absolutely dependent upon the species or assemblage of species to be preserved, and the

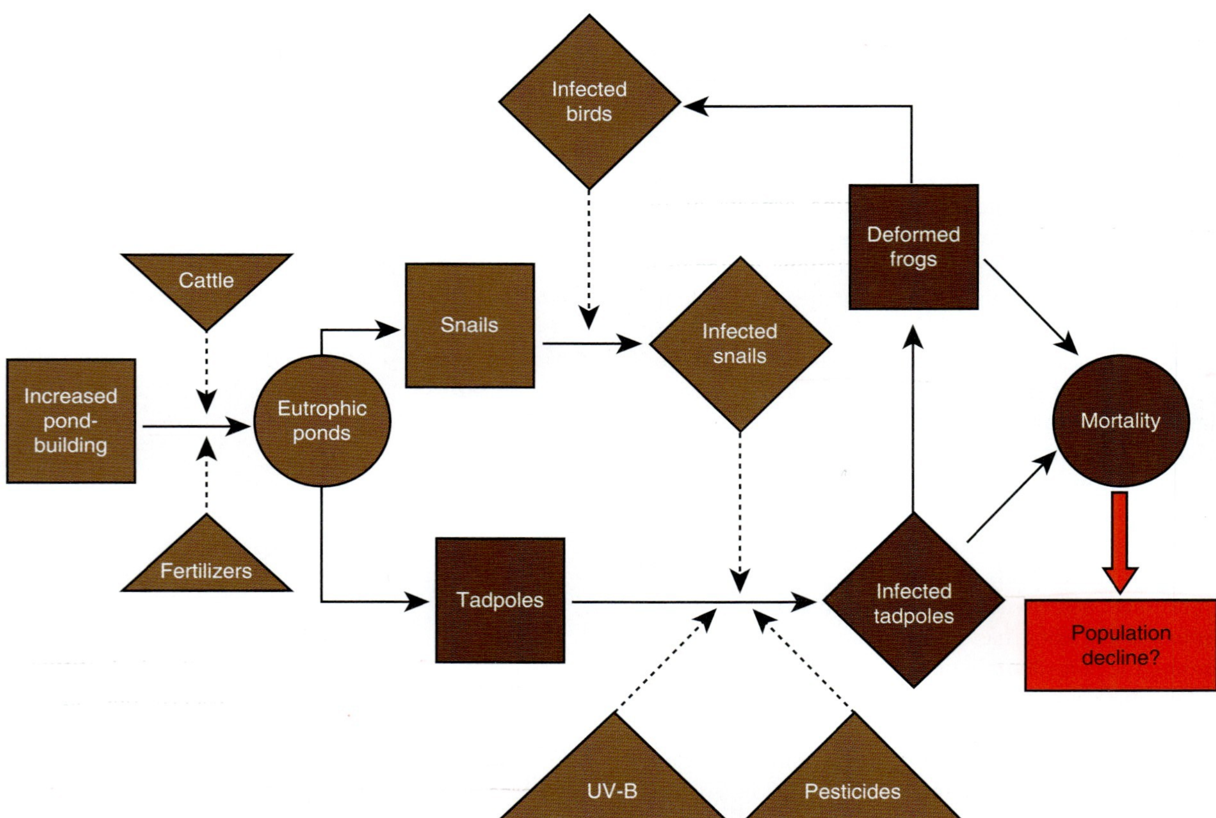

FIGURE 14.14 Relationships between amphibian declines and environmental factors are complex and undoubtedly vary among populations. This flow diagram shows how parasites, artificial pond eutrophication, UV radiation, and pesticides interact to cause declines. Adapted from Blaustein and Johnson, 2003.

identification of the necessary area for full protection requires a thorough knowledge of the natural history of the involved species, especially the habitat requirements of all life stages. As with all conservation issues, a tug-of-war exists regarding the amount of space needed according to biological conservationists and according to the local populace, business interests, and government. Further, ongoing scientific discussions are critical in determining areas in the world that are hot spots—areas of high biodiversity—that require immediate protection, the appropriate size and shape of reserves, the nearness of reserves to one another, and the length and shape of corridors connecting reserves. The issues are complex, so the following review can examine them only superficially.

In addition, a side issue is a philosophical debate on whether conservationists should target species or communities and ecosystems for preservation. Most biologists argue for the latter, but public and political support is required, and game and charismatic species, such as pandas or gorillas, identify the conservation need to a broader audience, thereby gaining the support necessary for reserve establishment. The advantage is that most of these charismatic species are large; hence, they require large areas and often diverse habitats. These requirements protect numerous other species and their communities as well. The major disadvantage is that single-species conservation can become so narrowly focused on the preservation of the target species that it loses sight of the necessity to preserve the entire habitat and ecosystem of the species.

Reserves are established to prevent the extinction of species, so the issues that determine the location, size, shape, and other aspects of a reserve relate to the survival of a population or species; this concept is referred to as the minimum viable population concept (MVP; see also the section "Habitat Modification, Fragmentation, and Loss"). The probability of extinction increases as population size decreases; this relationship gave rise to the minimum dynamic area (MDA) concept (Tables 14.2 and 14.5). MDA is the amount of habitat necessary for the maintenance of the MVP for a species. The important aspect of this definition is amount of habitat. Habitat represents the actual space used by a species and not just the amount of land or water area that theoretically should permit survival. The focus on habitat usage emphasizes the necessity of knowing all aspects of the natural history of an organism and the need for research into all facets of an organism's biology.

MVPs have been calculated for only a few species, mainly mammals, but not for any species of amphibian or reptile. Some recent studies examine aspects of the MVP for turtles and amphibians. These studies examine the conservation value of federal regulations designed to protect U.S. wetlands and their biota. Within the United States, any wetland larger than 0.4 hectare requires protection. One aspect of this protection is a requirement that a terrestrial buffer zone is established around a wetland to prevent development and encroachment into the natural area. Biologists at the Savannah River Ecology Laboratory have had several ponds completely enclosed by drift fences and pitfall traps for more than 2 decades in order to track the inward and outward movement of every individual of each amphibian and reptilian species living in the pond community. These studies have shown that some semiaquatic species spend considerable time in upland sites, for both nesting and hibernation. Researchers attached radio transmitters to 73 mud turtles (*Kinosternon subrubrum*), 14 Florida cooters (*Pseudemys floridana*), and 6 sliders (*Trachemys scripta*) to map the terrestrial movements of these species. They discovered that the federally mandated terrestrial buffer zone for wetlands protected none of the hibernation or nesting sites of these turtles (Fig. 14.15). Even the strictest state statutes protect less than 50% of these types of terrestrial sites. To encompass the total terrestrial area used by these turtles, the buffer zone must extend about 240 m beyond the outer edge of the federally mandated zone of protection.

A similar situation arises when pond-breeding amphibians are examined. Adults of many pond breeders are terrestrial except when breeding; only the larvae are aquatic. Using data from six species of *Ambystoma* salamanders, adults were on the average 125 m from the edge of their breeding ponds. This distance is a mean value, and salamanders often were even farther from ponds. Assuming that the area within the mean distance contains 50% of the population, a buffer zone would need to extend 164 m beyond the pond's edge to encompass the land-use activities of 95% of the sampled populations.

These studies highlight the difficulty of identifying and providing sufficient space to preserve one or a few

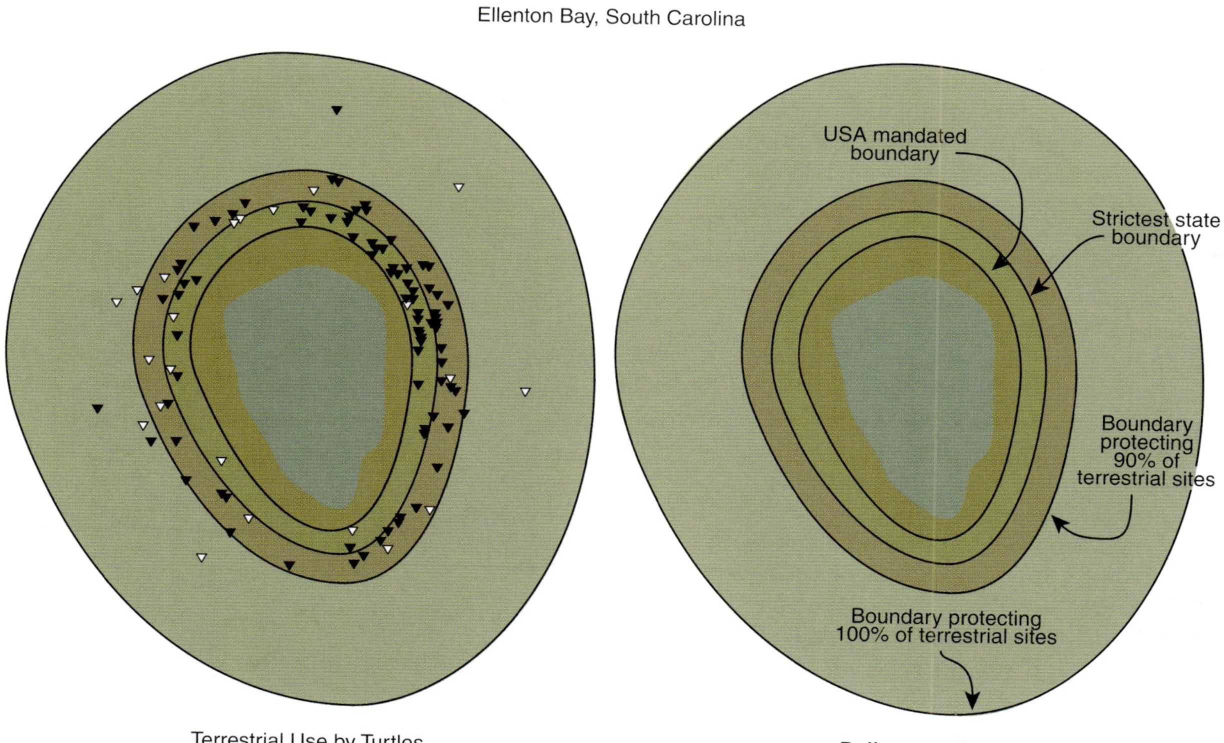

FIGURE 14.15 The inadequacy of the federally mandated terrestrial buffer zones to protect U.S. wetlands. The schematic diagram shows terrestrial use by three species of semiaquatic turtles living in a 10-ha pond in southeastern South Carolina. The left diagram maps the nesting sites (solid inverted triangles) and hibernation sites (open inverted triangles); the right diagram illustrates various buffer-zone boundaries. Modified from Burke and Gibbons, 1994.

components of a single community, and this space would certainly be inadequate to attain MDA requirements of any of the preceding species. Conceptually, MDA (= minimum reserve size) must encompass at a minimum the distribution of an entire metapopulation, and to expand the MDA concept to community–ecosystem preservation, the MDA must encompass the metapopulation of the species with the largest metapopulation distribution. For small mammals, minimum reserve size is estimated to range between 10,000 to 100,000 ha (100–1000 km^2). Small nonmigratory amphibians and reptiles likely require less area. Using extant species with limited distributions offers one method of estimating potential reserve size. The Shenandoah salamander *Plethodon shenandoah* presently occurs as three isolates in the George Washington National Forest, Virginia. The isolates contain 8 to 10 populations, none of which appears to be a sink population, and in total they occupy about 128 ha. Because of *P. shenandoah*'s close association with talus slopes that lie between 780–1150 m, and a known history of recent total defoliation of the forest and of earlier heavy logging and forest fires, we have evidence of the resilience of this species to ecocatastrophes. These data permit an estimate of a 36 km^2 MDA linking all the isolates in a single rectangular forest reserve with a broad buffer zone. A similar process estimates the total land area presently occupied by *Varanus komodoensis* in five islands of the Lesser Sunda group and yields an estimated MDA of roughly 1500 km^2. MDA increases greatly when migratory species are considered. The entire North Atlantic gyre, coastal Florida, and the Greater Antilles would define the MDA for the nesting populations of the green sea turtle *Chelonia mydas* on Florida's east coast.

The preceding MDAs are speculations and illustrate only one of many factors that must be evaluated when establishing a reserve. Among many factors, the most important one is defining goals for the reserve. Without precise goals, the critical decisions on size, shape, and other aspects of the reserve cannot be made and conservationists will not be able to develop a convincing case to win the support of the local community and the government. Reserves have been established for the protection of amphibians and reptiles. The Archie Carr National Wildlife Refuge was created to protect nesting sea turtles and their nests, and the Komodo National Park was created to protect the Komodo dragon. Both examples represent "single-species" reserves, and only the latter is sufficiently large to potentially meet the MDA criteria for the species it is meant to protect. The reserve does serve that function for many smaller lizard and snake species. Even though the Archie Carr National Wildlife Refuge does not meet MDA criteria, its establishment is essential to protect a major nesting beach for two sea turtle species on a coastline that is experiencing rapid and unwise development.

The high potential for extinction of small populations remains a constant threat as the world's natural habitats become increasingly fragmented. The recognition that reserves cannot be as large as conservationists desire, and as the populations of many species require, led to the biological corridor concept. If single large reserves cannot be established, could numerous small reserves linked by corridors of natural habitat serve as well? Conceptually, biological corridors seem to offer a satisfactory solution, although critics immediately began to identify potential problems, such as increased mortality along the corridors because of concentration of predators and the inability of species with low dispersal abilities to find and use narrow corridors. Nonetheless, conservation management groups broadly adopted the corridor concept before the efficacy of corridors was evaluated. Such research is only beginning to test the concept, and, while still limited in scope, the research findings largely support the critics' arguments that corridors are not effective for most species.

Two studies of corridor efficacy have used amphibians. In field experiments, *Ensatina eschscholtzii* used disturbed corridors that lacked surface litter and vegetation as frequently as corridors with natural cover. Salamanders traversing the disturbed corridors moved faster and more frequently than those in the natural corridors, suggesting that selection of corridors by humans likely will not match the habitat-specific dispersal requirements of many species in the communities being conserved. Similar results and conclusions derive from surveys of species that occur in a river floodplain corridor between two reserves in southern Illinois. The corridor is undisturbed floodplain forest flanked by upland deciduous forest on adjacent bluffs. A 2-year survey of the corridor revealed only 14 species of the 37 amphibian and reptilian species of the upland forest reserves. Even intense surveys do not locate all species, but the occurrence of only 38% of the species suggests that the corridor is not suitable for many species and may act as a sink potentially reducing population size for some species in the reserves. Corridors certainly serve some species; however, they cannot be assumed to be effective for all species in a community. Corridors appear to be ineffective for most species of small amphibians and reptiles.

Captive Management

Depending on goals and other factors, animals can be managed in captivity for relatively short periods of time or for much longer periods. Temporary captivity for short periods may protect one or more life stages in order to increase survivorship during a presumed critical period of life. Temporary captivity includes head-start programs and programs in which eggs are maintained in hatcheries. Long-term captivity for the duration of an individual's life, possibly for several generations, may be for commercial purposes and eventually result in death of the animals for their meat, hide, or some other commercially valuable product. Animals in captivity can produce offspring for translocation to replace extinct populations or to augment the size of an existing population with offspring raised in captivity.

Hatcheries and head-start programs are current tools in sea turtle conservation. The effectiveness of these conservation tools remains uncertain because they treat the symptom of population decline (fewer sea turtles) rather than addressing the cause of the decline. These techniques are incorporated into management plans without investigations supporting their efficacy. Both hatchery and head-start programs demonstrate potential dangers of the program in the short and long term; however, it is also essential to note that both types of programs appear to have some successes in increasing survivorship to adulthood. Sea turtle hatcheries were begun in the 1970s to protect eggs from terrestrial predators and eroding beaches. Initially, eggs were placed in Styrofoam containers filled with local beach sand. Hatching success was comparable to protected nests remaining on the beach; however, temperature-dependent sex determination was as yet unknown. Because the containers were typically maintained in shaded conditions and the boxes were insulated, the nest temperatures were commonly lower than natural nests and the hatcheries produced mainly males. Further, sea turtles are site-specific nesters, and adult sea turtles return to their natal beach for nesting. How the hatchlings imprint upon the beach so that they can relocate it as adults remains unknown. Whether disturbing and moving a nest of eggs or hand-releasing hatchlings at water's edge affect the hatchlings' ability to imprint on their future nesting site also remains unknown. The apparent return of some Kemp's ridley sea turtles reared in hatcheries to Padre Island, Texas, indicates that rearing and release techniques allow imprinting in some individuals (but see following text).

Head-start programs typically involve maintaining hatchlings in captivity for 6 to 12 months and feeding them a protein-rich diet to increase growth rate. The goal is to enable young turtles to attain a larger size before releasing them into the sea. Presumably, this enhances survival because the number of potential predators that can kill turtles decreases as the size of turtles increases. The success of such programs remains questionable. Head-start turtles survive for years after release and grow at natural rates; however, head starting can interfere with the ability of some turtles to locate their parental nesting beaches. This evidence is circumstantial; nonetheless, the nesting attempts by head-start Kemp's ridleys on beaches distant from their natural nesting beach are suggestive.

The most successful examples of long-term captive management are crocodylian farming and ranching (Fig. 14.16).

FIGURE 14.16 Typical crocodile farm. (P. Ryan).

FIGURE 14.17 Taxa involved in captive management and/or translocation programs: *Alytes muletensis obstetricans* (left; E. G. Crespo) and *Anaxyrus houstonensis* (right; D. B. Fenolio). The program has been successful for *A. muletensis* but unsuccessful for *A. houstonensis*.

The successes involve both alligators and crocodiles and have been motivated by commercial interest, principally production of hides for the leather trade. Crocodylians have a long history of captive maintenance, but it was only in the late 1960s and early 1970s that crocodylians began to be managed for production of skins as a result of the decline in wild populations from overharvesting. Initially, skin-production programs were done by ranching, in which eggs from wild nests or wild-caught hatchlings are brought into captivity and raised until they attain market size. In addition to the legal protection of large juvenile and adult crocodylians, ranching provides an economic incentive to rural communities to protect large individuals in order to have an annual crop of eggs and juveniles. Recently in some areas a shift from ranching to farming has occurred because of the difficulty of obtaining adequate numbers of eggs and juveniles. Farming involves maintaining breeding adults for production of a sufficient annual volume of eggs and hatchlings. Because of the demands for profitability, ranches and farms have determined the population size necessary to maximize reproduction and growth in the commercially valuable species. Their contribution to conservation is a reduced need to harvest wild animals; however, the attention from the public and government focused on success in ranching a few of the species. Declines of other species of crocodylians have not received the support necessary to insure their survival.

The captive breeding of pet-trade or hobbyist species has become a large-scale commercial enterprise during the 1990s. Aside from aquarium-raised African clawed frogs and the red-eared turtle farms of the south-central United States, amphibians and reptiles in the pet-trade market were taken almost exclusively from wild populations until the mid-1980s. Wild populations remain a major source for this industry; however, captive rearing of hobbyist species now supplies large numbers of amphibians and reptiles to this market. A controversy exists concerning whether hobbyist captive rearing has reduced the demand for wild-caught amphibians and reptiles or has only fueled a demand for the rarer and more threatened species. Hobbyist maintenance has definitely contributed to our knowledge of biology of many species, and it does provide a pool of amphibian and reptilian species from which hobbyists can obtain healthy animals that will be long-lived pets. In spite of arguments to the contrary, hobbyist captive rearing does not produce animals for translocation.

We are aware of only a few successful translocation or repatriation programs for amphibians and reptiles (Table 14.9). A number of zoos and wildlife sanctuaries have successfully maintained and bred threatened or endangered amphibians and reptiles, some for several generations, but few have reported the reestablishment of extirpated populations. Perhaps the most successful program has been the gharial (*Gavialis gangeticus*) program in India. This crocodylian has been reestablished in several river drainages from which it was extirpated. Nevertheless, these remarkable animals sit on the edge of extinction with only about 400 breeding pairs still in the wild. The gharial is now considered "critically endangered" and was added to the 2007 Red List of endangered species issued by the World Conservation Union. The giant tortoise program in the Galápagos has also reported successful repatriation and augmentation. In contrast to these successes most programs have failed. A number of reasons may be cited for this lack of success. The primary reason is the difficulty of solving the actual environmental problem that caused the original population decline. Usually the habitat has either disappeared or been drastically modified.

TABLE 14.9 Examples of Long-Term Captive Management and Translocation Programs for Amphibians and Reptiles.

Taxon	Life stage released	Status	Location
Amphibians			
Alytes muletensis, Majorcan midwife toad[1]	E-A	S	Majorca
Bufo houstonensis, Houston toad[1]	E-J	U	Texas
Peltophryne lemur, Puerto Rican crested toad[5]	L-A	U	Puerto Rico
Reptiles			
Geochelone hoodensis, Española Tortoise[2]	J	I	Española, Galápagos
Gavialis gangeticus, gharial[3]	J	S	Chambal River, India
Lacerta agilis, sand lizard[4]	J-A	S	Southeastern England

Note: These examples include only programs in which the adults have been maintained in captivity for 1 or more years and the released offspring were hatched or born in captivity. A successful translocation requires maturation of juveniles in situ, their reproduction, and the survival of their offspring. Abbreviations: Life stage released: E, eggs; L, larvae; J, juveniles; A, adults. Success of the translocation: U, unknown; I, indeterminate (some individual surviving and maturing); S, successful.
Sources: [1]Beebee, 1996; [2]Cayot and Morillo, 1998; [3]Choudhury and Choudhury, 1986; [4]Corbett, 1988; [5]Paine et al., 1989.

Predators, including humans, rats, feral house cats, or others that contributed to the population decline are still active. Genetic stochasticity and other factors that lead toward extinction of small populations make the maintenance of viable stock in captivity extremely difficult. Conservation-oriented programs actively address this difficulty; commercial and hobbyist programs seldom do. In fact, the latter industry emphasizes genetically aberrant lineages because these designer amphibians and reptiles are commercially more valuable.

Augmentation, Repatriation, and Introduction

Augmentation, repatriation, and introduction fall under the rubric of translocation, defined as the intentional release of individuals to establish or enlarge the population of a target species. The target species is typically a threatened or endangered species, although introductions include intentional and unintentional release of individuals of nonthreatened species into a locality or habitat foreign to that species. The latter sort of release and colonization is examined in the "Exotic Species" section. Repatriation is the release of individuals of a species into a locality from which the species was extirpated, and augmentation is the release of individuals of a species into a locality containing the same species. All three types of translocations are widely used conservation tools, although the use of each remains controversial. They have been used with varying success in the conservation of amphibians and reptiles. The following examples briefly address some positive and negative aspects of translocation conservation.

In 1976, a fire swept through an English sand dune nature preserve, largely destroying this isolated patch of heath vegetation. Concern that the plant community would recover but not quickly enough to allow the survival of *Lacerta agilis* led to the capture of all surviving sand lizards. The lizards were maintained in captivity where they prospered, and in 1978, all were transferred to an outdoor vivarium to establish a breeding colony. Sand lizards from the breeding colony were repatriated into the reserve, and by 1988, the heath community had completely regenerated and the lizard population was healthy. The breeding colony provided additional lizards for repatriation in other sand dune heath communities in southeastern England and elsewhere, as well as an introduction into the Inner Hebrides. Apparently most of the translocations have been successful, although a few populations were destroyed when fires destroyed translocation sites.

A similar success story applies to the Majorcan midwife toad *Alytes muletensis*. This toad was discovered first as a fossil and considered extinct on the Balearic Islands off the east coast of Spain. A few isolated populations were later discovered in the deep mountain gorges of Majorca. Apparently, many populations of this species were heavily preyed upon and driven extinct by the European water snake *Natrix maura*, which was introduced into the Balearics by the Romans for religious use. Once rediscovered in 1980, the toads were given legal protection, some nature reserves were established, and two zoos established breeding colonies, each from a few individuals. These breeding colonies have been used for repatriation; of eight repatriations, three populations now have begun to reproduce, one translocation failed, and the status of the other four is indeterminate.

Captive-breeding programs and translocations are not always successful. The endangered Houston toad (*Anaxyrus houstonensis*) is a resident of southeastern Texas in pine flat woods with sandy soils. Agriculture and other development have eliminated many populations and reduced this species to a few isolated populations. This toad is adaptable to captive breeding, and a breeding program at the local Houston zoo has provided adults and thousands of eggs and metamorphs for translocation. In spite of massive efforts to reestablish the toad at extirpated and new sites, no new populations have become established.

A similar lack of success is common in "mitigation," or relocation, projects that move animals from sites that are scheduled for destruction because of development. Thousands of gopher tortoises (*Gopherus polyphemus*) have been relocated in central Florida because of development projects. Typically these translocated individuals are placed in existing populations. Of the hundreds of relocations, only a few include short-term monitoring of the relocated individuals, showing that usually 50% or more of the relocated individuals disappear from the new site within 2 years. Most other relocation projects show similar results. A site in southeast England scheduled for development had a large population of slowworms (*Anguis fragilis*). Slowworms were captured and relocated at a natural site that lacked slowworms but was otherwise ecologically similar. The translocation failed; the slowworms at the relocation sites were clearly less robust than those from nearby populations and did not become established and reproduce. In general, relocations largely fail. Further, they mislead the public, developers, and government officials by suggesting that natural populations are conserved. Although these efforts result from good intentions, they lead to poor conservation and should be eliminated as a development tradeoff strategy.

Relocation and augmentation programs have real and potential dangers built into them. The introduction of disease into healthy populations is a real danger. For example, a disease of the upper respiratory tract has decimated populations of the desert tortoise (*Gopherus agassizii*) in the Mojave Desert of southern California and now appears to be spreading through the gopher tortoise populations of Florida. Another potential danger is outbreeding depression. Small isolated populations are often closely adapted to their local environment, and while subtle to the human eye, microenvironments can be quite different among nearby populations. Augmentation introduces individuals with new genetic constitutions into a genetically stable population, thereby changing the relationship of the resident populations to its local environment. Over time the new genetic pool can adapt to the specific local environment, but the initial response of the local population is likely lower survivorship. For a small population, this decline in numbers can push it to extinction. The essential message of translocation conservation is to anticipate failure and proceed cautiously with intense scrutiny to avoid causing further injury to the species or population in need of intervention.

An Afterword

Evidence is mounting that humans are spending less and less time engaged in nature-based recreation. A study in 2006 by Oliver Pergams and Patricia Zaradic showed a 16-year downturn in the number of visits made to U. S. National Parks beginning in 1987. Subsequent work by the same authors in 2008 showed a similar downturn for other activities in nature, including camping, fishing, and visits to national parks in other countries (Spain and Japan). This downturn was significantly correlated with an increase in indoor activities, particularly spending hours with video games and surfing the Internet. Children can no longer identify the most common animals in the area where they live, and people living in cities frequently lose sight of or do not understand the connection between their lives and ecosystem services such as clean air, water, food, and climate. People are most likely to care about what they know about. The challenge of how to get people, 48% of whom now live in cities, to connect or reconnect to nature is a daunting one. As Peter Kareiva points out, the disconnect between humans and nature may well be the world's greatest environmental threat.

QUESTIONS

1. First, describe in some detail, the causes of global amphibian declines. Second, explain the significance of this issue.
2. What is "cutaneous chytridiomycosis" and why should we worry about it?
3. What is meant by a "buffer zone" and why is an understanding of the movements of adult amphibians, turtles, or other reptiles and amphibians critical to developing workable conservation strategies?
4. How do snails, birds, tadpoles, and adult frogs fit together in a conservation-based parasite–host story?
5. Given what you now know about amphibian and reptile declines, hormones and other chemicals in the environment, environmental law (and enforcement), and the competing interests of growth-based global economies, can you construct an approach to amphibian and reptile conservation that will be sustainable, and if so, how?

ADDITIONAL READING

Beebee, T. J. C. (1996). *Ecology and Conservation of Amphibians.* Chapman & Hall, London.

Drury, W. H., Jr. (1998). *Chance and Change. Ecology for Conservationists.* University of California Press, Los Angeles.

Heyer, W. R., Donnelly, M. A., McDiarmid, R. W., Hayek, L.-A. C., and Foster, M. S. (1994). *Measuring and Monitoring Biological Diversity. Standard Methods for Amphibians.* Smithsonian Institute Press, Washington, D.C.

Jaffe, M. (1994). *and No Birds Sing. The Story of an Ecological Disaster in a Tropical Paradise.* Simon & Schuster, New York.

Mendelson, J. R., III, Lips K. R., Gagliardo, R. W., Rabb, G. B., Collins, J. P., Diffendorfer, J. E., Daszak, P., Ibanez, R., Zippel, K. C., Lawson, D. P., Wright, K. M., Stuart, S. N., Gascon, C., da Silva, H. R., Burrowes, P. A., Joglar, R. L., La Marca, E., Lotters, S., du Preez, L. H., Weldon, C., Hyatt, A., Rodriguez-Mahecha, J. V., Hunt, S., Robertson, H., Lock, B., Raxworthy, C. J., Frost, D. R., Lacy, R. C., Alford, R. A, Campbell, J. A., Parra-Olea, G., Bolanos, F., Domingo, J. J. C., Halliday, T., Murphy, J, B., Wake, M. H., Coloma, L. A., Kuzmin, S. L., Price, M. S., Howell, K. M., Lau M., Pethiyagoda, R., Boone, M., Lannoo, M. J., Blaustein, A. R., Dobson, R. A., Griffiths, L. A., Crump, M., Wake, D. B., and Brodie, E. D. (2006). Biodiversity -Confronting amphibian declines and extinctions. *Science* **313:** 48–48.

Mrosovsky, N. (1983). *Conserving Sea Turtles.* British Herpetological Society, London.

Phillips, K. (1994). *Tracking the Vanishing Frogs.* St. Martin's Press, New York.

Primack, R. B. (1995). *A Primer of Conservation Biology.* Sinauer Assoc. Inc., Sunderland, Mass.

Stebbins, R. C., and Cohen, N. W. (1995). *A Natural History of Amphibians.* Princeton University Press, Princeton.

REFERENCES

General Principles

Balmford and Cowling, 2006; Bauer, 1999; Carr, 1940; Day, 1981; Folke, 2006; Heyer, 1967; Meffe, 1994; Meffe and Carroll, 1994; Meffe et al., 2006; Miller, 2005; Papenfuss, 1986; Paquette, 2007; Pianka, 1986, 1994; Pitman et al., 2007; Raup, 1991; Saunders et al., 2006; Savage, 1995; Stocks et al., 2007; Valladares-Padua, 2006.

Impact on Amphibian and Reptile Communities

Habitat Modification, Fragmentation, and Loss

Ash and Bruce, 1994; Christain, 1975; Colli et al., 2002; Coman and Brunner, 1972; Corn and Bury, 1988; Congdon et al., 1994; Costa et al., 2007; Crouse et al., 1987; Demaynadier and Hunter, 1998; Denny, 1974; Derr, 2000; Fisher and Shaffer, 1996; Gibbs, 1998; Herbeck and Larsen, 1999; Iverson, 1978, 1979; Iverson and Mamula, 1989; Means et al., 1996; Mittermeier, 1972; Mittermeier et al., 2000; Myers et al., 2000; Nogueira et al., 2005; Oliveira and Marquis, 2002; Pearman, 1997; Petranka et al., 1994; Pianka, 1994; Pianka and Vitt, 2003; Sartorius et al., 1999; Shaffer, 1981; Vitt et al., 1998, 2007.

Pollution and Disease

Aguirre et al., 1998; Beebee, 1996; Berger et al., 1998; Crawshaw, 1997; Davis et al., 2007; Dunson et. al., 1992; George, 1996; Grillitsch and Chovanec, 1995; Guillette, 1995; Guillette et al., 1996; Hayes et al., 1997; Herbst, 1994; Herbst and Klein, 1995; Laurance, 1996; Laurance et al., 1996; Lips, 1998, 1999; Pough and Wilson, 1977; Richards et al., 1993.

Harvesting Amphibians and Reptiles

Brazaitis et al., 1998; Fitzgerald, 1994a,b; Frazer, 1992; Jennings and Hayes, 1985; Melisch, 1998; Pérez et al., 2004; Shi and Parham, 2001; Shi et al., 2004, 2007; Shine et al., 1996e; Thorbjarnarson, 1992, 1999; Ugarte et al., 2007.

Exotic Species

Allen et al., 2004; Crossland, 1998; Fritts, 1988; Godley, 1980; Greenlees et al., 2006, 2007; Hagman and Shine, 2006, 2007; Kraus and Campbell, 2002; Kraus et al., 1999; Mitchell and Beck, 1992; Petren et al., 1993; Phillips et al., 2006, 2007; Pimentel et al., 2000; Rodda and Fritts, 1992; Ryan, 1988; Savidge, 1987; Urban et al., 2007; Watling, 1986; Whitaker, 1973; Wilson and Porras, 1983; Zug and Zug, 1979.

Extinction

Declining Amphibians

Andreone et al., 2005; Beebee, 1996; Bell et al., 2004; Berger et al., 1998; Blaustein and Johnson, 2003; Channing et al., 2006; Delfino, 2005; Dodd, 1997; Drake et al., 2007; Gower and Wilkinson, 2005; Halliday and Heyer, 1997; Humraskar and Velho, 2007; Joglar and Burrowes, 1996; Knapp and Morgan, 2006; Lawler et al., 1999; Lehtinen and Skinner, 2006; Lips et al., 2003; Mendelson et al., 2006; Morgan et al., 2007; Pechmann and Wake, 1997; Pechmann and Wilbur 1994; Piha et al., 2006; Pimenta et al., 2005; Pounds and Crump, 1994; Pounds et al., 1997, 2006; Ranvestel et al., 2004; Sessions and Ruth, 1990; Stuart et al., 2004; Tyler, 1991b; Vertucci and Corn, 1996; Webb and Joss, 1997; Whitfield et al., 2007.

Preservation and Management

Reserves and Corridors—Saving Habitats

Burke and Gibbons, 1995; Gibbs, 1998; Rosenberg et al., 1997; Semlitsch, 1998.

Captive Management

Frazer, 1992; Gibbons, 1994; Mrosovsky, 1983; Thornjarnarson, 1999.

Augmentation, Repatriation, and Introduction

Beebee, 1996; Burke, 1991; Corbett, 1988, Dodd and Siegel, 1991; Reinert, 1991; Spellerberg, 1988; Templeton, 1986.

Afterword

Kareiva, 2008; Pergams and Zaradic, 2008.

Herp Regulations in NYS

1. Diamondback terrapin has open season, license required
2. Native turtles:
   Snapping turtle has open season, size and bag limits
   All other species open season: none
3. Native snakes:
   Open season: none
4. Native lizards:
   Open season: none
5. Native Frogs:
   Open season for all except:
   - Leopard frogs (n. or s.) on LI
   - N. cricket frogs + E. spadefoot toads anywhere

   NO size or bag limits
6. Native salamanders:
   Open season: none

Part VI

Classification and Diversity

Chapter 15

Caecilians

OVERVIEW

Caecilians occur worldwide in the tropics, except for Madagascar and Oceania. Only about 170 species are known, distributed among 33 genera and six families. Most caecilians are fossorial, living in moist soils usually adjacent to streams, lakes, and swamps; a few species are aquatic. Because of their secretive nature, their biology is largely unknown, and much of our knowledge comes from observations made during capture and from captive or museum specimens.

Caecilians (Gymnophiona; Apoda, stem-based name) are amphibians that look like earthworms. They have blunt, bullet-shaped heads, cylindrical, limbless bodies, and short tails. The bodies of all caecilians are distinctly segmented by encircling primary grooves called *annuli,* and usually each segment (= primary annulus) contains a single vertebra. In some taxa, the primary annuli are further divided into secondary and even tertiary annuli by additional encircling grooves. The blunt heads are digging tools for creating the burrows in which these animals live. A combination of serpentine and internal concertina locomotion is used to move through burrows, and serpentine or undulatory locomotion is used when on the surface. Heavier-bodied caecilians typically use internal concertina locomotion. In these caecilians, the muscular body is loosely attached to the skin. During movement, the body moves and bends within the inflexible skin and shifts forward. The skin contracts and moves the entire animal forward. Slender caecilians use only undulatory locomotion on the surface and in burrowing; hence, they are confined to more friable soils. Aquatic species often have dorsal and ventral fins posteriorly and a somewhat laterally compressed body; undulatory locomotion is used for forward movement.

All caecilians have internal fertilization. The male copulatory organ, the phallodeum, is an eversible part of the posterior cloaca. During mating, the male lies on or entwines about the female so that their cloacae are in apposition and the phallodeum can be everted into the female's cloaca. Offspring develop internally or externally, depending upon species, and if externally, development is indirect or direct; developmental mode is invariable within each species.

Many shared derived traits confirm monophyly of all living caecilians. The snout bears a retractable sensory tentacle on each side of the head between the nostril and the eye; the tentacles aid in location and identification of prey. Many structures that are part of the eye in other vertebrates have been preempted for the tentacle in caecilians. The upper jaw protrudes beyond the lower jaw; this position allows prey capture in narrow spaces yet does not interfere with the head's use in burrowing. The jaw-closing mechanism of caecilians is unique in having a muscle that attaches to a process on the dentary and when contracted causes the lower jaw to swing upward (Fig. 15.1). Dermal (bony) scales often are present, lying deep within the tissue of the annular grooves. The skull of adult caecilians is heavily ossified, enabling it to withstand the jarring forces of digging or burrowing. Some elements, such as the maxillary and palatine, are fused as single bones. The eyes are vestigial, represented only by small darkly pigmented areas that lie beneath the skin and, in some cases, beneath skull bones. External ear openings are absent. The limbs have been completely lost; not even a remnant of the pectoral or pelvic girdles remains in the body wall.

The predominantly subterranean existence of caecilians has made study of this group difficult. Although published works on caecilians extend back 250 years, most of

Herpetology, 3rd Ed.

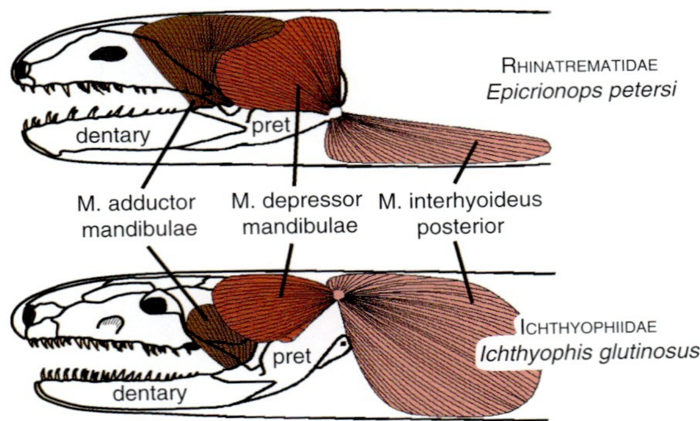

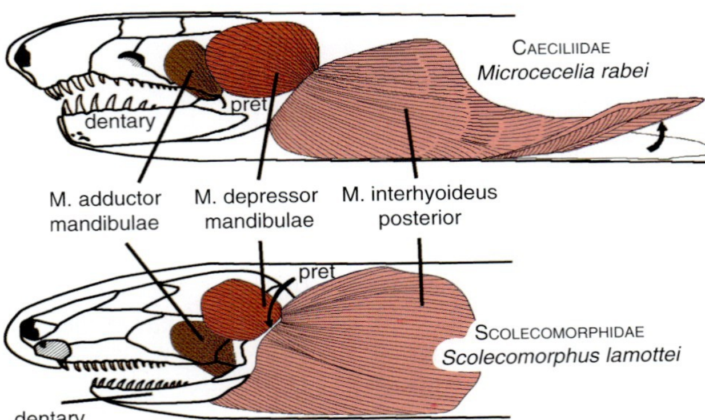

FIGURE 15.1 The unique dual jaw-closing mechanism present in all four major clades of caecilians consists of the mechanism ancestral in vertebrates, the masseter adductor mandibulae, a muscle that pulls up on the lower jaw, and a new mechanism, the masseter interhyoideus posterior, which pulls down on the processus retroarticularis (pret) an extension of the dentary bone. The result is that the lower jaw swings up. This mechanism is progressively more developed in more derived caecilian clades. Redrawn from Nussbaum, 1983.

this work dealt with how caecilians were related to other amphibians. Some authors during this time considered caecilians to be snakes! They existed as seldom-mentioned oddities, all lumped in a single family, the Caeciliidae, until Dr. Edward H. Taylor began a survey of these amphibians in the 1960s. In 1968, Dr. Taylor published a landmark monograph devoted to caecilians. His extensive work drew attention to caecilian diversity and how little was known about their systematics and life history. Three decades after their "rediscovery," we know as much about their phylogenetic relationships as we do for salamanders and frogs after a century of study; however, much of their biology remains unknown.

Taylor's 1968 monograph did not provide a phylogenetic analysis, although his partition of caecilians into three families, Caeciliidae, Ichthyophiidae, and Typhlonectidae, was an implicit hypothesis of relationships and monophyly. A year later, he proposed an additional family, Scolecomorphidae, and divided the largest family into two subfamilies, Caeciliinae and Dermophiinae, again

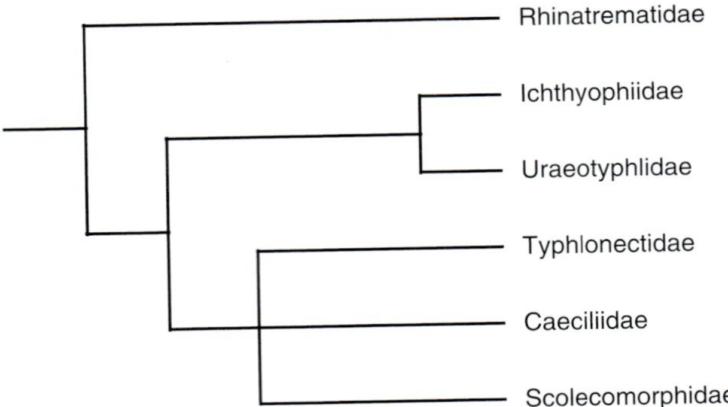

FIGURE 15.2 A cladogram depicting relationships among the families of extant caecilians. Based on Wilkinson and Nussbaum, 2006.

without phylogenetic analysis. The uniqueness of the genera *Rhinatrema* and *Epicrionops* prompted recognition of the family Rhinatrematidae in 1977. Two years later, *Uraeotyphlus*, which has close affinities to Ichthyophiidae, was transferred from Caeciliidae to its own subfamily within Ichthyophiidae and subsequently to its own family.

Subsequent studies have confirmed the basal divergence of rhinatrematids from the main gymnophionan lineage, then the divergence of ichthyophiids and uraeotyphlids, either independently or as a shared ancestor that subsequently diverged from the "higher" caecilians (Fig. 15.2). These confirmations were based on a variety of data sets, including immunology, RNA sequences, and morphology. In 2005, Wilkinson and Nussbaum retained the six families but considered Typhlonectidae, Caeciliidae, and Scolecomorphidae as an unresolved polytomy (Fig. 15.2). In addition, based on molecular data, Caeciliidae remained paraphyletic with respect to Typhlonectidae and Scolecomorphidae. Typhlonectids are a monophyletic clade, as are the three genera of caeciliids on the Seychelles. Scolecomorphids appear as a clade in some studies but not in others. Generally, the scolecomorphids are a sister group to the African–Seychelles–Asian caeciliids, and the typhlonectids are a sister group to the American caecilians. A recent phylogenetic analysis of all amphibians considered this polytomy to consist of one large family, the Caeciliidae. That analysis contained only a small number of caecilian species, and clades were supported by low bootstrap values. Thus, until more data are available, we continue to recognize six families.

TAXONOMIC ACCOUNTS

"Caeciliidae"

Tailless Caecilians

Classification: Amphibia; Lissamphibia; Gymnophiona.

Sister taxon: In part and separately, Typhlonectidae and Scolecomorphidae.

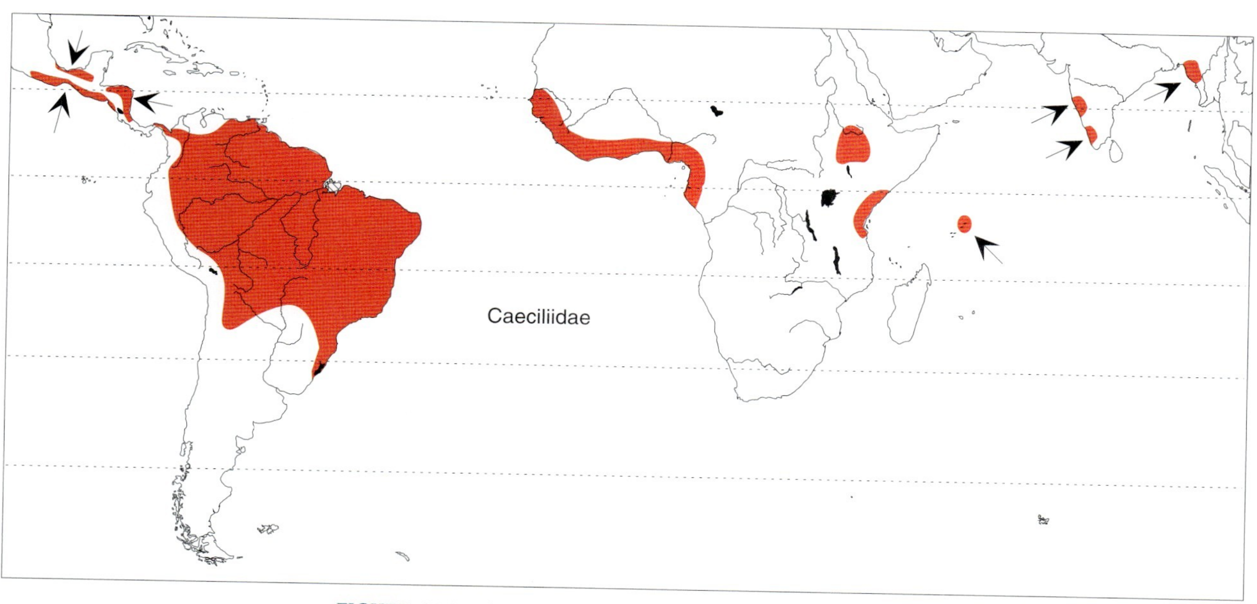

FIGURE 15.3 Geographic distribution of the extant Caeciliidae.

Content: Twenty-one genera, *Boulengerula, Brasilotyphlus, Caecilia, Dermophis, Gegeneophis, Geotrypetes, Grandisonia, Gymnopis, Herpele, Hypogeophis, Idiocranium, Indotyphlus, Luetkenotyphlus, Microcaecilia, Mimosiphonops, Oscaecilia, Parvicaecilia, Praslinia, Schistometopum, Siphonops, and Sylvacaecilia,* with 101 species.

Distribution: Tropical America, eastern and western equatorial Africa, Seychelles Archipelago, and India (Fig. 15.3).

Characteristics: All caeciliids have primary annuli, and some have the primary annuli divided by secondary grooves, but none has tertiary grooves (Fig. 15.4). Scales are present in the annular grooves of some genera and absent in others. The posterior end of the body is capped with a terminal shield but lacks a true tail, which has caudal vertebrae and myomeres. Eyes may or may not be visible externally; in some genera, eyes lie in bony sockets beneath the skin; in others, they lie beneath bone. Tentacles are variously positioned; in some taxa, the tentacles are adjacent to the nostrils, whereas in others, they are near the eyes. The middle ear contains a stapes.

Biology: Caeciliids range in size from the small *Idiocranium russeli* (98-104 mm adult total length, TL) to the large *Caecilia thompsoni* (1.5 m TL). Adults of most taxa range from 300 to 500 mm TL, although most adult *Oscaecilia* exceed 600 mm TL. All caeciliids are fossorial in moist soils, and most live in forests. Caeciliids have a variety of reproductive behaviors. Some are oviparous (e.g., *Grandisonia, Hypogeophis*), whereas others are viviparous (*Caecilia, Dermophis*). Some oviparous taxa lay eggs in or near water and have free-living larvae. Parents of other species (*Boulengerula taitanus, Hypogeophis rostratus*, and *I. russeli*) attend the eggs, which undergo direct development. Current evidence suggests that all direct-developing and many indirect-developing caecilians remain with their eggs until they hatch. The larval period usually extends from 10–12 months; metamorphosis occurs shortly before the next reproductive season begins (e.g., *Geotrypetes*). Development is faster in larvae with direct development. Reproduction appears to be seasonal or nearly continuous, depending largely on the climate in a particular area.

Ichthyophiidae

Asian Tailed Caecilians

Classification: Amphibia; Lissamphibia; Gymnophiona.

Sister taxon: Uraeotyphlidae.

Content: Two genera, *Caudacaecilia* and *Ichthyophis*, with 36 species.

Distribution: India, Sri Lanka, and Southeast Asia (Fig. 15.5).

Characteristics: Ichthyophiids have conspicuous primary annuli divided by secondary and tertiary grooves. Scales are present in most annular grooves but occasionally are absent from the anteriormost grooves. The body ends in a short, true tail that has caudal vertebrae and myomeres. The eyes are visible externally and lie in bony sockets beneath the skin. Each tentacle lies between the eye and the nostril, usually closer to the eye. The middle ear contains a stapes.

Biology: Ichthyophiids are moderate-sized caecilians, with adults of most species in the 200–300 mm TL size range; a few (e.g., *C. nigroflava, I. glutinosus,*

FIGURE 15.4 Representative caecilians. Clockwise from upper left: Sao Tome caecilian *Schistometopum thomense*, Caeciliidae (R. A. Nussbaum); monarch tailed caecilian *Ichthyophis monarchus*, Ichthyophiidae (L. L. Grismer); Bannan caecilian *Ichthyophis bannanicus*, Rhinatrematidae (E. D. Brodie, Jr.); African buried-eyed caecilian *Scolecomorphus vittatus*, Scolecomorphidae (L. W. Porras).

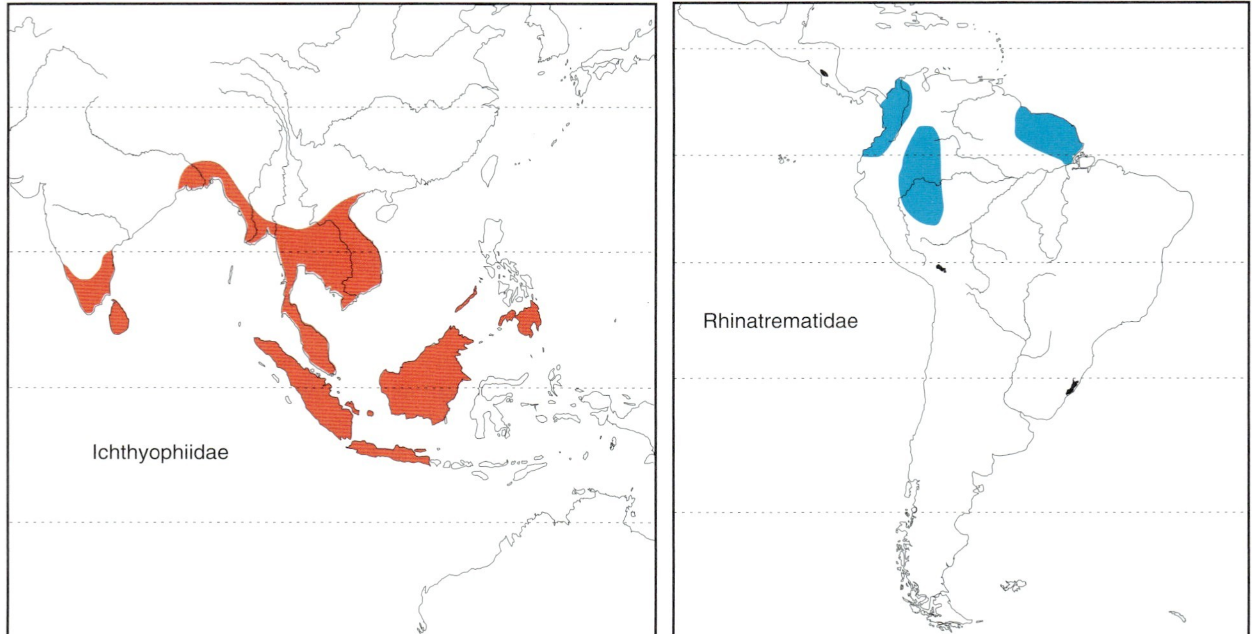

FIGURE 15.5 Geographic distributions of the extant Ichthyophiidae and Rhinatrematidae.

I. bombayensis) reach total lengths of 400–500 mm. All species in the two genera are oviparous. Development is indirect in the few known examples.

Ichthyophis deposits eggs in its burrows near water. The female remains with the eggs until the larvae hatch. Upon hatching, the gilled larvae exit the burrows and crawl overland to a nearby pond or stream. The entire developmental cycle from egg deposition to metamorphosis is about a year (*I. glutinosus, I. kohtaoensis*).

Ichthyophiids likely were present on the Indian plate before it collided with Laurasia. One or more dispersal events out of India produced South Asian ichthyophiids.

Rhinatrematidae

American Tailed Caecilians

Classification: Amphibia; Lissamphibia; Gymnophiona.

Sister taxon: The clade including all other gymnophionan families.

Content: Two genera, *Epicrionops* and *Rhinatrema*, with 8 and 1 species, respectively.

Distribution: Northern South America east of the Andes (Fig. 15.5).

Characteristics: Rhinatrematid caecilians have true tails in which the post-cloacal segment contains vertebrae, myomeres, and complete annuli in the skin. The primary annuli of the body are divided by secondary and tertiary grooves, and numerous scales are present in the primary annular grooves. The eyes are visible externally and lie beneath the skin in bony sockets. A tentacle arises near or at the anterior edge of each eye. The middle ear contains a stapes.

Biology: Rhinatrematids range from 200 to 330 mm adult TL. Both genera are presumably oviparous. Free-living larvae are known for *Epicrionops*, thereby indicating that this taxon deposits aquatic eggs.

Scolecomorphidae

Buried-Eyed Caecilians

Classification: Amphibia; Lissamphibia; Gymnophiona.

Sister taxon: Possibly a clade including Caeciliidae (possibly African) and Typhlonectidae.

Content: Two genera, *Crotaphatrema* and *Scolecomorphus*, each with 3 species.

Distribution: Eastern and western equatorial Africa (Fig. 15.6); as yet, no caecilians have been found in central Africa.

Characteristics: Scolecomorphids (Fig. 15.4) possess only primary annuli, and only a few vestigial scales occur in the posteriormost annuli. They lack a true tail. Bony orbits are absent, and the eyes lie beneath skull bones; however, because the eyes are attached to the tentacles, they move outward when the tentacles are extended. The middle ear lacks a stapes.

Biology: Scolecomorphids range from 150 to 360 mm adult TL, with most adults over 300 mm. They are mountain forest-floor residents, usually inhabiting areas adjacent to streams or other moist habitats. Three species of *Scolecomorphus* give birth to young. Their oviductal eggs are small, yet the developing embryos are many times the mass of the eggs; thus, a maternal–embryo nutrient transfer is likely. *Crotaphatrema* appears to be oviparous, because its oviductal eggs are much larger than those of *Scolecomorphus*.

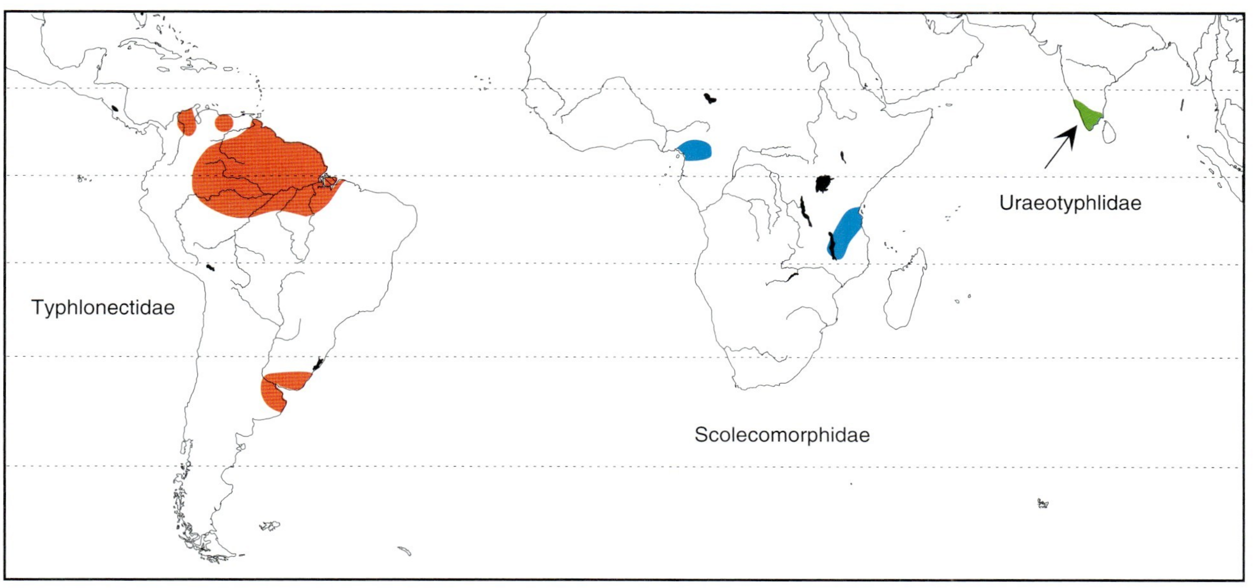

FIGURE 15.6 Geographic distributions of the extant Typhlonectidae, Scolecomorphidae, and Uraeotyphlidae.

FIGURE 15.7 Representative caecilians. From left: Water caecilian *Typhlonectes natans*, Typhlonectidae (C. Schwalbe); Kannan caecilian *Uraeotyphlus* cf. *narayana*, Uraeotyphlidae (D. J. Gower, The Natural History Museum, London).

Typhlonectidae

Aquatic Caecilians

Classification: Amphibia; Lissamphibia; Gymnophiona.

Sister taxon: Caeciliidae or a clade of Neotropical caeciliids.

Content: Five genera, including *Atretochoana, Chthonerpeton, Nectocaecilia, Potomotyphlus, and Typhlonectes,* with 14 species.

Distribution: South America east of the Andes (Fig. 15.6).

Characteristics: Typhlonectids (Fig. 15.7) have only primary annuli, although in a few species, some primary annuli are partially dissected by false secondary grooves. Dermal scales are absent in the grooves. They lack a true tail. Their eyes are moderately well developed and always visible in bony sockets beneath the skin. The sensory tentacles are small and usually closer to the nostrils than the eyes. Stapes are present.

Biology: Typhlonectids are the largest caecilians. Adult *Potomotyphlus* and *Typhlonectes* typically range from 300-600 mm adult TL, and *Atretochoana eiselti* reaches 800 mm TL. *Chthonerpeton* and *Nectocaecilia* are generally slender and range from 200 to 400 mm TL, although *C. viviparum* reaches a length of 560 mm. *Atretochoana, Potomotyphlus,* and *Typhlonectes* are aquatic, whereas *Chthonerpeton* and *Nectocaecilia* are semiaquatic. The bodies of the aquatic taxa are laterally compressed and bear a middorsal fold or fin, which presumably assists their undulatory swimming. *Atretochoana* is totally lungless and is the largest known lungless tetrapod, and *Chthonerpeton* has a rudimentary left lung. The other genera have well-developed left and right lungs. *Atretochoana* may have evolved in highly aerated waters, such as mountain streams, but only two specimens are known and they lack locality data. All typhlonectids are viviparous, and embryonic development can be divided into three primary stages. During the first stage, the embryo relies on yolk for nutrition. During the second stage, when yolk is depleted, the embryo is released into the uterus where it feeds on secretions and cells originating from the wall of the uterus. It may also feed on eggs or embryos of siblings. In the last stage, a pair of large gills surrounds the embryo and comes in contact with the uterine wall, where it serves as both a respiratory structure and a functional placenta.

Uraeotyphlidae

Kerala Caecilians

Classification: Amphibia; Lissamphibia; Gymnophiona.

Sister taxon: Ichthyophiidae.

Content: One genus, *Uraeotyphlus*, with 5 species.

Distribution: Southern peninsular India (Fig. 15.6).

Characteristics: *Uraeotyphlus* has primary annuli divided by secondary but not tertiary grooves, although none of the annular grooves completely encircles the body (Fig. 15.7). Scales are present in the annular grooves. The body ends in a short tail. The eyes are visible externally and lie in bony sockets beneath the skin. The tentacles are far forward beneath the nostrils. Each middle ear contains a stapes.

Biology: *Uraeotyphlus* range from 200 to 300 mm adult TL. Few individuals have been observed; hence, their biology is almost entirely unknown. All species are likely oviparous. Larvae of *U. oxyurus* possess a typical caecilian larval morphology; however, the structure of the mouth and throat suggests that the larvae are suction feeders and eat small prey.

QUESTIONS

1. What is the global distribution of caecilians?
2. Where do caecilians live and what morphological specializations allow them to live in these microhabitats?
3. What is the function of the tentacle of caecilians?

4. Describe the evolutionary relationships among caecilian families.
5. How has jaw and jaw muscle structure aided in understanding relationships among caecilian families?

REFERENCES

Overview

Cogger and Zweifel, 1998; Duellman and Trueb, 1986; Exbrayat, 2006a; Frost, 1985; Glaw and Köhler, 1998; Gower et al., 2002, 2005; Halliday and Adler, 1986; Hass et al., 1993; Hedges et al., 1993; Himstedt, 1996; Laurent, 1986; Lescure, 1986b; Lescure et al., 1986; Nussbaum and Wilkinson, 1989; Savage and Wake, 2001; San Mauro et al., 2004; Summers and O'Reilly, 1997; Taylor, 1968; Wake, 1993a; 1992b, 1977, 2006; Wilkinson 1989, 1992a, 1997; Wilkinson and Nussbaum, 1996, 2006; Zardoya and Meyer, 2000, 2001.

Taxonomic Accounts

"Caeciliidae"

Tailless Caecilians

Exbrayat, 1985, 2006b; Himstedt, 1996; Nussbaum, 1984; Nussbaum and Hinkel, 1994; M. Wake, 1977; Wilkinson, 1997; Wilkinson et al. 2002b, 2003.

Ichthyophiidae

Asian Tailed Caecilians

Breckenridge et al., 1987; Gower et al., 2002, 2005; Gundappa et al., 1981; Himstedt, 1996; Masood-Parveez and Nadkarni 1993; M. Wake, 1977; Wilkinson et al., 2002b.

Rhinatrematidae

American Tailed Caecilians

Gower et al., 2002; Nussbaum, 1977; Wilkinson, 1992a.

Scolecomorphidae

Buried-Eyed Caecilians

Loader et al., 2003; Nussbaum, 1985a; M. Wake, 1998; Wilkinson et al., 2003.

Typhlonectidae

Aquatic Caecilians

Billo et al., 1985; Exbrayat, 1984, 2006b; Exbrayat and Hraoui-Bloquet, 1992, 2006; Murphy et al., 1977; Nussbaum and Wilkinson, 1995; Wake, 1977; Wilkinson, 1989, 1996; Wilkinson and Nussbaum, 1997, 2006; Wilkinson et al., 1998.

Uraeotyphlidae

Kerala Caecilians

Bhatta, 1998; Exbrayat, 2006b; Nussbaum, 1979; Wilkinson, 1992b; Wilkinson and Nussbaum, 1996, 2006.

Chapter 16

Salamanders

OVERVIEW

Salamanders, the tailed amphibians, are largely a Northern Hemisphere (Holoarctic) group. All except the Plethodontidae are confined to temperate and subtropical areas of North America and/or Eurasia, including North Africa. Most terrestrial salamanders require moist habitats, typically forest, whereas aquatic salamanders may occur in vernal pools, spring seepages, streams, and large lakes and rivers. The only tropical salamanders are plethodontids, occurring mainly in mountains of Central America and a few species in the Amazon basin. These tropical invaders have been highly successful, representing more than 40% of the total extant salamander species, i.e., about 560 species.

Salamanders (Caudata; Urodela, stem-based name) have well-developed tails; cylindrical, often elongate, bodies; and distinct heads. Most also have well-developed limbs, frequently short relative to body length, although two clades have reduced or lost limbs. Salamander skulls are reduced by the loss of many elements, and other cranial elements are partly or totally cartilaginous (Fig. 2.16). Cartilaginous elements occur in the postcranial skeleton as well. Whether this cartilaginous condition reflects heterochrony is uncertain; however, heterochrony has occurred repeatedly in salamander evolution. Heterochrony (see the section "Development and Growth" in Chapter 2) involving paedomorphosis (interspecific) or paedogenesis (intraspecific) is recognized by the retention of larval traits in adults, such as gill slits and gills, and the absence of eyelids.

Derived lineages of salamanders have internal fertilization, although none has a copulatory organ. The basal lineages Cryptobranchoidei and Sirenidae have external fertilization. Internal fertilization occurs via a male-deposited spermatophore from which the female grasps a packet of sperm with her cloacal lips. With the exception of a few species, development occurs externally, either indirectly via a larval stage or directly into miniature salamanders. Salamanders have varied life histories. Although only 20–25% of known species have terrestrial adults that return to water to mate and deposit their eggs in water to hatch into gilled, aquatic larvae, this biphasic life history is considered the "typical" life history of salamanders. In contrast, most salamander species deposit eggs terrestrially in moist microhabitats, and these eggs hatch directly into fully formed juvenile salamanders. In a very few species, eggs hatch while still in the oviduct of the female.

Living salamanders share a suite of uniquely derived features that demonstrate the monophyly of salamanders. The major synapomorphies include the following: the ossification sequence of the skull, including the late appearance of the maxillae; a remodeling of the palate during metamorphosis; the absence of a middle ear; the origin of the jaw adductor muscle; and the presence of gill slits and external gills in aquatic larvae.

Relationships among the families of living salamanders have been the subject of many studies. Over a century ago, Cope grouped the nine families then recognized into two groups: Trematodera (Cryptobranchidae) and Amphiumoidea (all other salamanders); he also suggested the derivation of the caecilians from the amphiumas. By the 1930s, Noble had classified the eight families then recognized into five groups: Cryptobranchoidea (Cryptobranchidae, Hynobiidae); Ambystomoidea (Ambystomatidae);

Herpetology, 3rd Ed.

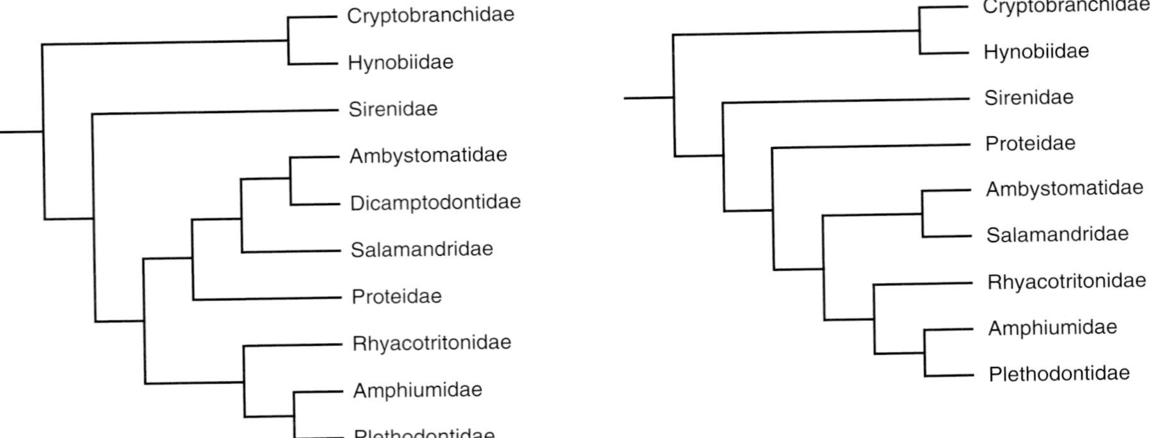

FIGURE 16.1 Cladograms depicting relationships among the families of extant salamanders. Left: Phylogenetic reconstruction based on combined morphological and molecular data for 32 amphibians, 28 of which were salamanders (Wiens et al., 2005). Right: Phylogenetic reconstruction based on 3747 unambiguously aligned base pairs of one mitochondrial and four nuclear genes sampled from 171 amphibians, 27 of which were salamanders (Roelants et al., 2007). Both cladograms redrawn from originals for uniformity, and only salamander families are included.

Salamandroidea (Amphiumidae, Plethodontidae, Salamandridae); Proteida (Proteidae); and Meantes (Sirenidae). His grouping became the accepted classification, and this consensus persisted into the middle 1960s. Three suprafamilial groups are now recognized, but differences of interpretation on the phylogenetic relationships among salamanders exist. A sampling of cladograms (Fig. 16.1) shows the range of differences among more recent investigators. Differences derive from the number of and the particular taxa used in producing the cladograms, the type and breadth of the data, and the manner of analysis used in each study. Although no taxa are invariably associated, cryptobranchids and hynobiids regularly appear as sister taxa near the base of the trees. In various studies, sirenids are the sister group to all other salamanders. No other pairings occur in more than one-half the cladograms, thus encouraging the various interpretations of relationships.

The previously accepted phylogenetic hypothesis of Larson and Dimmick combined morphological and molecular data; this hypothesis recognizes salamanders with internal fertilization as monophyletic. This cladogram yields three suprafamilial groups: Meantes—Sirenidae; Cryptobranchoidea—Cryptobranchidae, Hynobiidae; Salamandroidea—Ambystomatidae, Amphiumidae, Dicamptodontidae, Plethodontidae, Proteidae, Rhyacotritonidae, Salamandridae. Meantes is characterized by presumed external fertilization, a high number of microchromosomes, the absence of hindlimbs and pelvic girdle, and a derived spinal nerve morphology; Cryptobranchoidea has external fertilization, a high number of microchromosomes, and a presumed primitive spinal nerve morphology; and Salamandroidea has internal fertilization, a reduction in chromosome number, and usually a derived spinal nerve morphology.

Using the newest molecular techniques, most recent studies, e.g., that of Frost and colleagues in 2006, have verified many aspects of earlier studies. All have agreed that the cryptobranchids and hynobiids are sister taxa in the basal position. Some disagreement exists concerning whether sirenids are sister to proteids or to the remaining salamanders, but two recent studies by Wiens and colleagues and Roelants and colleagues agree on the latter relationship. Most all older and recent studies have agreed on the relationships among the remaining five families, with some disagreement over whether *Dicamptodon* should be placed in its own family; currently, it is considered to be an ambystomatid. Many recent studies have focused on the relationships of species within families, and many of these studies have provided new insights on the evolution of life history and morphology among salamanders. Undoubtedly, future studies will continue to provide many new surprises about salamander relationships.

TAXONOMIC ACCOUNTS

Cryptobranchidae

Asiatic Giant Salamanders and Hellbenders

Classification: Amphibia; Caudata; Cryptobrachoidei.

Sister taxon: Hynobiidae.

Content: Two genera, *Andrias* and *Cryptobranchus*, with 2 and 1 species, respectively.

Distribution: East-central China, Japan, and Appalachian and Ozark Mountains, U.S. (Fig. 16.2).

Characteristics: These giants are the largest living salamanders. The Japanese *Andrias japonicus* reaches 1.4 m

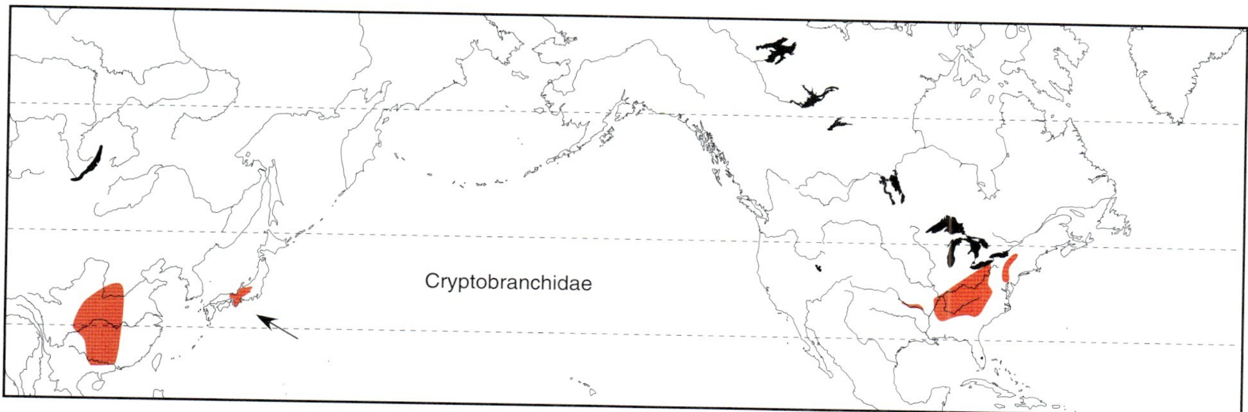

FIGURE 16.2 Geographic distribution of the extant Cryptobranchidae.

TL (total length), the Chinese *A. davidianus* 1.5 m TL, and the American *Cryptobranchus alleganiensis* (Fig. 16.3) 750 mm TL. All three are stout-bodied salamanders with four short, well-developed limbs and a heavy, laterally compressed tail. Cryptobranchids have a few paedomorphic traits, including a single pair of gill slits, open in *C. alleganiensis* and closed in *Andrias,* and the absence of eyelids. The lower jaw has separate angular and prearticular bones; the upper jaw has both premaxillae and maxillae, and the lacrimal is absent. Excluding the first spinal nerve, all pre- and postsacral spinal nerves exit intervertebrally. Costal grooves are lacking in the skin above the ribs, and nasolabial grooves are absent. Lungs are present, although vestigial and largely nonfunctional. Fertilization is external, and adult females lack spermathecae in the cloaca; both females and males have only ventral cloacal glands.

Biology: Cryptobranchids have extensively folded and wrinkled skin covering their dorsoventrally flattened bodies. The skin serves as a nearly exclusive respiratory surface because gills are absent and the small lungs are largely nonfunctional. All three species are confined to clear, cold mountain streams. Largely nocturnal, these salamanders hide beneath rocks and sunken logs during the day, sometimes emerging on heavily overcast days to forage or search for mates during the breeding season. Movement is typically by walking on the stream bottom, but undulatory locomotion is used for short-distance escapes to hiding places. These carnivores feed on a wide variety of invertebrate and vertebrate prey; crayfish are preferred by *C. alleganiensis*. In general, cryptobranchids lack the stereotypic courtship displays of the more derived families (see Chapter 9). During the breeding season, *C. alleganiensis* males excavate brooding sites beneath logs and wait for females to appear. When a female approaches, the male guides her into his nest chamber, where she remains until she has oviposited. The eggs, approximately 250 to 400, are laid in two gelatinous strings (one from each oviduct), and the male sheds seminal fluid containing sperm over them. A male may sequentially attract two or more females to his nest chamber, after which he guards the multiple egg clutches. During the entire year, whether breeding or not, adult males and females appear to defend specific rocks, logs, or other sites and drive away other individuals.

FIGURE 16.3 Representative cryptobranchid and hynobiid salamanders. From left: Hellbender *Cryptobranchus alleganiensis*, Cryptobranchidae (L. J. Vitt); Gensan salamander *Hynobius leechi*, Hynobiinae (L. L. Grismer).

Hynobiidae

Asiatic Giant Salamanders

Classification: Amphibia; Caudata; Cryptobrachoidei.

Sister taxon: Cryptobranchidae.

Content: Two subfamilies, Hynobiinae and Protohynobiinae.

Distribution: Asia, from the Urals to Japan, mainly above 40° N latitude (Fig. 16.4). Hynobiid genera segregate into a temperate group (e.g., *Batrachuperus*, *Liua*) predominantly north of and along the Himalayan axis and a cold-temperate and subarctic group (*Ranodon*, *Salamandrella*).

Characteristics: Hynobiids are heavy-bodied, thick-tailed salamanders with four short, well-developed limbs (Fig. 16.3). Most hynobiids are small (< 100 mm TL), although one species, *Ranodon sibiricus*, may reach 250 mm (TL). The lower jaw has separate angular and prearticular bones; the upper jaw has both premaxillae and maxillae, and the lacrimal is present. Excluding the first spinal nerve, all presacral and postsacral spinal nerves exit intervertebrally. Adults lack gills, gill slits, and nasolabial grooves; they have moveable eyelids. Costal grooves are present on the trunk. Lungs are usually well developed, although absent in *Onchodactylus*. Fertilization is external. Females lack spermathecae in the cloaca, and both females and adult males possess only ventral cloacal glands.

Biology: Hynobiids display little evidence of courtship. Most species are terrestrial except during the breeding season, when they migrate to breeding ponds or streams. Chemical communication may bring males and females together. The appearance of eggs extruding from females' vents also appears to be a visual signal that stimulates male *Hynobius*. One exception to this pattern is *Ranodon sibiricus*. In this species, males produce a rudimentary spermatophore, and the female lays eggs on the spermatophore instead of taking its sperm packet into her cloaca. In other hynobiids, females deposit eggs in a pair of gelatinous masses, one from each oviduct, and males then shed their sperm directly on these egg masses. Development in all species is indirect, with a free-living larval stage. Paedogenesis occurs in *Batrachuperus* and *Hynobius lichenatus*. Overall, the biology of the hynobiids remains poorly studied, with the exception of *Salamandrella*.

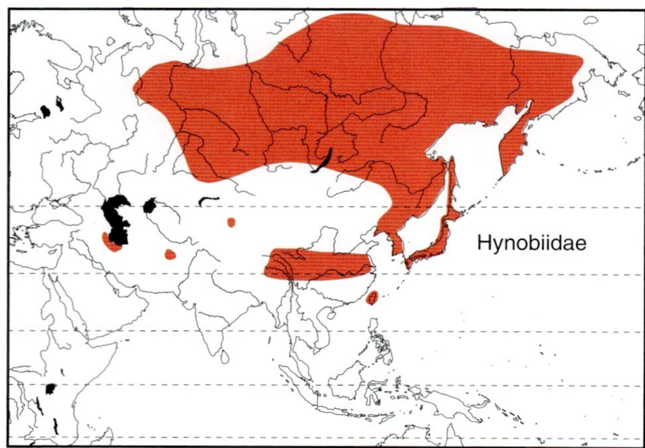

FIGURE 16.4 Geographic distribution of the extant Hynobiidae.

Hynobiinae

Sister taxon: Protohynobiinae.

Content: Eight genera, *Batrachuperus*, *Hynobius*, *Liua*, *Onychodactylus*, *Pachyhynobius*, *Paradactylodon*, *Pseudohynobius*, *Ranodon*, and *Salamandrella*, with 50 species.

Distribution: As for the family.

Biology: In general, hynobiids are either pond-breeding forms (*Salamandrella*) or stream-breeding forms (all other genera), with the exception that *Hynobius* has some species that use ponds and some that use streams for reproduction.

Protohynobiinae

Sister taxon: Hynobiinae.

Content: One genus, *Protohynobius*, with 1 species.

Distribution: Sichuan, China.

Characteristics: This species is known from only one specimen. It is placed as a separate subfamily based on a single morphological character, the presence of an internasal bone in the skull. Vomerine teeth are present. Tail is slender, and 13 costal grooves are present.

Sirenidae

Sirens and Dwarf Sirens

Classification: Amphibia; Caudata; Unnamed clade.

Sister taxa: Diadectosalamandroidei.

Content: Two genera, *Pseudobranchus* and *Siren*, each with 2 species.

Distribution: Coastal southeastern North America and the Mississippi River valley (Fig. 16.5).

Characteristics: Sirenids are moderately slender, eel-like salamanders with small forelimbs (Fig. 16.6); the hindlimbs and pelvic girdle are absent. Adult size ranges from 100 to 900 mm TL. The lower jaw has fused angular and prearticular bones; the upper jaw has premaxillae and small, floating maxillae, and the lacrimal is absent. All presacral, except the second and third, and all postsacral spinal nerves exit intravertebrally through foramina in the vertebrae. All sirenids are paedomorphic; adults have external gills and one or three pairs of gill slits; they have no eyelids. Costal grooves are present on the skin above the ribs, and nasolabial grooves are absent. Lungs are present, although small. The site of fertilization is unknown and is presumed to be external. Adult females lack spermathecae, and adult males lack reproductive glands in the cloaca. Sirenids are unlike any other salamanders in many aspects. For example,

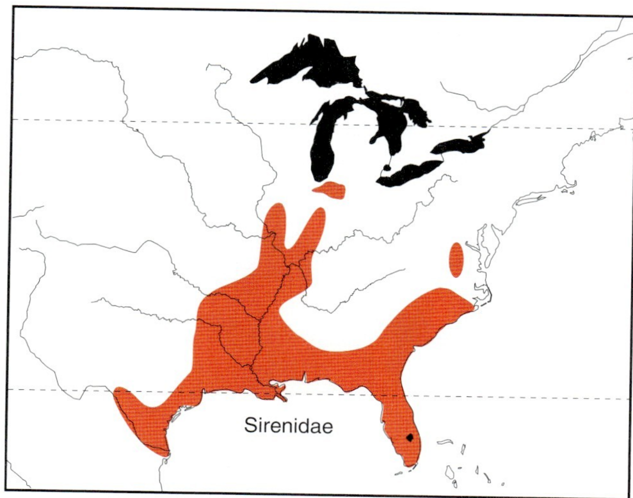

FIGURE 16.5 Geographic distribution of the extant Sirenidae.

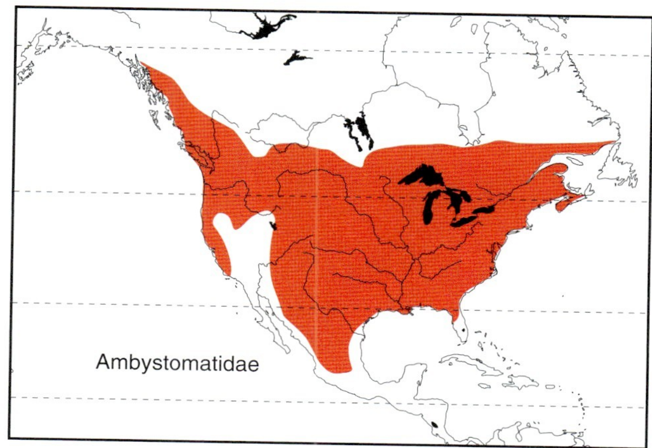

FIGURE 16.7 Geographic distributions of the extant Ambystomatidae (includes *Dicamptodon*).

FIGURE 16.6 Representative sirenid salamander. Greater siren *Siren lacertina* (L. J. Vitt).

a horny beak replaces premaxillary teeth, and an interventricular septum is present in the heart.

Biology: Sirenids typically live in heavily vegetated, slow-moving aquatic habitats, such as lakes, marshes, and swamps. They are active predators, preying on a variety of aquatic invertebrates, which they capture by suction feeding. The larger sirens readily capture crayfish; the dwarf sirens eat principally insect larvae and other small crustaceans and worms. In spite of their locally high abundance and widespread distribution, their biology is poorly known. Courtship behavior has not been observed; the oviductal morphology of *Siren* suggests external fertilization. Eggs are deposited singly or in small clusters attached to vegetation.

Ambystomatidae

Mole Salamanders and Pacific Mole Salamanders

Classification: Amphibia; Caudata; Diadectosalamandroidei; Treptobranchia.

Sister taxon: Salamandridae.

Content: Two genera, *Ambystoma* and *Dicamptodon*, with 33 and 4 species, respectively.

Distribution: North America to the southern rim of the Mexican Plateau (Fig. 16.7).

Characteristics: Ambystomatids are heavy-bodied, heavy-tailed salamanders with four short, well-developed limbs (Fig. 16.8). Adult size ranges from 80 to 550 mm, usually $>$ 160 mm TL. The lower jaw has fused angular and prearticular bones; the upper jaw has both premaxillae and maxillae, and the lacrimal is absent. All presacral spinal nerves, except the second, third, and fourth (*Ambystoma*), or first (*Dicamptodon*) and the postsacral spinal nerves exit intravertebrally through foramina in the vertebrae. Most adult ambystomatids lack gills and gill slits and have moveable eyelids, but the paedomorphic axolotl (*Ambystoma mexicanum*) and its relatives retain some larval traits. Within some species, i.e., *Ambystoma talpoideum*, some individuals retain larval traits such as gills, gill slits, and no eyelids. All ambystomatids have costal grooves on the skin above the ribs, well-developed and functional lungs, and no nasolabial grooves on the snout. Fertilization is internal, and adult females have spermathecae in the cloaca. Adult males have six sets of cloacal glands.

Biology: Most species are terrestrial during adulthood and return to water only for reproduction. Some species and/or populations have paedomorphic or paedotypic traits (see Heterochrony), e.g., the *Ambystoma tigrinum* complex (six species), *A. gracile*, and *A. talpoideum*. The *A. tigrinum* complex includes the axolotl (*A. mexicanum*). The ambystomatids occurring in the United States are predominantly winter breeders, migrating to ponds during brief midwinter warm rains, generally when air temperatures are greater than 10°C. The first wave of migrants is males, which await the females on subsequent nights. Courtship occurs in water; the males "dance" and nudge the females and then

FIGURE 16.8 Representative ambystomatid salamanders. Clockwise from upper left: Idaho giant salamander *Dicamptodon aterrimus* (W. Leonard); ringed salamander *Ambystoma annulatum* (J. P. Caldwell); small-mouthed salamander *Ambystoma texanum* (J. P. Caldwell); tiger salamander *Ambystoma rigrinum* (J. P. Caldwell).

deposit numerous spermatophores. Each female picks up one or more sperm packets from the spermatophores and, during the next several days, deposits eggs. The adults leave the ponds and return to their subterranean homes until the following year. *Ambystoma opacum* and *A. annulatum* deviate from this reproductive pattern by reproducing in late autumn. For most *Ambystoma* species, larval life lasts 3 to 4 months.

All four species of *Dicamptodon* (*D. aterrimus*, *D. copei*, *D. ensatus*, and *D. tenebrosus*) live in moist coastal forests. Metamorphosis occurs in all but *D. copei*, which is paedomorphic and permanently aquatic. Some populations of the other species are paedogenic. Postmetamorphic individuals of the other three species are predominantly terrestrial. Reproduction occurs in forest streams, and the terrestrial adults of the metamorphosed populations return to streams for all reproductive activity. Females typically deposit 50 or more eggs, depending upon body size, in water-filled chambers beneath logs and rocks within or beside streams. Females defend their eggs until they hatch, with incubation often as long as 6 months. *Dicamptodon* larvae are major invertebrate predators in the small forest streams. They forage mainly at night on the streambeds, and in many streams, *Dicamptodon* larvae are the major predator in abundance and biomass.

Salamandridae

Newts and European Salamanders

Classification: Amphibia; Caudata; Diadectosalamandroidei; Treptobranchia.

Sister taxon: Ambystomatidae.

Content: Two subfamilies, Pleurodelinae and Salamandrinae.

Distribution: Europe eastward to central Russia and southward into northeastern Africa, southeastern China and Japan, and eastern and western North America (Fig. 16.9); different generic groups occur in each region (e.g., *Notophthalmus* and *Taricha* in North America; *Pleurodeles*, *Salamandra*, and *Triturus* in Europe; *Cynops* and *Tylototriton* in Asia).

Characteristics: Body morphology of salamandrids ranges from moderately slender to robust; the four limbs are well developed and moderately short (Fig. 16.10). Most

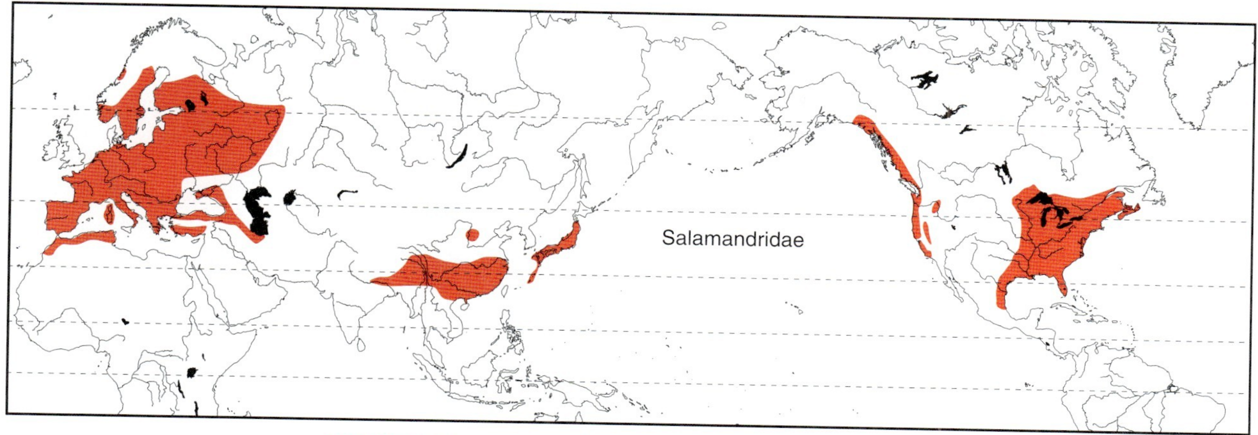

FIGURE 16.9 Geographic distribution of the extant Salamandridae.

FIGURE 16.10 Representative salamandrid salamanders. From left: Eastern newt *Notophthalmus viridescens*, Salamandrinae (L. J. Vitt); Himalayan newt *Tylotriton verrucosus*, Pleurodelinae (K. Nemuras).

adult salamandrids seldom exceed 200 mm TL, and even the larger taxa (e.g., European *Pleurodeles* and *Salamandra*) are less than 350 mm TL. The lower jaw has fused angular and prearticular bones; the upper jaw has both premaxillae and maxillae, and the lacrimal is absent. All presacral, except the second and third, and postsacral spinal nerves exit intravertebrally through foramina in the vertebrae. Adults lack gills and gill slits, except in the paedotypic populations of *Notophthalmus* and *Triturus*, and all have moveable eyelids. Costal grooves above the ribs and nasolabial grooves are absent. Lungs are present and functional. Fertilization is internal; adult females have spermathecae in the cloaca, and adult males possess five sets of cloacal glands.

Biology: Salamandrids typically have a granular or rugose skin because of numerous poison glands, and the secretions of these glands are the most toxic of all salamanders. In association with their high toxicity, many salamandrids are brightly colored, at least ventrally, and advertise their toxicity to potential predators. The bright coloration may be seasonal in appearance. All have courtship displays in which the male circles the female and nudges or rubs her, and in a few species, the male grasps the female and deposits his spermatophore in or near her cloaca. Three life cycles are evident among the taxa with aquatic larvae. In some species, e.g., *Cynops*, *Pleurodeles*, the larvae metamorphose into aquatic juveniles and all individuals remain aquatic throughout adult life. Others (*Taricha*, *Triturus*) have aquatic larvae; upon metamorphosis, the salamanders become terrestrial and return to water only to breed. Paedogenesis occurs in some populations of a few species, including *Notophthalmus viridescens*, *Triturus alpestris*, *T. cristatus*, and *T. helvaticus*.

Pleurodelinae

Sister taxon: Salamandrinae.

Content: Sixteen genera, *Calotriton*, *Cynops*, *Echinotriton*, *Euproctus*, *Lissotriton*, *Mesotriton*, *Neurergus*, *Notophthalmus*, *Ommatotriton*, *Pachytriton*, *Paramesotriton*, *Pleurodeles*, *Salamandrina*, *Taricha*, *Triturus*, and *Tylototriton*, with 59 species.

Distribution: As for the family.

Characteristics: Species in these genera are characterized as newts because of their rough, keratinized skin and their aquatic life history.

Biology: The majority deposit eggs in the water and have a free-living larval stage. The genus *Notophthalmus* (Fig. 16.10) has a triphasic life cycle: aquatic larvae, terrestrial juveniles called *efts,* and aquatic adults.

Salamandrinae

Sister taxon: Pleurodelinae.

Content: Four genera, *Chioglossa, Lyciasalamandra, Mertensiella,* and *Salamandra,* with 15 species.

Distribution: Southern and central Europe, northwest Africa, and western Asia.

Characteristics: Populations of *Salamandra atra* are highly variable plenotypically, e.g., in body coloration. Despite this high degree of phenotypic variation, they exhibit a limited amount of genetic variation.

Biology: Species in these genera are terrestrial salamanders that live in forested areas. At least four species in the genus *Salamandra* and all species of *Lyciasalamandra* are viviparous (see Reproductive Modes).

Proteidae

Olm, Mud Puppies, and Water Dogs

Classification: Amphibia; Caudata; Diadectosalamadroidei; Hydatinosalamandroidei.

Sister taxon: Treptobatrachia (clade containing Abystomatidae and Salamandridae).

Content: Two genera, *Necturus* and *Proteus*, with 5 and 1 species, respectively.

Distribution: Eastern half of North America and eastern Adriatic coast of Europe (Fig. 16.11).

Characteristics: Proteids are moderately robust salamanders with four short, well-developed limbs and large, laterally compressed tails (Fig. 16.12). Adults of three species

FIGURE 16.12 Lewis's water dog *Necturus lewisi*, Proteidae (R. W. Van Devender).

of *Necturus* and the more slender *Proteus* are 200 to 250 mm TL; *N. punctatus* is < 200 mm TL, and *N. maculosus* is the largest taxon, 250 to 350 mm but occasionally to 480 mm TL. The lower jaw of proteids has the angular and prearticular bones fused; the upper jaw has only premaxillae, and the lacrimal is absent. All pre- and postsacral spinal nerves, except the first one, exit intervertebrally. All proteids are paedomorphic; adults have external gills, two pairs of gill slits, and no eyelids. Costal grooves are present on the trunk, and nasolabial grooves are absent. Lungs are present, although small. Fertilization is internal; adult females have spermathecae, and adult males possess six sets of cloacal glands.

Biology: Both genera are totally aquatic, but the North American *Necturus* dwells in surface waters, whereas the European *Proteus anguinus* is a cave species. Superficially, *P. anguinus* appears more similar to the paedomorphic spelerpine plethodontids than to the *Necturus* species, because it has a slender body and limbs, reduced eyes beneath the skin, and a pigmentless skin. All species of *Necturus* prefer clear water and rocky, siltfree substrates. They are nocturnal foragers and eat a variety of prey with a preference for crayfish. *N. maculosus* courts in the autumn, but egg laying does

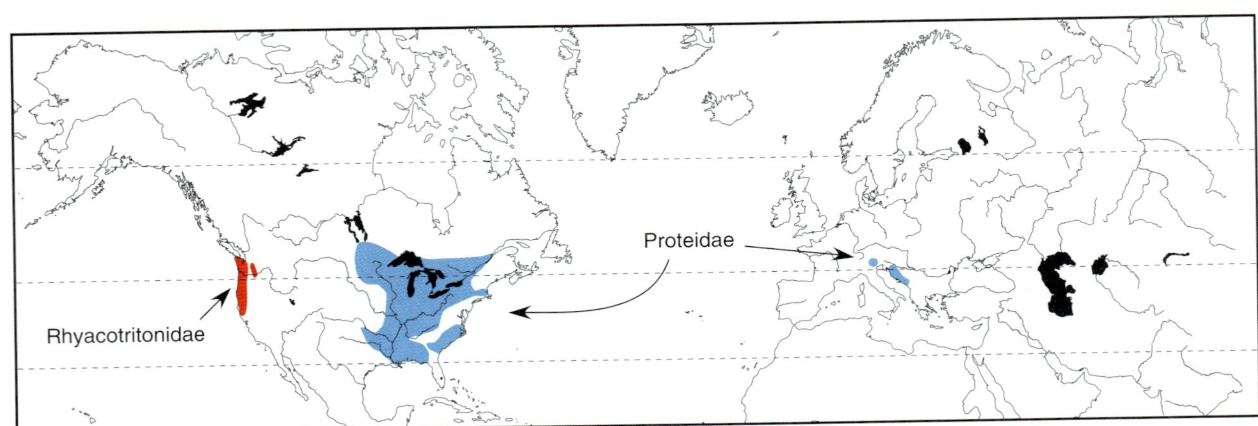

FIGURE 16.11 Geographic distributions of the extant Rhyacotritonidae and Proteidae.

not occur until the subsequent spring. Up to 50 eggs are attached to the roof of the female's shelter, and whether or not they receive active care, they are protected by her presence. Individuals of the cave-dwelling *P. anguinus* frequently aggregate in shelters under rocks or in fissures and use chemical signals to locate other individuals. *P. anguinus* commonly deposits up to 70 eggs in a season, but apparently warmer water temperatures may induce retention of eggs, resulting in the birth of two fully formed larvae.

Rhyacotritonidae

Torrent Salamanders

Classification: Amphibia; Caudata; Diadectosalamandroidei; Plethosalmandroidei.

Sister taxon: Xenosalamandroidei (clade containing Amphiumidae and Plethodontidae).

Content: One genus, *Rhyacotriton*, with 4 species.

Distribution: Pacific Northwest of United States (Fig. 16.11).

Characteristics: Rhyacotritonids are heavy-bodied, heavy-tailed salamanders with four short, well-developed limbs. Adult size ranges from 90 to 120 mm TL. The lower jaw has angular and prearticular bones fused; the upper jaw has both premaxillae and maxillae, and the lacrimal is present. All except the first presacral spinal nerve exit intervertebrally, and all postsacral nerves exit intravertebrally. Adults lack gills and gill slits. Eyelids are present and functional. Costal grooves are present on the skin above the ribs, and nasolabial grooves are absent. Small lungs are present. Fertilization is internal; adult females have spermathecae, and adult males possess six sets of cloacal glands and unique enlarged, rectangular vent glands.

Biology: These salamanders are semiaquatic residents of humid conifer forest. The larvae and transformed individuals (Fig. 16.13) live in shallow areas of rocky rubble in cold, well-aerated forest streams and spring seepages; occasionally they wander into deeper pools or adults forage on the forest floor during heavy rains. Courtship is presumed to occur on land or in the splash zone of streams. Fertilization is internal via spermatophores. Females deposit 3 to 15 eggs, each attached singly to the underside of rocks. The eggs hatch in 7 to 10 months, and larval development requires 3 to 5 years owing to the cold temperature of the aquatic nesting sites.

Amphiumidae

Amphiumas

Classification: Amphibia; Caudata; Diadectosalamandroidei; Plethosalmandroidei; Xenosalamandroidei.

Sister taxon: Plethodontidae.

Content: One genus, *Amphiuma*, with 3 species.

Distribution: Southeastern United States, including the southern half of the Mississippi River valley and along the coastal plain to Virginia (Fig. 16.14).

Characteristics: *Amphiuma* are heavy-bodied, eellike salamanders with four tiny, weakly developed limbs (Fig. 16.15). Adults can be large depending on species, occasionally exceeding 1 meter TL. The lower jaw has

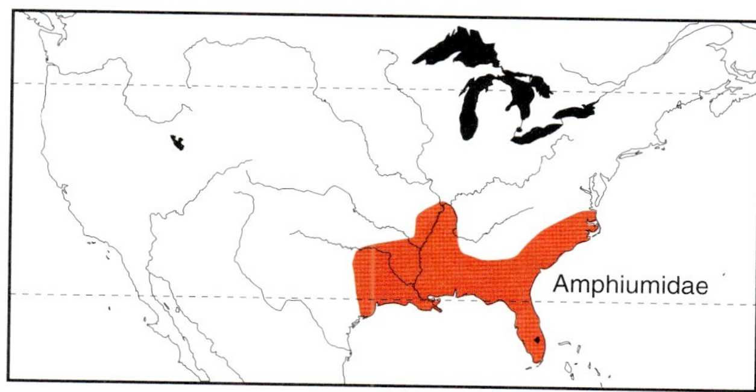

FIGURE 16.14 Geographic distribution of the extant Amphiumidae.

FIGURE 16.13 Representative rhyacotritonid salamander. Cascade torrent salamander *Rhyacotriton cascadae*, Rhyacotritonidae (W. Leonard).

FIGURE 16.15 Representative amphiumid salamander. Three-toed amphiuma *Amphiuma tridactylum*, Amphiumidae (R. W. Van Devender).

angular and prearticular bones fused; the upper jaw has both premaxillae and maxillae, but the lacrimal is absent. Excluding the first spinal nerve, all presacral and all, except the distalmost, postsacral spinal nerves exit intervertebrally. Amphiumids display some paedomorphic traits; adults have internal gills and a single pair of gill slits, and eyelids and a tongue are absent. They have costal grooves in the skin above the ribs and lack nasolabial grooves on the snout. Lungs are present. Fertilization is internal; adult females have spermathecae in the cloaca, and adult males possess five sets of cloacal glands, of which the posteriormost set has a unique morphology and histology.

Biology: Amphiumas lose their external gills during a partial metamorphosis. The limbs are greatly reduced, but the number of toes allows identification of the three species: *A. tridactylum* (Fig. 16.15) has three toes on each foot, *A. means* two toes, and *A. pholeter* one toe. The former two species are large salamanders with adult length exceeding 1 meter TL, whereas the latter species is considerably smaller, < 300 mm TL. All species are aquatic, although *A. means* has been found active on land during rainy nights. Field observations indicate that males court several females simultaneously or that multiple females contend for the attention of a single male. Since females in other closely related genera are passive or even rebuff the male's efforts, these observations require confirmation. Gender is not easily determined, and the observations may have consisted of several males vying for a single female. Courtship ends with the male depositing a spermatophore directly into the female's cloaca by means of cloacal apposition. In all species, females stay with and coil around their eggs, usually beneath logs, rocks, and other detritus at the water's edge. For at least *A. tridactylum*, females reproduce every 2 years and produce about 200 eggs each time.

Plethodontidae

Lungless Salamanders

Classification: Amphibia; Caudata; Diadectosalamandroidei; Plethosalmandroidei; Xenosalamandroidei.

Sister taxon: Amphiumidae.

Content: Four subfamilies, Bolitoglossinae, Hemidactyliinae, Plethodontinae, and Spelerpinae.

Distribution: Americas, occurring from southern Canada to southwestern Brazil, and disjunctly, central Mediterranean Europe and Korea (Fig. 16.16).

Characteristics: Plethodontids display a diversity of body shapes, but all have four limbs; some taxa are stocky and short limbed, and others are elongate and slender limbed; some have tails equal to body length, and in others, the tails are twice the length of the body (Fig. 16.17). Adult body size ranges from 25 to 30 mm TL in the diminutive *Thorius* to 320 mm TL in *Pseudoeurycea belli* (both Mexican bolitoglossines). The lower jaw has the angular and prearticular bones fused; the upper jaw has both premaxillae and maxillae, and the lacrimal is absent. All presacral, except the second, third, and fourth, and the postsacral spinal nerves exit intravertebrally through foramina in the vertebrae. Adults lack gills and gill slits and have moveable eyelids except in the paedomorphic taxa, e.g., *Eurycea*. Costal grooves are present on the trunk, and all species possess a pair of nasolabial grooves on the snout. Lungs are absent. Fertilization is internal; adult females have spermathecae in the cloaca, and adult males possess six sets of cloacal glands.

Bolitoglossinae

Sister taxon: Unnamed clade continuing Plethodontinae and Spelerpinae.

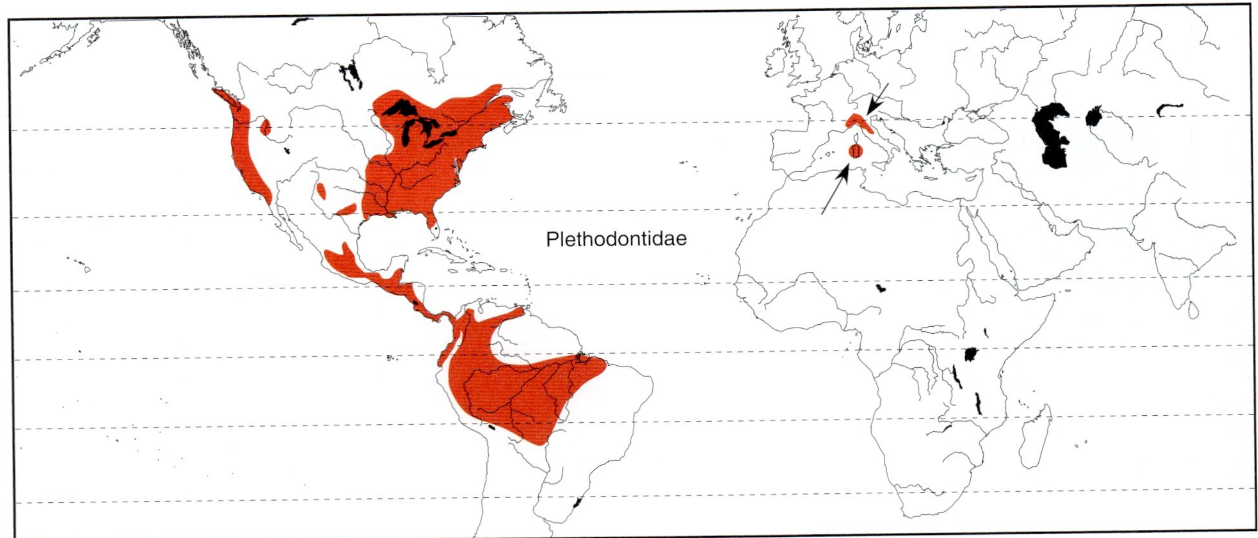

FIGURE 16.16 Geographic distribution of the extant Plethodontidae.

FIGURE 16.17 Representative plethodontid salamanders. Clockwise from upper left: Western slimy salamander *Plethodon albagula,* Plethodontinae (J. P. Caldwell); red salamander *Pseudotriton ruber*, Spelerpinae (L. J. Vitt); cave salamander *Eurycea lucifuga*, Spelerpinae (J. P. Caldwell); unnamed tropical salamander *Bolitoglossa* sp., Bollitoglossinae (J. P. Caldwell).

Content: *Batrachoseps*, *Bolitoglossa*, *Bradytriton*, *Chiropterotriton*, *Cryptotriton*, *Dendrotriton*, *Nototriton*, *Nyctanolis*, *Oedipina*, *Parvimolge*, *Pseudoeurycea*, and *Thorius* with 249 species.

Distribution: Western and southern North America to Brazil.

Characteristics: Bolitoglossines have the typical vertebrate jaw mechanism in which the skull remains rigid and the lower jaw swings downward; the occipital condyles are short and atlanto-mandibular ligaments are absent. Embryos and larvae have three pairs of gill slits.

Biology: The terrestrial bolitoglossines occupy a variety of habitats from forest-floor leaf litter and burrows to rock screes and cliffs; some species are arboreal, occurring high in trees. All species of bolitoglossines have direct development.

Hemidactylinae

Sister taxon: Unnamed clade, sister to all other plethodontids.

Content: One genus, *Hemidactylium*, with 1 species.

Distribution: Extreme southeastern Canada throughout eastern one-third of United States, with numerous disjunct populations throughout its range, the westernmost of which is in southeastern Oklahoma.

Characteristics: The single species, the four-toed salamander, has only four toes on the hind foot and a constriction at the base of the tail. The distinctive ventral coloration is white with black spots or blotches.

Biology: This species inhabits forested areas and breeds in swamps, bogs, vernal pools, and other types of nonmoving water. Clumps of moss at edges of streams are typical breeding habitat. Eggs are laid just above the waterline, and after hatching, larvae move into water. Females sometimes brood eggs communally.

Plethodontinae

Sister taxon: Spelerpinae.

Content: Seven genera, *Aneides*, *Desmognathus*, *Ensatina*, *Hydomates*, *Karsenia*, *Phaeognathus*, and *Plethodon*, with 93 species.

Distribution: United States and southern Canada, Mediterranean Europe, and Korean Peninsula.

Characteristics: Desmognathines have a unique jaw-opening mechanism in which the lower jaw is held stationary and the skull swings upward. The cranial and cervical skeleton and musculature have unique features associated with this behavior, including stalked occipital condyles and atlanto-mandibular ligaments. Embryos and larvae have four pairs of gill slits.

Biology: These salamanders are predominantly aquatic, although some species live streamside and forage along the stream or nearby. Other species (e.g., *D. apalachicolae*, *D. carolinensis*) are more terrestrial, but surface activity and habitat selection is driven by the requirement for high humidity. The large *Phaeognathus hubrichti* and the smallest desmognathine, *D. wrighti*, are terrestrial. The former lives in burrows and feeds at the burrow mouth, and *D. wrighti* lives under the forest-floor litter. While other taxa lay their eggs in wet situations from spring seepages to beneath rocks and leaf mats in streams, these two species lay their eggs terrestrially, and neither has an aquatic larval stage. *D. aeneus*, another diminutive species, deposits eggs in seepages, and although larvae hatch, they do not feed and quickly metamorphose. Most, if not all, plethodontines show parental care with females attending their eggs until they hatch.

Spelerpinae

Sister taxon: Plethodontinae.

Content: Four genera, *Eurycea*, *Gyrinophilus*, *Pseudotriton*, and *Stereochilus*, with 35 species.

Distribution: Eastern North America north of Mexico.

Characteristics: Larval periods may range from a few months to 2–3 years in *Eurycea*, and as long as 4 years or more in *Gyrinophilus porphyriticus*.

Biology: Paedomorphosis occurs only in the spelerpine salamanders, e.g., species of *Eurycea* of Edwards Plateau in Texas and *Gyrinophilus palleucus*. All these paedomorphs are subterranean aquatic or spring residents. In addition to incomplete metamorphosis and the retention of gills, most paedomorphs are slender-bodied and -limbed and have degenerate eyes and reduced skin pigmentation.

QUESTIONS

1. What is the global distribution of salamanders and how would you explain this distribution?
2. How are plethodontid salamanders distinguished from most other salamanders?
3. In what kinds of microhabitats would you expect to find sirens and amphiumas?
4. Which families of salamanders would you expect to find in Australia, the Seychelles, and Madagascar.
5. How can some salamanders have internal fertilization without a copulatory organ, and in which salamanders does this occur?

REFERENCES

General

Bruce, 2005; Cogger and Zweifel, 1992; Dowling and Duellman, 1974-1978; Duellman and Trueb, 1986; Edwards, 1976; Frost, 1985; Frost et al., 2006; Hairston, 1996; Halliday and Adler, 1986; Hay et al., 1995; Hillis, 1991; Larson, 1991; Larson and Dimmick, 1993; Macey, 2005; Petranka, 1998; Roelants et al., 2007; Simons and Brainerd, 1999; Weisrock et al., 2005; Wiens et al., 2005, 2006, 2007a; Weisrock et al., 2005, 2006a,b.

Taxonomic Accounts

Cryptobranchidae

Asiatic Giant Salamanders and Hellbenders

Nickerson and May, 1973; Sever, 1991b.

Hynobiidae

Asiatic Giant Salamanders

Adler and Zhao, 1990; Kuzmin, 1995; Larson et al., 2003; Tanaka, 1989; Vorobyeva, 1994, 1995; Weisrock et al., 1999; Zeng et al., 2006; Zhang et al., 2003, 2006; Zhao et al., 1988.

Hynobiinae

Zeng et al., 2006; Zhang et al., 2006.

Protohynobiinae

Fei and Ye, 2000.

Sirenidae

Sirens and Dwarf Sirens

Liu et al., 2004, 2006; Martof, 1972, 1974; Moler and Kezer, 1993; Petranka, 1998; Roelants et al., 2007; Sever, 1991; Sever et al., 1996b; Wiens et al., 2005.

Ambystomatidae

Mole Salamanders and Pacific Mole Salamanders

Carstens et al., 2005; Good, 1989; Good and Wake, 1992; Jones et al., 1993; Larson et al., 2003; Nussbaum 1976; Nussbaum et al., 1983; Parker, 1994; Pfingsten and Downs, 1989; Reilly and Brandon, 1994; Ryan and Semlitsch, 2003; Sever, 1991a; Sever and Kloepeer, 1993; Shaffer, 1993; Shaffer et al., 1991; Shaffer and Knight, 1996.

Salamandridae

Newts and European Salamanders

Griffiths, 1996; Frost et al., 2006; Hayashi and Matsui, 1989; Titus and Larsen, 1995; Twitty, 1966; Wake and Özeti, 1969; Weisrock et al., 2006.

Pleurodelinae

Carranza and Amat, 2005; Kozak et al., 2006; Titus and Larson, 1995; Veith et al., 2004; Weisrock et al., 2006.

Salamandrinae

Kozak et al., 2006; Riberon et al., 2004; Veith et al., 1998; Weisrock et al., 2006b.

Proteidae

Olm, Mud Puppies, and Water Dogs

Ashton, 1990; Engelmann et al., 1986; Gao and Shubin, 2001; Guillaume, 2000a,b; Pfingsten and Downs, 1989; Trontelj and Goricki, 2003; Wiens et al., 2005; Roelants et al., 2007.

Rhyacotritonidae

Torrent Salamanders

Good and Wake, 1992; Leonard et al., 1993; Nussbaum, 1976; Nussbaum et al., 1983; Petranka, 1998; Sever, 1992; Wagner et al., 2006; Welsch and Lind, 1996; Wiens et al., 2005.

Amphiumidae

Amphiumas

Fontenot, 1999; Larson and Dimmick, 1993; Means, 1996; Salthe, 1973a,b; Wiens et al., 2005.

Plethodontidae

Lungless Salamanders

Chippindale et al., 2004; Macey, 2005; Mueller et al., 2004; Parra-Olea et al., 2004; Petranka, 1998; Sever, 1994; Tilley and Bernardo, 1993; Titus and Larson, 1997.

Bolitoglossinae

Houck and Verrell, 1993; Jackman et al., 1997a; Jaeger and Forester, 1993; Lanza et al., 1995; Petranka, 1998; Wiens et al., 2007.

Hemidactylinae

Chippindale et al., 2004; Macey, 2005; Petranka, 1998; Tilley and Bernardo, 1993; Titus and Larson, 1997.

Plethodontinae

Min et al., 2005; Petranka, 1998; Tilley and Bernardo, 1993; Titus and Larson, 1997; Sever et al., 1990; Wake and Elias, 1983; Wiens et al., 2007.

Spelerpinae

Bruce, 2005; Chippindale et al., 2005; Macey, 2005.

E. troglodytes : Valdina Farms
- aquatic
- cave dweller
- eyes small
- Ext. gills retained
- subterranean intermittent pools/streams
- troglodyte
- lack pigmentation

E. latitans : Cascade Cavern
- Aquatic
- eyes small and lidless
- snout flattened
- tail fin conspicuous, elevated
- subterranean streams and pools
- cave dwelling

E. nana : San Marcos
- Aquatic
- Belly yellowish-white
- Neotenic
- restricted to site where it was first collected
- spring pool at source of San Marcos river

E. neotenes : Texas
- Aquatic
- External gills have long, bright red filaments
- short stout legs
- narrow tail fin
- Edward's Plateau
- rarely transforms to terrestrial form
- small cave streams, springs, seeps

E. tridentifera : Comal Blind
- subterranean aquatic
- Large head, flattened snout
- eyes very small
- underground waters of limestone caves
- transparent underside
- occurs w/ TX salamander

E. (T.) rathbuni : Texas Blind
- cave dweller
- long spindly legs
- larvae do not transform to adults
- Eyes reduced
- snout flattened
- subterranean
- Balcones Escarpment of Edwards Plateau vicinity of San Marcos, TX
- endangered
- one entrance to habitat, a national preserve
- eat invertebrates

Chapter 17

Frogs

OVERVIEW

Frogs and toads occur worldwide on all continents, except Antarctica, and on most continental islands. They are a diverse group with more than 5450 species. Frogs and toads live in most aquatic and terrestrial habitats from lowlands to mountaintops, although their inability to physiologically adapt to saltwater has largely excluded them from estuarine and marine habitats. Their highest diversity is in moist tropical sites; for example, about half of all known species live in the New World tropics. Nevertheless, frogs commonly occur in arid or cold-temperate localities.

Frogs and toads (Anura; Salientia, stem-based name) are unmistakable with their unique short, tailless bodies; broad, flat heads with big mouths; and long, muscular hindlimbs. This body form is associated with and likely evolved as an adaptation for saltatory (jumping) locomotion. The long hindlimbs extend synchronously and provide the propulsive force to lift and propel the frog forward. The short body provides a compact mass to be hurled forward, and the shortened vertebral column, robust pectoral girdle, and forelimbs readily absorb the shock of landing. Frogs regularly leap 2 to 10 times their body length; a few species are capable of prodigious leaps of 30 to 40 times their body length. Of course, not all frogs move by leaping. Some use a typical vertebrate walking gait, and frogs that normally leap walk when moving slowly or for a short distance.

With few exceptions, frogs have external fertilization. Males typically grasp (amplex) females in such a manner that their cloacae are juxtaposed, ensuring fertilization of the eggs as they are deposited. Indirect development of free-living larvae is common, although direct development is widespread. Larval (indirect) development of anurans is strikingly different from that of salamanders and caecilians. The anuran larva or tadpole is structurally, physiologically, ecologically, and behaviorally different from the fully developed froglet or adult. The shift from tadpole to froglet requires a major reorganization of anatomy and physiology as the larva metamorphoses. This contrasting body form and lifestyle may partially explain the lack of paedomorphosis and paedogenesis in anurans.

Living anurans share a suite of unique features attesting to their monophyly. All have greatly shortened vertebral columns, consisting of nine or fewer vertebrae; most

Herpetology, 3rd Ed.

clades have eight. All presacral vertebrae, except the atlas (first vertebra), have transverse processes, and dorsal ribs are absent (in most clades) or reduced, unicapitate, and usually confined to the second through fourth vertebrae in some primitive clades. Presacral vertebrae are firmly articulated, allowing only moderate lateral and dorsoventral flexure; postsacral vertebrae are fused into a rod-shaped urostyle lying within an elongated dorsopelvic pocket formed by the uniquely elongated and anteriorly oriented iliae. The epipodial elements of both fore- and hindlimbs are fused, at least at their ends, forming a robust radioulna and tibiofibula in each, respectively. The ankle is elongated and similarly consists of a pair of fused bones (fibulare or astragalus and tibiale or calcaneum) that form a sturdy strut. All frogs lack teeth on the dentary of the lower jaw, except for the hylid *Gastrotheca guentheri* (formerly *Amphignathodon*) and possess large subcutaneous lymph spaces beneath the skin. As previously noted, the anuran larva is structurally unlike that of the other extant amphibians; for example, the jaws are toothless, and keratinous jaw sheaths and labial teeth are usually present as functional but nonhomologous substitutes.

As with many groups of plants and animals, molecular and total evidence phylogenies are being produced rapidly, revealing new ideas about the number of families of frogs and their relationships. Aspects of these arrangements are controversial and have led to publication of competing hypotheses. Undoubtedly, refinement of the phylogeny will occur over a period of years. Viewed from a longer perspective, significant progress has been made in the last 30 years in terms of our knowledge of frog relationships. Looking back even further, Boulenger's 1882 *Catalogue of Batrachia Salientia* included about 1800 species classified in two suborders: Aglossa, with 2 families; Phaneroglossa, with 12 families divided into two series, Firmisternia and Arcifera. His classification, as all classifications of that era, was phenetic; nonetheless, some of his contemporaries and successors were broadly surveying anuran anatomy and recognizing character suites that still form the morphological core of present phylogenetic analyses.

In the early 1900s, G. K. Noble was first to attempt construction of an evolutionary classification of anurans. He examined a large spectrum of characters, drawing on the dentition and pectoral girdle characters of Cope and the vertebral characters of Nichols, and added his thigh-musculature characters to produce a dendrogram of relationships and a classification that was widely accepted into the 1960s. Problems with some of the characters and their interpretation were soon noted, and new character complexes were discovered that offered new insights into phylogenetic relationships. A new generation of systematists provided interpretations based on new analytical protocols, new characters, and character coding. The most recent large-scale attempt to classify frogs and understand their relationships was produced in 2006 and is adopted here with additional more recent modifications (Fig. 17.1).

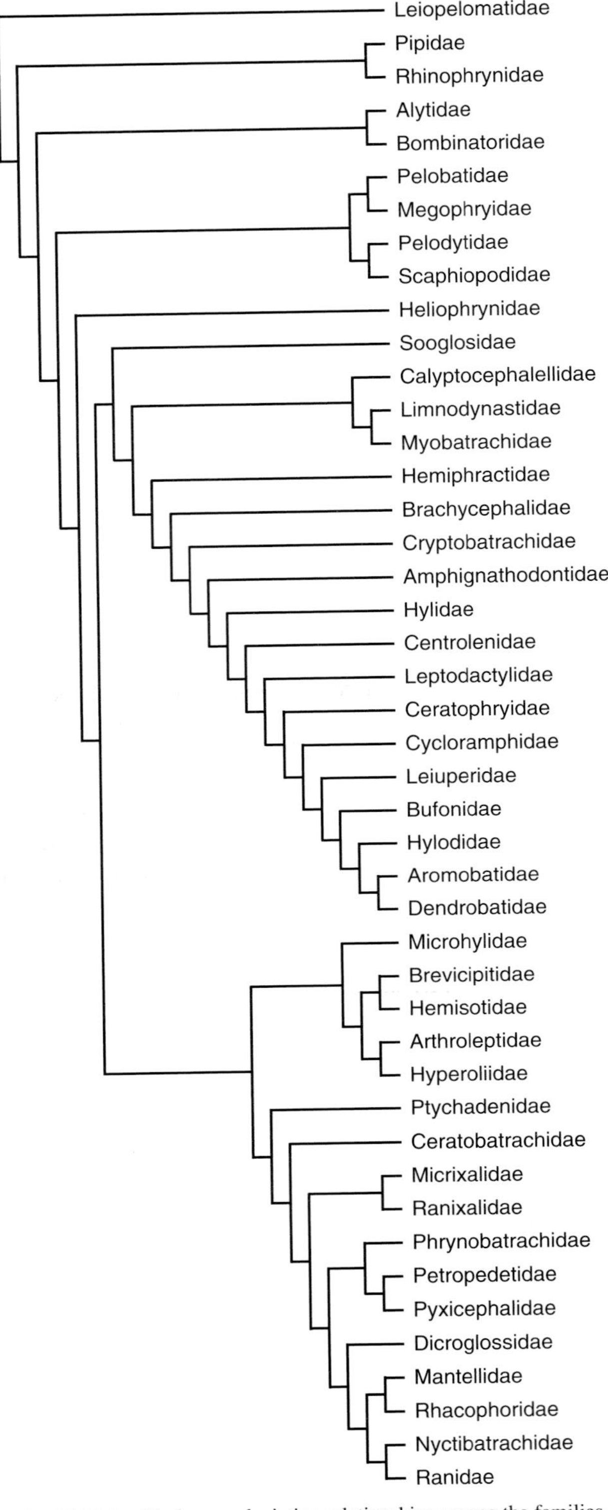

FIGURE 17.1 Cladogram depicting relationships among the families of extant frogs. The cladogram is reconstructed from a variety of recent studies (see text).

Amphibian taxonomy, and particularly anuran taxonomy, is in a state of rapid growth and change. Partly this is because of the explosive growth in the number of herpetologists all over the world, resulting in discovery and description of many heretofore unknown species. In addition, new molecular methods have generated great interest in studying relationships among anurans. Many large genera and families have been examined in detail in the last decade and are being partitioned into smaller groups. During this process, frogs that were once thought to be closely related (e.g., *Eleutherodactylus* and *Leptodactylus*; many genera of "ranids") are discovered to have entirely different relationships. The general goal of all systematists is to uncover evolutionary relationships among all amphibians. With the diversity and number of frog species already known and continuing to be discovered, this process will be ongoing for many years to come.

TAXONOMIC ACCOUNTS

Leiopelmatidae

Tailed Frogs and New Zealand Frogs

Classification: Amphibia; Lissamphibia; Anura.

Sister taxon: Lalagobatrachia, the clade containing all other living Anura.

Content: Two genera, *Ascaphus* and *Leiopelma*, with 2 and 4 species, respectively.

Distribution: Disjunct within northwestern North America (*Ascaphus*) and New Zealand (*Leiopelma*) (Fig. 17.2).

Characteristics: Frogs in the genus *Ascaphus* attain a body size of 35 to 50 mm SVL (snout-vent length). A unique modification of the cloaca and tail muscles produces an intromittent or copulatory organ in males (Fig. 17.3), one of two such structures for internal fertilization in anurans. Frogs in the genus *Leiopelma* are moderately small (30–49 mm adult SVL) and are unique among anurans in having ventral inscriptional ribs. In all leiopelmatids, the skull lacks palatines and has paired frontoparietals. The facial nerve exits through the facial foramen anterior to the auditory capsule; the trigeminal and facial nerve ganglia are separate. The vertebral column consists of nine presacral notochordal vertebrae, and all are amphicoelous. The transverse processes of the sacral vertebra are slightly expanded, and this vertebra has a cartilaginous connection to the urostyle. Adults have free dorsal ribs on the second through fourth, occasionally the fifth, presacral vertebrae. The pectoral girdle is arciferal with a distinct sternum. The fibulare and tibiale are fused at their proximal and distal ends. No intercalary cartilage occurs between the terminal and penultimate phalanges of each digit, and the tips of the terminal phalanges are generally blunt to pointed.

Tadpoles of *Ascaphus* have keratinized mouthparts and two small, fused spiracular tubes with a single anteromedial spiracle. The branchial chamber of tadpoles of *Leiopelma* does not close, so a spiracle does not form.

Biology: Tailed frogs are streamside residents of clear, cold, and unsilted mountain streams, living in forest from near sea level to over 2000 m elevation. They are largely nocturnal, active along streams at night or foraging in the forest on rainy evenings. During the day, they hide beneath stones and detritus at or near the stream's edge or in shallow areas within the stream. Courtship occurs in September and October; males are voiceless, and apparently visual cues are used by males and females to find one another. Amplexus is inguinal and copulation commonly occurs underwater. In addition to the rarity of internal fertilization among frogs, the female tailed frog stores sperm in her oviducts for nearly 9 months;

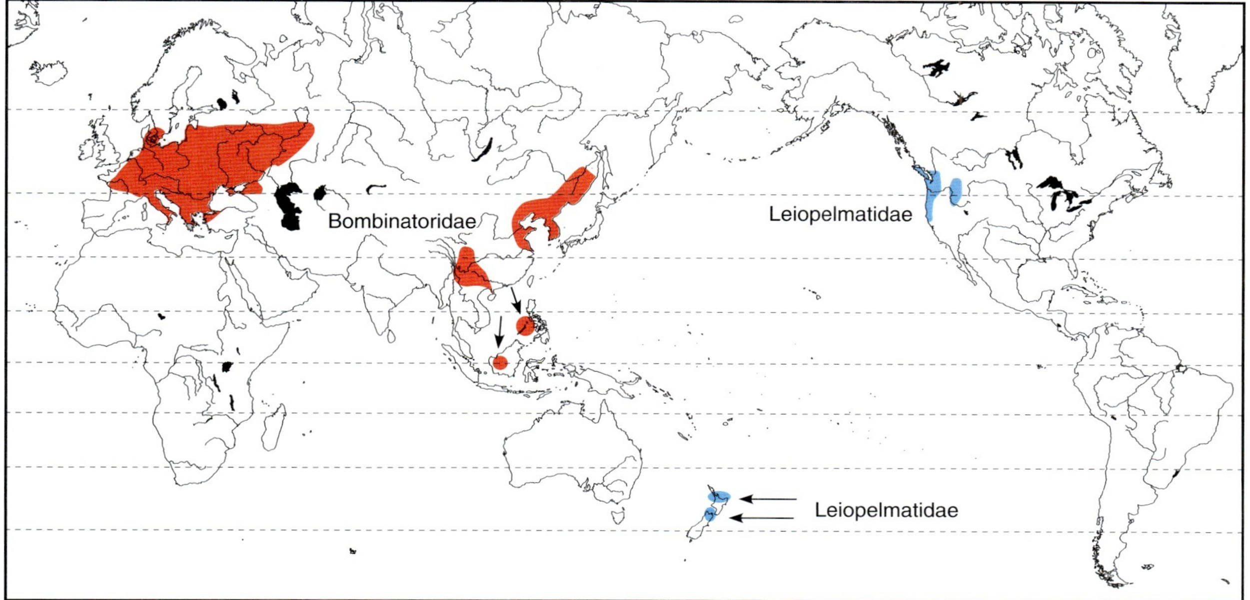

FIGURE 17.2 Geographic distributions of the extant Leioplematidae and Bombinatoridae.

FIGURE 17.3 Representative early frogs. Clockwise from upper left: Tailed frog *Ascaphus truei*, Ascaphidae (W. Leonard); Surinam toad *Pipa pipa*, Pipidae (J. P. Caldwell); Mesoamerican burrowing toad *Rhinophrynus dorsalis*, Rhinophrynidae (J. A. Campbell); midwife toad *Alytes obstetricans,* Alytidae (E. Crespo).

fertilization occurs at the time of ovulation and egg deposition from June to August. Females deposit 40 to 150 unpigmented eggs in small strings attached to undersides of boulders or in cobble in riffles or pools of fast-flowing, rocky streams. In the cold water (11°C), the eggs take about 6 weeks to hatch into streamlined tadpoles with reduced tail fins and suctorial oral discs. The latter structure permits the tadpoles to simultaneously cling upside down to the undersurface of rocks and feed on the algal crust in rapidly flowing streams. The larval phase lasts 2 to 3 years; metamorphosis usually occurs in late summer.

All species of *Leiopelma* are secretive frogs that survive in only a few areas along the borders of cool forest creeks, seepage areas, or open ridges. *L. hochstetteri* is semiaquatic and restricted to wet areas along streams compared with the other three species, which are terrestrial. Courtship occurs in spring and summer (September through January). Although they lack vocal sacs and tympana, males produce quiet chirping calls during sexual encounters; amplexus is inguinal. Tadpoles of all four species are endotrophic; they do not feed. The females deposit small clusters of 1 to 22 large, yolky eggs in small depressions beneath rocks or logs. Adults of *L. hochstetteri* remain near the eggs, but no obvious parental care occurs. Tadpoles of *L. hochstetteri* are nidicolous and remain near the site of oviposition until metamorphosis, although they are capable of swimming. In the other three species, *L. archeyi, L. hamiltoni*, and *L. pakeka*, males provide parental care by brooding the eggs; upon hatching, the exoviviparous tadpoles move onto the flanks and dorsum of the parent, where they complete their development. The presence of open gill slits, some intestinal looping, and rotation of the palatoquadrate early in development support the idea that the ancestor of leiopelmatids had a free-living, feeding, aquatic tadpole stage.

Pipidae

Platannas, African Clawed Frogs, and Surinam Toads

Classification: Amphibia; Lissamphibia; Anura; Lalagobatrachia; Xenoanura.

Sister taxon: Rhinophrynidae.

Content: Five genera, *Hymenochirus*, 4 species; *Pipa*, 7 species; *Pseudhymenochirus*, 1 species; *Silurana*, 2 species; and *Xenopus*, 17 species.

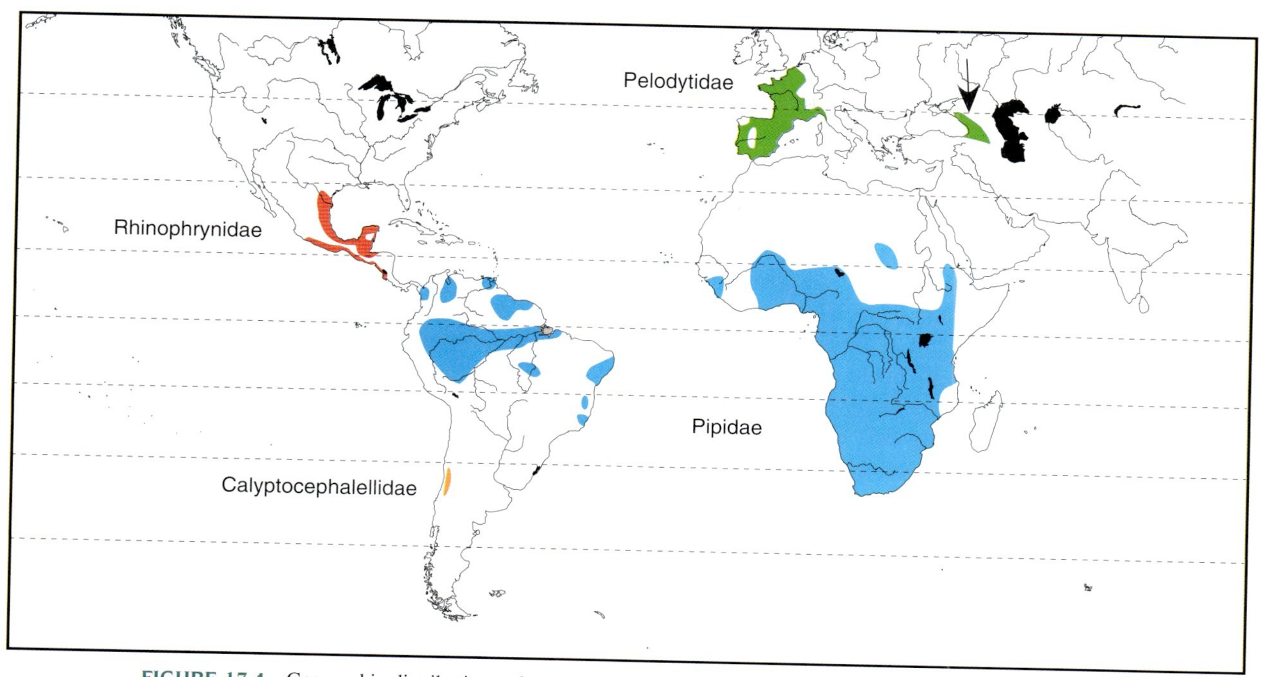

FIGURE 17.4 Geographic distributions of the extant Pipidae, Rhinophrynidae, Pelodytidae, and Calyptocephalellidae.

Distribution: Sub-Saharan Africa (*Hymenochirus, Pseudhymenochirus, Silurana*, and *Xenopus*) and tropical South America to Panama (*Pipa*) (Fig. 17.4).

Characteristics: Adult size is variable, ranging from the small *Pipa parva* (27–44 mm adult SVL) and *Hymenochirus* (25–33 mm SVL) to the larger *Xenopus laevis* (60–130 mm SVL) and *Pipa pipa* (105–170 mm SVL). All pipids are highly aquatic and possess dorsoventrally depressed bodies and large muscular hindlimbs and webbed feet. All lack tongues but retain the lateral line organs as adults. The pipid skull lacks palatines and has a single frontoparietal. The facial nerve exits through the anterior acoustic foramen in the auditory capsule; the trigeminal and facial nerve ganglia fuse to form a prootic ganglion. The vertebral column possesses six to eight presacral stegochordal vertebrae, and all are opisthocoelous. The transverse processes of the sacral vertebra are broadly expanded, and this vertebra has a bicondylar articulation with the urostyle. Postmetamorphs have dorsal ribs fused to the second through fourth presacral vertebrae. The pectoral girdle is arciferal, pseudofirmisternal in *Hymenochirus*, with a small sternum. The fibulare and tibiale are fused at their proximal and distal ends. No intercalary cartilage occurs between the terminal and penultimate phalanges of each digit, and the tips of the terminal phalanges are pointed. The larvae lack keratinized mouthparts, and the left and right branchial chambers are emptied by separate spiracles.

Hymenochirus, Pseudhymenochirus, and *Pipa* lack palpebral membranes (= nictitating membranes), nasolacrimal or subocular tentacles, and an epipubis. The tadpoles of these species do not have a sensory barbel at each corner of the mouth. *Silurana* and *Xenopus* have palpebral membranes, nasolacrimal or subocular tentacles, and an epipubis. Their tadpoles have long, thin sensory barbels at the corners of the mouth.

Biology: All pipids are aquatic frogs, occurring in a variety of habitats, usually still or slow-moving water among vegetation. *Xenopus* seemingly occurs in every freshwater habitat south of the Sahara, including roadside puddles. This broad distribution is largely that of the *X. laevis* complex, which occupies this entire area, and encompasses the much smaller distributions of *Silurana* and the other species of *Xenopus*. The genetic diversity of *Xenopus* is polyploid derived, and all or most extant species likely arose from interspecific hybridization events. Males of all pipids lack vocal cords and vocal sacs, but they attract females by producing a series of sharp clicking notes made from snapping the hyoid apparatus while underwater. In *Xenopus laevis*, sexually receptive females respond to male clicking with a rapping sound, resulting in a duet and allowing the pair to locate each other in dark, murky water where breeding occurs. Elaborate reproductive behavior, especially in *Pipa, Hymenochirus*, and *Pseudhymenochirus*, includes the performance by an amplexed pair of a series of aquatic somersaults (turnovers) that allows the male to fertilize the eggs prior to his rolling onto the female's back in *Pipa* and *Pseudhymenochirus* (Fig. 5.6) or being deposited at the water surface in *Hymenochirus*. In *Pipa*, amplexus lasts longer than 12 hours to allow morphological and physiological changes of dorsal skin. When the eggs roll onto the female's back, they sink into the skin and eventually become fully embedded. In *Pipa pipa* (Fig. 17.3) and *P. arrabali* development is direct, and toadlets "hatch" from their skin pockets; it is indirect in *P. carvalhoi*, *P. myersi*, and *P. parva*, with larvae emerging and completing their development as free-living tadpoles. Turnovers were previously thought not to occur in *Xenopus;* however, recent work on *Xenopus wittei* revealed that this species has an elaborate courtship that includes turnovers and deposition of eggs under floating vegetation at the surface of the water. Studies of the reproductive behavior of other species of *Xenopus* have been hampered because they typically breed at night in murky water.

Rhinophrynidae

Mexican Burrowing Toad

Classification: Amphibia; Lissamphibia; Anura; Lalagobatrachia; Xenoanura.

Sister taxon: Pipidae.

Content: Monotypic, *Rhinophyrnus dorsalis*.

Distribution: Tropical and subtropical lowlands of extreme southern Texas to Costa Rica (Fig. 17.4).

Characteristics: *Rhinophyrnus dorsalis* (Fig. 17.3) is a peculiar frog with a tiny, cone-shaped head and four short but robust limbs projecting from a large, somewhat flattened, globular body (75–85 mm SVL). Its skull lacks palatines and has a single frontoparietal. The facial nerve exits through the anterior acoustic foramen in the auditory capsule; the trigeminal and facial nerve ganglia fuse to form a prootic ganglion. The vertebral column possesses eight presacral notochordal vertebrae, and all are opisthocoelous. The transverse processes of the sacral vertebra are broadly expanded, and this vertebra has a bicondylar articulation with the urostyle. Postmetamorphs have no dorsal ribs on the presacral vertebrae. The pectoral girdle is arciferal and lacks a sternum. The fibulare and tibiale are fused at their proximal and distal ends. No intercalary cartilage occurs between the terminal and penultimate phalanges of the digits, and the tips of the terminal phalanges are blunt. The tadpole lacks keratinized mouthparts, and the left and right branchial chambers are emptied by separate spiracles.

Biology: The globular microcephalic body form of *R. dorsalis* denotes a fossorial existence and a diet of soft-bodied, subterranean arthropods such as termites and ant larvae. Numerous morphological features permit this frog to capture its prey in subterranean burrows. The snout is covered with an epidermal amour, and the lips have an unusual double closure that is sealed by secretions from submandibular glands. Muscles act to stiffen the tongue so that it can be projected straight out from the mouth, rather than flipped outward in typical frog fashion. It digs with the hindlimbs; its spades are on the inside edge of each hind foot. *Rhinophrynus dorsalis* breeds in temporary pools, where males call while floating. Amplexus is inguinal, and females deposit several thousand eggs that sink to the bottom. Duration of the tadpole stage is unknown; tadpoles swim in aggregations of 50 to several hundred individuals. Although the tadpoles are primarily filter feeders, the lower jaw develops early, allowing them to feed on larger prey. Some wild-caught individuals contained conspecific tadpoles in their intestines.

Alytidae

Midwife Toads and Painted Frogs

Classification: Amphibia; Lissamphibia; Anura; Lalagobatrachia; Sokolanura.

Sister taxon: Bombinatoridae.

Content: Two genera, *Alytes* and *Discoglossus*, with 5 and 6 species, respectively.

Distribution: Western and central Europe, northwestern Africa, Israel, and possibly Syria (Fig. 17.5).

Characteristics: Alytids are moderate-sized frogs, with adults ranging from 40 to 55 mm SVL in *Alytes* and 60 to 75 mm SVL in *Discoglossus*. The skull lacks palatines and has a pair of frontoparietals. The facial nerve exits through the facial foramen anterior to the auditory

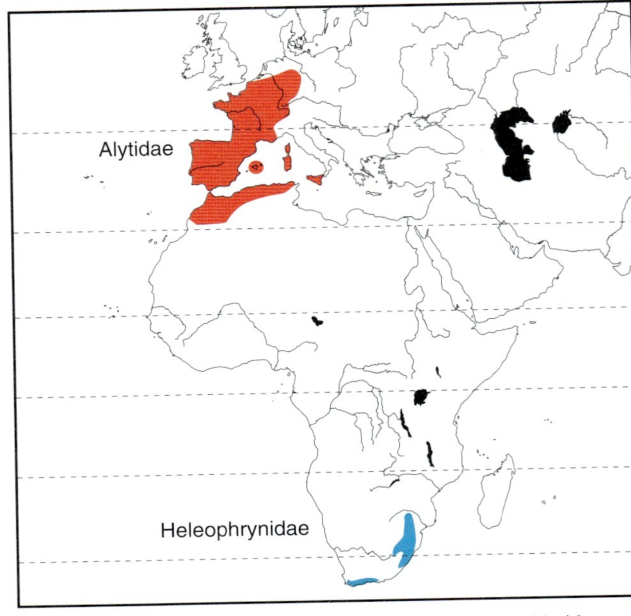

FIGURE 17.5 Geographic distributions of the extant Alytidae and Heleoprynidae.

capsule; the trigeminal and facial nerve ganglia are separate. The vertebral column possesses eight presacral stegochordal vertebrae, and all are opisthocoelous. The transverse processes of the sacral vertebra are broadly expanded, and this vertebra has a bicondylar articulation with the urostyle. Postmetamorphs have dorsal ribs on the second through fourth presacral vertebrae. The pectoral girdle is arciferal with a distinct sternum. The fibulare and tibiale are fused at their proximal and distal ends. No intercalary cartilage occurs between the terminal and penultimate phalanges of each digit, and the tips of the terminal phalanges are blunt to pointed. The larvae have keratinized mouthparts and two small, fused spiracular tubes with a single anteromedial spiracle.

Biology: The species of *Alytes* are fossorial and terrestrial frogs that live in wooded areas as well as more open habitats near ponds and streams. They are nocturnal and during the day hide beneath rocks and logs. They dig their own burrows, constructing a system of underground tunnels. They burrow using the forelimbs, and they sometimes do push-ups to pack the substrate against the tunnel with their heads. Forward burrowing is typical and presumably protects the egg strings wrapped around the hind legs of the male during parental care. *Discoglossus* is more aquatic and occurs mainly at the edge of fast-flowing streams with rocky substrates. Males of both genera have voices, and amplexus is inguinal. During one season, females of *Discoglossus pictus* deposit about 500 to 1000 eggs singly on vegetation or in small clusters on the stream bottom. Development to metamorphosis occurs in 3 to 8 weeks, depending on water temperature. In *Alytes*, males fertilize a clutch of 20 to 100 egg strings during amplexus, which are then wrapped around their hind legs. The eggs are carried by the male until the larvae are about to hatch (3 weeks in *A. cisternasii*; 4 to 5 weeks in *A. obstetricans*: Fig. 17.3), and then the male returns to water, allowing the tadpoles to swim free. The tadpoles overwinter and metamorphose in late spring and early summer.

Bombinatoridae

Fire-Bellied Toads

Classification: Amphibia; Lissamphibia; Anura; Lalagobatrachia; Sokolanura.

Sister taxon: Alytidae.

Content: Two genera, *Barbourula* and *Bombina*, with 2 and 8 species, respectively.

Distribution: Europe, southern China, Borneo, and Philippine Islands (Fig. 17.2).

Characteristics: *Bombina* contains moderate-sized (40–80 mm adult SVL) toadlike frogs; *Barbourula* is somewhat larger (60–100 mm adult SVL). The skull lacks palatines and has paired frontoparietals. The facial nerve exits through the facial foramen anterior to the auditory capsule; the trigeminal and facial nerve ganglia are separate. The vertebral column possesses eight presacral stegochordal vertebrae, and all are opisthocoelous. The transverse processes of the sacral vertebra are broadly expanded, and this vertebra has a bicondylar articulation with the urostyle. Postmetamorphs have dorsal ribs on the second through fourth presacral vertebrae, articulating with transverse processes in *Barbourula* and fused in *Bombina*. The pectoral girdle is arciferal with a distinct sternum. The fibulare and tibiale are fused at their proximal and distal ends. No intercalary cartilage occurs between the terminal and penultimate phalanges of each digit, and the tips of the terminal phalanges are blunt to pointed. The larvae have keratinized mouthparts and two small, fused spiracular tubes with a single anteromedial spiracle.

Biology: The fire-bellied toads, *Bombina*, are mainly diurnal and aquatic, spending much of their time in slow-moving waters of marshes and ponds. Although dark and camouflaged above, they are readily visible because they are active in open areas. Warty, glandular skin with toxic secretions protects them from many predators, and when attacked, they advertise their toxicity by an Unken reflex (Fig. 11.16). This arching reflex displays their bright undersides of black mottling on yellow, orange, or red backgrounds. European *Bombina* breeds from late April to midsummer; males call day and night, although most reproductive activity occurs in the early evening. Amplexus is inguinal. Females deposit 60 to 200 eggs in numerous small egg clusters that are attached to either vegetation or the substrate. The embryos hatch within 4 to 10 days, and the tadpoles develop rapidly, usually metamorphosing in 35 to 45 days except in cooler localities. Little is known of the biology of *Barbourula* (Fig. 17.6). They are cryptic and highly, although not exclusively, aquatic frogs. Juveniles and adults live in small, stone-bottomed streams in mountainous areas; juveniles live in shallow pools and seldom emerge, whereas adults occupy rock crevices or sit beneath rocks at the water–air interface. Their hands and feet are fully webbed. Females produce approximately 70 to 80 moderately large, weakly pigmented ova; presumably, the eggs are laid in the water beneath rocks.

Pelobatidae

Western Palearctic Spadefoot Toads

Classification: Amphibia; Lissamphibia; Anura; Sokolanura; Pelobatoidea.

Sister taxon: Megophryidae.

Content: One genus, *Pelobates*, with 4 species.

Distribution: Europe, western Asia, and northwestern Africa (Fig. 17.7).

Characteristics: Pelobatids are moderate-sized frogs (50–80 mm adult SVL) with squat toadlike bodies and

FIGURE 17.6 Representative frogs. Clockwise from upper left: Busuanga jungle frog *Barbourula busuangensis*, Bombinatoridae (R. M. Brown); common Eurasian spadefoot toad *Pelobates fuscus*, Pelobatidae (C. Mattison); long-footed toad frog *Megophrys longipes*, Megophryidae (L. L. Grismer); parsley frog *Pelodytes punctatus*, Pelodytidae (C. Raxworthy).

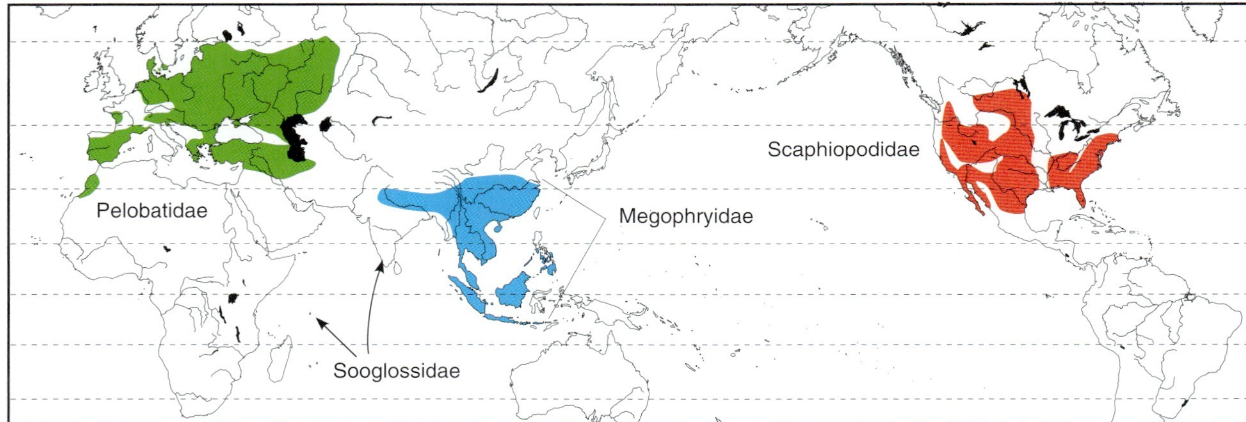

FIGURE 17.7 Geographic distributions of the extant Pelobatidae, Megophryidae, Scaphiopodidae, and Sooglossiade.

warty although soft skin (Fig. 17.6). The colloquial name is derived from the large, keratinous-edged, crescentic tubercle on the outer edge of each hind foot. The pelobatid skull lacks palatines and has a pair of frontoparietals. The facial nerve exits through the anterior acoustic foramen in the auditory capsule; the trigeminal and facial nerve ganglia fuse to form a prootic ganglion. The vertebral column possesses eight presacral stegochordal vertebrae, and all are amphicoelous. The transverse processes of the sacral vertebra are broadly expanded, and this vertebra has a bicondylar articulation with the urostyle. Postmetamorphs have no dorsal ribs on the presacral vertebrae. The pectoral girdle is arciferal with a distinct sternum. The fibulare and tibiale are fused at their proximal and distal ends. No intercalary cartilage occurs between the terminal and penultimate phalanges of each digit, and the tips of the terminal phalanges are blunt. The larvae have keratinized mouthparts, and the left and right branchial chambers fuse behind the heart and are emptied by a spiracle on the left side at midbody.

Biology: Spadefoots are fossorial (subterranean) and burrow backward with an alternating shuffling movement of the hindlimbs. Spadefoots spend much of their lives in burrows. *Pelobates* breeds in spring and has a 2- to 3-month larval period (*Pelobates syriacus*) or has slower development and overwinters for 1 to 3 years, depending on the local climate (*Pelobates fuscus*). This extended larval period results in large tadpoles (to 180 mm total length), the largest tadpoles among European anurans.

Megophryidae

Asian Toad Frogs

Classification: Amphibia; Lissamphibia; Anura; Sokolanura; Pelobatoidea.

Sister taxon: Pelobatidae.

Content: Ten genera, *Borneophrys*, *Brachytarsophrys*, *Leptobrachella*, *Leptobrachium*, *Leptolalax*, *Megophrys*, *Ophryophryne*, *Oreolalax*, *Scutiger*, and *Xenophrys*, with 137 species.

Distribution: Subtropical and tropical Asia from Nepal to the Philippines and Greater Sunda Islands (Fig. 17.7).

Characteristics: Megophryids are small to large frogs (15–120 mm adult SVL). The skull lacks palatines and has paired frontoparietals. The facial nerve exits through the anterior acoustic foramen in the auditory capsule; the trigeminal and facial nerve ganglia fuse to form a prootic ganglion. The vertebral column possesses eight presacral stegochordal vertebrae, and all are amphicoelous. The transverse processes of the sacral vertebra are moderately expanded, and this vertebra has a single condylar articulation or is fused with the urostyle. Postmetamorphs have no dorsal ribs on the presacral vertebrae. The pectoral girdle is arciferal with a distinct sternum. The fibulare and tibiale are fused at their proximal and distal ends. No intercalary cartilage occurs between the terminal and penultimate phalanges of each digit, and the tips of the terminal phalanges are blunt to pointed. The larvae have keratinized mouthparts, and the left and right branchial chambers fuse behind the heart and are emptied by a spiracle on the left side at midbody.

Biology: Adult megophryids primarily dwell in leaf litter of tropical forests, and they breed in flowing water of streams (Fig. 17.6). Amplexus is inguinal; eggs are laid in water and hatch into free-living larvae. Among the 10 genera of megophryids, tadpoles in 5 genera, *Leptobrachella, Leptolalax, Megophrys, Ophryophryne*, and *Xenophrys*, are unusual in having well-developed, supernumerary bony vertebrae in their tails. These tadpoles are typically fossiorial in streams and burrow into hard or rocky substrates to avoid fast-moving water. Attachment of muscle to the caudal skeletal elements lends extra strength to the tail, helping facilitate movement into the substrate. Tadpoles in other genera in the family have a more typical, globose body form. These tadpoles also live in streams, but they avoid rushing water and instead inhabit quiet pools or edges of the shore. Many of these tadpoles have funnel-shaped mouthparts. Several species of *Leptobrachium* are called "moustache frogs" because they develop large spines on the upper jaw during the breeding season. These males call beneath large stone underwater and appear to guard their egg clutches. The moustache spines may aid in warding off predators.

Pelodytidae

Parsley Frogs

Classification: Amphibia; Lissamphibia; Anura; Mesobatrachia; Pelodytoidea.

Sister taxon: Scaphiopodidae.

Content: One genus, *Pelodytes*, with 3 species.

Distribution: Southwestern Europe and the Caucasus Mountains in southwestern Asia (Fig. 17.4).

Characteristics: The three species of *Pelodytes* (Fig. 17.6) are moderately small frogs, 30 to 55 mm SVL. The skull lacks palatines and has a pair of frontoparietals. The facial nerve exits through the anterior acoustic foramen in the auditory capsule; the trigeminal and facial nerve ganglia fuse to form a prootic ganglion. The vertebral column possesses eight presacral stegochordal vertebrae, and all are amphicoelous. The transverse processes of the sacral vertebra are broadly expanded, and this vertebra has a bicondylar articulation with the urostyle. Postmetamorphs have no dorsal ribs on the presacral vertebrae. The pectoral girdle is arciferal with a distinct sternum. The fibulare and tibiale are fused along their entire lengths. No intercalary cartilage occurs between the terminal and penultimate phalanges of each digit, and the tips of the terminal phalanges are blunt to pointed. The larvae have keratinized mouthparts, and the left and right branchial chambers fuse behind the heart and are emptied by a spiracle on the left side at midbody.

Biology: Pelodytids are terrestrial, living in moist habitats from sea level to midmountain elevations (2300 m). They are nocturnal until the breeding season, when reproductive activity occurs throughout the day and night. *Pelodytes ibericus* prefers open areas and breeds in ponds and flooded fields. *Pelodytes punctatus* males call primarily from submerged positions. Amplexus is inguinal and occurs in the water. Females of *Pelodytes punctatus* lay 1000 to 1600 eggs, whereas females of *P. caucasicus* deposit about 40 eggs. In both, the eggs hatch quickly, and metamorphosis occurs in 75 to 80 days.

Scaphiopodidae

Nearctic Spadefoot Toads

Classification: Amphibia; Lissamphibia; Anura; Mesobatrachia; Pelodytoidea.

Sister taxon: Pelodytidae.

Content: Two genera, *Scaphiopus* and *Spea*, with 3 and 4 species, respectively.

Distribution: Southern Canada, western and central United States, to temperate southern Mexico (Fig. 17.7).

Characteristics: Pelobatids are moderate-sized frogs (50–80 mm adult SVL) with squat toadlike bodies and warty although soft skin. The colloquial name is derived from the large, keratinous-edged, crescentic tubercle on the outer edge of each hind foot. The scaphiopodid skull lacks palatines and has a pair of frontoparietals. The facial nerve exits through the anterior acoustic foramen in the auditory capsule; the trigeminal and facial nerve ganglia fuse to form a prootic ganglion. The vertebral column possesses eight presacral stegochordal vertebrae, and all are amphicoelous. The transverse processes of the sacral vertebra are broadly expanded, and this vertebra has a bicondylar articulation with the urostyle. Postmetamorphs have no dorsal ribs on the presacral vertebrae. The pectoral girdle is arciferal with a distinct sternum. The fibulare and tibiale are fused at their proximal and distal ends. No intercalary cartilage occurs between the terminal and penultimate phalanges of each digit, and the tips of the terminal phalanges are blunt. The larvae have keratinized mouthparts, and the left and right branchial chambers fuse behind the heart and are emptied by a spiracle on the left side at midbody.

Biology: Like Palearctic spadefoots, Nearctic spadefoots are fossorial (subterranean) and burrow backward with an alternating shuffling movement of the hindlimbs (Fig. 17.8). They also spend much of their lives in burrows, but contrary to the general misconception that Nearctic species emerge only for reproduction, they regularly forage on the surface in late spring and summer

FIGURE 17.8 Representative frogs. Clockwise from upper left: Eastern spadefoot toad *Scaphiopus holbrookii* (J. P. Caldwell); cape ghost frog *Heleophryne purcelli*, Heleophrynidae (J. Visser, courtesy of the Natural History Museum, The University of Kansas); Seychelles rock frog *Sooglossus thomasseti*, Sooglossidae (G. R. Zug); Seychelles frog *Sooglossus sechellensis*, Sooglossidae (G. R. Zug).

during damp evening hours. This misconception arises from the explosive reproductive habitats of *Scaphiophus* and *Spea*, and their generally drier habitat preference. In these species, reproduction can occur on any warm evening with heavy rains from early spring to late summer. As temporary ponds form, the males establish a raucous chorus and are soon joined by the females; inguinal amplexus and egg deposition soon follow. Most often, a local population's annual reproduction is completed in a single short period; individuals may call and breed even during the day after the first heavy rains of the season. The larval period can be as rapid as 6 to 8 days in *Scaphiopus couchii* but is usually 24 to 32 days.

Heleophrynidae

Ghost Frogs

Classification: Amphibia; Lissamphibia; Anura; Neobatrachia.

Sister taxon: Phthanobatrachia (clade containing all remaining Neobatrachia).

Content: One genus, *Heleophryne,* with 6 species.

Distribution: Mountainous areas of the Cape and Transvaal regions of South Africa (Fig. 17.5).

Characteristics: Heleophrynids are moderately small- to medium-sized (35–65 mm adult SVL) tree frog–like anurans with expanded digit tips (Fig. 17.8). Their skull has paired palatines and frontoparietals. The facial nerve exits through the anterior acoustic foramen in the auditory capsule; the trigeminal and facial nerve ganglia fuse to form a prootic ganglion. The vertebral column possesses eight presacral notochordal vertebrae, and all are amphicoelous. The transverse processes of the sacral vertebra are not expanded, and this vertebra has a bicondylar articulation with the urostyle. Postmetamorphs have no dorsal ribs on the presacral vertebrae. The pectoral girdle is arciferal with a distinct sternum. The fibulare and tibiale are fused at their proximal and distal ends. No intercalary cartilage occurs between the terminal and penultimate phalanges of each digit, and the tips of the terminal phalanges are blunt to slightly flared. The larvae have keratinized mouthparts but lack jaw sheaths, and the left and right branchial chambers fuse behind the heart and are emptied by a spiracle on the left side at midbody.

Biology: The six species of heleophrynids occur only in swift-flowing, rocky streams in isolated mountain gorges. This area is being converted to housing developments, thus threatening the frogs' habitats and survival. Adults (Fig. 17.8) are active mainly at night as sit-and-wait predators in the splash zone of the streams. Their expanded digital pads allow them to move easily and quickly over slippery rocks. The reproductive biology is largely unknown. Unlike many torrent-inhabiting frogs, the males call (*H. purcelli*) and inguinal amplexus is preceded by an elaborate courtship that includes tactile behavior between the male and female. A few large unpigmented eggs are attached beneath rocks in the streams. The tadpoles have a large oral disc, permitting them to cling to rock surfaces while feeding. Development is prolonged, and metamorphosis may occur 1 to 2 years after hatching.

Sooglossidae

Seychelles Frogs

Classification: Amphibia; Lissamphibia; Anura; Neobatrachia; Phthanobatrachia; Hyloides.

Sister taxon: Notogaeanura (clade containing all remaining hyloids).

Content: Three genera, *Nasikabatrachus, Sechellophryne,* and *Sooglossus,* with 1, 2, and 2 species, respectively.

Distribution: Granitic islands of the Seychelles Islands in the Indian Ocean (*Sechellophryne* and *Sooglossus*); Western Ghats of southern India (*Nasikabatrachus*) (Fig. 17.7).

Characteristics: Sooglossids range from tiny (10–14 mm adult SVL, *Sechellophryne*) to moderate-sized (45–55 mm in *Sooglossus thomasseti*; Fig. 17.8) to large (52–90 mm in *Nasikabatrachus*) terrestrial frogs. The skull has paired palatines and frontoparietals. The facial nerve exits through the anterior acoustic foramen in the auditory capsule; the trigeminal and facial nerve ganglia fuse to form a prootic ganglion. The vertebral column possesses eight presacral holochordal vertebrae, and all are procoelous. The transverse processes of the sacral vertebra are moderately expanded, and this vertebra has a bicondylar articulation with the urostyle. Postmetamorphs have no dorsal ribs on the presacral vertebrae. The pectoral girdle is pseudoarciferal to arciferal with a distinct sternum that is cartilaginous or ossified. The fibulare and tibiale are fused at their proximal and distal ends. No intercalary cartilage occurs between the terminal and penultimate phalanges of each digit, and the tips of the terminal phalanges are rounded to pointed. The larvae have keratinized mouthparts, and the left and right branchial chambers fuse behind the heart and are emptied by a spiracle on the left side at midbody.

Biology: On the Seyschelles Islands, sooglossids are inhabitants of moist forests and are nocturnal. *Sechellophryne gardineri* and *Sooglosssus sechellensis* live principally in the forest-floor litter, although they occasionally hide in axils of tree ferns. *Sooglossus thomasseti* is also a forest-floor resident and commonly is found along streams and in rivulets. Males call individually, not in choruses, but both females and males lack tympana. Amplexus is inguinal, and egg deposition is terrestrial. *Sechellophryne gardineri* females deposit eggs beneath leaves or rocks and stay with them for 3 to 4 weeks as they undergo direct development and hatch into tiny froglets. Females of *Sooglossus sechellensis* (Fig. 17.8) also deposit eggs beneath forest-floor debris and attend them for 2 to 3 weeks; the eggs hatch into nonfeeding tadpoles that wriggle onto the female's back, where they remain until metamorphosis. The reproductive behavior of *Sooglossus thomasseti* is similar in that females deposit eggs in a terrestrial nest, and nonfeeding tadpoles undergo direct development. *Nasikabatrachus sahyadrensis* is a burrowing frog, only recently discovered. This unusual frog apparently lives most of its life deep underground but emerges to breed during monsoon rains; pairs have inguinal amplexus. It has a robust body and short legs, typical of other but unrelated burrowing frogs. The small, basally attached tongue extends through a buccal grove and easily penetrates termite tunnels. The relationship of this frog to other sooglossids on the Seyschelles Islands indicates that this lineage may have been present prior to the breakup of Gondwanaland more than 130 million years ago.

Calyptocephalellidae

Chilean Toads

Classification: Amphibia; Lissamphibia; Anura; Neobatrachia; Australobatrachia.

Sister taxon: Myobatrachioidea.

Content: Two genera, *Calyptocephalella* and *Telmatobufo*, with 1 and 3 species, respectively.

Distribution: Mountains of central Chile (Fig. 17.4).

Characteristics: The sternum is cartilaginous. The presacral vertebrae lack a bony or cartilaginous shield; the transverse processes of the anterior presacral vertebrae are long and not expanded. The tips of the terminal phalanges are blunt, pointed, or T-shaped. Amplexus is axillary.

Biology: The species of *Telmatobufo* live in mountain streams. Their tadpoles have morphological adaptations for living in fast-moving water, including sucker-like mouths and muscular tails. *Calyptocephaella* breeds in ponds and lagoons, and its deep-bodied tadpoles have high dorsal and ventral fins.

Limnodynastidae

Australian Ground Frogs

Classification: Amphibia; Lissamphibia; Anura; Neobatrachia; Hyloides; Australobatrachia; Myobatrachoidea.

Sister taxon: Myobatrachidae.

Content: Eight genera, *Adelotus*, *Heleioporus*, *Lechriodus*, *Limnodynastes*, *Neobatrachus*, *Notaden*, *Opisthodon*, and *Philoria*, with 44 species.

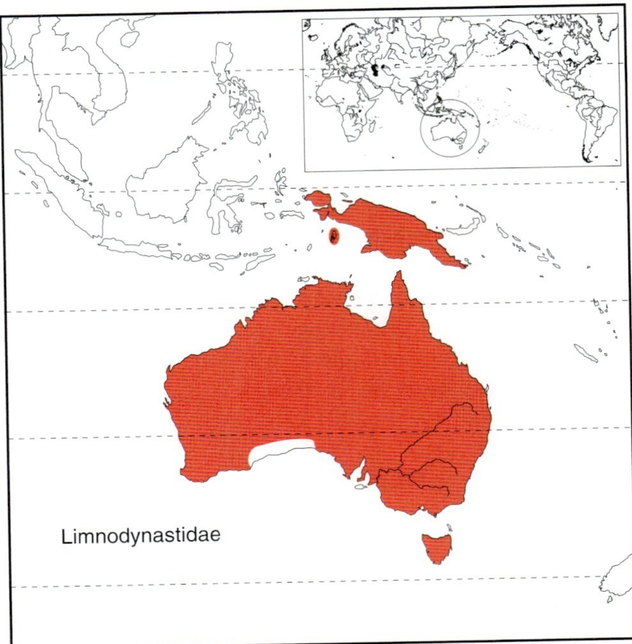

FIGURE 17.9 Geographic distribution of the extant Limnodynastidae.

Distribution: *Limnodynastes* and *Lechriodus* occur in both Australia and New Guinea (Fig. 17.9); all other genera occur in Australia.

Characteristics: Limnodynastids are toadlike terrestrial frogs that range from small to large (SVL, 100 mm in *Heleioporus australiacus*, the giant burrowing frog) in size (Fig. 17.10). The skull has paired palatines and frontoparietals. The facial nerve exits through the anterior acoustic foramen in the auditory capsule; the trigeminal and facial nerve ganglia fuse to form a prootic ganglion. The vertebral column possesses eight presacral holochordal vertebrae, and all are procoelous. The transverse processes of the sacral vertebra are cylindrical, and this vertebra has a bicondylar articulation with the urostyle. Postmetamorphs have no dorsal ribs on the presacral vertebrae. The pectoral girdle is arciferal with a distinct sternum. The fibulare and tibiale are fused at their proximal and distal ends. No intercalary cartilage occurs between the terminal and penultimate phalanges of each digit, and the tips of the terminal phalanges are blunt to pointed. The larvae have keratinized mouthparts, and the left and right branchial chambers fuse behind the heart and are emptied by a spiracle on the left side at midbody.

Biology: Limnodynastids live in a variety of habitats from dry scrub and savannas to marshes, stream or lake shores, and floor of the rain forest. All species are terrestrial, although individuals occasionally forage near the ground in the foliage of shrubs. The species living in the drier habitats use burrows to escape the heat and aridity of daytime and drought conditions. In the latter situation, the burrow is plugged and the frog estivates until rains arrive. In wet periods, the frogs emerge in the evening to

FIGURE 17.10 Representative frogs. Clockwise from upper left: Burrowing frog *Limnodynastes ornatus*, Limnodynastidae (S. J. Richards); common eastern froglet *Crinia signifera*, Myobatrachidae (S. Wilson); Spix's horned tree frog *Hemiphractus scutatus*, Hemiphractidae (J. P. Caldwell); Brazilian rain frog *Pristimantis* sp., Brachycephalidae (J. P. Caldwell).

feed. Reproduction is usually associated with heavy rains. Males attract females by vocalizing; *Heleioporus* and *Neobatrachus* lack vocal sacs yet produce loud calls. Amplexus is inguinal. *Neobatrachus* and *Notaden* produce strings of eggs in the water. The remainder of the limnodynastids deposits eggs in foam nests that are produced by cloacal secretions from the male and female. The foam nests, depending upon species, are deposited in burrows, on shorelines, or floating on the water.

Myobatrachidae

Australian Toadlets and Turtle Frogs

Classification: Amphibia; Lissamphibia; Anura; Neobatrachia; Hyloides; Australobatrachia; Myobatrachoidea.

Sister taxon: Limnodynastidae.

Content: Thirteen genera, *Arenophryne, Assa, Crinia, Geocrinia, Metacrinia, Mixophyes, Myobatrachus, Paracrinia, Pseudophyrne, Rheobatrachus, Spicospina, Taudactylus,* and *Uperoleia*, with 82 species.

Distribution: Australia and New Guinea (Fig. 17.11).

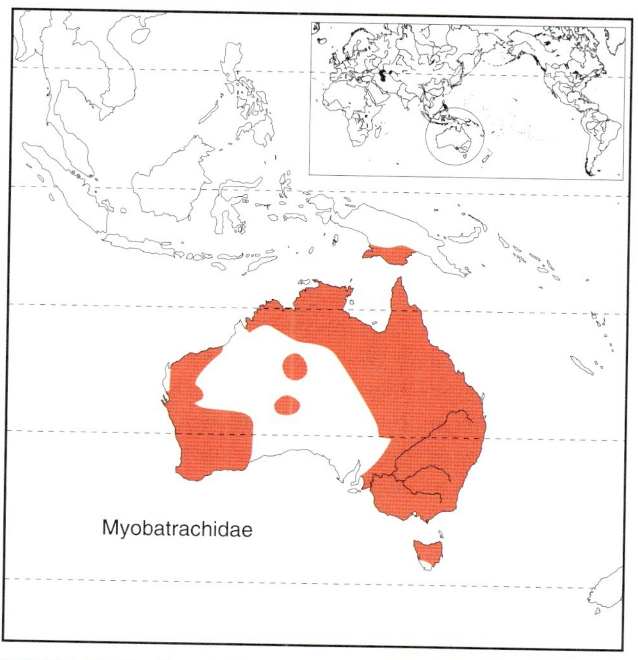

FIGURE 17.11 Geographic distribution of the extant Myobatrachidae.

Characteristics: Myobatrachids consist predominantly of small frogs (13–36 mm adult SVL), with the exception of *Myobatrachus* (34–50 mm SVL) and the enigmatic *Rheobatrachus* (33–79 mm). Among the small taxa, body form is either typical frog or toadlike, in contrast to the obese, molelike *Myobatrachus*. The myobatrachid skull has paired palatines and frontoparietals. The facial nerve exits through the anterior acoustic foramen in the auditory capsule; the trigeminal and facial nerve ganglia fuse to form a prootic ganglion. The vertebral column possesses eight presacral holochordal vertebrae, and all are procoelous. The transverse processes of the sacral vertebra are moderately expanded, and this vertebra has a bicondylar articulation with the urostyle. Postmetamorphs have no dorsal ribs on the presacral vertebrae. The pectoral girdle is arciferal with a distinct sternum. The fibulare and tibiale are fused at their proximal and distal ends. No intercalary cartilage occurs between the terminal and penultimate phalanges of each digit, and the tips of the terminal phalanges are usually blunt. Tadpoles of most species have keratinized mouthparts, and the left and right branchial chambers fuse behind the heart and are emptied by a spiracle on the left side at midbody.

Biology: Like the limnodynastids, all myobatrachids are terrestrial frogs that occupy a diverse set of habitats from dry grassland, scrub, and savannas to marshes, stream or lake shores, and the floor of the rain forest. *Uperoleia* is the most speciose taxon with 25 species and occurs in grassland and dry forest habitats around the periphery of Australia, although individual species have small geographic ranges. Reproductive data are unknown for most species of *Uperoleia,* but presumably all deposit eggs in water. Presumably all myobatrachids have inguinal amplexus. *Crinia* is also speciose, with 15 species, and broadly distributed but occurs mainly in moist habitats (Fig. 17.10). Some of the unusual reproductive behaviors of Australian anurans occur among the myobatrachids. For example, two species of *Geocrinia* deposit yolk-filled eggs in moist leaf litter or grass on land. Embryos develop to an advanced tadpole stage before hatching from the egg capsules, which occurs in response to flooding of the clutch. Tadpoles are washed into nearby pools where they continue development for several months before metamorphosis. *Arenophryne* and *Myobatrachus* burrow headfirst in sandy soils distant from water; both lay a few large eggs, buried deep in the soil, which undergo direct development and metamorphose into burrowing froglets. *Assa* lays 10 to 11 eggs in terrestrial but boggy situations; the male attends the developing egg mass. When the larvae hatch, he sits in the egg mass and the larvae wriggle onto him and into his inguinal tadpole pockets, emerging about 2 months later as froglets. Perhaps the strangest of all are the stomach- or gastric-brooding *Rheobatrachus.* After the eggs are fertilized, the female swallows the eggs or tadpoles (which stage remains unknown!). The eggs or embryos produce prostaglandin E_2, which blocks the production of stomach acids. The embryos develop in the female's stomach, and froglets emerge from the female's mouth in about 2 months.

Hemiphractidae

Horned Frogs

Classification: Amphibia; Lissamphibia; Anura; Neobatrachia; Hyloides; Nobleobatrachia.

Sister taxon: Meridianura (clade containing Brachycephalidae and Cladophrynia).

Content: One genus, *Hemiphractus*, with 6 species.

Distribution: Panama, through Colombia and Ecuador to Bolivia, including the northern Amazon basin in Brazil (Fig. 17.12).

Characteristics: The triangular skull is strongly ossified and exostosed. The pupils are horizontal. Of the superficial mandibular musculature, the interhyoideus does not extend posteriorly beyond the mandibles, and the intermandibular muscle is a single sheet bearing a large, median aponeurosis. This condition reflects the posterior position of the jaw articulation. The vertebrae have enlarged neural spines.

Biology: Frogs in the genus *Hemiphractus* are robust bodied and live on the ground, where they are inconspicuous in the leaf litter (Fig. 17.10). Individuals may be found less than a meter above ground perched on small shrubs at night. All species of *Hemiphractus* have direct development. Eggs are brooded openly on the dorsum of the female, not in pockets or brood chambers, and hatch as

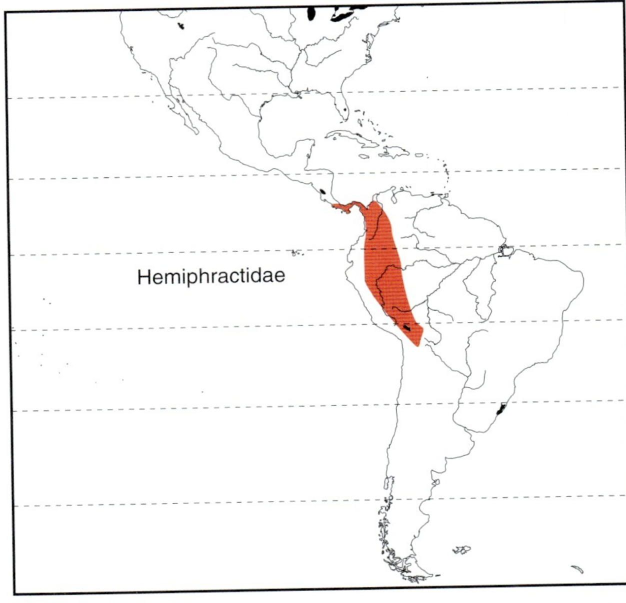

FIGURE 17.12 Geographic distribution of the extant Hemiphractidae.

froglets. Developing embryos are completely enclosed in large, membranous gills. Vocal apertures are lacking, although males produce calls.

Brachycephalidae

Rain Frogs, Three-Toed Toadlets, and Others

Classification: Amphibia; Lissamphibia; Anura; Neobatrachia; Hyloides; Nobleobatrachia; Meridianura.

Sister taxon: Cladophrynia (clade containing Cryptobatrachidae and Tinctanura).

Content: Seventeen genera, *Adelophryne, Atopophrynus, Barycholos, Brachycephalus, Craugastor, Dischidodactylus, Eleutherodactylus, Euparkerella, Geobatrachus, Holoaden, Ischnocnema, Limnophrys, Oreobates, Phrynopus, Phyllonastes, Phyzelaphryne,* and *Pristimantis*, with 823 species.

Distribution: Tropics and subtropics from southwestern United States and Antilles to southern South America (Fig. 17.13).

Characteristics: Brachycephalids range from tiny to medium-sized frogs. Among the smallest is *Brachycephalus*, in which adults are 12 to 18 mm SVL. The skull has a pair of palatines and frontoparietals. The facial nerve exits through the anterior acoustic foramen in the auditory capsule; the trigeminal and facial nerve ganglia fuse to form a prootic ganglion. The vertebral column possesses seven presacral holochordal vertebrae, and all are procoelous. The transverse processes of the sacral vertebra are moderately expanded, and this vertebra has a bicondylar articulation with the urostyle. Postmetamorphs have no dorsal ribs on the presacral vertebrae. The pectoral girdle is arciferal and lacks a sternum. The fibulare and tibiale are fused at their proximal and distal ends. No intercalary cartilage occurs between the terminal and penultimate phalanges of each digit, and the tips of the terminal phalanges are blunt to pointed. *Brachycephalus* has reduced numbers of digits: two or three functional fingers are present on each forefoot and three toes on the hind foot.

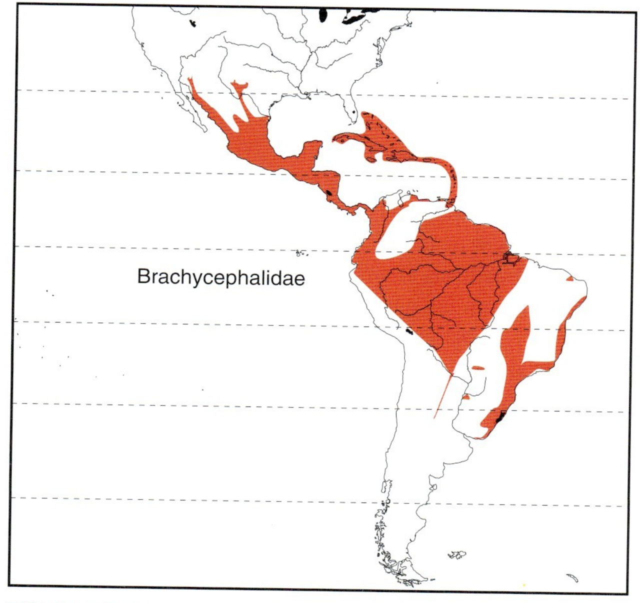

FIGURE 17.13 Geographic distribution of the extant Brachycephalidae.

Biology: Brachycephalids are leaf litter and arboreal inhabitants of rain forests (Fig. 17.10). *Brachycephalus ephippium* is a tiny, bright orange or yellow frog that lives in leaf litter of seasonal rain forests. The bright color is likely an aposematic warning of its toxic skin secretions, a tetrodotoxin-like compound. Males give a low buzzlike call from future nest sites beneath cover. Amplexus is initially inguinal in *B. ephippium* but shifts to a more axillary position as the female deposits the eggs, which she later coats with soil particles, perhaps for camouflage or to reduce desiccation. Most species of *Eleutherodactylus, Craugastor*, and *Pristimantis* are cryptically colored and frequently arboreal in rain forests, but other brachycephalids, e.g. *Ischnocnema*, are ground-dwelling frogs. As far as known, development is direct with few exceptions. Viviparity has evolved in the Puerto Rican species *Eleutherodactylus jasperi*, and it and *E. coqui* have internal fertilization. Presumably sperm transfer occurs via cloacal apposition.

Cryptobatrachidae

Stefanias and Others

Classification: Amphibia; Lissamphibia; Anura; Neobatrachia; Nobleobatrachia; Meridianura.

Sister taxon: Tinctanura (clade containing Amphignathodontidae and Athesphatanura).

Content: Two genera, *Cryptobatrachus* and *Stefania*, with 3 and 18 species, respectively.

Distribution: Northern South America including parts of Colombia, Venezuela, Guyana, and Brazil (Fig. 17.14).

Characteristics: The terminal phalanges are claw-shaped, and intercalary elements are present. Of the superficial mandibular musculature, in all *Cryptobatrachus* and two *Stefania* examined, the interhyoideus extends slightly posteriorly beyond the mandibles, and the intermandibular muscle is a single sheet bearing a large, median raphe. Another species of *Stefania* differs from this arrangement.

Biology: All species of *Cryptobatrachus* and *Stefania* have direct development. Eggs are brooded openly on the dorsum of the female, not in pockets or brood chambers, and hatch as froglets. Developing embryos are partially (*Cryptobatrachus*) or completely (*Stefania*) enclosed in large, membranous gills. Vocal apertures are lacking, and neither *Cryptobatrachus* nor *Stefania* are known to call.

FIGURE 17.14 Geographic distribution of the extant Cryptobatachidae.

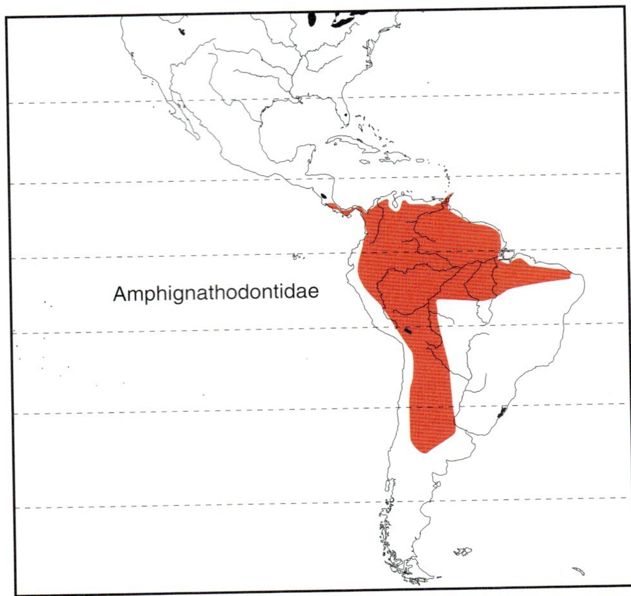

FIGURE 17.15 Geographic distribution of the extant Amphignathodontidae.

Amphignathodontidae

Marsupial Frogs

Classification: Amphibia; Lissamphibia; Anura; Neobatrachia; Nobleobatrachia; Tinctanura.

Sister taxon: Athesphatanura (clade containing Hylidae and Leptodactyliformes).

Content: Two genera, *Flectonotus* and *Gastrotheca*, with 5 and 53 species, respectively.

Distribution: Costa Rica, Panama, Trinidad, Tobago, and widespread throughout much of South America to northern Argentina (Fig. 17.15).

Characteristics: Of the superficial mandibular musculature, in all *Flectonotus* and many *Gastrotheca* examined, the interhyoideus extends slightly posteriorly beyond the mandibles, and the intermandibular muscle is a single sheet bearing a large, median raphe. In other species of *Gastrotheca*, the intermandibular muscle is differentiated by an apical element.

Biology: *Flectonotus* and *Gastrotheca* (Fig. 17.16) have a specialized dorsal pouch in which eggs are carried. Developing embryos are partially (*Flectonotus*) or completely (*Gastrothecaa*) enclosed in large, membranous gills. The gills develop an extensive capillary net that acts as a placenta for the maternal transfer of gases, water, and nutrients. In *Flectonotus* and some *Gastrotheca*, the eggs hatch as advanced tadpoles. In other *Gastrotheca*, the eggs are held in the pouch throughout the entire developmental period, and froglets hatch and emerge from the pouch after several months. All *Flectonotus* and *Gastrotheca* have openings to the vocal sacs, and males call.

Hylidae

Ameroaustralian Tree Frogs

Classification: Amphibia; Lissamphibia; Anura; Neobatrachia; Nobleobatrachia; Athesphatanura.

Sister taxon: Leptodactyliformes (clade containing Diphyabatrachia and Chthonobatrachia).

Content: Three subfamilies, Hylinae, Pelodryadinae, and Phyllomedusinae, with 844 species.

Distribution: North and South America, the West Indies, disjunctly in Eurasia, and the Australo–Papuan Region (Fig. 17.16).

Characteristics: Hylids range in size from tiny frogs (12–20 mm adult SVL, e.g., *Litoria microbelos*, *Pseudacris ocularis*) to giants (135–140 mm adult SVL, e.g., *Litoria infrafrenata*, *Hyla vasta*). Most are tree frogs in the sense of living in arboreal habitats, although some are ground dwelling. The hylid skull has paired palatines and frontoparietals. The facial nerve exits through the anterior acoustic foramen in the auditory capsule; the trigeminal and facial nerve ganglia fuse to form a prootic ganglion. The vertebral column possesses eight presacral holochordal vertebrae, and all are procoelous. The transverse processes of the sacral vertebra are slightly to moderately expanded, and this vertebra has a bicondylar articulation with the urostyle. Postmetamorphs have no dorsal ribs on the presacral vertebrae. The pectoral girdle is arciferal with a distinct sternum. The fibulare and tibiale are fused at their proximal and distal ends. An intercalary cartilage occurs between the terminal and penultimate phalanges of each digit, and the tips of the terminal

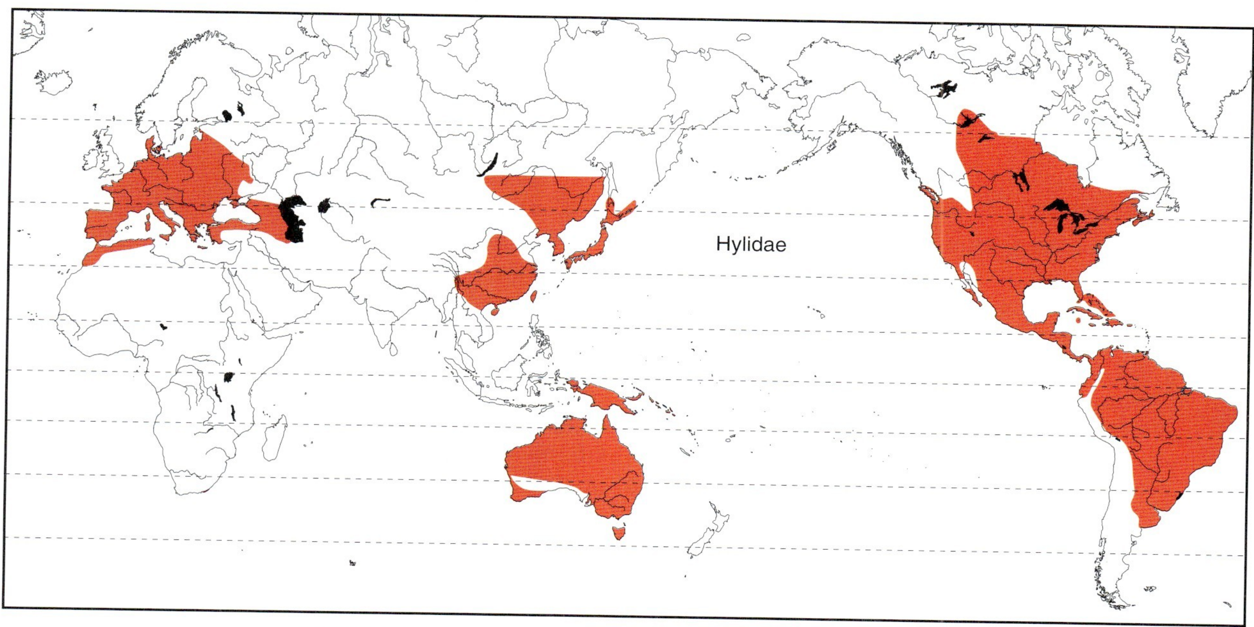

FIGURE 17.16 Geographic distribution of the extant Hylidae.

FIGURE 17.17 Representative hylid frogs. Clockwise from upper left: Walker's marsupial frog *Gastrotheca walkeri*, Amphignathodontidae (K.-H. Jungfer); Demerara Falls tree frog *Hypisboas cinerascens*, Hylinae (J. P. Caldwell); paradox frog *Pseudis paradoxa*, Hylinae (L. V. Vitt); mission golden-eyed tree frog *Trachycephalus resinifictrix*, Hylinae (J. P. Caldwell).

phalanges are pointed, occasionally claw shaped. The larvae have keratinized mouthparts, and the left and right branchial chambers fuse behind the heart and are emptied by a spiracle on the left side at midbody. All hylids have axillary amplexus.

Hylinae

Sister taxon: Unnamed taxon containing Phyllomedusinae and Pelodryadinae.

Content: Thirty-seven genera, *Acris*, *Anotheca*, *Aparasphenodon*, *Aplastodiscus*, *Argenteohyla*, *Bokermannohyla*, *Bromeliohyla*, *Charadrahyla*, *Corythomantis*, *Dendropsophus*, *Duellmanohyla*, *Ecnomiohyla*, *Exerodonta*, *Hyla*, *Hyloscirtus*, *Hypsiboas*, *Isthmohyla*, *Itapotihyla*, *Megastomatohyla*, *Myersiohyla*, *Nyctimantis*, *Osteocephalus*, *Osteopilus*, *Phyllodytes*, *Plectrohyla*, *Pseudacris*, *Pseudis*, *Ptychohyla*, *Scarthyla*, *Scinax*, *Smilisca*, *Sphaenorhynchus*, *Tepuihyla*, *Tlalocohyla*, *Trachycephalus*, *Triprion*, and *Xenohyla*, with 606 species.

Distribution: Disjunctly across Eurasia and throughout the Americas, and West Indies.

Characteristics: Ossification of the skull is variable, commonly lacking extensive fusion. The skin usually is not fused to roofing bones, although it is fused in the casque-headed taxa (e.g., *Osteopilus*, *Triprion*). The pupils are horizontal. Of the superficial mandibular musculature, the interhyoideus extends posteriorly beyond the lower jaw, and the intermandibular muscle usually is undifferentiated.

Biology: Hylines are predominantly arboreal frogs, although a few such as *Acris* and *Pseudacris* are terrestrial or live close to the ground on grasses and forbs; some species of *Smilisca* and *Triprion* are burrowers. Reproductive behavior includes male vocalization to attract females, and axillary amplexus stimulated by female contact with the male. Egg deposition occurs in water in sites ranging from tree holes and bromeliads to streams and lakes. Eggs hatch into free-swimming tadpoles and eventually metamorphose into froglets. Parental care is uncommon but occurs in the gladiator frogs (e.g., *Hypsiboas boans*, *H. faber*, *H. rosenbergi*). In these species, the males build nests in or adjacent to streams by pivoting around in sand or mud and pushing substrate with their feet. Males call to attract females to their nest for egg deposition and then guard the eggs and tadpoles. *Osteopilus brunneus* and some species of *Osteocephalus* deposit unfertilized eggs in bromeliads or tree holes to feed tadpoles developed from eggs previously deposited by the same parents. *Pseudis* is composed of highly aquatic frogs with fully webbed feet. Most live in permanent bodies of water, usually in lakes, marshes, or large ponds. In the Chaco region where streams and lakes dry up, *P. paradoxus* estivates in burrows in dry mud. *Pseudis* is paradoxical because they have giant tadpoles that metamorphose into strikingly smaller froglets. *P. paradoxa* has tadpoles that reach 220 mm total length and 98 g in mass, yet the tadpoles metamorphose into froglets one-third or less than the length of the tadpole.

Pelodryadinae

Sister taxon: Phyllomedusinae.

Content: One genus, *Litoria*, with 181 species.

Distribution: Mainly Australia and New Guinea, although present on a few southern Indonesian islands and with scattered introductions on Southwest Pacific islands.

Characteristics: Ossification of the skull is variable, commonly lacking extensive fusion, and the skin usually is not fused to roofing bones. The pupils are horizontal. Of the superficial mandibular musculature, the interhyoideus extends posteriorly beyond the lower jaw, and the intermandibular muscle has a separate apical element.

Biology: Pelodryadines are terrestrial to arboreal frogs (Fig. 17.18). A few species are semifossorial. The terrestrial *Litoria nasuta* is known to Australian children as the "rocket frog" because of its prodigious jumps of over 1 m. Reproductive behavior and development follows the typical anuran pattern. The male vocalizes to attract females, although a few species lack vocal sacs and are either voiceless or produce quiet calls. Amplexus is axillary and stimulated by female contact with the male. Eggs are deposited mainly in ephemeral pools or streams and lakes and hatch into free-swimming tadpoles. Parental care is unknown in pelodryadines.

Phyllomedusinae

Sister taxon: Pelodryadinae.

Content: Seven genera, *Agalychnis*, *Cruziohyla*, *Hylomantis*, *Pachymedusa*, *Phasmahyla*, *Phrynomedusa*, and *Phyllomedusa*, with 57 species.

Distribution: Southern Mexico to Argentina.

Characteristics: Ossification of the skull is variable, commonly lacking extensive fusion, and the skin usually is not fused to roofing bones. The pupils are vertical. Of the superficial mandibular musculature, the interhyoideus extends posteriorly beyond the lower jaw, and the intermandibular muscle has lateral accessory slips.

Biology: Most phyllomedusines are highly arboreal frogs (Fig. 17.18). Although capable jumpers, they usually walk slowly and methodically among branches to forage or search for resting sites. Some phyllomedusines (e.g., *Phyllomedusa hypochondrialis* and *P. sauvagii*) are uricotelic, having developed the ability to excrete uric acid rather than urea as a water-saving mechanism. Further, most species appear to have a lipid skin secretion that permits them to reduce water loss from the skin. *Phyllomedusa sauvagii* uses its hindlimbs in a contortionist-like manner to wipe its entire body with the secretion. This species is also able to tolerate excess heat loads without

FIGURE 17.18 Representative hylid and centrolenid frogs. Clockwise from upper left: West Sepik tree frog *Litoria leucova*, Pelodryadinae (S. Richards); splendid leaf frog *Cruziohyla calcarifer*, Phyllomedusinae (J. P. Caldwell); Tukeit Hill tree frog *Allophryne ruthveni*, Allophryninae (J. P. Caldwell); Amazonian glass frog *Hyalinobatrachium nouraguense*, Centroleninae (J. P. Caldwell).

resorting to increased skin evaporation loss to shed the excess heat. Phyllomedusines derive their colloquial name, leaf frogs, from their egg-laying behavior. Egg deposition typically occurs on leaves or branches overhanging water. While in amplexus with a male, the female selects a deposition site and deposits 100 to 150 eggs, which the male fertilizes. The female and male, still in amplexus, descend to the water so that the female can absorb water before returning to the original egg site to deposit more eggs. This sequence may be repeated several times. Not all leaf frogs deposit eggs in this manner; *Phrynomedusa marginata* hides its eggs in crevices.

Centrolenidae

Glass Frogs, Ruthven's Frog

Classification: Amphibia; Lissamphibia; Anura; Neobatrachia; Nobleobatrachia; Athesphatanura; Diphyabatrachia.

Sister taxon: Cruciabatrachia (clade containing Leptodactylidae and Chthonobatrachia).

Content: Two subfamilies, Allophryninae and Centroleninae, with 144 species.

Distribution: Southern Mexico to Panama, Andes from Venezuela to Bolivia, Amazon and Orinoco River basins, Guiana Shield, and Atlantic forests of southeastern Brazil and northeastern Argentina (Fig. 17.19).

Characteristics: Centrolenids (Fig. 17.18) vary in body size from small species (< 22 mm adult SVL), medium-sized species (22–35 mm), and large-sized species (35–55 mm) to a few giants (to 77 mm SVL in *Centrolene geckoideum*). The colloquial name refers to the transparent abdominal peritoneum and skin, through which the heart and other internal organs are visible. The skull has paired palatines and frontoparietals. The facial nerve exits through the anterior acoustic foramen in the auditory capsule; the trigeminal and facial nerve ganglia fuse to form a prootic ganglion. The vertebral column possesses eight presacral holochordal vertebrae, and all are procoelous. The transverse processes of the sacral vertebra are moderately expanded, and this vertebra has a bicondylar articulation with the urostyle. Postmetamorphs have no dorsal ribs on

FIGURE 17.19 Geographic distribution of the extant Centrolenidae.

the presacral vertebrae. The pectoral girdle is arciferal with a distinct sternum. The fibulare and tibiale are partially or completely fused. An intercalary cartilage occurs between the terminal and penultimate phalanges of each digit, and the tips of the terminal phalanges are T-shaped. The vermiform larvae have keratinized mouthparts, and the left and right branchial chambers fuse behind the heart and are emptied by a spiracle on the left side at midbody.

Biology: Centrolenids occur in a variety of forested habitats, including evergreen and semideciduous forests, rain forests, cloud forests, and páramos. They are typically found near streams and rivers, often in trees or other vegetation overhanging moving water. All species are nocturnal and deposit egg clutches on the upper surfaces of leaves overhanging water. Males of many species guard one or more clutches until hatching. Upon hatching, tadpoles drop into water below, where they are fossorial, living in leaf packs or in sandy or muddy substrate along the shoreline.

Allophryninae

Sister taxon: Centroleninae.

Content: Monotypic, *Allophyrne ruthveni.*

Distribution: Northern South America from Venezuela through Guyana, Suriname, and French Guiana to Amapá, Brazil, to Rondônia in western Brazil and Pará in eastern Brazil.

Characteristics: *Allophryne ruthveni* (Fig. 17.18) is a small species, 26 to 31 mm adult SVL. Its skull is strongly ossified dorsally, has paired palatines and frontoparietals, and toothless maxillae. The fibulare and tibiale are fused at their proximal and distal ends. No intercalary cartilage occurs between the terminal and penultimate phalanges of the digits, and the tips of the terminal phalanges are T-shaped. Tadpoles of *A. ruthveni* are unknown.

Biology: This species occurs in rain forests; individuals congregate in large breeding aggregations in trees and low vegetation along rivers as rising water begins flooding the forest during the wet season. Smaller choruses also occur in trees and shrubs at the edge of small ponds and flooded depressions in the forest.

Centroleninae

Sister taxon: Allophryninae.

Content: Four genera, *Centrolene*, *Cochranella*, *Hyalinobatrachium*, and *Nymphargus*, with 143 species.

Distribution: Southern Mexico to Panama, Andes from Venezuela to Bolivia, Amazon and Orinoco River basins, Guiana Shield, and Atlantic forests of southeastern Brazil and northeastern Argentina.

Characteristics: Centroleninae have an intercalary element between distal and penultimate phalanges, and T-shaped terminal phalanges. The tibiale and fibulare are partially or completely fused. A dilated process is present on the medial side of the third metacarpal. The ventral skin is usually transparent.

Biology: Centroleninae are tropical forest residents that spend their lives largely in the trees, except for the aquatic larval stage. Males typically call from vegetation over small to large streams and rivers. Unlike many arboreal frogs, the female does not descend to water after amplexing with the male but deposits her eggs on the leaf on which the male is calling. Parental care is common, and a male guards one to several clutches of 2 to 30 eggs (Fig. 5.15). Major predators of the eggs are various "frog flies" of the families Ephydridae and Drosophilidae that deposit their eggs on the frog egg mass and whose larvae will then consume the frog embryos. Vermiform tadpoles hatch and drop into the water below, where they complete their development, commonly living within leaf packs or burrowing into mud or sand substrate at the stream margin. Tadpoles that burrow are bright red because of dense capillary beds in the skin, which function in respiration in this low-oxygen environment. The large *Centrolene geckoideum* is an exception to arboreal breeding. It lives along small forest streams and attaches its eggs to rocks behind waterfalls; subsequently, the male parent attends the eggs.

Leptodactylidae

White-Lipped Frogs and Tropical Grass Frogs

Classification: Amphibia; Lissamphibia; Anura; Neobatrachia; Athesphatanura; Cruciabatrachia.

Sister taxon: Chthonobatrachia (clade containing Ceratophryidae and Hesticobatrachia).

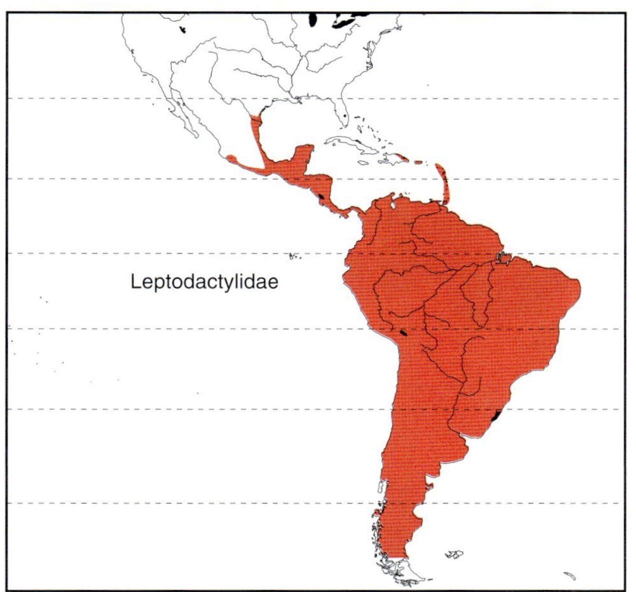

FIGURE 17.20 Geographic distribution of the extant Leptodactylidae.

Content: Four genera, *Hydrolaetare*, *Leptodactylus*, *Paratelmatobius*, and *Scythrophrys*, with 2, 81, 6, and 1 species, respectively.

Distribution: Southern North America, South America, and the West Indies (Fig. 17.20).

Characteristics: Most leptodactylids are moderate in size, but some are very large (e.g., 145–185 mm adult SVL, *Leptodactylus pentadactylus*). The skull has paired palatines and frontoparietals. The facial nerve exits through the anterior acoustic foramen in the auditory capsule; the trigeminal and facial nerve ganglia fuse to form a prootic ganglion. The vertebral column possesses eight presacral holochordal vertebrae, and all are procoelous. The transverse processes of the sacral vertebra are cylindrical, and this vertebra has a bicondylar articulation with the urostyle. Postmetamorphs have no dorsal ribs on the presacral vertebrae. The pectoral girdle is arciferal, rarely pseudofirmisternal, with a distinct sternum. The fibulare and tibiale are fused at their proximal and distal ends. No intercalary cartilage occurs between the terminal and penultimate phalanges of each digit, and the tips of the terminal phalanges are variable. The tadpoles have keratinized mouthparts, and the left and right branchial chambers fuse behind the heart and are emptied by a spiracle on the left side at midbody.

Biology: Most species of the large genus *Leptodactylus* are terrestrial (Fig. 17.21). Eggs are deposited in a foam nest produced from cloacal secretions in almost all species of *Leptodactylus* (Figs. 5.3 and 5.9). Location of foam nests varies among species. Foam nests may be located on the water surface of ponds, and tadpoles drop into the water below, or they may be deposited in depressions close to water, and tadpoles are washed into ponds during rains. In one group, the *Leptodactylus pentadactylus* group, eggs are deposited in foam nests in depressions or burrows, and tadpoles may be washed into water or may develop entirely in terrestrial nests. In *L. fallax*, the female may remain with the nest and deposit tropic eggs as food for the tadpoles. In *L. labryrinthicus*, only about 10% of the eggs are fertilized, and the developing tadpoles feed on the unfertilized eggs. Species formerly referred to as the genus *Adenomera* (now *Leptodactylus*) deposit eggs in foam nests, but some species have nonfeeding, endotrophic tadpoles (e.g., *L. marmoratus*), whereas others have aquatic tadpoles. Species of *Paratelmatobius* have brightly colored venters that they display by remaining upside down if disturbed. These frogs deposit large eggs either on pond bottoms (*P. cardosoi*) or adhering to rock surfaces by small rivulets (*P. poecilogaster*). Tadpoles are aquatic, living on the bottom of ponds until metamorphosis. Tadpoles of some species of *Leptodactylus* form large schools (Fig. 5.21).

Ceratophryidae

Horned Frogs, Water Frogs, and Others

Classification: Amphibia; Lissamphibia; Anura; Neobatrachia; Athesphatanura; Cruciabatrachia; Chthonobatrachia.

Sister taxon: Hesticobatrachia (clade containing Cycloramphidae and Calamitophrynia).

Content: Three subfamilies, Batrachylinae, Ceratophryinae, and Telmatobiinae, with 85 species.

Distribution: South America.

Characteristics: The skull has paired palatines and frontoparietals. The facial nerve exits through the anterior acoustic foramen in the auditory capsule; the trigeminal and facial nerve ganglia fuse to form a prootic ganglion. The vertebral column possesses eight presacral holochordal vertebrae, and all are procoelous. The transverse processes of the sacral vertebra are cylindrical, and this vertebra has a bicondylar articulation with the urostyle. The sternum is cartilaginous. In some species, a bony shield lies over the presacral vertebrae and fuses with the dermis; the transverse processes of the anterior presacral vertebrae are long and distally expanded. The tips of the terminal phalanges are knobbed. Amplexus is axillary.

Biology: Ceratophryids may be terrestrial or totally aquatic. Terrestrial species deposit eggs in ponds or other nonmoving water and have a tadpole stage. Body forms vary from the ceratophrynines, which have squat bodies with very large heads and wide mouths, to *Telmatobius culeus*, a large (250 mm SVL) aquatic frog that lives in Lake Titicaca in Peru and Bolivia. These frogs have highly folded skin and obtain oxygen primarily by cutaneous respiration. Tadpoles are carnivorous in *Ceratophrys* and *Lepidobatrachus*.

FIGURE 17.21 Representative leptodactylid and ceratophryid frogs. Clockwise from upper left: Moustached frog *Leptodactylus mystacinus*, Leptodactylidae (J. P. Caldwell); San Jose white-lipped frog *Leptodactylus stenodema*, Leptodactylidae (J. P. Caldwell); Martinez's tropical grass frog *Leptodactylus martinezi*, Leptodactylidae (J. P. Caldwell); Suriname horned frog *Ceratophrys cornuta*, Ceratophryinae (J. P. Caldwell).

Batrachylinae

Sister taxon: Unnamed clade consisting of Ceratophryinae and Telmatobiinae.

Content: Two genera, *Atelognathus* and *Batrachyla*, with 9 and 5 species, respectively.

Distribution: Central and southern Chile and adjacent Argentina.

Characteristics: *Atelognathus* has a large frontoparietal fontanelle, short palatine bones, and large nasals. It lacks quadratojugals, columellae, and tympanic annuli. The quadratojugal is reduced to a small spur in some *Batrachyla*, and the columellae are present. In both genera, the sternum is cartilaginous, and the pectoral girdle is arciferal. The eight presacral vertebrae are procoelous and lack a bony or cartilaginous shield; the transverse processes of the presacral vertebrae are short and not expanded.

Biology: Most species of *Atelognathus* are restricted to the Andean slopes or basaltic lagoons in Argentinian Patagonia. Amplexus is inguinal in some species of *Batrachyla*, and eggs are deposited on damp vegetation near water.

Ceratophryinae

Sister taxon: Telmatobiinae.

Content: Three genera, *Ceratophrys, Chacophrys*, and *Lepidobatrachus,* with 8, 1, and 3 species, respectively.

Distribution: South America.

Characteristics: The bones of the skull are co-ossified with the overlying skin, and the nonpedicellate teeth are fanglike. The sternum is cartilaginous. In some species, a bony shield lies over the presacral vertebrae and fuses with the dermis. The transverse processes of the anterior presacral vertebrae are long and distally expanded. The tips of the terminal phalanges are knobbed. Amplexus is axillary.

Biology: Ceratophryines are best known for the voracious predatory behavior of *Ceratophrys calcarata* and *C. cornuta* (Fig. 17.21). Their heads are large compared with their bodies, and their big mouths and fanglike teeth in the upper jaws enable them to capture and consume large prey, including lizards, other frogs, and small mammals, typically from ambush. The other two genera share the same body form and feeding behavior. All ceratophyrines are seasonal breeders, laying numerous small eggs in aquatic habitats; the eggs hatch into free-living tadpoles. *Chacophrys* and *Lepidobatrachus* are fossorial and inhabit arid areas, and all species possess well-developed metatarsal spades. *Ceratophrys ornata* produces a keratinous cocoon and remains in torpor during the driest part of the year; perhaps the other arid land species do also. Tadpoles of *Ceratophrys* and *Lepidobatrachus* are carnivorous, although the tadpoles of *Chacophrys* are typical grazers.

Telmatobiinae

Sister taxon: Ceratophryinae.

Content: Two genera, *Batrachophrynus* and *Telmatobius*, with 2 and 57 species, respectively.

Distribution: Andean South America, from Ecuador to Chile and Argentina.

Characteristics: The skin overlaying the skull is not co-ossified with the roofing bones. The sternum is cartilaginous. The presacral vertebrae lack a bony or cartilaginous shield; the transverse processes of the anterior presacral vertebrae are long and not expanded. The tips of the terminal phalanges are knobbed. Amplexus is axillary.

Biology: Frogs in the genus *Telmatobius* reach about 60 mm in SVL; one species of *Batrachophrynus* reaches 160 mm SVL. Many species of *Telmatobius* are totally aquatic, living in streams or lakes, whereas others are semiterrestrial.

Cycloramphidae

Mouth-Brooding Frogs, Smooth Horned Frogs, and Others

Classification: Amphibia; Lissamphibia; Anura; Neobatrachia; Athesphatanura; Cruciabatrachia; Hesticobatrachia.

Sister taxon: Calamitophrynia (clade containing Leiuperidae and Agastorophrynia).

Content: Twelve genera, *Alsodes, Crossodactylodes, Cycloramphus, Eupsophus, Hylorina, Limnomedusa, Macrogenioglottus, Odontophrynus, Proceratophrys, Rhinoderma, Thoropa*, and *Zachaenus*, with 95 species.

Distribution: Southern tropical and temperate South America (Fig. 17.22).

Characteristics: The skull has paired palatines and frontoparietals. The facial nerve exits through the anterior acoustic foramen in the auditory capsule; the trigeminal and facial nerve ganglia fuse to form a prootic ganglion. The vertebral column possesses eight presacral holochordal vertebrae, and all are procoelous. Transverse processes of the vertebrae are long or short, and the sacral diapophyses are slightly dilated. The larvae have keratinized mouthparts, and the left and right branchial chambers fuse behind the heart and are emptied by a spiracle on the left side at midbody.

Biology: Amplexus is axillary. Most genera have free-swimming tadpoles, and most deposit numerous small eggs in water. *Thoropa* has large eggs that are

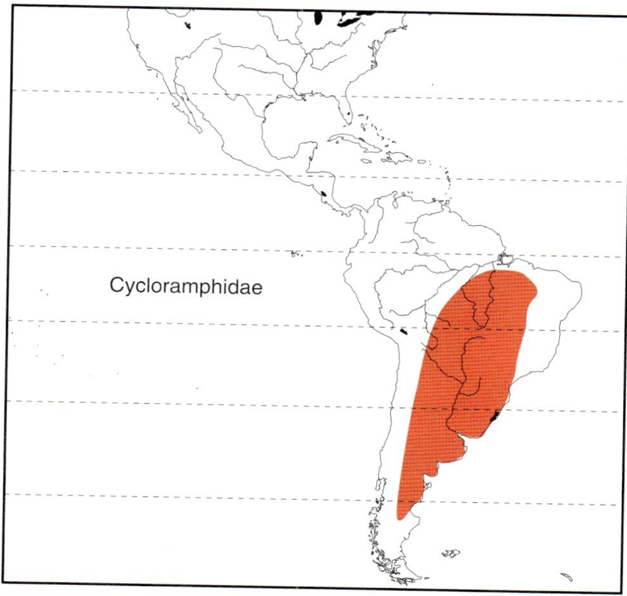

FIGURE 17.22 Geographic distribution of the extant Cycloramphidae.

deposited on rock ledges with dripping water. The tadpoles of *Thoropa* are flattened and vermiform, adaptations for living in torrential streams. Male *Rhinoderma* (mouth-brooding frogs) retain developing tadpoles in their vocal sacs, a unique behavior among anurans (Fig. 17.23). During courtship, the male calls to attract a female. After the eggs are deposited, the male attends them for about 20 days. When tadpoles are about to hatch and the egg mass is dissolving, the male gathers the hatchlings in his mouth. Male *R. rufrum* carry the larvae to water, where they complete their development. In contrast, *R. darwini* males manipulate the larvae into the vocal sacs, where the embryos undergo direct development and emerge as froglets about 50 days later. Both species are terrestrial residents of moist, temperate forests of Chile and adjacent Argentina. *Zachaenus* deposits a few large eggs in a moist, terrestrial habitat. The eggs develop in the gelatinous mass and remain in the mass until metamorphosis.

Leiuperidae

Foam-Nesting Frogs and Dwarf Frogs

Classification: Amphibia; Lissamphibia; Anura; Neobatrachia; Athesphatanura; Cruciabatrachia; Calamitophrynia.

Sister taxon: Agastorophrynia (clade containing Bufonidae and Nobleobatrachia).

Content: Seven genera, *Edalorhina, Engystomops, Eupemphix, Physalaemus, Pleurodema, Pseudopaludicola*, and *Somuncuria*, with 77 species.

Distribution: Mexico throughout Central and South America (Fig. 17.24).

FIGURE 17.23 Representative cycloramphid and leiuperid frogs. Clockwise from upper left: Goiás smooth-horned frog *Proceratophrys goyana*, Alsodinae (J. P. Caldwell); Darwin's mouth-breeding frog *Rhinoderma darwinii*, Rhinodermatidae (W. E. Duellman, courtesy of the Natural History Museum, The University of Kansas); Cuibá dwarf frog *Eupemphix nattereri*, Leiuperidae (J. P. Caldwell); Perez's snouted frog *Edalorhina perezi*, Leiuperidae (J. P. Caldwell).

FIGURE 17.24 Geographic distribution of the extant Leiuperidae.

Characteristics: The skull has paired palatines and frontoparietals. The facial nerve exits through the anterior acoustic foramen in the auditory capsule; the trigeminal and facial nerve ganglia fuse to form a prootic ganglion. The vertebral column possesses eight presacral holochordal vertebrae, and all are procoelous. The transverse processes of the sacral vertebra are cylindrical, and this vertebra has a bicondylar articulation with the urostyle. Postmetamorphs have no dorsal ribs on the presacral vertebrae. The pectoral girdle is arciferal, rarely pseudofirmisternal, with a distinct sternum. The fibulare and tibiale are fused at their proximal and distal ends. No intercalary cartilage occurs between the terminal and penultimate phalanges of each digit, and the tips of the terminal phalanges are variable. The larvae have keratinized mouthparts, and the left and right branchial chambers fuse behind the heart and are emptied by a spiracle on the left side at midbody.

Biology: Amplexus is axillary. Most genera deposit eggs in a foam nest on water; eggs hatch quickly and tadpoles undergo development in water. An exception is

Pseudopaludicola, in which small clutches of eggs are deposited in water. In some species of *Physalaemus*, particularly in drier regions, breeding events include thousands of frogs that call both at night and throughout the day (Fig. 17.23).

Bufonidae

True Toads, Harlequin Frogs, and Relatives

Classification: Amphibia; Lissamphibia; Anura; Neobatrachia; Athesphatanura; Cruciabatrachia; Calamitophrynia; Agastorophrynia.

Sister taxon: Nobleobatia (clade containing Hylodidae and Dendrobatoidea).

Content: Forty-five genera, *Adenomus, Altiphrynoides, Amietophrynus, Anaxyrus, Andinophryne, Ansonia, Atelopus, Bufo, Bufoides, Capensibufo, Churamiti, Crepidophryne, Dendrophryniscus, Didynamipus, Duttaphrynus, Epidalea, Frostius, Ingerophrynus, Laurentophryne, Leptophyrne, Melanophryniscus, Mertensophryne, Metaphryniscus, Nannophryne, Nectophryne, Nectophrynoides, Nimbaphrynoides, Ollotis, Oreophrynella, Osornophyrne, Parapelophryne, Pedostibes, Pelophryne, Peltophryne, Phrynoidis, Poyntonophrynus, Pseudepidalea, Pseudobufo, Rhaebo, Rhinella, Schismaderma, Truebella, Vandijkophrynus, Werneria,* and *Wolterstorffina*, with 505 species.

Distribution: Worldwide on all continents except Antarctica and Australia (Fig. 17.25). *Rhinella marina* has been introduced widely in the Caribbean, Oceania, Philippines, and Australia.

Characteristics: Bufonids vary greatly in size, ranging from the tiny *Dendrophryniscus carvalhoi, Mertensophryne micranotis*, and *Pelophryne brevipes* (< 20 mm adult SVL) to giants, such as *Rhinella marina*, which has a maximum length to 230 mm SVL (Fig. 17.26). Bufonids are the only anurans to possess a Bidder's organ in male tadpoles, and this organ persists in most adult males. All adults lack teeth in the upper jaw, thus bufonids are toothless amphibians. The bufonid skull has paired palatines and frontoparietals. The facial nerve exits through the anterior acoustic foramen in the auditory capsule; the trigeminal and facial nerve ganglia fuse to form a prootic ganglion. The vertebral column possesses five to eight presacral holochordal vertebrae, and all are procoelous. The transverse processes of the sacral vertebra are moderately expanded, and this vertebra has a bicondylar articulation with the urostyle. Postmetamorphs have no dorsal ribs on the presacral vertebrae. The pectoral girdle is arciferal, rarely pseudofirmisternal, with a distinct sternum. The fibulare and tibiale are fused at their proximal and distal ends. No intercalary cartilage occurs between the terminal and penultimate phalanges of each digit, and the tips of the terminal phalanges are blunt to pointed. The larvae have keratinized mouthparts, and the left and right branchial chambers fuse behind the heart and are emptied by a spiracle on the left side at midbody.

Biology: Bufonids have a diverse array of life histories. Although the majority is terrestrial to semifossorial, some (*Pseudobufo*) are aquatic and others (*Pedostibes*) are arboreal. Most have prominent skin glands, often with highly toxic skin secretions. Many species have a thick, warty, often spiny, skin and enlarged concentrations of glands in the temporal-neck area forming prominent

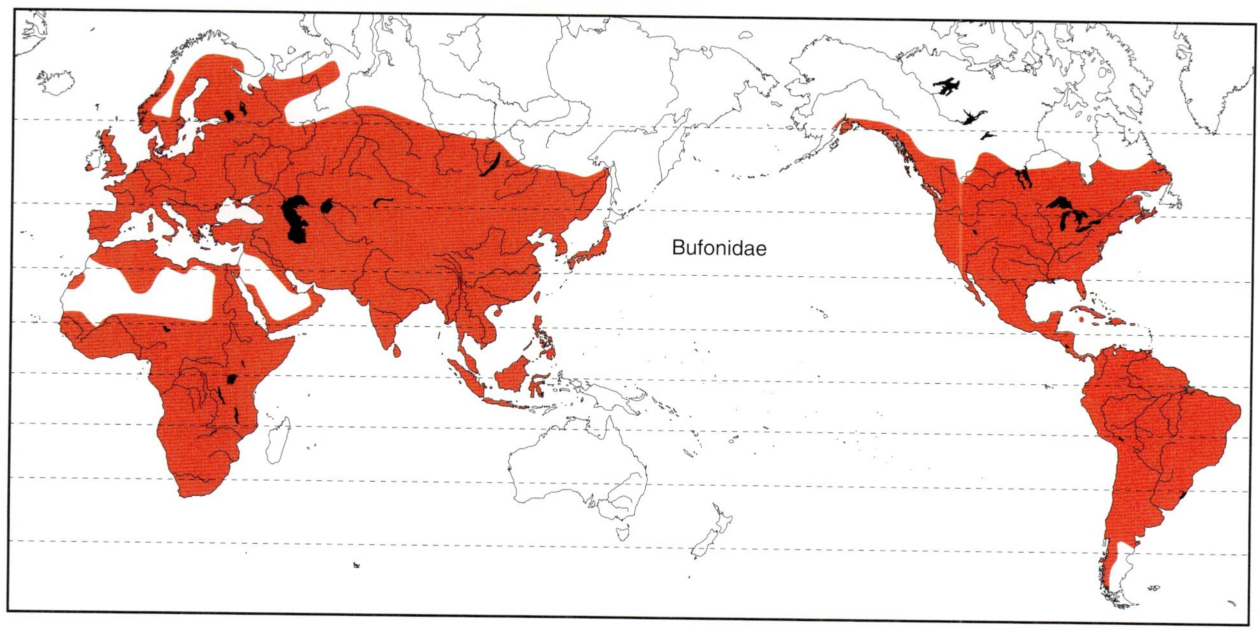

FIGURE 17.25 Geographic distribution of the extant Bufonidae.

FIGURE 17.26 Representative bufonids. Clockwise from upper left: Beautiful harlequin frog *Atelopus pulcher* (R. W. McDiarmid); ocellated toad *Rhinella ocellata* (J. P. Caldwell); *Dendrophryniscus minutus* (J. P. Caldwell); American toad *Anaxyrus americanus* (J. P. Caldwell).

parotoid glands. *Atelopus* (Fig. 17.26) lacks the prominent glandular swellings, but its skin secretions are more readily lethal to predators; its bright aposematic coloration of red, yellow, or orange markings on a black background advertise its toxicity to predators. *Bufo* and many other bufonids reproduce using axillary amplexus (rarely inguinal, e.g., *Osornophyrne*, *Ollotis fastidiosa*); they typically deposit strings of pigmented eggs in water, and these hatch into free-living tadpoles. The tadpoles develop quickly and generally metamorphose within 2 to 10 weeks of hatching. Other bufonids have terrestrial eggs and direct development (e.g., *Osornophryne*) and even internal fertilization and viviparity. Two species of *Nectophrynoides* and two of *Nimbaphrynoides* are viviparous.

Hylodidae

Stream-Dwelling Frogs

Classification: Amphibia; Lissamphibia; Anura; Neobatrachia; Athesphatanura; Cruciabatrachia; Agastorophrynia; Nobleobatia.

Sister taxon: Dendrobatoidea.

Content: Three genera, *Crossodactylus*, *Hylodes,* and *Megaelosia*, with 39 species.

Distribution: Northwestern to southern Brazil and adjacent Argentina (Fig. 17.27).

Characteristics: The sternum is cartilaginous, occasionally calcified in old adults. The presacral vertebrae lack a bony or cartilaginous shield; the transverse processes of the anterior presacral vertebrae are short and not expanded. The tips of the terminal phalanges are variable. Amplexus is axillary.

Biology: These streamside frogs are usually small (< 35 mm adult SVL), although adult *Megaelosia* may attain lengths of 120 mm. All are diurnal predators. *Crossodactylus* spends much of its time in the water, even as adults; the other taxa occur on rocks and vegetation along streams. All species deposit eggs in water and have a typical tadpole stage.

Aromobatidae

Cryptic Forest Frogs

Classification: Amphibia; Lissamphibia; Anura; Neobatrachia; Athesphatanura; Cruciabatrachia; Nobleobatia; Dendrobatoidea.

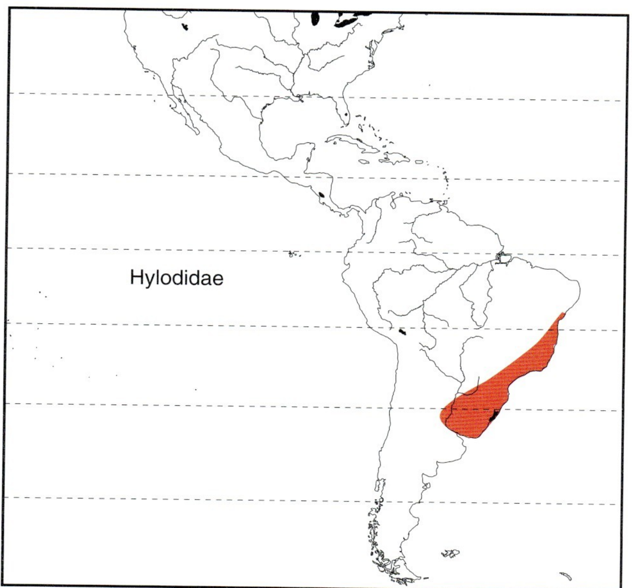

FIGURE 17.27 Geographic distribution of the extant Hylodidae.

Sister taxon: Dendrobatidae.

Content: Three subfamilies, Anomaloglossinae, Aromobatinae, and Allobatinae, with 96 species.

Distribution: Central America, South America, and the Lesser Antilles, with the highest diversity occurring on the eastern Andean slopes, the Amazon basin, and the Atlantic forest of Brazil (Fig. 17.28).

Characteristics: All aromobatids have supradigital scutes. The skull has paired palatines (absent in some groups) and frontoparietals. The facial nerve exits through the anterior acoustic foramen in the auditory capsule, and the trigeminal and facial nerve ganglia fuse to form a prootic ganglion. The vertebral column possesses eight presacral holochordal vertebrae, and all are procoelous. The transverse processes of the sacral vertebra are cylindrical, and this vertebra has a bicondylar articulation with the urostyle. Postmetamorphs have no dorsal ribs on the presacral vertebrae. The pectoral girdle is firmisternal with a distinct sternum. The fibulare and tibiale are fused at their proximal and distal ends. No intercalary cartilage occurs between the terminal and penultimate phalanges of each digit, and the tips of the terminal phalanges are usually T-shaped. Tadpoles have keratinized mouthparts, and the left and right branchial chambers fuse behind the heart and are emptied by a spiracle on the left side at midbody.

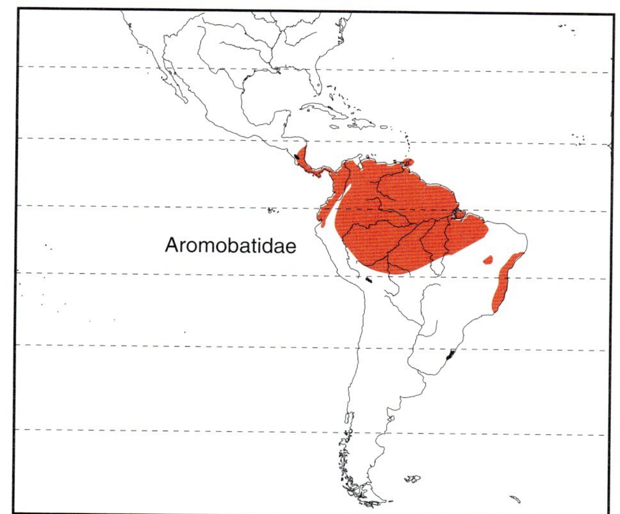

FIGURE 17.28 Geographic distribution of the extant Aromobatidae.

Biology: All aromobatids lack the ability to sequester alkaloids in their skin, in contrast to the sister taxon, Dendrobatidae. Although life histories vary considerably within aromobatids, many deposit relatively small clutches of eggs in a terrestrial location. After the eggs develop into tadpoles, one of the parent frogs transports the tadpoles on its back to a small forest pool or a backwater pool in a small stream, where the tadpoles remain until they metamorphose.

Anomaloglossinae

Sister taxon: Unnamed clade containing Aromobatinae and Allobatinae.

Content: Two genera, *Anomaloglossus* and *Rheobates*, with 20 and 2 species, respectively.

Distribution: Widespread from the Pacific slopes of the Andes in Colombia through the Amazon basin to the Atlantic forest in Brazil.

Characteristics: *Anomaloglossus* is characterized by having a median lingual process on the tongue, which is lacking in *Rheobates*. These brown, cryptically colored frogs are typically small and slender with minimal toe webbing or large and robust with moderate to extensive toe webbing.

Biology: Most species are cryptic brown or gray and live in leaf litter of tropical forests. Most deposit eggs in terrestrial nests, and tadpoles are transported by one of the parents to a forest pool or other small body of water, where they complete development. *Anomaloglossus beebei* breeds in giant terrestrial bromeliads, where four eggs are deposited above the waterline of the bromeliad tank. A pair-bonded male and female provide care to their offspring by moistening the eggs and transporting the tadpoles (male) and depositing trophic eggs to feed the tadpoles (female). Two other species, *A. stepheni* and *A. degranvillei*, have endotrophic tadpoles. Tadpoles of *A. stepheni* develop into froglets in a leaf nest on the forest floor, whereas those of *A. degranvillei* remain on the parent's back until metamorphosis.

Aromobatinae

Sister taxon: Allobatinae.

Content: Two genera, *Aromobates* and *Mannophryne*, with 12 and 15 species, respectively.

FIGURE 17.29 Representative aromobatids. From left: St. Teresa collared frog *Mannophryne oblitterata*, Aromobatinae (L. J. Vitt); brilliant-thighed frog *Allobates femoralis*, Allobatinae (J. P. Caldwell).

Distribution: Parts of Venezuela and Colombia; Trinidad and Tobago.

Characteristics: These small to moderate sized, cryptically colored frogs generally have a robust body form and have basal to extensive toe webbing. *Mannophryne* is a distinct clade in which all species have a dark throat collar (Fig. 17.29). *Aromobates* lacks the throat collar.

Biology: Many aromobatines live along high-elevation streams in cloud forests. The generic name *Aromobates* was given in reference to the original type species, *A. nocturnus*, which has a noxious, but not toxic, skin secretion with a mercaptan-like (skunklike) odor. The odiferous skin of this frog lacks alkaloids like those of the true poison frogs in the sister taxon, Dendrobatidae. *Aromobates nocturnus* is nocturnal unlike other aromobatids and usually found swimming or sitting in water.

Allobatinae

Sister taxon: Aromobatinae.

Content: One genus, *Allobates*, with 46 species.

Distribution: Widespread from Nicaragua through South America to Bolivia and Brazil; Martinique.

Characteristics: Dorsal coloration is cryptic in most species, although the *Allobates femoralis* group has bright dorsolateral stripes and flash colors. Toe webbing is basal in most species. The skull lacks palatine bones. This group of frogs is variable in terms of morphology and behavior.

Biology: Most species are terrestrial, living in leaf litter of tropical forests. Males call most frequently in early morning and late afternoon, and most breeding occurs during the rainy season. Calls of the widespread *A. femoralis* group vary geographically, perhaps in response to the calls of co-occurring frogs with similar calls (Fig. 17.29). *A. caeruleodactylus* has blue fingers, which may be used to signal the boundaries of its territory to intruding males. Two species, *A. nidicola* and *A. chalcopis*, have endotrophic tadpoles that develop to froglets in terrestrial nests.

Dendrobatidae

Poison Frogs

Classification: Amphibia; Lissamphibia; Anura; Neobatrachia; Athesphatanura; Cruciabatrachia; Nobleobatia; Dendrobatoidea.

Sister taxon: Aromobatidae.

Content: Three subfamilies, Colostethinae, Dendrobatinae, and Hyloxalinae, with 164 species.

Distribution: Southern Nicaragua to northern South America through the Amazonian Basin to Bolivia (Fig. 17.30).

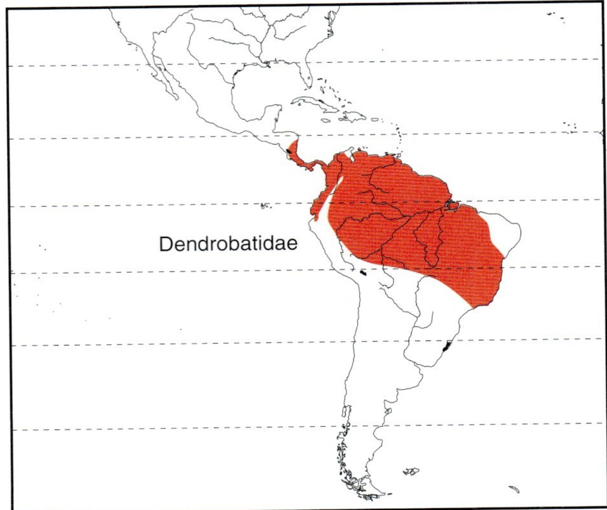

FIGURE 17.30 Geographic distribution of the extant Dendrobatidae.

Characteristics: All dendrobatids have supradigital scutes. The dendrobatid skull has paired palatines (absent in some) and frontoparietals. The facial nerve exits through the anterior acoustic foramen in the auditory capsule, and the trigeminal and facial nerve ganglia fuse to form a prootic ganglion. The vertebral column possesses six to eight presacral holochordal vertebrae, and all are procoelous. The transverse processes of the sacral vertebra are cylindrical, and this vertebra has a bicondylar articulation with the urostyle. Postmetamorphs have no dorsal ribs on the presacral vertebrae. The pectoral girdle is firmisternal with a distinct sternum. The fibulare and tibiale are fused at their proximal and distal ends. No intercalary cartilage occurs between the terminal and penultimate phalanges of each digit, and the tips of the terminal phalanges are usually T-shaped. Tadpoles have keratinized mouthparts, and the left and right branchial chambers fuse behind the heart and are emptied by a spiracle on the left side at midbody.

Biology: Dendrobatids are diurnal frogs and may occur in riparian, terrestrial, or semiarboreal microhabitats in tropical forests (Fig. 17.31). All species deposit eggs in terrestrial nests and transport tadpoles to various types of water bodies. Many species have lipophilic alkaloids in their skin, which is derived from their diet of ants. In general, the brighter or more boldly colored dendrobatids are most toxic; several hundred alkaloids have been identified from their skin.

Colostethinae

Sister taxon: Unnamed clade containing Dendrobatinae and Hyloxalinae.

Content: Four genera, *Ameerega, Colostethus, Epipedobates*, and *Silverstoneia*, with 26, 19, 5, and 3 species, respectively.

Distribution: Widespread from southwestern Costa Rica through most of northern and Amazonian South America.

Characteristics: Most species are moderate to large in size. Dorsal coloration varies from conspicuous bright orange, green, or deep red to cryptic brown or gray. One group of *Ameerega* is deep red with highly granulated skin. Lipophilic alkaloids are absent in *Colostethus* and

FIGURE 17.31 Representative dendrobatids. Clockwise from upper left: Three-striped poison frog *Ameerga trivittata*, Colostethinae (J. P. Caldwell); splashback poison frog *Adelphobates galactonotus*, Dendrobatinae (J. P. Caldwell); strawberry poison frog *Oophaga pumilio*, Dendrobatinae (J. P. Caldwell); green-striped poison frog *Hyloxalus chlorocraspedus*, Hyloxinae (J. P. Caldwell).

Silverstoneia but present in some *Ameerega* and *Epipedobates*. Two species of *Colostethus* have the alkaloid tetrodotoxin present in their skin.

Biology: Most species live in rain forest and are terrestrial. Many are riparian, living along the banks of streams. All are diurnal, and males may be heard calling throughout daylight hours.

Dendrobatinae

Sister taxon: Hyloxalinae.

Content: Six genera, *Adelphobates* (4 species), *Dendrobates* (5 species), *Minyobates* (1 species), *Oophaga* (9 species), *Phyllobates* (5 species), and *Ranitomeya* (27 species).

Distribution: Nicaragua throughout most of South America to Bolivia and Brazil.

Characteristics: The skin on the dorsum is smooth, and many species are brightly colored. Lipophilic alkaloids are present in the skin. Frogs in the genus *Phyllobates* have batrachotoxin in their skin, which is one of the most toxic alkaloids produced by any animal.

Biology: Life histories are elaborate in this group of frogs. Presumably all dendrobatids have parental care, and most often the male parent, but occasionally the female, attends the eggs. Males attract females by calling, although they do not form choruses. Amplexus is cephalic or absent, and eggs are laid among leaf litter on the forest floor, along streams, or in arboreal retreats. After the eggs hatch, the tadpoles then wriggle upward onto the back of the parent, who transports them to a nearby pool, tree hole, or bromeliad tank, where they complete development. One clade of *Ranitomeya* has biparental care: The pair-bonded male and female return periodically to feed trophic eggs to their tadpoles. In *Oophaga*, only the female cares for the tadpoles by returning to their leaf axil nurseries and depositing unfertilized eggs for them. Tadpoles in this group appear to be obligatorily oophagous. In *Adelphobates*, tadpoles are transported to Brazil nut capsules on the forest floor, where cannibalism is common if more than one tadpole is transported to the same capsule.

Hyloxalinae

Sister taxon: Dendrobatinae.

Content: One genus, *Hyloxalus*, with 57 species.

Distribution: Panama south through most of South America to Bolivia and Brazil.

Characteristics: One group within this genus, the *H. ramose* group, is characterized by the presence of a black, glandular band on the inner surface of the upper arm. Two species, *H. azuriventris* and *H. chlorocraspedus*, and possibly others, are united by a pale dorsolateral stripe that converges toward the posterior dorsum.

Biology: Most species deposit eggs in terrestrial nests and transport their tadpoles to forest pools or backwater pools in small streams. *H. chlorocraspedus* transports its tadpoles to pools formed in fallen trees; its tadpoles feed on detritus but are also predaceous on small invertebrates.

Microhylidae

Narrow-Mouth Frogs and Toads

Classification: Amphibia; Lissamphibia; Anura; Neobatrachia; Phthanobatrachia; Ranoides.

Sister taxon: Afrobatrachia (clade containing Xenosyneunitanura and Laurentobatrachia).

Content: Eleven subfamilies, Asterophryinae, Cophylinae, Dyscophinae, Gastrophryninae, Hoplophryninae, Kalophryninae, Melanobatrachinae, Microhylinae, Otophryninae, Phrynomerinae, and Scaphiophryinae, with 64 genera and 426 species.

Distribution: Worldwide on all continents, except Antarctica (Fig. 17.32).

Characteristics: Microhylids have a broad range of body forms from a pointed-headed, fossorial habitus to a tree frog habitus. Body size is equally broad and ranges from the tiniest of frogs, *Syncope* and *Stumpffia* (9–13 mm adult SVL) to large *Glyphoglossus molossus* females (78–88 mm SVL). The microhylid skull has paired palatines and frontoparietals. The facial nerve exits through the anterior acoustic foramen in the auditory capsule; the trigeminal and facial nerve ganglia fuse to form a prootic ganglion. The vertebral column possesses eight, rarely seven, presacral holochordal vertebrae, and the vertebrae are all procoelous except for a biconcave surface on the last presacral (i.e., diplasiocoelous). The transverse processes of the sacral vertebra are cylindrical to broadly expanded, and this vertebra has a bicondylar articulation with the urostyle. Postmetamorphs have no dorsal ribs on the presacral vertebrae. The pectoral girdle is firmisternal with a distinct sternum, although many microhylids show a reduction of clavicle and procoracoid. The fibulare and tibiale are fused at their proximal and distal ends. No intercalary cartilage occurs between the terminal and penultimate phalanges, except in the phyrnomerines; the tips of the terminal phalanges are blunt, pointed, or T-shaped. Tadpoles lack keratinized mouthparts (except *Scaphiophryne* and *Otophryne*), and a large spiracular chamber is emptied by a caudomedial spiracle.

Biology: Microhylids are a diverse group of frogs with fossiorial, terrestrial, and arboreal species. The fossorial species tend to be ant and termite specialists. Some groups have direct development (e.g., Asterophryinae), and others have endotrophic, nonfeeding tadpoles (e.g., Cophylinae). At least two subfamilies have some species with endotrophic tadpoles and some with exotrophic tadpoles (e.g., Microhylinae and Hoplophryninae).

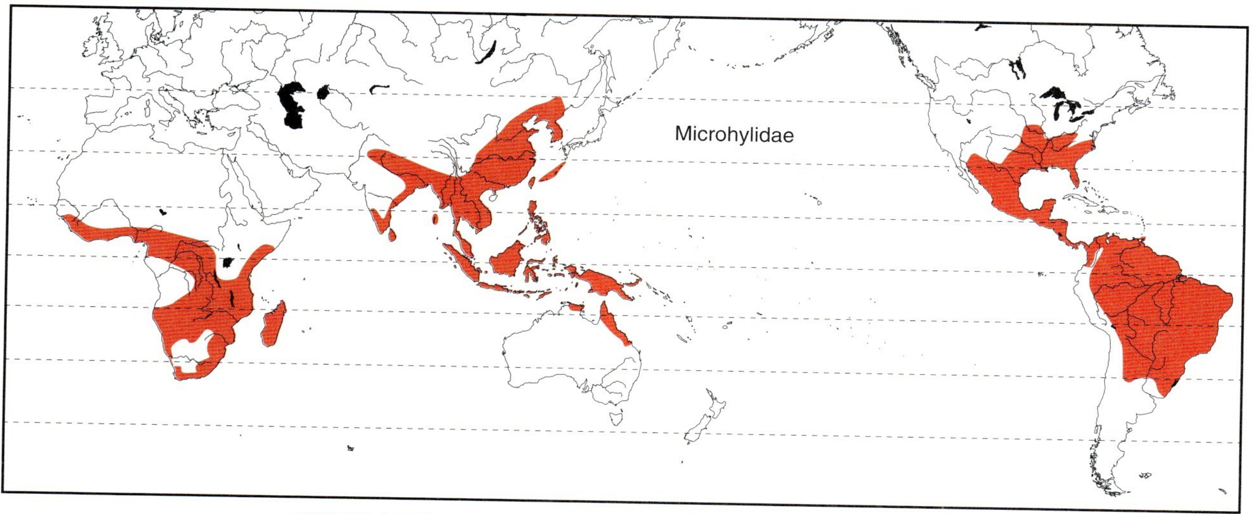

FIGURE 17.32 Geographic distribution of the extant Microhylidae.

Asterophryinae

Sister taxon: Microhylinae.

Content: Eighteen genera, *Albericus, Aphantophryne, Asterophrys, Austrochaperina, Bargenys, Callulops, Choerophryne, Cophixalus, Copiula, Genophryne, Hylophorbus, Liophryne, Mantophryne, Oreophryne, Oxydactyla, Pherohapsis, Sphenophryne*, and *Xenorhina,* with 218 species.

Distribution: Southern Philippines eastward to Indonesia and New Guinea, its adjacent islands, northern Australia, and the Molucca Islands.

Characteristics: The skull has paired ethmoids, a single large prevomer, and toothless maxillaries. The vertebral column is procoelous, and the pectoral girdle lacks clavicles and procoracoids.

Biology: Asterophyrines are terrestrial species that are fossorial (*Bargenys, Xenobatrachus*) or inhabit the forest floor (e.g., *Asterophrys, Callulops, Xenorhina*), and some species occur in grasslands and disturbed habitats. Asterophyrines range from small (20–24 mm adult SVL, *Hylophorbus rufescens*) to moderately large (60–80 mm SVL, *Callulops stictogaster*). Reproductive data are absent for most species, but where known, all species lay small clutches of large, well-yolked eggs in protected sites (forest floor or in trees), and a male is usually in attendance (Fig. 17.33), indicating that direct development may occur in some species.

Cophylinae

Sister taxon: Scaphiophryninae.

Content: Seven genera, *Anodonthyla, Cophyla, Madecassophryne, Platypelis, Plethodontohyla, Rhombophryne,* and *Stumpffia*, with 43 species.

Distribution: Madagascar.

Characteristics: The skull has paired ethmoids, paired prevomers, and usually teeth on the maxillaries. The vertebral column is procoelous, and the pectoral girdle usually has well-developed clavicles and procoracoids.

Biology: The cophylines consist of two clades with contrasting habits and reproductive behavior. *Anodonthyla, Cophyla*, and *Platypelis* are arboreal species; *Anodonthyla* occurs mainly on tree trunks or rocks, whereas the other two genera live on the branches and leaves of trees. All three taxa deposit small clutches of less than 100 eggs in tree holes or leaf axils (*Cophyla, Platypelis*) or in rock cavities (*Anodonthyla*). The eggs hatch into nonfeeding tadpoles that are attended by the male until metamorphosis. These three genera are mainly small-to-medium-sized frogs (16–40 mm SVL; *Platypelis grandis* 43–105 mm). The other genera include tiny (*Stumpffia pygmae* and *S. tridactyla,* 10–12 mm adult) to large (*Plethodontohyla inguinalis,* 55–100 mm) frogs with nearly exclusively terrestrial habits. These taxa typically lay eggs on the forest floor, either in cavities or in foam nests in the leaf litter. Their tadpoles are also nonfeeding and often have one parent in attendance.

Dyscophinae

Sister taxon: Unnamed clade containing Asterophryninae and Microhylinae.

Content: One genus, *Dyscophus*, with 3 species.

Distribution: Madagascar.

Characteristics: The skull has paired ethmoids, a single large prevomer, and teeth on the maxillaries. The vertebral column is diplasiocoelous. In the pectoral girdle, the clavicles and procoracoids range from well developed to vestigial.

Biology: Frogs in the genus *Dyscophus* are moderate to large (40–105 mm adult SVL) and inhabit the forest floors of Madagascar. The common name, tomato frog, is based on the frogs' bright red coloration. They usually breed in ephemeral pools or slow-moving backwaters of streams and swamps.

D. antongili lives around towns, where it burrows in sandy soil and breeds in sewage ditches. Eggs (1000+) are deposited on the water surface; they hatch within 36 hours, and the tadpoles grow moderately rapidly and metamorphose in 40 to 45 days.

Gastrophryninae

Sister taxon: Unnamed taxon.

Content: Nine genera, *Chiasmocleis*, *Ctenophryne*, *Dasypops*, *Dermatonotus*, *Elachistocleis*, *Gastrophryne*, *Hamptophryne*, *Hypopachus*, and *Nelsonophryne*, with 40 species.

Distribution: North and South America.

Characteristics: The skull usually has paired ethmoids, a single, generally small prevomer, and usually toothless maxillaries. The vertebral column is diplasiocoelous in most taxa, occasionally procoelous. In the pectoral girdle, the clavicles and procoracoids range from well developed to vestigial.

Biology: Gastrophrynines are predominantly terrestrial, stout-bodied, microcephalic frogs (Fig. 17.33). Most taxa are fossorial to semifossorial; they occupy a variety of habitats, from semiarid grasslands to scrub to rain forest. Most species deposit eggs that float on the surfaces of ponds, swamps, or other nonmoving water.

Hoplophryninae

Sister taxon: Uncertain.

Content: Two genera, *Hoplophryne* and *Parhoplophryne*, with 2 and 1 species, respectively.

Distribution: Usambara, Uluguru, and Magrotto Mountains, Tanzania.

Characteristics: The skull has a fused ethmoid and parasphenoid, the single prevomer is reduced anteriorly, and the maxillaries lack teeth. The vertebral column is procoelous, and the pectoral girdle has clavicles and procoracoids ranging from well developed to absent.

Biology: The three species are small (20–30 mm adult SVL) and toadlike. They are montane forest inhabitants and appear to be arboreal; they deposit eggs in holes in

FIGURE 17.33 Representative microhylids. Clockwise from upper left: Undescribed cross frog *Oreophryne* sp., Asterophryinae (S. J. Richards); Sumatra grainy frog *Kalophrynus pleurostigmus*, Kalophryninae (R. M. Brown); marbled rain frog *Scaphiophryne marmorata*, Scaphiophryninae (R. D. Bartlett); brown egg frog *Ctenophryne geayi*, Gastrophryninae (J. P. Caldwell).

the stems of bamboo or axils of bananas. The tadpoles are free living, and there is no evidence of parental care.

Kalophryninae

Sister taxon: Uncertain.

Content: One genus, *Kalophrynus,* with 14 species (Fig. 17.33).

Distribution: Southern China to Java and Philippines; Assam, India.

Characteristics: The skull usually has paired ethmoids, a single, generally small prevomer, and usually toothless maxillaries. The vertebral column is diplasiocoelous in most taxa, occasionally procoelous.

Biology: Some species breed in small pools, such as those formed in fallen logs or in pitcher plants, whereas others form nocturnal choruses in swampy areas or roadside ditches.

Melanobatrachinae

Sister taxon: Uncertain.

Content: One genus, *Melanobatrachus,* with 1 species.

Distribution: Southern India.

Characteristics: The skull has a fused ethmoid and parasphenoid, the single prevomer is reduced anteriorly, and the maxillaries lack teeth. The vertebral column is procoelous, and the pectoral girdle has clavicles and procoracoids ranging from well developed to absent.

Biology: *Melanobatrachus indicus* is a small frog (24–28 mm SVL) that lives along permanent forest streams of the Western Ghats. Little is known of its biology.

Microhylinae

Sister taxon: Asterophryninae.

Content: Nine genera, *Calluella, Chaperina, Glyphoglossus, Kaloula, Metaphrynella, Microhyla, Micryletta, Ramanella,* and *Uperodon,* with 68 species.

Distribution: Eastern Asia from India and Korea to the Greater Sunda Islands.

Characteristics: The skull usually has paired ethmoids, a single, generally small prevomer, and usually toothless maxillaries. The vertebral column is diplasiocoelous in most taxa, occasionally procoelous. In the pectoral girdle, the clavicles and procoracoids range from well developed to vestigial.

Biology: Microhylines are predominantly terrestrial, stout-bodied, microcephalic frogs. Most taxa are fossorial to semifossorial; they occupy a variety of habitats, from semiarid grasslands to scrub to rain forest. The widespread Asian *Kaloula* (15 species) has an elongate body, long limbs, and expanded toe tips; the species are active surface foragers, and some are arboreal and scansorial. Most microhylines have free-living tadpoles. The Asian *Calluella* are moderate-sized forest frogs (30–60 mm adult SVL) that are rarely observed because of their fossorial habitats.

Otophryninae

Sister taxon: Uncertain.

Content: One genus, *Otophryne*, with 2 species.

Distribution: Northern South America.

Characteristics: The skull has paired ethmoids, a pair of prevomers, and toothless maxillaries. The vertebral column is diplasiocoelous. In the pectoral girdle, the clavicles and procoracoids are well developed.

Biology: *Otophryne robusta,* the pe-ret' toad, is a moderate-sized (44–60 mm adult SVL), forest-floor frog. This diurnally active frog is a leaf mimic with a dorsal pattern that shows the pinnation of dry leaves and ranges in color from yellow through shades of dusky red to brown. It walks, rather than hops, among the leaf litter. Breeding occurs adjacent to forest streams, and the eggs are laid on land beneath wet leaves (perhaps in the water also, but that is not confirmed). The tadpoles have tiny needle-like, keratinized labial teeth and a long spiracular tube, apparently adaptations for their aquatic–fossorial habit of burrowing and feeding in the sand, either on stream banks or in the stream bottom. The length of tadpole stage is thought to be less than 1 year. Whether the keratinized mouthparts are basal and thus plesiomorphic, or adaptations to a fossorial lifestyle, requires further study.

Phrynomerinae

Sister taxon: Uncertain.

Content: One genus, *Phrynomantis,* with 5 species.

Distribution: Sub-Saharan Africa.

Characteristics: The skull has paired ethmoids, a single, anteriorly reduced prevomer, and toothless maxillaries. The vertebral column is diplasiocoelous. In the pectoral girdle, the clavicles and procoracoids are absent. An intercalary cartilage occurs between the terminal and penultimate phalanges of each digit.

Biology: *Phrynomantis* looks like an elongated, heavy-bodied *Dendrobates* with a similar skin texture and aposematic coloration. Skin secretions are toxic, at least in *P. bifasciatus*. Like *Dendrobates*, all species are diurnal and terrestrial and have a diet composed of ants. *Phrynomantis* contains mostly moderate-sized frogs (30–45 mm adult SVL), although in some populations, *P. bifasciatus* reaches 80 mm. They walk or run, seldom hopping, and often climb in the lower branches of shrubs. Because they are mainly savanna inhabitants, they are seldom seen except in the wet season when they reproduce in ephemeral ponds. The eggs are laid in small masses (100–1400) at the surface of the water, attached to vegetation or floating. The eggs hatch quickly; the tadpoles are filter feeders and remain suspended motionless, except for tiny vibrations of the tail, in the middle of the water column. Development is fairly rapid with metamorphosis occurring in 30–40 days.

Comment: *Phrynomantis* is a presumed senior synonym, hence it replaces *Phrynomerus* (Duellman, 1993).

Scaphiophryninae

Sister taxon: Cophylinae.

Content: Two genera, *Paradoxophyla* and *Scaphiophryne,* with 2 and 8 species, respectively.

Distribution: Madagascar.

Characteristics: The skull has a single large ethmoid, a single large prevomer, and toothless maxillaries. The vertebral column is diplasiocoelous. In the pectoral girdle, the clavicles and procoracoids are well developed.

Biology: Scaphiophryines are small- to moderate-sized frogs (20–50 mm adult SVL) (Fig. 17.33). They are predominantly terrestrial, occurring either in moist forest or grassland and scrublands. In the latter environments, they are burrowers and emerge for feeding and reproduction with the onset of the wet season. They usually breed explosively in ephemeral pools. Modest clutches of eggs (<1000) float on the surface of the water and hatch into free-living tadpoles. The forest dwellers feed nocturnally in the forest litter and have similar reproductive habits. The tadpole of *Scaphiophryne* differs from other microhylid tadpoles in having keratinized jaw sheaths and labial teeth, whereas *Paradoxophyla* has a typical microhylid tadpole that lacks these structures. Molecular data show a sister relationship between *Scaphiophryne* and *Paradoxophyla*, suggesting that the tadpole of *Scaphiophryne* is a reversal of the typical microhylid form.

Brevicipitidae

Rain Frogs

Classification: Amphibia; Lissamphibia; Anura; Neobatrachia; Ranoides; Allodapanura; Xinosyneunitanura.

Sister taxon: Hemisotidae.

Content: Five genera, *Balebreviceps, Breviceps, Callulina, Probreviceps*, and *Spelaeophryne*, with 1, 16, 2, 6, and 1 species, respectively.

Distribution: Sub-Saharan east and southern Africa (Fig. 17.34).

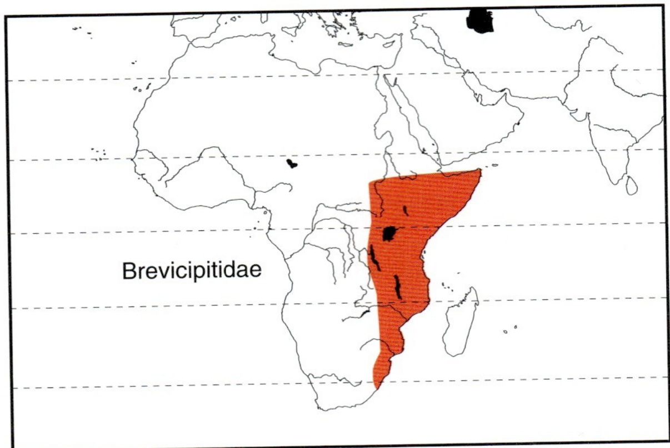

FIGURE 17.34 Geographic distribution of the extant Brevicipitidae.

Characteristics: The skull has no ethmoids, a single, anteriorly narrowed prevomer, and toothless maxillaries. The vertebral column is diplasiocoelous, and the pectoral girdle has well-developed clavicles and procoracoids. In addition, the middle ear is absent, and the urostyle and sacral vertebra may be fused.

Biology: The rain frog, *Breviceps*, and its relatives are nearly spherical in shape, with the head barely distinguishable from the body; limbs are short and robust (Fig. 17.35). Their globular appearance is further enhanced by a tendency to inflate the body when disturbed. These small-to-moderate-sized frogs (most 30–50 mm adult SVL) are backward burrowers and found from forest to near-desert habitats. As is common in many frogs with this body form, males are distinctly smaller than females; this size disparity and their short limbs prevent a typical amplexus. The problem is resolved by skin secretions that glue the male to the female's back for the duration of egg deposition. Small clutches of eggs are laid in subterranean nests, and the tadpoles have direct development. The female remains with the eggs until they hatch, usually in 4 to 5 weeks.

Hemisotidae

Shovel-Nosed Frogs

Classification: Amphibia; Lissamphibia; Anura; Neobatrachia; Ranoides; Allodapanura; Xinosyneunitanura.

Sister taxon: Brevicipitidae.

Content: One genus, *Hemisus*, with 9 species.

Distribution: Africa south of the Sahara (Fig. 17.36).

Characteristics: Hemisotids are small- to moderate-sized frogs (22–52 mm adult SVL, except *H. guttatum*, which reaches 75 mm) with stout bodies and small, pointed heads. The skull has paired palatines and frontoparietals. The facial nerve exits through the anterior acoustic foramen in the auditory capsule; the trigeminal and facial nerve ganglia fuse to form a prootic ganglion. The vertebral column possesses eight presacral holochordal vertebrae (first and second presacrals fused), and all are procoelous except for the biconcave surface of last presacral. The transverse processes of the sacral vertebra are moderately expanded, and this vertebra has a bicondylar articulation with the urostyle. Postmetamorphs have no dorsal ribs on the presacral vertebrae. The pectoral girdle is firmisternal with a distinct sternum. The fibulare and

FIGURE 17.35 Representative brevicipitid, arthroleptid, and hyperoliid frogs. Clockwise from upper left: Transvaal forest rain frog *Breviceps sylvestris,* Brevicipitidae (R. D. Bartlett); shovel-foot squeaker *Arthroleptis stenodactylus,* Arthroleptinae (L. W. Porras); forest tree frog *Leptopelis natalensis*, Leptopelinae (R. D. Bartlett); long reed frog *Hyperolius nasutus,* Hyperoliidae (R. D. Bartlett).

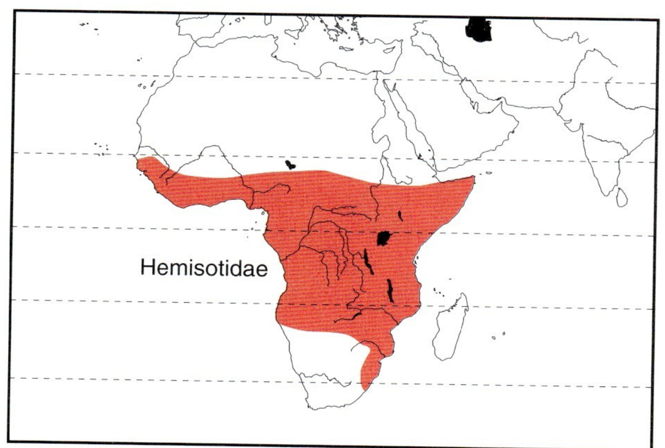

FIGURE 17.36 Geographic distribution of the extant Hemisotidae.

tibiale are fused at their proximal and distal ends. No intercalary cartilage occurs between the terminal and penultimate phalanges of each digit, and the tips of the terminal phalanges are blunt to pointed. The larvae have keratinized mouthparts, and the left and right branchial chambers fuse behind the heart and are emptied by a spiracle on the left side at midbody.

Biology: *Hemisus* species are burrowers, burrowing headfirst. They use the pointed and hardened snout as a ramming rod, moving the head up and down, throwing the soil to the rear with the forelimbs, and pushing forward with the hindlimbs. They are largely savanna inhabitants, although they also live in scrub and gallery forests. They appear to feed both beneath and on the surface, mainly on soft-bodied arthropods and worms. Reproduction begins with the earliest heavy rains of the wet season, or even before the rains arrive. The male calls from the ground; subsequently, a pair remains in inguinal amplexus while the female digs an incubation chamber near, but not in, an ephemeral pool. About 100 to 250 eggs in *H. marmoratus* and about 2000 in *H. guttatus* are fertilized in the chamber. The male burrows out of the chamber, and the female remains with eggs. When the eggs hatch, the female carries or guides her tadpoles to the nearby pool, digging an escape tunnel or a surface channel. The tadpoles are free living and metamorphose in approximately 3 to 4 weeks.

Arthroleptidae

Squeakers and Cricket Frogs

Classification: Amphibia; Lissamphibia; Anura; Neobatrachia; Ranoides; Allodapanura; Laurentobatrachia.

Sister taxon: Hyperoliidae.

Content: Two subfamilies, Arthroleptinae and Leptopelinae.

Distribution: Sub-Saharan Africa (Fig. 17.37).

Characteristics: Arthroleptids are mostly small frogs with pointed snouts and long limbs. The skull has paired palatines and frontoparietals. The facial nerve exits through the anterior acoustic foramen in the auditory capsule; the trigeminal and facial nerve ganglia fuse to form a prootic ganglion. The vertebral column possesses eight presacral holochordal vertebrae, and all are procoelous except for a biconcave surface on the last presacral. The transverse processes of the sacral vertebra are cylindrical, and this vertebra has a bicondylar articulation with the urostyle. Postmetamorphs have no dorsal ribs on the presacral vertebrae. The pectoral girdle is firmisternal with a distinct sternum. The fibulare and tibiale are fused at their proximal and distal ends. No intercalary cartilage occurs between the terminal and penultimate phalanges of each digit, and the tips of the terminal phalanges are blunt, pointed, or T-shaped. The larvae have keratinized mouthparts, and the left and right branchial chambers fuse behind the heart and are emptied by a spiracle on the left side at midbody.

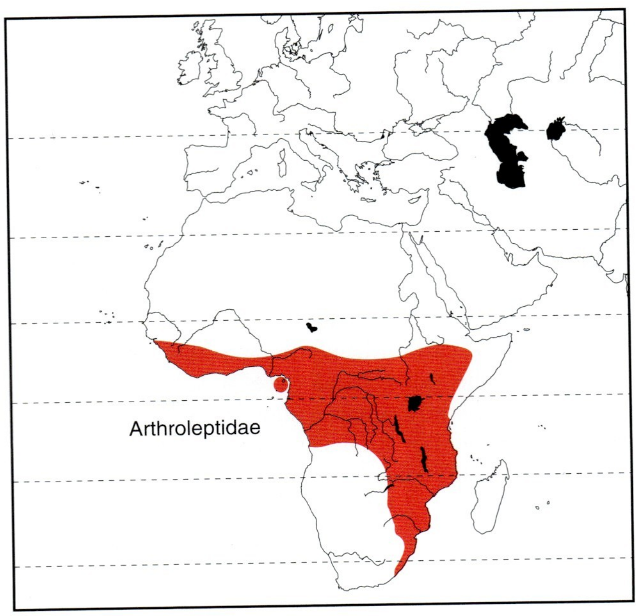

FIGURE 17.37 Geographic distribution of the extant Arthroleptidae.

Arthroleptinae

Sister taxon: Leptopelinae.

Content: Seven genera, *Arthroleptis, Astylosternus, Cardioglossa, Leptodactylodon, Nyctibates, Scotobleps,* and *Trichobatrachus*, with 81 species.

Distribution: Sub-Saharan Africa.

Characteristics: Arthroleptines have vertical or horizontal pupils, the vomerine bears teeth or is edentate, and the terminal phalanges on most species are T-shaped.

Biology: Arthroleptines are predominantly small (*Leptodactylon albiventris*, 20–21 mm adult SVL) to large (*Trichobatrachus robustus*, 80–130 mm SVL) frogs, although most are moderate in size (Fig. 17.36). Some are terrestrial, occurring in a variety of habitats from grassland to open forest, usually away from standing water. They typically breed after heavy summer rains. Males with high-pitched, insect-like squeaks form large, diffuse choruses. Amplexus is axillary, and small clutches (10–30) of large yolky eggs are laid in leaf litter of the forest floor. About 4 weeks after deposition, tiny froglets hatch. Other species are closely associated with running water, living either in the water or immediately adjacent to it. *Trichobatrachus robustus*, the hairy frog, is the most aquatic species. Dense patches of fine hair-like projections are located on the sides in adult males. These microvillae-like structures are heavily vascularized and likely associated with cutaneous respiration. *Cardioglossa* and other genera have a tadpole stage.

Leptopelinae

Sister taxon: Arthroleptinae.

Content: One genus, *Leptopelis*, with 51 species.

Distribution: Sub-Saharan Africa.

Characteristics: The vocal pouch and associated gular gland are absent. The forearm gland is well developed, but digital glands are absent. The aponeurosis of the palmar muscular is well developed and mobile.

Biology: *Leptopelis* consists mainly of arboreal forest species, with the greatest diversity in equatorial Africa; however, in more arid areas, the species are terrestrial to subfossorial and climb into the trees only for breeding (e.g., *L. bocagii*, Zimbabwe). Most species are medium-sized and range from 26 to 42 mm adult SVL, but *L. palmatus* ranges in size from 45 to 87 mm SVL. Many species use a gaping defense display in which the mouth is opened fully, the eyes are half-closed, and the body may be arched. Breeding is associated with heavy rains, usually at the beginning of the wet season. Males call solitarily. Amplexus is axillary, and eggs are deposited in various situations from ephemeral pools or backwaters of streams to holes in the ground. Parental care has not been reported for any species, and development includes a tadpole stage, except perhaps in *L. brevirostris*, in which the female has been reported to carry large eggs in her mouth.

Hyperoliidae

African Reed Frogs and Running Frogs

Classification: Amphibia; Lissamphibia; Anura; Neobatrachia; Ranoides; Allodapanura; Laurentobatrachia.

Sister taxon: Arthroleptidae.

Content: Seventeen genera, *Acanthixalus*, *Afrixalus*, *Alexteroon*, *Arequinus*, *Callixalus*, *Chlorolius*, *Chrysobatrachus*, *Cryptothylax*, *Heterixalus*, *Hyperolius*, *Kassina*, *Kassinula*, *Opisthothylax*, *Paracassina*, *Phlyctimantis*, *Semnodactylus*, and *Tachycnemis*, with 207 species.

Distribution: Sub-Saharan Africa, Madagascar, and the Seychelles (Fig. 17.38).

Characteristics: Most hyperoliids are tree frogs with expanded toe pads. The skull has paired palatines and frontoparietals. The facial nerve exits through the anterior acoustic foramen in the auditory capsule; the trigeminal and facial nerve ganglia fuse to form a prootic ganglion. The vertebral column possesses eight presacral holochordal vertebrae, and all are procoelous except for a biconcave surface on the last presacral (procoelous in *Acanthixalus* and *Callixalus*). The transverse processes of the sacral vertebra are cylindrical, and this vertebra has a bicondylar articulation with the urostyle. Postmetamorphs have no dorsal ribs on the presacral vertebrae. The pectoral girdle is firmisternal with a distinct sternum. Fibulare and tibiale are fused at their proximal and distal ends. An intercalary cartilage occurs between the terminal and penultimate phalanges of each digit; the tips of the terminal phalanges are blunt to pointed or T-shaped. Hyperoliids are unique in having a distinctive gular gland.

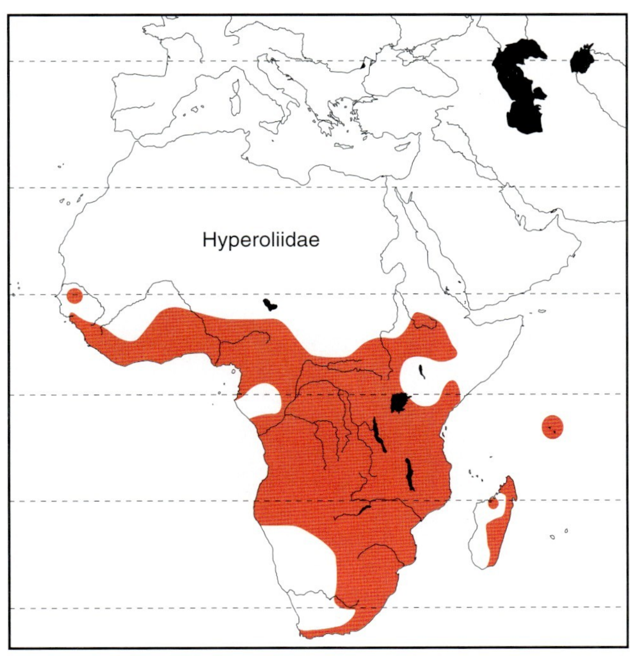

FIGURE 17.38 Geographic distribution of the extant Hyperoliidae.

The larvae have keratinized mouthparts, and the left and right branchial chambers fuse behind the heart and are emptied by a spiracle on the left side at midbody.

Biology: Hyperoliids range in size from small (17–22 mm adult SVL, e.g., *Afrixalus knysnae*, *Hyperolius pusillus*) to moderate (25–43 mm, e.g., *Hyperolius puncticulatus*) (Fig. 17.35). They occur in grasslands and marshes to full-canopied forest. Most are arboreal, whether living in grass and reeds or shrubs and trees. Males call from a variety of sites, usually in choruses. Amplexus is axillary, and eggs are laid in or over water and invariably attached to the vegetation. Development includes the tadpole stage (except in *Alexteroon obstetricans*); if the eggs are laid away from water, the tadpole must reach water upon hatching. *Tachycnemis seychellensis* is a moderate-sized tree frog (33–77 mm adult SVL) that lives in forests on the Seychelles Islands. Breeding appears to occur irregularly throughout the year. Males call in small choruses from low vegetation adjacent to ephemeral pools or forest streams. Eggs are usually deposited on vegetation overhanging the water, and the hatching tadpoles fall into the water. The duration of the tadpole phase is unknown.

Ptychadenidae

Grassland Frogs

Classification: Amphibia; Lissamphibia; Anura; Neobatrachia; Ranoides; Natatanura.

Sister taxon: Victoranura (clade containing Ceratobatrachidae and Telmatobatrachia).

Content: Three genera, *Hildebrandtia*, *Lanzarana*, and *Ptychadena*, with 3, 1, and 49 species, respectively.

Distribution: Sub-Saharan Africa (Fig. 17.39).

Characteristics: The skull lacks palatine bones, and the otic process of the squamosal is reduced or absent. The last presacral and sacral vertebrae are fused, and the clavicle is reduced and usually fused to the coracoid.

Biology: Most ptychadenids are slender, long-limbed frogs. Most species are moderate sized (40–60 mm adult SVL) and mainly inhabit savannas or grasslands. The ribbed or sharp-nosed frogs, *Ptychadena* (Fig. 17.40), are the most speciose (49 species) and define the distribution of the family. Owing to their semiarid habitats, they are most evident in the wet season and usually begin reproduction several weeks after the rains have begun. Males form noisy choruses in shallow, ephemeral pools, and females deposit modest-sized clutches of 200 to 500 eggs among the vegetation. The eggs hatch quickly, and the tadpoles usually metamorphose within 4 to 5 weeks. Two species of *Ptychadena* have unusual reproductive modes for the genus. *Ptychadena broadleyi* deposits eggs on moist rocks, and the tadpoles live in the film of water covering the rock face. Females of *Ptychadena*

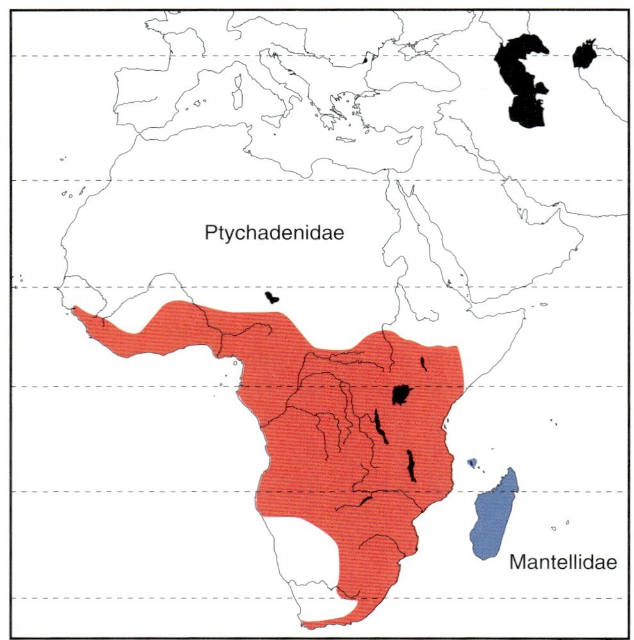

FIGURE 17.39 Geographic distribution of the extant Ptychadenidae and Mantellidae.

aequiplicata deposit eggs in communal masses of several hundred to a thousand eggs on the ground among vegetation. The communal masses are placed near dried ponds and undergo development for several weeks to Gosner Stage 28. When ponds fill, the eggs hatch almost immediately and tadpoles enter the ponds. Metamorphosis occurs within about 2 weeks after tadpoles enter the ponds.

Ceratobatrachidae

Triangle Frogs and Others

Classification: Amphibia; Lissamphibia; Anura; Neobatrachia; Ranoides; Natatanura; Victorana.

Sister taxon: Telmatobatrachia (clade containing Micrixalidae and Ametrobatrachia).

Content: Five genera, *Batrachylodes, Ceratobatrachus, Discodeles, Palmatorappia*, and *Platymantis*, with 8, 1, 5, 1, and 61 species, respectively.

Distribution: Malaysia, Philippines, Borneo, New Guinea, and Solomon Islands (Fig. 17.41).

Characteristics: Body form ranges from squat and toadlike (e.g., *Discodeles bufoniformis*) to tree frog-like

FIGURE 17.40 Representative ptychadenid, ceratobatrachid, dicroglossid, and mantellid frogs. Clockwise from upper left: Seychelles ribbed frog *Ptychadena mascareniensis*, Ptychadenidae (G. R. Zug); Papua wrinkled ground frog *Platymantis papuensis*, Ceratobatrachidae (E. R. Lingren); Burmese spadefoot *Sphaerotheca breviceps*, Dicroglossinae (G. R. Zug); black golden frog *Mantella cowanii*, Mantellinae (R. D. Bartlett).

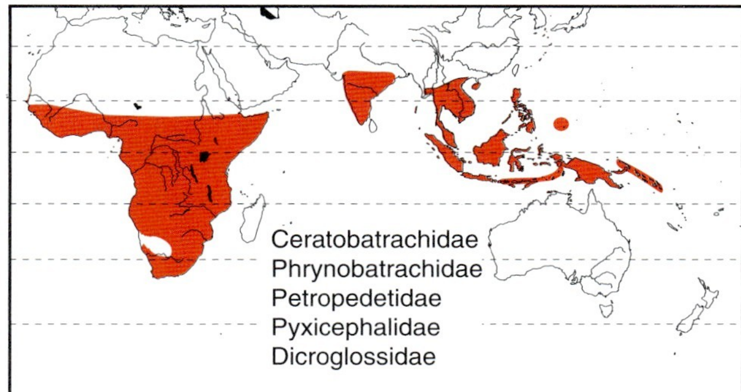

FIGURE 17.41 Map showing combined geographic distribution of the extant Ceratobatrachidae, Phrynobatrachidae, Petropedetidae, Pyxicephalidae, and Dicroglossidae (see text).

(some *Platymantis*). The pectoral girdle is firmisternal, and the condition of the sternum varies from not bifurcate to strongly bifurcate posteriorly. In the skull, the nasals are either reduced and do not touch or are generally in broad contact with one another and with the frontoparietals.

Biology: *Platymantis* and its relatives deposit terrestrial eggs that have direct development (Fig. 17.40). *Ceratobatrachus* has fangs (odontoids) on its lower jaw, which may be used to subdue large prey items.

Micrixalidae

Tropical Frogs

Classification: Amphibia; Lissamphibia; Anura; Neobatrachia; Ranoides; Natatanura; Victorana; Telmatobatrachia.

Sister taxon: Ranixalidae.

Content: One genus, *Micrixalus,* with 11 species.

Distribution: India.

Characteristics: The pectoral girdle is firmisternal, and the omosternum is not bifurcate. Tadpoles have only one row of labial teeth.

Biology: *Micrixalus saxicola* is a small torrent frog (25–35 mm SVL) that breeds in fast-flowing streams. Males are territorial and are often found on wet rocks along the stream. In addition to calling, males use a visual display, foot flagging, to challenge males that enter their territories. Eggs are deposited on the surface of rocks or along the shoreline, where they are bathed with water. *M. fuscus* was one of the most common amphibians in a study of a stream-breeding and rain forest floor communities in the Western Ghats, India.

Ranixalidae

Leaping Frogs

Classification: Amphibia; Lissamphibia; Anura; Neobatrachia; Ranoides; Natatanura; Victorana; Telmatobatrachia.

Sister taxon: Micrixalidae.

Content: One genus, *Indirana*, with 10 species.

Distribution: Central and southern India.

Characteristics: *Indirana* has a median lingual tubercle. Males have femoral glands of variable size and spicules around the margin of the jaw or on the chest region. The terminal phalanges are Y-shaped.

Biology: Most species are found in leaf litter or near streams in tropical moist deciduous or evergreen forests. Arboreal tadpoles are found on rocks adjacent to streams. They have elongate, dorsoventrally flattened bodies with low tail fins.

Phrynobatrachidae

Puddle Frogs

Classification: Amphibia; Lissamphibia; Anura; Neobatrachia; Ranoides; Natatanura; Victorana; Africanura.

Sister taxon: Pyxicephaloidea (clade containing Petropedetidae and Pyxicephalidae).

Content: One genus, *Phrynobatrachus,* with 76 species.

Distribution: Sub-Saharan Africa (Fig. 17.41).

Characteristics: *Phrynobatrachus* has a medial lingual tubercle. The pectoral girdle is firmisternal, and the omosternum is not bifurcate. Many species have T-shaped terminal digits but lack discs. They have a distinctive tarsal tubercle. Tadpoles have only one row of labial teeth.

Biology: Frogs of the genus *Phrynobatrachus* are among the most widespread and abundant amphibians in Africa. They are typically small, fast-moving frogs and occupy habitats from dry savannas to primary rain forests. Many savanna species breed in stagnant water, including small ponds, buffalo wallows, flooded rice paddies, and other types of turbid waters. Almost all species deposit many small eggs in a single-layered surface clutch on stagnant or slow-moving water and have exotrophic tadpoles. Two species, *P. guineensis* and *P. tokba*, have strikingly different reproductive modes. The former attaches a few large eggs above the waterline in water-filled tree holes, empty fruit capsules, or snail shells, and the latter has endotrophic (nonfeeding) tadpoles that develop in a terrestrial leaf nest in dense vegetation of secondary forests or savannas. Many species of *Phrynobatrachus* are known for their polymorphic colors and patterns, making identification of species in the field difficult.

Petropedetidae

African Water Frogs, Goliath Frog

Classification: Amphibia; Lissamphibia; Anura; Neobatrachia; Ranoides; Natatanura; Victorana; Africanura; Pyxicephaloidea.

Sister taxon: Pyxicephalidae.

Content: Two genera, *Conraua* and *Petropedetes*, with 6 and 10 species, respectively.

Distribution: Sub-Saharan Africa; central and southern India (Fig. 17.41).

Characteristics: Males of most species of *Petropedetes* have femoral glands, enlarged forearms, and a ring of papillae around the tympanum. *Petropedetes* and *Conraua* have spines on the chin and throat. *Petropedetes* has T-shaped terminal phalanges, whereas *Conraua* has simple terminal phalanges.

Biology: *Conraua goliath* is the largest frog in the world, reaching a size of 300 mm SVL and weighing as much as 3.3 kg. It is typically found in rapids in fast-moving, sandy-bottomed rivers in West African rain forest. Eggs are deposited in rocky areas of pools near rapids. Species of *Petropedetes* inhabit rocky streams in forested mountains. Tadpoles of *P. martiensseni* and *P. yakusini* are elongate and highly specialized for living in the water film of rock faces in streams.

Pyxicephalidae

African Bullfrogs and Others

Classification: Amphibia; Lissamphibia; Anura; Neobatrachia; Ranoides; Natatanura; Victorana; Africanura; Pyxicephaloidea.

Sister taxon: Petropedetidae.

Content: Two subfamilies, Cacosterninae and Pyxicephalinae.

Distribution: Sub-Saharan Africa (Fig. 17.41).

Characteristics: In the skull, the vomer is usually toothless or has only a small posterior patch of teeth. The tongue is notched. The vertebral column is composed of eight presacral vertebrae, the sacrum, and the urostyle. The first seven presacrals are procoelus, whereas the eighth presacral is amphicoelus. The terminal phalanges are usually T-shaped.

Biology: The pyxicelphalines are large, stocky bullfrogs that typically live in savanna-like habitats and breed in summer months. They have fanglike projections in the lower jaw and complex parental care. The cacosternines tend to be small frogs with generalized but varied ecology and reproduction. *Natalobatrachus* is a semiarboreal, rain forest dweller, and *Tomopterna* is a burrowing inhabitant of savannas.

Cacosterninae

Sister taxon: Pyxicephalinae.

Content: Eleven genera, *Amietia, Anhydrophryne, Arthroleptella, Cacosternum, Ericabatrachus, Microbatrachella, Natalobatrachus, Nothophryne, Poyntonia, Strongylopus,* and *Tomopterna*, with 59 species.

Distribution: Sub-Saharan Africa.

Characteristics: In the skull, the vomer is usually toothless or has only a small posterior patch of teeth. The tongue is notched. The terminal phalanges are usually T-shaped.

Biology: Cacosternines are mostly small (< 30 mm adult SVL), although some are medium-sized, typical ranid-like frogs. They are terrestrial or semiaquatic and generally live in moist habitats or rocky montane streams, although some live in savannas and emerge from subterranean retreats only with the arrival of the wet season. Some (e.g., *Cacosternum* and *Natalobatrachus*) deposit aquatic eggs that hatch into free-living tadpoles. Others have direct development. *Arthroleptella* lays small clutches of 20–40 eggs in damp cavities beneath moss or detritus; the eggs hatch into nonfeeding tadpoles that quickly metamorphose into tiny 3–4 mm froglets. *Anhydrophryne* deposits small clutches of 10–30 terrestrial eggs that hatch directly into froglets. *Tomopterna,* called pyxies, are pelobatid-like with short, robust bodies (30–60 mm adult SVL) and an enlarged, spade-shaped tubercle on each hind foot. These terrestrial frogs occur in dry habitats such as open forests, scrub, and grasslands. They are semifossorial, emerging on moist evenings to forage on the surface. Most are explosive breeders; they appear in great numbers after heavy rains, depositing eggs in ephemeral pools before returning to their terrestrial habitats. The tadpoles are free living and develop in 4–5 weeks.

Pyxicephalinae

Sister taxon: Cacosterninae.

Content: Two genera, *Aubria* and *Pyxicephalus*, with 2 and 3 species, respectively.

Distribution: Sub-Saharan Africa.

Characteristics: In *Pyxicephalus*, much of the skeleton is highly ossified. The frontoparietals are paired and highly exostosed. The vertebral column is composed of eight presacral vertebrae, the sacrum, and the urostyle. The first seven presacrals are procoelus, whereas the eighth presacral is amphicoelus. The neural arches bear dorsally projecting spines. The anterior presacral vertebrae have expanded sacral diapophyses. The urostyle has a bicondylar articulation with the sternum. The pectoral girdle is firmisternal, with an ossified omosternum and sternum. The tibiale and fibulare are fused their entire length.

Biology: Both pyxicephaline genera are moderately large and stocky frogs (*Aubria,* 50–100 mm; *Pyxicephalus,* 60–195 mm adult SVL). These frogs mainly occur in dry habitats, usually savannas. The African bullfrog (*P. adspersus*) occurs throughout much of the distribution of pyxicephalines and accounts for much of our knowledge of the clade. It has several geographic morphs, and these morphs vary in size from moderately large to very large. In general, *P. adspersus* is active only during the summer months, emerging when the summer rains occur and feeding voraciously. *P. adspersus* captures large prey, which it holds with two bony pseudoteeth (dentary tusks) in the front of the lower jaw. Reproduction occurs in ephemeral pools, and the 3000 to 4000 eggs are deposited. Males have been reported to individually guard a pool filled with eggs and tadpoles, and in some instances, to construct channels between bodies of water allowing tadpole schools to exit shallow pools. Although they occasionally eat tadpoles, no evidence exists indicating that they feed on their own tadpoles. *Aubria subsigillata* also deposits large egg clutches; when these hatch, the tadpoles form dense schools with the individual tadpoles appearing to touch one another.

Dicroglossidae

Fanged Frogs, Tiger Frogs, and Others

Classification: Amphibia; Lissamphibia; Anura; Neobatrachia; Ranoides; Natatanura; Victorana; Saukrobatrachia.

Sister taxon: Aglaioanura (clade containing Rhacophoroidea and Ranoidea).

Content: Two subfamilies, Dicroglossinae and Occidozyginae (Fig. 17.41).

Distribution: Sub-Saharan to central Africa, South Asia through the East Indies to the Philippines and New Guinea into the Southwest Pacific islands.

Characteristics: The pectoral girdle is firmisternal, and its sternum is moderately to strongly bifurcate posteriorly. In the skull, the nasals are generally in broad contact with one another and with the frontoparietals. Vomerine teeth are present in dicroglossines but absent in occidozygines. The tympanum varies from distinct to hidden. Digits are rounded or pointed and may have dorsal grooves in some species.

Biology: Tadpoles of dicroglossids show a wide range of morphological adaptations, from semiterrestrial forms that scrape algae from the surface film of rocks to carnivorous forms with specialized mouthparts.

Dicroglossinae

Sister taxon: Occidozyginae.

Content: Thirteen genera, *Allopaa, Annandia, Chrysopaa, Euphlyctis, Fejervarya, Hoplobatrachus, Limnonectes, Minervarya, Nannophrys, Nanorana, Ombrana, Quasipaa*, and *Sphaerotheca*, with 142 species.

Distribution: Widespread from Africa through parts of Asia and Indochina to Japan and the Philippines.

Characteristics: The pectoral girdle is firmisternal, and its sternum is moderately to strongly bifurcate posteriorly. In the skull, the nasals are generally in broad contact with one another and with the frontoparietals. Vomerine teeth are present. Some species of *Limnonectes* have enlarged heads and fangs in front of their mandibles.

Biology: Habitats vary from terrestrial (e.g., *Hoplobatrachus tigerinus*) to semiterrestrial (*Nannophrys ceylonensis*) to pond edges or paddy fields (*Euphlyctis*). Some species have direct development, whereas others have free-living tadpoles. Fanged frogs, exemplified by *Limnonectes kuhlii*, have an unusual suite of reproductive features, including enlarged heads and presence of fangs in males, parental care, nest building, absence of male call in some species, and presence of female call. In addition, pairs utilize a handstand-like posture for spawning. Shallow nests are constructed in forest streams in gravel or sand. *Nannophrys ceylonensis* is found near small streams along road cuts, where it inhabits boulders and vertical rock walls. Its tadpoles are adapted for semiterrestrial life in that the body is dorsoventrally flattened, the mouth is ventral, and the fins are reduced. The tadpoles scrape food from the surface film of rocks. Tadpoles of *Hoplobatrachus* are carnivorous.

Occidozyginae

Sister taxon: Dicroglossinae.

Content: Two genera, *Ingerana* and *Occidozyga*, with 10 and 12 species, respectively.

Distribution: Widespread through Asia and southern China; Greater and Lesser Sunda Islands.

Characteristics: Dicroglossines are a diverse group of frogs. Some are small (e.g., *Occidozyga baluensis*, 15–35 mm adult SVL). *Occidozyga* has the lateral-line system present in adults, entirely webbed feet, and pointed digits. The larvae are aquatic with reduced mouthparts.

Biology: Some species of *Occidozyga* have inguinal rather than axillary amplexus.

Mantellidae

Malagasy Poison Frogs and Others

Classification: Amphibia; Lissamphibia; Anura; Neobatrachia; Ranoides; Natatanura; Victorana; Saukrobatrachia; Rhacophoroidea.

Sister taxon: Rhacophoridae.

Content: Three subfamilies, Boophinae, Laliostominae, and Mantellinae, with 166 species.

Distribution: Madagascar and Mayotte Island (Fig. 17.39).

Characteristics: The skull has paired palatines and frontoparietals. The facial nerve exits through the anterior acoustic foramen in the auditory capsule; the trigeminal and facial nerve ganglia fuse to form a prootic ganglion. The vertebral column possesses eight presacral holochordal vertebrae, and all are procoelous except for a biconcave surface on the last presacral. The transverse processes of the sacral vertebra are cylindrical, and this vertebra has a bicondylar articulation with the urostyle. Postmetamorphs have no dorsal ribs on the presacral vertebrae. The pectoral girdle is firmisternal with a distinct sternum. Fibulare and tibiale are fused at their proximal and distal ends. Intercalary phalangeal elements are present. Tadpoles have keratinized mouthparts, and the left and right branchial chambers fuse behind the heart and are emptied by a spiracle on the left side at midbody.

Biology: Most mantellids are small- to medium-sized (15–50 mm adult SVL; *Mantidactylus guttulatus* reaches 100–120 mm) terrestrial or arboreal frogs; most species live in semiarid to wet forested habitats. Mantellinae (Fig. 17.40) is the most speciose (107 species) and behaviorally diverse group; many are cryptically colored in shades of green to brown. In contrast, *Mantella* (16 species) are commonly boldly, even gaudily, colored. Their bold and contrasting colors advertise their toxic skin secretions, containing lipophilic alkaloids. They share toxic skin secretions, advertising coloration, size, and habitus with some dendrobatids, but this similarity is because of convergence, not relationship. Reproductive behavior is diverse. Most, if not all, mantellids have male vocalization and axillary amplexus. In most genera, eggs appear to be laid away from water. For those with aquatic larvae, the hatching tadpoles drop into the water from clutches deposited in overhanging vegetation (e.g., *Guibemantis liber*) or are washed into streams or pools from terrestrial nests (e.g., *Mantidactylus betsileanus*). Other species have terrestrial or arboreal nonfeeding larvae (e.g., *Gephyromantis pseduodasper*), and direct development occurs in *Gephyromantis eiselti,* although this species does not have parental care. In *Mantella*, courtship is brief with no real amplexus; the male either lays on the head and shoulder of the female or loosely grasps her on the trunk. Eggs are deposited out of water.

Boophinae

Sister taxon: Unnamed clade containing Mantellinae and Laliostominae.

Content: One genus, *Boophis*, with 54 species.

Distribution: Madagascar and Mayotte Island.

Characteristics: Frogs in the genus *Boophis* are typically small, toxic, and arboreal.

Biology: These frogs inhabit tropical or subtropical lowland and montane forests. Tadpoles of several species of *Boophis* develop in high-altitude streams and have numerous rows of labial teeth. They attain a large size and may require more than a year to metamorphose.

Laliostominae

Sister taxon: Mantellinae.

Content: Two genera, *Aglyptodactylus* and *Laliostoma,* with 3 and 1 species, respectively.

Distribution: Madagascar.

Characteristics: These frogs are robust bodied, ranid-like in appearance.

Biology: *Laliostoma labrosum* is an abundant species that occurs in open areas, including rice fields and other agricultural sites. It is an explosive breeder in ponds and other still waters. Species of *Aglyptodactylus* are also explosive breeders, and their tadpoles may transform in as little as 12 days. Eggs are deposited as a single-layered surface film, and tadpoles of all species are exotrophic and morphologically similar.

Mantellinae

Sister taxon: Laliostominae.

Content: Eight genera, *Blommersia* (6 species), *Boehmantis* (1 species), *Gephyromantis* (34 species), *Guibemantis* (10 species), *Mantella* (16 species), *Mantidactylus* (29 species), *Spinomantis* (10 species), and *Wakea* (1 species).

Distribution: Madagascar and Mayotte Island.

Characteristics: Mantellines are a group of highly diverse, mostly small- to medium-sized frogs. Recently, the large genus *Mantidactylus* was partitioned into six genera. Species of *Mantella* are brightly colored and toxic. They are convergent in many characteristics with Neotropical dendrobatids.

Biology: Some of the species formerly considered part of the heterogenous *Mantidactylus* inhabit plant axils of the screw pine tree, *Pandanus*, whereas others are large, ground-dwelling species, some of which live around rain forest streams. *Mantella* are small, brightly colored, diurnal frogs and, like dendrobatids, have alkaloid toxins in their skin.

Rhacophoridae

Afroasian Tree Frogs

Classification: Amphibia; Lissamphibia; Anura; Neobatrachia; Ranoides; Natatanura; Victorana; Saukrobatrachia; Rhacophoroidea.

Sister taxon: Mantellidae.

Content: Two subfamilies, Buergeriinae and Rhacophorinae.

Distribution: Sub-Saharan Africa, Madagascar, and South Asia. (Fig. 17.42).

Characteristics: Rhacophorids are mainly tree frogs, ranging from small to large species. The skull has paired palatines and frontoparietals. The facial nerve exits through the anterior acoustic foramen in the auditory capsule; the trigeminal and facial nerve ganglia fuse to form a prootic ganglion. The vertebral column possesses eight presacral holochordal vertebrae, and all are procoelous except for a biconcave surface on the last presacral. The transverse processes of the sacral vertebra are cylindrical, and this vertebra has a bicondylar articulation with the urostyle. Postmetamorphs have no dorsal ribs on the presacral vertebrae. The pectoral girdle is firmisternal with a distinct sternum. Fibulare and tibiale are fused at their proximal

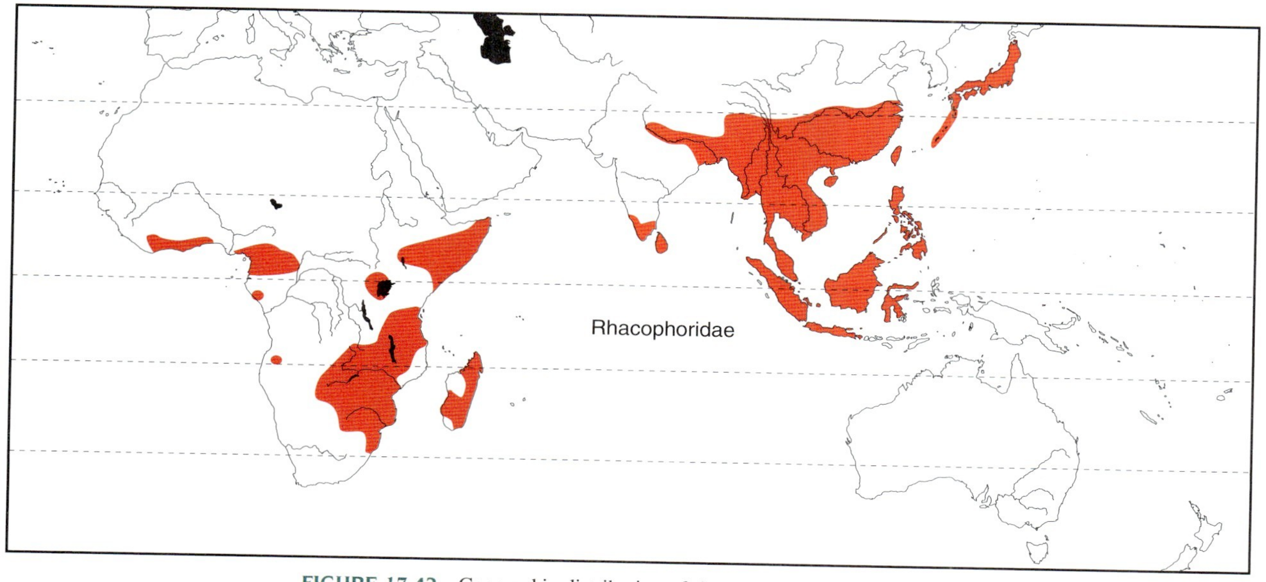

FIGURE 17.42 Geographic distribution of the extant Rhacophoridae.

and distal ends. An intercalary cartilage occurs between the terminal and penultimate phalanges of each digit, and the tips of the terminal phalanges are T-shaped and sometimes pointed. Tadpoles have keratinized mouthparts, and the left and right branchial chambers fuse behind the heart and are emptied by a spiracle on the left side at midbody.

Buergeriinae

Sister taxon: Rhacophorinae.

Content: One genus, *Buergeria*, with 4 species.

Distribution: Taiwan through Ryukyu Island to southern Japan.

Characteristics: The anterior horn of the hyoid is present but consists only of the medial arch. The sphenethmoid is a single bone.

Biology: Buergeriines (Fig. 17.34) are small- to moderate-sized frogs (25–70 mm adult SVL); females are larger than males. Although they are a tree frog-like, they are commonly found on the ground or in water, particularly in montane streams. They have an extended breeding season from early spring through summer. They do not form breeding aggregations or choruses; instead, the males establish and maintain territories along a stream, typically calling from the water's edge or in the water. Amplexus is axillary, and the eggs are deposited in the water. Tadpoles are free living and metamorphose in about 8 weeks.

Rhacophorinae

Sister taxon: Buergeriinae.

Content: Nine genera, *Aquixalus, Chiromantis, Feihyla, Kurixalus, Nyctixalus, Philautus, Polypedates, Rhacophorus*, and *Theloderma*, with 281 species.

Distribution: Tropical Africa and Asia to temperate China and Japan.

Characteristics: The anterior horn of the hyoid is absent, and the sphenethmoid is paired.

Biology: Rhacophorines are tree frogs. They range in size from small (30–45 mm adult SVL; Fig. 17.34) to large (e.g., *Rhacophorus dennysi*, 78–102 mm). The Asian *Philautus, Nyctixalus*, and *Theloderma* have tree hole egg deposition sites and nonfeeding tadpoles that have brief developmental periods. *Chiromantis, Polypedates, Rhacophorus*, and others deposit eggs in foam nests above water, mostly in shrubs and trees; upon hatching, the larvae drop into the water below and develop as free-living tadpoles. The foam nests often are created jointly by two or more amplectant pairs, and at least in *Chiromantis*, unpaired males may assist. The Malagasian taxa lay eggs directly in water and have a typical aquatic tadpole life cycle. The African *Chiromantis xerampelina* is an arid land species and has evolved special physiological and morphological adaptations to tolerate high temperature and reduce water loss (see the section "Thermoregulation" in Chapter 7).

Nyctibatrachidae

Robust Frogs

Classification: Amphibia; Lissamphibia; Anura; Neobatrachia; Ranoides; Natatanura; Victorana; Saukrobatrachia; Ranoidea.

Sister taxon: Ranidae.

Content: Two genera, *Lankanectes* and *Nyctibatrachus*, with 1 and 12 species, respectively.

Distribution: India and Sri Lanka.

Characteristics: Species of *Nyctibatrachus* are robust-bodied frogs. They range in size from small bodied

FIGURE 17.43 Representative rhacophorid and ranid frogs. Clockwise from upper left: Panther flying frog *Rhacophorus pardalis,* Rhacophorinae (R. M. Brown); tropical forest frog *Hylarana igorota,* Ranidae (R. M. Brown); crawfish frog, *Lithobates areolatus*, Ranidae (J. P. Caldwell); Amazon River frog, Ranidae *Lithobates palmipes*, Ranidae (J. P. Caldwell).

(14–15 mm SVL in males, 15–17 mm SVL in females of *N. beddomii*) to large (adult SVL to 84 mm, *N. karnatakaensis*). They have a median lingual process, concealed tympanum, dorsum with longitudinal skin folds, femoral glands, and expanded finger and toe discs. Maxillary teeth are present, and the omosternum and sternum have a bony style. *Lankanectes* lacks a median lingual process, digital discs, and femoral glands. It has a forked omosternum and the unusual presence of a functional lateral-line system in adults, a characteristic also present in two genera of dicroglossids. In addition, like some species of *Limnonectes* (a discoglossine genus), *Lankanectes* has fangs on its mandibles.

Biology: *Nyctibatrachus* occurs near streams in hilly evergreen forests. Males are territorial and amplexus is absent.

Ranidae

True Frogs

Classification: Amphibia; Lissamphibia; Anura; Neobatrachia; Ranoides; Natatanura; Victorana; Saukrobatrachia; Ranoidea.

Sister taxon: Nyctibatrachidae.

Content: Seventeen genera, *Amolops, Babina, Clinotarsus, Glanirana, Huia, Humerana, Hylarana, Lithobates, Meristogenys, Nasirana, Odorrana, Pelophylax, Pseudorana, Pterorana, Rana, Sanguirana,* and *Staurois,* with 315 species (Fig. 17.43).

Distribution: Cosmopolitan except for southern South America and most of Australia (Fig. 17.44).

Characteristics: The ranid skull has paired palatines and frontoparietals. The facial nerve exits through the anterior acoustic foramen in the auditory capsule; the trigeminal and facial nerve ganglia fuse to form a prootic ganglion. The vertebral column possesses eight presacral holochordal vertebrae, and all are procoelous except for a biconcave surface on the last presacral. The transverse processes of the sacral vertebra are cylindrical, and this vertebra has a bicondylar articulation with the urostyle. Postmetamorphs have no dorsal ribs on the presacral vertebrae. The pectoral girdle is firmisternal, rarely pseudoarciferal, with a distinct sternum. The fibulare and tibiale fused at their proximal and distal ends. No intercalary cartilage occurs between the terminal and penultimate phalanges of each digit; the

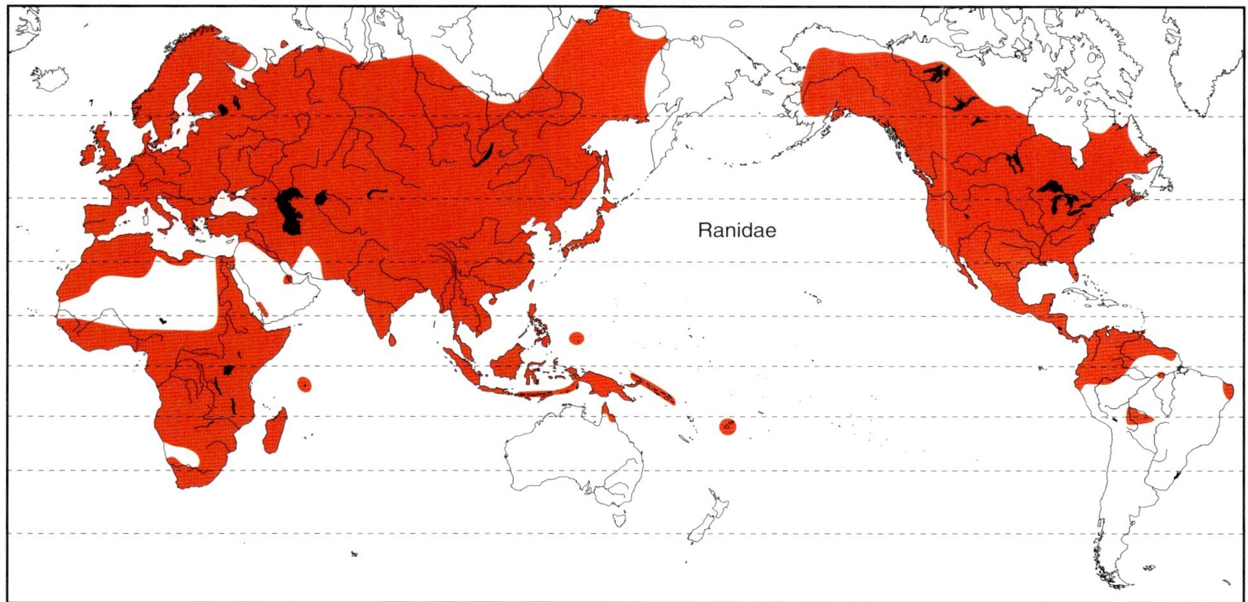

FIGURE 17.44 Geographic distribution of the extant Ranidae.

terminal phalanges are blunt, pointed, or T-shaped. The larvae have keratinized mouthparts, and the left and right branchial chambers fuse behind the heart and are emptied by a spiracle on the left side at midbody.

Biology: Most ranids are medium to large species (American *Lithobates catesbeianus*, 85–180 mm; New Guinean *Hylarana arfaki*, 90–160 mm). Many species are typical pond breeders that deposit eggs in clumps or as surface films. Tadpoles usually transform within several months, although some, such as *Lithobates catesbianus*, may take 1-2 years to reach metamorphosis. Species of *Amolops* breed in fast-moving streams and rivers. All tadpoles of *Amolops* are gastromyzophorous, possessing a large abdominal sucker that is used to attach to rocks in torrential streams. Species of *Odorrana* are called cascade frogs because they occur in forested, montane rivers where they call from boulders in areas with cascading waterfalls. Species of *Odorrana* have odoriferous, highly toxic skin.

QUESTIONS

1. In what kinds of habitats would you expect to find the most species of frogs and why?
2. What families of frogs occur in the Pacific Northwest of the United States?
3. Explain why, until recently, it has been so difficult to reconstruct phylogenetic relationships among frog families.
4. Based on what you now know about the diversity of frogs, describe two good examples of morphological or ecological convergence at the family level.

REFERENCES

Overview

Biju and Bossuyt, 2003; Bossuyt et al., 2006; Che et al., 2007; Cisneros-Heredia and McDiarmid, 2007; Cogger and Zweifel, 1998; Dubois, 1984, 1986; Duellman, 1993; Duellman and Trueb, 1986; Ford and Cannatella, 1993; Frost, 1985; Frost et al., 2006; Fu et al., 2007; Grant et al., 2006; Griffiths, 1963; Halliday and Adler, 1986; Haas, 2003; Hays et al., 1995; Hoegg et al., 2004; Inger, 1967; Jamieson, 2003; Laurent, 1986; Lynch, 1973; Maglia et al., 2001; Noble, 1922; Rao and Wilkinson, 2007; Roelants et al., 2007; Ruvinsky and Maxson, 1996; Smith et al., 2007; Stebbins and Cohen, 1995; Van Bocxlaer et al., 2006; Van der Meijden et al., 2005, 2007a,b; Veith et al., 2006; Vences et al., 2007; Vial, 1973; Wells, 2007; Zheng et al., 2007.

Taxonomic Accounts

Leiopelmatidae

Talied Frogs and New Zealand Frogs

Bell, 1978, 1985, 1994; Bell and Wassersug, 2003; Brown, H., 1975, 1989; Diller and Wallace, 1999; Green, 1994; Green and Campbell, 1984; Green and Cannatella, 1993; Green et al., 1989; Karraker et al, 2006; Leonard et al., 1993; Metter, 1967, 1968; Nielson et al, 2001; Worthy, 1987.

Pipidae

Platannas, African Clawed Frogs, and Surinam Toads

Báez and Pugener, 2003; Báez and Trueb, 1997; Cannatella and de Sá, 1993; Cannatella and Trueb, 1988a,b; de Sá and Hillis, 1990; Evans et al., 2004, 2005; Frost et al., 2006; Graf, 1996; Kobel et al., 1996; Measey and Tinsley, 1997; Rabb and Rabb, 1960, 1963a,b; Rabb and Snediger, 1960; Tinsley et al., 1996; Tobias et al, 1998; Trueb and Cannatella, 1986; Wassersug, 1996; Weygoldt, 1976.

Rhinophrynidae

Mexican Burrowing Toad

Foster and McDiarmid, 1982, 1983; Fouquette, 1969; Lee, 1996; Swart and de Sá, 1999; Trueb and Gans, 1983.

Alytidae

Midwife Toads and Painted Frogs

Arntzen and García-París, 1995; Berger and Michalowski, 1973; Bosch and Boyero, 2003; Brown and Crespo, 2000; Engelmann et al., 1986; Frost et al., 2006; Hemmer and Alcover, 1984; Márquez, 1992, 1995; Márquez and Bosch, 1997; Salvador, 1985; Zangari et al., 2006.

Bombinatoridae

Fire-Bellied Toads

Alcala, and Brown, 1987; Barandun, 1995; Engelmann et al., 1986; Frost et al., 2006; Grillitsch et al., 1983; Inger, 1954; Iskandar, 1995; M. Lang, 1989b.

Pelobatidae

Western Palearctic Spadefoot Toads

Bragg, 1965; Engelmann et al., 1986; Frost et al., 2006; Justus et al., 1977; Maglia, 1998; Nöllert, 1990; Sage et al, 1982; Salvador, 1985; Wiens and Titus, 1991.

Megophryidae

Asian Toad Frogs

Berry, 1975; Fu et al., 2007; Haas, 2005; Handrigan et al., 2007; Henrici, 1994; Inger, 1966; Lathrop 1997; Rao and Wilkinson, 2007; Rao and Yang, 1997; Rao et al., 2006; Zheng and Fu, 2007; Zheng et al., 2006, 2007.

Pelodytidae

Parsley Frogs

Borkin and Anissimova, 1987; Esteban et al., 2004; Henrici, 1994; Malkmus, 1995; Salvador, 1985; Sánchez-Herráiz et al., 2000; Veith et al., 2006.

Scaphiopodidae

Nearctic Spadefoot Toads

Bragg, 1965; Engelmann et al., 1986; Frost et al., 2006; Justus et al., 1977; Maglia, 1998; Nöllert, 1990; Sage et al., 1982; Salvador, 1985; Wiens and Titus, 1991.

Heleophrynidae

Ghost Frogs

Boycott, 1988; Boycott and de Villiers, 1986; Passmore and Carruthers, 1995; Poynton, 1964; Wager, 1986.

Sooglossidae

Seychelles Frogs

Biju and Bossuyt, 2003; Dodd, 1982; Frost et al., 2006; Gerlach and Willi, 2002; Mitchell and Altig, 1983; Nussbaum, 1980, 1985; Nussbaum and Wu, 2007; Nussbaum et al. 1982; Radhakrishnan et al, 2007; Roelants et al., 2007; Van der Meijden et al., 2007.

Calyptocephalellidae

Chilean Toads

Formas et al., 2001; Formas and Espinoza, 1975; Frost et al., 2006; Nuñez and Formas, 2000.

Limnodynastidae

Australian Ground Frogs

Barker et al., 1995; Frost et al., 2006; Heyer and Liem, 1976; Knowles et al., 2004; Lemckert and Brassil, 2003; Littlejohn et al., 1993; Tyler, 1985, 1989.

Myobatrachidae

Australian Toadlets and Turtle Frogs

Barker et al., 1995; Frost et al., 2006; Gollman and Gollman, 1991; Littlejohn et al., 1993; Read et al., 2001; Roberts et al., 1997; Roelants et al., 2007; Tyler, 1985, 1989.

Hemiphractidae

Horned Frogs

Darst and Cannatella, 2004; Faivovich et al., 2005; Frost et al., 2006; Sheil and Mendelson, 2001; Trueb, 1974; Tyler and Duellman, 1995; Wiens et al., 2005, 2006.

Brachycephalidae

Rain Frogs, Three-Toed Toadlets, and Others

Frost et al., 2006; Heinicke et al., 2007; Heyer et al., 1990; Pombal et al., 1994, 1998.

Cryptobatrachidae

Stefanias and Others

Darst and Cannatella, 2004; Faivovich et al., 2005; Frost et al., 2006; Tyler and Duellman, 1995; Wiens et al., 2005, 2006.

Amphignathodontidae

Marsupial Frogs

Darst and Cannatella, 2004; del Pino, 1980, 1989; Duellman et al., 1988; Faivovich et al., 2005; Frost et al., 2006; Tyler and Duellman, 1995; Weygoldt and de Carvalho Silva, 1991; Wiens et al., 2005, 2006.

Hylidae

Ameroaustralian Tree Frogs

Darst and Cannatella, 2004; Duellman, 1970; Duellman and Trueb, 1986; Faivovich et al., 2005; Frost et al., 2006; Schwartz and Henderson, 1991; Tyler and Davies, 1993; Wiens et al., 2005, 2006.

Hylinae

Aguiar et al., 2007; Cei, 1980; Cocroft, 1994; Darst and Cannatella, 2004; Duellman, 1970, 2001; Duellman and Wiens, 1992; Emerson, 1988; Faivovich et al., 2005; Frost et al., 2006; Gallardo, 1987; Garda and Cannatella, 2007; Geiger, 1995; Hedges, 1986; Heyer et al., 1990; Jungfer and Weygoldt, 1999; Kluge, 1981; Lemmon et al., 2007; Maeda and Matsui, 1989; Wiens et al., 2005, 2006, 2007b.

Pelodryadinae

Barker et al., 1995; Duellman, 2001; Faivovich et al., 2005; Frost et al., 2006; Pyke and Osborne, 1996; Tyler, 1985; Tyler and Davies, 1978, 1993.

Phyllomedusinae

Bagnara and Rastogi, 1992; Cruz, 1990; Shoemaker et al., 1989; Shoemaker and McClanahan, 1982; Weygoldt, 1991.

Centrolenidae

Glass Frogs, Ruthven's Frog

Darst and Cannetella, 2004; Cisneros-Heredia and McDiarmid, 2007; Faivovich et al., 2005; Frost et al., 2006; Grant et al., 2006; Lynch et al., 1983; McDiarmid, 1978; Ruiz-Carranza and Lynch, 1991; Villa, 1984; Wiens et al., 2005.

Allophryninae

Austin et al., 2002; Caldwell, 1996b; Caldwell and Hoogmoed, 1998; Duellman, 1975; Frost et al., 2006; Hoogmoed, 1969; Noble, 1931.

Centroleninae

Darst and Cannetella, 2004; Cisneros-Heredia and McDiarmid, 2007; Faivovich et al., 2005; Frost et al., 2006; Grant et al., 2006; Lynch et al., 1983; McDiarmid, 1978; Ruiz-Carranza and Lynch, 1991; Villa, 1984; Wiens et al., 2005.

Leptodactylidae

White-Lipped Frogs and Tropical Grass Frogs

Camargo et al., 2006; Frost et al., 2006; Grant et al., 2006; Heyer, 1975, 1979, 1994; Heyer and Crombie, 2005; Heyer et al., 1990; Kokubum and Giaretta, 2005; Lynch 1971; Pombal and Haddad, 1999; Prado et al., 2002; Shepard and Caldwell, 2005.

Ceratophryidae

Horned Frogs, Water Frogs, and Others

Frost et al., 2006.

Batrachylinae

Basso, 1998; Cei, 1984; Formas, 1997; Nussbaum, 1980.

Ceratophryinae

Cei, 1980; Maxson and Ruibal, 1988; Quinzio et al., 2006.

Telmatobiinae

Cei, 1980; Duellman, 1975; Gallardo, 1987; Hutchison, 1982; Schwartz and Henderson, 1991; Townsend et al., 1981; M. Wake, 1993b.

Cycloramphidae

Mouth-Brooding Frogs, Smooth Horned Frogs, and Others

Busse, 1970; Cei, 1980; Frost et al., 2006; Goicoechea et al., 1986; Grant et al., 2006; Lynch, 1971.

Leiuperidae

Foam-Nesting Frogs and Dwarf Frogs

Duellman and Morales, 1990; Grant et al., 2006; Lynch, 1971, 1978; Murphy, 2003a,b; Ron et al., 2006; Schlüter, 1990.

Bufonidae

True Toads, Harlequin Frogs, and Relatives

Berry, 1975; Blair, 1972; Channing and Stanley, 2002; Frost et al., 2006; Grant et al., 2006; Grandison and Ashe, 1983; Graybeal, 1997; Graybeal and Cannatella, 1995; Inger, 1966; Izecksohn, 1993; Lötters, 1996; Poynton, 1996; Poynton and Broadley, 1988; Pramuk, 2002, 2006; Pramuk et al., 2008; Tihen, 1965; M. Wake, 1993b; Xavier, 1986.

Hylodidae

Stream-Dwelling Frogs

Cei, 1980; Heyer et al., 1990.

Aromobatidae

Cryptic Forest Frogs

Grant et al., 2006.

Anomaloglossinae

Bourne et al., 2001; Caldwell and Lima, 2003; Grant et al., 2006.

Aromobatinae

Myers et al., 1991; Péfaur, 1993.

Allobatinae

Amezquita et al., 2006; Caldwell and Lima, 2003; Lima and Caldwell, 2001; Lima et al., 2002.

Dendrobatidae

Poison Frogs

Grant et al., 2006.

Colostethinae

Grant, 2007.

Dendrobatinae

Caldwell, 1996a,b; Caldwell and Oliveira, 1999.

Hyloxalinae

Caldwell, 1996a,b; Caldwell, 2005; Ford, 1993; Grant et al., 1997; Lötters et al., 2000; Myers and Daly, 1983; Myers et al., 1991.

Microhylidae

Narrow-Mouth Frogs and Toads

Bossuyt et al., 2006; Ford and Cannatella, 1993; Frost et al., 2006; Laurent, 1986; Parker, 1934; Van Bocxlaer et al., 2006; Van der Meijden et al., 2007; Vences et al., 2003a,b.

Asterophryinae

Burton, 1986; Burton and Zweifel, 1995; Johnston and Richards, 1993; Menzies, 1976; Zweifel, 1972, 1985.

Cophylinae

Blommers-Schlösser and Blanc, 1991; Glaw and Vences, 1994; Köhler et al., 1997.

Dyscophinae

Andreone et al., 2005; Berry, 1975; Blommers-Schlösser and Blanc, 1991; Glaw and Vences 1994; Inger and Steubing, 1997; Manthey and Grossman, 1997; Pintak, 1987.

Gastrophryninae

Wild, 1995.

Hoplophryninae

Daltry and Martin, 1997; Laurent, 1986; Parker, 1934; Van Bocxlaer et al., 2006.

Kalophryninae

Das and Haas, 2003; Matsui et al., 1996.

Melanobatrachinae

Daltry and Martin, 1997; Laurent, 1986; Parker, 1934.

Microhylinae

Donnelly et al., 1990; Dutta and Manamendra-Archchi, 1996; Inger and Steubing, 1997; Manthey and Grossman, 1997; Zweifel, 1986; Krügel and Richter, 1995.

Otophryninae

Parker, 1934; Wassersug and Pyburn, 1987; Van der Meijden et al., 2007; Wild, 1995.

Phrynomerinae

Lambiris, 1989; Passmore and Carruthers, 1995; Poynton and Broadley, 1985; Rödel, 1996; Stewart, 1967.

Scaphiophryninae

Blommers-Schlösser and Blanc, 1991; Ford and Cannatella, 1993; Glaw and Vences 1994; Grosjean et al., 2007; Van der Meijden et al., 2007.

Brevicipitidae

Rain Frogs

Channing, 1995; Lambiris, 1989; Passmore and Carruthers, 1995; Poynton and Broadley, 1985; Stewart, 1967; Van der Meijden et al., 2007; Wager, 1986.

Hemisotidae

Shovel-Nosed Frogs

Blommers-Schlosser, 1993; Channing, 1995; Emerson, 1976; Kaminsky et al., 1999; Laurent, 1986; Rödel, 1996; Rödel et al., 1995; Van der Meijden et al., 2007; Wager, 1986.

Arthroleptidae

Squeakers and Cricket Frogs

Bossuyt et al., 2006; Branch, 1991; Dubois, 1981; Frost et al., 2006; Laurent, 1986; Passmore and Carruthers, 1995; Poynton and Broadley, 1985; Scott, 2005.

Arthroleptinae

Bossuyt et al., 2006; Lambiris, 1989; Noble, 1925; Perret, 1966; Poyton and Broadley, 1985; Scott, 2005; Stewart, 1967; Vences et al., 2003; Wager, 1986.

Leptopelinae

Channing, 2001; Channing and Howell, 2006; Emerson et al., 2000; Lambiris, 1989; Perret, 1966; Poynton and Broadley, 1987; Schiotz, 1967, 1975; Van der Meijden et al., 2007; Wager, 1986.

Hyperoliidae

African Reed Frogs and Running Frogs

Blommers-Schlösser and Blanc, 1991; Frost et al., 2006; Channing, 1989; Drewes, 1984; Glaw and Vences, 1994; Herrmann, 1993; Laurent, 1986; Liem, 1970; Nussbaum, 1984: Nussbaum and Wu, 1995; Passmore and Carruthers, 1995; Perret, 1966; Poyton and Broadley, 1987; Rödel, 1996; Schiotz, 1967, 1975; Stewart, 1967; Vences et al., 2003c.

Ptychadenidae

Grassland Frogs

Channing, 2001; Channing and Howell, 2006; Lambiris, 1989; Lamotte and Ohler, 2000; Perret, 1966; Poynton and Broadley, 1985; Rödel, 1996; Rödel et al., 2002; Schiotz, 1963; Stewart, 1967; Vences et al., 2004; Wager, 1986.

Ceratobatrachidae

Triangle Frogs and Others

Chen et al., 2005; Fabrezi and Emerson, 2003.

Micrixalidae

Tropical Frog

Dubois et al., 2001; Dubois, 1986; Frost et al., 2006; Krishna and Krishna, 2006; Roelants et al., 2004; Vasudevan et al., 2001.

Ranixalidae

Leaping Frogs

Altig and Johnson, 1989; Van Bocxlear et al., 2006.

Phrynobatrachidae

Puddle Frogs

Bossuyt et al., 2006; Crutsinger et al., 2004; Frost et al., 2006; Howell, 2000; Largen, 2001; Rödel, 1998; Rödel and Ernst, 2002a,b; Stewart, 1974.

Petropedetidae

African Water Frogs, Goliath Frog

Bossuyt et al., 2006; Channing et al., 2002; Drewes et al., 1989; Roelants et al., 2004; Sabater-Pi, 1985; Scott, 2005.

Pyxicephalidae

African Bullfrogs and Others

Frost et al., 2006; Van der Meijden et al., 2005.

Cacosterninae

Frost et al., 1996; Lambiris, 1989; Largen, 1991; Laurent, 1986; Poynton and Broadley, 1985; Rödel, 1996, 1998; Van der Meijden et al., 2006; Wager, 1986.

Pyxicephalinae

Branch, 1991; Channing et al., 1994; Kok et al., 1989; Lambiris, 1989; Perret, 1966; Rödel, 1996; Schiotz, 1963; Sheil, 1999; Stewart, 1967.

Dicroglossidae

Fanged Frogs, Tiger Frogs, and Others

Dutta and Manamendra-Arachchi, 1996; Emerson and Berrigan, 1993; Frost et al., 2006; Inger, 1996a,b; Inger and Steubing, 1997; Jiang and Zhou, 2005; Olher and Dubois, 1999; Sabater-Pi, 1985.

Dicroglossinae

Delorme et al., 2004; Emerson et al., 2000; Evans et al., 2003; Grosjean et al., 2004; Ohler et al., 1999; Orlov, 1997; Tsuji and Lue, 1998; Wickramasinghe et al., 2007.

Occidozyginae

Dubois and Ohler, 2000; Marmayou et al., 2000.

Mantellidae

Malagasy Poison Frogs and Others

Blommers-Schlösser, 1993; Blommers-Schlösser and Blanc, 1991; Daly et al., 1996; Glaw and Vences, 1994, 2006; Vences and Glaw, 2002, 2005; Vences et al., 2002a,b, 2007.

Boophinae

Glaw and Vences, 2006; Thomas et al., 2005; Vences and Glaw, 2002, 2005; Vences et al., 2002, 2007.

Laliostominae

Glos, 2003; Glos and Linsenmair, 2004; Glaw et al., 1998; Nussbaum et al., 1999.

Mantellinae

Glaw and Vences, 2006; Lehtinen et al., 2007; Vences et al., 1999.

Rhacophoridae

Afroasian Tree Frogs

Channing, 1989; Herrmann, 1993; Liem, 1970; Richards and Moore, 1998.

Buergeriinae

Lue, 1991; Maeda and Matsui, 1989.

Rhacophorinae

Berry, 1975; Channing, 1989; Dutta and Manamendra-Archchi, 1996; Glaw and Vences, 1994; Inger and Stuebing, 1997; Liu and Hu, 1961; Maeda and Matsui, 1989; Poynton and Broadley, 1987; Shoemaker et al., 1989.

Nyctibatrachidae

Robust Frogs

Bossuyt and Milikovitch, 2000, 2001; Das and Kunte, 2005; Delorme et al., 2004; Dinesh et al., 2007; Dubois and Ohler, 2001; Roelants et al., 2004.

Ranidae

True Frogs

Bain et al., 2003; Bossuyt et al., 2006; Che et al., 2007; Dubois, 1985, 1992; Frost et al., 2006; Hillis and Wilcox, 2005; Scott, 2005; Van der Meijden et al., 2005.

Chapter 18

Turtles

OVERVIEW

About 285+ turtle species occur in cool-temperate to tropical habitats throughout much of the world. They are ecologically and morphologically diverse, including marine, freshwater, and terrestrial species, varying in size from small to giant. Morphological diversity is reflected in shell shapes that range from nearly spherical to nautically streamlined. Considerable physiological variation exists as well, allowing some marine species to dive to depths of over half a kilometer and some upland desert species to exist in habitats with less than 10 centimeters of rainfall each year. Turtles are renowned for a slow, plodding locomotion that is more imaginary than real. Turtle life histories are characterized by slow growth, late maturity, repeated reproduction, and long lives. Because of these life history traits, harvesting of adult turtles by humans has a major impact on turtle populations, thus many species and populations are declining toward extinction.

Turtles (Testudines) are reptilian tanks, armored above and below, and capable of withdrawing the head and neck, limbs, and tails either partially or fully within the shell. No other tetrapod has a bony shell that encloses both the pectoral and pelvic girdles. The upper shell, the carapace, is formed from fusion of the eight trunk vertebrae and ribs to an overlying set of dermal bones; the lower shell, the plastron, arises from the fusion of parts of the sternum and pectoral girdle with external dermal bones. The shell is robust in some taxa, such as in tortoises and mud turtles, with only small openings for the head and appendages. In other turtles, such as leatherback sea turtles and softshell turtles, the shell is lightly built and has reduced bony elements or has lost them completely. The neck of all turtles is extremely flexible and consists of eight cervical vertebrae. Extant turtles are divided into two clades based on the movement or retraction pattern of the neck. The Pleurodira or side-neck turtles retract the head and neck by laying them to the side; thus, the sides of the neck and head are exposed in the gap between the carapace and plastron (Fig. 18.1). The Cryptodira or hidden-neck turtles retract the neck into a medial slot within the body cavity; the neck forms a vertical S-shape when viewed laterally, and only the tip of the nose is exposed between the shielding forearms. In spite of the different mechanics of neck retraction, the structure of the cervical vertebrae in the two groups is very similar.

All turtles are oviparous. The number of eggs deposited by females of different species ranges from one to more than a hundred. The number of eggs in a clutch is generally positively associated with female size; small turtles lay one or two eggs, and larger turtles lay a dozen or more. Most turtles have a stereotypic nest-digging behavior. Egg chambers are dug with the hindlimbs, which work alternately to scoop out a flask-shaped chamber as deep as the hindlimbs can reach. Fertilization is internal, and because the shell surrounds the body in both sexes, males generally balance their plastron on top of the female's carapace during copulation. Males of many species have a slightly concave plastron to facilitate mating.

Living and extinct turtles share a large suite of unique characteristics. No one questions the monophyly of turtles, although the origin of turtles is controversial (see Chapter 3). In addition to the uniquely evolved carapace and plastron, all testudines share a special cranial architecture (see Fig. 2.18) that includes the presence of a maxillary, a

FIGURE 18.1 Side-neck turtles (Pleurodira), such as *Phrynops gibbus* (left), can withdraw their head and neck only within the outer margin of the shell, whereas hidden-neck turtles (Cryptodira), such as *Malaclemys terrapin* (right), withdraw the neck and head within the shell. (L. J. Vitt)

premaxillary, and a dentary without teeth and bearing a horny sheath; the absence of a postparietal, postfrontal, and ectopterygoid; a small or absent lacrimal; a large quadrate that abuts the squamosal to form the temporal surface of the skull; and a rodlike stapes without a foramen or processes. Some other features that distinguish turtles include the presence of a largely nonsensory but strongly secretory pineal organ; the absence of nasal conchae; the presence of a lower eyelid tendon; prominent epicondyles and an ectepicondylar foramen or groove on the humerus; and a subspherical and elevated femur head.

Turtles have always been recognized as a unique and natural group. Linnaeus included all turtles in *Testudo* and recognized 15 species in his 1766 edition of the *Systema Naturae*. The partitioning of turtle species into more genera began soon thereafter. In 1805, Brongniart subdivided turtles based on habitat into marine (*Chelonia*), freshwater (*Emys*), and land (*Testudo*) species. The first hierarchical arrangement appeared in 1806 when Duméril constructed a listing of sequentially indented pairs of diagnostic traits to differentiate the preceding three genera and a new one, *Chelus*. The recognition of new genera and species continues to the present time. Throughout the 19th and 20th centuries, biologists have attempted to recognize natural groups, but the relative stability of turtle classification is recent and based on a combination of a cladistic phylogenetic analysis of fossil and extant turtle morphology combined with molecular data (Table 3.5; Fig. 18.2).

The discovery of new fossil turtles and the use of molecular data support the basal division of extant turtles into the pleurodires and cryptodires. This divergence of turtle clades

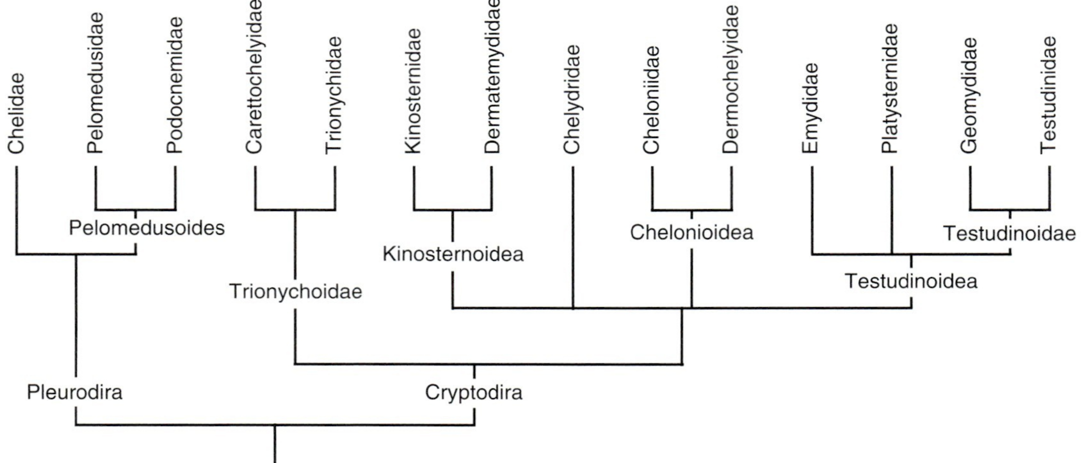

FIGURE 18.2 Cladogram depicting relationships among the families of extant turtles. The cladogram derives from Gaffney and Meylan, 1988, and is modified based on Meylan, 1996; Shaffer et al., 1997; Fujita et al., 2004; and Parham et al., 2006.

is ancient, occurring more than 220 mybp, and these two clades lived contemporaneously with the most primitive turtle, *Proganochelys* (see Chapter 3). The recognition of these two clades arose from their contrasting neck-retraction mechanics, but other characters support the monophyly of each. For example, pleurodiran turtles have the pelvic girdle fused to the plastron and a jaw closure mechanism with an articulation on the trochlear surface of the pterygoid; cryptodires have a flexible articulation of the pelvic girdle with the plastron and a jaw closure mechanism with an articulation on the trochlear surface of the otic capsule. Two clades of extant side-necks, Pelomedusidae and Chelidae, have been recognized for much of the 20th century. As fossils were incorporated into phylogenetic analyses, it became evident that pelomedusids were polyphyletic. Resolution of this problem has occurred with the recognition of the Pelomedusoides as the sister group to the Chelidae, and the classification of the Pelomedusoides into two fossil clades and the extant Pelomedusidae and Podocnemididae clades. Combined molecular and morphological data show that both the Australian chelids and the South American chelids are monophyletic.

The position of the Chelydridae remains uncertain (Fig. 18.2). It has been considered the sister group to all other extant families of cryptodiran turtles based on combined molecular and morphological data, although an alternative arrangement has placed the chelydrids as a sister group to all extant cryptodirans except a trionychid–carettochelyid clade. More recent analyses place the Chelydridae as the sister taxon to the Kinosternidae–Dermatemydidae clade. Because many issues remain uncertain, we place chelydrids in a polytomy with Kinosternoidea–Chelonioidea–Testudinoidea. Although previous studies indicated a sister-group relationship of snapping turtles to the big-headed turtle *Platysternon megacephalum*, chromosomal and molecular evidence indicates that *Platysternon* falls within the Testudinoidea.

Combined molecular and morphological data support the recognized groupings of Cheloniidae–Dermochelyidae, Trionychidae–Carettochelyidae, Kinosternidae–Dermatemydidae, and Emydidae–Platysternidae–Geomydidae–Testudinoidae, each as clades. Agassiz in his 1857 classification of turtles recognized the relationship of all extant sea turtles, but owing to the leatherback's extreme specializations, many subsequent biologists placed *Dermochelys* in a separate group (Atheca) equivalent to cryptodires and pleurodires. Fossil data suggest a long separation of the leatherback clade and hard-shelled sea turtles; nonetheless, sea turtles comprise a monophyletic group. The five genera of extant cheloniids are commonly divided into two subgroups; however, the inclusion of fossil taxa suggests otherwise. Molecular data proposes a different pattern of relationships among extant cheloniids but does not account for relationships among extant and fossil taxa.

The trionychid–carettochelyid and the kinosternid–dermatemydid clades have been recognized as sister groups of a larger clade (Chelomacryptodira). Although morphological characters continue to support this relationship, molecular data alone or combined with morphological data suggest trionychids–carettochelyids as the sister group to all other cryptodires. Fossil evidence also indicates trionychids and carettochelyids are sister taxa. Dermatemydids, however, are the sister group to extant kinosternids and several fossil genera. Staurotypines and kinosternines are sister taxa based on all evidence except karyotype.

Portions of the Emydidae–Platysternidae–Geomydidae–Testudinoidae clade (Testudinoidea) have a long history of recognition; however, the proposed relationships therein have been variable. Combined data indicate that the emydids are the sister group of the geomydid–testudinid clade (Testudinoidae). The monophyly of the emydids has strong support as do the clades Testudinoidae and Testudinidae; however, the monophyly of the Geomydidae is uncertain. Both the emydids and the testudinids have sets of shared derived characteristics that confirm their monophyly. The geomydids do not, and it is possible that testudinids arose from within the presently conceived geomydid group. Finally, recent molecular evidence ties *Platysternon* (the sole member of the Platysternidae) to emydids, thus expanding the Testudinoidae. Because the relationship of platysternids with emydids remains uncertain, we treat the Testudinoidea as an unresolved polytomy.

TAXONOMIC ACCOUNTS

Pleurodira

Chelidae

Australoamerican Side-Neck Turtles

Classification: Reptilia; Parareptilia; Testudines; Pleurodira.

Sister taxon: Pelomedusoides.

Content: Twelve genera, *Acanthochelys*, *Chelodina*, *Chelus*, *Elseya*, *Elusor*, *Emydura*, *Hydromedusa*, *Phrynops*, *Platyemys*, *Pseudemydura*, *Rheodytes*, and *Rhinemys*, with 50+ species.

Distribution: Australia, New Guinea, and South America (Fig. 18.3).

Characteristics: The Australoamerican side-necks range in adult CL from 12 to 14 cm (straight carapace length) for *Pseudemydura umbrina* to about 48 cm for *Chelodina expansa*; most chelid species range in CL from 20 to 35 cm (Fig. 18.4). As a group, they have flattened skulls and shells. The jaw closure mechanism articulates on a pterygoid trochlear surface that lacks a synovial capsule but contains a fluid-filled saclike duct from the buccal cavity. The skull lacks the epipterygoid but possesses an

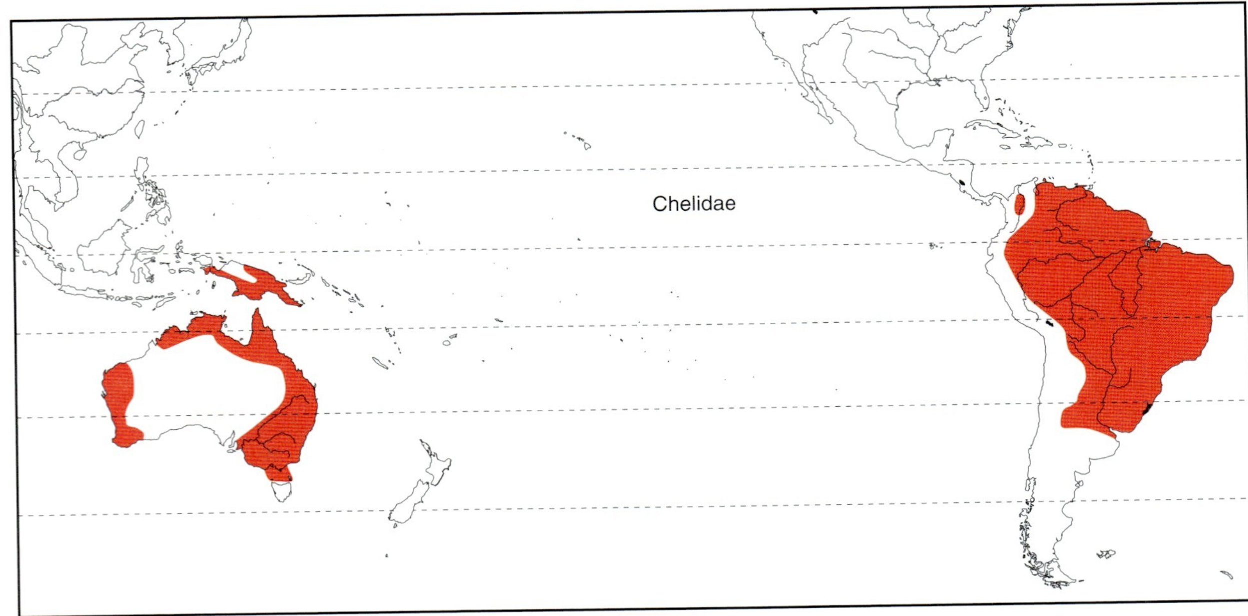

FIGURE 18.3 Geographic distribution of the extant Chelidae.

FIGURE 18.4 Representative chelid side-neck turtles. Clockwise from upper left: Juvenile of Geoffrey's side-necked turtle *Phrynops geoffroanus,* Chelidae (L. J. Vitt); mata mata, Chelidae (L. J. Vitt); narrow-breasted snake-neck turtle *Chelodina oblongata*, Chelidae (R. W. Barbour); northern Australian snake-neck turtle *Chelodina rugosa*, Chelidae (C. K. Dodd, Jr.).

internal carotid canal in the prootic and strong parietal–squamosal and postorbital–squamosal contact. The facial nerve has a hyomandibular branch. The plastron lacks a mesoplastron and has well-developed plastral buttresses that articulate with the costals on each side of the carapace; the carapace has 11 pairs of sutured peripherals around its margin and a nuchal without costiform processes. The neck withdraws horizontally, and this mechanism is reflected in an anteriorly oriented articular surface of the first thoracic vertebra; other vertebral traits are the inclusion of the 10th thoracic vertebra in the sacral complex and procoelous caudal vertebrae. The pelvic girdle is firmly fused to the plastron, and the ilium lacks a thelial process. The karyotype is 2N = 50 or 54.

Biology: Most chelids (Fig. 18.4) are predominantly aquatic, some highly so (e.g., *Elusor macrurus*, *Rheodytes leukops*), and they seldom leave the water except to deposit eggs. Species that live in seasonally drying marshes or ponds, such as *P. umbrina*, have extended aestivation–hibernation periods, during which individuals remained buried in the mud. The Neotropical *Platemys platycephalus* and *Phyrnops zuliae* are semiaquatic; they commonly leave the water to forage on the forest floor. Overall, chelids are opportunistic omnivores and take food ranging from filamentous algae and periphyton to arthropods, mollusks, and small vertebrates. *Chelus fimbriatus* and the species of *Chelodina* are carnivores that regularly catch fish and other active prey by a gape-suck mechanism. Their long necks are retracted until a prey approaches, and then rapidly extended; as the head nears the prey, the mouth opens and the buccal cavity is rapidly enlarged to create a vacuum that sucks the prey into the enlarging cavity. We have observed *Platemys platycepala* entering Amazonian rain forest ponds when large numbers of hylid frogs (*Osteocephalus*) were breeding. The turtles were feeding on frog egg masses as they were being deposited. Other taxa forage for small animal prey or carrion, or graze on aquatic vegetation. Seasonality of chelid reproduction varies considerably, and numerous patterns exist. They range from a "typical" spring or late dry season egg laying and hatching 8 to 10 weeks later, to egg deposition before the summer drought and eggs hatching about 180 days later (*P. umbrina*). In some Australian chelids, egg deposition occurs in late fall, and the eggs hatch about a year later (*Chelodina expansa*). In one species (*Chelodina rugosa*), eggs are deposited in submerged nests and hatch 9 to 10 months later at the beginning of the wet season.

Comment: A recent molecular study of chelid turtles proposed three subfamilies (Chelodininae, Chelidinae [sic], Hydromesinae). None of these groups was characterized morphologically, and the chelodinines are paraphyletic. Nevertheless, some structure is evident, with *Hydromedusa* as sister to other South American chelids and these as a group sister to the Australian chelids.

Pelomedusoides

Pelomedusidae

African Mud Terrapins

Classification: Reptilia; Parareptilia; Testudines; Pleurodira.

Sister taxon: Clade containing Podocnemididae and extinct Bothremydidae.

Content: Two genera, *Pelomedusa* and *Pelusios*, with 1 and 15+ species, respectively.

Distribution: Sub-Saharan Africa, Madagascar, and granitic Seychelles (Fig. 18.5).

Characteristics: The African mud terrapins are small (12 cm adult CL, *Pelusios nana*) to moderately large (46 cm CL, *Pelusios sinuatus*); most species are 20 to 30 cm CL. Most species have oblong, moderately high-domed carapaces, large plastra that are hinged in *Pelusios* and not hinged in *Pelomedusa*, and moderate-sized heads. The jaw closure mechanism articulates on a pterygoid trochlear surface that lacks a synovial capsule but contains a fluid-filled saclike duct from the buccal cavity. The skull lacks the epipterygoid and parietal–squamosal contact but possesses an internal carotid canal in the prootic and strong postorbital–squamosal contact. The facial nerve has a hyomandibular branch. The plastron has a mesoplastron and well-developed plastral buttresses that articulate with the costals on each side of the carapace; the carapace has 11 pairs of sutured peripherals around its margin and a nuchal without costiform processes. The neck withdraws horizontally, and this mechanism is reflected in an

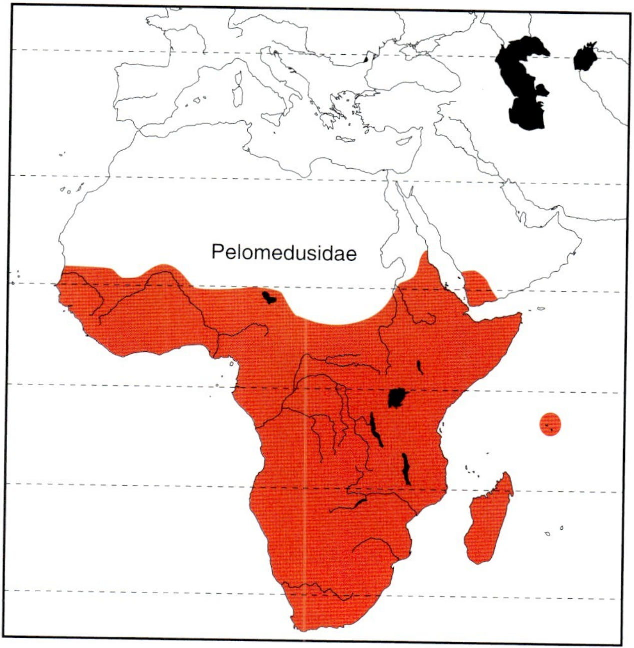

FIGURE 18.5 Geographic distribution of the extant Pelomedusidae.

anteriorly oriented articular surface of the first thoracic vertebra; other vertebral traits are the inclusion of the 10th thoracic vertebra in the sacral complex and procoelous caudal vertebrae. The pelvic girdle is firmly fused to the plastron, and the ilium lacks a thelial process. The karyotype is 2N = 34 or 36.

Biology: The mud terrapins (Fig. 18.6) are semi-aquatic or aquatic, bottom-walking turtles of slow-moving waters, principally lakes, swamps, and marshes and even ephemeral waterways. Their biology is little studied. They appear to be predominantly carnivorous, eating a variety of arthropods, worms, and other small animals, which they find by slow, methodical foraging on the bottom of their aquatic habitats. Species in seasonally dry waterways aestivate or hibernate in the bottom or on shore immediately adjacent to the drying habitat. Pelomedusids generally produce small to modest clutches of 6 to 18 eggs, depending upon female size. Egg deposition occurs in the more equitable season of the year, with known incubation periods ranging from 8 to 10 weeks.

Podocnemidae

Madagascan Big-Headed Turtles and American Side-Neck River Turtles

Classification: Reptilia; Parareptilia; Testudines; Pleurodira.

Sister taxon: Bothremydidae, a fossil clade.

Content: Three genera, *Erymnochelys*, *Peltocephalus*, and *Podocnemis*, with 1, 1, and 6 species, respectively.

Distribution: Madagascar and the northern half of South America east of the Andes (Fig. 18.7).

Characteristics: Podocnemids are moderately large turtles, ranging in adult CL from 20 to 25 cm (male *Podocnemis erythrocephala*) to 80 cm (female *Podocnemis expansa*). The jaw closure mechanism articulates on a pterygoid trochlear surface that lacks a synovial capsule but contains a fluid-filled saclike duct from the buccal cavity. The skull lacks the epipterygoid and parietal–squamosal contact but possesses an internal carotid canal in the prootic, and strong postorbital–squamosal contact. The

FIGURE 18.6 Representative pelomedusoid side-neck turtles. Clockwise from upper left: Adanson's mud terrapin *Pelusios adansoni*, Pelomedusidae (R. W. Barbour); helmet turtle *Pelomedusa subrufra*, Pelomedusidae (G. R. Zug); yellow-spotted river turtle *Podocnemis unifilis*, Podocnemidae (L. J. Vitt); red-headed river turtle *Podocnemis erythrocephala*, Podocnemidae (T. C. S. Avila-Pires).

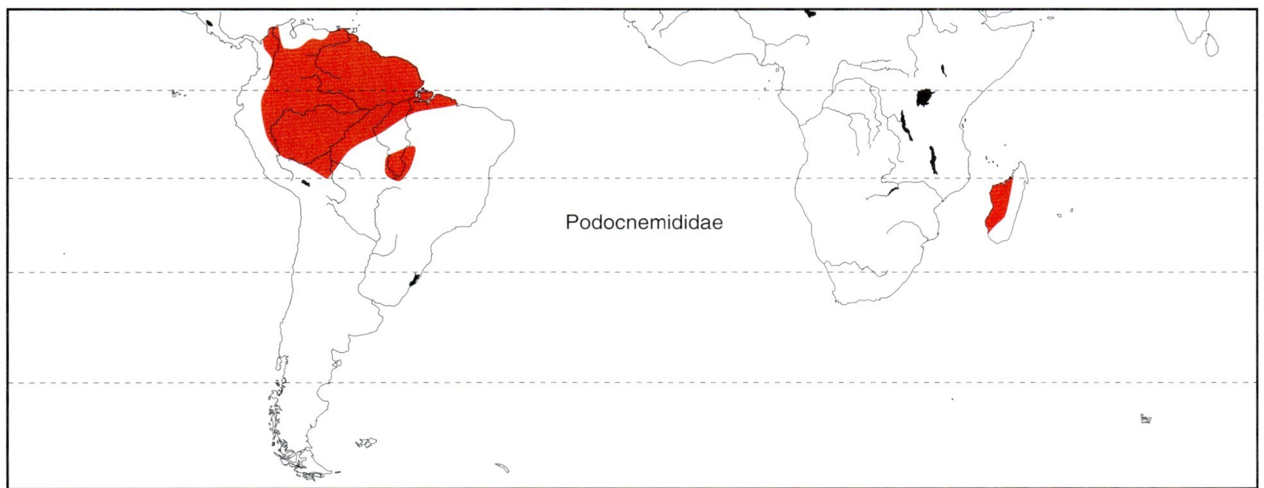

FIGURE 18.7 Geographic distribution of the extant Podocnemidae.

facial nerve has a hyomandibular branch. The plastron has a mesoplastron and well-developed plastral buttresses that articulate with the costals on each side of the carapace; the carapace has 11 pairs of sutured peripherals around its margin and a nuchal without costiform processes. The neck withdraws horizontally, and this mechanism is reflected in an anteriorly oriented articular surface of the first thoracic vertebra; other vertebral traits are the inclusion of the 10th thoracic vertebra in the sacral complex and procoelous caudal vertebrae. The pelvic girdle is firmly fused to the plastron, and the ilium lacks a thelial process. The karyotype is 2N = 28.

Biology: Podocnemids (Fig. 18.6) are mainly river turtles that have broad, domed, streamlined shells for active swimming in moderate currents. They feed on a variety of plant material, including aquatic vegetation and plant products that fall into the water; however, they are not strict herbivores and opportunistically catch and eat small, slow-moving animal prey and carrion. They nest predominantly on sandy riverbanks or sandbars. *P. expansa* nests en masse, and each female lays 60 to 120 eggs. Smaller species accordingly deposit smaller clutches, and most are solitary nesters. Incubation is variable. Eggs of *Podocnemis expansa* require 42 to 47 days to hatch, whereas those of *Podocnemis vogli* require 127 to 149 days.

Cryptodira

Chelydridae

Snapping Turtles

Classification: Reptilia; Parareptilia; Testudines; Cryptodira.

Sister taxon: Most likely the clade is composed of the Kinosternidae and the Dermatemydidae.

Content: Two genera, *Chelydra* and *Macrochelys*, with 1 species each.

Distribution: Southern two-thirds of North America east of the Rockies, portions of Mesoamerica, and southernmost Central America into Ecuador (Fig. 18.8).

Characteristics: Chelydrids range in adult CL from the giant *Macrochelys temminckii* at 80 cm (maximum) to the smaller *Chelydra serpentina* at a maximum of 47 cm. They are large headed and have broad, flattened carapaces with reduced plastra; they possess among the longest tails of all turtles. The jaw closure mechanism of chelydrids articulates on a trochlear surface of the otic capsule and is enclosed in a synovial capsule. An epipterygoid is present in the skull; the internal carotid canal lies in the pterygoid, and the parietal–squamosal and postorbital–squamosal are in strong contact. The facial nerve lacks a hyomandibular branch. The plastron lacks a mesoplastron, and the plastral buttresses articulate loosely or firmly with the costals of the carapace; the carapace has 11 pairs of sutured peripherals around its margin and a nuchal with large costiform processes. The neck withdraws vertically, and this mechanism is reflected in an anteroventrally oriented articular surface of the first thoracic vertebra; other vertebral traits are the exclusion of the 10th thoracic vertebra from the sacral complex, and amphicoelous and opisthocoelous caudal vertebrae. The pelvic girdle flexibly articulates with the plastron, and the ilium lacks a thelial process. The plastron is greatly reduced and cruciform, and the plastral bridge is rigid; the skull roof is strongly emarginated. The karyotype is 2N = 52.

Biology: *Chelydra* and *Macrochelys* are aquatic turtles (Fig. 18.9). *Macrochelys* rarely leave the water except to nest, whereas *Chelydra* commonly make terrestrial forays in addition to nesting on land. Feeding, mating, and hibernation occur in water, so these terrestrial movements seem to be related to nesting and dispersal. Chelydrids are opportunistic omnivores; *M. temminckii* has a wormlike lingual appendage with which to lure fish, but it also eats mollusks, other invertebrates, and plant matter. *Chelydra* catches prey

FIGURE 18.8 The two extant chelydrid turtles. From left: Alligator snapping turtle *Macrochelys temminckii* (L. J. Vitt); common snapping turtle *Chelydra serpentina* (L. J. Vitt). Note the moss growing on the back of the common snapping turtle.

from ambush and also actively searches for prey, which includes aquatic vertebrates, invertebrates, and plant material. *Macrochelys temminckii* usually lives in lakes and deep, slow-moving streams, although it often travels long distances and forages in smaller streams. *Chelydra serpentina* is mainly a shallow-water inhabitant and occurs in freshwater habitats. Egg laying is mainly spring and early summer for both, and clutch size is related to female body size; *M. temminckii* has the largest clutches (20–50 eggs). Clutches of *M. temminckii* are not as large as might be expected based on its body size, differing little from the clutch size seen in *Chelydra*.

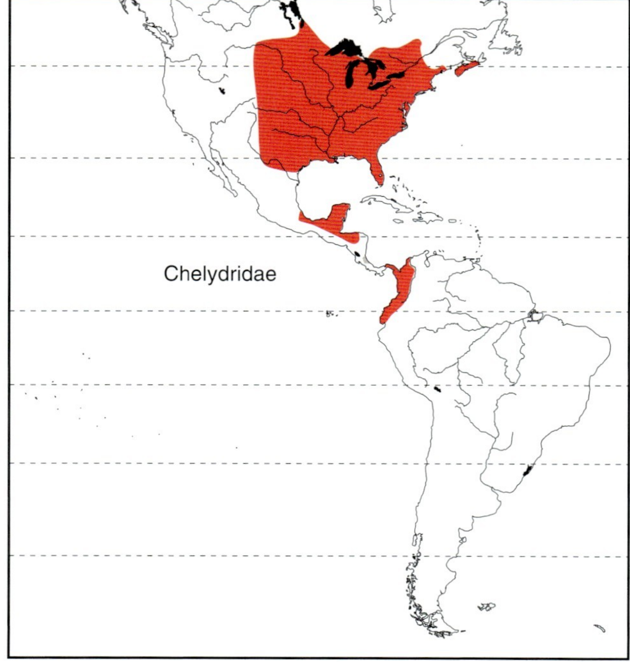

FIGURE 18.9 Geographic distribution of the extant Chelydridae.

Chelonioidea

Cheloniidae

Hard-Shelled Sea Turtles

Classification: Reptilia; Parareptilia; Testudines; Cryptodira.

Sister taxon: Dermochelyidae.

Content: Five genera, *Caretta*, *Chelonia*, *Eretmochelys*, *Lepidochelys,* and *Natator*, with 6 species.

Distribution: Worldwide in tropical and temperate seas (Fig. 18.10).

Characteristics: Cheloniid sea turtles are large, ranging in adult CL from about 60 cm (*Lepidochelys*) to 1.0–1.4 m (*Chelonia*). They have flattened, streamlined shells covered with epidermal scutes and forelimbs modified into large flippers. The jaw closure mechanism articulates on a trochlear surface of the otic capsule and is enclosed in a synovial capsule. An epipterygoid is present in the skull; the internal carotid canal lies in the pterygoid, and there is strong parietal–squamosal and postorbital–squamosal contact. The facial nerve lacks a hyomandibular branch. The plastron lacks a mesoplastron, and the plastral buttresses do not join into the costals of the carapace; the carapace has 11 or more pairs of sutured peripherals around its margin and a nuchal without costiform processes. The neck withdraws vertically, and this mechanism is reflected in an anteroventrally oriented articular surface of the first thoracic vertebra; other vertebral traits include the exclusion of the 10th thoracic vertebra from the sacral complex and procoelous caudal vertebrae. The pelvic girdle flexibly articulates with the plastron, and the ilium lacks a thelial process. The karyotype is 2N = 56.

Biology: Cheloniids are marine turtles, emerging on land only to nest and rarely to bask (Hawaiian and Galapagos *Chelonia mydas*). They swim via forelimb propulsion; the flippers move in a figure-eight stroke, just as in avian aerial flight but with forward thrust produced by both the

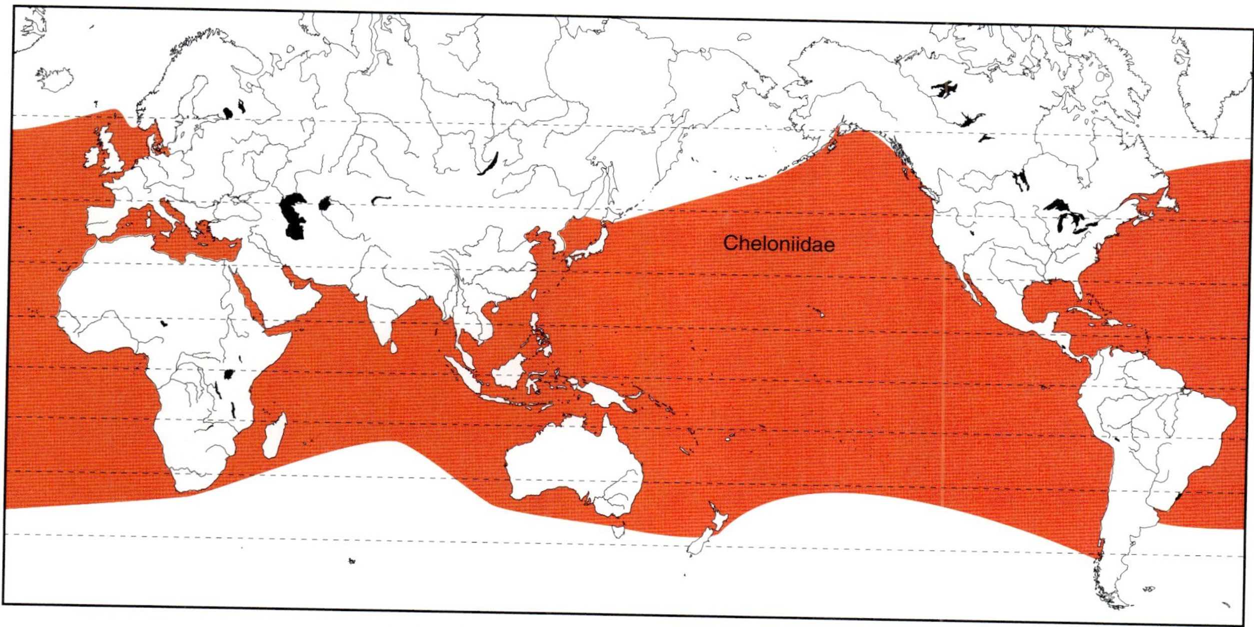

FIGURE 18.10 Geographic distribution of the extant Cheloniidae.

FIGURE 18.11 Representative chelonioid turtles. From left: Green sea turtle *Chelonia mydas*, Cheloniidae (G. R. Zug); leatherback sea turtle *Dermochelys coriacea*, Dermochelyidae (C. K. Dodd, Jr.).

up and down stroke; strongly webbed hind feet serve mainly as rudders. As adults, all cheloniids except *Lepidochelys olivacea* are near-shore or continental-slope residents. Cheloniids appear to have a pelagic stage from immediately after hatching for about 4 to 12 years. Although this aspect of juvenile biology is unknown for *Natator depressus*, presumably newly hatched juveniles are not pelagic. Cheloniids tend to be dietary specialists as adults; for example, *Chelonia mydas* (Fig. 18.11) eats mainly marine grasses or algae, *Caretta caretta* eats decapod crustaceans and mollusks, and *Eretmochelys imbricata*, sponges and soft corals. Most sea turtles (*Lepidochelys* excepted) require 25 or more years to reach reproductive maturity and have a multiyear reproductive cycle. During a reproductive season, a female typically deposits two to five clutches of eggs at approximately 2-week intervals. Clutch size is variable within a species, depending to some extent on the female's body size and nourishment; typically clutch size is more than 100 eggs. *E. imbricata* has the highest average clutch size, 130, and *N. depressus* the lowest, 52.

Dermochelyidae

Leatherback Sea Turtles

Classification: Reptilia; Parareptilia; Testudines; Cryptodira.

Sister taxon: Cheloniidae.

Content: Monotypic, *Dermochelys coriacea*.

Distribution: Worldwide in tropical to cold temperate seas (Fig. 18.12).

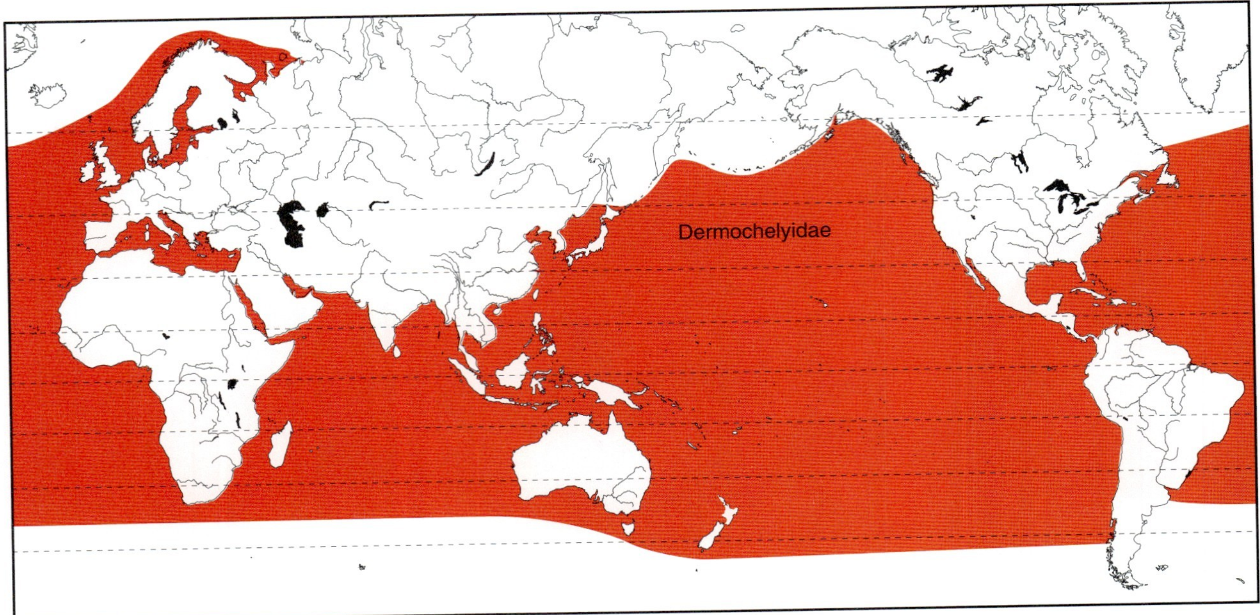

FIGURE 18.12 Geographic distribution of the extant Dermochelyidae.

Characteristics: Adult leatherbacks average from 1.34-1.67 m CL among different populations. They have broad, streamlined, ridged shells that lack epidermal scutes (Fig. 18.11). Their forelimbs are modified into large flippers, and their hindlimbs are typical for turtles but strongly webbed. The jaw closure mechanism articulates on a trochlear surface of the otic capsule and is enclosed in a synovial capsule. An epipterygoid is present in the skull; the internal carotid canal lies in the pterygoid, and there is strong contact between the parietal and squamosal and the postorbital and squamosal. The facial nerve lacks a hyomandibular branch. The plastron lacks a mesoplastron, and the plastral buttresses do not link into the costals of the carapace; the carapace has numerous atypical peripherals along the lateral margins and a nuchal without costiform processes. The neck withdraws vertically, and this mechanism is reflected in an anteroventrally oriented articular surface of the first thoracic vertebra; other vertebral traits are the exclusion of the 10th thoracic vertebra from the sacral complex and procoelous caudal vertebrae. The pelvic girdle flexibly articulates with the plastron, and the ilium lacks a thelial process. The karyotype is 2N = 56.

Biology: Leatherbacks are highly specialized, pelagic sea turtles (Fig. 18.11). They are unique among the living reptiles because they are inertial endotherms (see Chapter 7). They maintain body temperatures above ambient temperatures and do so even in the cooler waters of the north and south temperate zones. Body heat is generated by muscle activity, not by cellular metabolism as in avian reptiles. Heat loss is reduced by the large surface-to-body ratio and by the high insulation properties of an oil-laden skin. To further conserve body heat, the forelimbs have a circulatory counterflow system that transfers heat from the arteries to the veins and back to the body core. It is unknown at what stage juveniles shift from ectothermy to inertial endothermy, although the shift is probably size related owing to the physics of heat exchange associated with surface-to-volume ratio. Amazingly, leatherbacks support their endothermy on a diet of jellyfish, salps, and other gelatinous invertebrates, prey more liquid than solid but obviously highly nutritious. Leatherbacks are highly migratory, potentially crossing and recrossing the length and breadth of entire ocean basins. Their movements seem tied to the pursuit of jellyfish blooms and other aggregations of their prey. Like their sister group, the cheloniids, dermochelyids have a multiyear reproductive cycle. Females return to their nesting beaches, mainly onc biennial to triennial reproductive cycles and lay multiple clutches within one nesting season. Clutch size averages about 80 eggs (range, 46–160), and most clutches contain a moderate percentage of yolkless eggs, the function of which remains a mystery.

Chelomacryptodira, Trionychoidea

Carettochelyidae

Pig-Nose Turtles

Classification: Reptilia; Parareptilia; Testudines; Cryptodira.

Sister taxon: Trionychidae.

Content: Monotypic, *Carettochelys insculpta*.

Distribution: Southern New Guinea and northwestern Australia (Fig. 18.13).

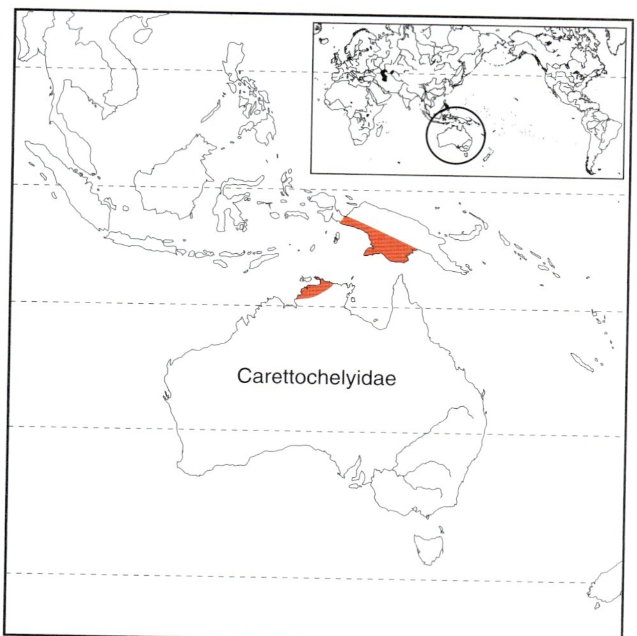

FIGURE 18.13 Geographic distribution of the extant Carettochelyidae.

Characteristics: Adults have heavy, moderately domed shells and range from 30 to 55 cm CL. The shell lacks epidermal scutes; instead, it is covered with a smooth epidermal skin. The forelimbs are modified flippers with two well-developed claws. The jaw closure mechanism articulates on a trochlear surface of the otic capsule and is enclosed in a synovial capsule. An epipterygoid is present in the skull; the internal carotid canal lies in the pterygoid, and the parietal but not the postorbital touches the squamosal. The facial nerve lacks a hyomandibular branch. The plastron lacks a mesoplastron, and the plastral buttresses form firm articulations with the costals of the carapace. The carapace has 11 pairs of sutured peripherals around its margin and a nuchal without costiform processes. The neck withdraws vertically, and this mechanism is reflected in an anteroventrally oriented articular surface of the first thoracic vertebra; other vertebral traits are the exclusion of the 10th thoracic vertebra from the sacral complex and procoelous caudal vertebrae. The pelvic girdle flexibly articulates with the plastron, and the ilium has a thelial process. The karyotype is 2N = 68.

Biology: *Carettochelys insculpta* (Fig. 18.14.) is a highly aquatic turtle that lives mainly in large rivers and

FIGURE 18.14 Representative trionychoid turtles. Clockwise from upper left: Pig-nose turtle *Carettochelys insculpta*, Carettochelyidae (R. W. Barbour); Indian softshell turtle *Nilssonia gangeticus*, Trionychinae (E. O. Moll); spiny softshell turtle *Apalone spinifera*, Trionychinae (L. J. Vitt); Burmese flap-shell turtle *Lissemys scutata*, Cyclanorbinae (G. R. Zug).

estuaries associated with rivers. As with sea turtles, the flipper-shaped forelimbs are the major locomotor appendages and propel the animal using a figure-eight stroke. This type of stroke mimics underwater flying in sea turtles and penguins and apparently is used predominantly for slow and moderate-speed locomotion; when pursued, the turtle reverts to the typical quadrapedal swimming gait of other aquatic turtles. The broadly webbed hindlimbs are typical of aquatic testudines. Pig-nose turtles emerge from the water only to lay eggs. Nesting occurs in the latter part of the dry season, mainly from August to October when the river sandbanks and bars are exposed. Clutch size is about 7 to 19 eggs, which hatch after an 8–10-week incubation. *C. insculpta* is an opportunistic omnivore; fruit, seeds, and leaves of riparian vegetation and submergent plants are commonly eaten, as are a variety of invertebrates and vertebrates.

Trionychidae

Softshell Turtles

Classification: Reptilia; Parareptilia; Testudines; Cryptodira.

Sister taxon: Carettochelyidae.

Content: Two subfamilies, Cyclanorbinae and Trionychinae.

Distribution: North America, Africa, and South and East Asia to New Guinea (Fig. 18.15).

Characteristics: Softshells are flattened, pancake-shaped turtles that have reduced bony carapaces and plastrons. The carapace and plastron are naked, lacking epidermal scutes, but are covered with a thick, leathery skin. The jaw closure mechanism articulates on the trochlear surface of the otic capsule and is enclosed in a synovial capsule. An epipterygoid is present in the skull; the internal carotid canal lies in the pterygoid, and the parietal but not the postorbital touches the squamosal. The facial nerve lacks a hyomandibular branch. The plastron lacks a mesoplastron, and plastral buttresses do not form. The flattened carapace lacks peripheral bones (except in *Lissemys*), and the nuchal lacks costiform processes. The neck withdraws vertically; this mechanism is reflected in an anteroventrally oriented articular surface of the first thoracic vertebra. Other vertebral traits are the exclusion of the 10th thoracic vertebra from the sacral complex and procoelous caudal vertebrae. The pelvic girdle flexibly articulates with the plastron, and the ilium has a thelial process. The karyotype is 2N = 66.

Cyclanorbinae

Sister taxon: Trionychinae.

Content: Three genera, *Cyclanorbis*, *Cycloderma*, and *Lissemys*, with 6 species.

Distribution: Sub-Saharan and northeastern Central Africa, and South Asia (Fig. 18.15).

Characteristics: The lattice-like plastral skeleton has bilaterally fused hyoplastral and hypoplastral bones, and externally the plastron has well-developed femoral flaps.

Biology: Flap-shelled softshells are small-to-moderate-sized turtles. The smallest taxon is *Lissemys* (maximum

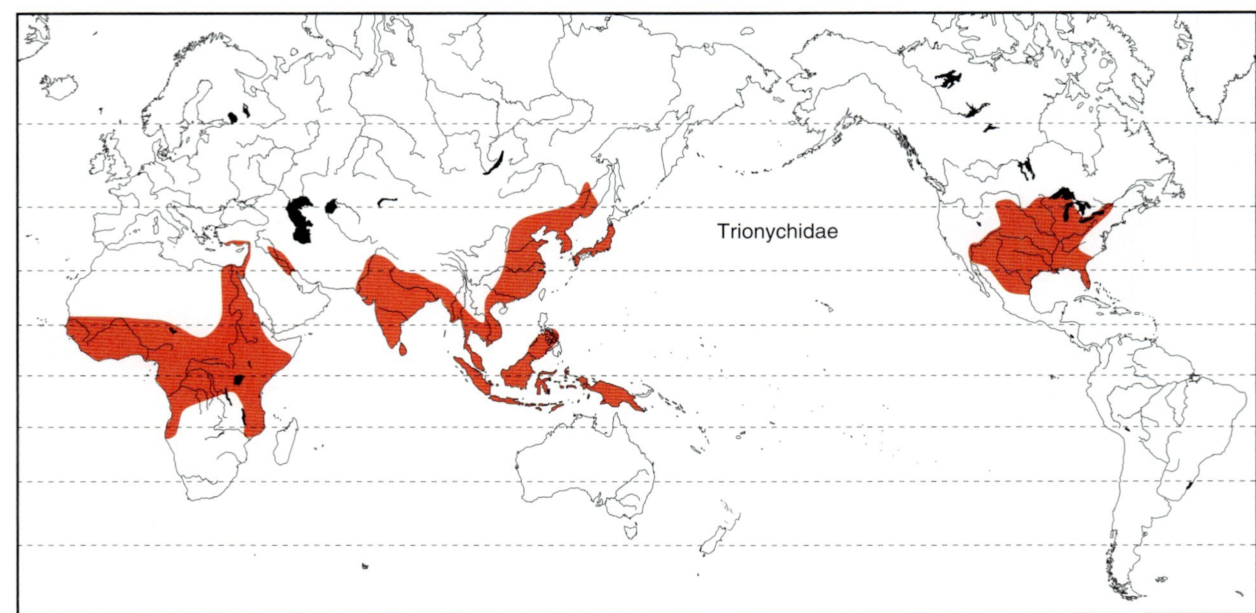

FIGURE 18.15 Geographic distribution of the extant Trionychidae.

adult CL, 37 cm) and the largest is *Cyclanorbis elegans* (to 60 cm). The biology of the African taxa, *Cycanorbis* and *Cycloderma*, is little studied; the South Asian *Lissemys* is somewhat better known. All cyclanorbines are probably bottom dwellers like trionychines. They actively forage and also lie partially hidden in the bottom silt or sand, waiting for passing prey. They are presumably opportunistic omnivores, eating invertebrates, small vertebrates, and occasional plant matter. Clutch size is small to modest; *L. punctata* deposits 2 to 14 eggs; good evidence indicates that clutch size varies geographically and that females produce multiple clutches each year. Incubation ranges from 30 to 40 days to more than 300 days.

Trionychinae

Sister taxon: Cyclanorbinae.

Content: Ten genera, *Amyda*, *Apalone*, *Chitra*, *Dogania*, *Nilssonia* (including *Aspideretes*), *Palea*, *Pelochelys*, *Pelodiscus*, *Rafetus,* and *Trionyx*, with 20+ species.

Distribution: Eastern North America, South Asia to Japan southward to New Guinea, and north-central sub-Saharan Africa into Southwest Asia (Fig. 18.15).

Characteristics: The lattice-like plastral skeleton has separate hyoplastral and hypoplastral bones on each side, and externally the plastron lacks femoral flaps.

Biology: Trionychine softshells are moderate to large turtles (Fig. 18.14). *Pelodiscus sinensis*, the Chinese softshell, is the smallest species (20–25 cm adult CL); *Pelochelys* and *Chitra* are much larger with shell lengths to more than a meter as adults, and the other genera range in adult CL from 40 to 60 cm. All are highly aquatic turtles, spending much of their time partially buried on the bottom waiting for prey. Their long necks and protruding, snorkel-like snouts permit them to extend their noses to the water surface to breathe; they also depend to some extent upon cutaneous respiration. They actively forage for prey, and their flattened hydrodynamically efficient habitus makes them excellent and fast swimmers. The three North American species of *Apalone* often occur at high densities. Softshells live primarily in rivers and lakes. *Dogania subplana* appears to be the only softshell that occurs in small mountain streams. All trionychines are predominantly carnivorous, although they likely feed on plant matter, particularly when animal prey is not readily available. Temperate and subtemperate species are predominately spring breeders, and tropical species lay eggs in the early dry season. Clutch size is small to moderate; for example, the three species of *Apalone* deposit 4 to 30 eggs, whereas the smaller *P. sinensis* lays 9 to 15 eggs per clutch. *Trionyx triunguis* reaches 95 cm CL and can deposit over 100 eggs, but more typically it produces half that number. Incubation generally requires 8 to 10 weeks, although in *Nilssonia gangeticus*, it is 36 to 42 weeks or as brief as 28 days in *Pelodiscus sinensis*.

Kinosternoidae

Dermatemydidae

Mesoamerican River Turtles

Classification: Reptilia; Parareptilia; Testudines; Cryptodira.

Sister taxon: Kinosternidae.

Content: Monotypic, *Dermatemys mawii*.

Distribution: Caribbean–Gulf drainage of Mesoamerica (Fig. 18.16).

Characteristics: *Dermatemys mawii* has an oblong, slightly domed carapace, a large plastron, and a moderately small head. Its jaw closure mechanism articulates on a trochlear surface of the otic capsule and is enclosed in a synovial capsule. An epipterygoid is present in the skull; the internal carotid canal lies in the pterygoid, and the parietal but not the postorbital touches the squamosal. The facial nerve lacks a hyomandibular branch. The plastron lacks a mesoplastron, and the plastral buttresses articulate with costals of the carapace; the carapace has 10 pairs of sutured peripherals around its margin and a nuchal with distinct costiform processes. The neck withdraws vertically, and this mechanism is reflected in an anteroventrally oriented articular surface of the first thoracic vertebra; other

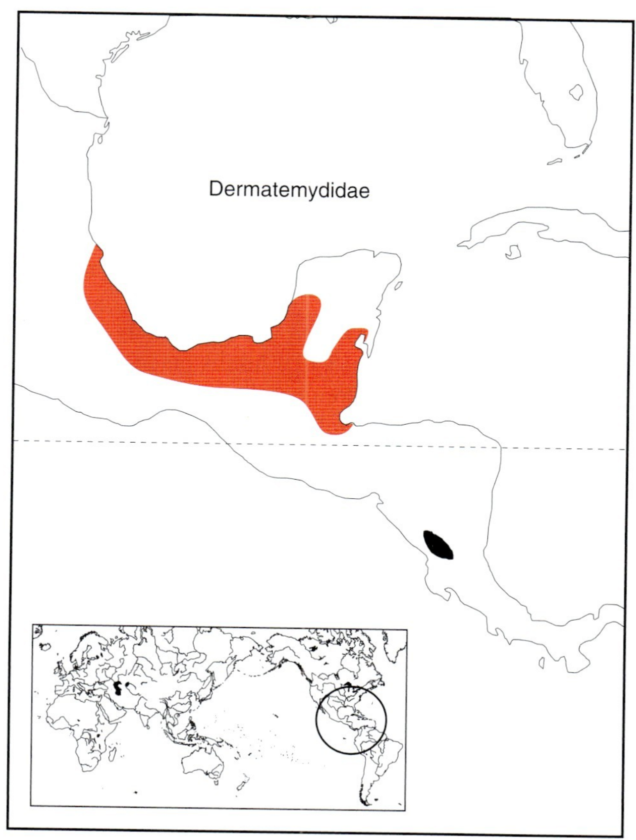

FIGURE 18.16 Geographic distribution of the extant Dermatemydidae.

FIGURE 18.17 Representative kinosternoid turtles. Clockwise from upper left: Mesoamerica river turtle *Dermatemys mawii*, Dermatemydidae (D. Moll); Tabasco mud turtle *Kinosternon acutum*, Kinosterninae (G. R. Zug); yellow mud turtle *Kinosternon flavescens*, Kinosterninae (L. J. Vitt); narrow-bridged musk turtle *Claudius angustatus*, Staurotypinae (R. W. Barbour).

vertebral traits are the exclusion of the 10th thoracic vertebra from the sacral complex and procoelous caudal vertebrae. The pelvic girdle flexibly articulates with the plastron, and the ilium has a thelial process. The karyotype is 2N = 56.

Biology: *Dermatemys mawii* is a moderately large and highly aquatic turtle (Fig. 18.17). Adults range in CL from 33 to 65 cm. They live predominantly in slow-moving areas of large rivers and lakes. Adults and juveniles are totally herbivorous; they eat a variety of aquatic plants and streamside vegetation, fruits, and seeds that fall into the water, particularly figs. Presumably, they are nocturnal, spending the day resting near the bottom or basking at the surface of the water; foraging occurs at night. The turtles court and mate from May to September; egg deposition (2–20 eggs in a clutch) occurs mainly from October to December, and a single individual will deposit eggs as many as four times. Females nest along streams. In Belize, the nesting occurs during the period with greatest rainfall and rising river levels; some early nests are submerged, but developmental arrest allows the embryos to survive. Incubation in these populations is 8 to 10 months; hatching occurs in June and July with the beginning of the rainy season.

Comment: This unique turtle is easily captured and prized as a local food item. Human exploitation has decimated and extirpated most populations, and it is now as endangered as many of the Asian turtles.

Kinosternidae

Mud Turtles and Musk Turtles

Classification: Reptilia; Parareptilia; Testudines; Cryptodira.

Sister taxon: Dermatemydidae.

Content: Two subfamilies, Kinosterninae and Staurotypinae.

Distribution: Eastern North America to the Amazon drainage of South America (Fig. 18.18).

Characteristics: Kinosternids have oblong, moderately domed carapaces and moderate to large heads. The plastron is commonly hinged and has 11 or fewer epidermal scutes. The jaw closure mechanism articulates on the trochlear surface of the otic capsule and is enclosed in a synovial capsule. An epipterygoid is present in the skull; the internal carotid canal lies in the pterygoid, and the parietal but not the postorbital touches the squamosal.

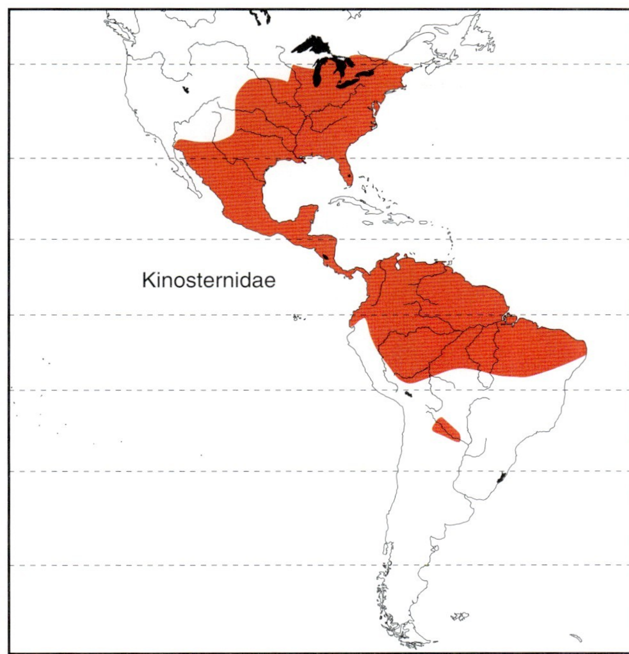

FIGURE 18.18 Geographic distribution of the extant Kinosternidae.

The facial nerve lacks a hyomandibular branch. The plastron lacks a mesoplastron, and the plastral buttresses do not form interdigitating articulations with costals of the carapace; the carapace has 10 pairs of sutured peripherals around its margin and a nuchal without costiform processes. The neck withdraws vertically, and this mechanism is reflected in an anteroventrally oriented articular surface of the first thoracic vertebra; other vertebral traits are the exclusion of the 10th thoracic vertebra free of the sacral complex and procoelous caudal vertebrae. The pelvic girdle flexibly articulates with the plastron, and the ilium has a thelial process. The karyotype is 2N = 54 or 56.

Kinosterninae

Sister taxon: Staurotypinae.

Content: Two genera, *Kinosternon* and *Sternotherus*, with 16+ and 4 species, respectively.

Distribution: Eastern North America to the Amazon drainage of South America (Fig. 18.18).

Characteristics: The well-developed plastron lacks an entoplastral bone and is usually hinged.

Biology: Mud turtles and musk turtles are small-to-moderate-sized turtles (Fig. 18.17), ranging in adult CL from 8 to 12 cm (e.g., *Sternotherus depressus*) and from 15 to 27 cm (*Kinosternon scorpioides*); most species have a maximum adult shell length less than 18 cm. They are generally aquatic species and live in various waterways, including ephemeral pools, marshes and swamps, and large rivers and lakes. All are bottom walkers and poor swimmers. They forage and mate in water; however, some species hibernate on land and others, particularly tropical species, appear to forage on land during wet weather. One species, *K. flavescens*, appears to spend much time on land in underground retreats. During summer rains, these turtles appear on roads and in temporary ponds, often in large numbers. Kinosternids have relatively small clutches, most commonly 1 to 4 eggs, although clutches of up to 16 eggs are deposited by larger species. Incubation is moderately long, usually 100 to 150 days. Aquatic invertebrates, small vertebrates, and carrion dominate the diets of kinosternines. *Sternotherus minor* eats mollusks, and they have proportionately larger heads resulting from large jaw muscles and broad jaw surfaces for crushing snails and bivalves. Large head size associated with eating mollusks occurs in other turtle clades.

Staurotypinae

Sister taxon: Kinosterninae.

Content: Two genera, *Claudius* and *Staurotypus* (Fig. 18.17), with 1 and 2 species, respectively.

Distribution: The Caribbean and Gulf of Mexico and Pacific drainage of Mesoamerica (Fig. 18.18).

Characteristics: The plastron has an entoplastral bone, and the plastron is either moderately reduced with a hinge (*Staurotypus*) or strongly reduced (cruciform) without a hinge (*Claudius*).

Biology: Staurotypines include the small species *C. angustatus* (9-15 cm adult CL) and the largest kinosternid species, *Staurotypus triporcatus* (30–38 cm CL). The biology of the three species is poorly known. *Claudius angustatus* occurs principally in seasonally flooded marshes or pastures and appears to be active only during the rainy season (June–February). Nesting occurs at the end of the wet season (November–February), and from one to five eggs are deposited beneath vegetation; the stereotypic nest digging does not occur. The natural incubation period is unknown, and captive incubation is long, about 100 to 200 days. The two *Staurotypus* inhabit slow- to fast-flowing waters of marshes to large rivers, and rarely occur in ephemeral waters. Reproduction in captive individuals suggests only a slightly larger clutch (3–10 eggs) for *Staurotypus*. All staurotypines are carnivorous, feeding on a variety of aquatic invertebrates and small vertebrates; *S. triporcatus* feeds heavily on snails year-round, and occasionally other turtles become a major prey.

Testudinoidea

Emydidae

Cooters, Sliders, American Box Turtles, and Allies

Classification: Reptilia; Parareptilia; Testudines; Cryptodira.

Sister taxon: Platysternidae or Testudinoidae.

FIGURE 18.19 Representative emydid and platysternid turtles. Clockwise from upper left: European pond turtle *Emys orbicularis*, Emydinae (R. W. Barbour); chicken turtle *Deirochelys reticulata*, Deirochelinae (L. J. Vitt); red-eared slider *Trachemys scripta,* Deirochelinae (L. J. Vitt); big-headed turtle *Platysternon megacephala*, Platysternidae (R. W. Van Devender).

Content: Two subfamilies, Emydinae and Deirochelinae (Fig. 18.19).

Distribution: Europe to Ural Mountains and North America southward to eastern Brazil (Fig. 18.20).

Characteristics: Emydids include small species such as *Calemys muhlenbergi* (8–11 cm adult CL) to moderate-sized species, such as *Pseudemys concinna* (35–40 cm CL). These turtles have oval to oblong and moderately domed carapaces; the plastron is large and occasionally hinged (Fig. 18.19). The jaw closure mechanism articulates on a trochlear surface of the otic capsule and is enclosed in a synovial capsule. An epipterygoid is present in the skull; the internal carotid canal lies in the pterygoid, and the parietal but not the postorbital touches the squamosal. The facial nerve lacks a hyomandibular branch. The plastron lacks a mesoplastron, and the plastral buttresses usually articulate with the costals of the carapace; the carapace has 11 pairs of sutured peripherals around its margin and a nuchal without costiform processes. The neck withdraws vertically, and this mechanism is reflected in an anteroventrally oriented articular surface of the first thoracic vertebra; other vertebral traits are the exclusion of the 10th thoracic vertebra from the sacral complex and procoelous caudal vertebrae. The pelvic girdle flexibly articulates with the plastron, and the ilium lacks a thelial process. The karyotype is 2N = 50.

Biology: Emydids include semiaquatic, aquatic, and terrestrial turtles (Fig. 18.19); they live in most permanent water habitats from marshes to large rivers and lakes. *Terrapene* is mainly a terrestrial group, whereas *Malaclemys terrapin* is largely estuarine and adapted to brackish water. With the exceptions of *Pseudemys* and female *Trachemys* and *Graptemys*, adult CL of emydids is less than 20 cm. Sexual dimorphism is common and often strikingly so in *Pseudemys* and *Graptemys*. In *Graptemys*, adult males are commonly half the size of adult females; for example, female *G. barbouri* are 17 to 26 cm CL and males only 9 to 13 cm. Most taxa are omnivores, and juveniles eat mainly animal prey; in contrast, the large *Pseudemys* are strongly herbivorous. These predominantly temperate turtles deposit eggs in spring; hatching occurs later in the summer, commonly with a 60–80-day incubation period. Hatchlings of some species, e.g., *Chrysemys picta*, regularly overwinter in the nest in the northern part of their distribution. Clutch size is small to modest; 2 to 10 eggs compose

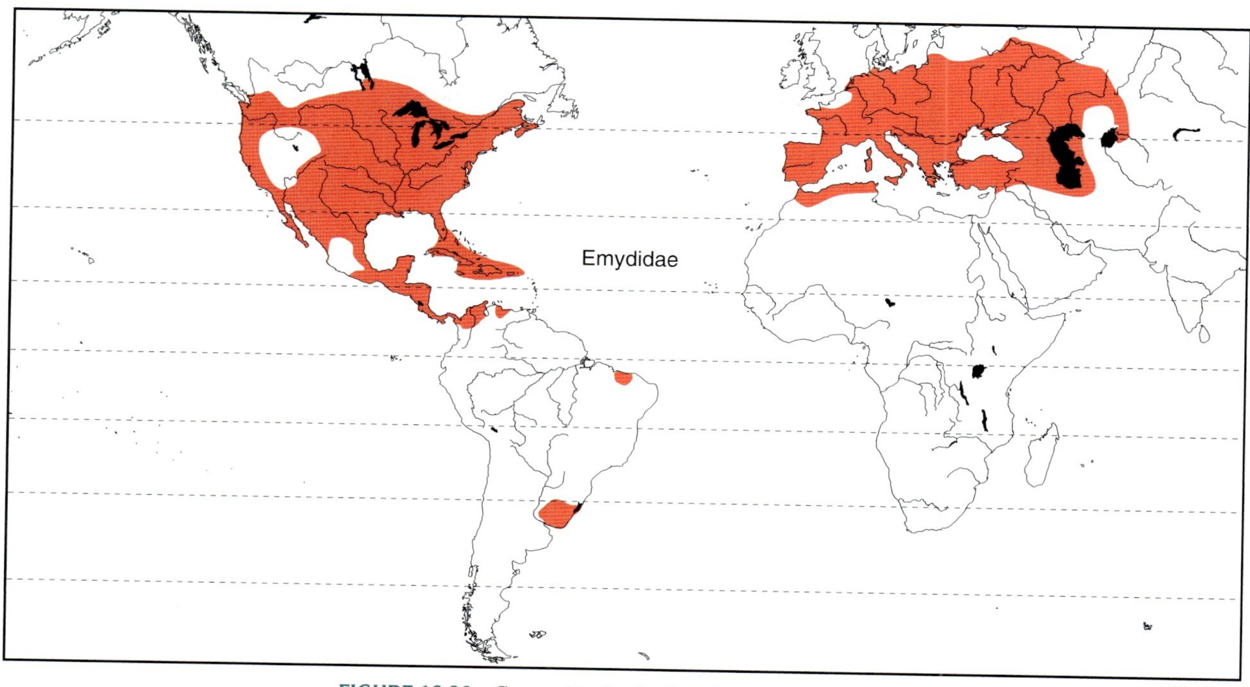

FIGURE 18.20 Geographic distribution of the extant Emydidae.

the average clutch for most emydids, although *Pseudemys* and *Trachemys* typically have larger clutches.

An interesting contrast exists ecologically and evolutionarily between the two subfamilies of emydids. The emydinae are conservative with respect to diets (omnivorous) but have diversified in habitat use (terrestrial, aquatic, and semiterrestrial species). On the other hand, deirochelines are conservative in habitat use (aquatic) but have diversified in terms of diets (herbivorous, omnivorous, and carnivorous species).

Plastral kinesis evolved independently in each emydid subfamily.

Emydinae

Content: Four genera, *Calemys, Clemmys*, *Emys* (includes *Emydoidea*), and *Terrapene*, with 11 species.

Distribution: Europe to the Ural Mountains, most of North America but not the central West (Fig. 18.20).

Characteristics: Palatine extends from the triturating surface, and the posterior palatine foramen is much larger than the foramen orbitale–nasal.

Biology: Some emydines are aquatic (*E. orbicularis* and *E. trinacris*), some are swamp dwellers (*Clemmys guttata*), and yet others are terrestrial (*Terrapene* and *Calemys insculpta*). Historically, these turtles were very common and, because most spend at least some time on land, have been severely impacted by humans. For example, *C. muhlenbergii* and *C. insculpta* have nearly disappeared in the eastern United States, and *Emys marmorata* has suffered severe declines in the western United States. Although box turtles (*T. carolina* and *T. ornata*) remain common in parts of their ranges, both have suffered population reductions and in some areas extirpation as the result of land-use changes and road mortality. Box turtles spend considerable time basking on and crossing paved roads during spring and often appear in large numbers following rainstorms during summer.

Comment: Relationships of emydine turtles have been problematical until recently. A recent molecular phylogeny restricts *Clemmys* to *C. guttata*, resurrects *Calemys* to include the former *Glyptemys muhlenbergii* and *G. insculpta* (formerly *Clemmys*), and expands *Emys* to include three species, *E. orbicularis, E. blandingii* (former *Emydoidea*), and *Emys marmorata*.

Deirochelyinae

Content: Six genera, *Chrysemys, Deirochelys, Graptemys, Malaclemys, Pseudemys,* and *Trachemys,* with 31 species.

Distribution: Most of North America, Caribbean Islands, Central America, northern South America, and two disjunct regions in northeastern and southeastern Brazil (Fig. 18.20).

Characteristics: The jugal contacts the palatine, epipubes do not ossify, and the foramen caroticopharyngeale is reduced or absent.

Biology: This is a truly interesting group of turtles. First, all species are aquatic, although *Deirochelys* spends considerable time on land. Some (*Deirochelys* and *Malaclemys*) feed on crustaceans, others (*Graptemys*) feed on mollusks, and yet others (*Pseudemys*) feed on plant material. *Chrysemys* and *Trachemys* are omnivorous but rely to a large extent on a variety of animal prey. *Trachemys scripta* is not only the most common North American turtle but also the best studied and has served as a model organism for studies on the ecology, behavior, and life histories of turtles in general.

Platysternidae

Big-headed Turtle

Sister taxon: Most likely, Emydidae, but possibly Testudinoidae.

Content: Monotypic, *Platysternon megacephalum*.

Distribution: Southern China southward into Thailand (Fig. 18.21).

Characteristics: The plastron is moderate-sized and the plastral bridge is flexible; the skull roof is complete.

Biology: *Platysternon megacephalum* is a relatively small turtle reaching about 18 cm in length (Fig. 18.19). It has been rarely studied in the wild, and its biology is known principally from captive animals. This species occurs in small, rocky streams in mountainous areas of Southeast Asia, mostly above 700 m elevation. Presumably it forages at night and spends the day hiding beneath rocks and logs in streams. In captivity, its eats a range of animal matter. It likely eats fish, frogs, and assorted invertebrates in the wild. Clutch size is one to three eggs. The karyotype is 2N = 54.

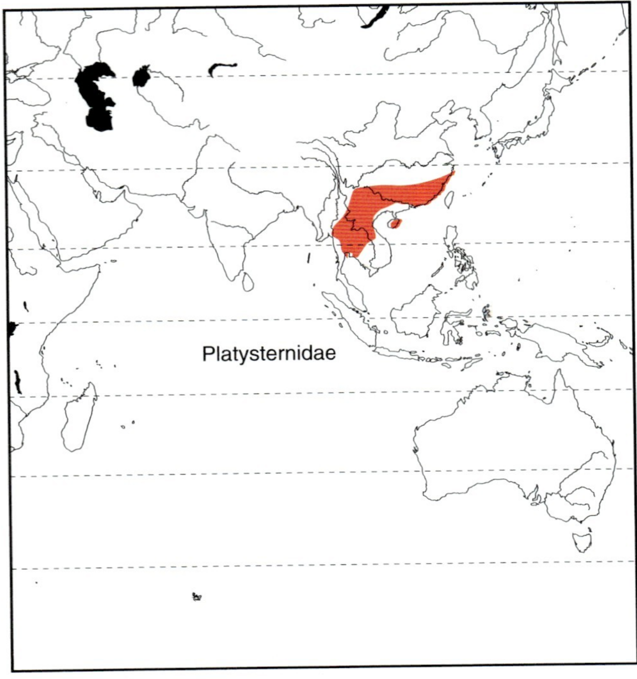

FIGURE 18.21 Geographic distribution of the extant Platysternidae.

Testudinoidea, Testudinoidae

Geomydidae

Asian River Turtles, Leaf and Roofed Turtles, Asian Box Turtles, and Allies

Classification: Reptilia; Parareptilia; Testudines; Cryptodira.

Sister taxon: Testudinidae.

Content: Twenty-three genera, *Batagur*, *Callagur*, *Chinemys*, *Cuora*, *Cyclemys*, *Geoclemys*, *Geoemyda*, *Hardella*, *Heosemys*, *Hieremys*, *Kachuga*, *Leucocephalon*, *Malayemys*, *Mauremys*, *Melanochelys*, *Morenia*, *Notochelys*, *Ocadia*, *Orlitia*, *Rhinoclemmys*, *Sacalia*, *Siebenrockiella*, and *Vijayachelys*, with 65+ species.

Distribution: Southern Europe to Japan and East Indies, and central and northern South America (Fig. 18.22).

Characteristics: Geomydids are small to large turtles with oval to oblong and moderately domed or flattened carapaces; the plastron is large and occasionally hinged (Fig. 18.23). The jaw closure mechanism articulates on a trochlear surface of the otic capsule and is enclosed in a synovial capsule. An epipterygoid is present in the skull; the internal carotid canal lies in the pterygoid, and the parietal but not the postorbital touches the squamosal. The facial nerve lacks a hyomandibular branch. The plastron lacks a mesoplastron, and the plastral buttresses usually articulate firmly with the costals of the carapace; the carapace has 11 pairs of sutured peripherals around its margin and a nuchal without costiform processes. The neck withdraws vertically, and this mechanism is reflected in an anteroventrally oriented articular surface of the first thoracic vertebra; other vertebral traits are the exclusion of the 10th thoracic vertebra from the sacral complex and procoelous caudal vertebrae. The pelvic girdle flexibly articulates with the plastron, and the ilium lacks a thelial process. The karyotype is 2N = 52.

Biology: Geomydids are a diverse group of turtles. They range in adult CL from the small *Geoemyda spengleri* and *Heosemys silvatica* (to 13 cm) to the large *Orlitia borneensis* (to 80 cm), and from totally terrestrial (*G. spengleri*, *H. silvatica*) to highly aquatic species that emerge on land only to lay eggs (*O. borneensis*, *Batagur baska*). Other species live in mountain streams (*Cyclemys dentata*, *Cuora trifasciata*) or estuaries (*B. baska*, *Callagur borneoensis*). Some taxa are specialized carnivores (aquatic snails—male *Malayemys subtrijuga*) to strict herbivores (*Kachuga smithi*). Within a single clade, habits and habitat preferences can be strikingly different; for example, the Neotropical *Rhinoclemys* has totally terrestrial species (e.g., *R. annulata*), semiaquatic species (*R. areolata*), and highly aquatic species (*R. nasuta*); the terrestrial and aquatic species are either herbivorous or omnivorous. Shell morphology is similarly diverse and includes high-domed to flattened species. Reproductive behavior is only beginning to be documented.

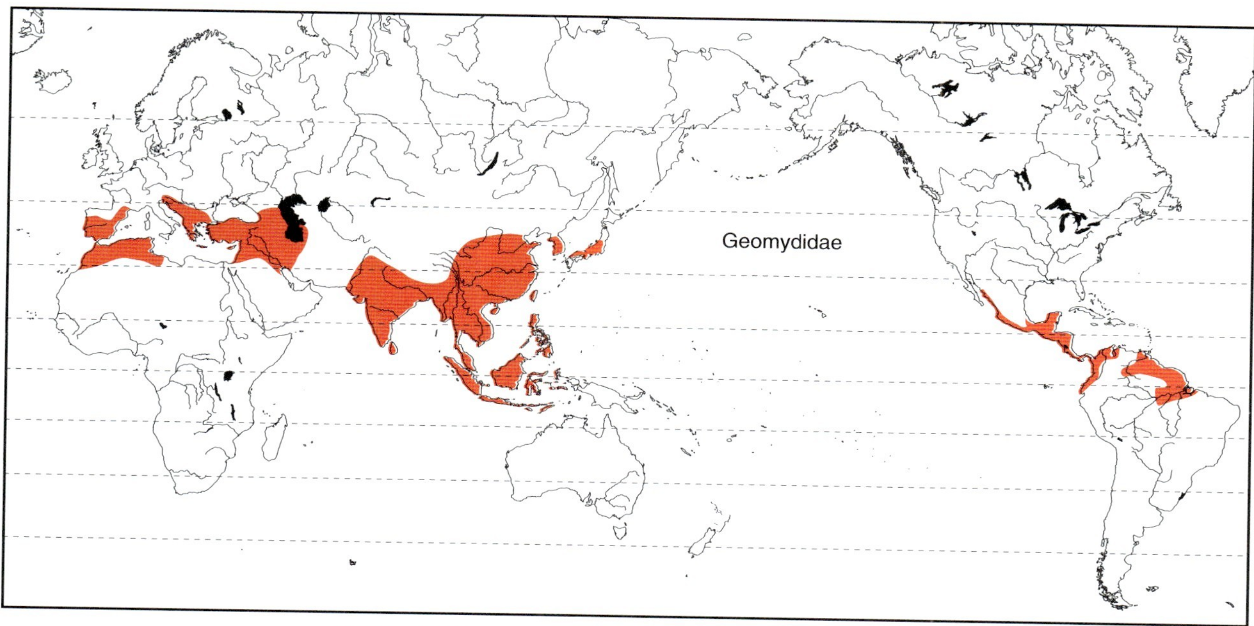

FIGURE 18.22 Geographic distribution of the extant Geomydidae.

FIGURE 18.23 Representative geomydid turtles. Clockwise from upper left: Yellow-headed box turtle *Cuora aureocapita*, Geomydidae (C. H. Ernst); giant Asian pond turtle *Heosemys grandis*, Geomydidae (G. R. Zug); Philippine forest turtle *Siebenrockiella leytensis* (R. Brown); South American wood turtle, *Rhinoclemys punctularia* (L. J. Vitt).

Most species produce fewer than 10 eggs per clutch, although many appear to have multiple clutches within a single reproductive season. The largest clutches occur in *Geoclemys hamiltoni* (18–30 eggs; female 30–40 cm CL) and *Kachuga dhongoka* (30–35 eggs; to 48 cm CL), yet the large *Batagur baska* (50–60 cm CL) averages 20 eggs per clutch, and the similar-sized *C. borneoensis* (50–60 cm CL) has clutches of 15 to 25 eggs. Incubation period is unknown for most species, but where known, it is commonly from 3 to 5 months.

Comment: Geomydids, formerly referred to as the Bataguridae, are the most species group of extant turtles. Because many species have small distributions and occur in the most densely human-populated part of the world, they are subjected to the highest levels of human predation. Because conservation of these turtles is largely ignored in Asia, many species will become extinct during the next decade. One species, *Heosemys leytensis*, was known from only two specimens collected in the 1980s, with another two from 1920 until it was rediscovered on Palawan and Dumaran Islands in the Philippines.

Testudinidae

Tortoises

Classification: Reptilia; Parareptilia; Testudines; Cryptodira.

Sister taxon: Geomydidae.

Content: Fifteen genera, *Angonoka, Astrochelys, Chelonoidis, Chersina, Dipsochelys, Geochelone, Gopherus, Homopus, Indotestudo, Kinixys, Malacochersus, Manouria, Psammobates, Pyxis,* and *Testudo*, with 45+ species.

Distribution: Southern North America to southern South America, circum-Mediterranean Euroafrica to Indomalaysia, sub-Saharan Africa, Madagascar, and some oceanic islands (Fig. 18.24).

Characteristics: With a single exception (*Malacochersus tornieri*), all tortoises have well-developed, high-domed shells, and without exception, all share unique columnar or elephantine hindlimbs. The jaw closure mechanism articulates on a trochlear surface of the otic capsule and is enclosed in a synovial capsule. An epipterygoid is present in the skull; the internal carotid canal lies in the pterygoid, and the parietal but not the postorbital touches the squamosal. The facial nerve lacks a hyomandibular branch. The plastron lacks a mesoplastron, and the plastral buttresses articulate firmly with the costals of the carapace; the carapace has 11 pairs of sutured peripherals around its margin and a nuchal without costiform processes. The neck withdraws vertically, and this mechanism is reflected in an anteroventrally oriented articular surface of the first thoracic vertebra; other vertebral traits are the exclusion of the 10th thoracic vertebra from the sacral complex and procoelous caudal vertebrae. The pelvic girdle flexibly articulates with the plastron, and the ilium lacks a thelial process. The karyotype is 2N = 52.

Biology: All tortoises are terrestrial (Fig. 18.25). They live in diverse habitats, including deserts, arid grasslands, and scrub (*Gopherus agassizii, Testudo kleinmanni*) to wet evergreen forests, (*Geochelone denticulata, Kinixys erosa*), and from sea level (*Geochelone gigantea*) to mountainsides (1000 m elevation; *Indotestudo forsteni*). Most species, however, occupy semiarid habitats. Adult CL ranges from 8.5 cm in the smallest tortoise, *Homopus*

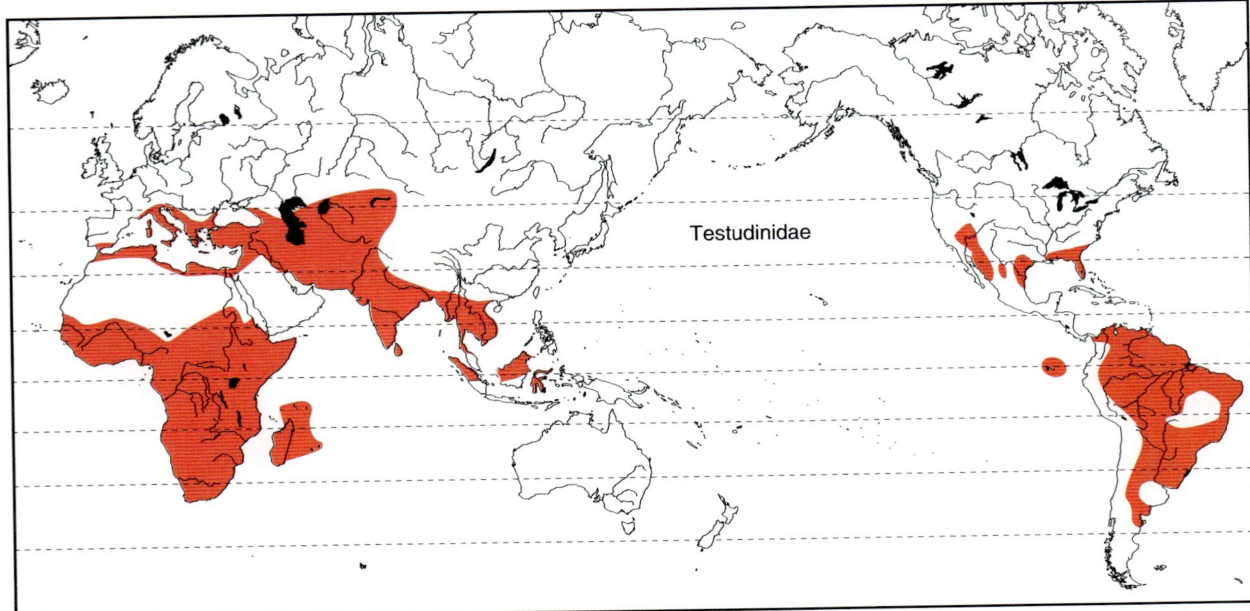

FIGURE 18.24 Geographic distribution of the extant Testudinidae.

FIGURE 18.25 Representative tortoises. Clockwise from upper left: Berlandier's tortoise *Gopherus berlandieri*, Testudinidae (R. W. Van Devender); Asian brown tortoise *Manouria emys*, Testudinidae (R. W. Barbour); yellow-footed tortoise, *Geochelone denticulata* (L. J. Vitt); red-footed tortoise, *Geochelone carbonaria,* in defensive posture (L. J. Vitt).

signatus, to 130 cm in the largest, *Geochelone elephantopus*. Most tortoises are herbivores and eat flowers, seeds, fruits, and foliage; a few species, such as *Geochelone carbonaria*, are opportunistic omnivores, eating what they can find on the forest floor. Most species lay small clutches, seldom exceeding 20 eggs (including the Galapagos and Aldabran giant tortoises), and many species have clutches of only 1 to 2 eggs. *Manouria* is the exception; *M. impressa* averages more than 30 eggs in a clutch. Incubation is characteristically long in tortoises; the average incubation periods are between 100 and 160 days for most species and supposedly as long as 18 months in *Geochelone pardalis*.

QUESTIONS

1. What are the differences between pleurodiran and cryptodiran turtles?
2. Turtles have been placed as a sister clade to all other "reptiles" by some authors and as a sister clade to crocodylians and birds by other authors. What evidence supports each of these placements of turtles?
3. What is the global distribution of the turtle family Chelydridae?
4. Among turtles, which family is represented by the most living species, and does this family also have the largest global distribution?
5. If you landed in Australia, what turtle families would you expect to find?

REFERENCES

General

Ananjeva et al., 1988; Bickham and Carr, 1983; Cogger and Zweifel, 1998; David, 1994; Dutton et al., 1996; Ernst and Barbour, 1989a; Ernst et al., 1994; Fujita et al., 2004; Gaffney, 1975, 1977, 1984; Gaffney and Meylan, 1988; Iwabe et al., 2005; Hirayama, 1994; Hutchinson, 1991; Iverson, 1992; Joyce et al., 2004; Laurin and Reisz, 1995; Legler and Georges, 1993; McDowell, 1964; Meylan, 1987, 1996; Meylan and Gaffney, 1989; Near et al., 2005; Parham and Fastovsky, 1998; Parham et al., 2006; Seddon et al., 1997; Shaffer et al., 1997; Williams, 1950.

Taxonomic Accounts

Chelidae

Australoamerican Side-Neck Turtles

Cann (1998); Fujita et al., 2004; Gaffney et al., 1991; Georges and Adams, 1992, 1996; Georges et al., 1993, 1998, 2002; Kennett et al., 1993a,b; Kuchling, 1999; Legler and Georges, 1993; Lieb et al., 1999; McCord and Thompson, 2002; McCord et al., 2001; Pritchard and Trebbau, 1984; Seddon et al., 1997; Shaffer et al., 1997, 1998.

Pelomedusidae

African Mud Terrapins

Ernst and Barbour, 1989a; Fujita et al., 2004; Gaffney et al., 1991; Georges et al., 1998; Meylan, 1996; Shaffer et al., 1997.

Podocnemidae

Madagascan Big-Headed Turtles and American Side-Neck River Turtles

Gaffney et al., 1991; Georges et al., 1998; Kuchling, 1999; Meylan, 1996; Pritchard and Trebbau, 1984; Shaffer et al., 1997.

Chelydridae

Snapping Turtles

Congdon et al., 1994; Ernst et al., 1994; Fujita et al., 2004; Gaffney et al., 1991; Roman et al., 1999; Shaffer et al., 1997; Sites and Crandall, 1997.

Cheloniidae

Hard-Shelled Sea Turtles

Bjorndal, 1996; Fujita et al., 2004; Gaffney et al., 1991; Limpus and Miller, 1993; Márquez, M., 1990; Miller, 1996; Shaffer et al., 1997.

Dermochelyidae

Leatherback Sea Turtles

Bjorndal, 1996; Fujita et al., 2004; Gaffney et al., 1991; Limpus, 1993; Márquez, M., 1990; Miller, 1996; Shaffer et al., 1997; Zug and Parham, 1996.

Carettochelyidae

Pig-Nose Turtles

Cann, 1998; Fujita et al., 2004; Gaffney et al., 1991; Georges and Wombey, 1993; Shaffer et al., 1997; Webb et al., 1986.

Trionychidae

Softshell Turtles

Fujita et al., 2004; Meylan, 1987; Shaffer et al., 1997.

Cyclanorbinae

Das, 1995; Ernst et al., 1994; Rashid and Swingland, 1998.

Trionychinae

Das, 1995; Ernst and Barbour, 1989a; Ernst et al., 1994; Kuchling, 1999; Plummer, 1977a,b; Rashid and Swingland, 1998; Vasudevan, 1997.

Dermatemydidae

Mesoamerican River Turtles

Ernst et al., 1994; Fujita et al., 2004; Gaffney et al., 1991; Moll, 1988; Polisar, 1996; Shaffer et al., 1997.

Kinosternidae

Mud Turtles and Musk Turtles

Fujita et al., 2004; Gaffney et al., 1991; Iverson, 1991a, 1998; Shaffer et al., 1997.

Kinosterninae

Ernst et al., 1994; Frazer et al., 1991; Fujita et al., 2004; Gibbons, 1970; Iverson, 1991b; Iverson et al., 1993; Moll, 1990; van Loben Sels et al., 1997.

Staurotypinae

Ernst and Barbour, 1989a; Flores-Villela and Zug, 1995; Moll, 1990; Vogt and Guzman G., 1988.

Emydidae

Cooters, Sliders, American Box Turtles, and Allies

Ernst et al., 1994; Feldman and Parham, 2002; Gaffney and Meylan, 1988; Gaffney et al., 1991; Gibbons, 1990; Shaffer et al., 1997; Stephens and Wiens, 2003, 2004.

Emydinae

Feldman and Parham, 2002; Fujita et al., 2004; Fritz et al., 2005.

Deirochelyinae

Gaffney and Meylan, 1988; Fujita et al., 2004; Gibbons, 1990; Stephens and Wiens, 2004.

Platysternidae

Big-headed Turtle

Cox et al., 1998; Ernst and Barbour, 1989a; Fujita et al., 2004; Gaffney et al., 1991; Parham et al., 2006; Roman et al., 1999; Shaffer et al., 1997.

Geomydidae

Asian River Turtles, Leaf and Roofed Turtles, Asian Box Turtles, and Allies

Cox et al., 1998; Das, 1995; Diesmos et al., 2004, 2005; Ernst and Barbour, 1989a; Fujita et al., 2004; Gaffney et al., 1991; Rashid and Swingland, 1998; Shaffer et al., 1997; Stuart and Parham, 2004.

Testudinidae

Tortoises

Crumly, 1982, 1984; Ernst and Barbour, 1989a; Fujita et al., 2004; Gaffney et al., 1991; Moll and Klemens, 1996; Shaffer et al., 1997; Swingland and Klemens, 1989.

Chapter 19

Crocodylians

OVERVIEW

Today's crocodylians represent only a small fraction of the species that have lived since their origin in the Late Triassic over 220 million years ago. Twenty-three species of extant crocodylians are distributed throughout the world's tropics and subtropics, even extending slightly into the temperate zone. These comprise the clade Crocodylia, which also includes some Tertiary and Late Cretaceous species. This restricted use of Crocodylia is recent and derives from cladistic analyses of diverse fossil crocodylians (see Chapter 3). The broader clade, Crocodyliformes, includes all fossil and extant taxa. Although the vernacular name *crocodilians* appears throughout popular literature, *crocodylians* is technically correct (the clade is Crocodylia).

All crocodylians share a similar elongated body with a robust skull, a long snout and strongly toothed jaws, a short neck, a robust cylindrical trunk extending without constriction into a thick laterally compressed tail, and short but strongly developed limbs. Bony plates (osteoderms) that are covered with thick keratinous skin provide armor to the neck, trunk, and tail. This body form is an ancient one, hence the frequent mislabeling of crocodylians as living fossils, which they are not. The body form is, however, ancient and persists owing at least partially to the functional success of an aquatic predator that ambushes prey in shallow water or at waterside. All modern crocodylians are semiaquatic and spend much of their life in water, although they regularly bask on the shoreline and construct terrestrial nests for the incubation of their eggs.

All crocodylians are oviparous, and fertilization is internal. Moderate-sized clutches average from 12 to 48 eggs. Clutch size generally increases with female body size among the various taxa; for example, the small caiman *Paleosuchus trigonatus* (1.3 m adult SVL) has an average of 15 eggs in a clutch, and *Crocodilus porosus* (2.7 m SVL) has an average of 48 eggs per clutch. Similarly, clutch size increases with female body size within species. Eggs usually are deposited in mounds of vegetation and other detritus near the shoreline or on floating vegetation mats in shallow water. Females construct the nests using the entire body to bulldoze available debris into a mound. If surface debris is inadequate, the female digs a nest cavity in the ground. Parental care as nest attendance is common. The guarding parent is usually the female, although a male may also attend the nest, as is the case in *Crocodylus novaeguineae*. Parental care extends beyond nest attendance. The female opens the nest, helps to break the eggshells and free hatchlings, and transports the hatchlings to the water. This level of parental care and crèche or juvenile guarding has not been reported for all species; in fact, the reproductive behavior of wild crocodylians is not fully documented for more than a third of the extant species.

Crocodylians are uniquely characterized by a shared set of skeletal features. These features include an earflap on the skull table; foramen magnum formed by the basioccipital and exoccipitals; dorsal skull sculpturing of pits and ridges; bony eustachian tubes; trunk covered with a dorsal shield of unfused osteoderms; and a unique rod-shaped pubic process on the ischium. All members of the Crocodylia have a scapula with nearly horizontal anterior and posterior edges.

In the 10th edition of the *Systema Naturae,* Linnaeus classified a single crocodylian as the lizard, *Lacerta crocodilus*, diagnosed as a four-legged animal with a compressed tail. Eighteenth-century naturalists recognized the existence of other crocodylians at that time, even though Linnaeus described only one species. Other species were soon formally described. In a later edition, Linnaeus adopted Gmelin's Crocodili (= Crocodylia) for the group and thereby recognized crocodylians as a natural group. Since then, their monophyly has not been questioned, although assorted higher-level group names have been applied to them.

Recent phylogenetic studies of crocodylians have yielded two competing hypotheses to explain the relationships of the extant genera and families (Fig. 19.1). The difference between the two cladograms rests on the phylogenetic position of *Gavialis*: Is it the sister taxon to all other crocodylians, or is it nested within other crocodylians? If it is nested within other crocodylians, is *Tomistoma*, the "false" gharial, its closest living relative? Morphology supports the former relationship and molecular data support the latter. Combined morphological and molecular data support the *Tomistoma–Gavialis* pair as a sister taxon to the Crocodylidae within crocodylians (Fig. 19.1), but fossil gavialoids are about 70 million years older than divergence times for the *Tomistoma–Gavialis* pair indicated by molecular data. The morphological tree (Fig. 19.1) retains *Gavialis* as the outgroup. The striking superficial similarity between snouts (long and slender) of *Gavialis* and *Tomistoma* add to the confusion, but slender snouts have evolved multiple times in the evolutionary history of crocodylians, and substantial differences exist in the arrangement of skull bones between *Gavialis* and *Tomistoma*. Molecular data cannot address relationships among the numerous extinct crocodylians, and coding of morphological data can affect phylogenetic reconstructions. Repeated molecular analyses tie *Tomistoma* to *Gavialis* with these as the sister taxon to Crocodylidae, which we adopt in the taxonomy that follows.

Among extant crocodylians, all data sets indicate a sister-group relationship between *Crocodylus* and *Osteolaemus*, and a sister-group relationship between *Alligator* and the caimans. In the latter group, the relationships among species are variable, but the phylogeny most used is (*Alligator* [*Paleosuchus* {*Caiman*, *Melanosuchus*}]). The later pairing indicates that *Caiman* is paraphyletic.

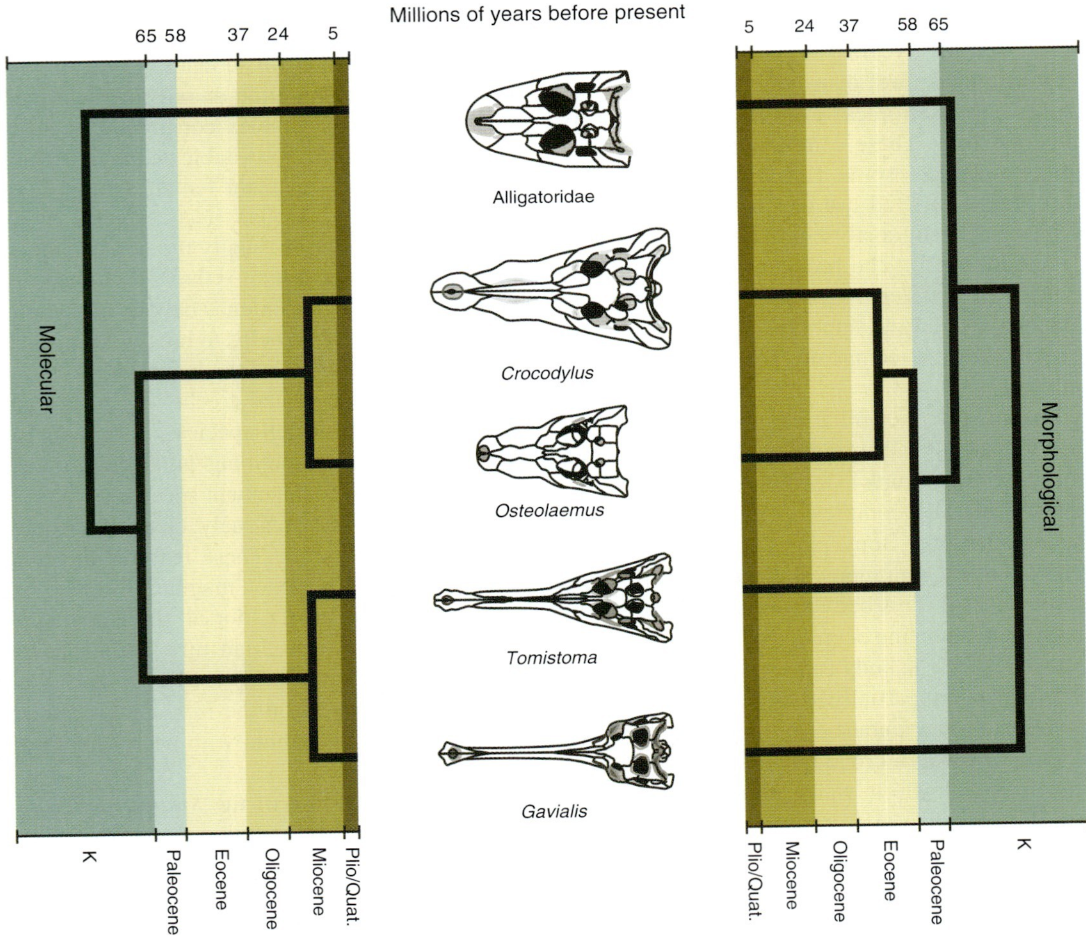

FIGURE 19.1 Cladograms depicting the two most likely relationship patterns among extant crocodylians. The cladograms are redrafted based on Brochu, 2003 (Fig. 8). Cladograms are redrawn from originals for uniformity.

TAXONOMIC ACCOUNTS

Gavialoidea

Gavialidae

Gharials and the "False" Gharial

Classification: Reptilia; Diapsida; Sauria; Archosauria; Crocodylia.

Sister taxon: Crocodylidae.

Content: Two genera, *Gavialis* and *Tomistoma*, with 1 species each.

Distribution: *Gavialis* occurs in South Asia, formerly in the upper portions of the Indus, Ganges, Brahmaputra, Bhima, Manahandi, and Ayeydrwady Rivers, but is now extinct in many areas (Fig. 19.2). *Tomistoma* occurs in freshwater streams of the Malaya Peninsula, Sumatra, and Borneo.

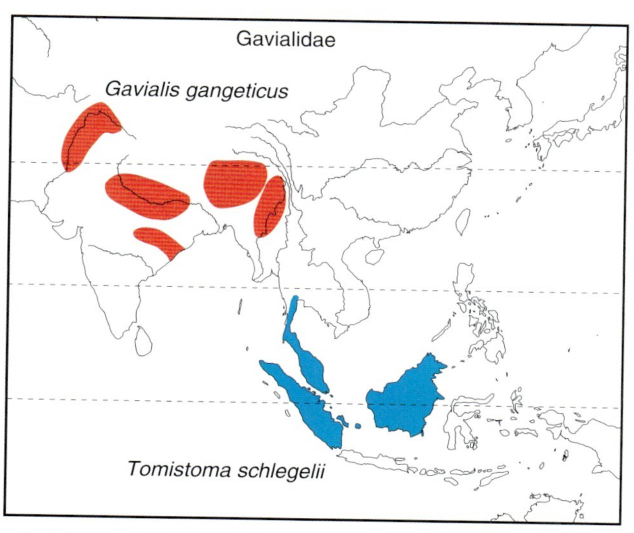

FIGURE 19.2 Geographic distribution of the extant Gavialidae.

Characteristics: *Gavialis gangeticus* attains a total length of 6.5 m. Among living crocodylians, *Gavialis* has the longest and narrowest jaws. All teeth in the anterior half of the upper and lower jaws lie outside the closed mouth, a characteristic unique to *Gavialis gangeticus*. The lower jaws are joined anteriorly by a long symphyseal articulation, and the anterior processes of the surangular have distinctly different lengths. In the skull, the ectopterygoid abuts the maxillary at its tooth row margin; the anterior process of the palatine is long and extends beyond the suborbital fenestra; and each parietal contains a sinus that opens into the cranial pneumatic system. The vertebral column contains a large, blocklike proatlas, a crested neural spine on the axis, and a slightly or unnotched axial hypapophysis. Lingual salt glands are absent or poorly developed and their exit pores are small; the surface of the tongue is not keratinized.

Tomistoma schlegelii (Fig. 19.3) attains 4 meters TL. It differs from all living crocodiles by having a narrow, elongate snout, a lower jaw with a long splenial symphysis, a postorbital that does not touch the quadrate or quadratojugal, and a suborbital fenestra with a distinct posterior notch. The jaw is not as narrow as that in *Gavialis*.

Biology: *Gavialis* is the most aquatic of living crocodylians and seldom moves far from water (Fig. 19.3). They prefer deep, fast-flowing rivers, where the adults congregate in deep holes at river bends and at the confluence of smaller streams. The juveniles select smaller side streams or river backwaters. As in other crocodylians, *Gavialis* regularly basks, particularly in winter when the water of its upstream habitats is cooler. The narrow, elongate, tooth-filled jaws are highly effective for catching fish, their primary food. *Gavialis* catches fish with a quick sideward snap of the jaws. With the fish impaled on the teeth, the head is lifted out of the water and backward, and then with a sideward head jerk, the fish drops head-first deep into the mouth. Frogs are also a common prey, and birds and mammals are eaten less frequently.

FIGURE 19.3 The two species of gavialids. From left: Gharial *Gavialis gangeticus* (C. A. Ross) and the "false" gharial *Tomistoma schlegelii* (G. Webb).

Male *Gavialis* reach maturity in about 15–18 years and at about 4 m TL; females mature earlier at about 7–8 years and at a smaller size (2.6–3 m TL). As males mature, they develop an irregular growth, the boss, on the tip of the snout. This boss grows progressively larger with age. Although its function is uncertain, it overlaps the nostrils and can cause a hissing and buzzing sound with each breath. Because this sound becomes part of the male's territorial defense behavior and may be important in courting, males with larger bosses having a social advantage. Today, most adult *Gavialis* are 4 m or less; an old record verified a maximum 6.45 m total length.

Nesting occurs in the late dry season (March–April), several months after mating. Females lay clutches consisting of 35 to 60 large eggs in nests typically dug on steep-sloped stream banks. The female guards her nest during an incubation period of 60+ days. When the eggs begin to hatch, the female assists the hatchlings as they emerge from the eggs. The hatchlings remain in a crèche with the female in attendance until the monsoon rains arrive. Flooding disrupts the nesting area and disperses the young.

Tomistoma is commonly incorrectly identified as a fish eater like the *Gavialis*; however, it appears to be mainly an ambusher of waterside prey. Mammals, birds, and in some areas, crab-eating macaques are a common prey. Its natural history is poorly known because its populations have been extirpated or reduced throughout its range. In captivity, female *Tomistoma* mature in 6 to 10 years at 2.5–3 m TL. Females construct large detritus nesting mounds and, typically in June and July, lay 20–40 eggs. Eggs of *T. schlegelii* are very large, each egg double or triple the mass of any other crocodylian egg. Eggs have a 10–12-week incubation period. Presumably hatchlings experience the same level of parental care as in other crocodylians but this is uncertain.

Comment: Morphological-based phylogenetic analyses place *Tomistoma* in the Crocodylidae and *Gavialis* as the sister taxon to all other extant crocodilians (Fig. 19.1). Molecular data place *Tomistoma* + *Gavialis* as sister to the Crocodylidae.

Alligatoroidea

Alligatoridae

Alligators and Caimans

Classification: Reptilia; Diapsida; Sauria; Archosauria; Crocodylia.

Sister taxon: Crocodylidae.

Content: Two subfamilies, Alligatorinae and Caimaninae, with 9 species.

Distribution: Eastern North America, Central and South America, and eastern China (Fig. 19.4).

Characteristics: Alligators and caimans (Fig. 19.5) commonly have broad, moderately long jaws. All teeth of the lower jaw lie inside the closed mouth. The lower jaws are joined anteriorly by a narrow symphyseal articulation, and the anterior processes of the surangular are subequal. In the skull, the ectopterygoid is broadly separated from the maxillary tooth row; the anterior process of the palatine is long and extends beyond the suborbital fenestra, and each parietal is solid. The vertebral column contains a moderate-sized and flattened proatlas, a crested neural spine on the axis, and a deeply notched axial hypapophysis. Lingual salt glands are absent, and the surface of the tongue is keratinized.

Alligatorinae

Sister taxon: Caimaninae.

Content: One genus, *Alligator*, with 2 species.

Distribution: Extant alligators are exclusively Holarctic. *Alligator mississippiensis* occurs in southeastern North America, and *A. sinensis* occurs in the lower reaches of the Yangtze River of eastern China.

Characteristics: Alligators are moderate-sized crocodylians, attaining lengths to 2.1 m TL (*A. sinensis*; Fig. 19.5) and 4 meters (*A. mississippiensis*). They have a narrow, parallel-sided dorsal horn on the hyoid plate, paired nasal foramina (Fig. 2.18), and a pointed anterior tip of angular extending dorsally to or beyond posterior intermandibular foramen.

Biology: *A. mississippiensis* lives in a wide range of habitats, including freshwater sloughs immediately behind coastal sand dunes, marshes and swamps, and large lakes and rivers. While seriously overharvested in the 1950s and 1960s, government protection, coupled with the alligator's high reproductive potential and relatively short generation time, has allowed populations in its core distribution area to rebound. American alligators have again assumed their role as a top predator of aquatic vertebrates in some regions. The situation for *A. sinensis* has improved but remains fragile. An effective breeding program has produced sufficient animals for reintroduction; unfortunately, no available protected areas exist for such releases. Populations in a single large reserve have increased in the areas of preferred habits; elsewhere populations are small or extirpated.

American alligators are opportunistic carnivores and eat a wide variety of animals. Vertebrates from fish to mammals (including other gators) are regularly eaten. In contrast, the Chinese alligator feeds heavily on mollusks, about 40-50% of its diet; they also eat a variety of small vertebrates.

Among crocodylians, only alligators live in areas where seasonal temperatures are below freezing. *Alligator mississippiensis* does not hibernate in cold weather; *A. sinensis* does. Large juvenile and adult *A. mississippiensis* select steep-sided shorelines where they can float with the tip of the snout above water and the body and tail in deeper,

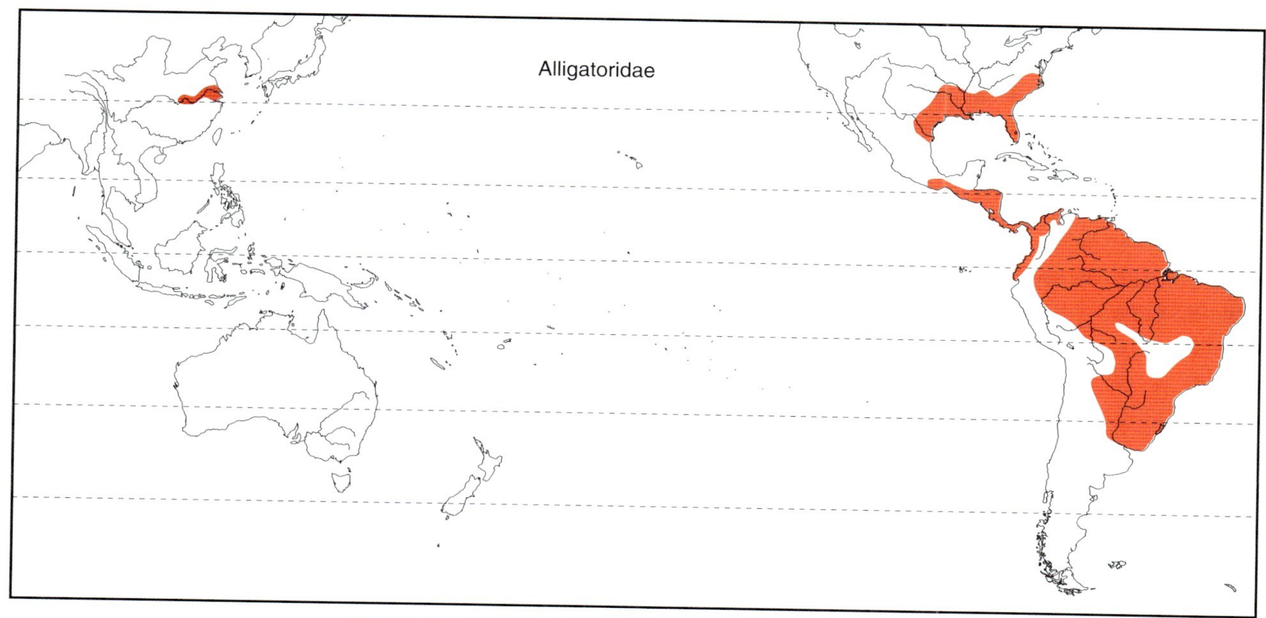

FIGURE 19.4 Geographic distribution of the extant Alligatoridae.

FIGURE 19.5 Representative alligatorids. Clockwise from upper left: Chinese alligator *Alligator sinensis*, Alligatorinae (C. K. Dodd, Jr.); spectacled caiman *Caiman crocodilus*, Caimaninae (J. P. Caldwell); Cuvier's dwarf caiman *Paleosuchus palpebrosus*, Caimaninae (J. P. Caldwell); black caiman *Melanosuchus niger*, Caimaninae (L. J. Vitt).

warmer water. If the shoreline water freezes, an alligator will maintain an ice-free hole around its snout in order to breathe. When possible, American alligators remain in burrows during cold days and emerge to bask during warm, sunny ones. *A. sinensis* digs extensive burrow systems and resides in them year-round; the burrow systems are complex, containing numerous tunnels and watered and dry chambers. These chambers are used for hibernation from about October through February; indeed, hibernation seems necessary to stimulate reproduction in Chinese alligators.

In the southern half of their distribution, *A. mississippiensis* reaches sexual maturity at about 2.0 m TL in 7 to 10 years. Because of their small relative size, young, sexually mature males are usually unable to compete with larger males for territories and females, so most males do not breed for the first time until they are 2.4-2.8 m TL and about 15–20 years old. Courtship begins 8 to 10 weeks before the eggs are deposited in mid-June, and females frequently mate with more than one male. Nesting begins with mound construction; the female heaps dirt and vegetation into a large mound, usually near the shoreline, although occasionally on a floating vegetation mat in shallow water. She digs a cavity in the mound and deposits an average 35 to 40 eggs, which will hatch in 65 to 70 days. Parental care includes guarding the nest and the crèche. *Alligator sinensis* has a similar reproductive pattern, but because it is a smaller species, it matures somewhat earlier. They mate May to June and begin nesting in July; they produce smaller clutches (average, 24 eggs).

Caimaninae

Sister taxon: Alligatorinae.

Content: Three genera, *Caiman*, *Melanosuchus,* and *Paleosuchus,* with 7 species.

Distribution: Central and southern Mexico to Ecuador and east of the Andes into Uruguay, Paraguay, and northern Argentina (Fig. 19.4).

Characteristics: Caimans are small to large crocodylians, to 1.7 m TL in *Paleosuchus* and 2 to 5 meters in *Caiman*. They have a broad flaring dorsal horn on hyoid plate, a large nasal foramen, and a blunt anterior tip of angular not extending to the posterior intermandibular foramen.

Biology: Caimans occur in a diversity of freshwater habitats throughout the lowlands of Central and South America. *Caiman crocodilus* and *C. fuscus* are the most widespread species and appear to be the most tolerant ecologically. They occupy the broadest range of habitats, preferring slow-moving backwaters of rivers and ponds and lakes. *Caiman crocodilus* is still fairly abundant in the llanos of Venezuela and elsewhere, but it is heavily harvested. Harvesting, habitat modifications, and lower reproductive potential of caimans continue to threaten their survival in many regions.

All species of caimans build nest mounds in which they deposit their eggs. Clutch size is related to body size, and smaller individuals and the smaller species lay fewer eggs. Clutch size is 10 to 15 eggs in *Paleosuchus trigonatus*, 15–40 in *C. crocodilus*, and 30–60 in *M. niger*. Evidence suggests that all caimans have parental care that includes guarding the nest and the crèche. Most crocodylians select open-canopy microhabitats adjacent to or marginally in forest, but *P. trigonatus* is a regular inhabitant of closed-canopied, small streams in the rain forests of the Amazon and Orinoco basins, although it also occurs in open areas. Because shallow streams offer little protection, adults often seek shelter in deep cavities under stream banks or in logs and debris away from the stream. The closed canopy does not permit sunlight to heat the nesting mounds, so females often place their nests adjacent to and partially on termite mounds to obtain additional heat generated by the termite nest chamber.

Crocodyloidea

Crocodylidae

Crocodiles and Dwarf Crocodiles

Classification: Reptilia; Diapsida; Sauria; Archosauria; Crocodylia.

Sister taxon: Gavialidae.

Content: Three genera, *Crocodylus, Mecistops,* and *Osteolaemus*, with 12, 1, and 1 species, respectively (Fig. 19.6).

Distribution: The genus *Crocodylus* is pantropic in distribution; *Mecistops* and *Osteolamus* occur in western and west-central Africa (Fig. 19.7).

Characteristics: Crocodiles range in total length from the small (2.5 m TL) dwarf crocodile *O. tetraspis* to the largest extant crocodylian *C. porosus* (to 7 m TL). Crocodiles differ from gharials by having broader snouts, lower jaws with short splenial symphyses, postorbitals touching the quadrates and quadratojugals, and suborbital fenestrae without a distinct posterior notch. Most crocodiles have moderately long and often broad jaws. A unique feature is that only the fourth mandibular tooth lies externally on each side of the mouth when it is closed. Occasionally, the first mandibular tooth perforates the upper jaw and its tip is visible as well when the mouth is closed. The lower jaws are usually joined anteriorly by a narrow symphyseal articulation, and the anterior processes of the surangular have distinctly different lengths. In the skull, the ectopterygoid abuts the maxillary at its tooth row margin; the anterior process of the palatine is short and does not extend beyond the suborbital fenestra; and each parietal is solid. The vertebral column contains a moderate-sized and flattened proatlas, an uncrested neural spine on the axis, and a deeply notched axial hypapophysis. Lingual salt glands are well developed and their exit pores are large; the surface of the tongue is not keratinized.

FIGURE 19.6 Representative crocodiles. Clockwise from upper left: Saltwater crocodile *Crocodylus porosus* (G. R. Zug); Saltwater crocodile *Crocodylus porosus* floating (R. Shine); Johnstone's crocodile *Crocodylus johnstoni* (R. Shine); dwarf crocodile *Osteolaemus tetraspis*, Crocodylinae (A. Britton).

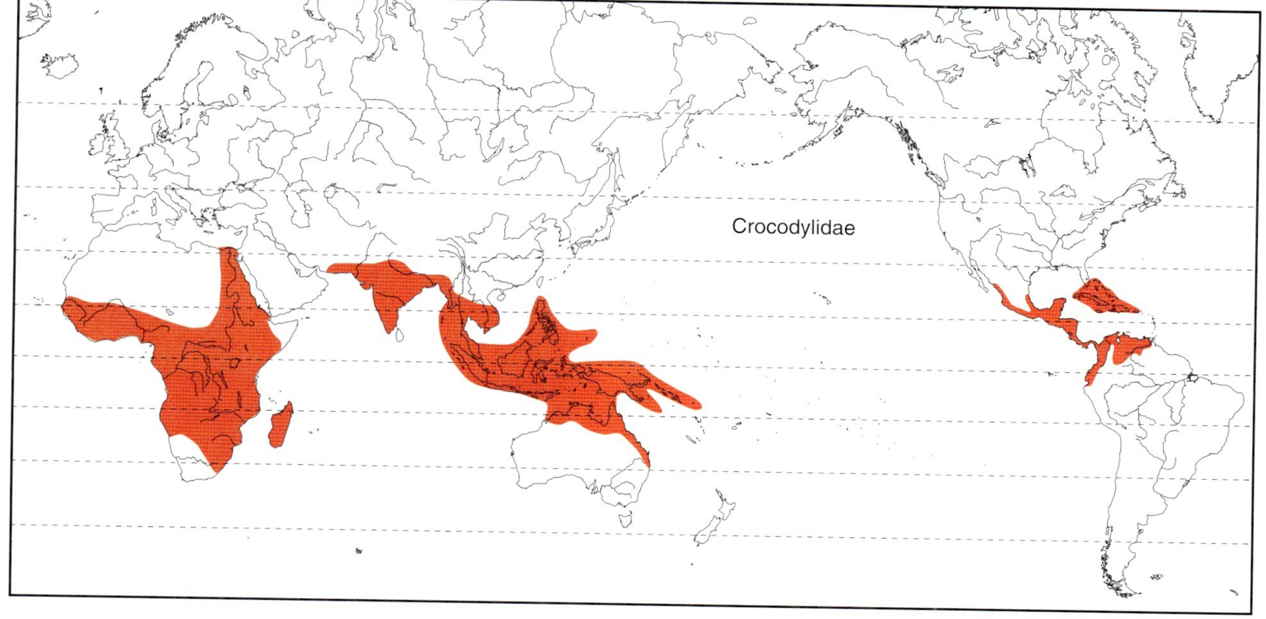

FIGURE 19.7 Geographic distribution of the extant Crocodylidae.

Biology: Species of *Crocodylus* occur mainly in aquatic habitats with open canopies, including freshwater marshes, the margins of large rivers and lakes, tidal marshes, and mangrove forests. *Mecistops cataphractus* and *Osteolaemus tetraspis* are exceptions, living in small to moderate-sized rain forest streams, often beneath a closed canopy. *Osteolaemus tetraspis* is exclusively nocturnal, differing from other crocodiles. Most other crocodiles hunt at night, but they are active diurnal predators as well. About one-third of the species, including *C. acutus*, *C. intermedius*, *C. niloticus*, and *C. porosus*, reach lengths greater than 4 meters TL; the others are mostly 2 to 3 meters, whereas *O. tetraspis* is seldom larger than 1.5 m.

All crocodiles appear to be mound builders and nesters if given the opportunity; if adequate vegetation and surface detritus are not available, the female digs a nest in sand or soil of the shoreline. Where data are available, all crocodiles have parental care that includes nest and crèche guarding. Clutch size is related to body size, and the small-bodied species (e.g., *O. tetraspis*, *M. cataphractus*) seldom lay more than 25 eggs, whereas the large species (e.g., *C. niloticus*, *C. porosus*) regularly lay more than 50 eggs.

QUESTIONS

1. Why has the phylogenetic position of *Gavialis* been so controversial among crocodylian systematists, and, is the issue finally resolved?
2. Can you provide a reasonable explanation why crocodylians have been such a successful group of reptiles and have remained relatively unchanged throughout their long evolutionary history?
3. The shape of jaws in crocodylians varies considerably among both living and extinct species. What are the functions of these different-shaped jaws?
4. Compare the crocodylian faunas of South America with those of Africa.

REFERENCES

General

Benton and Clark, 1988; Brochu, 1997a, 1999, 2001, 2003, 2004; Clark, 1994; Cogger and Zweifel, 1998; Densmore, 1983; Densmore and White, 1991; Greer, 1970a; Hass et al., 1992; Harshman et al., 2003; Norell, 1989; Poe, 1997; Ross, 1989; Tarsitano et al., 1989; Thorbjarnarson, 1992, 1996; Webb and Manolis, 1993; Webb et al., 1987.

Taxonomic Accounts

Gavialidae

Gharials and the "False" Gharial

Brochu, 1997a; J. Lang, 1989; Magnusson et al., 1989; Shine, 1988; Thorbjarnarson, 1990, 1996; Webb and Manolis, 1993; Willis et al., 2007; Whitaker and Basu, 1983.

Alligatoridae

Alligators and Caimans

Brochu, 1999; J. Lang, 1989; Magnusson et al., 1989; Thorbjarnarson, 1992.

Alligatorinae

Joanen and McNease, 1980; J. Lang, 1989; McIlhenny, 1935; Webb and Manolis, 1993; Webb and Vernon, 1992.

Caimaninae

Amato and Gatesy, 1994; Brazaitis et al., 1998; J. Lang, 1989; Magnusson et al., 1989; Ouboter and Nanhoe, 1987; Webb and Manolis, 1993.

Crocodylidae

Crocodiles and Dwarf Crocodiles

Brochu, 1999, 2003; Cott, 1961; Densmore and Owen, 1989; Graham and Beard, 1973; Lang, 1989; Magnusson et al., 1989; McAliley et al., 2006; Ouboter and Nanhoe, 1987; Thorbjarnarson, 1996; Webb and Manolis, 1989, 1993; Webb et al., 1987.

Chapter 20

Tuataras and Lizards

OVERVIEW—SPHENODONTIDA

Tuataras are relics of a formerly more diverse group, persisting only on small islets off the main islands of New Zealand. Even before the arrival of humans, tuataras were less diverse than lizards in the same area. Why have the tuataras dwindled to so few species? Only speculative answers are possible, and all hypotheses are likely to include a competitive component. Lizards were probably their main competitors, at least during the Tertiary, but the real answer is certainly more complex than just competition with lizards.

Sphenodontidans and squamates comprise the Lepidosauria. Lepidosaurs share numerous derived characteristics, including a transverse cloacal opening (the vent); tongue notched distally and used to capture prey (lingual prehension); full-body ecdysis; imperforate stapes; teeth attached superficially to the jaw bones; pelvic bones fused in adults; fracture planes or septa in the caudal vertebrae; and numerous other anatomical traits. The sphenodontidans and squamates apparently diverged in the early Late Triassic, and the sphenodontidans seemingly have always been a group with low diversity (see Chapter 3 for fossil history).

Sphenodontidans differ from squamates by the presence of gastralia; a narrow quadrate with greatly reduced or lateral concha; lower temporal fenestra enclosed or partially so; jugal in the mid-temporal arch touching the squamosal posteriorly; prominent coronoid process on the mandible; several anterior teeth of the palatine series enlarged; dentary and mandibular teeth generally enlarged, regionalized, and fused to dorsal margin of bone; and the premaxillary teeth replaced by chisel-shaped extensions of the premaxillary bones that have given rise to the tuatara's other vernacular name, half-beaks (see Fig. 3.19).

TAXONOMIC ACCOUNT

Sphenodontidae

Tuataras

Classification: Reptilia; Diapsida; Sauria; Lepidosauria; Sphenodontida.

Sister taxon: Clade containing extinct sapheosaurs and *Homeosaurus*.

Content: One extant genus, *Sphenodon*, with 2 species.

Distribution: New Zealand, but now restricted to small coastal islands; *S. punctatus* occurs on about 30 islands off the northeast coast of the North Island and western Cook Strait, and *S. guntheri* is restricted to a single island, the North Brother Island in the Cook Strait.

Characteristics: Tuataras are lizard-like, stout-bodied (19–28 cm adult SVL) reptiles with large heads and thick tails (Fig. 20.1). They have a chisel-beaked upper jaw overhanging the lower jaw, a series of erect spines on the nape and back, and rudimentary hemipenes. They lack a tympanum.

Biology: Adult tuataras forage principally at night, commonly at temperatures that range from 12 to 16°C.

FIGURE 20.1 Full body (left) and head (right) of the tuatara *Sphenodon punctatus* (P. Ryan).

They are not exclusively nocturnal animals and, in warm summer months, bask at their burrow entrances, retreating when they become too hot and emerging after they cool. Their prey consists predominantly of insects and other arthropods, although they occasionally eat skinks, geckos, and seabirds. *Sphenodon punctatus* is most numerous on those islands shared with nesting seabirds, an indication of a lack of or a reduction in rat predators. Bird nesting activities yield abundant arthropod prey for tuataras and burrows for daily shelter and winter hibernation. Islands with moderate to high rat populations have tuatara populations composed nearly exclusively of adults, because rats prey on the eggs and juveniles. Such populations persist only because tuataras are long lived, living up to 50-60 years.

Courtship and mating occur in January, but egg laying is delayed until October–December of the following year. Females produce clutches, on average, every 4 years, which includes a 3-year Vitellogenic cycle. The female digs a small nest cavity and deposits 5 to 15 eggs, returning over several nights to fill the cavity. Development is slow and stops during the winter, and hatching occurs 11 to 16 months after egg deposition. Optimal incubation temperatures in the laboratory are 18–22°C, the lowest known in living reptiles. The eggs absorb moisture during incubation, so the mass of the hatchlings is 1.2–1.3 times greater than the original egg mass.

OVERVIEW—SQUAMATA

The nearly 7200 species of lizards are the most diverse and speciose living clade of reptiles. Of course this total includes snakes, which are actually reduced-limbed or limbless lizards. Our vernacular recognizes lizards and snakes as two different groups, but our classificatory label, Squamata, denotes monophyletic status of these reptiles and requires use of the vernacular squamates. Squamates, thus, have a precise definition and diagnosis whereas lizards do not, and they require a shared perception between reader and writer. That perception is generally adequate, but it cannot be formalized with a name, such as Lacertilia, because that classification excludes snakes and creates a paraphyletic taxon. Because snakes are a monophyletic group arising from within a group of lizards, the taxon Serpentes and its definition delimit a monophyletic group. Herein, the term *lizard* represents our shared perception and excludes Serpentes, but only for purposes of communication.

This chapter is about lizards, but first, Squamata must be defined. Squamates have more than 50 shared–derived features attesting to their monophyly. Skeletal features include a single (fused) premaxillary and a single parietals; reduced nasals; no vomerine teeth; specialized ulnare–ulna and radiale–radius joints (wrist); a specialized ankle joint; and a hooked fifth metatarsal. Among soft anatomical structures, squamates have well-developed paired copulatory organs (hemipenes); saccular ovaries; a vomeronasal (Jacobson's) organ separated from the nasal capsule; a lacrimal duct joined to the vomeronasal duct; femoral and preanal glands; and no caruncle, but instead have an egg tooth.

Lizards and their snake descendents are the only living squamates. Excluding snakes, lizards are still the most speciose extant reptiles, with about 4450 species. Lizards occur on all continents except Antarctica, and on most tropical and subtropical oceanic islands. This widespread occurrence denotes their broad ecological, physiological, and behavioral adaptations to extremely hot to cold climates, to extremely arid to freshwater–marine habitats, and to lowland to high elevation regions. Their highest species diversity appears to be in semiarid habitats. For example, 53 species of lizards occur at one site in the Great Victoria Desert, Australia, and in some areas, particularly islands, densities can be greater than 3000 per ha^{-1} (*Emoia cyanura*).

Charles C. Camp's *Classification of Lizards* represents the first explicit attempt at an evolutionary analysis of squamate relationships. His dendrogram provides a series of dichotomous branches, and the overall pattern is not strikingly different from patterns seen in many recent phylogenetic (explicitly cladistic) studies. For example, his analysis recognized iguanians as the first branch of the dendrogram and geckos as the next branch. His results also suggest that varanoids and snakes are sister groups, although the sister-group concept was not adopted for reptilian classification until the 1960s. The first explicitly cladistic analysis of squamates appeared in 1988. This analysis, by Richard Estes and his colleagues, examined a wide representation of squamate genera and families and several hundred characters that were reduced to 148 useful ones. The resulting cladogram and other more recent ones are similar to Camp's. The classifications are also similar, although today's classifications recognize only monophyletic groups (Fig. 20.2).

The major branches of the Estes et al. cladogram and of most subsequent ones show Iguania as the sister group of all other squamates, the Scleroglossa. Scleroglossa then branches into geckos and allies and autarchoglossans. Other similarities include sister-group relationships between Teiidae and Gymnophthalmidae, between the latter pair and Lacertidae, between *Varanus* and *Lanthanotus*, and between the latter pair and *Heloderma*, but thereafter sister-group pairings do not match. One cause of dissimilarities is that the analyses compare different sets of taxa. This alone can account for different branching patterns. Additional differences arise from the size of the character data set and its diversity, which includes the level of interrelatedness of the characters (e.g., whether characters represent one functional unit or many). Because a consensus does not yet exist, our selection of a squamate cladogram is arbitrary (Fig. 20.2).

Iguania contains as few as 2 groups or as many as 12. Historically, Iguania consisted of Agamidae, Chamaeleonidae, and Iguanidae. Agamidae and Chamaeleonidae, which compose Acrodonta, are more closely related to one another than either is to Iguanidae. That generality is still supported by the majority of phylogenetic analyses. However, is Agamidae or Iguanidae monophyletic? One cladistic analysis by Darrel Frost and Richard Etheridge (1989) indicated that neither lineage was monophyletic and proposed a new classification that recognized numerous new families (Corytophanidae, Crotaphytidae, Hoplocercidae, Iguanidae, Opluridae, Phrynosomatidae, Polychrotidae, Tropiduridae) for the original Iguanidae, and a single family for agamids and chameleons. This classification has been adopted widely, although not unanimously. Another study that used molecular data and also reanalyzed the Frost–Etheridge data supported monophyly of Iguanidae, Acrodonta, and Chamaeleonidae but was unable to confirm or reject monophyly of Agamidae.

Membership of and relationships within Gekkota and its component families also differ among systematists. All agree that membership of geckos (gekkonoids) includes eublepharids, gekkonids, pygopods, and diplodactylines, although the analyses yield different branching patterns

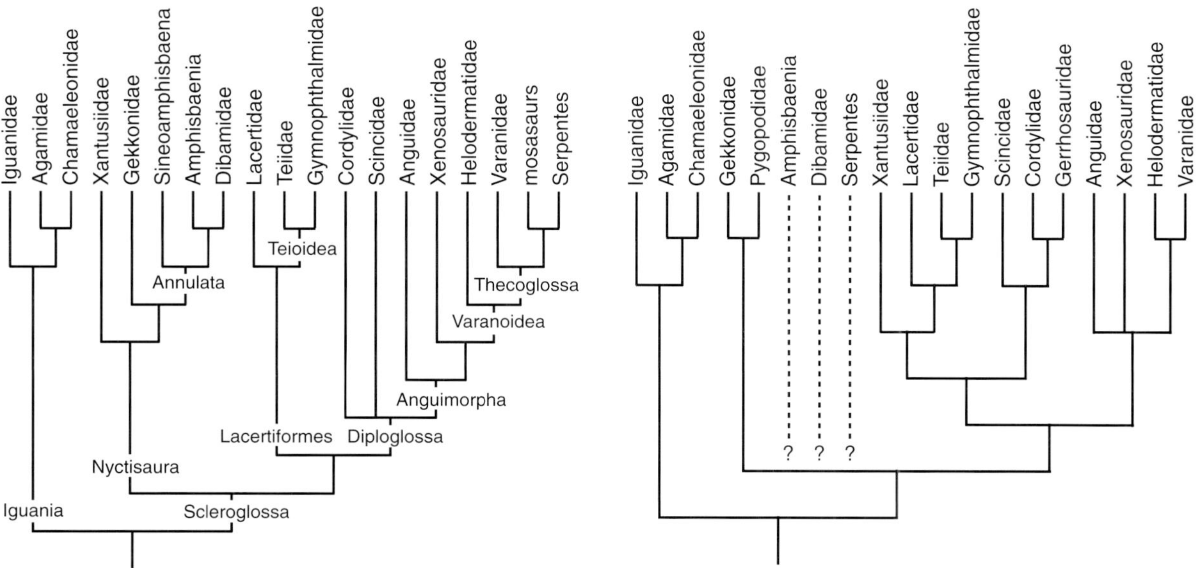

FIGURE 20.2 Two cladograms depicting relationships among extant taxa of squamates with emphasis on the phylogeny of lizards. The cladogram on the left derives mainly from Lee (1988) for the basic branching patterns, and the position of the Annulata clade derives from Wu and collaborators (1996). The cladogram on the right is a composite from Estes et al. (1988), Frost and Etheridge (1989), and additions from the University of Texas Herpetology Web site page. The key differences between these two "classical" phylogenies are positions of snakes, amphisbaenians, and dibamids, and resolution of some of the branching patterns. Cladograms redrawn from originals for uniformity.

and different assessments of monophyly of these groups. A multicharacter study by Arnold Kluge recognized monophyly of Gekkonidae and Pygopodidae including diplodactylines. Subsequently, other authors included pygopods in Gekkonidae, an arrangement that does not alter monophyly of the latter. More recently, a molecular study confirmed monophyly of Gekkonidae, Diplodactylinae, Pygopodinae, and Gekkoninae, and the sister-group relationship of diplodactylines and pygopodines. That study did not resolve relationships of gekkonines, eublepharines, or each of these to the diplodactyline–pygopodine clade. Indeed, it suggested that eublepharines are either a sister group of diplodactyline–pygopodines or of gekkonines rather than the sister group of all other geckos.

Although gekkonoid monophyly is well supported, its sister-group relationship is uncertain. A common interpretation is that gekkonoids are the same as Gekkota, which is the sister group of all other extant lizards (Autarchoglossa) except iguanians. Other relationships have been proposed. Proposed sister taxa are Annulata (= amphisbaenians, dibamids) and snakes; all other lizards excluding Iguania and Annulata; anguimorphs; scincomorphs; and Xantusiidae. These different hypotheses emphasize that our resolution of phylogenetic relationships among lizards is not firmly resolved and that several alternative classifications are equally likely at this time.

Some relationships persist among all or most analyses. Teioids (Gymnophthalmidae and Teiidae) are consistently paired and, in turn, usually linked to Lacertidae, forming Lacertiformes or lacertoids, although a teioid–amphisbaenian pairing has been suggested. The genera *Lanthanotus* and *Varanus* are another consistent pair that forms the sister group to *Heloderma*. Thereafter, relationships within scleroglossans or autarchoglossans are less certain, as indicated by the numerous proposals of Gekkota relationships. But in spite of these differing hypotheses on interfamilial relationships among squamates, neither snakes nor amphisbaenians currently are considered a basal sister group to all other squamates. Their origins (i.e., sister-group relationships) are to a subgroup of lizards. The two current competing hypotheses are that Serpentes is a sister group to varanids–varanoids or to a dibamid–amphisbaenid clade. Considerable discussion has occurred as to whether snakes arose from a mosasaur-like ancestor in a marine environment or whether snake ancestors were terrestrial–subterranean, possibly similar to present-day scolecophidian snakes (see Chapter 3 and Chapter 21). Both these hypotheses still place snakes within, not sister to, lizards.

Several recent nuclear gene–based squamate phylogenies present a strikingly different interpretation of relationships between iguanians, gekkotans, and autarchoglossans. An analysis by Dan Townsend and his colleagues (2004) places the family Dibamidae as sister to all other squamates, and Gekkonidae as sister to all remaining squamates, thus dissolving Scleroglossa, Scincomorpha sister

FIGURE 20.3 Two different interpretations of higher taxonomic relationships among squamate reptiles based on nuclear-gene studies. Both of these differ dramatically from the phylogenies illustrated in Figure 20.2 but are similar in that dibamids are the outgroup to all other squamates, and Iguania and snakes are deeply embedded in the phylogenies. Left from Townsend et al., 2004; right from Vidal and Hedges, 2005.

to remaining squamates, and Iguania in a clade with anguimorps + snakes (Fig. 20.3, left). Another nuclear-gene phylogeny produced similar patterns but with better resolution within clades (Fig. 20.3, right). Both of these move snakes out of anguimorphs. More importantly, if these relationships are corroborated with additional analyses and more complete data sets, then reinterpretation of fossil, morphological, ecological, physiological, and behavioral analyses based on phylogenies will be necessary.

TAXONOMIC ACCOUNTS

Iguania

Agamidae

Angleheads, Calotes, Dragon Lizards, and Allies

Classification: Reptilia; Diapsida; Sauria; Lepidosauria; Squamata.

Sister taxon: Chamaeleonidae.

Content: Two subfamilies, Agaminae and "Leiolepidinae."

Distribution: Africa, Asia, and Australia (Fig. 20.4).

Characteristics: Agamids are small to large lizards (45-350 cm adult SVL), covered dorsally and ventrally by overlapping scales or granular, juxtaposed scales. No osteoderms occur dorsally or ventrally on the trunk. All species are limbed, and the pectoral girdle has a T-shaped or cruciform interclavicle and curved rod-shaped clavicles. The tail is usually long to moderately long (from just less than SVL to 1.4 times SVL) and lacks fracture planes in caudal vertebrae (except in some *Uromastyx*). The tongue is covered dorsally with reticular papillae and lacks lingual scales; the foretongue is nonretractable. The skull

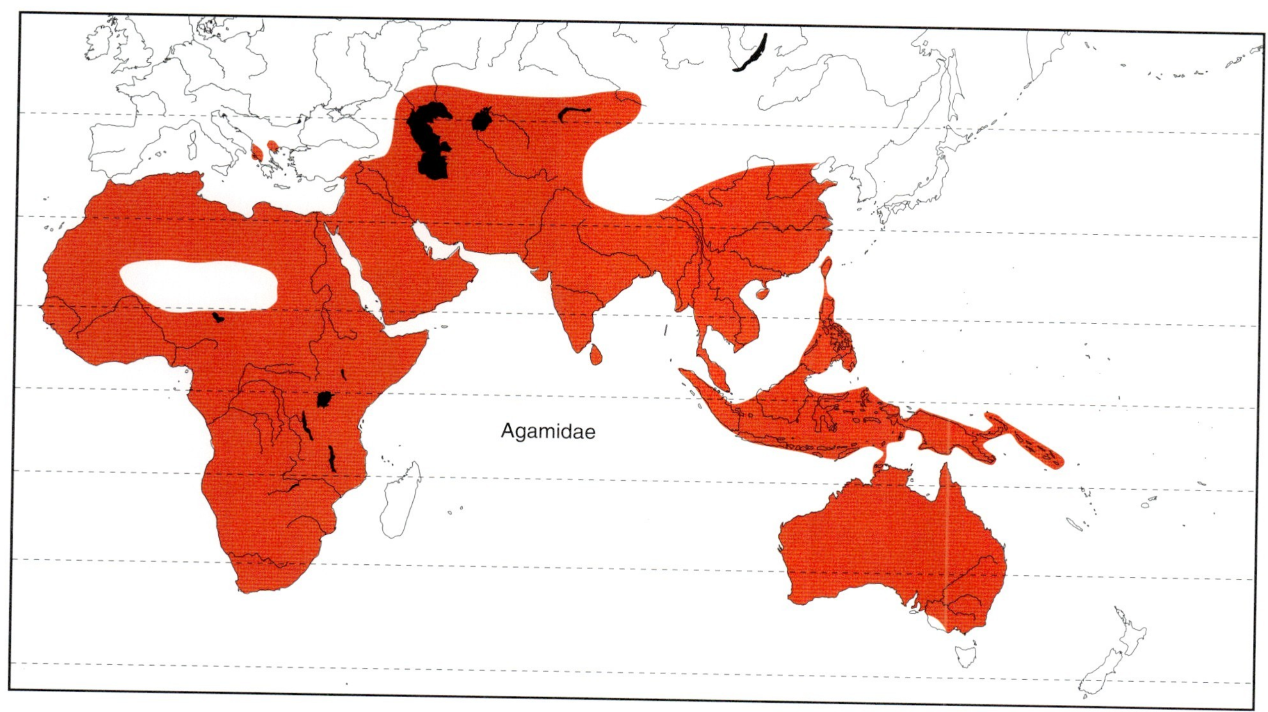

FIGURE 20.4 Geographic distribution of the extant Agamidae.

possesses paired nasals, postorbitals and squamosals, and a frontal and parietal; a parietal foramen usually perforates the frontoparietal suture. Attachment of the marginal dentition is acrodont, and the pterygoid lacks teeth.

Comment: Opinion varies on delineation of Leiolepidinae as a clade. Primitive morphological characters appear to link *Leiolepis* and *Uromastyx*, so it is possible that they are not sister groups.

Agaminae

Sister taxon: "Leiolepidinae."

Content: Fifty-two genera, *Acanthocercus, Acanthosaura, Agama, Amphibolurus, Aphaniotis, Brachysura, Bronchocoela, Bufoniceps, Caimanops, Calotes, Ceratophora, Chelosania, Chlamydosaurus, Cophotis, Coryphophylax, Cryptagama, Ctenophorus, Dendragama, Diporiphora, Draco, Gonocephalus, Harpesaurus, Hydrosaurus, Hypsicalotes, Hypsilurus, Japalura, Laudakia, Lophocalotes, Lophognathus, Lyriocephalus, Mantheyus, Mictopholis, Moloch, Oreodeira, Oriocalotes, Oriotarus, Otocryptis, Phoxophrys, Phrynocephalus, Physignathus, Pogona, Psammophilus, Pseudocalotes, Pseudotrapelus, Ptyctolaemus, Rankinia, Salea, Sitana, Thaumatorhynchus, Trapelus, Tympanocryptis,* and *Xenagama,* with ±400 species.

Distribution: Africa, Asia, and Australia.

Characteristics: Agamines have large lacrimal foramina and epiotic foramina.

Biology: Agamines are a diverse clade of predominantly terrestrial and semiarboreal lizards (Fig. 20.5); a few are highly arboreal, but none is fossorial. The diversity results in part from their extensive distribution in the Old World and independent adaptive radiations in Africa, Asia, and Australia. They range in size from the small *Cryptagama aurita* (40–45 mm adult SVL) to the large water dragon *Hydrosaurus amboinensis* (350 mm SVL, 1.1 m TL), and in body shape from stout-bodied, short-limbed taxa (e.g., *Moloch, Phrynocephalus*) to slender and long-limbed taxa (e.g., *Draco, Sitana, Diporiphora*). Agamines are usually diurnal, and most are heliotherms that regularly bask to maintain elevated body temperatures. Among the most spectacular ecologically are the gliding lizards in the genus *Draco*. Not only do they glide using dorsal skin flaps supported by elongate ribs, but they also are able to direct their glides. *Moloch* is another spectacular agamine that not only does not look like a lizard but also is an ant specialist. Agamines are predominantly carnivores, preying largely on arthropods by using a sit-and-wait foraging behavior. Most, perhaps all, agamines are oviparous, although reports suggest that some *Phrynocephalus* and *Cophotis ceylanica* are viviparous. Clutch size is generally correlated with body size within species; small-bodied taxa deposit smaller clutches (e.g., 2 eggs, *Ctenophorus fordi*), and larger-bodied species deposit larger clutches (e.g., 30–35 eggs, *Pogona*). Clutch size varies for most species from 4 to 10 eggs. Eggs are deposited in nests dug by the females, and incubation is commonly 6 to 8 weeks.

FIGURE 20.5 Representative agamid lizards. Clockwise from upper left: Spotted butterfly lizard *Leiolepis guttata*, Leiolepidinae (R. D. Bartlett); rhinocerus agama *Ceratophora tennentii*, Agaminae (C. Austin); Australian water dragon *Lophognathus longirostris* (E. R. Pianka); Dabbs mastigure *Uromastyx acanthinura*, Leiolepidinae (L. L. Grismer).

"Leiolepidinae"

Sister taxon: Agaminae.

Content: Two genera, *Leiolepis* and *Uromastyx*, with 7 and 14 species, respectively.

Distribution: Northern Africa eastward to Southeast Asia.

Characteristics: Leiolepidines have small lacrimal foramina and lack epiotic foramina.

Biology: All leiolepidine species are terrestrial (Fig. 20.5) and use burrows for daily and seasonal retreats. They can climb and occasionally forage in low shrubs. All are predominantly herbivorous, eating foliage, flowers, fruits, and seeds. Both *Leiolepis* and *Uromastyx* are oviparous. Clutch size is moderate in both taxa, ranging from 2 to 8 eggs in *Leiolepis* (110–150 mm adult SVL) and 8 to 20 eggs in *Uromastyx hardwickii* (340–400 mm adult TL). All species usually lay their eggs within the female's burrow system, either in late spring–early summer or at the beginning of the dry season. Incubation is approximately 8 to 10 weeks, and hatchlings appear to stay within the parent's burrow system for several weeks to several months before leaving to establish their own burrows.

Chamaeleonidae

Chameleons

Classification: Reptilia; Diapsida; Sauria; Lepidosauria; Squamata.

Sister taxon: Agamidae.

Content: Six genera, *Bradypodion, Brookesia, Calumma, Chamaeleo, Furcifer,* and *Rhampholeon,* with 24, 26, 31, 54, 20, and 16 species, respectively.

Distribution: Africa, the Middle East, Madagascar, southern Spain, Sri Lanka, and India (Fig. 20.6).

Characteristics: Chameleons are unique lizards that have strongly laterally compressed bodies, prehensile tails, head casques covering their necks (Fig. 20.7), zygodactylous feet (i.e., fusion of sets of two and three digits, forming opposable, two-digited mitten-like fore- and hind feet; manus fusion 1-2-3 and 4-5, pes 1-2 and 3-4-5), projectile tongues, and independently movable eyes with muffler-like lids. Most species have a skin of small, juxtaposed scales. No osteoderms occur dorsally or ventrally on the trunk. All species are limbed; the specialized pectoral girdle lacks an interclavicle and

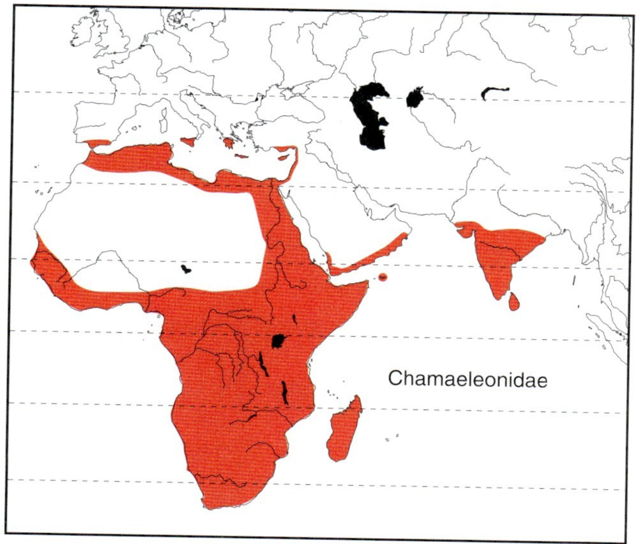

FIGURE 20.6 Geographic distribution of the extant Chamaeleonidae.

clavicles. The tail is moderately short (about two-thirds SVL) to long and usually prehensile; the caudal vertebrae lack fracture planes. The tongue is covered dorsally with reticular papillae and lacks lingual scales; the foretongue is nonretractable into the hind tongue. The skull has paired nasals (occasionally fused), postorbitals, squamosals, and a single frontal and a parietal; a parietal foramen, when present, perforates the frontal bone. Attachment of the marginal dentition is acrodont, and the pterygoid lacks teeth.

Biology: Chameleons vary greatly in adult SVL with some *Chamaeleo* reaching 70 to 630 mm, and *Brookesia* and *Rhampholeon* averaging 25 to 55 mm in SLV. Chameleons are largely although not exclusively arboreal. Many features of their morphology previously described are associated with a specialized arboreal existence and prey capture. They are stalkers, walking along narrow branches with a slow, somewhat jerky gait that suggests a leaf shaken by wind. After insect prey is located visually, locomotion is frozen, and the independently mobile eyes focus with the head adjusted to center the eyes binocularly on

FIGURE 20.7 Representative chamaeleonids. Clockwise from upper left: Cameroon stump-tailed chameleon *Rhampholeon spectrum* (C. Mattison); brown leaf chameleon *Brookesia superciliaris* (C. Mattison); cape dwarf chameleon *Bradypodium pumilis* (D. Hillis); south-central chameleon *Furcifer minor* (R. D. Bartlett).

the prey (but see Chapter 11); the tongue shoots forward—nearly the length of the body—and entraps the prey and recoils into the mouth. In addition to camouflaging their gait and other body movements, chameleons adjust their body colors to match their background to escape detection by visual-searching bird and mammal predators.

They live in diverse forest habitats from scrub to evergreen rain forest; some live high in the canopy, others in shrubs of the understory, and a few live mainly on the ground in grassy or scrub habitats. Both egg-laying and live-bearing taxa are known. Clutch and litter size generally correlate with body size; the smaller taxa generally produce 2 to 8 eggs or neonates, and the larger species typically deposit more than 20 and as many as 50 eggs, but litter size is generally 20 embryos or less, even for large females. Incubation time is variable and may reach 300 days for winter-nesting *Chamaeleo dilepis*.

The tiny leaf or stump-tailed chameleons (*Brookesia* and *Rhampholeon*) are often called leaf chameleons because a combination of their small body and a morphology resembling a small twig or leaf makes them highly cryptic when on the ground. Coloration of many resembles that of lichens. Only the distal end of the tail of these chameleons is prehensile. These tiny chameleons can be divided into two morphotypes, the leaf morphotype and the twig morphotype. Although most *Rhampholeon* have the leaf morphology, some *Brookesia* (e.g., *B. bekolosy*) do as well. Chameleons with the twig morphotype generally have elongate bodies with enlarged vertebral process on either side of the spinal column that project upward, giving the lizard a sawtooth aspect. In *Brookesia stumpffi*, which apparently lays eggs containing advanced embryos, the incubation period varies from 28 to 30 days. In some species, males ride on the female after mating and may ride her for several days. Whether this represents extended mate defense to ensure paternity remains unstudied.

Comment: Some authors consider *Brookesia* and *Rhampholeon* to be in a separate subfamily, the Brookesinae.

Iguanidae

Anoles, Iguanas, and Allies

Classification: Reptilia; Diapsida; Sauria; Lepidosauria; Squamata.

Sister taxon: Acrodonta (= Agamidae and Chamaeleonidae).

Content: Eight subfamilies, Corytophaninae, Crotaphytinae, Hoplocercinae, Iguaninae, Oplurinae, Phrynosomatinae, Polychrotinae, and Tropidurinae.

Distribution: Throughout the Americas, Madagascar, and west-central Pacific islands (Fig. 20.8).

Characteristics: Iguanids range from small (30 mm adult SVL, *Anolis ophiolepis*) to large (750 mm adult SVL, *Cyclura nubilia*) lizards, many covered dorsally and ventrally by large, keeled, overlapping scales and others with small, granular scales. No osteoderms occur dorsally or ventrally on the trunk. All species are limbed, and the pectoral girdle has a T-shaped interclavicle and curved rod-shaped clavicles. The tail is usually long to moderately long, and many iguanids have caudal autotomy with fracture planes variously located in the caudal vertebrae. The tongue is covered dorsally with reticular papillae and lacks lingual scales; the foretongue is nonretractable. The skull possesses paired nasals, postorbitals,

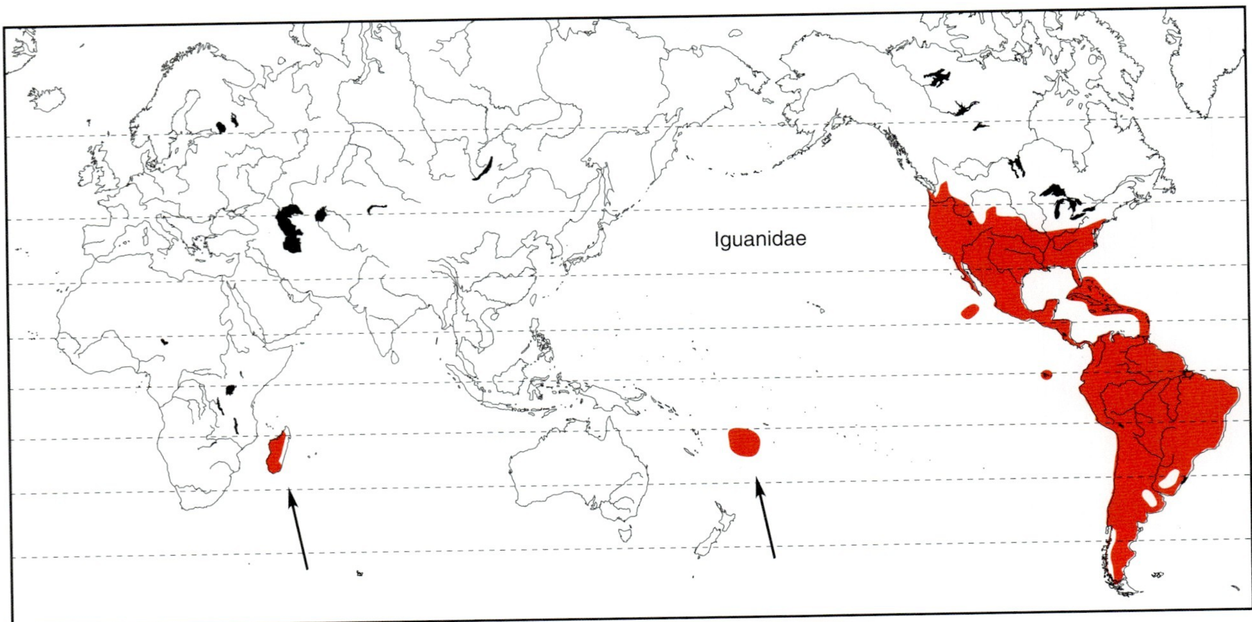

FIGURE 20.8 Geographic distribution of the extant Iguanidae.

squamosals, and single (fused) frontal and parietal bones; a parietal foramen is present in most species and perforates the frontal, frontoparietal suture, or parietal. Attachment of the marginal dentition is pleurodont, and teeth are present or absent on the pterygoids.

Corytophaninae

Sister taxon: Uncertain, possibly Polychrotinae.

Content: Three genera, *Basiliscus*, *Corytophanes*, and *Laemanctus*, with 4, 3, and 2 species, respectively.

Distribution: Southern Mexico to northern South America.

Characteristics: In the skull, the lacrimal foramen is not enlarged, the jugal and squamosal are in broad contact, the parietal foramen perforates the frontal (absent in *Laemanctus*), palatine teeth are absent, and pterygoid teeth are present. Meckel's groove in the mandible is usually fused. Males lack femoral pores, and spinuate scale organs are absent.

Biology: Corytophanines are largely arboreal lizards (Fig. 20.9), living in dry scrub forest to wet rain forest. They are casque-headed, slender-bodied, long-limbed, and long-tailed lizards, generally ranging from 90 to 200 mm adult SVL. Of the three genera, species of *Corytophanes* and *Laemanctus* are strongly arboreal and rarely ascend to the ground except to lay eggs. They are capable of rapid locomotion but typically use a slow, methodical gait, reminiscent of chameleons, and although not capable of rapid color change, they are cryptically camouflaged. In contrast, basilisks (*Basciliscus*) are low-level forest inhabitants, foraging largely on the ground but returning to trees to escape predators and sleep or bask. Basilisks are also capable of running bipedally and noted for their ability to run across the surface of water. All species with the exception of *C. pericarinata* are oviparous. Basilisks have 8 to 18 eggs per clutch, fewer (2–8 eggs) in the other two genera. The viviparous *C. pericarinata* produce an average litter of 7 neonates (3–10). Egg deposition likely occurs year-round in equitable habitats and from the early to the middle of the dry season in the more climatically extreme areas. Incubation is about 8 to 12 weeks.

FIGURE 20.9 Representative iguanid lizards I. Clockwise from upper left: Green basilisk *Basiliscus plumifrons*, Corytophaninae (L. J. Vitt); collared lizard (female) *Crotaphytus collaris*, Crotaphytinae (L. J. Vitt); Boulenger's dwarf iguana *Hoplocercus spinosus*, Hoplocercinae (L. J. Vitt); green iguana *Iguana iguana*, Iguaninae (L. J. Vitt).

Crotaphytinae

Sister taxon: Uncertain.

Content: Two genera, *Crotaphytus* and *Gambelia*, with 7 and 3 species, respectively.

Distribution: Southwestern United States and northern Mexico.

Characteristics: In the skull, the lacrimal foramen is not enlarged, the jugal and squamosal are not in broad contact, the parietal foramen perforates the frontoparietal suture, and palatine and pterygoid teeth are present. Meckel's groove in the mandible is open. Males have femoral pores, and spinuate scale organs are absent.

Biology: Crotaphytines are moderately large (100-140 mm adult SVL), stout-bodied lizards with long, strong limbs and long tails (Fig. 20.9). They are principally diurnal predators, frequently preying upon other lizards, although arthropods form a significant component of their diet. They occur predominantly in dry, open habitats and select a rock or other vantage point from which they can search for prey. When prey is sighted, they jump from their perch and chase their prey. They are fast and capable of bipedal running. They produce clutches of modest size, averaging three to eight eggs per female; clutch size increases as female size and age increase.

Hoplocercinae

Sister taxon: Uncertain, possibly Iguaninae.

Content: Three genera, *Enyalioides*, *Hoplocercus*, and *Morunasaurus*, with 7, 1, and 3 species, respectively.

Distribution: Disjunct, from the Isthmus of Panama to northern South America and in upland areas of the Amazon basin. *Hoplocercus* occurs in the Brazilian cerrados and the southern portion of the Amazon basin.

Characteristics: In the skull, the lacrimal foramen is enlarged, the jugal and squamosal are not in broad contact, the parietal foramen perforates frontoparietal suture, palatine teeth are absent, and pterygoid teeth are present. Meckel's groove in the mandible is open. Males have femoral pores, and spinuate scale organs are absent.

Biology: Hoplocercines are moderately large lizards (90-150 mm adult SVL), each genus with a different morphology. For example, *Hoplocercus* is a robust lizard, somewhat like a spiny iguana (Fig. 20.9), and *Enyalioides* is more slender with longer hindlimbs. This difference in body form is associated with more terrestrial habits in open habitats in the former and semiarboreal habits in forest habitats in the latter. All species are insectivorous. Reproductive behavior is little studied.

Iguaninae

Sister taxon: Uncertain, possibly Hoplocercinae.

Content: Eight genera, *Amblyrhynchus*, *Brachylophus*, *Conolophus*, *Ctenosaura*, *Cyclura*, *Dipsosaurus*, *Iguana*, and *Sauromalus*, with 29+ species.

Distribution: Americas from southwestern United States to Paraguay and southern Brazil, West Indies, Galapagos, and west-central Pacific islands.

Characteristics: In the skull, the lacrimal foramen is not enlarged, the jugal and squamosal are not in broad contact, the parietal foramen perforates the frontoparietal suture, palatine teeth are absent, and pterygoid teeth are present. Meckel's groove in the mandible is fused. Males have femoral pores, and spinuate scale organs are absent.

Biology: Iguanines (the iguanas) are typically large lizards; most species exceed 200 mm adult SVL, although some, such as the Fijian banded iguana (*B. fasciata*) and the desert iguana (*D. dorsalis*) attain sexual maturity at 140 to 160 mm SVL. Iguanas are predominantly terrestrial in mesic to xeric habitats. Only *Iguana* and *Brachylophus* (Fig. 20.9) display strong arboreality, rarely descending to the ground. They are predominantly to exclusively herbivores, feeding on a wide variety of plant parts, including flowers and fruits as well as foliage. *Amblyrhynchus cristatus* feeds exclusively on marine algae and grazes beneath the water even though it is not an exceptionally proficient swimmer. All iguanas are oviparous and produce moderately large clutches, ranging from 2 to 8 eggs in the small-bodied *D. dorsalis* and 12 to 88 eggs in the large-bodied *Cyclura* and *Iguana*. Nutrition is a significant factor in clutch size; large-bodied species in resource-poor environments produce fewer eggs. Several of the larger iguanas (e.g., *I. iguana*, *Conolophus pallidus*) migrate from their home ranges to special nesting sites to deposit eggs. For most iguanas, incubation is about 10 to 12 weeks but commonly requires more than 30 weeks in the two Fijian iguanas.

Oplurinae

Sister taxon: Uncertain, possibly Tropidurinae or Polychrotinae.

Content: Two genera, *Chalarodon* and *Oplurus*, with 1 and 6 species, respectively.

Distribution: Madagascar and the Comores.

Characteristics: In the skull, the lacrimal foramen is not enlarged, the jugal and squamosal are not in broad contact, the parietal foramen perforates the frontoparietal suture, palatine teeth are present or absent, and pterygoid teeth are present. Meckel's groove in the mandible is variably open or fused. Males lack femoral pores, and spinuate scale organs are present.

Biology: In looks and behavior, Madagascan oplurines share many features with phrynosomatines and tropidurines. They range from 60 to 90 mm adult SVL (*Chalarodon*) to 90 to 150 mm (*Oplurus*; Fig. 20.10). They include arboreal and terrestrial taxa. *Oplurus* lives mainly

FIGURE 20.10 Representative iguanid lizards II. Clockwise from upper left: *Oplurus grandiere*, Oplurinae (G. R. Zug); round-tailed horned lizard *Phrynosoma modestum*, Phrynosomatinae (L. J. Vitt); Brazilian bush anole *Polychrus acutirostris* (L. J. Vitt); harlequin tree runner *Plica umbra*, Tropidurinae (L. J. Vitt).

on rocks, and *Chalarodon* lives in sandy areas. For all, scrub to desert habitats are largely xeric. All species are oviparous; the smaller *C. madagascarensis* typically lays two eggs and the somewhat larger *Oplurus* deposits clutches of four to six eggs. Nests are regularly dug and eggs are deposited in the ground, but some rock dwellers deposit eggs in rock crevices.

Phrynosomatinae

Sister taxon: Uncertain, possibly Tropidurinae.

Content: Ten genera, *Callisaurus, Cophosaurus, Holbrookia, Petrosaurus, Phrynosoma, Sator, Sceloporus, Uma, Urosaurus,* and *Uta,* with 110+ species.

Distribution: Southern half of North America to western Panama.

Characteristics: In the skull, the lacrimal foramen is not enlarged, the jugal and squamosal are not in broad contact, the parietal foramen perforates the frontoparietal suture, and palatine and pterygoid teeth are absent. Meckel's groove in the mandible is open. Males have femoral pores, and spinuate scale organs are absent.

Biology: Phrynosomatines are the dominant iguanid lizards of North America and Mexico; species diversity of this clade declines southward through Central America. They are largely arid-adapted species (Fig. 20.10) and reach their greatest abundance in xeric habitats of the southwestern United States and the Mexican Plateau. *Sceloporus* is the most diverse genus with more than 70 species. The moderately robust, spiny-scaled body of many *Sceloporus* epitomizes the spiny lizard appearance shared with many other iguanid and agamid genera. This body form also largely characterizes a terrestrial–semiterrestrial, sit-and-wait foraging lizard that preys largely on insects and other arthropods. *Urosaurus*, *Holbrookia*, and their relatives are smaller-scaled, slender-bodied, and longer-limbed lizards, and although they and *Sceloporus* may not look like the pancake-bodied *Phrynosoma* (an ant specialist), they are all closely related. Phrynosomatines are predominantly moderate-sized lizards, and most species range from 50 to 100 mm adult SVL. A few species are larger but none exceeds 200 mm SVL. Phrynosomatines are predominantly oviparous. Clutches consist of 2 to 28 eggs, although most species produce less than 10 eggs per clutch. Several species of *Phrynosoma* and *Sceloporus* are live bearers, producing litters of 6 to 30 neonates. Most species occur in seasonal environments, hence reproduction is strongly seasonal. The first clutch is deposited in middle to late spring, and often a second clutch is produced a few weeks later. Incubation times generally range from 6 to 8 weeks. Some high-elevation viviparous *Sceloporus*, such as *S. jarrovi*, ovulate in fall, carry embryos during winter, with females basking on rock outcrops to gain heat, and give birth in early spring.

Polychrotinae

Sister taxon: Uncertain, possibly Corytophaninae or Oplurinae.

Content: Eight genera, *Anisolepis*, *Anolis*, *Diplolaemus*, *Enyalius*, *Leiosaurus*, *Polychrus*, *Pristidactylus,* and *Urostrophus*, with 320+ species.

Distribution: Southeastern United States through Central America and the West Indies to nearly the southern tip of South America.

Characteristics: The lacrimal foramen in the skull is not enlarged, the jugal and squamosal are not in broad contact, the parietal foramen usually perforates the frontoparietal suture (occasionally the parietal), palatine teeth are present or absent, and pterygoid teeth are present. Meckel's groove in the mandible is fused. Males lack femoral pores (except *Polychrus*), and spinuate scale organs are typically present.

Biology: Polychrotinae are the most speciose iguanid lizards (Fig. 20.10). One genus, *Anolis* (anoles), has about 200 species and has been divided into as many as three genera (*Anolis, Norops*, and *Dactyloa*). The adaptive radiation of West Indian *Anolis* within and among islands has provided a theoretical and experimental springboard for numerous evolutionary and ecological studies. Polychrotines, also known as anolines, are predominately arboreal species as indicated by their specialized foot morphology. They range in size from 30 mm SVL (*Anolis ophiolepis*) to greater than 180 mm SVL (*A. equestris* complex), although most species are within 40 to 80 mm SVL. Most species are sexually dimorphic with larger males. All polychrotines are diurnal, and most are sit-and-wait foragers on arthropod prey. Anoles have a unique reproductive physiology that includes continual egg production. Only 1 egg is laid at a time—in a terrestrial nest—but eggs are produced in rapid succession. Oogenetic maturation, ovulation, and egg shelling occur alternately between left and right ovaries and oviducts, and under ideal conditions, a female in good condition will lay an egg every 7 to 20 days. Continuous reproduction does not typically occur in the wild because most environments are climatically cyclic; thus food availability and quality are also cyclic. Other polychrotines have a typical lizard reproduction and produce clutches with variable numbers of eggs (range, 2–30) deposited in a terrestrial nest.

Comment: *Anolis* species are morphologically similar, and this similarity has made it difficult to determine relationships among various species groups. One solution recognized two major clades, *Anolis* and *Norops*, and several much smaller ones. This solution has been accepted by some researchers but rejected by others because their proposed taxa are likely paraphyletic. Molecular studies have identified monophyletic clades, but to date, no study has included all species.

Tropidurinae

Sister taxon: Uncertain, possibly Phrynosomatinae.

Content: Eleven genera, *Ctenoblepharys, Leiocephalus, Liolaemus, Microlophus, Phymaturus, Plesiomicrolophus, Plica, Stenocercus, Tropidurus, Uracentron,* and *Uranoscodon,* with 175+ species.

Distribution: West Indies from the Bahamas to the southern tip of South America.

Characteristics: In the skull, the lacrimal foramen is not enlarged, the jugal and squamosal are not in broad contact, the parietal foramen perforates the frontoparietal suture or is absent, palatine teeth are absent, and pterygoid teeth are usually present. Meckel's groove in the mandible is variably fused. Males lack femoral pores, and spinuate scale organs are absent.

Biology: Tropidurines are morphologically similar to sceloporines and include both spiny (Fig. 20.10) and smooth-scaled forms. They are generally a more diverse group of lizards, living in a broader range of habitats from mesic forest to deserts. Nonetheless, like phrynosomatines, they occur primarily in open habitats, and many are arid adapted. Tropidurines show three adaptive radiations: a leiocephaline clade in the West Indies, a tropidurine one in northern and central South America, and a liolaemine clade in grasslands and deserts of southern South America. These radiations are often assigned subfamily status (Leiocephalinae, Tropidurinae, and Liolaeminae); the former two clades have larger body size (> 65 mm adult SVL), and many have spiny lizard appearance. They also are typical insectivores. A number of species in the genera *Tropidurus*, *Plica*, and *Uracentron* eat large numbers of ants, and some even specialize on ants. Liolaemines are somewhat smaller and smoother-scaled lizards. Many species include significant amounts of plant matter in their diets, and the number of independent origins of herbivory within these lizards may exceed all other origins of herbivory in squamates. Most tropidurines are oviparous with clutches ranging from 1 to 14 eggs, with larger species generally producing larger clutches. Some populations of the southern latitude and high-elevation species, such as *Liolaemus magellanicus*, are viviparous.

Nyctisauria

Gekkonidae

Geckos and Pygopods

Classification: Reptilia; Diapsida; Sauria; Lepidosauria; Squamata.

Sister taxon: Uncertain; two main hypotheses are Annulata or Scleroglossa–Autarchoglossa.

Content: Four subfamilies, Diplodactylinae, Eublepharinae, Gekkoninae, and Pygopodinae.

Distribution: Pantropic on all landmasses (Fig. 20.11).

Characteristics: Gekkonoids are small (16–18 mm adult SVL, *Sphaerodactylus parthenopion*) to large (370 mm SVL, *Hoplodactylus delcourti*) lizards. Most species are covered dorsally and ventrally by small, granular scales that are occasionally interspersed with tubercles. No osteoderms occur dorsally on the trunk; they occur ventrally in some geckos. Most species are distinctly limbed with a pectoral girdle having a T-shaped or cruciform interclavicle and angular clavicles. The tail is usually moderately short to long (two-thirds to just longer than SVL). Pygopodines are snakelike with highly reduced rear limbs and elongate bodies and tails. Caudal autotomy is common, and a fracture plane occurs posterior to the transverse processes of each caudal vertebra. The tongue is covered dorsally with peglike papillae and lacks lingual scales; the foretongue is nonretractable. The skull has paired nasals and single or paired frontals and parietals, squamosals are present or absent, and postorbitals and a parietal foramen are absent. Attachment of the marginal dentition is pleurodont, and the pterygoid lacks teeth.

Comment: Eublepharines are regularly given familial status and considered the sister group of the clade containing all other geckos. However, most studies examining the relationships among geckos usually include only one or two outgroups of other lizards, and conversely, studies examining the relationships of squamates seldom treat eublepharines independently of other geckos. Pygopodines are also often considered a family sister to other gekkonoids. Two genera, *Teratoscincus* and *Aeluroscalabotes*, are placed in their own monotypic subfamilies by some authors.

Diplodactylinae

Sister taxon: Pygopodinae.

Content: Twenty genera, *Bavayia, Carphodactylus, Crenadactylus, Dierogekko, Diplodactylus, Eurydactylodes, Hoplodactylus, Naultinus, Nephrurus, Oedura, Phyllurus, Pseudothecadactylus, Rhacodactylus, Rhynchoedura, Saltuarius, Strophurus,* and *Underwoodisaurus*, with 110+ species.

Distribution: Australia, New Caledonia, and New Zealand.

Characteristics: Body is not elongate or snakelike; both fore- and hindlimbs are well developed. The skin is soft with numerous small, juxtaposed scales. The skull has paired premaxillaries, paired parietals, and an imperforate stapes, except in *Eurydactylodes*. The eye is covered by a spectacle and usually contains 20 or more sclerotic ossicles. The auditory meatus is fully encircled by a closure muscle, and the tectorial membrane is thickened medially.

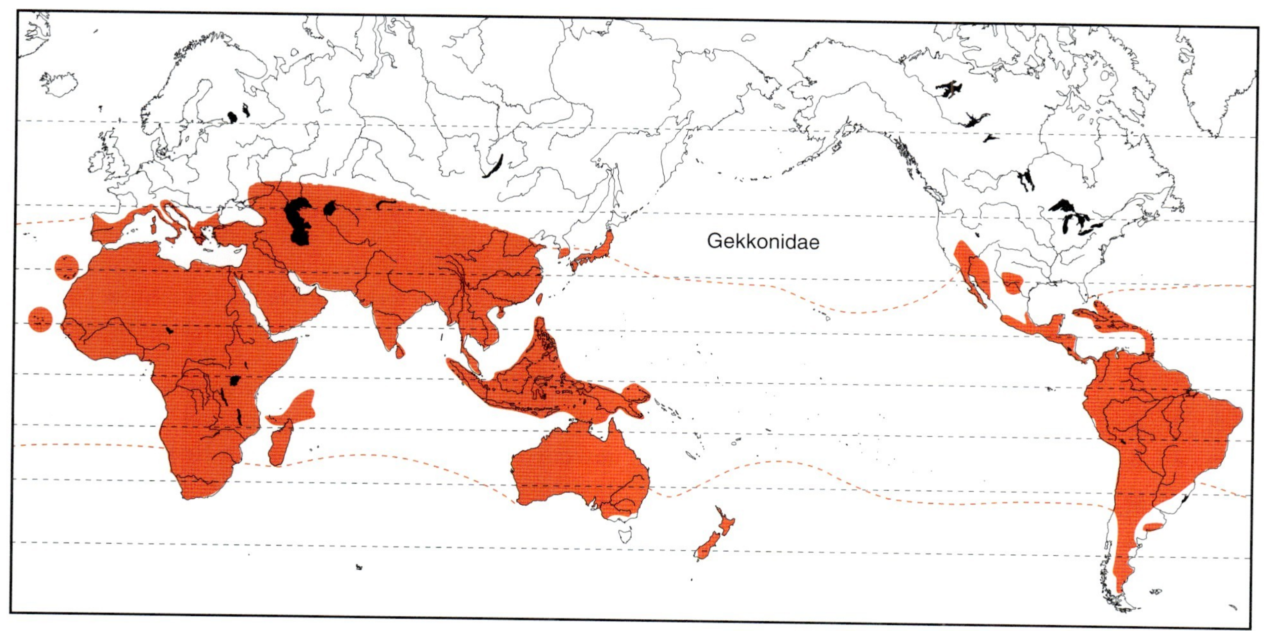

FIGURE 20.11 Geographic distribution of the extant Gekkonidae. Dashed red lines indicate envelope of oceanic islands containing gecko species.

FIGURE 20.12 Representative geckos. Clockwise from upper left: Beautiful gecko *Diplodactylus pulcher*, Diplodactylinae (B. Maryan); Texas banded gecko *Coleonyx brevis*, Eublepharinae (L. J. Vitt); cosmopolitan house gecko *Hemidactylus mabouia*, Gekkoninae (L. J. Vitt); southern pygopodid *Aprasia inaurita*, Pygopodinae (M. Kearney).

Biology: Diplodactylines are predominately moderate-sized geckos (60–110 mm adult SVL; Fig. 20.12), although *Hoplodactylus delcourti* attains a 370 mm SVL and *Rhacodactylus* adults commonly exceed 100 mm SVL. Diplodactylines occupy a wide range of habitats from cool, moist forest through dry scrub to desert. Most are nocturnal and many are arboreal, although a few, such as *Nephrurus,* are distinctly terrestrial, spending the day in burrows and foraging on the surface at night. Most are insectivorous, although the prehensile-tailed *Naultinus* and a few others are nectivores, or at least nectar and flowers form a significant portion of their diet. Most are oviparous and typically lay two eggs, which remain leathery through incubation. A few, such as *Hoplodactylus* and *Naultinus*, are viviparous and produce only two neonates.

Eublepharinae

Sister taxon: Uncertain, likely Gekkoninae–Diplodactylinae–Pygopodinae clade.

Content: Six genera, *Aeluroscalabotes*, *Coleonyx*, *Eublepharis*, *Goniurosaurus*, *Hemitheconyx,* and *Holodactylus*, with 25 species.

Distribution: Disjunct in southwestern North America and northern Central America, and sub-Saharan Africa and southern Asia.

Characteristics: Body is not elongate or snakelike; both fore- and hindlimbs are well developed. The skin is soft with numerous small, juxtaposed scales. The skull has paired premaxillaries, a single parietal, and a stapes perforated by a branch of the facial artery. The eye lacks a spectacle cover and usually contains 20 or more sclerotic ossicles. The auditory meatus has a semicircular closure muscle, and the tectorial membrane is uniform.

Biology: Eublepharines are moderate to large geckos, ranging from 45 to 155 mm adult SVL (Fig. 20.12). With the exception of Bornean *Aeluroscalabotes*, they are terrestrial geckos with narrow digits, and all are nocturnal insectivores. Their disjunct Northern Hemisphere distribution suggests an ancient lizard clade, and presently each regional occurrence denotes a separate center of diversification. The American radiation produced six species (*Coleonyx*), two living on the floor of mesic tropical forests and four in the Southwest deserts. The Asian *Eublepharis* consists of eight species in the Asian deserts from Iraq to northeastern peninsular India. The other Asian taxa are mainly forest inhabitants; *Goniurosaurus* lives on the forest floor or rock outcrops, and *Aeluroscalabotes* lives above the forest floor on logs and understory shrubs. The African radiation (*Hemitheconyx* and *Holodactylus*, two species each) occurs mainly in scrub and desert habitats. All eublepharines are surface foragers and have a fixed clutch size of two eggs.

Gekkoninae

Sister taxon: Uncertain, possibly the Diplodactylinae–Pygopodinae clade.

Content: Seventy-nine genera, *Afroedura, Afrogecko, Agamura, Ailuronyx, Alsophylax, Aristelliger, Asaccus, Asiocolotes, Blaesodactylus, Bogertia, Briba, Bunopus, Calodactylodes, Carinatogecko, Chondrodactylus, Christinus, Cnemaspis, Coleodactylus, Colopus, Cosymbotus, Crossobamon, Cryptactites, Cyrtodactylus, Cyrtopodion, Dixonius, Dravidogecko, Ebenavia, Elasmodactylus, Euleptes, Geckoella, Geckolepis, Geckonia, Gehyra, Gekko, Goggia, Gonatodes, Gonydactylus, Gymnodactylus, Haemodracon, Hemidactylus, Hemidactylus, Hemiphyllodactylus, Heteronotia, Homonota, Homopholis, Lepidoblepharis, Lepidodactylus, Luperosaurus, Lygodactylus, Matoatoa, Microscalabotes, Nactus, Narudasia, Pachydactylus, Palmatogecko, Paragehyra, Paroedura, Perochirus, Phelsuma, Phyllodactylus, Phyllopezus, Pristurus, Pseudogekko, Pseudogonatodes, Ptenopus, Ptychozoon, Ptyodactylus, Quedenfeldtia, Rhoptropus, Saurodactylus, Sphaerodactylus, Stenodactylus, Tarentola, Teratolepis, Teratoscincus, Thecadactylus, Tropiocolotes, Urocotyledon,* and *Uroplatus,* with 800+ species.

Distribution: Pantropic and temperate Eurasia.

Characteristics: Body is not elongate or snakelike; both fore- and hindlimbs are well developed. The skin is soft with numerous small, juxtaposed scales, except in *Teratoscincus*. The skull has a single premaxillary, a single parietal (paired in sphaerodactyl geckos), and an imperforate or perforated stapes. The eye is covered by a spectacle and contains 14 sclerotic ossicles. The auditory meatus has a semicircular closure muscle, and the tectorial membrane is uniform.

Biology: Gekkonines have the greatest species richness of all lizard groups. In addition to their high species richness, their ecologies and life histories are likely to be equally diverse. Varying greatly in morphology, particularly foot morphology, they nonetheless remain recognizable as geckos and most are nocturnal. Nevertheless, several genera, including *Phelsuma, Coleodactylus, Pseudogonatodes, Lepidoblepharis, Lygodactylus,* and most *Gonatodes* (Fig. 20.12) are diurnal. Most gekkonines are small-to-moderate-sized lizards, ranging from 35 to 100 mm adult SVL. A few, such as *Gekko*, commonly exceed 100 mm SVL as adults, and one clade, *Sphaerodactylus*, are typically tiny geckos (most less than 30 mm SVL). Most geckos are rupicolous or arboreal. Arid land species commonly occur on rock outcrops and cliffs, and forest species occupy a variety of elevated sites from low understory to high in the canopy. Other geckos are strictly terrestrial, living on leaf litter, in burrows, inside termite nests, or beneath surface detritus. Most are insectivorous. However, larger species commonly eat smaller geckos, and a few species at least supplement their diet with nectar, fruit, and sap. Some small species, such as *Coleodactylus amazonicus*, eat springtails and mites that they capture in leaf litter. All gekkonines are oviparous, typically depositing two eggs that have flexible shells when laid. The shells quickly harden and become resistant to water loss. The tiny *Sphaerodactylus, Coleodactylus, Pseudogonatodes,* and *Lepidoblepharis* deposit single eggs, which may reflect a morphological constraint on clutch volume because of small body size. One large species, *Thecadactylus rapicauda*, also has a clutch size of a single egg. About a dozen species of gekkonines are parthenogenetic, several of which have spread widely throughout the Indoaustralia and Pacific region via accidental transport by humans (e.g., *Hemidactylus garnotii, Lepidodactylus lugubris*).

Pygopodinae

Sister taxon: Diplodactylinae.

Content: Seven genera, *Aprasia, Delma, Lialis, Ophidiocephalus, Paradelma, Pletholax,* and *Pygopus,* with 35+ species.

Distribution: Australia and southern New Guinea.

Characteristics: Body is elongate and snakelike; external evidence of forelimbs is lacking, and hindlimbs are flaplike (Fig. 20.12). The skin is comprised of large, overlapping scales. The skull has paired premaxillaries, paired parietals (single in *Lialis*), and an imperforate stapes. The eye is covered by a spectacle and contains 11–19 sclerotic ossicles. The auditory meatus is fully encircled by a closure muscle, and the tectorial membrane is uniform.

Biology: The snakelike pygopods are moderate (59 mm adult SVL, *Delma australis*) to large (310 mm SVL, *Lialis jicari*) lizards. Most species are between 70 and 120 mm SVL as adults. Pygopods are largely but not entirely diurnal. They both search for and ambush prey, and most taxa are insectivorous. They eat a broad variety of arthropods, although a few appear to be dietary specialists, such as *Aprasia*, which feeds on ants. Large species prey occasionally on small vertebrates, and *Lialis* appears to prey only on lizards, especially skinks. *Lialis* has a highly flexible hinge in the middle of the skull, and this added flexibility permits them to tightly grasp the hard, slippery-scaled skinks. Pygopods typically lay two eggs, although nests of six or more pygopod eggs have been found, indicating communal nesting. The eggs retain a flexible shell throughout an 8–10-week incubation period.

Nyctisaura–Annulata

Dibamidae

Blind Skinks

Classification: Reptilia; Diapsida; Sauria; Lepidosauria; Squamata.

Sister taxon: Uncertain; two main hypotheses suggest Amphisbaenia or Scincomorpha.

Content: Two genera, *Anelytropsis* and *Dibamus*, with 1 and 9 species, respectively.

Distribution: Disjunct, Mexico and eastern Indochina to the East Indies (Fig. 20.13).

Characteristics: Dibamids are small-to-moderate-sized (50–200 mm adult SVL) snakelike lizards (Fig. 20.14). They lack forelimbs and have only flaplike hindlimbs. The body is cloaked in shiny, smooth, overlapping scales. No osteoderms occur dorsally or ventrally on the trunk. Forelimbs are absent, and neither limb nor pectoral girdle bones are present. The tail is short and autotomous with a fracture plane anterior to the transverse processes of each caudal vertebra. The tongue is covered dorsally with filamentous papillae and lacks lingual scales. The foretongue is nonretractable. The skull has paired nasals and frontals, the postorbitals and squamosals are present or absent, and the parietal bone is single (fused). A parietal foramen is absent. Attachment of the marginal dentition is pleurodont, and the pterygoid lacks teeth.

Biology: Dibamids are predominantly subsurface lizards, living beneath surface detritus and often in burrows and crevices in the ground. They apparently are not strict burrowers but depend upon burrows and other openings in soil, although they are capable of digging in loose humus or friable soils. *Dibamus* is a forest-floor inhabitant and requires moist soils; during dry season, it lives deep in the moisture shadow, beneath rocks and fallen trees. *Anelytropsis* is more arid adapted and lives in dry upland forest and scrub. Dibamids are insectivorous, and all are presumably oviparous. Limited evidence suggests that *Dibamus* lays a single egg but that it may lay multiple sequential clutches. After deposition, the eggshell hardens, forming a barrier to water loss as in gekkonines. Reproductive data are not available for *Anelytropsis*.

Nyctisaura–Annulata–Amphisbaenia

Amphisbaenidae

Worm Lizards

Classification: Reptilia; Diapsida; Sauria; Lepidosauria; Squamata.

Sister taxon: Uncertain, possibly Bipedidae or Rhineuridae.

Content: Seventeen genera, *Amphisbaena, Ancylocranium, Anops, Aulura, Baikia, Blanus, Bronia, Cercolophia, Chirindia, Cynisca, Dalophia, Geocalamus, Leposternon, Loveridgea, Mesobaena, Monopeltis,* and *Zygaspis*, with ±130 species.

Distribution: Greater Antilles, South America, and Africa (Fig. 20.15).

Characteristics: Amphisbaenids are wormlike, limbless lizards. They have an annulate appearance that results from rings of rectangular, juxtaposed scales encircling the body and tail. No osteoderms occur dorsally or ventrally on the trunk. The external limbless appearance is accompanied by total absence of fore- and hindlimb skeletons. Pelvic vestiges and occasionally sternal or pectoral vestiges persist. The tail is short and autotomous in most species, but regeneration does not occur if the tail is lost. Most appear to have a single fracture plane anterior to the transverse processes of a caudal vertebra. The position

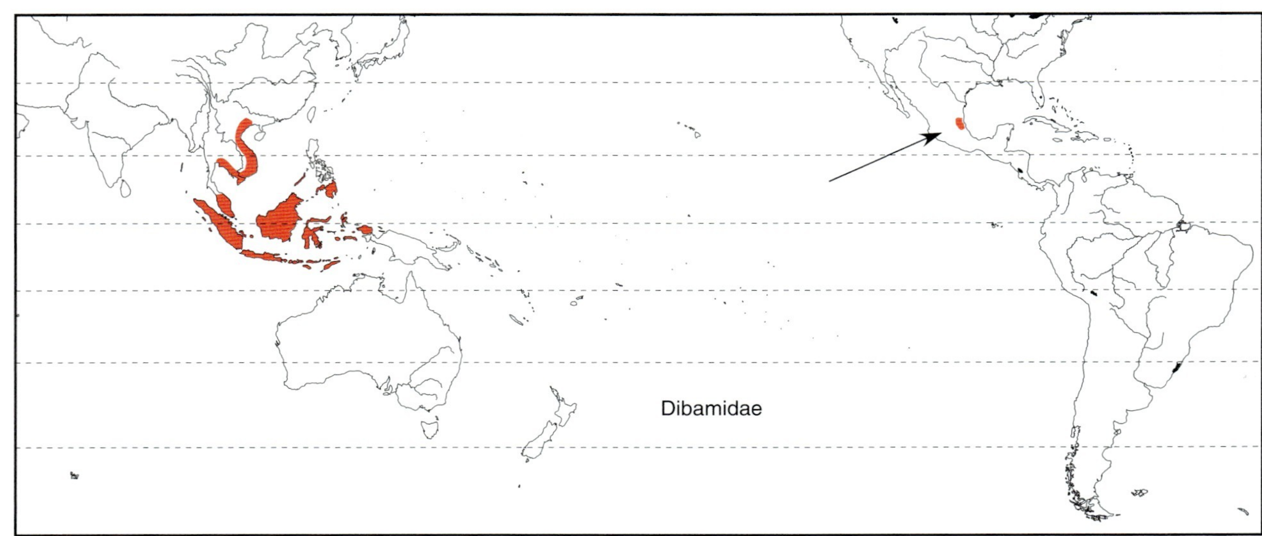

FIGURE 20.13 Geographic distribution of the extant Dibamidae.

FIGURE 20.14 Representative annulate lizards. Clockwise from upper left: Blind skink *Dibamis* sp., Dibamidae (R. W. Murphy); Bahia worm lizard *Leposternon polystegum*, Amphisbaenidae (L. J. Vitt); mole-limbed worm lizard *Bipes biporus*, Bipedidae (L. L. Grismer); Florida worm lizard *Rhineura floridana*, Rhineuridae (R. G. Tuck, Jr.).

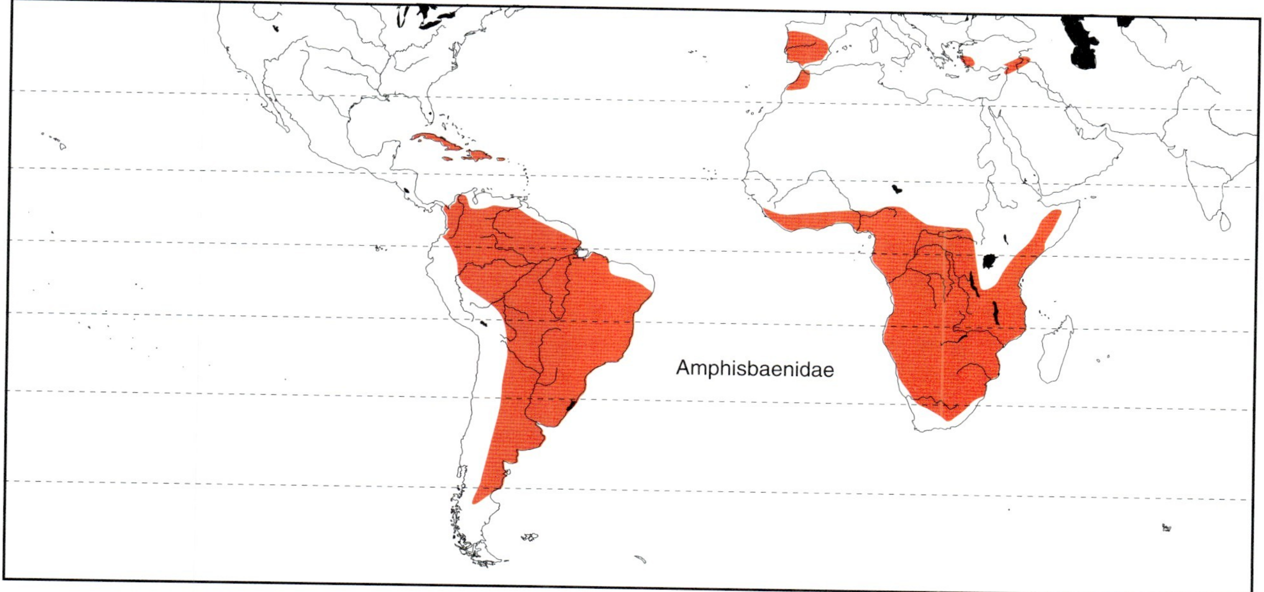

FIGURE 20.15 Geographic distribution of the extant Amphisbaenidae.

of the cleavage plane is often detectible as a dent that circles the proximal end of the tail. The tongue bears filamentous papillae and is covered dorsally with lingual scales arranged in diagonal rows; the posterior edges of these scales are smooth. The foretongue is nonretractable. The skull has paired nasals and frontals, a single, large premaxillary and parietal, and no postorbitals or squamosals. A parietal foramen is absent, except in *Monopeltis*. Attachment of the marginal dentition is pleurodont, and the pterygoid lacks teeth.

Biology: Amphisbaenids are moderate-sized worm lizards (Fig. 20.14); most range from 250 to 400 mm adult SVL, although a few species are larger or smaller. *Amphisbaena alba* reaches 720 mm TL, whereas a few smaller species, such as *Chirindia rondoense,* are only 90 to 120 mm SVL as adults. All are burrowers and create their own burrow systems. The blunt-cone or bullet-head taxa (e.g., *Amphisbaena*, *Blanus*, and *Zygaspis*) burrow by simple head ramming. The spade-snouted taxa (e.g., *Leposternon* and *Monopeltis*) tip the head downward, thrust forward, and then lift upward to compress soil to the roof of the burrow. The laterally compressed keeled-headed taxa (e.g., *Anops* and *Ancyclocranium*) ram the head forward and then alternately swing it to the left and right to compress the soil to the sides of the burrow. The ecology and life histories of most amphisbaenians are poorly studied. Most, if not all, feed on a variety of arthropods and other invertebrates. Among amphisbaenids, most species appear to be oviparous, although *Loveridgea ionidesii* and *Monopeltis capensis* are live bearers. Reproductive data are limited. Clutch size appears small, typically from two to four elongate eggs, and clutch size may be related to body size.

Bipedidae

Mole-Limbed Worm Lizards

Classification: Reptilia; Diapsida; Sauria; Lepidosauria; Squamata.

Sister taxon: Uncertain, possibly Amphisbaenidae.

Content: One genus, *Bipes*, with 3 species.

Distribution: Coastal southwestern Mexico and southern Baja California (Fig. 20.16).

Characteristics: *Bipes* is unique among amphisbaenians because it has large molelike forelimbs and forefeet (Fig. 20.14). The annulate appearance results from rings of rectangular, juxtaposed scales encircling the body and tail. No osteoderms occur dorsally or ventrally on the trunk. Bipedids lack only hindlimb elements; they have unassignable pelvic remnants and robust forelimb and pectoral girdle skeletons, although an interclavicle and clavicles are absent. The tail is short and autotomous, and regeneration does not occur. The fracture plane occurs anterior to the transverse processes of the caudal vertebra. The tongue bears filamentous papillae and is covered dorsally with lingual scales arranged in diagonal rows; the posterior edges of these scales are smooth. The foretongue is nonretractable. The skull has paired nasals and frontals and a single, large premaxillary and parietal. It lacks postorbitals, squamosals, and usually a parietal foramen. Attachment of marginal dentition is pleurodont, and the pterygoid lacks teeth.

Biology: *Bipes* are small-to-moderate-sized worm lizards, ranging from 120 to 240 mm adult SVL. They are blunt headed and burrow by head ramming in sandy desert soils. They prey mainly on arthropods, captured presumably in or immediately adjacent to the burrow tunnels. All three species are oviparous and lay small clutches of one to four eggs.

Rhineuridae

Florida Worm Lizard

Classification: Reptilia; Diapsida; Sauria; Lepidosauria; Squamata.

Sister taxon: Uncertain, possibly Amphisbaenidae.

Content: Monotypic, *Rhineura floridana*.

Distribution: Central Florida (Fig. 20.16).

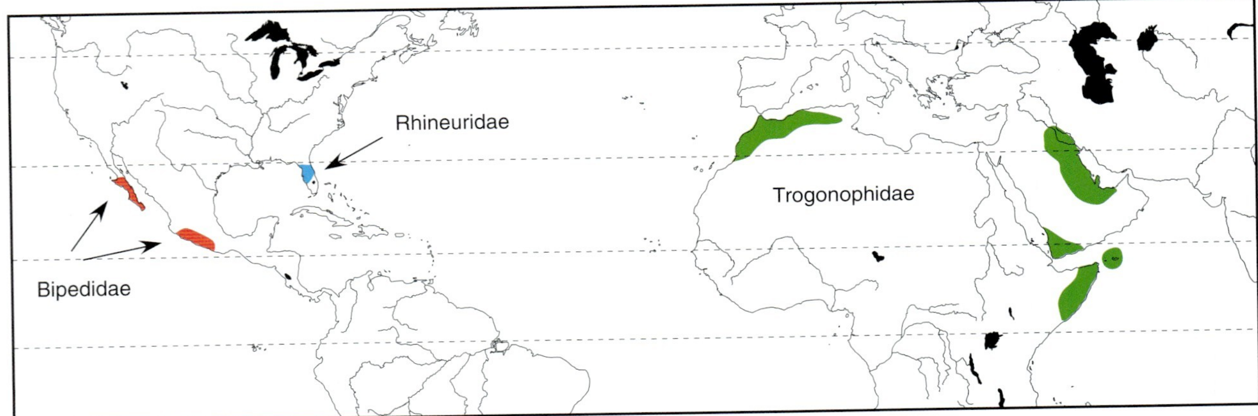

FIGURE 20.16 Geographic distributions of the extant Bipedidae, Rhineuridae, and Trogonophidae.

Characteristics: Rhineurids are limbless, wormlike lizards (Fig. 20.14). The annulate appearance results from rings of rectangular, juxtaposed scales encircling the body and tail. No osteoderms occur dorsally or ventrally on the trunk. The external limbless appearance is accompanied by the total absence of fore- and hindlimb skeletons. However, pelvic vestiges and occasionally sternal or pectoral vestiges persist. The tail is short and lacks autotomy. The tongue bears filamentous papillae and is covered dorsally with lingual scales arranged in diagonal rows; the posterior edges of these scales are smooth. The foretongue is nonretractable. The skull possesses paired nasals and frontals, and a single, large premaxillary and parietal. It lacks postorbitals, squamosals, and a parietal foramen. Attachment of the marginal dentition is pleurodont, and the pterygoid lacks teeth.

Biology: *Rhineura floridana* is a moderate-sized, burrowing lizard that ranges from 240 to 380 mm adult SVL. Although confined to sandy soils, it occurs in mesic hammock forest to xeric scrub forest. It preys largely on invertebrates, which it captures within the burrow system or on the surface near burrow openings. It is oviparous and usually lays a clutch of two eggs.

Trogonophidae

Spade-Headed Worm Lizards

Classification: Reptilia; Diapsida; Sauria; Lepidosauria; Squamata.

Sister taxon: Uncertain, possibly clade containing Amphisbaenidae, Bipedidae, and Rhineuridae.

Content: Four genera, *Agamodon*, *Diplometopon*, *Pachycalamus,* and *Trogonophis*, with 3, 1, 1, and 1 species, respectively.

Distribution: North Africa, Horn of Africa, and eastern Arabian Peninsula (Fig. 20.16).

Characteristics: Trogonophids are limbless, wormlike lizards. The annulate appearance results from rings of rectangular, juxtaposed scales encircling the body and tail. No osteoderms occur dorsally or ventrally on the trunk. The external limbless appearance is accompanied by the total absence of limb and girdle skeletons. The tail is short and lacks caudal autotomy. The tongue bears filamentous papillae and is covered dorsally with lingual scales arranged in diagonal rows; the posterior edges of these scales are smooth. The foretongue is nonretractable. The skull has paired nasals and frontals, a large premaxillary, and a large parietal; it lacks postorbitals, squamosals, and a parietal foramen. Attachment of the marginal dentition is acrodont, which easily distinguishes trogonophids from other amphisbaenians (pleurodont dentition). The pterygoid lacks teeth. Also, the body of trogonophids is triangular in cross section, whereas the body of other amphisbaenians is round.

Biology: Trogonophids are the most divergent amphisbaenians. Accentuating the peculiarity of a wormlike morphology, they have shorter, heavier bodies and strongly flattened snouts with slightly upturned edges. They live in dry sandy soils. Unlike other amphisbaenians, they dig with an oscillating head movement followed by an upward or side-to-side sweep. They create their burrows by an alternating rotational movement of the head that simultaneously shaves off the sides of the tunnel and compacts the walls. Feeding apparently occurs mainly in the burrow or immediately adjacent to it. All trogonophids are small to moderate in size, ranging from 80 to 240 mm SVL. They are oviparous, except the live-bearing *T. wiegmanni*, which produces about five neonates in a litter.

Nyctisaura

Xantusiidae

Night Lizards

Classification: Reptilia; Diapsida; Sauria; Lepidosauria; Squamata.

Sister taxon: Uncertain, two main hypotheses suggest Annulata or Lacertiformes.

Content: Three genera, *Cricosaura*, *Lepidophyma*, and *Xantusia*, with 1, 17, and 6 species, respectively.

Distribution: Western United States and eastern Mexico through Central America to northern South America. *Cricosaura typica* occurs at Cabo Cruz, Cuba (Fig. 20.17).

Characteristics: Xantusiids are small lizards, less than 100 mm adult SVL. Dorsally, they bear small, granular scales and ventrally large, juxtaposed scales (Fig. 20.18). No osteoderms occur dorsally or ventrally on the trunk. All species are limbed, and the pectoral girdle has a cruciform interclavicle and angular clavicles. The tail is usually long and autotomous. A fracture plane occurs anterior to the transverse processes of each caudal vertebra. The tongue is covered dorsally with peglike papillae and lacks lingual scales. The foretongue is nonretractable. The skull has paired nasals, squamosals, and either single or paired frontal and parietal bones. Postorbitals are absent, but if present, a parietal foramen perforates the parietal bone. Attachment of the marginal dentition is pleurodont, and the pterygoid lacks teeth.

Biology: Night lizards are extremely secretive lizards. Although their elliptical pupils suggest that they are nocturnal, they are diurnal to crepuscular but seldom venture into the open. Rather, they forage slowly in and under ground litter, in rock crevices, or beneath a canopy of low, dense vegetation. Whether desert or forest inhabitants, all are probably sedentary and may have home ranges of only a few square meters. All appear to be insectivores and to consume a large variety of arthropods. *Cricosaura typica* has reduced limbs, and its movements are predominantly serpentine. Although limbs of other taxa are also short, they use walking and running gaits. All xantusiids are live bearers, producing one to eight

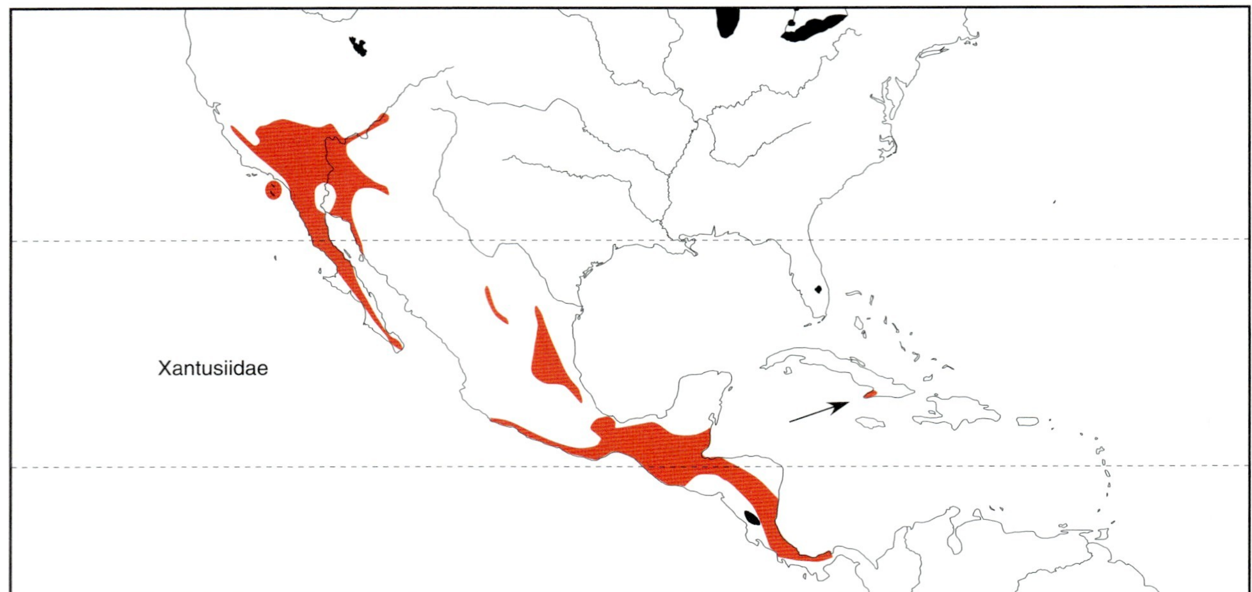

FIGURE 20.17 Geographic distribution of the extant Xantusiidae.

FIGURE 20.18 Representative xantusiid lizards. From left: Granite night lizard *Xantusia henshawi*, Xantusiidae (L. J. Vitt); yellow-spotted night lizard *Lepidophyma flavimaculatum*, Xantusiidae (L. J. Vitt).

young each year or biennially. *Xantusia* that have been studied in the field are long lived and late maturing, which is unusual in small-bodied lizards. Because adult males and females are found together over extended time periods, the possibility exists that these lizards have relatively long-term pair bonds. Tropical forest-dwelling *Lepidophyma* consists of unisexual (parthenogenetic) and bisexual species. Otherwise, little is known of their biology, although one cave-dwelling species, *L. smithii*, feeds mainly on figs that fall into their retreats.

Lacertiformes

Lacertidae

Wall Lizards, Rock Lizards, and Allies

Classification: Reptilia; Diapsida; Sauria; Lepidosauria; Squamata.

Sister taxon: Teioidea.

Content: Thirty-one genera, *Acanthodactylus, Adolfus, Algyroides, Australolacerta, Darevskia, Eremias, Gallotia, Gastropholis, Heliobolus, Holaspis, Iberolacerta, Ichnotropis, Lacerta, Latastia, Meroles, Mesalina, Nucras, Omanosaura, Ophisops, Parvilacerta, Pedioplanis, Philochortus, Podarcis, Poromera, Psammodromus, Pseuderemias, Takydromus, Teira, Timon, Tropidosaura*, and *Zootoca*, with 220+ species (see Comment).

Distribution: Most of Africa, Europe, and Asia southward into the northern East Indies (Fig. 20.19).

Characteristics: Lacertids are small to large lizards, ranging from 40 to 260 mm adult SVL. Body scalation is variable; dorsal and lateral body scales range from large, overlapping smooth or keeled scales to small, granular scales; rectangular ventral scales are juxtaposed or overlapping (Fig. 20.20). No osteoderms occur dorsally or ventrally on the trunk. All species are limbed; the pectoral

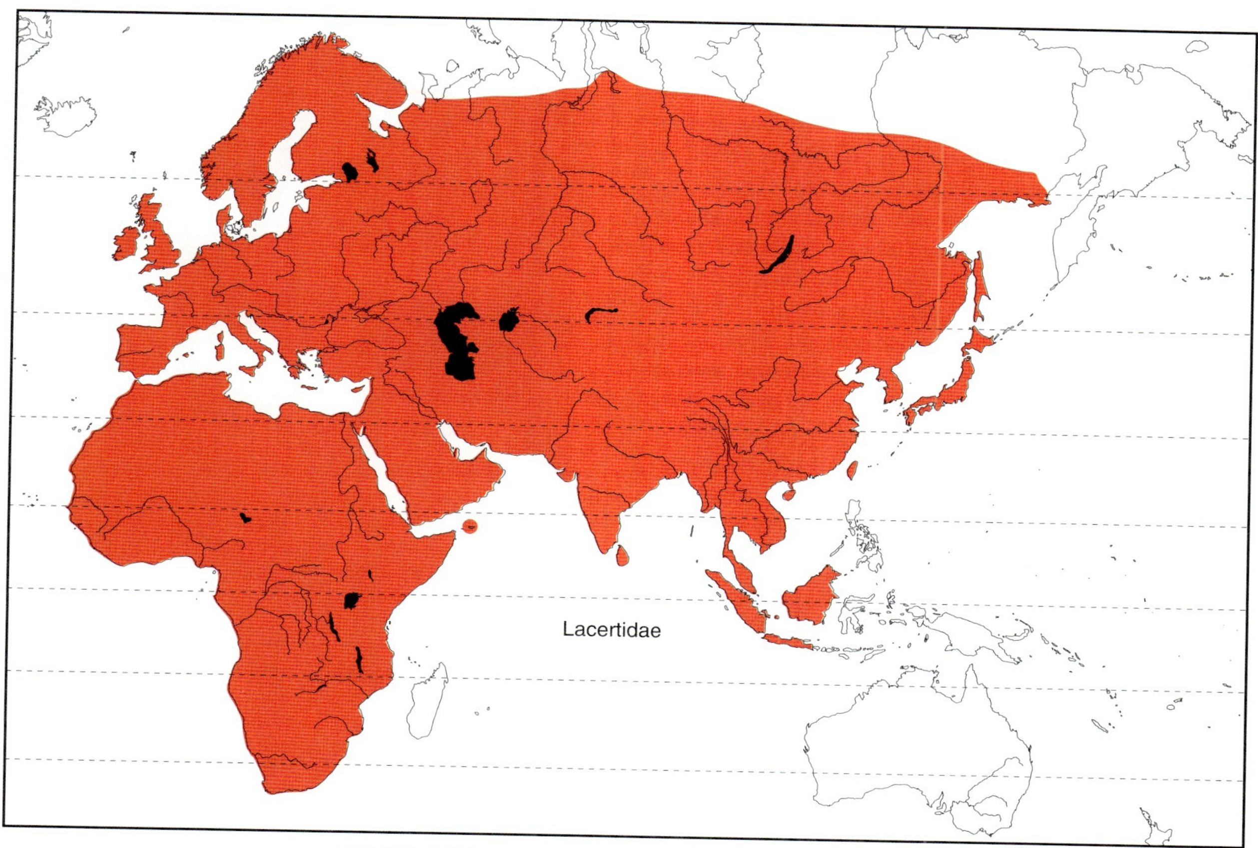

FIGURE 20.19 Geographic distribution of the extant Lacertidae.

FIGURE 20.20 Representative lacertid lizards. From left: Saw tail lizard *Holaspis guentheri*, Lacertidae (L. W. Porras); Italian wall lizard *Podarcis sicula*, Lacertidae (L. J. Vitt).

girdle has a cruciform interclavicle and angular clavicles. The tail is autotomous, usually long, and more than two times longer than SVL in *Takydromus*. Each caudal vertebra has a fracture plane anterior to the transverse processes. The tongue bears filamentous papillae and lingual scales, arranged in alternating rows dorsally. The posterior edges of the lingual scales are smooth, and the foretongue is nonretractable. The skull has paired nasals, postorbitals, squamosals, and most often a parietal and a frontal. A parietal foramen is absent. Attachment of the marginal dentition is pleurodont, and pterygoid teeth are present or absent.

Biology: Lacertids and teiids are sometimes referred to as Old and New World ecological equivalents. This generality roughly fits the behavioral, ecological, and reproductive similarities between *Lacerta* and *Cnemidophorus*, which are terrestrial, occur in arid landscapes, and have some parthenogenetic populations. The elongate, whiptail morphology is common to most other lacertids as well. Most lacertids are terrestrial, although a few are arboreal, such as *Holaspis guentheri*, which is known for its parachuting behavior even though it appears similar morphologically to other taxa. Lacertids range in size from less than 40 mm adult SVL (*Algyroides fitzingeri*) to nearly 260 mm SVL (*Gallotia stehlini*). Adults of most species are less than 120 mm SVL. Lacertids are largely insectivores and forage on the ground or low in shrubs and on bases of trees. *Meroles anchietae* regularly eats seeds, an uncommon food for lizards. Some *Lacerta* and *Australolacerta* are strongly saxicolous and are efficient, speedy climbers of rock surfaces. Some *Takydromus* spend more time off the ground than on, usually in thick grass or shrubs; *Gastropholis* and *Holaspis* are strongly arboreal, often high in trees and seldom on the ground. Most lacertids are oviparous and produce modest clutches, usually less than 10 eggs; however, clutch size is related to body size and large species, such as *L. leipida* (180–200 mm SVL) lay 20 or more eggs. Populations of the viviparous *L. vivipara* occur in areas of northern Europe, which has 6 months of freezing temperatures. Females carry from 4 to 11 embryos for 3 or 4 months. Birth occurs from late July to early October. Spanish *L. vivipara* reportedly are oviparous.

Comment: Some authors recognize three subfamilies, Laertinae, Eremianae, and Gallotianae. Although the "tropical" Afroasia taxa form a clade, relationships of the Palearctic genera are less easily resolved, and "*Lacerta*" is clearly a paraphyletic group. One study suggested that several subgenera (*Omanosaura*, *Teira*, *Timon*, and *Zootoa*) should be elevated to generic status; this recommendation has not received wide support.

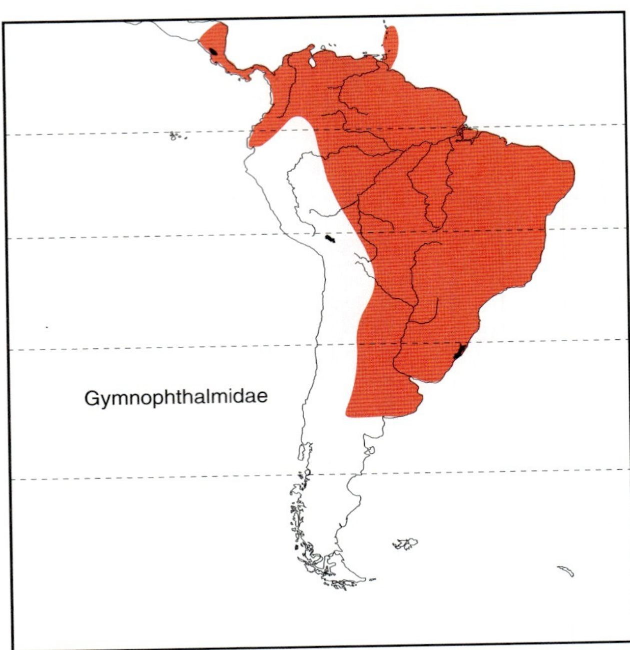

FIGURE 20.21 Geographic distribution of the extant Gymnophthalmidae.

Teioidea

Gymnophthalmidae

Gymnophthalmids

Classification: Reptilia; Diapsida; Sauria; Lepidosauria; Squamata.

Sister taxon: Teiidae.

Content: Thirty-eight genera, *Alopoglossus, Amapasaurus, Anadia, Anotosaura, Arthrosaura, Bachia, Calyptommatus, Cercosaura, Colobodactylus, Colobosaura, Colobosauroides, Echinosaura, Ecpleopus, Euspondylus, Gymnophthalmus, Heterodactylus, Iphisa, Leposoma, Macropholidus, Micrablepharus, Neusticurus, Nothobachia, Opipeuter, Petracola, Pholidobolus, Placosoma, Potamites, Procellosaurinus, Proctoporus, Psilophthalmus, Ptychoglossus, Rhachisaurus, Riama, Riolama, Stenolepis, Teuchocercus, Tretioscincus,* and *Vanzosaura,* with 160+ species.

Distribution: Southern Central America to southern South America east of the Andes (Fig. 20. 21).

Characteristics: Gymnophthalmids are mostly small lizards, less than 60 mm adult SVL. Their scalation and overall morphology are highly variable. Dorsal and lateral scales are small (some *Cercosaura*) to large (*Iphisa*), and smooth (*Bachia*) to strongly keeled (*Arthrosaura*). In some, small and large scales are interspersed and overlapping. Ventral scales are usually larger than dorsal scales and smooth or keeled. No osteoderms occur dorsally or ventrally on the trunk. Most species have limbs, and the limbs are usually small but well developed (reduced in *Bachia*, absent in *Calyptommatus*). The pectoral girdle has a cruciform interclavicle and angular clavicles. The tail varies from moderately short to long and is autotomous. A fracture plane occurs anterior to the transverse processes of each caudal vertebra. The tongue bears filamentous papillae and lingual scales arranged in diagonal rows on the dorsal surface. The posterior edges of the lingual scales are smooth, and the foretongue is nonretractable. The skull has paired nasals, postorbitals, squamosals, and single (fused) frontal and parietal bones. The parietal foramen is absent. Attachment of the marginal dentition is pleurodont, and pterygoid teeth are present or absent.

Biology: Gymnophthalmids, often referred to as "microteiids," are generally small lizards (Fig. 20.22). Ecologically, gymnophthalmids are highly diverse. For the most part, they are diurnal, but some species have been observed foraging at night. They occur in lowland rain forests, high-elevation habitats in the Andes, the savanna-like cerrados, and semiarid habitats, including relictual sand dunes of the Rio São Francisco in northeastern Brazil. Many are terrestrial and forage within forest-floor detritus. *Tretioscincus* forages above the ground on tree trunks. *Alopoglossus angulatus* and *Potamites* occur in swampy areas and along streams and readily diving into water and swimming away like salamanders. *Bachia* and *Calyptommatus* are subterranean. All are insectivores and a few (e.g., *Calyptommatus*) feed on large numbers of termites. Reproductive biology is known only for a few

FIGURE 20.22 Representative gymnophthalmid and teiid lizards. Clockwise from upper left: Rain forest bachia *Bachia flavescens*, Gymnophthalmidae (L. J. Vitt); red-tailed gymnophthalmid *Vanzosaura rubricauda*, Gymnophthalmidae (L. J. Vitt); keeled jungle runner *Kentropyx pelviceps*, Teiinae (L. J. Vitt); golden tegu *Tupinambis teguixin*, Tupinambinae (L. J. Vitt).

species, all of which are oviparous. Depending on species, clutch size is one or two eggs. Occasionally, nests are found with more than two eggs, suggesting communal nesting (e.g., *Proctoporus raneyi*). Populations of most species have both females and males, although parthenogenetic species occur in the genera *Gymnophthalmus* and *Leposoma*.

Comment: Gymnophthalmids have been divided into as many as five subfamilies (Rhachisaurinae, Gymnophthalminae, Ecpleopinae, Cercosaurinae, Alopoglossinae).

Teiidae

Whiptail Lizards, Tegus, and Allies

Classification: Reptilia; Diapsida; Sauria; Lepidosauria; Squamata.

Sister taxon: Gymnophthalmidae.

Content: Two extant, Teiinae and Tupinambinae, and two extinct clades, Chamopsiinae and "Polyglyphanodontinae."

Distribution: Americas, from northern United States to Chile and Argentina (Fig. 20.23).

Characteristics: Teiids are small (55 mm adult SVL, *Aspidoscelis inornatus*) to large (400 mm adult SVL, *Tupinambis merianae*) lizards. The dorsal and lateral body scales are usually small and granular, whereas the rectangular ventral scales are larger, juxtaposed, and arranged in transverse rows. No osteoderms occur dorsally or ventrally on the trunk. All species have well-developed limbs. The pectoral girdle has a T-shaped interclavicle and angular clavicles. The tail is autotomous, usually long, and a fracture plane occurs anterior to the transverse processes of each caudal vertebra. The tongue bears filamentous papillae and dorsal lingual scales, arranged in diagonal rows. The posterior edges of the lingual scales are smooth, and the foretongue is nonretractable. The skull has paired nasals, postorbitals, squamosals, and a single frontal (occasionally paired) and parietal bones. A parietal foramen is often present and perforates the parietal. Attachment of marginal dentition is pleurodont, and pterygoid teeth are present or absent.

Teiinae

Sister taxon: Tupinambinae.

Content: Six genera, *Ameiva, Aspidoscelis, Cnemidophorus, Dicrodon, Kentropyx,* and *Teius,* with 110+ species.

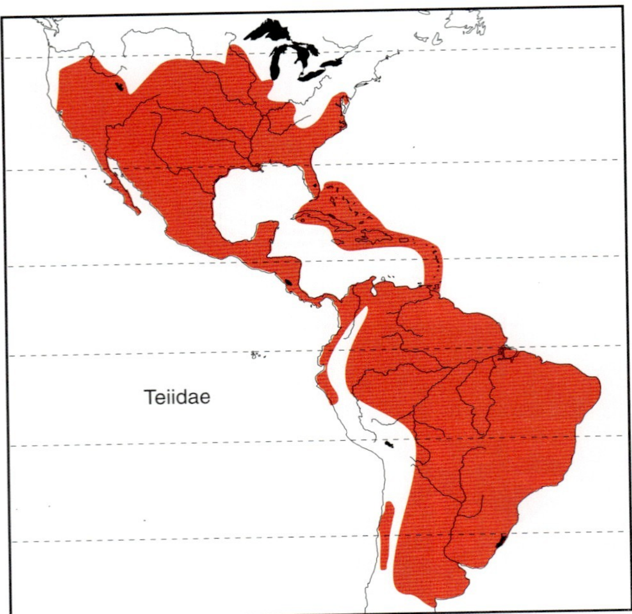

FIGURE 20.23 Geographic distribution of the extant Teiidae.

Distribution: Southern North America to northern Argentina.

Characteristics: The skull has the anteromedial edge of the supratemporal fenestra formed by the postfrontal and/or postorbital bones, a medially expanded quadrate with sliding articulation with the pterygoid, and a nasal process on the maxillary. The retroarticular process of the mandible bears a dorsal pit or sulcus.

Biology: Teiines share a strikingly similar overall morphology, with a streamlined body, long whiplike tail, and long hindlimbs (Fig. 20.22). *Ameiva* (ca. 45–200 mm adult SVL), *Aspidoscelis*, and *Cnemidophorus* (55–150 mm) are the best-known teiines because of their widespread occurrence and often moderately high population densities. All teiines appear to be active at relatively high body temperatures, often approaching 40°C. Nearctic species of *Aspidoscelis* remain inactive and in their burrows until environmental temperatures reach about 24°C; once active, they maintain body temperatures at 36°C or higher. Their thermal physiology generally limits the amount of time active each day and determines their total yearly activity period. Most teiine genera are active foragers and use a combination of vision and chemical cues to detect prey. Individuals often dig prey from under the surface, and some break into termite tunnels. They feed on a variety of arthropods, often consuming large numbers of termites. They appear to avoid insects with chemical defenses such as beetles and most ants. Three species, *Cnemidophorus arubensis, C. murinus,* and *Dicrodon guttulatum,* are herbivorous. *Aspidoscelis* is predominantly a temperate-zone, arid land taxon, whereas *Cnemidophorus* is primarily tropical occurring from Central America to northern Argentina. *Ameiva* and others are predominantly tropical and often abundant in tropical dry forest, cerrado, semiarid regions, or clearings and other open areas in tropical rain forest. The large-bodied *A. ameiva* and the smaller-bodied *C. lemniscatus* are abundant in Amazonian areas where disturbance of the forest provides open habitats that facilitate colonization. All teiines are oviparous, and clutch size is associated with lizard body size. *Aspidoscelis* and *Cnemidophorus* have clutches ranging from two to six eggs, but one species, *C. arubensis*, produces a single egg. *Ameiva* tend to be slightly larger and its average clutch size is slightly larger (four to seven eggs). *Dicrodon*, *Kentropyx*, and *Teius* are similar in body and clutch size with *Aspidoscelis* and *Cnemidophorus*. *Aspidoscelis, Cnemidophorus, Teius,* and *Kentropyx* contain unisexual and bisexual species, and nearly one-third of the 45± species of *Aspidoscelis* are parthenogenetic. The diversity of South American teiine lizards is just beginning to be appreciated, as new species are being described on a regular basis.

Tupinambinae

Sister taxon: Teiinae.

Content: Four genera, *Callopistes, Crocodilurus, Dracaena,* and *Tupinambis,* with 2, 1, 2, and 7 species, respectively.

Distribution: South America east of the Andes to central Argentina and Chile northward in interandean valleys.

Characteristics: The skull has the anteromedial edge of the supratemporal fenestra formed by the parietal, an unexpanded quadrate without a pterygoid articulation, and a maxillary without a nasal process. The retroarticular process of the mandible is smooth dorsally.

Biology: Tupinambines are a much less speciose group than the teiines, but overall they are more diverse in habits and habitat preference. They range in size from the smaller *Callopistes maculatus* (120–170 mm adult SVL) to the larger tupinambines, *Crocodilurus* (to 220 mm SVL), *Dracaena* (to 360 mm SVL) and *Tupinambis* (250–420 mm SVL). The seven tegu species (*Tupinambis*; Fig. 20.22) occur in a wide range of habitats from Amazon rain forest to grasslands and semiarid areas. Tegus are opportunistic omnivores as adults that feed on a combination of invertebrates, vertebrates and their eggs, and a variety of fruits. Even though *Tupinambis* are large and conspicuous lizards, new species continue to be discovered, and some described species, such as *T. teguixin*, likely represent several taxa. *Callopistes maculatus* lives in arid habitats of coastal and piedmont Chile. It preys largely on other lizards. *Crocodilurus amazonicus* is semiaquatic, living along edges of streams, lagoons, or lakes that are bordered by forest. *C. amazonicus* basks on trees above water and forages on riverbanks and in water. It feeds on a variety of arthropods and small vertebrates, including mollusks, crabs, and fish. When approached from land, it escapes into water by swimming in a serpentine fashion. *Dracaena guianensis* is also a semiaquatic resident of forest steams and lakes. It is caiman-like in appearance and spends more time in water than *C. amazonicus*. Its head is broad and heavily muscled, and its molariform teeth crush snails, its major food. Reproductive data are limited for these large lizards. Clutch size undoubtedly is associated with body size, although *D. guianensis* apparently has small clutches of 2 to 4 eggs. Tegus have large clutches, from 4 to 32 eggs for *T. teguixin*, and presumably the incubation period is moderately long, from 3 to 4 months. *Tupinambis* often deposit clutches of eggs inside termite nests, including arboreal nests of *Nasutitermes*. The lizards dig a cavity in the nest and deposit eggs; termites cover the opening, sealing the eggs in the nest. Parthenogenesis has not been reported in tupinambines.

Diploglossa

Cordylidae

Crag, Girdled, and Plated Lizards

Classification: Reptilia; Diapsida; Sauria; Lepidosauria; Squamata.

Sister taxon: Uncertain, possibly Scincidae.

Content: Two subfamilies, Cordylinae and Gerrhosaurinae (see Comment).

Distribution: Sub-Saharan Africa and Madagascar (Fig. 20.24).

Characteristics: Cordylids are small to moderately large lizards that range in adult SVL from 60 to 300 mm and are typically heavily armored (Fig. 20.25). The scales may abut or overlap and frequently are strongly keeled. Rectangular osteoderms underlie the scales dorsally and ventrally on the trunk; they are thicker and stronger dorsally. A longitudinal ventrolateral groove or fold separates the dorsal and ventral scale armor. All species have limbs, and the pectoral girdle has a T-shaped or cruciform interclavicle and either curved rodlike or angular clavicles. The tail is moderately short to long and autotomous. A fracture plane occurs anterior to the transverse processes of each caudal vertebra. The tongue bears filamentous papillae and dorsal lingual scales arranged in alternating rows. The posterior edges of the lingual scales are serrate, and the foretongue is nonretractable. The skull has paired nasals, postorbitals, and squamosals, and a single parietal and frontal (occasionally paired); there is no evidence of a parietal foramen. Attachment of marginal dentition is pleurodont, and pterygoid teeth are present or absent.

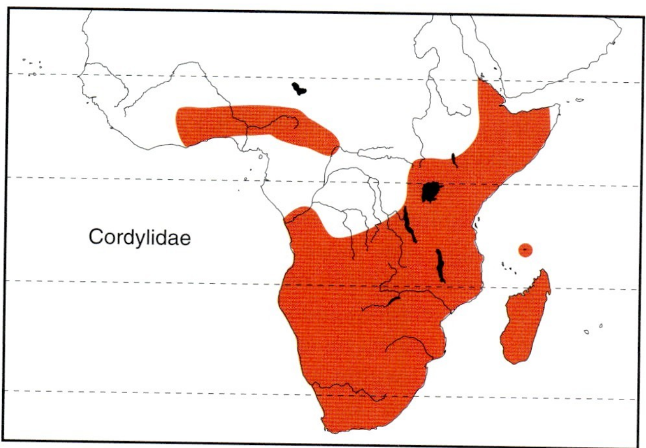

FIGURE 20.24 Geographic distribution of the extant Cordylidae.

Comment: Cordylid phylogeny is currently in a state of flux. The two subfamilies that we recognize have been repeatedly considered families or subfamilies during the last 17 years. The most recent molecular analysis once again considers them subfamilies.

Cordylinae

Sister taxon: Gerrhosaurinae.

Content: Three genera, *Chamaesaura*, *Cordylus*, and *Platysaurus*, with 3, 37, and 15 species, respectively.

Distribution: Sub-Saharan Africa.

Characteristics: The head has four parietal scales, and each nostril is enclosed in nasal or nasal and postnasal scales; cycloid scales are present on the throat. The skull has a slitlike supratemporal fossa on each side, the anteroventral border of the orbit formed by maxillary and jugal, and lacrimals not visible externally. The zygosphene and zygantra of opposing vertebrae form a strong articulation.

Biology: Cordylines, the girdled lizards, live in semiarid and arid habitats, and not unexpectedly, all are diurnal and heliophilic. *Cordylus* and *Platysaurus* (Fig. 20.25) are mainly rock dwellers that forage in surrounding grassland or scrub. They are agile climbers of rock surfaces and typically perch on rock crevices and enter crevices for escape. Their rough scaly bodies armor them from predator attacks in these crevices as well as when they are exposed. *Chamaesaura* is a clade of elongate, reduced-limb lizards. They live in grasslands and use undulatory locomotion, an especially effective locomotor pattern in thick grass. Although somewhat snakelike because of small limbs, elongate body, and tail, which may be two times body length, *Chamaesaura* retains the cordyline armored appearance. All cordylines are predominantly insectivores, although small vertebrates and plant material can be regular dietary items. Cordylines are viviparous, except for the oviparous *Platysaurus* species. The larger taxa typically give birth to litters of one to four neonates, and somewhat surprisingly, the smaller-bodied *Chamaesaura* commonly has four to nine young in a litter. *Platysaurus* produces clutches of only two eggs.

FIGURE 20.25 Representative cordylids. From left: Black-lined plated lizard *Gerrhosaurus nigrolineatus*, Gerrhosaurinae (L. J. Vitt); black spiny tail lizard *Cordylus niger*, Cordylinae (D. Bauwens).

Gerrhosaurinae

Sister taxon: Cordylinae.

Content: Five genera, *Cordylosaurus*, *Gerrhosaurus*, *Tetradactylus*, *Tracheloptychus*, and *Zonosaurus*, with 1, 7, 7, 2, and 17 species, respectively.

Distribution: Sub-Saharan Africa and Madagascar.

Characteristics: The head has two parietal scales, and each nostril is enclosed in three or four scales, including an infralabial scale; cycloid scales are lacking. The skull has lost the supratemporal fossae; the anteroventral border of the orbit is formed by the jugal, and the lacrimals are visible externally. The zygosphene and zygantra of opposing vertebrae do not articulate.

Biology: Gerrhosaurines, the plated lizards, are more diverse ecologically than the cordylines. Although predominantly residents of semiarid and arid habitats, some Madagascan taxa are forest residents, and *Z. maximus* is possibly semiaquatic. All are diurnal and most are heliothermic. *Gerrhosaurus* (Fig. 20.25) and *Cordylosaurus* are stout, scale-armored, mainly rock-dwelling lizards. *Tetradactylus* has a variety of body forms from a strong-limbed morphology to an elongate, reduced-limb body form, and in some species, forelimbs are lost and hindlimbs are tiny. The elongate taxa live in grasslands and use lateral undulation locomotion. *Gerrhosaurus skoogi* is a sand diver or sand swimmer, living largely in sand dune habitats. It is also an omnivore and regularly eats foliage. The Madagascan *Tracheloptychus* and *Zonosaurus* are less heavily armored and appear skinklike. They live in habitats from sand dunes to dry forest. Plated lizards are generally omnivores. Although insects and other arthropods are the major prey, plant matter is commonly eaten. Larger species often prey on small vertebrates. Gerrhosaurines are oviparous. Clutch size is small, varying from two to six eggs per clutch. Clutch size appears not to be associated with body size. The largest gerrhosaurine, *G. validus*, averages only four eggs (range, two to five) per clutch.

Scincidae

Skinks

Classification: Reptilia; Diapsida; Sauria; Lepidosauria; Squamata.

Sister taxon: Uncertain, likely Cordylidae.

Content: Two subfamilies, Acontinae and "Scincinae" (see Comment).

Distribution: Nearly worldwide (Fig. 20.26).

Characteristics: Skinks are small to large lizards (27-350 mm adult SVL). They are nearly always covered dorsally and ventrally by overlapping scales (Fig. 20.27). Osteoderms underlie the scales dorsally and ventrally on the trunk. Body form ranges from strong limbed to no external limbs; in strongly reduced limbed taxa, the interclavicle is absent or cruciform; the clavicles are angular. Tails are long to moderately long. Caudal autotomy is common but not universal in skinks; autotomous caudal vertebrae have a fracture plane anterior to the transverse processes. The tongue bears filamentous papillae and dorsal lingual scales arranged in alternating rows. The posterior edges of the lingual scales are serrate, and the foretongue is nonretractable. The skull has paired nasals and squamosals, either single or paired postorbitals and frontals, and a single fused parietal. A parietal foramen is present or absent, and when present, perforates the parietal. Attachment of marginal dentition is pleurodont, and pterygoid teeth are present or absent.

Comment: Although the four-subfamily taxonomy (Acontinae, Feylininae, Lygosominae, and Scincinae) has dominated the literature since it was proposed by Alan Greer in 1970 and appears in other herpetology textbooks (including our last edition) and virtually all popular literature and Web sites, recent molecular analyses indicate that feylinines are nested within a group of sub-Saharan scincines, lygosomines are nested within a global group of scincines, and acontines may also be nested within global scincines, although support for the latter is weak.

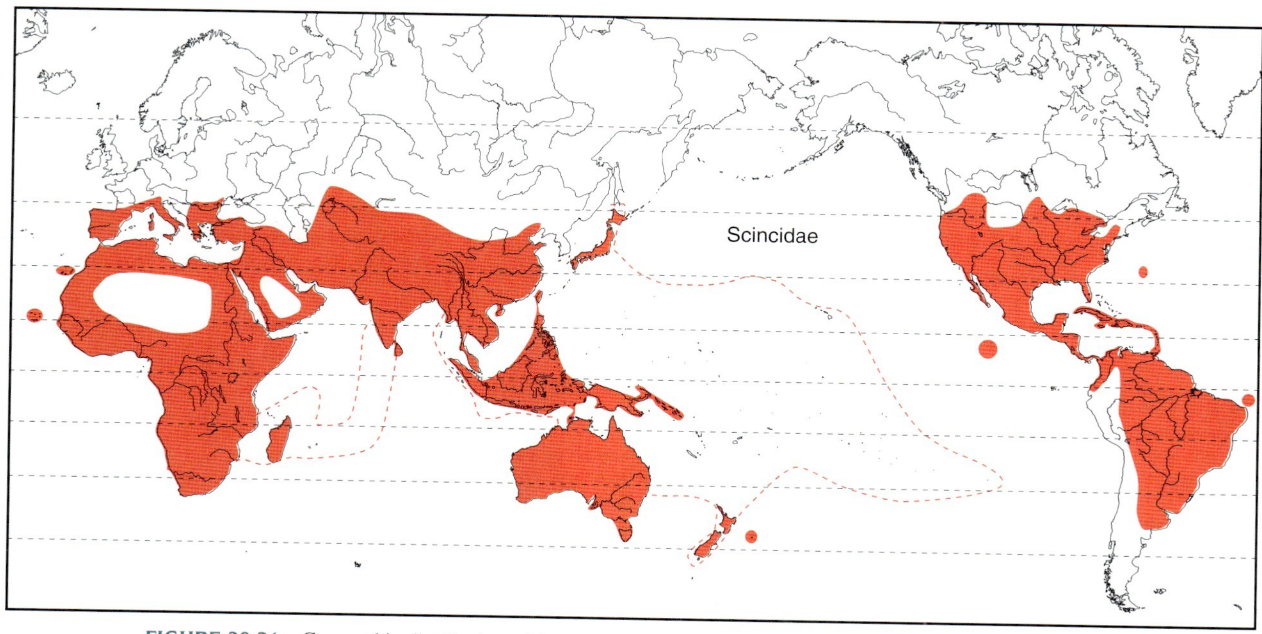

FIGURE 20.26 Geographic distribution of the extant Scincidae. Dashed red lines indicate envelope of oceanic islands containing skink species.

FIGURE 20.27 Representative skinks. Clockwise from upper left: Pygmy saw-tailed skink *Egernia depressa* (B. Maryan); black-spotted mabuya *Mabuya nigropunctata* (L. J. Vitt); eastern sand skink *Scincus mitrans* (R. D. Bartlett); Great Plains skink *Plestiodon obsoletus* (L. J. Vitt).

Scincidae, Acontinae, and Lygosominae appear to be monophyletic, but the proportion of "lygosomine" species used to support monophyly is so low that additional work will be necessary to sort out this large and complex group of skinks. Synonomies and subdivision of former genera have resulted in numerous taxonomic changes within skinks. Several large genera (e.g., *Mabuya* and *Scincella*) are highly complex and speciose and will undoubtedly be further dissected. Consequently, we consider only two subfamilies of skinks to be supported, the Acontinae and the Scincinae. We recognize that the "Scincinae" is a temporary fix for the evolving and complex problem of sorting out skink relationships and is, in itself, unresolved.

Acontinae

Sister taxon: Undetermined.

Content: Four genera, *Acontias*, *Acontophiops*, *Microacontias*, and *Typhlosaurus*, with 27, 1, 4, and 9 species, respectively.

Distribution: Southern Africa.

Characteristics: The skull has paired frontal bones, left and right palatines not touching medially on the palate, a complete supratemporal arch, and a small posttemporal fenestra. Limbs are absent.

Biology: Acontines are limbless, fossorial skinks. Unlike the feylines, these skinks retain large head scales and the general limbless anguimorph morphology. All are large skinks, ranging from about 110 mm in the smaller species to about 550 mm TL in *Acontias plumbeus*. They are predominantly arid land species, burrowing in sandy soils or living within bunchgrass. All are viviparous and produce litters of 1 to 4 neonates in the smaller species and 10 to 14 young in the larger species.

"Scincinae"

Sister taxon: Unresolved.

Content: One-hundred and thirty-three genera, *Ablepharus, Afroablepharus, Amphiglossus, Androngo, Anomalopus, Apterygodon, Asymblepharus, Ateuchosaurus, Barkudia, Bartleia, Bassiana, Brachymeles, Caledoniscincus, Calyptotis, Carlia, Cautula, Celatiscincus, Chabanaudia, Chalcides, Chalcidoseps, Chioninia, Coeranoscincus, Coggeria, Cophoscincopus, Corucia, Cryptoblepharus, Cryptoscincus, Ctenotus, Cyclodina, Cyclodomorphus, Dasia, Davewakeum, Egernia, Emoia, Eremiascincus, Eroticoscincus, Eugongylus, Eulamprus, Eumeces, Eumecia, Euprepes, Eurylepis, Eutropis, Feylinia, Fojia, Geomyersia, Geoscincus, Glaphyromorphus, Gnypetoscincus, Gongylomorphus, Gongylus, Graciliscincus, Haackgreerius, Hakaria, Hemiergis, Hemisphaeriodon, Isopachys, Janetaescincus, Kanakysaurus, Lacertaspis, Lacertoides, Lamprolepis, Lampropholis, Lankascincus, Larutia, Leiolopisma, Leptoseps, Leptosiaphos, Lerista, Lioscincus, Lipinia, Lobulia, Lygisaurus, Lygosoma, Mabuya, Macroscincus, Marmorosphax, Melanoseps, Menetia, Mesoscincus, Mochlus, Morethia, Nangura, Nannoscincus, Nessia, Niveoscincus, Notoscincus, Novoeumeces, Oligosoma, Ophiomorus, Ophioscincus, Pamelaescincus, Panaspis, Papuascincus, Paracontias, Paralipinia, Parvoscincus, Phoboscincus, Plestiodon, Prasinohaema, Proablepharus, Proscelotes, Pseudoacontias, Pseudemoia, Pygomeles, Riopa, Ristella, Saiphos, Saproscincus, Scelotes, Scincella, Scincopus, Scincus, Scolecoseps, Sepsina, Sepsophis, Sigaloseps, Simiscincus, Sirenoscincus, Sphenomorphus, Sphenops, Tachygyia, Tiliqua, Trachylepis, Tribolonotus, Tropidophorus, Tropidoscincus, Typhlacontias, Vietnascincus,* and *Voeltzkowia*, with about 1200 species.

Distribution: Nearly worldwide but not extending much above 60° N latitude, and absent from Antarctica.

Characteristics: The skull has single or paired frontal bones, left and right palatines either touching or separated medially on the palate, a complete supratemporal arch, and a post-temporal fenestra. Limbs are usually present, although limb reduction has evolved independently many times.

Biology: This is a highly diverse group, taxonomically, ecologically, behaviorally, and in terms of reproductive diversity. They can be found in tropical rain forests, seasonal savannas, arid deserts, and coniferous forests. They live in and on the ground, in shrubs, on tree trunks, on rocks, and along margin of watercourses. On oceanic islands, some species occur at remarkable densities. They range in size from 27 mm adult SVL in some of the smaller species (e.g., *Menetia greyi*) to 350 mm SVL in large species such as *Corucia zebrata*. Most have a cylindrical body and tail, a more or less conical head, well-developed moderately short limbs, and shiny, smooth scales (Fig. 20.27). Others are short and robust, with heavily keeled scales. Most species are diurnal, but some are nocturnal and many are crepuscular. Some are highly active whereas others are slow and sluggish. Behavioral diversity is great, with many species appearing to be nonterritorial whereas others are strongly territorial. Tail autotomy occurs in most species, and regeneration is often rapid and nearly complete. Foraging behavior and diets also vary considerably. Many appear to be active foragers, searching for prey nearly continuously while active. Others use a sit-and-wait strategy to find and capture prey. Although most feed on a diversity of arthropods, small mollusks, and other invertebrates, larger carnivorous species frequently feed on small vertebrates, including other lizards. Some species are herbivorous, and many carnivorous species occasionally eat fruits. Reproductively, these skinks are diverse as well. Most species are oviparous, depositing from 1 to as many as 18 or more eggs. Although most abandon their clutches, some attend or guard eggs until they hatch. Many species deposit their clutches in isolation whereas others have communal nesting, with several females depositing clutches in the same location. Many of these skinks are viviparous, and the degree of matrotrophy varies from none to the most extreme known in reptiles (see Chapter 4). Long-term pair bonds and long-term parental care are known in a few species.

Anguimorpha

Anguidae

Alligator Lizards, Galliwasps, Glass Lizards, and Allies

Classification: Reptilia; Diapsida; Sauria; Lepidosauria; Squamata.

Sister taxon: Uncertain, possibly clade containing Xenosauridae and Varanoidea.

Content: Four subfamilies, Anguinae, Anniellinae, Diploglossinae, and Gerrhonotinae.

Distribution: Disjunct, Americas, Europe, Southwest Asia, and southern Asia (Fig. 20.28).

Characteristics: Anguids are small (55–70 mm adult SVL, *Elgaria parva*) to very large (500–520 mm SVL and 1.4 m maximum TL, *Ophisaurus apodus*) limbed to limbless lizards. All are heavily armored with largely nonoverlapping scales. Osteoderms underlie these scales dorsally and ventrally on the trunk, and a longitudinal ventrolateral groove or fold separates this dorsal and ventral scale armor in some taxa. The fold allows body expansion for breathing, feeding, and reproduction yet maintains the shield effect of the scale armor. Body form ranges from strong limbed to no external limbs; in strongly reduced-limbed taxa, the interclavicle is absent or cruciform; the clavicles are angular. Tails are short to very long. Caudal autotomy is common but not universal among anguids; autotomous caudal vertebrae have a fracture plane anterior to the transverse processes. The tongue bears filamentous papillae and lacks lingual scales. The foretongue retracts into the hind tongue. The skull possesses paired nasals and postorbitals, present or absent paired squamosals, single or paired frontals, and a single (fused) parietal. A parietal foramen is present and perforates the parietal. Attachment of the marginal dentition is pleurodont, and pterygoid teeth are present or absent.

Anguinae

Sister taxon: Uncertain, possibly clade containing Diploglossinae and Gerrhonotinae.

Content: Three genera, *Anguis, Ophisaurus*, and *Pseudopus*, with 2, 13, and 1 species, respectively.

Distribution: North America and Eurasia.

Characteristics: Anguines are robust, elongate, limbless lizards. The tail is long, typically twice the length of the body. A ventrolateral fold is well developed in *Ophisaurus* and *Pseudopus* (Fig. 20.29) but indistinct in *Anguis*. The skull possesses paired frontals, pterygoid teeth, more than 15 teeth on the dentaries, and unicuspid posterior teeth. The frontoparietal scales are very small and widely separated.

Biology: *Anguis fragilis,* the slowworm, is a moderately abundant resident of scrub and open habitats with dense ground coverage. Slowworms are largely diurnal, occasionally basking. They can be easily found by turning

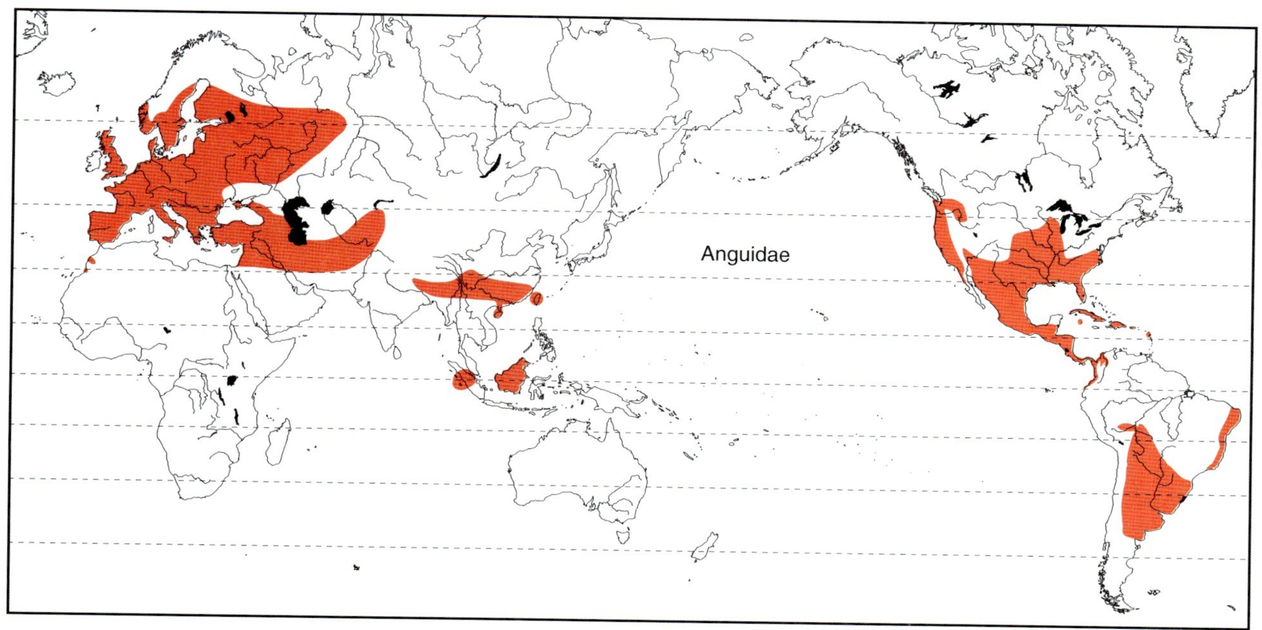

FIGURE 20.28 Geographic distribution of the extant Anguidae.

FIGURE 20.29 Representative anguid lizards. Clockwise from upper left: California legless lizard *Anniella pulchra*, Anniellinae (L. J. Vitt); eastern glass lizard *Ophisaurus ventralis*, Anguinae (L. J. Vitt); Bocourt's arboreal alligator lizard *Abronia vasconcelosii*, Gerrhonotinae (J. A. Campbell); banded galliwasp *Diploglossus fasciatus*, Diploglossinae (O. A. V. Marques).

rocks exposed to sun, under which they subsurface bask to gain heat. They are most often observed on the surface slowly searching for snails and slugs. They also eat arthropods and small vertebrates. They are generally active from early spring to late fall with mating occurring in late spring. Slowworms are viviparous and have a gestation period that lasts 8 to 12 weeks. Litters of 4 to 28 but usually less than 12 young are born in late August and early September. If mating occurs late or if a cool summer occurs, females retain embryos over winter. The other two genera are typically referred to as *glass lizards* because their long tails not only can autotomize, but at least in *Ophisaurus*, the tail often breaks into several pieces. *Ophisaurus* is more speciose with a broader size range (150–300 mm adult SVL) and lives in a greater variety of habitats. These are usually found in open habitats with heavy ground cover. They can often be observed along the edge of forest patches in the southern United States in morning or late afternoon. *Ophisaurus* preys more heavily on arthropods, although it eats a broad array of small semifossorial and terrestrial animals. *Pseudopus* is much larger than other glass lizards, reaching 400 mm SVL. In contrast to *Anguis*, *Ophisaurus* and *Pseudophus* are oviparous and deposit clutches of 4 to 20 eggs; females appear to remain with the eggs during the 8–10-week incubation period.

Anniellinae

Sister taxon: Uncertain, possibly Anguinae.

Content: One genus, *Anniella*, with 2 species.

Distribution: California and Baja, California.

Characteristics: Anniellines are small, elongate, limbless lizards. The tail is short, less than two-thirds body length. A ventrolateral fold is absent. The skull possesses paired frontals, pterygoid teeth, fewer than 15 teeth on the dentaries, and unicuspid posterior teeth. The frontoparietal scales are small and separated.

Biology: *Anniella* are 150 to 165 mm in adult SVL and have a snakelike morphology, except that their tails comprise a larger portion of their total length than that of most snakes (Fig. 2.29). They inhabit coastal sand dunes and valleys from sea level to 1600 m elevation and are largely confined to friable soils that retain some moisture. They obtain their highest abundance in sandy soils with moderate plant cover. *Anniella gerominensis* tends to be more restricted to coastal dunes than *A. pulchra*, and in areas where they occur together, *A. pulchra* is much more common. These lizards spend much of their time underground, emerging on the surface early in the morning, apparently for short time periods. While underground, they are usually at the base of plants, and it is likely that some if not a considerable amount of foraging takes place under leaf litter and other debris associated with shrubs. They are susceptible to desiccation and able to "drink" interstitial water from the soil when soil moisture content exceeds 7%. They eat a broad variety of small arthropods. Mating occurs in spring or early summer, and live offspring are born in late summer and early fall. Litter size is usually two and the offspring are large.

Diploglossinae

Sister taxon: Gerrhonotinae.

Content: Three genera, *Celestus*, *Diploglossus*, and *Ophiodes*, with 28, 18, and 4 species, respectively.

Distribution: West Indies, Central America, and central South America.

Characteristics: Diploglossines are elongate lizards that generally have small but well-developed limbs and long, easily autotomized tails (Fig. 2.29). The limbs are greatly reduced in some taxa. For example, the South American *Ophiodes* are similar ecologically and morphologically to *Ophisaurus* (Anguinae). A ventrolateral fold is generally lacking. The skull has paired frontals, no pterygoid teeth, more than 15 teeth on the dentaries, and bicuspid posterior teeth. The frontoparietal scales are small and separated.

Biology: Galliwasps contain some of the smallest anguid taxa (60 mm adult SVL, *Celestus macrotus*) and some large taxa (280 mm SVL, *Diploglossus anelpistus*). Depending on species, diploglossines can be terrestrial or fossorial. Most live in forested habitats, although some live in more arid grassland or scrub habitats. Most activity occurs during the day, but some have been observed at dawn, dusk, and after dark, indicating that at least some activity may take place at night. All studied species prey mainly on arthropods and other invertebrates. Both oviparity (some *Diploglossus*) and viviparity (*Celestus* and some species of *Diploglossus*) occur. Clutch size or number of offspring is correlated with body size. Small species such as *D. delasagra* lay 2 eggs, others such as *C. curtissi* give birth to 2–5 neonates, and larger species such as *D. warreni* bear 8 to 27 neonates. Some, and possibly all, oviparous species attend their eggs until they hatch.

Gerrhonotinae

Sister taxon: Diploglossinae.

Content: Six genera, *Abronia*, *Barisia*, *Coloptychon*, *Elgaria*, *Gerrhonotus,* and *Mesaspis*, with 27, 4, 1, 6, 4, and 6 species, respectively.

Distribution: Northwestern United States southward to western Panama.

Characteristics: Gerrhonotines are stout-bodied lizards with short, well-developed limbs (Fig. 2.29). The tail is usually longer than the body length. A well-developed ventrolateral fold is present on the lateral body surfaces. The skull has a frontal, pterygoid teeth, more than 15 teeth on the dentaries, and bicuspid posterior teeth. The frontoparietal scales are nearly or barely touching on the dorsal midline.

Biology: Alligator lizards derive their name from heavy armoring on the head, body, and tail, and strong broad jaws. None is aquatic, although some occur in moist habitats from tropical upland forests to coastal and montane forests of western North America. In the Pacific Northwest of the United States and southern Canada, *E. coerulea* bask along rock crevices exposed to afternoon sun during sunny days in winter. A few species live in oak savannas and deserts. The tropical genus *Abronia* is the most speciose of the gerrhonotines and is arboreal, even possessing a prehensile tail. The gerrhonotines are mostly moderate-sized lizards, less than 110 mm adult SVL, although *G. liocephalus* attains 200 mm SVL. They are carnivorous, feeding mainly on arthropods, other invertebrates, and small vertebrates. Oviparous and viviparous species appear to mate in spring and produce eggs or offspring in late summer or early fall. Some oviparous species may produce more than a single clutch in a season. Oviparous species usually deposit 2–40 eggs, and viviparous species bear 2–15 young. Incubation of eggs normally requires at least 8 to 10 weeks, and gestation of live-born young takes 8 to 12 weeks. Some, and possibly all, oviparous species attend their eggs until they hatch, and some (e.g., *E. multicarinata*) may share egg-laying sites.

Xenosauridae

Knob-Scaled Lizards

Classification: Reptilia; Diapsida; Sauria; Lepidosauria; Squamata.

Sister taxon: Uncertain, two main hypotheses, Anguidae or the varanoid–thecoglossan clade.

Content: Two genera, *Shinisaurus* and *Xenosaurus*, with 1 and 6 species, respectively (see Comment).

Distribution: Disjunct, southern China and eastern Mexico into Guatemala (Fig. 20.30).

Characteristics: Xenosaurids are moderate-sized lizards (100–150 mm adult SVL). They are covered dorsally and ventrally by granular, juxtaposed scales and large keeled tubercles. Ventrally, the trunk contains small, nonarticulate osteoderms, but none is present dorsally.

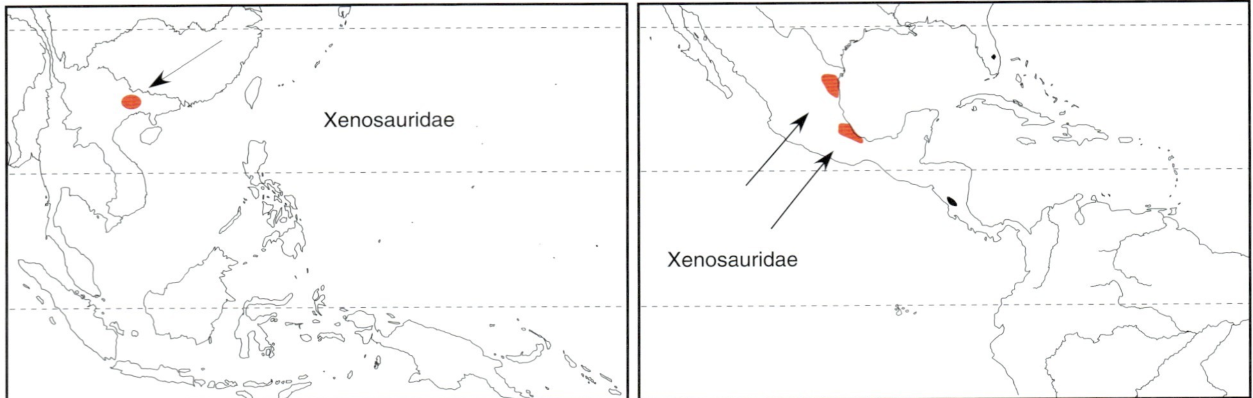

FIGURE 20.30 Geographic distribution of the extant Xenosauridae.

Limbs are well developed. The pectoral girdle has a T-shaped or cruciform interclavicle and angular clavicles. The tail is about 1.2 times body length. Caudal autotomy is present or absent; when present, a fracture plane occurs anterior to the transverse processes of each caudal vertebra. The tongue bears filamentous papillae and lacks lingual scales. The foretongue retracts into the hind tongue. The skull has paired nasals, postorbitals and squamosals, single or paired frontals, and a parietal. A parietal foramen is present and perforates the frontoparietal suture. Attachment of marginal dentition is pleurodont, and pterygoid teeth are present or absent.

Biology: Both Asian and American taxa appear stenohydric, generally requiring moist surroundings and losing water rapidly in dry conditions. The Asian *Shinisaurus crocodilurus* occurs in moist montane forests along streams. It is semiaquatic and during the day forages in mountain streams for fish, tadpoles, and other animal prey. At night, the lizards rest on branches overhanging water. Knob-scaled lizards were previously reported to be nocturnal, but field observations in China and Mexico show them to be diurnal. The American *Xenosaurus* (Fig. 20.31) contains terrestrial species that occur in moist cloud to dry scrub forest, most commonly associated with rock outcrops where they live in narrow crevices. Compared with *Shinisaurus, Xenosaurus* are dorsoventrally flattened and can be difficult to remove from crevices. Individuals defend their crevices from other individuals. *Xenosaurus* preys mainly on arthropods, particularly beetles, grasshoppers, and crickets, although they occasionally feed on other prey as well. Xenosaurids are live bearers. *S. crocodilurus* may produce litters of 15 neonates, usually many fewer, and has a gestation period of 8 to 14 months. *Xenosaurus* bears litters of two to eight young, most often two, and gestation requires 11 to 12 months. Postnatal parental care possibly occurs in *X. newmanorum*.

Comment: *Shinisaurus* and *Xenosaurus* are considered in separate subfamilies by some authors and clearly have been separated evolutionarily for a long time.

FIGURE 20.31 Representative extant xenosaurids. From left: Chinese crocodile lizard *Shinisaurus crocodilurus* (L. W. Porras); flathead knob-scaled lizard *Xenosaurus platyceps*, Xenosauridae (L. J. Vitt).

Varanoidea

Helodermatidae

Gila Monster and Mexican Beaded Lizard

Classification: Reptilia; Diapsida; Sauria; Lepidosauria; Squamata.

Sister taxon: Thecoglossa or Varanidae.

Content: One genus, *Heloderma*, with 2 species.

Distribution: Southwestern North America, from the Sonoran Desert southward along the Mexican Pacific coast to Guatemala (Fig. 20.32).

Characteristics: Helodermatids are large lizards (300-500 mm adult SVL). They are the only lizards with well-developed venom glands. They have broad, somewhat flattened heads, robust bodies, short well-developed limbs, and heavy tails. They have a thick skin with rows of rounded scales circling the body, giving them a beaded appearance (Fig. 20.33). Scales are somewhat tuberculated dorsally and laterally and are slightly larger and squarish ventrally. Ventrally the trunk contains small, nonarticulate osteoderms, but none is present dorsally. The pectoral girdle has a T-shaped interclavicle and angular clavicles. The tail is moderately short, about two-thirds body length. Caudal autotomy does not occur. The tongue bears filamentous papillae and lacks lingual scales. The foretongue retracts into the hind tongue. The skull has paired nasals, frontals, and squamosals, no postorbitals, and a parietal. A parietal foramen is absent. Attachment of the marginal dentition is pleurodont, and pterygoid teeth occur.

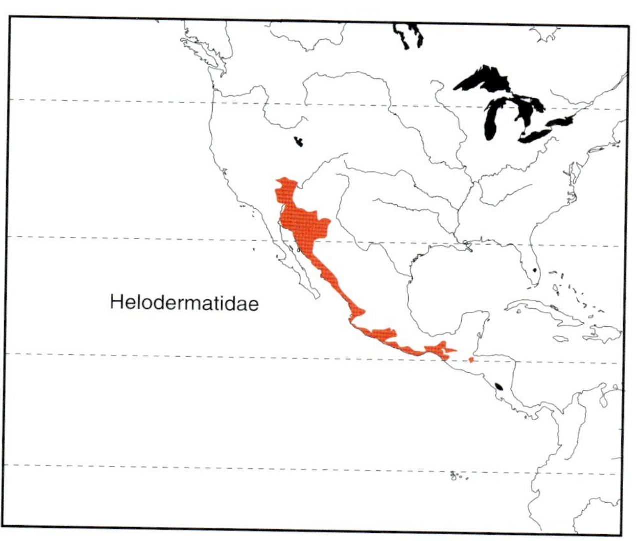

FIGURE 20.32 Geographic distribution of the extant Helodermatidae.

Biology: The two species of *Heloderma* are active during much of the year, and daily activity patterns vary with season. During spring and fall, activity occurs mostly in late morning and late afternoon, with a shift toward earlier and later activity during summer, when some individuals can be active at night. Surface (and thus observable) activity appears associated with finding mates and food, but there is also a tendency for increased surface activity independent from mate searching and food associated with wet periods. When not foraging or mate searching, *Heloderma* seek refuge in underground burrows and similar retreats. Diets of both species are highly varied, consisting of a variety of mammals, birds, reptiles, and their eggs, and some invertebrates. *H. horridum* will climb trees and other shrubs to take bird eggs and nestlings, and both species will dig nests of reptiles to feed on eggs. They can consume huge prey. For example, one *H. suspectum* was observed to swallow a juvenile rabbit that weighed 33% as much as the lizard. Unlike snakes, which have no pectoral girdle, *Heloderma* must force whatever they swallow through the pectoral girdle. Nevertheless, they eat some remarkably large prey items, which they swallow whole. Home ranges average 21–58 hectares, depending on species and locality, and individuals can easily move more than 1.5 km per day when active on the surface. *Heloderma* likely produce clutches every other year. Mating occurs in spring in *H. suspectum*. An average of

FIGURE 20.33 Representative extant Helodermatids. From left: Gila monster *Heloderma suspectum*, Helodermatidae (L. J. Vitt); Mexican beaded lizard *Heloderma horridum* (C. Schwalbe).

about 6 (2–12) eggs are laid in mid-July to mid-August. It is unclear whether eggs overwinter, but hatchings have been observed to emerge from nesting sites in May, suggesting that either eggs overwintered or they hatched in fall and hatchlings remained in the nest until the following year.

Thecoglossa

Varanidae

Monitors, Goannas, and Earless Monitors

Classification: Reptilia; Diapsida; Sauria; Lepidosauria; Squamata.

Sister taxon: Helodermatidae or clade containing Serpentes and extinct thecoglossans.

Content: Two subfamilies, Lanthanotinae and Varaninae.

Distribution: Warm temperate and tropical Africa, Asia, and Australia (Fig. 20.34).

Characteristics: Varanids are generally large lizards. They have thick skin with numerous rows of small, rounded scales circling the body. The ventral scales are slightly larger than dorsal scales. Dorsally the trunk lacks osteoderms; ventrally small, nonarticulate osteoderms are present in some species. Monitors have well-developed limbs. The pectoral girdle has a T-shaped or cruciform interclavicle and angular clavicles. The tail is long to very long and lacks caudal autotomy. The tongue bears filamentous papillae and lacks lingual scales. The foretongue retracts into the hind tongue. The skull has paired frontals and squamosals, no postorbitals, and a nasal and parietal. A parietal foramen is usually present and perforates the parietal. Attachment of marginal dentition is pleurodont, and pterygoid teeth are present or absent.

Lanthanotinae

Sister taxon: Varaninae.

Content: Monotypic, *Lanthanotus borneensis*.

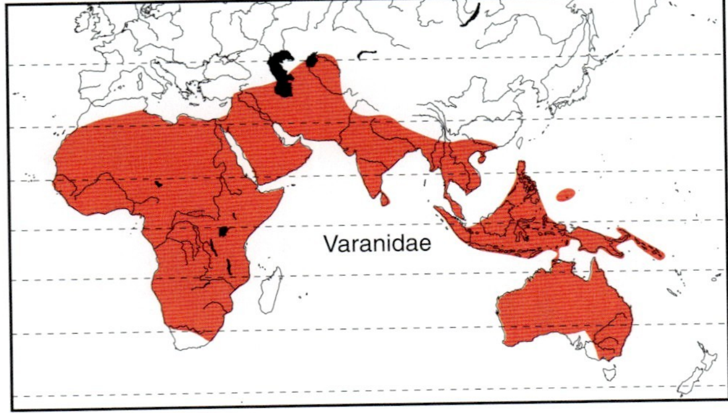

FIGURE 20.34 Geographic distribution of the extant Varanidae.

Distribution: Borneo.

Characteristics: *Lanthanotus borneensis*, the earless monitor, has pterygoid teeth and lacks a parietal eye and a hemibaculum.

Biology: *Lanthanotus borneensis* is poorly studied, partially owing to its preferred habitat and habits (Fig. 20.35). Most information derives from captive individuals. Adults (309–440 mm TL) appear to be restricted to forested habitats. Presumably, adults and juveniles are semiaquatic and live in or adjacent to forest streams and swamps. They forage at night, on land and in water; they eat invertebrates and small vertebrates. During the day, they rest in burrows that can be partially flooded. They are oviparous and produce small clutches (two to six eggs).

Varaninae

Sister taxon: Lanthanotinae.

Content: One genus, *Varanus*, with 63 species.

Distribution: Sub-Saharan Africa eastward through Asia to Australia and islands in the southwestern Pacific.

Characteristics: Varanines have no pterygoid teeth, have a parietal eye, and have a hemibaculum (i.e., a cartilaginous strut in each hemipenis).

Biology: Monitors are distinct lizards with relatively small heads and long necks, long and robust bodies, well-developed limbs, and long, muscular tails (Fig. 20.35). They range in size from the pygmy goanna *V. brevicauda* at a maximum 120 mm adult SVL (230 mm TL) to the largest known lizard, *V. komodensis* (3.1 m maximum TL), weighing more than 200 kg. Almost all monitors are active predators and have strong jaws and sharp, conically recurved teeth. They are alert, active lizards and, partially as a result of the large size of some species (e.g., *V. komodoensis*), have captured the imagination of humans, likely forming the substance of most dragon myths. The smaller species prey mainly on arthropods and small vertebrates. They catch live prey but also are scavengers. With increasing body size, prey shift to larger vertebrates, including mammals, birds, and reptiles and their eggs. Komodo dragons, for example, may have preyed on pygmy elephants before they were driven to extinction by man. The Philippine butaan (*V. olivaceus*) seasonally eats fruit. Most species are terrestrial to semiarboreal, although a few species (e.g., *V. doreanus*) are strongly arboreal. The Australian bulliwallah (*V. mertensi*) is seldom more than a meter from water and commonly feeds and escapes into water. All varanids are oviparous, and none shows evidence of parental care. One species (*V. komodensis*) may be facultatively parthenogenetic, at least in captivity (see Chapter 4). Clutch size is generally associated with body size. The smaller Australian species have 2 to 4 eggs in a clutch, and the larger species, such as *V. bengalensis*, deposit 5 to 42 eggs, although clutches of the largest monitor (*V. komodoensis*)

FIGURE 20.35 Representative extant varanids. From left: Earless monitor *Lanthanotus borneensis*, Lanthanotinae (L. W. Porras); Gould's goanna *Varanus gouldii*, Varaninae (E. R. Pianka).

average only 16 eggs (range 2–30). Eggs are typically buried and have a moderately long incubation, seldom less than 100 days to nearly 1 year.

Comment: *Varanus* contains nine morphologically distinct subgroups that contain one or more species. These subgroups have formal subgeneric names that are occasionally used as generic names. An area cladogram recognizes two African clades, two Asian clades, and one Australian clade.

QUESTIONS

1. Why is a snake a lizard?
2. Loss of limbs has arisen independently in which clades of squamate reptiles?
3. Where does the Tuatara live and how does it differ from squamates, morphologically, behaviorally, and ecologically?
4. Alternative hypotheses exist for the relationships of major clades of lizards and recent nuclear gene data suggest that the taxonomy that we use in this book may change in the near future. What are these differing relationships and what kinds of evidence support each one?

REFERENCES

Overview—Sphenodontida

Cogger and Zweifel, 1998; Daughtery et al., 1990; Evans et al., 2001; Gauthier et al., 1988a; Gill and Whitaker, 1996; Robb, 1986.

Taxonomic Account

Sphenodontidae

Tuataras

Castanet et al., 1988; Newman, 1987, 1988; Newman and McFannden, 1990; Newman and Watson, 1985; Newman et al., 1994; Tyrrell, 2001.

Overview—Squamata

Ananjeva et al., 1988; Camp, 1923; Cogger and Zweifel, 1998; Donnellan et al., 1999; Estes et al., 1988; Frost and Etheridge, 1989; Fry et al., 2005; Gao and Norell, 1998; Gauthier et al., 1988a; Hallermann, 1994, 1998; Harris et al., 2001; Kluge, 1987, 1989; Lee, 1998, 2005; Macey et al., 1997; Pianka and Vitt, 2003; Rieppel, 1994; Russell, 1988; Schwenk, 1988, 1994b; Townsend et al., 2004; Vidal and Hedges, 2002a, 2002b, 2005; Williams, 1988; Wu et al., 1996.

Taxonomic Accounts

Agamidae

Angleheads, Calotes, Dragon Lizards, and Allies

Estes et al., 1988; Frost and Etheridge, 1989; Hallermann, 1998; Joger, 1991; Macey et al., 1997; McGuire and Heang, 2001; Schwenk, 1988; Witten, 1993.

Agaminae

Böhme, 1981; Borsuk-Bailynicka and Moody, 1984; Cox et al., 1998; Daniel, 1983; Fitch, 1970; Greer, 1989; Pianka, 1986; Witten, 1993; McGuire and Heang, 2001; Wilms and Bohme, 2000.

"Leiolepidinae"

Cox et al., 1998; Daniel, 1983; Geniez et al., 2000, 2004; Peters, 1971; Pianka and Vitt, 2003.

Chamaeleonidae

Chameleons

Bauer, 1997; Böhme, 1981; Branch, 1988; Burrage, 1973; Cox et al., 1998; Daniel, 1983; Estes et al., 1988; Fitch, 1970; Glaw and Vences, 1994; Greer, 1989; Hallermann, 1998; Hofman et al., 1991; Klaver and Böhme, 1986, 1997; Martin, 1992; Müller et al., 2004; Necas, 2004; Necas and Schmidt, 2004; Pianka and Vitt, 2003; Raxworthy and Nussbaum, 1995; Raxworthy et al., 2002; Rieppel and Crumly, 1997; Schwenk, 1988; Tolley et al., 2004; Witten, 1993.

Iguanidae

Anoles, Iguanas, and Allies

Estes et al., 1988; Frost and Etheridge, 1989; Hallermann, 1994, 1998; Macey et al., 1997; Schwartz and Henderson, 1991; Schwenk, 1988, 1994b; Williams, 1988.

Corytophaninae

Frost and Etheridge, 1989; Lang, 1989a; McCoy, 1968; Pianka and Vitt, 2003; Van Devender, 1982.

Crotaphytinae

Frost and Etheridge, 1989; McGuire, 1996: Pianka and Vitt, 2003.

Hoplocercinae

Avila-Pires, 1995; Duellman, 1978; Frost and Etheridge, 1989; Vitt and de la Torre, 1996.

Iguaninae

Alberts 2004; Burghardt and Rand, 1982; Frost and Etheridge, 1989; Frost et al., 2001a; Grismer, 2002; Norell and de Queiroz, 1991; Pianka and Vitt, 2003; Schulte et al., 2003; Wiens and Hollingsworth, 2000; Wiewandt, 1982.

Oplurinae

Blanc, 1977; Frost and Etheridge, 1989; Glaw and Vences, 1994; Pianka and Vitt, 2003; Titus and Frost, 1996.

Phrynosomatinae

Dunham et al., 1988; Fitch, 1970; Frost and Etheridge, 1989; Grismer, 2002; Leache and McGuire, 2006; Pianka and Vitt, 2003; Reeder, 1995; Reeder and Montanucci, 2001; Reeder and Wiens, 1996; Sherbrooke, 1981; Sites et al., 1992; Wiens and Reeder, 1997.

Polychrotinae

Avila-Pires, 1995; Etheridge and Williams, 1991; Frost and Etheridge, 1989; Frost et al., 2001; Guyer and Savage, 1987; Irschick et al., 1997; Jackman et al., 1997b; Losos, 1994; Nicholson, 2002; Pianka and Vitt, 2003; Poe, 1998, 2004; Roughgarden, 1995; Savage and Guyer, 1989; Schoener, 1974; Vitt and Lacher, 1981; Vitt et al., 1995, 1996; Williams, 1983.

Tropidurinae

Avila-Pires, 1995; Espinoza et al., 2004; Etheridge, 1995; Etheridge and Espinoza, 2000; Frost, 1992; Frost and Etheridge, 1989; Frost et al., 2001b; Jaksíc and Schwenk, 1983; Pianka and Vitt, 2003; Schulte et al., 2000; Torres, 2007; Torres et al., 2006; Vitt, 1991; Vitt and Goldberg, 1983; Vitt and Zani, 1996d; Vitt et al., 1997.

Gekkonidae

Geckos and Pygopods

Bauer, 1994; Bauer and Lamb, 2005; Bauer et al., 1997; Donnellan et al., 1999; Estes et al., 1988; Hallermann, 1998; Kluge, 1987, 1991; Rösler, 1995; Russell et al., 1997; Schwenk, 1988.

Diplodactylinae

Bauer, 1990; Donnellan et al., 1999; Gill and Whitaker, 1996; Greer, 1989; King and Horner, 1993; Pianka and Vitt, 2003.

Eublepharinae

Dial and Grismer, 1992, 1994; Dial and Schwenk, 1996; Grismer, 1988; Grismer et al., 1994; Inger and Lian, 1996; Ota et al., 1999; Pianka and Vitt, 2003.

Gekkoninae

Bauer and Lamb, 2005; Caranza and Arnold, 2006; Donnellan et al., 1999; Good et al., 1997; Greer, 1989; Ineich, 1992; King and Horner, 1993; Pianka and Vitt, 2003; Schwartz and Henderson, 1991; Vitt and Zani, 1997.

Pygopodinae

Donnellan et al., 1999; Jennings et al., 2003; Greer, 1989; Kluge, 1974, 1976; Patchell and Shine, 1986a,b; Pianka and Vitt, 2003; Saint et al., 1998; Shea, 1993.

Dibamidae

Blind Skinks

Estes et al., 1988; Greer, 1985; Hallermann, 1998; Schwenk, 1988.

Amphisbaenidae

Worm Lizards

Broadley, 1997b; Colli and Zamboni, 1999; Estes et al., 1988; Gans, 1974, 1978; Gans and Kraklau, 1989; Hallermann, 1998; Kearney, 2003; Macey et al., 2004; Pianka and Vitt, 2003; Schwenk, 1988.

Bipedidae

Mole-Limbed Worm Lizards

Estes et al., 1988; Gans, 1978; Grismer, 2002; Hallermann, 1998; Kearney, 2003; Macey et al., 2004; Papenfuss, 1982; Saint et al., 1998; Schwenk, 1988.

Rhineuridae

Florida Worm Lizard

Estes et al., 1988; Gans, 1978; Hallermann, 1998; Schwenk, 1988.

Trogonophidae

Spade-Headed Worm Lizards

Estes et al., 1988; Gans, 1974, 1978; Hallermann, 1998; Kearney, 2003; Schwenk, 1988.

Xantusiidae

Night Lizards

Bezy, 1988, 1989; Bezy and Camarillo, 2002; Crother and Presch, 1994; Estes et al., 1988; Fellers and Drost, 1991; Grismer, 2002; Hallermann, 1998; Hedges et al., 1991; Ramírez-Bautista et al., 2008; Schwartz and Henderson, 1991; Schwenk, 1988; Vicario et al., 2003; Zweifel and Lowe, 1966.

Lacertidae

Wall Lizards, Rock Lizards, and Allies

Arnold, 1989, 1993, 1998, 2000; Braña and Bea, 1987; Darevskii, 1978; Englemann et al., 1986; Estes et al., 1988; Fu, 1998; Hallermann, 1998; Mayer and Benyr, 1994; Mayer and Bishoff, 1996; Peréz-Mellado et al., 2004; Schwenk, 1988.

Gymnophthalmidae

Gymnophthalmids

Avila-Pires, 1995; Castoe et al., 2004; Cole et al., 1990; Doan, 2003; Doan and Castoe, 2005; Duellman, 1978; Estes et al., 1988; Hallermann, 1998; Kizirian, 1996; Pellegrino et al., 2001; Schwenk, 1988; Vitt and Avila-Pires, 1998; Vitt and de la Torre, 1996; Vitt et al., 1998b, 2007.

Teiidae

Whiptail Lizards, Tegus, and Allies

Colli et al., 2002, 2003; Denton and O'Neill, 1995; Estes et al., 1988; Fitzgerald et al., 1999; Giugliano et al., 2007; Hallermann, 1998; Presch, 1974, 1988; Reeder et al., 2002; Schwenk, 1988.

Teiinae

Avila et al., 1992; Avila-Pires, 1995; Cole et al., 1995; Colli et al., 2002, 2003; Dias et al., 2002; Reeder et al., 2002; Rocha et al., 2000; Sartorius et al., 1999; Schwartz and Henderson, 1991; Vitt and Breitenbach, 1993; Vitt and de la Torre, 1996; Vitt and Zani, 1996c; Vitt et al., 1995a, 2001; Wright, 1993; Wright and Vitt, 1993.

Tupinambinae

Avila-Pires, 1995; Cei, 1993; Colli et al., 1998; Donoso-Barros, 1966; Fitzgerald et al., 1999; Giugliano et al., 2007; Manzani and Abe, 2002; Martins, 2006; Mesquita et al., 2006; Perés and Colli, 2004; Sullivan and Estes, 1997.

Cordylidae

Crag, Girdled, and Plated Lizards

Estes et al., 1988; Frost et al., 2001; Hallermann, 1998; Lang, 1991; Odierna et al., 2002; Schwenk, 1988; Scott et al., 2004.

Cordylinae

Branch, 1988; Broadley and Branch, 2002; Lang, 1991; Mouton, 1997; Scott et al., 2004.

Gerrhosaurinae

Branch, 1988; Glaw and Vences, 1994; Lamb et al., 2003; Lang, 1991; Odierna et al., 2003; van Dyke, 1997.

Scincidae

Skinks

Bauer, 2003; Brandley et al., 2005; Brandley et al., 2005; Donnellan et al., 2002; Estes et al., 1988; Greer, 1970b, 1974, 1989; Griffith et al., 2000; Greer and Biswas, 2004; Griffith et al., 2000; Honda et al., 1999, 2000; Reeder, 2003; Schmitz et al., 2005.

Acontinae

Branch, 1988; Broadley, 1997a; Daniels et al., 2006; Huey et al., 1974.

"Scincinae"

Blackburn, 1993b; Branch, 1988; Brandley et al., 2005; Broadley, 1997a; Brygoo and Roux-Estève, 1983; Bull and Pamula, 1998; Glaw and Vences, 1994; Greer, 1970b, 1989; Huey and Pianka, 1981; Hutchinson, 1992, 1993; Grismer, 2002; Pianka and Vitt, 2003; Stewart and Thompson, 1996.

Anguidae

Alligator Lizards, Galliwasps, Glass Lizards, and Allies

Estes et al., 1988; Gauthier, 1982; Hallermann, 1998; Macey et al., 1999; Schwenk, 1988; Vidal and Hedges, 2005.

Anguinae

Fitch, 1989; Frazer, 1989; Gauthier, 1982; Mitchell, 1994; Pianka and Vitt, 2003; Somma, 2003.

Anniellinae

Bell et al., 1995; Fusari, 1985; Gauthier, 1982; Goldberg and Miller, 1985; Grismer, 2002; Pianka and Vitt, 2003; Stebbins, 1954.

Diploglossinae

Cei, 1993; Gauthier, 1982; Greene et al., 2006; Grismer, 2002; Hedges, 1996; Hedges et al., 1992; Pianka and Vitt, 2003; Savage and Lips, 1993; Schwartz and Henderson, 1991; Vitt, 1985.

Gerrhonotinae

Campbell and Frost, 1993; Fitch, 1970; Gauthier, 1982; Good, 1994; Greene et al., 2006; Grismer, 2002; Pianka and Vitt, 2003; Shine, 1994; Somma, 2003; Stebbins, 1954.

Xenosauridae

Knob-Scaled Lizards

Ballinger et al., 1995; Estes et al., 1988; Hallermann, 1998; Lemos-Espinal et al., 1997a, 1997b, 1998, 2003a, 2003b; Macey et al., 1999; Mägdefrau, 1997; Pianka and Vitt, 2003; Schwenk, 1988.

Helodermatidae

Gila Monster and Mexican Beaded Lizard

Beck, 1990, 2005; Beck and Jennings, 2003; Beck and Lowe, 1991; Bogert, 1993; Estes et al., 1988; Hallermann, 1998; Lowe et al., 1986; Norell and Gao, 1997; Pianka and Vitt, 2003; Pregill et al., 1986; Schwenk, 1988.

Varanidae

Monitors, Goannas, and Earless Monitors

Baverstock et al., 1993; Estes et al., 1988; Hallermann, 1998; Lee, 1997; Pianka and Vitt, 2003; Pianka et al., 2004; Schwenk, 1988.

Lanthanotinae

Manthey and Grossman, 1997; Proud, 1978.

Varaninae

Auffenberg, 1981, 1988, 1994; Bennett, 1998; Greer, 1989; Jennings and Pianka, 2004; King and Green, 1993; Pianka, 1994b, 1995; Pianka and Vitt, 2003; Pianka and King 2004; Zeigler and Böhme, 1997.

Chapter 21

Snakes

OVERVIEW

Snakes are the second most speciose group of living reptiles, with over 2900 species (see http://www.reptile-database.org/). Like lizards, they occur on all continents except Antarctica. They have had a more successful marine radiation than lizards, yet they have been less successful than lizards in dispersing onto the world's oceanic islands. All have elongate, "limbless" morphology, but this morphology exists in some other squamate clades as well (e.g., Pygopodidae, Anguidae). Nevertheless, snakes exhibit a diversity of shapes, sizes, and surface textures. This diversity in morphology reflects diverse behavioral, ecological, and physiological diversity. As a group, snakes eat a wide variety of prey, all are carnivores, and diets of many species are highly specialized.

Snakes (Serpentes; Ophidia, stem-based name) are limbless or nearly so. The pectoral girdle and forelimbs are totally absent; where present, the pelvic girdle and hindlimbs are rudimentary and visible externally as small horny "spurs," one on each side of the cloaca opening. Elongation of the body is accomplished by an increase in number of vertebrae, which typically range between 120 and 240, although the number can be more than 500. The numerous vertebrae, each with a pair of ribs in the neck and trunk, create a remarkably flexible body, and this flexibility permits effective undulatory locomotion in water, on and underground, and in bushes and trees. The body is covered with epidermal scales, the number, size, and arrangement of which are often species specific. In most snakes, the venter (underside) has a series of large, transversely rectangular scales (scutes) that extend from the throat onto the tail. In many snakes, the number of large ventral scales equals the number of vertebrae.

Without limbs, snakes capture, manipulate, and consume their prey using only the body and mouth. Some capture prey with their mouth and simply swallow them, some hold their prey with portions of their body and their mouth, some constrict prey, and yet others inject highly toxic venoms that disable or kill prey. Major modifications of cranial anatomy aid in subduing and swallowing prey. Some of these modifications are unique to snakes, including the exclusion of the supraoccipital from the margin of the foramen magnum by exoccipitals and a flexible ligamentous symphysis between the dentaries.

Other unique traits have no apparent connection to feeding, such as the absence of ciliary-body muscles in the eyes and the presence of a tracheal lung. Some characteristics occur in both snakes and one or more taxa of reduced-limbed or limbless lizards; these features include no squamosal, no epipterygoid, no sclerotic ossicles in the eyes (each eye is covered by a transparent scale called a *spectacle*), and the absence of the tympanum and the eustachian tube. The limbless condition results in body modifications as well, including the presence of more than 30 presacral (trunk and neck) vertebrae; the left lung is absent or greatly reduced and the right lung is dominant.

The early classification of snakes was based on extant species in museum collections. Consequently, classification was based entirely on external appearance. In 1758, Linnaeus recognized snakes as Serpentes, a class distinct from reptiles, with three genera and nearly 200 species. His successors recognized additional species and began to divide them into groups based on similarity of external form. Only in the mid-19th century did C. Duméril depart from tradition and include characteristics of the skull and its dentition in snake classification. Subsequently, E. D. Cope began the search for snake relationships by examining a greater variety of internal structures, including vertebral, lung, and hemipenial morphology. His posthumously published classification in 1900 recognized five suborders: Epanodonta (Typhlopidae), Catodonta (Leptotyphlopidae), Tortricina (Aniliidae, Cylindrophiidae, Uropeltidae), Colubroidea (all other snakes, exclusive of vipers, divided into four divisions), and Solenoglypha (Viperidae). Cope's groups were well defined by a variety of characteristics in addition to the aforementioned ones. While Cope's was an innovative classification, Boulenger's classification in 1893 was simpler and won wide acceptance, being used into the middle of the 20th century. The Boulenger classification began at the familial level with no higher-level groupings; however, it did divide the Colubridae into series (Aglypha, Opisthoglypha, and Proteroglypha), each with one or more subfamilies.

Hoffstetter's (1955, 1962) classification in the mid-1900s began the effort to reflect evolution by incorporating fossils; however, our modern approach to snake classification owes much to G. Underwood's controversial paper *A Contribution to the Classification of Snakes*, published in 1967. His broad selection and intimate examination of characters and his willingness to cleave the larger poly- and paraphyletic taxa into monophyletic ones provide the foundation for most modern studies. His classification is the only recent one to broadly survey the morphological spectrum of representatives of all groups of snakes. It uses Hoffstetter's groups and divides snakes into three major groups (Scolecophidia, Henophidia, and Caenophidia) and most suprageneric taxa currently recognized, although not necessarily now at the same taxonomic level. His study just preceded the use of cladistic analysis in herpetology and lacks dendrograms of snake relationships.

All recent analyses indicate that snakes diverged early into scolecophidians (blind snakes) and alethinophidians (Fig. 21.1; Table 3.7). The blind snakes contain three major clades: Anomalepididae, Leptotyphlopidae, and Typhlopidae. Monophyly of the scolecophidans has strong support and is based on numerous shared–derived characteristics, including the absence of an artery through the trigeminal foramen; the mandible less

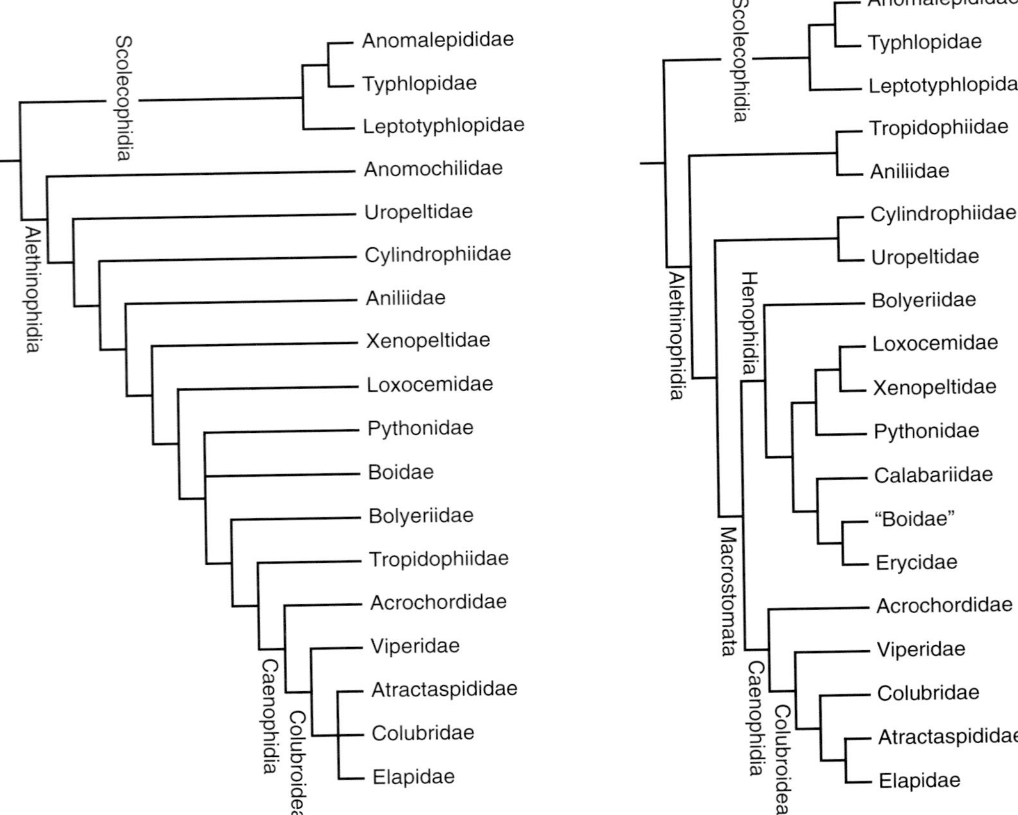

FIGURE 21.1 Two alternative hypotheses for relationships among the families of extant snakes. The cladogram on the left is a composite from Cundall et al., 1994, and Cadle, 1987. The cladogram on the right was redrawn from Vidal and Hedges, 2002, to reorder snake families similar to the order on the left.

than half the length of the jaw; vestigial pelvis and hindlimb within the body wall; thymus paired; epidermal lipid glands on the anteriormost head shields; undifferentiated smooth, glossy, cycloid body scales; and the absence of enlarged ventral scales. Within the scolecophidians, anomalepidids and typhlopids are each other's closest relatives and the sister clade to the leptotyphlopids.

The branching pattern and sister-group relationships are less certain in the alethinophidians, although some sister-group and clade relationship hypotheses appear regularly from different data sets and analyses. A colubroid clade of elapids, atractaspidids, colubrids, and viperids, and a colubroid–Acrochordidae sister group (clade Caenophidia) appear in most phylogenies. The proposed relationships among the other clades and to these robust clades are much more variable. The henophidian (primitive snakes, excluding caenophidians) and booid (boids, pythonids, bolyerids, tropidiophids) groups of Underwood and earlier systematists are paraphyletic in the classical taxonomy (Fig. 21.1, left) and resolved much differently in the more recent phylogeny based on gene sequence analysis (Fig. 21.1, right). Alethinophidia is the sister clade to Scolecophidia and, based on molecular data, includes one clade containing Aniliidae and Tropidophiidae and another clade, Caenophidia, containing remaining snake families. The Caenophidia consists of two large clades, the Colubroidea (Acrochordidae, Viperidae, Elapidae, Colubridae, and Atractaspidae) and the Henophidia (remaining families). Key differences in these two taxonomies include placement of bolyeriids and tropidophiids and better resolution of sister-clade relationships in the more recent molecular phylogeny. Skeletal and molecular data suggest an early divergence in the alethinophidians, with one lineage giving rise to the glossy snakes (Aniliidae, Tropidophiidae, Uropeltidae, and Cylindrophiidae) and the other to remaining snakes, although how this occurred differs between the two kinds of data (see Fig. 21.1). Morphological and molecular data also indicate that Aniliidae is the sister group to the *Cylindrophis*–Uropeltidae clade, but molecular data tie Tropidophiidae to Aniliidae whereas morphological data place Tropidophiidae as sister to Colubroidea. The uncertainties of relationships occur in part because most studies do not include all or most representatives of these "primitive" snakes, and different analyses often use different outgroups to root the tree. Consequently, results and interpretations differ among studies.

Acrochordus is considered to be a caenophidian and the sister group to the colubroids (Viperidae (Atractaspididae (Colubridae, Elapidae))). With similar strong support, vipers are the sister group to the other colubroids (Fig. 21.1). The relationships among the other three clades are less certain, and this uncertainty has yielded multiple interpretations. The atractaspidids have been placed as either a clade within the Colubridae, as the sister group to "aparallactine" colubrids or included with them, or more recently as a separate clade containing the fanged *Atractaspis* and the nonfanged aparallactines. The atractaspidids appear to be the sister group of elapids, although viperids and other African colubrids have been proposed as sister groups. The elapids have long been recognized as a monophyletic group, although the placement of a few African snakes (e.g., *Homoroselaps*) in elapids remains unclear. Sea snakes and terrestrial elapids were not separate clades as represented historically. Sea snakes arose from within the Australian radiation of terrestrial elapids. The present interpretation suggests that the terrestrial elapids of Africa, the Americas, and Asia (i.e., Elapinae) represent an unresolved group, and the sea snakes, sea kraits, and terrestrial Papuaustralian elapids (i.e., Hydrophiinae) form a monophyletic group. Further, *Bungarus* is the likely sister group to the hydrophiine clade. *Laticauda* has affinities to a divergent group of Papuan elapids, and affinities of the viviparous sea snakes (i.e., formerly hydrophiines) are within the Australian elapids. Not all these relationships have been confirmed by independent study.

Viperidae also has been proposed as the sister group of the Atractaspididae; however, evidence points to viperids as the sister group to the other three colubroid clades or as sister to the clade containing only atrachtaspids and colubrids. Relationships in and among the colubrids are far less resolved and seemingly change with each new data set and analysis, largely because it is a huge and complex group. Because the elapids and atractaspidids, and possibly the viperids, likely arose within the colubrids, the colubrid group is paraphyletic and perhaps even polyphyletic. A few groups appear to be robustly monophyletic, such as Boodontinae, Pseudocyrhophinae, and Xenodermatinae. Others, e.g., Colubrinae and Xenodontinae, are also likely clades (when membership is restricted). A growing dilemma within colubrid systematics is the constant redefinition of groups so that the species content of one author's group usually is different from the same-named group of another author. Also, because of the high species diversity of many of the colubrid genera and higher taxa, it is difficult for researchers to include all representatives of ingroups and appropriate outgroups in their studies. Colubrid relationships remain a challenge and likely will not be resolved in the near future.

Considering the current state of flux in our understanding of squamate phylogeny, we retain our family list as in the second edition of this book, with the exception that we move *Anomochilus* into the Cylindrophidae thus eliminating the Anomochilidae based on recent molecular analyses. We also retain Boinae and Erycinae in the Boidae, recognizing that "Boidae" is paraphyletic. We comment in each family account on likely sister relationships based on recent molecular analyses.

TAXONOMIC ACCOUNTS

Scolecophidia

Anomalepididae

Dawn Blind Snakes

Classification: Reptilia; Diapsida; Sauria; Lepidosauria; Squamata.

Sister taxon: Typhlopidae.

Content: Four genera, *Anomalepis*, *Helminthophis*, *Liotyphlops*, and *Typhlophis*, with 4, 3, 8, and 2 species, respectively.

Distribution: Disjunct in Central and South America (Fig. 21.2).

Characteristics: Anomalepidids are thin-bodied blind snakes. Most range in adult TL between 150 and 300 mm, and none is larger than 400 mm. Cranially, these snakes have two common carotid arteries, edentulous premaxillaries, longitudinally oriented maxillaries with solid teeth, and optic foramina that perforate the frontals. The mandible has a coronoid bone, and each dentary has one to three teeth. They lack cranial infrared receptors in pits or surface indentations. No limb vestiges are evident externally, although pelvic remnants occur in the trunk musculature. Intracostal arteries arise from the dorsal aorta at nearly every trunk segment. The left lung is absent, a tracheal lung is present, and the left oviduct is usually well developed, although variously reduced in *Anomalepis*.

Biology: The anomalepidids are fossorial snakes that are usually associated with subterranean ant and termite nests. We know little of their biology because of their cryptozoic lifestyle. Presumably, they are like other scolecophidians and prey on soft-bodied invertebrates and the larvae and eggs of these animals. Termites and early life history stages of ants (eggs, larvae) are likely a major food. One species, *Typhlophis squamosus* (Fig. 21.3), can be easily found in termite nests inside rotted logs on the forest floor in rain forest of the southern Amazon. Based on the limited reproductive data available, all are oviparous and lay small clutches that consist of 2 to 13 eggs. When captured, *Typhlophis* thrashes the body and repeatedly jabs its sharp tail in defense.

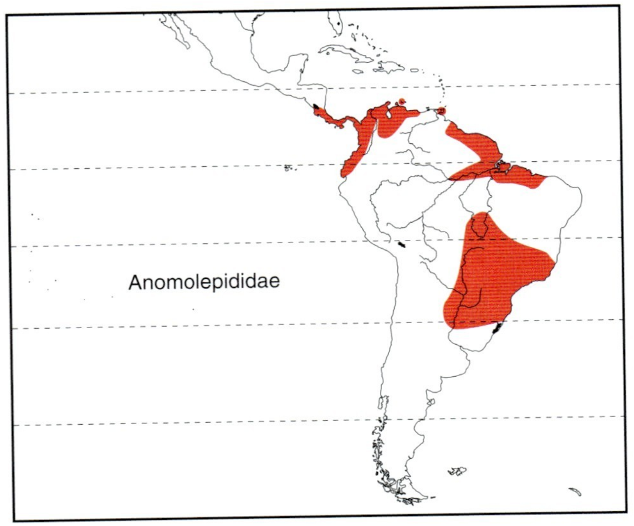

FIGURE 21.2 Geographic distribution of the extant Anomalepididae.

Leptotyphlopidae

Slender Blind Snakes, Thread Snakes

Classification: Reptilia; Diapsida; Sauria; Lepidosauria; Squamata.

Sister taxon: Anomalepididae and Typhlopidae clade.

Content: Two genera, *Leptotyphlops* and *Rhinoleptus*, with 103 and 1 species, respectively.

Distribution: Tropics and subtropics of Africa and the Americas, and Southwest Asia (Fig. 21.4).

Characteristics: Of the scolecophidians, leptotyphlopids are typically the thinnest-bodied members (Fig. 21.3). They reach a maximum of 400 mm adult SVL, but most are 150 to 250 mm SVL. Cranially, these snakes have two common carotid arteries, edentulous premaxillaries, longitudinally oriented maxillaries lacking teeth, and optic foramina that perforate the frontals. The mandible has a coronoid bone, and each dentary has four or five teeth. They lack cranial infrared receptors in pits or surface indentations. No limb vestiges are evident externally, although pelvic remnants occur in the trunk musculature. Intracostal arteries arise from the dorsal aorta at nearly every trunk segment. They lack a left lung, a tracheal lung, and a left oviduct.

Biology: Slender blind snakes are fossorial and occur in a variety of habitats from semidesert to forest. They feed on soft-bodied invertebrates, although termites appear to be the primary food of some species. Unlike many termite predators, they are capable of living in termite nests and are permanent residents within termite galleries. They have evolved a secretion that averts the attack of the soldier termites and ants, possibly by deceiving the potential attackers into considering them as nest mates as they move freely through tunnels in social insect nests. They are occasionally observed in barn owl nests, most likely brought there by the owls to feed their young. Some of the snakes apparently escape in the owl nest, where they survive feeding on insects and their larvae in the owl nest. Some species, such as *L. macrolepis* in the Amazon rain forest, have been observed on rainy nights nearly 2 meters aboveground, wrapped around small tree trunks with the head and neck extending perpendicular to the trunk and moving back and forth. They may climb trees to locate termite nests by detecting airborne chemical cues associated with the release of termite alates. The Texas thread snake, *L. dulcis*, can often be found under surface objects in early spring in clusters of as many as 18 individuals, nearly all of which are males. Leptotyphlopids are oviparous, laying 1 to 12 small, elongate eggs. Females of Texas thread snakes coil around their eggs,

FIGURE 21.3 Representative scolecophidian snakes. Clockwise from upper left: Trinidad blind snake *Typhlophis squamosus*, Anomalepididae (L. J. Vitt); reticulated blind snake *Typhlops reticulatus*, Typhlopidae (L. J. Vitt); Texas thread snake *Leptotyphlops dulcis*, Leptotyphlopidae (Buddy Brown); seven-line thread snake *Leptotyphlops septemstriatus*, Leptotyphlopidae (L. J. Vitt).

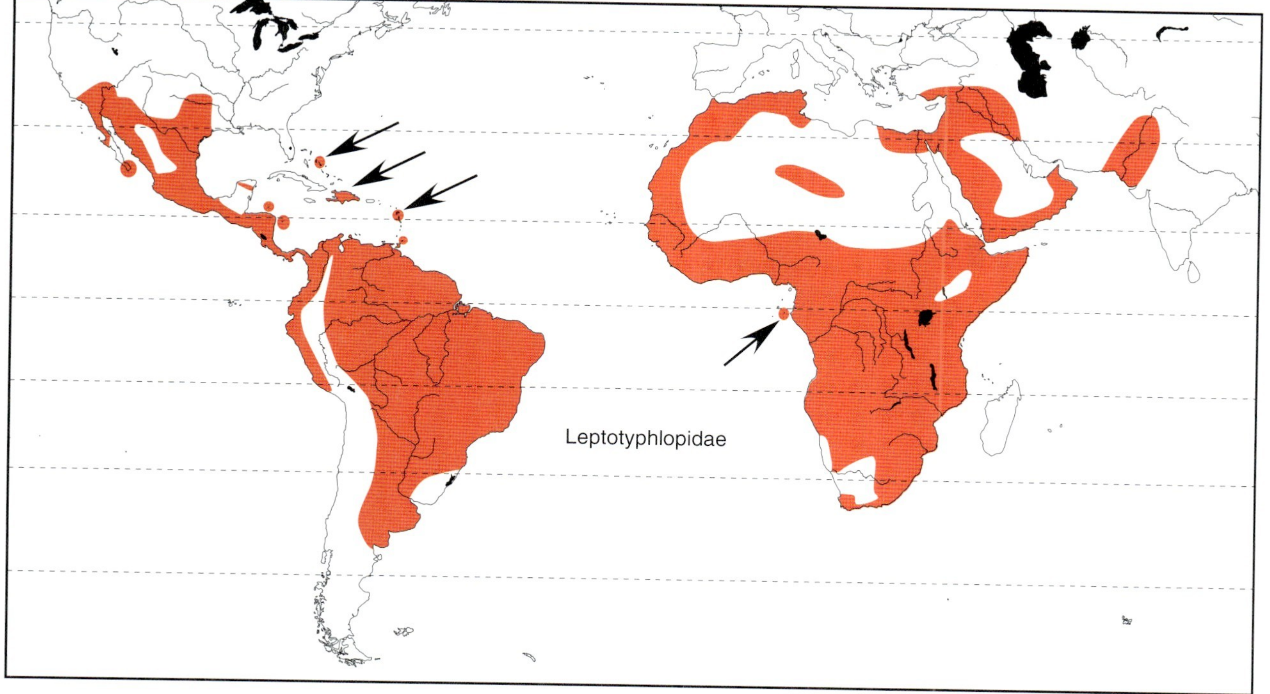

FIGURE 21.4 Geographic distribution of the extant Leptotyphlopidae.

possibly exhibiting parental care. This behavior may occur in other species. Surprisingly little is known about slender blind snakes, considering how common they often are.

Typhlopidae

Blind Snakes

Classification: Reptilia; Diapsida; Sauria; Lepidosauria; Squamata.

Sister taxon: Anomalepididae.

Content: Seven genera, *Acutotyphlops*, *Cyclotyphlops*, *Grypotyphlops*, *Ramphotyphlops*, *Rhinotyphlops*, *Typhlops*, and *Xenotyphlops*, with 4, 1, 1, 60, 15, 341, and 2 species, respectively.

Distribution: Cosmopolitan in tropical regions (Fig. 21.5).

Characteristics: Blind snakes range from small (140-180 mm TL, *Ramphotyphlops braminus*) to large (950 mm maximum TL, *Rhinotyphlops schlegelii*). Cranially, these snakes have two common carotid arteries, edentulous premaxillaries, transversally oriented maxillaries with solid teeth, and optic foramina that perforate the frontals. The mandible has a coronoid bone and lacks teeth on the dentary. They lack cranial infrared receptors in pits or surface indentations. No limb vestiges are evident externally, although pelvic remnants occur in the trunk musculature. Intracostal arteries arise from the dorsal aorta at nearly every trunk segment. The left lung is vestigial or absent, and the tracheal lung is multichambered; the left oviduct is absent.

Biology: Typhlopids are the most speciose blind snakes (Fig. 21.3) and occupy a variety of habitats from near desert to rain forest. All are subterranean, but some have been observed in arboreal situations, presumably having followed a termite trail or a termite gallery tunnel to climb a tree. The possibility exists that, like some leptotyphlopids, they climb to position themselves aboveground to orient on chemical cues originating from social insect nests during alate (winged males and females) releases. Termites, ants, and their larvae and eggs appear to be the major food, although blind snakes consume other soft-bodied arthropods. Reproductive data are unavailable for most species. Of the known species, all are oviparous, with the possible exception of one report in which embyros were observed in *Typhlops diardi*; however, this observation may represent delayed egg deposition, not viviparity. Clutch size varies with body size, ranging from 2 to 7 eggs (*Ramphotyphlops braminus*) to 40 to 60 eggs (*Rhinotyphlops schlegelii*). Eggs are deposited shortly after fertilization and incubated typically for 6 to 10 weeks, or they can be held within the oviducts and laid only a week or so before hatching (*Typhlops bibronii*). To date, the Brahminy blind snake (*Ramphotyphlops braminus*) is the only known unisexual species of snake. It is triploid, no doubt of hybrid origin, and parental species remain undetermined (see Chapter 4). Because a single individual can start a new population, and because it is small and lives in soil, it has become the most widely dispersed snake species. It now occurs in all continental and many insular tropical areas, apparently arriving as a stowaway in the root mass of exotic "potted" plants. These introduced snakes are now common in many parts of the southeastern United States, especially Florida.

Alethinophidia

"Tropidophiidae"

Dwarf Boas

Classification: Reptilia; Diapsida; Sauria; Lepidosauria; Squamata.

Sister taxon: Aniliidae.

Content: Three subfamilies, Tropidophiinae, Ungaliophiinae, and Xenophidioninae.

Distribution: Malaysia and tropical America (Fig. 21.6).

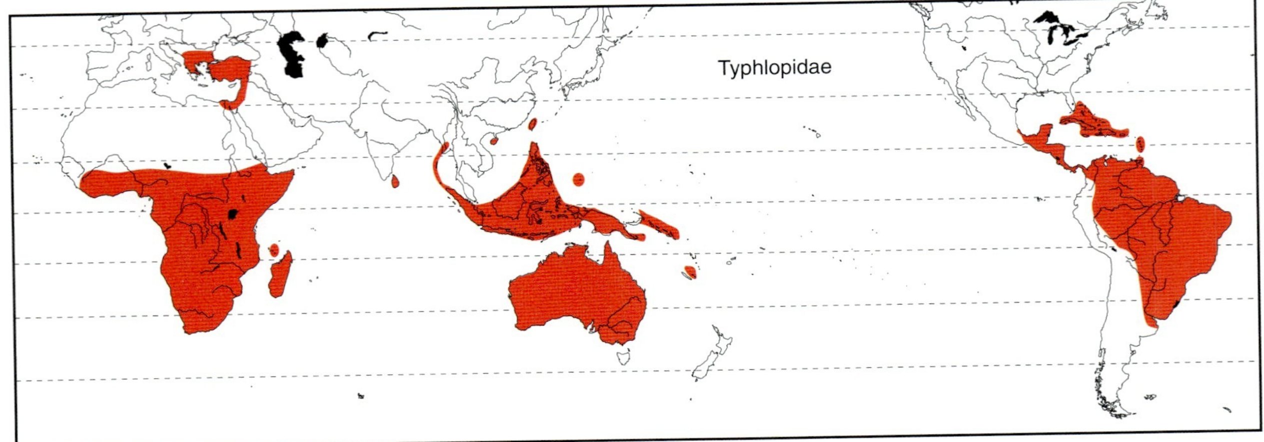

FIGURE 21.5 Geographic distribution of the extant Typhlopidae.

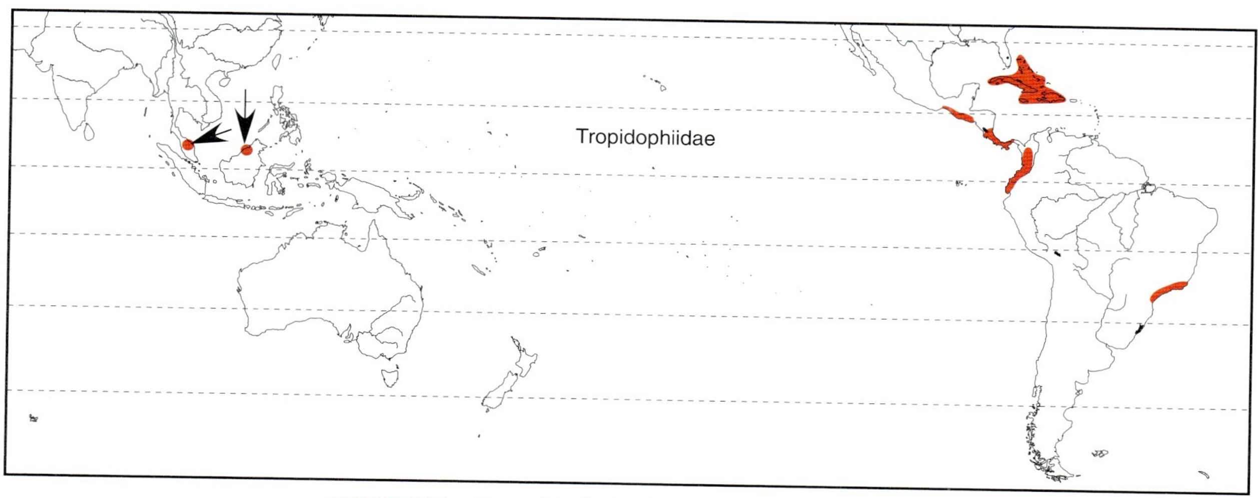

FIGURE 21.6 Geographic distribution of the extant Tropidophiidae.

Characteristics: These small-to-moderate-sized snakes share traits with both "booid" and colubroid snakes. Cranially, they have two common carotid arteries, edentulous premaxillaries, longitudinally oriented maxillaries with solid teeth, and optic foramina that perforate the frontal–parietal sutures. The coronoid is reduced or absent, and the dentary bears numerous teeth. Tropidophiids lack cranial infrared receptors in pits or surface indentations. Hindlimb vestiges appear externally as cloacal spurs in males, and pelvic remnants occur in the trunk musculature, except in *Xenophidion*. Intracostal arteries arise from the dorsal aorta at nearly every trunk segment or at intervals of several segments. The left lung is absent (occasionally vestigial in *Ungaliophis*), and a tracheal lung is well developed; both left and right oviducts are well developed.

Comment: The Tropidophiidae as presented here is problematical. A recent molecular phylogeny places tropidophiines distant from ungaliophiines but doesn't include xenophidionines. Until further resolution is available we consider the three traditional subfamilies of the Tropidophiidae, recognizing that the group is likely polyphyletic.

Tropidophiinae

Sister taxon: Traditionally, Ungaliophiinae but more likely the Caenophidia.

Content: Two genera, *Trachyboa* and *Tropidophis*, with 2 and 21 species, respectively.

Distribution: West Indies, Central and South America.

Characteristics: The dentary lacks an anterior canine-like tooth; the hyoid horns are parallel; and pelvic remnants are present.

Biology: *Trachyboa* and *Tropidophis* (Fig. 21.7) range in adult TL from 200 mm to 1 m, but most species and individuals are less than 600 mm. They are mainly forest inhabitants and are terrestrial to semiarboreal foragers. They feed mainly on small vertebrates, predominantly lizards. All are viviparous, and litter size is typically 10 or fewer young.

Ungaliophiinae

Sister taxon: Traditionally, Tropidophiinae but possibly Erycinae.

Content: Two genera, *Exiliboa* and *Ungaliophis*, with 1 and 2 species, respectively.

Distribution: Disjunct, from southern Mexico to northern Columbia.

Characteristics: The dentary lacks an anterior canine-like tooth; the hyoid horns are semiparallel; and pelvic remnants are present.

Biology: Ungaliophiines are moderately small snakes (< 760 mm adult TL) that occur in wet to dry forested habitats. *Ungaliophis* is purportedly arboreal or semiarboreal; *Exiliboa placata* is terrestrial, preferring rocky areas. They are secretive snakes, likely nocturnal foragers, and they prey mainly on amphibians and lizards. Ungaliophiines are live bearers, and *E. placata* bears 8 to 13 neonates in September and October.

Xenophidioninae

Sister taxon: Unknown, given uncertainties of the relationship of Ungaliophiinae to Tropidophiinae.

Content: One genus, *Xenophidion* (Fig. 21.7), with two species.

Distribution: Malaysia.

Characteristics: The dentary has a large, anterior canine-like tooth, the hyoid horns are strongly divergent, and pelvic remnants are absent.

FIGURE 21.7 From top to bottom: Haitian dwarf boa *Tropidophis haitianus*, Tropidophiinae (L. L. Grismer); Schaefer's spine-jaw snake *Xenophidion schaeferi*, Xenophidioninae (W. Grossman); false coral snake *Anilius scytale*, Aniliidae (L. J. Vitt).

Biology: Presently, the two species of *Xenophidion* are each known from a single specimen. Both are rain forest–floor inhabitants. They are small snakes, likely not exceeding 300 mm SVL as adults. A single mature female of *Xenophidion acanthagnathus* collected at 600 M in Borneo in a selectively logged forest contained several large-shelled eggs. Although the diet is unknown, a large tooth on the front of the lower jaw suggests that small vertebrates capable of struggling are prey.

Aniliidae

False Coral Snakes, South American Pipe Snakes

Classification: Reptilia; Diapsida; Sauria; Lepidosauria; Squamata.

Sister taxon: Tropidophiidae.

Content: Monotypic, *Anilius scytale*.

Distribution: Northern South America (Fig. 21.8).

Characteristics: *Anilius scytale* is another smooth, shiny-scaled snake (Fig. 21.7) with a very short tail and ventral scales barely larger than the dorsal ones. Cranially, *A. scytale* has two common carotid arteries, teeth on the premaxillaries, short longitudinally oriented maxillaries with solid teeth, and optic foramina that perforate the frontal–parietal sutures. The mandible has a coronoid bone, and the dentary bears teeth. *Anilius scytale* lacks cranial infrared receptors in pits or surface indentations. The small eyes are covered by a large scale. Hindlimb vestiges appear externally as cloacal spurs, and pelvic remnants occur in the trunk musculature. Intracostal arteries arise from the dorsal aorta at nearly every trunk segment. The left lung is reduced but present, and a tracheal lung is absent; both left and right oviducts are well developed.

Biology: This false coral snake receives its name from its striking red and black ringed pattern. Adults are typically less than 600 mm TL, although occasionally they exceed 1 meter TL. *Anilius* is generally fossorial or at least spends the daylight hours beneath forest-floor litter. The authors have captured these snakes in surface traps at night, and one was found foraging underwater in a small stream in the morning. Although predominantly a forest inhabitant, *Anilius* occasionally occurs in cultivated areas and other human-disturbed habitats. Adults prey on fish,

FIGURE 21.8 Geographic distribution of the extant Aniliidae.

amphisbaenians, and other snakes. Sexual maturity occurs at about 350 mm TL, and females give birth to 7 to 15 neonates, typically early in the wet season. These snakes have a defensive display in which they flatten the body and tail and raise the tail off the ground, waving it around as they either crawl off or tighten their body into a ball.

Cylindrophiidae

Pipe Snakes and Dwarf Pipe Snakes

Classification: Reptilia; Diapsida; Sauria; Lepidosauria; Squamata.

Sister taxon: Uropeltidae.

Content: Two genera, *Anomochilus* with 2 species and *Cylindrophis* with 8 species.

Distribution: Disjunct, Sri Lanka and Southeast Asia through the East Indies for *Cylindrophis;* and Malaya Peninsula, Sumatra, and Borneo for *Anomochilus* (Fig. 21.9).

Characteristics: *Anomochilus* are small snakes (250-350 mm adult SVL) with short tails and have previously been placed in their own family, the Anomochiliidae (Fig. 21.10). *Cylindrophis* species are moderate to large, thick-bodied, short-tailed snakes. Both genera have smooth, shiny scales, ventral scales that are only slightly larger than dorsal scales, two common carotid arteries, edentulous premaxillaries, optic foramina that perforate the frontal–parietal sutures, mandible with a coronoid bone, and teeth present or absent on the dentary. They lack cranial infrared receptors in pits or surface indentations. Hindlimb vestiges appear externally as cloacal spurs, and pelvic remnants occur in the trunk musculature. Intracostal arteries arise from the dorsal aorta at nearly every trunk segment. The small left lung is present (reduced in *Anomochilus*), and a tracheal lung is absent; both left and right oviducts are well developed. *Anomochilus* has diagonally oriented maxillaries with solid teeth, whereas *Cylindrophis* has longitudinally oriented maxillaries with solid teeth.

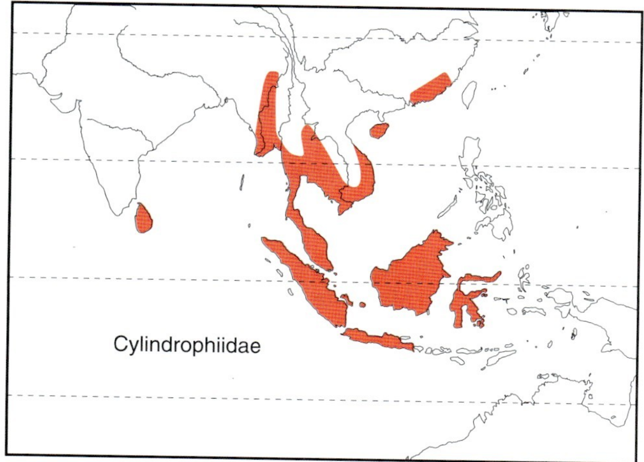

FIGURE 21.9 Geographic distribution of the extant Cylindrophiidae.

FIGURE 21.10 Representative Cylindrophid and Xenodontid snakes. *Anomocheilus leonardi*, Cylindrophiidae (I. Das); red-tailed pipe snake *Cylindrophis ruffus*, Cylindrophiidae (R. W. Murphy); Drummond–Hays shield-tail *Rhinophis drummondhayi*, Uropeltidae (I. Das).

Biology: *Anomochilus* is known from fewer than 10 specimens, none of which is accompanied by detailed observations, so the biology of these species is largely conjecture. Head and body morphology suggests fossorial habits. The snakes likely use preexisting tunnels or burrow through forest-floor detritus or friable soils. The mouth is small and the unique upper and lower jaw apparatus suggests a diet of small, generally soft-bodied prey. *Anomochilus* is oviparous. A female from Borneo contained four eggs. *Cylindrophis* species range in adult size from about 300 to 900 mm TL. All are fossorial snakes. The two best-known species, *C. maculatus* (Sri Lanka) and *C. ruffus* (Indochina; Fig. 21.10) are moderately abundant and occur widely from suburban gardens to forest, and occasionally even in mats of floating vegetation. They prefer moist, friable soils and apparently dig and create their own burrow systems. *Cylindrophis* regularly uses a tail head–mimicry display as a defense mechanism and uncommonly a death-feigning behavior. The pipe snakes are principally nocturnal foragers, searching on the surface in leaf litter for a variety of invertebrate and vertebrate prey, including earthworms, eels, caecilians, and other snakes. All are viviparous and produce litters of 2 to 12 neonates.

Uropeltidae

Shield-Tail Snakes

Classification: Reptilia; Diapsida; Sauria; Lepidosauria; Squamata.

Sister taxon: Cylindrophiidae.

Content: Eight genera, *Brachyophidium*, *Melanophidium*, *Platyplectrurus*, *Plectrurus*, *Pseudotyphlops*, *Rhinophis*, *Teretrurus,* and *Uropeltis*, with 1, 3, 2, 6, 1, 15, 1, and 23 species, respectively.

Distribution: Sri Lanka and southern India (Fig. 21.11).

Characteristics: Shield-tails have cone-shaped heads, often with a strongly keratinized tip, and a uniquely enlarged and roughened scale on the end of a short tail. Cranially, these snakes have two common carotid arteries, edentulous premaxillaries, longitudinally oriented maxillaries with solid teeth, and optic foramina that perforate the frontals. The mandible has a coronoid bone, and the dentary bears teeth. Shield-tails lack cranial infrared receptors in pits or surface indentations. Girdle and limb vestiges do not occur externally or internally. Intracostal arteries arise from the dorsal aorta at nearly every trunk segment. The left lung is usually present but small, and a tracheal lung is absent; both left and right oviducts are well developed.

Biology: The shield-tails are fossorial snakes. Much of their morphology from head to tail and the smooth, glossy scale covering appear associated with burrowing (Fig. 21.10). They are almost exclusively forest inhabitants, occurring in open areas only where the soil is friable, permitting them to burrow deeply and avoid high soil-surface temperatures. They seldom appear on the surface unless uncovered by surface predators (e.g., jungle fowl) or forced to the surface by flooded soils. When exposed, uropeltids hide their heads beneath body coils or debris and present the armored tail to the attacking predator; this behavior allows them to begin burrowing. The conical head and heavily muscled anterior quarter of the body facilitates digging. Digging begins with the head embedded in the tunnel wall and the muscular body folded into a series of loops within the skin envelope. The head is driven forward by straightening the body loops; the head then anchors the body, and the trunk is pulled forward as well as formed into loops within the skin. This concertina-style burrowing is effective in moist and friable soils, and shield-tails quickly disappear within a self-created hole while the tail shield plugs the hole and protects the escaping snake.

Shield-tails range in size from the very small *Platyplectrurus trilineatus* (100 to 130 mm adult SVL) to the moderate-sized *Uropeltis* (e.g., 420 mm maximum TL, *U. myhendrae*). Diet is unknown, but because the snakes are totally subterranean, their diet likely consists principally of earthworms and other soft-bodied invertebrates, and perhaps small burrowing vertebrates. Uropeltids appear to be exclusively viviparous, but data are limited. Litter size is small, three to nine embryos (usually four), and pregnancy may be confined to a single oviduct–uterus.

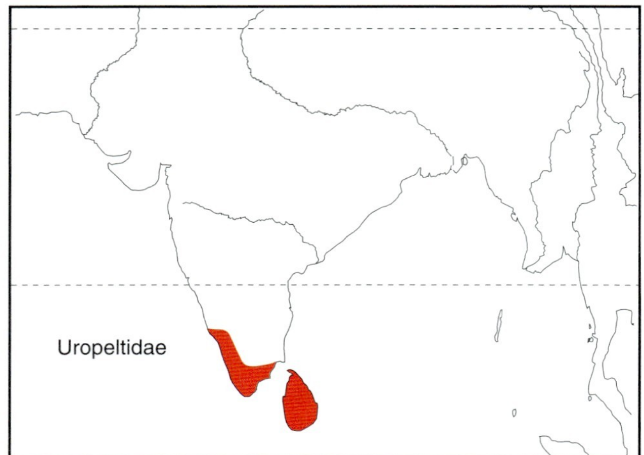

FIGURE 21.11 Geographic distribution of the extant Uropeltidae.

Bolyeriidae

Mascarene or Split-Jaw Boas

Classification: Reptilia; Diapsida; Sauria; Lepidosauria; Squamata.

Sister taxon: Clade containing all other members of the Henophidia.

Content: Two monotypic genera, *Bolyeria multocarinata* and *Casarea dussumieri*.

Distribution: Mauritius and northern islets (Fig. 21.12).

Characteristics: Bolyerines are unique among snakes because they possess a maxillary that is divided and hinged into anterior and posterior elements. They are slender boalike snakes (800 mm to 1.38 m TL) without cloacal spurs. Cranially, these snakes have two common carotid

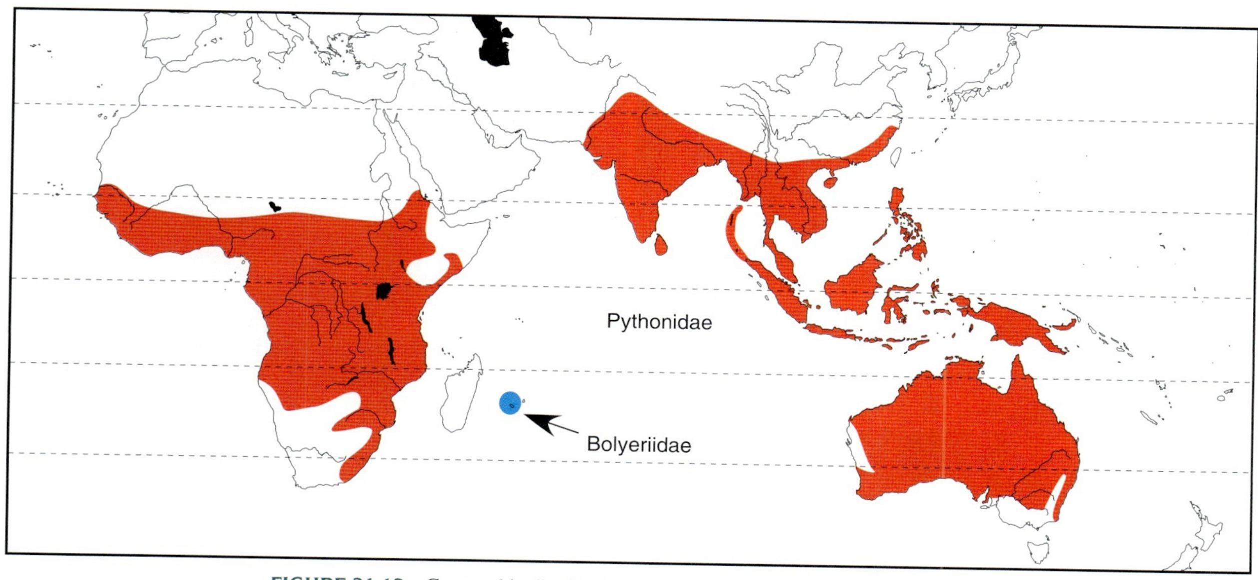

FIGURE 21.12 Geographic distribution of the extant Pythonidae and Bolyeriidae.

arteries, edentulous premaxillaries, longitudinally oriented maxillaries with solid teeth, and optic foramina that perforate the frontal–parietal sutures. The mandible has a coronoid bone, and the dentary lacks teeth. They lack cranial infrared receptors in pits or surface indentations. Girdle and limb elements are entirely absent. Intracostal arteries arise from the dorsal aorta at nearly every trunk segment. The left lung is greatly reduced, and there is no tracheal lung; both left and right oviducts are well developed.

Biology: The bolyeriids consist of two recently extant taxa. *Bolyeria multocarinata* was known from the northern islets near Mauritius, but it is now presumably extinct, as none has been seen since 1975 in spite of extensive searching. *Casarea dussumieri* previously occurred on Mauritius and still occurs today on Round Island (Fig. 21.13). The hinged lower jaw appears to be an adaptation to catch and hold hard, slippery-scaled skinks. Other squamates have evolved similar cranial adaptations for durophagous prey. Field observations indicate that *C. dussumieri* is nocturnal and approaches prey slowly with raised head and anterior trunk and strikes only when within a few millimeters of the prey. Once grasped, the skink or gecko might be constricted. *C. dussumieri* is oviparous; reproduction in *B. multocarinata* is unknown.

Loxocemidae

Mesoamerican Python

Classification: Reptilia; Diapsida; Sauria; Lepidosauria; Squamata.

Sister taxon: Xenopeltidae.

Content: Monotypic, *Loxocemus bicolor*.

Distribution: Southern Mexico to Costa Rica (Fig. 21.14).

FIGURE 21.13 Dussumier's split-jaw boa *Casarea dussumieri*, Bolyeriidae (Suzanne L. Collins, The Center for North American Amphibians and Reptiles).

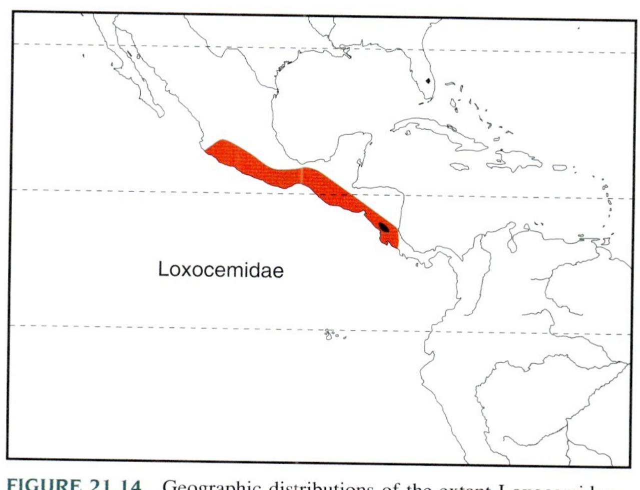

FIGURE 21.14 Geographic distributions of the extant Loxocemidae.

FIGURE 21.15 Representative loxocemid, xenopeltid, and pythonid snakes. Clockwise from upper left: Mesoamerican python *Loxocemus bicolor*, Loxocemidae (J. A. Campbell); sunbeam snake *Xenopeltis unicolor*, Xenopeltidae (G. R. Zug); carpet python *Morelia spilota* (L. J. Vitt); Burmese python *Python molurus*, Pythonidae (R. W. Murphy).

Characteristics: *Loxocemus bicolor* has supraorbital (postfrontal) bones, a cranial feature of primitive snakes (Fig. 21.15). In addition, this taxon has two common carotid arteries, teeth on the premaxillaries, longitudinally oriented maxillaries with solid teeth, and optic foramina that perforate the frontal–parietal sutures. The mandible has a coronoid bone, and the dentary bears teeth. They lack cranial infrared receptors in pits or surface indentations. Hindlimb vestiges appear externally as cloacal spurs, and pelvic remnants occur in the trunk musculature. Intracostal arteries arise from the dorsal aorta at nearly every trunk segment. The left lung is large, but no tracheal lung occurs; both left and right oviducts are well developed.

Biology: *Loxocemus bicolor* (Fig. 21.15) attains an adult SVL of 1.4 m, although most adults are less than 1 m. They are relatively uncommon or infrequently seen throughout their distribution; hence, their biology is incompletely known. Although labeled as burrowers, they appear to be more secretive than fossorial, and they generally live in tropical or subtropical dry forests. Apparently, they forage only at night, eating a variety of small terrestrial vertebrates (reptiles and mammals) and even sea turtle and iguana eggs. They are oviparous, laying small clutches of four relatively large eggs.

Xenopeltidae

Sunbeam Snakes

Classification: Reptilia; Diapsida; Sauria; Lepidosauria; Squamata.

Sister taxon: Loxocemidae.

Content: One genus, *Xenopeltis*, with 2 species.

Distribution: Southeast Asia, from Burma through East Indies to the Philippines (Fig. 21.16).

Characteristics: Sunbeam snakes obtain their name from the iridescent glow reflected from their smooth, shiny scales. They have blunt heads, cylindrical bodies, and short tails but large ventral scales. Cranially, these snakes have two common carotid arteries, teeth on the premaxillaries, longitudinally oriented maxillaries with solid teeth, and optic foramina that perforate the frontal–parietal sutures. The mandible lacks a coronoid bone, and the dentary bears numerous small teeth. They lack cranial infrared receptors in pits or surface indentations. Girdle and

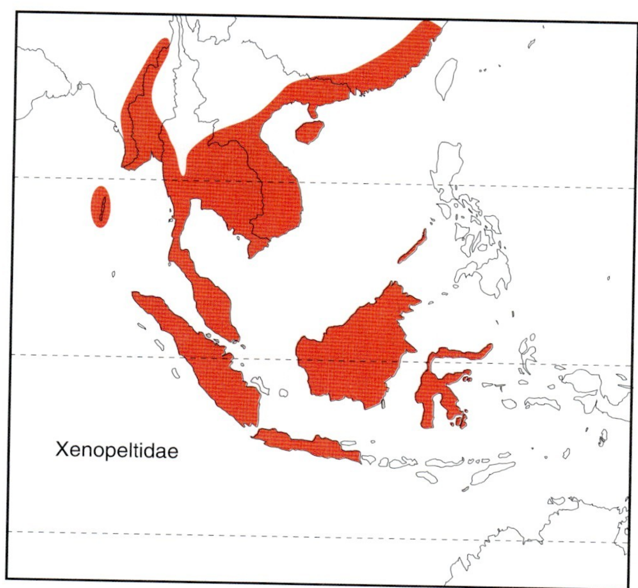

FIGURE 21.16 Geographic distribution of the extant Xenopeltidae.

limb vestiges are absent. Intracostal arteries arise from the dorsal aorta at nearly every trunk segment. The left lung is large, and the tracheal lung is absent. Both left and right oviducts are well developed.

Biology: Of the two species, *X. unicolor* (Fig. 21.15) has the widest distribution, and because it is moderately abundant, its biology is reasonably known. Adult *X. unicolor* attain total lengths to 1.3 m; however, most individuals do not exceed 800 mm TL. They are secretive snakes and associated with low coastal areas where they occur in lowland rain forest, rice fields, and other habitats. They can be found as much as 700 meters above sea level. These terrestrial snakes burrow in mud. Even though often described as nocturnal, they appear to forage during the day also, or at least diurnally during some seasons. The diet is broad and includes frogs, lizards, and snakes. They have a broad ecological tolerance and, although commonly associated with water, they occur widely from urban gardens to low montane forest and scrub forest. They are oviparous and can lay as many as 17 eggs in a clutch, but clutch size is usually smaller.

Pythonidae

Pythons

Classification: Reptilia; Diapsida; Sauria; Lepidosauria; Squamata.

Sister taxon: The clade containing Loxocemidae +Xenopeltidae.

Content: Eight genera, *Aspidites, Antaresia, Apodora. Bothrochilus, Leiopython, Liasis, Morelia,* and *Python*, with 2, 4, 1, 1, 1, 3, 13, and 12 species, respectively.

Distribution: Sub-Saharan Africa, South and Southeast Asia to Australia (Fig. 21.12).

Characteristics: Pythons are large to giant snakes (Fig. 21.15). Cranially, they have two common carotid arteries, teeth on the premaxillaries (except in *Aspidites*) without ascending processes, longitudinally oriented maxillaries with solid teeth, paired supraorbitals, optic foramina that perforate the frontal–parietal sutures, and a low or no supraoccipital crest. The mandible has a coronoid bone. Many pythons have cranial infrared receptors in interlabial pits. Hindlimb vestiges appear externally as cloacal spurs, and pelvic remnants occur in the trunk musculature. Intracostal arteries arise from the dorsal aorta at nearly every trunk segment. The left lung is usually large, and a tracheal lung is absent; both left and right oviducts are well developed.

Biology: Adult pythons range from the Australian pygmy python *Liasis childreni* (350–600 mm adult TL) to the giant reticulated python *Python reticulatus* (2.5–10 m TL); adults of most species are less than 4 meters. Pythons occur in a wide range of habitats from desert to rain forest. Forest and scrub species forage on and above the ground for vertebrate prey; mammals and birds become the food of the larger individuals. Some species are semiaquatic, e.g., *Liasis fuscus,* but birds and mammals are still the major prey. All pythons are oviparous, and in most (if not all) species, females coil about the eggs. In some, such as *P. molurus*, parental care is true brooding; the female maintains an elevated body temperature to aid incubation (see Chapters 5 and 7). Clutch size is associated with body size. Smaller and/or the more slender species have clutches of about 5 to 16 eggs, and the larger-bodied species have clutches of 30 to 60 eggs, occasionally over 100 eggs, as reported for *P. reticulatus*.

Boidae

Boas

Classification: Reptilia; Diapsida; Sauria; Lepidosauria; Squamata.

Sister taxon: Uncertain, possibly the clade (Pythonidae (Loxocemidae+Xenopeltidae)).

Content: Two subfamilies, "Boiinae" and Erycinae (considered separate families by some authors; see Fig. 21.1).

Distribution: Western North America to southern subtropical South America, West Indies, central Africa to South Asia, Madagascar, and Southwest Pacific islands (Fig. 21.17).

Characteristics: The "true" boas are small to large snakes. Cranially, they share two common carotid arteries, edentulous premaxillaries with ascending processes, longitudinally oriented maxillaries with solid teeth, optic foramina that perforate the frontal–parietal sutures, and a strongly developed supraoccipital crest. The mandible has a coronoid bone. Most boids have cranial infrared receptors in interlabial pits. Hindlimb vestiges appear

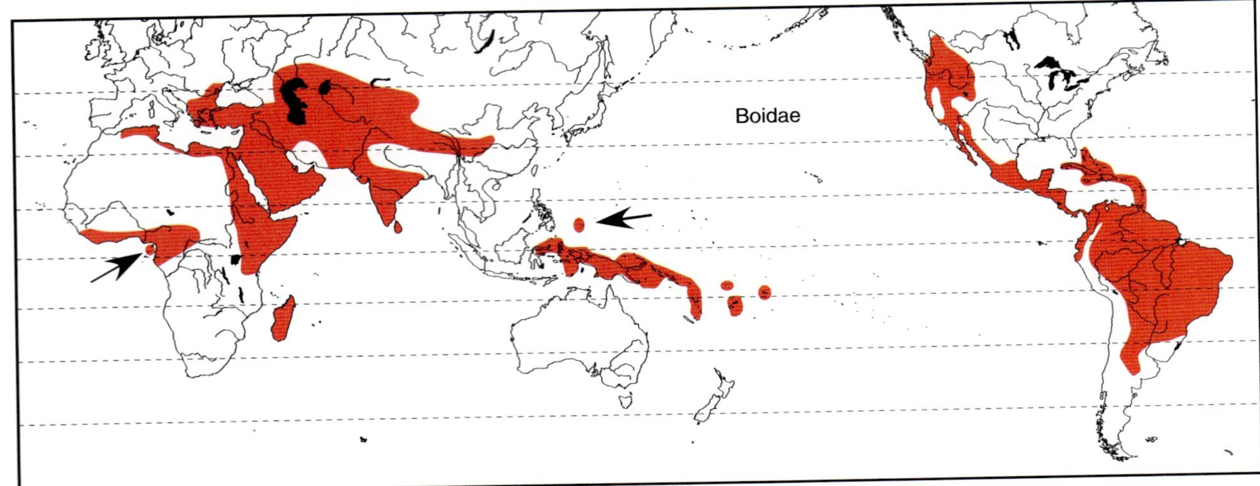

FIGURE 21.17 Geographic distribution of the extant Boidae.

externally as cloacal spurs, and pelvic remnants occur in the trunk musculature. Intracostal arteries arise from the dorsal aorta at nearly every trunk segment. The left lung is moderately to well developed, and a tracheal lung is absent; both left and right oviducts are well developed.

"Boinae"

Sister taxon: Erycinae.

Content: Seven genera, *Acrantophis, Boa, Candoia, Corallus, Epicrates, Eunectes,* and *Sanzinia,* with 2, 20, 3, 6, 11, 4, and 1 species, respectively.

Distribution: Disjunct, tropical Americas including the West Indies, Madagascar, and islands of Southwest Pacific.

Characteristics: Prefrontals touch medially or nearly so, labial sensory pits occur, and caudal vertebrae have simple neural arches.

Biology: Boines (Fig. 21.18) range from moderate-sized species (e.g., 600–900 mm adult SVL, *Candoia aspera*) to truly giant snakes (e.g., *Eunectes*, at least to 8 m maximum TL and possibly to 11.5 m). The small to large species are mostly arboreal snakes, although they are regularly found on ground; the largest-bodied clade, *Eunectes*, is aquatic. Contrary to what was depicted in the movie *Anaconda*, *Eunectes* are slow moving on land. Many are bird and mammal predators and are largely nocturnal. Some, such as *Corallus hortulanus*, use their infrared heat-sensing organs to locate sleeping prey, such as birds. Others, such as *Corallus caninus*, position themselves low on trunks of small trees in the forest with the head oriented down, apparently waiting for small mammals such as marsupial mice and rats to move within striking range. *Eunectes* often eat large prey such as caimans and capabaras, after which they lie in water for days with the large lump in their stomach floating high on the water. Others, such as *Candoia*, capture some endotherms but appear to eat mostly lizards and frogs. All are viviparous. Different populations within the *Candoia carinata* complex have litters ranging from 4 to 6 neonates in some to 40 and 50 neonates in others. Thus, litter number is not strongly associated with body size. The large *B. constrictor* and *Eunectes* can produce as many as 60 to 70 young, but they usually produce many fewer. *Epicrates cenchria* present spectacular displays when disturbed in the field, in which they raise the coiled tail off the ground, swing it back and forth, and rapidly crawl off, often entering burrows. The combination of smooth, reflective scales and waving motion of the coiled tail produces brilliant flashes of bluish light that easily distract the attention of a would-be pursuer. A remarkable amount of information exists on the biology of tree boas in the genus *Corallus*, assembled by Robert W. Henderson. As a result, these are the best-known boids ecologically and behaviorally.

Comment: The karyotypic differences between American *Boa* and Madagascan boids, as well as their long independent evolutionary histories, argue for the recognition of *Sanzinia* as distinct from *Boa*. The "Boinae" is likely polyphyletic.

Erycinae

Sister taxon: Boinae.

Content: Four genera, *Calabaria, Charina, Eryx,* and *Gongylophis,* with 1, 2, 11, and 3 species, respectively.

Distribution: Disjunct, western North America and central Africa eastward through Asia to western China.

Characteristics: Prefrontals are widely separated medially, labial sensory pits are absent, and caudal vertebrae have forked neural arches.

Biology: The sand boas (*Eryx*), rosy boas, and rubber boas (*Charina*; Fig. 21.19) are semifossorial snakes, usually living in semiarid to arid habitats. *Charina bottae* is an exception occurring in moist, montane conifer forests. All

FIGURE 21.18 Representative boid snakes in the subfamily Boinae. Clockwise from upper left: Boa constrictor *Boa constrictor* (L. J. Vitt); Brazilian rainbow boa (L. J. Vitt); garden tree boa (L. J. Vitt); juvenile emerald tree boa (L. J. Vitt).

are moderate-sized snakes, typically less than 700 mm TL. They have robust, cylindrical bodies, short tails, blunt heads, and small eyes. In the Pacific Northwest of the United States, rubber boas can be very common and easily found in early summer under surface objects exposed to sun. It is not uncommon to find several under a single surface item that has small mammal burrows, suggesting that they may overwinter in mammal burrows. They are predominantly nocturnal or crepuscular foragers and prey mainly on small reptiles and mammals. All are viviparous with litter size usually less than 10 neonates. When disturbed in the field, they often roll into a tight ball and expose the blunt tail as a head mimic. Although the blunt tails appear scarred, they are born with the blunt tail with irregular scales.

FIGURE 21.19 Rosy boa *Charina trivirgata*, Erycinae (L. J. Vitt).

Comment: The erycines have been placed differently by different authors. The most recent analyses based on gene sequence data place them sister to a likely polyphyletic "Boidae" (which, in the context of the taxonomy that we use, would place them sister to a likely polyphyletic "Boinae").

Caenophidia

Acrochordidae

Wart Snakes or File Snakes

Classification: Reptilia; Diapsida; Sauria; Lepidosauria; Squamata.

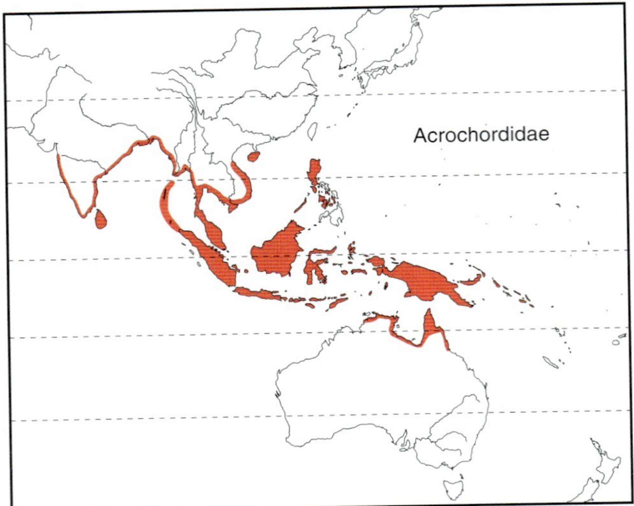

FIGURE 21.20 Geographic distribution of the extant Acrochordidae.

Sister taxon: Clade containing Viperidae, Atractaspididae, Colubridae, and Elapidae.

Content: One genus, *Acrochordus*, with 3 species.

Distribution: South and Southeast Asia to the Philippines and northern Australia (Fig. 21.20).

Characteristics: Acrochordids are small-headed and thick-bodied snakes; the skin is baggy, appearing several sizes too large for the body. The skin is covered dorsally and ventrally by numerous small, nonoverlapping, granular scales that have numerous bristle-tipped tubercles arising from the interscalar skin. Cranially, acrochordids have only a left carotid artery, edentulous premaxillaries, longitudinally oriented maxillaries with solid teeth, and optic foramina that perforate the parietal. The mandible lacks a coronoid bone, and the dentary bears numerous teeth. No cranial infrared receptors occur in pits or surface indentations. Girdle and limb elements are absent externally and internally. Intracostal arteries arise from the dorsal aorta at intervals of several trunk segments. The left lung is absent, and a tracheal lung is well developed; the left and right oviducts are well developed.

Biology: Acrochordids are large snakes, ranging in adult TL from about 800 mm to 1 m (*A. granulatus*) to 1.9 to 2.7 m (*A. javanicus*; Fig. 21.21); adult males are always significantly smaller than females. All three species are aquatic and largely incapable of terrestrial locomotion. *Acrochordus granulatus* is a brackish and marine species, *A. arafurae* is a freshwater resident, and *A. javanicus* occurs in both fresh and saltwater. All three feed principally on fish, and *A. arafurae* apparently exclusively so. Prey capture usually requires the fish to touch the anterior part of the snake's body, which triggers the snake to trap the fish in body loops and coils using the bristly tubercles for adhesion. The snake quickly shifts the fish forward in a wavelike action of the skin folds and rapidly swallows it. Acrochordids are viviparous, and litters range from 4 to 40 young, all born in the water. Clutch size is correlated with body size, and *A. arafurae* and *A. javanicus* are the most fecund.

Viperidae

Vipers and Pit Vipers

Classification: Reptilia; Diapsida; Sauria; Lepidosauria; Squamata.

Sister taxon: Uncertain, possibly a clade containing Atractaspididae, Colubridae, and Elapidae.

Content: Three subfamilies, Azemiopinae, Crotalinae, and Viperinae.

Distribution: Worldwide, except Papuaustralia and oceanic islands (Fig. 21.22).

Characteristics: Viperids are venomous snakes; a rotating fang apparatus allows the development of long fangs that are erected when biting and folded against the palate when the mouth is closed. Most viperids have robust bodies and distinctly triangular heads. Cranially, viperids

FIGURE 21.21 From left to right, little file snake *Acrochordus granulatus*, Acrochordidae (C. Siler); juvenile Arafura file snake *Acrochordus arafurae*, Acrochordidae (D. Nelson).

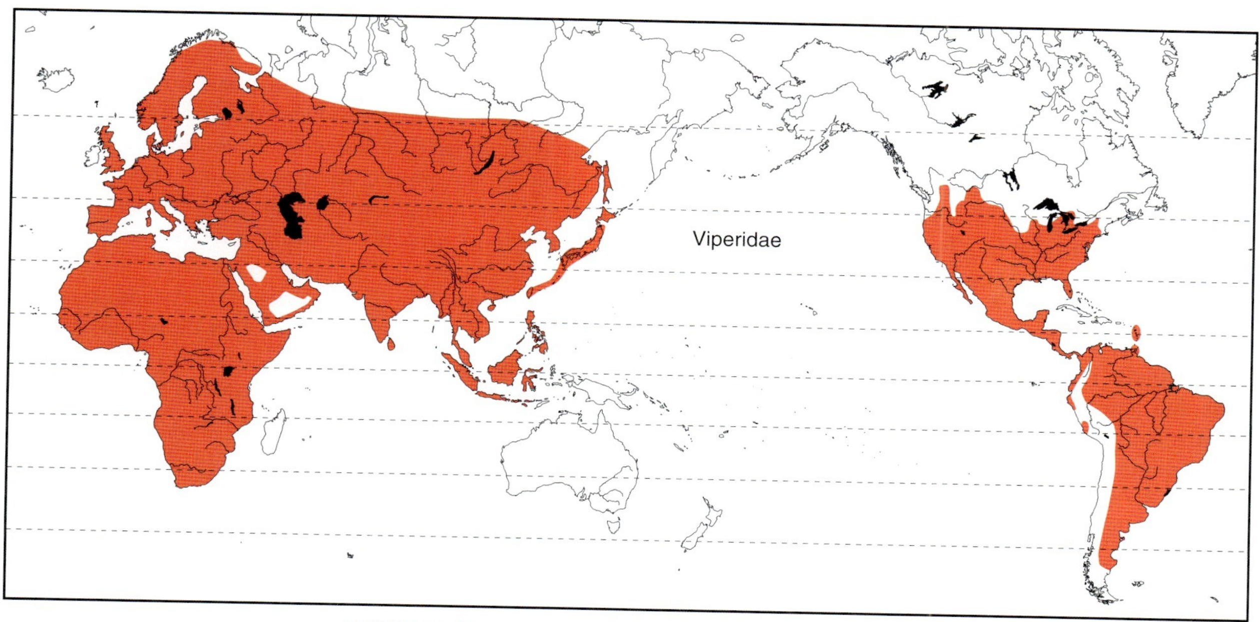

FIGURE 21.22 Geographic distribution of the extant Viperidae.

have only a left carotid artery, edentulous premaxillaries, blocklike rotating maxillaries with tubular teeth, and optic foramina that perforate the frontal–parietal or frontal–parietal–parasphenoid sutures. The mandible lacks a coronoid bone, and the dentary bears teeth. Cranial infrared receptors occur in loreal pits in crotalines or beneath scale surfaces in other taxa. Girdle and limb elements are absent externally and internally. Intracostal arteries arise from the dorsal aorta at intervals of several trunk segments. The left lung is usually absent or vestigial when present, and a tracheal lung is usually present; left and right oviducts are well developed.

Comment: The most recent molecular analysis suggests separating *Causus* into a separate subfamily, the Causinae. The subfamilial phylogeny would become (Causinae (Viperinae (Azemiopinae (Crotalinae)))).

FIGURE 21.23 Fea's viper *Azemiops feae*, Azemiopinae (R. W. Murphy).

Azemiopinae

Sister taxon: Clade containing Crotalinae and Viperinae.

Content: Monotypic, *Azemiops feae*.

Distribution: South-central China and adjacent areas of Burma and Vietnam.

Characteristics: *Azemiops* lacks a loreal pit, has a distinct choanal process on the palatine, has a large posteromedial orbital process on the prefrontal, and lacks a tracheal lung.

Biology: Fea's vipers are moderate-sized snakes (600-720 mm adult TL). They live at mid-elevations (800-1000 m) in moist montane forest, particularly among clumps of bamboo and tree ferns (Fig. 21.23). *Azemiops feae* appears to be semifossorial and lives in the forest litter in high humidity localities. The snakes spend the day beneath objects, often in wet situations, and emerge late at night to forage on and in the surface litter. Mammals are likely the major prey. *Azemiops* appears to dehydrate rapidly even in moderately dry conditions. Whether reproduction is oviparous or viviparous remains unknown.

Crotalinae

Sister taxon: Viperinae.

Content: Twenty-six genera, *Agkistrodon, Atropoides, Bothriechis, Bothriopsis, Bothrocophias, Bothrops, Calloselasma, Cerrophidion, Crotalus, Cryptelytrops, Deinagkistrodon, Gloydius, Hypnale, Lachesis, Ophryacus, Ovophis, Parius, Popeia, Porthidium, Protobothrops, Sistrurus, Triceratolepidophis, Trimeresurus, Tropidolaemus, Viridovipera*, and *Zhaoermia*, with more than 160 species.

Distribution: Southwest and southern Asia and the Americas.

Characteristics: Crotalines have a well-developed loreal pit for infrared receptors, have a small choanal process on the palatine, lack a posteromedial process on the prefrontal, and have a tracheal lung, except in *Lachesis*.

Biology: Crotalines are small to large snakes, ranging in adult TL from 300 to 660 mm in *Crotalus pricei* to a maximum 3.75 m in *Lachesis muta*. They are predominantly nocturnal snakes, and they use (not exclusively) their heat-sensory apparatus to locate prey. They prey mainly on vertebrates, usually birds and mammals in the larger crotaline species and amphibians and reptiles in the smaller ones; semiaquatic taxa eat fish and frogs. They occur in numerous habitats from deserts to cool mountain forests and wet tropical lowlands. Crotalines are mainly terrestrial, but a few taxa are semiaquatic, and 20+ tropical Asian and American species are arboreal (Fig. 21.24). In general, most appear to be long-lived species, maturing slowly and reproducing in 2-to-3-year cycles, except for the species in habitats with high prey density. Most crotalines are viviparous, although a few, such as *Calloselasma*, some *Trimeresurus*, and *Lachesis*, are oviparous and commonly attend eggs, suggesting some parental care. Litter or clutch size is generally associated with body size. Smaller species typically produce fewer eggs or young than larger ones; however, even the largest taxon, *L. muta*, produces only about a dozen eggs, and the much smaller *Sistrurus catenatus* averages nearly 12 neonates. Overall, crotalines produce about 10 eggs or neonates per reproductive event, and viviparous species tend to produce more offspring than oviparous ones of equivalent size.

Comment: The paraphyletic *Bothrops* has been divided into several genera, and other divisions have been proposed subsequently (e.g., *Trimeresurus*). We generally recognize all proposed genera; however, species content and even the recognition of the various genera continue to be actively investigated.

Viperinae

Sister taxon: Crotalinae.

Content: Thirteen genera, *Adenorhinos, Atheris, Bitis, Causus, Cerastes, Daboia, Echis, Eristocophis, Macrovipera, Monatatheris, Proatheris, Pseudocerastes,* and *Vipera,* with more than 65 species.

FIGURE 21.24 Representative viperids. Clockwise from upper left: Prairie rattlesnake *Crotalus viridis*, Crotalinae (L. J. Vitt); speckled forest pit viper, *Bothriopsis taeniata*, Crotalinae (L. J. Vitt); Brazilian lance-head pit viper *Bothrops moojeni*, Crotalinae (L. J. Vitt); ottoman viper *Vipera xanthina*, Viperinae (R. W. Barbour).

Distribution: Africa, Europe, and Asia.

Characteristics: Viperines lack a loreal pit, a choanal process on the palatine, and a posteromedial process on the prefrontal; all have a tracheal lung, except for *Bitis atropos*.

Biology: Viperines are modest-sized snakes; none is known to exceed 2 m SVL, and most taxa are less than 1 m adult SVL (Fig. 21.24). *Bitis* contains the largest species (*B. arietans*, *B. gabonica*, and *B. nasicornis*, all with maximum adult SVLs of 1.4 m or larger) and some of the smallest species (*B. peringueyi*, 300 mm maximum adult SVL). Most viperines are terrestrial, although a few forage low in bushes, and *Atheris* is arboreal. They occur in forest to desert habitats and from equatorial to subarctic regions. Although viperines are commonly labeled as diurnal species, many forage nocturnally; the activity patterns of most taxa are associated with climate and seasonal temperature regimes. For example, the European *Vipera* is diurnal and the desert *Cerastes* is nocturnal. Viperines prey mainly on small vertebrates. Viperines include oviparous taxa (e.g., *Causus*, *Echis coloratus*) and viviparous taxa (e.g., *Bitis*, *Echis carinatus*, most *Vipera*). Clutch or litter size is moderate in most taxa, usually not exceeding 10 eggs or neonates, but the large-bodied species of *Bitis* produces 40 to 100 neonates.

"Colubridae"

Colubrids

Classification: Reptilia; Diapsida; Sauria; Lepidosauria; Squamata.

Sister taxon: The clade containing Atractaspidae and Elapidae.

Content: Seven subfamilies, Lamprophiinae, Colubrinae, Homalopsinae, Natricinae, Pareatinae, Xenodermatinae, and Xenodontinae.

Distribution: Worldwide, except Antarctica and oceanic islands (Fig. 21.25).

Characteristics: Colubrids represent the most structurally diverse group of snakes and include aglyphous, opisthoglyphous, and proteroglyphous taxa. Cranially, colubrids have only a left carotid artery, edentulous premaxillaries, usually longitudinally oriented maxillaries with solid or grooved teeth, and optic foramina that usually perforate the frontal–parietal–parasphenoid sutures. The mandible lacks a coronoid bone, and the dentary bears teeth. No cranial infrared receptors occur in pits or surface indentations. Girdle and limb elements are absent externally and internally. Intracostal arteries arise from the dorsal aorta at intervals of several trunk segments. The left lung is greatly reduced or more often absent, and a tracheal lung can be present or absent; left and right oviducts are well developed.

Comment: Taxonomy and relationships among colubrid snakes has been and will continue to be in a state of flux. Colubridae contains all snakes that cannot be easily placed in one of the other major colubroid families (Elapidae and Atractaspidae), which amounts to more than about 60% of all known snake species! Because no synapomorphy defines the family, it is clearly paraphyletic. The number of subfamilies recognized varies from at least 13 to as few as 6. The following subfamilies represent only those colubrid groups that are widely recognized by herpetologists. The generic content, especially of the larger groups,

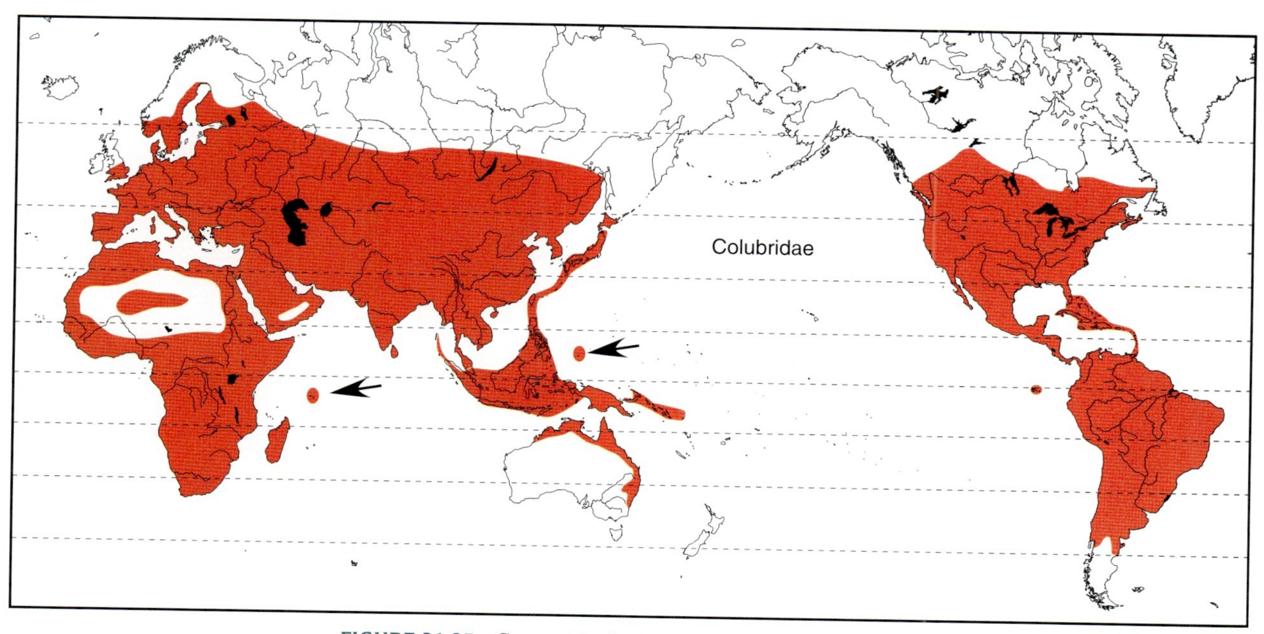

FIGURE 21.25 Geographic distribution of the extant Colubridae.

also attempts to reflect a general, certainly not unanimous, consensus. We have relied primarily on Zaher's (1999) list of colubrid genera and subfamilial assignments. Because we recognize fewer subfamilies, we assigned his incertae sedis taxa to our different subfamilies. We make no assumptions about the monophyly of the subfamilies that we recognize, and by necessity, our lists of genera are incomplete.

"Colubrinae"

Sister taxon: Uncertain.

Content: More than 100 genera, including *Aeluroglena, Ahaetulla, Argyrogena, Arizona, Blythia, Bogertophis, Boiga, Calamaria, Calamorhabdium, Cemophora, Cercaspis, Chilomeniscus, Chionactis, Chironius, Chrysopelea, Collorhabdium, Coluber, Conopsis, Coronella, Crotaphopeltis, Cryptophidion, Cyclocorus, Dasypeltis, Dendrelaphis, Dendrophidion, Dinodon, Dipsadoboa, Dispholidus, Dryadophis, Drymarchon, Drymobius, Drymoluber, Dryocalamus, Dryophiops, Eirenis, Elachistodon, Elaphe, Elapoidis, Entechinus, Etheridgeum, Ficimia, Geagras, Gongylosoma, Gonyophis, Gonyosoma, Gyalopion, Haplocercus, Hapsidophrys, Helophis, Hemirhagerrhis, Hierophis, Ithycyphus, Lampropeltis, Leptodrymus, Leptophis, Lepturophis, Liochlorophis, Liopeltis, Lycodon, Lytorhynchus, Macrocalamus, Macroprotodon, Malpolon, Masticophis, Mastigodryas, Meizodon, Myersophis, Oligodon, Opheodrys, Oreocalamus, Oxybelis, Pantherophis, Philothamnus, Phyllorhynchus, Pituophis, Poecilopholis, Prosyma, Psammophis, Psammophylax, Pseudocyclophis, Pseudoficimia, Pseudorabdion, Pseustes, Ptyas, Rabdion, Rhabdops, Rhamphiophis, Rhamnophis, Rhinobothryum, Rhinocheilus, Rhynchocalamus, Rhynchophis, Salvadora, Scaphiodontophis, Scaphiophis, Scolecophis, Senticolis, Sibynophis, Simophis, Sonora, Spalerosophis, Spilotes, Stegonotus, Stenorrhina, Stilosoma, Symphimus, Sympholis, Tantilla, Tantillita, Telescopus, Tetralepis, Thelotornis, Thermophis, Thrasops, Toluca, Trachischium, Trimorphodon, Xenelaphis,* and *Zaocys,* with ±650 species.

Distribution: Worldwide, as the family.

Biology: Colubrines are highly diverse in body form, and in ecology and behavior (Fig. 21.26). They range from small (160–190 mm TL, *Tantilla relicta*) to very large (e.g., 3.7 m TL, *Ptyas carinatus*). Body forms may be slender (*Sibynophis*), elongate viperine (*Boiga*), racer-like

FIGURE 21.26 Representative colubrids. Clockwise from upper left: Scarlet snake *Cemophora coccinae*, Colubrinae (L. J. Vitt); yellow-belly water snake *Enhydris plumbae*, Homalopsinae (D. R. Karns); Madagascaran hognose snake *Lioheterodon madagascariensis*, Lamprophiinae (H. I. Uible); banded water snake *Nerodia fasciata*, Natricinae (L. J. Vitt).

(*Masticophis*), or muscular serpentine (*Chironius*), as well as many others. Colubrines occur from brackish water habitats to high montane forest; some are desert inhabitants, whereas others are aquatic. Some are burrowers, many are terrestrial or semiarboreal, and others are arboreal. Species may be diet generalists or specialists. Generalists often prey on small vertebrates and occasionally invertebrates; specialists may eat only orthopteran insects (*Opheodrys*) or birds (*Thelotornis*). Colubrines are predominantly oviparous; the few viviparous species are usually small snakes. Clutch size generally is associated with body size. The small-bodied *Tantilla gracilis* produces clutches of 1 egg, and the much larger *Pantherophis obsoleta* has clutches to 40 eggs; however, most species produce clutches of 10 or fewer eggs.

Comment: Generic content derives primarily from Zaher's (1999) Colubrinae, Calamariinae, Psammophiinae, and Colubridae *incertae sedis*.

Homalopsinae

Sister taxon: Uncertain, possibly Pareatinae.

Content: Eleven genera, *Bitia*, *Brachyorrhos*, *Cantoria*, *Cerberus*, *Enhydris*, *Erpeton*, *Fordonia*, *Gerarda*, *Heurnia*, *Homalopsis*, and *Myron*, with 35+ species.

Distribution: Southern Asia from India to China and south to northern Australia.

Characteristics: Homalopsines are distinguished from other colubrids by valvular, crescentic, dorsal nostrils; small, dorsally oriented eyes (eye diameter less than vertical distance from bottom of orbit to mouth); nasal scales usually larger than internasals; and the last two or three maxillary teeth enlarged and grooved with well-developed venom (Duvernoy) glands.

Biology: Homalopsines are aquatic snakes and live in a variety of freshwater, brackish, and marine habitats, typically in shallow water (Fig. 21.26). Envenomation is an important aspect of prey capture for all taxa. Prey is bitten and held; a chewing action introduces the venom into the prey, and once subdued, the prey is swallowed. Most freshwater homalopsines eat fish or fish and frogs. *Fordonia leucobalia* is a crab specialist; it first pins the crab beneath a body loop and then bites and envenomates it. For all taxa, foraging appears to be mainly nocturnal, and most actively search for prey. Homalopsines are small (200–380 mm adult SVL, *Myron richardsonii*) to large (1.4 m maximum TL, *Homalopsis buccata*). All homalopsines are viviparous. Litter size is modest, from 5 to 15 neonates in most species, but larger individuals and larger species can have 20 to 30 young.

"Lamprophiinae"

Sister taxon: Uncertain.

Content: Forty-four genera, *Alluaudina*, *Boaedon*, *Bothrolycus*, *Bothrophthalmus*, *Brygophis*, *Buhoma*, *Chamaelycus*, *Compsophis*, *Cryptolycus*, *Dendrolycus*, *Dipsina*, *Ditypophis*, *Dromicodryas*, *Dromophis*, *Duberria*, *Exallodontophis*, *Gastropyxis*, *Geodipsas*, *Gonionotophis*, *Grayia*, *Heteroliodon*, *Hormonotus*, *Ithycyphus*, *Lamprophis*, *Langaha*, *Lioheterodon*, *Liophidium*, *Liopholidophis*, *Lycodonomorphus*, *Lycodryas*, *Lycophidion*, *Madagascarophis*, *Mehelya*, *Micropisthodon*, *Mimophis*, *Montaspis*, *Pararhadinaea*, *Polemon*, *Pseudaspis*, *Pseudoboodon*, *Pseudoxyrhopus*, *Pythonodipsas*, *Scaphiophis*, and *Stenophis*, with 205+ species.

Distribution: Sub-Saharan Africa and Madagascar.

Biology: Lamprophiines are a moderately diverse group (Fig. 21.26). They are mainly terrestrial to semifossorial, but a few (e.g., *Langaha*, *Stenophis*) are arboreal. The majority of species are less than 1 m TL, although some genera (e.g., *Brygophis*, *Dromophis*, *Leioheterodon*, *Mehelya*) have a maximum TL of 1.0 to 1.5 m, and *Grayia smythii* reaches 2.6 m. Body form ranges from typical terrestrial racer-habitus to blunt-headed, cylindrical-bodied burrowers and also includes big-headed, thin-bodied arboreal forms. Most taxa prey upon vertebrates, and none appears to be a dietary specialist. Lamprophiines are nearly exclusively oviparous, although *Stenophis* has both oviparous and viviparous species. Clutch size tends to be small, commonly less than 10 eggs per clutch even in some large-bodied species, but the large *Scaphiophis* has clutches up to 48 eggs.

Comment: Generic content is derived from Zaher's (1999) Boodontinae, Boodontinae *incertae sedis*, and Pseudoxyrhophiinae.

"Natricinae"

Sister taxon: Uncertain.

Content: Thirty-eight genera, *Adelophis*, *Afronatrix*, *Amphiesma*, *Amphiesmoides*, *Amplorhinus*, *Anoplohydrus*, *Aspidura*, *Atretium*, *Balanophis*, *Clonophis*, *Geodipsas*, *Haplocercus*, *Helophis*, *Hologerrhum*, *Hydrablabes*, *Hydraethiops*, *Iguanognathus*, *Limnophis*, *Natriciteres*, *Natrix*, *Nerodia*, *Opisthotropis*, *Parahelicops*, *Pararhabdophis*, *Plagiopholis*, *Psammodynastes*, *Pseudagkistrodon*, *Pseudoxenodon*, *Regina*, *Rhabdophis*, *Seminatrix*, *Sinonatrix*, *Storeria*, *Thamnophis*, *Tropidoclonion*, *Tropidonophis*, *Virginia*, and *Xenochrophis*, with 195+ species.

Distribution: North America to northern Central America, Africa, and Eurasia through the East Indies.

Biology: Natricines are small (160–250 mm adult SVL, *Virginia striatula*) to large (1.4–2.0 m maximum TL, *Natrix*, *Nerodia*, and *Xenochrophis*). Many species are labeled aquatic, and though these natricines feed and hide in water, they regularly exit the water for basking and reproduction in contrast to the aquatic homalopsines or acrochordids. The aquatic species are with few exceptions freshwater inhabitants, and the exceptions, such as *Nerodia fasciata* (Fig. 21.26), have some populations behaviorally and physiologically adapted to saltwater. Most other natricines are terrestrial to semifossorial, the majority of which live in moist habitats from marsh to forest. The aquatic species prey predominantly on fish and amphibians, but a few, like the crayfish-eating *Regina septemvittata*, are dietary specialists. Other species, generally the smaller ones or juveniles of larger species, eat slugs, snails, earthworms, and soft-bodied arthropods. American natricines are exclusively viviparous, whereas the Old World taxa are largely, but not exclusively, oviparous. Clutch size tends to be modest (2–20 eggs) in the oviparous taxa and even in the large-bodied taxa (e.g., 10–40 eggs, *Xenochrophis*). Litter size is somewhat larger in equivalent-sized viviparous species, although the prodigious 80 to 100 fetuses reported for *Nerodia cyclopion* is uncommon.

Comment: Kraus and Brown (1998) suggest that natricines are diphyletic; the New World thamnophines are monophyletic, whereas the Old World natricines are paraphyletic. Generic content is derived from Zaher's (1999) Natricinae, Natricinae *incertae sedis*, and Pseudoxenodontinae.

Pareatinae

Sister taxon: Uncertain, possibly Homalopsinae.

Content: Three genera, *Aplopeltura*, *Internatus*, and *Pareas*, with 18+ species.

Distribution: Southeast Asia from eastern India to China and southward to Java, Borneo, and Minanao.

Characteristics: Pareatines have a blunt snout, lack a mental groove, and have no teeth on the anterior part of the maxillary.

Biology: Pareatines are called slug-eating snakes because of their specialized diet of slugs and snails. The long slender body and oversized head are convergent with morphology seen in New World snail specialists. This morphology is an adaptation for slow arboreal searching on the slender twigs and branches at the ends of limbs and for traversing wide gaps. All taxa are moderate sized and have adults that range between 450 to 900 mm TL, although they appear small because of their slender body form. They forage at night, and upon finding a snail, they slide their lower jaw beneath the snail and the shell and bite the body. They use their teeth and independent jawbones in a ratchet-like fashion to exert a continuous pulling pressure on the snail's body, which eventually relaxes and is then ripped from its shell attachment. All pareatines are oviparous and have small clutches of two to eight eggs.

Xenodermatinae

Sister taxon: Uncertain, possibly all other Colubridae.

Content: Six genera, *Achalinus*, *Fimbrios*, *Oxyrhabdium*, *Stoliczkaia*, *Xenodermus*, and *Xylophis*, with 15 species.

Distribution: Disjunct, Assam, northern Indochina and adjacent China to Japan, and peninsular Malaysia, Sumatra, Java, and Borneo.

Characteristics: Xenodermatines have small orbits, from which the optic nerve exits between the parietal and frontal; the ophthalmic nerve exits through a foramen in the parietal, a unique characteristic, and they have numerous (> 20) small maxillary teeth.

Biology: Xenodermatines are a small group of peculiar snakes, generally living in moist forest habitats. They are small-to-moderate-sized, slender-bodied snakes; the maximum TL is less than 800 mm, but most individuals and species are less than 550 mm TL. All are secretive snakes, probably nocturnal, and either forest-floor or low arboreal foragers. The little dietary data available suggest that they are opportunistic carnivores and that vertebrates are their major prey. Limited reproductive data indicate that all are oviparous and have small clutch size, reportedly four or fewer eggs.

Xenodontinae

Sister taxon: Uncertain, possibly all other Colubridae exclusive of Xenodermatinae.

Content: Over 90 genera, including *Adelphicos, Alsophis, Amastridium, Antillophis, Apostolepis, Arrhyton, Atractus, Boiruna, Calamodontophis, Carphophis, Cercophis, Chersodromus, Clelia, Coniophanes, Conophis, Contia, Crisantophis, Cryophis, Darlingtonia, Diadophis, Diaphorolepis, Dipsas, Ditaxodon, Drepanoides, Echinanthera, Elapomorphus, Emmochiliopis, Enuliophis, Enulius, Eridiphas, Erythrolamprus, Farancia, Geophis, Gomesophis, Helicops, Heterodon, Hydrodynastes, Hydromorphus, Hydrops, Hypsiglena, Hypsirhynchus, Ialtris, Imantodes, Leptodeira, Lioheterophis, Liophis, Lystrophis, Manolepis, Ninia, Nothopsis, Opisthoplus, Oxyrhopus, Parapostolepis, Phalotris, Philodryas, Phimophis, Pliocercus, Pseudablabes, Pseudoboa, Pseudoeryx, Pseudoleptodeira, Pseudotomodon, Psomophis, Ptycophis, Rhachidelus, Rhadinaea, Rhadinophanes, Saphenophis, Sibon, Sibynomorphus, Siphlophis, Sordellinia, Synophis, Tachymenis, Taeniophallus, Tantalophis, Thalesius, Thamnodynastes, Tomodon, Tretanorhinus, Trimetopon, Tripanurgos, Tropidodipsas, Tropidodryas, Umbrivaga, Uromacer, Uromacerina, Urotheca, Waglerophis, Xenodon, Xenopholis,* and *Xenoxybelis,* with 540+ species.

Distribution: Americas.

Biology: Xenodontines are highly diverse in body form, ecology, and behavior (Fig. 21.27). Most xenodontines are small-to-moderate-sized snakes (less than 800 mm adult TL); less than a dozen genera have adults greater than 1 m SVL, e.g., *Alsophis, Clelia, Farancia, Hydrodynastes,* and *Uromacer*. Body form ranges from small and slender (*Diadophis*) through heavy bodied (*Xenodon*) to racer-like (*Philodryas*). Arboreal xenodontines display two body forms. Diurnal hunters have long, muscular bodies and elongate, pointed heads (e.g., *Uromacer*), whereas nocturnal searchers are slender bodied

FIGURE 21.27 Representative colubrids. Clockwise from upper left: Gomes's pampas snake *Phimophis eglasiasi*, Xenodontinae (L. J. Vitt); Aesculapiam false coral snake *Erythrolamprus aesculapii*, Xenodontinae (L. J. Vitt); keel-back water snake *Helicops angulatus*, Xenodontinae (L. J. Vitt); southern hognose snake *Heterodon simus*, Xenodontinae (L. J. Vitt).

and have blunt oversized heads (e.g., *Dipsas*, *Imantodes*). They occur in all habitats but marine ones, although some taxa are aquatic in freshwater. Some species burrow, while others are terrestrial or arboreal. A majority of the species appears to be generalists or dietary opportunists that eat predominantly small vertebrates. Some species are prey specialists, such as the snail- and slug-eating *Dipsas* and *Sibon*. Xenodontines are predominantly oviparous. Clutch size generally has a direct association with body size and ranges from small clutches of 1 to 3 eggs (*Imatodes cenchoa*) to over 100 eggs (*Farancia abacura*).

Comment: Generic content is derived from Zaher's (1999) Dipsadinae, Dipsadinae *incertae sedis*, Xenodontinae, and Xenodontinae *incertae sedis*.

Atractaspididae

Stiletto Vipers or Mole Vipers

Classification: Reptilia; Diapsida; Sauria; Lepidosauria; Squamata.

Sister taxon: Elapidae.

Content: Two subfamilies, Aparallactinae and Atractaspidinae.

Distribution: Sub-Saharan Africa and the bordering coast of Arabian Peninsula including Israel and Jordan (Fig. 21.28).

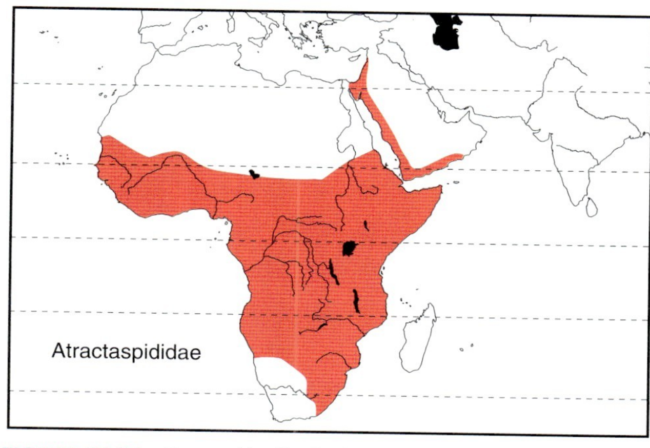

FIGURE 21.28 Geographic distribution of the extant Atractaspididae.

Characteristics: Atractaspidids are slender, cylindrical-bodied snakes with short, usually blunt-snouted head, small eyes, and a short tail (Fig. 21.29). All taxa are often labeled venomous and considered to possess "venom glands." However, *Atractaspis* is the only taxon capable of a fatal envenomation of humans, although *Macrelaps* is potentially lethal. Cranially, atractaspidids have only a left carotid artery, edentulous premaxillaries, longitudinally oriented maxillaries with enlarged, tubular anterior teeth, and optic foramina that perforate the frontal–parietal–parasphenoid sutures. The mandible lacks a coronoid bone, and the dentary bears teeth. No cranial infrared receptors occur in pits or surface indentations. Girdle and limb elements are absent externally and internally. Intracostal arteries arise from the dorsal aorta at intervals of several trunk segments. The left lung is greatly reduced or absent, and a tracheal lung is also present or absent; left and right oviducts are well developed.

Aparallactinae

Sister taxon: Atractaspidinae.

Content: Ten genera, *Amblyodipsas*, *Aparallactus*, *Brachyophis*, *Chilorhinophis*, *Elapotinus*, *Hypoptophis*, *Macrelaps*, *Micrelaps*, *Polemon*, and *Xenocalamus*, with 40+ species.

Distribution: Sub-Saharan Africa.

Characteristics: Enlarged, grooved teeth occur on each maxilllary, either posteriorly (opisthoglyphous) or anteriorly (proteroglyphous); a tracheal lung is present, and the left lung is often absent.

Biology: Aparallactines are small (200–300 mm adult SVL, *Aparallactus nigriceps*) to large (about 1.1 m maximum TL, *Macrelaps microlepidotus*). All are terrestrial to semifossorial snakes, occurring in a variety of habitats from grassland to moist forest. *Aparallactus* is generally a centipede specialist; the other taxa prey mainly on small vertebrates that live in or on the surface litter. These snakes include oviparous and viviparous species; clutch or litter size is small, usually less than 10 eggs or young.

FIGURE 21.29 Bibron's mole viper *Atractaspis bibronii*, Atractaspidinae (W. R. Branch).

Atractaspidinae

Sister taxon: Aparallactinae.

Content: Two genera, *Atractaspis* and *Homoroselaps*, with 19 and 2 species, respectively.

Distribution: As for the family (Fig. 21.28).

Characteristics: Each maxillary bears a large semierect fang anteriorly, a tracheal lung is present or absent, and usually the left lung is present but small.

Biology: *Atractaspis* is a venomous, highly fossorial taxon. All species are blunt headed, apparently capable of using their heads in burrowing, although they are likely dependent upon the burrows of their mammalian prey. Because they live and feed subterraneanly, they cannot use the typical snake strike to achieve envenomation. Instead, they crawl alongside their prey (mainly newborn rodents and burrowing reptiles), depress their lower jaw and shift it toward the opposite side thereby exposing their exceptionally long fangs, and with a backward stab, envenomate the prey. *Atractaspis* is oviparous and lays small clutches of 2 to 11 eggs. For most species, adults range from 400 to 600 mm TL. *Homoroselaps* was formerly placed in the Elapidae.

Elapidae

Cobras, Kraits, Sea Snakes, Death Adders, and Allies

Classification: Reptilia; Diapsida; Sauria; Lepidosauria; Squamata.

Sister taxon: Uncertain, possibly Colubridae or Atractaspididae.

Content: Two subfamilies, Elapinae and Hydrophiinae.

Distribution: Southern North America to southern South America, Africa, southern Asia to southern Australia, and the tropical Indian and Pacific Oceans (Fig. 21.30).

Characteristics: Elapids are venomous snakes that have an erect fang anteriorly on each maxillary bone. Cranially, they have only a left carotid artery, edentulous premaxillaries, longitudinally oriented, shortened maxillaries with anterior teeth that are large and tubular, and optic foramina usually perforate the frontal–parietal–parasphenoid sutures. The mandible lacks a coronoid bone, and the dentary bears teeth. No cranial infrared receptors occur in pits or surface indentations. The girdle and limb elements are absent externally and internally. Intracostal arteries arise from the dorsal aorta at intervals of several trunk segments. The left lung is greatly reduced or absent; a tracheal lung is commonly present in the marine taxa and absent in terrestrial ones. Left and right oviducts are well developed.

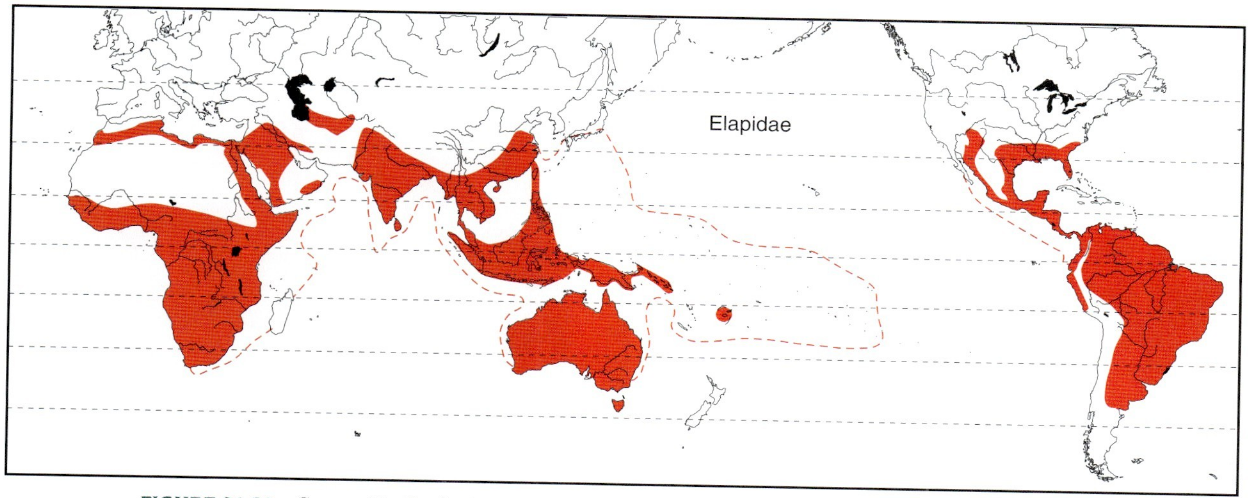

FIGURE 21.30 Geographic distribution of the extant Elapidae. Dashed red lines indicate distribution within the oceans of the Hydrophinae.

Comment: Relationships among elapids remain unresolved with each new phylogeny, suggesting different sister-group relationships and different subfamily content. Most recent analyses agree that the two marine snake clades (sea kraits or *Laticauda*; *Hydrophis* and other sea snakes) arose independently from within the Australian terrestrial elapids. However, the possibility exists of three independent origins of marine snakes.

"Elapinae"

Sister taxon: Hydrophiinae or the clade containing hydrophiines and *Bungarus*.

Content: Seventeen genera, *Aspidelaps*, *Boulengerina*, *Bungarus*, *Calliophis*, *Dendroaspis*, *Elapsoidea*, *Hemachatus*, *Hemibungarus*, *Homoroselaps*, *Maticora*, *Micruroides*, *Micrurus*, *Naja*, *Ophiophagus*, *Paranaja*, *Pseudohaje*, and *Walterinnesia*, with 130+ species.

Distribution: Americas, Africa, and Eurasia through the East Indies to the Philippines.

Characteristics: The palatine bones have choanal processes, except in *Dendroaspis*.

Biology: Elapines comprise a moderately diverse group of predominantly terrestrial snakes (Fig. 21.31). Of the 17 genera, only two African taxa (mambas, *Dendroaspis*; tree cobras, *Pseudohaje*) are arboreal, and only the African *Boulengerina* is aquatic. The remainder of the taxa are semifossorial (*Aspidelaps*, *Calliopsis*, *Micrurus*) and surface foragers (*Bungarus*, *Naja*). The semifossorial or surface-litter foragers are commonly brightly patterned (aposematic), presumably to alert potential predators of their venomous bite, and most are called coral snakes. Adult size in elapines ranges from small (less than 500 mm TL) for some of the semifossorial taxa to the very large king cobra, *Ophiophagus*, which attains lengths to 5.8 m. Adults of most species are less than 4 m TL. The kraits and various cobras commonly range in adult TL from 1 to 2 m. All elapines are mainly vertebrate predators; for example, *Boulengerina* eats exclusively fish, *Micrurus* mainly snakes and lizards, *Hemachatus* anurans, and *Dendroaspis* birds and mammals. Elapines are mostly oviparous, but a few species are viviparous (e.g., *Hemachatus*). Clutch size is generally associated with body size; smaller species tend to produce 10 or fewer eggs, and the larger species commonly produce more than 20 eggs.

Hydrophiinae

Sister taxon: Elapinae.

Content: Forty-three genera, *Acalyptophis*, *Acanthophis*, *Aipysurus*, *Aspidomorphus*, *Astrotia*, *Austrelaps*, *Cacophis*, *Demansia*, *Densonia*, *Drysadalia Echiopsis*, *Elapognathus*, *Emydocephalus*, *Enhydrina*, *Ephalophis*, *Furina*, *Hemiaspis*, *Hoplocephalus*, *Hydrelaps*, *Hydrophis*, *Kerilia*, *Kolphophis*, *Lapemis*, *Laticauda*, *Loveridgelaps*, *Micropechis*, *Notechis*, *Ogmodon*, *Oxyuranus*, *Parahydrophis*, *Parapistocalamus*, *Pelamis*, *Praescutata*, *Pseudechis*, *Pseudonaja*, *Rhinoplocephalus*, *Salomonelaps*, *Simoselaps*, *Suta*, *Thalassophis*, *Toxicocalamus*, *Tropidechis*, and *Vermicella*, with 165+ species.

Distribution: Papuaustralia and tropical Indian and Pacific Oceans.

Characteristics: The palatine bones lack choanal processes.

Biology: The hydrophiines contain terrestrial and aquatic taxa (Fig. 21.31). The aquatic group contains two clades, both of which are marine and show distinct adaptations for an aquatic existence. The terrestrial species include semifossorial and surface foragers. A few surface foragers (e.g., *Tropidechis*) occasionally climb low in shrubs or trees, but none is truly arboreal; the same situation exists for the taxa that forage in or near water (e.g., *Notechis ater*),

FIGURE 21.31 Representative elapid snakes. Clockwise from upper left: Cerrado coral snake *Micrurus brasiliensis* (L. J. Vitt); Phillipine karait *Hemibungarus calligaster*, Elapinae (R. M. Brown); yellow-lip sea krait *Laticauda colubrina*, Hydrophiinae (G. R. Zug); curl snake *Suta suta*, Hydrophiinae (T. Schwaner).

for they are at best semiaquatic. These terrestrial taxa range from small snakes (200–400 mm adult SVL, *Drysadalia*) to large ones (0.8–2.2 m SVL, *Oxyuranus*). Their prey is composed nearly exclusively of vertebrates and lizards. Terrestrial hydrophiines include both oviparous and viviparous species.

The "true" sea snakes include a diverse array of genera (14; e.g., *Aipysurus*, *Hydrophis*, *Thalassophis*). All are totally aquatic; their laterally compressed bodies, paddle-like tail, and loss of enlarged ventral scales and associated muscular links make them incapable of terrestrial locomotion. Most species are 750 mm to 1.5 m SVL (maximum to 2.7 m TL, *Hydrophis spiralis*). Even though they eat mostly fish, they are often specialists, eating only certain types of fish or fish of a limited size range. Surprisingly, they largely avoid invertebrates. All sea snakes are viviparous, and birth occurs in the water. Litter size varies from 1 to 30, but most species produce litters of fewer than 10 neonates.

The sea kraits, *Laticauda*, occupy the middle ground between the terrestrial hydrophiines and the sea snakes. Aside from less reduced ventral scales than the sea snakes, they regularly come ashore and have good terrestrial locomotion. As egg layers, they must lay their eggs on land, and *L. colubrina* seemingly always comes ashore to digest its food, mainly eels. In spite of large adult size (0.8–1.0 m SVL), they produce moderate-sized clutches of 1 to 10 eggs.

QUESTIONS

1. Which family of snakes would you expect to be the dominant family in Australia and why do think that family has been able to dominate the snake fauna?
2. What characteristics suggest that "snakes" are a monophyletic group of limbless lizards?
3. Describe some of the major differences (morphologically and ecologically) between scolecophidian and alethinophidian snakes.
4. Describe, in as much detail as possible, morphological and ecological diversity within colubrid snakes.

REFERENCES

General

Ananjeva et al., 1988; Bauchot, 1994; Broadley, 1983; Cogger and Zweifel, 1998; David and Vogel, 1996; Dixon and Werler, 2000; Ernst and Ernst, 2003; Ernst and Zug, 1996; Gibbons and Dorcas, 2004; Grandison,

1977; Greene, 1997; Greer, 1997; Mattison, 2007; Rossman et al., 1996; Seigel and Collins, 1993; Seigel et al., 2002; Shine, 1991b.

Systematic

Burbrink et al., 2000; Cadle, 1987, 1992; Cundall et al., 1993; David and Ineich, 1999; Dowling et al., 1996; Gower et al., 2005; Gravlund, 2001; Heise et al., 1995; Hoffstetter, 1955, 1962; Kelly et al., 2003; Keogh, 1998; Kluge, 1991; Kraus and Brown, 1998; Lawson et al., 2005; Lee and Scanlon, 2002; Liem et al., 1971; McCarthy, 1985; McDiarmid et al., 1999; McDowell, 1969, 1974, 1987; Rieppel, 1988; Scanlon and Lee, 2004; Slowinski and Keogh, 2000; Slowinski and Lawson, 2002, 2005; Smith et al., 1977; Underwood, 1967; Underwood and Kochva, 1993; Vidal and Hedges, 2002a,b, 2004, 2005; Wallach and Günther, 1998; Wilcox et al., 2002; Zaher, 1994a,b, 1999.

Taxonomic Accounts

Anomalepididae

Dawn Blind Snakes

Dixon and Kofron, 1983; Kofron, 1988; Lancini and Kornacker, 1989; Martins and Oliveira, 1998; McDiarmid et al., 1999; Underwood, 1967; Wallach, 1998; Wallach and Günther, 1997.

Leptotyphlopidae

Slender Blind Snakes, Thread Snakes

Gelbach and Balbridge, 1987; Hahn and Wallach, 1998; Martins and Oliveira, 1998; McDiarmid et al., 1999; Underwood, 1967; Wallach, 1998; Watkins, et al., 1969; Webb et al., 2000.

Typhlopidae

Blind Snakes

Branch, 1988; Broadley, 1983; Ehmann and Bamford, 1993; Fitch, 1970; Greer, 1997; Martins and Oliveira, 1998; McDiarmid et al., 1999; Underwood, 1967; Wallach, 1993, 1998; Wynn et al., 1987.

"Tropidophiidae"

Dwarf Boas

Lawson et al., 2004; McDiarmid et al., 1999; Underwood, 1967; Gower et al., 2005; Wallach, 1998; Wallach and Günther, 1998; Wilcox et al., 2002; Zaher, 1994a.

Tropidophiinae

Schwartz and Henderson, 1991; Tolson and Henderson, 1993.

Ungaliophiinae

Campbell and Camarillo R., 1992; Villa and Wilson, 1990; Wilcox et al., 2002.

Xenophidioninae

Manthey and Grossmann, 1997; Steubing and Inger, 1999; Wallach, 1998; Wallach and Günther, 1998.

Aniliidae

False Coral Snakes, South American Pipe Snakes

Gower et al., 2005; Martins and Oliveira, 1998; McDiarmid et al., 1999; Starace, 1998; Underwood, 1967.

Cylindrophiidae

Pipe Snakes and Dwarf Pipe Snakes

Adler et al., 1992; Cadle et al., 1990; Cox et al., 1998; Cundall, 1995; Cundall and Rossman, 1993; David and Vogel, 1996; de Lang and Vogel, 2005; Gower et al., 2005; Manthley and Grossmann, 1997; McDiarmid et al., 1999; Nanayakkara, 1988; Smith, 1943; Steubing, 1994; Steubing and Inger, 1999; Underwood, 1967; Wallach, 1998.

Uropeltidae

Shield-Tail Snakes

Cadle et al., 1990; Gans, 1976, 1979, 1986; Gower et al., 2005; McDiarmid et al., 1999; Rajendran, 1985; Underwood, 1967; Wallach, 1998.

Bolyeriidae

Mascarene or Split-Jaw Boas

Bullock, 1986; Cundall and Irish, 1986, 1989; Grandison, 1977; McDiarmid et al., 1999; Underwood, 1967; Vidal and Hedges 2005; Wallach and Günther, 1998.

Loxocemidae

Mesoamerican Python

Greene, 1997; McDiarmid et al., 1999; Odinchenko and Latyshev, 1996; Orlov, 2001; Underwood, 1967; Wallach, 1998.

Xenopeltidae

Sunbeam Snakes

Cox et al., 1998; David and Vogel, 1996; McDiarmid et al., 1999; Orlov, 2000; Steubing and Inger, 1999; Underwood, 1967; Wallach, 1998.

Pythonidae

Pythons

Barker and Barker, 1994; Ehrmann, 1993; Ernst and Zug, 1996; Gower et al., 2005; Kluge, 1993a; McDiarmid et al., 1999; O'Shea, 1996; Shine, 1991b; Shine and Slip, 1990; Underwood, 1967; Underwood and Stimson, 1990.

Boidae

Boas

Burbrink, 2005; Henderson and Powell, 2007; Kluge, 1991, 1993b; McDiarmid et al., 1999; McDowell, 1987; Noonan and Chippindale, 2006; Underwood, 1967; Vences et al., 2001; Wilcox et al., 2002.

"Boinae"

Burbrink, 2005; Ehmann, 1993; Fitch, 1970; Henderson, 1993, 2002; Kluge, 1991; Noonan and Chippindale, 2006; O'Shea, 1996; Tolson and Henderson, 1993; Vences et al., 2001; Wilcox et al., 2002.

Erycinae

Burbrink, 2005; Kluge, 1993b; Noonan and Chippindale, 2006; Tokar, 1989, 1996. Wilcox et al., 2003.

Acrochordidae

Wart Snakes or File Snakes

Cox et al., 1998; Manthey and Grossmann, 1997; McDiarmid et al., 1999; Shine, 1986a,b; Shine and Houston, 1993; Shine and Lambeck, 1985; Underwood, 1967; Wallach, 1998; Wallach and Günther, 1998.

Viperidae

Vipers and Pit Vipers

Castoe et al., 2006; McDiarmid et al., 1999; Thorpe et al., 1997; Underwood, 1967; Wallach, 1998.

Azemiopinae

Liem et al., 1971;

Crotalinae

Beaupre and Duvall, 1998; Brown, 1993; Cadle, 1992; Campbell and Brodie, 1992; Campbell and Lamar, 1989; Castoe et al., 2006; Gloyd and Conant, 1990; Klauber, 1982; Ripa, 1999; Rubio, 1998; Zamudio and Greene, 1997.

Viperinae

Ashe and Marx, 1988; Brodmann, 1987; Castoe et al., 2006; Hermann and Joger, 1997; Saint Girons, 1992; Seigel and Ford, 1987; Sprawl and Branch, 1995.

"Colubridae"

Colubrids

Brown and Krauss, 1998; Underwood, 1967; Vidal and Hedges, 2002; Wallach, 1998; Zaher, 1999.

"Colubrinae"

Cox et al., 1998; Ernst and Barbour, 1989b; Seigel and Ford, 1987; Shine et al., 2006c.

Homalopsinae

Cox et al., 1998; Greer, 1997; Gyi, 1970; Murphy and Voris, 1994.

"Lamprophiinae"

Branch, 1988; Cadle, 1994; Glaw and Vences, 1994; Pitman, 1974; Shine et al., 1996.

"Natricinae"

Cox et al., 1998; Engelmann et al., 1986; Ernst and Barbour, 1989b; Manthey and Grossmann, 1997; Rossman et al., 1996.

Pareatinae

Cox et al., 1998; Manthey and Grossmann, 1997; Steubing and Inger, 1999; Rao and Yang, 1992.

Xenodermatinae

Manthey and Grossmann, 1997; Smith, 1943; Zhao and Adler, 1993.

Xenodontinae

Cadle, 1984; Ernst and Barbour, 1989b; Lancini and Kornacker, 1989; Pérez-Santos and Moreno, 1991; Starace, 1998; Zaher, 1999.

Atractaspididae

Stiletto Vipers or Mole Vipers

Cadle, 1994; Nagy et al., 2005; Underwood, 1967; Underwood and Kochva, 1993; Wallach, 1998; Zaher, 1999.

Aparallactinae

Branch, 1988; Villers, 1975.

Atractaspidinae

Branch, 1988; David and Ineich, 1999; Sprawl and Branch, 1995; Underwood and Kochva, 1993; Villers, 1975.

Elapidae

Cobras, Kraits, Sea Snakes, Death Adders, and Allies

David and Ineich, 1999; Grandison, 1977; Keogh, 1998; Ramussen, 1997; Slowinski et al., 1997, 2001; Underwood, 1967; Wallach, 1985; 1998.

"Elapinae"

Branch, 1988; Cox et al., 1998; Manthey and Grossmann, 1997; Pitman, 1974; Roze, 1996.

Hydrophiinae

Greer, 1997; Heatwole, 1999; Heatwole and Cogger, 1993; Heatwole and Guinea, 1993; Rasmussen, 1997; Shea et al., 1993; Voris and Voris, 1995.

Part VII

Data Sources

Bibliography

Adams, C. S., and Cooper, W. E., Jr. (1988). Oviductal morphology and sperm storage in the keeled earless lizard, *Holbrookia propinqua*. *Herpetologica* **44:** 190–197.

Adams, M., Foster, R., Hutchinson, M. N., Hutchinson, R. G., and Donnellan, S. C. (2003). The Australian scincid lizard *Menetia greyii*: A new instance of widespread vertebrate parthenogenesis. *Evolution* **57:** 2619–2627.

Adler, K. (1970). The role of extraoptic photoreceptors in amphibian rhythms and orientation: A review. *Journal of Herpetology* **4:** 99–112.

Adler, K. (1976). Extraocular photoreception in amphibians. *Photochemistry and Photobiology* **23:** 275–298.

Adler, K. (2003). Salamander classification and reproductive biology: An historical overview. In *Reproductive Biology and Phylogeny of Urodela*. D. M. Sever (Ed.). Pp. 1–29. Science Publishers, Inc., Nefield, NH.

Adler, K., and Phillips, J. B. (1991). Orientation in a desert lizard (*Uma notata*): Time compensated compass movement and polarotaxis. *Journal of Comparative Physiology A: Neuroethology, Sensory, Neural, and Behavioral Physiology* **156:** 547–552.

Adler, K., and Taylor, D. H. (1973). Extraocular perception of polarized light by orienting salamanders. *Journal of Comparative Physiology A: Neuroethology, Sensory, Neural, and Behavioral Physiology* **87:** 203–212.

Adler, K., and Zhao, E. (1990). Studies on hynobiid salamanders, with description of a new genus. *Asiatic Herpetological Research* **3:** 37–45.

Adler, K., Zhao, E., and Darevsky, I. S. (1992). First records of the pipe snake (*Cylindrophis*) in China. *Asiatic Herpetological Research* **4:** 37–41.

Adolf, E. F. (1932). The vapor tension relations of frogs. *Biological Bulletin* **62:** 112–125.

Agassiz, L. (1857). *Contributions to the Natural History of the United States of America. Vol. I. Part I. Essay on Classification. Part II. North American Testudinata. Vol. II., Part III. Embryology of the Turtle*. Little, Brown and Co., Boston, Mass.

Aguiar, O., Bacci, M., Lima, A. P., Rossa-Feres, D. C., Haddad, C. F. B., and Recco-Pimentel, S. M. (2007). Phylogenetic relationships of *Pseudis* and *Lysapsus* (Anura, Hylidae, Hylinae) inferred from mitochondrial and nuclear gene sequences. *Cladistics* **23:** 455–463.

Aguirre, A. A., Spraker, T. R., Balazs, G. H., and Zimmerman, B. (1998). Spirorchidiasis and fibropapillomatosis in green turtles from the Hawaiian Islands. *Journal of Wildlife Diseases* **34:** 91–98.

Ahlberg, A., Arndorff, L., and Guy-Ohlson, D. (2002). Onshore climate change during the Late Triassic marine inundation of the Central European Basin. *Terra Nova* **14:** 241–248.

Ahlberg, P. E. (1995). *Elginerpeton pancheni* and the earliest tetrapod clade. *Nature* **373:** 420–425.

Ahlberg, P. E., and Clack, J. A. (2006). A firm step from water to land. *Nature* **440:** 747–749.

Ahlberg, P. E., and Milner, A. R. (1994). The origin and the early diversification of tetrapods. *Nature* **368:** 507–514.

Ahlberg, P. E., Friedman, M., and Blom, H. (2005). New light on the earliest known tetrapod jaw. *Journal of Vertebrate Paleontology* **25:** 720–724.

Ahlberg, P. E., Luksevics, E., and Mark-Kurik, E. (2000). A near-tetrapod from the Baltic Middle Devonian. *Palaeontology* **43:** 533–548.

Alberts, A. C. (1992). Pheromonal self-recognition in the desert iguana. *Copeia* **1992:** 229–232.

Alberts, A. C. (2004). *Iguanas: Biology and Conservation*. University of California Press, Berkeley.

Albino, A. M. (1996). The South American fossil Squamata (Reptilia: Lepidosauria). In *Contributions of Southern South America to Vertebrate Paleontology*. G. Arratia (Ed.). Pp. 185–202. *Müncher Geowiss. Abh.*

Alcala, A. C., and Brown, W. C. (1987). Notes on the microhabitat of the Philippine discoglossid frog *Barbourula*. *Silliman Journal* **34:** 12–17.

Alcala, A. C., Ross, C. A., and Alcala, E. L. (1988). Observations on reproduction and behavior of captive Philippine crocodiles (*Crocodylus mindorensis* Schmidt). *Silliman Journal* [1987] **34:** 18–28.

Alcock, J. (1998). *Animal Behavior*. Sinauer Associates, Inc., Sunderland, MA.

Aleksiuk, M., and Gregory, P. T. (1974). Regulation of seasonal mating behavior in *Thamnophis sirtalis parietalis*. *Copeia* **1974:** 681–689.

Alexy, K. J., Brunjes, K. J., Gassett, J. W., and Miller, K. V. (2003). Continuous remote monitoring of gopher tortoise burrow use. *Wildlife Society Bulletin* **31:** 1240–1243.

Alibardi, L. (2003). Adaptation to the land: The skin of reptiles in comparison to that of amphibians and endotherm amniotes. *Journal of Experimental Zoology Part B-Molecular and Developmental Evolution* **298B:** 12–41.

Alifanov, V. R. (1993). New lizards of Marocephalosauridae from the Upper Cretaceous of Mongolia, with some critical comments to the classification of Teiidae (sensu Estes, 1983). *Paleontologicheskii Zhurnal* **1993:** 57–74. [in Russian]

Allen, C. R., Epperson, D. M., and Garmestani, A. S. (2004). Red imported fire ant impacts on wildlife: A decade of research. *American Midland Naturalist* **152:** 88–103.

Allsop, D. J., Warner, D., Lankilde, T., Du, W., and Shine, R. (2006). Do operational sex ratios influence sex allocation in viviparous lizards with temperature-dependent sex determination? *Journal of Evolutionary Biology* **19:** 1175–1182.

Al-Sorkhy, M. K., and Amr, Z. (2003). Platyhelminth parasites of some amphibians in Jordan. *Turkish Journal of Zoology* **27:** 89–93.

Altig, R., and Johnston, G. F. (1986). Major characteristics of free-living anuran tadpoles. *Smithsonian Herpetological Information Service* **67:** 1–75.

Altig, R., and Johnston, G. F. (1989). Guilds of anuran larvae: relationships among developmental modes, morphologies, and habitats. *Herpetological Monographs* **3:** 81–109.

Amato, G., and Gatesy, J. (1994). PCR assays of variable nucleotide sites for identification of conservation units. In *Ecology and Evolution: Approaches and Applications*. B. Schierwater, B. Streit, G. P. Wagner, and R. DeSalle (Eds.). Pp. 215–226. Birkhauser Verlag, Basel.

Amezquita, A., Hodl, W., Lima, A. P., Castellanos, L., Erdtmann, L., and De Araujo, M. C. (2006). Masking interference and the evolution of the acoustic communication system in the amazonian dendrobatid frog *Allobates femoralis*. *Evolution* **60:** 1874–1887.

Ananjeva, N. B., Borkin, L. Y., Darevsky, I. S., and Orlov, N. L. (1988). *Dictionary of Animal Names in Five Languages. Amphibians and Reptiles*. Russky Yazyk Publishers, Moscow.

Anderson, J. D. (1967). A comparison of the life histories of coastal and montane populations of *Ambystoma macrodactylum* in California. *American Midland Naturalist* **77:** 323–355.

Anderson, R. A. (1993). An analysis of foraging in the lizard, *Cnemidophorus tigris*. In *Biology of Whiptail Lizards (Genus* Cnemidophorus*)*. J. W. Wright and L. J. Vitt (Eds.). Pp. 83–116. Oklahoma Museum of Natural History, Norman, OK.

Anderson, R. A. (2007). Food acquisition modes and habitat use in lizards: Questions from an integrative perspective. In *Lizard Ecology: The Evolutionary Consequences of Foraging Mode*. S. M. Reilly, L. D. McBrayer, and D. B. Miles (Eds.). Pp. 450–490. Cambridge University Press, Cambrdge UK.

Anderson, R. A., and Karasov, W. H. (1981). Contrasts in energy intake and expenditure in sit-and-wait and widely foraging lizards. *Oecologia* **49:** 67–72.

Anderson, R. A., and Vitt, L. J. (1990). Sexual selection versus alternative causes of sexual dimorphism in teiid lizards. *Oecologia* **84:** 145–157.

Andreas B. Sichert, A. B., Friedel, P., and van Hemmen, J. L. (2006). Snake's perspective on heat: Reconstruction of input using an imperfect detection system. *Physical Review Letters* **97:** 068105.

Andreone, F., Cadle, J. E., Cox, N., Glaw, F., Nussbaum, R. A., Raxworthy, C. J., Stuart, S. N., Vallan, D., and Vences, M. (2005). Species review of amphibian extinction risks in Madagascar: Conclusions from the global amphibian assessment. *Conservation Biology* **19:** 1790–1802.

Andrewartha, H. G., and Birch, L. C. (1954). *Distribution and Abundance of Animals*. University of Chicago Press, Chicago, IL.

Andrews, R. M. (1976). Growth rate in island and mainland anoline lizards. *Copeia* **1976:** 477–482.

Andrews, R. M. (1979). Reproductive effort of female *Anolis limifrons* (Sauria: Iguanidae). *Copeia* **1979:** 620–626.

Andrews, R. M. (1982a). Patterns of growth in reptiles. In *Biology of the Reptilia, Volume 13. Physiology D*. C. Gans and F. H. Pough (Eds.). Pp. 273–320. Academic Press, London.

Andrews, R. M. (1982b). Spatial variation in egg mortality of the lizard *Anolis limifrons*. *Herpetologica* **38:** 165–171.

Andrews, R. M., and Rand, A. S. (1974). Reproductive effort in anoline lizards. *Ecology* **55:** 1317–1327.

Apesteguia, S., and Zaher, H. (2006). A Cretaceous terrestrial snake with robust hindlimbs and a sacrum. *Nature* **440:** 1037–1040.

Aranzabal, M. C. U. (2003a). The ovary and oogenesis. *Reproductive Biology and Phylogeny of Urodela* **1:** 135–150.

Aranzabal, M. C. U. (2003b). The testes, spermatogenesis and mate reproductive ducts. *Reproductive Biology and Phylogeny of Urodela* **1:** 183–202.

Aresco, M. J. (1999). Habitat structures associated with juvenile gopher tortoise burrows on pine plantations in Alabama. *Chelonian Conservation and Biology* **3:** 507–509.

Arnold, E. N. (1984). Evolutionary aspects of tail shedding in lizards and their relatives. *Journal of Natural History* **18:** 127–169.

Arnold, E. N. (1986). Mite pockets of lizards, a possible means of reducing damage by ectoparasites. *Biological Journal of the Linnean Society* **29:** 1–21.

Arnold, E. N. (1988). Caudal autotomy as a defense. In *Biology of the Reptilia. Volume 16, Ecology B. Defense and Life History*. C. Gans and R. B. Huey (Eds.). Pp. 235–273. Alan R. Liss, Inc., New York.

Arnold, E. N. (1989). Towards a phylogeny and biogeography of the Lacertidae: Relationships within an Old-World family of lizards derived from morphology. *Bulletin of the British Museum of Natural History (Zoology)* **55:** 209–257.

Arnold, E. N. (1993). Phylogeny and the Lacertidae. In *Lacertids of the Mediterranean Region: A Biological Approach*. E. D. Valakos, W. Böhme, V. Pérez-Mellado, and P. Maragou (Eds.). Pp. 1–16. Hellenic Zoological Society, Athens.

Arnold, E. N. (1998). Structural niche, limb morphology and locomotion in lacertid lizards (Squamata, Lacertidae): A preliminary survey. *Bulletin of the British Museum of Natural History (Zoology)* **64:** 63–89.

Arnold, E. N. (2000). The gender of *Podarcis* and the virtues of stability, a reply to W. Böhme. *Bonner Zoologische Beiträge* **49:** 71–74.

Arnold, S. J. (1982). A quantitative approach to antipredator performance: Salamander defenses against snake attack. *Copeia* **1982:** 247–253.

Arnold, S. J., and Duvall, D. (1994). Animal mating systems: A synthesis based on selection theory. *American Naturalist* **143:** 317–348.

Arntzen, J. W., and García-París, N. (1995). Morphological and allozyme studies of midwife toads (genus *Alytes*), including the description of two new taxa from Spain. *Contributions to Zoology* **65:** 5–34.

Asana, J. J. (1931). The natural life history of *Calotes versicolor* (Boulenger), the common blood sucker. *Journal of the Bombay Natural History Society* **34:** 1041–1047.

Ash, A. N., and Bruce, R. C. (1994). Impact of timber harvesting on salamanders. *Conservation Biology* **8:** 300–301.

Ashe, J. S., and Marx, H. (1988). Phylogeny of the viperine snakes (Viperinae): Part II. Cladistic analysis and major lineages. *Fieldiana: Zoology* **52:** 1–23.

Ashton, R. E. (1975). A study of movement, home range, and winter behavior of *Desmognathus fuscus*. *Journal of Herpetology* **9:** 85–91.

Ashton, R. E., Jr. (1990). *Necturus lewisi* Brimley. *Catalogue of American Amphibians and Reptiles* **456:** 1–2.

Atsatt, S. R. (1939). Color changes as controlled by temperature and light in the lizards of the desert regions of southern California. *University of California Publications in Biological Science, Los Angeles* **1:** 237–276.

Auffenberg, W. (1965). Sex and species discrimination in two sympatric South American tortoises. *Copeia* **1965:** 335–342.

Auffenberg, W. (1978). Social and feeding behavior in *Varanus komodoensis*. In *Behavior and Neurology of Lizards*. N. Greenberg and P. D. MacLean (Eds.). Pp. 301–331. National Insittute of Mental Health, Poolesville, MD.

Auffenberg, W. (1981). *The Behavioral Ecology of the Komodo Dragon*. University Presses Florida, Gainesville, FL.

Auffenberg, W. (1982). Feeding strategy of the Caicos ground iguana, *Cyclura carinata*. In *Iguanas of the World*. G. M. Burghart and A. S. Rand (Eds.). Pp. 84–116. Noyes Publications, Park Ridge, NJ.

Auffenberg, W. (1988). *Gray's Monitor Lizard*. University Presses of Florida, Gainesville, FL.

Auffenberg, W. (1994). *The Bengal Monitor*. University Press Florida, Gainesville, FL.

Auffenberg, W., and Franz, R. (1982). The status and distribution of the gopher tortoise (*Gopherus polyphemus*). In *North American Tortoises: Conservation and Ecology*. Pp. 95–126. Wildlife Research Report No. 12, United States Fish and Wildlife Service.

Auffenberg, W., and Weaver, W. G. (1969). *Gopherus berlandieri* in southeastern Texas. *Bulletin of the Florida State Museum* **13:** 141–203.

Augé, M. (2005). Evolution des le´zards du Paléogéne en Europe. Mémoires du Muséum National d'Histoire Naturelle **192:** 1–369.

Austin, J. D., Lougheed, S. C., Tanner, K., Chek, A. A., Bogart, J. P., and Boag, P. T. (2002). A molecular perspective on the evolutionary affinities of an enigmatic neotropical frog, *Allophryne ruthveni*. *Zoological Journal of the Linnean Society* **134:** 335–346.

Autumn, K. (1999). Secondarily diurnal geckos return to cost of locomotion typical of diurnal lizards. *Physiological and Biochemical Zoology* **72:** 339–351.

Autumn, K., and Han, B. (1989). Mimicry of scorpions by juvenile lizards, *Teratoscincus roborowskii* (Gekkonidae). *Chinese Herpetological Research* **2:** 60–64.

Autumn, K., Farley, C. T., Emshwiller, M., and Full, R. J. (1997). Low cost of locomotion in the banded gecko: A test of the nocturnality hypothesis. *Physiological Zoology* **70:** 660–669.

Autumn, K., Jindrich, D., DeNardo, D., and Mueller, R. (1999). Locomotor performance at low temperature and the evolution of nocturnality in geckos. *Evolution* **53:** 580–599.

Autumn, K., Weinstein, R. B., and Full, R. J. (1994). Low cost of locomotion increases performance at low temperature in a nocturnal lizard. *Physiological Zoology* **67:** 238–262.

Averianov, A., and Danilov, I. (1996). Agamid lizards (Reptilia, Sauria, Agamidae) from the Early Eocene of Kyrgyzstan. *Neues Jahrbuch für Geologie und Paläontologie, Monatshefte* **1996:** 739–750.

Avila, L. J., Acosta, J. C., and Marotri, R. A. (1992). Composición, variación, anual y estacional de la dieta de *Teius suquiensis* (Sauria: Teiidae) en la provincia de Córdoba (Argentina). *Cuadernos de Herpetologica* **7:** 5–13.

Avila-Pires, T. C. S. (1995). Lizards of Brazilian Amazonia (Reptilia: Squamata). *Zoologische Verhandelingen, Leiden* **299:** 1–706.

Avise, J. C. (2000). *Phylogeography: The History and Formation of Species*. Harvard University Press, Cambridge, Mass.

Ayala, S. C. (1977). *Plasmodia* of reptiles. In *Parasitic Protozoa. Volume 3: Gregarines, Haemogregarines, Coccidia, Plasmodia, and Haemoprotids*. J. P. Kreier (Ed.). Pp. 269–309. Academic Press, New York.

Primack, R. B. (1995). *A Primer of Conservation Biology*. Sinauer Associates Inc., Sunderland, Mass.

Bachman, M. D. (1984). Defensive behavior of brooding female red-backed salamanders, *Plethodon cinereus*. *Herpetologica* **40:** 436–443.

Backwell, P. R. Y., and Passmore, N. (1990). Polyandry in the leaf-folding frog, *Afrixalus delicatus*. *Herpetologica* **46:** 7–10.

Badenhorst, A. (1978). The development and the phylogeny of the organ of Jacobson and the tentacular apparatus of *Ichthuphis glutinosus* (Linné). *Annals of the Stellenbosch University* **1:** 1–26.

Báez, A. M. (1996). The fossil record of the Pipidae. In *The Biology of Xenopus*. R. C. Tinsley and H. R. Kobel (Eds.). Pp. 329–347. Zoological Society of London, London.

Báez, A. M., and Basso, N. G. (1996). The earliest known frogs of the Jurassic of South America: Review and cladistic appraisal of their relationships. In *Contributions of Southern South America to Vertebrate Paleontology*. G. Arratia (Ed.). Pp. 131–158. *Müncher Geowiss. Abh.* **A 30**.

Báez, A. M., and Pugener, L. A. (1998). A new Paleogene pipid frog from northwestern Patagonia. *Journal of Vertebrate Paleontology* **18:** 511–524.

Baez, A. M., and Pugener, L. A. (2003). Ontogeny of a new Palaeogene pipid frog from southern South America and xenopodinomorph evolution. *Zoological Journal of the Linnean Society* **139:** 439–476.

Báez, A. M., and Trueb, L. (1997). Redescription of the Paleogene *Shelania pascuali* from Patagonia and its bearing on the relationships of fossil and Recent pipoid frogs. *Scientific Papers, Natural History Museum, University of Kansas* **4:** 1–41.

Bagnara, J. T. (1986). Pigment cells. In *Biology of the Integument. 2. Vertebrates*. J. Bereiter-Hahn, A. G. Matoltsy, and K. S. Richards (Eds.). Pp. 136–149. Springer Verlag, Berlin.

Bagnara, J. T., and Rastogi, R. K. (1992). Reproduction in the Mexican leaf frog, *Pachymedusa dacnicolor*.

In *Reproductive Biology of South American Vertebrates.* W. C. Hamlett (Ed.). Pp. 98–111. Springer-Verlag, New York.

Bain, R. H., Lathrop, A., Murphy, R. W., Orlov, N. L., and Cuc, H. T. (2003). Cryptic species of a cascade frog from Southeast Asia: Taxonomic revisions and descriptions of six new species. *American Museum Novitates* **3417:** 1–60.

Baird, I. L. (1970). The anatomy of the reptilian ear. In *Biology of the Reptilia. Volume 2. Morphology B.* C. Gans and T. S. Parsons (Eds.). Pp. 193–275. Academic Press, London.

Balinsky, J. B. (1981). Adaptations of nitrogen metabolism to hyperosmotic environments in Amphibia. *Journal of Experimental Zoology* **215:** 335–350.

Ballinger, R. E. (1977). Reproductive strategies: Food availability as a source of proximal variation in a lizard. *Ecology* **58:** 628–635.

Ballinger, R. E., Lemos-Espinal, J., Sanoja-Sarabia, S., and Coady, N. R. (1995). Ecological observations of the lizard, *Xenosaurus grandis* in Cuautlapán, Veracruz, México. *Biotropica* **27:** 128–132.

Balmford, A., and Cowling, R. M. (2006). Fusion or failure? The future of conservation biology. *Conservation Biology* **20:** 692–695.

Barandun, J. (1995). *Reproductive Ecology of Bombina variegata (Amphibia).* Inaugural-Dissertation der Universität Zürich, Zürich.

Barbault, R. (1972). Les peuplements d'amphibiens des savanes de Lamto (Cote-d'Ivoire). *Annals of the University d'Abidjan,* **5:** 59–142.

Barbault, R. (1975a). Les peuplements de lézards des savanes de Lamto (Cote-d'Ivoire). *Annals of the University d'Abidjan* **8:** 147–221.

Barbault, R. (1975b). Observations écologiques sur la reproduction des lézards tropicaux: Les stratégies de ponte en forêt en an savane. *Bulletin de la Société Zoologique de France* **100:** 153–168.

Barbault, R. (1976). Population dynamics and reproductive patterns of three African skinks. *Copeia* **1976:** 483–490.

Barbault, R. (1977). Structure et dynamique d'une herpetocenose de savane (Lamto, Cote d'Ivoire). *Geo-Eco-Trop* **4:** 309–334.

Barbault, R., and Mou, Y.-P. (1988). Population dynamics of the common wall lizard, *Podarcis muralis,* in southwestern France. *Herpetologica* **44:** 38–47.

Barbault, R., and Rodrigues, M. T. (1978). Observations sur la reproduction et la dynamique des populations de quelques anoures tropicaux. II. *Phrynobatrachus plicatus. Geo-Eco-Trop* **2:** 455–466.

Barbault, R., and Rodrigues, M. T. (1979). Observations sur la reproduction et la dynamique des populations de quelques anoures tropicaux III. *Arthroleptis poecilonotus. Tropical Ecology* **20:** 64–77.

Barbeau, T. R., and Lillywhite, H. B. (2005). Body wiping behaviors associated with cutaneous lipids in hylid tree frogs of Florida. *Journal of Experimental Biology* **208:** 2147–2156.

Barbour, R. W., Harvey, M. J., and Hardin, J. W. (1969). Home range, movements, and activity of the eastern worm snake, *Carphophis amoenus amoenus. Ecology* **50:** 470–476.

Barker, D. G., and Barker, T. M. (1994). *Pythons of the World. Volume 1, Australia.* The Herpetocultural Library, Lakeside, CA.

Barker, J., Grigg, G., and Tyler, M. (1995). *A Field Guide to Australian Frogs.* Surrey Beatty & Sons Pty Ltd, Chipping Norton Pty Ltd, N.S.W.

Barthalmus, G. T., and Bellis, E. D. (1972). Home range, homing and the homing mechanism of the salamander, *Desmognathus fuscus. Copeia* **1974:** 632–642.

Bartholomew, G. A. (1982). Physiological control of body temperature. In *Biology of the Reptilia. Volume 12. Physiology C. Physiological Ecology.* C. Gans and F. H. Pough (Eds.). Pp. 167–211. Academic Press, New York.

Basso, N. G. (1998). A new telmatobiine leptodactylid frog of the genus *Atelognathus* from Patagonia. *Herpetologica* **54:** 44–52.

Bauchot, R. Ed. (1994). *Les Serpents.* Bordas, Paris.

Bauer, A. M. (1990). Phylogenetic systematics and biogeography of the Carphodactylini (Reptilia: Gekkonidae). *Bonner Zoologische Monographien* **30:** 1–218.

Bauer, A. M. (1994). Gekkonidae (Reptilia, Sauria) Part I Australia and Oceania. *Das Tierreich* **109:** 1–306.

Bauer, A. M. (1997). Pertioneal pigmentation and generic allocation in the Chamaeleonidae. *African Journal of Herpetology* **46:** 117–123.

Bauer, A. M. (1999). Twentieth century amphibian and reptile discoveries. *Cryptozoology* [1997–1998] **13:** 1–17.

Bauer, A. M. (2003). Geckos and pygopods (Gekkonidae). In M. Hutchins, J. B. Murphy, and N. Schlager (Eds.). Pp. 259–270. *Grzimek's Animal Life Encyclopedia,* (2nd Ed.) *Vol. 7, Reptiles.* Gale Group, Farmington Hills, MI.

Bauer, A. M. (2007). The foraging biology of the Gekkota: Life in the middle. In *Lizard Ecology: The Evolutionary Consequences of Foraging Mode.* S. M. Reilly, L. D. McBrayer, and D. B. Miles (Eds.). Pp. 371–404. Cambridge University Press, Cambrdge UK.

Bauer, A. M., Good, D. A., and Branch, W. R. (1997). The taxonomy of the Southern African leaf-toed geckos (Squamata: Gekkonidae), with a review of Old World "*Phyllodactylus*" and the description of five new genera [*Haemodracon, Dixonius, Cryptactites, Goggia, Afrogecko*: Reinstatement of Euleptes]. Proceedings of the California Academy of Sciences **49:** 447–497.

Bauer, A. M., Russell, A. P., and Dollahon, N. R. (1990). Skin folds in the gekkonid genus *Rhacodactylus*: A natural test of the damage limitation hypothesis of mite pocket function. *Canadian Journal of Zoology* **68:** 1196–1201.

Bauer, A. M., Russell, A. P., and Dollahon, N. R. (1993). Function of the mite pockets of lizards: A reply to E. N. Arnold. *Canadian Journal of Zoology* **71:** 865–868.

Bauer, A. M., Russel, A. P., and Shadwick, R. E. (1989). Mechanical properties and morphological correlates of fragile skin in gekkonid lizards. *Journal of Experimental Biology* **145:** 79–102.

Bauer, A. M., and Lamb, T. (2005). Phylogenetic relationships of southern African geckos in the *Pachydactylus* group (Squamata: Gekkonidae). *African Journal of Herpetology* **54:** 105–129.

Bauwens, D., Garland, T. J., Castilla, A. M., and Damme, R. V. (1995). Evolution of sprint speed in lacertid lizards: Morphological, physiological, and behavioral covariation. *Evolution* **49:** 848–863.

Baverstock, P. R., King, D., King, M., Birrell, J., and Krieg, M. (1993). The evolution of species of the Varanidae: Microcomplement fixation analysis of serum albumins. *Australian Journal of Zoology* **41:** 621–638.

Beaupre, S. J., and Duvall, D. J. (1998). Integrative biology of rattlesnake. Contributions to biology and evolution. *BioScience* **48:** 531–538.

Bechtel, H. B., and Bechtel, E. (1991). Scaleless snakes and a breeding report of scaleless *Elaphe obsoleta linheimeri*. *Herpetological Review* **22:** 12–14.

Beck, D. D. (1990). Ecology and behavior of the Gila monster in southwestern Utah. *Journal of Herpetology* **24:** 54–68.

Beck, D. D. (2005). *Biology of Gila Monsters and Beaded Lizards*. University of California Press, Berkeley.

Beck, D. D., and Lowe, C. H. (1991). Ecology of the beaded lizard, *Heloderma horridum*, in a tropical dry forest in Jalisco, México. *Journal of Herpetology* **25:** 395–406.

Beck, D. D., and Jennings, R. D. (2003). Habitat use by Gila monsters: The importance of shelters. *Herpetological Monographs* **17:** 111–129.

Beebee, T. J. C. (1996). *Ecology and Conservation of Amphibians*. Chapman & Hall, London.

Bell, B. D. (1978). Observations on the ecology and reproduction of the New Zealand leiopelmid frogs. *Herpetologica* **34:** 340–354.

Bell, B. D. (1985). Development and parental-care in the endemic New Zealand frogs. In *Biology of Australasian Frogs and Reptiles*. G. Grigg, R. Shine, and H. Ehmann (Eds.). Pp. 269–278. Surrey Beatty & Sons Pty Ltd., Chipping Norton Pty Ltd., Sydney, Australia.

Bell, B. D. (1994). A review of the status of New Zealand *Leiopelma* species (Anura: Leiopelmatidae), including a summary of demographic studies in Coromandel and Maud Island. *New Zealand Journal of Zoology* **21:** 341–349.

Bell, B. D., and Wassersug, R. J. (2003). Anatomical features of *Leiopelma* embryos and larvae: Implications for anuran evolution. *Journal of Morphology* **256:** 160–170.

Bell, B. D., Carver, S., Mitchell, N. J., and Pledger, S. (2004). The recent decline of a New Zealand endemic: How and why did populations of Archey's frog *Leiopelma archeyi* crash over 1996–2001? *Biological Conservation* **120:** 189–199.

Bell, C. J., Mead, J. I., and Fay, L. P. (1995). Neogene history of *Anniella* Gray, 1852 (Squamata, Anniellidae) with comments on postcranial osteology. *Copeia* **1995:** 719–726.

Bell, E. L. (1955). An aggregation of salamanders. *Proceedings of the Pennsylvania Academy of Science* **24:** 265–66.

Bell, G. (1980). The costs of reproduction and their consequences. *American Naturalist* **116:** 45–76.

Bellairs, A. d' A. (1969). *The Life of Reptiles. Volumes I & II.* Weidenfeld & Nicolson, London.

Bellairs, A. d' A., and Kamal, A. M. (1981). The chondrocranium and the development of the skull in Recent reptiles. In *Biology of the Reptilia. Volume 11. Morphology F.* C. Gans and T. S. Parsons (Eds.). Pp. 1–263. Academic Press, London.

Benabib, M. (1994). Reproduction and lipid utilization of tropical populations of *Sceloporus variabilis*. *Herpetological Monographs* **8:** 160–180.

Benabib, M., and Congdon, J. D. (1992). Metabolic and water-flux rates of free-ranging tropical lizards *Sceloporus variabilis*. *Physiological Zoology* **65:** 788–802.

Bennett, A. F. (1980). The metabolic foundations of vertebrate behavior. *BioScience* **30:** 452–456.

Bennett, A. F. (1982). The energetics of reptilian activity. In *Biology of the Reptilia. Volume 13. Physiology D. Physiological Ecology*. C. Gans and F. H. Pough (Eds.). Pp. 155–199. Academic Press, New York.

Bennett, A. F. (1991). The evolution of activity capacity. *Journal of Experimental Biology* **160:** 1–23.

Bennett, A. F., and Gleeson, T. T. (1979). Metabolic expenditure and the cost of foraging in the lizard *Cnemidophorus murinus*. *Copeia* **1979:** 573–577.

Bennett, A. F., and Licht, P. (1975). Evaporative water loss in scaleless snakes. *Comparative Biochemistry and Physiology* **52A:** 213–215.

Bennett, D. (1998). *Monitor Lizards. Natural History, Biology & Husbandry*. Edition Chimaira, Frankfurt am Main.

Bentivegna, F. (2002). Intra-Mediterranean migrations of loggerhead sea turtles (*Caretta caretta*) monitored by satellite telemetry. *Marine Biology* **141:** 795–800.

Bentley, P. J. (1966). The physiology of the urinary bladder of Amphibia. *Biological Reviews* **41:** 275–316.

Bentley, P. J., and Blumer, W. F. C. (1962). Uptake of water by the lizard, *Moloch horridus*. *Nature* **194:** 699–700.

Benton, M. J. (1990). Phylogeny of the major tetrapod groups: Morphological data and divergence dates. *Journal of Molecular Evolution* **30:** 409–424.

Benton, M. J. (1991). Amniote phylogeny. In *Origins of the Higher Groups of Tetrapods. Controversy and Consensus*. H.-P. Schultze and L. Trueb (Eds.). Pp. 317–330. Comstock Publishing Associates, Ithaca, NY.

Benton, M. J. (1994). Late Triassic to Middle Jurassic extinctions among continental tetrapods: Testing the pattern. In *In the Shadow of Dinosaurs. Early Mesozoic Tetrapods*. N. C. Fraser and H.-D. Sues (Eds.). Pp. 366–397. Cambridge University Press, New York.

Benton, M. J. (1997a). *Vertebrate Paleontology*. Second edition. Chapman & Hall, London.

Benton, M. J. (1997b). Origin and early evolution in dinosaurs. In *The Complete Dinosaur*. J. O. Farlow and M. K. Brett-Surman (Eds.). Pp. 204–215. Indiana University Press, Bloomington.

Benton, M. J. (1997c). Models for the diversification of life. *Trends in Ecology & Evolution* **12:** 490–495.

Benton, M. J. (2000). Stems, nodes, crown clades, and rank-free lists: Is Linnaeus dead? *Biological Reviews* **75:** 633–648.

Benton, M. J., and Clark, J. M. (1988). Archosaur phylogeny and the relationships of Crocodylidae. In *The Phylogeny and Classification of the Tetrapods. Vol. 1: Amphibians, Reptiles, Birds*. M. J. Benton (Ed.). Pp. 295–338. Clarendon Press, Oxford.

Bentz, S., Combes, C., Euzet, L., Riutord, J. J., and Verneau, O. (2003). Evolution of monogenean parasites across vertebrate hosts illuminated by the phylogenetic position of *Euzetrema* Combes, 1965 within the Monopisthocotylea. *Biological Journal of the Linnean Society* **80:** 727–734.

Berger, L. (1977). Systematics and hybridization in the *Rana esculenta* complex. In *The Reproductive Biology of Amphibians*. D. H. Taylor and S. I. Guttman (Eds.). Pp. 367–410. Plenum Press, New York.

Berger, L., and Michalowski, J. (1973). *Keys for the Identification of Vertebrates of Poland. Part 2. Amphibia*. Institute of Systematic Zoology, Cracow.

Berger, L., Speare, R., Daszak, P., Green, D. E., Cunningham, A. A., Goggin, C. L., Slocombe, R., Ragan, M. A., Hyatt, A. D., McDonald, K. R., Hines, H. B., Lips, K. R., Marantelli, G., and Parkes, H. (1998). Chytridiomycosis causes amphibian mortality associated with population declines in the rain forests of Australia and Central America. *Proceedings of the National Academy of Sciences (USA)* **95:** 9031–9036.

Berry, J. F., and Shine, R. (1980). Sexual size dimorphism and sexual selection in turtles (Order Testudines). *Oecologia* **44:** 185–191.

Berry, P. Y. (1975). *The Amphibian Fauna of Peninsular Malaysia*. Tropical Press, Kuala Lumpur.

Bezy, R. L. (1988). The natural history of the night lizards, family Xantusiidae. In *Proceedings of the Conference on California Herpetology*. H. F. De Lisle, et al. (Eds.). Pp. 1–12. Southwest. Herpetological Society, Van Nuys, CA.

Bezy, R. L. (1989). Morphological differentiation in unisexual and bisexual xantusiid lizards of the genus *Lepidophyma* in Central America. *Herpetological Monographs* **3:** 6–80.

Bezy, R. L., and Camarillo, J. L. (2002). Systematics of xantusiid lizards of the genus *Lepidophyma*. Los Angeles County Museum Contributions in Science **493:** 1–41.

Bhatta, G. (1998). A field guide to the caecilians of the Western Ghats, India. *Journal of Biosciences* **23:** 73–85.

Bi, K., and Bogart, J. P. (2006). Identification of intergenomic recombinations in unisexual salamanders of the genus *Ambystoma* by genomic in situ hybridization (GISH). *Cytogenetic and Genome Research* **112:** 307–312.

Bi, K., Bogart, J. P., and Fu, J. (2007). Intergenomic translocations in unisexual salamanders of the genus *Ambystoma* (Amphibia, Caudata). *Cytogenetic and Genome Research* **116:** 289–297.

Bickford, D. (2002). Animal behaviour- Male parenting of New Guinea froglets. *Nature* **418:** 601–602.

Bickham, J. W., and Carr, J. L. (1983). Taxonomy and phylogeny of the higher categories of cryptodiran turtles based on a cladistic analysis of chromosomal data. *Copeia* **1983:** 918–932.

Biju, S. D., and Bossuyt, F. (2003). New frog family from India reveals an ancient biogeographical link with the Seychelles. *Nature* **425:** 711–714.

Billo, R. R., Straub, J. O., and Senn, D. G. (1985). Vivipare Apode (Amphibia: Gymnophiona), *Typhlonectes compressicaudus* (Duméril & Bibron, 1841): Kopulation, Tragziet und Geburt. *Amphibia-Reptilia* **6:** 1–9.

Billo, R., and Wake, M. H. (1987). Tentacle development in *Dermophis mexicanus* (Amphibia, Gymnophiona) with an hypothesis of tentacle origin. *Journal of Morphology* **192:** 101–111.

Birkhead, T. (2000). *Promiscuity: An Evolutionary History of Sperm Competition*. Faber and Faber Ltd., London.

Birkhead, T. R., and Møller, A. P. (1995). Extra-pair copulation and extra-pair paternity in birds. *Animal Behaviour* **49:** 843–848.

Bishop, S. C. (1941). The salamanders of New York. *New York State Museum Bulletin* **324:** 1–365.

Bjorndal, K. A. (1980). Nutrition and grazing behavior of the green turtle, *Chelonia midas*. *Marine Biology* **56:** 147–154.

Bjorndal, K. A. (1987). Digestive efficiency in a temperate herbivorous reptile, *Gopherus polyphemus*. *Copeia* **1987:** 714–720.

Bjorndal, K. A. (1989). Flexibility of digestive responses in two generalist herbivores, the tortoises *Geochelone carbonaria* and *Geochelone denticulata*. *Oecologia* **78:** 317–321.

Bjorndal, K. A. (1996). Foraging ecology and nutrition of sea turtles. In *The Biology of Sea Turtles*. P. L. Lutz and J. A. Musick (Eds.). Pp. 199–231. CRC Press, Boca Raton, FL.

Blackburn, D. (1982). Evolutionary origins of viviparity in the Reptilia. I. Sauria. *Amphibia-Reptilia* **3:** 185–205.

Blackburn, D. (1985). Evolutionary origins of viviparity in the Reptilia. II. Serpentes, Amphisbaenia, and Ichthyosauria. *Amphibia-Reptilia* **6:** 259–291.

Blackburn, D. G. (1994). Standardized criteria for the recognition of embryonic nutritional patterns in squamate reptiles. *Copeia* **1994:** 925–935.

Blackburn, D. G. (1993a). Standardized criteria for the recognition of reproductive modes in squamate reptiles. *Herpetologica* **49:** 118–132.

Blackburn, D. G. (1993b). Chorioallantoic placentation in squamate reptiles: Structure, function, development, and evolution. *Journal of Experimental Zoology* **266:** 414–430.

Blackburn, D. G. (1998a). Structure, function, and evolution of the oviducts of squanate reptiles, with special reference to vivipartiy and placentation. *The Journal of Experimental Zoology* **282:** 560–617.

Blackburn, D. G. (1998b). Reconstructing the evolution of viviparity and placentation. *Journal of Theoretical Biology* **192:** 183–190.

Blackburn, D. G. (1999a). Viviparity and oviparity: evolution and reproductive strategies. In *Encyclopedia of Reproduction* (Volume 4). E. Knobil and J. D. Neill (Eds.). Pp. 994–1003. Academic Press, London.

Blackburn, D. G. (1999b). Placenta and placenta analogs in reptiles and amphibians. *Encyclopedia of Reproduction* **3:** 840–847.

Blackburn, D. G. (1999c). Are viviparity and egg-guarding evolutionarily labile? *Herpetologica* **55:** 556–573.

Blackburn, D. G. (2000a). Classification of the reproductive patterns of amniotes. *Herpetological Monographs* 371–377.

Blackburn, D. G. (2000b). Reptilian viviparity: Past research, future directions, and appropriate models. *Comparative Biochemistry and Physiology A: Molecular Integrative Physiology* **127:** 391–409.

Blackburn, D. G. (2005). Evolutionary origins of viviparity in fishes. In *Viviparity in Fishes*. H. Grier and M. Uribe (Eds.). Pp. 303–317. New Life Publications, Homestead, Florida.

Blackburn, D. G. (2006a). Evolution of viviparity in reptiles: introduction to the symposium. *Herpetological Monographs* **20:** 129–130.

Blackburn, D. G. (2006b). Squamate reptiles as model organisms for the evolution of viviparity. *Herpetological Monographs* **20:** 131–146.

Blackburn, D. G., and Flemming, A. F. (2008). Morphology, development, and evolution of fetal membranes and placentation in squamate reptiles. *Journal of Experimental Zoology*: in press.

Blackburn, D. G., and Vitt, L. J. (1992). Reproduction in viviparous South American lizards of the genus *Mabuya*. In *Reproductive Biology of South American Vertebrates*. W. C. Hamlett (Ed.). Pp. 150–164. Springer-Verlag, New York.

Blackburn, D. G., and Vitt, L. J. (2002). Specializations of the chorioallantoic placenta in the Brazilian scincid lizard, *Mabuya heathi*: A new placental morphotype for reptiles. *Journal of Morphology* **254:** 121–131.

Blackburn, D. G., Evans, H. E., and Vitt, L. J. (1985). The evolution of fetal nutritional adaptations. *Fortschritte der Zoologie* **30:** 437–439.

Blackburn, D. G., Vitt, L. J., and Beuchat, C. A. (1984). Eutherian-like reproductive specializations in a viviparous reptile. *Proceedings of the National Academy of Sciences (USA)* **81:** 4860–4863.

Blair, W. F. (1960). *The Rusty Lizard: A Population Study*. University of Texas Press, Austin, TX.

Blair, W. F., (Ed.). (1972). *Evolution in the Genus* Bufo. University of Texas Press, Austin, TX.

Blanc, C. P. (1977). Reptiles, Sauriens Iguanidae. *Faune Madagascar* **45:** 1–195.

Blaustein, A. R., and Johnson, P. T. (2003). The complexity of deformed amphibians. *Frontiers in Ecology and the Environment* **1:** 87–94.

Blaylock, L. A., Ruibal, R., and Platt-Aloia, K. (1976). Skin structure and wiping behavior of phyllomedusine frogs. *Copeia* **1976:** 283–295.

Blom, H. (2005). Taxonomic revision of the Late Devonian tetrapod *Ichthyostega* from East Greenland. *Palaeontology* **48:** 111–134.

Blommers-Schlösser, R. M. A. (1975). Observations on the larval development of some Malagasay frogs, with notes on their ecology and biology (Anura, Scaphiophryninae and Cophylinae). *Beufortia* **24:** 7–26.

Blommers-Schlösser, R. M. A. (1993). Systematic relationships of Mantellinae Laurent 1946 (Anura Ranoidea). *Ethology, Ecology & Evolution* **5:** 199–218.

Blommers-Schlösser, R. M. A., and Blanc, Ch. P. (1991). Amphibiens (première partie). *Faune de Madagascar* **75:** 1–379.

Blouin, M. S. (1990). Evolution of palatability differences between closely-related treefrogs. *Journal of Herpetology* **24:** 309–311.

Blouin-Demers, G., Kissner, K. J., and Weatherhead, P. J. (2000). Plasticity in preferred body temperature of young snakes in response to temperature during development. *Copeia* **2000:** 841–845.

Bodie, J. R., and Semlitsch, R. D. (2000). Spatial and temporal use of floodplain habitats by lentic and lotic species of aquatic turtles. *Oecologia* **122:** 138–146.

Boersma, P. D. (1982). The benefits of sleeping aggregations in marine iguanas, *Amblyrhynchus cristatus*. In *Iguanas of the World*. G. M. Burghardt and A. S. Rand (Eds.). Pp. 292–299. Noyes Publ., Park Ridge, NJ.

Bogart, J. P. (2003). Genetics and systematics of hybrid species. *Reproductive Biology and Phylogeny of Urodela* **1:** 109–134.

Bogart, J. P., Bi, K., Fu, J. Z., Noble, D. W. A., and Niedzwiecki, J. (2007). Unisexual salamanders (genus *Ambystoma*) present a new reproductive mode for eukaryotes. *Genome* **50:** 119–136.

Bogart, J. P., Elinson, R. P., and Licht, L. E. (1989). Temperature and sperm incorporation in polyploidy salamanders. *Science* **246:** 1032–1034.

Bogert, C. M. (1993). Preface to The Gila Monster and Its Allies. In *The Gila Monster and Its Allies* pp. v–xv. Society for the Study of Amphibians and Reptiles, Ithaca, NY.

Boglioli, M. D., Michener, W. K., and Guyer, C. (2000). Habitat selection and modification by the gopher tortoise, *Gopherus polyphemus*, in Georgia longleaf pine forest. *Chelonian Conservation and Biology* **3:** 699–705.

Böhmme, W. (Ed.) (1981). *Handbuch der Reptilien und Amphibien Europas. Band 1. Echsen (Sauria) I*. Akademie Verlagsges, Wiesbaden.

Böhme, W., Hutterer, R., and Bings, W. (1985). Die Stimme der Lacertidae, speziell der Canareneidechsen. *Bonner Zoologische Beiträge* **36:** 337–354.

Boiko, V. P. (1984). The participation of chemoreception in the organization of reproductive behavior in *Emys orbicularis* (Testudines, Emydidae). *Zoologotjeskij Zhurnal* **63:** 584–589. (English abstract).

Bolt, J. R. (1991). Lissamphibian origins. In *Origins of the Higher Groups of Tetrapods. Controversy and Consensus*. H.-P. Schultze and L. Trueb (Eds.). Pp. 194–222. Comstock Publishing Associates, Ithaca, NY.

Bonine, K. E. (2007). Physiological correlates of foraging mode. In *Lizard Ecology: The Evolutionary Consequences of Foraging Mode*. S. M. Reilly, L. D. McBrayer, and D. B. Miles (Eds.). Pp. 94–119. Cambridge University Press, Cambrdge UK.

Borkin, L. J., and Anissimova, E. V. (1987). The vertebral structure and vocal sacs in the Caucasian parsley frog (*Pelodytes caucasicus*) and its taxonomic position. *Proceedings of the Zoological Institute, Leningrad* **158:** 59–76. [in Russian].

Bosch, J., Boyero, L., and Rohlf, F. J. (2003). Precopulatory behaviour and the evolutionary relationships of Discoglossidae. *Journal of Zoological Systematics and Evolutionary Research* **41:** 145–151.

Bossuyt, F., and Milinkovitch, M. C. (2000). Convergent adaptive radiations in Madagascan and Asian ranid frogs reveal covariation between larval and adult traits. *Proceedings of the National Academy of Sciences (USA)* **97:** 6585–6590.

Bossuyt, F., and Milinkovitch, M. C. (2001). Amphibians as indicators of Early Tertiary 'Out-of-India' dispersal of vertebrates. *Science* **292:** 93–95.

Bossuyt, F., Brown, R. M., Hillis, D. M., Cannatella, D. C., and Milinkovitch, M. C. (2006). Phylogeny and biogeography of a cosmopolitan frog radiation: Late cretaceous diversification resulted in continent-scale endemism in the family ranidae. *Systematic Biology* **55:** 579–594.

Bouchard, S. S., and Bjorndal, K. A. (2006). Ontogenetic diet shifts and digestive constraints in the omnivorous freshwater turtle *Trachemys scripta*. *Physiological and Biochemical Zoology* **79:** 150–158.

Bour, R. (1984). Les tortues terrestres géantes des iles de océean indien occidental: données géographiques, taxinomiques et phylogénétiques. *Stvdia Geologica Salmanticensia* **1:** 17–76.

Bourn, D., and Coe, M. (1978). The size, structure and distribution of the giant tortoise population of Aldabra. *Philosophical Transactions of the Royal Society of London, series B: Biological Sciences* **282:** 139–175.

Bourne, G. R., Collins, A. C., Holder, A. M., and McCarthy, C. L. (2001). Vocal communication and reproductive behavior of the frog *Colostethus beebei* in Guyana. *Journal of Herpetology* **35:** 272–281.

Bourquin, O., and Sowler, S. G. (1980). The vertebrates of Vernon Crookes Nature Reserve: 1. *Lammergeyer* **28:** 20–32.

Bowen, B. W., and Avise, J. C. (1996). Conservation genetics of marine turtles. In *The Biology of Sea Turtles*. P. L. Lutz and J. A. Musick (Eds.). Pp. 190–237. CRC Press, Boca Raton, FL.

Bowen, B. W., and Karl, S. W. (1996). Population genetics, phylogeography, and molecular evolution. In *The Biology of Sea Turtles*. P. L. Lutz and J. A. Musick (Eds.). Pp. 29–50. CRC Press, Boca Raton, FL.

Boycott, R. C. (1988). Evidence of tactile communication during courtship in *Heleophryne* (Anura: Heleophrynidae). *Journal of Herpetological Association of Africa* **35:** 12–13.

Boycott, R. C., and de Villiers, A. L. (1986). The status of *Heleophyrne rosei* Hewitt (Anura: Leptodactylidae) on Table Mountain and recommendations for its conservation. *South African Journal of Wildlife Research* **16:** 129–134.

Bragg, A. N. (1965). *Gnomes of the Night*. University of Pennsylvania Press, Philadelphia.

Brainerd, E. L., and Owerkowicz, T. (2006). Functional morphology and evolution of aspiration breathing in tetrapods. *Respiratory Physiology & Neurobiology* **154:** 73–88.

Braña, F., and Bea, A. (1987). Bimodalite de la reproduction chez *Lacerta vivipara* (Reptilia, Lacertidae). *Bulletin de la Societe Herpetologique de France* **44:** 1–5.

Branch, B (1988). *Field Guide to the Snakes and Other Reptiles of Southern Africa*. Ralph Curtis Books, Sanibel Island, FL.

Branch, B. (1991). *Everyone's Guide to Snakes of Southern Africa*. CNA, Johannesburg.

Branch, L. C. (1983). Social behavior of the tadpoles of *Phyllomedusa vaillanti*. *Copeia* **1983:** 420–428.

Brandão, C. R. F., and Vanzolini, P. E. (1985). Notes on the incubatory inquilinism between Squamata (Reptilia) and the Neotropical fungus-growing ant genus *Acromyrmex* (Hymenoptera: Formicidae). *Papéis Avulsos Zoologia (São Paulo)* **36:** 31–36.

Brandley, M. C., Schmitz, A., and Reeder, T. W. (2005). Partitioned Bayesian analysis, partition choice, and the phylogenetic relationships of scincid lizards. *Systematic Biology* **54:** 373–390.

Brattstrom, B. H. (1963). A preliminary review of the thermal requirements of amphibians. *Ecology* **44:** 238–255.

Brattstrom, B. H. (1965). Body temperature of reptiles. *American Midland Naturalist* **73:** 376–422.

Brattstrom, B. H. (1979). Amphibian temperature regulation studies in the field and laboratory. *American Zoologist* **19:** 345–356.

Brazaitis, P., Watanabe, M. E., and Amato, G. (1998). The caiman trade. *Scientific American* **278:** 52–58.

Breckenridge, W. R., Nathanael, S., and Pereira, L. (1987). Some aspects of the biology and development of *Ichthyophis glutinosus* (Amphibia: Gymnophiona). *Journal of Zoology, London* **211:** 437–449.

Breitenbach, G. L. (1982). The frequency of communal nesting and solitary brooding in the salamander, *Hemidactylium scutatum*. *Journal of Herpetology* **16:** 341–346.

Breitenbach, G. L., Congdon, J. D., and Van Loben Sels, R. C. (1984). Winter temperatures of *Chrysemys picta* nest in Michigan: Effects on hatchling survival. *Herpetologica* **40:** 76–81.

Brenowitz, E. A., and Rose, G. J. (1999). Female choice and plasticity of male calling behaviour in the Pacific treefrog. *Animal Behaviour* **57:** 1337–1342.

Briggs, J. L., and Storm, R. M. (1970). Growth and population structure of the cascade frog, *Rana cascadae* Slater. *Herpetologica* **26:** 283–300.

Britson, C. A. (1998). Predatory responses of largemouth bass (*Micropterus salmoides*) to conspicuous and cryptic hatchling turtles: A comparative experiment. *Copeia* **1998:** 383–390.

Britson, C. A., and Gutzke, W. (1993). Antipredator mechanisms of hatchling freshwater turtles. *Copeia* **1993:** 435–440.

Broadley, D. G. (1974). Reproduction in the genus *Platysaurus* (Sauria: Cordylidae). *Herpetologica* **30:** 379–380.

Broadley, D. G. (1978). A revision of the genus *Platysaurus* A. Smith (Sauria: Cordylidae). *Occasional Papers of the National Museums and Monuments of Rhodesia* **6:** 129–185.

Broadley, D. G. (1983). *Fitzsimons' Snakes of Southern Africa*. Delta Books, Johannesburg.

Broadley, D. G. (1997a). Family Scincidae. In *Proceedings of the Fitzsimons Commemorative Symposium*. J. H. van Dyke (Ed.). Pp. 66–75. Herpetol. Assoc. Africa, Matieland, S. Afr.

Broadley, D. G. (1997b). Sub-order Amphisbaenia. In *Proceedings of the Fitzsimons Commenorative Symposium*. J. H. van Dyke (Ed.). Pp. 83–86. Herpetol. Assoc. Africa, Matieland, S. Afr.

Broadley, D. G., and Branch, W. R. (2002). A review of the small east African *Cordylus* (Sauria: Cordylidae), with the description of a new species. *African Journal of Herpetology* **51:** 9–34.

Brochu, C. A. (1997a). Morphology, fossils, divergence timing, and the phylogenetic relationships of *Gavialis*. *Systematic Biology* **46:** 479–522.

Brochu, C. A. (1997b). A review of "*Leidyosuchus*" (Crocodyliformes, Eusuchia) from the Cretaceous through Eocene of North America. *Journal of Vertebrate Paleontology* **17:** 679–697.

Brochu, C. A. (1999). Phylogenetics, taxonomy, and historical biogeography of Alligatoroidea. *Society of Vertebrate Paleontology Memoir* **6:** 9–100.

Brochu, C. A. (2001). Crocodylian snouts in space and time: Phylogenetic approaches toward adaptive radiation. *American Zoologist* **41:** 564–585.

Brochu, C. A. (2003). Phylogenetic approaches toward crocodylian history. *Annual Review of Earth and Planetary Sciences* **31:** 357–397.

Brochu, C. A. (2004). Calibration age and quartet divergence date estimation. *Evolution* **58:** 1375–1382.

Brockmann, H. J. (1993). Parasitizing conspecifics: Comparisons between Hymenoptera and birds. *Trends in Ecology & Evolution* **8:** 2–4.

Brodie, E. D., III. (1993). Differential avoidance of coral snake banded patterns by free-ranging avian predators in Costa Rica. *Evolution* **47:** 227–235.

Brodie, E. D., III., and Brodie, E. D., Jr. (1999a). Costs of exploiting poisonous prey: Evolutionary trade-offs in a predator-prey arms race. *Evolution* **53:** 626–631.

Brodie, E. D., III., and Brodie, E. D., Jr. (1999b). Predator-prey arms races: Asymmetrical selection on prey may be reduced when prey are dangerous. *Bioscience* **49:** 557–568.

Brodie, E. D., III., and Janzen, F. J. (1995). Experimental studies of coral snake mimicry: Generalized avoidance of ringed snake patterns by free-ranging avian predators. *Functional Ecology* **9:** 186–190.

Brodie, E. D., III., and Ridenhour, B. J. (2003). Reciprocal selection at the phenotypic interface of coevolution. *Integrative and Comparative Biology* **43:** 408–418.

Brodie, E. D., Jr. (1968). Investigation on the skin toxin of the red-spotted newt, *Notophthalmus viridescens viridescens*. *American Midland Naturalist* **80:** 276–280.

Brodie, E. D., Jr., and Brodie, E. D., III. (1980). Differential avoidance of mimetic salamanders by free-ranging birds. *Science* **208:** 181–183.

Brodie, E. D., Jr., and Smatresk, N. J. (1990). The antipredator arsenal of fire salamaders: Spraying of secretions from highly pressurized dorsal skin glands. *Herpetologica* **46:** 1–7.

Brodie, E. D., Jr., Hensel, J. L., and Johnson, J. A. (1974). Toxicity of urodele amphibians *Taricha, Notophthalmus, Cynops*, and *Paramesotriton* (Salamandridae). *Copeia* **1974:** 506–511.

Brodie, E. D., Jr., Johnson, J. A., and Dodd, C. K., Jr. (1974). Immobility as a defensive behavior in salamanders. *Herpetologica* **30:** 79–85.

Brodie, E. D., Jr., Nussbaum, R. A., and Storm, R. M. (1969). An egg-laying aggregation of five species of Oregon reptiles. *Herpetologica* **25:** 223–227.

Brodie, E. D., Jr., Ridenhour, B. J., and Brodie, E. D., III. (2002). The evolutionary response of predators to dangerous prey: Hotspots and coldspots in the geographic mosaic of coevolution between garter snakes and newts. *Evolution* **56:** 2067–2082.

Brodmann, P. (1987). *Die Giftschlangen Europas und die Gattung Vipera in Afrika und Asien*. Kümmerly+Frey, Bern.

Brown, G. P., and Shine, R. (2004). Effects of reproduction on the antipredator tactics of snakes (*Tropidonophis mairii*, Colubridae). *Behavioral Ecology and Sociobiology* **56:** 257–262.

Brown, G. P., and Shine, R. (2006). Why do most tropical animals reproduce seasonally? Testing alternative hypotheses on the snake *Tropidonophis mairii* (Colubridae). *Ecology* **87:** 133–143.

Brown, G. P., Shilton, C. M., and Shine, R. (2006). Do parasites matter? Assessing the fitness consequences of haemogregarine infection in snakes. *Canadian Journal of Zoology* **84:** 668–676.

Brown, H. A. (1975). Temperature and development of the tailed frog, *Ascaphus truei*. *Comparative Biochemistry and Physiology* **50A:** 397–405.

Brown, H. A. (1989). Developmental anatomy of the tailed frog (*Ascaphus truei*): A primitive frog with large eggs and slow development. *Journal of Zoology, London* **217:** 525–537.

Brown, J. H., and Lomolino, M. V. (1998). *Biogeography*. (2nd Ed.). Sinauer Associates, Inc., Sunderland MA.

Brown, J. L., and Orians, G. H. (1970). Spacing patterns in mobile animals. *Annual Review of Ecology and Systematics* **1:** 239–262.

Brown, L. E., and Crespo, E. G. (2000). Burrowing behavior of the midwife toad *Alytes cisternasii* and *Alytes obstetricans* (Anura, Discoglossidae). *Alytes* **17:** 101–113.

Brown, W. M., and Krauss, F. (1998). Phylogenetic relationships of colubroid snakes based on mitochondrial DNA sequences. *Zoological Journal of the Linnean Society* **122:** 455–487.

Brown, W. S. (1993). Biology, status, and management of the timber rattlesnake (*Crotalus horridus*): A guide for conservation. *Herpetological Circular* **22:** 1–78.

Brown, W. S., and Parker, W. S. (1984). Growth, reproduction and demography of the racer, *Coluber constrictor mormon*, in northern Utah. *Museum of Natural History, University of Kansas Specical Publication* **10:** 13–40.

Bruce, R. C. (1988a). An ecological life table for the salamander *Eurycea wilderae*. *Copeia* **1988:** 15–26.

Bruce, R. C. (1988b). Life history variation in the salamander *Desmognathus quadramaculatus*. *Herpetologica* **44:** 218–227.

Bruce, R. C. (2005). Theory of complex life cycles: Application in plethodontid salamanders. *Herpetological Monographs* 180–207.

Bruce, R. C., Jaeger, R., and Houck, L. D. (2000). *The Biology of Plethodontid Salamanders*. Kluwer Academic/Plenum Publishers, New York.

Brust, D. G. (1993). Maternal brood care by *Dendrobates pumilio*: A frog that feeds its young. *Journal of Herpetology* **27:** 96–98.

Brygoo, É. R., and Roux-Estève, R. (1983). *Feylinia*, genre de lézards africains de la famille des Scincidae, sous-famille des Feyliniinae. *Bulletin du Muséum national d'Histoire naturelle de Paris 4e série* **5:** 307–341.

Buchholz, D. R., Singamsetty, S., Karadge, U., Williamson, S., Langer, C. E., and Elinson, R. P. (2007). Nutritional endoderm in a direct developing frog: A potential parallel to the evolution of the amniote egg. *Developmental Dynamics* **236:** 1259–1272.

Buckley, D., Alcobendas, M., Garcia-Paris, M., and Wake, M. H. (2007). Heterochrony, cannibalism, and the evolution of viviparity in *Salamandra salamandra*. *Evolution & Development* **9:** 105–115.

Buffetaut, E. (1989). Evolution. In *Crocodiles and Alligators*. C. A. Ross (Ed.). Pp. 26–41. Golden Press, Silverwater, N.S.W.

Bull, C. M. (1988). Mate fidelity in an Australian lizard *Trachydosaurus rugosus*. *Behavioral Ecology and Sociobiology* **23:** 45–49.

Bull, C. M. (1994). Population dynamics and pair fidelity in sleepy lizards. In *Lizard Ecology: Historical and Experimental Perspectives*. L. J. Vitt and E. R. Pianka (Eds.). Pp. 159–174. Princeton University Press, Princeton, NJ.

Bull, C. M., and Bagurst, B. C. (1998). Home range overlap of mothers and their offspring in the sleepy lizard, *Tiliqua rugosa*. *Behavioral Ecology and Sociobiology* **42:** 357–362.

Bull, C. M., and Burzacott, D. (1993). The impact of tick load on the fitness of their lizard hosts. *Oecologia* **96:** 415–419.

Bull, C. M., and Pamula, Y. (1998). Enhanced vigilance in monogamous pairs of the lizard, *Tiliqua rugosa*. *Behavioral Ecology* **9:** 452–455.

Bull, C. M., Cooper, S. J. B., and Baghurst, B. C. (1998). Social monogamy and extra-pair fertilization in an Australian lizard, *Tiliqua rugosa*. *Behavioral Ecology and Sociobiology* **44:** 63–72.

Bull, C. M., Griffin, C. L., Bonnett, M., Gardner, M. G., and Cooper, S. J. B. (2001). Discrimination between related and unrelated individuals in the Australian lizard *Egernia striolata*. *Behavioral Ecology and Sociobiology* **50:** 173–179.

Bull, C. M., Griffin, C. L., Lanham, E. J., and Johnston, G. R. (2000). Recognition of pheromones from group members in a gregarious lizard, *Egernia stokesii*. *Journal of Herpetology* **34:** 92–99.

Bull, J. J. (1980). Sex determination in reptiles. *Quarterly Review of Biology* **55:** 3–21.

Bull, J. J. (1987). Evolution of phenotypic variance. *Evolution* **41:** 303–315.

Bullock, D. J. (1986). The ecology and conservation of reptiles on Round Island and Gunner's Quoin, Mauritius. *Biological Conservation* **37:** 135–156.

Bunnell, P. (1973). Vocalizations in the territorial behavior of the frog, *Dendrobates pumilio*. *Copeia* **1973:** 277–284.

Burbrink, F. T. (2005). Inferring the phylogenetic position of *Boa constrictor* among the Boinae. *Molecular Phylogenetics and Evolution* **34:** 167–180.

Burbrink, F. T., Lawson, R., and Slowinski, J. B. (2000). Molecular phylogeography of the North American rat snake (*Elaphe obsoleta*): A critique of the subspecies concept. *Evolution* **54:** 2107–2114.

Burger, J., and Zappalorti, R. T. (1986). Nest site selection by pine snakes, *Pituophis melanoleucus*, in the New Jersey Pine Barrens. *Copeia* **1986:** 116–121.

Burggren, W. W. (1987). Form and function in reptilian circulation. *American Zoologist* **27:** 5–19.

Burghardt, G. M., and Rand, A. S. (1982). *Iguanas of the World: Their Behavior, Ecology, and Conservation*. Noyes Publications, Park Ridge, N.J.

Burke, A. C. (1989). Development of the turtle carapace: Implications for the evolution of a novel bauplan. *Journal of Morphology* **199:** 363–378.

Burke, R. L. (1991). Relocations, repatriations, and translocations of amphibians and reptiles: Taking a broader view. *Herpetologica* **47:** 350–357.

Burke, R. L., Ewert, M. A., McLemore, J. B., and Jackson, D. R. (1996). Temperature-dependent sex determination and hatching success in the gopher tortoise (*Gopherus polyphemus*). *Chelonian Conservation and Biology* **2:** 86–88.

Burke, V. J., and Gibbons, J. W. (1995). Terrestrial buffer zones and wetland conservation: A case study of freshwater turtles in a Carolina bay. *Conservation Biology* **9:** 1365–1369.

Burke, V. J., Nagle, R. D., Osentoski, M., and Congdon, J. D. (1993). Common snapping turtles' association with ant mounds. *Journal of Herpetology* **27:** 114–115.

Burns, G., Ramos, A., and Muchlinski, A. (1996). Fever response in North American snakes. *Journal of Herpetology* **30:** 133–139.

Burrage, B. R. (1973). Comparative ecology and behavior of *Chamaeleo pumilus pumilus* (Gmelin) & *C. namaquensis* A. Smith (Sauria: Chamaeleonidae). *Annals of the South African Museum* **61:** 1–158.

Burrowes, P. A. (2000). Parental care and sexual selection in the Puerto Rican cave-dwelling frog, *Eleutherodactylus cooki*. *Herpetologica* **56:** 375–386.

Burton, T. C. (1986). A reassessment of the Papuan subfamily Asterophryinae (Anura: Microhylidae). *Records of the South Australian Museum* **19:** 405–450.

Burton, T. C., and Zweifel, R. G. (1995). A new genus of genyophrynine microhylid frog from New Guinea. *American Museum Novitates* **3129:** 1–7.

Bury, R. B., and Corn, P. S. (1988). Douglas-fir forests in the Oregon and Washington Cascades: Abundance of terrestrial herpetofauna. *United States Department of Agriculture Forest Service Technical Report* **RM166:** 11–22.

Bury, R. B., and Whelan, J. A. (1984). Ecology and management of the bullfrog. *United States Fish and Wildlife Services Resource Publications* **155:** 1–23.

Busack, S. D. (1977). Zoogeography of amphibians and reptiles in Cádiz Province, Spain. *Annals of the Carnegie Museum* **46:** 285–316.

Busse, K. (1970). Care of the young by male *Rhinoderma darwini*. *Copeia* **1970:** 395.

Bustard, H. R. (1967). Gekkonid lizards adapt fat storage to desert environments. *Science* **158:** 1197–1198.

Bustard, H. R. (1970). The role of behavior in the natural regulation of numbers in the gekkonid lizard *Gehyra variegata*. *Ecology* **51:** 724–728.

Buth, D. G. (1984). The application of electrophoretic data in systematic studies. *Annual Review of Ecology and Systematics* **15:** 501–522.

Butler, J. A., Bowman, R. D., Hull, T. W., and Sowell, S. (1995). Movements and home range of hatchling and yearling gopher tortoises, *Gopherus polyphemus*. *Chelonian Conservation and Biology* **1:** 173–180.

Byrne, P. G., and Roberts, J. D. (2004). Intrasexual selection and group spawning in quacking frogs (*Crinia georgiana*). *Behavioral Ecology* **15:** 872–882.

Cadle, J. E. (1984). Molecular systematics of Neotropical xenodontine snakes. III. Overview of xenodontine phylogeny and the history of New World snakes. *Copeia* **1984:** 641–652.

Cadle, J. E. (1987). Geographic distribution: Problems in phylogeny and zoogeography. In *Snakes: Ecology and Evolutionary Biology*. R. A. Seigel, J. T. Collins, and S. S. Novak (Eds.). Pp. 77–105. Macmillian Publishing Co., New York.

Cadle, J. E. (1992). Phylogenetic relationships among vipers: Immunological evidence. In *Biology of the Pitvipers*. J. A. Campbell and E. D. Brodie, Jr. (Eds.). Pp. 41–48. Selva, Tyler, TX.

Cadle, J. E. (1994). The colubrid radiation in Africa (Serpentes: Colubridae): Phylogenetic relationships and evolutionary paterns based on immunological data. *Biological Journal of the Linnean Society* **110:** 103–140.

Cadle, J. E., and Greene, H. W. (1993). Phylogenetic patterns, biogeography, and the ecological structure of Neotropical snake assemblages. In *Species Diversity in Ecological Communities: Historical and Geographical Perspectives*. R. E. Ricklefs and D. Schluter (Eds.). Pp. 281–293. Chicago University Press, Chicago.

Cadle, J. E., Dessauer, H. C., Gans, C., and Gartside, D. F. (1990). Phylogenetic relationships and molecular evolution in uropeltid snakes (Serpentes: Uropeltidae): Allozymes and albumin immunology. *Biological Journal of the Linnean Society* **40:** 293–320.

Caldwell, J. P. (1982). Disruptive selection: A tail color polymorphism in *Acris* tadpoles in response to differential predation. *Canadian Journal of Zoology* **60:** 2818–2827.

Caldwell, J. P. (1987). Demography and life history of two species of chorus frogs (Anura: Hylidae) in South Carolina. *Copeia* **1987:** 114–127.

Caldwell, J. P. (1989). Structure and behavior of *Hyla geographica* tadpole schools, with comments on classification of group behavior in tadpoles. *Copeia* **1989:** 938–950.

Caldwell, J. P. (1992). Diversity of reproductive modes in anurans: Faculatative nest construction in gladiator frogs. In *Reproductive Biology of South American Vertebrates*. W. C. Hamlett (Ed.). Pp. 85–97. Springer-Verlag, New York.

Caldwell, J. P. (1993). Brazil nut fruit capsules as phytotelmata: Interactions among anuran and insect larvae. *Canadian Journal of Zoology* **71:** 1193–1201.

Caldwell, J. P. (1994). Natural history and survival of eggs and early larval stages of *Agalychnis calcarifer* (Anura: Hylidae). *Herpetological Natural History* **2:** 57–66.

Caldwell, J. P. (1996a). The evolution of myrmecophagy and its correlates in poison frogs (family Dendrobatidae). *Journal of Zoology, London* **240:** 75–101.

Caldwell, J. P. (1996b). Diversity of Amazonian anurans: The role of systematics and phylogeny in identifying macroecological and evolutionary patterns. In *Neotropical Biodiversity and Conservation*. A. Gibson (Ed.). Pp. 73–88. *Occasional Papers of the Mildred E. Mathias Botanical Garden* **1**.

Caldwell, J. P. (1997). Pair bonding in spotted poison frogs. *Nature* **385:** 211.

Caldwell, J. P., and Hoogmoed, M. S. (1998). Allophrynidae, *Allophryne, A. ruthveni*. *Catalogue of American Amphibians and Reptiles* **661:** 1–3.

Caldwell, J. P., and Lima, A. P. (2003). A new Amazonian species of *Colostethus* (Anura: Dendrobatidae) with a nidicolous tadpole. *Herpetologica* **59:** 219–234.

Caldwell, J. P., and Lopez, P. T. (1989). Foam-generating behavior in tadpoles of *Leptodactylus mystaceus*. *Copeia* **1989:** 498–502.

Caldwell, J. P., and Oliveira, V. R. L. (1999). Determinants of biparental care in the spotted poison frog, *Dendrobates vanzolinii* (Anura: Dendrobatidae). *Copeia* **1999:** 565–575.

Caldwell, J. P., and Shepard, D. B. (2007). Calling site fidelity and call structure of a Neotropical toad, *Rhinella ocellata* (Anura: Bufonidae). *Journal of Herpetology* **41:** 611–621.

Caldwell, J. P., and Vitt, L. J. (1999). Dietary asymmetry in leaf litter frogs and lizards in a transitional northern Amazonian rain forest. *Oikos* **84:** 383–397.

Caldwell, J. P., Thorpe, J. H., and Jervey, T. O. (1980). Predator-prey relationships among larval dragonflies, salamanders, and frogs. *Oecologia* **46:** 285–289.

Caldwell, M. W. (1996). Ichthyosauria: A preliminary phylogenetic analysis of diapsid affinities. *Neues Jahrbuch für Geologie und Paläontologie, Abhandlungen* **200:** 361–386.

Caldwell, M. W., and Lee, M. S. Y. (1997). A snake with legs from the marine Cretaceous of the Middle East. *Nature* **386:** 705–709.

Callaway, J. M. (1997). Introduction. In *Ancient Marine Reptiles*. J. M. Callaway and E. L. Nicholls (Eds.). Pp. 3–16. Academic Press, San Diego.

Camargo, A., de Sá, R. O., and Heyer, W. R. (2006). Phylogenetic analyses of mtDNA sequences reveal three cryptic lineages in the widespread Neotropical frog *Leptodactylus fuscus* (Schneider, 1799) (Anura, Leptodactylidae). *Biological Journal of the Linnean Society* **87:** 325–341.

Camp, C. L. (1923). Classification of the lizards. *Bulletin of American Museum of Natural History* **48:** 289–481.

Campbell, J. A., and Brodie, E. D., Jr., (Eds.). (1992). *Biology of the Pitvipers*. Selva, Tyler, TX.

Campbell, J. A., and Camarillo, R., J. A. (1992). The Oaxacan dwarf boa, *Exiliboa placata* (Serpentes: Tropidiophiidae): Descriptive notes and life history. *Caribbean Journal of Science* **28:** 17–20.

Campbell, J. A., and Frost, D. R. (1993). Anguid lizards of the genus *Abronia*: Revisionary notes, descriptions of four new species, a phylogenetic analysis, and key. *Bulletin of American Museum of Natural History* **216:** 1–121.

Campbell, J. A., and Lamar, W. W. (1989). *The Venomous Reptiles of Latin America*. Comstock Publishing Associates, Ithaca, N.Y.

Campbell, J. A., and Quinn, H. R. (1975). Reproduction in a pair of Asiatic cobras, *Naja naja* (Serpentes, Elapidae). *Journal of Herpetology* **9:** 229–233.

Cann, J. (1998). *Australian Freshwater Turtles*. Beaumont Publishers Pte Ltd, Singapore.

Cannatella, D. C., and de Sá, R. O. (1993). *Xenopus laevis* as a model organism. *Systematic Biology* **42:** 476–507.

Cannatella, D. C., and Trueb, L. (1988a). Evolution of pipoid frogs: Intergeneric relationships of the aquatic frog family Pipidae (Anura). *Zoological Journal of the Linnean Society* **94:** 1–38.

Cannatella, D. C., and Trueb, L. (1988b). Evolution of pipoid frogs: Morphology and phylogenetic relationships of *Pseudhymenochirus*. *Journal of Herpetology* **22:** 439–456.

Capranica, R. R. (1986). Morphology and physiology of the auditory system. In *Biology of the Integument. 2. Vertebrates*. J. Bereiter-Hahn, A. G. Matoltsy, and K. S. Richards (Eds.). Pp. 551–575. Springer Verlag, Berlin.

Carpenter, C. C. (1953). A study of hibernacula and hibernating associations of snakes and amphibians in Michigan. *Ecology* **34:** 74–80.

Carpenter, C. C. (1957). Hibernation, hibernacula and associated behavior of the three-toed box turtle (*Terrapene carolina triunguis*). *Copeia* **1957:** 278–282.

Carpenter, C. C. (1961). Patterns of social behavior in the desert iguana, *Dipsosaurus dorsalis*. *Copeia* **1961:** 396–405.

Carpenter, C. C. (1977). Communication and displays of snakes. *American Zoologist* **17:** 217–223.

Carpenter, G. C., and Duvall, D. (1995). Fecal scent marking in the western banded gecko (*Coleonyx variegatus*). *Herpetologica* **51:** 33–38.

Carr, A. F., Jr. (1940). A contribution to the herpetology of Florida. *University of Florida Publications, Biological Sciences* **3:** 1–118.

Carranza, S., and Amat, F. (2005). Taxonomy, biogeography and evolution of *Euproctus* (Amphibia: Salamandridae), with the resurrection of the genus *Calotriton* and the description of a new endemic species from the Iberian Peninsula. *Zoological Journal of the Linnean Society* **145:** 555–582.

Carranza, S., and Arnold, E. N. (2006). Systematics, biogeography, and evolution of *Hemidactylus* geckos (Reptilia: Gekkonidae) elucidated using mitochondrial DNA sequences. *Molecular Phylogenetics and Evolution* **38:** 531–545.

Carranza, S., Arnold, E. N., Mateo, J. A., and Lopez-Jurado, L. F. (2000). Long-distance colonization and radiation in gekkonid lizards, *Tarentola* (Reptilia: Gekkonidae), revealed by mitochondrial DNA sequences. *Proceedings of the Royal Society of London Series B, Biological Sciences* **267:** 637–649.

Carroll, R. L. (1988). *Vertebrate Paleontology and Evolution*. W.H. Freeman & Co, New York.

Carroll, R. L. (1991). The origin of reptiles. In *Origins of the Higher Groups of Tetrapods. Controversy and Consensus*. H.-P. Schultze and L. Trueb (Eds.). Pp. 331–353. Comstock Publishing Associates, Ithaca, NY.

Carroll, R. L. (1995). Problems of the phylogenetic analysis of Paleozoic choanates. *Bulletin du Muséum national d'Histoire naturelle de Paris* **7:** 389–445.

Carroll, R. L. (2007). The palaeozoic ancestry of salamanders, frogs and caecilians. *Zoological Journal of the Linnean Society* **150:** 1–140.

Carroll, R. L., and Baird, D. (1972). Carboniferous stem reptiles of the family Romeriidae. *Bulletin of the Museum of Comparative Zoology* **143:** 321–361.

Carroll, R. L., and Wild, R. (1994). Marine members of the Sphenodontia. In *In the Shadow of the Dinosaurs. Early Mesozoic Tetrapods*. N. C. Fraser and H.-D. Sues (Eds.). Pp. 70–83. Cambridge University Press, New York.

Carroll, R. L., Kuntz, A., and Albright, K. (1999). Vertebral development and amphibian evolution. *Evolution & Development* **1:** 36–48.

Carstens, B. C., Degenhardt, J. D., Stevenson, A. L., and Sullivan, J. (2005). Accounting for coalescent stochasticity in testing phylogeographical hypotheses: Modelling Pleistocene population structure in the Idaho giant salamander *Dicamptodon aterrimus*. *Molecular Ecology* **14:** 255–265.

Cartledge, V. A., Withers, P. C., McMaster, K. A., Thompson, G. G., and Bradshaw, S. D. (2006). Water balance of field-excavated aestivating Australian desert frogs, the cocoon-forming *Neobatrachus aquilonius* and the non-cocooning *Notaden nichollsi* (Amphibia: Myobatrachidae). *Journal of Experimental Biology* **209:** 3309–3321.

Case, T. J. (1976). Seasonal aspects of thermoregulatory behavior in the chuckwalla, *Sauromalus obesus*, (Reptilia, Lacertilia: Iguanidae). *Journal of Herpetology* **10:** 85–95.

Case, T. J., Bolger, D. T., and Richman, A. D. (1992). Reptilian extinctions: The last ten thousand years. In *Conservation Biology*. P. L. Fledler and S. K. Jain (Eds.). Pp. 91–125. Chapman and Hall, N.Y.

Castaneda, L. E., Sabat, P., Gonzalez, S. P., and Nespolo, R. F. (2006). Digestive plasticity in tadpoles of the Chilean giant frog (*Caudiverbera caudiverbera*): Factorial effects of diet and temperature. *Physiological and Biochemical Zoology* **79:** 919–926.

Castanet, J., and Baez, M. (1991). Adaptation and evolution in *Gallotia* lizards from the Canary Islands: Age, growth, maturity and longevity. *Amphibia-Reptilia* **12:** 81–102.

Castanet, J., Newman, D. G., and Saint Girons, H. (1988). Skeletochronological data on the growth, Age, and population structure of the tuatara, *Sphenodon punctatus*, on Stephens and Lady Alice Islands, New Zealand. *Herpetologica* **44:** 25–37.

Castoe, T. A., and Parkinson, C. L. (2006). Bayesian mixed models and the phylogeny of pitvipers (Viperidae: Serpentes). *Molecular Phylogenetics and Evolution* **39:** 91–110.

Castoe, T. A., Doan, T. A., and Parkinson, C. L. (2004). Data partitions and complex models in Bayesian analysis: The phylogeny of gymnophthalmid lizards. *Systematic Biology* **53:** 448–469.

Catton, W. T. (1976). Cutaneous mechanoreceptor. In *Frog Neurobiology. A Handbook*. R. Llinás and W. Precht (Eds.). Pp. 629–642. Springer-Verlag, Berlin.

Caughley, G., and Gunn, A. (1996). *Conservation Biology in Theory and Practice*. Blackwell Science, Cambridge, Mass.

Cecil, S. G., and Just, J. J. (1979). Survival rate, population density and development of a naturally occurring anuran larvae (*Rana catesbeiana*). *Copeia* **1979:** 447–453.

Cei, J. M. (1980). Amphibians of Argentina. *Monitore zoologico italiano, (N. S.) Monographia* **2:** 1–609.

Cei, J. M. (1984). A new leptodactylid frog, genus *Atelognathus*, from Southern Patagonia. *Herpetologica* **40:** 47–51.

Cei, J. M. (1993). Reptiles del noroeste, nordeste y este de la Argentina. Herpetofauna de las selvas subtropicales, puna y pampas. *Museu Regionale di Scienze Naturali – Torino Monographías* **14:** 1–949.

Censky, E. J. (1995). Mating strategy and reproductive success in the teiid lizard, *Ameiva plei*. *Behaviour* **132:** 529–557.

Censky, E. J. (1996). The evolution of sexual size dimorphism in the teiid lizard *Ameiva plei*: A test of alternative hypotheses. In *Contributions to West Indian Herpetology: A tribute to Albert Schwartz*. R. Powell and R. W. Henderson (Eds.). Pp. 277–289. Society for the Study of Amphibians and Reptiles, Ithaca, N.Y.

Censky, E. J. (1997). Female mate choice in the non-territorial lizard *Ameiva plei* (Teiidae). *Behavioral Ecology and Sociobiology* **40:** 221–225.

Channing, A. (1989). A re-evaluation of the phylogeny of Old World treefrogs. *South African Journal of Zoology* **24:** 116–131.

Channing, A. (1995). The relationship between *Breviceps* (Anura: Microhylidae) and *Hemisus* (Hemisotidae) remains equivocal. *Journal of the Herpetological Association of Africa* **44:** 55–57.

Channing, A., and Howell, K. M. (2006). *Amphibians of East Africa*. Cornell. University Press, Ithaca, New York.

Channing, A., du Preez, L., and Passmore, N. (1994). Status, vocalization and breeding biology of two species of African bullfrogs (Ranidae: *Pyxicephalus*). *Journal of Zoology, London* **234:** 141–148.

Channing, A., Finlow-Bates, K. S., Haarklau, S. E., and Hawkes, P. G. (2006). The biology and recent history of the critically endangered Kihansi spray toad *Nectophrynoides asperginis* in Tanzania. *Journal of East African Natural History* **95:** 117–138.

Chapple, D. G. (2003). Ecology, life-history, and behavior in the Australian Scincid genus *Egernia*, with comments on the evolution of complex sociality in lizards. *Herpetological Monographs* **17:** 145–180.

Che, J., Pang, J. F., Zhao, H., Wu, G. F., Zhao, E. M., and Zhang, Y. P. (2007a). Molecular phylogeny of the Chinese ranids inferred from nuclear and mitochondrial DNA sequences. *Biochemical Systematics and Ecology* **35:** 29–39.

Che, J., Pang, J. F., Zhao, H., Wu, G. F., Zhao, E. M., and Zhang, Y. P. (2007b). Phylogeny of Raninae (Anura: Ranidae) inferred from mitochondrial and nuclear sequences. *Molecular Phylogenetics and Evolution* **43:** 1–13.

Chen, L. Q., Murphy, R. W., Lathrop, A., Ngo, A., Orlov, N. L., Ho, C. T., and Somorjai, I. L. M. (2005). Taxonomic chaos in Asian ranid frogs: An initial phylogenetic resolution. *Herpetological Journal* **15:** 231–243.

Chippindale, P. T., Bonett, R. M., Baldwin, A. S., and Wiens, J. J. (2004). Phylogenetic evidence for a major reversal of life history evolution in plethodontid salamanders. *Evolution* **58:** 2809–2822.

Chong, G., Heatwole, H., and Firth, B. T. (1973). Panting thresholds of lizards-II. Diel variation in panting threshold of *Amphibolurus muricatus*. *Comparative Biochemistry and Physiology* **46:** 827–829.

Chou, L. M., Leong, C. F., and Choo, B. L. (1988). The role of optic, auditory and olfactory senses in prey hunting by two species of geckos. *Journal of Herpetology* **22:** 349–351.

Christian, K. A. (1996). Solar absorptance of some Australian lizards and its relationship to temperature. *Australian Journal of Zoology* **44:** 59–67.

Christian, K. A., and Bedford, G. (1993). High reproductive expenditure per progeny in geckos relative to other lizards. *Journal of Herpetology* **27:** 351–354.

Christian, K. A., and Parry, D. (1997). Reduced rates of water loss and chemical properties of skin secretions of the frogs *Litoria caerulea* and *Cyclorana australis*. *Australian Journal of Zoology* **45:** 13–20.

Christian, K. A., and Weavers, B. W. (1996). Thermoregulation of monitor lizards in Australia: An evaluation of methods in thermal biology. *Ecological Monographs* **66:** 139–157.

Christian, K. A., Corbett, L. K., Green, B., and Weavers, B. W. (1995). Seasonal activity and energetics of two species of varanid lizards in tropical Australia. *Oecologia* **103:** 349–357.

Christian, K. A., Tracy, C. R., and Porter, W. P. (1983). Seasonal shifts in body temperature and use of microhabitats by Galapagos land iguanas (*Conolophus pallidus*). *Ecology* **64:** 463–468.

Christian, K., and Bedford, G. S. (1995). Physiological consequences of filarial parasites in the frillneck lizard, *Chlamydosaurus kingii*, in northern Australia. *Canadian Journal of Zoology* **73:** 2302–2306.

Christian, K., Green, B., and Kennett, R. (1996). Some physiological consequences of estivation by freshwater crocodiles, *Crocodylus johnstoni*. *Journal of Herpetology* **30:** 1–9.

Christian, K. A., Tracy, C. R., and Tracy, C. R. (2006). Evaluating thermoregulation in reptiles: An appropriate null model. *American Naturalist* **168:** 421–430.

Christiansen, D. G., Fog, K., Pedersen, B. V., and Boomsma, J. J. (2005). Reproduction and hybrid load in all-hybrid populations of *Rana esculenta* water frogs in Denmark. *Evolution* **59:** 1348–1361.

Christiansen, J. L., and Dunham, A. E. (1972). Reproduction of the yellow mud turtle (*Kinosternon flavescens flavenscens*) in New Mexico. *Herpetologica* **28:** 130–137.

Cifelli, R. L., and Nydam, R. L. (1995). Primitive, helodermatid-like platynotan from the Early Cretaceous of Utah. *Herpetologica* **51:** 286–291.

Ciofi, C., and Chelazzi, G. (1994). Analysis of homing pattern in the colubrid snake *Coluber viridiflavus*. *Journal of Herpetology* **28:** 477–484.

Cisneros-Heredia, D. F., and McDiarmid, R. W. (2007). Revision of the characters of Centrolenidae (Amphibia: Anura: *Athesphatanura*), with comments on its taxonomy and the description of new taxa of glassfrogs. *Zootaxa* **1572:** 3–82.

Clack, J. A. (1998). A new Early Carboniferous tetrapod with a mélange of crown-group characters. *Nature* **394:** 66–69.

Clack, J. A. (2006). The emergence of early tetrapods. *Palaeogeography Palaeoclimatology Palaeoecology* **232:** 167–189.

Clark, D. R. J. (1971). The strategy of tail-autotomy in the ground skink, *Lygosoma laterale*. *Journal of Experimental Zoology* **176:** 295–302.

Clark, J. M. (1994). Patterns of evolution in Mesozoic Crocodyliformes. In *In the shadow of the dinosaurs. Early Mesozoic Tetrapods*. N. C. Fraser and H.-D. Sues (Eds.). Pp. 84–97. Cambridge University Press, New York.

Clark, R. W. (2004). Kin recognition in rattlesnakes. *Proceedings of the Royal Society of London Series B-Biological Sciences* **271:** S243–S245.

Clark, V. C., Raxworthy, C. J., Rakotomalala, V., Sierwald, P., and Fisher, B. L. (2005). Convergent evolution of chemical defense in poison frogs and arthropod prey between Madagascar and the Neotropics. *Proceedings of the National Academy of Sciences (USA)* **102:** 11617–11622.

Clarke, B. T. (1997). The natural history of amphibian skin secretions, their normal functioning and potential medical applications. *Biological Reviews* **72:** 365–379.

Clevenger, A. P., Wierzchowski, J., Chruszcz, B., and Gunson, K. (2002). GIS-generated, expert-based models for identifying wildlife habitat linkages and planning mitigation passages. *Conservation Biology* **16:** 503–514.

Clobert, J., Massot, M., Lecomte, J., Sorci, G., Fraipont, M., and Barbault, R. (1994). Determinants of dispersal behavior: The common lizard as a case study. In *Lizard Ecology: Historical and Experimental Perspectives*. L. J. Vitt and E. R. Pianka (Eds.). Pp. 183–206. Princeton University Press, Princeton, NJ.

Clos, L. M. (1995). A new species of *Varanus* (Reptilia: Sauria) from the Miocene of Kenya. *Journal of Vertebrate Paleontology* **15:** 254–267.

Clough, M., and Summers, K. (2000). Phylogenetic systematics and biogeography of the poison frogs: Evidence from mitochondrial DNA sequences. *Biological Journal of the Linnean Society* **70:** 515–540.

Coates, M. I., and Clack, J. A. (1995). Romer's gap: Tetrapod origins and terrestriality. *Bulletin du Muséum national d'Histoire naturelle de Paris* **7:** 373–388.

Coates, M. I., Jeffery, J. E., and Ruta, M. (2002). Fins to limbs: What the fossils say. *Evolution & Development* **4:** 390–401.

Coates, M. I., Ruta, M., and Milner, A. R. (2000). Early tetrapod evolution. *Trends in Ecology & Evolution* **15:** 327–328.

Cock Buning, T. (1985). Thermal sensitivity as a specialization for prey capture and feeding in snakes. *American Zoologist* **23:** 363–75.

Cocroft, R. B. (1994). A cladistic analysis of chorus frog phylogeny (Hylidae: *Pseudacris*). *Herpetologica* **50:** 420–437.

Cody, M. L. (1996). Introduction to long-term community ecological studies. In *Long-term Studies of Vertebrate Communities*. M. L. Cody and J. A. Smallwood (Eds.). Pp. 1–15. Academic Press, San Diego.

Cody, M. L., and Smallwood, J. A. (Eds.). (1996). *Long-term Studies of Vertebrate Communities*. Academic Press, San Diego.

Cogger, H. G., and Zweifel, R. G. (1992). *Reptiles and Amphibians: A comprehensive illustrated guide*. Publishing, New York.

Cogger, H. G. and Zweifel, R. G. (Eds.). (1998). *Encyclopedia of Reptiles & Amphibians*. Academic Press, San Diego.

Colbert, E. H. (1967). Adaptations for gliding in the lizard *Draco*. *American Museum Novitates* **2283:** 1–20.

Colbert, E. H. (1970). The Triassic gliding reptile *Icarosaurus*. *Bulletin of the American Museum of Natural History* **143:** 84–142.

Cole, C. J. (1984). Unisexual lizards. *Scientific American* **250:** 94–100.

Cole, C. J., and Townsend, C. R. (1983). Sexual behavior in unisexual lizards. *Animal Behaviour* **31:** 724–728.

Cole, C. J., and Townsend, C. R. (1990). Parthenogenetic lizards as vertebrate systems. *Journal of Experimental Zoology* **4:** 174–176.

Cole, C. J., Dessauer, H. C., Townsend, C. R., and Arnold, M. G. (1990). Unisexual lizards of the genus *Gymnophthalmus* (Reptilia: Teiidae) in the Neotropics: Genetics, origin, and systematics. *American Museum Novitates* **2994:** 1–29.

Cole, C. J., Dessauer, H. C., Townsend, C. R., and Arnold, M. G. (1995). *Kentropyx borckiana* (Squamata: Teiidae): A unisexual lizards of hybrid origin in the Guiana region, South America. *American Museum Novitates* **3145:** 1–23.

Colinvaux, P. A. (1996). Quaternary environmental history and forest diversity in the Neotropics. In *Evolution and Environment in Tropical America*. J. B. C. Jackson, A. F. Budd, and A. J. Coates (Eds.). Pp. 359–405. University of Chicago Press, Chicago, IL.

Colli, G. R., and Zamboni, D. S. (1999). Ecology of the worm-lizard *Amphisbaena alba* in the Cerrado of central Brazil. *Copeia* **1999:** 733–742.

Colli, G. R., Bastos, R. P., and Araujo, A. F. B. (2002). The character and dynamics of the Cerrado herpetofauna. In*: The Cerrados of Brazil: Ecology and Natural History of a Neotropical Savanna*. P. S. Oliveira and R. J. Marquis (Eds.). Pp. 223–241. Columbia University Press, NY.

Colli, G. R., Péres, A. K., and Cunha, H. J. (1998). A new species of *Tupinambis* (Squamata: Teiidae) from central Brazil, with an analysis of morphological and genetic variation in the genus. *Herpetologica* **54:** 477–492.

Colli, G. R., Caldwell, J. P., Costa, G. C., Gainsbury, A. M., Garda, A. A., Mesquita, D. O., Filho, C. M. M. R., Soares, A. H. B., Silva, V. N., Valdujo, P. H., Vieira, G. H. C., Vitt, L. J., Werneck, F. P., Wiederhecker, H. C., and Zatz, M. G. (2003). A new species of *Cnemidophorus* (Squamata, Teiidae) from the Cerrado Biome in central Brazil. *Occasional Papers of the Sam Noble Oklahoma Museum of Natural History* **14:** 1–14.

Coman, B. J., and Brunner, H. (1972). Food habits of the feral house cat in Victoria. *Journal of Wildlife Management* **36:** 848–853.

Conant, R., and Collins, J. T. (1991). *A Field Guide to Reptiles and Amphibians. Eastern and Central North America*. Second edition. Houghton Mifflin Co., Boston.

Conant, R., and Collins, J. T. (1998). *A Field Guide to Reptiles and Amphibians. Eastern and Central North America*. Third edition. Houghton Mifflin Co, Boston.

Congdon, J. D., and Gibbons, J. W. (1987). Morphological constraint on egg size: A challenge to optimal egg size theory? *Proceedings of the National Academy of Sciences (USA)* **84:** 4145–4147.

Congdon, J. D., and Gibbons, J. W. (1996). Structure and dynamics of a turtle community over two decades. In *Long-term Studies of Vertebrate Communities*. M. L. Cody and J. A. Smallwood (Eds.). Pp. 137–159. Academic Press, San Diego.

Congdon, J. D., Ballinger, R. E., and Nagy, K. A. (1979). Energetics, temperature and water relations in winter aggregated *Sceloporus jarrovi* (Sauria: Iguanidae). *Ecology* **60:** 30–35.

Congdon, J. D., Breitenbach, G. L., van Loben Sels, R. C., and Tinkle, D. W. (1987). Reproduction and nesting ecology of snapping turtles (*Chelydra serpentina*) in southeastern Michigan. *Herpetologica* **43:** 39–54.

Congdon, J. D., Dunham, A. E., and Tinkle, D. W. (1982). Energy budgets and life histories of reptiles. In *Biology of the Reptilia. Volume 13. Physiology D. Physiological Ecology*. C. Gans and F. H. Pough (Eds.). Pp. 233–271. Academic Press, New York.

Congdon, J. D., Dunham, A. E., and van Loben Sels, R. C. (1994). Demographics of common snapping turtles (*Chelydra serpentina*): Implications for conservation and management of long-lived organisms. *American Zoologist* **34:** 397–408.

Congdon, J. D., Tinkle, D. W., Breitenbach, G. L., and van Loben Sels, R. C. (1983). Nesting ecology and hatchling success in the turtle *Emydoidea blandingi*. *Herpetologica* **39:** 417–429.

Congdon, J. D., Vitt, L. J., and King, W. W. (1974). Geckos: Adaptive significance and energetics of tail autotomy. *Science* **184:** 1379–1380.

Congdon, J. D., Vitt, L. J., van Loben Sels, R. C., and Ohmart, R. D. (1982). The ecological significance of water flux rates in arboreal desert lizards of the genus *Urosaurus*. *Physiological Zoology* **55:** 317–322.

Congdon, J. D., Dunham, A. E., and van Loben Sels, R. C. (1993). Delayed sexual maturity and demographics of Blanding's turtles (*Emydoidea blandingii*): Implications for conservation and management of long-lived organisms. *Conservation Biology* **7:** 826–833.

Cooper, W. E., Jr. (1984). Female secondary sexual coloration and sex recognition in the keeled earless lizard, *Holbrookia propinqua*. *Animal Behaviour* **32:** 1142–1150.

Cooper, W. E., Jr. (1985). Female residency and courtship intensity in a territorial lizard, *Holbrookia propinqua*. *Amphibia-Reptilia* **6:** 63–69.

Cooper, W. E., Jr. (1994a). Prey chemical discrimination, foraging mode, and phylogeny. In *Lizard Ecology: Historical and Experimental Perspectives*. L. J. Vitt and E. R. Pianka (Eds.). Pp. 95–116. Princeton University Press, Princeton, NJ.

Cooper, W. E., Jr. (1994b). Chemical discrimination by tongue-flicking in lizards: A review with hypotheses on its origin and its ecological and phylogenetic relationships. *Journal of Chemical Ecology* **20:** 439–487.

Cooper, W. E., Jr. (1995). Foraging mode, prey chemical discrimination, and phylogeny in lizards. *Animal Behaviour* **50:** 973–985.

Cooper, W. E., Jr. (1996). Chemosensory recognition of familiar and unfamiliar conspecifics by the Scincid lizard *Eumeces laticeps*. *Ethology* **102:** 1–11.

Cooper, W. E., Jr. (2000a). Effect of temperature on escape behaviour by an ectothermic vertebrate, the keeled earless lizard (*Holbrookia propinqua*). *Behaviour* **137:** 1299–1315.

Cooper, W. E., Jr. (2000b). Tradeoffs between predation risk and feeding in a lizard, the broad-headed skink (*Eumeces laticeps*). *Behaviour* **137:** 1175–1189.

Cooper, W. E., Jr. (2003a). Risk factors affecting escape behavior by the desert iguana, *Dipsosaurus dorsalis*: Speed and directness of predator approach, degree of cover, direction of turning by a predator, and temperature. *Canadian Journal of Zoology* **81:** 979–984.

Cooper, W. E., Jr. (2003b). Sexual dimorphism in distance from cover but not escape behavior by the keeled earless lizard *Holbrookia propinqua*. *Journal of Herpetology* **37:** 374–378.

Cooper, W. E., Jr. (2003c). Shifted balance of risk and cost after autotomy affects use of cover, escape, activity, and foraging in the keeled earless lizard (*Holbrookia propinqua*). *Behavioral Ecology and Sociobiology* **54:** 179–187.

Cooper, W. E., Jr. (2003d). Social behavior and antipredatory defense in lizards. In *Lizard Social Behavior*. S. F. Fox, J. K. McCoy, and T. A. Baird (Eds.). Pp. 107–141. Johns Hopkins University Press, Baltimore, MD.

Cooper, W. E., Jr. (2003e). Effect of risk on aspects of escape behavior by a lizard, *Holbrookia propinqua*, in relation to optimal escape theory. *Ethology* **109:** 617–626.

Cooper, W. E., Jr. (2005). When and how do predator starting distances affect flight initiation distances? *Canadian Journal of Zoology* **83:** 1045–1050.

Cooper, W. E., Jr. (2006). Risk factors affecting escape behavior in Puerto Rican *Anolis* lizards. *Canadian Journal of Zoology* **84:** 495–504.

Cooper, W. E. (2007a). Foraging modes as suites of coadapted movement traits. *Journal of Zoology, London* **272:** 45–56.

Cooper, W. E., Jr. (2007b). Lizard chemical senses, chemosensory behavior, and foraging mode. In *Lizard Ecology: The Evolutionary Consequences of Foraging Mode*. S. M. Reilly, L. D. McBrayer, and D. B. Miles (Eds.). Pp. 237–270. Cambridge University Press, Cambrdge UK.

Cooper, W. E., and Alberts, A. C. (1990). Responses to chemical food stimuli by an herbivorous actively foraging lizard, *Dipsosaurus dorsalis*. *Herpetologica* **46:** 259–266.

Cooper, W. E., Jr., and Crews, D. (1988). Sexual coloration, plasma concentrations of sex steroid hormones, and responses to courtship in the female keeled earless lizard (*Holbrookia propinqua*). *Hormones and Behavior* **22:** 12–25.

Cooper, W. E., and Greenberg, N. (1992). Reptilian coloration and behavior. In *Biology of the Reptilia. Volume 18, Physiology. Hormones, Brain, and Behavior*. C. Gans and D. Crews (Eds.). Pp. 298–422. University of Chicago Press, Chicago.

Cooper, W. E., Jr., and Lemos-Espinal, J. A. (2001). Coordinated ontogeny of food preference and responses to chemical food stimuli by a lizard *Ctenosaura pectinata* (Reptilia: Iguanidae). *Ethology* **107:** 639–653.

Cooper, W. E., Jr., and Perez-Mellado, V. (2004). Tradeoffs between escape behavior and foraging opportunity by the Balearic lizard (*Podarcis lilfordi*). *Herpetologica* **60:** 321–324.

Cooper, W. E., Jr., and Vitt, L. J. (1985). Blue tails and autotomy: Enhancement of predation avoidance in juvenile skinks. *Zeitschrift für Tierpsychologie* **70:** 265–276.

Cooper, W. E., Jr., and Vitt, L. J. (1987a). Intraspecific and interspecific aggression in lizards of the scincid genus *Eumeces*: Chemical detection of conspecific sexual competitors. *Herpetologica* **43:** 7–14.

Cooper, W. E., Jr., and Vitt, L. J. (1987b). Deferred agonistic behavior in a long-lived scincid lizard *Eumeces laticeps*. *Oecologia* **72:** 321–326.

Cooper, W. E., Jr., and Vitt, L. J. (1991). Influence of detectability and ability to escape on natural selection of conspicuous autotomous defenses. *Canadian Journal of Zoology* **69:** 757–764.

Cooper, W. E., Jr., and Vitt, L. J. (2002a). Optimal escape and emergence theories. *Comments on Theoretical Biology* **7:** 283–294.

Cooper, W. E., Jr., and Vitt, L. J. (2002b). Distribution, extent, and evolution of plant consumption by lizards. *Journal of Zoology, London* **257:** 487–517.

Cooper, W. E., Jr., Caldwell, J. P., Vitt, L. J., Pérez-Mellado, V., and Baird, T. A. (2002). Food chemical discriminations and correlated evolution between plant diet and plant-chemical discrimination in lacertiform lizards. *Canadian Journal of Zoology* **80:** 655–663.

Cooper, W. E., Jr., Pérez-Mellado, V., and Hawlena, D. (2006). Magnitude of food reward affects escape behavior and acceptable risk in Balearic lizards, *Podarcis lilfordi*. *Behavioral Ecology* **17:** 554–559.

Cooper, W. E., Jr., Vitt, L. J., Hedges, R., and Huey, R. B. (1990). Locomotor impairment and defense in gravid lizards (*Eumeces laticeps*): Behavioral shift in activity may offset costs of reproduction in an active forager. *Behavioral Ecology and Sociobiology* **27:** 153–157.

Corben, C. J., Ingram, G. J., and Tyler, M. J. (1974). Gastric brooding—unique form of parental care in an Australian frog. *Science* **186:** 946–947.

Corbett, K. F. (1989). *Conservation of European Reptiles and Amphibians*. Christopher Helm, London.

Corbett, K. F. (1988). Conservation strategy for the sand lizard (*Lacerta agilis agilis*) in Britain. *Mertensiella* **1:** 101–109.

Cordes, J. E., and Walker, J. A. (2006). Evolutionary and systematic implications of skin histocompatibility among parthenogenetic teiid lizards: Three color pattern classes of *Aspidoscelis dixoni* and one of *Aspidoscelis tesselata*. *Copeia* **2006:** 14–26.

Corn, P. S., and Bury, R. B. (1988). Logging in western Oregon: responses of headwater habitats and stream amphibians. *Forest Ecology and Management* **29:** 39–57.

Costa, G. C., Nogueira, C., Machado, R. B., and Colli, G. R. (2007). Squamate richness in the Brazilian Cerrado and its environmental-climatic associations. *Diversity and Distributions* **13:** 714–724.

Costa, G. C., Vieira, G. H. C., Teixeira, R. D., Garda, A. A., Colli, G. R., and Báo, S. N. (2004). An ultrastructural comparative study of the sperm of *Hyla pseudopseudis, Scinax rostratus*, and *S. squalirostris* (Amphibia: Anura: Hylidae). *Zoomorphology* **123:** 191–197.

Costa, G. C., Wolfe, C., Shepard, D. B., Caldwell, J. P., and Vitt, L. J. (2008). Detecting the influence of climatic variables on species distributions: A test using GIS niche-based models along a steep longitudinal environmental gradient. *Journal of Biogeography* **35:** 637–646.

Cote, S., Carroll, R., Cloutier, R., and Bar-Sagi, L. (2002). Vertebral development in the Devonian Sarcopterygian fish *Eusthenopteron foordi* and the polarity of vertebral evolution in non-amniote tetrapods. *Journal of Vertebrate Paleontology* **22:** 487–502.

Cott, H. B. (1961). Scientific results of an inquiry into the ecology and economic status of the Nile crocodile (*Crocodylus niloticus*) in Uganda and Northern Rhodesia. *Transactions of the Zoological Society of London* **29:** 211–356.

Coulson, K. L. (1974). The polarization of light in the environment. In *Planets, Stars, and Nebula: Studied with Photopolarimetry*. T. Grebels (Ed.). Pp. 444–471. University of Arizona Press, Tucson.

Coulson, R. A., and Hernandez, T. (1983). Alligator Metabolism. Studies on Chemical Reactions In Vivo. *Comparative Biochemistry and Physiology* **74B:** 1–182.

Cowles, R. B., and Bogert, C. M. (1944). A preliminary study of the thermal requirements of desert reptiles. *Bulletin of the American Museum of Natural History* **83:** 261–296.

Cox, C. B., and Moore, P. D. (1985). *Biogeography: An Ecological and Evolutionary Approach (4th Ed.)*. Blackwell Scientific Publications, Oxford.

Cox, M. J., van Dijk, P. P., Nabhitabhata, J., and Thirakhupt, K. (1998). *A Photographic Guide to Snakes and Other Reptiles of Peninsular Malaysia, Singapore and Thailand*. Ralph Curtis Books, Sanibel Island, FL.

Crawford, E. C., Jr., and Barber, B. J. (1974). Effects of core, skin, and brain temperature on panting in the lizard *Sauromalus obesus*. *American Journal of Physiology* **226:** 569–573.

Crawford, E. C., Jr., and Kampe, G. (1971). Physiological responses of the lizard *Sauromalus obesus* to changes in ambient temperature. *American Journal of Physiology* **220:** 1256–1260.

Crawford, E. C., Palomeque, J., and Barber, B. J. (1977). Physiological basis for head–body temperature differences in a panting lizard. *Comparative Biochemistry and Physiology* **56A:** 161–163.

Crawshaw, G. J. (1997). Disease in Canadian amphibian populations. In *Amphibians in Decline. Canadian Studies of a Global Problem*. D. M. Green (Ed.). Pp. 258–270. *Herpetological Conservation* **1**.

Cree, A. (1988). Water-balance responses of the hylid frog *Litoria aurea*. *Journal of Experimental Zoology* **247:** 119–125.

Cree, A., Thompson, M. B., and Daugherty, C. H. (1995). Tuatara sex determination. *Nature* **375:** 543.

Crews, D., and Moore, M. C. (1993). Psychobiology of reproduction of unisexual whiptail lizards. In *Biology of Whiptail Lizards (Genus* Cnemidophorus*)*. J. W. Wright and L. J. Vitt (Eds.). Pp. 257–282. Oklahoma Museum of Natural History, Norman, OK.

Crews, D., Bergeron, J. M., Bull, J. J., Flores, D., Tousignant, A., Skipper, J. K., and Wibbles, T. (1994). Temperature dependent sex determination in reptiles: Proximate mechanisms, untimate outcomes, and practical applications. *Developmental Genetics* **15:** 297–312.

Crossland, M. R. (1998). A comparison of cane toad and native tadpoles as predators of native anuran eggs, hatchlings and larvae. *Wildlife Research* **25:** 373–381.

Crother, B. I., and Presch, W. F. (1994). Xantusiid lizards, concern for analysis, and the search for a best estimate of phylogeny: Further comments. *Molecular Phylogenetics and Evolution* **3:** 272–274.

Crouse, D. T., Crowder, L. B., and Caswell, H. (1987). A stage-based population model for loggerhead sea turtles and implications for conservation. *Ecology* **68:** 1412–1423.

Crumly, C. R. (1982). A cladistic analysis of *Geochelone* using cranial osteology. *Journal of Herpetology* **16:** 215–234.

Crumly, C. R. (1984). A hypothesis for the relationships of land tortoise genera (family Testudinidae). *Studia Geologica Salmanticensia* **1:** 115–124.

Crump, M. L. (1986). Homing and site fidelity in a Neotropical frog, *Atelopus varius* (Bufonidae). *Copeia* **1986:** 438–444.

Crump, M. L. (1995). Parental care. In *Amphibian Biology. Volume 2, Social Behaviour*. H. Heatwole and B. K. Sullivan (Eds.). Pp. 518–567. Surrey Beatty & Sons Pty Ltd., Chipping Norton Pty Ltd, N.S.W.

Crutsinger, G., Pickersgill, M., Channing, A., and Moyer, D. (2004). A new species of *Phrynobatrachus* (Anura: Ranidae) from Tanzania. *African Zoology* **39:** 19–23.

Cruz, C. A. G. da (1990). Sobre as relaçoes intergenénericas de Phyllomedusinae da foresta atlântica (Amphibia, Anura, Hylidae). *Revista Brasileira de Biologia* **50:** 709–726.

Cuéllar, O. (1976). Intraclonal histocompatibility in a parthenogenetic lizard: Evidence of genetic homogeneity. *Science* **193:** 150–153.

Cundall, D. (1995). Feeding behaviour in *Cylindrophis* and its bearing on the evolution of alethinophidian snakes. *Journal of Zoology, London* **237:** 353–376.

Cundall, D., and Irish, F. J. (1986). Aspects of locomotor and feeding behaviour in the Round Island boa *Casarea dussumieri*. *Dodo, Journal of the Jersey Wildlife Preservation Trust* **23:** 108–111.

Cundall, D., and Irish, F. J. (1989). The function of the intramaxillary joint in the Round boa, *Casarea dussumieri*. *Journal of Zoology, London* **217:** 569–598.

Cundall, D., and Rossman, D. A. (1993). Cephalic anatomy of the rare Indonesian snake *Anomochilus weberi*. *Zoological Journal of the Linnean Society* **109:** 253–273.

Cundall, D., Lorenz-Elwood, J., and Groves., J. D. (1987). Asymmetric suction feeding in primitive salamanders. *Experientia* **43:** 1229–1231.

Cundall, D., Wallach, V., and Rossman, D. A. (1993). The systematic relationships of the snake genus *Anomochilus*. *Zoological Journal of the Linnean Society* **109:** 275–299.

Cunningham, J. D. (1960). Aspects of the ecology of the Pacific slender salamander, *Batrachoseps pacificus*, in southern California. *Ecology* **41:** 88–99.

Czopek, J. (1961). Vascularization of respiratory surfaces in *Ambystoma mexicanum* (Cope) in ontogeny. *Zoologica Poloniae* **8:** 131–149.

Czopek, J. (1965). Quantitative studies on the morphology of respiratory surfaces in amphibians. *Acta Anatomica (Basel)* **62:** 296–323.

Daeschler, E. B., Shubin, N. H., and Jenkins, F. A. (2006). A Devonian tetrapod-like fish and the evolution of the tetrapod body plan. *Nature* **440:** 757–763.

Daltry, J. C., and Martin, G. N. (1997). Rediscovery of the black narrow-mouth frog, *Melanobatrachus indicus* Beddome, 1978. *Hamadryad* **22:** 57–58.

Daly, J. W. (1998). Thirty years of discovering arthropod alkaloids in amphibian skin. *Journal of Natural Products* **61:** 162–172.

Daly, J. W., Andriamaharavo, N. R., Andriantsiferana, M., and Myers, C. W. (1996). Madagascan poison frogs (*Mantella*) and their skin alkaloids. *American Museum Novitates* **3177:** 1–34.

Daly, J. W., Garraffo, H. M., Hall, G. S. E., and Cover, J. F. (1997). Absence of skin alkaloids in captive-raised Madagascan mantelline frogs (*Mantella*) and sequestration of dietary alkaloids. *Toxicon* **35:** 1131–1135.

Daly, J. W., Garraffo, H. M., Jain, P., Spande, T. F., Snelling, R. R., Jaramillo, C., and Rand, A. S. (2000). Arthropod-frog connection: Decahydroquinoline and pyrrolizidine alkaloids common to microsympatric myrmicine ants and dendrobatid frogs. *Journal of Chemical Ecology* **26:** 73–85.

Daly, J. W., Garraffo, H. M., Spande, T. F., Clark, V. C., Ma, J. Y., Ziffer, H., and Cover, J. F. (2003). Evidence for an enantioselective pumiliotoxin 7-hydroxylase in dendrobatid poison frogs of the genus *Dendrobates*. *Proceedings of the National Academy of Sciences (USA)* **100:** 11092–11097.

Daly, J. W., Garraffo, H. M., Spande, T. F., Jaramillo, C., and Rand, A. S. (1994). Dietary source for skin alkaloids of poison frogs (Dendrobatidae). *Journal of Chemical Ecology* **20:** 943–955.

Daly, J. W., Kaneko, T., Wilham, J., Garraffo, H. M., Spande, T. F., Espinosa, A., and Donnelly, M. A. (2002). Bioactive alkaloids of frog skin: Combinatorial bioprospecting reveals that pumiliotoxins have an arthropod source. *Proceedings of the National Academy of Sciences (USA)* **99:** 13996–14001.

Daly, J. W., Myers, C. W., and Whittaker, N. (1987). Further classification of skin alkaloids from Neotropical poison frogs (Dendrobatidae), with a general survey of toxic/noxious substances in the Amphibia. *Toxicon* **25:** 1023–1095.

Daniel, J. C. (1983). *The Book of Indian Reptiles*. Bombay Natural History Society, Bombay.

Daniels, C. B. (1983). Running: An escape strategy enhanced by autotomy. *Herpetologica* **3:** 162–165.

Daniels, S. R., Heideman, N. J. L., Hendricks, M. G. J., and Crandall, K. A. (2006). Taxonomic subdivisions within the fossorial skink subfamliy Acontinae (Squamata; Scincidae) reconsidered: A multilocus perspective. *Zoologica Scripta* **35:** 353–362.

Darevsky, I. S. (1958). Natural parthenogenesis in certain subspecies of rock lizard, *Lacerta saxicola* Eversmann. *Doklady Akademii Nauk SSSR, Biological Science* **122:** 877–879.

Darevsky, I. S. (1978). *Rock Lizards of the Caucasus*. Indian National Science Documentation Centre, New Delhi.

Darevsky, I. S. (1992). Evolution and ecology of parthenogenesis in reptiles. In *Herpetology: Current Research on the Biology of Amphibians and Reptiles*. K. Adler (Ed.). Pp. 21–39. Society for Study of Amphibians and Reptiles, Oxford, Ohio.

Darst, C. R., and Cannatella, D. C. (2004). Novel relation-ships among hylold frogs inferred from 12S and 16S mitochondrial DNA sequences. *Molecular Phylogenetics and Evolution* **31:** 462–475.

Darst, C. R., Menendez-Guerrero, P. A., Coloma, L. A., and Cannatella, D. C. (2005). Evolution of dietary specialization and chemical defense in poison frogs (Dendrobatidae): A comparative analysis. *American Naturalist* **165:** 56–69.

Das, I. (1995). *Turtles and Tortoises of India*. Oxford University Press, Bombay.

Das, I. (1996). Folivory and seasonal changes in diet in *Rana hexadactyla* (Anura: Ranidae). *Journal of Zoology, London* **238:** 785–794.

Das, I., and Haas, A. (2003). A new species of *Kalophrynus* (Anura: Microhylidae) from the Highlands of north-central Borneo. *Raffles Bulletin of Zoology* **51:** 109–113.

Das, I., and Kunte, K. (2005). New species of *Nyctibatrachus* (Anura: Ranidae) from Castle Rock, Karnataka State, Southwest India. *Journal of Herpetology* **39:** 465–470.

Daugherty, C. H., Cree, A., Hay, J. M., and Thompson, M. B. (1990). Neglected taxonomy and continuing existence of tuatara (*Sphenodon*). *Nature* **347:** 177–179.

David, P. (1994). Liste des reptiles actuels du monde I. Chelonii. *Dumerilia* **1:** 7–127.

David, P., and Ineich, I. (1999). Les serpents venimeux du monde: Systématique et répartition. *Dumerilia* **3:** 3–499.

David, P., and Vogel, G. (1996a). *The Snakes of Sumatra*. Edition Chimaira, Frankfurt am Main.

David, P., and Vogel, G. (1996b). *The Snakes of Sumatra. An Annotated Checklist and Key with Natural History Notes*. Bücher Kreth, Frankfurt/M.

David, R. S., and Jaeger, R. G. (1981). Prey location through chemical cues by a terrestrial salamander. *Copeia* **1981:** 435–440.

Davis, A. K., Yabsley, M. J., Keel, M. K., and Maerz, J. C. (2007). Discovery of a novel alveolate pathogen affecting southern leopard frogs in Georgia: Description of the disease and host effects. *Ecohealth* **4:** 310–317.

Dawson, W. R. (1975). On the physiological significance of the preferred body temperatures of reptiles. In *Perspectives in Biophysical Ecology*. D. M. Gates and R. B. Schmerl (Eds.). Pp. 443–473. Springer-Verlag, New York.

Day, D. (1981). *The Doomsday Book of Animals. A Natural History of Vanished Species*. Viking Press, New York.

De Blij, H. J., Muller, P. O., and Williams, R. S. (1995). *Physical Geography: The Global Environment*. (1995). Oxford University Press, New York, NY.

de Lang, R., and Vogel, G. (2005). The snakes of Sulawesi. A field guide to the land snakes of Sulawesi with identification keys. Frankfurter Beiträge zur Naturkunde, 25, Edition Chimaira, Frankfurt am Main.

de Queiroz, K. (1997). The Linnean hierarchy and the evolutionization of taxonomy, with emphasis on the problem of nomenclature. *Aliso* **15:** 125–144.

de Queiroz, K., and Gauthier, J. (1992). Phylogenetic taxonomy. *Annual Review of Ecology and Systematics* **23:** 449–480.

de Queiroz, K., and Gauthier, J. (1994). Toward a phylogenetic system of biological taxonomy. *Trends in Ecology and Evolution* **9:** 27–31.

de Sá, R. O., and Hillis, D. M. (1990). Phylogenetic relationships of the pipid frogs *Xenopus* and *Silurana*: An integration of ribosomal DNA and morpholog. *Molecular Biology and Evolution* **7:** 365–376.

deBraga, M., and Rieppel, O. (1997). Reptile phylogeny and the interrelationships of turtles. *Zoological Journal of the Linnean Society* **120:** 281–354.

Deeming, D. C. (2004). Prevelance of TSD in crocodilians. In *Temperature-Dependent Sex Determination in Vertebrates*. N. Valenzuela and V. Lance (Eds.). Pp. 33–41. Smithsonian Books, Washington, D. C.

del Pino, E. M. (1989). Marsupial frogs. *Scientific American* **160:** 76–84.

Delêtre, M., and Measey, G. J. (2004). Sexual selection vs ecological causation in a sexually dimorphic caecilian, *Schistometopum thomense* (Amphibia Gymnophiona Caeciliidae). *Ethology Ecology & Evolution* **16:** 243–253.

Delfino, M. (2005). The past and future of extant amphibians. *Science* **308:** 49–50.

Delorme, M., Dubois, A., Kosuch, J., and Vences, M. (2004). Molecular phylogenetic relationships of *Lankanectes corrugatus* from Sri Lanka: Endemism of South Asian frogs and the concept of monophyly in phylogenetic studies. *Alytes* **22:** 53–64.

Delpino, E. M. (1980). Morphology of the Pouch and Incubatory Integument in marsupial frogs (Hylidae). *Copeia* **1980:** 10–17.

Delsol, M., Extrayat, J.-M., Flatin, M., J., and Gueydan-Baconnier, M. (1986). Nutrition embronnaire chez *Typhlonectes compressicaudus* (Duméril et Bibron, 1841), Amphibien Apode vivipare. *Memoires Societe Zoologique de France* **43:** 39–54.

Delsol, M., Flatin, J., Extrayat, J.-M., and Bons, M. (1981). Développement de *Typhlonectes compressicaudus*. *Comptes Rendus des Séances de l'Académie des Sciences (Paris)* **293:** 281–285.

Delsol, M., Flatin, J., Extrayat, J.-M., and Bons, M. (1983). Développement embronnaire de *Typhlonectes compressicaudus* (Duméril et Bibron, 1841). Constitution d'un ectotrophoblaste. *Bulletin de la Société Zoologique de France* **108:** 680–681.

Demaynadier, P. G., and Hunter, M. L., Jr. (1998). Effects of silvicultural edges on the distribution and abundance of amphibians in Maine. *Conservation Biology* **8:** 300–301.

Denny, R. N. (1974). The impact of uncontrolled dogs on wildlife and livestock. *Transactions of the North American Wildlife and Natural Resources Conference* **39:** 257–291.

Densmore, L. D. (1983). Biochemical and immunological systematics of the order Crocodilia. *Evolutionary Biology* **16:** 397–465.

Densmore, L. D., and Owen, R. D. (1989). Molecular systematics of the Order Crocodilia. *American Zoologist* **29:** 831–841.

Densmore, L. D., and White, P. S. (1991). The systematics and evolution of the Crocodilia as suggested by restriction endonuclease analysis of mitochondrial and nuclear ribosomal DNA. *Copeia* **1991:** 602–615.

Denton, R. K., Jr., and O'Neill, R. C. (1995). *Prototeius stageri* gen. et sp. nov., a new teiid lizard from the Upper Cretaceous Marshalltown Formation of New Jersey, with a preliminary phylogenetic revision of the Teiidae. *Journal of Vertebrate Paleontology* **15:** 235–253.

Denver, R. J. (1997). Proximate mechanisms of phenotypic plasticity in amphibian metamorphosis. *American Zoologist* **37:** 172–184.

Derickson, W. K. (1976). Ecological and physiological aspects of reproductive strategies in two lizards. *Ecology* **57:** 445–458.

Derr, M. (2000). Editorial. In *Caribbean, Endangered Iguanas Get Their Day. New York Times Science Section*, October 10.

Dial, B. E. (1981). Function and energetics of autotomized tail thrashing in *Lygosoma laterale* (Sauria: Scincidae). *American Zoologist* **21:** 1001.

Dial, B. E., and Fitzpatrick, L. C. (1981). The energetic costs of tail autotomy to reproduction in the lizard *Coleonyx brevis* (Sauria: Gekkonidae). *Oecologia* **51:** 310–317.

Dial, B. E., and Fitzpatrick, L. C. (1983). Lizard tail autotomy: function and energetics of post autotomy tail movement in *Scincella lateralis*. *Science* **219:** 391–393.

Dial, B. E., and Grismer, L. L. (1992). A phylogenetic analysis of physiological-ecological character evoution in the lizard genus *Coleonyx* and its implications for historical biogeographic reconstruction. *Systematic Biology* **4:** 178–195.

Dial, B. E., and Grismer, L. L. (1994). Phylogeny and physiology: Evolution of lizards in the genus *Coleonyx*. In *Herpetology of the North American Deserts: Proceedings of a Symposium*. P. R. Brown and J. W. Wright (Eds.). Pp. 239–254. Southwestern Herpetological Society Special Publication No. 5, Van Nuys, CA.

Dial, B. E., and Schwenk, K. (1996). Olfaction and predator detection in *Coleonyx brevis* (Squamata: Eublepharidae), with comments on the functional significance of buccal pulsing in geckos. *Journal of Experimental Zoology* **276:** 415–424.

Dias, E. J. D., Rocha, C. F. D., and Vrcibradic, D. (2002). New *Cnemidophorus* (Squamata: Teiidae) from Bahia State northeastern Brazil. *Copeia* **2002:** 1070–1077.

Dicke, U., Roth, G., Kaas, J. H., and Bullock, T. H. (2007). Evolution of the amphibian nervous system. *Evolution of Nervous Systems: A Comprehensive Reference, Volume 2, Nonmammalian Vertebrates* 61–124.

Diego-Rasilla, F. J. (2003). Homing ability and sensitivity to the geomagnetic field in the alpine newt, *Triturus alpestris*. *Ethology Ecology & Evolution* **15:** 251–259.

Diemer, J. E. (1992). Home range and movements of the *tortoise Gopherus polyphemus* in northern Florida. *Journal of Herpetology* **26:** 158–165.

Diesmos, A. C., Gee, G. V. A., Diesmos, M. L., Brown, R. M., Widmann, P. J., and Dimalibot, J. C. (2004). Rediscovery of the Philippine forest turtle, *Heosemys leytensis* (Chelonia; Bataguridae), from Palawan Island, Philippines. *Asiatic Herpetological Research* **10:** 22–27.

Diesmos, A. C., Parham, J. F., Stuart, B. L., and Brown, R. M. (2005). The phylogenetic position of the recently rediscovered Philippine forest turtle (Bataguridae: *Heosemys leytensis*). *Proceedings of the California Academy of Sciences* **56:** 31–41.

Diller, L. V., and Wallace, R. L. (1999). Distribution and habitat of *Ascaphus truei* in streams on managed, young growth forests in north coastal California. *Journal of Herpetology* **33:** 71–79.

Dinesh, K. P., Radhakrishnan, C., Reddy, A. H. M., and Gurnraja, K. V. (2007). *Nyctibatrachus karnatakaensis* nom. nov., a replacement name for the giant wrinkled frog from the Western Ghats. *Current Science* **93:** 246–250.

Dingle, H. (1996). *Migration.*. Oxford University Press, Oxford.

Dixon, J. R., and Kofron, C. P. (1983). The Central and South American Anomalepid snakes of the genus *Liotyphlops*. *Amphibia-Reptilia* **4:** 241–264.

Dixon, J. R., and Werler, J. E. (2000). *Texas Snakes, Identification, Distribution, and Natural History*. University of Texas Press, Austin, TX.

Doan, T. M. (2003). A new phylogentic classification for the gymnophthalmid genera *Cercosaura, Pantadoctylus* and *Prionodactylus* (Reptilia: Squamata). *Zoological Journal of the Linnean Society* **137:** 101–115.

Doan, T. M., and Castoe, T. A. (2005). Phylogenetic taxonomy of the Cercosaurini (Squamata: Gymnophthalmidae), with new genera for species of *Neusticurus* and *Proctoporus*. *Zoological Journal of the Linnean Society* **143:** 405–416.

Dobzhansky, T. (1950). Evolution in the tropics. *American Scientist* **38:** 209–221.

Dodd, C. K., Jr. (1994). The effects of drought on population structure, activity, and orientation of toads (*Bufo quercicus* and *B. terrestris*) at a temporary pond. *Ethology Ecology & Evolution* **6:** 331–349.

Dodd, C. K., Jr. (1982). Immobility in *Sooglossus gardineri* (Anura: Sooglossidae). *Herpetological Review* **13:** 33–34.

Dodd, C. K., Jr. (1989). Status of the Red Hills salamander is reassessed. *Endangered Species Technical Bulletin* **14:** 10–11.

Dodd, C. K., Jr. (1997). Imperiled amphibians. A historical perspective. In *Aquatic Fauna in Peril: The Southeastern Perspective*. G. W. Benz and D. E. Collins (Eds.). Pp. 165–200. Special Publication No. 1, Southeast Aquatic Research Institute, Lenz Design & Communications, Decator, GA.

Dodd, C. K., Jr., and Byles, R. (2003). Post-nesting movements and behavior of loggerhead sea turtles (*Caretta caretta*) departing from east-central Florida nesting beaches. *Chelonian Conservation and Biology* **4:** 530–536.

Dodd, C. K., Jr., and Cade, R. S. (1998). Movement patterns and the conservation of amphibians breeding in small temporary wetlands. *Conservation Biology* **12:** 331–339.

Dodd, C. K., Jr., and Seigel, R. A. (1991). Relocation, repatriation, and translocation of amphibians and reptiles: Are they conservation strategies that work? *Herpetologica* **47:** 336–350.

Dole, D. W., Rose, B. B., and Tachiki, K. H. (1981). Western toads (*Bufo boreas*) learn odor of prey insects. *Herpetologica* **37:** 63–68.

Donnellan, S. C., Hutchinson, M. N., and Saint, K. M. (1999). Molecular evidence for the phylogeny of Australian gekkonoid lizards. *Biological Journal of the Linnean Society* **67:** 97–118.

Donnellan, S. C., Hutchinson, M. N., Dempsey, P., and Osborne, W. (2002). Systematics of the *Egernia whitii* species group (Lacertilia: Scincidae) in southeastern Australia. *Australian Journal of Zoology* **50:** 439–59.

Donnelly, M. A., and Guyer, C. (1994). Patterns of reproduction and habitat use in an assemblage of Neotropical hylid frogs. *Oecologia* **98:** 291–302.

Donnelly, M. A., de Sá, R., and Guyer, C. (1990). Description of the tadpoles of *Gastrophyrne pictiventris* and *Nelsonophryne aterrima* (Anura: Microhylidae) with a review of morphological variation in free-swimming microhylid larvae. *American Museum Novitates* **2976:** 1–19.

Donoso-Barros, R. (1966). *Reptiles de Chile*. Edic. University Chile, Santiago de Chile.

Doody, J. S., and Young, J. E. (1995). Temporal variation in reproduction and clutch mortality of leopard frogs (*Rana utricularia*) in south Mississippi. *Journal of Herpetology* **29:** 614–616.

Doody, J. S., Young, J. E., and Georges, A. (2002). Sex differences in activity and movements in the pig-nosed turtle, *Carettochelys insculpta*, in the wet-dry tropics of Australia. *Copeia* **2002:** 93–103.

Doughty, P., Sinervo, B., and Burghardt, G. M. (1994). Sex-biased dispersal in a polygynous lizard, *Uta stansburiana*. *Animal Behaviour* **47:** 227–229.

Dowling, H. G., and Duellman, W. E. (1974–1978). *Systematic Herpetology. A Synopsis of Families and Higher Categories*. HISS Publ., New York.

Dowling, H. G., Hass, C. A., Hedges, S. B., and Highton, R. (1996). Snake relationships revealed by slow-evolving proteins: A preliminary survey. *Journal of Zoology, London* **240:** 1–28.

Downes, S., and Shine, R. (2001). Why does tail loss increase a lizard's later vulnerability to snake predators? *Ecology* **82:** 1293–1303.

Downie, J. R. (1984). How *Leptodactylus fuscus* tadpoles make foam, and why. *Copeia* **1984:** 778–780.

Drake, D. L., Altig, R., Grace, J. B., and Walls, S. C. (2007). Occurrence of oral deformities in larval anurans. *Copeia* **2007:** 449–458.

Drewes, R. C., Altig, R., and Howell, K. M. (1989). Tadpoles of three frog species endemic to the forests of the Eastern Arc Mountains, Tanzania. *Amphibia-Reptilia* **10:** 435–443.

Drewes, R. C. (1984). A phylogenetic analysis of the Hyperoliidae (Anura): treefrogs of Africa, Madagascar, and the Seychelles Islands. *Occasional Papers of the California Academy of Sciences* **139:** 1–70.

Drury, W. H., Jr. (1998). *Chance and Change. Ecology for Conservationists.* University of California Press, Los Angeles.

Dubois, A. (1981). Liste de genres et sous-genres nominaux de Ranoidea (amphibien anoures) du monde, avec identification de leurs espèces-types: conséquences nomenclaturales. *Monitore zoologico italiano (nuova serie) supplemento.* **15:** 225–284.

Dubois, A. (1984). La nomenclature supragénérique des amphibiens anoures. *Mémoires du Museum national d'Histoire naturelle, series A, Zoologie* **131:** 1–64.

Dubois, A. (1985). Diagnose préliminaire d'un nouveau genre de Ranoidea (Amphibiens, Anoures) du sud de l'Inde. *Alytes* **4:** 113–118.

Dubois, A. (1986). Miscellanea taxinomica batrachologica (I). *Alytes* **5:** 7–95.

Dubois, A. (1992). Notes sur la classification des Ranidae (amphibiens anoures). *Bulletin Mensuel de la Société Linnéenne de Lyon* **61:** 305–352.

Dubois, A., and Ohler, A. (2000). Systematics of *Fejervarya limnocharis* (Gravenhorst, 1829) (Amphibia, Anura, Ranidae) and related species. 1. Nomenclatural status and type-specimens of the nominal species *Rana limnocharis* Gravenhorst, 1829. *Alytes* **18:** 15–50.

Dubois, A., and Ohler, A. (2001). A new genus for an aquatic ranid (Amphibia, Anura) from Sri Lanka. *Alytes* **19:** 81–106.

Dubois, A., Ohler, A., and Biju, S. D. (2001). A. new genus and species of Ranidae (Amphibia, Anura) from south-western India. *Alytes* **19:** 53–79.

Dudley, R., Byrnes, G., Yanoviak, S. P., Borrell, B., Brown, R. M., and McGuire, J. A. (2007). Gliding and the functional origins of flight: Biomechanical novelty or necessity? *Annual Review of Ecology Evolution and Systematics* **38:** 179–201.

Duellman, W. E. (1970). The frogs of Middle America. Volume 1. *Monograph of the Museum of Natural History, University of Kansas* **1:** 1–427.

Duellman, W. E. (1975). The biology of an equatorial herpetofauna in Amazonian Ecuador. *Miscellaneous Publications of the Museum of Natural History, University of Kansas* **42:** 1–14.

Duellman, W. E. (1978). The biology of an equatorial herpetofauna in Amazonian Ecuador. *Miscellaneous Publications of the Museum of Natural History, University of Kansas* **65:** 1–352.

Duellman, W. E. (1993). Amphibian species of the world: additions and corrections. *University of Kansas Museum of Natural History Special Publication* **21:** 1–372.

Duellman, W. E. (2001). The hylid frogs of Middle America (Volume 1). *Contributions to Herpetology* **18:** 1–694.

Duellman, W. E. (2003). An overview of anuran phylogeny, classification and reproductive modes. In *Reproductive Biology and Phylogeny of Anura.* B. G. M. Jamieson (Ed.). Pp. 1–18. Science Publishers, Inc., Nefield, NH.

Duellman, W. E., and Trueb, L. (1986). *Biology of Amphibans.* McGraw-Hill Book Co., New York.

Duellman, W. E., and Trueb, L. (1994). *Biology of Amphibians.* Johns Hopkins University Press, Baltimore, MD.

Duellman, W. E., and Wiens, J. J. (1992). The status of the hylid frog genus *Ololygon* and the recognition of *Scinax* Wagler, 1830. *Occasional Papers of the Museum of Natural History, University of Kansas* **151:** 1–23.

Duellman, W. E., Maxson, L. R., and Jesiolowski, C. A. (1988). Evolution of marsupial frogs (Hylidae: Hemiphractinae): Immunological evidence. *Copeia* **1988:** 527–543.

Dundee, H. A., and Miller, M. C. (1968). Aggregative behavior and habitat conditioning by the prairie ringneck snake, *Diadophis punctatus arnyi. Tulane Studies in Zoology and Botany* **15:** 41–58.

Dunham, A. E. (1978). Food availability as a proximate factor influencing individual growth rates in the iguanid lizard *Sceloporus merriami. Ecology* **59:** 770–778.

Dunham, A. E. (1993). Realized niche overlap, resource abundance, and the intensity of interspecific competition. In *Lizard Ecology: Studies of a Model Organism.* R. B. Huey, E. R. Pianka, and T. W. Schoener (Eds.). Pp. 261–280. Harvard University Press, Cambridge, Mass.

Dunham, A. E., and Miles, D. B. (1985). Patterns of covariation in life history traits of squamate reptiles: The effects of size and phylogeny reconsidered. *American Naturalist* **126:** 231–257.

Dunham, A. E., Grant, B. W., and Overall, K. L. (1989). Interfaces between biophysical and physiological ecology and the population ecology of terrestrial vertebrate ectotherms. *Physiological Zoology* **62:** 335–355.

Dunham, A. E., Miles, D. B., and Reznick, D. N. (1988). Life history patterns in squamate reptiles. In *Biology of the Reptilia. Volume 16. Ecology B. Defense and Life History. C.* Gans and R. B. Huey (Eds.). Pp. 441–552. A. R. Liss, New York.

Dunson, W. A. (1970). Some aspects of electrolyte and water balance in three estuarine reptiles, the diamondback terrapin, American alligator, and "salt water" crocodile. *Comparative Biochemistry and Physiology* **32:** 161–174.

Dunson, W. A., Wyman, R. L., and Corbett, E. S., covenors (1992). A symposium on amphibian declines and habitat acidification. *Journal of Herpetology* **26:** 349–442.

Dupre, R. K., Hicks, J. W., and Wood, S. C. (1986). The effect of hypoxia on evaporative cooling thresholds of lizards. *Journal of Thermal Biology* **11:** 223–227.

Durand, J. F. (2004). Major African contributions to Palaeozoic and Mesozoic vertebrate palaeontology. *Journal of African Earth Sciences* **43:** 53–82.

During, M. von, and Miller, M. R. (1979). Sensory nerve endings of the skin and deeper structures. In *Biology of the Reptilia. Volume 9. Neurology A.* C. Gans, A., R. G. Northcutt, and P. Ulinski (Eds.). Pp. 407–441. Academic Press, London.

Durtsche, R. D. (1999). The Ontogeny of Diet in the Mexican Spiny-tailed Iguana, *Ctenosaura pectinata:* Physiological Mechanisms and Ecological Consequences. Ph. D. Thesis, University of Oklahoma, Norman.

Durtsche, R. D. (2000). Ontogenetic plasticity of food habits in the Mexican spiny-tailed iguana, *Ctenosaura pectinata. Oecologia* **124:** 185–195.

Dutta, S. K., and Manamendra-Archchi, K. (1996). *The Amphibian Fauna of Sri Lanka*. Wildlife Heritage Trust of Sri Lanka, Colombo.

Dutton, P. H., Davis, S. K., Guerra, T., and Owens, D. (1996). Molecular phylogeny for marine turtles based on sequences of the ND4–Leucine tRNA and control regions of mitochondrial DNA. *Molecular Phylogenetics and Evolution* **5:** 511–521.

Duvall, D., Arnold, S. J., and Schuett, G. W. (1992). Pitviper mating systems: ecological potential, sexual selection and microevolution. In *Biology of Pitvipers*. J. A. Campbell and E. D. Brodie, Jr. (Eds.). Pp. 321–336. Selva, Tyler, TX.

Duvall, D., King, M. B., and Gutzwiller, K. J. (1985). Behavioral ecology of the prairie rattlesnake. *National Geographic Research* **1:** 80–111.

Duvall, D., Schuett, G. W., and Arnold, S. J. (1993). Ecology and evolution of snake mating systems. In *Snakes- Ecology and Behavior*. R. A. Seigel and J. T. Collins (Eds.). Pp. 165–200. McGraw-Hill, Inc., New York.

Dyke, G. J., Nudds, R. L., and Rayner, J. M. V. (2006a). Limb disparity and wing shape in pterosaurs. *Journal of Evolutionary Biology* **19:** 1339–1342.

Dyke, G. J., Nudds, R. L., and Rayner, J. M. V. (2006b). Flight of *Sharovipteryx*: the world's first delta-winged glider. *Journal of Evolutionary Biology* **19:** 1040–1043.

Echelle, A. A., Echelle, A. F., and Fitch, H. S. (1971). A comparative analysis of aggressive display in nine species of Costa Rican *Anolis*. *Herpetologica* **27:** 271–288.

Edmund, A. G. (1969). Dentition. In *Biology of the Reptilia. Volume 1. Morphology A.* C. Gans, A. d'A. Bellairs, and T. S. Parsons (Eds.). Pp. 117–200. Academic Press, London.

Edwards, J. L. (1976). Spinal nerves and their bearing on salamander phylogeny. *Journal of Morphology* **148:** 305–328.

Edwards, J. L. (1989). Two perspectives on the evolution of the tetrapod limb. *American Zoologist* **29:** 235–254.

Eernisse, D. J., and Kluge, A. G. (1993). Taxonomic congruence versus total evidence, and amniote phylogeny inferred from fossils, molecules, and morphology. *Molecular Biology and Evolution* **10:** 1170–1195.

Ehmann, H. (1993). Family Boidae. In *Fauna of Australia. Volume 2A. Amphibia & Reptilia.* C. J. Gasby, G. J. B. Ross, and P. L. Beesley (Eds.). Pp. 284–289. Australian Government Publishing Service, Canberra.

Ehmann, H., and Bamford, M. J. (1993). Family Typhlopidae. In *Fauna of Australia. Volume 2A. Amphibia & Reptilia.* C. J. Gasby, G. J. B. Ross, and P. L. Beesley (Eds.). Pp. 280–283. Australian Government Publishing Service, Canberra.

Ehmann, H., and Swan, G. (1985). Reproduction and development in the marsupial frog, *Assa darlingtoni* (Leptodactylidae, Anura). In *Biology of Australasian Frogs and Reptiles*. G. Grigg, R. Shine, and H. Ehmann (Eds.). Pp. 279–285. Surrey Beatty & Sons Pty Ltd., Chipping Norton Pty Ltd, Sydney, Australia.

Elias, H., and Shapiro, J. (1957). Histology of the skin of some toads and frogs. *American Museum Novitates* **1819:** 1–27.

Elinson, R. P. (1987). Changes in developmental patterns: embryos of amphibians with large eggs. In *Development as an Evolutionary Process*. R. A. Faff and E. C. Raff (Eds.). Pp. 1–21. Alan R. Liss, Inc., New York.

Elinson, R. P., and Fang, H. (1998). Secondary coverage of the yolk by the body wall in the direct developing frog, *Eleutherodactylus coqui*: An unusual process for amphibian embryos. *Development Genes and Evolution* **208:** 457–466.

Elphick, M. J., and Shine, R. (1998). Longterm effects of incubation temperatures on the morphology and locomotor performance of hatchling lizards (*Bassiana duperreyi*, Scincidae). *Biological Journal of the Linnean Society* **63:** 429–447.

Elton, C. (2001). *Animal Ecology*. University of Chicago Press, Chicago IL.

Emerson, S. B. (1975). Burrowing in frogs. *Journal of Morphology* **149:** 437–458.

Emerson, S. B. (1988). The giant tadpole of *Pseudis paradoxa*. *Biological Journal of the Linnean Society* **34:** 93–104.

Emerson, S. B., and Berrigan, D. (1993). Systematics of Southeast Asian ranids: Multiple origins of voicelessness in the subgenus *Limnonectes* (Fitzinger). *Herpetologica* **49:** 22–31.

Emerson, S. B., and Koehl, M. A. R. (1990). The interaction of behavioral and morphological change in the evolution of a novel locomotor type: Flying frogs. *Evolution* **44:** 1931–1946.

Emerson, S. B., Richards, C., Drewes, R. C., and Kjer, K. M. (2000). On the relationships among ranoid frogs: A review of the evidence. *Herpetologica* **56:** 209–230.

Emlen, S. T., and Oring, L. W. (1977). Ecology, sexual selection, and the evolution of mating systems. *Science* **197:** 215–223.

Endler, J. A. (1978). A predator's view of animal color patterns. In *Evolutionary Biology. Volume 11.* M. K. Hecht, W. C. Steere, and B. Wallace (Eds.). Pp. 319–364. Plenum Press, New York.

Endler, J. A. (1986). *Natural Selection in the Wild.*. Princeton University Press, Princeton, NJ.

Endler, J. A. (1997). *Geographic Variation, Speciation, and Clines*. Princeton University Press, Princeton, NJ.

Engbretson, G. A. (1992). Neurobiology of the lacertilian parietal eye system. *Ethology, Ecology and Evolution* **4:** 89–107.

Engelmann, W.-E., Fritsche, J., Günther, R., and Obst, F. J. (1986). *Lurche und Kriechtiere Europas.*. F. Enke Verlag, Stuttgart.

Epperson, D. M., and Heise, C. D. (2003). Nesting and hatchling ecology of gopher tortoises (*Gopherus polyphemus*) in southern Mississippi. *Journal of Herpetology* **37:** 315–324.

Epple, A., and Brinn, J. E. (1986). Pancreatic islets. In *Vertebrate Endocrinology: Fundamentals and Biomedical Implications. Vol. 1. Morphological Considerations*. P. K. T. Pang and M. P. Schreibman (Eds.). Pp. 279–317. Academic Press, Orlando, FL.

Ernst, C. H. (1971). Population dynamics and activity cycles of *Chrysemys picta* in southeastern Pennsylvania. *Journal of Herpetology* **5:** 151–160.

Ernst, C. H., and Barbour, R. W. (1989a). *Turtles of the World.* Smithsonian Institute Press, Washington, DC.

Ernst, C. H., and Barbour, R. W. (1989b). *Snakes of Eastern North America.* George Mason University Press, Fairfax, Va.

Ernst, C. H., and Ernst, E. M. (2003). *Snakes of the United States and Canada.* Smithsonian Institute Press, Washington, DC.

Ernst, C. H., and Zug, G. R. (1996). *Snakes In Question.* Smithsonian Institution Press, Washington, D.C.

Ernst, C. H., Lovich, J. E., and Barbour, R. W. (1994). *Turtles of the United States and Canada.* Smithsonian Institute Press, Washington, D.C.

Erspamer, V. (1994). Bioactive secretions of the amphibian integument. In *Amphibian Biology. Volume 1, The Integument.* H. Heatwole and G. T. Barthalmus (Eds.). Pp. 178–350. Surrey Beaty & Sons, Chipping Norton Pty Ltd, N.S.W.

Espinoza, R. E., Wiens, J. J., and Tracy, C. R. (2004). Recurrent evolution of herbivory in small, cold-climate lizards: Breaking the ecophysiological rules of reptilian herbivory. *Proceedings of the National Academy of Sciences, USA* **101:** 16819–16824.

Esteban, M., Sanchez-Herraiz, M. J., Barbadillo, L. J., and Castanet, J. (2004). Age structure and growth in an isolated population of *Pelodytes punctatus* in northern Spain. *Journal of Natural History* **38:** 2789–2801.

Estes, R. (1981). *Gymnophiona, Caudata. Handbuch der Paläoherpetologie, 2/2.* Gustav Fischer Verlag, Stuttgart.

Estes, R. (1983a). *Sauria terrestria, Amphisbaenia. Handbuch der Paläoherpetologie, 10A.* Gustav Fischer Verlag, Stuttgart.

Estes, R. (1983b). The fossil record and early distribution of lizards. In *Advances in Herpetology and Evolutionary Biology.* A. G. J. Rhodin and K. Miyata (Eds.). Pp. 365–398. Harvard University Press, Cambridge.

Estes, R., and Hoffstetter, R. (1976). Les Urodèles du Miocéne de la Grive-Saint-Alban (Isère, France). Musem national d'Histoire naturelle *(Sci. Terre)* **57:** 297–343.

Estes, R., and Reig, O. A. (1973). The early fossil record of frogs. A review of the evidence. In *Evolutionary Biology of the Anurans.* J. L. Vial (Ed.). Pp. 11–63. University of Missouri Press, Columbia.

Estes, R., de Queiroz, K., and Gauthier, J. (1988). Phylogenetic relationships within Squamata. In *Phylogenetic Relationships of the Lizard Families. Essay Commemorating Charles L. Camp.* R. Estes and G. Pregill (Eds.). Pp. 119–281. Stanford University Press, Stanford, CA.

Etheridge, K. (1990). Water balance in estivating sirenid salamanders (*Siren lacertina*). *Herpetologica* **46:** 400–406.

Etheridge, R. (1995). Redescription of *Ctenoblepharys adspersa* Tschudi, 1845, and the taxonomy of Liolaeminae (Reptilia: Squamata: Tropiduridae). *American Museum Novitates* **3142:** 1–34.

Etheridge, R., and Espinoza, R. E. (2000). Taxonomy of the Liolaeminae (Squamata: Iguania: Tropiduridae) and a semi-annotated bibliograpy. *Smithsonian Herpetological Information Service* **126:** 2–64.

Etheridge, R., and Williams, E. E. (1991). A review of the South American lizard genera *Urostrophus* and *Anisolepis* (Squamata: Iguania: Polychridae). *Bulletin of the Museum of Comparative Zoology* **152:** 317–361.

Eubanks, J. O., Michener, W. K., and Guyer, C. (2003). Patterns of movement and burrow use in a population of gopher tortoises (*Gopherus polyphemus*). *Herpetologica* **59:** 311–321.

Evans, B. J., Brown, R. M., McGuire, J. A., Supriatna, J., Andayani, N., Diesmos, A., Iskandar, D., Melnick, D. J., and Cannatella, D. C. (2003). Phylogenetics of fanged frogs: Testing biogeographical hypotheses at the interface of the Asian and Australian faunal zones. *Systematic Biology* **52:** 794–819.

Evans, B. J., Kelley, D. B., Melnick, D. J., and Cannatella, D. C. (2005). Evolution of RAG-1 in polyploid clawed frogs. *Molecular Biology and Evolution* **22:** 1193–1207.

Evans, B. J., Kelley, D. B., Tinsley, R. C., Melnick, D. J., and Cannatella, D. C. (2004). A mitochondrial DNA phylogeny of African clawed frogs: Phylogeography and implications for polyploid evolution. *Molecular Phylogenetics and Evolution* **33:** 197–213.

Evans, C. M., and Brodie, E. D., Jr. (1994). Adhesive strength of amphibian skin secretions. *Journal of Herpetology* **28:** 499–502.

Evans, S. E. (1988). The early history and relationships of the Diapsida. *Systematics Association Special Volume Series* 221–260.

Evans, S. E. (1993). Jurassic lizard assemblages. *Revue Paléobiologie*: 55–65.

Evans, S. E. (1994). A new anguimorph lizard from the Jurassic and Lower Cretaceous of England. *Palaeontology* **37:** 33–49.

Evans, S. E. (1995). Lizards: Evolution, early radiation and biogeography. In *Sixth Symposium on Mesozoic Terrestrial Ecosystems and Biota, Short Papers.* A. Sun and Y. Wang (Eds.). Pp. 51–55. China Ocean Press, Beijing, China.

Evans, S. E., and Barbadillo, L. J. (1998). An unusual lizard (Reptilia: Squamata) from the Early Cretaceous of Las Hoyas, Spain. *Zoological Journal of the Linnean Society* **124:** 235–265.

Evans, S. E., and Chure, D. C. (1998). Paramacellodid lizard skulls from the Jurassic Morrison Formation at Dinosaur National Monument, Utah. *Journal of Vertebrate Paleontology* **18:** 99–114.

Evans, S. E., and Milner, A. R. (1993). Frogs and salamanders from the Upper Jurassic Morrison Formation (Quarry Nine, Como Bluff) of North America. *Journal of Vertebrate Paleontology* **13:** 24–30.

Evans, S. E., and Milner, A. R. (1994). Middle Jurassic microvertebrate assemblages from the British Isles. In *the shadow of the dinosaurs. Early Mesozoic Tetrapods.* N. C. Fraser and H.-D. Sues (Eds.). Pp. 23–37. Cambridge University Press, New York.

Evans, S. E., and Milner, A. R. (1996). A metamorphosed salamander from the early Cretaceous of Las Hoyas, Spain. *Philosophical Transactions of the Royal Society of London. Series B: Biological Sciences* **351:** 627–646.

Evans, S. E., Milner, A. R., and Mussett, F. (1988). The earliest known salamander (Amphibia, Caudata): A record from the Middle Jurassic of England. *Geobios* **21:** 539–552.

Evans, S. E., Milner, A. R., and Mussett, F. (1990). A discoglossid frog from the Middle Jurassic of England. *Palaeontology* **33:** 299–311.

Evans, S. E., Milner, A. R., and Werner, C. (1996). Sirenid salamanders and a gymnophionan amphibian from the Cretaceous of the Sudan. *Palaeontology* **39:** 77–95.

Evans, S. E., Prasad, G. V. R., and Manhas, B. K. (2001). Rhynchocephalians (Diapsida: Lepidosauria) from the Jurassic Kota Formation of India. *Zoological Journal of the Linnean Society* **133:** 309–334.

Ewert, M. A. (1985). Embryology of turtles. In *Biology of the Reptilia. Volume 14, Development A.* C. Gans, F. Billett, and P. F. A. Maderson (Eds.). Pp. 75–267. John Wiley & Sons, New York.

Ewert, M. A., Echtberger, C. R., and Nelson, C. E. (2004). Turtle sex-determining modes and TSD patterns, and some TSD pattern correlates. In *Temperature-Dependent Sex Determination in Vertebrates.* N. Valenzuela and V. Lance (Eds.). Pp. 21–32. Smithsonian Books, Washington, D. C.

Exbrayat, J.-M. (1984). Cycle sexuel et reproduction chez un amphibien apode: *Typhlonectes compressicaudus* (Duméril et Bibron, 1841). *Bulletin de la Société herpétolgique de France* **32:** 31–35.

Exbrayat, J.-M. (1985). Le cycle de reproduction chez un amphibien apodes influence des variations sasonnieres. *Bulletin de la Société zoologique de France* **110:** 301–305.

Exbrayat, J.-M. (2006a). *Reproductive Biology and Phylogeny of Gymnophiona (Caecilians).* Science Publishers, Enfield, NH.

Exbrayat, J.-M. (2006b). Modes of parity and oviposition. In *Reproductive Biology and Phylogeny of Gymnophiona (Caecilians).* J.-M. Exbrayat (Ed.). Pp. 303–323. Science Publishers, Enfield, NH.

Exbrayat, J- M. (Ed.) (2006c). *Reproductive Biology and Phylogeny of Gymnophiona (Caecilians).* Volume 5 of the series *Reproductive Biology and Phylogeny*, series Editor, B. G. M. Jamieson. Science Publishers, Inc., Enfield, NH, USA.

Exbrayat, J.-M., and Hraoui-Bloquet, S. (1992). La nutrition embryonnaire et les relations foeto-maternelles chez *Typhlonectes compressicaudus* amphibien gymnophione vivipare. *Bulletin de la Société herpétolgique de France* **61:** 53–61.

Exbrayat, J.-M., and Hraoui-Bloquet, S. (2006). Viviparity in *Typhlonectes compressicauda.* In *Reproductive Biology and Phylogeny of Gymnophiona (Caecilians).* J.-M. Exbrayat (Ed.). Pp. 325–357. Science Publishers, Enfield, NH.

Ezaz, T., Stiglec, R., Veyrunes, F., and Graves, J. A. M. (2006). Relationships between vertebrate ZW and XY sex chromosome systems. *Current Biology* **16:** R736–R743.

Fabrezi, M., and Emerson, S. B. (2003). Parallelism and convergence in anuran fangs. *Journal of Zoology, London* **260:** 41–51.

Faivovich, J., Haddad, C. F. B., Garcia, P. C. A., Frost, D. R., Campbell, J. A., and Wheeler, W. C. (2005). Systematic review of the frog family Hylidae, with special reference to Hylinae: phylogenetic analysis and taxonomic revision. *Bulletin of the American Museum of Natural History* **294:** 1–240.

Farley, C. T., and Emshwiller, M. (1993). Does maximum power output limit running speed? *American Zoologist* **33:** 141A.

Farley, C. T. (1997). Maximum speed and mechanical power output in lizards. *Journal of Experimental Biology* **200:** 2189–2195.

Farley, C. T., and Emshwiller, M. (1996). Efficiency of uphill locomotion in nocturnal and diurnal lizards. *Journal of Experimental Biology* **199:** 587–592.

Farlow, J. O. and Brett-Surman, M. K., Ed. (1997). *The Complete Dinosaur.* Indiana University Press, Bloomington.

Farrell, A. P., Gamperl, A. K., and Francis, E. T. B. (1998). Comparative aspects of heart morphology. In *Biology of the Reptilia. Volume 19, Morphology G.* C. Gans and A. S. Gaunt (Eds.) Pp. 375–424. Society for the Study of Amphibians and Reptiles, Ithaca, NY.

Farrell, T. M., May, P. G., and Pilgrim, M. A. (1995). Reproduction in the rattlesnake, *Sistrurus miliarius barbouri*, in central Florida. *Journal of Herpetology* **29:** 21–27.

Feder, M. E., and Burggren, W. W. (1985). Cutaneous gas-exchange in vertebrates: Design, patterns, control and implications. Cutaneous gas-exchange in vertebrates - design, patterns, control and implications. *Biological Reviews of the Cambridge Philosophical Society* **60:** 1–45.

Feder, M. E., and Lynch, J. F. (1982). Effects of latitude, season, elevation, and microhabitat on field body temperatures of neotropical and temperate zone salamanders. *Ecology* **63:** 1657–1664.

Feduccia, A. (1999). 1,2,3 = 2,3,4: Accommodating the cladogram. *Proceedings of the National Academy of Sciences (USA)* **96:** 4740–4742.

Feduccia, A., and Wild, R. (1993). Birdlike characters in the Triassic archosaur *Megalancosaurus. Naturwissenschaften* **80:** 564–566.

Fei, L., and Ye, C. (2000). A new hynobiid subfamily with a new genus and new species of Hynobiidae (Amphibia: Caudata) from West China. *Cultum Herpectologica Sinica* **8:** 64–70.

Feldman, C. R., and Parham, J. F. (2002). Molecular phylogenetics of emydine turtles: Taxonomic revision and the evolution of shell kinesis. *Molecular Phylogenetics and Evolution* **22:** 388–398.

Fellers, G. M., and Drost, C. A. (1991a). Ecology of the island night lizard, *Xantusia riversiana*, on Santa Barbara Island, California. *Herpetological Monographs* **5:** 28–78.

Fellers, G. M., and Drost, C. A. (1991b). *Xantusia riversianna. Catalogue of American Amphibians and Reptiles* **518:** 1–4.

Felsenstein, J. (2004). *Inferring Phylogenies.* Sinauer Associates, Sunderland, MA.

Ferguson, D. E. (1967). Sun-compass orientation in anurans. In *Animal Orientation and Navigation.* R. E. Storm (Ed.). Pp. 21–34. Oregon State University Press, Corvalis.

Ferguson, D. E. (1971). The sensory basis of orientation in amphibians. *Annals of the New York Academy of Sciences* **188:** 30–36.

Ferguson, M. W. J., and Joanen, T. (1982). Temperature of egg incubation determines sex in *Alligator mississippiensis. Nature* **297:** 850–853.

Fernandez, M., and Gasparini, Z. (2000). Salt glands in a Tithonian metriorhynchid crocodyliform and their physiological significance. *Lethaia* **33:** 269–276.

Firth, B. T., and Heatwole, H. (1976). Panting thresholds of lizards: Role of pineal complex in panting responses in an agamid, *Amphibolurus muricatus. General and Comparative Endocrinology* **29:** 388–401.

Fisher, R. A. (1930). *The Genetical Theory of Natural Selection.* Oxford University Press, Oxford.

Fisher, R. N., and Shaffer, H. B. (1996). The decline of amphibians in the California's Great Central Valley. *Conservation Biology* **10:** 1387–1397.

Fitch, H. S. (1954). Life history and ecology of the five-lined skink, *Eumeces fasciatus. Miscellaneous Publications of the Museum of Natural History University of Kansas* **8:** 1–156.

Fitch, H. S. (1955). Habits and adaptations of the great plains skink (*Eumeces obsoletus*). *Ecological Monographs* **25:** 59–83.

Fitch, H. S. (1960). Autecology of the copperhead. *Museum of Natural History, University of Kansas Publications* **13:** 85–288.

Fitch, H. S. (1965). The University of Kansas Natural History Reserve in 1965. *University of Kansas, Museum of Natural History Miscellaneous Publications* **42:** 1–60.

Fitch, H. S. (1970). Reproductive cycles in lizards and snakes. *Miscellaneous Publications of the Museum of Natural History University of Kansas* **52:** 1–247.

Fitch, H. S. (1975). A demographic study of the ringneck snake (*Diadophis punctatus*) in Kansas. *University of Kansas, Museum of Natural History Miscellaneous Publications.* **62:** 1–53.

Fitch, H. S. (1982). Reproductive cycles in tropical reptiles. *Occasional Papers of the Museum of Natural History University of Kansas* **96:** 1–53.

Fitch, H. S. (1985). Variation in clutch and litter size in New World reptiles. *Miscellaneous Publications of the Museum of Natural History University of Kansas* **76:** 1–76.

Fittkau, E.-J. (1970). Role of caimans in the nutrient regime of mouth-lakes of Amazon affluents (an hypothesis). *Biotropica* **2:** 138–142.

Fitzgerald, L. A. (1994a). *Tupinambis* lizards and people: A sustainable use approach to conservation and development. *Conservation Biology* **8:** 12–16.

Fitzgerald, L. A. (1994b). The interplay between life history and environmental stochasticity: Implications for the management of exploited lizard populations. *American Zoologist* **34:** 371–381.

Fitzgerald, L. A., Cook, J. A., and Aquino, A. L. (1999). Molecular phylogenetics and conservation of *Tupinambis* (Sauria: Teiidae). *Copeia* **1999:** 894–905.

Fitzgerald, L. A., Cruz, F. B., and Perotti, G. (1993). The reproductive cycle and the size at maturity of *Tupinambis rufescens* (Sauria: Teiidae) in the dry Chaco of Argentina. *Journal of Herpetology* **27:** 70–78.

Fleishman, L. J. (1992). The influence of the sensory system and the environment on motion patterns in the visual displays of anoline lizards and other vertebrates. *American Naturalist* **139:** S36–S61.

Fleming, A. F. (1994). Male and female reproductive cycles of the viviparous lizard, *Mabuya capensis* (Sauria: Scincidae) from South Africa. *Journal of Herpetology* **28:** 334–341.

Fleming, A. F., and van Wyk, J. H. (1992). The female reproductive cycle of the lizard *Cordylus p. polyzonus* (Sauria: Cordylidae) in the southwestern Cape Province, South Africa. *Journal of Herpetology* **26:** 121–127.

Flintoft, L., and Skipper, M. (2006). Model organisms: *Xenopus tropicalis* goes genetic. *Nature Reviews Genetics* **7:** 588–589.

Flores-Villela, O. A., and Zug, G. R. (1995). Reproductive biology of the chopontil, *Claudius angustatus*, (Testudines: Kinosternidae), in southern Veracruz, México. *Chelonian Conservation and Biology* **1:** 181–186.

Folke, C. (2006). The economic perspective: Conservation against development versus conservation for development. *Conservation Biology* **20:** 686–688.

Follett, B. K., and Redshaw, M. R. (1974). The physiology of vitellogenesis. In *Physiology of the Amphibia, Volume 2.* B. Lofts (Ed.). Pp. 219–308. Academic Press, New York.

Fontenot, C. L., Jr. (1999). Reproductive biology of the aquatic salamander *Amphiuma tridactylum* in Louisiana. *Journal of Herpetology* **33:** 100–105.

Ford, L. S. (1993). The phylogenetic position of the dart-poison frogs (Dendrobatidae) among anurans: An examination of the competing hypotheses and their characters. *Ethology Ecology & Evolution* **5:** 219–231.

Ford, L. S., and Cannatella, D. C. (1993). The major clades of frogs. *Herpetological Monographs* **7:** 94–117.

Ford, N. B. (1982). Courtship behavior of the queen snake *Regina septemvittata. Herpetological Review* **13:** 72.

Ford, N. B., and Low, J. R. (1984). Sex-pheromone source location by garter snakes: A mechanism for detection of direction in nonvolatile trails. *Journal of Chemical Ecology* **10:** 1193–1199.

Ford, N. B., and O'Bleness, M. L. (1986). Species and sexual specificity of pheromone trails of the checkered garter snake, *Thamnophis marcianus. Journal of Herpetology* **20:** 259–262.

Ford, N. B., and Schofield, C. W. (1984). Species specificity of sex pheromone trails in the plains garter snake, *Thamnophis radix. Herpetologica* **40:** 51–55.

Ford, N. B., and Seigel, R. A. (1989). Phenotypic plasticity in reproductive traits: evidence from a viviparous snake. *Ecology* **70:** 1768–1774.

Forester, D. C., and Czarnowsky, R. (1985). Sexual selection in the spring peeper, *Hyla crucifer* (Amphibia, Anura): The role of the advertisement call. *Behaviour* **92:** 112–128.

Formanowicz, D. R., Jr., Brodie, E. D., Jr., and Wise, S. C. (1989). Foraging behavior of matamata turtles: The effects of prey density and the presence of a conspecific. *Herpetologica* **45:** 61–67.

Formas, J. R. (1997). A new species of *Batrachyla* (Anura: Leptodactylidae) from southern Chile. *Herpetologica* **53:** 6–13.

Formas, J. R., Nunez, J. J., and Brieva, L. M. (2001). Osteología, taxonomía y relaciones filogenéticas de las ranas del género *Telmatobufo* (Leptodactylidae). *Revista chilena de historia natural* **74:** 365–387.

Formas, J. R., and Espinoza, N. D. (1975). Karyological relationships of frogs of the genus *Telmatobufo* (Anura, Leptodactylidae). *Herpetologica* **31:** 429–432.

Forsman, A., and Shine, R. (1995). Sexual size dimorphism in relation to frequency of reproduction in turtles (Testudines: Emydidae). *Copeia* **1995:** 727–729.

Foster, M. S., and McDiarmid, R. W. (1982). Study of aggregative behavior of *Rhinophrynus dorsalis* tadpoles: design and analysis. *Herpetologica* **38:** 395–404.

Foster, M. S., and McDiarmid, R. W. (1983). *Rhinophrynus dorsalis* (alma de vaca, sapo borracho, Mexican burrowing toad). In *Costa Rican Natural History*. D. H. Janzen (Ed.). Pp. 419–421. University of Chicago Press, Chicago.

Fouquette, M. J., Jr., and Delahoussaye, A. J. (1977). Sperm morphology in the *Hyla rubra* group (Amphibia, Anura, Hylidae), and its bearing on generic status. *Journal of Herpetology* **11:** 387–396.

Fouquette, M. J., Jr. (1969). Rhinophrynidae, *Rhinophrynus, R. dorsalis*. *Catalogue of American Amphibians and Reptiles* **78:** 1–2.

Fox, H. (1977). The urogenital system of reptiles. In *Biology of the Reptilia. Volume 6. Morphology E*. C. Gans and T. S. Parsons (Eds.). Pp. 1–157. Academic Press, London.

Fox, H. (1983). *Amphibian Morphogenesis*. Humana Press, Clifton, N.J.

Fox, H. (1985). Changes in amphibian skin during larval development and metamorphosis. In *Metamorphosis*. M. Balls and M. Bownes (Eds.). Pp. 59–87. Clarendon Press, Oxford.

Fox, H. (1986). Epidermis. Dermis. Dermal glands. In *Biology of the Integument. 2. Vertebrates*. J. Bereiter-Hahn, A. G. Matoltsy, and K. S. Richards (Eds.). Pp. 78–110, 111–115, 16–135, respectively, Springer Verlag, Berlin.

Fox, S. F., and Rostker, M. A. (1982). Social cost of tail loss in *Uta stansburiana*. *Science* **218**: 692–693.

Fox, S. F., Heger, N. A., and Delay, L. S. (1990). Social cost of tail loss in *Uta stansburiana*: Lizard tails as status-signalling badges. *Animal Behaviour* **39:** 549–554.

Fox, S. F., McCoy, J. K., and Baird, T. A. (Eds.) (2003). *Lizard Social Behavior*. John Hopkins University Press, Cambridge, Mass.

Foxon, G. E. H. (1964). Blood and respiration. In *Physiology of the Amphibia*. J. A. Moore (Ed.). Pp. 151–209. Academic Press, New York.

Frair, W., Ackman, R. G., and Mrosovsky, N. (1972). Body temperature of *Dermochelys coriacea*: Warm turtle from cold water. *Science* **177:** 191.

Francis, E. T. B. (1934). *The Anatomy of the Salamander*. Clarendon Press, Oxford.

Frazer, D. (1989). *Reptiles and Amphibians in Britian*. Bloomsbury Books, London.

Frazer, N. B. (1992). Sea turtle conservation and halfway technology. *Conservation Biology* **6:** 179–184.

Frazer, N. B., Gibbons, J, W., and Greene, J. L. (1991). Life history and demography of the common mud turtle *Kinosternon subrubrum* in South Carolina, USA. *Ecology* **72:** 2218–2231.

Frazer, N. B., Gibbons, J. W., and Greene, J. L. (1990). Life tables of a slider turtle population. In *Life History and Ecology of the Slider Turtle*. J. W. Gibbons (Ed.). Pp. 183–200. Smithsonian Institution Press, Washington.

Frazier, J., and Peters, G. (1981). The call of the Aldabra tortoise (*Geochelone gigantea*). *Amphibia-Reptilia* **2:** 165–179.

Frazzetta, T. H. (1986). The origin of amphikinesis in lizards. A problem in functional morphology and the evolution of adaptive systems. *Evolutionary Biology* **20:** 419–461.

Freake, M. J. (1998). Variation in homeward orientation performance in the sleepy lizard (*Tiliqua rugosa*): Effects of sex and reproductive period. *Behavioral Ecology and Sociobiology* **43:** 339–344.

Freake, M. J. (1999). Evidence for orientation using the e-vector direction of polarised light in the sleepy lizard *Tiliqua rugosa*. *Journal of Experimental Biology* **202:** 1159–1166.

Frey, E., Sues, H.-D., and Munk, W. (1997). Gliding mechanism in the Late Permian reptile *Coelurosauravus*. *Science* **275:** 1450–1452.

Friedl, T. W., and Klump, G. M. (1997). Some aspects of population biology in the European treefrog, *Hyla arborea*. *Herpetologica* **53:** 321–330.

Fritts, T. H. (1988). The brown tree snake, *Boiga irregularis*, a threat to Pacific islands. *United States Department of Interior, Fish and Wildlife Service, Biological Report* **88:** 1–36.

Fritz, U., Fattizzo, T., Guicking, D., Tripepi, S., Pennisi, M. G., Lenk, P., Joger, U., and Wink, M. (2005). A new cryptic species of pond turtle from southern Italy, the hottest spot in the range of the genus *Emys* (Reptilia, Testudines, Emydidae). *Zoologica Scripta* **34:** 351–371.

Fritzsch, B. (1990). Evolution of tetrapod hearing. *Nature* **344:** 116.

Fritzsch, B., and Wake, M. H. (1988). The inner ear of gymnophione amphibians and its nerve supply: A comparative study of regressive events in a complex sensory system. *Zoomorphology* **108:** 201–217.

Frolich, L. M. (1997). The role of the skin in the origin of amniotes: Permeability barrier, protective covering and mechanical support. In *Amniote Origins. Completing the Transition to Land*. S. S. Sumida and K. L. M. Martin (Eds.). Pp. 327–352. Academic Press, San Diego.

Frost, D. R. (1992). Phylogenetic analysis and taxonomy of the *Tropidurus* group of lizards (Iguania: Tropiduridae). *American Museum Novitates* **3033:** 1–68.

Frost, D. R. (Ed.) (1985). *Amphibian Species of the World*. Allen Press and The Association of Systematics Collections, Lawrence, Kansas.

Frost, D. R., and Etheridge, R. (1989). A phylogenetic analysis and taxonomy of iguanian lizards (Reptilia: Squamata). *Miscellaneous Publications, University of Kansas, Museum of Natural History* **81:** 1–65.

Frost, D. R., Etheridge, R., Janies, D., and Titus, T. A. (2001b). Total evidence, sequence alignment, evolution of polychrotid lizards, and a reclassification of the Iguania (Squamata: Iguania). *American Museum Novitates* **3343:** 1–38.

Frost, D. R., Grant, T., Faivovich, J., Bain, R. H., Haas, A., Haddad, C. F. B., De Sa, R. O., Channing, A., Wilkinson, M., Donnellan, S. C., Raxworthy, C. J., Campbell, J. A., Blotto, B. L., Moler, P., Drewes, R. C., Nussbaum, R. A., Lynch, J. D., Green, D. M., and Wheeler, W. C. (2006). The amphibian tree of life. *Bulletin of the American Museum of Natural History* **297:** 1–370.

Frost, D. R., Rodrigues, M. T., Grant, T., and Titus, T. A. (2001). Phylogenetics of the lizard genus Tropidurus (Squamata: tropiduridae: tropidurinae): Direct optimization, descriptive efficiency, and sensitivity analysis of congruence between molecular data and morphology. *Molecular Phylogenetics and Evolution* **21:** 352–371.

Fry, B. G., Vidal, N., Norman, J. A., Vonk, F. J., Scheib, H., Ramjan, S. F. R., Kuruppu, S., Fung, K., Hedges, S. B., Richardson, M. K., Hodgson, W. C., Ignjatovic, V., Summerhayes, R., and Kochva, E. (2005). Early evolution of the venom system in lizards and snakes. *Nature* **439:** 584–588.

Fu, J. (1998). Toward the phylogeny of the family Lacertidae: Implications from mitochondrial DNA 12S and 16S gene sequences (Reptilia: Squamata). *Molecular Phylogenetics and Evolution* **9:** 118–30.

Fu, J., Weadick, C. J., and Bi, K. (2007). A phylogeny of the high-elevation Tibetan megophryid frogs and evidence for the multiple origins of reversed sexual size dimorphism. *Journal of Zoology, London* **273:** 315–325.

Fujita, M. K., Engstrom, T. N., Starkey, D. E., and Shaffer, H. B. (2004). Turtle phylogeny: insights from a novel nuclear intron. *Molecular Phylogenetics and Evolution* **31:** 1031–1040.

Full, R. J. (1986). Locomotion without lungs: Energetics and performance of a lungless salamander. *American Journal of Physiology* **251:** 775–780.

Funk, W. C., Caldwell, J. P., Peden, C. E., Padial, J. M., De la Riva, I., and Cannatella, D. C. (2007). Tests of biogeographic hypotheses for diversification in the Amazonian forest frog, *Physalaemus petersi. Molecular Phylogenetics and Evolution* **44:** 825–837.

Fusari, M. H. (1985). Drinking of soil water by the California legless lizard, *Anniella pulchra. Copeia* **1985:** 981–986.

Futuyma, D. J. (1998). *Evolutionary Biology*. (3rd Edition). Sinauer Associates, Sunderland, Mass.

Gabe, M. (1970). The adrenal. In *Biology of the Reptilia. Volume 3. Morphology C.* C. Gans and T. S. Parsons (Eds.). Pp. 263–318. Academic Press, London.

Gaffney, E. S. (1975). A phylogeny and classification of the higher categories of turtles. *Bulletin of the American Museum of Natural History* **155:** 387–436.

Gaffney, E. S. (1977). The side-necked turtle family Chelidae: A theory of relationships using shared derived characters. *American Museum Novitates* **2620:** 1–28.

Gaffney, E. S. (1979). Comparative cranial morphology of Recent and fossil turtles. *Bulletin of the American Museum of Natural History* **164:** 65–376.

Gaffney, E. S. (1980). Phylogenetic relationships of the major groups of amniotes. In *The Terrestrial Environment and the Origin of Land Vertebrates.* A. L. Panchen (Ed.). Pp. 593–610. Academic Press, London.

Gaffney, E. S. (1984). Historical analysis of theories of chelonian relationship. *Systematic Zoology* **33:** 283–301.

Gaffney, E. S. (1986). Triassic and Early Jurassic turtles. In *The Beginnings of the Age of Dinosaurs.* K. Padian (Ed.). Pp. 183–187. Cambridge University Press, Cambridge.

Gaffney, E. S. (1990). The comparative osteology of the Triassic turtle *Proganochelys. Bulletin of the American Museum of Natural History* **194:** 1–263.

Gaffney, E. S. (1991). The fossil turtles of Australia. In *Vertebrate Palaeontology of Australasia*. P. Vickers-Rich, J. M. Monaghan, R. F. Baird, and T. H. Rich (Eds.). Pp. 704–719. Pioneer Design Studio, Lilydale, Australia.

Gaffney, E. S., and Forster, C. A. (2003). Side-Necked Turtle Lower Jaws (Podocnemididae, Bothremydidae) from the Late Cretaceous Maevarano Formation of Madagascar. *American Museum Novitates* **3397:** 1–13.

Gaffney, E. S., and Kitching, J. W. (1994). The most ancient African turtle. *Nature*. **369:** 55–58.

Gaffney, E. S., and Meylan, P. (1988). A phylogeny of turtles. In *The Phylogeny and Classification of the Tetrapods. Vol. 1: Amphibians, Reptiles, Birds.* M. J. Benton (Ed.). Pp. 157–219. Clarendon Press, Oxford.

Gaffney, E. S., Ashok, S., Herman, S., Swarn Deep, S., and Rahul, S. (2003). *Sankuchemys*, a new side-necked turtle (Pelomedusoides, Bothremydidae) from the late Cretaceous of India. *American Museum Novitiates* **3405:** 1–10.

Gaffney, E. S., Hutchinson, J. H., Jenkins, F. A., Jr., and Meeker, L. J. (1987). Modern turtle origins: the oldest known cryptodire. *Science* **237:** 289–291.

Gaffney, E. S., Meylan, P. A., and Wyss, A. R. (1991). A computer assisted analysis of the relationships of the higher categories of turtles. *Cladistics* **7:** 313–335.

Gaffney, E., and Schleich, H. H. (1994). New reptile material from the German Tertiary. 16. On *Chelydropsis murchisoni* (Bell, 1892) from the Middle Miocene Locality of Unterwohlbach/South Germany. *Courier Forschungsinstitut Senckenberg* **173:** 197–213.

Galbraith, D. A., Chandler, M. W., and Brooks, R. J. (1987). The fine-structure of home ranges of male *Chelydra serpentina*: are snapping turtles territorial. *Canadian Journal of Zoology* **65:** 2623–2629.

Gallardo, J. M. (1987). *Anfibios Argentinos*. Bibliotecha Mosaico, Buenos Aires.

Gamble, T., Simons, A. M., Colli, G. R., and Vitt, L. J. (2008). Tertiary climate change and the diversification of the Amazonian gecko genus *Gonatodes* (Sphaerodactylidae, Squamata). *Molecular Phylogenetics and Evolution* **46:** 269–277.

Gans, C. (1974). *Biomechanics: An Approach to Vertebrate Biology*. J. B. Lippincott Co, Philadelphia.

Gans, C. (1976). Aspects of the biology of uropeltid snakes. In *Morphology and Biology of Reptiles*. A. d'A. Bellairs and C. B. Cox (Eds.). Pp. 191–205. Linnean Society of London, London.

Gans, C. (1978). The characteristics and affinities of the Amphisbaenia. *Transactions of the Zoological Society of London* **34:** 347–416.

Gans, C. (1979). A subterranean snake with a funny tail. *Natural History* **88:** 70–75.

Gans, C. (1986). Automimicry and Batesian mimicry in uropeltid snakes: Pigment pattern, proportions, and behavior. *Journal of the Bombay Natural History Society* **83:** 152–158.

Gans, C., and coeditors. (1969 et seq). *Biology of the Reptilia* several publishers. Volumes 1–4, 6. 11, 19.

Gans, C., and Gorniak, G. C. (1982). Functional morphology of lingual protrusion in marine toad (*Bufo marinus*). *Journal of Morphology* **163:** 195–222.

Gans, C., and Kraklau, D. M. (1989). Studies on amphisbaenians (Reptilia) 8. Two genera of small species from East Africa (*Geocalamus* and *Loveridgea*). *American Museum Novitates* **2944:** 1–28.

Gans, C., and Maderson, P. F. A. (1973). Sound producing mechanisms in recent reptiles: A review and comment. *American Zoologist* **13:** 1195–1203.

Gans, C., Bellairs, A. d., and Parsons, T. S. (1969). Biology of the Reptilia Volume 1. Morphology A. *Biology of the Reptilia Volume 1 Morphology A*: p. vii–xv, 1–373.

Gans, C., Darevski, I., and Tatarinov, L. P. (1987). *Sharovipteryx*, a reptilian glider? *Paleobiology* **13:** 415–426.

Gans, C., Gillingham, J. C., and Clark, D. L. (1984). Courtship, combat and male combat in tuatara. *Journal of Herpetology* **18:** 194–197.

Gao, K. (1997). *Sineoamphisbaena* phylogenetic relationships discussed: reply. *Canadian Journal of Earth Sciences* **34:** 886–889.

Gao, K. Q., and Shubin, N. H. (2003). Earliest known crown-group salamanders. *Nature* **422:** 424–428.

Gao, K., and Hou, L. (1995). Late Cretaceous fossil record and paleobiogeography of iguanian squamates. In *Sixth Symposium on Mesozoic Terrestrial Ecosystems and Biota, Short Papers*. A. Sun and Y. Wang (Eds.). Pp. 47–50. China Ocean Press, Beijing.

Gao, K., and Nessov, L. A. (1998). Early Cretaceous squamates from the Kyzylkum Desert, Uzbekistan. *Neues Jahrbuch fuer Geologie und Palaeontologie Abhandlungen* **207:** 289–309.

Gao, K., and Norell, M. A. (1998). Taxonomic revision of *Carusia* (Reptilia: Squamata) from the Late Cretaceous of the Gobi Desert and phylogenetic relationships of anguimorphan lizards. *American Museum Novitates* **3230:** 1–51.

Gao, K., and Shubin, N. (2001). Late Jurassic salamanders from Northern China. *Nature* **410:** 574–577.

Garda, A. A., and Cannatella, D. C. (2007). Phylogeny and biogeography of paradoxical frogs (Anura, Hylidae, Pseudae) inferred from 12S and 16S mitochondrial DNA. *Molecular Phylogenetics and Evolution* **44:** 104–114.

Garda, A. A., Costa, G. C., Colli, G. R., and Báo, S. N. (2004). Spermatozoa of Pseudinae (Amphibia, Anura, Hylidae), with a test of the hypothesis that sperm ultrastructure correlates with reproductive modes in anurans. *Journal of Morphology* **261:** 196–205.

Gardner, J. D., Russell, A. P., and Brinkman, D. B. (1995). Systematics and taxonomy of soft-shelled turtles (Family Trionychidae) from the Judith River Group (mid-Campanian) of North America. *Canadian Journal of Earth Sciences* **32:** 631–643.

Gardner, M. G., Bull, C. M., Cooper, S. J. B., and Duffield, G. A. (2001). Genetic evidence for a family structure in stable social aggregations of the Australian lizard *Egernia stokesii*. *Molecular Ecology* **10:** 175–183.

Garland, T. J., Jr., Martin, K. L. M., and Diaz-Uriarte, R. (1997). Reconstructing ancestral trait values using squared-change parsimony: Plasma osmolarity at the amniote transitio. In *Amniote Origins: Completing the Transition to Land*. S. S. Sumida and K. L. M. Martin (Eds.). Pp. 425–501. Academic Press, San Diego.

Garland, T. J., Jr. (1994). Phylogenetic analysis of lizard endurance capacity in relation to body size and body temperature. In *Lizard Ecology: Historical and Experimental Perspectives*. L. J. Vitt and E. R. Pianka (Eds.). Pp. 237–259. Princeton University Press, Princeton, NJ.

Garland, T. J., Jr., and Losos, J. B. (1994). Ecological morphology of locomotor performance in squamate reptiles. In *Ecological Morphology: Integrative Organismal Biology*. P. C. Wainwright and S. M. Reilly (Eds.). Pp. 204–302. University of Chicago Press, Chicago.

Garrick, L. D., and Lang, J. W. (1977). Social signals and behaviors of adult alligators and crocodiles. *American Zoologist* **17:** 225–239.

Garrick, L. D., Lang, J. W., and Herzog, J. (1978). Social signals of adult American alligators. *Bulletin of the American Museum of Natural History* **160:** 153–192.

Garstka, J. W., and Crews, D. (1981). Female sex pheromones in the skin and circulation of a garter snake. *Science* **214:** 681–683.

Garstka, J. W., and Crews, D. (1986). Pheromones and reproduction in garter snakes. In *Chemical Signals in Vertebrates, IV. Ecology, Evolution, and Comparative Biology*. D. Duvall, D. Müller-Schwarze, and R. M. Silverstein (Eds.). Pp. 243–260. Plenum Press, New York.

Gasc, J.-P. (1981). Axial musculature. In *Biology of the Reptilia. Volume 11. Morphology F*. C. Gans and T. S. Parsons (Eds.). Pp. 355–435. Academic Press, London.

Gascon, C., Lougheed, S. C., and Bogart, J. P. (1996). Genetic and morphological variation in *Vanzolinius discodactylus*: A test of the river hypothesis of speciation. *Biotropica* **28:** 376–387.

Gascon, C., Lougheed, S. C., and Bogart, J. P. (1998). Patterns of genetic population differentiation in four species of Amazonian frogs: A test of the riverine barrier hypothesis. *Biotropica* **30:** 104–119.

Gascuel, O. (2007). Advances and limitations of maximum likelihood phylogenetics. http://www.newton.ac.uk/webseminars/pg+ws/2007/plg/plgw01/0904/gascuel/001.html.

Gasparini, Z. (1996). Biogeographic evolution of the South American crocodilians. In *Contributions of Southern South America to Vertebrate Paleontology*. G. Arratia (Ed.), Pp. 159–184. *Münchner Geowissenschaftliche Abhandlungen* **A 30**.

Gauthier, J. A. (1986). Saurischian monophyly and the origin of birds. In *The Origin of Birds and the Evolution of Flight*. K. Padian (Ed.). Pp. 1–55. Memoirs of the California Academy of Science, San Francisco.

Gauthier, J. A. (1982). Fossil xenosaurid and anguid lizards from the Early Eocene Wasatch Formation, southeast Wyoming, and a revision of the Anguioidea. *Contributions to Geology, University of Wyoming* **21:** 7–54.

Gauthier, J. A., Kluge, A. G., and Rowe, T. (1988c). The early evolution of the Amniota. In *The Phylogeny and Classification of the Tetrapods. Vol. 1: Amphibians, Reptiles, Birds*. M. J. Benton (Ed.). Pp. 103–155. Clarendon Press, Oxford.

Gauthier, J. A., Cannatella, D., de Queiroz, K., Kluge, A. G., and Rowe, T. (1989). Tetrapod phylogeny. In *The Hierarchy of Life*. B. Fernholm, K. Bremer, and H. Jörnwall (Eds.). Pp. 337–353. Elsevier Science Publisher B. V., Amsterdam.

Gauthier, J. A., Estes, R., and de Queiroz, K. (1988a). A phylogenetic analysis of Lepidosauromorpha. In *Phylogenetic Relationships of the Lizard Families. Essay Commemorating Charles L. Camp*. R. Estes and G. Pregill (Eds.). Pp. 15–98. Stanford University Press, Stanford, CA.

Gauthier, J. A., Kluge, A. G., and Rowe, T. (1988b). Amniote phylogeny and the importance of fossils. *Cladistics* **4:** 105–209.

Gauthreaux, S. A., Jr. (Ed.). (1980). *Animal Migration, Orientation, and Navigation*. Academic Press, London.

Geffeney, S. L., Fujimoto, E., Brodie, E. D., III., Brodie, E. D., Jr., and Ruben, P. C. (2005). Evolutionary diversification of TTX-resistant sodium channels in a predator-prey interaction. *Nature* **434:** 759–763.

Geffeney, S., Brodie, E. D., Jr., Ruben, P. C., and Brodie, E. D., III, (2002). Mechanisms of adaptation in a predator-prey arms race: TTX-resistant sodium channels. *Science* **297:** 1336–1339.

Gehlbach, F. R., and Walker, B. (1970). Acoustic behavior of the aquatic salamander, *Siren intermedia*. *Bioscience* **20:** 1107–1108.

Gehlbach, F. R., Kimmel, J. R., and Weems, W. A. (1969). Aggregations and body water relations in tiger salamanders (*Ambystoma tigrinum*) from the Grand Canyon rims, Arizona. *Physiological Zoology* **42:** 173–182.

Geiger, A. (1995). Der Laubfrosch (*Hyla arborea*) Ökologie und Artenschutz. *Mertensiella* **6:** 1–196.

Geise, W., and Linsenmair, K. E. (1986). Adaptations of the reed frog *Hyperolius viridiflavus* (Amphibia, Anura, Hyperoliidae) to its arid environment: 2. Some aspects of the water economy of *Hyperolius viridiflavus nitidulus* under wet and dry season conditions. *Oecologia* **68:** 542–548.

Gehlbach, F. R., and Balbridge, R. S. (1987). Live blind snakes (*Leptotyphlops dulcis*) in eastern screech owl (*Oyus osio*) nests: A novel commensalism. *Oecologia* **71:** 560–563.

Geniez, P., Mateo, J. A., and Bons, J. (2000). A checklist of amphibians and reptiles of Western Sahara. *Herpetozoa* **13:** 149–163.

Geniez, P., Mateo, J. A., Geniez, M., and Pether, J. (2004). *The Amphibians and Reptiles of the Western Sahara: An Atlas and Field Guide*. Edition Chimaira, Frankfurt am Main.

Gensel, P. G., and Andrews, H. N. (1987). The evolution of early land plants. *American Scientist* **75:** 478–489.

George, R. H. (1996). Health problems and diseases of sea turtles. In *The Biology of Sea Turtles*. P. L. Lutz and J. A. Musick (Eds.). Pp. 363–385. CRC Press, Boca Raton, FL.

Georges, A. G., and Adams, M. (1992). A phylogeny for Australian chelid turtles based on allozyme electrophoresis. *Australian Journal of Zoology* **40:** 453–476.

Georges, A. G., and Adams, M. (1996). Electrophoretic delineation of species boundaries within the short-necked freshwater turtles of Australia (Testudines: Chelidae). *Zoological Journal of the Linnean Society* **118:** 241–260.

Georges, A. G., and Wombey, J. C. (1993). Family Carettochelyidae. In *Fauna of Australia. Volume 2A, Amphibia & Reptilia*. C. J. Glasby, G. J. B. Ross, and P. I. Beesley (Eds.). Pp. 153–156. Australian Government Public Service, Canberra.

Georges, A. G., Limpus, C. J., and Parmenter, C. J. (1993). Natural history of the Chelonia. In *Fauna of Australia. Volume 2A, Amphibia & Reptilia*. C. J. Glasby, G. J. B. Ross, and P. I. Beesley (Eds.). Pp. 120–128. Australian Government Public Service, Canberra.

Georges, A. G., Adams, M., and McCord, W. (2002). Electrophoretic delineation of species boundaries within the genus *Chelodina* (Testudines: Chelidae) of Australia, New Guinea and Indonesia. *Zoological Journal of the Linnean Society* **134:** 401–421.

Georges, A. G., Birrell, J., Saint, K. M., McCord, W., and Donnellan, S. C. (1998). A phylogeny for side-necked turtles (Chelonia: Pleurodira) based on mitochondrial and nuclear gene sequence variation. *Biological Journal of the Linnean Society* **67:** 213–246.

Gerlach, J., and Willi, J. (2002). A new species of frog, genus *Sooglossus* (Anura, sooglossidae) from Silhouette island, Seychelles. *Amphibia-Reptilia* **23:** 445–458.

Gerritsen, J., and Strickler, J. R. (1977). Encounter probabilities and community structure in zooplankton: A mathematical model. *Journal of Fisheries Research Board of Canada* **34:** 73–82.

Gibbons, J. W. (1968). Carapacial algae in a population of the painted turtle, *Chrysemys picta*. *American Midland Naturalist* **79:** 517–519.

Gibbons, J. W. (1970). Terrestrial activity and the population dynamics of aquatic turtles. *American Midland Naturalist* **83:** 404–414.

Gibbons, J. W. (1972). Reproduction, growth, and sexual dimorphism in the canebrake rattlesnake (*Crotalus horridus atricaudatus*). *Copeia* **1972:** 222–226.

Gibbons, J. W. (1994). Reproductive patterns of reptiles and amphibians: considerations for captive breeding and conservation. In *Captive Management and Conservation of Amphibians and Reptiles*. J. B. Murphy, K. Adler, and J. T. Collins (Eds.). Pp. 119–123. Society for the Study of Amphibians and Reptiles, Ithaca, NY.

Gibbons, J. W., and Bennett, D. H. (1974). Determination of anuran terrestrial activity patterns by a drift fence method. *Copeia* **1974:** 236–243.

Gibbons, J. W., and Dorcas, M. E. (2004). *North American Watersnakes: A Natural History*. University of Oklahoma Press, Norman, OK.

Gibbons, J. W., and Lovich, J. E. (1990). Sexual dimorphism in turtles with emphasis on the slider turtle (*Trachemys scripta*). *Herpetological Monographs* **4:** 1–29.

Gibbons, J. W., and Nelson, D. H. (1978). The evolutionary significance of delayed emergence from the nest by hatchling turtles. *Evolution* **32:** 297–303.

Gibbons, J. W., and Semlitsch, R. D. (1991). *Guide to the Reptiles and Amphibians of the Savannah River Site*. University of Georgia Press, Athens, Ga.

Gibbons, J. W., and Smith, M. H. (1968). Evidence of orientation by turtles. *Herpetologica* **24:** 331–333.

Gibbons, J. W. (Ed.) (1990). *Life History and Ecology of the Slider Turtle*. Smithsonian Institute Press, Washington. D.C.

Gibbons, J. W., Greene, J. L., and Congdon, J. D. (1990). Temporal and spatial movement patterns of sliders and other turtles. In *Life History and Ecology of the Slider Turtle*. J. W. Gibbons (Ed.). Pp. 201–215. Smithsonian Institution Press, Washington.

Gibbs, J. P. (1998). Distribution of woodland amphibians along a forest fragmentation gradient. *Landscape Ecology* **13:** 263–268.

Gibson, R. C., and Buley, K. R. (2004). Maternal care and obligatory oophagy in *Leptodactylus fallax*: A new reproductive mode in frogs. *Copeia* **2004:** 128–135.

Gill, B., and Whitaker, T. (1996). *New Zealand Frogs & Reptiles*. David Bateman, Auckland.

Gill, D. E. (1978). The metapopulation ecology of the red-spotted newt, *Notophthalmus viridescens*. *Ecological Monographs* **48:** 145–166.

Gill, D. E. (1979). Density dependence and homing behavior in adult red-spotted newts *Notophthalmus viridescens* (Rafinesque). *Ecology* **60:** 800–813.

Gillingham, J. C. (1987). Social behavior. In *Snakes: Ecology and Evolutionary Biology*. R. A. Seigel, J. T. Collins, and

S. S. Novak (Eds.). Pp. 184–209. Macmillian Publishing Co., New York.

Gillingham, J. C., Carmichael, C., and Miller, T. (1995). Social behavior of the tuatara, *Sphenodon punctatus*. *Herpetological Monographs* **9:** 5–16.

Gillis, R., and Ballinger, R. E. (1992). Reproductive ecology of red-chinned lizards (*Sceloporus undulatus erythrocheilus*) in southcentral Colorado: Comparisons with other populations of a wide-ranging species. *Oecologia* **89:** 236–243.

Gilmore, C. W. (1928). Fossil Lizards of North America. *Memoirs of the National Academy of Science, Washington* **22**: Pp. 1–201.

Giugliano, L. G., Collevatti, R. G., and Colli, G. R. (2007). Molecular dating and phylogenetic relationships among Teiidae (Squamata) inferred by molecular and morphological data. *Molecular Phylogenetics and Evolution* **45:** 168–179.

Glaw, F., and Köhler, J. (1998). Amphibian species diversity exceeds that of mammals. *Herpetological Review* **29:** 11–12.

Glaw, F., and Vences, M. (1994). *A Field Guide to the Amphibians and Reptiles of Madagascar*. M. Vences and F. Glaw, Verlag GbR, Köln.

Glaw, F., and Vences, M. (2006). Phylogeny and genus-level classification of mantellid frogs (Amphibia, Anura). *Organisms Diversity & Evolution* **6:** 236–253.

Glaw, F., Vences, M., and Böhme, W. (1998). Systematic revision of the genus *Aglyptodactylus* Boulenger, 1919 (Amphibia: Ranidae), and analysis of its phylogenetic relationships to other Madagascan ranid genera (*Tomopterna, Boophis, Mantidactylus*, and *Mantella*). *Journal of Zoological Systematics and Evolutionary Research* **36:** 17–37.

Glaw, F., Vences, M., and Gossmann, V. (2000). A new species of *Mantidactylus* (subgenus *Guibemantis*) from Madagascar, with a comparative survey of internal femoral gland structure in the genus (Amphibia: Ranidae: Mantellinae). *Journal of Natural History* **34:** 1135–1154.

Glodek, G. S., and Voris, H. K. (1982). Marine snake diets: Prey composition, diversity and overlap. *Copeia* **1982:** 661–666.

Glor, R. E., Vitt, L. J., and Larson, A. (2001). A molecular phylogenetic analysis of diversification in Amazonian *Anolis* lizards. *Molecular Ecology* **10:** 2661–2668.

Glos, J. (2003). The amphibian fauna of the Kirindy dry forest in western Madagascar. *Salamandra* **39:** 75–90.

Glos, J., and Linsenmair, K. E. (2004). Descriptions of the tadpoles of *Aglyptodactylus laticeps* and *Aglyptodactylus securifer* from Western Madagascar, with notes on life history and ecology. *Journal of Herpetology* **38:** 131–136.

Gloyd, H. K., and Conant, R. (1990). *Snakes of the Agkistrodon Complex: A Monographic Review*. Society for the Study of Amphibians and Reptiles, Ithaca, NY.

Godley, B. J., Richardson, S., Broderick, A. C., Coyne, M. S., Glen, F., and Hays, G. C. (2002). Long-term satellite telemetry of the movements and habitat utilisation by green turtles in the Mediterranean. *Ecography* **25:** 352–362.

Godley, J. S. (1980). Foraging ecology of the stripped swamp snake, *Regina alleni*, in southern Florida. *Ecological Monographs* **50:** 411–436.

Godley, J. S. (1983). Observations on the courtship, nests and young of *Siren intermedia* in southern Florida. *American Midland Naturalist* **110:** 215–219.

Goicoechea, O., Carrido, O, and Jorquera, B. (1986). Evidence for a trophic paternal-larval relationship in the frog *Rhinoderma darwinii*. *Journal of Herpetology* **20:** 168–178.

Goin, C. J. (1960). Amphibians, pioneers of terrestrial breeding habits. *Smithsonian Report* **1959:** 427–455.

Goldberg, S. R. (1971). Reproductive cycle of the ovoviviparous iguanid lizard *Sceloporus jarrovi* Cope. *Herpetologica* **27:** 123–131.

Goldberg, S. R., and Miller, C. M. (1985). Reproduction of the silvery legless lizard, *Anniella puchra pulchra* (Anniellidae), in southern California. *Southwestern Naturalist* **30:** 617–619.

Gollmann, B., and Gollmann, G. (1991). Embryonic development of the myobatrachine frogs *Geocrinia laevis, Geocrinia victoriana*, and their natural hybrids. *Amphibia-Reptilia* **12:** 103–110.

Gomez, N. A., Acosta, M., Zaidan, F., and Lillywhite, H. B. (2006). Wiping behavior, skin resistance, and the metabolic response to dehydration in the arboreal frog *Phyllomedusa hypochondrialis*. *Physiological and Biochemical Zoology* **79:** 1058–1068.

Good, D. A. (1989). Hybridization and cryptic species In: *Dicamptodon* (Caudata: Dicamptodontidae). *Evolution* **43:** 728–744.

Good, D. A. (1994). Species limits in the genus *Gerrhonotus* (Squamata: Anguidae). *Herpetological Monographs* **8:** 180–202.

Good, D. A., and Wake, D. B. (1992). Geographic variation and speciation in the torrent salamanders of the genus *Rhyacotriton* (Caudata: Rhyacotritonidae). *University of California Publications, Zoology* **126:** i–xii, 1–91.

Good, D. A., Bauer, A. M., and Sadlier, R. A. (1997). Allozyme evidence for the phylogeny of giant New Caldeonian geckos (Squamata: Diplodactylidae: *Rhacodactylus*), with comments on the status of *R. leachianus henkeli*. *Australian Journal of Zoology* **45:** 317–330.

Goodman, M., Miyamoto, M., and Czelusniak, J. (1987). Patterns and process in vertebrate phylogeny revealed by coevolution of molecules and morphologies. In *Molecules and Morphology in Evolution: Conflict or Compromise*. C. Patterson (Ed.). Pp. 141–176. Cambridge University Press, Cambridge.

Goodrich, E. S. (1930). *Studies on the Structure and Development of Vertebrates. Volume I, 1958 reprint*. Dover Publishers, Inc. New York.

Goodwin, T. M., and Marion, W. R. (1978). Aspects of the nesting ecology of American alligators (*Alligator mississippiensis*) in north-central Florida. *Herpetologica* **34:** 43–47.

Goodyear, C. P. (1971). Y-axis orientation of the oak toad, *Bufo quercicus*. *Herpetologica* **27:** 320–323.

Goodyear, C. P., and Altig, R. (1971). Orientation of bullfrogs, (*Rana catesbeiana*), during metamorphosis. *Copeia* **1971:** 362–364.

Gosner, K. L. (1960). A simplified table for staging anuran embryos and larvae with notes on identification. *Herpetologica* **16:** 183–190.

Gower, D. J., and Wilkinson, M. (1996). Is there any consensus on basal archosaur phylogeny? *Proceedings of the Royal Society of London* **B 263:** 1399–1406.

Gower, D. J., and Wilkinson, M. (2005). Conservation biology of caecilian amphibians. *Conservation Biology* **19:** 45–55.

Gower, D. J., Bahir, M. M., Mapatuna, Y., Pethiyagoda, R., Raheem, D., and Wilkinson, M. (2005). Molecular phylogenetics of Sri Lankan *Ichthyophis* (Amphibia: Gymnophiona: Ichthyophiidae), with discovery of new species of caecilian amphibians. *Raffles Bulletin of Zoology,* Supplement **12:** 153–161.

Gower, D. J., Kupfer, A., Oommen, O. V., Himstedt, W., Nussbaum, R. A., Loader, S. P., Presswell, B., Müller, H., Krishna, S. B., Boistel, R., and Wilkinson, M. (2002). A molecular phylogeny of ichthyophiid caecilians (Amphibia: Gymnophiona: Ichthyophiidae): Out of India or out of South East Asia? *Proceedings of the Royal Society, Biological Sciences* **269:** 1563–1569.

Gower, D. J., Vidal, N., Spinks, J. N., and McCarthy, C. J. (2005). The phylogenetic position of Anomochilidae (Reptilia: Serpentes): First evidence from DNA sequences. *Journal of Zoological Systematics and Evolutionary Research* **43:** 315–320.

Gradstein, S. R., and Equihua, C. (1995). An epizoic bryophyte and algae growing on the lizard *Corytophanes cristatus* in Mexican rain forest. *Biotropica* **27:** 265–268.

Grace, M. S., Woodward, O. M., Church, D. R., and Calisch, G. (2001). Prey targeting by the infrared-imaging snake *Python molurus*: Effects of experimental and congenital visual deprivation. *Behavioural Brain Research* **119:** 23–31.

Graf, J.-D. (1996). Molecular approaches to the phylogeny of *Xenopus*. In *The Biology of Xenopus*. R. C. Tinsley and H. R. Kobel (Eds.). Pp. 379–401. Zool. Soc. London, London.

Graham, A., and Beard, P. (1973). *Eyelids of Morning. The Mingled Destinies of Crocodiles and Men.* New York Graphic Soc., Ltd., Greenwich, Conn.

Graham, C. H., Ron, S. R., Santos, J. C., Schneider, C. J., and Moritz, C. (2004). Integrating phylogenetics and environmental niche models to explore speciation mechanisms in dendrobatid frogs. *Evolution* **58:** 1781–1793.

Graham, J. B. (1997). *Air-Breathing Fishes. Evolution, Diversity, and Adaptation.* Academic Press, San Diego.

Grandison, A. G. C. (1977). *H. W. Parker Snakes a Natural History*. Second Editon. British Museum of Natural History, London.

Grandison, A. G. C., and Ashe, S. (1983). The distribution, behavioural ecology and breeding strategy of the pygmy toad, *Mertensophryne micranotis* (Lov.). *Bulletin of the British Museum of Natural History Zoology* **45:** 85–93.

Grant, B. W. (1990). Trade-offs in activity time and physiological performance for thermoregulating desert lizards, *Sceloporus merriami*. *Ecology* **71:** 2323–2333.

Grant, D., Anderson, O., and Twitty, V. C. (1973). Homing orientation by olfaction in newts (*Taricha rivularis*). *Science* **160:** 1354–1355.

Grant, T. (2007). A new, toxic species of *Colostethus* (Anura: Dendrobatidae: Colostethinae) from the Cordillera Central of Colombia. *Zootaxa* **1555:** 39–51.

Grant, T., Frost, D. R., Caldwell, J. P., Gagliardo, R., Haddad, C. F. B., Kok, P. J. R., Means, D. B., Noonan, B. P., Schargel, W. E., and Wheeler, W. C. (2006). Phylogenetic systematics of dart-poison frogs and their relatives (Amphibia: Athesphatanura: Dendrobatidae). *Bulletin of the American Museum of Natural History* **299:** 1–262.

Grant, T., Humphrey, E. C., and Myers, C. W. (1997). The median lingual process of frogs: A bizarre character of Old World ranoids discovered in South American dendrobatids. *American Museum Novitates* **3212:** 1–40.

Grassman, M. A., Owens, D. W., McVey, J. P., and M., R. Marquez, M. (1984). Olfactory-based orientation in artificially imprinted sea turtles. *Science* **224:** 83–84.

Gravlund, P. (2001). Radiation within the advanced snakes (Caenophidia) with special emphasis on African opistoglyph colubrids, based on mitochondrial sequence data. *Zoological Journal of the Linnean Society* **72:** 99–114.

Gray, E. M. (1995). DNA fingerprinting reveals a lack of genetic variation in northern populations of the western pond turtle (*Clemmys marmorata*). *Conservation Biology* **9:** 1244–1255.

Gray, R. D. (1987). Faith and foraging: A critique of the 'paradigm argument from design'. In *Foraging Behavior*. A. C. Kamil, J. R. Krebs, and H. R. Pulliam (Eds.). Pp. 69–140. Plenum Press, New York.

Graybeal, A. (1997). Phylogenetic relationships of bufonid frogs and tests of alternate marcoevolutionary hypotheses characterizing their radiation. *Zoological Journal of the Linnean Society* **119:** 297–338.

Graybeal, A., and Cannatella, D. C. (1995). A new taxon of Bufonidae from Peru, with descriptions of two new species and a review of the phylogenetic status of supraspecific bufonid taxa. *Herpetologica* **51:** 105–131.

Green, D. M. (1994). Genetic and cytogenetic diversity in Hochstetter's frog, *Leiopelma hochstetteri*, and its importance for conservation management. *New Zealand Journal of Zoology* **21:** 417–424.

Green, D. M., and Campbell, R. W. (1984). *The Amphibians of British Columbia*. Royal British Columbia Museum Handbook, Victoria.

Green, D. M., and Sessions, S. K. (Eds.). (1991). *Amphibian Cytogenetics and Evolution*. Academic Press, New York.

Green, D. M., Sharbel, T. F., Hitchmough, R. A., and Daugherty, G. H. (1989). Genetic variation in the genus *Leiopelma* and relationships to other primitive frogs. *Zeitschrift für zoologische Systematik und Evolutionsforschung* **27:** 65–79.

Green, D. M., and Cannatella, D. C. (1993). Phylogenetic significance of the amphicoelous frogs, Ascaphidae and Leiopelmatidae. *Ethology Ecology & Evolution* **5:** 233–245.

Greenberg, G., and Noble, G. K. (1944). Social behavior of the American chameleon (*Anolis carolinensis* Voigt). *Physiological Zoology* **17:** 392–439.

Greene, H. W. (1982). Dietary and pheonotypic diversity in lizards: Why are some organisms specialized? *Environmental Adaptation and Evolution, a Theoretical and Empirical Approach.*. D. Mossakowski and G. Roth (Eds.). Pp. 107–128. Gustav Fischer Verlag, Stuttgart.

Greene, H. W. (1986). Natural history and evolutionary biology. In *Predator-prey Relationships: Perspectives and Approaches from the Study of Lower Vertebrates*. M. E. Feder and G. V. Lauder (Eds.). Pp. 99–108. University of Chicago Press, Chicago.

Greene, H. W. (1988). Antipredator mechanisms in reptiles. In *Biology of the Reptilia. Volume 16: Ecology B. Defense and Life History.* C. Gans and R. B. Huey (Eds.). Pp. 1–152. Alan R. Liss, Inc., New York.

Greene, H. W. (1997). *Snakes. The Evolution of Mystery in Nature*. University of California Press, Berkeley.

Greene, H. W., and McDiarmid, R. W. (1981). Coral snake mimicry: Does it occur? *Science* **213:** 1207–1212.

Greene, H. W., May, P. G. D. L., Hardy, S., Sciturro, J. M., and Farrell, T. M. (2002). Parental behavior by vipers. In *Biology of the Vipers*. G. W. Schuett, M. Hoggren, M. E. Douglas, and H. W. Greene (Eds.). Pp. 179–225. Eagle Mountain Publications, Eagle Mountain, Utah.

Greene, H. W., Sigala Rodriguez, J. J., and Powell, B. J. (2006). Parental behavior in anguid lizards. *South American Journal of Herpetology* **1:** 9–19.

Greenlees, M. J., Brown, G. P., Webb, J. K., Phillips, B. L., and Shine, R. (2006). Effects of an invasive anuran (the cane toad, *Bufo marinus*) on the invertebrate fauna of a tropical Australian floodplain. *Animal Conservation* **9:** 431–438.

Greenlees, M. J., Brown, G. P., Webb, J. K., Phillips, B. L., and Shine, R. (2007). Do invasive cane toads (*Chaunus marinus*) compete with Australian frogs (*Cyclorana australis*)? *Austral Ecology* **32:** 900–907.

Greer, A. E. (1970a). Evolutionary and systematic significance of crocodilian nesting habits. *Nature* **227:** 523–524.

Greer, A. E. (1970b). A subfamilial classification of scincid lizards. *Bulletin of the Museum of Comparative Zoology* **139:** 151–183.

Greer, A. E. (1974). The generic relationships of the scincid lizard genus *Leiolopisma* and its relatives. *Australian Journal of Zoology, Supplementary Series* **31:** 1–67.

Greer, A. E. (1975). Clutch size in crocodilians. *Journal of Herpetology* **9:** 319–322.

Greer, A. E. (1976). A most successful invasion: The diversity of Australia's skinks. *Australian Natural History* **18:** 428–433.

Greer, A. E. (1985). The relationships of the lizard genera *Anelytropsis* and *Dibamus*. *Journal of Herpetology* **19:** 116–156.

Greer, A. E. (1989). *The Biology and Evolution of Australian Lizards*. Surrey Beatty & Son, Chipping Norton Pty Ltd, N.S.W.

Greer, A. E., and Biswas, S. (2004). A generic diagnosis for the Southeast Asian scincid lizard Genus *Tropidophorus* Dumeril & Bibron, 1839 with some additional comments on morphology and distribution. *Journal of Herpetology* **38:** 426–430.

Gregory, P. T. (1974). Patterns of spring emergence of red-sided garter snake (*Thamnophis sirtalis parietalis*) in interlake region of Manitoba. *Canadian Journal of Zoology* **52:** 1063–1069.

Gregory, P. T. (1982). Reptilian hibernation. In *Biology of the Reptilia. Volume 13, Physiology D, Physiological Ecology*. C. Gans and F. H. Pough (Eds.). Pp. 53–154. Academic Press, New York.

Gregory, P. T. (1984). Communal denning in snakes. In *Vertebrate Ecology and Systematics: A Tribute to Henry S. Fitch*. R. A. Seigel, L. E. Hunt, J. L. Knight, L. Malaret, and N. L. Zuschlag (Eds.). Pp. 57–75. University of Kansas Museum of Natural History, Special Publication, Lawrence, KS.

Gregory, W. K. (1951). *Evolution Emerging. Vol. II*. Macmillian Co., New York.

Griffith, H., Ngo, A., and Murphy, R. W. (2000). A cladistic evaluation of the cosmopolitan genus *Eumeces* Wiegmann (Reptilia, Squamata, Scincidae). *Russian Journal of Herpetology* **7:** 1–16.

Griffiths, I. (1963). The phylogeny of the Salientia. *Biological Reviews* **38:** 241–292.

Griffiths, R. A. (1996). *Newts and Salamanders of Europe*. Academic Press, London.

Grillitsch, B., and Chovanec, A. (1995). Heavy metals and pesticides in anuran spawn and tadpoles, water, and sediment. *Toxicological and Environmental Chemistry* **50:** 131–155.

Grillitsch, B., Grillitsch, H., Häupl, M., and Tiedemann, F. (1983). *Lurche und Kriechtiere Niederösterreichs.*. Facultas Verlag, Wien.

Grinnell, J. (1917). Field tests of theories concerning distributional control. *American Naturalist* **51:** 115–128.

Grismer, L. L. (1988). Phylogeny, taxonomy, classification, and biogeography of eublepharid geckos. In *Phylogenetic Relationships of the Lizard Families. Essay Commemorating Charles L. Camp*. R. Estes and G. Pregill (Eds.). Pp. 369–469. Stanford University Press, Stanford, CA.

Grismer, L. L. (2002). *Amphibians and Reptiles of Baja California, Including Its Pacific Islands and the Islands in the Sea of Cortés*. University of California Press, Berkeley, CA.

Grismer, L. L., Ota, H., and Tanaka, S. (1994). Phylogeny, classification, and biogeography of *Goniurosaurus kuroiwae* (Squamata: Eublepharidae) from the Ryukyu Archipelago, Japan, with description of a new subspecies. *Zoological Science* **11:** 319–335.

Groot, T. V. M., Bruins, E., and Breeuwer, J. A. J. (2003). Molecular genetic evidence for parthenogenesis in the Burmese python, *Python molurus bivittatus*. *Heredity* **90:** 130–135.

Grosjean, S., Glos, J., Teschke, M., Glaw, F., and Vences, M. (2007). Comparative larval morphology of Madagascan toadlets of the genus *Scaphiophryne*: Phylogenetic and taxonomic inferences. *Zoological Journal of the Linnean Society* **151:** 555–576.

Grosjean, S., Vences, M., and Dubois, A. (2004). Evolutionary significance of oral morphology in the carnivorous tadpoles of tiger frogs, genus *Hoplobatrachus* (Ranidae). *Biological Journal of the Linnean Society* **81:** 171–181.

Gross, M. R., and Shine, R. (1981). Parental care and mode of fertilization in ectothermic vertebrates. *Evolution* **35:** 775–793.

Groves, J. D. (1981). Observations and comments on the post-parturient behavior of some tropical boas of the genus *Epicrates*. *British Journal of Herpetology* **6:** 89–91.

Grubb, P. (1971). The growth, ecology and population structure of giant tortoises on Aldabra. *Philosophical Transactions of the Royal Society of London series B* **260:** 327–372.

Guard, C. I. (1979). The reptilian digestive system: General characteristics. In *Comparative Physiology: Primitive Mammals*. K. Schmidt-Nielsen, L. Bolis, and C. R. Taylor (Eds.). Pp. 43–51. Cambridge University Press, New York.

Guillaume, O. (2000). Role of chemical communication and behavioural interactions among conspecifics in the choice of shelters by the cave-dwelling salamander *Proteus anguinus* (Caudata, Proteidae). *Canadian Journal of Zoology* **78:** 167–173.

Guillaume, O. (2000). The ventral skin glands, new additional cloacal glands in *Proteus anguinus* (Caudata, Proteidae). I. Female. *Acta Zoologica* **81:** 213–221.

Guillette, L. J., Jr. (1987). The evolution of viviparity in fishes, amphibians and reptiles: An endocrine approach. In *Hormones and Reproduction in Fishes, Amphibians, and Reptiles*. D. O. Norris and R. E. Jones (Eds.). Pp. 523–569. Plenum, New York.

Guillette, L. J., Jr. (1995). Endocrine disrupting environmental contaminants and developmental abnormalities in embryos. *Human and Ecological Risk Assessment* **1995:** 25–36.

Guillette, L. J., Jr., and Gongora, G. L. (1986). Notes on oviposition and nesting in the high elevation lizard, *Sceloporus aeneus*. *Copeia* **1986:** 232–233.

Guillette, L. J., Jr., and Méndez-de la Cruz, F. (1993). The reproductive cycle of the viviparous Mexican lizard *Sceloporus torquatus*. *Journal of Herpetology* **27:** 168–174.

Guillette, L. J., Jr., Arnold, S. F., and Mclachlan, J. A. (1996). Ecoestrogens and embryos—Is there a scientific basis for concern? *Animal Reproduction Science* **42:** 13–24.

Guillette, L. J., Jr., Jones, R. E., Fitzgerald, K. T., and Smith, H. M. (1980). Evolution of viviparity in the lizard genus *Sceloporus*. *Herpetologica* **36:** 201–215.

Guimond, R. W., and Hutchison, V. H. (1972). Pulmonary, brachial, and cutaneous gas exchange in the mudpuppy, *Necturus maculosus maculosus* (Rafinesque). *Comparative Biochemisty and Physiology* **42A:** 367–392.

Guisan, A., and Hofer, U. (2003). Predicting reptile distributions at the mesoscale: relation to climate and topography. *Journal of Biogeography* **30:** 1233–1243.

Gundappa, K. R., Balakrishna, R. T. A., and Shakuntala, K. (1981). Ecology of *Ichthyophis glutinosus* (Linn.) (Apoda, Amphibia). *Current Science, Bangalore* **50:** 480–483.

Guttman, S. I. (1985). Biochemical studies of anuran evolution. *Copeia* **1985:** 292–309.

Guyer, C. (1991). Orientation and homing behavior as a measure of affinity for the home range in two species of iguanid lizards. *Amphibia-Reptilia* **12:** 373–384.

Guyer, C., and Savage, J. M. (1987). Cladistic relationships among anoles (Sauria: Iguanidae). *Systematic Zoology* **35:** 509–531.

Gyi, K. K. (1970). A revision of colubrid snakes of the subfamily Homalopsinae. *University of Kansas Publications, Musuem of Natural History* **20:** 47–223.

Haas, A. (2003). Phylogeny of frogs as inferred from primarily larval characters (Amphibia: Anura). *Cladistics* **19:** 23–89.

Haas, G. (1973). Muscles of the jaws and associated structures in the Rhynchocephalia and Squamata. In *Biology of the Reptilia Volume 4. Morphology* D. C. Gans and T. S. Parsons (Eds.). Pp. 285–400. Academic Press, London.

Haddad, C. F. B., and Giaretta, A. A. (1999). Visual and acoustic communication in the Brazilian torrent frog, *Hylodes asper* (Anura: Leptodactylidae). *Journal of Herpetology* **55:** 324–333.

Haddad, C. F. B., and Prado, C. P. A. (2005). Reproductive modes in frogs and their unexpected diversity in the Atlantic Forest of Brazil. *Bioscience* **55:** 724–724.

Haddad, C. F. B., and Sawaya, R. J. (2000). Reproductive modes of Atlantic forest hylid frogs: A general overview and the description of a new mode. *Biotropica* **32:** 862–871.

Haffer, J. (1969). Speciation in Amazonian birds. *Science* **165:** 131–137.

Haffer, J. (1979). Quaternary biogeography of tropical lowland South America. In *The South American herpetofauna: Its origin, evolution, and dispersal*. W. E. Duellman (Ed.). Pp. 107–140. Museum of Natural Historty, University of Kansas Monograph 7.

Haffer, J. (1974). Avian speciation in tropical South America. *Publications of the Nuttall Ornithological Club* **14:** 1–390.

Hagman, M., and Shine, R. (2006). Spawning-site selection by feral cane toads (*Bufo marinus*) at an invasion front in tropical Australia. *Austral Ecology* **31:** 551–558.

Hagman, M., and Shine, R. (2007). Effects of invasive cane toads on Australian mosquitoes: Does the dark cloud have a silver lining? *Biological Invasions* **9:** 445–452.

Hahn, D. E., and Wallach, V. (1998). Comments on the systematics of Old World *Leptotyphlops* (Serpentes: Leptotyphlopidae), with description of a new species. *Hamadryad* **23:** 50–62.

Hahn, W. E. (1967). Estradiol-induced vitellinogenesis and concomitant fat mobilization in the lizard *Uta stansburiana*. *Comparative Biochemistry and Physiology* **23:** 83–93.

Hahn, W. E., and Tinkle, D. W. (1965). Fat body cycling and experimental evidence for its adaptive significance to ovarian follicle development in the lizard *Uta stansburiana*. *Journal of Experimental Zoology* **158:** 79–86.

Hairston, N. G. (1983). Growth, survival and reproduction of *Plethodon jordani*: Trade-offs between selective pressures. *Copeia* **1983:** 1024–1035.

Hairston, N. G., Sr. (1996). Predation and competition in salamander communities. In *Long-term Studies of Vertebrate Communities*. M. L. Cody and J. A. Smallwood (Eds.). Pp. 161–189. Academic Press, San Diego.

Hall, B. K. (2007). Homoplasy and homology: Dichotomy or continuum? *Journal of Human Evolution* **52:** 473–479.

Hall, D. H., and Steidl, R. J. (2007). Movements, activity, and spacing of sonoran mud turtles (*Kinosternon sonoriense*) in interrupted mountain streams. *Copeia* **2007:** 403–412.

Hallermann, J. (1994). Zur Morphologie der Ethmodialregion der Iguania (Squamata) – eine vergleichend-anatomische Untersuchung. *Bonner Zoologische Monographien* **35:** 1–133.

Hallermann, J. (1998). The ethmoidal region of *Dibamus taylori* (Squamata: Dibamidae), with a phylogenetic hypothesis on dibamid relationships within Squamata. *Zoological Journal of the Linnean Society* **122:** 385–426.

Halliday, T. R. (1977). The courtship of European newts: An evolutionary perspective. In *The* Reproductive Biology of Amphibians. D. H. Taylor and S. I. Guttman (Eds.). Pp. 185–232. Plenum Press, New York.

Halliday, T. R., and Heyer, W. R. (1997). The case of the vanishing frogs. *MIT's Technology Review* May/June: 56–63.

Halliday, T. R., and Adler, K. (Eds.) (1986). *The Encyclopedia of Reptiles and Amphibians* Facts on File Inc., New York.

Halliday, T. R., and Tejedo, M. (1995). Intrasexual selection and alternative mating behaviour. In *Amphibian Biology. Volume 2, Social Behaviour*. H. Heatwole and B. K. Sullivan (Eds.).

Pp. 419–468. Surrey Beatty & Sons Pty Ltd, Chipping Norton Pty Ltd, N.S.W.

Halpern, M. (1992). Nasal chemical senses in reptiles: Structure and function. In *Biology of the Reptilia. Volume 18, Physiology. Hormones, Brain, and Behavior*. C. Gans and D. Crews (Eds.). Pp. 422–523. University of Chicago Press, Chicago.

Hamilton, J., and Coe, M. (1982). Feeding, digestion and assimilation of a population of giant tortoises (*Geochelone gigantea*) on Aldabra atoll. *Journal of Arid Environments* **5:** 127–144.

Handrigan, G. R., Haas, A., and Wassersug, R. J. (2007). Bony-tailed tadpoles: The development of supernumerary caudal vertebrae in larval megophryids (Anura). *Evolution & Development* **9:** 190–202.

Hanken, J. (1986). Developmental evidence for amphibian origins. *Evolutionary Biology* **20:** 389–417.

Hanken, J. (1989). Development and evolution in amphibians. *American Scientist* **77:** 336–343.

Hanken, J. and Hall, B. K. (Eds.). (1993). *The Skull. Volumes 1–3*. University of Chicago Press, Chicago.

Hardy, D. L. (1994). A re-evaluation of suffocation as the cause of death during constriction by snakes. *Herpetological Review* **25:** 45–47.

Harlow, P. S. (2004). Temperature-dependent sex determination in lizards. In *Temperature-Dependent Sex Determination in Vertebrates*. N. Valenzuela and V. Lance (Eds.). Pp. 42–58. Smithsonian Books, Washington, D. C.

Harris, D. J., Marshall, J. C., and Crandall, K. A. (2001). Squamate relationships based on C-mos nuclear DNA sequences: Increased taxon sampling improves bootstrap support. *Amphibia-Reptilia* **22:** 235–242.

Harshman, J., Huddleston, C. J., Bollback, J. P., Parsons, T. J., and Braun, M. J. (2003). True and false gharials: A nuclear gene phylogeny of Crocodylia. *Systematic Biology* **52:** 386–402.

Hartline, P. H. (1967). The unbelievable fringed turtle. *International Turtle and Tortoise Society Journal* **1:** 24–29.

Hartmann, M. T., Marques, O. A. V., and Alameida-Santos, S. M. (2005). Reproducive biology of the southern Brazilian pitviiper *Bothrops neuwiedi pubescens* (Serpentes, Viperidae). *Amphibia-Reptilia* **25:** 77–85.

Hass, C. A., Hoffman, M. A., Densmore, L. D., and Maxson, L. R. (1992). Crocodilian evolution: Insights from immunological data. *Molecular Phylogenetics and Evolution* **1:** 193–201.

Hass, C. A., Nussbaum, R. A., and Maxson, L. R. (1993). Immunological insights into the evolutionary history of caecilian (Amphibia: Gymnophiona): Relationships of the Seychellean caecilians and a preliminary report on family-level relationships. *Herpetological Mongraphs* **7:** 56–63.

Hay, J. M., Ruvinsky, I., Hedges, S. B., and Maxson, L. R. (1995). Phylogenetic Relationships of amphibian families inferred from DNA sequences of mitochondrial 12S and 16S ribosomal RNA genes. *Molecular Biology and Evolution* **12:** 928–937.

Hayashi, T., and Matsui, M. (1989). Preliminary study of phylogeny in the family Salamandridae: allozyme data. In *Current Herpetology in East Asia*. M. Matsui, T. Hikida, and R. C. Goris (Eds.). Pp. 157–167. Herpetological Society of Japan, Kyoto.

Hayes, M. P. (1983). Predation on the adults and prehatching stages of glass frogs (Centrolenidae). *Biotropica* **15:** 74–76.

Hayes, T. B. (1997a). Amphibian metamorphosis: An integrative approach. *American Zoologist* **37:** 121–123.

Hayes, T. B. (1997b). Steroids as potential modulators of thyroid hormone activity in anuran metamorphosis. *American Zoologist* **37:** 185–194.

Hayes, T. B., Wu, T. H., and Gill, T. N. (1997). DDT-like effects as a result of coricosterone treatment in an anuran amphibian: Is DDT a corticoid mimic or a stressor? *Environmental Toxicology and Chemistry* **16:** 1948–1953.

Hays, J., Hoffman, P. D., and Blaustein, A. R. (1995). Differences in DNA-repair activity and sensitivity to UV-B light among eggs and oocytes of declining and stable amphibian species. *Journal of Cellular Biochemistry* 296.

Hetherington, T. E. (1985). Role of the opercularis muscle in seismic sensitivity in the bullfrog *Rana catesbeiana*. *Journal of Experimental Zoology* **235:** 27–34.

Heatwole, H. (1999). *Sea Snakes. Second Edition*. Krieger Publ. Co., Malabar, FL.

Heatwole, H., and Barthalmus, G. T. (1994). *Amphibian Biology. Volume 1. The Integument*. Surrey Beatty & Sons Pty Ltd, Chipping Norton Pty Ltd, N.S.W.

Heatwole, H., and Cogger, H. G. (1993). Family Hydrophiidae. In *Fauna of Australia. Volume 2A. Amphibia & Reptilia*. C. J. Gasby, G. J. B. Ross, and P. L. Beesley (Eds.). Pp. 310–318. Australian Government Publishing Service, Canberra.

Heatwole, H., and Guinea, M. L. (1993). Family Laticaudinae. In *Fauna of Australia. Volume 2A. Amphibia & Reptilia*. C. J. Gasby, G. J. B. Ross, and P. L. Beesley (Eds.). Pp. 319–321. Australian Government Publishing Service, Canberra.

Heatwole, H. and Sullivan, B. K. (Eds.) (1995). *Amphibian Biology, Vol. 2, Social Behaviour*. Surrey Beaty & Sons, Chipping Norton, NSW.

Heatwole, H., Firth, B. T., and Stoddart, H. (1975). Influence of season, photoperiod and thermal acclimation on panting threshold of *Amphibolurus muricatus*. *Journal of Experimental Zoology* **191:** 183–192.

Heatwole, H., Firth, B. T., and Webb, G. J. W. (1973). Panting thresholds of lizards I. some methodological and internal influences on panting threshold of an agamid, *Amphibolurus muricatus*. *Comparative Biochemistry and Physiology* **46:** 799–826.

Hedges, S. B. (1986). An electrophoretic analysis of holarctic hylid frog evolution. *Systematic Zoology* **35:** 1–21.

Hedges, S. B. (2001b). Afrotheria: Plate tectonics meets genomics. *Proceedings of the National Academy of Sciences (USA)* **98:** 1–2.

Hedges, S. B. (2001a). Molecular evidence for the early history of living vertebrates. *Major Events in Early Vertebrate Evolution: Palaeontology, Phylogeny, Genetics and Development* **61:** 119–134.

Hedges, S. B. (1996). The origin of West Indian amphibians and reptiles. In *Contributions to West Indian Herpetology: A tribute to Albert Schwartz*. R. Powell and R. W. Henderson

(Eds.). Pp. 95–128. Society for the Study of Amphibians and Reptiles, Ithaca, NY.

Hedges, S. B., and Maxson, L. R. (1996). Re: Molecules and morphology in amniote phylogeny. *Molecular Phylogenetics and Evolution* **6:** 312–314.

Hedges, S. B., and Poling, L. L. (1999). A molecular phylogeny of reptiles. *Science* **283:** 998–1001.

Hedges, S. B., Bezy, R. L., and Maxson, L. R. (1991). Phylogenetic relationships and biogeography of xantusiid lizards, inferred from mitochondrial DNA sequences. *Molecular Biology and Evolution* **8:** 767–780.

Hedges, S. B., Hass, C. A., and Maxson, L. R. (1992). Caribbean biogeography: Molecular evidence for dispersal in West Indian terrestrial vertebrates. *Proceedings of the National Acadademy of Sciences (USA)* **89:** 1909–1913.

Hedges, S. B., Nussbaum, R. A., and Maxson, L. R. (1993). Caecilian phylogeny and biogeography inferred from mitochondrial DNA sequences of the 12S rRNA and 16S rRNA genes (Amphibia: Gymnophiona). *Herpetological Monographs* **7:** 64–76.

Heinicke, M. P., Duellman, W. E., and Hedges, S. B. (2007). Major Caribbean and Central American frog faunas originated by ancient oceanic dispersal. *Proceedings of the National Academy of Sciences (USA)* **104:** 10092–10097.

Heise, P. J., Maxson, L. R., Dowling, H. G., and Hedges, S. B. (1995). Higher-level snake phylogeny inferred from mitochondrial DNA sequences of 12S rRNA and 16S rRNA genes. *Molecular Biology and Evolution* **12:** 259–265.

Hemmer, H. and Alcover, J. A. (Eds.). (1984). *Història Biològica del Ferreret. (Life History of the Mallorcan Midwife Toad).* Editorial Moll, Palma de Mallorca.

Henderson, R. W. (1993). *Corallus. Catalogue of American Amphibians and Reptiles* **572:** 1–2.

Henderson, R. W. (2002). *Neotropical Treeboas, Natural History of the* Corallus hortulanus *Complex.* Kreiger Publishing Company, Malabar, FL.

Henderson, R. W., and Powell, R. (2007). *Biology of the Boas and Pythons.* Eagle Mountain Publishing, LC, Eagle Mountain, Utah.

Hennig, W. (1966). *Phylogenetic Systematics.* University of Chicago Press, Chicago.

Henrici, A. C. (1994). *Tephrodytes brassicarvalis*, new genus and species (Anura: Pelodytidae), from the Arikareean Cabbage Patch Beds of Montana, USA, and pelodytid-pelobatid relationships. *Annals of the Carnegie Museum of Natural History* **63:** 155–183.

Herbeck, L. A., and Larsen, D. R. (1999). Plethodontid salamander response to silvicultural practices in Missouri Ozark forest. *Conservation Biology* **13:** 623–632.

Herbst, L. H. (1994). Fibropapillomatosis of marine turtles. *Annual Review of Fish Diseases* **4:** 389–425.

Herbst, L. H., and Klein, P. A. (1995). Green turtle fibropapillomatosis: Challenges to assessing the role of environmental cofactors. *Environmental Health Perspectives* **103, suppl. 4:** 27–30.

Hermann, H. -W., and Joger, U. (1997). Evolution of viperine snakes. In *Venomous Snakes, Ecology, Evolution and Snakebite.* R. S. Thorpe, W. Wüster, and A. Malhotra (Eds.). Pp. 43–61. Zoological Society of London.

Herrel, A., Reilly, S. M., McBrayer, L. D., and Miles, D. B. (2007). Herbivory and foraging mode in lizards. In *Lizard Ecology: The Evolutionary Consequences of Foraging Mode.* S. M. Reilly, L. D. McBrayer, and D. B. Miles (Eds.). Pp. 209–236. Cambridge University Press, Cambrdge UK.

Herrmann, H. -J. (1993). *Ruder- und Riedfrösche. Baumfrösche mit interessantem Verhalten für attraktive Terriarien.* Tetra-Verlag, Melle, Germany.

Hertz, P. E., Huey, R. B., and Nevo, E. (1982). Fight versus flight: body temperature influences defensive responses of lizards. *Animal Behaviour* **30:** 676–679.

Hertz, P. E., Huey, R. B., and Stevenson, R. D. (1993). Evaluating temperature regulation by field-active ectotherms: The fallacy of the inappropriate question. *American Naturalist* **142:** 796–818.

Hetherington, T. E. (1985). Role of the opercularis muscle in seismic sensitivity in the bullfrog *Rana catesbeiana. Journal of Experimental Zoology* **235:** 27–34.

Hetherington, T. E., Jaslow, A. P., and Lombard, R. E. (1986). Comparative morphology of the amphibian opercularis system: I. General design features and functional interpretation. *Journal of Morphology* **190:** 43–61.

Heulin, B. (1990). comparative-study on eggshell membrane of oviparous and parous populations of the lizard *Lacerta vivipara. Canadian Journal of Zoology* **68:** 1015–1019.

Heyer, W. R. (1967). A herpetofaunal study of an ecological transect through the Cordillera de Tilarán, Costa Rica. *Copeia* **1967:** 259–271.

Heyer, W. R. (1975). A preliminary analysis of the intergeneric relationships of the frog family Leptodactylidae. *Smithsonian Contributions to Zoology* **199:** 1–55.

Heyer, W. R. (1979). Systematics of the *pentadactylus* species group of the frog genus *Leptodactylus* (Amphibia: Leptodactylidae). *Smithsonian Contributions from Zoology* **301:** 1–43.

Heyer, W. R. (1994). *Hyla benitezi* (Amphibia, Anura, Hylidae) - first record for Brazil and its biogeographical significance. *Journal of Herpetology* **28:** 497–499.

Heyer, W. R., and Crombie, R. I. (2005). *Leptodactylus lauramiriamae*, a distinctive new species of frog (Amphibia: Anura: Leptodactylidae) from Rondônia, Brazil. *Proceedings of the Biological Society of Washington* **118:** 590–595.

Heyer, W. R., and Liem, D. S. (1976). Analysis of the intergeneric relationships of the Australian frog family Myobatrachidae. *Smithsonian Contributions from Zoology* **233:** 1–29.

Heyer, W. R., and Maxon, L. R. (1982). Distribution, relationships, and zoogeography of lowland frogs. The *Leptodactylus* complex in South America, with special reference to Amazonia. In *Biological diversification in the tropics.* G. T. Prance (Ed.). Columbia University Press, New York, NY.

Heyer, W. R., Donnelly, M. A., McDiarmid, R. W., Hayek, L.-A. C., and Foster, M. S. (1994).In *Measuring and Monitoring Biological Diversity. Standard Methods for Amphibians* Smithsonian Institute Press, Washington, D.C.

Heyer, W. R., Rand, A. S., Cruz, C. A. G. da, Peixoto, O. L., and Nelson, C. E. (1990). Frogs of Boracéia. *Arquivos de Zoologia* **31:** 231–410.

Hille, B. (1992). *Ionic Channels of Excitable Membranes* Sinauer Associates, Sunderland, MA.

Hillis, D. M. (1991). The phylogeny of amphibians: Current knowledge and the role of cytogenetics. In *Amphibian Cytogenetics and Evolution*. D. M. Green and S. K. Sessions (Eds.). Pp. 7–31. Academic Press, San Diego.

Hillis, D. M. (2007). Making evolution relevant and exciting to biology students. *Evolution* **61:** 1261–1264.

Hillis, D. M., and Green, D. M. (1990). Evolutionary changes of heterogametic sex in the phylogenetic history of amphibians. *Journal of Evolutionary Biology*. **3:** 49–64.

Hillis, D. M., and Moritz, C. (1990). An overview of applications of molecular systematics. In *Molecular Systematics, Second Edition*. D. Hillis and C. Moritz (Eds.). Pp. 502–515. Sinauer Associates, Sunderland, Mass.

Hillis, D. M., and Wilcox, T. P. (2005). Phylogeny of the New World true frogs (*Rana*). *Molecular Phylogenetics and Evolution* **34:** 299–314.

Hillis, D. M., Mable, B. K., Larson, A., Davis, S. K., and Zimmer, E. A. (1996). Nucleic acid IV: Sequencing and cloning. In *Molecular Systematics, Second Edition*. D. M. Hillis, C. Moritz, and B. K. Mable (Eds.). Pp. 321–381. Sinauer Associates, Inc., Sunderland, Mass.

Hillis, D. M., Moritz, C., and Mable, B. K. (Eds.) (1996). *Molecular Systematics, Second Edition*. Sinauer Associates, Inc., Sunderland, Mass.

Hillyard, S. D. (1999). Behavioral, molecular and integrative mechanisms of amphibian osmoregulation. *Journal of Experimental Zoology* **283:** 662–674.

Hillyard, S. D., Hoff, K. V., and Propper, C. (1998). The water absorption response: A behavioral assay for physiological processes in terrestrial amphibians. *Physiological Zoology* **71:** 127–138.

Himstedt, W. (1996). *Die Blindwühlen*. Westarp Wissenschaften, Magdeburg.

Hirayama, R. (1994). Phylogenetic systematics of chelonioid sea turtles. *The Island Arc* **3:** 270–284.

Hirayama, R. (1997). Distribution and diversity of Cretaceous chelonioids. In *Ancient Marine Reptiles*. J. M. Callaway and E. L. Nicholls (Eds.). Pp. 225–241. Academic Press, San Diego.

Hirayama, R. (1998). Oldest known sea turtle. *Nature* **392:** 705–708.

Hirshfield, M. F., and Tinkle, D. W. (1975). Natural selection and the evolution of reproductive effort. *Proceedings of the National Academy of Sciences (USA)* **72:** 2227–2231.

Hödl, W. (1990). Reproductive diversity in Amazonian lowland frogs. *Fortschritte der Zoologie* **38:** 41–60.

Hoegg, S., Vences, M., Brinkmann, H., and Meyer, A. (2004). Phylogeny and comparative substitution rates of frogs inferred from sequences of three nuclear genes. *Molecular Biology and Evolution* **21:** 1188–1200.

Hoff, K. S., and Hillyard, S. D. (1991). Angiotension II stimulates cutaneous drinking in the toad, *Bufo punctatus*: A new perspective on the evolution of thirst in terrestrial vertebrates. *Physiological Zoology* **64:** 1165–1172.

Hoffstetter, R. (1955). Rhynchocephalia, Squamata. Squamata de types modernes. In *Traité de Paléontologie*. V. J. Piveteau (Ed.). Pp. 556–605, 606–662, respectively. Masson et Cie, Édit, Paris.

Hoffstetter, R. (1962). Revue des récentes acquisitions concernant l'historire et la systématique des squamates. *Colloques Internationaux du Centre National de la Recherche Scientifique* **104:** 243–279.

Hoffstetter, R., and Gasc, J.-P. (1969). Vertebrae and ribs of modern reptiles. In *Biology of the Reptilia. Volume 1. Morphology A*, C. Gans, d'A. Bellairs, and T. S. Parsons (Eds.). Pp. 201–310. Academic Press, London.

Hofman, A., Maxson, L. R., and Arntzen, J. W. (1991). Biochemical evidence pertaining to the taxonomic relationships within the family Chamaeleonidae. *Amphibia-Reptilia* **12:** 245–265.

Höggren, M., and Tegelström, H. (1995). DNA fingerprinting shows within-season multiple paternity in the adder (*Vipera berus*). *Copeia* **1995:** 271–277.

Holder, M., and Lewis, P. O. (2003). Phylogeny estimation: Traditional and Bayesian approaches. *Nature Reviews Genetics* **4:** 275–284.

Holman, J. A. (1995). *Pleistocene Amphibians and Reptiles in North America*. Oxford University Press, New York.

Holmberg, A. R. (1957). Lizard hunts on the north coast of Peru. *Fieldiana Anthropology, Chicago Natural History Museum* **36:** 203–220.

Holmstrom, W. F. (1978). Preliminary observations on prey herding in the matamata turtle, *Chelus fimbriatus* (Reptilia, Testudines, Chelidae*). Journal of Herpetology* **12:** 573–674.

Honda, M., Ota, H., Kobayashi, Nabhitabhata, J., M., Yong, H.-S., and Hikida, T. (1999). Phylogenetic relationships of the flying lizards, genus *Draco* (Reptilia: Agamidae). *Zoological Science* **16:** 535–549.

Honda, M., Ota, H., Kobayashi, Nabhitabhata, J., M., Yong, H. -S., Sengpku, S., and Hikida, T. (2000). Phylogenetic relationships of the family Agamidae (Reptilia: Iguania) inferred from mitochondrial DNA sequences. *Zoological Science* **17:** 527–537.

Hoogmoed, M. S. (1969). Notes on the herpetofauna of Surinam. II. On the occurrence of *Allophryne ruthveni* Gaige (Amphibia, Salientia, Hylidae) in Surinam. *Zoologische Mededelingen* **44:** 75–81.

Hoogmoed, M. S., and Cadle, J. E. (1990). Natural history and distribution of *Agalychnis craspedopus* (Funkhouser, 1957) (Amphibia: Anura: Hylidae). *Zoologische Mededelingen* **65:** 129–142.

Horne, E. A., and Jaeger, R. G. (1988). Territorial pheromones of female red-backed salamanders. *Ethology* **78:** 143–152.

Hotton, N., MacLean, P. D., Roth, J. J., and Roth, E. C. (1986). *The Ecology and Biology of Mammal-like Reptiles*. Smithsonian Institute Press, Washington, D.C.

Hotton, N., Olson, E. C., and Beerbower, R. (1997). Amniote origins and the discovery of herbivory. In *Amniote Origins. Completing the Transition to Land*. S. S. Sumida and K. L. M. Martin (Eds.). Pp. 207–264. Academic Press, San Diego.

Houck, L. D., and Arnold, S. J. (2003). Courtship and mating behavior. In *Reproductive Biology and Phylogeny of Urodela*. D. M. Server (Ed.). Pp. 383–424. Science Publishers, Inc. Enfield, NH.

Houck, L. D., and Sever, D. M. (1994). Role of the skin in reproduction and behaviour. In *Amphibian Biology. Volume 1, The Integument*. H. Heatwole and G. T. Barthalmus (Eds.). Pp. 351–381. Surrey Beaty & Sons, Chipping Norton Pty Ltd, N.S.W.

Houck, L. D., and Verrell, P. A. (1993). Studies of courtship behavior in plethodontid salamanders: A review. *Herpetologica* **49:** 175–184.

Hourdry, J., and Beaumont, A. (1985). *Les Métamorphoses des Amphibien*. Masson, Paris.

Howard, R. D. (1978a). The influence of male-defended oviposition sites on early embryo mortality in bullfrogs. *Ecology* **59:** 789–798.

Howard, R. D. (1978b). The evolution of mating strategies in bullfrogs, *Rana catesbeiana*. *Evolution* **32:** 850–871.

Howard, R. R., and Brodie, E. D., Jr. (1973). A Batesian mimetic complex in salamanders: Responses of avian predators. *Herpetologica* **29:** 33–41.

Howell, K. M. (2000). An overview of East African amphibian studies, past, present and future: A view from Tanzania. *African Journal of Herpetology* **49:** 147–164.

Howes, B. J., Lindsay, B., and Lougheed, S. C. (2006). Range-wide phylogeography of a temperate lizard, the five-lined skink (*Eumeces fasciatus*). *Molecular Phylogenetics and Evolution* **40:** 183–194.

Howland, J. M. (1992). Life history of *Cophosaurus texanus* (Sauria: Iguanidae): Environmental correlates and interpopulational variation. *Copeia* **1992:** 82–93.

Howland, J. M., Vitt, L. J., and Lopez, P. T. (1990). Life on the edge: The ecology and life history of the tropidurine iguanid lizard *Uranoscodon superciliosum*. *Canadian Journal of Zoology* **68:** 1366–1373.

Hua, S., and Buffetaut, E. (1997). Part V: Crocodylia. Introduction. In *Ancient Marine Reptiles*. J. M. Callaway and E. L. Nicholls (Eds.). Pp. 357–374. Academic Press, San Diego.

Huang, W.-S. (2006). Parental care in the long-tailed skink, *Mabuya longicaudata*, on a tropical Asian island. *Animal Behaviour* **72:** 791–795.

Huang, W.-S. (1997a). Reproductive cycle of the skink, *Sphenomorphus taiwanensis*, in central Taiwan. *Journal of Herpetology* **31:** 287–290.

Huang, W.-S. (1997b). Reproductive cycle of the oviparous lizard *Japalura brevipes* (Agamidae: Reptilia) in Taiwan, Republic of China. *Journal of Herpetology* **31:** 22–29.

Huang, W.-S. (1998). Reproductive cycles of the grass lizard, *Takydromus hsuehshanensis*, with comments on reproductive patterns of lizards from the central high elevation area of Taiwan. *Copeia* **1998:** 866–873.

Huey, R. B. (1982). Temperature, physiology, and the ecology of reptiles. In *Biology of the Reptilia. Volume 12, Physiology C. Physiological Ecology*. C. Gans and F. H. Pough (Eds.). Pp. 25–91. Academic Press, London.

Huey, R. B., and Pianka, E. R. (1981a). Ecological consequences of foraging mode. *Ecology* **62:** 991–999.

Huey, R. B., and Pianka, E. R. (1981b). Natural selection for juvenile lizards mimicking noxious beetles. *Science* **195:** 201–203.

Huey, R. B., and Slatkin, M. (1976). Costs and benefits of lizard thermoregulation. *The Quarterly Review of Biology* **51:** 363–384.

Huey, R. B., and Stevenson, R. D. (1979). Integrating thermal physiology and ecology of ectotherms: A discussion of approaches. *American Zoologist* **19:** 357–366.

Huey, R. B., Bennett, A. F., John-Alder, H., and Nagy, K. A. (1984). Locomotor capacity and foraging behaviour of Kalahari lacertid lizards. *Animal Behaviour* **32:** 41–50.

Huey, R. B., Niewiarowski, P. H., Kaufmann, J., and Herron, J. C. (1989b). Thermal biology of nocturnal ectotherms: Is sprint performance of geckos maximal at low body temperatures? *Physiological Zoology* **62:** 488–504.

Huey, R. B., Peterson, C. R., Arnold, S. J., and Porter, W. P. (1989a). Hot rocks and not-so-hot rocks: Retreat-site selection by garter snakes and its thermal consequences. *Ecology* **70:** 931–944.

Huey, R. B., Pianka, E. R., and Schoener, T. W. (Eds.) (1983). *Lizard Ecology. Studies of a Model Organism*. Harvard University Press, Cambridge, Mass.

Huey, R. B., Pianka, E. R., Egan, M. E., and Coons, L. W. (1974). Ecological shifts in sympatry: Kalahari fossorial lizards (*Typhlosaurus*). *Ecology* **55:** 304–316.

Hughes, G. M. (Ed.) (1976). *Respiration of Amphibious Vertebrates*. Academic Press, London.

Humphries, C. J., and Parenti, L. R. (1999). *Cladistic Biogeography: Interpreting Patterns of Plant and Animal Distributions*. (2nd Edition). Oxford University Press, Oxford.

Humphries, R. L. (1956). An unusual aggregation of *Plethodon glutinosus* and remarks on its subspecific status. *Copeia* **1956:** 122–123.

Humraskar, D., and Velho, N. (2007). The need for studies on amphibians in India. *Current Science* **92:** 1032.

Hunt, J. A., and Farrar, E. S. (2003). Phenotypic plasticity of the digestive system of tadpoles in response to diet. *Integrative and Comparative Biology* **43:** 934.

Hunt, R. H., and Ogden, J. J. (1991). Selected aspects of the nesting ecology of American alligators in Okefenokee Swamp. *Journal of Herpetology* **24:** 448–453.

Hutchinson, G. E. (1957). Concluding remarks. *Cold Spring Harbor Symposia on Quantitative Biology* **22:** 415–427.

Hutchinson, J. H. (1991). Early Kinosterninae (Reptilia: Testudines) and their phylogenetic significance. *Journal of Vertebrate Paleontology* **11:** 145–167.

Hutchinson, J. H. (1992). Western North American reptile and amphibian record across the Eocene/Oligocene boundary and its climatic implications. In *Eocene-Oligocene Climatic and Biotic Evolution*. D. R. Prothero and W. A. Berggen (Eds.). Pp. 451–463. Princeton University Press, Princeton.

Hutchinson, M. N. (1992). Origins of the Australian scincid lizards: A preliminary report on the skinks of Riversleigh. *The Beagle* **9:** 61–70.

Hutchinson, M. N. (1993). Family Scincidae. In *Fauna of Australia. Volume 2A. Amphibia & Reptilia*. C. J. Gasby, G. J. B. Ross, and P. L. Beesley (Eds.). Pp. 261–279. Australian Government Publishing Service, Canberra.

Hutchison, V. H., and Dupré, R. K. (1992). Thermoregulation. In *Environmental Physiology of the Amphibians*. M. E. Feder and W. W. Burggren (Eds.). Pp. 206–249. University of Chicago Press, Chicago.

Hutchison, V. H., and Larimer, J. (1960). Reflectivity of the integuments of some lizards from different habitats. *Ecology* **41:** 199–209.

Hutchison, V. H., Dowling, H. G., and Vinegar, A. (1966). Thermoregulation in a brooding female Indian python, *Python molurus bivittatus*. *Science* **151:** 694–696.

Hutchison, V. H., Haines, H. B., and Engbretson, G. A. (1976). Aquatic life at high altitude: Respiratory adaptations in the Lake Titicaca frog, *Telmatobius culeus*. *Respiration Physiology* **27:** 115–129.

Hutton, J. M. (1989). Production efficiency at crocodile rearing stations in Zimbabwe. In *Crocodiles. Proceedings of the 8th Working Meeting of the Crocodile Specialist Group, IUCN* Pp. 1–15. The World Conservation Union, Gland, Switzerland.

Huyghe, K., Vanhooydonck, B., Scheers, H., Molina-Borja, M., and Van Damme, R. (2005). Morphology, performance and fighting capacity in male lizards, *Gallotia galloti*. *Functional Ecology* **19:** 800–807.

Ineich, I. (1992). La parthenogenese chez les Gekkonidae (Reptilia, Lacertilia): Origine et evolution. *Bulletin de la Société zoologique de France* **117:** 253–266.

Inger, R. F. (1954). Systematics and zoogeography of Philippine Amphibia. *Fieldiana: Zoology* **33:** 181–531.

Inger, R. F. (1966). The systematics and zoogeography of the Amphibia of Borneo. *Fieldiana: Zoology* **52:** 1–402.

Inger, R. F. (1967). The development of a phylogeny of frogs. *Evolution* **21:** 369–384.

Inger, R. F. (1980). Densities of floor-dwelling frogs and lizards in lowland forests of southeast Asia and Central America. *American Naturalist* **115:** 761–770.

Inger, R. F. (1986). Diets of tadpoles living in a Bornean rain forest. *Alytes* **5:** 153–164.

Inger, R. F. (1996). Commentary on a proposed classification of the family Ranidae. *Herpetologica* **52:** 241–246.

Inger, R. F. (1999). Distribution patterns of amphibians in southern Asia and adjacent islands. In *Patterns of Distribution of Amphibians*. W. E. Duellman (Ed.). Pp. 445–482. Johns Hopkins University Press, Baltimore, MD.

Inger, R. F., and Bacon, J. P. (1968). Annual reproduction and clutch size in rain forest frogs from Sarawak. *Copeia* **3:** 602–606.

Inger, R. F., and Colwell, R. K. (1977). Organization of contiguous communities of amphibians and reptiles in Thailand. *Ecological Monographs* **47:** 229–253.

Inger, R. F., and Greenberg, B. (1966). Annual reproductive patterns of lizards from a Bornean rain forest. *Ecology* **47:** 1007–1021.

Inger, R. F., and Lian, T. F. (1996). *The Natural History of Amphibians and Reptiles in Sabah*. Natural History Publications (Borneo) Sdn. Bhd., Kota Kinabalu, Sabah.

Inger, R. F., Shaffer, H. B., Koshy, M., and Bakde, R. (1984). A report on a collection of amphibians and reptiles from the Ponmudi, Kerala, South India. *Journal of the Bombay Natural History Society* **81:** 551–570.

Inger, R. R., and Stuebing, R. B. (1997). *A Field Guide to the Frogs of Borneo*. Natural History Publ. (Borneo) Sdn. Bhd., Kota Kinabalu, Sabah.

Iordansky, N. N. (1973). The skull of the Crocodilia. In *Biology of the Reptilia. Volume 4. Morphology* D. C. Gans and T. S. Parsons (Eds.). Pp. 201–262. Academic Press, London.

Irish, F. J., Williams, E. E., and Seling, E. (1988). Scanning electron microscopy of changes in epidermal structure occurring during the shedding cycle in squamate reptiles. *Journal of Morphology* **197:** 105–126.

Irschick, D. J., Vitt, L. J., Zani, P. A., and Losos, J. B. (1997). A comparison of evolutionary radiations in mainland and Caribbean *Anolis* lizards. *Ecology* **78:** 2191–2203.

Irwin, W. P., Amy, J. H., and Lohmann, K. J. (2004). Magnetic field distortions produced by protective cages around sea turtle nests: Unintended consequences for orientation and navigation? *Biological Conservation* **118:** 117–120.

Iskandar, D. T. (1995). Notes on the second specimen of *Barbourula kalimantanensis* (Amphibia: Anura: Discoglossidae). *Raffles Bulletin of Zoology* **43:** 309–311.

Iverson, J. B. (1978). Impact of feral cats and dogs on populations of West-Indian rock iguana, *Cyclura carinata*. *Biological Conservation* **14:** 63–73.

Iverson, J. B. (1979). Behavior and ecology of the rock iguana *Cyclura carinata*. *Bulletin of the Florida State Museum, Biological Sciences* **24:** 175–358.

Iverson, J. B. (1982). Adaptations to herbivory in iguanine lizards. In *Iguanas of the World: Their Behavior, Ecology, and Conservation*. G. M. Burghardt and A. S. Rand (Eds.). Pp. 60–67. Noyes Publications, Park Ridge, NJ.

Iverson, J. B. (1990). Nesting and parental care in the mud turtle, *Kinosternon flavescens*. *Canadian Journal of Zoology* **68:** 230–233.

Iverson, J. B. (1991a). Phylogenetic hypotheses for the evolution of modern kinosternine turtles. *Herpetological Monographs* **5:** 1–27.

Iverson, J. B. (1991b). Life history and demography of the yellow mud turtle, *Kinosternon flavescens*. *Herpetologica* **47:** 373–395.

Iverson, J. B. (1992). *A Revised Checklist with Distribution Maps of the Turtles of the World*. Privately Printed, Richmond, Ind.

Iverson, J. B. (1998). Molecules, morphology, and mud turtle phylogenetics. *Chelonian Conservation Biology* **3:** 113–117.

Iverson, J. B., and Mamula, M. R. (1989). Natural growth in the Bahamian iguana *Cyclura cychlura*. *Copeia* **1989:** 502–505.

Iverson, J. B., Balgooyen, C. P., Byrd, K. K., and Lyddan, K. K. (1993). Latidtudinal variation in egg and clutch size in turtles. *Canadian Journal of Zoology* **71:** 2448–2461.

Iverson, J. B., Higgins, H., Sirulnik, A., and Griffiths, C. (1997). Local and geographic variation in the reproductive biology of the snapping turtle (*Chelydra serpentina*). *Herpetologica* **53:** 96–117.

Iwabe, N., Hara, Y., Kumazawa, Y., Shibamoto, K., Saito, Y., Miyata, T., and Katoh, K. (2005). Sister group relationship of turtles to the bird-crocodilian clade revealed by nuclear DNA-coded proteins. *Molecular Biology and Evolution* **22:** 810–813.

Izecksohn, E. (1993). Três novas espécies de *Dendrophryniscus* Jiménez de la Espada das regioes sudeste e sul do Brasil (Amphibia, Anura, Bufonidae). *Revista Brasileira de Zoologia* **10:** 473–488.

Jackman, T. R., Applebaum, G., and Wake, D. B. (1997a). Phylogenetic relationships of bolitoglossine salamanders: A demonstration of the effects of combining morphological and molecular data sets. *Molecular Biology and Evolution* **14:** 883–891.

Jackman, T. R., Larson, A., de Queiroz, K., and Losos, J. B. (1999). Phylogenetic relationships and tempo of early diversification in *Anolis* lizards. *Systematic Biology* **48:** 254–285.

Jackman, T., Losos, J. B., Larson, A., and de Queiroz, K. (1997b). Phylogenetic studies of convergent adaptive radiations in Caribbean *Anolis* lizards. In *Molecular Evolution and Adaptive Radiation*. T. J. Givnish and K. J. Sytsma (Eds.). Pp. 535–557. Cambridge University Press, New York.

Jackson, C. G., Jr., and Davis, J. D. (1972). Courtship display behavior of *Chrysemys concinna suwanniensis*. *Copeia* **1972:** 385–387.

Jacobson, E. R., and Whitford, W. G. (1971). Physiological responses to temperature in the patch-nosed snake *Salvadora hexalepis*. *Herpetologica* **27:** 289–295.

Jacobson, S. K. (1985). Reproductive behavior and male mating success in two species of glass frogs (Centrolenidae). *Herpetologica* **41:** 396–404.

Jaeger, C. B., and Hillman, D. E. (1976). Morphology of gustatory organs. In *Frog Neurobiology. A Handbook*. R. Llinás and W. Precht (Eds.). Pp. 588–606. Springer-Verlag, Berlin.

Jaeger, R. G. (1976). A possible prey-call window in anuran auditory perception. *Copeia* **1976:** 833–834.

Jaeger, R. G. (1981). Dear enemy recognition and the costs of aggression between salamanders. *American Naturalist* **117:** 962–974.

Jaeger, R. G. (1986). Pheromonal markers as territorial advertisement by terrestrial salamanders. In *Chemical Signals in Vertebrates 4*. D. Duvall, D. Müller-Schwarze, and R. M. Silverstein (Eds.). Pp. 191–203. Plenum Press, New York.

Jaeger, R. G., and Forester, D. C. (1993). Social behavior of plethodontid salamanders. *Herpetologica* **49:** 163–175.

Jaeger, R. G., Goy, J., Tarver, M., and Marquez, C. (1986). Salamander territoriality: Pheromonal markers as advertisement by males. *Animal Behaviour* **34:** 860–864.

Jaeger, R. G., Kalvarsky, D., and Shimizu, N. (1982). Territorial behaviour of the red-backed salamander: Expulsion of intruders. *Animal Behaviour* **30:** 490–496.

Jaffe, M. (1994). *and No Birds Sing. The Story of an Ecological Disaster in a Tropical Paradise*. Simon & Schuster, New York.

Jaksíc, F. M., and Schwenk, K. (1983). Natural history observations on *Liolaemus magellanicus*, the southernmost lizard in the world. *Herpetologica* **39:** 457–461.

James, C. D. (1991). Annual variation in reproductive cycles of scincid lizards (*Ctenotus*) in central Australia. *Copeia* **1991:** 744–760.

James, C. D., and Shine, R. (1985). The seasonal timing of reproduction: A tropical-temperate comparison in Australian lizards. *Oecologia* **67:** 464–474.

James, C. D., and Shine, R. (1988). Life-history strategies of Australian lizards: A comparison between the tropics and the temperate zone. *Oecologia* **75:** 307–316.

James, F. C., and McCullough, C. E. (1985). Data analysis and the design of experiments in ornithology. In *Current Ornithology. Vol. 2*. R. F. Johnston (Ed.). Pp. 1–63.

James, F. C., and McCullough, C. E. (1990). Multivariate analysis in ecology and systematics: Panacea or Pandora's box? *Annual Review of Ecology and Systematics* **21:** 129–166.

Jameson, D. L. (1955). The population dynamics of the cliff frog, *Syrrhophus marnocki*. *American Midland Naturalist* **54:** 342–381.

Janzen, F. J., and Krenz, J. G. (2004). Phylogenetics: Which was first, TSD or GSD? In *Temperature-Dependent Sex Determination in Vertebrates*. N. Valenzuela and V. Lance (Eds.). Pp. 121–130. Smithsonian Books, Washington, D. C.

Janzen, F. J., and Paukstis, G. (1991). Environmental sex determination in reptiles: Ecology, evolution, and experimental design. *Quarterly Review of Biology* **66:** 149–179.

Janzen, F. J., and Phillips, P. C. (2006). Exploring the evolution of environmental sex determination, especially in reptiles. *Journal of Evolutionary Biology* **19:** 1775–1784.

Janzen, F. J., Krenz, J. G., Haselkorn, T. S., Brodie, E. D. Jr., and Brodie, E. D., III. (2002). Molecular phylogeography of common garter snakes (*Thamnophis sirtalis*) in western North America: Implications for regional historical forces. *Molecular Ecology* **11:** 1739–1751.

Jenkins, F. A., and Walsh, D. M. (1993). An Early Jurassic caecilian with limbs. *Nature* **365:** 246–249.

Jennings, M. R., and Hayes, M. P. (1985). Pre-1900 overharvest of California red-legged frogs (*Rana aurora draytonii*): The inducement for bullfrog (*Rana catesbeiana*) introduction. *Herpetologica* **41:** 94–103.

Jennings, W. B., and Pianka, E. R. (2004). Tempo and timing of the Australian *Varanus* radiation. In *Varanoid Lizards of the World*. E. R. Pianka and D. R. King (Eds.). Indiana University Press, Bloomington, IN.

Jennings, W. B., Pianka, E. R., and Donnellan, S. (2003). Systematics of lizards in the family Pygopodidae with implications for the diversification of Australian temperate biotas. *Systematic Biology* **52:** 757–780.

Jenssen, T. A. (1970a). The ethoecology of *Anolis nebulosus* (Sauria: Iguanidae). *Journal of Herpetology* **4:** 1–38.

Jenssen, T. A. (1970b). Female response to filmed displays of *Anolis nebulosus* (Sauria, Iguanidae). *Animal Behaviour* **18:** 640–647.

Jenssen, T. A., and Hover, E. L. (1976). Display analysis of the signature display of *Anolis limifrons* (Sauria: Iguanidae). *Behaviour* **57:** 227–240.

Jenssen, T. A., and Nunez, S. C. (1998). Spatial and breeding relationships of the lizard *Anolis carolinensis*: Evidence of intrasexual selection. *Behaviour* **135:** 981–1003.

Jenssen, T. A., Decourcy, K. R., and Congdon, J. D. (2005). Assessment in contests of male lizards (*Anolis carolinensis*): How should smaller males respond when size matters? *Animal Behaviour* **69:** 1325–1336.

Jiang, J. P., and Zhou, K. Y. (2005). Phylogenetic relationships among Chinese ranids inferred from sequence data set of 12S and 16S rDNA. *Herpetological Journal* **15:** 1–8.

Joanen, T. (1969). Nesting ecology of alligators in Louisiana. *Proceedings of the Annual Conference of the Southeastern Association of Game and Fish Commissioners* **23:** 141–151.

Joanen, T., and McNease, L. (1980). Reproductive biology of the American alligator in soutwest Lousiana. In *Reproductive Biology and Disease of Captive Reptiles*. J. B. Murphy and J. T. Collins (Eds.). Pp. 153–159. Society for the Study of Amphibians and Reptiles, Athens, Ohio.

Joger, U. (1991). A molecular phylogeny of agamid lizards. *Copeia* **1991:** 616–622.

Joglar, R. L., and Burrowes, P. A. (1996). Declining amphibian populations in Puerto Rico. In *Contribution to West Indian Herpetology. A Tribute to Albert Schwartz*. R. Powell and R. W. Henderson (Eds.). Pp. 371–380. Society for the Study of Amphibians and Reptiles, Ithaca, NY.

Joglar, R. L., Burrowes, P. A., and Rios, N. (1996). Biology of the Purto Rican cave-dwelling frog, *Eleutherodactylus cooki*, and some recommendations for conservation. In *Contribution to West Indian Herpetology. A Tribute to Albert Schwartz*. R. Powell and R. W. Henderson (Eds.). Pp. 251–258. Society for the Study of Amphibians and Reptiles, Ithaca, NY.

Johnson, C. R. (1969). Aggregation as a means of water conservation in juvenile *Limnodynastes* from Australia. *Herpetologica* **25:** 275–276.

Johnson, S. A. (2003). Orientation and migration distances of a pond-breeding salamander (Salamandridae, *Notophthalmus perstriatus*). *Alytes* **21:** 3–22.

Johnston, G. R., and Richards, S. J. (1993). Observations on the breeding biology of a microhylid frog (genus *Oreophryne*) from New Guinea. *Transactions of the Royal Society of South Australia* **117:** 105–107.

Joly, J. (1968). Données écologiques sur la salamandre tachetée *Salamandra Salamandra*. *Annalles Sciences Naturelles Zoologiques* **12:** 301–366.

Jones, D. R. (1996). The crocodilian central circulation: Reptilian or avian? *Verhandlungen der Deutschen Zoologischen Gesellschaft* **89:** 209–218.

Jones, T. R., Kluge, A. G., and Wolf, A. J. (1993). When theories and methodology clash: A phylogenetic reanalysis of the North American ambystomatid salamanders (Caudata: Ambystomatidae). *Systematic Biology* **42:** 92–102.

Jorgensen, C. B. (1992). Growth and reproduction. In *Environmental Physiology of the Amphibians*. M. E. Feder and W. W. Burggren (Eds.). Pp. 439–466. University of Chicago Press, Chicago.

Jorgensen, C. B. (1997). 200 years of amphibian water economy: From Robert Townson to the present. *Biological Reviews of the Cambridge Philosophical Society* **72:** 153–237.

Joyce, W. G., Parham, J. F., and Gauthier, J. A. (2004). Developing a protocol for the conversion of rank-based taxon names to phylogenetically defined clade names, as exemplified by turtles. *Journal of Paleontology* **78:** 989–1013.

Juncá, F. A. (1998). Reproductive biology of *Colostethus stepheni* and *Colostethus marchesianus* (Dendrobatidae), with the description of a new anuran mating behavior. *Herpetologica* **54:** 377–387.

Juncá, F. A., Altig, R., and Gascon, C. (1994). Breeding biology of *Colostethus stepheni*, a dendrobatid frog with a nontransported nidicolous tadpole. *Copeia* **3:** 747–750.

Jungfer, K.-H. (1996). Reproduction and parental care of the coronated treefrog, *Anotheca spinosa* (Steindachner, 1864) (Anura: Hylidae). *Herpetologica* **52:** 25–32.

Jungfer, K.-H., and Schiesari, L. C. (1995). Description of a central Amazonian and Guianan tree frog, genus *Osteocephalus* (Anura, Hylidae), with oophagous tadpoles. *Alytes* **13:** 1–13.

Jungfer, K.-H., and Weygoldt, P. (1999). Biparental care in the tadpole-feeding Amazonian treefrog *Osteocephalus oophagus*. *Amphibia-Reptilia* **20:** 235–249.

Junqueira, L. C. U., Jared, C., and Antoniazzi, M. M. (1999). Structure of the caecilian *Siphonops annulatus* (Amphibia, Gymnophiona): General aspect of the body, disposition of the organs and structure of the mouth, oesophagus and stomach. *Acta Zoologica* **80:** 75–84.

Justus, J. T., Sandomir, M., Urquhart, T., and Ewan, B. O. (1977). Developmental rates of two species of toads from the desert Southwest. *Copeia* **1977:** 592–594.

Kachigan, S. K. (1982). *Multivariate Statistical Analysis. A Conceptual Introduction*. Radius Press, New York.

Kalb, H. J., and Zug, G. R. (1990). Age estimates for a population of American toads, *Bufo americanus* (Salientia: Bufonidae), in northern Virginia. *Brimleyana* **16:** 79–86.

Kaminsky, S. D., Linsenmair, K. E., and Grafe, T. U. (1999). Reproductive timing, nest construction and tadpole guidance in the African pig-nosed frog, *Hemisus marmoratus*. *Journal of Herpetology* **33:** 119–123.

Kardong, K. V. (1992). Proximate factors affecting guidance of the rattlesnake strike. *Zoologische Jahrbuecher Abteilung fuer Anatomie und Ontogenie der Tiere* **122:** 233–244.

Kardong, K. V. (1995). *Vertebrates: Comparative Anatomy, Function, Evolution*. William C. Brown Publishers, Dubuque, IA.

Kardong, K. V. (1997). Squamates at geological boundaries: Unbounded evolution. *Zoology* **100:** 152–163.

Kardong, K. V. (1998). *Vertebrates. Comparative Anatomy, Function, Evolution*. Second Edition. McGraw-Hill, New York.

Kardong, K. V. (2006). *Vertebrates. Comparative Anatomy, Function, Evolution*. Fourth Edition. McGraw Hill, Boston, MA.

Kardong, K. V., and Bels, V. L. (1998). Rattlesnake strike behavior: Kinematics. *Journal of Experimental Biology* **201:** 837–850.

Kareiva, P. (2008). Ominous trends in nature recreation. *Proceedings of the National Academy of Sciences (USA)* **105:** 2757–2758.

Karraker, N. E., Pilliod, D. S., Adams, M. J., Bull, E. L., Corn, P. S., Diller, L. V., Dupuis, L. A., Hayes, M. P., Hossack, B. R., Hodgson, G. R., Hyde, E. J., Lohman, K., Norman, B. R., Ollivier, L. M., Pearl, C. A., and Peterson, C. R. (2006). Taxonomic variation in oviposition by tailed frogs (*Ascaphus* spp.). *Northwestern Naturalist* **87:** 87–97.

Kats, L. B., Petranka, J. W., and Sih, A. (1988). Antipredator defenses and the persistence of amphibian larvae with fishes. *Ecology* **69:** 1865–1870.

Katz, U. (1989). Strategies of adaptation to osmotic stress in anuran Amphibia under salt and burrowing conditions. *Comparative Biochemistry and Physiology* **93A:** 499–503.

Kearney, M. (2003a). Diet in the amphisbaenian *Bipes biporus*. *Journal of Herpetology* **37:** 404–408.

Kearney, M. (2003b). Systematics and evolution of the Amphisbaenia: A phylogenetic hypothesis based on morphological evidence from fossil and recent forms. *Herpetological Monographs* **17:** 1–75.

Kearney, M., and Shine, R. (2004). Developmental success, stability, and plasticity in closely related parthenogenetic and sexual lizards (*Heteronotia*, Gekkonidae). *Evolution* **58:** 1560–1572.

Kearney, M., and Stuart, B. L. (2004). Repeated evolution of limblessness and digging heads in worm lizards revealed by DNA from old bones. *Proceedings of the Royal Society of London Series B-Biological Sciences* **271:** 1677–1683.

Kearney, M., Blacket, M. J., Strasburg, J. L., and Moritz, C. (2006). Waves of parthenogenesis in the desert: Evidence for the parallel loss of sex in a grasshopper and a gecko from Australia. *Molecular Ecology* **15:** 1743–1748.

Kelly, C. M. R., Barker, N. P., and Villet, M. H. (2003). Phylogenetics of advanced snakes (Caenophidia) based on four mitochondrial genes. *Systematic Biology* **52:** 439–459.

Kennett, R., Christian, K., and Pritchard, D. (1993a). Underwater nesting of the tropical freshwater turtle, *Chelodina rugosa*. *Australian Journal of Zoology* **41:** 47–52.

Kennett, R., Georges, A., and Palmer-Allen, M. (1993b). Early develomental arrest during immersion of eggs of a tropical freshwater turtle, *Chelodina rugosa* (Testudinata: Chelidae), from northern Australia. *Australian Journal of Zoology* **41:** 37–45.

Keogh, J. S. (1998). Molecular phylogeny of elapid snakes and a consideration of their biogeographic history. *Biological Journal of the Linnean Society* **63:** 177–203.

Kiester, A. R., Gorman, G. C., and Arroyo, D. C. (1975). Habitat selection behavior of three species of *Anolis* lizards. *Ecology* **56:** 220–225.

King, D., and Green, B. (1993). *Goanna. The Biology of Varanid Lizards*. New S. Wales University Press, Kensington, N.S.W.

King, M., and Horner, P. (1993). Family Gekkonidae. In *Fauna of Australia. Volume 2A. Amphibia & Reptilia*. C. J. Gasby, G. J. B. Ross, and P. L. Beesley (Eds.). Pp. 221–233. Australian Government Publishing Service, Canberra.

Kinkead, R. (1997). Episodic breathing in frogs: Converging hypotheses on neural control of respiration in air breathing vertebrates. *American Zoologist* **37:** 31–40.

Kitimasak, W., Thirakhupt, K., and Moll, D. L. (2003). Eggshell structure of the Siamese narrow-headed softshell turtle *Chitra chitra* Nupthand, 1986 (Testudines: Trionychidae). *ScienceAsia* **29:** 95–98.

Kizirian, D. A. (1996). A review of Ecuadorian *Proctoporus* (Squamata: Gymnophthlamidae) with descriptions of nine new species. *Herpetological Monographs* **10:** 85–155.

Klauber, L. M. (1982). *Rattlesnakes. Their Habits, Life Histories, and Influence on Mankind*. Abridged Edition. University of California Press, Berkeley.

Klaver, C. J. J., and Böhme, W. (1986). Phylogeny and classification of the Chamaeleonidae (Sauria) with special reference to hemipenis morphology. *Bonner Zoologische Monographien* **22:** 1–64.

Klaver, C. J. J., and Böhme, W. (1997). Chamaeleonidae. *Das Tierreich* **109:** 1–306.

Kluge, A. G. (1974). A taxonomic revision of the lizard family Pygopodidae. *Miscellaneous Publications, Museum of Zoology, University of Michigan* **147:** 1–221.

Kluge, A. G. (1976). Phylogenetic relationships in the lizard family Pygopodidae: An evaluation of theory, methods and data. *Miscellaneous Publications, Museum of Zoology, University of Michigan* **152:** 1–72.

Kluge, A. G. (1981). The life history, social organization, and parental behavior of *Hyla rosenbergi* Boulenger, a nest-building gladiator frog. *Miscellaneous Publications, Museum of Zoology, University of Michigan* **160:** 1–170.

Kluge, A. G. (1987). Cladistic relationships among the Gekkonoidea (Squamata, Sauria). *Miscellaneous Publications, Museum of Zoology, University of Michigan* **173:** 1–54.

Kluge, A. G. (1989). Progess in squamate classification. *Herpetologica* **45:** 368–379.

Kluge, A. G. (1991a). Checklist of gekkonoid lizards. *Smithsonian Herpetological Information Service* **85:** 1–35.

Kluge, A. G. (1991b). Boine snake phylogeny and research cycles. *Miscellaneous Publications, Museum of Zoology, University of Michigan* **178:** 1–58.

Kluge, A. G. (1993a). *Aspidites* and the phylogeny of pythonine snakes. *Records of the Australian Museum, supplement* **19:** 1–77.

Kluge, A. G. (1993b). *Calabaria* and the phylogeny of erycine snakes. *Zoological Journal of the Linnean Society* **107:** 293–351.

Kluge, A. G., and Farris, J. S. (1969). Quantitative phyletics and the evolution of anurans. *Systematic Zoology* **18:** 1–32.

Knapp, R. A., and Morgan, J. A. T. (2006). Tadpole mouthpart depigmentation as an accurate indicator of chytridiomycosis, an emerging disease of amphibians. *Copeia* **2006:** 188–197.

Knowles, R., Mahony, M., Armstrong, J., and Donnellan, S. (2004). Systematics of sphagnum frogs of the genus *Philoria* (Anura: Myobatrachidae) in eastern Australia, with the description of two new species. *Records of the Australian Museum* **56:** 57–74.

Kobel, H. R., Loumont, C., and Tinsley, R. C. (1996). The extant species. In *The Biology of Xenopus*. R. C. Tinsley and H. R. Kobel (Eds.). Pp. 9–33. Zoological Society of London, London.

Kochva, E. (1978). Phylogeny of oral glands in reptiles as related to the origin and evolution of snakes. In *Toxins: Animal, Plant and Microbial*. P. Rosenberg (Ed.). Pp. 29–37. Pergamon Press, Oxford.

Kofron, C. P. (1983). Female reproductive cycle of the Neotropical snail-eating snake *Sibon sanniola* in northern Yucatan, Mexico. *Copeia* **1983:** 963–969.

Kofron, C. P. (1988). The Central and South American blindsnakes of the genus *Anomalepis*. *Amphibia-Reptilia* **9:** 7–14.

Köhler, J., Glaw, F., and Vences, M. (1997). Notes on the reproduction of *Rhombophryne* (Anura: Microhylidae) at Nosy Be, northen Madagascar. *Revue francaise de Aquariologie (Herpetologie)* **24:** 1–2.

Kok, D., du Preez, L. H., and Channing, A. (1989). Channel construction by the African bullfrog: Another anuran parental care strategy. *Journal of Herpetology* **23:** 435–437.

Kokubum, M. N. D., and Giaretta, A. A. (2005). Reproductive ecology and behaviour of a species of *Adenomera* (Anura, Leptodactylinae) with endotrophic tadpoles: Systematic implications. *Journal of Natural History* **39:** 1745–1758.

Komnick, H. (1986). Chloride cells and salt glands. In *Biology of the Integument*. G. Matoltsy and K. S. Richards (Eds.). Pp. 499–516. Springer Verlag, Berlin.

Kozak, K. H., Blaine, R. A., and Larson, A. (2006). Gene lineages and eastern North American palaeodrainage basins: phylogeography and speciation in salamanders of the

Eurycea bislineata species complex. *Molecular Ecology* **15:** 191–207.

Kraus, F., and Brown, W. M. (1998). Phylogenetic relationships of colubroid snakes based on mitochondrial DNA sequences. *Zoological Journal of the Linnean Society* **122:** 455–487.

Kraus, F., and Campbell, E. W. (2002). Human-mediated escalation of a formerly eradicable problem: The invasion of Caribbean frogs in the Hawaiian Islands. *Biological Invasions* **4:** 327–332.

Kraus, F., Campbell, E. W., Allison, A., and Pratt, T. (1999). *A Field Guide to Florida Reptiles and Amphibians*. Gulf Publishing Company, Houston, TX.

Krishna, S. N., and Krishna, S. B. (2006). Visual and acoustic communication in an endemic stream frog, *Micrixalus saxicolus* in the Western Ghats, India. *Amphibia-Reptilia* **27:** 143–147.

Kronauer, D. J. C., Bergmann, P. J., Mercer, J. M., and Russell, A. P. (2005). A phylogeographically distinct and deep divergence in the widespread Neotropical turnip-tailed gecko, *Thecadactylus rapicauda*. *Molecular Phylogenetics and Evolution* **34:** 431–437.

Kropach, C. (1971). Sea snake (*Pelamis platurus*) aggregations on slicks in Panama. *Herpetologica* **27:** 131–135.

Krügel, P., and Richter, S. (1995). *Syncope antenori*—a bromeliad breeding frog with free-swimming, nonfeeding tadpoles (Anura, Microhylidae). *Copeia* **1995:** 955–963.

Kuchling, G. (1998). Managing the last survivors: Integration of *in situ* and *ex situ* conservation of *Pseudemydura umbrina*. In *Proceedings: Conservation, Restoration, and Management of Tortoises and Turtles – An International Conference [1997]*. J. V. Abbema (Ed.). Pp. 339–344. New York Turtle and Tortoise Society, New York.

Kuchling, G. (1999). *The Reproductive Biology of the Chelonia*. Springer-Verlag, Berlin.

Kunte, K. (2004). Natural history and reproductive behavior of *Nyctibatrachus* cf. *humayuni (Anura: Ranidae)*. *Herpetological Review* **35:** 137–140.

Kupfer, A., Muller, H., Antoniazzi, M. M., Jared, C., Greven, H., Nussbaum, R. A., and Wilkinson, M. (2006). Parental investment by skin feeding in a caecilian amphibian. *Nature* **440:** 926–929.

Kupfer, A., Nabhitabhata, J., and Himstedt, W. (2005). Life history of amphibians in the seasonal tropics: Habitat, community and population ecology of a caecilian (genus *Ichthyophis*). *Journal of Zoology, London* **266:** 237–247.

Kusano, T. (1980). Breeding and egg survival of a population of a salamander, *Hynobius nebulosus tokyoensis* Tago. *Researches on Population Ecology* **21:** 181–196.

Kuzmin, S. L. (1995). *The Clawed Salamanders of Asia*. Westarp Wissenschaften, Magdeburg.

Lailvaux, S. P., and Irschick, D. J. (2006). No evidence for female association with high-performance males in the green anole lizard, *Anolis carolinensis*. *Ethology* **112:** 707–715.

Lailvaux, S. P., and Irschick, D. J. (2007). The evolution of performance-based male fighting ability in Caribbean *Anolis* lizards. *American Naturalist* **170:** 573–586.

Lamb, T., Meeker, A. M., Bauer, A. M., and Branch, W. R. (2003). On the systematic status of the desert plated lizard (*Angolosaurus skoogi*): Phylogenetic inference from DNA sequence analysis of the African Gerrhosauridae. *Biological Journal of the Linnean Society* **78:** 253–261.

Lambiris, A. J. L. (1989). The frogs of Zimbabwe. *Museo Regionale di Scienze Naturali Monographia* **10:** 1–247.

Lamotte, M., and Ohler, A. (2000). Revision of the *Ptychadena stenocephala* species group (Amphibia, Anura). *Zoosystema* **22:** 569–583.

Lampo, M., and De Leo, G. A. (1998). The invasion ecology of the toad *Bufo marinus*: From South America to Australia. *Ecological Applications* **8:** 388–396.

Lancaster, J. R., Wilson, P., and Espinoza, R. E. (2006). Physiological benefits as precursors of sociality: Why banded geckos band. *Animal Behaviour* **72:** 199–207.

Lance, V. A., Elsey, R. M., and Lang, J. W. (2000). Sex ratios of American alligators (Crocodylidae): Male or female biased? *Journal of Zoology, London* **252:** 71–78.

Lancini, A. R., and Kornacker, P. M. (1989). *Die Schlangen von Venezuela*. Verlag Armitano Ed. C. A, Carcas.

Lande, R. (1980). Sexual dimorphism, sexual selection, and adaptation in polygenic characters. *Evolution* **34:** 292–305.

Landmann, L. (1975). The sense organs in the skin of the head of Squamata (Reptilia). *Israel Journal of Zoology* **24:** 99–135.

Landmann, L. (1986). Epidermis and dermis. In *Biology of the Integument*. J. Vertebrates, A. G. Bereiter-Hahn, K. S. Matoltsy, and Richards (Eds.). Pp. 150–187. Springer Verlag, Berlin.

Lang, J. W. (1989). Social behavior. In *Crocodiles and Alligator*. C. A. Ross (Ed.). Pp. 102–117. Golden Press Pty Ltd, Silverwater, N.S.W.

Lang, J. W., and Andrews, H. V. (1994). Temperature-dependent sex determination in crocodilians. *Journal of Experimental Zoology* **270:** 28–44.

Lang, M. (1989a). Phylogenetic and biogeographic patterns of basiliscine iguanians (Reptilia: Squamata: "Iguanidae"). *Bonner Zoologische Monographien* **28:** 1–172.

Lang, M. (1989b). Notes on the genus *Bombina* Oken (Anura: Bombinatoridae): III. Anatomy, systematics, hybrization, fossil record and biogeography. *British Herpetological Society Bulletin* **28:** 43–49.

Lang, M. (1991). Generic relationships within Cordyliformes (Reptilia: Squamata). *Bulletin de l'Institut Royal des Sciences Naturelles de Belgique* **61:** 121–188.

Lannoo, M. J. (1987a). Neuromast topography in anuran amphibians. *Journal of Morphology* **191:** 115–129.

Lannoo, M. J. (1987b). Neuromast topography in urodele amphibians. *Journal of Morphology* **191:** 247–263.

Lanza, B., Caputo, V., Nascetti, G., and Bullini, L. (1995). Morphologic and genetic studies of the European plethodontid salamanders: taxonomic inferences (genus *Hydromantes*). *Museo Regionale di Scienza Naturali (Torino) Monographias* **14:** 1–366.

Lappin, A. K., and Husak, J. F. (2005). Weapon performance, not size, determines mating success and potential reproductive output in the collared lizard (*Crotaphytus collaris*). *American Naturalist* **166:** 426–436.

Largen, M. J. (1991). A new genus and species of petropedetine frog (Amphibia Anura Ranidae) from high altitude in the mountains of Ethiopia. *Tropical Zoology* **4:** 139–152.

Larson, A. (1991). A molecular perspective on the evolutionary relationships of the salamander families. *Evolutionary Biology* **25:** 211–277.

Larson, A., and Dimmick, W. W. (1993). Phylogenetic relationships of the salamander families: An analysis of congruence among morphological and molecular characters. *Herpetological Monographs* **7:** 77–93.

Larson, A., Weisrock, D. W., and Kozak, K. H. (2003). Phylogenetic systematics of salamanders (Amphibia: Urodela), a review. In *Reproductive Biology and Phylogeny of Urodela (Amphibia)*. D. M. Sever (Ed.). Pp. 31–108. Science Publishers, Inc. Enfield, New Hampshire.

Larson, J. H., and Guthrie, D. J. (1985). The feeding system of terrestrial tiger salamanders (*Ambystoma tigrinum melanostictum* Baird). *Journal of Morphology* **147:** 137–154.

Lasiewski, R. C., and Bartholomew, G. A. (1969). Condensation as a mechanism for water gain in nocturnal desert poikilotherms. *Copeia* **1969:** 405–407.

Lathrop, A. (1997). Taxonomic review of the megophryid frogs (Anura: Pelobatoidea). *Asiatic Herpetological Research* **7:** 68–79.

Lauder, G., and Gillis, G. B. (1997). Origin of the amniote feeding mechanism: Experimental analyses of outgroup clades. In *Amniote Origins. Completing the Transition to Land.* S. S. Sumida and K. L. M. Martin (Eds.). Pp. 169–206. Academic Press, San Diego.

Laurance, W. F. (1996). Catastrophic declines of Australian rainforest frogs: Is unusual weather responsible? *Biological Conservation* **77:** 203–212.

Laurance, W. F., McDonald, K. R., and Speare, R. (1996). Epidemic disease and the catastrophic decline of Australian rainforest frogs. *Conservation Biology* **10:** 406–413.

Laurent, R. F. (1954). Aperçu de la biogéographie des batraciens et des reptiles de la région des grands lacs. *Bulletin de la Société Zoologique de France* **79:** 290–310.

Laurent, R. F. (1979). Herpetofaunal relationships between Africa and South America. In *The South American Herpetofauna: Its Origin, Dispersal, and Evolution*. W. E. Duellman (Ed.). Pp. 55–71. Museum of Natural History, University of Kansas, Monograph No. 7, Lawrence, KS.

Laurent, R. F., and Teran, E. M. (1981). Lista de los anfibios y reptiles de la provincia de Tucumán. *Miscelanea de la Fundación Miguel Lillo* **71:** 1–15.

Laurin, M. (2002). Tetrapod phylogeny, amphibian origins, and the definition of the name tetrapoda. *Systematic Biology* **51:** 364–369.

Laurin, M., and Reisz, R. R. (1995). A re-evaluation of early amniote phylogeny. *Zoological Journal of the Linnean Society* **113:** 165–223.

Laurin, M., and Reisz, R. R. (1997). A new perspective on tetrapod phylogeny. In *Amniote Origins. Completing the Transition to Land.* S. S. Sumida and K. L. M. Martin (Eds.). Pp. 9–59. Academic Press, San Diego.

Lawler, S. P. (1989). Behavioural responses to predators and predation risk in four species of larval anurans. *Animal Behaviour* **38:** 1039–1047.

Lawler, S. P., Dritz, D., Strange, T., and Holyoak, M. (1999). Effects of introduced mosquitofish and bullfrogs on the threatened California red-legged frog. *Conservation Biology* **13:** 613–622.

Lawson, R., Slowinski, J. B., and Burbrink, F. T. (2004). A molecular approach to discerning the phylogenetic placement of the enigmatic snake *Xenophidion schaeferi* among the Alethinophidia. *Journal of Zoology, London* **263:** 285–294.

Lawson, R., Slowinski, J. B., Crother, B. I., and Burbrink, F. T. (2005). Phylogeny of the Colubroidea (Serpentes): New evidence from mitochondrial and nuclear genes. *Molecular Phylogenetics and Evolution* **37:** 581–601.

Leache, A. D., and McGuire, J. A. (2006). Phylogenetic relationships of horned lizards (*Phrynosoma*) based on nuclear and mitochondrial data: Evidence for a misleading mitochondrial gene tree. *Molecular Phylogenetics and Evolution* **39:** 628–644.

Leary, C. J., Fox, D. J., Shepard, D. B., and Garcia, A. M. (2005). Body size, age, growth and alternative mating tactics in toads: Satellite males are smaller but not younger than calling males. *Animal Behaviour* **70:** 663–671.

Leary, C. J., Garcia, A. M., and Knapp, R. (2006). Elevated corticosterone levels elicit non-calling mating tactics in male toads independently of changes in circulating androgens. *Hormones and Behavior* **49:** 425–432.

Leary, C. J., Jessop, T. S., Garcia, A. M., and Knapp, R. (2004). Steroid hormone profiles and relative body condition of calling and satellite toads: implications for proximate regulation of behavior in anurans. *Behavioral Ecology* **15:** 313–320.

Lebedev, O. A. (1997). Palaeontology: Fins made for walking. *Nature* **390:** 21–22.

Lee, J. C. (1975). The autecology of *Xantusia henshawi henshawi* (Sauria: Xantusiidae). *Transactions of the San Diego Soceity of Natural History* **17:** 259–278.

Lee, J. C. (1996). *The Amphibians and Reptiles of the Yucatán Peninsula*. Comstock Publishing Associates, Ithaca, N.Y.

Lee, M. S. Y. (1993). The origin of the turtle body plan: Bridging a famous morphological gap. *Science* **261:** 1716–1720.

Lee, M. S. Y. (1995). Historical burden in systematics and the interrelationships of 'parareptiles. *Biological Reviews* **70:** 459–547.

Lee, M. S. Y. (1997). Pareisaur phylogeny and the origin of turtles. *Zoological Journal of the Linnean Society* **120:** 197–280.

Lee, M. S. Y. (1997). The phylogeny of varanoid lizards and the affinities of snakes. *Philosophical Transactions of the Royal Society of London Series B-Biological Sciences* **352:** 53–91.

Lee, M. S. Y. (1998). Convergent evolution and character correlation in burrowing reptiles: Towards a resolution of squamate relationships. *Biological Journal of the Linnean Society* **65:** 369–453.

Lee, M. S. Y. (2005). Squamate phylogeny, taxon sampling, and data congruence. *Organisms, Diversity & Evolution* **5:** 25–45.

Lee, M. S. Y., and Scanlon, J. D. (2002). Snake phylogeny based on osteology, soft anatomy and ecology. *Biological Review* **77:** 333–401.

Lee, M. S. Y., Scanlon, J. D., and Caldwell, M. W. (2000). Snake origins. *Science* **288:** 1343–1344.

Legler, J. M., and Fitch, H. S. (1957). Observations on hibernation and nests of the collared lizard, *Crotaphytus collaris*. *Copeia* **1957:** 305–307.

Legler, J. M., and Georges, A. (1993). Family Chelidae. In *Fauna of Australia. Volume 2A, Amphibia & Reptilia*. C. J. Glasby, G. J. B. Ross, and P. I. Beesley (Eds.). Pp. 120–128. Australian Government Public Service, Canberra.

Lehtinen, R. M., and Skinner, A. A. (2006). The enigmatic decline of Blanchard's Cricket Frog (*Acris crepitans blanchardi*): A test of the habitat acidification hypothesis. *Copeia* **2006:** 159–167.

Lehtinen, R. M., Nussbaum, R. A., Richards, C. M., Cannatella, D. C., and Vences, M. (2007). Mitochondrial genes reveal cryptic diversity in plant-breeding frogs from Madagascar (Anura, Mantellidae, *Guibemantis*). *Molecular Phylogenetics and Evolution* **44:** 1121–1129.

Lemaster, M. P., and Mason, R. T. (2002). Variation in a female sexual attractiveness pheromone controls male mate choice in garter snakes. *Journal of Chemical Ecology* **28:** 1269–1285.

LeMaster, M. P., Moore, I. T., and Mason, R. T. (2001). Conspecific trailing behaviour of red-sided garter snakes, *Thamnophis sirtalis parietalis*, in the natural environment. *Animal Behaviour* **61:** 827–833.

Lemckert, F., and Brassil, T. (2003). Movements and habitat use by the giant burrowing frog, *Heleioporus australiacus*. *Amphibia-Reptilia* **24:** 207–211.

Lemmon, E. M., Lemmon, A. R., Collins, J. T., Lee-Yaw, J. A., and Cannatella, D. C. (2007). Phylogeny-based delimitation of species boundaries and contact zones in the trilling chorus frogs (*Pseudacris*). *Molecular Phylogenetics and Evolution* **44:** 1068–1082.

Lemos-Espinal, J. A., Smith, G. R., and Ballinger, R. E. (1998). Thermal ecology of the crevice-dwelling lizard, *Xenosaurus newmanorum*. *Journal of Herpetology* **32:** 141–144.

Lemos-Espinal, J. A., Smith, G. R., and Ballinger, R. E. (2003a). Ecology of *Xenosaurus grandis agrenon*, a knob-scaled lizard from Oaxaca, México. *Journal of Herpetology* **37:** 192–196.

Lemos-Espinal, J., Smith, G. R., and Ballinger, R. E. (1997a). Natural history of the Mexican knob-scaled lizard, *Xenosaurus rectocollaris*. *Herpetological Natural History* [1996] **4:** 151–154.

Lemos-Espinal, J., Smith, G. R., and Ballinger, R. E. (1997b). Neonate-female associations in *Xenosaurus newmanorum*: A case of parental care in a lizard? *Herpetological Review* **28:** 22–23.

Lena, J. P., and de Fraipont, M. (1998). Kin recognition in the common lizard. *Behavioral Ecology and Sociobiology* **42:** 341–347.

Lenk, P., Eidenmueller, B., Staudter, H., Wicker, R., and Wink, M. (2005). A parthenogenetic *Varanus*. *Amphibia-Reptilia* **26:** 507–514.

Leonard, W. P., Brown, H. A., Jones, L. L. C., McAllister, K. R., and Storm, R. M. (1993). *Amphibians of Washington and Oregon*. Seattle Audubon Soc, Seattle.

Lescure, J. (1968). Le comportement social des Batraciens. *Revue Comparatif Animal* **2:** 1–33.

Lescure, J. (1979). Étude taxonomique et éco-éthologique d'un ampihibien des petites Antilles *Leptodactylus fallax* Muller, 1926 (Leptodactylidae). *Bulletin du Muséum National d'Histoire Naturelle, Paris* **1:** 757–774.

Lescure, J. (1986a). La vie sociale de l'adulte. In *Traité de Zoologie. Anatomie, Systématique, Biologie. Tome XIV. Batraciens*. P.-P. Grassé and M. Delsol (Eds.). Pp. 525–537. Masson, Paris.

Lescure, J. (1986b). Histoire de la classification des cecilies (Amphibia: Gymnophiona). *Mémoires Société Zoologique de France* **43:** 11–19.

Lescure, J., Renous, S., and Gasc, J.-P. (1986). Proposition d'une nouvelle classification des amphibiens gymnopiones. *Mémoires Société Zoologique de France* **43:** 145–177.

Lieb, C. S., Sites, J. W., and Archie, J. W. (1999). The use of isozyme characters in systematic studies of turtles: preliminary data for Australian Chelids. *Biochemical Systematics and Ecology* **27:** 157–183.

Liebman, P. A., and Granda, A. M. (1971). Microspectrophotometric measurements of visual pigments in two species of turtle, *Trachemys scripta* and *Chelonia mydas*. *Vision Research* **11:** 105–114.

Liem, K. F., Marx, H., and Rabb, G. R. (1971). The viperid snake *Azemiops*: Its comparative cephalic anatomy and phylogenetic position in relation to Viperinae and Crotalinae. *Fieldiana: Zoology* **59:** 1–126.

Liem, S. S. (1970). The morphology, systematics, and evolution of the Old World treefrogs (Rhacophoridae and Hyperoliidae). *Fieldiana: Zoology* **57:** 1–145.

Ligon, D. B., and Stone, P. A. (2003). Radiotelemetry reveals terrestrial estivation in Sonoran Mud Turtles (*Kinosternon sonoriense*). *Journal of Herpetology* **37:** 750–754.

Lillywhite, H. B. (1970). Behavioral thermoregulation in the bullfrog. *Copeia* **1970:** 158–168.

Lillywhite, H. B. (1987). Temperature, energetics, and physiological ecology. In *Snakes: Ecology and Evolutionary Biology*. R. A. Seigel, J. T. Collins, and S. S. Novak (Eds.). Pp. 422–477. Macmillian Publishing Co, New York.

Lillywhite, H. B., and Maderson, P. F. A. (1982). Skin structure and permeability. In *Biology of the Reptilia. Volume 12. Physiology C*. C. Gans and F. H. Pough (Eds.). Pp. 397–442. Academic Press, London.

Lima, A. P., and Caldwell, J. P. (2001). A new Amazonian species of *Colostethus* with sky blue digits. *Herpetologica* **57:** 180–189.

Lima, A. P., Caldwell, J. P., and Biavati, G. M. (2002). Territorial and reproductive behavior of an Amazonian dendrobatid frog, *Colostethus caeruleodactylus*. *Copeia* **2002:** 44–51.

Limpus, C. J. (1993). Family Dermochelyidae. In *Fauna of Australia. Volume 2A, Amphibia & Reptilia*. C. J. Glasby, G. J. B. Ross, and P. I. Beesley (Eds.). Pp. 139–141. Australian Government Public Service, Canberra.

Limpus, C. J., and Miller, J. D. (1993). Family Cheloniidae. In *Fauna of Australia. Volume 2A, Amphibia & Reptilia*. C. J. Glasby, G. J. B. Ross, and P. I. Beesley (Eds.). Pp. 133–138. Australian Government Public Service, Canberra.

Lindquist, E. D., and Hetherington, T. E. (1996). Field studies of visual and acoustic signaling in the "earless" Panamanian golden frog, *Atelopus zeteki*. *Journal of Herpetology* **30:** 347–354.

Lips, K. R. (1998). Decline of a tropical montane amphibian fauna. *Conservation Biology* **12:** 106–117.

Lips, K. R. (1999). Mass mortality and population declines of anurans at an upland site in western Panama. *Conservation Biology* **13:** 117–125.

Lips, K. R., Green, D. E., and Papendick, R. (2003). Chytridiomycosis in wild frogs from southern Costa Rica. *Journal of Herpetology* **37:** 215–218.

Little, C. (1990). *The Terrestrial Invasion. An Ecophysiological Approach to the Origins of Land Animals*. Cambridge University Press, Cambridge.

Littlejohn, M. J., Roberts, J. D., Watson, G. F., and Davies, M. (1993). The Myobatrachidae. In *Fauna of Australia. Volume 2A. Amphibia & Reptilia*. C. J. Gasby, G. J. B. Ross, and P. L. Beesley (Eds.). Pp. 41–57. Australian Government Publishing Service, Canberra.

Litzgus, J. D., Costanzo, J. P., Brooks, R. J., and Lee, R. E. (1999). Phenology and ecology of hibernation in spotted turtles (*Clemmys guttata*) near the northern limit of their range. *Canadian Journal of Zoology* **77:** 1348–1357.

Liu, C.-Z., and Hu, S.-Q. (1961). *The Anura of China*. Science Press, Bejing. [in Chinese].

Liu, F.-G. R., Moler, P. E., Whidden, H. P., Miyamoto, M. M., 2004. Allozyme variation in the salamander genus *Pseudobranchus*: Phylogeographic and taxonomic signifcance. *Copeia* **2004:** 136–144.

Liu, F- G. R., Moler, P. E., and Miyamoto, M. M. (2006). Phylogeography of the salamander genus *Pseudobranchus* in the southeastern United States. *Molecular Phylogenetics and Evolution* **39:** 149–159.

Llinás, R., and Precht, W. (Eds.) (1976). *Frog Neurobiology. A Handbook*. Springer-Verlag, Berlin.

Lloyd, M., Inger, R. F., and King, F. W. (1968). On the diversity of reptile and amphibian species in a Bornean rain forest, *American Naturalist* **102:** 497–515.

Loader, S. P., Pisani, D., Cotton, J. A., Gower, D. J., Day, J. J., and Wilkinson, M. (2007). Relative time scales reveal multiple origins of parallel disjunct distributions of African caecilian amphibians. *Biology Letters* **3:** 505–508.

Loader, S. P., Wilkinson, M., Gower, D. J., and Msuya, C. A. (2003). A remarkable young *Scolecomorphus vittatus* (Amphibia: Gymnophiona: Scolecomorphidae) from the North Pare Mountains, Tanzania. *Journal of Zoology, London* **259:** 93–101.

Lohmann, K. J. (1992). How sea turtles navigate. *Scientific American* **266:** 100–106.

Lohmann, K. J., Cain, S. D., Dodge, S. A., and Lohmann, C. M. F. (2001). Regional magnetic fields as navigational markers for sea turtles. *Science* **294:** 364–366.

Lohmann, K. J., Lohmann, C. M. F., and Putman, N. F. (2007). Magnetic maps in animals: Nature's GPS. *Journal of Experimental Biology* **210:** 3697–3705.

Lohmann, K. J., Lohmann, C. M. F., Ehrhart, L. M., Bagley, D. A., and Swing, T. (2004). Geomagnetic map used in sea-turtle navigation. *Nature* **428:** 909–910.

Lohmann, K. J., Witherington, B. E., Lohmann, C. M. F., and Salmon, M. (1966). Orientation, navigation, and natal beach homing in sea turtles. In *The Biology of Sea Turtles*. P. L. Lutz and J. A. Musick (Eds.). Pp. 107–135. CRC Press, Boca Raton, FL.

Lombard, R. E., and Sumida, S. S. (1992). Recent progress in understanding early tetrapods. *American Zoologist* **32:** 609–622.

Lombard, R. E., and Wake, D. B. (1976). Tongue evolution in the lungless salamanders, family Plethodontidae I. Introduction, theory, and a general model of dynamics. *Journal of Morphology* **148:** 265–286.

Lombard, R. E., and Wake, D. B. (1977). Tongue evolution in the lungless salamanders, family Plethodontidae II. Function and evolutionary diversity. *Journal of Morphology* **153:** 39–79.

Lombard, R. E., and Wake, D. B. (1986). Tongue evolution in the lungless salamanders, family Plethodontidae. 4. Phylogeny of plethodontid salamanders and the evolution of feeding dynamics. *Systematic Zoology* **35:** 532–551.

Lopez, C. H., and Brodie, E. D., Jr. (1977). The function of costal grooves in salamanders (Amphibia, Urodela). *Journal of Herpetology* **11:** 372–374.

López, P., Salvador, A., and Cooper, W. E., Jr. (1997). Discrimination of self from other males by chemosensory cues in the amphisbaenian (*Blanus cinereus*). *Journal of Comparative Psychology* **111:** 105–109.

Losos, J. B. (1992). The evolution of convergent structure in Caribbean *Anolis* communities. *Systematic Biology* **41:** 403–420.

Losos, J. B. (1994). Integrative approaches to evolutionary ecology: *Anolis* lizards as model systems. *Annual Review of Ecology and Systematics* **25:** 467–493.

Losos, J. B., Jackman, T. R., Larson, A., de Queiroz, K., and Rodríguez-Schettino, L. (1998). Contingency and determinism in replicated adaptive radiations of island lizards. *Science* **279:** 2115–2118.

Lötters, S. (1996). *The Neotropical Toad Genus Atelopus. Checklist-Biology-Distribution*. M. Vences and F. Glaw Verlags GbR, Köln.

Lötters, S., Jungfer, K.-H., and Widmer, A. 2000. A new genus of aposematic poison frog (Amphibia: Anura: Dendrobatidae) from the upper Amazon basin, with notes on its reproductive behaviour and tadpole morphology. *Jahreshefte der Gesellschaft für Naturkunde in Württemberg* **156:** 233–243.

Louw, G. N., and Holm, E. (1972). Physiological, morphological and behavioural adaptations of the ultrapsammophilous, Namib Desert lizard *Aporosaura anchietae* (Bocage). *Madoqua, series II* **1:** 67–81.

Lovern, M. B., and Passek, K. M. (2002). Sequential alternation of offspring sex from successive eggs by female green anoles, *Anolis carolinensis*. *Canadian Journal of Zoology* **80:** 77–82.

Lovern, M. B., and Wade, J. (2003). Yolk testosterone varies with sex in eggs of the lizard. *Anolis carolinensis*. *Journal of Experimental Zoology* **295A:** 206–210.

Lowe, C. H., Schwalbe, C. R., and Johnson, T. B. (1986). *The Venomous Reptiles of Arizona*. Arizona Fish and Game Department, Phoenix.

Lue, K. Y. (1991). *The Amphibians of Taiwan*. Taiwan Prov. Dept. Education, Tapei.

Luiselli, L., Capula, M., and Shine, R. (1996). Reproductive output, costs of reproduction, and ecology of the smooth snake, *Coronella austriaca*, in the eastern Italian Alps. *Oecologia* **106:** 100–110.

Luiselli, L., Capula, M., and Shine, R. (1997). Food habits, growth rates, and reproductive biology of grass snakes, *Natrix natrix* (Colubridae) in the Italian Alps. *Journal of Zoology, London* **241:** 371–380.

Luppa, H. (1977). Histology of the digestive tract. In *Biology of the Reptilia. Volume 6. Morphology E.* C. Gans and T. S. Parsons (Eds.). Pp. 225–313. Academic Press, London.

Luthardt, G., and Roth, G. (1979). The relationship between stimulus orientation and stimulus movement pattern in the prey catching behavior of *Salamandra salamandra. Copeia* **1979:** 442–447.

Lutz, P. L. and Musick, J. A. (Eds.) (1996). *The Biology of Sea Turtles.* CRC Press, Boca Raton, FL.

Lynch, J. D. (1971). Evolutionary relationships, osteology, and zoogeography of leptodactyloid frogs. *Miscellaneous Publications of the Museum of Natural History, University of Kansas* **53:** 1–238.

Lynch, J. D. (1973). The transition from archaic to advanced frogs. In *Evolutionary Biology of the Anurans.* J. L. Vial (Ed.). Pp. 133–182. University of Missouri Press, Columbia.

Lynch, J. D. (1978). A new eleutherodactyline frog from the Andes of northern Colombia (leptodactylidae). *Copeia* **1978:** 17–21.

Lynch, J. D., Ruiz, P. M., and Rueda, J. V. (1983). Notes on the distribution and reproductive biology of *Centrolene geckoideum* Jimenez de la Espada in Colombia and Ecuador (Amphibia: Centrolenidae). *Studies on Neotropical Fauna and Environment* **18:** 239–243.

Lynn, W. G. (1970). The thryoid. In *Biology of the Reptilia. Volume 3. Morphology C.* C. Gans, and T. S. Parsons (Eds.). Pp. 201–234. Academic Press, London.

MacArthur, R. H. (1984). *Geographical Ecology: Patterns in the Distribution of Species.* Princeton University Press, Princeton, NJ.

MacArthur, R. H., and Pianka, E. R. (1966). On optimal use of a patchy environment. *American Naturalist* **100:** 603–609.

MacArthur, R. H., and Wilson, E. O. (1967). *The Theory of Island Biogeography.* Princeton University Press, Princeton, N.J.

Macdonald, L. A., and Mushinsky, H. R. 1988. Foraging ecology of the gopher tortoise, *Gopherus polyphemus*, in a sandhill habitat. *Herpetologica* **44:** 345–353.

Macey, J. R. (2005). Plethodontid salamander mitochondrial genomics: A parsimony evaluation of character conflict and implications for historical biogeography. *Cladistics* **21:** 194–202.

Macey, J. R. J. A., Schulte, Larson, A., Tuniyev, B. S., Orlov, N., and Papenfuss, T. J. (1999). Molecular phylogenetics, tRNA evolution, and historical biogeography in anguid lizards and related taxonomic families. *Molecular Phylogenetics and Evolution* **12:** 250–272.

Macey, J. R., Larson, A., Ananjeva, N. B., and Papenfuss, T. J. (1997). Evolutionary shifts in three major structural features of the mitochondrial genome among iguanian lizards. *Journal of Molecular Evolution* **44:** 660–674.

Macey, J. R., Papenfuss, T. J., Kuehl, J. V., Fourcade, H. M., and Boore, J. L. (2004). Phylogenetic relationships among amphisbaenian reptiles based on complete mitochondrial genomic sequences. *Molecular Phylogenetics and Evolution* **33:** 22–31.

Maderson, P. F. A. (1965). The structure and development of the squamate epidermis. In *Biology of the Skin and Hair Growth.* A. G. Lyne and B. F. Short (Eds.). Pp. 129–153. Angus & Robertson, Sydney.

Maderson, P. F. A., Rabinowitz, T., Tandler, B., and Alibardi, L. (1998). Ultrastructural contributions to an understanding of the cellular mechanisms involved in lizard skin shedding with comments on the function and evolution of a unique lepidosaurian phenomenon. *Journal of Morphology* **236:** 1–24.

Madison, D. M. (1969). Homing behavior of the red-cheeked salamander, *Plethodon jordani. Animal Behaviour* **17:** 25–39.

Madison, D. M., and Shoop, C. R. (1970). Homing behavior, orientation, and home range of salamanders tagged with Tantalum-182. *Science* **168:** 1484–1487.

Madsen, T. (1984). Movements, home range size and habitat use of radio-tracked grass snakes (*Natrix natrix*) in southern Sweden. *Copeia* **1984:** 707–713.

Madsen, T., and Shine, R. (1993). Male mating success and body size in European grass snakes. *Copeia* **1993:** 561–564.

Madsen, T., and Shine, R. (1994). Components of lifetime reproductive success in adders, *Vipera berus. Journal of Animal Ecology* **63:** 561–568.

Madsen, T., and Shine, R. (1996). Seasonal migration of predators and prey—a study of pythons and rats in tropical Australia. *Ecology* **77:** 149–156.

Madsen, T., Shine, R., Loman, J., and Hakansson, T. (1992). Why do female adders copulate so frequently? *Nature* **355:** 440–441.

Madsen, T., Ujvari, B., and Olsson, M. (2005). Old pythons stay fit: Effects of haematozoan infections on life history traits of a large tropical predator. *Oecologia* **142:** 407–412.

Maeda, N., and Matsui, M. (1989). *Frogs and Toads of Japan.* Bun-Ichi Sogo Shuppan Co., Tokyo.

Mägdefrau, H. (1997). Biologie, Haltung und Zucht der Krokodilschwanz-Höckerechse (*Shinisaurus crocodilurus*). *Zeitschrift des Kölner Zoo* **2:** 55–60.

Maglia, A. M. (1996). Ontogeny and feeding ecology of the red-backed salamander, *Plethodon cinereus. Copeia* **1996:** 576–586.

Maglia, A. M. (1998). Phylogenetic relationships of extant pelobatoid frogs (Anura: Pelobatoidea): Evidence from adult morphology. *Scientific Papers of the Natural History Museum, University of Kansas* **10:** 1–19.

Maglia, A. M., Pugener, L. A., and Trueb, L. (2001). Comparative development of anurans: Using phylogeny to understand ontogeny. *American Zoologist* **41:** 538–551.

Magnusson, W. E. (1980). Hatching and creche formation by *Crocodylus porosus. Copeia* **1980:** 359–362.

Magnusson, W. E., Vliet, K. A., Pooley, A. C., and Whitaker, R. (1989). Reproduction. In *Crocodiles and Alligators.* C. A. Ross (Ed.). Pp. 118–135. Golden Press Pty Ltd, Silverwater, N.S.W.

Mahmoud, I. Y., and Klicka, J. (1979). Feeding, drinking, and excretion. In *Turtles, Perspectives and Research.* M. Harless and H. Morlock (Eds.). Pp. 229–243. John Wiley and Sons, New York.

Main, A. R., and Bull, C. M. (1996). Mother-offspring recognition in two Australian lizards, *Tiliqua rugosa* and *Egernia stokesii*. *Animal Behaviour* **52:** 193–200.

Maiorana, V. C. (1977). Tail autotomy, functional conflicts and their resolution by a salamander. *Nature* **265:** 533–535.

Malkmus, R. (1995). *Die Amphibien und Reptilien Portugals, Madeiras und der Azoren.* Westarp Wissenschaften, Magdeburg.

Manley, G. A. (1990). *Peripheral Hearing Mechanisms in Reptiles and Birds.* Springer-Verlag, Berlin.

Manthey, U., and Grossmann, W. (1997). *Amphibien & Reptilien Südostasiens.* Natur und Tier Verlag, Münster, Germany.

Manzani, P. R., and Abe, A. S. (2002). A new species of *Tupinambis* Daudin, 1803 from southeastern Brazil. *Arquivos do Museu Nacional, Rio de Janeiro* **60:** 295–302.

Markwick, P. J. (1998). Fossil crocodilians as indicators of Late Cretaceous and Cenozoic climates: Implications for using palaeontological data in reconstructing palaeoclimate. *Palaeogeography, Palaeoclimatology, Palaeoecology* **137:** 205–271.

Marmayou, J., Dubois, A., Ohler, A., Pasquet, E., and Tillier, A. (2000). Phylogenetic relationships in the Ranidae. Independent origin of direct development in the genera *Philautus* and *Taylorana*. *Comptes Rendus De L Academie Des Sciences Serie Iii-Sciences De La Vie-Life Sciences* **323:** 287–297.

Márquez, R. (1992). Terrestrial paternal care and short breeding seasons: reproductive phenology of the midwife toads *Alytes obstetricans* and *A. cisternasii*. *Ecography* **15:** 279–288.

Márquez, R. (1995). Female choice in the midwife toads (*Alytes obstetricans* and *A. cisternasii*). *Behaviour* **132:** 151–161.

Márquez, M. R. (1990). Sea turtles of the world. An annotated and illustrated catalogue of sea turtles species known to date. *Foof and Agriculture Organization of the United Nations Fisheries Synopsis* **125/11:** 1–81.

Marquez, R., and Bosch, J. (1997). Female preference in complex acoustical environments in the midwife toads *Alytes obstetricans* and *Alytes cisternasii*. *Behavioral Ecology* **8:** 588–594.

Marquis, R. J., Donnelly, M. A., and Guyer, C. (1986). Aggregations of calling males of *Agalychnis calcifer* Boulenger (Anura: Hylidae) in a Costa Rican lowland wet forest. *Biotropica* **18:** 173–175.

Martin, J. (1992). *Masters of Disguise: A Natural History of Chameleons.* Facts on File, Inc, New York.

Martin, J. A., and Salvador, A. (1993). Tail loss reduces mating success in the Iberian rock-lizard, *Lacerta monticola*. *Behavioral Ecology and Sociobiology* **32:** 185–189.

Martín, J., and Lopéz, P. (1999). When to come out from a refuge: Risk-sensitive and state-dependent in an alpine lizard. *Behavioral Ecology* **10:** 487–492.

Martin, K. L. M., and Nagy, K. A. (1997). Water balance and the physiology of the amphibian to amniote transition. In *Amniote Origins. Completing the Transition to Land.* S. S. Sumida and K. L. M. Martin (Eds.). Pp. 399–423. Academic Press, San Diego.

Martin, K. L. M., and Sumida, S. S. (1997). An integrated approach to the origin of amniotes: Completing the transition to land. In *Amniote Origins: Completing the Transition to Land.* S. S. Sumida and K. L. M. Martin (Eds.). Pp. 1–8. Academic Press, San Diego, CA.

Martin, L. D. (1991). Mesozoic birds and the origin of birds. In *Origins of the Higher Groups of Tetrapods. Controversy and Consensus.* H. P. Schultze and L. Trueb (Eds.). Pp. 485–540. Comstock Publishing Associates, Ithaca, NY.

Martin, L. D., and Rothschild, B. M. (1989). Paleopathology and diving mosasaurs. *American Scientist* **77:** 460–467.

Martin, W. F., and Gans, C. (1972). Muscular control of the vocal tract during release signaling in the toad *Bufo valliceps*. *Journal of Morphology* **137:** 1–28.

Martins, E. P. (1993). Contextual use of the push-up display by the sagebrush lizard, *Sceloporus graciosus*. *Animal Behaviour* **45:** 25–36.

Martins, E. P. (1994). Phylogenetic perspectives on the evolution of lizard territoriality. In *Lizard Ecology: Historical and Experimental Perspectives.* L. J. Vitt and E. R. Pianka (Eds.). Pp. 118–144. Princeton University Press, Princeton, NJ.

Martins, E. P., Labra, A., Halloy, M., and Thompson, J. T. (2004). Large-scale patterns of signal evolution: An interspecific study of *Liolaemus* lizard headbob displays. *Animal Behaviour* **68:** 453–463.

Martins, M. (2006). Life in the water: Ecology of the jacarerana lizard, *Crocodilurus amazonicus*. *Herpetological Journal* **16:** 171–177.

Martins, M., and Oliveira, M. E. (1998). Natural history of snakes in forests of the Manaus region, central Amazonia, Brazil. *Herpetological Natural History* **6:** 78–150.

Martof, B. S. (1953). Home range and movements of the green frog, *Rana clamitans*. *Ecology* **34:** 529–543.

Martof, B. S. (1972). *Pseudobranchus* Gray. *Catalogue of American Amphibians and Reptiles* **118:** 1–4.

Martof, B. S. (1974). *Siren* Linneaus. *Catalogue of American Amphibians and Reptiles* **152:** 1–2.

Marvin, G. A. (1996). Life history and population characteristics of the salamander *Plethodon kentucki* with a review of *Plethodon* life histories. *American Midland Naturalist* **136:** 385–400.

Maslin, T. P. (1950). The production of sound in caudate Amphibia. *University of Colorado Studies, Series in Biology No* **1:** 29–45.

Mason, R. T. (1992). Reptilian pheromones. In *Biology of the Reptilia. Volume 18, Physiology. Hormones, Brain, and Behavior.* C. Gans and D. Crews (Eds.). Pp. 114–228. University of Chicago Press, Chicago.

Mason, R. T., and Crews, D. (1985). Female mimicry in garter snakes. *Nature* **316:** 59–60.

Mason, R. T., and Crews, D. (1986). Pheromone mimicry in garter snakes. In *Chemical Signals in Vertebrates. Volume 4. Ecology, Evolution, and Comparative Biology.* D. Duvall, D. Müller-Schwarze, and R. M. Silverstein (Eds.). Pp. 279–283. Plenum Press, New York.

Masood-Parveez, U., and Nadkarni, V. B. (1993). The ovarian cycle in an oviparous gymnophione amphibian *Ichthyophis beddomei* (Peters). *Journal of Herpetology* **27:** 59–63.

Massey, A. (1988). Sexual interactions in red-spotted newt populations. *Animal Behaviour* **36:** 205–210.

Masta, S. E., Sullivan, B. K., Lamb, T., and Routman, E. J. (2002). Molecular systematics, hybridization, and phylogeography of the *Bufo americanus* complex in Eastern North America. *Molecular Phylogenetics and Evolution* **24:** 302–314.

Mathis, A. (1991). Territories of male and female terrestrial salamanders: Costs, benefits, and intersexual spatial associations. *Oecologia* **86:** 433–440.

Mathis, A., Jaeger, R. G., Keen, W. H., Ducey, P. K., Walls, S. C., and Buchanan, B. W. (1995). Aggression and territoriality by salamanders and a comparison with the territorial behaviour of frogs. In *Amphibian Biology. Volume 2, Social Behaviour*. H. Heatwole and B. K. Sullivan (Eds.). Pp. 633–676. Surrey Beatty & Sons Pty Ltd., Chipping Norton Pty Ltd, N.S.W.

Matsui, M., ChanArd, T., and Nabhitabhata, J. (1996). Distinct specific status of *Kalophrynus pleurostigma interlineatus* (Anura, Microhylidae). *Copeia* **1996:** 440–445.

Mattison, C. (2007). *The New Encyclopedia of Snakes*. Cassell Illustrated, London.

Mautz, W. J. (1982). Patterns of evaporative water loss. In *Biology of Reptilia Volume 12, Physiology C: Physiological Ecology*. C. Gans and F. H. Pough (Eds.). Pp. 443–481. Academic Press, New York, NY.

Mautz, W. J., and Lopezforment, W. (1978). Observations on activity and diet of cavernicolous lizard *Lepidophyma smithii* (Sauria: Xantusiidae). *Herpetologica* **34:** 311–313.

Mautz, W. J., and Nagy, K. A. (1987). Ontogenetic changes in diet, field metabolic-rate, and water flux in the herbivorous lizard *Dipsosaurus dorsalis*. *Physiological Zoology* **60:** 640–657.

Maxson, L. R., and Maxson, R. D. (1990). Proteins II: immunological techniques. In *Molecular Systematics*, Second Edition. D. M. Hillis, C. Moritz, and B. K. Mable (Eds.). Pp. 127–155. Sinauer Associates, Inc, Sunderland, MA.

Maxson, L. R., and Ruibal, R. (1988). Relationships of frogs in the leptodactylid subfamily Ceratophryinae. *Journal of Herpetology* **22:** 228–231.

Maxson, R. D., and Maxson, L. R. (1986). Micro-complement fixation: A quantitative estimator of protein evolution. *Molecular Biology and Evolution* **3:** 375–388.

Mayer, W., and Benyr, G. (1994). Albumin-Evolution und Phylogenese in der Familie Lacertidae (Reptilia: Sauria). *Annalen des Naturhistorischen Museums in Wien* **96B:** 621–648.

Mayer, W., and Bishoff, W. (1996). Beiträge zur taxonomischen Revision der Gattung *Lacerta* (Reptilia: Lacertidae). Teil 1: *Zootoca, Omanosaura, Timon* and *Teira* als eigenständige Gattungen. *Salamandra* **32:** 163–170.

Mayr, E., and Ashlock, S. D. (1991). *Principles of Systematic Zoology*. McGraw-Hill Book Co, New York.

Mazzotti, F. J., and Dunson, W. A. (1989). Osmoregulation in crocodilians. *American Zoologist* **29:** 903–920.

McAliley, L. R., Willis, R. E., Ray, D. A., White, P. S., Brochu, C. A., and Densmore, L. D. (2006). Are crocodiles really monophyletic? Evidence for subdivisions from sequence and morphological data. *Molecular Phylogenetics and Evolution* **39:** 16–32.

McBrayer, L. D., and Reilly, S. M. (2002). Testing amniote models of prey transport kinematics: A quantitative analysis of mouth opening patterns in lizards. *Zoology* **105:** 71–81.

McBrayer, L. D., Corbin, C. E., Reilly, S. M., and Miles, D. B. (2007). Patterns of head shape variation in lizards: Morphological correlates of foraging mode. In *Lizard Ecology: The Evolutionary Consequences of Foraging Mode*. S. M. Reilly, L. D. McBrayer, and D. B. Miles (Eds.). Pp. 271–301. Cambridge University Press, Cambrdge UK.

McCarthy, C. J. (1985). Monophyly of elapid snakes (Serpentes: Elapidae), an assessment of the evidence. *Zoological Journal of the Linnean Society* **83:** 79–93.

McCay, M. G. (2001). Aerodynamic stability and maneuverability of the gliding frog *Polypedates dennysi*. *Journal of Experimental Biology* **204:** 2817–2826.

McCay, M. G. (2003). Winds under the rain forest canopy: The aerodynamic environment of gliding tree frogs. *Biotropica* **35:** 94–102.

McClanahan, L. L. Jr., and Baldwin, R. (1969). Rate of water uptake through the integument of the desert toad, *Bufo punctatus*. *Comparative Biochemistry and Physiology* **28:** 381–389.

McClanahan, L. L. Jr., Ruibal, R., and Shoemaker, V. H. (1983). Rate of cocoon formation and its physiological correlates in a ceratophryd frog. *Physiological Zoology* **56:** 430–435.

McClanahan, L. L. Jr., Shoemaker, V. H., and Ruibal, R. (1976). Structure and function of the cocoon of a ceratophryd frog. *Copeia* **1976:** 179–185.

McClanahan, L. L. Jr., Stinner, J. N., and Shoemaker, V. H. (1978). Skin lipids, water loss, and metabolism in a South American tree frog (*Phyllomedusa sauvagei*). *Physiological Zoology* **51:** 179–187.

Mccord, W. P., and Thomson, S. A. (2002). A new species of *Chelodina* (Testudines: Pleurodira: Chelidae) from northern Australia. *Journal of Herpetology* **36:** 255–267.

McCord, W. P., Ouni, M. J., and Lamar, W. W. (2001). A taxonomic reevaluation of *Phrynops* (Testudines: Chelidae) with the description of two new genera and a new species of *Batrachemys*. *Revista de Biologı́a Tropical, San Jose´* **49:** 715–764.

McCoy, C. J. (1968). Reproductive cycles and viviparity in Guatemalan *Corythophanes percarinatus* (Reptilia: Iguanidae). *Herpetologica* **24:** 175–178.

McCue, M. D. (2004). General effects of temperature on animal biology. In *Temperature-Dependent Sex Determination in Vertebrates*. N. Valenzuela and V. Lance (Eds.). Pp. 71–78. Smithsonian Books, Washington, D. C.

McDiarmid, R. W. (1978). Evolution of parental care in frogs. In *The Development of Behavior: Comparative and Evolutionary Aspects*. G. M. Burghardt and M. Bekoff (Eds.). Pp. 127–147. Garland STPM Press, New York.

McDiarmid, R. W., Campbell, J. A., and Touré, T. A. (1999).*Snake Species of the world. A Taxonomic and Geographic Reference* Volume 1. The Herpetologists' League, Washington, D.C.

McDonald, K. R., and Tyler, M. J. (1984). Evidence of gastric brooding in the Australian leptodactylid frog *Rheobatrachus vitellinus*. *Transactions of the Royal Society of South Australia* **108:** 226.

McDowell, S. B. (1969). Notes on the Australian seas-snake *Ephalophis greyi* M. Smith (Serpentes: Elapidae, Hydrophiinae) and the origin and classification of sea-snakes. *Zoological Journal of the Linnean Society* **48:** 333–349.

McDowell, S. B. (1974). A catalogue of the snakes of New Guinea and the Solomons, with special reference to those in the Bernice P. Bishop Museum, part I. Scolecophidia. *Journal of Herpetology* **8:** 1–128.

McDowell, S. B. (1987). Systematics. In *Snakes: Ecology and Evolutionary Biology*. R. A. Seigel, J. T. Collins, and S. S. Novak (Eds.). Pp. 3–50. MacMillan, New York.

McDowell, S. B., Jr. (1964). Partition of the genus *Clemmys* and related problems in the taxonomy of the aquatic Testudinidae. *Proceedings of the Zoological Society of London* **143:** 239–279.

McGowan, C. (1991). *Dinosaurs, Spitfires, and Sea Dragons*. Harvard University. Press, Cambridge, Mass.

McGowan, G., and Evans, S. E. (1995). Albanerpetontid amphibians from the Cretaceous of Spain. *Nature* **373:** 143–145.

McGregor, J. H., and Teska, W. R. (1989). Olfaction as an orientation mechanism in migrating *Ambystoma maculatum*. *Copeia* **1989:** 779–781.

McGuire, J. A. (1996). Phylogenetic systematics of crotaphytid lizards (Reptilia: Iguania: Crotaphytidae). *Bulletin of the Carnegie Museum of Natural History* **32:** 1–143.

McGuire, J. A. (2003). Allometric prediction of locomotor performance: An example from Southeast Asian flying lizards. *American Naturalist* **161:** 337–349.

McGuire, J. A., and Dudley, R. (2005). The cost of living large: Comparative gliding performance in flying lizards (Agamidae: *Draco*). *American Naturalist* **166:** 93–106.

McGuire, J. A., and Heang, K. B. (2001). Phylogenetic systematics of southeast Asian flying lizards (Iguania: Agamidae: *Draco*) as inferred from mitochondrial DNA sequences. *Biological Journal of the Linnean Society* **72:** 203–229.

McGuire, J. A., Linkem, C. W., Koo, M. S., Hutchison, D. W., Lappin, A. K., Orange, D. I., Lemos-Espinal, J., Riddle, B. R., and Jaeger, J. R. (2007). Mitochondrial introgression and incomplete lineage sorting through space and time: Phylogenetics of crotaphytid lizards. *Evolution* **61:** 2879–2897.

McIlhenny, E. A. (1935). *The Alligator's Life History*. Christopher Publ. House, Boston.

McIntyre, N. E., and Wiens, J. A. (1999). Interactions between landscape structure and animal behavior: The roles of heterogeneously distributed resources and food deprivation on movement patterns. *Landscape Ecology* **14:** 437–447.

McPherson, R. J., and Marion, K. R. (1981). The reproductive biology of female *Sternotherus odoratus* in an Alabama population. *Journal of Herpetology* **15:** 389–396.

McVey, M. E., Zahary, R. G., Perry, D., and MacDougal, J. (1981). Territoriality and homing behavior in the poison dart frog (*Dendrobates pumilio*). *Copeia* **1981:** 1–8.

Means, D. B. (1996). *Amphiuma pholeter* Neill. *Catalogue of American Amphibians and Reptiles* **622:** 1–2.

Means, D. B., Palis, J. G., and Baggett, M. (1996). Effects of slash pine silviculture on a Florida population of flatwoods salamanders. *Conservation Biology* **10:** 426–437.

Measey, G. J., and Tinsley, R. C. (1997). Mating behavior of *Xenopus wittei* (Anura: Pipidae). *Copeia* **1997:** 601–609.

Medica, P. A., and Turner, F. B. (1984). Natural longevity of lizards in southern Nevada. *Herpetological Review* **15:** 34–35.

Medica, P. A., Bury, R. B., and Luckenbach, R. A. (1980). Drinking and construction of water catchments by the desert tortoise, *Gopherus agassizii*, in the Mojave Desert. *Herpetologica* **36:** 301–304.

Meffe, G. K. (1994). Human population control: The missing awareness. *Conservation Biology* **8:** 310–313.

Meffe, G. K. and Carroll, C. R. (Eds.) (1994). *Principles of Conservation Biology*. Sinauer Associates, Inc., Sunderland, MA.

Meffe, G. K., Carroll, C. R., and Pimm, S. L. (1994). Community-level conservation: species interactions, disturbance regimes, and invading species. In *Principles of Conservation Biology*. G. K. Meffe and C. R. Carroll (Eds.). Pp. 209–236. Sinauer Associates, Inc, Sunderland, MA.

Meffe, G. K., Ehrenfeld, D., and Noss, R. F. (2006). Conservation biology at twenty. *Conservation Biology* **20:** 595–596.

Meffe, G. K., Ehrlich, A. H., and Ehrenfeld, D. (1993). Human-population control: The missing agenda. *Conservation Biology* **7:** 1–3.

Melisch, R. (1998). Wildlife trade, sustainable use, and conservation implications from a WWF and TRAFFIC perspective. In *Conservation, Trade and Sustainable Use of Lizards and Snakes in Indonesia*. W. Erdelen (Ed.). Pp. 1–8. Mertensiella 7, Rheinbach, Germany.

Mendelson, J. R., III, Lips, K. R., Gagliardo, R. W., Rabb, G. B., Collins, J. P., Diffendorfer, J. E., Daszak, P., Ibanez, R., Zippel, K. C., Lawson, D. P., Wright, K. M., Stuart, S. N., Gascon, C., da Silva, H. R., Burrowes, P. A., Joglar, R. L., La Marca, E., Lotters, S., du Preez, L. H., Weldon, C., Hyatt, A., Rodriguez-Mahecha, J. V., Hunt, S., Robertson, H., Lock, B., Raxworthy, C. J., Frost, D. R., Lacy, R. C., Alford, R. A., Campbell, J. A., Parra-Olea, G., Bolanos, F., Domingo, J. J. C., Halliday, T., Murphy, J. B., Wake, M. H., Coloma, L. A., Kuzmin, S. L., Price, M. S., Howell, K. M., Lau, M., Pethiyagoda, R., Boone, M., Lannoo, M. J., Blaustein, A. R., Dobson, A., Griffiths, R. A., Crump, M. L., Wake, D. B., and Brodie, E. D. (2006). Biodiversity—Confronting amphibian declines and extinctions. *Science* **313:** 48.

Menezes, V. A., Rocha, C. F. D., and Dutra, D. F. (2004). Reproductive ecology of the parthenogenetic whiptail lizard *Cnemidophorus nativo* in a Brazilian restinga habitat. *Journal of Herpetology* **38:** 280–282.

Menin, M., Rodrigues, D. d. J., and de Azevedo, C. S. (2005). Predation on amphibians by spiders, (Arachnida, Araneae) in the Neotropical region. *Phyllomedusa* **4:** 39–47.

Menzies, J. I. (1976). *Handbook of Common New Guinea Frogs*. Wau Ecological Institute Handbook No. 1, Wau.

Mertens, R., and Wermuth, H. (1960). *Die Amphibien und Reptilien Europas*. Verlag Waldemar Kramer, Frankfurt am Main.

Mesquita, D. O., Colli, G. R., França, F. G. R., and Vitt, L. J. (2006). Ecology of a Cerrado lizard assemblage in the Jalapão region of Brazil. *Copeia* **2006:** 460–471.

Metter, D. E. (1967). Variation in the Ribbed Frog *Ascaphus truei* Stejneger. *Copeia* **1967:** 634–649.

Metter, D. E. (1968). *Ascaphus* and *A. truei*. *Catalogue of American Amphibians and Reptiles* **69:** 1–2.

Meyers, J. J., O'Reilly, J. C., Monroy, J. A., and Nishikawa, K. C. (2004). Mechanism of tongue protraction in microhylid frogs. *Journal of Experimental Biology* **207:** 21–31.

Meylan, P. A. (1987). The phylogenetic relationships of soft-shelled turtles (Family Trionychidae). *Bulletin of the American Museum of Natural History* **186:** 1–110.

Meylan, P. A. (1996). Skeletal morphology and relationships of the Early Cretaceous side-necked turtle, *Araripemys barretoi* (Testudines: Pelomedusoides: Araripemydidae), from the Santana Formation of Brazil. *Journal of Vertebrate Paleontology* **16:** 20–33.

Meylan, P. A., and Gaffney, E. S. (1989). The skeletal morphology of the Cretaceous cryptodiran turtle, *Adocus*, and the relationships of the Trionychoidea. *American Museum Novitates* **2941:** 1–60.

Miaud, C., Andreone, F., Riberon, A., De Michelis, S., Clima, V., Castanet, J., Francillon-Vieillot, H., and Guyetant, R. (2001). Variations in age, size at maturity and gestation duration among two neighbouring populations of the alpine salamander (*Salamandra lanzai*). *Journal of Zoology, London* **254:** 251–260.

Middendorf, G. A., and Sherbrooke, W. C. (1992). Canid elicitation of blood squirting in a horned lizard. *Copeia* **1992:** 519–527.

Milius, S. (2006). No-dad dragons: Komodos reproduce without males. *Science News* **170:** 403.

Miller, J. D. (1996). Reproduction in sea turtles. In *The Biology of Sea Turtles*. P. L. Lutz and J. A. Musick (Eds.). Pp. 51–81. CRC Press, Boca Raton, FL.

Miller, J. R. (2005). Biodiversity conservation and the extinction of experience. *Trends in Ecology & Evolution* **20:** 430–434.

Miller, M. R. (1954). Further observations on reproduction in the lizard *Xantusia vigilis*. *Copeia* **1954:** 38–40.

Miller, M. R., and Lagios, M. D. (1970). The pancreas. In *Biology of the Reptilia. Volume 3. Morphology C*. C. Gans and T. S. Parsons (Eds.). Pp. 319–346. Academic Press, London.

Mills, M. S., Hudson, C. L., and Berna, H. J. (1995). Spatial ecology and movements of the brown water snake (*Nerodia taxispilota*). *Herpetologica* **51:** 412–423.

Milner, A. C. (1988). The relationships and origin of living amphibians. In *The Phylogeny and Classification of the Tetrapods. Vol. 1: Amphibians, Reptiles, Birds*. M. J. Benton (Ed.). Pp. 59–102. Clarendon Press, Oxford.

Milner, A. C. (1980). A review of them Nectridea (Amphibia). In *The Terrestrial Environment and the Origin of Land Vertebrates*. A. L. Panchen (Ed.). Pp. 377–405. Academic Press, London.

Milner, A. C. (1994). Late Triassic and Jurassic amphibians: Fossil record and phylogeny. In *the shadow of the dinosaurs. Early Mesozoic Tetrapods*. N. C. Fraser and H.-D. Sues (Eds.). Pp. 23–37. Cambridge University Press, New York.

Milner, A. R. (1993). The Paleozoic relatives of lissamphibians. *Herpetological Monographs* **7:** 8–27.

Milner, A. R., Smithson, T. R., Milner, A. C., Coates, M. I., and Rolfe, W. D. I. (1986). The search for early tetrapods. *Modern Geology* **10:** 1–28.

Milton, T. H., and Jenssen, T. A. (1979). Description and significance of vocalizations by *Anolis grahami*. *Copeia* **1979:** 481–489.

Min, M. S., Yang, S. Y., Bonett, R. M., Vieites, D. R., Brandon, R. A., and Wake, D. B. (2005). Discovery of the first Asian plethodontid salamander. *Nature* **435:** 87–90.

Minnich, J. E. (1979). Reptiles. In *Comparative Physiology of Osmoregulation in Animals*. G. M. Hughes (Ed.). Pp. 391–641. Academic Press, London.

Minnich, J. E. (1982). The use of water. In *Biology of the Reptilia* (Volume 12): *Physiology C (Physiological Ecology)*. C. Gans and F. H. Pough (Eds.). Pp. 325–395. Academic Press, London.

Mitchell, J. C. (1988). Population ecology and life histories of the freshwater turtles, *Chrysemys picta* and *Sternotherus odoratus* in an urban lake. *Herpetological Monographs* **2:** 40–61.

Mitchell, J. C. (1994). *The Reptiles of Virginia*. Smithsonian Institute Press, Washington, D.C.

Mitchell, J. C., and Beck, R. A. (1992). Free-ranging domestic cat predation on native vertebrates in rural and urban Virginia. *Virginia Journal of Science* **43:** 197–207.

Mitchell, S. L., and Altig, R. (1983). The feeding ecology of *Sooglossus gardineri*. *Journal of Herpetology* **17:** 281–282.

Mittermeier, R. A. (1972). Jamaica's endangered species. *Oryx* **11:** 258–262.

Mittermeier, R. A., Myers, N., and Mittermeier, C. G. 2000. *Hotspots: Earth's biologically richest and most endangered terrestrial ecoregions*. CEMEX, Conservation International.

Mlynarski, M. (1976). *Testudines. Handbuch der Paläoherpetologie, 7*. Gustav Fischer Verlag, Stuttgart.

Modha, M. L. (1967). The ecology of the Nile crocodile (*Crocodylus niloticus* Laurenti) on Central Island, Lake Rudolf. *East African Wildlife Journal* **5:** 74–95.

Moler, P. E., and Kezer, J. (1993). Karyology and systematics of the salamander genus *Pseudobranchus* (Sirenidae). *Copeia* **1993:** 39–47.

Moll, D. (1980). Natural history of the river terrapin, *Batagar baska* (Gray) in Malaysia (Testudines: Emydidae). *Malayasian Journal of Science* **6:** 23–62.

Moll, D. (1989). Food and feeding behavior of the turtle, *Dermatemys mawei*, in Belize. *Journal of Herpetology* **23:** 445–447.

Moll, D. (1990). Population sizes and foraging ecology in a tropical freshwater stream turtle community. *Journal of Herpetology* **24:** 48–53.

Moll, D., and Klemens, M. W. (1996). Ecological characteristics of the pancake tortoise, *Malacochersus tornieri*, in Tanzania. *Chelonian Conservation Biology* **2:** 26–35.

Moll, E. O., and Legler, J. M. (1971). The life history of a Neotropical slider turtle, *Pseudemys scripta* (Schoepff) in Panama. *Bulletin of the Los Angeles County Museum of Science* **11:** 1–102.

Monagas, W. R., and Gatten, R. E., Jr. (1983). Behavioural fever in the turtles *Terrapene carolina* and *Chrysemys picta*. *Journal of Thermal Biology* **8:** 285–288.

Montanucci, R. R. (1974). Convergence, polymorphism or introgressive hybridization: Analysis of interaction between *Crotaphytus collaris* and *Crotaphytus reticulatus* (Sauria: Iguanidae). *Copeia* **1974:** 87–101.

Mook, C. C. (1921). Notes on the postcranial skeleton in Crocodilia. *Bulletin of the American Museum of Natural History* **44:** 67–100.

Moon, B. R. (2000). The mechanics and muscular control of constriction in gopher snakes (*Pituophis melanoleucus*) and a king snake (*Lampropeltis getula*). *Journal of Zoology, London* **252:** 83–98.

Morafka, D. J. (1994). Neonates: Missing links in the life histories of North American tortoises. *U S Fish and Wildlife Service Fish and Wildlife Research* **13:** 161–173.

Morales, V. R., and McDiarmid, R. W. (1996). Annotated checklist of the amphibians and reptiles of Pakita, Manu National Park Reserve Zone, with comments on the herpetofauna of Madre de Dios. In *Manu. The Biodiversity of Southeastern Peru*. D. E. Wilson and A. Sandoval (Eds.). Pp. 503–522. Smithsonian Institute, Washington, D. C.

Morey, S. R. (1990). Microhabitat selection and predation in the Pacific treefrog, *Pseudacris regilla*. *Journal of Herpetology* **24:** 292–296.

Morgan, J. A. T., Vredenburg, V. T., Rachowicz, L. J., Knapp, R. A., Stice, M. J., Tunstall, T., Bingham, R. E., Parker, J. M., Longcore, J. E., Moritz, C., Briggs, C. J., and Taylor, J. W. (2007). Population genetics of the frog-killing fungus *Batrachochytrium dendrobatidis*. *Proceedings of the National Academy of Sciences (USA)* **104:** 13845–13850.

Mori, A. (1996). A comparative study of the development of prey handling behavior in young rat snakes, *Elaphe quadrivirgata* and *E. climacophora*. *Herpetologica* **52:** 313–322.

Morin, P. J. (1987). Predation, breeding asunchrony, and the outcome of competition among treefrog tadpoles. *Ecology* **68:** 675–683.

Moritz, C., and Hillis, D. M. (1996). Molecular systematics: context and controversies. In *Molecular Systematics*. Second Edition. D. M. Hillis, C. Moritz, and B. K. Mable (Eds.). Pp. 1–13. Sinauer Associates, Inc, Sunderland, MA.

Moritz, C., Patten, J. L., Schneider, C. J., and Smith, T. B. (2000). Diversification of rainforest faunas: an integrated molecular approach. *Annual Reviews of Ecology and Systematics* **31:** 533–563.

Morreale, S. J., Ruiz, G. J., Spotila, J. R., and Standora, E. A. (1982). Temperature-dependent sex determination: Current practices threaten conservation of sea turtles. *Science* **216:** 1245–1247.

Mosher, H. S., Fischer, H. G., Fuhrman, F. A., and Buchwald, H. D. (1964). Tarichatoxin-tetrodotoxin - potent neurotoxin: Nonprotein substance isolated from California newt is same as toxin from puffer fish. *Science* **144:** 1100–1110.

Moskovits, D. K., and Bjorndal, K. (1990). Diet and Food Preferences of the tortoises *Geochelone carbonaria* and *G. denticulata* in northwestern Brazil. *Herpetogica* **46:** 207–218.

Motychak, J. E., and Brodie, E. D. (1999). Evolutionary response of predators to dangerous prey: Preadaptation and the evolution of tetrodotoxin resistance in garter snakes. *Evolution* **53:** 1528–1535.

Mouton, P. le F. N. (1997). Family Cordylidae. In *Proceedings of the Fitzsimons Commemorative Symposium*. J. H. van Dyke (Ed.). Pp. 19–23. Herpetological Association of Africa, Matieland, S. Afr.

Mrosovsky, N. (1983). *Conserving Sea Turtles*. British Herpetological Society, London.

Mrosovsky, N. (1994). Sex ratios in sea turtles. *Journal of Experimental Zoology* **270:** 16–27.

Mueller, H. (2006). Ontogeny of the skull, lower jaw, and hyobranchial skeleton of *Hypogeophis rostratus* (Amphibia: Gymnophiona: Caeciliidae) revisited. *Journal of Morphology* **267:** 968–986.

Mueller, R. L., Macey, J. R., Jaekel, M., Wake, D. B., and Boore, J. L. (2004). Morphological homoplasy, life history evolution, and historical biogeography of plethodontid salamanders inferred from complete mitochondrial genomes. *Proceedings of the National Academy of Sciences (USA)* **101:** 13820–13825.

Müller, R., Lutzmann, N., and Walbröl, U. (2004). *Furcifer pardalis—Das Pantherchamäleon*. Natur und Tier Verlag, Münster, Germany.

Murphy, J. B., Quinn, H, and Campbell, J. A. (1977). Observations on the breeding habits of the aquatic caecilian *Typhlonectes compressicaudus*. *Copeia* **1977:** 66–69.

Murphy, J. C., and Voris, H. K. (1994). A key to the homalopsine snakes. *The Snake* **26:** 123–133.

Murphy, P. J. (2003a). Context-dependent reproductive site choice in a Neotropical frog. *Behavioral Ecology* **14:** 626–633.

Murphy, P. J. (2003b). Does reproductive site choice in a Neotropical frog mirror variable risks facing offspring? *Ecological Monographs* **73:** 45–67.

Murphy, R. W., Sites, J. W., Buth, D. G., and Haufler, C. H. (1996). Proteins: Isozyme electrophoresis. In *Molecular Systematics* (Second Edition). D. M. Hillis, C. Moritz, and B. K. Mable (Eds.). Pp. 51–120. Sinauer Associates, Inc., Sunderland, MA.

Mushinsky, H. R., Hebrard, J. J., and Vodopich, D. S. (1982). Ontogeny of water snake foraging ecology. *Ecology* **63:** 1624–1629.

Myers, C. W., and Daly, J. W. (1983). Dart-poison frogs. *Scientific American* **248:** 120–133.

Myers, C. W., and Rand, A. S. (1969). Checklist of amphibians and reptiles of Barro Colorado Island, Panama, with comments on faunal change and sampling. *Smithsonian Contributions to Zoology* **10:** 1–11.

Myers, C. W., Paolillo, O. A., and Daly, J. W. (1991). Discovery of a defensively malodorous and nocturnal frog in the family Dendrobatidae: phylogenetic significance of a new genus and species from the Venezuelan Andes. *American Museum Novitates* **3022:** 1–33.

Myers, E. M., and Zamudio, K. R. (2004). Multiple paternity in an aggregate breeding amphibian: the effect of reproductive skew on estimates of male reproductive success. *Molecular Ecology* **13:** 1951–1963.

Myers, N., Mittermeier, R. A., Mittermeier, C. G., da Fonesca, G. A. B., and Kent, J. (2000). Biodiversity hotspots for conservation priorities. *Nature* **403:** 853–858.

Nagy, K. A. (1972). Water and electrolyte budgets of a free-living desert lizard, *Sauromalus obesus*. *Journal of Comparative Physiology B* **79:** 39–62.

Nagy, K. A. (1982). Field studies of water relations. In *Biology of the Reptilia* (Volume 12): *Physiology C (Physiological Ecology)*. C. Gans and F. H. Pough (Eds.) Pp. 483–501. Academic Press, London.

Nagy, K. A. (1983a). The doubly labeled water ($^3H^{18}O$) method: a guide to its use. *University of California, Los Angeles, Publication* **12–1417:** 1–45.

Nagy, K. A. (1983b). Ecological energetics. In *Lizard Ecology: Studies of a Model Organism*. R. B. Huey, E. R. Pianka, and T. W. Schoener (Eds.). Pp. 25–54. Harvard University Press, Cambridge, Mass.

Nagy, K. A., and Medica, P. A. (1986). Physiological ecology of desert tortoises in southern Nevada. *Herpetologica* **42:** 73–92.

Nagy, K. A., Huey, R. B., and Bennett, A. F. (1984). Field energetics and foraging mode of Kalahari lacertid lizards. *Ecology* **65:** 588–596.

Nagy, Z. T., Vidal, N., Vences, M., Branch, W. R., Pauwels, O. S. G., Wink, M., and Joger, U. (2005). Molecular systemaics of African Colubroidea (Squamata: Serpentes). In *African Biodiversity: Molecules, Organisms, Ecosystems, Proceedings of the 5th International Symposium on Tropical Biology*. B. A. Huber, B. J. Sinclair, and e Lampe K. H. (eds.). Pp. 221–228. Museum Koenig, Bonn.

Nanayakkara, G. L. A. (1988). An analysis of the defensive behaviour of the Sri Lankan pipe snake, *Cylindrophis maculatus* (Linne, 1758). *The Snake* **20:** 81–83.

National Research Council (1990). *Decline of the Sea Turtles*. National Academy Press, Washington, D.C.

Narahashi, T. (2001). Pharmacology of tetrodotoxin. *Journal of Toxicology-Toxin Review* **20:** 67–84.

Navas, C. A., and Chaui-Berlinck, J. G. (2007). Respiratory physiology of high-altitude anurans: 55 years of research on altitude and oxygen. *Respiratory Physiology & Neurobiology* **158:** 307–313.

Near, T. J., Meylan, P. A., and Shaffer, H. B. (2005). Assessing concordance of fossil calibration points in molecular clock studies: An example using turtles. *American Naturalist* **165:** 137–146.

Necas, P. (2004). *Chameleons, Nature's Hidden Jewels*. Edition Chimaira, Frankfurt am Main.

Necas, P., and Schmidt, W. (2004). *Stump-tailed Chameleons. Miniature Dragons of the Rainforest*. Edition Chimaira, Frankfurt, Germany.

Nei, M., and Kumar, S. (2000). *Molecular Evolution and Phylogenetics*. Oxford University Press, New York.

Nelsen, O. E. (1953). *Comparative Embryology of the Vertebrates*. Blakiston Co., Inc, New York.

Nelson, N. J., Thompson, M. B., Pledger, S., Keall, S. N., and Daugherty, C. H. (2004). Egg mass determines hatchling size, and incubation temperature influences post-hatching growth, of tuatara *Sphenodon punctatus*. *Journal of Zoology, London* **263:** 77–87.

Newman, D. G. (1987). Burrow use and population densities of tuatara (*Sphenodon punctatus*) and how they are influenced by fairy prions (*Pachyptila turtur*) on Stephens Island, New Zealand. *Herpetologica* **43:** 336–344.

Newman, D. G., and McFadden, I. (1990). Status of the tuatara, *Sphenodon punctatus*, on Hongiora and Ruamahua-Iti islands, aldermen group, New Zealand. *New Zealand Journal of Zoology* **17:** 153–156.

Newman, D. G., Watson, P. R., and McFadden, I. (1994). Egg-production by tuatara on Lady Alice and Stephens Island, New-Zealand. *New Zealand Journal of Zoology* **21:** 387–398.

Newman, D. G., Watson, P. R., Grigg, G., Shine, R., and Ehmann, H. (1985). The contribution of radiography to the study of the reproductive ecology of the tuatara, *Sphenodon punctatus*. In *Biology of Austrasian Frogs and Reptiles*. G. Grigg, R. Shine, and H. Ehmann (Eds.). Pp. 7–10. Surrey Beatty & Sons Pty Ltd, Chipping Norton Pty Ltd., Sydney, Australia.

Newton, W. D., and Trauth, S. E. (1992). Ultrastructure of the spermatozoon of the lizard *Cnemidophorus sexlineatus* (Sauria: Teiidae). *Herpetologica* **48:** 330–343.

Nicholson, K. E. (2002). Phylogenetic analysis and a test of the current infrageneric classification of *Norops* (beta *Anolis*). *Herpetological Monographs* **16:** 93–120.

Nickerson, M. A., and May, C. E. (1973). The hellbenders: North American giant salamanders. *Milwaukee Public Museum, Publications in Biology and Geology* **1:** 1–106.

Nielsen, R. (2005). Molecular signatures of natural selection. *Annual Review of Genetics* **39:** 197–218.

Nielson, M., Lohman, K., and Sullivan, J. (2001). Phylogeography of the tailed frog (*Ascaphus truei*): Implications for the biogeography of the Pacific Northwest. *Evolution* **55:** 147–160.

Niewiarowski, P. H. (1994). Understanding geographic life-history variation in lizards. In *Lizard Ecology: Historical and Experimental Perspectives*. L. J. Vitt and E. R. Pianka (Eds.). Pp. 31–49. Princeton University Press, Princeton NJ.

Niewiarowski, P. H., Congdon, J. D., Dunham, A. E., Vitt, L. J., and Tinkle, D. W. (1997). Tales of lizard tails: Effects of tail autotomy on subsequent survival and growth of free-ranging hatchling *Uta stansburiana*. *Canadian Journal of Zoology* **75:** 542–548.

Noble, G. K. (1922). The phylogeny of the Salientia. I. The osteology and the thigh musculature; their bearing on classification and phylogeny. *Bulletin of the American Museum of Natural History* **46:** 1–87.

Noble, G. K. (1925). The integumentary, pulmonary, and cardiac modifications correlated with increased cutaneous respiration in the Amphibia: A solution of the 'hairy frog' problem. *Journal of Morphology and Physiology,* **40:** 341–416.

Noble, G. K. (1931). *The Biology of the Amphibia*. McGraw-Hill Book Co, New York, NY.

Noble, G. K., and Bradley, H. T. (1933). The mating behavior of lizards; Its bearing on the theory of sexual selection. *Annals of the New York Academy of Sciences* **35:** 25–100.

Noble, G. K., and Clausen, H. J. (1936). The aggregation behavior of *Storeria dekayi* and other snakes, with especial reference to the sense organs involved. *Ecological Monographs* **6:** 271–316.

Noble, G. K., and Mason, E. R. (1933). Experiments on the brooding habits of the lizards *Eumeces* and *Ophisaurus*. *American Museum Novitates* **619:** 1–29.

Nogueira, C, Paula H. Valdujo, P. H., and França, F. G. R. Habitat variation and lizard diversity in a Cerrado area of Central Brazil. *Studies on Neotropical Fauna and Environment* **40:** 105–112.

Nokhbatolfoghahai, A., and Downie, J. R. (2007). Amphibian hatching gland cells: Pattern and distribution in anurans. *Tissue & Cell* **39:** 225–240.

Nöllert, A. (1990). *Die Knoblauchkröte*. A. Ziemsen Verlag, Wittenberg Lutherstadt.

Noonan, B. P. (2000). Does the phylogeny of pelomedusoid turtles reflect vicariance due to continental drift? *Journal of Biogeography* **27:** 1245–1249.

Noonan, B. P., and Chippindale, P. T. (2006). Vicariant origin of Malagasy reptiles supports late cretaceous antarctic land bridge. *American Naturalist* **168:** 730–741.

Norell, M. A. (1989). The higher level relationships of the extant Crocodylia. *Journal of Herpetology* **23:** 325–335.

Norell, M. A., and de Queiroz, K. (1991). The earliest iguanine lizard (Reptilia: Squamata) and its bearing on iguanine phylogeny. *American Museum Novitates* **2997:** 1–16.

Norell, M. A., and Gao, K. (1997). Braincase and phylogenetic relationships of *Estesia mongoliensis* from the Late Cretaceous of the Gobi Desert and the recognition of a new clade of lizards. *American Museum Novitates* **3211:** 1–25.

Norell, M. A., Clark, J.M, and Hutchison, J. H. (1994). The Late Cretaceous alligatoroid *Brachychamps Montana* [sic] (Crocodylia): New material and putative relationships. *American Museum Novitates* **3116:** 1–26.

Nores, M. (1999). An alternative hypothesis for the origin of Amazonian bird diversity. *Journal of Biogeography* **26:** 475–485.

Normark, B. B., Judson, O. P., and Moran, N. A. (2003). Genomic signatures of ancient asexual lineages. *Biological Journal of the Linnean Society* **79:** 69–84.

Norris, K. S., and Dawson, W. R. (1964). Observations on the water economy and electrolyte excretion of chuckwallas (Lacertilia, *Sauromalus*). *Copeia* **1964:** 638–646.

Nuñez, J. J., and Formas, J. R. (2000). Evolutionary history of the Chilean frog genus *Telmatobufo* (Leptodactylidae): an immunological approach. *Amphibia-Reptilia* **21:** 351–356.

Nussbaum, R. A. (1976). Geographic variation and systematics of salamanders of the genus *Dicamptodon* Strauch (Ambystomatidae). *Miscellaneous Publications of the Museum of Zoology, University of Michigan* **149:** 1–94.

Nussbaum, R. A. (1977). Rhinatrematidae: a new family of caecilians (Amphibia: Gymnophiona). *Occasional Papers of the Museum of Zoology, University of Michigan* **682:** 1–30.

Nussbaum, R. A. (1979). The taxonomic status of the caecilian genus *Uraeotyphlus* Peters. *Occasional Papers of the Museum of Zoology, University of Michigan* **687:** 1–20.

Nussbaum, R. A. (1980). Phylogenetic implications of amplectic behavior in sooglossid frogs. *Herpetologica* **36:** 1–5.

Nussbaum, R. A. (1983). The evolution of a unique dual jaw-closing mechanism in caecilians (Amphibia: Gymnophiona) and its bearing on caecilian ancestry. *Journal of Zoology, London* **199:** 545–554.

Nussbaum, R. A. (1984). Amphibians of the Seychelles. In *Biogeography and Ecology of the Seychelles Islands*. D. R. Stoddart (Ed.). Pp. 378–415. W. Junk, The Hague.

Nussbaum, R. A. (1985a). Systematics of caecilians (Amphibia: Gymnophiona) of the family Scolecomorphidae. *Occasional Papers of the Museum of Zoology, University of Michigan* **713:** 1–49.

Nussbaum, R. A. (1985b). Amphibian fauna of the Seychelles Archipelago. *National Geographic Society Research Reports* **1977:** 53–62.

Nussbaum, R. A., and Hinkel, H. (1994). Revision of East African caecilians of the genera *Afrocaecilia* Taylor and *Boulengerula* Tornier (Amphibia: Gymnophiona: Caeciliaidae). *Copeia* **1994:** 750–760.

Nussbaum, R. A., and Tait, C. K. (1977). Aspects of the life history and ecology of the Olympic salamander, *Rhyacotriton olympicus* (Gaige). *American Midland Naturalist* **98:** 176–199.

Nussbaum, R. A., and Wilkinson, M. (1989). On the classification and phylogeny of caecilians (Amphibia: Gymnophiona), a critical review. *Herpetological Monographs* **3:** 1–42.

Nussbaum, R. A., and Wu, S. H. (1995). Distribution, variation, and systematics of the Seychelles treefrog, *Tachycnemis seychellensis* (Amphibia: Anura: Hyperoliidae). *Journal of Zoology, London* **236:** 383–406.

Nussbaum, R. A., and Wu, S. H. (2007). Morphological assessments and phylogenetic relationships of the Seychellean frogs of the family Sooglossidae (Amphibia: Anura). *Zoological Studies* **46:** 322–335.

Nussbaum, R. A., Brodie, E. D., Jr., and Storm, R. M. (1983). *Amphibians and Reptiles of the Pacific Northwest*. University Press Idaho, Moscow, Idaho.

Nussbaum, R. A., Jaslow, A., and Watson, J. (1982). Vocalization in frogs of the family Sooglossidae. *Journal of Herpetology* **16:** 198–204.

Nussbaum, R. A., Raxworthy, C. J., Raselimanana, A. P., and Ramanamanjato, J.-B. (1999). Amphibians and reptiles of the Reserve Naturelle Integrale d'Andohahela, Madagascar. *Fieldiana Zoology* **94:** 155–173.

Nussbauum, R. A., and Wilkinson, M. (1995). A new genus of lungless tetrapod: A radically divergent caecilian (Amphibiia: Gymnophiona). *Proceedings of the Royal Society B-Biological Sciences* **261:** 331–335.

Nussear, K. E., Esque, T. C., Haines, D. F., and Tracy, C. R. (2007). Desert Tortoise hibernation: Temperatures, timing, and environment. *Copeia* **2007:** 378–386.

O'Shea, M. (1996). *A Guide to the Snakes of Papua New Guinea*. Independent Publishing, Port Moresby.

O'Connor, D. E., and Shine, R. (2004). Parental care protects against infanticide in the lizard *Egernia saxatilis* (Scincidae). *Animal Behaviour* **68:** 1361–1369.

Odierna, G., Canapa, A., Andreone, F., Aprea, G., and Barucca, M. (2002). A Phylogenetic Analysis of Cordyliformes (Reptilia: Squamata): Comparison of Molecular and Karyological Data. *Molecular Phylogenetics and Evolution* **23:** 37–42.

Odinchenko, V. I., and Latyshev, V. A. (1996). Keeping and breeding in captivity the Mexican burrowing python *Loxocemus bicolor* (Cope, 1861) at Moscow Zoo. *Russian Journal of Herpetology* **3:** 95–97.

Odum, E. P. (1993). *Ecology and Our Endangered Life-Support Systems*. Sinauer Associates, Inc., Sunderland MA.

Oelrich, T. M. (1956). The anatomy of the head of *Ctenosaura pectinata* (Iguanidae). *Miscellaneous Publications of the Museum of Zoology, University of Michigan* **94:** 1–122.

Ohler, A., Grosjean, S., and Hoyos, J. M. (1999). Observation of nest constructing in *Taylorana hascheana* (Anura: Ranidae). *Revue Francaise d'Aquariologie Herpetologie* **26:** 67–70.

Olalla-Tarraga, M. A., and Rodriguez, M. A. (2007). Energy and interspecific body size patterns of amphibian faunas in Europe and North America: anurans follow Bergmann's rule, urodeles its converse. *Global Ecology and Biogeography* **16:** 606–617.

Oliveira, P. S., and Marquis, R. J. (Eds.) (2002). *The Cerrados of Brazil: Ecology and Natural History of a Neotropical Savanna*. Columbia University Press, New York.

Olsson, M., and Madsen, T. (1996). Costs of mating with infertile males selects for late emergence in females and lizards (Lacerta agilis L.). *Copeia* **1996:** 462–464.

Olsson, M., and Madsen, T. (1998). Sexual selection and sperm competition in reptiles. In *Sexual Selection and Sperm Competition*. T. Birkhead and A. P. Moller (Eds.). Pp. 503–577. Academic Press, London.

Olsson, M., and Shine, R. (1997). The limits of reproductive output: Offspring size versus number in the sand lizard (*Lacerta agilis*). *American Naturalist* **149:** 179–188.

Olsson, M., Madsen, T., and Shine, R. (1997). Is sperm really so cheap? Costs of reproduction in male adders, *Vipera berus*. *Proceedings of the Royal Society of London B* **264:** 455–459.

Olsson, M., Shine, R., and Madsen, T. (1996). Sperm selection by females. *Nature* **383:** 585.

Oosterbrock, P. (1987). More appropriate definitions of paraphyly and polyphyly, with a comment on the Farris 1974 model. *Systematic Zoology* **36:** 103–108.

Organ, J. A. (1961). Studies of the local distribution, life history, and population dynamics of the salamander genus *Desmognathus* in Virginia. *Ecological Monographs* **31:** 189–220.

Orlov, N. L. (2000). Distribution, comparative morphology and biology of snakes of *Xenopeltis* (Serpentes: Macrostomata: Xenopeltidae) in Vietnam. *Russian Journal of Herpetology* **7:** 103–114.

Orlov, N. L. (1997). Viperid snakes (Viperidae Bonaparte, 1840) of Tam-Dao mountain ridge (Vinh-Phu and Bac-Thai Provinces, Vietnam). *Russian Journal of Herpetology* **4:** 64–74.

Ortega, C. E., Stranc, D. S., Casal, M. P., Hallman, G. M., and Muchlinski, A. E. (1991). A positive fever response in *Agama agama* and *Sceloporus orcutti* (Reptilia, Agamidae and Iguanidae). *Journal of Comparatvie Physiology B* **161:** 377–381.

Orton, G. L. (1953). The systematics of vertebrate larvae. *Systematic Zoology* **2:** 63–75.

Osborne, L., and Thompson, M. B. (2005). Chemical composition and structure of the eggshell of three oviparous lizards. *Copeia* **2005:** 683–692.

Ostrom, J. H. (1991). The question of the origin of birds. In *Origins of the Higher Groups of Tetrapods. Controversy and Consensus*. H.-P. Schultze and L. Trueb (Eds.). Pp. 467–484. Comstock Publishing Associates, Ithaca, NY.

Ota, H., Honda, M., Kobayashi, M., Sengoku, S., and Hikida, T. (1999). Phylogenetic relationships of eublepharid geckos (Reptilia: Squamata): A molecular approach. *Zoological Science* **16:** 659–666.

Oterino, J., Toranzo, G. S., Zelarayan, L., Ajmat, M. T., Bonilla, F., and Buhler, M. I. (2006). Behaviour of the vitelline envelope in *Bufo arenarum* oocytes matured in vitro in blockade to polyspermy. *Zygote* **14:** 97–106.

Ottaviani, G., and Tazzi, A. (1977). The lymphatic system. In *Biology of the Reptilia. Volume 6. Morphology E*. C. Gans, and T. S. Parsons (Eds.). Pp. 315–462. Academic Press, London.

Ouboter, P. E., and Nanhoe, L. M. R. (1989). Notes on nesting and parental care in *Caiman crocodilus crocodilus* in northern Suriname and an analysis of crocodilian nesting habitats. *Amphibia-Reptilia* **8:** 331–348.

Overall, K. L. (1994). Lizard egg environments. In *Lizard Ecology: Historical and Experimental Perspectives*. L. J. Vitt and E. R. Pianka (Eds.). Pp. 51–72. Princeton University Press, Princeton, NJ.

Packard, G. C., and Kilgore, D. L., Jr. (1968). Internal fertilization: Adaptive value to early amniotes. *Journal of Theoretical Biology* **20:** 245–248.

Packard, G. C., and Packard, M. J. (1980). Evolution of the cleidoic egg among reptilian antecedents of birds. *American Zoologist* **20:** 351–362.

Packard, G. C., and Packard, M. J. (1988). The physiological ecology of reptilian eggs and embryos. In *Biology of the Reptilia. Volume 16, B, Defense and Life History*. C. Gans and R. B. Huey (Eds.). Pp. 523–605. Alan R. Liss, Inc, New York.

Packard, G. C., Packard, M. J., and Benigan, L. (1991). Sexual differentiation, growth, and hatching success by embryonic painted turtles incubated in wet and dry environments at fluctuation temperatures. *Herpetologica* **47:** 125–132.

Packard, G. C., Taigen, T. L., Packard, M. J., and Boardman, T. J. (1981). Changes in mass of eggs of softshell turtles (*Trionyx spiniferus*) incubated under hydric conditions simulating those of natural nests. *Journal of Zoology, London* **193:** 81–90.

Packard, G. C., Tracy, C. R., and Roth, J. J. (1977). The physiological ecology of reptilian eggs and embryos, and the evolution of viviparity within the class Reptilia. *Biological Reviews* **52:** 71–105.

Packard, M. J. (1994). Patterns of mobilization and deposition of calcium in embryos of oviparous, amniotic vertebrates. *Israel Journal of Zoology* **40:** 481–492.

Packard, M. J., and DeMarco, V. G. (1991). Eggshell structure and formation in eggs of oviparous reptiles. In *Egg Incubation: Its Effects on Embryonic Development in Birds and Reptiles*. D. C. Deeming and M. W. J. Ferguson (Eds.). Pp. 53–69. Cambridge University Press, Cambridge.

Packard, M. J., and Packard, G. C. (1985). Effect of variation in water-balance of eggs on calcium mobilizaton and growth by embryonic painted turtles (*Chrysemys picta*). *American Zoologist* **25:** A134–A134.

Packard, M. J., and Seymour, R. S. (1997). Evolution of the amniote egg. In *Amniote Origins. Completing the Transition to Land*. S. S. Sumida and K. L. M. Martin (Eds.). Pp. 265–290. Academic Press, San Diego.

Packard, M. J., Packard, G. C., and Boardman, T. J. (1982). Structure of eggshells and water relations of reptilian eggs. *Herpetologica* **38:** 136–155.

Paladino, F. V., O'Connor, M. P., and Spotila, J. R. (1990). Metabolism of leatherback turtles, gigantothermy, and thermoregulation of dinosaurs. *Nature* **344:** 858.

Palmer, B. D., Demarco, V., and Guillette, L. J., Jr. (1993). Oviductal morphology and eggshell formation in the lizard, *Sceloporus woodi*. *Journal of Morphology* **217:** 205–217.

Palumbi, S. R. (1996). Nucleic acids II: The polymerase chain reaction. In *Molecular Systematics, Second Edition*. D. M. Hillis, C. Moritz, and B. K. Mable (Eds.). Pp. 205–247. Sinauer Associates, Inc., Sunderland, MA.

Panchen, A. L. (1991). The early tetrapods: Classification and the shape of cladograms. In *Origins of the Higher Groups of Tetrapods. Controversy and Consensus*. H.-P. Schultze and L. Trueb (Eds.). Pp. 110–144. Comstock Publishing Associates, Ithaca, NY.

Panchen, A. L., and Smithson, T. R. (1988). The relationships of the earliest tetrapods. In *The Phylogeny and Classification of the Tetrapods. Vol. 1: Amphibians, Reptiles, Birds*. M. J. Benton (Ed.). Pp. 1–32. Clarendon Press, Oxford.

Pang, P. K. T. and Schreibman, M. P. (Eds.). (1986). *Vertebrate Endocrinology: Fundamentals and Biomedical Implications. Vol. 1. Morphological Considerations*. Academic Press, Orlando, FL.

Papenfuss, T. J. (1982). The ecology and systematics of the amphisbaenian genus *Bipes*. *Occasional Papers of the California Academy of Sciences* **136:** 1–42.

Papenfuss, T. J. (1986). Amphibian and reptile diversity along elevational transects in the White-Inyo Range. In *Natural History of the White-Inyo Range, Eastern California and Western Nevada and High Altitude Physiology*. C. A. Hall, Jr., and D. J. Young (Eds.). Pp. 129–136. University of California White Mountain Research Station Symposium, Bishop, CA.

Parham, J. F., and Fastovsky, D. E. (1998). The phylogeny of cheloniid sea turtles revisited. *Chelonian Conservation Biology* [1997]. **2:** 548–554.

Parham, J. F., Feldman, C. R., and Boore, J. L. (2006a). The complete mitochondrial genome of the enigmatic bigheaded turtle (*Platysternon*): Description of unusual genomic features and the reconciliation of phylogenetic hypotheses based on mitochondrial and nuclear DNA. *Bmc Evolutionary Biology* **6**.

Parham, J. F., Macey, J. R., Papenfuss, T. J., Feldman, C. R., Turkozan, O., Polymeni, R., and Boore, J. (2006b). The phylogeny of Mediterranean tortoises and their close relatives based on complete mitochondrial genome sequences from museum specimens. *Molecular Phylogenetics and Evolution* **38:** 50–64.

Parham, J. F., Simison, W. B., Kozak, K. H., and Feldman, C. R. (2000). A reassessment of some recently described Chinese turtles. *American Zoologist* **40:** 1163.

Parham, J. F., Simison, W. B., Kozak, K. H., Feldman, C. R., and Shi, H. T. (2001). New Chinese turtles: Endangered or invalid? A reassessment of two species using mitochondrial DNA, allozyme electrophoresis and known-locality specimens. *Animal Conservation* **4:** 357–367.

Parker, E. D., and Selander, R. K. (1976). The organization of genetic diversity in the parthenogenetic lizard *Cnemidophorus tesselatus*. *Genetics* **84:** 791–805.

Parker, H. W. (1934). *A Monograph of the Frogs of the Family Microhylidae*. The British Museum, London.

Parker, M. S. (1994). Feeding ecology of stream-dwelling Pacific giant salamander larvae. (*Dicamptodon tenebrosus*). *Copeia* **1994:** 705–718.

Parker, W. S., and Brown, W. S. (1980). Comparative ecology of two colubrid snakes, *Masticophis t. taeniatus* and *Pituophis melanoleucus deserticola*, in northern Utah. *Milwaukee Public Museum, Publications in Biology and Geology* **7:** 1–104.

Parmenter, C. J. (1985). Reproduction and survivorship of *Chelodina longicollis* (Testudinata: Chelidae). In *Biology of Australasian Frogs and Reptiles*. G. Grigg, R. Shine, and H. Ehmann (Eds.). Pp. 53–61. Surrey Beatty & Sons., Chipping Norton Pty Ltd., Sydney, Australia.

Parmenter, C. J., and Heatwole, H. (1975). Panting thresholds of lizards. 4. Effect of dehydration on panting threshold of *Amphibolurus barbatus* and *Amphibolurus muricatus*. *Journal of Experimental Zoology* **191:** 327–332.

Parmley, D., and Holman, J. A. (1995). Hemphillian (Late Miocene) snakes from Nebraska, with comments on Arikareea through Blancan snakes of midcontinental North America. *Journal of Vertebrate Paleontology* **15:** 79–95.

Parra-Olea, G., Garcia-Paris, M., and Wake, D. B. (2004). Molecular diversification of salamanders of the tropical American genus *Bolitoglossa* (Caudata: Plethodontidae) and its evolutionary and biogeographical implications. *Biological Journal of the Linnean Society* **81:** 325–346.

Parrish, J. M. (1997). Evolution of the archosaurs. In *The Complete Dinosaur*. J. O. Farlow and M. K. Brett-Surman (Eds.). Pp. 191–203. Indiana University Press, Bloomington.

Parsons, T. S. (1970). The nose and Jacobson's organ. In *Biology of the Reptilia. Volume 2. Morphology B*. C. Gans and T. S. Parsons (Eds.). Pp. 99–191. Academic Press, London.

Passmore, N., and Carruthers, S. V. (1995). *South African Frogs. A Complete Guide. Revised edition*. Southern Books & Witwatersrand University Press, Johannesburg.

Patchell, F. C., and Shine, R. (1986a). Feeding mechanisms in pygopodid lizards: How can *Lialis* swallow such large prey? *Journal of Herpetology* **20:** 59–64.

Patchell, F. C., and Shine, R. (1986b). Food habitas and reproductive biology of the Australian legless lizards (Pygopodidae). *Copeia* **1986:** 30–39.

Patten, J. L., da Silva, M. N. F., and Malcolm, J. R. (2000). Mammals of the Rio Juruá and the evolutionary and ecological diversification of Amazonia. *Bulletin of the American Museum of Natural History* **244:** 1–306.

Pauly, G. B., Bernal, X. E., Rand, A. S., and Ryan, M. J. (2006). The vocal sac increases call rate in the tungara frog *Physalaemus pustulosus*. *Physiological and Biochemical Zoology* **79:** 708–719.

Pearman, P. B. (1997). Correlates of amphibian diversity in an altered landscape of Amazonian Ecuador. *Conservation Biology* **11:** 1211–1225.

Pearson, O. P. (1977). Effect of substrate and of skin color on thermoregulation of a lizard. *Comparative Biochemistry and Physiology* **58A:** 353–358.

Pearson, O. P., and Bradford, D. F. (1976). Thermoregulation of lizards and toads at high altitudes in Peru. *Copeia* **1976:** 155–169.

Pechmann, J. H. K., and Semlitsch, R. D. (1986). Diel activity patterns in the breeding migrations of winter-breeding anurans. *Canadian Journal of Zoology* **64:** 1116–1120.

Pechmann, J. H. K., and Wake, D. B. (1997). Declines and disappearances of amphibian populations. In *Principles of Conservation Biology*. G. K. Meffe and C. R. Carroll (Eds.). Pp. 135–137. Sinauer Associates, Inc., Sunderland, MA.

Pechmann, J. H. K., and Wilbur, H. M. (1994). Putting declining amphibian populations in perspective: Natural fluctuations and human impacts. *Herpetologica* **50:** 65–84.

Pechmann, J. H. K., Scott, D. E., Gibbons, J. W., and Semlitsch, R. D. (1989). Influence of wetland hydroperiod on diversity and abundance of metamorphosing juvenile amphibians. *Wetlands Ecology and Management* **1:** 3–11.

Pechmann, J. H. K., Scott, D. E., Semlitsch, R. D., Caldwell, J. P., Vitt, L. J., and Gibbons, J. W. (1991). Declining amphibian populations: The problem of separating human impacts from natural fluctuations. *Science* **253:** 892–895.

Pefaur, J. E. (1993). Description of a new *Colostethus* (Dendrobatidae) with some natural history comments on the genus in Venezuela. *Alytes* **11:** 88–96.

Pefaur, J. E., and Duellman, W. E. (1980). Community structure in high Andean herpetofaunas. *Transactions of the Kansas Academy of Science* **83:** 45–65.

Pellegrino, K. C. M., Rodrigues, M. T., Yonenaga-Yassuda, Y., and Sites, J. J. W. (2001). A molecular perspective on the evolution of microteiid lizards (Squamata, Gymnophthalmidae), and a new classification for the family. *Biological Journal of the Linnean Society* **74:** 315–338.

Péres, A. K., and Colli, G. R. (2004). The taxonomic status of *Tupinambis rufescens* and *T. duseni* (Squamata: Teiidae), with a redescription of the two species. *Occasional Papers of the Sam Noble Oklahoma Museum of Natural History* **15:** 1–12.

Perez, I., Gimenez, A., Sanchez-Zapata, J. A., Anadon, J. D., Martinez, M., and Esteve, M. A. (2004). Non-commercial collection of spur-thighed tortoises (*Testudo graeca graeca*): A cultural problem in southeast Spain. *Biological Conservation* **118:** 175–181.

Perez-Higareda, G., Rangel-Rangel, A., Smith, H. M., and Chiszar, D. (1989). Comments on the food and feeding habits of Morelet's crocodile. *Copeia* **1989:** 1039–1041.

Pérez-Mellado, V., Riera, N., and Perera, A. (Eds.) (2004). *The Biology of Lacertid Lizards: Evolutionary and Ecological Perspectives*. Institut Menorquí d'Estudis, Maó, Menorca.

Pérez-Santos, C., and Moreno, A. G. (1991). Serpientes de Ecuador. *Museo Regionale di Scienze Naturali – Torino Monografie* **11:** 1–538.

Pergams, O. R. W., and Zaradic, P. A. (2008). Evidence for a fundamental and pervasive shift away from nature-based recreation. *Proceedings of the National Academy of Sciences (USA)* **105:** 2295–2300.

Perret, J.- L. (1966). Les amphibiens du Cameroun. *Zoologische Jahrbücher Abtheelungen Systematik, Ökologie und Geographie der Tiere* **93:** 289–464.

Perry, G. (1999). The evolution of search modes: Ecological versus phylogenetic perspectives. *American Naturalist* **153:** 98–109.

Perry, G. (2007). Movement patterns in lizards: Measurement, modality, and behavioral correlates. In *Lizard Ecology: The Evolutionary Consequences of Foraging Mode*. S. M. Reilly, L. D. McBrayer, and D. B. Miles (Eds.). Pp. 13–48. Cambridge University Press, Cambridge, UK.

Perry, G., and Pianka, E. R. (1997). Animal foraging: Past, present, and future. *Trends in Ecology & Evolution* **12:** 360–364.

Perry, G., Lampl, I., Lerner, A., Rothenstein, D., Shani, E., Sivan, N., and Werner, Y. L. (1990). Foraging mode in lacertid lizards: Variation and correlates. *Amphibia-Reptilia* **11:** 373–384.

Perry, S. F. (1983). Reptilian lungs. Functional anatomy and evolution. *Advances in Anatomy, Embryology and Cell Biology* **29:** 1–81.

Perry, S. F. (1998). Lungs: Comparative anatomy, functional morphology, and evolution. In *Biology of the Reptilia. Volume 19, Morphology G*. C. Gans and A. S. Gaunt (Eds.). Pp. 1–92. Society for the Study of Amphibians and Reptiles, Ithaca, NY.

Peters, G. (1971). Die intragenerischen Gruppen und die Phylogenese der Schetterlingsagamen (Agamidae: *Leiolepis*). *Zoologische Jahrbücher für Systematik* **98:** 11–130.

Peterson, C. C. (1996). Anhomeostasis: seasonal water and solute relations in two populations of the desert tortoise (*Gopherus agassizii*) during chronic drought. *Physiological Zoology* **69:** 1324–1358.

Peterson, C. C., and Stone, P. A. (2000). Physiological capacity for estivation of the Sonoran Mud Turtle, *Kinosternon sonoriense*. *Copeia* **2000:** 684–700.

Peterson, C. L., Wilkinson, R. F., Topping, M. S., and Metter, D. E. (1983). Age and growth of the Ozark hellbender (*Cryptobranchus alleganiensis bishopi*). *Copeia* **1983:** 225–231.

Petranka, J. W. (1998). *Salamanders of the United States and Canada*. Smithsonian Institute Press, Washington, D.C.

Petranka, J. W., Brannon, M. P., Hopey, M. E., and Smith, C. K. (1994). Effects of timber harvesting on low elevation populations of southern Applachian salamanders. *Forest Ecology and Management* **67:** 135–147.

Petranka, J. W., Kats, L. B., and Sih, A. (1987). Predator-prey interactions among fish and larval amphibians: Use of chemical cues to detect predatory fish. *Animal Behaviour* **35:** 420–425.

Petren, K., Bolger, D., and Case, T. J. (1993). Mechanisms in the competitive success of an invading sexual gecko over an asexual native. *Science* **259:** 354–358.

Petriella, S., Reboreda, J. C., Otero, M., and Segura, E. T. (1989). Antidiuretic responses to osmotic cutaneous stimulation in the toad, *Bufo arenarum*. *Journal of Comparative Physiology* **159B:** 91–95.

Pfennig, D. W., and Murphy, P. J. (2000). Character displacement in polyphenic tadpoles. *Evolution* **54:** 1738–1749.

Pfennig, D. W., Harcombe, W. R., and Pfennig, K. S. (2000). Coral snake mimics are protected only when they occur with their model. *American Zoologist* **40:** 1170.

Pfennig, D. W., Harper, G. R., Brumo, A. F., Harcombe, W. R., and Pfennig, K. S. (2007). Population differences in predation on Batesian mimics in allopatry with their model: Selection against mimics is strongest when they are common. *Behavioral Ecology and Sociobiology* **61:** 505–511.

Pfingsten, R. A., and Downs, F. L. (Eds.). (1989). Salamanders of Ohio. *Bulletin of the Ohio Biological Survey* **7:** 1–315.

Phillips, B. L., and Shine, R. (2004). Adapting to an invasive species: Toxic cane toads induce morphological change in Australian snakes. *Proceedings of the National Academy of Sciences (USA)* **101:** 17150–17155.

Phillips, B. L., Brown, G. P., Greenlees, M., Webb, J. K., and Shine, R. (2007). Rapid expansion of the cane toad (*Bufo marinus*) invasion front in tropical Australia. *Austral Ecology* **32:** 169–176.

Phillips, B. L., Brown, G. P., Webb, J., and Shine, R. (2006). Runaway toads: An invasive species evolves speed and thus spreads more rapidly through Australia. *Nature* **439:** 803.

Phillips, J. B., and Borland, S. C. (1994). Use of a specialized magnetoreception system for homing by the eastern red-spotted newt *Notophthalmus viridescens*. *Journal of Experimental Biology* **188:** 275–291.

Phillips, J. B., Deutschlander, M. E., Freake, M. J., and Borland, S. C. (2001). The role of extraocular photoreceptors in newt magnetic compass orientation: Parallels between light-dependent magnetoreception and polarized light detection in vertebrates. *Journal of Experimental Biology* **204:** 2543–2552.

Phillips, K. (1994). *Tracking the Vanishing Frogs*. St. Martin's Press, New York.

Pianka, E. R. (1966). Convexity, desert lizards and spatial heterogeneity. *Ecology* **47:** 1055–1059.

Pianka, E. R. (1970). On '*r*' and '*K*' selection. *American Naturalist* **104:** 592–597.

Pianka, E. R. (1973). The structure of lizard communities. *Annual Review of Ecology and Systematics* **4:** 53–74.

Pianka, E. R. (1975). Niche relations of desert lizards. In *Ecology and Evolution of Communities*. M. Cody and J. Diamond (Eds.). Pp. 292–314. Harvard University Press, Cambridge, Mass.

Pianka, E. R. (1985). Some intercontinental comparisons of desert lizards. *National Geographic Research* **1:** 490–504.

Pianka, E. R. (1986). *Ecology and Natural History of Desert Lizards. Analyses of the Ecological Niche and Community Structure*. Princeton University Press, Princeton.

Pianka, E. R. (1988). *Evolutionary Ecology*. Harper & Row, New York.

Pianka, E. R. (1992). The state of the art in community ecology. In *Herpetology: Current Research on the Biology of Amphibians and Reptiles*. K. Adler (Ed.). Pp. 141–162. Society for the Study of Amphibians and Reptiles, Oxford Ohio.

Pianka, E. R. (1994a). Biodiversity of Australian desert lizards. In *Biodiversity and Terrestrial Ecosystems*. C. I. Peng and C. H. Chou (Eds.). Pp. 259–281. Institute of Botany, Academia Sinica Monograph No. 14, Taipei.

Pianka, E. R. (1994b). Comparative ecology of *Varanus* in the Great Victoria Desert. *Australian Journal of Ecology* **19:** 395–408.

Pianka, E. R. (1995). Evolution of body size: Varanid lizards as a model system. *American Naturalist* **146:** 398–414.

Pianka, E. R. (1996). Long-term changes in lizard assemblages in the Great Victoria Desert: Dynamic habitat mosaics in response to wildfires. In *Long-term Studies of Vertebrate Communities*. M. L. Cody and J. A. Smallwood (Eds.). Pp. 191–215. Academic Press, San Diego, CA.

Pianka, E. R. and King, D. R. (Eds.). (2004). *Varanoid Lizards of the World*. Indiana University Press, Bloomingrton, Indiana.

Pianka, E. R., and Parker, W. S. (1975). Ecology of horned lizards: A review with special reference to *Phrynosoma platyrhinos*. *Copeia* **1975:** 141–162.

Pianka, E. R., and Vitt, L. J. (2003). *Lizards. Windows to the Evolution of Diversity*. University of California Press, Brekeley, CA.

Pieau, C. (1971). Sur la proportion sexuelle chez les embryons de deux chéloniens (*Testudo graeca* L. et *Emys orbicularis* L.) issus d'oeufs incubés artificiellement. *Comptes Rendus de l'Académie des Sciences, Paris* **274:** 3071–3074.

Pieau, C. (1975). Temperature and sex differentiation in embryos of two chelonians, *Emys orbicularis* L., and *Testudo graeca* L. In *Intersexuality in the Animal Kingdom*. R. Reinboth (Ed.). Pp. 332–339. Springer-Verlag, New York, NY.

Piha, H., Pekkonen, M., and Merila, J. (2006). Morphological abnormalities in amphibians in agricultural habitats: A case study of the common frog *Rana temporaria*. *Copeia* **2006:** 810–817.

Pike, D. A. (2006). Movement patterns, habitat use, and growth of hatchling tortoises, *Gopherus polyphemus*. *Copeia* **2006:** 68–76.

Pilliod, D. S., Peterson, C. R., and Ritson, P. I. (2002). Seasonal migration of Columbia spotted frogs (*Rana luteiventris*) among complementary resources in a high mountain basin. *Canadian Journal of Zoology* **80:** 1849–1862.

Pilorge, T. (1987). Density, size structure, and reproductive characteristics of three populations of *Lacerta vivipara*. *Herpetologica* **43:** 345–356.

Pimenta, B. V. S., Haddad, C. F. B., Nascimento, L. B., Cruz, C. A. G., and Pombal, J. P. (2005). Comment on "Status and trends of amphibian declines and extinctions worldwide." *Science* **309:** 1999b.

Pimentel, D., Lach, L., Zuniga, R., and Morrison, D. (2000). Environmental and economic costs of nonindigenous species in the United States. *Bioscience* **50:** 53–65.

Pintak, T. (1987). Zur Kenntnìs des tomatenfrosches *Dyscophus antongili* (Erandidier, 1877) (Anura: Microhylidae). *Salamandra* **23:** 106–121.

Pitman, C. R. S. (1974). *A Guide to the Snakes of Uganda. Revised Edition*. Wheldon & Wesley, Ltd., Codicote, U.K.

Pitman, N. C. A., Azaldegui, M. D. L., Salas, K., Vigo, G. T., and Lutz, D. A. (2007). Written accounts of an Amazonian landscape over the last 450 years. *Conservation Biology* **21:** 253–262.

Pledge, N. S. (1992). The Curramulka local fauna: A new Late Tertiary fossil assemblage from Yorke Peninsula, South Australia. *The Beagle* **9:** 115–142.

Plummer, M. V. (1976). Some aspects of nesting success in the turtle, *Trionyx muticus*. *Herpetologica* **32:** 353–359.

Plummer, M. V. (1977a). Reproduction and growth in the turtle *Trionyx muticus*. *Copeia* **1977:** 440–447.

Plummer, M. V. (1977b). Activity, habitat and population structure in the turtle, *Trionyx muticus*. *Copeia* **1977:** 431–440.

Plummer, M. V. (1985). Growth and maturity in green snakes (*Opheodrys aestivus*). *Herpetologica* **41:** 28–33.

Plummer, M. V. (2004). Seasonal inactivity of the Desert Box Turtle, *Terrapene ornata luteola*, at the species' southwestern range limit in Arizona. *Journal of Herpetology* **38:** 589–593.

Plummer, M. V., and Shirer, H. W. (1975). Movement patterns in a river population of the softshell turtle, *Trionyx muticus*. *Occasional Papers of the Museum of Natural History, University of Kansas* **43:** 1–26.

Poe, S. (1997). Data set incongruence and the phylogeny of crocodilians. *Systematic Biology* [1996] **45:** 393–414.

Poe, S. (1998). Skull characters and the cladistic relationships of the Hispaniolan dwarf twig *Anolis*. *Herpetological Monographs* **12:** 192–236.

Poe, S. (2004). Phylogeny of anoles. *Herpetological Monographs* **18:** 37–89.

Poinar, G. R., Jr., and Cannatella, D. C. (1987). An Upper Eocene frog from the Dominican Republic and its implication for Caribbean biogeography. *Science* **237:** 1215–1216.

Polcyn, M. J., Jacobs, L. L., and Haber, A. (2005). A morphological model and CT assessment of the skull of *Pachyrhachis problematicus* (Squamata, Serpentes), a 98 million year old snake with legs from the Middle East. Palaeontologia Electronica **8:** Article number 26A.

Polisar, J. (1996). Reproductive biology of a flood-season nesting freshwater turtle of the northern Neotropics: *Dermatemys mawii* in Belize. *Chelonian Conservation Biology* **2:** 13–25.

Pombal, J. P., and Haddad, C. F. B. (1999). Frogs of the genus *Paratelmatobius* (Anura: Leptodactylidae) with descriptions of two new species. *Copeia* **1999:** 1014–1026.

Pombal, J. P., Jr., Wistuba, E. M., and Bornshein, M. R. (1998). A new species of brachycephalid (Anura) from the Atlantic rain forest of Brazil. *Journal of Herpetology* **32:** 70–74.

Pooley, A. C. (1974). Parental care in the Nile crocodile: A preliminary report on behaviour of a captive female. *Lammergeyer* **3:** 22–44.

Pooley, A. C. (1977). Nest opening response of the Nile crocodile *Crocodylus niloticus*. *Journal of Zoology, London* **182:** 17–26.

Porter, W. P., and Gates, D. M. (1969). Thermodynamic equilibria of animals with environment. *Ecological Monographs* **39:** 227–244.

Pough, F. H. (1988). Mimicry and related phenomenona. In *Biology of the Reptilia, Vol. 16, Ecology B. Defense and Life History*. C. Gans and R. B. Huey (Eds.). Pp. 153–234. Alan R. Liss, Inc, New York, NY.

Pough, F. H., and Taigen, T. L. (1990). Metabolic correlates of the foraging and social behaviour of dart-poison frogs. *Animal Behaviour* **39:** 145–155.

Pough, F. H., and Wilson, R. E. (1977). Acid precipitation and reproductive success of *Ambystoma* salamanders. *Water Air Soil Pollution* **7:** 307–316.

Pough, F. H., Andrews, R. M., Cadle, J. E., Crump, M. L., Savitzky, A. H., and Wells, K. D. (1998). *Herpetology*. Prentice Hall, Upper Saddle River, NJ.

Pounds, J. A., and Crump, M. L. (1994). Amphibian declines and climate disturbance: The case of the golden toad and the harlequin frog. *Conservation Biology* **8:** 72–85.

Pounds, J. A., Bustamante, M. R., Coloma, L. A., Consuegra, J. A., Fogden, M. P. L., Foster, P. N., La Marca, E., Masters, K. L., Merino-Viteri, A., Puschendorf, R., Ron, S. R., Sanchez-Azofeifa, G. A., Still, C. J., and Young, B. E. (2006). Widespread amphibian extinctions from epidemic disease driven by global warming. *Nature* **439:** 161–167.

Pounds, J. A., Fogden, M. P. L., Savage, J. M., and Gorman, G. C. (1997). Test of null models for amphibian declines on a tropical mountain. *Conservation Biology* **11:** 1307–1322.

Poynton, J. C. (1964). The Amphibia of southern Africa: A faunal study. *Annals of the Natal Museum* **17:** 1–334.

Poynton, J. C. (1996). Diversity and conservation of African bufonids (Anura): Some preliminary findings. *African Journal of Herpetology* **45:** 1–7.

Poynton, J. C., and Broadley, D. G. (1985–1991). Amphibia Zambesiaca. Parts 1–5 *Annals of the Natal Museum* **26:** 503–553, **27:** 115–181, **28:** 161–229, **29:** 447–490, **32:** 221–277.

Poynton, J. C., and Broadley, D. G. (1987). Amphibia Zambesiaca 3. Rhacophoridae and Hyperoliidae. *Annals of the Natal Museum* **28:** 161–229.

Poynton, J. C., and Broadley, D. G. (1988). Amphibia Zambesiaca 4. Bufonidae. *Annals of the Natal Museum* **29:** 447–490.

Prado, C. P. A., Abdalla, F. C., Silva, A. P. Z., and Zina, J. (2004). Late gametogenesis in *Leptodactylus labyrinthicus* (Amphibia, Anura, Leptodactylidae) and some ecological considerations. *Brazilian Journal of Morphological Science* **21:** 177–184.

Prado, C. P. A., Toledo, L. F., Zina, J., and Haddad, C. F. B. (2005). Trophic eggs in the foam nests of *Leptodactylus labyrinthicus* (Anura, Leptodactylidae): An experimental approach. *Herpetological Journal* **15:** 279–284.

Prado, C. P. A., Uetanabaro, M., and Haddad, C. F. B. (2002). Description of a new reproductive mode in *Leptodactylus* (Anura, Leptodactylidae), with a review of the reproductive specialization toward terrestriality in the genus. *Copeia* **2002:** 1128–1133.

Pramuk, J. B. (2002). Combined evidence and cladistic relationships of West Indian toads (Anura: Bufonidae). *Herpetological Monographs* **16:** 121–151.

Pramuk, J. B. (2006). Phylogeny of South American *Bufo* (Anura: Bufonidae) inferred from combined evidence. *Zoological Journal of the Linnean Society* **146:** 407–452.

Pramuk, J. B., Robertson, T., Sites, J. W., and Noonan, B. P. (2008). Around the world in 10 million years: Biogeography of the nearly cosmopolitan true toads (Anura: Bufonidae). *Global Ecology and Biogeography* **17:** 72–83.

Prance, G. T. (1973). Phytogeographic support for the theory of Pleistocene refuges in the Amazon Basin, based on evidence from distributional patterns in Caryocaraceae, Diphapetalaceae and Lecythidaceae. *Acta Amazonica* **3:** 5–28.

Prance, G. T. (1982). A review of the phytogeographic evidences for Pleistocene climate changes in the Neotropics. *Annals of the Missouri Botanical Garden* **69:** 594–624.

Prasad, G. V. R., and Rage, J.- C. (1991). A discoglossid frog in the latest Cretaceous (Maastrichtian) of India. Further evidence for a terrestrial route between India and Laurasia in the latest Cretaceous. *Comptes Rendus de l'Académie des Sciences, Paris* **313:** 273–278.

Pregill, G. K., Gauthier, J. A., and Greene, H. W. (1986). The evolution of helodermatid squamates, with description of a new taxon and an overview of Varanoidea. *Transactions of the San Diego Society of Natural History* **21:** 167–202.

Presch, W. (1974). Evolutionary relationships and biogeography of the macroteiid lizards (Family Teiidae, Subfamily Teiinae). *Bulletin of the Southern California Academy of Sciences* **73:** 23–32.

Presch, W. (1988). Phylogenetic relationships of the Scincomorpha. In *Phylogenetic Relationships of the Lizard Families. Essay Commemorating Charles L. Camp.* R. Estes and G. Pregill. (Eds.). Pp. 471–492. Stanford University Press, Stanford, CA.

Price, T. D. (1984). The evolution of sexual size dimorphism in Darwin's finches. *American Naturalist* **123:** 500–518.

Pritchard, P. C. H., and Trebbau, P. (1984). *The Turtles of Venezuela.* Society for the Study of Amphibians and Reptiles, Athens, Ohio.

Proud, K. R. S. (1978). Some notes on a captive earless monitor lizard *Lanthanotus borneensis*. *Sarawak Museum Journal* **26:** 235–242.

Pyke, G. H., and Osborne, W. S. (Eds.). (1996). The green & golden bell frog *Litoria aurea*: Biology and conservation. *Australian Zoologist* **30:** 132–258.

Qualls, C. P. (1996). Influence of the evolution of viviparity on eggshell morphology in the lizard, *Lerista bougainvillii*. *Journal of Morphology* **228:** 119–125.

Qualls, C. P., and Shine, R. (1995). Maternal body-volume as a constraint on reproductive output in lizards: Evidence from the evolution of viviparity. *Oecologia* **103:** 73–78.

Quay, W. B. (1972). Integument and the environment: Glandular composition, function, and evolution. *American Zoologist* **12:** 95–108.

Quay, W. B. (1979). The parietal eye-pineal complex. In *Biology of the Reptilia. Volume 9. Neurology A.* C. Gans, R. G. Northcutt, and P. Ulinski (Eds.). Pp. 245–406. Academic Press, London.

Quay, W. B. (1986). Glands. In *Biology of the Integument. 2. Vertebrates.* J. Bereiter-Hahn, A. G. Matoltsy, and K. S. Richards (Eds.). Pp. 188–193. Springer Verlag, Berlin.

Quinzio, S. I., Fabrezi, M., and Faivovich, J. (2006). Redescription of the tadpole of *Chacophrys pierottii* (Vellard, 1948) (Anura, Ceratophryidae). *South American Journal of Herpetology* **1:** 202–209.

Rabb, G. B., and Rabb, M. S. (1960). On the mating and egg-laying behavior of the Surinam toad, *Pipa pipa*. *Copeia* **1960:** 271–276.

Rabb, G. B., and Rabb, M. S. (1963a). Additional observations on breeding behavior of the Surinam toad, *Pipa pipa*. *Copeia* **1963:** 636–642.

Rabb, G. B., and Rabb, M. S. (1963b). On the behavior and breeding biology of the African pipid frog *Hymenochirus boettgeri*. *Zeitschrift für Tierpsychologie* **20:** 215–241.

Rabb, G. B., and Snediger, R. (1960). Observations on breeding and development of the Surinam toad, *Pipa pipa*. *Copeia* **1960:** 40–44.

Radder, R. S., Saidapur, S. K., Shine, R., and Shanbhag, B. A. (2006). The language of lizards: Interpreting the function of visual displays of the Indian rock lizard, *Psammophilus dorsalis* (Agamidae). *Journal of Ethology* **24:** 275–283.

Radhakrishnan, C., Gopi, K. C., and Palot, M. J. (2007). Extension of range of distribution of *Nasikabatrachus sahyadrensis* Biju & Bossuyt (Amphibia: Anura: Nasikabatrachidae) along Western Ghats, with some insights into its bionomics. *Current Science* **92:** 213–216.

Radtkey, R. R., Donnellan, S. C., Fisher, R. N., Moritz, C., Hanley, K. A., and Case, T. J. (1995). When species collide: The origin and spread of an asexual species of gecko. *Proceedings of the Royal Society of London* **B 259:** 145–152.

Rage, J.-C. (1984). *Serpentes. Handbuch der Paläoherpetologie, 11.* Gustav Fischer Verlag, Stuttgart.

Rage, J.-C. (1986). Le plus ancien amphibien apode (Gymnophiona) fossile. Remarques sur la répartiition et l'histoire paléobiogéographique des gymnophiones. *Comptes Rendus de l'Académie des Sciences, Paris* **302:** 1033–1036.

Rage, J.- C. (1987). Fossil history. In *Snakes: Ecology and Evolutionary Biology.* R. A. Seigel, J. T. Collins, and S. S. Novak (Eds.). Pp. 51–76. Macmillian Publ.ishing Co., New York.

Rage, J.-C. (1996). Les Madsoiidae (Reptilia, Serpentes) du Crétacé supérieur d'Europe: Témoins gondwaniens d'une dispersion transtéthysienne. *Comptes Rendus de l'Académie des Sciences, Paris, serie II a* **322:** 603–608.

Rage, J. C., and Albino, A. M. (1989). *Dinilysia patagonica* (Reptilia, Serpentes): Materiel vertebral additionel du Cretace superieur d'Argentine. Etude complementaire des vertebres, variations intraspecifiques et intracolumnaires. *Neues Jahrbuch fuer Geologie und Palaeontologie Monatshefte* **1989:** 433–447.

Rage, J.-C., and Richter, A. (1994). A snake from the Lower Cretaceous (Barremian) of Spain: The oldest known snake. *Nues Jarbuch für Geologie und Paläontologie, Monatshefte* **1994:** 561–565.

Rage, J.-C., and Rocek, Z. (1989). Redescription of *Triadobatrachus massinoti* (Piveteau, 1936) an anuran from the Early Triassic. *Palaeontographica Abteilung A* **206:** 1–16.

Rage, J.-C., Marshall, L. G., and Gayet, M. (1993). Enigmatic Caudata (Amphibia) from the Upper Cretaceous of Gondwana. *Geobios* **26:** 515–519.

Ragghianti, M., Bucci, S., Marracci, S., Casola, C., Mancino, G., Hotz, H., Guex, G. D., Plotner, J., and Uzzell, T. (2007). Gametogenesis of intergroup hybrids of hemiclonal frogs. *Genetical Research* **89:** 39–45.

Rajendran, M. V. (1985). *Studies in Uropeltid Snakes.* Publication Division, Madurai Kamaraj University, Madurai, India.

Ralph, C. L. (1983). Evolution of pineal control of endocrine function in lower vertebrates. *American Zoologist* **23:** 597–605.

Ramírez-Bautista, A., Vitt, L. J., Ramírez-Hernández, A., Quijano, F. M., and Smith, G. R. (2008). Reproduction and sexual dimorphism of *Lepidophyma sylvaticum* (Squamata: Xantusiidae), a tropical night lizard from Tlanchinol, Hidalgo, Mexico. *Amphibia-Reptilia* **29:** 1–10.

Rand, A. S. (1982). Clutch and egg size in Brazilian iguanid lizards. *Herpetologica* **38:** 171–178.

Rand, A. S., Dugan, B. A., Monteza, H., and Vianda, D. (1990). The diet of a generalized folivore: *Iguana iguana* in Panama. *Journal of Herpetology* **24:** 211–214.

Rand, W. M., and Rand, A. S. (1976). Agonistic behavior in nesting iguanas: A stochastic analysis of dispute settlement dominated by the minimization of energy cost. *Zeitschrift für Tierpsychologie* **40:** 279–299.

Ranvestel, A. W., Lips, K. R., Pringle, C. M., Whiles, M. R., and Bixby, R. J. (2004). Neotropical tadpoles influence stream benthos: Evidence for the ecological consequences of decline in amphibian populations. *Freshwater Biology* **49:** 274–285.

Rao, D.-Q., and Wilkinson, J. A. (2007). A new species of *Amolops* (Anura: Ranidae) from Southwest China. *Copeia* **2007:** 913–919.

Rao, D.-Q., and Yang, D.-T. (1992). Phylogenetic systematics of Pareinae (Serpentes) of southeastern Asia and adjacent islands with relationship between it and the geology change. *Acta Zoologica Sinica* **38:** 139–150.

Rao, D.-Q., and Yang, D.-T. (1997). The karyotypes of Megophryidae (Pelobatoidea) with a discussion on their classification and phylogenetic relationships. *Asiatic Herpetological Research* **7:** 93–102.

Rao, D.-Q., Wilkinson, J. A., and Zhang, M. W. (2006). A new species of the genus *Vibrissaphora* (Anura: Megophryidae) from Yunnan Province, China. *Herpetologica* **62:** 90–95.

Räsänen, M. E., Linna, A. M., Santos, J. C. R., and Negri, F. R. (1995). Late Miocene tidal deposits in the Amazon foreland basin. *Science* **269:** 386–390.

Rashid, S. M. A., and Swingland, I. R. (1998). On the ecology of some freshwater turtles in Bangladesh. In *Proceedings: Conservation, Restoration, and Management of Tortoises and Turtles – An International Conference [1997].* J. Van Abbema (Ed.). Pp. 224–242. New York Turtle & Tortoise Society, Purchase, N.Y.

Rasmussen, A. R. (1997). Systematics of sea snakes: A critical review. In *Venomous Snakes, Ecology, Evolution and Snakebite.* R. S. Thorpe, W. Wüster, and A. Malhotra (Eds.). Pp. 15–30. Zoological Society of London, Oxford.

Raup, D. M. (1991). *Extinction: Bad Genes or Bad Luck?* W. W. Norton & Co., New York.

Raxworthy, C. J., and Nussbaum, R. A. (1995). Systematics, speciation and biogeography of the dwarf chameleons (*Brookesia*; Reptilia, Squamata, Chamaeleontidae) of northern Madagascar. *Journal of Zoology, London* **235:** 525–558.

Raxworthy, C. J., Forstner, M. R. J., and Nussbaum, R. A. (2002). Chameleon radiation by oceanic dispersal. *Nature* **415:** 784–787.

Raxworthy, C. J., Martinez-Meyer, E., Horning, N., Nussbaum, R. A., Schneider, G. E., Ortega-Huerta, M. A., and Peterson, A. T. (2003). Predicting distributions of known and unknown reptile species in Madagascar. *Nature* **426:** 837–841.

Read, K., Keogh, J. S., Scott, I. A. W., Roberts, J. D., and Doughty, P. (2001). Molecular phylogeny of the Australian frog genera *Crinia, Geocrinia,* and allied taxa (Anura: Myobatrachidae). *Molecular Phylogenetics and Evolution* **21:** 294–308.

Reagan, D. P. (1991). The response of *Anolis* lizards to hurricane-induced habitat changes in a Puerto Rican rain forest. *Biotropica* **23:** 468–474.

Rebouças-Spieker, R., and Vanzolini, P. E. (1978). Parturition in *Mabuya macrorhyncha* Hoge, 1946 (Sauria, Scincidae), with a note on the distribution of maternal behavior in lizards. *Papéis Avulsos de Zoologia (São Paulo)* **32:** 95–99.

Reeder, T. W. (1995). Phylogenetic relationships among phrynosomatid lizards as inferred from mitochondrial ribosomal DNA sequences: Substitutional bias and information content of transitions relative to transversions. *Molecular Phylogenetics and Evolution* **4:** 203–222.

Reeder, T. W. (2003). A phylogeny of the Australian *Sphenomorphus* group (Scincidae: Squamata) and the phylogenetic placement of the crocodile skinks (*Trilobonotus*): Bayesian approaches to assessing congruence and obtaining confidence in maximum likelihood inferred relationships. *Molecular Phylogenetics and Evolution* **27:** 384–397.

Reeder, T. W., and Montanucci, R. R. (2001). Phylogenetic analysis of the horned lizards (Phrynosomatidae: *Phrynosoma*): Evidence from mitochondrial DNA and morphology. *Copeia* **2001:** 309–323.

Reeder, T. W., and Wiens, J. J. (1996). Evolution of the lizard family Phrynosomatidae as inferred from diverse types of data. *Herpetological Monographs* **10:** 43–84.

Reeder, T. W., Cole, C. J., and Dessauer, H. C. (2002). Phylogenetic relationships of whiptail lizards of the genus *Cnemidophorus* (Squamata: Teiidae): A test of monophyly, reevaluation of karyotypic evolution, and review of hybrid origins. *American Museum Novitates* **3365:** 1–61.

Reid, K. J. (1955). Reproduction and development in the northern diamondback terrapin, *Malaclemys terrapin terrapin. Copeia* **1955:** 310–311.

Reilly, S. M., and Brandon, R. A. (1994). Partial paedomophosis in the Mexican stream ambystomatids and the taxonomic status of the genus *Rhyacosiredon* Dunn. *Copeia* **1994:** 656–662.

Reilly, S. M., McBrayer, L. B., and Miles, D. B. (Eds.) (2007a). *Lizard Ecology: The Evolutionary Consequences of Foraging Mode.* Cambridge University Press, Cambridge, MA.

Reilly, S. M., McBrayer, L. D., and Miles, D. B. (2007b). Prey capture and prey processing behavior and the evolution of lingual and sensory characteristics: Divergences and convergences in lizard feeding biology. *Lizard Ecology: The Evolutionary Consequences of Foraging Mode.* S. M. Reilly, L. D. McBrayer, and D. B. Miles (Eds.). Pp. 302–333. Cambridge University Press, Cambridge, MA.

Reilly, S. M., Wiley, E. O., and Meinhardt, D. J. (1997). An integrative approach to heterochrony: The distinction between interspecific and intraspecific phenomena. *Biological Journal of the Linnean Society* **60:** 119–143.

Reinert, H. K. (1991). Translocations as a conservation strategy for amphibians and reptiles: Some comments, concerns, and observations. *Herpetologica* **47:** 357–363.

Reisz, R. R., and Laurin, M. (1991). *Owenetta* and the origin of turtles. *Nature* **349:** 324–326.

Reno, H. W., Gehlbach, F. R., and Turner, R. A. (1972). Skin and aestivational cocoon of the aquatic amphibian, *Siren intermedia* Le Conte. *Copeia* **1972:** 625–631.

Rhen, T., and Lang, J. W. (2004). Phenotypic effects of incubation temperature in reptiles. In *Temperature-Dependent Sex Determination in Vertebrates.* N. Valenzuela and V. Lance (Eds.). Pp. 91–98. Smithsonian Books, Washington, D. C.

Riberon, A., Miaud, C., Guyetant, R., and Taberlet, P. (2004). Genetic variation in an endemic salamander, *Salamandra atra*, using amplified fragment length polymorphism. *Molecular Phylogenetics and Evolution* **31:** 910–914.

Richards, C. M., and Moore, W. S. (1998). A molecular phylogenetic study of the Old World treefrog family Rhacophoridae. *Herpetological Journal* **8:** 41–46.

Richards, S. J., McDonald, K. R., and Alford, R. A. (1993). Declines in populations of Australia's endemic tropical rainforest frogs. *Pacific Conservation Biology* **1:** 66–77.

Richardson, B. J., Baverstock, P. R., and Adams, M. (1986). *Allozyme Electrophoresis. A Handbook for Animal Systematics and Population Studies*. Academic Press, Sydney.

Ricklefs, R. E. and Schluter, D. (Eds.). (1994). *Species Diversity in Ecological Communities*. University of Chicago Press, Chicago, IL.

Riemer, W. J. (1957). The snake *Farancia abacura*: An attended nest. *Herpetologica* **13:** 31–32.

Rieppel, O. (1988). A review of the origin of snakes. *Evolutionary Biology* **22:** 37–130.

Rieppel, O. (1993). Euryapsid relationships: A preliminary analysis. *Nues Jahrbuch für Geologie und Paläontologie, Abhandlungen* **188:** 241–264.

Rieppel, O. (1994). The Lepidosauromorpha: An overview with special emphasis on the Squamata. In *the shadow of the dinosaurs. Early Mesozoic Tetrapods*. N. C. Fraser and H.-D. Sues (Eds.). Pp. 23–37. Cambridge University Press, New York.

Rieppel, O., and Crumly, C. (1997). Paedomorphosis and skull structure in Malagasy chamaeleons (Reptilia: Chamaeleoninae). *Journal of Zoology, London* **243:** 351–380.

Rieppel, O., and deBraga, M. (1996). Turtles as diapsid reptiles. *Nature* **384:** 453–455.

Riley, J., Stimson, A. F., and Winch, J. M. (1985). A review of Squamata ovipositing in ant and termite nests. *Herpetological Review* **16:** 38–43.

Ripa, D. (1999). Keys to understanding the bushmasters (genus *Lachesis* Daudin, 1803). *Bulletin of the Chicago Herpetological Society* **34:** 45–92.

Robb, J. (1986). *New Zealand Amphibians and Reptiles In Colour. Revised*. Collins, Auckland.

Roberts, J. D., Horwitz, P., Wardell-Johnson, G., Maxson, L. R., and Mahony, J. (1997). Taxonomy, relationships and conservation of a new genus and species of myobatrachid frog from the high rainfall region of southwestern Australia. *Copeia* **1997:** 373–381.

Robertson, A. V., Ramsden, C., Niedzwiecki, J., Fu, J. Z., and Bogart, J. P. (2006). An unexpected recent ancestor of unisexual *Ambystoma*. *Molecular Ecology* **15:** 3339–3351.

Robinson, M. D., and Cunningham, A. B. (1978). Comparative diet of two Namib Desert sand lizards (Lacertidae). *Madoqua* **11:** 41–54.

Robinson, P. L. (1962). Gliding lizards from the Upper Keuper of Great Britain. *Proceedings of the Geological Society of London* **1601:** 137–146.

Rocek, Z. (1994a). A review of the fossil Caudata of Europe. *Abhandlungen und Berichte für Naturkunde, Magdeburg* **17:** 51–56.

Rocek, Z. (1994b). Taxonomy and distribution of Tertiary discoglossids (Anura) of the genus *Latonia* V. Meyer, 1843. *Geobios* **27:** 717–751.

Rocek, Z., and Lamaud, P. (1995). *Thaumastosaurus bottii* de Stefano, 1903, an anuran with Gondwanan affinities from the Eocene of Europe. *Journal of Vertebrate Palaeontology* **15:** 506–515.

Rocek, Z., and Nessov, L. A. (1993). Cretaceous anurans from Central Asia. *Palaeontographica Abteilung* **A 226:** 1–54.

Rocha, C. F. D. (1990). Reproductive effort in the Brazilian sand lizard *Liolaemus lutzae* (Sauria: Iguanidae). *Ciência e Cultura (São Paulo)* **42:** 1203–1206.

Rocha, C. F. D., and Bergallo, H. G. (1990). Thermal biology and flight distance of *Tropidurus oreadicus* (Sauria Iguanidae) in an area of Amazonian Brazil. *Ethology, Ecology & Evolution* **2:** 263–268.

Rocha, C. F. D., Araújo, A. F. B., Vrcibradic, D., and Costa, E. M. M. d. (2000). New *Cnemidophorus* (Squamata; Teiidae) from coastal Rio de Janeiro state, southeastner Brazil. *Copeia* **2000:** 501–509.

Rodda, G. H. (1984). The orientation and navigation of juvenile alligators: Evidence of magnetic sensitivity. *Journal of Comparative Physiology* **154A:** 649–658.

Rodda, G. H. (1985). Navigation in juvenile alligators. *Zeitschrift für Tierpsychologie* **68:** 65–77.

Rodda, G. H., and Fritts, T. H. (1992). The impact of the introduction of the colubrid snake *Boiga irregularis* on Guam's lizards. *Journal of Herpetology* **26:** 166–174.

Rodda, G. H., and Phillips, J. B. (1992). Navigational systems develop along similar lines in amphibians, reptiles, and birds. *Ethology, Ecology & Evolution* **4:** 43–51.

Rodda, G. H., Fritts, T. H., and Campbell, E. W. (1999). The feasibility of controlling the Brown Treesnake in small plots. *Problem Snake Management*: 468–477.

Rodda, G. H., Fritts, T. H., and McCoid, M. J. (1999). An overview of the biology of the Brown Treesnake* (*Boiga irregularis*), a costly introduced pest on Pacific Islands. In *Problem Snake Management: The Habu and the Brown Treesnake*. G. H. Rodda, Y. Sawai, D. Chizar, and H. Tanaka (Eds.). Pp. 469–477. Cornell University Press, Ithaca, NY.

Rodda, G. H., McCoid, M. J., Fritts, T. H., and Campbell, E. W. (1999). Population trends and limiting factors in *Boiga irregularis*. In *Problem Snake Management: The Habu and the brown treesnake*. G. H. Rodda, Y. Sawai, D. Chiszar, and H. Tanaka (Eds.). Pp. 236–253. Cornell University Press, Ithaca, NY.

Rödel, M. -O., Grafe, T. U., Rudolf, V. H. W., and Ernst, R. (2002). A review of west African spotted *Kassina*, including a description of *Kassina schioetzi* sp nov (Amphibia: Anura: Hyperoliidae). *Copeia* **2002:** 800–814.

Rödel, M.-O. (1996). *Amphibien der westafrikanischen Savanne*. Edition Chimaira, Frankfurt am Main.

Rödel, M.-O. (1998). A reproductive mode so far unknown in African ranids: *Phrynobatrachus guineensis* Guibé & Lamotte, 1961 breeds in tree holes (Anura: Ranidae). *Herpetozoa* **11:** 19–26.

Rödel, M.-O., Spieler, M., Grabow, K., and Böckheler, C. (1995). *Hemisus marmoratus* (Peters, 1854) (Anura: Hemisotidae), Fortpflanzungsstrategien eines Savannenfrosches. *Bonner Zoologische Beiträge* **45:** 191–207.

Rodriguez, L. B., and Cadle, J. E. (1990). A preliminary overview of the herpetofauna of Cocha Cashu, Manu National Park, Peru. In *Four Neotropical Rainforests*. A. H. Gentry (Ed.). Pp. 410–425. Yale University Press, New Haven, CT.

Roelants, K., Gower, D. J., Wilkinson, M., Loader, S. P., Biju, S. D., Guillaume, K., Moriau, L., and Bossuyt, F. (2007).

Global patterns of diversification in the history of modern amphibians. *Proceedings of the National Academy of Sciences (USA)* **104:** 887–892.

Roelants, K., Jiang, J. P., and Bossuyt, F. (2004). Endemic ranid (Amphibia: Anura) genera in southern mountain ranges of the Indian subcontinent represent ancient frog lineages: Evidence from molecular data. *Molecular Phylogenetics and Evolution* **31:** 730–740.

Rogers, C. L., and Kerstetter, R. E. (1974). *The Ecosphere: Organisms, Habitats, and Disturbances.* Harper and Row, New York.

Roman, J., Santhuff, S. D., Moler, P. R., and Bowen, B. W. (1999). Population structure and cryptic evolutionary units in the alligator snapping turtle. *Conservation Biology* **13:** 135–142.

Rome, L. C., Stevens, D. E., and John-Adler, H. B. (1992). The influence of temperature and thermal acclimation on physiological function. In *Environmental Physiology of the Amphibians.* M. E. Feder and W. W. Burggren (Eds.). Pp. 183–205. University of Chicago Press, Chicago.

Romer, A. S. (1956). *Osteology of the Reptiles.* University of Chicago Press, Chicago.

Romer, A. S. (1960). Vertebrate-bearing continental Triassic strata in Mendoza region, Argentina. *Geological Society of America Bulletin* **71:** 1279–1293.

Romer, A. S. (1966). *Vertebrate Paleontology.* (Third Edition). University of Chicago Press, Chicago.

Ron, S. R., Santos, J. C., and Cannatella, D. C. (2006). Phylogeny of the tungara frog genus *Engystomops* (=*Physalaemus pustulosus* species group; Anura: Leptodactylidae). *Molecular Phylogenetics and Evolution* **39:** 392–403.

Rose, B. (1982). Lizard home ranges: Methodology and functions. *Journal of Herpetology* **16:** 253–269.

Rose, G. J., and Gooler, D. M. (2007). Function of the amphibian central auditory system. *Hearing and Sound Communication in Amphibians* **28:** 250–290.

Rose, W. (1962). *The Reptiles and Amphibians of Southern Africa.* Maskew Miller, Ltd., Cape Town, South Africa.

Rosenberg, D. K., Noon, B. R., and Meslow, E. C. (1997). Biological corridors, form, function, and efficacy. *BioScience* **47:** 677–687.

Rosenberg, H. I., and Russell, A. P. (1980). Structural and functional aspects of tail squirting: A unique defense mechanism of *Diplodactylus* (Reptilia: Gekkonidae). *Canadian Journal of Zoology* **58:** 865–881.

Rosenberg, H. I., Russell, A. P., and Kapoor, M. (1984). Preliminary characterization of the defensive secretion of *Diplodactylus* (Reptilia: Gekkonidae). *Copeia* **1984:** 1025–1028.

Rösler, H. (1995). *Geckos der Welt. Alle Gattungen.* Urania-Verlag, Leipzig.

Ross, C. A. (1990). *Crocodylus raninus* S. Müller and Schlegel, a valid species of crocodile (Reptilia: Crocodylidae) from Borneo. *Proceedings of the Biological Society of Washington* **103:** 955–961.

Ross, C. A. (Ed.) (1989). *Crocodiles and Alligators.* Golden Press Pty. Ltd, Silverwater, N.S.W.

Rossi, J. V. (1983). The use of olfactory cues by *Bufo marinus. Journal of Herpetology* **17:** 72–73.

Rossman, D. A., and Willliams, K. L. (1966). Defensive behavior of the South American colubrid snakes *Pseustes sulphureus* (Wagler) and *Spilotes pullatus* (Linnaeus). *Proceedings of the Louisiana Academy of Sciences* **29:** 152–156.

Rossman, D. A., Ford, N. B., and Seigel, R. A. (Eds.). (1996). *The Garter Snakes: Evolution and Ecology.* University of Oklahoma Press, Norman, OK.

Roth, G. (1986). Neural mechanisms of prey recognition: an example in amphibians. In *Predator-prey Relationships: Perspectives and Approaches from the Study of Lower Vertebrates.* M. E. Feder and G. V. Lauder (Eds.). Pp. 42–68. University of Chicago Press, Chicago.

Roth, G. (1987). *Visual Behavior in Salamanders.* Springer-Verlag, New York.

Roughgarden, J. (1995). Anolis. *Lizards of the Caribbean: Ecology, Evolution, and Plate Tectonics.* Oxford University Press, New York.

Rowe, C. L., and Dunson, W. A. (1995). Impacts of hydroperiod on growth and survival of larval amphibians in temporary ponds of central Pennsylvania, USA. *Oecologia* **102:** 397–403.

Roze, J. A. (1996). *Coral Snakes of the Americas: Biology, Identification, and Venoms.* Krieger Publishing Company, Malabar, FL.

Ruben, J. A., Hillenius, W. J., Geist, N. R., Leitch, A., Jones, T. D., Currie, P. J., Horner, J. R., and Espe, G. (1996). The metabolic status of some Late Cretaceous dinosaurs. *Science* **273:** 1204–1207.

Rubio, M. (1998). *Rattlesnakes: Protrait of a Predator.* Smithsonian Institute Press, Washington, DC.

Ruby, D. E. (1978). Seasonal changes in the territorial behavior of the iguanid lizard, *Sceloporus jarrovi. Copeia* **1978:** 430–438.

Ruby, D. E. (1981). Phenotypic correlates of male reproductive success in the lizard, *Sceloporus jarrovi.* In *Natural Selection and Social Behavior.* R. D. Alexander and D. W. Tinkle (Eds.). Pp. 96–107. Chiron Press, New York.

Ruby, D. E. (1984). Male breeding success and differential access to females in *Anolis carolinensis. Herpetologica* **40:** 272–280.

Ruby, D. E., and Dunham, A. E. (1987). Variation in home range size along an elevational gradient in the iguanid lizard *Sceloporus merriami. Oecologia* **71:** 473–480.

Ruby, D. E., and Niblick, H. A. (1994). A behavioral inventory of the desert tortoise: Development of an ethogram. *Herpetological Monographs* **8:** 88–102.

Rugh, R. (1951). *The Frog. Its Reproduction and Development.* McGraw-Hill Book Co, New York.

Ruibal, R., and Shoemaker, V. (1984). Osteoderms in anurans. *Journal of Herpetology* **18:** 313–328.

Ruiz-Carrnaza, P. M., and Lynch, J. D. (1991). Ranas Centrolenidae de Colombia. I. Propuesta de una nueva clasificación genérica. *Lozania* **57:** 1–30.

Runkle, L. S., Wells, K. D., Robb, C. C., and Lance, S. L. (1994). Individual, nightly, and seasonal variation in calling behavior of the gray treefrog, *Hyla versicolor*: Implications for energy expenditure. *Behavioral Ecology* **5:** 318–325.

Russell, A. P. (1988). Limb muscles in relation to lizard systematics: a reappraisal. In *Phylogenetic Relationships of the Lizard Families. Essay Commemorating Charles L. Camp.* R. Estes and G. Pregill (Eds.). Pp. 493–568. Stanford University Press, Stanford, CA.

Russell, A. P., Bauer, A. M., and Johnson, M. K. (2005). Migration in amphibians and reptiles: An overview of patterns and orientation mechanisms in relation to life history strategies. In *Migration of Organisms: Climate, Geography, Ecology* A. M. T Elewa (Ed.). Springer Publishing, New York.

Russell, A. P., and Wu, X.-C. (1997). The Crocodylomorpha at and between geological boundaries: The Baden-Powell approach to change? *Zoology* **100:** 164–182.

Russell, A. P., Bauer, A. M., and Laroiya, R. (1997). Morphological correlates of the secondarily symmetrical pes of gekkotan lizards. *Journal of Zoology, London* **241:** 767–790.

Ruta, M., and Coates, M. I. (2007). Dates, nodes and character conflict: Addressing the lissamphibian origin problem. *Journal of Systematic Palaeontology* **5:** 67–122.

Ruta, M., Coates, M. I., and Quicke, D. L. J. (2001). Early tetrapod relationships revisited. *American Zoologist* **41:** 1574.

Ruta, M., Coates, M. I., and Quicke, D. L. J. (2003). Early tetrapod relationships revisited. *Biological Reviews* **78:** 251–345.

Ruta, M., Jeffery, J. E., and Coates, M. I. (2003). A supertree of early tetrapods. *Proceedings of the Royal Society of London Series B-Biological Sciences* **270:** 2507–2516.

Ruta, M., Wagner, P. J., and Coates, M. I. (2006). Evolutionary patterns in early tetrapods. I. Rapid initial diversification followed by decrease in rates of character change. *Proceedings of the Royal Society B-Biological Sciences* **273:** 2107–2111.

Ruvinsky, I., and Maxson, L. R. (1996). Phylogenetic relationships among bufonoid frogs (Anura: Neobatrachia) inferred from mitochondrial DNA sequences. *Molecular Phylogenetics and Evolution* **5:** 533–547.

Ryan, M. J. (1985). *The Túngara Frog. A Study in Sexual Selection and Communication.* University of Chicago Press, Chicago.

Ryan, M. J., and Rand, A. S. (1995). Female responses to ancestral advertisement calls in túngara frogs. *Science* **269:** 390–392.

Ryan, M. J. (Ed.) (2001). *Anuran Communication.* Smithsonian Institution Press, Washington, D. C.

Ryan, M. J., Tuttle, M. D., and Rand, A. S. (1982). Bat predation and sexual advertisment in a Neotropical frog. *American Naturalist* **119:** 136–139.

Ryan, M. J., Tuttle, M. D., and Taft, L. K. (1981). The costs and benefits of frog chorusing behavior. *Behavioral Ecology and Sociobiology* **8:** 273–278.

Ryan, P. (1988). *Fiji's Natural Heritage.* Southwestern Publ. Co, Auckland.

Ryan, T. J., and Semlitsch, R. D. (2003). Growth and the expression of alternative life cycles in the salamander *Ambystoma talpoideum* (Caudata: Ambystomatidae). *Biological Journal of the Linnean Society* **80:** 639–646.

Sabater-Pi, J. (1985). Contribution to the biology of the giant frog (*Conraua goliath*, Boulenger). *Amphibia-Reptilia* **6:** 143–153.

Sage, R. D., Prager, E. M., and Wake, D. B. (1982). A Cretaceous divergence time between pelobatid frogs (*Pelobates* and *Scaphiopus*): Immunological studies of serum albumin. *Journal of Zoology, London* **198:** 481–494.

Saint Girons, H. (1970). The pituitary gland. In *Biology of the Reptilia. Volume 3. Morphology C.* C. Gans and T. S. Parsons (Eds.). Pp. 135–199. Academic Press, London.

Saint Girons, H. (1988). Les glandes céphaliques exocrines des reptiles. I. Données anatomiques et histologiques. *Annales des Sciences Naturelles, Zoologie, Paris 13e serie* **9:** 221–255.

Saint Girons, H. (1992). Strategies reproductrices des Viperidae dans les zones temperees fraiches et froides. *Bulletin de la Société Zoologique de France* **117:** 267–278.

Saint Girons, M.-C. (1970). Morphology of the circulating blood cells. In *Biology of the Reptilia. Volume 3. Morphology C.* C. Gans and T. S. Parsons (Eds.). Pp. 73–91. Academic Press, London.

Saint, K. M., Austin, C. C., Donnellan, S. C., and Hutchinson, M. N. (1998). C-mos, a nuclear marker useful for squamate phylogenetic analysis. *Molecular Phylogenetics and Evolution* **10:** 259–263.

Saint-Aubain, M. L. de (1985). Blood flow patterns of the respiratory systems in larval and adult amphibians: Functional morphology and phylogenetic significance. *Zeitschrift für Zoologische Systematik und Evolutionsforschung* **23:** 229–240.

Sakaluk, S. K., and Belwood, J. J. (1984). Gecko phonotaxis to cricket calling song: A case of satellite predation. *Animal Behaviour* **32:** 659–662.

Salemi, M., and Vandamme, A.-M. (2003). *The Phylogenetic Handbook: A Practical Approach to DNA and Protein Phylogeny.* Sinauer Associates, Sunderland, Mass.

Salthe, S. N. (1969). Reproductive modes and the numbers and sizes of ova in the urodeles. *American Midland Naturalist* **81:** 467–490.

Salthe, S. N. (1973a). *Amphiuma means* Garden. *Catalogue of American Amphibians and Reptiles* **148:** 1–2.

Salthe, S. N. (1973b). *Amphiuma tridactylum* Cuvier. *Catalogue of American Amphibians and Reptiles* **149:** 1–2.

Salthe, S. N., and Duellman, W. E. (1973). Quantitative constraints associated with reproductive mode in anurans. In *Evolutionary Biology of the Anurans.* J. L. Vial (Ed.). Pp. 229–249. University of Missouri Press, Columbia.

Salthe, S. N., and Mecham, J. S. (1974). Reproduction and courtship patterns. In *Physiology of the Amphibia, Volume 2.* B. Lofts (Ed.). Pp. 309–521. Academic Press, New York.

Salvador, A. (1985). *Guia de Campo de los Anfibios y Reptiles de la Peninsula Iberica, Islas Baleares y Canarias.* Privately printed, Madrid.

Salvador, A., Martín, J., López, P., and Veiga, J. P. (1996). Long-term effect of tail loss on home-range size and access to females in male lizards (*Psammodromus algirus*). *Copeia* **1996:** 208–209.

Salvidio, S. (1993). Life history of the European plethodontid salamander *Speleomantes ambrosii* (Amphibia, Caudata). *Herpetological Journal* **3:** 55–59.

San Mauro, D., Gower, D. J., Oommen, O. V., Wilkinson, M., and Zardoya, R. (2004). Phylogeny of caecilian amphibians (Gymnophiona) based on complete mitochondrial genomes and nuclear RAG1. *Molecular Phylogenetics and Evolution* **33:** 413–427.

Sanchez-Herraiz, M. J., Barbadillo, L. J., Machordom, A., and Sanchez, B. (2000). A new species of pelodytid frog from the Iberian Peninsula. *Herpetologica* **56:** 105–118.

Sanchiz, B., and Rocek, Z. (1996). An overview of the anuran fossil record. In *The Biology of Xenopus.* R. C. Tinsley and H. R. Kobel (Eds.). Pp. 317–328. Zool. Soc. London, London.

Santos, J. C., Coloma, L. A., and Cannatella, D. C. (2003). Multiple, recurring origins of aposematism and diet specialization in poison frogs. *Proceedings of the National Academy of Sciences (USA)* **100:** 12792–12797.

Saporito, R. A., Donnelly, M. A., Hoffman, R. L., Garraffo, H. M., and Daly, J. W. (2003). A siphonotid millipede (*Rhinotus*) as the source of spiropyrrolizidine oximes of dendrobatid frogs. *Journal of Chemical Ecology* **29:** 2781–2786.

Saporito, R. A., Donnelly, M. A., Norton, R. A., Garraffo, H. M., Spande, T. F., and Daly, J. W. (2007). Oribatid mites as a major dietary source for alkaloids in poison frogs. *Proceedings of the National Academy of Sciences (USA)* **104:** 8885–8890.

Saporito, R. A., Garraffo, H. M., Donnelly, M. A., Edwards, A. L., Longino, J. T., and Daly, J. W. (2004). Formicine ants: An arthropod source for the pumiliotoxin alkaloids of dendrobatid poison frogs. *Proceedings of the National Academy of Sciences (USA)* **101:** 8045–8050.

Sartorius, S., Vitt, L. J., and Colli, G. R. (1999). Use of naturally and anthropogenically distrurbed habitats in Amazonian rainforest by the teiid lizard *Ameiva ameiva*. *Biological Conservation* **90:** 91–101.

Saunders, C. D., Brook, A. T., and Myers, O. E. (2006). Using psychology to save biodiversity and human well-being. *Conservation Biology* **20:** 702–705.

Savage, J. M. (1995). Systematics and the biodiversity crisis. *BioScience* **45:** 673–679.

Savage, J. M., and Guyer, C. (1989). Infrageneric classification and species composition of the anole genera, *Anolis, Ctenonotus, Dactyloa, Norops*, and *Semiurus* (Sauria: Iguanidae). *Amphibia-Reptilia* **10:** 105–116.

Savage, J. M., and Lips, K. R. (1993). A review of the status and biogeography of the lizard genera *Celestus* and *Diploglossus* (Squamata: Anguidae), with description of two new species from Costa Rica. *Revista de Biología Tropical* **41:** 817–842.

Savage, J. M., and Wake, M. H. (2001). Reevaluation of the status of taxa of Central American caecilians (Amphibia: Gymnophiona), with comments on their origin and evolution. *Copeia* **2001:** 52–64.

Savidge, J. A. (1987). Extinction of an island forest avifauna by an introduced snake. *Ecology* **68:** 660–668.

Savitsky, A. H. (1983). Coadapted character complexes among snakes: Fossoriality, piscivory, and durophagy. *American Zoologist* **23:** 397–409.

Sazima, I. (1991). Caudal luring in two Neotropical pitvipers, *Bothrops jararaca* and *B. jararacussu*. *Copeia* **1991:** 245–248.

Sazima, I., and Abe, A. S. (1991). Habits of five Brazilian snakes with coral-snake pattern, including a summary of defensive tactics. *Studies on Neotropical Fauna and Environment* **26:** 159–164.

Scalia, F. (1976). Structure of the olfactory and accessory olfactory systems. In *Frog Neurobiology. A Handbook*. R. Llinás and W. Precht (Eds.). Pp. 213–233. Springer-Verlag, Berlin.

Scanlon, J. D. (1992). A new large madtsoiid snake from the Miocene of the Northern Territory. *The Beagle* **9:** 49–60.

Scanlon, J. D., and Lee, M. S. Y. (2004). Phylogeny of Australasian venomous snakes (Colubroidea, Elapidae, Hydrophiinae) based on phenotypic and molecular evidence. *Zoologica Scripta* **33:** 335–366.

Schabetsberger, R., Jehle, R., Maletzky, A., Pesta, J., and Sztatecsny, M. (2004). Delineation of terrestrial reserves for amphibians: Post-breeding migrations of Italian crested newts (*Triturus c. carnifex*) at high altitude. *Biological Conservation* **117:** 95–104.

Schaeffel, R., and de Queiroz, A. (1990). Alternative mechanisms of enhanced underwater vision in the garter snake *Thamnophis melanogaster* and *T. couchii*. *Copeia* **1990:** 50–58.

Schaffner, F. (1998). The liver. In *Biology of the Reptilia. Volume 19, Morphology G*. C. Gans and A. S. Gaunt (Eds.). Pp. 485–531. Society for the Study of Amphibians and Reptiles, Ithaca, NY.

Schall, J. J. (1983). Lizard malaria: Parasite-host ecology. In *Lizard Ecology. Studies of a Model Organism*. R. B. Huey, E. R. Pianka, and T. W. Schoener (Eds.). Pp. 84–100. Harvard University Press, Cambridge, Mass.

Schall, J. J. (1992). Parasite-mediated competition in *Anolis* lizards. *Oecologia* **92:** 58–64.

Schall, J. J., and Pianka, E. R. (1978). Geographical trends in numbers of species. *Science* **201:** 679–686.

Schall, J. J., and Pianka, E. R. (1980). Evolution of escape behavior diversity. *American Naturalist* **115:** 551–566.

Schauinsland, H. (1903). Beiträge zur Entwicklungsgechichte und Anatomie der Wirbeltiere. I. *Sphenodon, Callorhynchus, Chamäleo. Abh. Ges. Zool* **39:** 1–95.

Scheltinga, D. M., and Jamieson, B. G. M. (2003a). Spermatogenesis and the mature spermatozoon: Form function and phylogenetic implications. *Reproductive Biology and Phylogeny of Anura* **2:** 119–251.

Scheltinga, D. M., and Jamieson, B. G. M. (2003b). The mature spermatozoon. *Reproductive Biology and Phylogeny of Urodela* **1:** 203–274.

Scheltinga, D. M., and Jamieson, B. G. M. (2006). Ultrastructure and phylogeny of caecilian spermatozoa. In *Reproductive Biology and Phylogeny of Gymnophiona (Caecilians)*. J.-M. Exbrayat (Ed.). Pp. 247–274. Science Publishers, Enfield, NH.

Schiesari, L. C., Grillitsch, B., and Vogl, C. (1996). Comparative morphology of phytotelmonous and pond-dwelling larvae of four Neotropical treefrog species (Anura, Hylidae, *Osteocephalus oophagus, Osteocephalus taurinus, Phrynohyas resinifictrix, Phrynohyas venulosa). Alytes* **13:** 109–139.

Schiotz, A. (1963). The Amphibia of Nigeria. *Videnskabelige Meddelelser fra Dansk naturhistorik Forening* **125:** 1–92.

Schiotz, A. (1967). The treefrogs (Rhacophoridae) of West Africa. *Spolia zoologica Musei hauniensis* **25:** 1–346.

Schiotz, A. (1975). *The Treefrogs of East Africa*. Steenstrupia, Copenhagen.

Schluter, A. (1990). Reproduction and tadpole of *Edalorhina perezi* (Amphibia, Leptodactylidae). *Studies on Neotropical Fauna and Environment* **25:** 49–56.

Schmalhausen, I. I. (1968). *The Origin of Terrestrial Vertebrates*. Academic Press, New York.

Schmitt, M. (1989). Claims and limits of phylogenetic systematics. *Zeitschrift fur Zoologische Systematik und Evolutionsforschung* **27:** 181–190.

Schmitz, A., Brandley, M. C., Mausfeld, P., Vences, M., Glaw, F., Nussbaum, R. A., and Reeder, T. W. (2005). Opening the black box: Phylogenetics and morphological evolution of the malagasy fossorial lizards of the subfamily "Scincinae. *Molecular Phylogenetics and Evolution* **34:** 118–133.

Schmuck, R., and Linsemair, K. E. (1988). Adaptations of the reed frog, *Hyperolius viridiflavus* (Amphibia, Anura, Hyperoliidae) to its environment: III. Aspects of nitrogen metabolism and osmoregulation in the reed frog, *Hyperolius viridiflavus taeniatus*, with special reference to the role of iridophores. *Oecologia* **75:** 354–361.

Schoener, T. W. (1974). Resource partitioning in ecological communities. *Science* **185:** 27–38.

Schoener, T. W. (1971). Theory of feeding strategies. *Annual Review of Ecology and Systematics* **2:** 369–404.

Schoener, T. W. (1979). Inferring the properties of predation and other injury-producing agents from injury frequencies. *Ecology* **60:** 1110–1115.

Schoener, T. W. (1986). Overview: Kinds of ecological communities—ecology becomes pluralistic. In *Community Ecology*. J. Diamond and T. J. Case (Eds.). Pp. 467–479. Harper and Row, New York.

Schoener, T. W., and Schoener, A. (1982). The ecological correlates of survival in some Bahamian *Anolis* lizards. *Oikos* **39:** 1–16.

Schreibman, M. P. (1986). Pituitary gland. In *Vertebrate Endocrinology: Fundamentals and Biomedical Implications. Vol. 1. Morphological Considerations*. P. K. T. Pang and M. P. Schreibman (Eds.). Pp. 11–55. Academic Press, Orlando, FL.

Schubauer, J. P., Gibbons, J. W., and Spotila, J. R. (1990). Home range and movement patterns of slider turtles inhabiting Par Pond. In *Life History and Ecology of the Slider Turtle*. J. W. Gibbons (Ed.). Pp. 223–232. Smithsonian Institution Press, Washington.

Schulte, J. A., Losos, J., Cruz, F. B., and Nuñez, H. (2004). The relationship between morphology, escape behavior and microhabitat occupation in the lizard clade *Liolaemus* (Iguanidae: Tropidurinae: Liolaemini). *Journal of Evolutionary Biology* **17:** 408–420.

Schulte, J. A., II., Macey, J. R., Espinoza, R. E., and Larson, A. (2000). Phylogenetic relationships in the iguanid lizard genus *Liolaemus*: Multiple origins of viviparous reproduction and evidence for recurring Andean vicariance and dispersal. *Biological Journal of the Linnean Society* **69:** 75–102.

Schulte, J. A., Valladares, J. P., and Larson, A. (2003). Phylogenetic relationships within iguanidae inferred using molecular and morphological data and a phylogenetic taxonomy of iguanian lizards. *Herpetologica* **59:** 399–419.

Schultz, J. (2005). *The Ecozones of the World*. (2nd Ed.). Springer, Berlin.

Schultze, H.-P. (1991). A comparison of controversial hypotheses on the origin of tetrapods. In *Origins of the Higher Groups of Tetrapods. Controversy and Consensus*. H.-P. Schultze and L. Trueb (Eds.). Pp. 29–67. Comstock Publishing Associates, Ithaca, NY.

Schultze, H.-P. (1994). Comparison of hypotheses on the relationships of sarcopterygians. *Systematic Biology* **43:** 155–173.

Schultze, H.-P. and Trueb, L. (Eds.) (1991). *Origins of the Higher Groups of Tetrapods. Controversy and Consensus*. Comstock Publishing Associates, Ithaca, NY.

Schwartz, A., and Henderson, R. W. (1991). *Amphibians and Reptiles of the West Indies: Descriptions, Distributions, and Natural History*. University of Florida Press, Gainesville, FL.

Schwartz, E. R., Schwartz, C. W., and Kiester, A. R. (1984). The three-toed box turtle in central Missouri, part II: a nineteen-year study of home range, movements and population. *Missouri Department of Conservation Terrestrial series* **12:** 1–28.

Schwenk, K. (1985). Occurrence, distribution and functional significance of taste buds in lizards. *Copeia* **1985:** 91–101.

Schwenk, K. (1986). Morphology of the tongue in the tuatara, *Sphenodon punctatus*, (Reptilia: Lepidosauria), with comments on function and phylogeny. *Journal of Morphology* **188:** 129–156.

Schwenk, K. (1988). Comparative morphology of the lepidosaur tongue and its revelance to squamate phylogeny. In *Phylogenetic Relationships of the Lizard Families. Essay Commemorating Charles L. Camp*. R. Estes and G. Pregill (Eds.). Pp. 569–598. Stanford University Press, Stanford, CA.

Schwenk, K. (1993). The evolution of chemoreception in squamate reptiles: A phylogenetic approach. *Brain Behavior and Evolution* **41:** 124–137.

Schwenk, K. (1994a). Why snakes have forked tongues. *Science* **263:** 1573–1577.

Schwenk, K. (1994b). Systematics and subjectivity: The phylogeny and classification of iguanian lizards revisited. *Herpetological Review* **25:** 53–57.

Schwenk, K. (1995). Of tongues and noses: Chemoreception in lizards and snakes. *Trends in Ecology and Evolution* **10:** 7–12.

Schwenk, K. (2000). Feeding: Form, Function, and Evolution in Tetrapod Vertebrates. Academic Press, San Diego, CA.

Schwenk, K., and Greene, H. W. (1987). Water collection and drinking in *Phrynocephalus helioscopus*: A possible condensation mechanism. *Journal of Herpetology* **21:** 134–139.

Schwenk, K. (Ed.) (2000). *Feeding, Form, Function, and Evolution in Tetrapod Vertebrates*. Academic Press, San Diego, CA.

Scott, A. C. (1980). The ecology of some Upper Palaeozoic floras. In *The Terrestrial Environment and the Origin of Land Vertebrates*. A. L. Panchen (Ed.). Pp. 87–115. Academic Press, London.

Scott, D. E. (1990). Effects of larval density in *Ambystoma opacum*: an experiment in large-scale field enclosures. *Ecology* **71:** 296–306.

Scott, D. E. (1994). The effect of larval density on adult demographic traits in *Ambystoma opacum*. *Ecology* **75:** 1383–1396.

Scott, E. (2005). A phylogeny of ranid frogs (Anura: Ranoidea: Ranidae), based on a simultaneous analysis of morphological and molecular data. *Cladistics* **21:** 507–574.

Scott, I. A. W., Keogh, J. S., and Whiting, M. J. (2004). Shifting sands and shifty lizards: molecular phylogeny and biogeography of African flat lizards (*Platysaurus*). *Molecular Phylogenetics and Evolution* **31:** 618–629.

Scott, N. J., Jr. (1976). The abundance and diversity of the herpetofaunas of tropical forest litter. *Biotropica* **8:** 41–58.

Seben, K. P. (1987). The ecology of indeterminate growth in animals. *Annual Review of Ecology and Systematics* **18:** 371–407.

Secor, S. M. (1994). Ecological significance of movements and activity range for the sidewinder, *Crotalus cerastes*. *Copeia* **1994:** 631–645.

Secor, S. M., and Nagy, K. A. (1994). Bioenergetic correlates of foraging for the snakes *Crotalus cerastes* and *Masticophis flagellum*. *Ecology* **75:** 1600–1614.

Secor, S. M., Jayne, B. C., and Bennett, A. F. (1992). Locomotor performance and energetic cost of sidewinding by the snake *Crotalus cerastes*. *Journal of Experimental Biology* **163:** 1–14.

Seddon, J. M., Baverstock, P. R., and Georges, A. (1998). The rate of mitochondrial 12S rRNA gene evolution is similar in freshwater turtles and marsupials. *Journal of Molecular Evolution* **46:** 460–464.

Seddon, J. M., Georges, A., Baverstock, P. R., and McCord, W. (1997). Phylogenetic relationships of chelid turtles (Pleurodira: Chelidae) based on mitochondrial 12S rRNA gene sequence variation. *Molecular Phylogenetics and Evolution* **7:** 55–61.

Seebacher, F., and Shine, R. (2004). Evaluating thermoregulation in reptiles: The fallacy of the inappropriately applied method. *Physiological and Biochemical Zoology* **77:** 688–695.

Seigel, R. A. (1983). Natural survival of eggs and tadpoles of the wood frog, *Rana sylvatica*. *Copeia* **1983:** 1096–1098.

Seigel, R. A. and Collins, J. T. (Eds.) (1993). *Snakes: Ecology and Behavior*. McGraw-Hill, Inc, New York.

Seigel, R. A. and Collins, J. T. (Eds.) (2002). *Snakes: Ecology and Behavior*. McGraw-Hill, Inc, New York.

Seigel, R. A., and Ford, N. B. (1991). Phenotypic plasticity in the reproductive characteristics of an oviparous snake, *Elaphe guttata*: Implications for life history studies. *Herpetologica* **47:** 301–307.

Seigel, R. A., and Ford, N. B. (1987). Reproductive ecology. In *Snakes: Ecology and Evolutionary Biology*. R. Seigel, J. T. Collins, and S. S. Novak (Eds.). Pp. 210–252. Macmillan Publishing Company, New York, NY.

Seigel, R. A., Fitch, H. S., and Ford, N. B. (1986). Variation in relative clutch mass on snakes among and within species. *Herpetologica* **42:** 179–185.

Seigel, R. A., Huggins, M. M., and Ford, N. B. (1987). Reduction in locomotor ability as a cost of reproduction in gravid snakes. *Oecologia* **73:** 481–485.

Seigel, R. A., Loraine, R. K., and Gibbons, J. W. (1995). Reproductive cycles and temporal variation in fecundity in the black swamp snake, *Seminatrix pygaea*. *American Midland Naturalist* **134:** 371–377.

Seigel, R. A., Smith, R. B., Demuth, J., Ehrhart, L. M., and Snelson, F. F. (2002). Amphibians and reptiles of the John F. Kennedy Space Center, Florida: A long-term assessment of a large protected habitat (1975–2000). *Florida Scientist* **65:** 1–12.

Selcer, K. W. (1990). Egg-size relationships in a lizard with fixed clutch size: Variation in a population of the Mediterranean gecko. *Herpetologica* **46:** 15–21.

Semlitsch, R. D. (1980). Geographic and local variation in population parameters of the slimy salamander *Plethodon glutinosus*. *Herpetologica* **36:** 6–16.

Semlitsch, R. D. (1981). Effects of implanted tantalum-182 wire tags on the mole salamander, *Ambystoma talpoideum*. *Copeia* **1981:** 735–737.

Semlitsch, R. D. (1985a). Analysis of climatic factors influencing migrations of the salamander *Ambystoma talpoideum*. *Copeia* **1985:** 477–489.

Semlitsch, R. D. (1985b). Reproductive strategy of a facultatively paedomorphic salamander *Ambystoma talpoideum*. *Oecologia* **65:** 305–313.

Semlitsch, R. D. (1987). Relationship of pond drying to the reproductive success of the salamander *Ambystoma talpoideum*. *Copeia* **1987:** 61–69.

Semlitsch, R. D. (1998). Biological delineation of terrestrial buffer zones for pond-breeding salamanders. *Conservation Biology* **12:** 1113–1119.

Semlitsch, R. D., and Gibbons, J. W. (1988). Fish predation in size-structured populations of treefrog tadpoles. *Oecologia* **75:** 321–326.

Semlitsch, R. D., Scott, D. E., and Pechmann, J. H. K. (1988). Time and size at metamorphosis related to adult fitness in *Ambystoma talpoideum*. *Ecology* **69:** 184–192.

Semlitsch, R. D., Scott, D. E., Pechmann, J. H. K., and Gibbons, J. W. (1996). Structure and dynamics of an amphibian community: Evidence from a 16–year study of a natural pond. In *Long-term Studies of Vertebrate Communities*. M. L. Cody and J. A. Smallwood (Eds.). Pp. 217–248. Academic Press, San Diego.

Semple, C., and Steel, M. (2003). *Phylogenetics*. Oxford Lecture Series in Mathematics and Its Applications, Oxford University Press, Oxford, UK.

Sessions, S. K. (1996). Chromosomes: Molecular cytogenetics. In *Molecular Systematics, Second Edition*. D. M. Hillis, C. Moritz, and B. K. Mable (Eds.). Pp. 121–168. Sinauer Associates, Inc., Sunderland, Mass.

Sessions, S. K., and Ruth, S. B. (1990). Explanation for naturally occurring supernumerary limbs in amphibians. *Journal of Experimental Zoology* **254:** 38–47.

Sever, D. M. (1991a). Comparative anatomy and phylogeny of the cloacae of salamanders (Amphibia: Caudata). I. Evolution at the family level. *Herpetologica* **47:** 165–193.

Sever, D. M. (1991b). Comparative anatomy and phylogeny of the cloacae of salamanders (Amphibia: Caudata). II. Cryptobranchidae, Hynobiidae, and Sirenidae. *Journal of Morphology* **207:** 283–301.

Sever, D. M. (1992). Comparative anatomy and phylogeny of the cloacae of salamanders (Amphibia: Caudata). VI. Ambystomatidae and Dicamptodontidae. *Journal of Morphology* **212:** 305–322.

Sever, D. M. (1994). Observations on regionalization of secretory activity in the spermathecae of salamanders and comments on phylogeny of sperm storage in female amphibians. *Herpetologica* **50:** 383–397.

Sever, D. M. (2003). Courtship and mating glands. *Reproductive Biology and Phylogeny of Urodela* **1:** 323–381.

Sever, D. M. (Ed.) (2003). *Reproductive Biology and Phylogeny of Anura*. Volume 2 of the series *Reproductive Biology and Phylogeny*, series Editor, B. G. M. Jamieson. Science publishers, Inc., Enfield, NH, USA.

Sever, D. M., and Kloepeer, N. M. (1993). Spermathecal cytology of *Ambystoma opacum* (Amphibia: Ambystomatidae) and the phylogeny of sperm storage organs in female salamanders. *Journal of Morphology* **217:** 114–127.

Sever, D. M., Heinz, E. A., Lempart, P. A., and Taghon, M. S. (1990). Phylogenetic significance of the cloacal anatomy of female bolitoglossine salamanders (Plethodontidae, tribe Bolitoglossini). *Herpetologica* **46:** 431–446.

Sever, D. M., Rania, L. C., and Krenz, J. D. (1996a). Annual cycle of sperm storage in spermathecae of the red-spotted newt,

Notophthalmus viridescens (Amphibia: Salamandridae). *Journal of Morphology* **227:** 155–170.

Sever, D. M., Rania, L. C., and Krenz, J. D. (1996b). Reproduction of the salamander *Siren intermedia* Le Conte with especial reference to oviducal anatomy and mode of fertilization. *Journal of Morphology* **227:** 335–348.

Sexton, O. J. (1959). Spatial and temporal movements of a population of the painted turtle *Chrysemys picta marginata* (Agassiz). *Ecological Monographs* **29:** 113–140.

Sexton, O. J., Ortleb, E. P., Hathaway, L. M., and Ballinger, R. E. (1971). Reproductive cycles of three species of anoline lizards from the Isthmus of Panama. *Ecology* **52:** 201–215.

Seymour, R. S. (1982). Physiological adaptations to aquatic life. *Biology of Reptilia* **13:** 1–51.

Seymour, R. S., and Bradford, D. F. (1995). Respiration of amphibian eggs. *Physiological Zoology* **68:** 1–25.

Shaffer, H. B. (1993). Phylogenetics of model organisms: the laboratory axolotl, *Ambystoma mexicanum*. *Systematic Biology* **42:** 508–522.

Shaffer, H. B., and McKnight, M. L. (1996). The polytypic species revisited: Genetic differentiation and molecular phylogenetics of the tiger salamander *Ambystoma tigrinum* (Amphibia: Caudata) complex. *Evolution* **50:** 417–433.

Shaffer, H. B., Meylan, P., and McKnight, M. L. (1997). Tests of turtle phylogeny: molecular, morphological, and paleontological approaches. *Systematic Biology* **46:** 235–268.

Shaffer, H. B., Clark, J. M., and Kraus, F. (1991). When molecules and morphology clash: A phylogenetic analysis of the North American ambystomatid salamanders (Caudata: Ambystomatidae). *Systematic Zoology* **40:** 284–303.

Shaffer, M. L. (1981). Minimum population sizes for species conservation. *BioScience* **31:** 131–134.

Shah, B., Shine, R., Hudson, S., and Kearney, M. (2003). Sociality in lizards: Why do thick-tailed geckos (*Nephrurus milii*) aggregate? *Behaviour* **140:** 1039–1052.

Shaldybin, S. L. (1981). Wintering and number of amphibians and reptiles of Lazo State Nature Reserve. In *Herpetological Investigations in Siberia and the Far East*. L. J. Borkin (Ed.). Pp. 123–124. Academy of Science USSR, Zoology Institute, Leningrad.

Shea, G. M. (1993). Family Pygopodidae. In *Fauna of Australia. Volume 2A. Amphibia & Reptilia*. C. J. Gasby, G. J. B. Ross, and P. L. Beesley (Eds.). Pp. 234–239. Australian Government Publishing Service, Canberra.

Shea, G., Shine, R., and Covacevich, J. C. (1993). Family Elapidae. In *Fauna of Australia. Volume 2A. Amphibia & Reptilia*. C. J. Gasby, G. J. B. Ross, and P. L. Beesley (Eds.). Pp. 295–309. Australian Government Publishing Service, Canberra.

Sheil, C. A. (1999). Osteology and skeletal development of *Pyxicephalus adspersus* (Anura: Ranidae: Raninae). *Journal of Morphology* **240:** 49–75.

Sheil, C. A., and Mendelson, J. R. (2001). A new species of *Hemiphractus* (Anura: Hylidae: Hemiphractinae), and a redescription of *H. johnsoni*. *Herpetologica* **57:** 189–202.

Sheldahl, L. A., and Martins, E. P. (2000). The territorial behavior of the western fence lizard, *Sceloporus occidentalis*. *Herpetologica* **56:** 469–479.

Shepard, D. B., and Caldwell, J. P. (2005). From foam to free-living: Ecology of larval *Leptodactylus labyrinthicus*. *Copeia* **2005:** 803–811.

Sherbrooke, W. C. (1981). *Horned Lizards. Unique Reptiles of Western North America*. Southwestern Parks & Monument Association, Globe, AZ.

Sherbrooke, W. C. (1990). Rain-harvesting in the lizard, *Phrynosoma cornutum*: Behavior and integumental morphology. *Journal of Herpetology* **24:** 302–308.

Sherbrooke, W. C., and Montanucci, R. R. (1988). Stone mimicry in the round-tailed horned lizard, *Phrynosoma modestum* (Sauria: Iguanidae). *Journal of Arid Environments* **14:** 275–284.

Sherbrooke, W. C., Scardino, A. J., de Nys, R., and Schwarzkopf, L. (2007). Functional morphology of scale hinges used to transport water: Convergent drinking adaptations in desert lizards (*Moloch horridus* and *Phrynosoma cornutum*). *Zoomorphology* **126:** 89–102.

Shi, H. T., and Parham, J. F. (2001). Preliminary observations of a large turtle farm in Hainan Province, People's Republic of China. *Turtle and Tortoise Newsletter* **3:** 4–6.

Shi, H. T., Parham, J. F., Lau, M., and Tien-Hsi, C. (2007). Farming endangered turtles to extinction in China. *Conservation Biology* **21:** 5–6.

Shi, H., Fan, Z., Yin, F., and Yuan, Z. (2004). New data on the trade and captive breeding of turtles in Guangxi Province, south China. *Asiatic Herpetological Research* **10:** 126–128.

Shine, R. (1978). Sexual size dimorphism and male combat in snakes. *Oecologia* **33:** 269–277.

Shine, R. (1979). Sexual selection and sexual dimorphism in the Amphibia. *Copeia* **1979:** 297–306.

Shine, R. (1980a). Ecology of the Australian death adder *Acanthophis antarcticus* (Elapidae): Evidence for convergence with the Viperidae. *Herpetologica* **36:** 281–289.

Shine, R. (1980b). Costs" of reproduction in reptiles. *Oecologia* **46:** 92–100.

Shine, R. (1980c). Reproduction, feeding and growth in the Australian burrowing snake *Vermicella annulata*. *Journal of Herpetology* **14:** 71–77.

Shine, R. (1983). Reptilian viviparity in cold climates: Testing the assumptions of an evolutionary hypothesis. *Oecologia* **57:** 397–405.

Shine, R. (1985a). The reproductive biology of Australian reptiles: a search for general patterns. In *Biology of Australasian Frogs and Reptiles*. G. Grigg, R. Shine, and H. Ehmann (Eds.). Pp. 297–303. Surrey Beatty & Sons Pty Ltd., Chipping Norton Pty Ltd., Sydney, Australia.

Shine, R. (1985b). The evolution of viviparity in reptiles: an ecological analysis. In *Biology of the Reptilia. Volume 15, Developmental Biology B*. C. Gans and F. Billett (Eds.). Pp. 677–680. John Wiley & Sons, Inc, New York.

Shine, R. (1986a). Ecology of a low-energy specialist: Food habits and reproductive biology of the Arafura filesnake (Acrochordidae). *Copeia* **1986:** 424–437.

Shine, R. (1986b). Sexual differences in morphology and niche utilization in an aquatic snake, *Acrochordus arafurae*. *Oecologia* **69:** 260–267.

Shine, R. (1987). The evolution of viviparity: Ecological correlates of reproductive mode within a genus of Australian snakes (*Pseudechis*: Elapidae). *Copeia* **1987:** 551–563.

Shine, R. (1988). Parental care in reptiles. In *Biology of the Reptilia, Vol. 16, Ecology B. Defense and Life History*. C. Gans and R. D. Huey (Eds.). Pp. 275–329. Alan R. Liss, Inc., New York.

Shine, R. (1991a). Intersexual dietary divergence and the evolution of sexual dimorphism in snakes. *American Naturalist* **138:** 103–122.

Shine, R. (1991b). *Australian Snakes. A Natural History*. Cornell University Press, Ithaca, N.Y.

Shine, R. (1994). Sexual dimorphism in snakes revisited. *Copeia* **1994:** 326–346.

Shine, R. (2002). An empirical test of the 'predictability' hypothesis for the evolution of viviparity in reptiles. *Journal of Evolutionary Biology* **15:** 553–560.

Shine, R. (2003a). Locomotor speeds of gravid lizards: Placing 'costs of reproduction' within an ecological context. *Functional Ecology* **17:** 526–533.

Shine, R. (2003b). Reproductive strategies in snakes. *Proceedings of the Royal Society of London B* **270:** 995–1004.

Shine, R. (2004). Seasonal shifts in nest temperature can modify the phenotypes of hatchling lizards, regardless of overall mean incubation temperature. *Functional Ecology* **18:** 43–49.

Shine, R. (2005). Life-history evolution in reptiles. *Annual Review of Ecology Evolution and Systematics* **36:** 23–46.

Shine, R., and Bull, J. J. (1979). The evolution of life-bearing in lizards and snakes. *American Naturalist* **113:** 905–923.

Shine, R., and Fitzgerald, M. (1996). Large snakes in a mosaic rural landscape: The ecology of carpet pythons *Morelia spilota* (Serpentes: Pythonidae) in coastal eastern Australia. *Biological Conservation* **76:** 113–122.

Shine, R., and Harlow, P. S. (1996). Maternal manipulation of offspring phenotypes via nest-site selection in an oviparous lizard. *Ecology* **77:** 1808–1817.

Shine, R., and Houston, D. (1993). Family Acrochordidae. In *Fauna of Australia. Volume 2A. Amphibia & Reptilia*. C. J. Gasby, G. J. B. Ross, and P. L. Beesley (Eds.). Pp. 322–324. Australian Government Publishing Service, Canberra.

Shine, R., and Lambeck, R. (1985). A radiotelemetric study of movements, thermoregulation and habitat utilization of Arafura filesnakes (Serpentes: Acrochordidae). *Herpetologica* **41:** 351–361.

Shine, R., and Mason, R. T. (2005). Does large body size in males evolve to facilitate forcible insemination? A study on garter snakes. *Evolution* **59:** 2426–2432.

Shine, R., and Olsson, M. (2003). When to be born? Prolonged pregnancy or incubation enhances locomotor performance in neonatal lizards (Scincidae). *Journal of Evolutionary Biology* **16:** 823–832.

Shine, R., and Schwaner, T. (1985). Prey constriction by venomous snakes: A review, and new data on Australian species. *Copeia* **1985:** 1067–1071.

Shine, R., and Schwarzkopf, L. (1992). The evolution of reproductive effort in lizards and snakes. *Evolution* **46:** 62–75.

Shine, R., and Slip, D. J. (1990). Biological aspects of the adaptive radiation of Australasian pythons (Serpentes: Boidae). *Herpetologica* **46:** 283–290.

Shine, R., Branch, W. R., Harlow, P. S., and Webb, J. K. (1996a). Sexual dimorphism, reproductive biology, and food habits of two species of African filesnakes (*Mehelya*, Colubridae). *Journal of Zoology, London* **240:** 327–340.

Shine, R., Elphick, M. J., and Barrott, E. G. (2003). Sunny side up: Lethally high, not low, nest temperatures may prevent oviparous reptiles from reproducing at high elevations. *Biological Journal of the Linnean Society* **78:** 325–334.

Shine, R., Elphick, M. J., and Harlow, P. S. (1997a). The influence of natural incubation environments on the phenotypic traits of hatchling lizards. *Ecology* **78:** 2559–2568.

Shine, R., Haagner, G. V., Branch, W. R., Harlow, P. S., and Webb, J. K. (1996b). Natural history of the African shieldnose snake *Aspidelaps scutatus* (Serpentes, Elapidae). *Journal of Herpetology* **30:** 361–366.

Shine, R., Harlow, P. S., and Keogh, J. S. (1996e). Commercial harvesting of giant lizards: The biology of water monitors *Varanus salvator* in southern Sumatra. *Biological Conservation* **77:** 125–134.

Shine, R., Harlow, P. S., Branch, W. R., and Webb, J. K. (1996c). Life on the lowest branch: Sexual dimorphism, diet, and reproductive biology of an African twig snake, *Thelotornis capensis* (Serpentes, Colubridae). *Copeia* **1996:** 290–299.

Shine, R., Langkilde, T., and Mason, R. T. (2004a). Courtship tactics in garter snakes: How do a male's morphology and behavior influence his mating success? *Animal Behaviour* **67:** 477–483.

Shine, R., Langkilde, T., Wall, M. D., and Mason, R. T. (2005b). Alternative male mating tactics in garter snakes. *Animal Behaviour* **70:** 387–396.

Shine, R., Madsen, T. R. L., Elphick, M. J., and Harlow, P. S. (1997). The influence of nest temperatures and maternal brooding on hatchling phenotypes of water pythons. *Ecology* **78:** 1713–1721.

Shine, R., O'Connor, D., LeMaster, M. P., and Mason, R. T. (2001a). Pick on someone your own size: Ontogenetic shifts in mate choice by male garter snakes result in size-assortative mating. *Animal Behaviour* **61:** 1133–1141.

Shine, R., O'Connor, D., and Mason, R. T. (2000a). Female mimicry in garter snakes: Behavioural tactics of "she-males" and the males that court them. *Canadian Journal of Zoology* **78:** 1391–1396.

Shine, R., O'Connor, D., and Mason, R. T. (2000b). Sexual conflict in the snake den. *Behavioral Ecology and Sociobiology* **48:** 392–401.

Shine, R., O'Donnell, R. P., Langkilde, T., Wall, M. D., and Mason, R. T. (2005a). Snakes in search of sex: The relationship between mate-locating ability and mating success in male garter snakes. *Animal Behaviour* **69:** 1251–1258.

Shine, R., Phillips, B., Langkilde, T., Lutterschmidt, D., Waye, H., and Mason, R. T. (2004b). Mechanisms and consequences of sexual conflict in garter snakes (*Thamnophis sirtalis*, Colubridae). *Behavioral Ecology* **15:** 654–660.

Shine, R., Phillips, B., Waye, H., LeMaster, M., and Mason, R. T. (2001b). Advantage of female mimicry to snakes. *Nature* **414:** 267.

Shine, R., Schwarzkopf, L., and Caley, M. J. (1996d). Energy, risk, and reptilian reproductive effort: A reply to Niewiarowski and Dunham. *Evolution*. **50:** 2111–2114.

Shine, R., Sun, L., Fitzgerald, M., and Kearney, M. 2002. Behavioral responses of free-ranging pit-vipers (*Gloydius shedaoensis*, Viperidae) to approach by a human. *Copeia* **2002:** 834–850.

Shine, R., Langkilde, T., Wall, M., and Mason, R. T. (2005c). The fitness correlates of scalation asymmetry in garter snakes (*Thamnophis sirtalis parietalis*). *Functional Ecology* **19:** 306–314.

Shine, R., T. Madsen, R. L., Elphick, M. J., and Harlow, P. S. (1997b). The influence of nest temperatures and maternal brooding on hatchling phenotypes of water pythons. *Ecology* **78:** 1713–1721.

Shoemaker, V. H. (1992). Exchange of water, ions, and respiratory gases in terrestrial amphibians. In *Environmental Physiology of the Amphibians*. M. E. Feder and W. W. Burggren (Eds.). Pp. 125–150. University of Chicago Press, Chicago.

Shoemaker, V. H., and Bickler, P. E. (1979). Kidney and bladder function in a uricotelic treefrog (*Phyllomedusa sauvagei*). *Journal of Comparative Physiology* **133:** 211–218.

Shoemaker, V. H., and McClanahan, L. L. (1980). Nitrogen excretion and water balance in amphibians of Borneo. *Copeia* **1980:** 446–451.

Shoemaker, V. H., and McClanahan, L. L. (1982). Enzymatic correlates and uricotelism in tree frogs of the genus *Phyllomedusa*. *Journal of Experimental Zoology* **220:** 163–169.

Shoemaker, V. H., and Nagy, K. A. (1977). Osmoregulation in amphibians and reptiles. *Annual Review Physiology* **39:** 449–471.

Shoemaker, V. H., Baker, M. A., and Loveridge, J. P. (1989). Effect of water balance on thermoregulation in waterproof frogs (*Chiromantis* and *Phyllomedusa*). *Physiological Zoology* **62:** 133–146.

Shoop, C. R. (1960). The breeding habits of the mole salamander, *Ambystoma talpoideum* (Holbrook), in southeastern Louisiana. *Tulane Studies in Zoology* **8:** 65–82.

Shoop, C. R. (1968). Migratory orientation of *Ambystoma maculatum*: Movements near breeding ponds and displacements of migrating individuals. *Biological Bulletin* **135:** 230–238.

Shubin, N. H., and Jenkins, F. A., Jr. (1995). An Early Jurassic jumping frog. *Nature* **377:** 49–52.

Shubin, N. H., Daeschler, E. B., and Coates, M. I. (2004). The early evolution of the tetrapod humerus. *Science* **304:** 90–93.

Shubin, N. H., Daeschler, E. B., and Jenkins, F. A. (2006). The pectoral fin of *Tiktaalik roseae* and the origin of the tetrapod limb. *Nature* **440:** 764–771.

Simon, C. A. (1975). The influence of food abundance on territory size in the iguanid lizard *Sceloporus jarrovi*. *Ecology* **56:** 993–998.

Simon, M. P. (1983). The ecology of parental care in a terrestrial breeding frog from New Guinea. *Behavioral Ecology and Sociobiology* **14:** 61–67.

Simon, M. P., and Toft, C. A. (1991). Diet specialization in small vertebrates: Mite eating in frogs. *Oikos* **61:** 263–278.

Simons, R. S., and Brainerd, E. L. (1999). Morphological variation of hypaxial musculature in salamanders (Lissamphibia: Caudata). *Journal of Morphology* **241:** 153–164.

Simpson, G. G. (1961). *Principles of Animal Taxonomy*. Columbia University Press, New York.

Sinch, U. (1990). Migration and orientation in anuran amphibians. *Ethology, Ecology & Evolution* **2:** 65–79.

Sinervo, B., and Zamudio, K. R. (2001). The evolution of alternative reproductive strategies: Fitness differential, heritability, and genetic correlation between the sexes. *Journal of Heredity* **92:** 198–205.

Sinsch, U. (1989). Behavioral thermoregulation of the Andean toad (*Bufo spinulosus*) at high altitudes. *Oecologia* **80:** 32–38.

Sites, J. W., Jr, and Crandall, K. A. (1997). Testing species boundaries in biodiversity studies. *Conservation Biology* **11:** 1289–1297.

Sites, J. W., Jr, Davis, S. K., Guerra, T., Iverson, J. B., and Snell, H. L. (1996). Character congruence and phylogenetic signal in molecular and morphological data sets: A case study in the living iguanas (Squamata, Iguanidae). *Molecular Biology and Evolution* **13:** 1087–1105.

Sites, J. W., Jr., Archie, J. W., Cole, C. J., and Flores Villela, O. (1992). A review of phylogenetic hypotheses for lizards of the genus *Sceloporus* (Phrynosomatidae): Implications for ecological and evolutionary studies. *Bulletin of the American Museum of Natural History* **213:** 1–110.

Slade, N. A., and Wassersug, R. J. (1975). On the evolution of complex life cycles. *Evolution* **29:** 568–571.

Slowinski, J. B., and Keogh, J. S. (2000). Phylogenetic relationships of elapid snakes based on Cytochrome b mtDNA sequences. *Molecular Phylogenetics and Evolution* **15:** 157–164.

Slowinski, J. B., and Lawson, R. (2002). Snake phylogeny: Evidence from nuclear and mitochondrial genes. *Molecular Phylogenetics and Evolution* **24:** 194–202.

Slowinski, J. B., Boundy, J., and Lawson, R. (2001). The phylogenetic relationships of Asian coral snakes (Elapidae: *Calliophis* and *Maticora*) based on morphological and molecular characters. *Herpetologica* **57:** 233–245.

Slowinski, J. B., Knight, A., and Rooney, A. P. (1997). Inferring species trees from gene trees: A phylogenetic analysis of the Elapidae (Serpentes) based on the amino acid sequences of venom proteins. *Molecular Phylogenetics and Evolution* **8:** 349–362.

Slowinski, J. B., and Lawson, R. (2005). Elapid relationships. In *Ecology and Evolution in the Tropics: Essays in tribute to J.M. Savage*. M. A. Donnelly, B. I. Crother, C. Guyer, M. H. Wake, and M. E. White (Eds.). Pp. 174–189. University of Chicago Press, Chicago, IL.

Smatresk, N. J. (1994). Respiratory control in the transition from water to air breathing in vertebrates. *American Zoologist* **34:** 264–279.

Smith, C. C., and Fretwell, S. D. (1974). The optimal balance between size and number of offspring. *American Naturalist* **108:** 499–506.

Smith, H. M., Smith, R. B., and Sawin, H. L. (1977). A summary of snake classification (Reptilia, Serpentes). *Journal of Herpetology* **11:** 115–121.

Smith, M. A. (1943). *The Fauna of British India, Ceylon and Burma. Vol. III. Serpentes*. Taylor and Francis, London.

Smith, R. B., Breininger, D. R., and Larson, V. L. (1997). Home range characteristics of radiotagged gopher tortoises on Kennedy Space Center, Florida. *Chelonian Conservation and Biology* **2:** 358–362.

Smith, S. A., Arif, S., de Oca, A. N. M., and Wiens, J. J. (2007). A phylogenetic hot spot for evolutionary novelty in middle American treefrogs. *Evolution* **61:** 2075–2085.

Smith, S. M. (1975). Innate recognition of coral snake pattern by a possible avian predator. *Science* **187:** 759–760.

Smith, S. M. (1977). Coral-snake pattern recognition and stimulus generalisation by naive great kiskadees (Aves: Tyrannidae). *Nature* **265:** 535–536.

Smith, T. L., Povel, G. D. E., and Kardong, K. V. (2002). Predatory strike of the tentacled snake (*Erpeton tentaculatum*). *Journal of Zoology, London* **256:** 233–242.

Snell, H. L., and Tracy, C. R. (1985). Behavioral and morphological adaptations by Galapagos land iguanas (*Conolophus subcristatus*) to water and energy requirements of eggs and neonates. *American Zoologist* **25:** 1009–1018.

Sokal, R. R. (1986). Phenetic taxonomy: Theory and methods. *Annual Review of Evology and Systematics* **17:** 423–442.

Som, C., and Reyer, H. U. (2007). Hemiclonal reproduction slows down the speed of Muller's ratchet in the hybridogenetic frog *Rana esculenta*. *Journal of Evolutionary Biology* **20:** 650–660.

Somma, L. A. (2003). *Parental Behavior in Lepidosaurian and Testudinian Reptiles*. Krieger Publishing Company, Malabar, Florida.

Somma, L. A., and Fawcett, J. D. (1989). Brooding behaviour of the prairie skink, *Eumeces septentrionalis*, and its relationship to the hydric environment of the nest. *Zoological Journal of the Linnean Society* **95:** 245–256.

Spellerberg, I. F. (1988). Ecology and management of *Lacerta agilis* L. populations in England. *Mertensiella* **1:** 113–121.

Spinar, Z. V. (1972). *Tertiary Frogs from Central Europe*. W. Junk N. V, The Hague.

Spotila, J. R., Dunham, A. E., Leslie, A. J., Steyermark, A. C., Plotkin, P. T., and Paladino, F. V. (1998). Worldwide population decline of *Dermochelys coriacea*: Are leatherback seaturtles going extinct? *Chelonian Conservation Biology* **2:** 209–222.

Spotila, J. R., O'Connor, M. P., and Bakken, G. S. (1992). Biophysics of heat and mass transfer. In *Environmental Physiology of the Amphibians*. M. E. Feder and W. W. Burggren (Eds.). Pp. 59–80. University of Chicago Press, Chicago.

Spotila, J. R., O'Connor, M. P., and Paladino, F. V. (1997). Thermal biology. In *The Biology of Sea Turtles*. P. L. Lutz and J. A. Musick (Eds.). Pp. 297–314. CRC Press, Boca Raton, FL.

Spotila, J. R., Terpin, K. M., and Dodson, P. (1977). Mouth gaping as an effective thermoregulatory device in alligators. *Nature* **265:** 235–236.

Sprawl, S., and Branch, B. (1995). *The Dangerous Snakes of Africa*. Ralph Curtis Books, Sanibel Island, FL.

Spray, D. C. (1976). Pain and temperature receptors of anurans. In *Frog Neurobiology. A Handbook*. R. Llinás and W. Precht (Eds.). Pp. 607–628. Springer-Verlag, Berlin.

Stammer, D. (1976). Reptiles. In *Around Mt. Isa. A Guide to the Flora and Fauna*. H. Horton (ed.) University of Queensland, St Lucia, Queensland.

Stamps, J. A. (1977). Social behavior and spacing patterns in lizards. In *Biology of the Reptilia. Volume 7. Ecology and Behavior*. C. Gans and D. W. Tinkle (Eds.). Pp. 265–334. Academic Press, New York.

Stamps, J. A. (1978). A field study on the ontogeny of social behavior in the lizard *Anolis aeneus*. *Behaviour* **64:** 1–31.

Stamps, J. A. (1983). Sexual selection, sexual dimorphism, and territoriality. In *Lizard Ecology. Studies of a Model Organism*. R. B. Huey, E. R. Pianka, and T. W. Schoener (Eds.). Pp. 169–204. Harvard University Press, Cambridge, Mass.

Stamps, J. A., Krishnan, V. V., and Andrews, R. M. (1994). Analyses of sexual size dimorphism using null growth-based models. *Copeia* **1994:** 598–613.

Stamps, J. A., Mangel, M., and Phillips, J. A. (1998). A new look at relationships between size at maturity and asymptotic size. *American Naturalist* **152:** 470–479.

Stamps, J., and Krishnan, V. V. (1997). Sexual bimaturation and sexual size dimorphism in animals with asymptotic growth after maturity. *Evolutionary Ecology* **11:** 21–39.

Starace, F. (1998). *Guide des Serpents et Amphisbènes de Guyana Française*. Ibis Rouge Editions, Guadeloupe.

Stearns, S. C. (1976). Life-history tactics: A review of the ideas. *Quarterly Review of Biology* **51:** 3–47.

Stearns, S. C. (1977). The evolution of life history traits: A critique of the theory and a review of the data. *Annual Review of Ecology and Systematics* **8:** 145–171.

Stearns, S. C. (1992). *The Evolution of Life Histories*. Oxford University Press, Oxford.

Stebbins, R. C. (1954). *Amphibians and Reptiles of Western North America*. McGraw-Hill Book Co., Inc, New York.

Stebbins, R. C., and Cohen, N. W. (1995). *A Natural History of Amphibians*. Princeton University Press, Princeton.

Steel, R. (1973). *Crocodylia. Handbuch der Paläoherpetologie, 16*. Gustav Fischer Verlag, Stuttgart.

Stein, K., Palmer, C., Gill, P., and Benton, M. (in press). The aerodynamics of the Late Triassic British Kuehneosauridae. *Paleontology* in press.

Stephens, P. R., and Wiens, J. J. (2003). Ecological diversification and phylogeny of emydid turtles. *Biological Journal of the Linnean Society* **79:** 577–610.

Stephens, P. R., and Wiens, J. J. (2004). Convergence, divergence, and homogenization in the ecological structure of emydid turtle communities: The effects of phylogeny and dispersal. *American Naturalist* **164:** 244–254.

Steubing, R. (1994). A new species of *Cylindrophis* (Serpentes: Cylindrophiidae) from Sarawak, western Borneo. *Raffles Bulletin of Zoology* **42:** 967–973.

Stewart, J. R. (1989). Facultative placentotrophy and the evolution of squamate placentation: Quality of eggs and neonates in *Virginia striatula*. *American Naturalist* **133:** 111–137.

Stewart, J. R., and Blackburn, D. G. (1988). Reptilian placentation: structural diversity and terminology. *Copeia* **1988:** 839–852.

Stewart, J. R., and Thompson, M. B. (1996). Evolution of reptilian placentation: development of extraembryonic membranes of the Australian scincid lizards, *Bassiana duperreyi* (oviparous) and *Pseudemoia entrecasteauxii* (viviparous). *Journal of Morphology* **227:** 349–370.

Stewart, J. R., and Thompson, M. B. (2000). Evolution of placentation among squamate reptiles: Recent research and future directions. *Comparative Biochemistry and Physiology Part A* **127:** 411–431.

Stewart, J. R., and Thompson, M. B. (2003). Evolutionary transformations of the fetal membranes of viviparous reptiles: A case study of two lineages. *Journal of Experimental Zoology* **299A:** 13–32.

Stewart, M. M. (1967). *Amphibians of Malawi*. State University of New York Press, New York.

Stewart, M. M. (1974). Parallel pattern polymorphism in the genus *Phrynobatrachus* (Amphibia: Ranidae). *Copeia* **1974:** 823–832.

Stewart, M. M., and Pough, F. H. (1983). Population density of tropical forest frogs: Relation to retreat sites. *Science* **221:** 570–571.

Steyn, W. (1963). *Angolocaurus* [sic] *skoogi* (Andersson)—a new record from south west Africa. *Cimbebasia* **6:** 8–11.

Stickel, L. F. (1950). Populations and home range relationships of the box turtle *Terrapene c. carolina* (Linnaeus). *Ecological Monographs* **20:** 351–378.

Stickel, L. F. (1989). Home range behavior among box turtles (*Terrapene c. carolina*) of a bottomland forest in Maryland. *Journal of Herpetology* **23:** 40–44.

Stocks, A., McMahan, B., and Taber, P. (2007). Indigenous, colonist, and government impacts on Nicaragua's Bosawas reserve. *Conservation Biology* **21:** 1495–1505.

Stockwell, D., and Peters, D. (1999). The GARP modelling system: Problems and solutions to automated spatial prediction. *International Journal of Geographical Information Science* **13:** 143–158.

Stone, P. A. (2001). Movements and demography of the Sonoran mud turtle, *Kinosternon sonoriense*. *Southwestern Naturalist* **46:** 41–53.

Storrs, G. W. (1993). Function and phylogeny in sauropterygian (Diapsida) evolution. *American Journal of Science* **293–A:** 63–90.

Strüssmann, C., Ribeiro, R. A. K., Ferreira, V. L., and Béda, A. de F. (2007). Herpetofauna do Pantanal brasiliero. In *Herpetologia no Brasil II*. L. B. Nascimento and M. E. Oliveira (Eds.). Pp. 66–84. Sociedade Brasileira de Herpetologica, Belo Horizonte.

Stuart, B. L., and Parham, J. F. (2004). Molecular phylogeny of the critically endangered Indochinese box turtle (*Cuora galbinifrons*). *Molecular Phylogenetics and Evolution* **31:** 164–177.

Stuart, S. N., Chanson, J. S., Cox, N. A., Young, B. E., Rodrigues, A. S. L., Fischman, D. L., and Waller, R. W. (2004). Status and trends of amphibian declines and extinctions worldwide. *Science* **306:** 1783–1786.

Stuebing, R. B., and Inger, R. F. (1999). *A Field Guide to the Snakes of Borneo*. Natural History Publications (Borneo), Kota Kinabalu.

Sues, H.-D., Clark, J. M., and Jenkins, F. A., Jr. (1994). A review of the Early Jurassic tetrapods from the Glen Canyon Group of the American Southwest. In *In the shadow of the dinosaurs. Early Mesozoic Tetrapods*. N. C. Fraser and H.-D. Sues (Eds.). Pp. 284–294. Cambridge University Press, New York.

Sues, H.-D., Clark, J. M., and Jenkins, F. A. (1994). A review of the Early Jurassic tetrapods from the Glen Canyon Group of the American Southwest. In *In the Shadow of the Dinosaurs: Early Mesozoic Tetrapods*. N. C. Fraser and H.-D. Sues (Eds.). Pp. 284–294. Cambridge University Press, New York.

Sullivan, B. K. (1989). Desert environments and the structure of anuran mating systems. *Journal of Arid Environments* **17:** 175–183.

Sullivan, B. K., Ryan, M. J., and Verrell, P. A. (1995). Female choice and mating system structure. In *Amphibian Biology. Volume 2, Social Behaviour*. H. Heatwole and B. K. Sullivan (Eds.). Pp. 468–517. Surrey Beatty & Sons Pty Ltd., Chipping Norton Pty Ltd, N.S.W.

Sullivan, R. M., and Estes, R. (1997). A reassessment of the fossil Tupinambinae. In *Vertebrate Paleontology in the Neotropics. The Miocene Fauna of La Venta, Colombia*. R. F. Kay, R. H. Madden, R. L. Cifelli, and J. J. Flynn (Eds.). Pp. 100–112. Smithsonian Institute Press, Washington, DC.

Sullivan, R. M., and Holman, J. A. (1996). Squamata. In *The Terrestrial Eocene-Oligocene Transition in North America*. D. R. Prothero and R. J. Emry (Eds.). Pp. 354–372. Cambridge University Press, New York.

Sumida, S. S. (1997). Locomotor features of taxa spanning the origin of amniotes. In *Amniote Origins. Completing the Transition to Land*. S. S. Sumida and K. L. M. Martin (Eds.). Pp. 353–398. Academic Press, San Diego.

Sumida, S. S., and Martin, K. L. M. (2000). *Amniote Origins. Completing the Transition to Land*. Academic Press, San Diego, CA.

Summers, A. P., and O'Reilly, J. C. (1997). A comparative study of locomotion in the caecilians *Dermophis mexicanus* and *Typhlonectes natans* (Amphibia: Gymnophiona). *Zoological Journal of the Linnean Society* **121:** 65–76.

Summers, K. (2003). Convergent evolution of bright coloration and toxicity in frogs. *Proceedings of the National Academy of Sciences (USA)* **100:** 12533–12534.

Summers, K., and Amos, W. (1997). Behavioral, ecological, and molecular genetic analyses of reproductive strategies in the Amazonian dart-poison frog, *Dendrobates ventrimaculatus*. *Behavioral Ecology* **8:** 260–267.

Summers, K., and Clough, M. E. (2001). The evolution of coloration and toxicity in the poison frog family (Dendrobatidae). *Proceedings of the National Academy of Sciences (USA)* **98:** 6227–6232.

Summers, K., and Earn, D. J. D. (1999). The cost of polygyny and the evolution of female care in poison frogs. *Biological Journal of the Linnean Society* **66:** 515–538.

Summers, K., and McKeon, C. S. (2004). The evolutionary ecology of phytotelmata use in Neotropical poison frogs. *Miscellaneous Publications Museum of Zoology University of Michigan* **193:** 55–73.

Swan, G. (1990). *A Field Guide to the Snakes and Lizards of New South Wales*. Three Sisters Productions, Sydney, New South Wales, Australia.

Swart, C. C., and de Sa, R. O. (1999). The chondrocranium of the Mexican burrowing toad, *Rhinophrynus dorsalis*. *Journal of Herpetology* **33:** 23–28.

Swingland, I. R. (1977). Reproductive effort and life history strategy of the Aldabran giant tortoise. *Nature* **269:** 402–404.

Swingland, I. R., and Klemens, M. W. (Eds.). (1989). The conservation biology of tortoises. *Occasional Papers of the Internation Union for Conservation of Nature Species Survival Commission* **1:** 1–202.

Swofford, D. L., Olsen, G. J., Waddell, P. J., and Hillis, D. M. (1996). Phylogenetic inference. In *Molecular Systematics, Second Edition*. D. M. Hillis, C. Moritz, and B. K. Mable (Eds.). Pp. 321–381. Sinauer Associates, Inc., Sunderland, Mass.

Sylber, C. K. (1988). Feeding habits of the lizard *Sauromalus varius* and *S. hispidus* in the Gulf of California. *Journal of Herpetology* **22:** 413–424.

Taigen, T. L., Pough, F. H., and Stewart, M. M. (1984). Water balance of terrestrial anuran (*Eleutherodactylus coqui*) eggs: Importance of parental care. *Ecology* **65:** 248–255.

Taigen, T. L., Wells, K. D., and Marsh, R. L. (1985). The enzymatic basis of high metabolic rates in calling frogs. *Physiological Zoology* **58:** 719–726.

Tanaka, K. (1989). Mating strategy of male *Hynobius nebulosus*. In *Current Herpetology in East Asia*. M. Matsui et al. (Eds.). Pp. 437–448. Herpetological Society of Japan, Koyoto.

Taplin, L. E., Grigg, G. C., Harlow, P., Ellis, T. M., and Dunson, W. A. (1982). Lingual salt glands in *Crocodylus acutus* and *Crocodylus johnstoni* and their absence from *Alligator mississipiensis* and *Caiman crocodilus*. *Journal of Comparative Physiology* **149:** 43–47.

Tarsitano, S. F., Frey, E., and Riess, J. (1989). The evolution of the Crocodilia: A conflict between morphological and biochemical data. *American Zoologist* **29:** 843–856.

Tarsitano, S., and Hecht, M. K. (1980). A reconsideration of the reptilian relationships of *Archeopteryx*. *Zoological Journal of the Linnean Society* **69:** 149–182.

Tarsitano, S., and Riess, J. (1982). Plesiosaur locomotion — underwater flight versus rowing. *Nues Jahrbuch für Geologie und Paläontologie, Abhandlungen* **164:** 188–192.

Tattersall, G. J., and Gerlach, R. M. (2005). Hypoxia progressively lowers thermal gaping thresholds in bearded dragons, *Pogona vitticeps*. *Journal of Experimental Biology* **208:** 3321–3330.

Tattersall, G. J., Cadena, V., and Skinner, M. C. (2006). Respiratory cooling and thermoregulatory coupling in reptiles. *Respiratory Physiology & Neurobiology* **154:** 302–318.

Taylor, D. H., and Adler, K. (1973). Spatial orientation by salamanders using plane polarized light. *Science* **181:** 285–287.

Taylor, E. H. (1968). *The Caecilians of the World. A Taxonomic Review*. University of Kansas Press, Lawrence, KS.

Tchernov, E. (1986). *Évolution des Crocodiles en Afrique du Nord et de l'Est*. Édit. Centre National de la Recherche Scientifique, Paris.

Teather, K. L. (1991). The relative importance of visual and chemical cues for foraging on newborn blue-striped garter snakes (*Thamnophis sirtalis similis*). *Behaviour* **117:** 255–261.

Temple, S. A. (1991). Conservation biology: New goals and new patterns for managers of biological resources. In *"Challenges in the Conservation of Biological Resources: A Practioner's Guide"* D. J. Decker, M. E. Krasney, G. R. Goff, C. R. Smith, and D. W. Goss (Eds.), Pp. 45–54. Westview Press, Boulder, Co.

Templeton, A. L. (1986). Coadaptation and outbreeding depression. In *Conservation Biology: The Science of Scarcity and Diversity*. M. E. Soulé (Ed.). Pp. 105–116. Sinauer Associates, Sunderland, MA.

Thomas, M., Raharivololoniaina, L., Glaw, F., Vences, M., and Vieites, D. R. (2005). Montane tadpoles in Madagascar: Molecular identification and description of the larval stages of *Mantidactylus elegans*, *Mantidactylus madecassus*, and *Boophis laurenti* from the Andringitra Massif. *Copeia* **2005:** 174–183.

Thomas, R. (1965). A congregation of the blind snake, *Typhlops richardi*. *Herpetologica* **21:** 309.

Thompson, M. B., and Speake, B. K. (2006). A review of the evolution of viviparity in lizards: Structure, function and physiology of the placenta. *Journal of Comparative Physiology B-Biochemical Systemic and Environmental Physiology* **176:** 179–189.

Thomson, K. S. (1980). The ecology of Devonian lobe-finned fishes. In *The Terrestrial Environment and the Origin of Land Vertebrates*. A. L. Panchen (Ed.). Pp. 87–122. Academic Press, London.

Thomson, K. S. (1993). The origin of the tetrapods. *American Journal of Science* **293–A:** 33–62.

Thorbjarnarson, J. (1992). *Crocodiles: An Action Plan for Their Conservation*. IUCN, Gland.

Thorbjarnarson, J. (1996). Reproductive characteristics of the order Crocodylia. *Herpetologica* **52:** 8–24.

Thorbjarnarson, J. (1999). Crocodile tears and skins: International trade, economic constraints, and limits to the sustainable use of crocodilians. *Conservation Biology* **13:** 465–470.

Thorbjarnarson, J. B. (1990). Notes on the feeding behavior of the gharial (*Gavialis gangeticus*) under semi-natural conditions. *Journal of Herpetology* **24:** 99–100.

Thorpe, R. S. (1976). Biometric analysis of geographic variation and racial affinities. *Biological Review* **51:** 407–452.

Thorpe, R. S. (1987). Geographic variation: A synthesis of cause, data, pattern and congruence in analysis and phylogenesis. *Bollettino di Zoologia* **54:** 3–11.

Thorpe, R. S., Wüster, W., and Malhotra, A. (Eds.) (1997). *Venomous Snakes. Ecology, Evolution and Snakebite*. Zoological Society of London, London.

Tiebout, H. M., and Cary, J. R. (1987). Dynamic spatial ecology of the water snake, *Nerodia sipedon*. *Copeia* **1987:** 1–18.

Tiedemann, F. (1990). *Lurche und Kriechtiere Wiens*. J & V Editions, Wien.

Tihen, J. A. (1965). Evolutionary trends in frogs. *American Zoologist* **5:** 309–318.

Tilley, S. G. (1980). Life histories and comparative demography of two salamander populations. *Copeia* **1980:** 806–812.

Tilley, S. G., and Bernardo, J. (1993). Life history evolution in plethodontid salamanders. *Herpetologica* **49:** 154–163.

Tilley, S. G., Lundrigan, B. L., and Brower, L. P. (1982). Erythrism and mimicry in the salamander *Plethodon cinereus*. *Herpetologica* **38:** 409–417.

Tinkle, D. W. (1967). The life and demography of the side-blotched lizard, *Uta stansburiana*. *Miscellaneous Publications of the Museum of Zoology, University of Michigan* **132:** 1–182.

Tinkle, D. W., and Gibbons, J. W. (1977). The distribution and evolution of viviparity in reptiles. *Miscellaneous Publications of the Museum of Zoology, University of Michigan* **154:** 1–55.

Tinkle, D. W., and Hadley, N. F. (1973). Reproductive effort and winter activity in the viviparous montane lizard *Sceloporus jarrovi*. *Copeia* **1973:** 272–277.

Tinkle, D. W., and Hadley, N. F. (1975). Lizard reproductive effort: caloric estimates and comments on its evolution. *Ecology* **56:** 427–434.

Tinkle, D. W., Congdon, J. D., and Rosen, P. C. (1981). Nesting frequency and success: Implications for the demography of painted turtles. *Ecology* **62:** 1426–1432.

Tinkle, D. W., Wilbur, H. M., and Tilley, S. G. (1970). Evolutionary strategies in lizard reproduction. *Evolution* **24:** 55–74.

Tinsley, R. C. (1990). The influence of parasite infection on mating success in spadefoot toads, *Scaphiopus couchii. American Zoologist* **30:** 313–324.

Tinsley, R. C. and Kobel, H. R. (Eds.) (1996). *The Biology of Xenopus*. Zoological Society of London, London.

Tinsley, R. C., Loumont, C., and Kobel, H. R. (1996). Geographic distribution and ecology. In *The Biology of Xenopus*. R. C. Tinsley and H. R. Kobel (Eds.). Pp. 34–59. Zool. Soc. London, London.

Titus, T. A., and Frost, D. R. (1996). Molecular homology assessment and phylogeny in the lizard family Opluridae (Squamata: Iguania). *Molecular Phylogenetics and Evolution* **6:** 49–62.

Titus, T. A., and Larson, A. (1995). A molecular phylogenetics perspective on the evolutionary radiation of the salamander family Salamandridae. *Systematic Biology* **44:** 125–151.

Titus, T. A., and Larson, A. (1997). Molecular phylogenetics of desmognathine salamanders (Caudata: Plethodontidae): A reevaluation of evolution in ecology, life history, and morphology. *Systematic Biology* [1996] **45:** 451–472.

Tobias, M. L., Viswanathan, S. S., and Kelley, D. B. (1998). Rapping, a female receptive call, initiates male-female duets in the South African clawed frog. *Proceedings of the National Academy of Sciences (USA)* **95:** 1870–1875.

Toft, C. A. (1980). Feeding ecology of thirteen syntopic species of anurans in a seasonal tropical environment. *Oecologia* **45:** 131–134.

Toft, C. A. (1995). Evolution of diet specialization in poison-dart frogs (Dendrobatidae). *Herpetologica* **51:** 202–216.

Tokar, A. A. (1989). A revision of the genus *Eryx* (Serpentes, Boidae), based upon osteological data. *Vestnik Zoologii* **1989:** 46–55.

Tokar, A. A. (1996). Taxonomic revision of the genus *Gongylophis* Wagler 1830: *G. colubrinus* (L. 1758) (Serpentes Boidae). *Tropical Zoology* **9:** 1–17.

Tokarz, R. R. (1992). Male mating preferences for unfamiliar females in the lizard *Anolis sagrei. Animal Behaviour* **44:** 843– 849.

Tokarz, R. R. (1995). Mate choice in lizards: A review. *Herpetological Monographs* **9:** 17–40.

Toledo, L. F. (2005). Predation of juvenile and adult anurans by invertebrates: Current knowledge and perspectives. *Herpetological Review* **36:** 395–400.

Toledo, L. F., Ribeiro, R. S., and Haddad, C. F. B. (2007). Anurans as prey: An exploratory analysis and size relationships between predators and their prey. *Journal of Zoology, London* **271:** 170–177.

Tolley, K., A., Tilbury, C. R., Branch, W. R., and Matthee, C. A. (2004). Phylogenetics of the southern African dwarf chameleons, *Bradypodion* (Squamata: Chamaeleonidae). *Molecular Phylogenetics and Evolution* **30:** 354–365.

Tolson, P. J., and Henderson, R. W. (1993). *The Natural History of West Indian Boas*. R&A Publishing, Ltd, Taunton, U.K.

Tome, M. A., and Pough, F. H. (1982). Responses of amphibians to acid precipitation. In *Acid Rain/Fisheries. Proceedings of an International Symposium on Acidic Precipitation and Fishery Impacts in Northeastern North America*. R. E. Johnson (Ed.). Pp. 245–254. American Fisheries Society, Bethesda, Md.

Tong, H., Gaffney, E. S., and Buffetaut, E. (1998). *Foxemys*, a new side-neck turtle (Bothremydidae: Pelomedusoides) from the Late Cretaceous of France. *American Museum Novitates* **3251:** 1–19.

Toranzo, G. S., Oterino, J., Zelarayan, L., Bonilla, F., and Buhler, M. I. (2007). Spontaneous and LH-induced maturation in *Bufo arenarum* oocytes: Importance of gap junctions. *Zygote* **15:** 65–80.

Torres-Carvajal, O. (2007). Phylogeny and biogeography of a large radiation of Andean lizards (Iguania, *Stenocercus*). *Zoologica Scripta* **36:** 311–326.

Torres-Carvajal, O., Schulte, J. A., and Cadle, J. E. (2006). Phylogenetic relationships of South American lizards of the genus *Stenocercus* (Squamata: Iguania): A new approach using a general mixture model for gene sequence data. *Molecular Phylogenetics and Evolution* **39:** 171–185.

Touchon, J. C., Gomez-Mestre, I., and Warkentin, K. M. (2006). Hatching plasticity in two temperate anurans: Responses to a pathogen and predation cues. *Canadian Journal of Zoology* **84:** 556–563.

Townsend, D. S., and Stewart, M. M. (1985). Direct development in *Eleutherodactylus coqui* (Anura: Leptodactylidae): A staging table. *Copeia* **1985:** 423–436.

Townsend, D. S., Stewart, M. M., Pough, F. H., and Brussard, P. F. (1981). Internal fertilization in an oviparous frog. *Science* **212:** 469–471.

Townsend, T. M., Larson, A., Louis, E., and Macey, J. R. (2004). Molecular phylogenetics of Squamata: The position of snakes, Amphisbaenians, and Dibamids, and the root of the Squamate tree. *Systematic Biology* **53:** 735–757.

Tracy, C. R. (1971). Evidence for the use of celestial cues by dispersing immature California toads (*Bufo boreas*). *Copeia* **1971:** 145–147.

Tracy, C. R., and Christian, K. A. (2005). Preferred temperature correlates with evaporative water loss in hylid frogs from northern Australia. *Physiological and Biochemical Zoology* **78:** 839–846.

Tracy, C. R., and Dole, J. W. (1969). Orientation of displaced California toads, *Bufo boreas*, to their breeding sites. *Copeia* **1969:** 693–700.

Tracy, C. R., Nussear, K. E., Esque, T. C., Dean-Bradley, K., DeFalco, L. A., Castle, K. T., Zimmerman, L. C., Espinoza, R. E., and Barber, A. M. (2006). The importance of physiological ecology in conservation biology. *Integrative and Comparative Biology* **46:** 1191–1205.

Trauth, S. E., Cooper, J., Vitt, L. J., and Perrill, S. A. (1987). Cloacal anatomy of the broad-headed skink, *Eumeces laticeps*, with a description of a female pheromonal gland. *Herpetologica* **43:** 458–466.

Trivers, R. L. (1972). Parental investment and sexual selection. In *Sexual Selection and the Descent of Man*. B. Campbell (Ed.). Pp. 136–179. Aldine, Chicago.

Trivers, R. L. (1976). Sexual selection and resource-accruing abilities in *Anolis garmani*. *Evolution* **30:** 253–269.

Trontelj, P., and Goricki, S. (2003). Monphyly of the family Proteidae (Amphibia: Caudata) tested by phylogenetic analysis of mitochondrial 12s rDNA sequences. *Natura Croatica* **12:** 113–120.

Troyer, K. (1984a). Structure and function of the digestive tract of an herbivorous lizard *Iguana iguana*. *Physiological Zoology* **57:** 1–8.

Troyer, K. (1984b). Behavioral acquisition of the hindgut fermentation system by hatchling *Iguana iguana*. *Behavioral Ecology and Sociobiology* **14:** 189–193.

Trueb, L. (1973). Bones, frogs, and evolution. In *Evolutionary Biology of the Anurans*. J. L. Vial (Ed.). Pp. 65–132. University of Missouri Press, Columbia.

Trueb, L. (1974). Systematic relationships of Neotropical horned frogs, genus *Hemiphractus* (Anura: Hylidae). *Occasional Papers of the Museum of Natural History, University of Kansas* **29:** 1–60.

Trueb, L., and Cannatella, D. C. (1986). Systematics, morphology, and phylogeny of genus *Pipa* (Anura: Pipidae). *Herpetologica* **42:** 412–449.

Trueb, L., and Cloutier, R. (1991a). Toward an understanding of the amphibians: Two centuries of systematic history. In *Origins of the Higher Groups of Tetrapods. Controversy and Consensus*. H.-P. Schultze and L. Trueb (Eds.). Pp. 176–193. Comstock Publishing Associates, Ithaca, NY.

Trueb, L, and Cloutier, R. (1991b). A phylogenetic investigation of the inter- and intrarelationships of the Lissamphibia (Amphibia: Temnospondylii). In *Origins of the Higher Groups of Tetrapods. Controversy and Consensus*. H.-P. Schultze and L. Trueb (Eds.). Pp. 223–313. Comstock Publishing Associates, Ithaca, NY.

Trueb, L., and Gans, C. (1983). Feeding specializations of the Mexican burrowing toad, *Rhinophrynus dorsalis* (Anura: Rhinophrynidae). *Journal of Zoology, London* **199:** 189–208.

Tsuji, H., and Lue, K. Y. (1998). Temporal aspects of the amplexus and oviposition behavior of the fanged frog *Rana kuhlii* from Taiwan. *Copeia* **1998:** 769–773.

Tu, M.-C., Chu, C.-W., and Lue, K.-Y. (1999). Specific gravity and mechanisms for its control in tadpoles of three anuran species from different water strata. *Zoological Studies* **38:** 76–81.

Tuberville, T. D., Gibbons, J. W., and Greene, J. L. (1996). Invasion of new aquatic habitats by male freshwater turtles. *Copeia* **1996:** 713–715.

Turner, F. B. (1977). The dynamics of populations of squamates, crocodilians and rhynchocephalians. In *Biology of the Reptilia. Volume 7. Ecology and Behavior*. C. Gans and D. W. Tinkle (Eds.). Pp. 157–264. Academic Press, New York.

Turner, F. B., Medica, P. A., Lannom, J. R., Jr., and Hoddenbach, G. A. (1969). A demographic analysis of fenced populations of the whiptailed lizard, *Cnemidophorus tigris*, in southern Nevada. *Southwestern Naturalist* **14:** 189–202.

Tuttle, M. D., and Ryan, M. J. (1981). Bat predation and the evolution of frog vocalization in the Neotropics. *Science* **214:** 677–678.

Twitty, V. C. (1966). *Of Scientists and Salamanders*. W. H. Freeman & Co, San Francisco.

Tyler, M. J. (1985). Reproductive modes in Australian Amphibia. In *Biology of Australasian Frogs and Reptiles*. G. Grigg, R. Shine, and H. Ehmann (Eds.). Pp. 265–267. Surrey Beatty & Sons Pty Ltd, Chipping Norton Pty Ltd., Sydney, Australia.

Tyler, M. J. (1989). *Australian Frogs*. Viking O'Neil, Ringwood, Australia.

Tyler, M. J. (1991a). *Kyarranus* Moore (Anura, Leptodactylidae) from the Tertiary of Queensland. *Proceedings of the Royal Society of Victoria* **103:** 47–51.

Tyler, M. J. (1991b). Hylid frogs from the mid-Miocene Camfield Beds of northern Australia. *The Beagle* **11:** 141–144.

Tyler, M. J. (1991c). Declining amphibian populations – a global phenomenon? An Australian perspective. *Alytes* **9:** 43–50.

Tyler, M. J., and Davies, M. (1978). Phylogenetic relationships of Australian hyline and neotropical phyllomedusine frogs of the family Hylidae. *Herpetologica* **34:** 219–224.

Tyler, M. J., and Davies, M. (1993). The Hylidae. In *Fauna of Australia. Volume 2A. Amphibia & Reptilia*. C. J. Gasby, G. J. B. Ross, and P. L. Beesley (Eds.). Pp. 58–63. Australian Government Publishing Service, Canberra.

Tyler, M. J., and Duellman, W. E. (1995). Superficial mandibular musculature and vocal sac structure in hemiphractine hylid frogs. *Journal of Morphology* **224:** 65–71.

Tyrrell, C. (2001). Completion of research on reproductive ecology of tuatara on northern islands, including Aorangi, Taranga and the Marotere Islands. 141. University of Otago, Dunedin.

Udvardy, M. D. F. (1975). *A classification of the Biogeographical Provinces of the World*. IUCN Occasional Paper no. 18. Morges, Switzerland.

Uetz, P. (1999). The EMBL reptile database. www.embl-heildberg.de/~uetz/LivingReptiles.html.

Ugarte, C. A., Rice, K. G., and Donnelly, M. A. (2007). Comparison of diet, reproductive biology, and growth of the pig frog (*Rana grylio*) from harvested and protected areas of the Florida everglades. *Copeia* **2007:** 436–448.

Ultsch, G. R., Bradford, D. F., and Freda, J. (1999). Physiology: Coping with the environment. In *Tadpoles: The Biology of Anuran Larvae*. R. W. McDiarmid and R. Altig (Eds.). Pp. 189–214. University of of Chicago Press, Chicago.

Underwood, G. (1967). *A Contribution to the Classification of Snakes*. British Museum of Natural History, London.

Underwood, G. (1970). The eye. In *Biology of the Reptilia. Volume 2. Morphology B*. C. Gans and T. S. Parsons (Eds.). Pp. 1–97. Academic Press, London.

Underwood, G., and Kochva, E. (1993). On the affinities of the burrowing asps *Atractaspis* (Serpentes: Atractaspididae). *Zoological Journal of the Linnean Society* **107:** 3–64.

Underwood, G., and Stimson, A. F. (1990). A classification of pythons (Serpentes, Pythoninae). *Journal of Zoology, London* **221:** 565–603.

Urban, M., Phillips, B. L., Skelly, D. K., and Shine, R. (2007). The cane toad's (*Chaunus marinus*) increasing ability to invade Australia is revealed by a dynamically updated range model. *Proceedings of the Royal Society London Series B* **274:** 1413–1419.

Valenzuela, N. (2004). Evolution and maintenance of temperature-dependent sex determination. In *Temperature-Dependent Sex Determination in Vertebrates*. N. Valenzuela and V. Lance (Eds.). Pp. 131–147. Smithsonian Books, Washington, D. C.

Valenzuela, N. and Lance, V. (Eds.) (2004). *Temperature-dependent Sex Determination in Vertebates*. Smithsonian Books, Washington D. C.

Valladares-Padua, C. (2006). Importance of knowledge-intensive economic development to conservation of biodiversity in developing countries. *Conservation Biology* **20:** 700–701.

Van Berkum, F. H., Huey, R. B., and Adams, B. A. (1986). Physiological consequences of thermoregulation in a tropical lizard (*Ameiva festiva*). *Physiological Zoology* **59:** 464–472.

van Beurden, E. K. (1984). Survival strategies of the Australian water-holding frog, *Cyclorana platycephalus*. In *Arid Australia*. H. G. Cogger and E. E. Cameron (Eds.). Pp. 223–234. Australian Museum, Sydney.

Van Bocxlaer, I., Roelants, K., Biju, S. D., Nagaraju, J., and Bossuyt, F. (2006). Late Cretaceous vicariance in Gondwanan amphibians. *PLos ONE* **1:** e74.

Van Damme, R., Bauwens, D., Brana, F., and Verheyen, R. F. (1992). Incubation temperature differentially affects hatching time, egg survival, and hatchling performance in the lizard *Podarcis muralis*. *Herpetologica* **48:** 220–228.

van den Elzen, P. (1975). Contribution á la connaissance de *Pelodytes punctatus* étudié en Camarque. *Bulletin de la Sociétéde Zoologique de France* **100:** 691–692.

van der Meijden, A., Boistel, R., Gerlach, J., Ohler, A., Vences, M., and Meyer, A. (2007a). Molecular phylogenetic evidence for paraphyly of the genus *Sooglossus*, with the description of a new genus of Seychellean frogs. *Biological Journal of the Linnean Society* **91:** 347–359.

van der Meijden, A., Vences, M., Hoegg, S., and Meyer, A. (2005). A previously unrecognized radiation of ranid frogs in Southern Africa revealed by nuclear and mitochondrial DNA sequences. *Molecular Phylogenetics and Evolution* **37:** 674–685.

van der Meijden, A., Vences, M., Hoegg, S., Boistel, R., Channing, A., and Meyer, A. (2007b). Nuclear gene phylogeny of narrow-mouthed toads (Family: Microhylidae) and a discussion of competing hypotheses concerning their biogeographical origins. *Molecular Phylogenetics and Evolution* **44:** 1017–1030.

Van Devender, R. W. (1982). Comparative demography of the lizard *Basiliscus basciliscus*. *Herpetologica* **38:** 189–208.

van Dyke, J. H. (1997). Family Gerrhosauridae. In *Proceedings of the Fitzsimons Commemorative Symposium*. (J. H. van Dyke Ed.), pp. 44–51. Herpotological Association of Africa, Matieland, S. Afr.

van Loben Sels, R. C., Congdon, J. D., and Austin, J. T. (1997). Life history and ecology of the Sonoran mud turtle (*Kinosternon sonoriense*) in southeastern Arizona: A preliminary report. *Chelonian Conservation Biology* **2:** 338–344.

Van Mieriop, L. H. S., and Barnard, S. M. (1978). Further observations on thermoregulation in the brooding female *Python molurus bivittatus* (Serpentes: Boidae). *Copeia* **1978:** 615–621.

Vanhooydonck, B., Herrel, A., and Van Damme, R. (2007). Interactions between habitat use, behavior, and trophic niche of lacertid lizards. In *Lizard Ecology: The Evolutionary Consequences of Foraging Mode*. S. M. Reilly, L. D. McBrayer, and D. B. Miles (Eds.). Pp. 427–449. Cambridge University Press, Cambrdge UK.

Vanzolini, P. E. (1973). Distribution and differentiation of animals along the coast and in continental inslands of the state of S. Paulo, Brasil. I. Introduction to the area and problems. *Papéis Avulsos Zoologia (São Paulo)* **26:** 281–294.

Vanzolini, P. E. (2002). A second note on the geographical differentiation of *Amphisbaena fuliginosa* L., 1758 (Squamata, Amphisbaenidae), with a consideration of the forest refuge model of speciation. *Anais da Academia Brasileira de Ciências* **74:** 609–648.

Vanzolini, P. E., and Williams, E. E. (1970a). South American anoles: The geographic differentiation and evolution of the *Anolis chrysolepis* species group (Sauria, Iguanidae). Part 1. *Arquivos de Zoologia* **19:** 1–124.

Vanzolini, P. E., and Williams, E. E. (1970b). South American anoles: The geographic differentiation and evolution of the *Anolis chrysolepis* species group (Sauria, Iguanidae). Part 2. *Arquivos de Zoologia* **19:** 125–298.

Vanzolini, P. E., and Williams, E. E. (1981). The vanishing refuge: a mechanism for ecogeographic speciation. *Papéis Avulsos Zoologia (São Paulo)* **34:** 251–255.

Vanzolini, P. E., Ramos-Costa, A. M. M., and Vitt, L. J. (1980). *Repteis das Caatingas*. Academia Brasileira de Ciências, Rio de Janeiro.

Vasudevan, K. (1996/97). Reproductive ecology of *Aspideretes gangeticus* (Cuvier) in north India. *International Congress on Chelonian Conservation 1995*, Pp. 21–28.

Vasudevan, K., Kumar, A., and Chellam, R. (2001). Structure and composition of rainforest floor amphibian communities in Kalakad-Mundanthurai Tiger Reserve. *Current Science* **80:** 406–412.

Vaughn, L. K., Bernheim, H. A., and Kluger, M. J. (1974). Fever in the lizard *Dipsosaurus dorsalis*. *Nature* **252:** 473–474.

Vaz-Ferreira, R., and Gehrau, A. (1975). Epimeletic behavior of the South American common frog *Leptodactylus ocellatus* (L.) (Amphibia, Leptodactylidae). I. Brood care and related feeding and aggressive activities. *Physis Buenos Aires* **34:** 1–14.

Vaz-Ferreira, R., Covelo de Zolessi, L., and Achaval, F. (1970). Oviposicion y desarollo de ofidios y lacertilios en hormigueros de *Acromyrmex*. *Physis* **79:** 431–459.

Veith, M., and Steinfartz, S. (2004). When non-monphyly results in taxonomic cosequences – the case of *Mertensiella* with the Salamandridae (Amphibia: Urodela). *Salamandra, Rheinbach* **40:** 67–80.

Veith, M., Fromhage, L., Kosuch, J., and Vences, M. (2006). Historical biogeography of Western Palaearctic pelobatid and pelodytid frogs: A molecular phylogenetic perspective. *Contributions to Zoology* **75:** 109–120.

Veith, M., Steinfartz, S., Zardoya, R., Seitz, A., and Meyer, A. (1998). A molecular phylogeny of 'true' salamanders (family Salamandridae) and the evolution of terrestriality of reproductive modes. *Journal of Zoological Systematics and Evolutionary Research* **36:** 7–16.

Vences, M., and Glaw, F. (2002a). Molecular phylogeography of *Boophis* tephraeomystax: A test case for east-west vicariance in Malagasy anurans (Amphibia, Anura, Mantellidae). *Spixiana* **25:** 79–84.

Vences, M., and Glaw, F. (2002b). Two new treefrogs of the *Boophis rappiodes* group from eastern Madagascar (Amphibia Mantellidae). *Tropical Zoology* **15:** 141–163.

Vences, M., and Glaw, F. (2005a). A new cryptic frog of the genus *Boophis* from the northwestern rainforests of Madagascar. *African Journal of Herpetology* **54:** 77–84.

Vences, M., and Glaw, F. (2005b). A new species of *Mantidactylus* from the east coast of Madagascar and its molecular phylogenetic relationships within the subgenus *Guibemantis*. *Herpetological Journal* **15:** 37–44.

Vences, M., Andreone, F., Glaw, F., Kosuch, J., Meyer, A., Schaefer, H. C., and Veith, M. (2002). Exploring the potential of life-history key innovation: Brook breeding in the radiation of the Malagasy treefrog genus *Boophis*. *Molecular Ecology* **11:** 1453–1463.

Vences, M., Freyhof, J., Sonnenberg, R., Kosuch, J., and Veith, M. (2001). Reconciling fossils and molecules: Cenozoic divergence of cichlid fishes and the biogeography of Madagascar. *Journal of Biogeography* **28:** 1091–1099.

Vences, M., Glaw, F., and Boehme, W. (1999). A review of the genus *Mantella* (Anura, Ranidae, Mantellinae): Taxonomy, distribution and conservation of Malagasy poison frogs. *Alytes (Paris)* **17:** 3–72.

Vences, M., Glaw, F., and Böhme, W. (1997/98). Evolutionary correlates of microphagy in alkaloid-containing frogs (Amphibia: Anura). *Zoologischer Anzeiger* **236:** 217–230.

Vences, M., Glaw, F., Kosuch, J., Boehme, W., and Veith, M. (2001). Phylogeny of South American and Malagasy boine snakes: Molecular evidence for the validity of *Sanzinia* and *Acrantophis* and biogeographic implications. *Copeia* **2001:** 1151–1154.

Vences, M., Kosuch, J., Glaw, F., Bohme, W., and Veith, M. (2003c). Molecular phylogeny of hyperoliid treefrogs: Biogeographic origin of Malagasy and Seychellean taxa and re-analysis of familial paraphyly. *Journal of Zoological Systematics and Evolutionary Research* **41:** 205–215.

Vences, M., Kosuch, J., Rodel, M. O., Lotters, S., Channing, A., Glaw, F., and Bohme, W. (2004). Phylogeography of *Ptychadena mascareniensis* suggests transoceanic dispersal in a widespread African-Malagasy frog lineage. *Journal of Biogeography* **31:** 593–601.

Vences, M., Raxworthy, C. J., Nussbaum, R. A., and Glaw, F. (2003b). A revision of the *Scaphiophryne marmorata* complex of marbled toads from Madagascar, including the description of a new species. *Herpetological Journal* **13:** 69–79.

Vences, M., Raxworthy, C. J., Nussbaum, R. A., and Glaw, F. (2003a). New microhylid frog (*Plethodontohyla*) from Madagascar, with semiarboreal habits and possible parental care. *Journal of Herpetology* **37:** 629–636.

Vences, M., Wahl-Boos, G., Hoegg, S., Glaw, F., Oliveira, E. S., Meyer, A., and Perry, S. (2007). Molecular systematics of mantelline frogs from Madagascar and the evolution of their femoral glands. *Biological Journal of the Linnean Society* **92:** 529–539.

Vences, M., Wanke, S., Odierna, G., Kosuch, J., and Veith, M. (2000). Molecular and karyological data on the South Asian Ranid genera *Indirana, Nyctibatrachus* and *Nannophrys* (Anura: Ranidae). *Hamadryad* **25:** 75–82.

Vermeij, G. J. (1982). Unsuccessful predation and evolution. *American Naturalist* **120:** 701–720.

Verrell, P. A. (1982). The sexual behaviour of the red-spotted newt, *Notopthalmus viridescens* (Amphibia: Urodela: Salamandridae). *Animal Behaviour* **30:** 1224–1236.

Verrell, P. A. (1983). The influence of the ambient sex ratio and intermale competition on the sexual behavior of the red-spotted newt, *Notophthalmus viridescens* (Amphibia, Urodela: Salamandridae). *Behavioral Ecology and Sociobiology* **13:** 307–313.

Verrell, P. A. (1989). The sexual strategies of natural populations of newts and salamanders. *Herpetologica* **45:** 265–282.

Verrell, P. A., and Halliday, T. (1985). Reproductive dynamics of a population of smooth newts, *Triturus vulgaris*, in southern England. *Herpetologia* **41:** 386–395.

Vertucci, F. A., and Corn, P. S. (1996). Evaluation of episodic acidification and amphibian declines in the Rocky Mountains. *Ecological Applications* **6:** 449–457.

Vial, J. L. (1968). The ecology of the tropical salamander, *Bolitoglossa subpalmata*, in Costa Rica. *Revista de Biologiá Tropical* **15:** 13–115.

Vial, J. L., Berger, T. J., and McWilliams, W. T. (1977). Quantitative demography of copperheads, *Agkistrodon contortrix*. *Researches on Population Ecology* **18:** 223–234.

Vial, J. L. (Ed.) (1973). *Evolutionary Biology of the Anurans*. University of Missouri Press, Columbia.

Vicario, S., Caccone, A., and Gauthier, J. (2003). Xantusiid "night" lizards: A puzzling phylogenetic problem revisited using likelihood-based Bayesian methods on mtDNA sequences. *Molecular Phylogenetics and Evolution* **26:** 243–261.

Vidal, N., and Hedges, S. B. (2002a). Higher-level relationships of snakes inferred from four nuclear and mitochondrial genes. *Comptes Rendus Biologies* **325:** 977–985.

Vidal, N., and Hedges, S. B. (2002b). Higher-level relationships of snakes inferred from four nuclear and mitochondrial genes. *Comptes Rendus Biologies* **325:** 987–995.

Vidal, N., and Hedges, S. B. (2004). Molecular evidence for a terrestrial origin of snakes. *Proceedings of the Royal Society of London Series B-Biological Sciences* **271:** S226–S229.

Vidal, N., and Hedges, S. B. (2005). The phylogeny of squamate reptiles (lizards, snakes, and amphisbaenians) inferred from nine nuclear protein-coding genes. *Comptes Rendus Biology* **328:** 1000–1008.

Vidal, N., Azvolinsky, A., Cruaud, C., and Hedges, S. B. (2008). Origin of tropical American burrowing reptiles by transatlantic rafting. *Biology Letters* **4:** 115–118.

Vieites, D. R., Min, M. S., and Wake, D. B. (2007). Rapid diversification and dispersal during periods of global warming by plethodontid salamanders. *Proceedings of the National Academy of Sciences (USA)* **104:** 19903–19907.

Vieites, D. R., Nieto-Roman, S., Barluenga, M., Palanca, A., Vences, M., and Meyer, A. (2004). Post-mating clutch piracy in an amphibian. *Nature* **431:** 305–308.

Viertel, B., and Richter, S. (1999). Anatomy: Viscera and endocrines. In *Tadpoles: The Biology of Anuran Larvae*. R. W. McDiarmid and R. Altig (Eds.). Pp. 92–148. University of of Chicago Press, Chicago.

Viets, B. E., Ewert, M. A., Talent, L. G., and Nelson, C. E. (1994). Sex-determining mechanisms in squamate reptiles. *Journal of Experimental Zoology* **270:** 45–56.

Villa, J. (1984). Biology of a neotropical glass frog, *Centrolenella fleischmanni* (Boettger), with special reference to its frogfly associates. *Milwaukee Public Museum Contributions in Biology and Geology* **55:** 1–60.

Villa, J. D., and Wilson, L. D. (1990). *Ungaliophis, U. continentalis, U. panamensis*. *Catalogue of American Amphibians and Reptiles* **480:** 1–4.

Villecco, E. I., Aybar, M. J., Riera, A. N. S., and Sanchez, S. S. (1999). Comparative study of vitellogenesis in the anuran amphibians *Ceratophrys cranwelli* (Leptodactilidae) and *Bufo arenarum* (Bufonidae). *Zygote* **7:** 11–19.

Villiers, A. (1975). *Les Serpents de l'Ouest Africain*. 3rd Ed. Les Nouvelles Éditions Africaines, Dakar.

Vincent, S. E., Shine, R., and Brown, G. P. (2005). Does foraging mode influence sensory modalites for prey detection? A comparison between males and female filesnakes (*Acrochordus arafurae*, Acrochordidae). *Animal Behavior* **70:** 715–721.

Vinegar, A., Hutchison, V., and Dowling, H. (1970). Metabolism, energetics, and thermoregulation during brooding of snakes of the genus *Python* (Reptilia. Boidae). *Zoologica* **55:** 19–48.

Vitt, L. J. (1981). Lizard reproduction: habitat specificity and constraints on relative clutch mass. *American Naturalist* **117:** 506–514.

Vitt, L. J. (1983). Ecology of an anuran-eating guild of terrestrial tropical snakes. *Herpetologica* **39:** 52–66.

Vitt, L. J. (1985). On the biology of the little known anguid lizard, *Diploglossus lessonae* in northeast Brazil. *Papeis Avulsos de Zoologia (São Paulo)* **36:** 69–76.

Vitt, L. J. (1986). Reproductive tactics of sympatric gekkonid lizards with a comment on the evolutionary and ecological consequences of invariant clutch size. *Copeia* **1986:** 773–786.

Vitt, L. J. (1991). Ecology and life history of the scansorial arboreal lizard *Plica plica* (Iguanidae) in Amazonian Brazil. *Canadian Journal of Zoology* **69:** 504–511.

Vitt, L. J. (1992a). Diversity of reproduction strategies among Brazilian lizards and snakes: The significance of lineage and adaptation. In *Reproductive Biology of South American Vertebrates*. W. C. Hamlett (Ed.). Pp. 135–149. Springer-Verlag, New York.

Vitt, L. J. (1992b). Mimicry of millipedes and centipedes by elongate terrestrial vertebrates. *National Geographic Reseach and Exploration* **8:** 76–95.

Vitt, L. J. (1993). Ecology of isolated open-formation *Tropidurus* (Reptilia: Tropiduridae) in Amazonian lowland rain forest. *Canadian Journal of Zoology* **71:** 2370–2390.

Vitt, L. J., and Avila-Pires, T. C. S. (1998). Ecology of two sympatric species of *Neusticurus* (Sauria: Gymnophthalmidae) in the western Amazon of Brazil. *Copeia* **1998:** 570–582.

Vitt, L. J., and Breitenbach, G. L. (1993). Life histories and reproductive tactics among lizards in the genus *Cnemidophorus* (Sauria: Teiidae). In *Biology of Whiptail Lizards (Genus Cnemidophorus)*. J. W. Wright and L. J. Vitt (Eds.). Pp. 211–243. Oklahoma Museum of Natural History, Norman, OK.

Vitt, L. J., and Colli, G. R. (1994). Geographical ecology of a neotropical lizard: *Ameiva ameiva* (Teiidae) in Brazil. *Canadian Journal of Zoology* **72:** 1986–2008.

Vitt, L. J., and Congdon, J. D. (1978). Body shape, reproductive effort, and relative clutch mass in lizards: Resolution of a paradox. *American Naturalist* **112:** 595–608.

Vitt, L. J., and Cooper, W. E., Jr. (1985). The evolution of sexual dimorphism in the skink *Eumeces laticeps*: An example of sexual selection. *Canadian Journal of Zoology* **63:** 995–1002.

Vitt, L. J., and Cooper, W. E., Jr. (1986a). Foraging and diet of a diurnal predator (*Eumeces laticeps*) feeding on hidden prey. *Journal of Herpetology* **20:** 408–415.

Vitt, L. J., and Cooper, W. E., Jr. (1986b). Tail loss, tail color, and predator escape in *Eumeces* (Lacertilia: Scincidae): Age-specific differences in costs and benefits. *Canadian Journal of Zoology* **64:** 583–592.

Vitt, L. J., and Cooper, W. E., Jr. (1988). Feeding responses of broad-headed skinks (*Eumeces laticeps*) to velvet ants (*Dasymutilla occidentalis*). *Journal of Herpetology* **22:** 485–488.

Vitt, L. J., and de la Torre, S. (1996). *Guia para la Investigacion de las Lagartijas de Cuyabeno—A Research Guide to the Lizards of Cuyabeno*. Centro de Biodiversidad y Ambiente, Pontificia Universidad Católica del Ecuador, Quito, Ecuador.

Vitt, L. J., and Goldberg, S. R. (1983). Reproductive ecology of two tropical iguanid lizards: *Tropidurus torquatus* and *Platynotus semitaeniatus*. *Copeia* **1983:** 131–141.

Vitt, L. J., and Lacher, T. E., Jr. (1981). Behavior, habitat, diet, and reproduction of the iguanid lizard *Polychrus acutirostris* in the Caatinga of northeastern Brazil. *Herpetologica* **37:** 53–63.

Vitt, L. J. and Pianka, E. R. (Eds.) (1994). *Lizard Ecology: Historical and Experimental Perspectives*. Princeton University Press, Princeton, N.J.

Vitt, L. J., and Pianka, E. R. (2005). Deep history impacts present-day ecology and biodiversity. *Proceedings of the National Academy of Sciences* **102:** 7877–7881.

Vitt, L. J., and Pianka, E. R. (2007). Feeding ecology in the natural world. In *Lizard Ecology: The Evolutionary Consequences of Foraging Mode*. S. M. Reilly, L. D. McBrayer, and D. B. Miles (Eds.). Pp. 141–172. Cambridge University Press, Cambrdge UK.

Vitt, L. J., and Price, H. J. (1982). Ecological and evolutionary determinants of relative clutch mass in lizards. *Herpetologica* **38:** 237–255.

Vitt, L. J., and Zani, P. A. (1996a). Organization of a taxonomically diverse lizard assemblage in Amazonian Ecuador. *Canadian Journal of Zoology* **74:** 1313–1335.

Vitt, L. J., and Zani, P. A. (1996b). Behavioural ecology of *Tropidurus hispidus* on isolated rock outcrops in Amazonia. *Journal of Tropical Ecology* **12:** 81–101.

Vitt, L. J., and Zani, P. A. (1996c). Ecology of the South American lizard *Norops chrysolepis* (Polychrotidae). *Copeia* **1996:** 56–68.

Vitt, L. J., and Zani, P. A. (1996d). Ecology of the lizard *Ameiva festiva* (Teiidae) in southeastern Nicaragua. *Journal of Herpetology* **30:** 110–117.

Vitt, L. J., and Zani, P. A. (1996e). Ecology of the elusive tropical lizard *Tropidurus* [=*Uracentron*] *flaviceps* (Tropiduridae) in lowland rain forest of Ecuador. *Herpetologica* **52:** 121–132.

Vitt, L. J., and Zani, P. A. (1997). Ecology of the nocturnal lizard *Thecadactylus rapicauda* (Sauria: Gekkonidae) in the Amazon region. *Herpetologica* **53:** 165–179.

Vitt, L. J., Congdon, J. D., and Dickson, N. A. (1977). Adaptive strategies and energetics of tail autotomy in lizards. *Ecology* **58:** 326–337.

Vitt, L. J., Zani, P. A., and Durtsche, R. D. (1995a). Ecology of the lizard *Norops oxylophus* (Polychrotidae) in lowland forest of southeastern Nicaragua. *Canadian Journal of Zoology* **73:** 1918–1927.

Vitt, L. J., Zani, P. A., Caldwell, J. P., and Carrillo, E. O. (1995b). Ecology of the lizard *Kentropyx pelviceps* (Sauria: Teiidae) in lowland forest of Ecuador. *Canadian Journal of Zoology* **73:** 691–703.

Vitt, L. J., Zani, P. A., and Caldwell, J. P. (1996b). Behavioural ecology of *Tropidurus hispidus* on isolated rock outcrops in Amazonia. *Journal of Tropical Ecology* **12:** 81–101.

Vitt, L. J., Avila-Pires, T. C. S., and Zani, P. A. (1996a). Observations on the ecology of the rare Amazonian lizard, *Enyalius leechii* (Polychrotidae). *Herpetological Natural History* **4:** 77–82.

Vitt, L. J., Zani, P. A., and Avila-Pires, T. C. S. (1997a). Ecology of the arboreal lizard *Tropidurus* (*=Plica*) *umbra* in the Amazonian region. *Canadian Journal of Zoology* **75:** 1876–1882.

Vitt, L. J., Zani, P. A., and Barros, A. A. M. (1997b). Ecological variation among populations of the gekkonid lizard *Gonatodes humeralis* in the Amazon Basin. *Copeia* **1997:** 32–43.

Vitt, L. J., Avila-Pires, T. C. S., Caldwell, J., and Oliveira, V. R. L. (1998a). The impact of individual tree harvesting on thermal environments of lizards in Amazonian rain forest. *Conservation Biology* **12:** 655–663.

Vitt, L. J., Sartorius, S. S., Avila-Pires, T. C. S., and Espósito, M. C. (1998b). Use of time, space, and food by the gymnophthlamid lizard *Prionodactylus eigenmanni* from the western Amazon of Brazil. *Canadian Journal of Zoology* **76:** 1681–1688.

Vitt, L., Zani, P., and Espósito, M. C. (1999). Historical ecology of Amazonian lizards: Implications for community ecology. *Oikos* **87:** 286–294.

Vitt, L. J., Sartorius, S. S., Avila-Pires, T. C. S., and Espósito, M. C. (2001). Life at River's Edge: Ecology of *Kentropyx altamazonica* in the Brazilian Amazon. *Canadian Journal of Zoology* **79:** 1855–1865.

Vitt, L. J., Pianka, E. R., Cooper, W. E., and Schwenk, K. (2003). History and the global ecology of squamate reptiles. *American Naturalist* **162:** 44–60.

Vitt, L. J., Caldwell, J. P., Sartorius, S. S., Cooper, W. E., Baird, T. A., Baird, T. D., and Perez-Mellado, V. (2005a). Pushing the edge: Extended activity as an alternative to risky body temperatures in a herbivorous teiid lizard (*Cnemidophorus murinus*: Squamata). *Functional Ecology* **19:** 152–158.

Vitt, L. J., Sartorius, S. S., Avila-Pires, T. C. S., Zani, P. A., and Esposito, M. C. (2005b). Small in a big world: Ecology of leaf-litter geckos in new world tropical forests. *Herpetological Monographs* 137–152.

Vitt, L. J., Shepard, D. B., Caldwell, J. P., Vieira, G. H. C., Franca, F. G. R., and Colli, G. R. (2007a). Living with your food: Geckos in termitaria of Cantão. *Journal of Zoology, London* **272:** 321–328.

Vitt, L. J., Colli, G. R., Caldwell, J. P., Mesquita, D. O., Garda, A. A., and Franca, F. G. R. (2007b). Detecting variation in microhabitat use in low-diversity lizard assemblages across small-scale habitat gradients. *Journal of Herpetology* **41:** 654–663.

Vitt, L. J., Ávila-Pires, T. C. S., Espósito, M. C., Sartorius, S. S., and Zani, P. A. (2007c). Ecology of *Alopoglossus angulatus* and *A. atriventris* in western Amazonia. *Phyllomedusa* **6:** 11–21.

Vogt, R. C., and Flores-Villela, O. (1992). Effects of incubation temperature on sex determination in a community of Neotropical freshwater turtles in southern Mexico. *Herpetologica* **48:** 265–270.

Vogt, R. C., and Guzman, S. (1988). Food partitioning in a Neotropical freshwater turtle community. *Copeia* **1988:** 37–47.

Volsøe, H. (1944). Structure and seasonal variation of the male reproductive organs of *Vipera berus* (L.). *Spoila Zoologica Musei Huaniensis* **5:** 1–157.

Vorburger, C. (2006). Geographic parthenogenesis: Recurrent patterns down under. *Current Biology* **16:** R641–643.

Voris, H. K. (1985). Population size estimates for a marine snake (*Enhydrina schistosa*) in Malaysia. *Copeia* **1985:** 955–961.

Voris, H. K., and Karns, D. R. (1996). Habitat utilization, movements, and activity patterns of *Enhydris plumbea* (Serpentes: Homalopsinae) in a rice paddy wetland in Borneo. *Herpetological Natural History* **4:** 111–126.

Voris, H. K., and Voris, H. H. (1983). Feeding strategies in marine snakes: An analysis of evolutionary, morphological, behavioral and ecological relationships. *American Zoologist* **23:** 411–425.

Voris, H. K., and Voris, H. H. (1995). Commuting on the tropical tides: the life of the yellow-lipped sea krait. *Ocean Realm* **1995:** 57–61.

Vorobyeva, E. I. (Ed.) (1994). *The Siberian Newt (Salamandrella keyserlingii Dybowski, 1870). Zoogeography, Systematics, Morphology*. Nauka, Moscow. [in Russian].

Vorobyeva, E. I. (Ed.) (1995). *The Siberian Newt (Salamandrella keyserlingii Dybowski, 1870). Ecology, Behavior, Conservation*. Nauka, Moscow. [in Russian].

Vorobyeva, E., and Schultze, H.-P. (1991). Description and systematics of panderichthyid fishes with comments on their relationship to tetrapods. In *Origins of the Higher Groups of Tetrapods. Controversy and Consensus*. H.-P. Schultze and L. Trueb (Eds.). Pp. 68–109. Comstock Publishing Associates, Ithaca, NY.

Vrijenhoek, R. C., Dawley, R. M., Cole, C. J., and Bogart, J. P. (1989). A list of the known unisexual vertebrates. In *Evolution and Ecology of Unisexual Vertebrates*. R. M. Dawley and J. P. Bogart (Eds.). Pp. 19–23. New York State Museum, Albany, NY.

Wager, V. A. (1986). *Frogs of South Africa. Their Fascinating Life Stories*. 2nd edition. Delta Books, Craighall, South Africa.

Wagner, P. J., Ruta, M., and Coates, M. I. (2006). Evolutionary patterns in early tetrapods. II. Differing constraints on available character space among clades. *Proceedings of the Royal Society B-Biological Sciences* **273:** 2113–2118.

Wainwright, P. C., and Bennett, A. F. (1992a). The mechanism of tongue projection in chameleons. II. Electromyographic tests of functional hypotheses. *Journal of Experimental Biology* **168:** 1–21.

Wainwright, P. C., and Bennett, A. F. (1992b). The mechanism of tongue projection in chameleons. II. Role of shape change in muscular hydrostat. *Journal of Experimental Biology* **168:** 23–40.

Wainwright, P. C., Kraklau, D. M., and Bennett, A. F. (1991). Kinematics of tongue projection in *Chamaeleo oustaleti*. *Journal of Experimental Biology* **159:** 109–133.

Wakahara, M., and Yamaguchi, M. (2001). Erythropoiesis and conversion of RBCs and hemoglobins from larval to adult type during amphibian development. *Zoological Science* **18:** 891–904.

Wake, D. B., and Dresner, I. G. (1967). Functional morphology and evolution of tail autotomy in salamanders. *Journal of Morphology* **122:** 265–306.

Wake, D. B., and Elias, P. (1983). New genera and a new species of Central American salamanders with a review of the tropical genera (Amphibia, Caudata, Plethodontidae). *Contributions in Science (Los Angeles)* **345:** 1–19.

Wake, D. B., and Hanken, J. (1996). Direct development in the lungless salamanders: What are the consequences for developmental biology, evolution and phylogenesis? *International Journal of Developmental Biology* **40:** 859–869.

Wake, D. B., and Özeti, N. (1969). Evolutionary relationships in the family Salamandridae. *Copeia* **1969:** 124–137.

Wake, M. H. (1977). The reproductive biology of caecilians: An evolutionary perspective. In *The Reproductive Biology of Amphibians*. D. H. Taylor and S. I. Guttman (Eds.). Pp. 73–101. Plenum Press, New York.

Wake, M. H. (1978). The reproductive biology of *Eleutherodactylus jasperi* (Amphibia, Anura, Leptodactylidae), with comments on the evolution of live-bearing systems. *Journal of Herpetology* **12:** 121–133.

Wake, M. H. (1980). Reproduction, growth, and population structure of the Central American caecilian *Dermophis mexicanus*. *Herpetologica* **36:** 244–256.

Wake, M. H. (1982). Diversity within a framework of constraints: Reproductive modes in the Amphibia. In *Environmental Adaptation and Evolution, a Theoretical and Empirical Approach*. D. Mossakowski and G. Roth (Eds.). Pp. 87–106. Gustav Fischer Verlag, Stuttgart.

Wake, M. H. (1989). *Hyman's Comparative Vertebrate Anatomy*. Third Edition. University of Chicago Press, Chicago.

Wake, M. H. (1992a). Reproduction in caecilians. In *Reproductive Biology of South American Vertebrates*. W. C. Hamlett (Ed.). Pp. 112–120. Springer-Verlag, New York.

Wake, M. H. (1992b). Biogeography of Mesoamerican caecilians (Amphibia: Gymnophiona). *Tulane Studies in Zoology and Botany, Supplementary Publication* **1:** 321–325.

Wake, M. H. (1993a). Non-traditional characters in the assessment of caecilian phylogenetic relationships. *Herpetological Monographs* **7:** 42–55.

Wake, M. H. (1993b). Evolution of oviductal gestation in amphibians. *Journal of Experimental Zoology* **266:** 394–413.

Wake, M. H. (1998). Cartilage in the cloaca: Phallodeal spicules in caecilians (Amphibia: Gymnophiona). *Journal of Morphology* **237:** 177–186.

Wake, M. H. (2006). A brief history of research on gymnophionian reproductive biology and development. In *Reproductive Biology and Phylogeny of Gymnophiona (Caecilians)*. J.-M. Exbrayat (Ed.). Pp. 1–37. Science Publishers, Enfield, NH.

Wake, M. H., and Dickie, R. (1998). Oviduct structure and function and reproductive modes in amphibians. *Journal of Experimental Zoology* **282:** 477–506.

Walker, A. D. (1961). Triassic reptiles from Elgin area: *Stagonolepis, Dasygnathus*, and their allies. *Philosophical Transactions of the Royal Society B-Biological Sciences* **244:** 103–204.

Wallach, V. (1985). A cladistic analysis of the terrestrial Australian Elapidae. In *Biology of Australasian Frogs and Reptiles*. G. Grigg, R. Shine, and H. Ehmann (Eds.). Pp. 223–253. Surrey Beatty & Sons, Chipping Norton Pty Ltd., Sydney, Australia.

Wallach, V. (1993). Presence of a left lung in the Typhlopidae (Reptilia: Serpentes). *Journal of the Herpetological Association of Africa* **42:** 32–33.

Wallach, V. (1998). The lungs of snakes. In *Biology of the Reptilia. Volume 19, Morphology G*. C. Gans and A. S. Gaunt (Eds.), Pp. 93–295. Society for the Study of Amphibians and Reptiles, Ithaca, NY.

Wallach, V., and Günther, R. (1997). Typhlopidae vs. Anomalepididae: The identity of *Typhlops mutilatus* Werner (Reptilia: Serpentes). *Mitt. Zool. Mus. Berlin* **73:** 333–342.

Wallach, V., and Günther, R. (1998). Visceral anatomy of the Malaysian snake genus *Xenophidion*, including a cladistic analaysis and allocation of a new family (Serpentes: Xenophidiidae). *Amphibia-Reptilia* **19:** 385–404.

Warburg, M. R., Lewinson, D., and Rosenberg, M. (1994). Structure and function of *Salamandra* skin and gills. *Mertensiella* **4:** 423–452.

Warkentin, K. M. (1995). Adaptive plasticity in hatching age: A response to predation risk trade-offs. *Proceedings of the National Academy of Sciences (USA)* **92:** 3507–3510.

Warkentin, K. M. (2000). Wasp predation and wasp-induced hatching of red-eyed treefrog eggs. *Animal Behaviour* **60:** 503–510.

Warkentin, K. M. (2005). How do embryos assess risk? Vibrational cues in predator-induced hatching of red-eyed treefrogs. *Animal Behaviour* **70:** 59–71.

Warner, D. A., and Shine, R. (2005). The adaptive significance of temperature-dependent sex determination: Experimental tests with a short-lived lizard. *Evolution* **59:** 2209–2221.

Warner, D. A., and Shine, R. (2006). Morphological variation does not influence locomotor performance within a cohort of hatchling lizards (*Amphibolurus muricatus*, Agamidae). *Oikos* **114:** 126–134.

Warner, D. A., and Shine, R. (2007). Reproducing lizards modify sex allocation in response to operational sex ratios. *Biology Letters* **3:** 47–50.

Warner, D. A., and Shine, R. (2008). The adaptive significance of temperature-dependent sex determination in a reptile. *Nature* **451:** 566–568.

Warner, D. A., Lovern, M., and Shine, R. (2007). Maternal nutrition affects reproductive output and sex allocation in a lizard with environmental sex determination. *Proceedings of the Royal Society London Series B* **274:** 883–890.

Wassarman, P. M. (1987). The biology and chemistry of fertilization. *Science* **235:** 553–560.

Wassersug, R. (1996). The biology of *Xenopus* tadpoles. In *The Biology of Xenopus*. R. C. Tinsley and H. R. Kobel (Eds.). Pp. 195–211. Zoological Society of London, London.

Wassersug, R. J. (1971). On the comparative palatability of some dry-season tadpoles from Costa Rica. *American Midland Naturalist* **86:** 101–109.

Wassersug, R. J. (1972). The mechanism of ultraplanktonic entrapment in anuran larvae. *Journal of Morphology* **137:** 279–288.

Wassersug, R. J. (1973). Aspects of social behavior in anuran larvae. In *Evolutionary Ecology of the Anurans*. J. L. Vial, (Ed.). Pp. 273–297. University of Missouri Press, Columbia.

Wassersug, R. J. (1974). Evolution of anuran life cycles. *Science* **185:** 377–378.

Wassersug, R. J. (1976). Internal oral features in *Hyla regilla* (Anura: Hylidae) larvae: An ontogenetic study. *Occasional Papers of the Museum of Natural History, University of Kansas* **49:** 1–24.

Wassersug, R. J. (1980). Internal oral features of larvae from eight anuran families: Functional, systematic, ecolutionary and ecological considerations. *Miscellaneous Publications of the Museum of Natural History, University of Kansas* **68:** 1–146.

Wassersug, R. J., and Pyburn, W. F. (1987). The biology of the pe-ret' toad, *Otophryne robusta* (Microhylidae), with special consideration of its fossorial larva and systematic relationships. *Zoological Journal of the Linnean Society* **91:** 137–169.

Wassersug, R. J., Lum, A. M., and Patel, M. J. (1981). An analysis of school structure for tadpoles (Anura: Amphibia). *Behavioral Ecology and Sociobiology* **9:** 15–22.

Watkins, J. F., Gehlbach, F. R., and Kroll, J. C. (1969). Attractant-repellant secretions in blind snakes (*Leptotyphlops dulcis*) and army ants (*Neivamyrex nigrescens*). *Ecology* **50:** 1098–1102.

Watling, D. (1986). *Mai Veikau. Tales of Fijian Wildlife*. Privately printed, Suva, Fiji.

Watrous, L. E., and Wheeler, Q. D. (1981). The outgroup comparison method of character analysis. *Systematic Zoology* **30:** 1–11.

Watts, P. C., Buley, K. R., Boardman, W., Ciofi, C., and Gibson, O. R. (2006). Parthenogenesis in komodo dragons. *Nature* **444:** 1021–1022.

Webb, C., and Joss, J. (1997). Does predation by the fish *Gambusia holbrooki* (Atheriniformes: Poeciliidae) contribute to declining frog populations? *Australian Zoologist* **30:** 316–324.

Webb, G. J. W. (1977). The natural history of *Crocodylus porosus*. In *Australian Animals and Their Environments*. H. Messel and S. Butler (Eds.). Pp. 239–312. Shakespeare Head Press, Sydney.

Webb, G. J. W., and Manolis, S. C. (1989). *Crocodiles of Australia*. Reed, Chatswood, N.S.W.

Webb, G. J. W., and Manolis, S. C. (1993). *Crocodiles of Australia. Revised Edtion*. Reed, Chatswood, N.S.W.

Webb, G. J. W., and Vernon, B. (1992). Crocodile management in the People's Republic of China – a review with recommendations. In *Crocodile Conservation Action (Crocodile Specialist Group, SSC)* Pp. 1–27. IUCN, Glands.

Webb, G. J. W., Buckworth, R., and Manolis, S. C. (1983). *Crocodylus johnstoni* in the McKinlay River, N. T. VI. Nesting biology. *Australian Wildlife Research* **10:** 607–637.

Webb, G. J. W., Choquenot, D., and Whitehead, P. J. (1986). Nests, eggs, and embryonic development of *Carettochelys insculpta* (Chelonia: Carettochelidae [sic]) from northern Australia. *Journal of Zoology, London* **1B:** 521–550.

Webb, G. J. W., Manolis, S. C., and Whitehead, P. J. (1987). *Wildlife Management: Crocodiles and Alligators*. Surrey Beatty & Sons, Chipping Norton Pty Ltd, N.S.W.

Webb, G. J. W., Messel, H., and Magnusson, W. (1977). The nesting of *Crocodylus porosus* in Arnhem Land, northern Australia. *Copeia* **1977:** 238–249.

Webb, J. K., and Shine, R. (1997). A field study of spatial ecology and movements of a threatened snake species, *Hoplocephalus bungaroides*. *Biological Conservation* **82:** 203–217.

Webb, J. K., and Shine, R. (1998). Ecological characteristics of a threatened snake species, *Hoplocephalus bungaroides* (Serpentes, Elapidae). *Animal Conservation* **1:** 185–193.

Webb, J. K., Shine, R., and Christian, K. A. (2006). The adaptive significance of reptilian viviparity in the tropics: Testing the maternal manipulation hypothesis. *Evolution* **60:** 115–122.

Weekes, H. C. (1935). A review of placentation among reptiles with particular regard to the function and evolution of the placenta. *Proceedings of the Zoological Society of London* **3:** 625–645.

Wegener, A. (1968). *The Origin of Continents and Oceans*. Methuen, London.

Weisrock, D. W., Harmon, L. J., and Larson, A. (2005). Resolving deep phylogenetic relationships in salamanders: Analyses of mitochondrial and nuclear genomic data. *Systematic Biology* **54:** 758–777.

Weisrock, D. W., Papenfuss, T. J., Macey, J. R., Litvinchuk, S. N., Polymeni, R., Ugurtas, I. H., Zhao, E., Jowkar, H., and Larson, A. (2006a). A molecular assessment of phylogenetic relationships and lineage accumulation rates within the family Salamandridae (Amphibia, Caudata). *Molecular Phylogenetics and Evolution* **41:** 368–383.

Weisrock, D. W., Shaffer, H. B., Storz, B. L., Storz, S. R., and Voss, S. R. (2006b). Multiple nuclear gene sequences identify phylogenetic species boundaries in the rapidly radiating clade of Mexican ambystomatid salamanders. *Molecular Ecology* **15:** 2489–2503.

Weiss, S. L. (2001). The effect of reproduction on food intake of a sit-and-wait foraging lizard, *Sceloporus virgatus*. *Herpetologica* **57:** 138–146.

Weldon, P. J. (1990). Responses by vertebrates to chemicals from predators. In *Chemical Signals in Vertebrates 5*. D. W. Macdonald, D. Müller-Schwarze, and S. E. Natynczuk (Eds.). Pp. 500–521. Oxford University Press, Oxford.

Wells, K. D. (1977). The social behaviour of anuran amphibians. *Animal Behaviour* **15:** 666–693.

Wells, K. D. (1980). Social behavior and communication of a dendrobatid frog (*Colostethus trinitatis*). *Herpetologica* **36:** 189–199.

Wells, K. D. (1981). Parental behavior of male and female frogs. In *Natural Selection and Social Behavior*. R. D. Alexander and D. W. Tinkle (Eds.). Pp. 184–197. Chiron Press, New York.

Wells, K. D. (1988). The effect of social interactions on anuran vocal behavior. In *The Evolution of the Amphibian Auditory System*. B. Fritzsch, M. J. Ryan, W. Wilczynski, T. E. Hetherington, and W. Walkowiak (Eds.). Pp. 433–454. Wiley, New York.

Wells, K. D. (2007). *The Ecology and Behavior of Amphibians*. University of Chicago Press, Chicago.

Wells, K. D., and Bard, K. M. (1988). Parental behavior of an aquatic-breeding tropical frog, *Leptodactylus bolivanus*. *Journal of Herpetology* **22:** 361–364.

Wells, K. D., and Wells, R. A. (1976). Patterns of movement in a population of the slimy salamander, *Plethodon glutinosus*, with observations on aggregations. *Herpetologica* **32:** 156–162.

Welsch, H. H., Jr., and Lind, A. J. (1996). Habitat correlates of the southern torrent salamander, *Rhyacotriton variegatus* (Caudata: Rhyacotritonidae), in northwestern California. *Journal of Herpetology* **30:** 385–398.

Werner, D. I., Baker, E. M., Gonzales, E. C., and Sosa, I. R. (1987). Kinship recognition and grouping in hatchling green iguanas. *Behavioral Ecology and Sociobiology* **21:** 83–89.

Werner, E. E. (1986). Amphibian metamorphosis: Growth rate, predator risk, and the optimal size at transformation. *American Naturalist* **128:** 319–341.

Werner, E. E., Skelly, D. K., Relyea, R. A., and Yurewicz, K. L. (2007a). Amphibian species richness across environmental gradients. *Oikos* **116:** 1697–1712.

Werner, E. E., Yurewicz, K. L., Skelly, D. K., and Relyea, R. A. (2007b). Turnover in an amphibian metacommunity: The role of local and regional factors. *Oikos* **116:** 1713–1725.

Werner, Y. (1997). Observations and comments on active foraging in geckos. *Russian Journal of Herpetology* **4:** 34–39.

Werner, Y. L., Okada, S., Ota, H., Perry, G., and Tokunaga, S. (1997). Varied and fluctuating foraging modes in nocturnal lizards of the family Gekkonidae. *Asiatic Herpetological Research* **7:** 153–165.

West, N. H., and Van Vliet, B. N. (1983). Open-loop analysis of the pulmocutaneous baroreflex in the toad *Bufo marinus*. *American Journal of Physiology - Regulatory, Integrative, and Comparative Physiology* **245:** 642–650.

Wever, E. G. (1978). *The Reptile Ear. Its Structure and Function*. Princeton University Press, Princeton.

Wever, E. G. (1985). *The Amphibian Ear*. Princeton University Press, Princeton.

Wevers, E. (1988). Enige opmerkingen over de pijlgifkikker *Dendrobates parvulus*. *Lacerta* **46:** 51–53.

Weygoldt, P. (1976). Beobachtungen zur Biologie und Ethologie von *Pipa* (*Hemipipa*) *carvalhoi* Mir. RiB. 1937 (Anura, Pipidae). *Zeitschrift für Tierpsychologie* **40:** 80–99.

Weygoldt, P. (1991). Zur Biologie und zum Verhalten von *Phyllomedusa marginata* Izeckshohn & Da Cruz, 1976 im Terrarium. *Salamandra* **27:** 83–96.

Weygoldt, P., and de Carvalho e Silva, S. P. (1991). Observations on mating, oviposition, egg sac formation and development in the egg-brooding frog, *Fritziana goeldii*. *Amphibia-Reptilia* **12:** 67–80.

Whimster, I. W. (1990). Neural induction of epidermal sensory organs in gecko skin. In *The Skin of Vertebrates*. R. I. C. Spearman and P. A. Riley (Eds.). Pp. 161–167. Linnean Society of London, London.

Whitaker, A. H. (1968). The lizards of the Poor Knights Islands. *New Zealand Journal of Science* **11:** 623–651.

Whitaker, A. H. (1973). Lizard populations on islands with and without Polynesian rats, *Rattus exulans* (Peale). *Proceedings of the New Zealand Ecological Society* **20:** 121–130.

Whitaker, R., and Basu, D. (1983). The gharial (*Gavialis gangeticus*): A review. *Journal of the Bombay Natural History Society* **79:** 531–548.

White, J. (1976). Reptiles of the Corruna Hills. *Herpetofauna* **8:** 21–23.

Whitear, M. (1977). A functional comparison between the epidermis of fish and of amphibians. *Symposium of the Zoological Society of London* **39:** 291–313.

Whitfield, S. M., Bell, K. E., Philippi, T., Sasa, M., Bolaños, F., Chaves, G., Savage, J. M., and Donnelly, M. A. (2007). Amphibian and reptile declines over 35 years at La Selva, Costa Rica. *Proceedings of the National Academy of Sciences (USA)* **104:** 8352–8356.

Whitford, W. G. (1973). The effects of temperature on respiration in the Amphibia. *American Zoologist* **13:** 505–512.

Whitford, W. G., and Bryant, M. (1979). Behavior of a predator and its prey: The horned lizard (*Phrynosoma cornutum*) and harvester ants (*Pogonomyrmex* spp.). *Ecology* **60:** 686–694.

Whiting, M. J., Dixon, J. R., and Murray, R. C. (1993). Spatial distribution of a population of Texas horned lizards (*Phrynosoma cornutum*: Phrynosomatidae) relative to habitat and prey. *Southwestern Naturalist* **38:** 150–154.

Whiting, M. J., Nagy, K. A., and Bateman, P. W. (2003). Evolution and maintenance of social status-signaling badges: Experimental manipulations in lizards. In *Lizard Social Behavior*. S. F. Fox, J. K. McCoy, and T. A. Baird (Eds.). Pp. 47–82. Johns Hopkins University Press, Baltimore, MD.

Whittier, J. M., and Crews, D. (1987). Seasonal reproduction: Patterns and control. In *Hormones and Reproduction in Fishes, Amphibians, and Reptiles*. D. O. Norris and R. E. Jones (Eds.). Pp. 385–409. Plenum Press, New York.

Wibbels, T., Bull, J. J., and Crews, D. (1994). Temperature dependent sex determination: A mechanistic approach. *Journal of Experimental Zoology* **270:** 71–78.

Wickramasinghe, D. D., Oseen, K. L., and Wassersug, R. J. (2007). Ontogenetic changes in diet and intestinal morphology in semi-terrestrial tadpoles of *Nannophrys ceylonensis* (Dicroglossidae). *Copeia* **2007:** 1012–1018.

Wiens, J. J. (1999). Phylogenetic evidence for multiple losses of a sexually selected character in phrynosomatid lizards. *Proceedings of the Royal Society of London Series B-Biological Sciences* **266:** 1529–1535.

Wiens, J. J., and Donoghue, M. J. (2004). Historical biogeography, ecology and species richness. *Trends in Ecology & Evolution* **19:** 639–644.

Wiens, J. J., and Hollingsworth, B. D. (2000). War of the iguanas: Conflicting molecular and morphological phylogenies and long-branch attraction in iguanid lizards. *Systematic Biology* **49:** 143–159.

Wiens, J. J., and Reeder, T. W. (1997). Phylogeny of the spiny lizards (*Sceloporus*) based on molecular and morphological evidence. *Herpetological Monographs* **11:** 1–101.

Wiens, J. J., and Titus, T. A. (1991). A phylogenetic analysis of *Spea* (Anura: Pelobatidae). *Herpetologica* **47:** 21–28.

Wiens, J. J., Bonett, R. M., and Chippindale, P. T. (2005). Ontogeny discombobulates phylogeny: Paedomorphosis and higher-level salamander relationships. *Systematic Biology* **54:** 91–110.

Wiens, J. J., Engstrom, T. N., and Chippindale, P. T. (2006). Rapid diversification, incomplete isolation, and the "speciation clock" in North American salamanders (Genus *Plethodon*): Testing the hybrid swarm hypothesis of rapid radiation. *Evolution* **60:** 2585–2603.

Wiens, J. J., Kuczynski, C. A., Duellman, W. E., and Reeder, T. W. (2007a). Loss and re-evolution of complex life cycles in marsupial frogs: Does ancestral trait reconstruction mislead? *Evolution* **61:** 1886–1899.

Wiens, J. J., Parra-Olea, G., Garcia-Paris, M., and Wake, D. B. (2007b). Phylogenetic history underlies elevational biodiversity patterns in tropical salamanders. *Proceedings of the Royal Society B-Biological Sciences* **274:** 919–928.

Wiewandt, T. A. (1982). Evolution of nesting patterns in iguanine lizards. In *Iguanas of the World: Their Behavior, Ecology, and Conservation*. G. M. Burghardt, and A. S. Rand (Eds.). p. 117–149. Noyes Publications, Park Ridge, N.J.

Wilbur, H. M. (1972). Competition, predation, and the structure of the *Ambystoma-Rana sylvatica* community. *Ecology* **53:** 3–21.

Wilbur, H. M. (1975a). A growth model for the turtle *Chrysemys picta*. *Copeia* **1975:** 337–343.

Wilbur, H. M. (1975b). The evolutionary and mathematical demography of the turtle *Chrysemys picta*. *Ecology* **56:** 64–77.

Wilbur, H. M. (1980). Complex life cycles. *Annual Review of Ecology and Systematics* **11:** 67–93.

Wilbur, H. M. (1987). Regulation of structure in complex systems: Experimental temporary pond communities. *Ecology* **68:** 1437–1452.

Wilbur, H. M. (1997). Experimental ecology of food webs: Complex systems in temporary ponds. *Ecology* **78:** 2279–2302.

Wilbur, H. M., and Collins, J. P. (1973). Ecological aspects of amphibian metamorphosis. *Science* **182:** 1305–1314.

Wilbur, H. M., and Fauth, J. E. (1990). Experimental aquatic food webs: Interactions between two predators and two prey. *American Naturalist* **135:** 176–204.

Wilcox, T. P., Zwickl, D. J., Heath, T. A., and Hillis, D. M. (2002). Phylogenetic relationships of the dwarf boas and a comparison of Bayesian and bootstrap measures of phylogenetic support. *Molecular Phylogenetics and Evolution* **25:** 361–371.

Wild, E. R. (1995). New genus and species of Amazonian microhylid frog with a phylogenetic analysis of New World genera. *Copeia* **1995:** 837–849.

Wiley, E. O. (1981). *Phylogenetics. The Theory and Practice of Phylogenetic Systematics*. John Wiley & Sons, New York.

Wiley, E. O. (1988). Vicariance biogeography. *Annual Review of Ecology and Systematics* **19:** 513–542.

Wiley, E. O., Siegel-Causey, D., Brooks, D. R., and Funk, V. A. (1991). The Compleat Cladist. A Primer of Phylogenetic Procedures. *University of Kansas Museum of Natural History, Special Publication* **19:** 1–158.

Wilhoft, D. C. (1963). Reproduction in the tropical Australian skink, *Leiolopisma rhomboidalis*. *American Midland Naturalist* **70:** 442–461.

Wilkinson, M. (1989). On the status of *Nectocaecilia fasciata* Taylor with a discussion of the phylogeny of the Typhlonectidae (Amphibia: Gymnophiona). *Herpetologica* **45:** 23–36.

Wilkinson, M. (1992a). The phylogenetic position of the Rhinatrematidae (Amphibia: Gymnophiona): Evidence from the larval lateral line system. *Amphibia-Reptilia* **13:** 74–79.

Wilkinson, M. (1992b). On the life history of the caecilian genus *Uraeotyphlus* (Amphibia: Gymnophiona). *Herpetological Journal* **2:** 121–124.

Wilkinson, M. (1996). Resolution of the taxonomic status of *Nectocaecilia hyadee* (Roze) and a revised key to the genera of the Typhlonectidae. *Journal of Herpetology* **30:** 413–415.

Wilkinson, M. (1997). Characters, congruence and quality: A study of neuroanatomical and traditional data in caecilian phylogeny. *Biological Reviews* **72:** 423–470.

Wilkinson, M., and Nussbaum, R. A. (1996). On the phylogenetic position of the Uraeotyphlidae (Amphibia: Gymnophiona). *Copeia* **1996:** 550–562.

Wilkinson, M., and Nussbaum, R. A. (1997). Comparative morphology and evolution of the lungless caecilian *Atretochoana eiselti* (Taylor) (Amphibia: Gymnophiona: Typhlonectidae). *Biological Journal of the Linnean Society* **62:** 39–109.

Wilkinson, M., and Nussbaum, R. A. (2006). Caecilian phylogeny and classification. In *Reproductive Biology and Phylogeny of Gymnophiona (Caecilians)*. J.-M. Exbrayat (Ed.). Pp. 39–78. Science Publishers, Enfield, NH.

Wilkinson, M., Loader, S. P., Gower, D. J., Sheps, J. A., and Cohen, B. L. (2003). Phylogenetic relationships of African caecilians (Amphibia; Gymnophiona): Insights from mitochondrial rRNA gene sequences. *African Journal of Herpetology* **52:** 83–92.

Wilkinson, M., Richardson, M. K., Gower, D. J., and Oommen, O. V. (2002a). Extended embryo retention, caecilian oviparity and amniote origins. *Journal of Natural History* **36:** 2185–2198.

Wilkinson, M., Sebben, A., Schwartz, E. N. F., and Schwartz, C. A. (1998). The largest lungless tetrapod: Report on a second specimen of *Atretochoana eiselti* (Amphibia: Gymnophiona: Typhlonectidae) from Brazil. *Journal of Natural History* **32:** 617–627.

Wilkinson, M., Sheps, J. A., Oommen, O. V., and Cohen, B. L. (2002b). Phylogenetic relationships of Indian caecilians (Amphibia: Gymnophiona) inferred from mitochondrial rRNA gene sequences. *Molecular Phylogenetics and Evolution* **23:** 401–407.

Willi, U. B., Bronner, G. N., and Narins, P. A. (2006). Middle ear dynamics in response to seismic stimuli in the Cape golden mole (*Chrysochloris asiatica*). *Journal of Experimental Biology* **209:** 302–313.

Williams, E. E. (1950). Variation and selection in the cervical central articulations of living turtles. *Bulletin of the American Museum of Natural History* **94:** 505–561.

Williams, E. E. (1969). The ecology of colonization as seen in the zoogeography of anoline lizards on small islands. *Quarterly Review of Biology* **44:** 345–389.

Williams, E. E. (1972). The origin of faunas. Evolution of lizard congeners in a complex island fauna: a trial analysis. *Evolutionary Biology* **6:** 47–89.

Williams, E. E. (1983). Ecomorphs, faunas, island size, and diverse end points in island radiations of *Anolis*. In *Lizard Ecology: Studies of a Model Organism*. R. B. Huey, E. R. Pianka, and T. W. Schoener (Eds.). Pp. 326–370. Harvard University Press, Cambridge, Mass.

Williams, E. E. (1988). A new look at the Iguania. In Proceedings of a Workshop on Neotropical Distribution Patterns. W. R. Heyer and P. E. Vanzolini (Eds.). Pp. 429–486. Academia Brasileira de Ciências, Rio de Janeiro.

Williams, G. C. (1966). *Adaptation and Natural Selection: A Critique of some Current Evolutionary Thought*. Princeton University Press, Princeton, NJ.

Willis, P. M. A. (1997). Review of fossil crocodilians from Australasia. *Australian Zoologist* **30:** 287–298.

Willis, R. E., McAliley, L. R., Neeley, E. D., and Densmore, L. D. (2007). Evidence for placing the false gharial (*Tomistoma schlegelii*) into the family Gavialidae: Inferences from nuclear gene sequences. *Molecular Phylogenetics and Evolution* **43:** 787–794.

Wilms, T., and Böhme, W. (2000). Revision of the *Uromastyx acanthinura* species group, with description of a new species from the central Sahara (Reptilia: Sauria: Agamidae). *Zool. Abh. Staatl. Mus. Tierk.* Dresden **51:** 73–104.

Wilson, D. S. (1998). Nest-site selection: Microhabitat variation and its effects on the survival of turtle embryos. *Ecology* **79:** 1884–1892.

Wilson, D. S., Mushinsky, H. R., and McCoy, E. D. (1994). Home range, activity, and use of burrows of juvenile gopher tortoises in central Florida. *U S Fish and Wildlife Service Fish and Wildlife Research* **13:** 147–160.

Wilson, D. S., Nagy, K. A., Tracy, C. R., Morafka, D. J., and Yates, R. A. (2001). Water balance in neonate and juvenile desert tortoises, *Gopherus agassizii*. *Herpetological Monographs* **15:** 158–170.

Wilson, E. O., and Bossert, W. H. (1971). *A Primer of Population Biology*. Sinauer Associates, Stamford, CT.

Wilson, J. (2001). A review of the modes of orientation used by amphibians during breeding migration. *Journal of the Pennsylvania Academy of Sciences* **74:** 61–66.

Wilson, L. D., and Porras, L. (1983). *The Ecological Impact of Man on the South Florida Herpetofauna. University of Kansas Musem of Natural History Special Publication* **7,** Lawrence, KS.

Winemiller, K. O., and Pianka, E. R. (1990). Organization in natural assemblages of desert lizards and tropical fishes. *Ecological Monographs* **60:** 27–55.

Winter, D. A., Zawada, W. M., and Johnson, A. A. (1986). Comparison of the symbiotic fauna of the family Plethodontidae in the Ouachita Mountains of western Arkansas. *Proceedings of the Arkansas Academy of Science* **40:** 82–85.

Wirot, N. (1979). *The Turtles of Thailand*. Siamfarm Zoological Garden, Bangkok, Thailand.

Withers, P. C. (1992). *Comparative Animal Physiology*. Saunders College Publishing, New York.

Withers, P. C. (1998). Urea: Diverse functions of a 'waste' product. *Clinical and Experimental Pharmacology and Physiology* **25:** 722–727.

Withers, P. C., and Richards, S. J. (1995). Cocoon formation by the treefrog *Litoria alboguttata* (Amphibia: Hylidae): A 'waterproof' taxonomic tool? *Journal of the Royal Society of Western Australia* **78:** 103–106.

Withers, P. C., Hillman, S. S., Drewes, R. C., and Sokol, O. M. (1982). Water loss and nitrogen excretion in sharp-nosed reed frogs (*Hyperolius nasutus:* Anura, Hyperoliidae). *Journal of Experimental Biology* **97:** 335–343.

Witmer, L. M. (1991). Perspectives on avian origins. In *Origins of the Higher Groups of Tetrapods. Controversy and Consensus*. H.-P. Schultze and L. Trueb (Eds.). Pp. 427–466. Comstock Publishing Associates, Ithaca, NY.

Witten, G. J. (1993). Family Agamidae. In *Fauna of Australia. Volume 2A. Amphibia & Reptilia*. C. J. Gasby, G. J. B. Ross, and P. L. Beesley (Eds.). Pp. 240–252. Australian Government Publishing Service, Canberra.

Woinarski, J. C. Z. (1989). The vertebrate fauna of broombush *Melaleuca uncinata* vegetation in northwestern Victoria. *Australian Wildlife Research* **16:** 217–238.

Wone, B., and Beauchamp, B. (2003). Movement, home range, and activity patterns of the horned lizard, *Phrynosoma mcallii*. *Journal of Herpetology* **37:** 679–686.

Woolbright, L. L. (1983). Sexual selection and size dimorphism in anuran Amphibia. *American Naturalist* **121:** 110–119.

Woolbright, L. L. (1991). The impact of Hurricane Hugo on forest frogs in Puerto Rico. *Biotropica* **23:** 462–467.

Worthy, T. H. (1987). Osteology of *Leiopelma* (Amphibia: Leiopelmatidae) and descriptions of three new subfossil *Leiopelma* species. *Journal of the Royal Society of New Zealand* **17:** 201–251.

Wright, J. W. (1993). Evolution of whiptail lizards (*Cnemidophorus*). In *Biology of Whiptail Lizards* (Genus *Cnemidophorus*). J. W. Wright and L. J. Vitt (Eds.). Pp. 27–81. Oklahoma Museum of Natural History, Norman, OK.

Wright, J. W. and Vitt, L. J. (Eds.) (1993). *Biology of Whiptail Lizards (Genus* Cnemidophorus*)*. Oklahoma Museum of Natural History, Norman, OK.

Wu, X.-C., Brinkman, D. B., and Russell, A. R. (1996a). A new alligator from the Upper Cretaceous of Canada and the relationships of early eusuchians. *Palaeontology* **39:** 351–375.

Wu, X.-C., Brinkman, D. B., and Russell, A. R. (1997a). *Sineoamphisbaena hexatabularis*, an amphisbaenian (Diaspsida: Squamata) from the Upper Cretaceous redbeds at Bayan Mandahu (Inner Mongolia, People's Republic of China), and comments on the phylogenetic relationships of the Amphisbaenia. *Canadian Journal of Earth Sciences* **33:** 541–577.

Wu, X.-C., Russell, A. R., and Brinkman, D. B. (1997b). Phylogenetic relationships of *Sineoamphisbaena hexatabularis*: Further considerations. *Canadian Journal of Earth Sciences* **34:** 883–885.

Wu, X.-C., Sues, H. -D., and Dong, Z. -M. (1996b). *Sichuanosuchus shuhanensis*, a new Early Cretaceous protosuchian (Archosauria: Crocodyliformes) from Sichuan (China), and the monophyly of Protosuchia. *Journal of Vertebrate Paleontology* **17:** 89–103.

Wynn, A. H., Cole, C. J., and Gardner, A. L. (1987). Apparent triploidy in the unisexual Brahminy blind snake, *Ramphotyphlops braminus*. *American Museum Novitates* **2868:** 1–7.

Xavier, F. (1986). La reproduction des *Nectophrynoides*. In *Traité de Zoologie. Vol XIV, fasc. I-B. Amphibiens*. P. P. Grasse and M. Delsol (Eds.). Pp. 497–513. Massons, Paris.

Ydenberg, R. C., and Dill, L. M. (1986). The economics of fleeing from predators. *Advances in the Study of Behavior* **16:** 229–249.

Yeomans, S. R. (1995). Water-finding in adult turtles: Random search or oriented behavior? *Animal Behaviour* **49:** 977–987.

Young, B. A. (1988). The cephalic vascular anatomy of three species of sea snakes. *Journal of Morphology* **196:** 195–204.

Young, B. A. (1997). On the absence of taste buds in monitor lizards (*Varanus*) and snakes. *Journal of Herpetology* **31:** 130–137.

Young, J. E., Christian, K. A., Donnellan, S., Tracy, C. R., and Parry, D. (2005). Comparative analysis of cutaneous evaporative water loss in frogs demonstrates correlation with ecological habits. *Physiological and Biochemical Zoology* **78:** 847–856.

Young, J. E., Tracy, C. R., Christian, K. A., and McArthur, L. J. (2006). Rates of cutaneous evaporative water loss of native Fijian frogs. *Copeia* **2006:** 83–88.

Zaher, H. (1994a). Les Tropidopheoidea (Serpentes; Alethinophidia) sont-ils réellement monophylétiques? Arguments en faveur de leur polyphylétisme. *Comptes Rendus de l'Academie des Sciences Paris, Serie III – Sciences de la Vie* **317:** 471–478.

Zaher, H. (1994b). Comments on the evolution of the jaw adductor musculature of snakes. *Zoological Journal of the Linnean Society* **111:** 339–384.

Zaher, H. (1999). Hemipenial morphology of the South American xenodontine snakes, with a proposal for a monophyletic Xenodontinae and a reappraisal of colubroid hemipenes. *Bulletin of the American Museum of Natural History* **240:** 1–168.

Zamudio, K. R., and Greene, H. W. (1997). Phylogeography of the bushmaster (*Lachesis muta*: Viperidae): Implications for Neotropical biogeography, systematics, and conservation. *Biological Journal of the Linnean Society* **62:** 421–442.

Zangari, F., Cimmaruta, R., and Nascetti, G. (2006). Genetic relationships of the western Mediterranean painted frogs based on allozymes and mitochondrial markers: Evolutionary and taxonomic inferences (Amphibia, Anura, Discoglossidae). *Biological Journal of the Linnean Society* **87:** 515–536.

Zardoya, R., and Meyer, A. (1998). Complete mitochondrial genome suggests diapsid affinities of turtles. *Proceedings of the National Academy of Sciences (USA)* **95:** 14226–14231.

Zardoya, R., and Meyer, A. (2000). Mitochondrial evidence on the phylogenetic position of caecilians (Amphibia: Gymnophiona). *Genetics* **155:** 765–775.

Zardoya, R., and Meyer, A. (2001). On the origin of and phylogenetic relationships among living amphibians. *Proceedings of the National Academy of Sciences (USA)* **98:** 7380–7383.

Zeigler, T., and Böhme, W. (1997). Genital structures and mating biology in squamate reptiles, especially the Platynota, with comments on systematics. *Mertensiella* **8:** 1–207.

Zeng, X. M., Fu, J. Z., Chen, L. Q., Tian, Y. Z., and Chen, X. H. (2006). Cryptic species and systematics of the hynobiid salamanders of the Liua-Pseudohynobius complex: Molecular and phylogenetic perspectives. *Biochemical Systematics and Ecology* **34:** 467–477.

Zhang, P., Chen, Y. Q., Liu, Y. F., Zhou, H., and Qu, L. H. (2003). The complete mitochondrial genome of the Chinese giant salamander, *Andrias davidianus* (Amphibia: Caudata). *Gene* **311:** 93–98.

Zhang, P., Chen, Y. Q., Zhou, H., Liu, Y. F., Wang, X. L., Papenfuss, T. J., Wake, D. B., and Qu, L. H. (2006). Phylogeny, evolution, and biogeography of Asiatic Salamanders (Hynobiidae). *Proceedings of the National Academy of Sciences (USA)* **103:** 7360–7365.

Zhao, E., and Adler, K. (1993). *Herpetology of China*. Society for the Study of Amphibians and Reptiles, Ithaca, NY.

Zhao, E., Hu, Q., Jiang, Y., and Yang, Y. (1988). *Studies on Chinese Salamanders*. Society for the Study of Amphibians and Reptiles, Oxford, Ohio.

Zheng, Y. C., and Fu, J. Z. (2007). Making a doughnut-shaped egg mass: Oviposition behaviour of *Vibrissaphora boringiae* (Anura: Megophryidae). *Amphibia-Reptilia* **28:** 309–311.

Zhu, M., Ahlberg, P. E., Zhao, W. J., and Jia, L. T. (2002). Palaeontology - First Devonian tetrapod from Asia. *Nature* **420:** 760–761.

Zimmerman, K., and Heatwole, H. (1990). Cutaneous photoreception: A new sensory mechanism for reptiles. *Copeia* **1990:** 860–862.

Zug, G. R. (1991). The lizards of Fiji: Natural history and systematics. *Bishop Museum Bulletin in Zoology* **2:** 1–136.

Zug, G. R., and Mitchell, J. C. (1995). Amphibians and reptiles of the Royal Chitwan National Park, Nepal. *Asiatic Herpetological Research* **6:** 172–180.

Zug, G. R., and Parham, J. F. (1996). Age and growth in leatherback turtles, *Dermochelys coriacea* (Testudines: Dermochelyidae): A skeletochronological analysis. *Chelonian Conservation Biology* **2:** 244–249.

Zug, G. R., and Zug, P. B. (1979). The marine toad, *Bufo marinus*: A natural history resumé of native populations. *Smithsonian Contributions to Zoology* **284:** 1–58.

Zweifel, R. G. (1972). Results of the Archbold Expeditions. No. 97. A revision of the frogs of the subfamily Asterophryinae, family Microhylidae. *Bulletin of the American Museum of Natural History* **148:** 411–546.

Zweifel, R. G. (1985). Australian frogs of the family Microhylidae. *Bulletin of the American Museum of Natural History* **182:** 265–388.

Zweifel, R. G. (1986). A new genus and species of microhylid frog from the Cerro de la Neblina region of Venezuela and a discussion of relationships among New World microhylid genera. *American Museum Novitates* **2863:** 1–24.

Zweifel, R. G. (1989). Long-term ecological studies on a population of painted turtles, *Chrysemys picta*, on Long Island, New York. *American Museum Novitates* **2952:** 1–55.

Zweifel, R. G., and Lowe, C. H. (1966). The ecology of a population of *Xantusia vigilis*, the desert night lizard. *American Museum Novitates* **2247:** 1–57.

Glossary

This glossary does not attempt completeness; rather we include potentially unfamiliar words that are not defined when they first appear in the text. Abbreviations: adj, adjective; n, noun; pl, plural; v, verb.

A

abiotic [adj] All nonliving components of the environment, e.g., weather and geology.

Age-Specific Mortality [n, adj, n] Proportion of individuals in any age group (cohort) that do not survive to reach the next age group.

alate [adj] State of having wings; also used as a noun in reference to the winged, mating stage of ants and termites.

allele See chromosome.

Allopatric [adj] Refers to species or populations that are geographically isolated from one another.

Allopatric speciation [adj, n] Process by which one species differentiates into two or more species as the result of a physical barrier, such as a river or mountain range. Also referred to as geographic speciation.

amniote [n] A tetrapod that arises developmentally from an amniotic egg, e.g., reptiles, birds, and mammals.

amphicoelous See vertebral structure.

amplexus [n], amplex [v], amplectant [adj], amplectic [adj] The "copulatory" behavior of frogs in which the male sits on the female's back and grasps her with his forelimbs; amplexus can be inguinal (forefeet grasping body immediately in front of hindlimbs), axillary (immediately behind forelimbs), cephalic (on head or neck), straddled (male sits on shoulders of female while frogs are vertical and sperm flows down the female's back), or glued (male is attached to females back by adhesive substance). In amplexus, the cloacae of the male and female are adpressed and sperm and eggs are extruded simultaneously. Amplexus is absent in some frogs.

anamniote [n] A tetrapod that lacks an amniotic egg in its development, e.g., amphibians.

anosmic [adj] Unable to smell; absence of the olfactory sense.

anterior [adj] See body location.

arciferal [adj] The anuran pectoral girdle architecture with the epicoracoids of the left and right side fused anteriorly and free and overlapping posteriorly.

auditory meatus [n] The car canal, either external or internal.

aufwuchs [n] The aquatic community of microorganisms living on the surface of submerged objects. Aufwuchs form a coating, often slimy, on which numerous animals, such as tadpoles, graze.

Australian See biogeographic realms.

autopod See limb segments.

axilla [n], axillary [adj] See body location.

B

Bayesian Inference [adj, n] An iterative process in which the degree of belief in a hypothesis is updated as evidence accumulates. Prior probabilities are continually updated, and posterior probabilities are then calculated based on new evidence. For examples, see http://en.wikipedia.org/wiki/Bayesian_inference.

Bidder's organ A band or cap of ovarian tissue on the testis of male bufonids.

biogeographic realms The major divisions of the world's terrestrial areas, based on shared endemism of plants and animals.

- **Australian [adj, n]** The biogeographic area of New Guinea and adjacent islands, and Australia and adjacent islands.
- **Ethiopian [adj, n]** The biogeographic area of Saharan and sub-Saharan Africa and the southern half of the Arabian Peninsula.
- **Gondwana** The southern continent arising from the breakup of Pangaea consisting of the future Antarctica, South America, Africa, Australia, and New Zealand.
- **Holarctic [adj, n]** The biogeographic area composed of the Nearctic and Palearctic.
- **Laurasia** The northern continent arising from the breakup of Pangaea consisting of the future North America, Greenland, and Eurasia.
- **Nearctic [adj, n]** The biogeographic area of North America including the Mexican Plateau.
- **Neotropical [adj, n]** The biogeographic area of Central America (excluding the Mexican Plateau), South America, and the Greater and Lesser Antilles.
- **Oriental [adj, n]** Southern Asia, south of the Himalayan mountains and their east and west neighboring

mountain ranges from the Indus Valley eastward through southern China and southward to the Seram-Halmahera seas.

Palearctic [adj, n] The biogeographic area of Europe, Africa north of the Sahara, and Asia north of the Himalayan mountains and their east and west neighboring mountain ranges.

Pangaea The megacontinent of the Paleozoic period containing all the continental blocks that would become our present continents. Pangaea began to break up in the early Mesozoic.

biota [n], biotic [adj] All living components of the environment.

bipedal See locomotion.

body location

anterior [adj] The front or head end of an animal.
axilla [n], axillary [adj] At the forelimb insertion.
distal [adj] Toward the tip of an extremity, i.e., most distant from the body.
dorsum [n], dorsal [adj] The top or upper surface of an animal.
inguen [n], inguinal [adj] At the hindlimb insertion.
lateral [adj] The side of an animal.
posterior [adj] The rear or tail end of an animal.
proximal [adj] Toward the origin of an extremity, i.e., closest to the body.
venter [n], ventral [adj] The underside or lower surface of an animal.

C

carnivore See diet.

Carolina Bay [adj, n] These ponds form in elliptical depressions and are distributed across the Atlantic seaboard states. They are typically rich in amphibians and reptiles. see http://en.wikipedia.org/wiki/Carolina_bays for more details.

chromosomes

alleles [n] The different forms of a gene occurring at the same position on different, homologous chromosomes.
diploid [adj] Possessing the typical number of chromosomes following the fusion of the sperm and ovum pronuclei, i.e., a pair each of homologous chromosomes is present. Symbol, 2N.
haploid [adj] Possessing one-half of the homologous chromosomes; the condition obtained by meiotic division to produce sex gametes. Symbol, 1N.
heterozygosity [n], heterozygous [adj] The genetic state in which two different alleles occur at the same position or locus on homologous chromosomes.
homozygosity [n], homozygous [adj] The genetic state in which two identical alleles occur at the same position or locus on homologous chromosomes.
karyotype [n], karyotypic [adj] The chromosome set of an organism and its structural characteristics.
polyploid [adj] Possessing more than two sets of homologous chromosomes.
triploid [adj] Possessing three sets of homologous chromosomes. Symbol, 3N.

clade [n] A group of organisms containing an ancestor and all its descendants.

Cladogenesis [n] see Macroevolution.

classification

node name or node-based name This classification category name labels a clade stemming from the immediate common ancestor of two or more designated descendants.
sister group [n] The taxon sharing the most recent common ancestor with another taxon. A pair of taxa sharing the same common ancestor.
stem name or stem-based name This classification category name labels a clade of all taxa that are more closely related to a specified set of descendants than to any other taxa.

congeners [n], congeneric [adj] Individuals, populations, or species of the same genus.

conspecifics [n], conspecific [adj] Individuals or populations of the same species.

crèche [n] Nest chamber.

D

deme See population.

detritovore See diet.

development

direct A developmental pattern in which an egg hatches into a miniature adult body form; no larval stage occurs and development is complete or nearly so prior to hatching.
indirect A developmental pattern in which an egg hatches into a larva; the larva is free-living and grows and develops further prior to metamorphosing into a miniature adult body form.

diet

carnivore [n], carnivorous [adj] A flesh-eating organism.
detritovore [n], detritivorous [adj] A detritus-eating organism.
durophagous [adj] Eating hard-bodied prey; often used in herpetology for snakes and lizards preying on skinks or related lizards armored with osteoderms beneath scales.
folivore [n], folivorous [adj] A foliage-eating organism.
frugivore [n], frugivorous [adj] A fruit-eating organism.
herbivore [n], herbivorous [adj] A plant-eating organism.
insectivore [n], insectivorous [adj] An insect-eating organism, although commonly used for eating any arthropod.
molluscivore [n], molluscivorous [adj] A mollusk-eating organism.
nectivore [n], nectivorous [adj] A nectar-eating organism.
omnivore [n], omnivorous [adj] An organism that consumes a variety of plant and animal matter.

diplasiocoelous See vertebral structure.

diploid See chromosomes.

distal See body location.

diverse [adj] Having numerous, different aspects, such as body forms, courtship behaviors, or temperature or habitat tolerances.

dorsum See body location.

durophagous [adj] See diet.

E

Ecomorph [n] A predictable morphology based on habitat use. For example, the twig ecomorph of *Anolis* lizards is thin-bodied with a long tail. Unrelated species of *Anolis* on different islands have converged on various ecomorphs.

edentate, edentulous [adj] Lacking teeth.

epipodium See limb segments.

Ethiopian See biogeographic realms.

exaptation [n] A structure, behavior, or physiological feature of an organism that serves one function in an ancestor but serves a new and different function in a descendant. A replacement word for the situation previously called pre-adaptation.

exostosis [n] The condition of a bone having a rugose surface, commonly arising from the fusion of bone and dermis or osteoderms.

extant [adj, n] The state of a population or species of being alive now; not extinct.

F

fertilization [n] The penetration of the ovum's cell membrane by the sperm and the fusion of the sperm and ovum pronuclei to reestablish a diploid state.

- **external** The condition when the sperm and ovum come in contact external to the reproductive tract or cloaca of a female.
- **internal** The condition when the sperm and ovum come in contact within the reproductive tract or cloaca of a female.
- **firmisternal [adj]** The anuran pectoral girdle architecture with the left and right epicoracoids fused anteriorly and posteriorly.
- **folivore** See diet.
- **fossorial [adj]** Living underground; not all fossorial animals are burrowers but instead may use preexisting holes and cavities in the earth.
- **frugivore** See diet.

Fertility Rate [adj, n] The average number of offspring that an organism can produce in its lifetime. Fertility rate is calculated by summing the average number of offspring produced at each age. For example, a turtle might produce 10 eggs at age 1, 30 at age 2, 35 at age 3, and so on. See also "net reproductive rate."

G

gait See locomotion.

Geographic speciation [adj, n] See Allopatric speciation.

Gondwana See biogeographic realms.

grade [n] A group of organisms that possess a similar adaptative level of organization.

H

habitus [n] The body shape or form of an organism, i.e., its general appearance.

haploid See chromosomes.

hatchling [n] An animal recently hatched from an egg. The duration of the hatchling state is variable, although its end in reptiles might be fixed by the disappearance of the yolk-sac scar.

heliophilic [adj] Sun-loving.

heliothermic [adj] Deriving heat from the sun.

herbivore See diet.

heterozygosity See chromosome.

Holarctic See biogeographic realms.

holochordal See vertebral structure.

homozygosity See chromosome.

hydroperiod [n] A cycle characterized by a period of dryness; often used in amphibian biology in reference to the period when an ephemeral pond has water.

I

inguen [n], inguinal [adj] See body location.

insectivore See diet.

K

karyotype See chromosomes.

L

lateral See body location.

Laurasia See biogeographic realms.

limb segments

- **autopod [n]** The distal part of the limb, including the mesopodium, metapodium, and the phalanges.
- **epidpodium [n], epipodial [adj]** The second segment of the limb, including either the radius and ulna or the tibia and fibula. Zeugopod is a synonym.

mesopodium [n] The third segment of the limb, including either the wrist bones (carpus) or the ankle bones (tarsus).

metapodium [n] A distal segment of the limb, including either the metacarpal or the metatarsal elements.

propodium [n], propodial [adj] The most proximal segment of the limb, including either the humerus or the femur. Stylopod is a synonym.

locomotion

bipedal [adj] Moving on two limbs.

gait [n] The pattern of limb movement.

quadrupedal [adj] Moving on four limbs.

rectilinear locomotion [n] A mode of limbless locomotion dependent upon a wavelike pattern of rib movement to move the animal forward.

saltatory [adj] Moving by jumping, either bipedally or quadrupedally.

serpentine [adj] A mode of limbless, undulatory locomotion in which all portions of the body pass along the same path and use the same frictional surfaces for pushing the body forward.

sidewinding [adj] A specialized mode of serpentine locomotion in which only two parts of the body touch the ground simultaneously.

undulatory [adj] A group of limbless locomotion patterns in which the body moves through a series of curves.

M

Macroevolution [n] Any evolutionary change occurring at or above the species level. At the very least, macroevolution results in the splitting of one species into two. The splitting of one species into two or the splitting of higher order clades is often called cladogenesis.

manus [n] Hand or forefoot.

meiosis [n], meiotic [adj] Gametic cell division in which the number of chromosomes in a sex cell is halved.

mesic [adj] Habitat with moderate moisture level or water availability; adapted to moist conditions.

Mesoamerica [n] The portion of Central America from central Mexico to Nicaragua.

mesopodium See limb segments.

metapodium See limb segments.

metapopulation See population.

Microevolution [n] Evolution that results from small changes in allele frequencies within a population. It occurs below the species level.

mitosis [n], mitotic [adj] Regular, nongametic cell division in which each homologous chromosome duplicates itself; when the cell and nucleus divide, the sister cells retain their original ploidy or number of chromosomes.

molluscivore See diet.

monoestrous [adj] Having a single gametogenic cycle within a single reproductive season. See also polyestrous.

monophyly [n], monophyletic [adj] A taxonomic group whose members share the same ancestor. See also clade, paraphyly, and polyphyly.

morph [n] A particular body form or colored group of individuals. Morph is used regularly in discussion of polymorphism and variation of individuals within a population or species.

morphology [n], morphological [adj] The study of an organism's form or shape, or the shape of one or more of an organism's parts.

N

Nearctic See biogeographic realms.

nectivore See diet.

neonate [n] An animal recently born, i.e., it has emerged from the female's reproductive tract.

Neotropical See biogeographic realms.

Net Reproductive Rate [adj, adj, noun] Number of offspring produced by a female during its lifetime taking into consideration not only the fertility rate, but also age-specific mortality rates.

nictitating membrane Same as palpebral membrane.

node-based names See classification.

notochordal See vertebral structure.

O

omnivore See diet.

opisthocoelous See vertebral structure.

Oriental See biogeographic realms.

oviposit [v] To lay eggs.

P

palpebral membrane [n] A transparent "eyelid" that lies beneath the true eyelids and can extend horizontally from its resting position in the inner corner of the eye to the outer corner.

Palearctic See biogeographic realms.

Pangaea See biogeographic realms.

panmixis [n], panmictic [adj] Random and unrestricted mating within a population, thereby allowing the interchange of genes among all parts of a population.

paraphyly [n], paraphyletic [adj] A taxonomic group containing most but not all taxa derived from the same ancestor. See also monophyly and polyphyly.

perennibranchiate [adj] The retention of external (larval) gills as an adult.

periphyton [n] A synonym of aufwuchs; see above.

pes [n] Foot, specifically the hindfoot.

pheromone [n] A chemical signal secreted by one animal that conveys specific information to another animal, usually a conspecific, and often elicits a specific behavioral and/or physiological response.

Philopatry [n] Refers to individual animals that return to a specific location, usually to breed or feed.

phylogenesis [n], phylogenetic [adj] The evolutionary history of a taxon.

phytotelma, phytotelmata [pl, n] Small bodies of water within or on plants, e.g., pools in bromeliads.

polyestrous [adj] Having two or more gametogenic cycles within a single reproductive season. See also monoestrous.

polyphyly [n], polyphyletic [adj] A taxonomic group whose members do not share the same ancestor. See also grade, monophyly, and paraphyly.

polyploid See chromosomes.

population

deme [n] A small local population, panmictic in concept if not in actuality.

metapopulation [n] A population of several to many smaller populations or demes in the same geographic area; the smaller populations potentially exchange members by migration.

population [n] All individuals of the same species within a prescribed area.

posterior See body location.

postmetamorph [n] An amphibian that has recently completed metamorphosis, or the entire life stage following metamorphosis, in contrast to the larval or premetamorphic stage.

primitive [adj, n] A character or condition that is the same as an ancestral character or condition.

procoelous See vertebral structure.

propodium See limb segments.

proximal See body location.

Q

quadrupedal See locomotion.

R

rectilinear locomotion See locomotion.

rupicolous [adj] Living on walls or rocks.

S

salps [n, pl] Free-swimming, oceanic tunicates in the genus *Salpa* with transparent, fusiform bodies.

saltatory See locomotion.

saxicolous [adj] Living on or among rocks.

serpentine See locomotion.

sidewinding See locomotion.

sister group See classification.

speciose [adj] A taxon with many species.

spermatheca [n] A chamber for storing spermatozoa, usually multibranched, in the wall of some female salamanders.

spermatophore [n] A mucoid pedestal to support the sperm packets of some male salamanders; it is produced in the cloaca.

stegochordal See vertebral structure.

stem name See classification.

supraciliary [adj] Above the eye; eyebrow area.

SVL [n] Snout-vent length; straight-line distance from the tip of the snout to the anterior edge of the vent.

Sympatric [adj] Refers to species or populations that occur together in the same geographic area.

Sympatric Speciation [adj., n] Refers to a process by which a species differentiates into two or more species with no physical barriers isolating the populations.

T

taxon, taxa [pl, n] All members of a taxonomic group of organisms, e.g., *Anolis,* all members of all species classified in this particular genus.

tectorial membrane [n] A membrane in the inner ear covering a patch of sensory hairs.

TL [n] Various; used for Tail Length or Total Length. For tail length, it is distance from posterior edge of the vent to the tip of the tail, and for total length, distance from tip of snout to tip of tail.

trackway [n] A fossilized trail of footprints.

triploid See chromosomes.

tympanum, tympana [pl, n] Eardrum.

U

undulatory See locomotion.

urticating hairs [n] Defensive hairlike structures that break off the surface of an organism and cause irritation to the attacking herbivore or predator.

V

venter, ventral See body location.

vertebral structure

amphicoelous [adj] A vertebra in which the centrum is concave on both the anterior and the posterior surface.
diplasiocoelous [adj] The condition of the vertebral column with seven procoelous presacral vertebrae, the eighth presacral vertebra is biconcave, and the sacral vertebra is biconvex posteriorly.
holochordal [adj] Structurally, a centrum in which the notochord has been totally replaced.
notochordal [adj] Structurally, a centrum in which a small remnant of the notochord remains in the center of the centrum.
opisthocoelous [adj] A vertebra in which the centrum is convex on the anterior surface and concave on the posterior surface.
procoelous [adj] A vertebra in which the centrum is concave on the anterior surface and convex on the posterior surface.
stegochordal [adj] Structurally, a flattened centrum in which only the dorsal portion of the notochordal sheath has ossified.

X

xeric [adj] Habitat with low moisture level or water availability; adapted to dry or arid conditions.

REFERENCES

Lincoln, R. J., et al. (1982). "A Dictionary of Ecology, Evolution and Systematics." Cambridge Univ. Press, Cambridge.
Lincoln, R. J., and Boxshall, G. A.(1987). "The Cambridge Illustrated Dictionary of Natural History." Cambridge Univ. Press, Cambridge.
Peters, J. A. (1964). "Dictionary of Herpetology." Hafner Publ. Co., New York.

Taxonomic Index

Note: Page Numbers followed by f indicate figures; t, tables.

B

C

D

E

F

G

H

M

P

Q

R

S

T

Subject Index

X

Y

Z